GEOCHIMICA ET COSMOCHIMICA ACTA

SUPPLEMENT 3

PROCEEDINGS

OF THE

THIRD LUNAR SCIENCE CONFERENCE

Houston, Texas, January 10–13, 1972

GEOCHIMICA ET COSMOCHIMICA ACTA

Journal of The Geochemical Society and The Meteoritical Society

SUPPLEMENT 3

PROCEEDINGS

OF THE

THIRD LUNAR SCIENCE CONFERENCE

Houston, Texas, January 10–13, 1972

Sponsored by
The Lunar Science Institute

VOLUME 3
PHYSICAL PROPERTIES

Edited by
DAVID R. CRISWELL
The Lunar Science Institute, Houston, Texas

THE MIT PRESS

Cambridge, Massachusetts, and London, England

It was set in Monotype Times Roman
by Wolf Composition Company, Inc.
Printed and bound by The Colonial Press Inc.
in the United States of America.

Library of Congress Cataloging in Publication Data

Lunar Science Conference, 3d, Houston, Tex., 1972. Proceedings.

(Geochimica et cosmochimica acta. Supplement 3)
"Sponsored by the Lunar Science Institute."
Includes bibliographies.
CONTENTS: v. 1. Mineralogy and petrology, edited by Elbert
A. King, Jr.—v. 2. Chemical and isotope analyses, organic
chemistry, edited by Dieter Heymann.—v. 3. Physical properties,
edited by David R. Criswell.
1. Lunar geology—Congresses. 2. Moon—Congresses. I. King,
Elbert A., ed. II. Heymann, Dieter, ed. III. Criswell, David R., ed.
IV. Lunar Science Institute. V. Series.
QB592.L85   1972   559.9′1   72–5496
ISBN 0–262–12060–7 (vol. 1)
      0–262–12062–3 (vol. 2)
      0–262–12063–1 (vol. 3)
      0–262–12064–X (3 vol. set)

Supplement 3
GEOCHIMICA ET COSMOCHIMICA ACTA
Journal of The Geochemical Society and The Meteoritical Society

Preface

THESE VOLUMES contain the *Proceedings of the Third Lunar Science Conference*, which was held 10–13 January 1972 at the Manned Spacecraft Center near Houston, Texas. The conference was held under the joint auspices of the National Aeronautics and Space Administration and The Lunar Science Institute, Houston, Texas, which has also sponsored publication of all the *Proceedings* sets. The two previous conferences dealt with analyses of lunar samples and geophysical information returned by Apollos 11 and 12 and Luna 16. Naturally, refinements and extensions of these studies appear in this set. However, the bulk of the new material deals with Apollo 14 and 15 samples and the voluminous supply of new, critically important data obtained by the complement of detectors housed in the scientific instrument bay (SEM-Bay) of the Apollo 15 Service Module. Geophysical, geochemical, and photographic data on a global basis were produced by these lunar orbiting experiments.

In Volume 3 the papers are arranged by specialties. Wherever applicable, papers on ALSEP (Apollo Lunar Surface Experiment Package) or SEM-Bay geophysical experiments open the section. Alphabetical ordering by first author usually applies thereafter. Three cross-reference sections have been added to the *Proceedings* to enhance their usefulness. Two "Lunar Sample Cross References" are presented (Volumes 2 and 3) that list for a given sample number every article in all three *Proceedings* sets in which that sample number is mentioned. A complete and annotated "Sample Inventory" is presented for Apollos 11, 12, 14, and 15 samples (Volume 1). The subject index for this *Proceedings* set has been expanded considerably by including many key words suggested by the authors. The subject index as well as an author index is included in each volume. It is a pleasure to acknowledge the assistance provided by J. L. Warner (Manned Spacecraft Center) in the preparation of the indices, sample cross references, and sample inventories.

Approximately 256 papers were presented at the Third Lunar Science Conference whereas 228 papers are published in the *Proceedings*. The conference document *Lunar Science—III* (Revised Abstracts of the Papers Presented at the Third Lunar Science Conference, Houston, 10–13 January 1972), editor C. Watkins, Lunar Science Institute Contribution No. 88 (Library Congress No. 79-189790) is available at the publication cost of $15 from The Lunar Science Institute. Useful general descriptions of the missions and initial scientific results are presented in "NASA (1972) Apollo 14 Preliminary Science Report, SP-272" and "NASA (1972) Apollo 15 Preliminary Science Report, SP-289." These are available through the U.S. Government Printing Office, Washington, D.C. 24042 for $3 and $8, respectively.

On behalf of The Lunar Science Institute, sponsor of these *Proceedings*, we express our thanks to NASA, the lunar Principal Investigators and their associates, the Associate Editors and referees for their cooperation and support. Special thanks are due Mmes Ann Geisendorff and Olene Edwards for their secretarial assistance.

Acknowledgment is due Pergamon Press, Ltd. for their generosity in permitting use of the name of *Geochimica et Cosmochimica Acta* to be associated with these *Proceedings* (as Supplements 3) with The MIT Press as the publisher.

David R. Criswell

The Lunar Science Institute
Houston, Texas
May 1972

Contents

Contents

Contents

Contents

Contents

A first look at the lunar orbital gamma-ray data

A. E. Metzger

Jet Propulsion Laboratory
Pasadena, Calif., 91103

J. I. Trombka

Goddard Space Flight Center
Greenbelt, Maryland 20771

L. E. Peterson, R. C. Reedy, and J. R. Arnold

University of Calif., San Diego
La Jolla, Calif., 92037

THE COLOR PLATES OVERLEAF give a relief map of radioactivity of the regions of the moon overflown by the Apollo 15 CSM. The data are the count rates in the energy band from 0.55 to 2.75 MeV, corrected for altitude and averaged over 5 degree $\times$ 5 degree areas. Analysis has shown that differences in counting rate in this energy region are chiefly due to differences in the concentration of the radioactive elements Th, U, and K. The remaining counts in this energy region are due to other elements and to continuum processes.

The series of colors from blue to red represents increasing radioactivity, from the low levels characteristic of the backside highlands (probably ≤ 1 ppm Th) to the highest levels in the western mare regions, on the order of 10 ppm Th, with corresponding amounts of U and K. The ranges of counting rate, with differences roughly proportional to Th and U content, are deep blue <74 cps, blue 74–78 cps, dark green 78–81 cps, light green 81–84 cps, yellow 84–88 cps, orange 88–91 cps, and red >91 cps. The background outline map is freely adapted from one furnished through the courtesy of Dr. Farouk El-Baz.

The most significant conclusions are the following:

(1) The major concentrations of radioactivity in the area covered are in the region of Mare Imbrium and Oceanus Procellarum. All levels observed in this region are higher than all levels outside it. (Preliminary results from the Apollo 16 experiment show the same pattern at near-equatorial latitudes.)

(2) The highest regions are in the area of Aristarchus in Oceanus Procellarum and a region in eastern Mare Imbrium.

(3) The highland regions show low radioactivity, except at the edge of the western maria where some lateral mixing seems to have occurred and in the backside area near Van de Graaff.

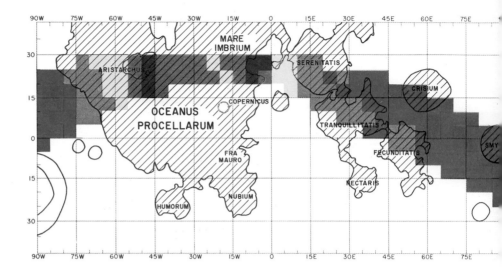

Radioactivity of the overflown regions of the moon, as measured by the total count rate in the region 0.55–2.75 MeV. The key is as follows:

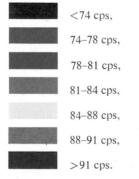

<74 cps,

74–78 cps,

78–81 cps,

81–84 cps,

84–88 cps,

88–91 cps,

>91 cps.

See text for fuller description.

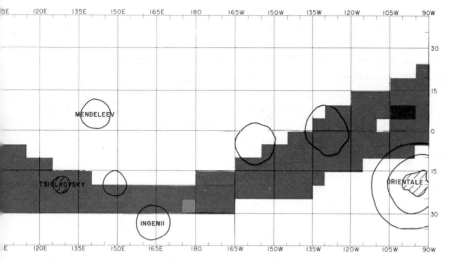

(4) While there are indications of activity in the eastern mare regions, it is at a much lower level than in the west. Some of this also may be due to mixing from the west. Small rises occur in some eastern maria; nothing is visible in the crater Tsiolkovsky. These small increases may be due to other elements, and should be interpreted with caution.

(5) The K/U ratio appears to be lower everywhere than that characteristic of the earth.

(6) Taken together with the results of the x-ray fluorescence experiment of Adler *et al.* (1972) and many measurements on surface samples, these data are most simply interpreted in terms of three major components in the lunar regolith. The high radioactivity component is identified with KREEP (Hubbard and Gast, 1971). The low radioactivity components are low-Al mare basalt, and highlands material with a composition rich in Ca and Al, like that observed by Patterson *et al.* (1970) at the Surveyor VII site.

Lines due to cosmic ray activation of O, Si, Fe, Mg and other elements have been measured. Significant results have also been obtained concerning the gamma ray flux in interstellar space. A fuller account of these results will appear elsewhere.

References

Hubbard N. J. and Gast P. W. (1971) *Proc. Second Lunar Sci. Conf., Geochim. Cosmochim. Acta*, Suppl. 2, Vol. 2, pp. 999–1020, M.I.T. Press.

Patterson J. H., Turkevich A. L., Franzgrote E. J., Economou T. E., and Sowinski K. P. (1970) *Science* **163**, 825–28.

Adler I., Trombka J. I., Gerard J., Lowman P., Yin L., Blodgett H., Gorenstein P., and Bjorkholm P. (1972) *Proc. Third Lunar Sci. Conf., Geochim. Cosmochim. Acta*, Suppl. 3, Vol. 3, M.I.T. Press.

GEOCHIMICA ET COSMOCHIMICA ACTA

Supplement 3

PROCEEDINGS

OF THE

THIRD LUNAR SCIENCE CONFERENCE

Houston, Texas, January 10–13, 1972

Proceedings of the Third Lunar Science Conference
(Supplement 3, *Geochimica et Cosmochimica Acta*)
Vol. 3, pp. 2157–2178
The M.I.T. Press, 1972

The Apollo 15 x-ray fluorescence experiment

I. Adler and J. Gerard

NASA Goddard Space Flight Center and National Academy of Science

J. Trombka, R. Schmadebeck, P. Lowman, H. Blodget,
L. Yin, E. Eller, and R. Lamothe

NASA Goddard Space Flight Center

and

P. Gorenstein, P. Bjorkholm, B. Harris, and H. Gursky

American Science and Engineering

Abstract—The prime purpose of the x-ray fluorescence spectrometer carried in the Scientific Instrument Module (SIM) of the Command-Service Module (CSM) was to map the lunar surface with respect to its chemical composition. Results are presented for Al, Mg, and Si as Al/Si and Mg/Si ratios for the various features overflow by the spacecraft. The lunar surface measurements involved observations of the intensity and characteristic energy distribution of the secondary or fluorescent x-rays produced by the interaction of solar x-rays with the lunar surface. The results showed that the highlands and maria are chemically different, with the highlands having considerably more Al and less Mg than the maria. The mare-highland contact is quite sharp and puts a limit on the amount of horizontal transport of material. The x-ray data suggests that the dominant rock type of the lunar highlands is a plagioclase-rich pyroxene bearing rock probably anorthositic gabbro or feldspathic basalt. Thus the moon appears to have a widespread differentiated crust (the highlands) systematically richer in Al and lower in Mg than the maria. This crust is pre-mare and may represent the first major internal differentiation of the moon.

Introduction

In the experiment being described, the production of characteristic x-rays follows the interaction of solar x-rays with the lunar surface. The situation is summarized in Fig. 1. Tabulated within the figure are the absorption edge energies E_k which are required to ionize the atoms in the K shell, and $E(K\alpha)$, the energies of the resulting characteristic x-ray lines. To produce the characteristic x-rays, an incident energy in excess of the binding energy of the electrons is necessary.

The results of numerous calculations indicate that the solar x-ray spectrum typical of the sun's active regions is energetically capable of producing measureable amounts of characteristic x-rays from all the abundant elements with atomic number 14 (Si) or less. During brief periods of more intense solar activity, excitation of higher atomic number elements will also occur.

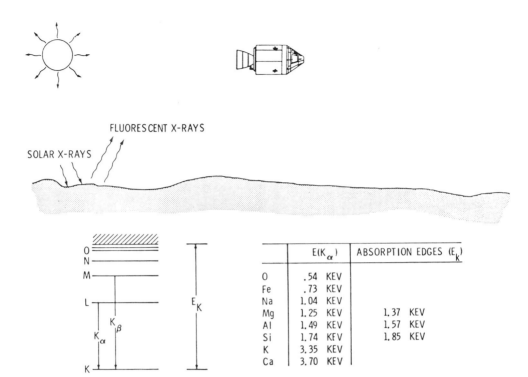

	E(K$_\alpha$)	ABSORPTION EDGES (E$_K$)
O	.54 KEV	
Fe	.73 KEV	
Na	1.04 KEV	
Mg	1.25 KEV	1.37 KEV
Al	1.49 KEV	1.57 KEV
Si	1.74 KEV	1.85 KEV
K	3.35 KEV	
Ca	3.70 KEV	

INNER ELECTRON TRANSITIONS GIVING
RISE TO CHARACTERISTIC K
X-RAY SPECTRA

Fig. 1. Production of fluorescent x-rays at the lunar surface.

The solar x-ray flux is known to vary on a time scale of minutes to hours. Thus, in the final analysis of the data, such features of the solar x-ray flux as intensity and spectral distribution must be considered in detail.

The solar x-ray flux, observed with low-resolution instruments such as proportional counters, decreases with increasing energy. If a strictly thermal mechanism of production is assumed, variable coronal temperatures are found to be somewhere between 10^6 to 10^{7}°K. Such variations in temperature produce changes in both flux and spectral composition. In observing the lunar surface, changes must be expected not only in fluorescent intensities but in the relative intensities from the various elements being observed. For example, if the solar spectrum hardens (larger fluxes of higher energies) or if there is an increase in solar characteristic line intensities on the high-energy side of the absorption edge of the heavier element, an enhancement of the intensities from the heavier elements relative to the lighter ones would be observed (see Fig. 2).

An x-ray monitor was used to follow the possible variation in solar x-ray intensity and spectral shape. In addition, detailed simultaneous measurements of the solar

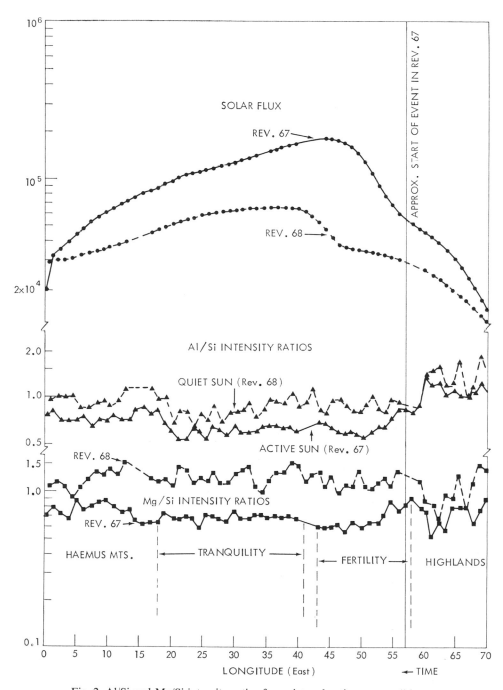

Fig. 2. Al/Si and Mg/Si intensity ratios for quiet and active sun conditions.

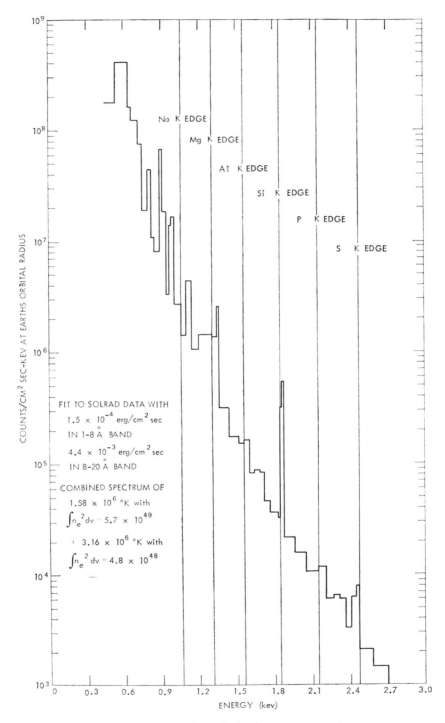

Fig. 3. An estimate of a typical quiet sun x-ray spectrum.

x-ray spectrum were obtained in flight during the mission from the various Explorer satellites that measure solar radiation.

An estimated representative solar x-ray spectrum based on Solrad data taken during the flight period is shown in Fig. 3. An attempt has been made to fit the observations with two temperatures, one for the quiet corona and the other for the more active regions. Superimposed on this curve along the energy axis are the K shell absorption edges for Mg, Al, Si, P, and S. Only the solar x-rays with energies on the high side of the absorption edges are capable of exciting those elements and to a degree depending on the incident flux and ionization cross sections. Therefore, under quiet sun conditions, the solar flux is most suitable for exciting the light elements, including the major rock-forming elements Si, Al, and Mg.

DESCRIPTION OF THE INSTRUMENTATION

The instrumentation which has been described in detail elsewhere (Adler *et al.*, 1971) will be only briefly described here.

The x-ray fluorescence and alpha particle experiment is shown in functional configuration in Fig. 4. The spectrometer consists of three main subsystems:

(1) Three large area proportional counters that have state-of-the-art energy resolution and 0.0025 cm thick beryllium (Be) windows.

(2) A set of large area filters for energy discrimination among the characteristic x-rays of Al, Si, and Mg.

(3) A data handling system for count accumulation, sorting into eight pulse-height channels, and finally, for relaying the data to the spacecraft telemetry.

The x-ray detector assembly consists of three proportional counters and two x-ray filters, mechanical collimators, an inflight calibration device, temperature monitors, and associated electronics. The detector assembly senses x-rays emitted from the lunar surface and converts them to voltage pulses proportional to their energies which are then processed in the x-ray processor assembly. Provisions for inflight calibration

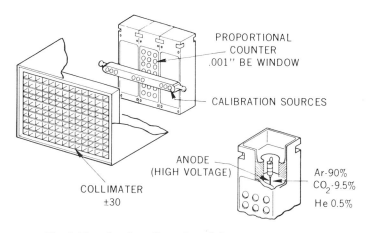

Fig. 4. Functional configuration of the x-ray spectrometer.

are made by means of programmed calibration sources, which, upon internal command, assume a position in front of the three detectors for calibration of gain, resolution, and efficiency.

The three proportional counters are identical, each having an effective window area of approximately 25 cm². The window consists of 0.0025 cm thick Be. The proportional counters are filled to a pressure of 1 atm with the standard P-10 mixture of 90% argon, 9.5% carbon dioxide, and 0.5% helium. To change the wavelength response, filters are mounted across the Be window aperture on two of the proportional counters. The filters consist of a Mg foil and an Al foil 5.08×10^{-4} to 1.27×10^{-3} cm thick. The third counter does not contain a filter. A single collimator assembly is used to define the field of view (FOV) of the three proportional counters as a single unit. The collimator consists of multicellular baffles that combine a large sensitive area and high resolution but are restricted in FOV. The FOV determines the total flux recorded from the lunar surface and the spatial resolution. The FOV is specified as $\pm 30°$ full width, half maximum (FWHM), in two perpendicular directions. The FWHM is the total angular width at which the collimator drops to one-half of its peak response. At the surface this translates into an instantaneous viewing area of 60×60 nautical miles at a spacecraft height of 60 nautical miles above the surface.

The inflight calibration device consists of a calibration rod with radioactive sources (Mg and Mn K radiation) that normally face away from the proportional counters. Upon internal command from the x-ray processor assembly, the rod is rotated 180° by a solenoid driver, thereby positioning the sources to face the proportional counters. Magnetically sensitive reed relays provide feedback signals indicating when the rod is fully in the calibration mode or fully in the noncalibration mode. These feedback signals are flag bits in the data telemetry output. The calibration command signal is generated in the x-ray processor assembly. The calibration cycle repeats every 16 min and continues for 64 sec.

In order to see higher energy radiation, if the sun was more active than anticipated, the bare detector (unfiltered) was programmed to operate in two modes, a high gain (normal) mode covering from about 0.75–2.75 keV and an alternate attenuated mode covering from 1.5–5.5 keV. During flight this detector alternately operates for six hours in the normal mode and two hours in the attenuate mode. The filtered detectors as well as the solar monitor continuously cover the energy range from 0.75–2.75 keV.

ANALYTICAL METHODS FOR DATA REDUCTION

The geochemical data reported below involving the ratios of magnesium and aluminum to silicon have been derived by an analysis of the pulse-height spectra observed with the three detectors. The following assumptions were made:

(1) All three proportional counters have identical characteristics.

(2) The detectors are 100% efficient for the radiation transmitted through the detector window.

(3) All three detectors cover the same energy region (in the normal mode). The intensity of x-rays observed in the three detectors are made up of the characteristic lines of the Si, Al, and Mg, the scattered solar flux and background x-rays, and high-

energy radiation from various sources, galactic and otherwise. The methods of data reduction used were required to extract the characteristic line intensities from the total flux. This is summed up in the following expression:

$$G_i = \int_{0.6 \text{ keV}}^{3.6 \text{ keV}} I(E)T_{\text{Be}}(E)F_i(E)\varepsilon_i(E) \, dE \,, \tag{1}$$

where G_i is the total count in detector i where $i = 1$ is the bare detector; $i = 2$ is the detector with the Mg filter and $i = 3$ is the detector with the Al filter; $T_{\text{Be}}(E)$ is the x-ray transmission efficiency of the detector window as a function of energy (since each detector has the same beryllium window (Be) and thickness, this transmission is independent of the detector); $F_i(E)$ is the x-ray transmission efficiency of the filters as a function of energy; $\varepsilon_i(E)$ is the detector efficiency as a function of energy after the x-rays have been transmitted through the window (this is reasonably assumed as 100% for all the wavelengths in this energy range); $I(E)$ (as indicated above) consists of the characteristic lines, the scattered radiation, and general background.

The background intensity was determined from measures on the dark side of the moon. The observations showed the background to be quite constant. Thus corrections could be simply made by subtracting these background values.

The problem of scatter was treated as follows: The scattered flux was divided into three energy regions covering the ranges 0.6–1.36 keV, 1.36–1.61 keV, and 1.61–2.6 keV. Each interval was chosen to include the absorption edges of Mg, Al, and Si respectively. Because of the nature of the solar spectrum the scattered flux in the region 1.36–2.6 is small compared to the fluorescence* (see Fig. 3).

In the low-energy region from 0.6–1.3 keV, if one combines the solar spectral distribution with the filtering effect of the Be window one gets a transmitted energy distribution which has an average energy of about 1.2 keV, very close to the Mg Kα line. Thus one can consider the scattered energy in this region as being essentially monoenergetic and use the appropriate mass absorption coefficient. As a consequence, the calculated flux of x-rays in this region is a combination of scattered and fluorescent x-rays.

Thus the integral in equation (1) reduces to

$$G_i = \sum_{j=1}^{3} I_j^T T_j^{\text{Be}} F_{ij} \,.$$

I_j^T is the total x-ray flux at energy j, where j is respectively the energy for the Mg, Al, and Si K-lines, and where $I_j^T = I_j^C + I_j^S$. I_j^C is the intensity of the characteristic line for energy j, and I_j^S is the scattered intensity for the energy j. $I_2^S = I_3^S = 0$. T_j^{Be} is the transmissions of the Be window for the energy j. F_{ij} is a filter factor for detector i and energy group j.

We thus have three simultaneous equations with three unknowns.

These equations can be solved by simple matrix inversion, but we selected a least

---

* The Al Si concentrations are high and most of the exciting energy is involved in photoelectric ionization.

Table 1. Intensity ratios and corresponding concentration ratios of Al/Si and Mg/Si for the various features overflown during the Apollo 15 remote sensing geochemical mapping experiment.

Feature	Intensity ratios			Concentration ratios	
	Al/Si ± 1σ	Mg/Si ± 1σ	N*	Al/Si ± 1σ	Mg/Si ± 1σ
West of Diophantus and Delisle, north-east of Schröters Valley	0.63 ± 0.28	0.96 ± 0.28	15	0.26 ± 0.13	0.21 ± 0.06
Mare Serenitatis	0.68 ± 0.16	1.12 ± 0.31	217	0.28 ± 0.08	0.21 ± 0.06
Diophantus and Delisle area	0.69 ± 0.20	1.16 ± 0.35	15	0.29 ± 0.10	0.26 ± 0.07
Archimedes Rille area	0.69 ± 0.16	0.86 ± 0.22	106	0.29 ± 0.08	0.19 ± 0.05
Mare Imbrium	0.71 ± 0.21	0.95 ± 0.28	159	0.30 ± 0.10	0.21 ± 0.06
Mare Tranquillitatis	0.81 ± 0.13	1.15 ± 0.21	170	0.34 ± 0.06	0.25 ± 0.04
Mare east of Littrow (Maraldi)	0.83 ± 0.16	0.99 ± 0.21	29	0.35 ± 0.08	0.22 ± 0.04
Palus Putredinus	0.83 ± 0.18	1.37 ± 0.16	8	0.35 ± 0.09	0.30 ± 0.03
Mare Fecunditatis	0.84 ± 0.12	1.13 ± 0.17	101	0.36 ± 0.06	0.25 ± 0.03
Apennine Mountains	0.84 ± 0.18	1.06 ± 0.22	55	0.36 ± 0.09	0.23 ± 0.05
Haemus Mountains, west border of Serenitatis	0.89 ± 0.21	1.12 ± 0.22	49	0.38 ± 0.10	0.25 ± 0.05
Mare Crisium	0.90 ± 0.16	1.19 ± 0.25	43	0.39 ± 0.08	0.26 ± 0.05
Tsiolkovsky	0.90 ± 0.24	0.78 ± 0.09	7	0.39 ± 0.11	0.18 ± 0.02
Haemus Mountains, south-south-west of Serenitatis	0.94 ± 0.15	1.16 ± 0.20	53	0.40 ± 0.07	0.26 ± 0.04
Littrow area	0.98 ± 0.22	1.13 ± 0.30	51	0.42 ± 0.10	0.25 ± 0.06
Mare Smythii	1.04 ± 0.13	1.22 ± 0.27	30	0.45 ± 0.06	0.27 ± 0.06
Taruntius area, between Tranquillitatis and Fecunditatis	1.06 ± 0.14	1.17 ± 0.14	45	0.45 ± 0.07	0.26 ± 0.02
Langrenus area, east of Fecunditatis to 62.5° E	1.10 ± 0.24	1.07 ± 0.27	37	0.48 ± 0.11	0.24 ± 0.06
Highlands between Crisium and Smythii (Mare Spumans and Mare Undarum area)	1.17 ± 0.13	0.99 ± 0.22	53	0.51 ± 0.06	0.22 ± 0.05
Highlands east of Fecunditatis, Kapteyn area 68–73°E 7.5–15°S	1.17 ± 0.21	0.98 ± 0.22	52	0.51 ± 0.10	0.22 ± 0.05
Highlands west of Crisium	1.17 ± 0.20	1.05 ± 0.25	80	0.51 ± 0.10	0.23 ± 0.05
Highlands east of Fecunditatis 62.5–68°E, 4–12.5°S	1.18 ± 0.20	0.98 ± 0.26	51	0.52 ± 0.10	0.22 ± 0.05
West border of Smythii to 4–5° out from Rim	1.19 ± 0.13	0.97 ± 0.16	43	0.52 ± 0.06	0.22 ± 0.03
South of Crisium, Apollonius area, to Fecunditatis, 50–60°E	1.20 ± 0.13	1.04 ± 0.13	42	0.53 ± 0.06	0.23 ± 0.03
East border of Crisium out to 6° from Rim	1.22 ± 0.19	0.99 ± 0.19	44	0.54 ± 0.09	0.22 ± 0.04
Tsiolkovsky—Rim	1.23 ± 0.25	0.70 ± 0.09	7	0.54 ± 0.12	0.16 ± 0.02
Highlands between Crisium and Smythii, 2.5°S 69°E, 5°S 76°E, 12°N 80°E, 10°N 83°E	1.24 ± 0.13	0.99 ± 0.17	48	0.55 ± 0.06	0.22 ± 0.03
Highlands west of Tsiolkovsky, 110–124°E to 9–21°S	1.29 ± 0.23	0.85 ± 0.19	89	0.57 ± 0.11	0.19 ± 0.04
Highland east of Fecunditatis 73–85°E, 10–19°S	1.30 ± 0.28	0.95 ± 0.24	70	0.58 ± 0.13	0.21 ± 0.05
South and south-west of Sklodowska, 86–101°E, 18–23°S	1.33 ± 0.29	0.86 ± 0.33	59	0.59 ± 0.14	0.19 ± 0.07
Pirquet, 135–145°E, 18–23°S	1.33 ± 0.32	0.72 ± 0.23	46	0.59 ± 0.15	0.16 ± 0.05
East border of Smythii, out to 4–5°	1.34 ± 0.20	0.92 ± 0.17	37	0.60 ± 0.10	0.21 ± 0.03
Pasteur Hilbert highlands area 101.5–110°E, 7–18°S	1.35 ± 0.21	0.79 ± 0.21	48	0.60 ± 0.10	0.18 ± 0.04
Hirayama, highlands east of Smithyii, 89°E 12°S, 100°E 15°S, 98°E 2°S, 103°E 5°S	1.39 ± 0.15	0.86 ± 0.18	63	0.62 ± 0.07	0.19 ± 0.04
Highlands around Tsiolkovsky	1.40 ± 0.26	0.71 ± 0.30	39	0.62 ± 0.12	0.15 ± 0.06
South part of Gagarin, 144–153°E, 21–23°S	1.45 ± 0.50	0.69 ± 0.22	22	0.65 ± 0.24	0.14 ± 0.05

Table 1 (*continued*)

Feature	Intensity ratios			Concentration ratios	
	Al/Si ± 1σ	Mg/Si ± 1σ	N*	Al/Si ± 1σ	Mg/Si ± 1σ
Al/Si and Mg/Si concentration ratios of selected lunar samples					
Apollo 12, Oceanus Procellarum average of type AB rocks[a]				0.22	0.22
Apollo 15, Hadley Apennines average of rocks[b]				0.22	0.27
Apollo 12, Oceanus Procellarum, type B rocks, average[a]				0.22	0.37
Apollo 11, Mare Tranquillitatis, high K rocks—average[c]				0.23	0.24
Apollo 12, Occanus Procellarium, type A rocks—average[a]				0.24	0.31
Rock 12013[a]				0.24–0.30	0.20
Apollo 11, Mare Tranquillitatis, average of low K rocks[c]				0.29	0.23
Dark of rock 12013[d,e,f]				0.33	0.22
Apollo 12, Oceanus Procellarium average of soils[a]				0.33	0.29
Surveyor VI, Sinus Medii, regolith[g,h]				0.34	0.20
Apollo 15, Hadley-Apennines, soils[b]				0.34	0.30
Surveyor V, Mare Tranquillitatis, regolith[i,h]				0.35	—
Luna 16, Mare Fecunditatis, rocks[i]				0.35	0.21
Apollo 11, Marc Tranquillitatis, bulk soils average[c]				0.37	0.24
Apollo 14, Fra Mauro, average of rocks[k]				0.38	0.26
KREEP average[d,e,f]				0.39	0.21
Apollo 14, Fra Mauro, soils[k]				0.41	0.26
Norite material, average[l,m]				0.42	0.20
Luna 16, Mare Fecunditatis, bulk soils[j]				0.42	0.27
Surveyor VII, rim of Tycho, regolith[n,h]				0.55	0.20
Anorthositic gabbros, Apollo 11 and 12[l,m]				0.64	0.21
Rock 15418, Apollo 15, gabbroic anorthosite[b]				0.67	0.15
Gabbroic anorthosites, Apollo 11 and 12[l,m]				0.82	0.074
Anorthosites, Apollo 11 and 12[l,m]				0.89	0.038
Rock 15415, Apollo 15, anorthosite genesis rock[b]				0.91	0.003

N* is the number of individual data points used to determine the average Al/Si and Mg/Si values.
±1 standard deviation and was obtained from the various passes over each feature.

[a] *Proc. Second Lunar Sci. Conf.* (1971)
[b] LSPET (1972).
[c] *Proc. Apollo 11 Lunar Sci. Conf.* (1970).
[d] McKay *et al.* (1971).
[e] Meyer *et al.* (1971a).
[f] Meyer *et al.* (1971b).
[g] Turkevich *et al.* (1968a).
[h] Mason and Melson (1970).
[i] Turkevich *et al.* (1967).
[j] Vinogradov (1971).
[k] LSPET (1971).
[l] Wood *et al.* (1971).
[m] Marvin *et al.* (1971).
[n] Turkevich *et al.* (1968b).

square inversion method which automatically calculates the statistical variance and covariance, the standard error due to counting statistics and background subtraction, and also interferences due to the fact that the filters are not ideal.

Furthermore an attempt has been made to arrive at actual concentration ratios for Al/Si. The approach to determining these concentrations is in part theoretical and in part empirical. The theoretical calculations are based on the assumption of a quiet sun and a coronal temperature of $4 \times 10^{6\circ}$K. These conditions give an x-ray energy distribution consisting of both a continuum and characteristic lines which is consistent with our solar monitor observations. Using this distribution and various compositions of soils as determined from the analysis of lunar samples we have been able to calculate a relationship between Al/Si and Mg/Si intensity ratios as a function of chemical concentration ratios.

Empirically, we have used soil values from the Apollo 11 site at Tranquility Base, Luna 16 values from Fecunditatis, and Apollo 15 values from Palus Putredinus to be ground truth values (see Table 1). With these values and the theoretically calculated slopes we have been able to determine the values of Al/Si concentrations and Mg/Si concentrations for various parts of the moon along the various ground tracks.

OPERATION OF THE EXPERIMENT

The x-ray experiment began to function 84 h into the flight, during the third revolution around the moon. From 84 to 102 h ground elapsed time (GET), the orbit was approximately 8 by 60 n. miles. After 102 h, the orbit was circularized and maintained at approximately 60 n. miles until transearth coast. During the orbital period, more than 100 h of surface measurements were made. The solar-monitor detector was used for simultaneous monitoring of solar x-ray flux. Lunar-far-side data were recorded on magnetic tape and telemetered on the near side. The data from the experiment were displayed in almost real time as numeric readout on a cathode ray tube monitor. The data were displayed in the form of running sums for the eight energy channels for each of the four detectors. The data obtained during the flight were used to plot compositional maps while the CSM was still orbiting the moon. These quick look values were for one-minute integration intervals resulting in some degradation of spatial resolution (there is a 3° longitudinal displacement along the ground track due to the spacecraft motion). The prime data now being processed and reported here will be on the basis of 24-sec integration invervals and a one-degree surface displacement (60 × 80 sq mile area).

As we have stated in the experiment description, the flux and energy distribution of the solar x-rays were expected to have a significant effect on the intensity of the fluorescent x-rays measured. This is very well demonstrated in Fig. 2. Our solar monitor indicated increasing solar activity during part of orbit 67 as well as 73. The intensities for part of orbit 67 are compared to orbit 68 for the same longitudes (there was some small displacement of latitude). Also shown in this figure are Al/Si and Mg/Si intensity ratios. It can be seen that for the more active periods the Al/Si and Mg/Si ratios were depressed relative to the more normal orbit 68. This is due to a hardened solar x-ray spectrum which more efficiently excites the silicon relative to the aluminum and magnesium. This observation points out the great importance of monitoring the solar x-rays simultaneously with the surface x-rays. Figure 5 is a plot of the integrated intensities registered by the solar monitor for the approximate period corresponding to the surface measurements. These values were taken at the subsolar point. Also shown are the corresponding surface measurements. With the exception of such orbits as 67 and 73, the solar flux was fairly stable, varying less than $\pm 30\%$ of the mean value. This stability was likewise reflected in the surface data, which indicate a relatively stable incident flux as well as a stable spectral distribution.

OBSERVATIONS

This report is based on the reduction of data recorded on the prime mission tapes. There are thousands of data points based on 24-sec integration times. Most of the time we will be using intensity ratios rather than concentrations for convenience's sake; however, these values just as effectively indicate important trends. The data presented here include Al/Si and Mg/Si relationships to terrain and returned lunar samples. The x-ray experiment provided data from that part of the lunar ground track illuminated by the sun. An examination of the ground tracks from east to west

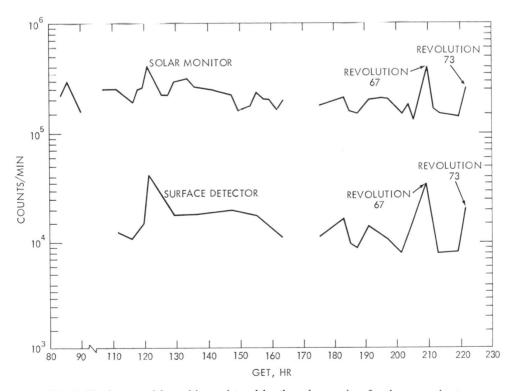

Fig. 5. The integrated intensities registered by the solar monitor for the approximate period corresponding to the surface measurements.

indicates that the spacecraft passed over such features as the craters Gagarin, Tsiol-kovsky, the far side and eastern limb highlands, the mare areas such as Smythii, Crisium, Fecunditatis, Tranquillitatis, Serenitatis, Imbrium, and Oceanus Procel-larum, and Haemus mountains and the Apennines. The Al/Si intensity ratios varied by more than a factor of 2 between the eastern limb highlands and the mare areas. The Mg/Si intensity ratios varied by about a factor of 2. The situation is clearly shown in Fig. 6. The part of the moon that was overflown has been divided into a number of areas, some representing distinct lunar features. The upper figures in each segment are Al/Si and the lower figures are Mg/Si intensity ratios. These values are averages, taken from a substantial number of ground tracks. The situation is further detailed in Table 1, where N is the number of individual data points used to calculate the averages. The table further indicates the scatter of observed values for each feature. Figures 7 and 8 show the Al/Si and Mg/Si intensity ratios and concentration ratios for north and south tracks versus longitude. In addition, the right-hand side of the figures shows the Al/Si and Mg/Si concentration ratios of selected lunar samples for reference.

The following observations can be made:

(1) The Al/Si intensity values are lowest over the mare areas and highest over the nonmare areas. The extremes vary from about 0.63 in Imbrium to 1.45 near Gagarin. The Mg/Si ratios tend to be lowest in the highlands and highest in the mare areas. The Mg/Si intensity ratios vary from about 0.69 at Gagarin to about 1.4 in Palus Putredinus.

(2) The Al/Si values for the Apennines and Haemus mountains is about 0.9, intermediate between Imbrium and the eastern limb highlands. On the other side of the Apennine mountains in the Archimedes Rille area, the observed values are about 0.70.

(3) An examination of the Al/Si coordinate plot shows that the values tend to increase from the western mare areas to the eastern limb highlands.

An interesting correlation has been observed between Al/Si ratios and optical albedo values, as is shown in Figs. 9 and 10. Generally, higher Al/Si ratios correspond to higher albedo values. There are occasional deviations from this relationship caused by surface features. The composition of areas with different albedos can be inferred by using the x-ray data. It is possible, for example, to state whether albedo variations are related to chemical differences or to the nature and, perhaps, age of a given feature.

GEOLOGIC INTERPRETATION

The x-ray data provide considerable insight to the moon's geology, when viewed in the context of returned sample analysis and the other orbital science experiments. The following interpretations are considerably more specific than those previously published (Adler *et al.*, 1972). However, they are still subject to revision; it should be stressed that although a wide range of lunar terrain was covered, the area for which we have data is less than 15% of the moon's total surface.

The most general conclusion from the measured x-ray intensities and shown

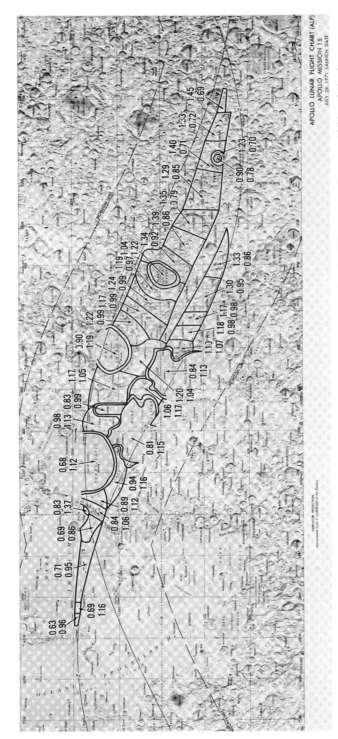

Fig. 6. Al/Si and Mg/Si intensity ratios for specific areas along the Apollo 15 groundtracks. The upper values are Al/Si and the lower values are the Mg/Si.

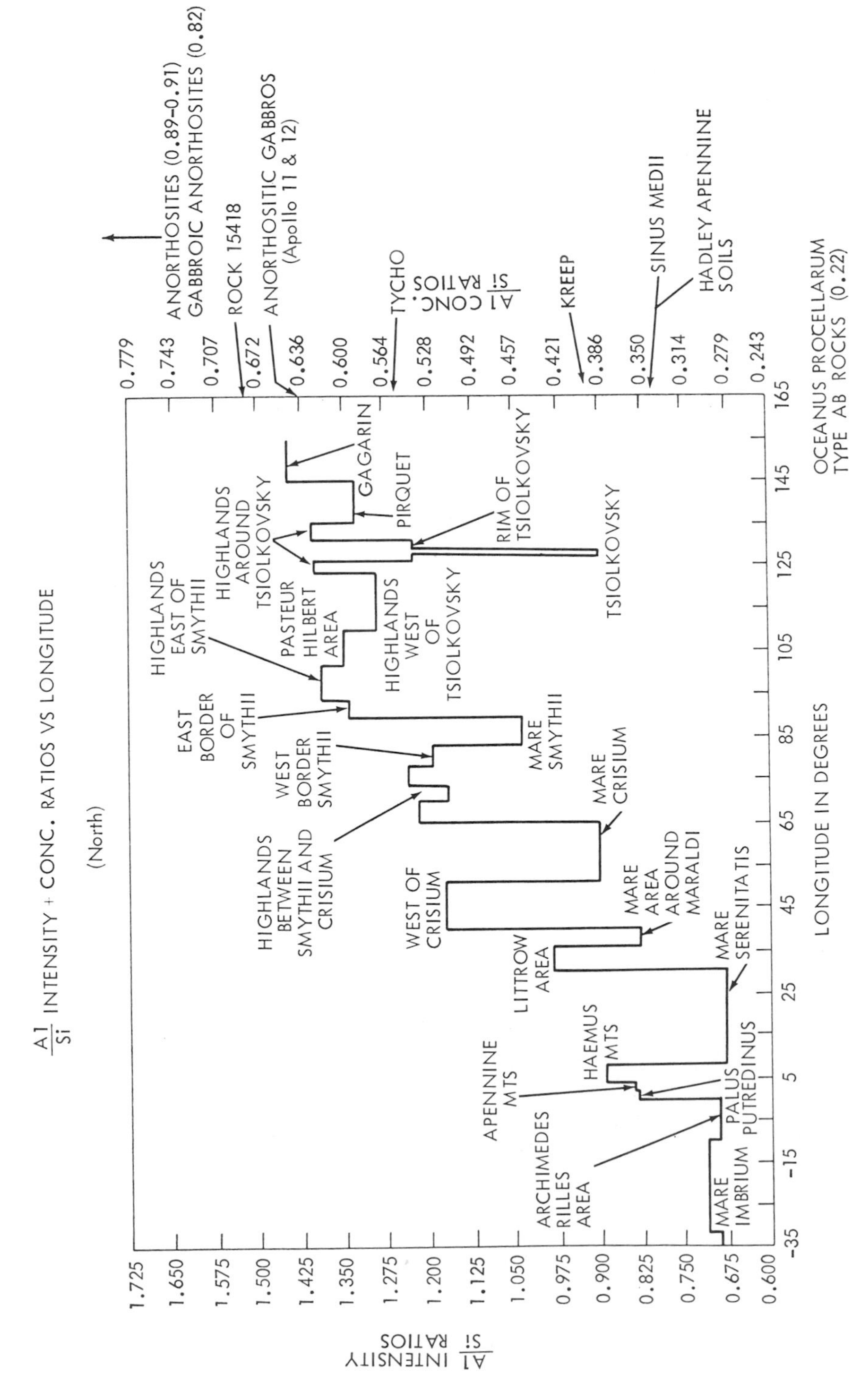

Fig. 7a. Al/Si ratios for a northerly track.

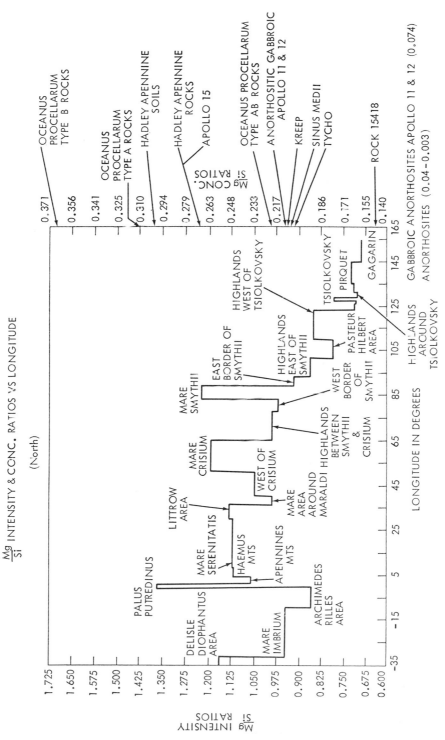

Fig. 7b. Mg/Si ratios for a northerly track.

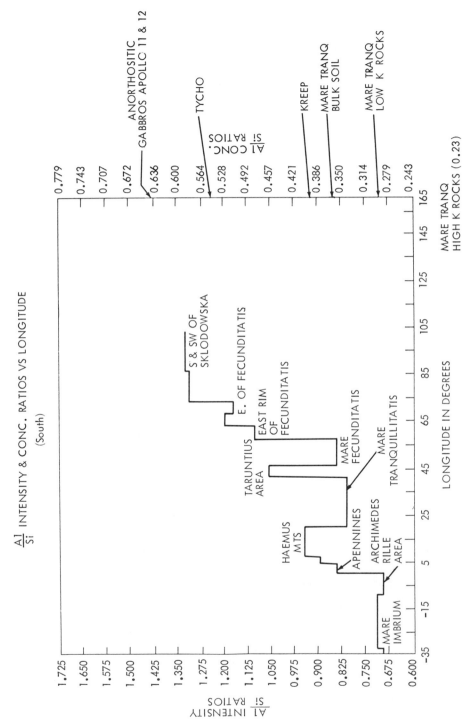

Fig. 8a. Al/Si ratios for a southerly track.

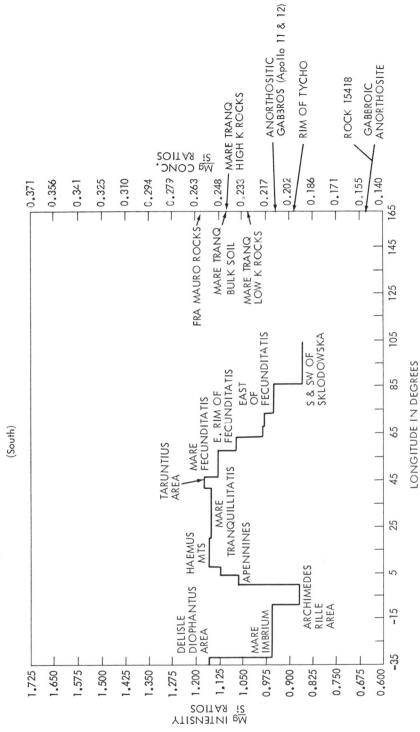

Fig. 8b. Mg/Si ratios for a southerly track.

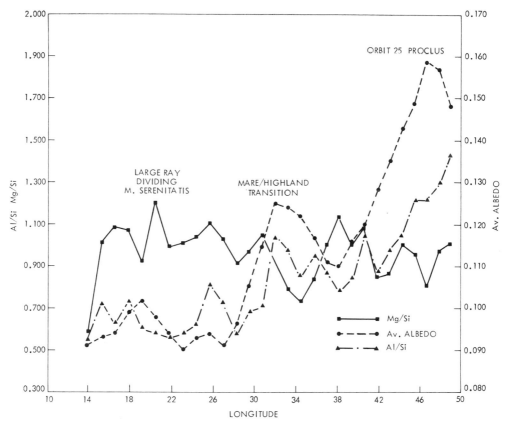

Fig. 9. Intensity ratios compared to optical albedos along the ground track corresponding
to orbit 25.

above is that the highlands and maria are chemically different, with the highlands having considerably more aluminum and less magnesium than the maria.
These differences, and the sharpness of the mare/highland contact shown by changes
in x-ray intensity, argue against the theory that the maria are essentially dust eroded
from the highlands (Gold, 1966) and electrostatically transported horizontally to low
spots. Such a process would tend to obliterate original chemical contacts, and there
is no obvious reason why it should produce chemical differentiation between highlands and maria. Further evidence against the dust theory in which the supposed
absence of far-side mare material is explained by the fact that the far side is shielded
from the earth's magnetic tail (Gold, 1972) is provided by the evidence for a marelike
composition of the dark material in the floor of Tsiolkovsky. The relative freshness
of the surrounding ejecta blanket indicates that this crater is probably Eratosthenian
in age, and probably formed after the moon's rotation became locked; thus it cannot
be argued that the far side was exposed to the earth's magnetic tail when Tsiolkovsky
was filled.

A number of specific conclusions can be drawn about the composition of the lunar
highlands. First, the x-ray intensity ratios, when reduced to composition ratios,

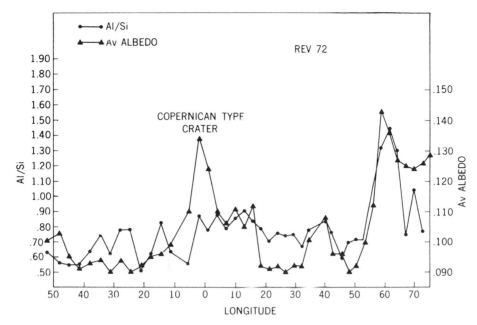

Fig. 10. Intensity ratios compared to optical albedos along the ground track corresponding
to orbit 72.

suggest that the highland crust is broadly similar over very large areas. There are regional gradients apparent within the highlands: in particular, the near side highlands are somewhat lower in aluminum and higher in magnesium than those of the far side. This can be explained by at least two phenomena. First, there has undoubtedly been some ballistic transport of mare material into the near side highlands. Second, there are many small patches of mare material, mapped as Imbrium or Eratosthenian mare by Wilhelm and McCauley (1971), in the front side highlands along the flight path.

The higher proportion of aluminum in the highlands is probably due largely to a greater proportion of plagioclase. This conclusion is based on correlative data from other orbital experiments and from examination of thin sections of samples from Apollo 14 and 15. The gamma-ray spectroscopy experiment shows that there is far too little ^{40}K in the highlands for the aluminum to be accounted for by alkali feldspar. Plagioclase, on the other hand, is abundant in most breccia samples from the Apennine Front and the Apollo 14 landing site, and widespread though scarce in the mare regolith. The higher albedo of the highlands appears to correlate well with aluminum content (Figs. 9 and 10). The television pictures from Surveyor VII by Shoemaker *et al.* (1968) also showed several blocks on the north slope of Tycho that were originally interpreted as feldspar. Taken together, these several lines of evidence strongly indicate plagioclase-rich highlands. However, the composition cannot be due only to enrichment in plagioclase, relative to the mare rocks, at the expense of other minerals, because the magnesium content does not go down as aluminum content goes up. Petrologically, this suggests that the highlands were not formed by simple crystal fractionation of a basaltic magma whose residue was later erupted to form the maria (Wood et al, 1970).

The question of what specific rock types comprise the highlands cannot be answered fully at this time, but some inferences can be drawn. First, it is clear that the highlands are not dominantly anorthosite. This is shown by Table 1; the Al/Si ratio of the highlands is too low and the Mg/Si ratio too high. There are, of course, almost certainly local concentrations of anorthosite, such as that from which sample 15415 came; Hollister et al. (1972) suggest that this rock was formed by crystal fractionation of a gabbroic anorthosite magma. The relation of KREEP to the lunar highlands is still not clear. The Al/Si ratio of KREEP is not as high as that of most parts of the highlands (Figs. 7 and 8). However, it does come close to that measured over the Apennine Mountains, which is of considerable interest in view of Myer's (1972) suggestion that the lower part of the highland crust (presumably exposed in the Apennines) is KREEP. Further speculation seems unwarranted until more highland samples are available.

When viewed with all other available information (summarized by Lowman, 1972), the x-ray data suggest that the dominant rock type of the lunar highlands is a plagioclase-rich pyroxene-bearing rock, probably anorthositic gabbro or feldspathic basalt. Local occurrences of anorthosite are probable; small areas of granitic (in a chemical sense) rock may also occur. KREEP, as previously noted, may be a significant constituent.

Turning to the maria, we see that the refined x-ray data do not support our earlier conclusion (Adler et al., 1972) that there are systematic chemical differences between the irregular and circular maria. This does not invalidate Soderblom's (1970) work, but simply suggests that the differences, if real, are not such as can be detected by the S-161 experiment.

The Al/Si ratios of the maria as calculated from the x-ray intensities appear systematically higher, and the Mg/Si ratios lower, than those of analyzed mare basalts. This effect is probably due to the fact that the x-ray fluorescence experiment is essentially a surface analyzer; it sees chiefly soils, and these have been shown to contain a substantial foreign component, perhaps from the highlands. Schnetzler et al. (1972), for example, show that soil 15531 could be formed by 88% 15555 basalt, 6% KREEP, and 6% plagioclase.

In summary, the main geologic conclusion from the S-161 experiment is that the moon appears to have a widespread differentiated crust (the highlands) systematically richer in aluminum and poorer in magnesium than the maria. The major rock type is probably chemically equivalent to anorthositic gabbro, with local areas of anorthosite and possibly KREEP and granitic rocks. This crust is pre-mare, probably between 4 and 4.6 b.y. old. Its formation may represent the first major internal differentiation of the moon.

REFERENCES

Adler I., Trombka J., Gerard J., Schmadebeck R., Lowman P., Blodget H., Yin L., Eller E., Lamothe R., Gorenstein P., Bjorkholm P., Harris B., and Gursky H. (1971) X-ray fluorescence experiment. In Apollo 15 Preliminary Science Report, NASA SP-289, Chap. 17.

Adler I., Trombka J., Gerard J , Lowman P., Schmadebeck R., Blodget H., Eller E., Yin L., Lamothe R., Gorenstein P., and Bjorkholm P. (1972) Apollo 15 geochemical x-ray fluorescence experiment: Preliminary report. *Science* **175**, 436–440.

Gold T. (1966) The moon's surface. In *The Nature of the Lunar Surface* (editors W. N. Hess, D. H. Menzel, and J. A. O'Keefe), pp. 107–121, Johns Hopkins Press.

Gold T., (1972) The depth of the lunar dust layer (abstract). In *Lunar Science—III.* (editor C. Watkins), pp. 321–322, Lunar Science Institute Contr. No. 88.

Hollister L., Trzcienski W. Jr., Dymek R., Kulick C., Weigand P., and Hargraves R. (1972) Igneous fragment 14310,21 and the origin of the mare basalts (abstract). In *Lunar Science—III* (editor C. Watkins), p. 386, Lunar Science Institute Contr. No. 88.

Lowman P. D. Jr. (1972) The geologic evolution of the moon. *J. Geol.* (in press).

LSPET (Lunar Sample Preliminary Examination Team) (1971) Preliminary examination of lunar samples from Apollo 14. *Science* **173**, 681–693.

LSPET (Lunar Sample Preliminary Examination Team) (1972) The Apollo 15 lunar samples: A preliminary description. *Science* **175**, 363–375.

Marvin U. B., Wood J. A., Taylor G. J., Reid J. B. Jr., Powell B. N., Dickey J. S. Jr., and Bower J. F. (1971) Relative proportions and probable sources of rock fragments in the Apollo 12 soil samples. *Proc. Second Lunar Sci. Conf., Geochim. Cosmochim. Acta* Suppl. 2, Vol. 1, pp. 679–699. MIT Press.

Mason B. H. and Melson W. G. (1970) *The Lunar Rocks*, p. 11, Wiley.

McKay D. S , Morrison D. A., Clanton U. S., Ladle G. H., and Lindsay J. F. (1971) Apollo 12 soil and breccia. *Proc. Second Lunar Sci. Conf., Geochim. Cosmochim. Acta* Suppl. 2, Vol. 1, pp. 755–773. MIT Press.

Meyer C. Jr. (1972) Mineral assemblages of lithic fragments of non-mare lunar rock types (abstract). In *Lunar Science—III* (editor C. Watkins), pp. 542–544, Lunar Science Institute Contr. No. 88.

Meyer C. Jr., Aitken F. K., Brett R., McKay D. S., and Morrison D. A. (1971a) Rock fragments and glasses rich in K, REE, P in Apollo 12 soils: Their mineralogy and origin. Unpublished proceedings of the Second Lunar Sci. Conf., Houston, January, 1971.

Meyer C. Jr., Brett R., Hubbard N. J., Morrison D. A. McKay D. S., Aitken F. K., Takeda H., and Schonfeld E. (1971b) Mineralogy, chemistry, and origin of the KREEP component in soil samples from the Ocean of Storms. *Proc. Second Lunar Sci. Conf., Geochim. Cosmochim. Acta* Suppl. 2, Vol. 1, pp. 393–411. MIT Press.

Proc. Apollo 11 Lunar Sci. Conf. (1970) *Geochim. Cosmochim. Acta* Suppl. 1, Vol. 1–3. Pergamon.

Proc. Second Lunar Sci. Conf. (1971) *Geochim. Cosmochim. Acta* Suppl. 2, Vol. 1–3. MIT Press.

Schnetzler C. C., Philpotts J. A., Nava D. F., Schuhmann S., and Thomas H. H. (1972) Geochemistry of Apollo 15 basalt 15555 and soil 15531. *Science* **175**, 426–428.

Shoemaker E. M., Batson R. M., Holt H. E., Morris E. C., Rennilson, J. J., and Whitaker E. A. (1968) Television observations from Surveyor VII. In *Surveyor VII: A Preliminary Report*, NASA SP-173, pp. 13–81. Office of Tech. Utilization, Washington, D.C.

Soderblom L. A. (1970) The distribution and relative ages of regional lithologies in the lunar maria (abstract). Abstracts with programs **2**, No. 70, Geological Society of America, pp. 690–691, 1970 annual meeting.

Turkevich A. L., Franzgrote E. J., and Patterson J. H. (1967) Chemical analysis of the moon at the Surveyor 5 landing site: Preliminary results. *Science* **158**, 635–637.

Turkevich A. L , Patterson J. H., and Franzgrote E. J. (1968a) Chemical analysis of the moon at the Surveyor 6 landing site. *Science* **160**, 1108–1110.

Turkevich A. L., Franzgrote E. J., and Patterson J. H. (1968b) Chemical analysis of the moon at the Surveyor 7 landing site: Preliminary results. *Science* **162**, 117–118.

Vinogradov A. P. (1971) Preliminary data on lunar ground brought to Earth by automatic probe "Luna-16." *Proc. Second Lunar Sci. Conf., Geochim. Cosmochim. Acta* Suppl. 2, Vol. 1, pp. 1–16. MIT Press.

Wilhelms D. E. and McCauley J. F. (1971) Geologic map of the front site of the moon, U.S. Geological Survey, Washington, D.C.

Wood J. A., Dickey J. S. Jr., Marvin U. B., and Powell B. N. (1970) Lunar anorthosites and a geophysical model of the moon. *Proc. Apollo 11 Lunar Sci. Conf., Geochim. Cosmochim. Acta* Suppl. 1, Vol. 1, pp. 965–988. Pergamon.

Wood J. A., Marvin U., Reid J. B., Taylor G. J., Bowen J. F., Powell B. N., and Dickey J. S. Jr. (1971) Relative proportions of rock types and nature of the light-colored lithic fragments in Apollo 12 soil samples. Unpublished proceedings of the Second Lunar Sci. Conf., January, 9171.

Proceedings of the Third Lunar Science Conference
(Supplement 3, *Geochimica et Cosmochimica Acta*)
Vol. 3, pp. 2179–2187
The M.I.T. Press, 1972

Observation of lunar radon emanation with the Apollo 15 alpha particle spectrometer

PAUL GORENSTEIN and PAUL BJORKHOLM

American Science and Engineering, 955 Massachusetts Avenue,
Cambridge, Massachusetts 02139

Abstract—The alpha particle spectrometer, a component of the orbital Sim Bay group of "geochemistry" experiments on Apollo 15, was designed to detect alpha particles emitted during the decay of isotopes of radon gas and her daughter products. The purpose was to measure the gross activity of radon on the lunar surface and to find possible regions of increased local activity. Results are presented from a partial analysis of Apollo 15 data. For the Moon as a whole, ^{220}Rn was not observed and the upper limit on its decay rate above the lunar surface is 3.8×10^{-4} disintegrations/cm²-sec. ^{222}Rn was marginally observed, but until further analysis can be carried out, we report the result as an upper limit of 10^{-3} disintegrations/cm²-sec. Possible variations of radon activity on the lunar surface are being investigated. ^{210}Po (a daughter product of ^{222}Rn) has been detected in a broad region from west of Mare Crisium to the Van de Graaff–Orlov region. The observed count rate is $(4.6 \pm 1.4) \times 10^{-3}$ disintegrations/cm²-sec. The observed level of ^{210}Po activity is in excess of the amount that would be in equilibrium with ^{222}Rn by about an order of magnitude. This implies that larger levels of radon emanation have occurred on the Moon within a time scale of 10^{1}-10^{2} years.

INTRODUCTION

ANALYSIS OF returned lunar samples has revealed significant concentration of uranium and thorium in lunar surface material. Both elements are unstable against radioactive decay and are the first members of two distinct highly complex decay series which terminate in stable isotopes of lead. Unstable isotopes of radon gas are produced as intermediate products of these series. Uranium produces ^{222}Rn and thorium produces ^{220}Rn. Radon is a rather special component of the decay series because it is a noble gas. There is a possibility that the radon will diffuse above the lunar surface where it remains trapped in an exceedingly rare atmosphere by the moon's gravity. As a result, the radioactive decay of the radon isotopes and their daughter products would have the effect of enhancing the radioactivity levels upon the surface of the moon.

Radon emanation from the soil is a well-known terrestrial phenomenon. Various effects promote the diffusion of easily detectable activity levels of radon into the atmosphere. Across the surface of the earth there exists gross differences in the amount of radon emanation that reflect local differences in concentrations of uranium and thorium and the ability of radon to diffuse through the soil. Generally speaking, there is a high degree of atmospheric radon activity where there is a high concentration of uranium and thorium. Volcanic activity and the evolution of volatiles from the ground is generally accompanied by radon emanation. Hence, a radon emanation map of the earth would be exceedingly nonuniform.

It is not unreasonable to expect that analogous effects are taking place across the surface of the moon. Significant concentrations of uranium and thorium, comparable to terrestrial values, are found. However, conditions on the moon are quite different from the earth. Some of the differences which may retard the diffusion of radon on the lunar surface are a lack of an atmosphere, lack of water vapor, and the grain size of the soil. Since the alpha spectrometer can only detect radon and her daughter products at or above the lunar surface, any retardation of the emanation and diffusion of radon will reduce the observed signal.

It is extremely difficult to determine a priori how the very high vacuum conditions in the soil affects the diffusion of radon. Previous measurements of lunar radon activity (to be discussed below) lead us to conclude that it retards diffusion in general. However, it might be expected that the presence of crevices or fissures in local regions that increase the amount of exposed surface would enhance the quantity of radon evolved into the atmosphere. It is quite reasonable to expect that volcanic or thermal sources of ordinary volatiles such as water vapor or carbon dioxide, should they exist on the moon, would also be sources of radon as they are on the earth. The movement of these common gases through rocks and material that contain uranium and thorium would very likely sweep radon to the surface. Because the uncertainties in this process are so large it is not possible to make a quantitative estimate of the amount of radon reaching the surface.

An early estimate by Kraner, Schroeder, Davidson, and Carpenter (1966) assumed terrestrial conditions for the diffusion coefficient and concentrations. When the actual concentrations of uranium and thorium are used their model predicts a rate of two disintegrations per sec cm^2 for ^{222}Rn and about 10^{-2} disintegrations per sec per cm^2 for ^{220}Rn. Actual observations of alpha emission from the moon have indicated that if the radon is present, the activity levels are considerably smaller than this. A measurement by Yeh and Van Allen (1969) from lunar orbiting Explorer 35 found no indication of alpha particle emission and set an upper limit that was about one-tenth of the value predicted by Kraner *et al.* (1966). Turkevich *et al.* (1970) reporting on background data obtained in the Surveyor 5, 6, and 7 alpha backscatter experiments cited evidence for a radioactive deposit at Mare Tranquilitatis (Surveyor 5) with an intensity of 0.09 $\pm$ 0.03 alpha disintegrations per sec per cm^2. Their instrument was deployed looking at and close to the lunar surface, well below the several kilometer scale height of any radon atmosphere. Thus, the Surveyor instrument was sensitive to only a small fraction of the total radon atmosphere. At the other two sites, Sinus Medii (Surveyor 6) and rim of Tycho (Surveyor 7) they report only upper limits to the alpha activity that are about a factor of two or three lower than Mare Tranquilitatis.

There are two other indirect measurements of alpha activity that look for the active deposit on returned samples that have been exposed to lunar radon. Lindstrom, Evans, Finkel, and Arnold (1971) looked for an excess of the radon daughter ^{210}Pb in Apollo 11 samples. They fail to find an excess to within 3%, which implies that the effect of the active deposit is less than 10^{-4} predicted by Kraner *et al.* This is the most pessimistic of all the experiments. However, there is a possibility that all or nearly all of the active deposit which resides entirely in the first micron of surface material could have been blown away by the action of the LM descent engine. A similar measurement was made by Economou and Turkevich (1971) upon the Surveyor 3 camera visor which was returned to earth from Oceanus Procellarum by the Apollo

12 astronauts. They found no evidence for the deposit and can set an upper limit that is about six times smaller than the value reported by Turkevich *et al.* for Mare Tranquilitatis. Here again one must remember that the slightest amount of abrasion or erosion could remove most of the active deposit. The net result from these pre-Apollo orbital measurements is that the active deposit on the lunar surface is probably several hundred times smaller than terrestrial diffusion rates would predict.

An interesting question concerns the degree to which radon remains localized. Any radon atoms reaching the lunar surface will move in ballistic trajectories. Emitted at thermal velocities of about 0.15 km/sec, at 300°K, they are decelerated by the gravitational pull of the moon. Typically, they reach a maximum altitude of about 10 km and fall back upon the surface. Essentially, no atoms have sufficient velocity to escape. It is evident that most of the shorter lived isotope ^{220}Rn ($T_{1/2} = 55$ sec) decays on its first ballistic trajectory. The alpha emission from ^{220}Rn and its daughter is confined to a region with a radius of 10 km around the point of emanation, thus preserving the localization to a very high degree. On the other hand, ^{222}Rn ($T_{1/2} = 3.8$ days) has sufficient time to migrate a considerable distance prior to decay. The largest uncertainty is the accommodation time or the elapsed time between the return to the surface of a freely falling radon atom and its reemission on a new trajectory. If we assume for an average ^{222}Rn atom a thermal velocity (300°K) and an emission angle of 45°, it impacts about 5 km from its point of emission and the process requires about 32 seconds. If the accommodation time is zero, then 2×10^4 bounces are possible during one mean life of ^{222}Rn. Hence, the displacement from the original point of emission is $\sqrt{2 \times 10^4} \times 5$ km or 700 km. For either nonzero accommodation time or lower temperatures, there will be a smaller spread of the activity. In any case some degree of localization may be preserved. Heyman and Yaniv (1971) have described a theoretical model for the displacement of ^{222}Rn in which they predict a pile up of ^{222}Rn at the sunrise terminator of the moon.

Detection of radon is, in the Apollo experiment, based on the fact that alpha particles are produced in its decay. Tables 1 and 2 list the kinetic energies of the alpha particles that are emitted by the radon isotopes of the uranium and thorium series plus the energies of the alphas from their subsequent daughter products. Alpha

Table 1. Uranium (Radium) series showing in detail the origins, half-lives energies and relative intensities of the main α-groups starting with radon (^{222}Rn). The relative intensities are normalized to 100 decays of ^{222}Rn above the lunar surface.

$$^{238}_{92}U \xrightarrow{\alpha} {}^{234}_{90}Th \xrightarrow{\beta^-} {}^{234}_{91}Pa \xrightarrow{\beta^-} {}^{234}_{92}U \xrightarrow{\alpha} {}^{230}_{90}Th \xrightarrow{\alpha} {}^{226}_{88}Ra \xrightarrow{\alpha} {}^{222}_{86}Rn$$

Isotopes	Half-lives	α-Energies (MeV)	Relative intensities
^{222}Rn	3.823 days	5.490	100
^{218}Po	3.05 min	6.002	50
^{214}Pb	26.8 min	$\beta-$	—
^{214}Bi	19.7 min	$\beta-$	—
^{214}Po	164×10^{-6} sec	7.687	50
^{210}Pb	21 yr	$\beta-$	—
^{210}Bi	5.01 days	$\beta-$	—
^{210}Po	138.4 days	5.305	50
^{206}Pb	Stable		

Table 2. Thorium series showing in detail the origins, half-lives, energies, and relative intensities of the main α-groups, starting with Thoron (^{220}Rn). The relative intensities are normalized to 100 decays of ^{222}Rn above the lunar surface and assuming at 7:1 ratio for the ^{222}Rn/^{220}Rn, as reported by Turkevich *et al.* (1970).

$$^{232}_{90}\text{Th} \xrightarrow{\alpha} {}^{228}_{88}\text{Rn} \xrightarrow{\beta^-} {}^{228}_{84}\text{Ac} \xrightarrow{\beta^-} {}^{228}_{90}\text{Th} \xrightarrow{\alpha} {}^{224}_{88}\text{Ra} \xrightarrow{\alpha} {}^{220}_{86}\text{Rn}$$

Isotopes	Half-lives	α-Energies (MeV)	Relative intensities
^{220}Rn	55 sec	6.287	14
^{216}Po	0.158 sec	6.777	7
^{212}Pb	10.64 hr	β−	—
^{212}Bi	60.0 min	6.090	0.7
		6.051	1.8
^{212}Po	0.304×10^{-6} sec	8.785	4.5
^{208}Tl	3.10 min	β−	—
^{208}Pb	Stable	—	—

particles from the decay of radon above the lunar surface will be seen at their full energy for there can be no significant slowing down in the lunar atmosphere. When the radon decay is such that an alpha particle is emitted upward a recoil nucleus with a kinetic energy of about one hundred kiloelectron volts is deposited on the lunar surface. The distance in which the heavy recoil nucleus is brought to rest is very much smaller than the range of the alpha particles that will be emitted subsequently. Hence the active deposit is itself a source of monoenergetic alpha particles. On the other hand, no alpha particles will reach the surface from radon which decays at a depth exceeding the alpha particle range. Typically, this is about 10 μ. Thus, alphas which are emitted between 0 and 10 μ are degraded in energy. Hence, the intensity of monoenergetic alpha particle emission is highly dependent on the effectiveness of the diffusion process.

EQUIPMENT

The spectrometer is designed to measure the energy of any incident alpha particles even in the presence of other energetic charged particles. The sensing elements are ten totally depleted silicon surface barrier detectors. They are each approximately 100 μ thick, 3 cm^2 active area, have a 90° field of view, and operate at -50 V bias. Additional gold, aluminum, and nickel layers were used at the contacts to assure light tight performance. The thickness of the detectors was chosen so that any background protons (deuterons or tritons) would give an output pulse of less than that for a 5 MeV alpha particle while the output for alpha particles up to 12 MeV would be linearly proportional to energy. This precludes the necessity for discriminating against other particles in any other way.

The ten detector pre-amplifier outputs are merged in a single summing amplifier and processed by a single analog to digital converter (ADC). While the use of one ADC minimizes the complexity of the hardware, it also means that the noise from all ten preamps is summed, resulting in a resolution degradation of about a factor of three. To circumvent this, each preamplifier has a bias offset of approximately 350 keV. This effectively removes the noise and allows the use of a single ADC without resolution degradation.

The ADC converts the energy pulse into a 9-bit digital signal. If the most significant bit is a 1, the ADC is disabled and the digital signal held until the next telemetry readout (every 100 milliseconds). If the most significant bit is zero, the ADC is reset and the next pulse is processed. This allows the instrument to digitize to a 9-bit accuracy and only transmit 8 bits.

This means that only the upper half of the digitized energy range is telemetered. Physically this is reasonable since the alpha energies of interest range from 5.3 to 8.8 MeV and it also prevents the usage of telemetry time by any low energy background. The actual telemetered energy range of the instrument was from 4.7 to 9.1 MeV. Parallel circuitry generates an analog signal from 0.25 to 4.75 V, in steps of 0.5 V, which identifies the detector which originated any given pulse.

Since the digital telemetry is limited to 80 bits/second (10 counts/second), an additional circuit is used which generates an analog signal proportional to the time from the end of one telemetry read cycle to the sensing of the first pulse with energy greater than 4.7 MeV. This allows the dead time correction of the data. Exclusive of housekeeping, the output consists of an 8-bit energy word, an analog voltage identifying the detector, and an analog voltage exponentially proportional to the count rate.

Five of the detectors had energy calibration sources in their field of view. The sources were ^{208}Po, alpha energy 5.114 MeV. The count rate of these sources was approximately 0.1 counts/second. An additional energy calibration comes from a small amount of ^{210}Po that was accidentally deposited on the detector surface during testing. This contamination was on all ten detectors in varying amounts. The worst case was approximately 0.047 counts/second and the best had an undetectable quantity at launch.

The spectrometer was turned on at 15:47 GMT, July 29, and remained on until 12:43 GMT, August 7, except for short periods during major burns, water and urine dumps. The spectrometer functioned as expected during this period except for occasional bursts of noise in two detectors.

RESULTS

An early examination of the data indicated that the amount of alpha activity observed was very small with no obviously high signal regions. Thus to increase sensitivity, the data were overlayed with the orbital period and examined within time intervals consistent with the field of view of the instrument on the lunar surface. The data examined to date were only post-LM descent to avoid any possible extra background due to the radioactive thermal generator attached to the LM. There are three distinct types of signals that can be searched for. First would be alpha particles detected at energies consistent with the decay of ^{222}Rn and her daughter products. The ratio of intensities in the ^{222}Rn line and the daughter product lines (except ^{210}Po) is predictable and provides a consistency check for any observed signal. Since the production of ^{210}Po is held up by the 21 year half-life of ^{210}Pb, the ratio of alpha particles from the decays of ^{222}Rn and ^{210}Po will depend on the time history of the radon evolution. The second type of signal that could be expected would be from ^{220}Rn and her daughter products. Again, since the daughter products come to equilibrium with ^{220}Rn within about 20 hours, the relative intensities of all the alpha lines can be reasonably predicted.

The third type of signal would be alphas from ^{210}Po only. These events indicate radon evolution which has occurred days to years previously. The radon and all daughters (except ^{210}Po) have died out and only ^{210}Po whose production is held up by the 21 yr half-life of ^{210}Pb remains.

At this point in time the results of our analysis are preliminary. The results will eventually be more complete and precise so this paper is essentially a report of our current progress.

^{222}Rn

We have used two methods to look for an excess in the number of counts whose energies correspond to alpha particles from the decay of ^{222}Rn. The first method is comparing the total number of counts obtained with the detector in a lunar orientation with that of a deep space orientation and restricting the comparison to the appropriate energy channels. The second method consists of examining the total energy spectrum in a lunar orientation and looking for an increase in those energy channels when alphas from ^{222}Rn are expected as compared to a background level that is given by neighboring channels. In the absence of radon emanation essentially all counts are due to cosmic ray interactions and the spectrum is expected to be uniform and featureless. With the first method we find for the region 40°E to 180°E an excess of $(2.9 \pm 1.1) \times 10^{-3}$ counts/sec in our detector corresponding to a lunar rate of $(1.32 \pm 0.50) \times 10^{-3}$ dis/cm^2-sec. From the second method we find for the same region a rate of $(0.92 \pm 0.25) \times 10^{-3}$ dis/cm^2-sec. Thus on the basis of the statistical significance there is evidence for the existence of ^{222}Rn over a large part of the moon. However, pending a more thorough examination of systematic errors such as uncertainties in the precise energy calibration of the detectors and possible live time corrections, we present these values as an upper limit rather than a finite result.

^{220}Rn

The second method, as described above, is more powerful in the case of looking for an indication of ^{220}Rn. No excess counts are observed in the energy region of ^{220}Rn disintegration in an energy spectrum consisting of 20 hours of data taken over all the lunar surface. The 3σ upper limit to the average decay rate of ^{220}Rn is 3.8×10^{-4} decays/cm^2-sec. This does not preclude the possibility of finding local concentrations of ^{220}Rn that exceed this limit.

^{210}Po

The result of the analysis does provide some evidence for a nonuniform distribution of ^{210}Po on the surface of the moon. An example is the region of the moon overflown by the CSM during revolutions 18–33 of Apollo 15.

Our sensitivity for detecting lunar surface concentrations of ^{210}Po was reduced by the fact that the detectors were slightly contaminated by an external source of ^{210}Po during a calibration procedure. However, it is still possible to look for variations of the total count rate of ^{210}Po with lunar longitude. The contamination level will be constant with position on the moon so real changes in count rate in the appropriate energy range can be interpreted as a lunar component.

A lunar region is within the field of view of the instrument for a multiple number of orbits, the exact number of orbits depending on lunar latitude. Thus, to look for local concentrations we have combined data from a number of orbits by folding over the orbital period of the Apollo spacecraft. The folded data from revolutions 18–33 were grouped in bins of 20° of lunar longitude. The count rates in various energy bands of the spectrum were then examined as a function of lunar longitude. Because the amount of contamination varied from detector to detector, those five detectors with the least contamination were examined separately and provide most of the sensitivity.

For those five detectors the count rate from 5.1 to 5.5 MeV tended to be systematically higher in the region of 40°E to 180°E. To increase the statistics all data from 32 consecutive hours were added together. Figure 1 shows the count rate in the region from 5.1 to 5.5 MeV as a function of longitude for revolutions 18–33. The data between 40°E and 180°E which extends from the western edge of Mare Crisium to the Van de Graaff–Orlov region are systematically higher than that in other longitude bins. The average count rate between 40°E and 180°E is 0.072 ± 0.002 counts/sec and between 0–40°E plus 180°W–0°W is 0.062 ± 0.001 counts/sec. These average count rates are shown as dashed lines on the figure. The excess between 40°E and 180°E corresponds to $(4.6 \pm 1.4) \times 10^{-3}$ dis/cm^2-sec.

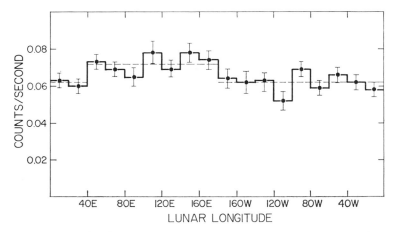

Fig. 1. Count rate of alpha particles with energy from 5.1 to 5.5 MeV for the five least contaminated detectors as a function of lunar longitude. This energy range includes ^{210}Po, a descendent of ^{222}Rn. This includes all available data from revolutions 18–33. The dashed lines indicate the average value of the count rate over two sections of longitude as indicated.

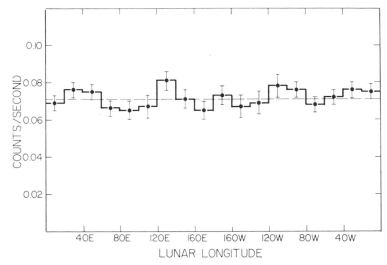

Fig. 2. Count rate of alpha particles with energy from 4.7 to 5.1 or 5.5 to 9.0 MeV for the five least contaminated detectors as a function of lunar longitude. This includes all available data from revolutions 18–33. The dashed line indicates the average count rate.

To verify that this is a true lunar signal and not a systematic live-time variation, all the data outside this energy bin were treated in a similar fashion. The results are shown in Fig. 2. These data do not show this variation. The count rate between 40°E and 180°E is 0.070 ± 0.002, and between 0–40°E plus 180°W–0°W it is 0.073 ± 0.001 counts/sec. If the variation seen in Fig. 1 were due to a live-time variation, both of these energy bins should be affected identically. Since they are not, we conclude that the excess in Fig. 1 is due to a true lunar signal. The energy spectra of the counts indicates that the excess is due to ^{210}Po only.

Summary and Discussion

A partial analysis of Apollo 15 orbital data from the alpha particle spectrometer indicates that a nonuniform concentration of ^{210}Po is found on the surface of the moon. For an area extending from west of Mare Crisium to the Van de Graaff–Orlov region we find an excess ^{210}Po activity of $(4.6 \pm 1.4) \times 10^{-3}$ dis/cm²-sec. This amount is approximately one-ninth of the value reported by Turkevich et al. (1970) for the Surveyor 5 landing site and about a factor of ten higher than the upper limit reported by Lindstrom et al. (1971) for a rock sample returned by Apollo 11.

Our value for the activity of ^{222}Rn, the progenitor of ^{210}Po, is at most 10^{-3} dis/cm²-sec, about a factor of nine smaller than the quality of ^{222}Rn that would be in equilibrium with the observed amount of ^{210}Po. This implies that the ^{210}Po activity at the time of Apollo 15 was a result of increased radon emanation from the moon within a time scale comparable to the 21 yr required for ^{222}Rn to decay to ^{210}Po. Referring to earth analogies, this is suggestive of transient radon emanation from the moon that would be promoted by transient lunar emission of more common volatiles or by volcanic activity.

Acknowledgment—This research was supported by The Manned Spacecraft Center under NASA Contract NAS9-9982.

REFERENCES

Economou T. E. and Turkevich A. L. (1971) Examination of returned Surveyor III camera visor for alpha radioactivity. *Proc. Second Lunar Sci. Conf., Geochim. Cosmochim. Acta* Suppl. 2, Vol. 3, pp. 2699–2703. MIT Press.

Heymann D. and Yaniv A. (1971) Distribution of radon-222 on the surface of the moon. *Nature Phys. Sci.* **233,** 27.

Kraner H. W., Schroeder G. L., Davidson G., and Carpenter J. W. (1966) Radioactivity of the lunar surface. *Science* **152,** 1235.

Lindstrom R. M., Evans J. C., Finkel R. C., and Arnold J. R. (1971) Radon emanation from the lunar surface. *Earth Planet. Sci. Lett.* **11,** 254.

Turkevich A. L., Paterson J. H., Franzgrote E. J., Sowinski K. P., and Economou T. E. (1970) Alpha radioactivity of the lunar surface at the landing sites of Surveyors 5, 6, and 7. *Science* **167,** 1722.

Yeh R. S. and Van Allen J. A. (1969) Alpha-particle emissivity of the moon: An observed upper limit. *Science* **166,** 370.

Proceedings of the Third Lunar Science Conference
(Supplement 3, *Geochimica et Cosmochimica Acta*)
Vol. 3, pp. 2189–2204
The M.I.T. Press, 1972

Analysis and interpretation of lunar laser altimetry

W. M. Kaula, G. Schubert, and R. E. Lingenfelter

University of California, Los Angeles

W. L. Sjogren

Jet Propulsion Laboratory, Pasadena, California

and

W. R. Wollenhaupt

Manned Spacecraft Center, Houston, Texas

Abstract—About $4\frac{1}{2}$ revolutions of laser altimetry were obtained by Apollo 15. This altimetry indicates a 2 km displacement of the center of mass from the center of figure toward the earthside. The terrae are quite rough, with frequent changes of 1 km or more in successive altitudes at about 33 km intervals. The mean altitude of terrae above maria is about 3 km with respect to the center of mass, indicating a thickness of about 24 km for a high-alumina crust. The maria are extremely level, with elevations varying not more than ± 150 m about the mean over some stretches of 200 to 600 km. However, different maria have considerably different mean elevations.

The largest unanticipated feature found is a 1400 km wide depression centered at about 180° longitude, and 2 km deep with respect to a 1737 km sphere (about 6 km deep with respect to the surrounding terrae). This basin has the appearance of typical terrae, though there are indications of a ring structure of about 600 km radius in the Orbiter photography. Altitudes across circum-Orientale features suggest that Mare Orientale is also a deep basin.

The data appear to corroborate a model of early large-scale differentiation of a crust, followed a considerable time later by short intense episodes of mare filling with low viscosity lavas.

Introduction

THE INCLUSION OF the laser altimeter in the Apollo 15 orbital science payload was originally motivated by the need to scale the metric photography taken by the three-inch mapping camera. However, it was soon apparent that the laser altimeter alone could obtain valuable information regarding topographic elevations on the moon, almost 2 orders of magnitude better than that previously available from earth-based photography. It therefore was proposed that an alternative mode of operation be adopted, in which the laser altimeter was fired regardless of whether photography could be taken.

The altimeter was designed and constructed by RCA Aerospace Systems Division of Burlington, Mass., under the leadership of Jason Woodward. It is a Q-switched ruby laser, mounted in the same frame as the three-inch mapping camera and operating in an altitude range of approximately 74 to 148 km. The pulse length is 10 nsec, and the pulse energy is 0.25 J. The beam width is 300 μrad, thus illuminating an area of about 30 m in diameter. The altitude accuracy is ± 2 m. Attitude is controlled within $\pm 2°$. When the laser is operated in synchronization with the mapping camera, the firing interval may vary from 16 to 32 sec, depending on spacecraft altitude. In separate operation the laser is fired every 20 sec, equivalent to about 33 km intervals at the moon's surface.

In the Apollo 15 flight, the laser operated intermittently, and eventually failed. About four and a half revolutions of partially overlapping data were obtained.

ALTITUDE DETERMINATION

In order to interpret the altimeter range measurements, an accurate estimate of the spacecraft position with respect to the moon at each measurement time is necessary. This is accomplished by use of Doppler frequency shift measurements made by the Manned Space Flight Network (MSFN) radar stations that track the spacecraft whenever it is in line of sight of the station. The location of the tracking stations and the earth-orbital geometry are such that the spacecraft, when not occulted by the moon, is in simultaneous view of at least two stations. The Doppler data are processed using a weighted least-squares technique to determine the moon-centered Cartesian components of the spacecraft orbit at a specified time. This determination is made with respect to the center of mass of the moon. The station observations are corrected for all known systematic errors, such as refraction effects, and timing differences between universal time and ground station time. A five-element spherical harmonic model, known as the L-1 model, is used to describe the lunar gravitational potential for both data processing and trajectory prediction. Coefficients of this model are $J_{20} = 2.07108 \times 10^{-4}$, $J_{30} = -0.21 \times 10^{-4}$, $C_{22} = 0.20716 \times 10^{-4}$, $C_{31} = 0.34 \times 10^{-4}$, and $C_{33} = 0.02583 \times 10^{-4}$. This model does not completely account for the observed gravitational effect and is the dominant error source in the orbital and prediction computations.

After the spacecraft orbit has been determined, the vehicle position with respect to the center of mass at altimeter measurement times is determined by using Cowell's method of numerical integration to integrate both the equations of motion and variational equations. The estimated altitude above the lunar surface is then computed with reference to a sphere having a radius of 1738 km. The altimeter slant range measurements are then converted to altitudes above the surface by accounting for the altimeter pointing angle deviations from the spacecraft local vertical. This correction has not been applied to the data presented in this paper; therefore, the uncertainties are larger than those expected for the final data reduction. The current uncertainties associated with the selenographic position of the laser altimeter points and the absolute radius values of these points are estimated to be 0.2° for both latitude and longitude and 640 m for radius. The uncertainty associated with the relative altitudes, on the lunar surface, derived from the altimeter measurement is 10 m.

A test of the analysis techniques and of the validity of the altimeter measurements was made by comparing the altimeter-derived radius values with those obtained by independent methods. Radius values were available for three relatively small craters located in Mare Smythii, Mare Serenitatis, and Palus Putredinis (Wollenhaupt et al., 1972). The altimeter-derived radius values and the crater values agreed to within 100 m. This can only be considered as a gross test because the laser measurements were not made on exactly the same features.

The numerical values for the altimeter measurement subvehicle points and radius deviations ΔR from a spherical moon of 1738 km radius about the center-of-mass, are presented in Table 1 for revolution 15–16, for which the most complete data is available (Wollenhaupt and Sjogren, 1972). A plot of the ground track or trace of the sub-vehicle point is shown in Figs. 1a, b. The altitude profile and the radius deviations from a 1738 km sphere are graphically presented in Figs. 2a, b.

ANALYSIS

"Broad" scale

A harmonic analysis of the data of Table 1 with respect to east longitude λ obtains for elevation h in kilometers:

$$h = -0.8 + 2.1 \cos (\lambda - 215°) + 0.7 \cos 2(\lambda - 90°). \qquad (1)$$

The constant term in equation (1) indicates that the mean lunar radius along this orbital track is 1737.2 km. A 2.1 km displacement of the center of mass from the center of figure in a direction 35° E of the earth-moon line is indicated by the first

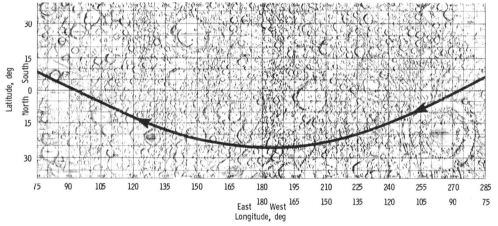

Fig. 1a. Laser altimeter groundtrack for revolution 15–16: lunar farside.

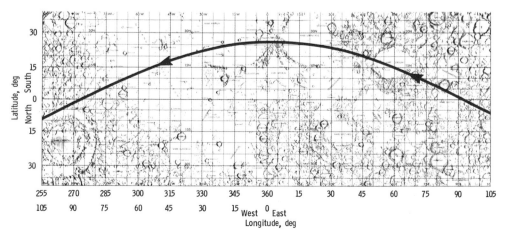

Fig. 1b. Laser altimeter groundtrack for revolution 15–16: lunar nearside.

W. M. KAULA *et al.*

Table 1. Laser altimeter subvehicle points and radius deviations from spherical moon of radius 1738 km.

Latitude (deg)	Longitude (deg)	ΔR (km)	Latitude (deg)	Longitude (deg)	ΔR (km)
8.5	291.4	−2.17	−20.8	221.8	2.40
8.0	290.4	−2.26	−21.1	220.6	2.79
7.5	289.2	−1.70	−21.4	219.4	3.36
7.0	288.1	−2.40	−21.8	218.1	3.41
6.5	287.0	−1.55	−22.0	216.9	1.70
6.0	285.9	−2.02	−22.3	215.6	3.38
5.4	284.8	−1.74	−22.6	214.4	2.76
5.0	283.7	−0.34	−22.9	213.0	2.23
4.4	282.6	−1.32	−23.1	211.8	1.78
3.9	281.5	0.17	−23.4	210.4	4.01
3.4	280.4	0.86	−23.6	209.2	1.64
2.9	279.3	−0.76	−23.8	208.0	0.34
2.4	278.2	−0.64	−24.0	206.7	3.96
1.9	277.2	−0.32	−24.2	205.5	−2.23
1.3	276.1	0.35	−24.4	204.3	−1.73
0.8	275.1	0.42	−24.5	203.2	−0.59
0.3	273.9	0.81	−24.7	201.6	−0.55
−0.3	272.8	0.01	−24.9	200.2	0.70
−0.9	271.8	0.84	−25.1	198.9	−0.52
−1.3	270.7	−0.27	−25.2	197.7	−0.82
−1.9	269.6	1.33	−25.3	196.4	−1.27
−2.4	268.5	2.45	−25.4	195.1	−3.04
−2.9	267.4	2.35	−25.5	193.8	−2.62
−3.4	266.3	1.96	−25.6	192.6	−2.29
−3.9	265.3	0.41	−25.7	191.2	−1.86
−4.4	264.2	4.13	−25.8	189.9	−3.31
−4.9	263.1	1.30	−25.8	188.7	−3.83
−5.5	262.0	1.70	−25.8	187.4	−1.20
−6.0	260.9	1.04	−25.9	186.0	−4.40
−6.5	259.8	0.97	−25.9	184.7	−4.86
−7.0	258.7	1.97	−25.9	183.4	−1.13
−7.5	257.6	0.94	−25.9	182.0	−4.62
−8.0	256.5	1.65	−25.8	178.2	−1.82
−8.5	255.4	1.90	−25.8	177.8	−4.17
−9.0	254.3	2.73	−25.7	176.3	−2.14
−9.5	253.2	4.30	−25.6	174.8	−4.86
−10.0	252.1	3.59	−25.6	173.5	−4.69
−10.5	251.0	2.29	−25.5	172.0	−3.93
−10.9	249.9	2.09	−25.4	170.6	−0.12
−11.4	248.8	2.48	−25.3	169.2	0.24
−11.9	247.6	2.25	−25.1	167.8	−2.15
−12.4	246.5	3.37	−24.9	166.4	−1.91
−12.8	245.3	3.76	−24.7	165.0	0.60
−13.3	244.2	2.68	−24.5	163.6	0.45
−13.7	243.1	2.87	−24.3	162.2	−2.92
−14.2	241.9	2.15	−24.1	160.9	1.47
−14.6	240.8	2.78	−23.9	159.5	−0.90
−15.1	239.6	5.20	−23.7	158.1	−3.09
−15.5	238.4	4.89	−23.4	156.7	2.21
−15.9	237.3	2.80	−23.2	155.4	2.79
−16.3	236.2	2.73	−22.9	154.1	4.12
−16.8	235.0	4.49	−22.7	152.8	2.38
−17.2	233.8	4.73	−22.4	151.4	−2.35
−17.6	232.6	3.57	−22.1	150.1	−0.91
−17.9	231.5	2.65	−21.8	148.7	−0.68
−18.3	230.2	2.88	−21.5	147.4	−0.35
−18.7	229.1	2.31	−21.2	146.1	0.28
−19.1	227.9	4.10	−20.8	144.9	4.75
−19.5	226.7	1.91	−20.5	143.6	1.78
−19.9	225.4	2.94	−20.1	142.2	0.90
−20.1	224.2	2.48	−19.7	141.0	2.57
−20.4	223.0	2.58	−19.4	139.6	1.44

Table 1. (continued).

Latitude (deg)	Longitude (deg)	ΔR (km)	Latitude (deg)	Longitude (deg)	ΔR (km)
−19.0	138.4	2.38	12.2	66.6	−4.34
−18.6	137.3	0.31	12.7	65.5	−3.73
−18.2	135.8	−1.61	13.1	64.3	−4.53
−17.8	134.6	2.43	13.6	63.2	−4.67
−17.4	133.3	1.46	14.0	62.0	−4.83
−17.0	132.1	2.27	14.4	60.8	−4.78
−16.5	130.8	1.46	14.9	59.7	−4.75
−16.1	129.6	2.46	15.3	58.5	−4.59
−15.6	128.4	1.64	15.7	57.4	−4.60
−15.2	127.1	2.09	16.2	56.3	−4.70
−14.7	125.9	1.87	16.6	55.1	−4.69
−14.3	124.7	0.25	17.0	53.9	−4.66
−13.8	123.5	0.38	17.4	52.7	−4.65
−13.3	122.3	0.80	17.8	51.6	−4.29
−12.8	121.1	1.19	18.2	50.4	−4.18
−12.3	119.9	−0.74	18.5	49.2	−1.44
−11.8	118.7	1.18	18.9	48.0	−2.65
−11.4	117.5	3.40	19.3	46.8	−1.55
−10.9	116.3	0.63	19.6	45.6	−1.34
−10.4	115.2	1.11	19.9	44.4	−1.56
−9.9	114.0	1.24	20.3	43.3	−1.97
−9.3	112.8	2.16	20.6	42.1	−1.78
−8.8	111.7	0.46	20.9	40.9	−4.21
−8.3	110.5	−1.00	21.2	39.6	−2.32
7.8	109.4	−1.22	21.5	38.4	−2.33
−7.2	108.2	0.82	21.8	37.1	−2.33
−6.7	107.0	0.74	22.1	35.9	−2.50
−6.2	105.8	0.57	22.4	34.6	−2.65
−5.6	104.7	2.23	22.7	33.4	−1.54
−5.1	103.6	1.41	22.9	32.1	−1.90
−4.6	102.4	1.01	23.2	30.8	−1.60
−4.0	101.2	1.04	23.4	29.6	−3.55
−3.4	100.1	0.04	23.6	28.3	−3.88
−2.8	98.9	0.15	23.9	27.0	−3.81
−2.3	97.8	−3.52	24.1	25.8	−3.68
−1.8	96.6	−2.18	24.3	24.5	−3.57
−1.1	95.4	−0.54	24.5	23.2	−3.49
−0.5	94.3	−1.04	24.7	21.9	−3.46
−0.1	93.2	−3.09	24.8	20.6	−3.34
0.4	92.1	−4.63	24.9	19.4	−3.47
0.9	91.0	−4.78	25.1	17.9	−3.46
1.5	89.9	−4.90	25.2	16.6	−3.42
2.0	88.8	−4.87	25.3	15.3	−3.36
2.5	87.6	−4.75	25.4	14.0	−3.31
3.0	86.5	−4.94	25.5	12.7	−3.27
3.5	85.4	−4.81	25.6	11.4	−3.30
4.1	84.3	−2.57	25.7	10.1	−3.32
4.7	83.1	−1.68	25.8	8.7	−3.22
5.2	82.0	−0.98	25.9	7.4	−2.11
5.8	80.8	−4.12	25.9	6.0	−0.76
6.3	79.7	−3.70	25.9	4.7	0.43
6.9	78.5	−2.07	26.0	3.4	−2.30
7.5	77.3	−2.84	26.0	2.1	−2.61
8.0	76.1	−3.07	26.0	0.8	−2.40
8.5	75.0	−0.68	25.9	359.4	−2.45
9.1	73.8	0.06	25.9	358.1	−2.43
9.6	72.6	−1.71	25.8	356.8	−2.22
10.1	71.5	−0.86	25.8	355.4	−1.17
10.6	70.3	−1.07	25.7	354.1	−0.54
11.1	69.1	−3.24	25.6	352.8	−2.14
11.7	67.8	−2.17	25.5	351.4	−2.24

Table 1. (continued).

Latitude (deg)	Longitude (deg)	ΔR (km)	Latitude (deg)	Longitude (deg)	ΔR (km)
25.4	350.2	−2.72	14.4	304.4	−1.24
25.3	348.8	−2.54	14.0	303.4	−1.59
25.2	347.5	−2.70	13.6	302.4	−1.60
25.0	346.2	−2.74	13.2	301.4	−1.95
24.9	344.9	−2.70	12.8	300.4	−2.09
24.7	343.6	−2.61	12.4	299.5	−2.02
24.5	342.3	−2.48	12.0	298.5	−1.97
24.3	341.0	−2.38	11.6	297.6	−1.91
24.1	339.7	−2.27	11.2	296.6	−1.85
23.9	338.4	−2.26	10.8	295.7	−1.96
23.6	337.1	−2.22	10.4	294.8	−1.80
23.4	335.8	−2.07	10.0	293.8	−1.80
23.2	334.5	−1.93	9.6	292.8	−1.84
22.9	333.3	−1.83	9.2	291.9	−1.81
22.7	332.0	−1.74	8.8	291.0	−1.53
22.4	330.8	−1.66	8.3	290.0	−1.36
22.1	329.4	−1.57	7.9	289.1	−0.75
21.8	328.2	−1.68	7.4	288.1	−1.74
21.5	327.0	−1.67	7.0	287.1	−1.38
21.2	325.8	−1.66	6.5	286.2	−0.92
20.9	324.5	−1.70	6.1	285.3	−1.20
20.6	323.2	−1.72	5.7	284.4	−0.32
20.3	322.0	−1.63	5.3	283.4	0.01
19.9	320.8	−1.77	4.9	282.5	−0.54
19.5	319.6	−1.68	4.4	281.5	1.15
19.2	318.4	−1.83	4.0	280.5	1.13
18.8	317.2	−1.82	3.6	279.6	0.34
18.4	316.0	−1.83	3.1	278.7	−0.04
18.0	314.8	−1.88	2.7	277.8	−0.60
17.6	313.5	−1.85	2.2	276.9	0.12
17.2	312.2	−1.96	1.7	275.9	0.21
16.8	311.0	−1.98	1.3	275.0	0.57
16.4	309.8	−1.93	0.8	274.1	1.22
16.0	308.6	−1.21	0.3	273.1	0.32
15.5	307.5	−1.45	−0.1	272.2	0.51
15.2	306.4	−1.41	−0.5	271.3	1.07
14.8	305.4	−1.46	−1.0	270.3	0.49

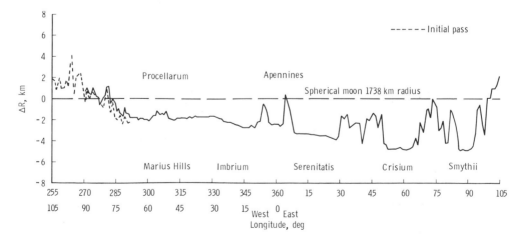

Fig. 2a. Altitude profile and radius deviations from a spherical moon: lunar farside.

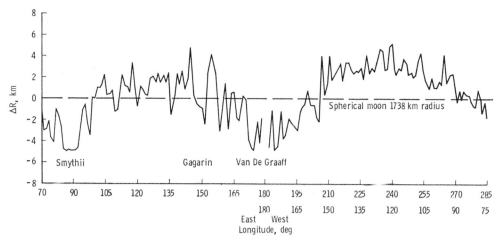

Fig. 2b. Altitude profile and radius deviations from a spherical moon: lunar nearside.

harmonic term. The second harmonic is equivalent to a flattening in the direction of the earth of about 0.7 km.

A number of investigations carried out prior to the availability of the Apollo 15 laser altimetry data have also indicated that the center of mass of the moon is displaced from the center of figure toward the earth by approximately 2 km. From the analysis of the impact times of the Ranger vehicles (Sjogren, 1967), it was determined that the lunar surface was about 2–3 km closer to the center of mass at the impact point than earth-based telescope models had indicated. Similar results have been obtained from earth-based radar bounce data (Shapiro *et al.*, 1968), the triangulations done on the Lunar Orbiter photographs (U.S. Army Topocommand, 1967–1968, 1970; U.S. Air Force A.C.I.C., 1967–1968, 1969–1970), elevations from Apollo landmark tracking data (Wollenhaupt and Ransford, 1971), data from the V/H sensor on Lunar Orbiter I (Wells, 1971), lunar laser retroreflector data (Garthwaite *et al.*, 1970), and gravity measurements at the sites of Apollo 11, 12, and 14 (Nance, 1971).

The data of Table 1, classified according to terrain type (see the following section), show that the terrae are elevated approximately 3 km with respect to the maria. Baldwin (1949, 1963) and Runcorn and Shrubsall (1968) have noted that separate analyses of the figures of the maria and terrae are important in understanding the past and present physical state of the moon's interior. We have therefore computed best-fit ellipses to the data of Table 1 for maria and terrae separately. Unfortunately, we are limited in data, especially for the maria points. In addition to the frontside maria, only a small number of altimeter readings over mare-type material on the farside are available. These include several points over Tsiolkovsky and Van de Graaff. However, it is not even clear that the isolated patches of mare-type material on the farside should be classified with the great mare basins of the frontside.

The values of the ellipse parameters that best fit the maria are sensitive to which of the farside maria points are included in the data (e.g., we have fitted ellipses to data that do not include any farside points and to data that include Tsiolkovsky and Van de Graaff both separately and together). In particular, the radii in the earth-moon direction and the center-of-mass, center-of-figure offsets in this direction are highly variable and strongly correlated. These offsets are observed to change sign according to which farside points are included. Certain characteristics of these ellipses are reasonably well behaved. The radii in a direction perpendicular to the earth-moon line are approximately 1734.7 km and the centers of mass are displaced to the east of the centers of figure by about 1.7 km. The ellipses are in fact elongated along the earth-moon line although the fit to the frontside maria plus several points in Van de Graaff is nearly circular.

The significance of these maria fits is a matter that requires considerable additional investigation. The results to date indicate that the maria do not lie on a present equipotential surface.

"Fine" scale

Table 2 gives a summary of areas measured by terrain type, with the limiting longitudes for revolution 15–16. The lateral spread of data about revolution 15–16 from other revolutions varies from 0 km at longitudes 0° and 180° to about 360 km at longitudes 90°W and 240 km at longitude 95°E.

The terrain types in Table 2 are based on the usual visual indicators: albedo; crater size, frequency, and freshness; hummock amplitude, wavelength, and direction; etc. One exception is that the dividing line between two terrae regions at longitude 206° is based on the altimetry itself.

When the refined data including attitude corrections are available, it is planned to refer altitudes to an equipotential rather than a sphere, to map and classify it in greater detail, and to characterize the variability of elevations by power spectra based on

Table 2. Summary of areas measured.

Region	Terrain type	Longitudes (deg)	Rev. 15–16 altitudes (km) resp. 1737 sphere			Points	No. other revs.
			Low	Mean	High		
Apennines	Circum-mare ring	4–8	−1.11	+0.19	+1.43	3	1
Serenitatis	Smooth mare	8–30	−2.88	−2.47	−2.22	17	1
Littrow-Römer vic.	Terrae	30–35	−1.65	−0.92	−0.54	4	1
Tranquillitatis	Mare	35–40	−1.50	−1.37	−1.32	4	1
Macrobius vic.	Terrae	40–50	−3.21	−1.06	−0.34	8	1
Crisium	Smooth mare	50–65	−3.83	−3.61	−3.18	13	1
Condorcet vic.	Terrae	65–85	−3.34	−1.28	+1.06	17	1
Smythii	Smooth mare	85–93	−3.94	−3.81	−3.63	7	1
Farside Highlands	Rough terrae	93–144	−2.52	+1.72	+4.40	43	6
Gagarin	Crater rim & floor	144–153	−1.35	+1.45	+5.75	7	6
Barbier-Paracelsus vic.	Rough terrae	153–171	−2.09	+1.06	+5.12	13	8
Van de Graaff	Crater floor	171–175	−3.86	−3.49	−2.93	3	8
Nassau-Plummer vic.	Depression	175–206	−3.86	−1.20	+1.70	22	8
Farside Highlands	Rough terrae	206–240	+1.34	+4.00	+6.20	28	7
Orientale Ejecta	Ejecta	240–252	+3.09	+3.67	+4.76	10	4
Montes Cordillera	Circum-mare ring	252–269	+1.41	+3.09	+5.30	16	4
Orientale Ejecta	Ejecta	269–283	+0.40	+1.49	+2.33	5	4
Olbers-Cavalerius vic.	Terrae-mare trans.	283–291	−0.74	−0.02	+1.01	9	2
Oceanus Procellarum	Mare	291–302	−1.09	−0.91	−0.80	11	2
Marius Hills	Domed & wrinkled	302–309	−0.60	−0.42	−0.21	7	2
Oceanus Procellarum	Mare	309–329	−0.98	−0.79	−0.63	16	2
Outer Imbrium	Mare	329–351	−1.74	−1.26	−0.57	17	1
Archimedes Ejecta	Ejecta	351–357	−1.24	−0.66	−0.46	5	1
Palus Putredinis	Smooth mare	357–4	−1.61	−1.44	−1.30	5	1

covariance analyses, etc. However, even in the preliminary data, there are significant variations with terrain type. We select two measures of these variations: "levelness" and "roughness."

It is evident from the altitudes in Table 2 that there are both considerable differences in mean elevations between regions and variations about the mean within a region. The eight regions listed as of maria type in Table 2 range from −0.79 to −3.81 km in mean elevation with respect to the 1737 km sphere. However, with the exception of outer Imbrium, none of these eight regions contains an elevation differing more than 0.5 km from its mean. In several places, the altitudes do not differ more than ±0.2 km from their means over stretches of some hundreds of kilometers; see, for example, Fig. 3. Conversely, the terrae regions virtually all contain elevations differing 2 km or more from their means.

From Table 2 we obtain a mean elevation of the terrae above the maria of 3 km. This value applies for elevations with respect to the center of mass, which should be used for calculating isostatic implications, etc. For elevations with respect to the center of figure, the mean elevation is 1.5 km, in fortuitously good agreement with the result of Runcorn and Shrubsall (1968).

Another distinct difference between regions is the magnitude of the change in elevation over the minimum distance sensed by the altimeter, 33 km. For want of a

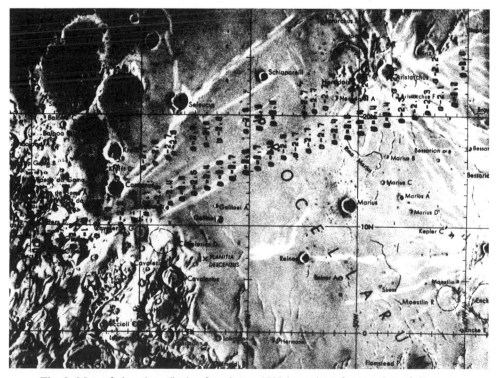

Fig. 3. Map of elevations (km) referred to a 1738 km sphere for a region of Oceanus Procellarum. The more northerly track represents data from revolution 33–34, the other track is for revolution 21–22.

Table 3. "Roughness." Percent frequency of altitude changes in successive points 20 sec (about 33 km) apart, by region and terrain type. A "reversal" is an altitude change of opposite sign to the previous one in an arbitrary sequence; a "continuation," one of same sign.

| | | Percent | | | | | | | |
| | | Reversals | | | | Continuations | | | |
Type	Longitudes (deg)	>1.5	1.5–0.6	0.5–0.2	≤0.16	≤0.16	0.2–0.5	0.6–1.5	>1.5
Maria	329–93			4	27	67	2		
Terrae	30–85	22	22	26			13	9	9
High terrae	93–171	24	18	10	8	2	13	10	15
Low terrae	171–206	35	9	9		4	26	13	4
High terrae	206–240	31	15	15	4	8	12	12	4
Orientale structures	240–291	6	33	14	4	4	17	17	4
Maria	291–329				65	35			

better word, we call this change "roughness." Table 3 is a summary divided into seven groups of regions, in order to have statistically significant samples. Here the maria are again remarkably level, nearly all of the changes in successive points are 160 m or less. Conversely, the terrae are rougher, having 60% or more of their changes greater than 600 m. In addition the terrae have a predominance of "reversals," whereas the maria, with the exception of Oceanus Procellarum (291°–329°), have more "continuations."

INTERPRETATION

"Broad" scale

The most prominent feature of the broad variations expressed by equation (1) is the offset of more than 2 km of the center of mass from the center of figure toward the earth. This offset cannot be accounted for in terms of the frontside mascons however. The total excess mass associated with the ringed maria Imbrium, Serenitatis, Crisium, Nectaris, and Humorum is about 1.5×10^{21} gm (Kaula, 1969). This excess mass could produce a center-of-mass, center-of-figure offset of about $(1.5 \times 10^{21}/7.35 \times 10^{25}) \times (1738)$ km or 35 m only.

The magnitude of the observed offset suggests that the differentiation of high alumina terrae (Adler et al., 1972) which is believed to have occurred within the first 10^8 years on petrological and radiochronological grounds (Papanastassiou and Wasserburg, 1971; Ringwood, 1970; Wetherill, 1971; Wood, 1970) entailed rather deep and broad overturns and produced a crust of variable thickness, thicker on the farside than on the nearside. The second and higher harmonic residuals in the gravitational field of these overturns have not persisted, because they cause greater stresses than a comparable first degree harmonic mass distribution. In fact, a first-degree harmonic surface load on an incompressible sphere will cause no stress at all but merely offset the center of mass (Jeffreys, 1970). The variable thickness crust essentially floats on the underlying mantle. The situation is not unlike what we find on the earth in the nonuniform distribution of continents over the globe and the average difference in elevation between continents and ocean basins.

The simplest analytical model that explains the offset is shown in Fig. 4, where the sketch has been exaggerated for clarity of presentation. The model consists of a spherical moon with an offset spherical "core." The "crust" has density ρ_0 and the "core" has density $\rho_0 + \Delta\rho$. With $\bar{x}$ the center-of-mass, center-of-figure offset, $\bar{\rho}$ the average density of the model, δ the distance separating the centers of the spheres, and r_0 and r_i the radii of the outer sphere and core, respectively, we find that

$$\bar{x} = \delta \frac{\Delta\rho}{\bar{\rho}} \left(\frac{r_i}{r_0}\right)^3. \tag{2}$$

In terms of the frontside crustal thickness t_f and the backside crustal thickness t_b, it follows that

$$\delta = (t_b - t_f)/2, \tag{3}$$

$$\frac{r_i}{r_0} = 1 - \frac{(t_b + t_f)}{2r_0}, \tag{4}$$

and the offset $\bar{x}$ is

$$\bar{x} = \frac{\Delta\rho}{\bar{\rho}} \left(\frac{t_b - t_f}{2}\right) \left(1 - \frac{t_b + t_f}{2r_0}\right)^3. \tag{5}$$

For the case $(t_b + t_f)/2r_0 \ll 1$, we find

$$t_b - t_f \approx \frac{2\bar{x}}{\Delta\rho/\bar{\rho}}, \tag{6}$$

if $\bar{x} = 2.1$ km, $\Delta\rho = 0.4$ gm/cm^3, and $\bar{\rho} = 3.34$ gm/cm^3 then the difference between the backside and frontside crustal thicknesses $t_b - t_f$ is 35 km. The thicker backside crustal material is, of course, to be associated with the extensive farside highlands and their roots for isostatic compensation.

Another estimate of crustal thickness is obtainable from the mean difference in elevation of the terrae and the maria, 3 km with respect to the center of mass. The absence of a significant positive gravity anomaly in the terrae with respect to the maria (see Muller and Sjogren, 1968) requires that the terrae be isostatically compensated. The terrae crustal thickness required t_T is

$$t_T = \frac{h\rho_M}{\rho_M - \rho_T}, \tag{7}$$

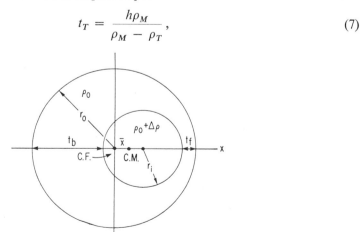

Fig. 4. A simple offset "core" model to explain the center-of-mass, center-of-figure separation.

where h is the mean maria-terrae elevation difference, ρ_M is the density of the mantle, and ρ_T is the density of the terrae. If $h = 3$ km, $\rho_M = 3.34$ gm/cm³, and $\rho_T = 2.94$ gm/cm³, then t_T is 25 km. This figure is higher than the mean of 17.5 km obtained from $(t_b - t_f)/2$ because t_f is not necessarily zero, and because the first degree harmonic does not represent all of the geographic distribution of terrae with respect to maria. Ransford and Sjogren (1972) suggest an alternate model.

The angle of 90° between the long axis of the ellipse fitting the figure and the dynamical principal axis is merely another manifestation of isostasy: The less dense terrae are concentrated near the limbs and the more dense maria near the principal axis. Ellipses fitted to the maria alone do have their long axes in the approximate direction of the earth, which suggests some sort of filling to a past tidal equipotential. Aside from the problems enumerated in the previous section, a serious difficulty with this interpretation is the lack of coincidence of the maria ellipses with the present center of mass. This would have required a shift of center of mass since the time of filling of the mare basins, in turn entailing an extent of asymmetric internal redistribution of mass quite inconsistent with the geophysical indications of present quiescence (Sonett *et al.*, 1971; Latham *et al.*, 1971).

Particular features

The most remarkable new feature is the great depression on the farside, extending from about longitude 160°E to 155°W, with depths up to 6 km with respect to a 1737 km sphere. These depths are as great as any found for the earthside maria, yet the terrain has the roughness and albedo characteristic of terrae (see Fig. 5). The evident suggestion is that the region is a basin which did not get flooded, as did the earthside maria. Baldwin (1969), Hartmann and Wood (1971) on the basis of Lunar Orbiter photographs, have pointed out a number of extremely large, ancient farside craters. However, none of the features they identified correlate with the large depression on the backside recorded by the laser altimeter. With the aid of our knowledge of the altimeter track and the farside topography along this track, we have examined the farside Orbiter photographs for visual evidence of this unfilled basin. We believe that such a basin may be defined by the remains of a ring structure best seen on Orbiter photo II-34M. Nearly an entire semicircle of the ring can be seen on this photograph. The arc follows a course from a point about midway between Hayford and Krasovsky, through the northwest rim of Stein, through the eastern rim of Coriolis, and thence to just east of Heaviside. The ring appears to be centered at about 10°S and 170°W and has a radius of about 600 km. The basin defined by this partial ring would be consistent with the depression found in the laser altimeter data along the orbital track.

On the other hand the depression indicated by the altimetry data may be just the outer edge of a much larger basin suggested by Zond-6 limb photography (Rodionov *et al.*, 1971). This basin has a radius of roughly 30° and is centered at about 60°S and 180°W.

Altimetry data from several passes across the northwest edge of Mare Orientale (Fig. 6) provide quantitative information on this major basin structure. These elevations show a general slope toward the center of Orientale, plus superimposed variations of a kilometer or two clearly associated with features in the rings around

Fig. 5. Map of elevations (km) referred to a 1738 km sphere for the great farside depression.

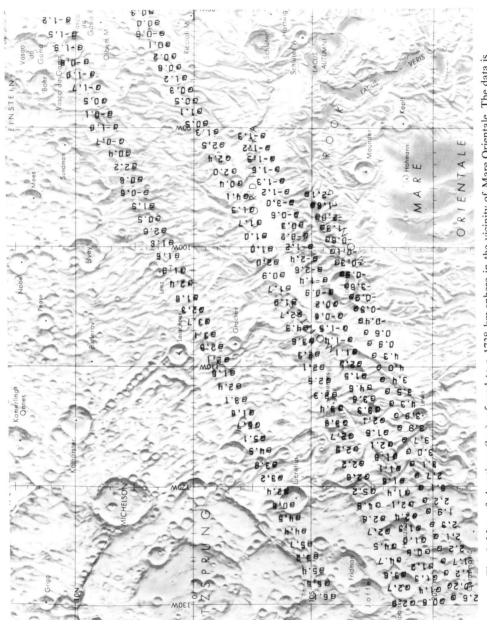

Fig. 6. Map of elevations (km) referred to a 1738 km sphere in the vicinity of Mare Orientale. The data is from revolutions 3–4, 7–8, 15–16, 33–34. The westerly precession of the orbit identifies the different revolutions.

Orientale. The scarp on the inner side of Montes Cordillera appear to be about 3 km in height. The crater Lowell is a depression of about 2 km. Montes Rook appears to be rougher but not higher than the ring outside it.

The earthside ringed maria which were crossed fairly close to their centers, Serenitatis, Crisium, and Smythii, show the expected depressions. The three maria have depths inversely correlated with their widths (as well as directly correlated with distance from the subearth point). There is also an inverse correlation with the sizes of the associated mass concentrations. These correlations are suggestive of the isostatic uplift explanation for crater depth-diameter relationships of Baldwin (1963). However, the uplift must be carried past isostatic balance to account for the mass concentrations: presumably a consequence of contraction of the lithosphere relative to the interior, resulting from the lunar thermal evolution (Kaula, 1971). The other portions of maria crossed, Oceanus Procellarum, Outer Imbrium, Palus Putredinis, and a small area of Tranquillitatis, are all somewhat shallower than Serenitatis. The fact that these areas are isostatically compensated, while shallower, suggest that they are composed of basalts less dense and less basic, and hence from a shallower source than the ringed maria basalts.

"Fine" scale

One of the more remarkable findings of the altimetry is the smoothness and levelness of the maria. The changes for some stretches, such as the portion of Oceanus Procellarum shown in Fig. 3, are small enough that they could arise largely from the neglected attitude corrections and variations in equipotential level. Given the evidences of relatively little movement of matter since filling of the maria (for example, the x-ray fluorescence (Adler *et al.*, 1972) and the cosmic-ray exposure ages (Eberhardt *et al.*, 1970; Wetherill, 1971)), this levelness must come from the properties of the lava that filled the mare. The lava must have had a low viscosity, as has been suggested by Murase and McBirney (1970) on the grounds of the low silica content and other petrological properties. Furthermore, the filling of each mare probably occurred in one epoch of limited duration.

The greater number of "reversals" rather than "continuations" in the terrae (Table 3) suggests that the events causing elevation differences on the moon have lateral scales predominantly smaller than the 33 km sampling distance. Since the terrae are saturated with craters of diameter comparable to the sampling distance the high frequency of "reversals" > 1.5 km may reflect the typical depths of these craters (a few kilometers).

Acknowledgment—This paper is publication No. 1018, Institute of Geophysics and Planetary Physics, University of California, Los Angeles, California, 90024.

REFERENCES

Adler I., Trombka J., Gerard J., Lowman P., Schmadebeck R., Blodget H., Eller E., Yen L., Lamothe R., Gorenstein P., and Bjorkholm P. (1972) Apollo 15 geochemical x-ray fluorescence experiment: Preliminary report. *Science* **175**, 436–440.
Baldwin R. B. (1949) *The Face of the Moon.* Univ. of Chicago Press, pp. 178–199.
Baldwin R. B. (1963) *The Measure of the Moon.* Univ. of Chicago Press, pp. 212–247.

Baldwin R. B. (1969) Ancient giant craters and the age of the lunar surface. *Astron. J.* **74,** 570–571.

Eberhardt P., Geiss J., Graf H., Grögler N., Krähenbühl U., Schwaller H., Schwarzmüller J., and Stettler A. (1970) Correlation between rock type and irradiation history of Apollo 11 igneous rocks. *Earth Planet. Sci. Lett.* **10,** 67–72.

Garthwaite K., Holdridge D. B., and Mulholland J. D. (1970) A preliminary special perturbation theory for the lunar motion. *Astron. J.* **75,** 1133–1139.

Hartmann W. K. and Wood C. A. (1971) Moon: Origin and evolution of multi-ring basins. *The Moon* **3,** 3–78.

Jeffreys H. (1970) *The Earth,* 5th ed. Cambridge University Press, pp. 471–484.

Kaula W. M. (1969) Interpretation of lunar mass concentrations. *Phys. Earth Planet. Interiors* **2,** 123–137.

Kaula W. M. (1971) Interpretation of the lunar gravitational field. *Phys. Earth Planet. Interiors* **4,** 185–192.

Latham G., Ewing M., Dorman J., Lammlein D., Press F., Toksöz N., Sutton G., Duennebier F., and Nakamura Y. (1971) Moonquakes. *Science* **174,** 687–692.

Muller P. M. and Sjogren W. L. (1968) Mascons: Lunar mass concentrations. *Science* **161,** 680–684.

Murase T. and McBirney A. R. (1970) Viscosity of lunar lavas. *Science* **167,** 1491–1493.

Nance R. L. (1971) Gravity measured at the Apollo 14 landing site. *Science* **174,** 1022–1023.

Papanastassiou D. A. and Wasserburg G. J. (1971) Lunar chronology and evolution from Rb–Sr studies of Apollo 11 and 12 samples. *Earth Planet. Sci. Lett.* **11,** 37–62.

Ransford G. A. and Sjogren W. L. (1972) Moon model—an offset core. *Nature* (in press).

Ringwood A. E. (1970) Petrogenesis of Apollo 11 basalts and implications for lunar origin. *J. Geophys. Res.* **75,** 6453–6479.

Rodionov B. N., Isavnina I. V., Avdeev Yu. F., Blagov V. D., Dorofeev A. S., Dunaev B. S., Zimon Ya. L., Kiselev V. V., Krosikov V. A., Lebedev O. N., Mikhailovskii A. B., Tiskchenko A. P., Nepoklonov B. V., Samoilov V. K., Truskov F. M., Chesnokov Yu. M., and Fivenskii Yu. I. (1971) New data on the moon's figure and relief based on results from the reduction of Zond-6 photographs. *Kosm. Issl.* **9,** 450–458.

Runcorn S. K. and Shrubsall M. H. (1968) The figure of the moon. *Phys. Earth Planet. Interiors* **1,** 317–325.

Shapiro A., Uliana E. A., Yaplee B. S., and Knowles S. H. (1968) Lunar radius from radar measurements. In *Moon and Planets II* (editor A. Dollfus), pp. 34–46. North-Holland.

Sjogren W. L. (1967) Estimate of four topographic lunar radii. In *Measure of the Moon* (editors Z. Kopal and C. L. Goudas), pp. 341–343. D. Reidel.

Sonett C. P., Schubert G., Smith B. F., Schwartz K., and Colburn D. S. (1971) Lunar electrical conductivity from Apollo 12 magnetometer measurements: Compositional and thermal inferences. *Proc. Second Lunar Sci. Conf., Geochim. Cosmochim. Acta* Suppl. 2, Vol. 3, pp. 2415–2431. MIT Press.

U.S. Air Force Aeronautical Chart and Information Center (ACIC) (1967–1968) Lunar 1:100,000 topomaps II P-8, III P-9, III P-12; (1969–1970) Lunar 1:250,000 topographic photomaps III 5-18—Mosting C, III 5-23—Fra Mauro—2nd ed.

U.S. Army Topocommand (1967–1968) Lunar 1:100,000 topomaps II P-2, II P-6, II P-13, III P-11; (1970) Lunar 1:250,000 topographic photomaps V 23.1—Rima Hyginus (2nd ed.), V 24—Hipparchus.

Wells W. D. (1971) Selenodetic data determined from Lunar Orbiter satellites. AAS/AISS Astrodynamics Specialists Conference (unpublished proceedings).

Wetherill G. W. (1971) Of time and the moon. *Science* **173,** 383–392.

Wollenhaupt W. R. and Ransford G. A. (1971) Lunar radii values from Apollo missions (abstract). *Bull. Am. Astron. Soc.* **3,** 272.

Wollenhaupt W. R. and Sjogren W. L. (1972) Comments on the figure of the moon based on preliminary results from laser altimetry. *The Moon* (in press).

Wollenhaupt W. R., Osburn R. K., and Ransford G. A. (1972) Comments on the figure of the moon from Apollo landmark tracking. *The Moon* (in press).

Wood J. A. (1970) Petrology of the lunar soil and geophysical implications. *J. Geophys. Res.* **75,** 6497–6513.

Proceedings of the Third Lunar Science Conference
(Supplement 3, *Geochimica et Cosmochimica Acta*)
Vol. 3, pp. 2205–2216
The M.I.T. Press, 1972

Lunar orbital mass spectrometer experiment

J. H. HOFFMAN and R. R. HODGES, JR.

The University of Texas at Dallas, Dallas, Texas

and

D. E. EVANS

Manned Spacecraft Center, Houston, Texas

Abstract—One of the Orbital Science experiments on Apollo 15 was a mass spectrometer designed to measure the composition and distribution of the lunar atmosphere. It operated for nearly 90 hours producing spectra of an unexpectedly complex nature, indicating many complex gas molecules exist in the vicinity of the spacecraft. The most plausible explanation is that there was continual vaporization of frozen or liquid drops of water, fuel or other matter that had been ejected from the spacecraft with small relative velocity so that these particles remained in nearby orbits. The search for naturally occurring gases in these spectra involves a statistical analysis of the data which has not been completed to date. A theoretical prediction on the possibilities of detecting lunar volcanism from orbit is included.

INTRODUCTION

THE MASS SPECTROMETER flown on Apollo 15 as part of the orbital science payload measured the concentration of gas molecules it encountered both in lunar orbit and during transearth coast for the purpose of studying the lunar atmosphere—its sources, sinks, and transport mechanisms. Nearly 40 hours of operation in lunar orbit and 50 additional hours in transearth coast produced over 5,000 mass spectra of an unexpectedly complex nature, particularly in lunar orbit, showing that the instrument was engulfed in a gas mixture containing many complex molecules. The origin and maintenance of this gas in lunar orbit is the subject of some interest.

In this paper we shall present first a brief description of the mass spectrometer and its performance characteristics, then a discussion of the preliminary results obtained to date, and finally a theoretical prediction on the possibilities of detecting lunar volcanism.

INSTRUMENT

The instrument, a sector-field, dual collector mass spectrometer (Hoffman *et al.*, 1971) was mounted on a boom stowed in the SIM bay of the Apollo Service Module which was capable of extending the instrument to a distance of 7.3 meters from the spacecraft. The purpose of the boom mount was to remove the instrument a reasonable distance from the spacecraft in the hope that it would be beyond the interacting cloud of outgassing molecules from the spacecraft, and in a collisionless, outwardly free streaming region. The instrument package is a rectangular box, 30 × 32 × 23 cm, weighing 11 kg, bisected by a baseplate, with the electronics portion on one side and the mass analyzer on the other. A plenum, in the form of a scoop, is mounted on the outboard side of the package and directed along the $-X$ axis of the spacecraft (i.e., opposite the Command Module end).

Control of the experiment functions, as well as the boom extension and retraction, is provided by a set of five switches in the Command Module (CM) operated by a crew member according to the mission flight plan or by request from the Mission Control Center.

The plenum contains the mass spectrometer ion source (a Nier type) employing redundant tungsten (with 1% rhenium) filaments mounted on either side of the ionization chamber. An emission control circuit activated by the ion source switch (ON position) in the command module powers the filaments. Two small heaters, consisting of ceramic blocks with imbedded resistors, are mounted on the sides of the ionization chamber. In order to outgas the ion source during flight, these heaters are activated by the ion source switch (STANDBY position). The ion source temperature reaches 300°C in 15 minutes. Several outgassing periods during the flight maintained the ion source in a reasonably outgassed state.

The mass analyzer is a single-focussing permanent magnet with second order angle focusing achieved by circular exit field boundaries, giving a mass resolution of better than 1% valley at mass 40 amu. Two collector systems permit simultaneous

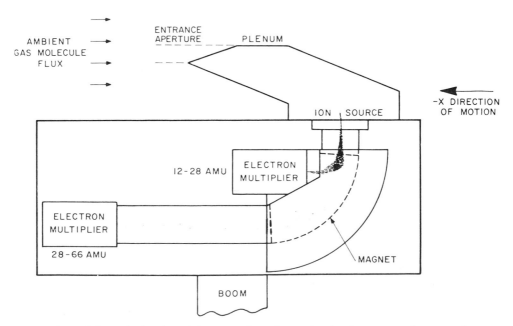

Fig. 1. Schematic drawing of the magnetic analyzer, showing ion source, plenum, and magnetic analyzer with two ion beam trajectories. The entrance aperture points in the $-X$ direction. When the spacecraft velocity is $-X$, ambient gas molecules would be directed into the plenum, as shown by the arrows, due to the high spacecraft velocity with respect to the ambient average molecular speeds.

scanning of two mass ranges, 12 to 28 amu and 28 to 66 amu. Figure 1 is a schematic drawing of the analyzer.

Voltage scan is employed utilizing a stepping high-voltage power supply. The ion accelerating voltage sweep is generated by varying the sweep high voltage in a series of 590 steps from 620 volts to 1,560 volts with a dwell time of 0.1 second per step. Between each sweep, thirty additional steps at zero volts are used to determine background counting rate, and to apply an internal calibration frequency. A sweep start flag indicates data or background, and serves as a marker for the start of each sweep. Mass number of the ion being detected is determined by the voltage step number at which the peak is detected. This step number is advanced by an enable pulse from the data handling system. Minimum number of steps between adjacent mass peaks below mass 54 is 12.

The detector systems employ electron multipliers, preamplifiers, and discriminators, which count the number of ions that pass through each collector slit on each of the sweep voltage steps. The ion count numbers are stored in 21-bit accumulators (one for each channel) until sampled by the scientific data system, a 64 kilobit/sec telemetry link to earth. Just prior to sampling, each data word is compressed pseudo-logarithmically into a 10-bit word consisting of a 6-bit mantissa and a 4-bit exponent. This system maintains 7-bit accuracy throughout the 21-bit range of data counts.

Instrument parameters, such as certain internal voltages, electron emission in the ion source, filament currents (to determine which filament was operating), multiplier voltages, sweep voltages, temperatures, multiplier and discriminator settings, and instrument current are monitored by a housekeeping circuit.

Initial calibration of the mass spectrometers, performed in a high vacuum chamber at The University of Texas at Dallas, verified that the proper mass ranges were scanned, and tested the resolution, linearity, mass discrimination, and dynamic range of the analyzer. Neon was introduced into the vacuum chamber with isotopic partial pressures ranging from 10^{-11} to 10^{-7} Torr. As is shown in Fig. 2, the instrument response was linear up to 1×10^{-8} Torr where the onset of saturation of the data-counting system occurred. The sensitivity of the instrument was verified to be greater than 3×10^{-5} A/Torr enabling the instrument to measure partial pressures down to 10^{-13} Torr. This is an end to end sensitivity that is essentially independent of electron multiplier gain provided the gain is sufficient to produce charge pulses large compared to the discriminator threshold. The sensitivity quoted includes the efficiency of the ion counting system. The uncertainties in the introduction of gases into the chamber, in the pressure measurement and in the wall effects precluded the determination of the absolute sensitivity in the University of Texas at Dallas chamber. The absolute calibration was performed in the Langley Research Center Molecular Beam Facility (MBF) as reported by Yeager et al. (1972).

The mass spectrometer was mounted in the MBF with the electronics package in the guard vacuum of the system and the plenum aligned with the axis of the chamber. A molecular beam, formed in the cryo-pumped chamber from a molecular furnace, impinged on the entrance aperture of the plenum. An externally controlled mechanical linkage allowed the plenum-beam angle to be varied from 0° to 40° with reference to

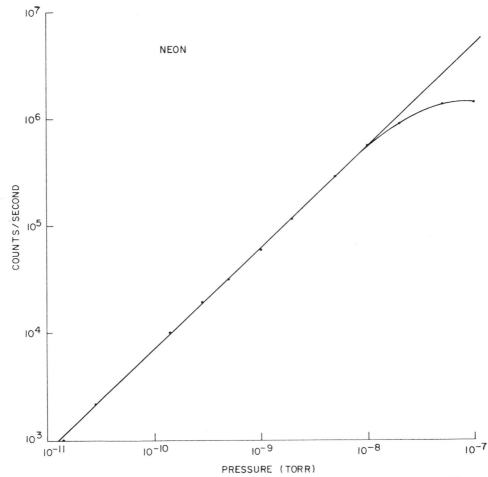

Fig. 2. Dynamic range of the mass spectrometer for neon, including neon-21 and neon-22 peaks to extend the pressure range.

a horizontal axis perpendicular to the beam axis (the spacecraft yaw direction). Pitch angles of $-5°$, 0, and $+5°$ (with reference to a vertical axis perpendicular to the beam axis) could be set manually with the system open. Separate tests were conducted with a combination of the three pitch angles and various yaw angles from $0°$ to $40°$. The mass spectrometer inlet was completely enclosed by a $4.2°$K extension copper tube so that the back scattering of molecules into the inlet was essentially eliminated. The $4.2°$K extension tube was enclosed by a $77°$K wall of the guard system.

A large amount of data was generated from these tests using three flight instruments and one qualification model. Figure 3 shows a typical set of curves of the output counting rate for neon and argon as a function of MBF beam flux; 10^{10} molecules per cm^2-sec is equivalent to 6.5×10^{-12} Torr of argon in the plenum. It can be seen that pressures below 10^{-13} Torr in the plenum are readily measurable. The data show a linear response well within the molecular beam accuracy (6%).

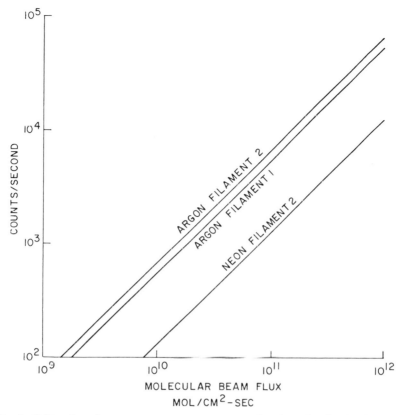

Fig. 3. Calibration of mass spectrometer output as a function of molecular beam flux for neon and argon. Combining these results with Fig. 2 shows that the instrument has a dynamic range of nearly 6 decades.

DISCUSSION OF RESULTS

The scientific objective of the orbital mass spectrometer experiment is to measure the composition and distribution of the ambient lunar atmosphere with a view to understanding its sources, sinks, and transport mechanisms. Light gases, such as hydrogen, helium, and neon probably originate from neutralization of solar wind ions at the surface of the moon, while Ar^{40} is most likely due to radioactive decay of K^{40}, and Ar^{36} and Ar^{38} may be expected as spallation products of cosmic ray interactions with surface materials. Molecular gases, such as carbon monoxide, hydrogen sulfide, ammonia, sulphur dioxide, and water vapor may be produced by lunar volcanism.

Light gases, such as hydrogen and helium are lost mainly by thermal escape. However, the lifetime against thermal escape for heavier gases is so long (~ 3000 yr for neon) that another process, photoionization and acceleration by the electric field associated with the motion of the solar wind, becomes the dominant loss process. Lifetimes against photoionization are of the order of 10^7 sec (Johnson, 1971). As a result, the dominant gas of solar origin to be expected on the moon is neon. Lighter gases should be less prevalent because of their more rapid thermal escape and heavier gases should be less prevalent because of their lesser abundances in the solar wind.

Hodges and Johnson (1968) have shown that light gases with negligible production and loss rates tend to be distributed at the lunar surface as the inverse 5/2 power of temperature, while heavier gases are influenced by the rotation of the moon. Neon falls into the former category, and its concentration on the antisolar side should be about 32 times that on the sunlit side. Its scale height on the dark side is about one-fourth that on the sunlit side, and thus, at a satellite altitude of 100 km, the diurnal fluctuation of neon concentration should be less than a factor of 2. Argon, being a heavier gas, is expected to be noticeably influenced by the rotation of the moon. It has a slightly less diurnal variation than neon and a longitudinal shift of its maximum towards sunrise, resulting in a concentration at sunrise that is approximately twice that at sunset.

Water vapor and other condensable gases probably exist in the lunar atmosphere but not on the dark side where the surface temperature is below $100°K$ and adsorption removes particles that come in contact with the surface. The rotation of the moon transports these adsorbed gases into sunlight where they are released into the atmosphere. Since surface heating occurs rapidly, this release probably occurs entirely within a few degrees longitude from the sunrise terminator, creating a pocket of gas (Hodges and Johnson, 1968).

In operation, the Apollo 15 lunar orbit mass spectrometer was deployed 7.3 meters from the spacecraft on a boom. The gas inlet of this instrument was directed so as to exclude the entire spacecraft from its field of view. When the spacecraft was flown with the SPS engine forward, the mass spectrometer gas inlet faced the direction of the velocity vector, and native lunar gas molecules (which have small thermal speeds as compared to the spacecraft velocity) were rammed into the inlet. When the direction of flight was opposite, i.e., with the Command Module forward, the inlet was in the wake of the plenum, excluding native gases. Therefore, the ram-wake effect should have been observed in the data. At a daytime temperature of $400°K$, the ram enhancement of the number density of gas molecules in the plenum should be a factor of 9.5 (assuming the native atmosphere is neon) and the wake depletion factor should be 0.0013 giving an expected ram-wake ratio of 10^4. Lower surface temperatures or heavier particles would cause the factor to be even larger.

Gases of spacecraft origin must have roughly the same mean velocity as the spacecraft and hence should not exhibit a ram-wake effect. Thus, if the source of contamination were constant, the difference between ram and wake data would be attributable to native lunar gases. In fact, very little ram-wake effect was observed, indicating the major source of the gases observed in lunar orbit was the CSM.

A rather complex spectrum was obtained in lunar orbit showing peaks at every mass number. Figure 4 is an example of a daytime spectrum. The upper channel is the high-mass range from 67 to 27 amu; the lower channel is the low-mass range from 28 to 12 amu. Mass 18, water vapor has saturated the counting system as has mass 17 (OH). During transearth coast (TEC), however, the amplitude of all peaks dropped by a factor of from 5 to 10 indicating the major source of gas observed in lunar orbit was removed. Figure 5 is a typical TEC spectrum. Also during TEC, a

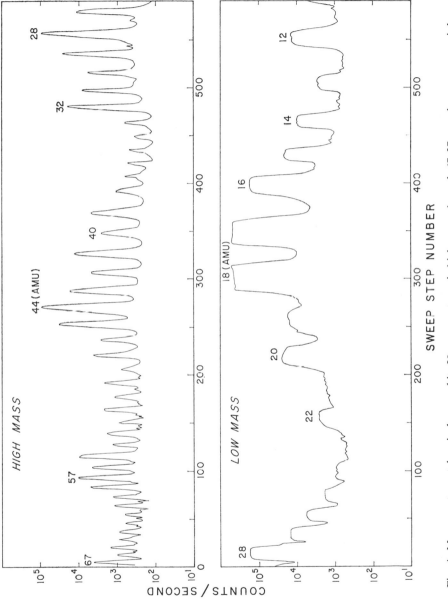

Fig. 4. Mass spectrum taken in lunar orbit. Upper spectrum is high mass channel, 67–27 amu; lower spectrum is low mass channel, 28–12 amu. Peak amplitudes (log scale) are plotted against sweep voltage step number that is equivalent to atomic mass number.

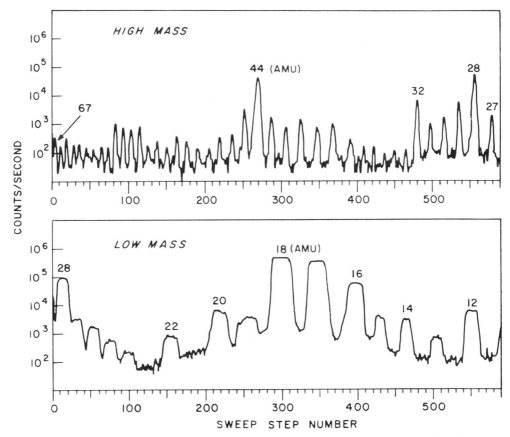

Fig. 5. Mass spectrum taken in transearth coast. Note decreased amplitudes of peaks with respect to those in Fig. 4.

boom retraction test, reducing the boom length in 4 steps to 1.25 meters showed no increase in any gas constituent. This implies that the mass spectrometer plenum is very effective in preventing molecules originating at the CSM from entering the ion source. Such outgassing molecules form essentially a collisionless gas within 1.25 meters from the CSM surface.

One of the unexpected results in the Apollo 15 data, that the levels of gases presumedly of spacecraft origin were significantly greater in lunar orbit than in the transearth coast, implies that orbital mechanics played an important role in the maintenance of the higher contaminant levels in lunar orbit. Several explanations of this effect have been proffered and all but one have been proved to be unrealistic.

The intuitive idea that the spacecraft was surrounded by a co-orbiting gas cloud can be attacked qualitatively on the ground that an extremely low electron temperature would be required to keep these molecules in orbits with perilunes above the surface of the moon. Quantitatively, a collisionless, stable cloud in lunar orbit of 10^5 tons of water at approximately one degree Kelvin would be required to produce the detected water vapor level. A collisional cloud, with molecules escaping continually from its periphery, would probably have monotonic radial decrease in concentration; however, during retraction of the mass spectrometer boom from 7.3 meters to 1.25 meters, no noticeable change in contaminant levels was detected. Therefore a cloud of contaminant gas, surrounding the spacecraft, is not a reasonable model.

The only plausible explanation to emerge is that the major part of the contamination came from the continual vaporization of frozen or liquid drops of water, fuel, or other matter that had been ejected from the spacecraft with small relative velocity, so that these particles remained in nearby orbits. It has been estimated that less than 10 grams of ice particles spread throughout a sphere of 100 meter radius about the spacecraft could account for all of the contaminant water vapor. This is quite realistic in terms of the amount of water dumped into orbit and the evaporative lifetimes of ice particles.

Figure 6 shows the diurnal variation of several of the major peaks in the spectrum. Water vapor abundance is determined by the mass 16 peak which is mainly the oxygen fragmentary ion from water. The diurnal variation of these peaks follows what might be expected if their origin is the vaporization of particulate matter in lunar orbit. The maximum concentration of these gases occurs slightly past the

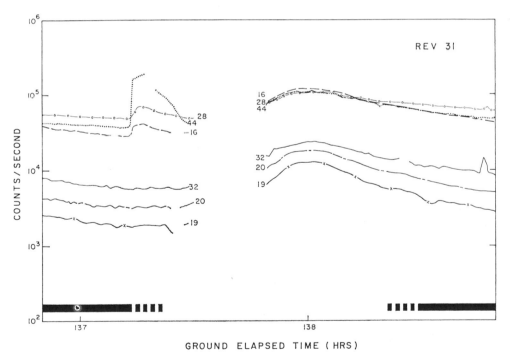

Fig. 6. Diurnal variation of major mass peaks observed in lunar orbit. Peak amplitude plotted as function of Apollo mission time (GET). Black line at bottom indicates spacecraft in darkness. Dashed line end locates terminators. Large amplitude excursion of mass 44 peak near terminator appears to be caused by sunlight entering plenum. Gap in plot due to data loss.

subsolar point where the equilibrium temperature of the particles would maximize due to the infrared heating from the lunar surface. From that point on through the night there is a continual decrease in gas concentrations due to the evaporative cooling of the particles and the decreasing heat input. Gas molecule lifetime, the transit time from leaving the solid particle to reaching the mass spectrometer, is assumed to be the direct flight path length divided by the molecule velocity (several hundred meters/sec). Therefore, photo-dissociation of these molecules is unlikely and cracking patterns observed in the mass spectrometer should be representative of the parent molecule. The mass 19 and 20 peaks are probably due in part to the O^{18} isotope. Also the reaction in the ion source

$$H_2O^+ + H_2O \rightarrow H_3O^+ + OH$$

gives a mass 19 peak which makes the 19/18 ratio H_2O pressure dependent. Similarly, the mass 20 peak is due to the O^{18} isotope with a possible addition of the Ne^{20} from the lunar atmosphere. It will require a statistical analysis of much of the data (which has not been reduced from magnetic tapes to date) to identify the neon in the mass 20 peak, or to set an upper bound on its abundance.

Mass 44, 28, and 16 exhibit a large amplitude excursion near the eastern terminator (on the left-hand side of Fig. 6) when the sunlight enters the plenum. There is evidence from laboratory studies of flashing photoflash bulbs into a vacuum chamber that a sudden increase in light intensity releases CO_2 from the chamber walls but has no effect on water vapor, O_2 or N_2. The 44, 28, and 16 peak increases in orbit are probably due to this same effect, the 28 and 16 being in part fragmentary ions from CO_2. Mass 22 follows the 44 peak very closely. This phenomenon is observed frequently in flight and occurs at the opposite terminator, where sunlight enters the plenum, when the spacecraft is flown in $+X$ orientation.

At the present stage of analysis of these data from Apollo 15, it is premature to draw any definite conclusions on the nature of the lunar atmosphere other than to reaffirm the general thoughts that it is indeed very rarefied as theoretical studies indicate.

Lunar Volcanism

If lunar volcanism is presently extant it is possible that its effects on the lunar atmosphere will be detected by the atmospheric sensors carried to the moon in the Apollo program. The currently operating cold cathode ionization gauges at the Apollo 14 and 15 landing sites provide nearly continuous monitoring of the atmospheric concentration and should indicate the occurrence of transient volcanic releases of gas near these sites. A mass spectrometer that is to be placed on the moon during the Apollo 17 mission should provide more complete information on the characteristics of such events, if any occur near enough to the instrument to produce a detectable transient. A wider search for active volcanic sources is being pursued through measurements with orbital mass spectrometers carried on the Apollo 15 and 16 service modules (Hodges et al., 1972).

The tenuous nature of the lunar atmosphere suggests that the volcanic release of gases is significantly less than on earth. Early results from the Apollo 14 cold cathode gauge (Johnson et al., 1971) indicate that native gases have a daytime concentration of less than 10^7 cm^{-3} at the surface. If the dominant loss process of these particles is photoionization, then their mean lifetime is of the order of 10^7 sec (Johnson, 1971). Since photoionization takes place only over the sunlit half of the moon, the average volcanic flux into the lunar atmosphere is half the product of daytime surface concentration, scale height, molecular mass, and photoionization rate; for an average daytime surface temperature of $360°$K, the Apollo 14 upper bound on concentration suggests that the maximum average volcanic flux is about 1.5×10^{-16} gm cm^{-2} sec^{-1}, which is at least 4 orders of magnitude less than the average flux for earth.

A more rapid removal mechanism than photoionization would increase this flux limit; however, there are no apparent candidate processes. Thermal escape is quite slow for the gases involved. In addition, their removal by chemical reaction with the surface is probably negligible, because over geologic time the chemical removal of volcanic gases at a rate only equal to the photoionization limit would have fixed about 20 gm cm^{-2} of these gases in the surface materials, which has not been detected in analyses of returned lunar samples. Thus the limit of 1.5×10^{-16} gm cm^{-2} sec^{-1} is probably a realistic one for the present rate of emission of volcanic gases. This amounts to a total lunar (global) release rate of about 60 gm sec^{-1}. It is evident that with such a slow rate of release of gases from the interior of the moon, the detection of an active volcanic source must require a fortuitous coincidence of event and experiment.

Figure 7 shows the mass dependence of the minimum expected time between detected events τ, based on a 10% perturbation of the ambient gas concentrations detected in the flight of Apollo 15, and on a global volcanic budget of 60 grams/sec (Hodges et al., 1972). This represents a most optimistic lower bound and may underestimate actual event repetition times by an order of magnitude or more, depending on the likelihood that the duration of typical gas releases may differ greatly from that necessary to establish a steady state perturbation (about 10 minutes).

The time required for lunar orbit is about 2 hours (7.2×10^3 sec) which in Fig. 5 is significantly less than the expected time between detected events τ for all mass numbers. In the 16 orbits of data obtained during Apollo 15, corresponding to about

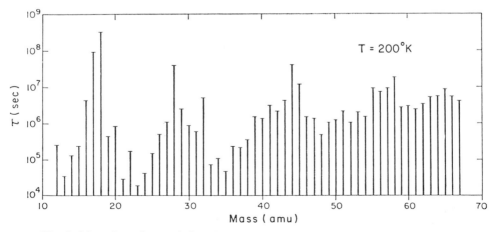

Fig. 7. Mass dependence of the minimum expected time between detected volcanic events in Apollo 15 orbital data.

1.2×10^5 sec, the situation is somewhat better. However the mass numbers corresponding to the smallest values of τ are not generally expected to correspond to volcanic gases. If the composition of lunar degassing were similar to that on earth, the following compounds would be included: H_2O, CO_2, N_2, H_2S, and SO_2. In Fig. 7 it can be noted that H_2S (34 amu) and SO_2 (64 amu) have the best chances of being detected in orbit.

Summary

The Apollo 15 lunar orbital mass spectrometer measured the composition and number density of neutral gas molecules along the flight path of the CSM as part of the Scientific Instrument Module (SIM) bay payload. The instrument, mounted on a boom extended from the SIM bay, observed a large number of gas molecules of many different species in the vicinity of the spacecraft in lunar orbit. This gas cloud seemed to be moving with the CSM, because its measured density was largely independent of the angle of attack of the mass spectrometer plenum, suggesting that the CSM is its most probable source. Many interesting variations of this gas cloud were observed as a function of orbital parameters. However, during TEC the amplitudes of all peaks in the spectra decreased by a factor of 5 to 10 and a boom retraction test showed no increase in gas densities down to a distance of 1.25 meters from the CSM surface. Molecules emanating from the spacecraft appear to be a collisionless, outwardly streaming gas beyond this point.

The orbital source of gas appears to be a continual vaporization of frozen or liquid drops of water, fuel, or other matter that had been ejected from the spacecraft into lunar orbit near the CSM.

Acknowledgments—The diligent efforts of the many people at the University of Texas at Dallas in the design and construction of this experiment is greatly appreciated. Through the efforts of one of us (D.E.E.), MSC supplied the thermal structure of the instrument. The work was monitored by MSC personnel.

Special thanks are due the crew of Apollo 15, especially Maj. Al Worden, Command Module pilot, for the excellent job in operating the experiment.

The NASA/Langley Research Center Molecular Beam Facility calibrations were made possible through the efforts of Paul Yeager and Al Smith.

This work was supported by NASA Contract NAS 9-10410.

References

Hodges R. R. and Johnson F. S. (1968) Lateral transport in planetary exospheres. *J. Geophys. Res.* **73,** 7307.

Hodges R. R., Hoffman J. H., Yeh T. T. J., and Chang G. K. (1972) Orbital search for lunar volcanism *J. Geophys. Res.*, in press.

Hoffman J. H., Hodges R. R., and Evans D. E. (1972) Lunar orbital mass spectrometer experiment. Apollo 15 Preliminary Science Report, NASA Report SP-289.

Johnson F. S., Evans D. E., and Carroll J. M. (1971) Cold cathode gauge experiment. Apollo 14 Preliminary Science Report, NASA Report SP-272.

Johnson F. S. (1971) Lunar atmosphere. *Rev. Geophys. Space Phys.* **9,** 813.

Yeager P., Smith A., Jackson J. J., and Hoffman J. H. (1972) Absolute calibration of Apollo lunar orbital mass spectrometer. (Submitted to *J. Vac. Sci.*).

Proceedings of the Third Lunar Science Conference
(Supplement 3, *Geochimica et Cosmochimica Acta*)
Vol. 3, pp. 2217–2230
The M.I.T. Press, 1972

Water vapor, whence comest thou?

J. W. Freeman, Jr., H. K. Hills, and R. R. Vondrak

Department of Space Science, Rice University,
Houston, Texas 77001

Abstract—During a 14-hour period on March 7, 1971, the Apollo 14 ALSEP suprathermal ion detector experiment (SIDE) observed an intense (maximum of $\sim 10^7$ ions/cm² sec sr), prolonged series of bursts of 48.6 eV ions at the lunar surface. The SIDE mass analyzer showed the mass per unit charge of these ions to be characteristic of water vapor if singly ionized. The event was also observed by the SIDE total ion detectors (TIDs) at the Apollo 14 site and at Apollo 12 (located 183 km to the west). The TID data from SIDE 14 indicate that the energy spectrum was narrower than the 20 eV interval between energy channels. Ion spectra due to the LM exhaust gases are shown to be readily identified by the SIDE and are distinctly different in character from the spectra obtained on March 7. Detailed consideration of other possible sources of water, including the Apollo 14 CSM, leads to the conclusion that the water vapor did not come from a man-made source. Also, it is estimated that the event may have involved a quantity of water much greater than that which has been artificially introduced into the lunar environment. Consequently, it appears to be of lunar origin.

Introduction

THE APOLLO 14 ALSEP Suprathermal Ion Detector Experiment (SIDE) observed ions of mass 18 amu/q during an event of approximately 14 hours duration on March 7, 1971. Because of their mass per unit charge, these ions are believed to be constituents of the water vapor group. Several gas sources associated with the Apollo 14 mission have been suggested as possible sources of this water vapor. However, all of these appear to us to be inadequate to explain the observed phenomena.

The SIDE consists of two positive-ion detectors: the mass analyzer (MA) and the total ion detector (TID). The conceptual design of the MA is illustrated in Fig. 1. The instrument is basically a neutral gas mass spectrometer for the lunar exosphere where photoionization by solar uv and charge exchange processes with the solar wind are allowed to play the role of the ionizer. The interplanetary electric field associated with the motion of the solar wind plays the role of the ion acceleration field in a conventional mass spectrometer (see Manka and Michel, 1970; and Manka *et al.*, 1972, for analysis of this acceleration mechanism and comparison with observational data). The MA then contains simply an ion velocity filter (Wien filter) and energy per unit charge filter in tandem followed by an ion detection system consisting of a post-analysis acceleration field and a channel-electron-multiplier operated as an ion counter (see Fig. 2). The velocity filter and energy/q filter are sequenced through a series of steps to provide coverage of twenty mass/q ranges at each of 6 energy/q steps: 0.2, 0.6, 1.8, 5.4, 16.2, and 48.6 eV/q. The efficacy of this system has been demonstrated by numerous observations by the SIDE of mass spectra associated with the LM rocket exhaust gases (Freeman, *et al.*, 1972), LM cabin venting (Hills and Freeman, 1971), and gas emission from the S-IVB and LM lunar impact events

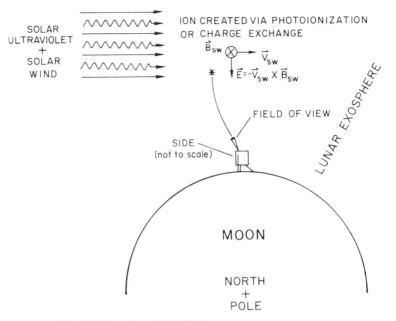

Fig. 1. Operation of the SIDE as a mass spectrometer.

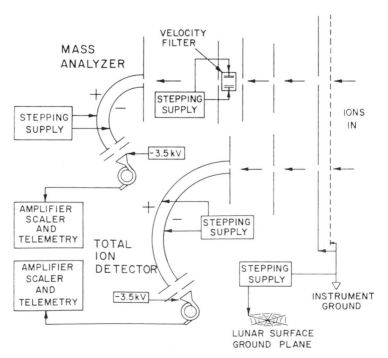

Fig. 2. Schematic diagram of the SIDE.

(Freeman *et al.*, 1971). For preliminary examination of the data the sensitivity of the instrument at 18 amu/q and the 48.6 eV/q energy level is approximately 0.05 ions/cm^3 and the resolution ($\Delta M/M$) is approximately 0.1. Figure 2 also depicts the TID, which is an ion energy/q spectrometer that records ions without mass discrimination in 20 differential energy channels from 3500 eV/q down to 10 eV/q. The lowest energy channels are at 10, 20, 30, 50, 70, and 100 eV/q; thus the TID lower energy channel data can be used as a check on the MA data.

EXHAUST GAS OBSERVATIONS

The exhaust gases from the LM engines are expected to exhibit a multi-component mass spectrum (Simoneit *et al.*, 1969). Figure 3 shows the spectrum as deduced from the Apollo 12 SIDE MA measurements 14 hours after the Apollo 12 lunar landing, together with one derived from the laboratory results of Simoneit *et al.* (1969). An amount of N_2 equal to the amount of H_2O has been added to the laboratory measurements as the best estimate of the amount of N_2 present (Aronowitz, *et al.*, 1968; Milford and Pomilla, 1967), since the laboratory results did not include a determination of the amount of N_2. Note the broad range of masses present and the good agreement of the two spectra. However, the observed spectrum contains a significant amount of mass 20 (neon), which is not expected to be a component of the engine exhaust. This may be neon liberated from the lunar surface by heat from the descent engine.

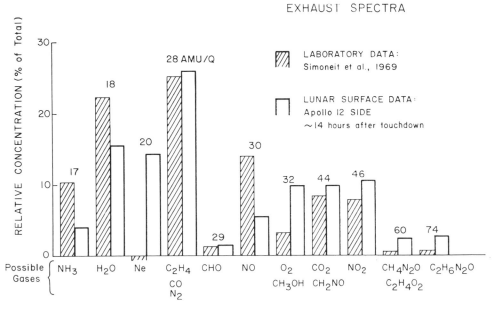

Fig. 3. Mass spectra of Lunar Module exhaust gases. The laboratory data are four trap averages from Simoneit *et al.* (1969) with N_2 added in amount equal to H_2O. The Apollo 12 SIDE mass spectrum is a 5 spectrum average measured 14 hours after LM touchdown.

The decay of intensity of the exhaust gas is demonstrated in Fig. 4, which shows the spectrum obtained approximately two months after the Apollo 12 landing compared with the spectrum obtained 14 hours after landing. After two months the intensities have decreased generally by an order of magnitude, but the spectrum is still very broad and readily identifiable as due to exhaust gases. There are several changes in the spectrum due to the operation of ionization, loss, dissociation, and recombination processes over the two months. Most striking is the disappearance of mass 18 ions and the appearance of a significant quantity of mass 16 ions. These mass 16 ions presumably come from the dissociation of the more massive molecules in the original spectrum; principally water vapor. For comparison, the maximum water vapor ion intensity observed by the Apollo 14 SIDE during the March 7 event is included in Fig. 4. It is approximately *three orders of magnitude* greater than the intensity observed by the Apollo 12 SIDE 14 hours after lunar landing.

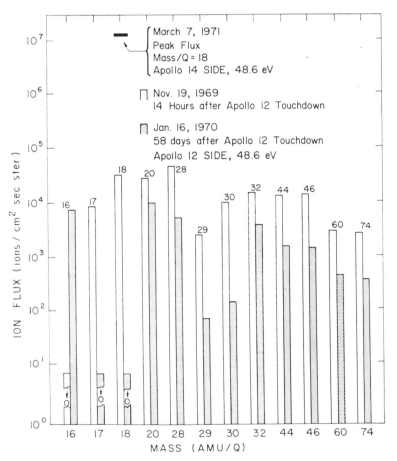

Fig. 4. Observed decrease in exhaust gas fluxes over a 58 day period. For comparison, the peak flux seen in the March 7th water vapor event is shown.

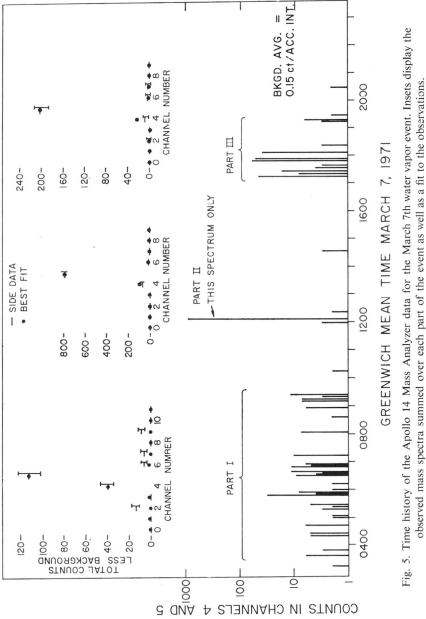

Fig. 5. Time history of the Apollo 14 Mass Analyzer data for the March 7th water vapor event. Insets display the observed mass spectra summed over each part of the event as well as a fit to the observations.

Water Vapor Observations, March 7, 1971

The Apollo 14 MA data from the March 7 event are shown in Fig. 5. Laboratory calibrations indicate that water vapor ions are recorded primarily in channel 5 of this instrument, with channel 4 recording approximately 12% as many counts due to water vapor as channel 5. Consequently, to show the principal features of the March 7 event, the total counts in adjacent channels 4 and 5 at 48.6 eV are displayed in the lower portion of the figure. Ions were observed during the March 7 event only at the highest energy level, 48.6 eV, and not at the next lower level, 16.2 eV. Twenty-channel mass spectra at 48.6 eV were obtained in 24 seconds every 2.58 minutes. The time history of the event leads to a separation of the data into three parts as indicated. The mass spectra accumulated in each part are shown in the upper portion of the figure, together with the best fit to the observations, based on laboratory calibration data. In Part I, the accumulated spectrum can be fit by a mixture of 85% mass 18 amu/q and 15% mass 16 amu/q while in Parts II and III the data fits well with 100% mass 18 amu/q. For Part I a fit including a percentage of mass 17 amu/q is also possible but presses the resolution of the experiment. The fit shown for Part I is primarily to illustrate an apparent change in the mass spectrum during the event. This change is suggestive of the injection of fresh gas and thereby a dynamic rather than a static event.

The TID recorded data consistent with that of the MA, showing sporadic bursts of low energy ions throughout the event. Figure 6 shows one energy spectrum obtained with the TID. The broad high energy spectrum at 250 eV to 1000 eV is due to ions in the earth's magnetosheath, and is not of interest in this event. The important point here is the high mono-energetic flux of low-energy ions (70 eV here). These low energy ions were observed sporadically and with various intensities during the event. The narrow low-energy portion of the spectrum occurred at various energies, but usually was observed in only one channel at a time (at 30, 50, or 70 eV/q). The width of a TID energy channel is approximately 10% FWHM.

The SIDE MA at the Apollo 12 site (183 km to the west) was quite noisy at the time, so no useable mass spectra were obtained. There was, however, good correlation between the TID responses of the two instruments during the event. The ion energies observed at the Apollo 12 site early in the event were higher than those at Apollo 14, being generally in the 100 eV/q and 250 eV/q channels, whereas the energies at Apollo 14 were in the range 30 eV/q to 70 eV/q. During Part II of the event the energies observed by the Apollo 12 TID decreased to a level matching those observed at the Apollo 14 site. Again we see the suggestion of a dynamic rather than a static event.

Discussion of Possible Sources

The mass spectra of LM exhaust shown in Fig. 4 indicate that the exhaust spectrum is readily identifiable by its broad mass range, and does not resemble the spectrum obtained in the March 7 event. This fact, together with the unprecedented intensity of the latter event, indicates that the March 7 event was not due to LM exhaust gases.

The maximum ion flux observed at the Apollo 14 SIDE during this event was

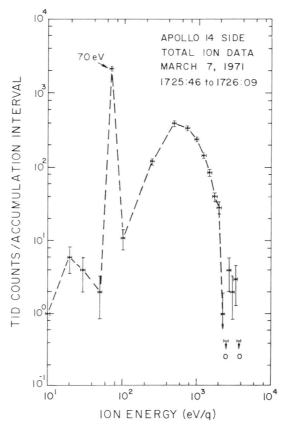

Fig. 6. Total ion detector data from the Apollo 14 SIDE at one time during the March 7th event. The peak at 70 eV corresponds to the water vapor ions seen in the mass analyzer. The broad spectrum peaking near 500 eV is that of magnetosheath ions and occurs independently of the water vapor ions.

approximately 1.5×10^7 ions/cm^2 sec sr. For comparison, when the LM cabin was vented during the Apollo 14 mission prior to the 2nd EVA, the SIDE observed a short-lived (approximately one minute) flux of 4×10^5 ions/cm^2 sec sr at 50 eV. However, during the venting the neutral gas pressure measured by the cold cathode gauge experiment (CCGE) went to a high value (Johnson *et al.*, 1971). This is an important point, for it establishes that the CCGE would have seen a large neutral gas pressure rise if enough gas were liberated *locally* to cause the high counting rate seen by the SIDE on March 7. There was no such increase corresponding to the SIDE observation (Johnson *et al.*, 1972); therefore the source must have been further away than the LM or the other debris left on the surface nearby. A gas source at a great distance could lead to the production of ions that subsequently reach the SIDE, while the neutral gas density at the instrument might never be high enough to be observed by the CCGE. The simultaneous observation of the event by the Apollo 12 SIDE further implies a distant or at least wide-spread source.

Perhaps the most attractive source, at first glance, for the observed water vapor ions, is the 43 kg of waste water (mostly excess H_2O from the fuel cells) dumped from the Command/Service Module (CSM) in lunar orbit. The attractive feature of such a source is the presumed narrow altitude distribution of gas which would result, thus leading to a narrow energy spectrum after acceleration through an electric field. However, we note that a narrow altitude distribution is not a prerequisite for the narrow energy spectrum observed. The relatively strong surface magnetic fields (Dyal *et al.*, 1971) result in an ion gyroradius (approximately 50 km) that precludes simple vertical acceleration from a high altitude. Furthermore, a narrow energy spectrum seems to be a general feature of heavy mass ion events *regardless of the source*. The Apollo 14 cabin vent for the 2nd EVA (Hills and Freeman, 1971) is one such event. Another example of a narrow energy spectrum peaked at 70 eV/q (and later at 20 eV/q) and associated with the Apollo 12 LM exhaust gases has been presented by Freeman *et al.* (1970, Fig. 6-5). A third example that occurred when the moon was outside the earth's magnetosheath is shown in Fig. 7. These examples illustrate the fact that monoenergetic fluxes of approximately 50 eV energy are frequently observed near the lunar surface, either in or out of the magnetosheath, and without regard to the temporal proximity of Apollo missions. The mechanism for the narrow energy spectrum is one of several features of the interaction between the solar wind and neutral gas which is not fully understood.

Most of the water dumped from the CSM will freeze promptly and then vaporize by sublimation. Exposure of this water vapor to the solar ultraviolet and the solar wind particles results in the production of H_2O^+ ions as well as the dissociation of H_2O into H and OH. Later the production of OH^+, and finally that of O^+, will

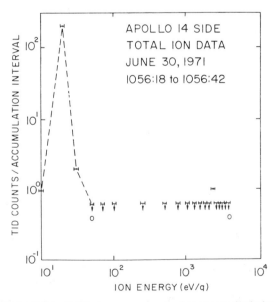

Fig. 7. A sample total ion detector measurement of monoenergetic ions that occurred when the moon was outside the magnetosheath.

dominate as further dissociation and ionization takes place and as the quantity of H_2O becomes depleted. Figure 8 shows the calculated production rates for these three ions as a function of ultraviolet exposure time. The curves shown are obtained from solution of the differential equations describing the reaction chain shown in the figure. The appropriate reaction rates were obtained from Hinteregger (1960) and Werner *et al.* (1967). The ionizing effect of solar wind charge exchange reactions is approximately the same as that of solar ultraviolet, but these processes can operate on a lunar gas cloud only during the local daytime. Thus the calculations shown give a reasonable indication of the total ion production rates to be expected from a cloud of water vapor in the lunar atmosphere. If the event of March 7 were due to water ejected from the Apollo 14 CSM a month earlier, then ions of masses 16 and 17 would be more abundantly produced than ions of mass 18, contrary to observation. In fact, in Part I of the event there is an indication of mass 16 and possibly mass 17 ions, but only a small amount. The later parts of the event showed only mass 18 ions. This change of spectrum is suggestive of a possible injection of fresh gas during the event.

An important quantity in regard to the CSM as a source is the amount of water dumped and its subsequent behavior. During three dumps a total of approximately 43 kg of water was ejected (C. Staresinich, private communication). Of this, a few percent would be immediately vaporized, freezing the remainder. The observed time history of the ion event indicates that if the source gas was in orbit, then it must have been dispersed relatively uniformly throughout the orbital path. This is implied by the fact that the counting rates in Part I were present for longer than the 2-hour orbital period of the CSM. We can estimate, as follows, the effect if all of the ice had become vapor within a few days, then been exposed to the solar uv for the remainder of the time until March 7. At the end of this time we take the appropriate ion production rate from Fig. 8, together with the volume over which the original gas has become dispersed, to find the average volume rate of production of ions. The integration over the narrow altitude range in which the observed ions originate (a small part of the total altitude distribution of the gas cloud) leads to the rate at which the observed ions are produced in a cm^2 column over the detector. A reasonable estimate of the ion flux expected at the SIDE is obtained as simply the total number of ions formed in one second anywhere along the cm^2 column out from the detector along its look direction, since these ions could move to the detector very rapidly once formed. When this is done, using a *minimum* volume of approximately 4×10^8 km^3, a fractional production rate of 10^{-7} sec^{-1}, and a source region altitude thickness of 1 km, the resulting upper limit on the expected counting rate of the SIDE is a few tenths of a count per second, more than three orders of magnitude below the observed peak rate, and a factor of at least 20 below the probable average observed rate.

For another calculation, consider the approximately 10^4 kg of fuel burned near the moon by the LM and CSM. Of this, approximately 2×10^3 kg is H_2O. With this much material in the lunar environment, there was still only an average of approximately 2 counts/sec observed by the Apollo 12 SIDE 14 hours after the landing, and only approximately 0.2 counts/sec 58 days later. Thus one can hardly expect counting rates of 2 to 700 counts/sec due to 43 kg of CSM water widely dispersed.

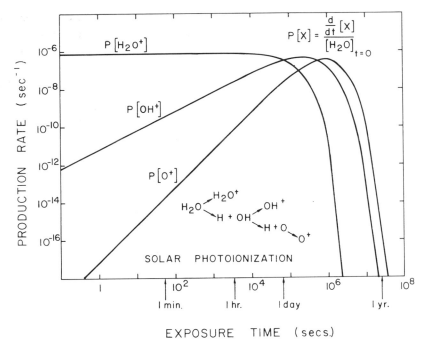

Fig. 8. Production rates of ions resulting from exposure of water to solar ultraviolet at the moon.

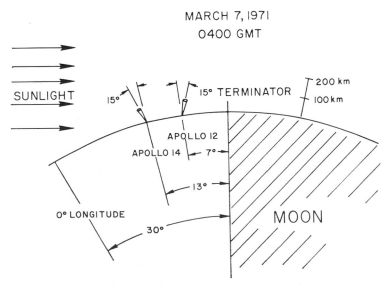

Fig. 9. Schematic of the geometry of the March 7th event. Each SIDE has a field of view about 6° square.

As mentioned, the March 7 event was also observed by the Apollo 12 SIDE, located some 183 km to the west of the Apollo 14 site. However, the two SIDE's are deployed such that their look directions are 36° apart, as shown in Fig. 9, which shows the geometry of the situation at the time of the event. It is interesting to note from the figure that the solar wind motional electric field may have been in the proper direction to accelerate ions into the Apollo 12 detector, but that any simple model doesn't account for the ions observed by both instruments. Observation of the ions by both instruments indicates complicated ion trajectories. Again, the surface fields almost certainly play an important role in determining these trajectories.

Although there were seismic signals ("swarms", not due to impact events) observed by the passive seismic experiment during this event, such seismic signals have been observed many times without simultaneous observations of high ion intensities (G. Latham, private communication). Thus there may be no meaningful relation between the seismic signals and the water vapor event. However, there is a possibility that such seismic signals are linked to a release of gas, but that the probability for the gas to be eventually detected as ions by the SIDE is quite low. For example, the solar wind motional electric field can lie parallel to the SIDE's look direction only near sunrise and sunset (assuming the typical solar wind flow directions). During either of these periods, there is a further requirement that the magnetic field point out of the ecliptic plane in order to have the solar wind motional field point in the desired direction. Since this requirement is met infrequently, conditions are usually unfavorable for acceleration of ions into the instrument by this mechanism. In addition, the gas source must be large enough and near enough to produce observable fluxes of ions. Thus the probability that a random venting of gas will result in ions observed by the SIDE is quite low if the solar wind motional electric field is the only accelerating mechanism. It should be noted that the solar wind motional electric field undergoes rapid temporal fluctuations in both magnitude and direction. Consequently, temporal variations in the ion observations may be the result of changes in the solar wind parameters rather than changes in the source of the ions.

We emphasize that the March 7 event was an intense event of extended duration and we have not yet seen another event like it, although there exist numerous smaller scale events with less definitive properties whose frequency appears to be independent of time. It would of course be comforting to find another large event displaced further in time from an Apollo mission. Because of a switchover in our data-processing program, examination of the Apollo 14 and 15 data has been delayed and only a fraction of the data from these missions has been examined. It is expected that a thorough analysis of all of the data will clarify the preliminary results presented herein.

QUANTITY OF WATER

It is of course desirable to have some estimate of the total quantity of water represented in this event. A lack of knowledge of the location of the source and the details of the ionization and acceleration mechanisms precludes a precise quantitative determination. However, we have looked for a similar event where the quantity of gas released is known in order to "calibrate" the water vapor event. Only one other event has been found that has yielded MA counting rates as high as those seen during the March 7 event and that was the detection of the Apollo 14 LM ascent engine exhaust gases by the Apollo 12 SIDE during the ascent trajectory burn (Hills and Freeman, 1971). The Apollo 14 SIDE was not operating at that time. The peak counts for this event in both the TID and MA were within a factor of 2 of those for the March 7 event and the ion energies were comparable. However, the ion flux duration was only about 2 minutes compared with some 14 hours (intermittent) for the March 7 event.

The flight path of the Apollo 14 LM as it left the Fra Mauro site was, at closest approach, approximately 22 km north of the Apollo 12 site at an altitude of approximately 15 km. Ions from the LM were recorded starting 1 minute after the LM had passed by the minimum slant range of 27 km. The total gas released by the LM was approximately 2200 kg and the burn time 431 seconds with an average emission rate of 5.2 kg/second. Therefore, if we assume equivalent ion accelerating conditions for the two events and if the water vapor source was no closer than 27 km the required emission rate is of the order of magnitude of at least 5 kg/sec and must have been sustained intermittently over a period of 14 hours. If these assumptions are valid, a total water emission of the order of 10^5 kg is implied for the March 7 event. The CCGE data indicate a distant source, but the source characteristics and accelerating conditions (including ion trajectories) may be greatly different in the two events. Consequently, this estimate of the source strength may be in error by several orders of magnitude if the two events are not completely equivalent.

Conclusion

None of the suggested artificial sources for the observed water vapor ions appears to us to be capable of producing the observed ion fluxes. Table 1 lists the prominent source candidates and objections to each. The LM and other mission-related items at the Apollo 14 site can be ruled out because they can't produce such an intense event over such a long time period, observable at both sites, and because the CCGE

Table 1.

Source	Difficulties
LM	1. Event duration and intensity 2. Simultaneous observation at two sites 3. Not observed by CCGE
Residual exhaust gases	1. Event intensity 2. Exhaust spectrum maintains identity for at least 58 days, with H_2O dissociation 3. Time history suggests impulsive injection of fresh gas during event
CSM waste water	1. Event intensity and duration 2. Narrow altitude distribution not required in order to explain observed narrow energy spectrum 3. Time history suggests injection of fresh gas during event
Moon	? ? ? ?

data indicate that these ions did not originate from a nearby gas source. The engine exhaust gases are also ruled out, primarily because the mass spectrum of the exhaust gases has been shown to maintain its identity for at least 58 days, and is readily distinguishable from the spectrum observed on March 7. The CSM waste water dump is unlikely to be the source, because it has been shown to lead to a counting rate much smaller than observed, and because consideration of the observation of the Apollo 14 LM ascent indicates a much greater quantity of water as the source. Also the appeal of the CSM as a source is diminished when it is recognized that a narrow altitude distribution for the gas is not required. Furthermore, the time history of the event suggests an impulsive injection of fresh gas during the event, long after the CSM has left the moon. It should be noted that the LM exhaust gases and items local to the Apollo site can be firmly excluded as possible sources because of the character of the observations. The CSM water dump is eliminated as a source due to the discrepancy between calculation and observation.

It has been brought to our attention at this conference (K. P. Florenskiy, private communication) that the Russian Lunokhod research vehicle (located at 35°W, 38°N, approximately 1400 km from Apollo 14) occasionally vents a "few liters" of water from its cooling system. We hope to obtain further information concerning the ventings but they do not appear to be a large enough source to account for the March 7 event.

Having examined all of the available artificial sources we conclude that the probable explanation is a natural source. There have been numerous suggestions of subsurface water on the moon, trapped below an ice layer beneath the lunar surface (Gold, 1960; Green, 1964; and Schubert et al., 1970) and liberated occasionally by the flexing or fracture of the lunar crust. The so-called lunar transient phenomena have been shown (Moore, 1971) to occur with a preference for times near lunar perigee passage and at likely positions for crustal stress. It seems likely that lunar venting is still taking place.

Acknowledgments—Dr. Hans Balsiger and Mr. Robert Lindeman were helpful in various portions of the data analysis. Discussion with members of Rice University Space Science Department and the staff of the Royal Institute of Technology in Stockholm were also useful in the interpretation of the data. Discussion at the Apollo 15 Investigators' Symposium, Houston, 1971 is also acknowledged. One of us (J.W.F.) was a Visiting Scientist at the Royal Institute of Technology in Stockholm during the preparation of this paper. This work was supported by NASA Contract NAS 9-5911.

References

Aronowitz L., Koch F., Scanlon J., and Sidran M. (1968) Contamination of lunar surface samples by the lunar module exhaust. *J. Geophys. Res.* **73**, 3231–3238.

Dyal P., Parkin C. W., Sonett C. P., DuBois R. L., and Simmons G. (1971) Lunar portable magnetometer experiment. Apollo 14 Preliminary Science Report, NASA SP-272, 227–237.

Freeman J. W. Jr., Balsiger H., and Hills H. K. (1970) Suprathermal ion detector experiment. Apollo 12 Preliminary Science Report, NASA SP-235, 83–92.

Freeman J. W. Jr., Hills H. K., and Fenner M. A. (1971) Some results from the Apollo 12 suprathermal ion detector. *Proc. Second Lunar Sci. Conf., Geochim. Cosmochim. Acta* Suppl. 2, Vol. 3, pp. 2093–2102. MIT Press.

Freeman J. W. Jr., Fenner M. A., Hills H. K., Lindeman R. A., Medrano R., and Meister J. (1972) Suprathermal ions near the moon. *XVth General Assembly IUGG* (Moscow, 1971). *Icarus* **16,** in press.

Gold T. (1960) Processes on the lunar surface. *Proc. IAU Symp. 14* (Leningrad, 1960). Academic Press, London.

Green J. (1964) Lunar defluidization and volcanism. North American Aviation Report SID 64-1340, 1964.

Hills H. K. and Freeman J. W. Jr. (1971) Suprathermal ion detector experiment. Apollo 14 Preliminary Science Report, NASA SP-272, 175–183.

Hinteregger H. (1960) Interplanetary ionization by solar extreme ultraviolet radiation. *Astrophys. J.* **132,** 801–811.

Johnson F. S., Evans D. E., and Carroll J. M. (1971) Cold-cathode-gage experiment. Apollo 14 Preliminary Science Report, NASA SP-272, 185–191.

Johnson F. S., Evans D. E., and Carroll J. M. (1972) Lunar atmosphere (abstract). In *Lunar Science—III* (editor C. Watkins), pp. 436–438, Lunar Science Institute Contr. No. 88.

Manka R. H. and Michel F. C. (1970) Lunar atmosphere as a source of argon-40 and other lunar surface elements. *Science* **169,** 278–280.

Manka R. H., Michel F. C., Freeman J. W. Jr., Dyal P., Parkin C. W., Colburn D. S., and Sonett C. P. (1972) Evidence for acceleration of lunar ions (abstract). In *Lunar Science—III* (editor C. Watkins), pp. 504–506, Lunar Science Institute Contr. No. 88.

Milford S. N. and Pomilla F. R. (1967) A diffusion model for the propagation of gases in the lunar atmosphere. *J. Geophys. Res.* **72,** 5433–5545.

Moore P. (1971) Transient phenomena on the moon. *XVth General Assembly IUGG* (Moscow, 1971).

Schubert G., Lingenfelter R., and Peale S. (1970) The morphology, distribution, and origin of lunar sinuous rilles. *Rev. Geophys. Space Sci.* **8,** 199–224.

Simoneit B. R., Burlingame A. L., Flory D. A., and Smith I. D. (1969) Apollo lunar module engine exhaust products. *Science* **166,** 733–738.

Werner M., Gold T., and Harwit M. (1967) On the detection of water on the moon. *Planet. Space Sci.* **15,** 771–774.

Proceedings of the Third Lunar Science Conference
(Supplement 3, *Geochimica et Cosmochimica Acta*)
Vol. 3, pp. 2231–2242
The M.I.T. Press, 1972

Lunar atmosphere measurements

FRANCIS S. JOHNSON and JAMES M. CARROLL

The University of Texas at Dallas, Dallas, Texas 75230

and

DALLAS E. EVANS

NASA Manned Spacecraft Center, Houston, Texas 77058

Abstract—Cold cathode ionization gauges were left on the lunar surface during Apollo Missions 14 and 15 to measure the amount of lunar gas. The observed nighttime concentration is very low, about 2×10^5 cm^{-3}, which is less than the neon concentration that might be expected from the solar wind. This suggests that the lunar surface is not saturated with solar wind neon, and hence that less neon is being released from the surface than impinges upon it. The low nighttime concentration shows that contaminant gases from the Apollo operations freeze out at night or become adsorbed on the cold lunar surface. Observed daytime concentrations have been two orders of magnitude greater than the nighttime values and appear to be due mainly to contamination in the landing area. The rate at which the contamination is decreasing is characterized by a time constant of a few months. Gas clouds have been seen at times and these appear to have been released from Apollo hardware left on the lunar surface.

INTRODUCTION

COLD CATHODE GAUGES were included in the Apollo lunar surface experiment packages (ALSEPs) to measure the amount of gas on the moon. The gauges were conventional magnetrons in stainless steel envelopes; they were supplied by the Norton Research Corporation. The gauge is shown schematically in Fig. 1. A permanent magnet provides a field of about 900 G, and $+4500$ V is applied to the anode of the gauge. The gauge is surrounded with a magnetic shield to reduce any possible effects on other instruments stowed near the gauge. The gauge aperture was covered but not sealed, and the cover was removed by remote command after deployment on the lunar surface.

SOLAR WIND SOURCE

The solar wind impinges upon the lunar surface with very little disturbance of the flow (Ness *et al.*, 1967; Sonett and Colburn, 1967). The solar wind ions must imbed themselves in the surface materials, but once the surface is saturated with a given constituent of the solar wind, that constituent must be released from the surface at the same average rate as it is brought to the lunar surface by the solar wind. After release from the surface, the neutralized particles constitute a lunar atmosphere. Those particles with sufficiently high velocity will escape from the moon just by virtue of their thermal motions, and this of course is most important for the lighter constituents.

The composition of the solar wind is reasonably approximated by cosmic abundances. The measured abundances are variable, making generalizations difficult.

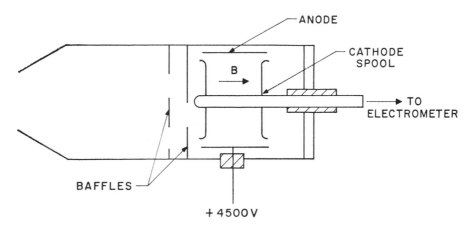

Fig. 1. Schematic representation of cold cathode gauge. B indicates the magnetic field.

Bame *et al.* (1970) give the following relative abundances based upon solar-wind observations in spacecraft on 6 July 1969: H, 5000; He, 150; O, 1; Si, 0.21; and Fe, 0.17. These compare fairly well with cosmic abundances given by Cameron (1971): H, 5000; He, 410; O, 4.5; Si, 0.21; and Fe, 0.17. Helium and oxygen abundances in the solar wind on this day were depressed relative to hydrogen, silicon, and iron, but the relative abundances of the latter agree well with cosmic abundances. Many other measurements have been made of the He/H ratio and an average value of 0.045 is accepted by Hundhausen (1970) based on measurements in several spacecraft programs.

Actual measurements of solar wind composition at the lunar surface by trapping in an aluminum foil give a He^4/Ne^{20} ratio of 550 (Buehler *et al.*, 1972) and a Ne^{20}/A^{36} ratio of 36 (Geiss *et al.*, 1971). Measurements in lunar surface materials generally indicate less He and Ne, presumably due to diffusive escape of these gases. Ilmenite samples show the least loss, and a Ne^{20}/A^{36} ratio of 33 and a He^4/A^{36} ratio of 7600 have been determined (Eberhardt *et al.*, 1970), thus indicating a He^4/Ne^{20} ratio of 230, less than measured in the aluminum foil. The A^{36}/Kr^{84} ratio was 2350 and the A^{36}/Xe^{132}, 16,000. All ratios were given as lower limits because of the possibility of diffusive escape favoring the loss of the lighter constituent. Combining the foil and ilmenite data, the relative abundances for He^4, Ne^{20}, A^{36}, Kr^{84}, and Xe^{132} are 1, 1.8×10^{-3}, 5×10^{-5}, 2.15×10^{-8}, and 3.1×10^{-9}, respectively. These compare with cosmic abundances (Cameron, 1971) 1, 10^{-3}, 9×10^{-5}, 1.7×10^{-8}, and 0.9×10^{-9}. Corrected for presence of other isotopes, the relative abundances of He, Ne, A, Kr, and Xe indicated by the analyses of lunar foils and samples are 1, 2×10^{-3} 6×10^{-5}, 3.8×10^{-8}, and 1.2×10^{-8}.

Buehler *et al.* (1972) report the following He^4 fluxes (cm^{-2} sec^{-1}) in the solar wind: 21 July 1969, 6.2×10^6; 19 November 1969, 8.1×10^6; 5 February 1971, 4.2×10^6; and 31 July 1971, 17.7×10^6. The solar wind fluxes measured in Vela satellites on these same days were 1.5×10^8, 1.3×10^8, 1.3×10^8, and 10^8 (Solar Geophysical Data, Dept. of Commerce) indicating an average flux for these four days of about 40% of the average value of 3×10^8 for the interval July 1965 to June 1967 (Hundhausen *et al.*, 1970). The He/H ratios are clearly quite variable, as can be seen from the above figures, and cannot be used to improve the average value of 0.045 quoted previously. We therefore for this discussion accept 0.045 as the best value for the He/H ratio and accept the relative abundances quoted in the previous paragraph for the noble gases. The He/H ratio of 0.045 and an average solar wind flux of $3 \times 10^8 \text{ cm}^{-2} \text{ sec}^{-1}$ yields a helium flux somewhat greater than the average of the He determinations from the foil experiment as listed above, but not as much greater as would be expected on the basis of the average solar wind proton flux compared with that for the days in question.

Thermal escape from the moon is very rapid for light gases such as hydrogen and helium, but the escape time rises exponentially for heavier gases and is quite long (approximately 200 yr) for gases as heavy as neon. Another escape process dominates for those particles whose thermal escape is slow, and this process is ionization followed by interaction with the magnetic and electric fields of the solar wind (Manka and Michel, 1971; Manka *et al.*, 1972). When a particle in the lunar atmosphere becomes photoionized, it immediately accelerates in response to the electric field of the solar wind. The electric field direction is perpendicular to both the solar wind velocity and the magnetic field. As the particle picks up velocity, it begins to react to the magnetic field and finally traces out a cycloidal path whose average direction is the same as that component of the solar wind velocity perpendicular to the magnetic field. What is important with regard to loss from the lunar atmosphere is that the newly ionized particle accelerates in the direction of the solar wind electric field and its motion is not much deviated by the magnetic field until it has moved a large fraction of a gyro radius (as observed in a coordinate system moving with the solar wind). As the radii of gyration for all but the lightest ions are large compared with the moon, the heavier ions are lost, either by acceleration to space or back into the lunar surface, in a time shorter than the ion angular gyro period. As the ion gyro periods are very short compared with the times required to photoionize gas particles, it is the ionization time that controls the loss process; this time is about 10^7 sec for most particles, although 10^6 sec is a better figure for argon. The loss times may be taken to be twice the ionization times to account for the fact that about half the ions are accelerated to the lunar surface and hence are not lost from the moon.

The distance over which the magnetic and electric fields in the solar wind are more or less uniform is very large compared with the size of the moon, approximately 0.01 AU (Jokipii, 1971).

Another loss process results from collisions between solar wind particles and atmospheric particles. For the most part, atmospheric particles undergoing such collisions are driven back to the lunar surface and are not lost from the moon. Only around the terminator are the particles lost to space. Overall, this loss rate is small compared with that due to ionization, and it is neglected here.

Table 1 shows the expected amounts of lunar atmosphere due to the solar wind source, assuming thermal escape for hydrogen and helium and ionization loss for heavier constituents. The total amount of gas in a unit column is the product of the incident flux and the lifetime, and the gas concentration is obtained by dividing the

Table 1. Expected amounts of lunar atmosphere near the subsolar point due to solar wind impingement on the lunar surface.

	H	He	Ne	Ar	Kr	Xe
Solar wind flux, atoms cm^{-2} sec^{-1}	3×10^8	1.3×10^7	2.7×10^4	8×10^2	0.5	0.16
Escape time, sec	3.5×10^3	10^4	2×10^7	2×10^6	2×10^7	2×10^7
Total gas, atoms cm^{-2}	10^{12}	1.3×10^{11}	5.4×10^{11}	1.6×10^9	10^7	3.2×10^6
Scale height, km	2000	500	100	55	25	15
Surface concentration, molecules cm^{-3}	5×10^3	3×10^3	5×10^4	3×10^2	4	2
Surface pressure, Torr	1.4×10^{-13}	8×10^{-14}	1.5×10^{-12}	8×10^{-15}	10^{-16}	6×10^{-17}

total amount by the scale height, also shown in Table 1. Owing to the highly variable temperature over the lunar surface, the gas concentrations cannot be expected to be uniform, and the figures given in Table 1 apply to the hot portion of the lunar surface, roughly that quarter of the surface for which the solar zenith angle is less than 60°. Note that neon is expected to be the principal constituent of such an atmosphere, lighter constituents being less plentiful because their thermal escape is so rapid and heavier constituents less plentiful because of their lower abundances in the solar wind.

The diffusion of gas over the lunar surface is a function of temperature. Gases that do not escape rapidly and that do not condense or adsorb on the cold nighttime surface distribute themselves according to a $T^{-5/2}$ concentration law, where T is the temperature of the lunar surface (Hodges and Johnson, 1968). This leads to concentrations on the nightside about 26 times greater than over the hot portion of the dayside. This effect increases the overall time constant for escape, as there is a substantial nighttime reservoir that does not participate in the escape processes. Table 2 shows concentrations of argon and neon to be expected on the day and night sides at the surface and at altitudes of 10 and 100 km, based on the solar wind source and the $T^{-5/2}$ concentration law.

INTERNAL RELEASE

Another possible source of lunar atmosphere is gas release from internal sources. The earth and presumably other terrestrial planets have acquired their atmospheres in this fashion. Table 3 shows the average rates of release over geologic time of the principal gases arising from the earth's interior (Johnson, 1971). Table 4 shows the concentrations to be expected in the lunar atmosphere if the release rate on the moon were the same per unit mass as on earth. As for Table 1, the total gas in a unit column is the product of the lifetime and the release rate, and the concentration is obtained by dividing by the scale height. The release rates per unit mass could be several orders

Table 2. Expected concentrations per cubic centimeter of neon and argon at the lunar surface and at 10 km and 100 km above the lunar surface due to the solar wind source.

	Neon		Argon	
	Day	Night	Day	Night
Surface	5×10^4	1.3×10^6	3×10^2	8×10^3
10 km	4×10^4	9×10^5	2.5×10^2	4×10^3
100 km	1.8×10^4	2.3×10^4	50	6

Table 3. Average rates of release over geologic time of gases
from the Earth's interior.

	10^{11} molecules cm^{-2} sec^{-1}
H_2O	6×10^9
CO_2	2×10^8
N_2	3×10^2
Ne	2×10^6
Ar	2×10^2
Kr	1.5×10
Xe	

Table 4. Expected amounts of lunar atmosphere near the subsolar point if the release rate per unit mass is same as for earth.

	H_2O	CO_2	N_2	Ne	Ar	Kr	Xe
Release rate, molecules cm^{-2} sec^{-1}	1.6×10^{10}	10^9	3.5×10^7	50	3.5×10^5	35	2.5
Lifetime, sec	3×10^4	10^7	2×10^7	2×10^7	2×10^6	2×10^7	2×10^7
Total gas, molecules cm^{-2}	5×10^{14}	10^{16}	7×10^{14}	10^9	7×10^{11}	7×10^8	5×10^7
Scale height, km	111	45	70	100	50	25	15
Surface concentration, molecules cm^{-3}	5×10^7	2×10^9	10^8	10^7	1.4×10^5	3×10^2	30
Surface pressure, Torr	1.3×10^{-9}	5×10^{-8}	3×10^{-9}	3×10^{-15}	4×10^{-12}	8×10^{-15}	10^{-15}

of magnitude lower on the moon than on earth and still be important compared with the solar wind source. Release from the lunar interior is apt to be sporadic and hence gas release events might be expected to appear with considerable prominence for restricted periods of time.

RESULTS FROM APOLLO 14 AND 15

The most important result so far obtained from the ALSEP cold cathode gauges is that the nighttime atmospheric concentration is very low, about 2×10^5 particles cm^{-3}, corresponding to 10^{-12} Torr. This is much lower than the expected concentration of neon from the solar wind. Thus, it appears that less neon is being released from the lunar surface than is impinging upon it, and hence that the lunar surface is not saturated with neon. The contaminant gases from the Apollo operations, and perhaps some natural lunar gases as well, are very completely adsorbed on the lunar surface at night and hence do not show up in the nighttime observations. Neon and argon should not adsorb in this manner and their nighttime concentrations are presumably much greater than their daytime values, although both daytime and nighttime concentrations are apparently less than the values indicated in Tables 1 and 2.

The lunar module has proved to be an intermittent gas source, especially near the time of the first lunar sunset. Figure 2 shows conditions just after the first lunar sunset for Apollo 14. The concentration had reached the low value typical of lunar night when a sudden increase occurred that lasted for several days. Two other small increases were superimposed on the big increase; these were at first thought possibly to be releases of gas from the lunar interior because their rise times were longer than those characteristic of impulsive releases of gas from the lunar module. However, the frequency of such events has decreased with time and it now seems more reasonable to associate them with the LM than with the moon. Just after the first sunset on

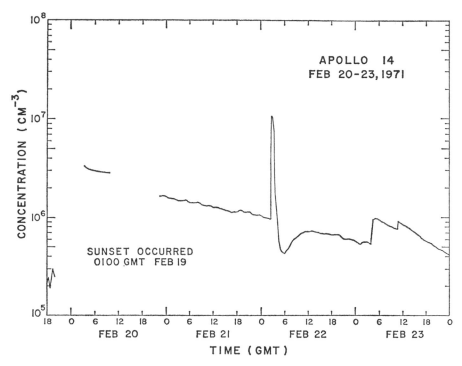

Fig. 2. Gas concentration at Apollo 14 site following first lunar sunset.

Apollo 15, a large release of gas occurred, less long lasting than on Apollo 14 but larger in magnitude, as shown in Fig. 3. Two short bursts of gas followed the large increase, reminiscent of the two small increases on Apollo 14. Apollo 15 exhibited a similar gas release at the time of its second lunar sunset, as shown in Fig. 4.

A short burst of gas occurred with amazing regularity on Apollo 14, following each lunar sunset by about two hours. The effect persisted for at least eight months with slowly decreasing amplitude. The source of this recurrent burst is almost surely associated with the LM.

The gas concentration during the daytime has been larger than at night by about two orders of magnitude. Figure 5 shows data for Apollo 14 for several lunar days along with the monthly temperature cycle for the gauge. The lunar days are indicated by the labels on the curves; no data were obtained for day 1, the day of the landing. Until day 9, data were not obtained during the middle of the day because high voltage arcing problems were feared at the higher temperatures prevalent then. To avoid crowding, data are not shown for days 6, 7, and 8, but they fall between the curves for days 5 and 9. The concentration has tended to decrease on each successive lunar day with a time constant of about three or four months. The decay in gas concentration does not appear to be entirely uniform and the difference in concentrations near midday on days 9 and 10 should not be used with confidence as an indication of decay rate; note for example the irregularity about six earth days after sunrise on day 10. Gas concentrations at the same temperature forenoon and afternoon on the same day are slightly lower during the afternoon.

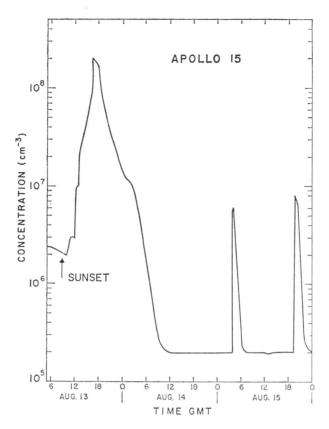

Fig. 3. Gas concentration at Apollo 15 site at time of first lunar sunset.

The combination of gas concentration and time constant suggests the adsorption on the lunar surface of a surprisingly large fraction of the LM exhaust gas, perhaps 10%. The initial gas concentration at the gauge was about 10^8 cm^{-3}. Assuming an outflow velocity from the landing site of 10^4 cm sec^{-1} and assuming uniform outflow over a hemisphere of radius 2×10^4 cm (the approximate distance from the LM to the gauge), the initial rate of gas loss was 2.4×10^{21} atoms sec^{-1}, or 0.1 g sec^{-1} for a molecular weight near 30. The estimate of total loss is obtained by multiplying the initial loss rate by the time constant, about 10^7 sec, but this must be reduced by about a factor of 4 since loss is important only over approximately the hottest quarter of the day, giving a total of about 2.5×10^5 g. The amount of exhaust gas directed at the landing site is about 3 tons, or 3×10^6 g.

First Lunar Sunrise after Apollo Landing

The gas concentrations during the first lunar sunrise on both Apollo 14 and 15 were different from the following sunrises. The degree of contamination was undoubtedly higher at that time than at later sunrises, as a considerable release of gas occurred after the first sunset and much of this must have become adsorbed near the

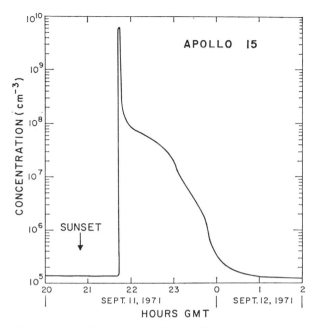

Fig. 4. Gas concentration at Apollo 15 site following second lunar sunset.

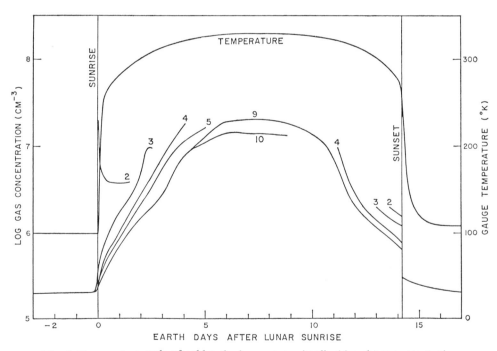

Fig. 5. Temperature cycle of cold cathode gauge on Apollo 14 and gas concentration data for several lunar days. The lunar days are numbered taking the day of the landing to be day 1.

landing site. The gas concentration rose about two orders of magnitude in about half an hour at sunrise and then fell slightly or held steady, as shown in Fig. 6 for Apollo 14 and Fig. 7 for Apollo 15.

The release of gas shown in Fig. 6 must be considered a possible source of the water vapor reportedly seen by the suprathermal ion detector experiment (SIDE) at this time (Freeman, 1971; Freeman *et al.*, 1972). Nothing unusual can be seen in the record during the interval 19.5 to 33 hours after sunrise, shown by the vertical lines in Fig. 6, during which the water vapor ions were observed. However, this is not necessarily inconsistent, as the SIDE is sensitive only to ions, and these may have come from a distant location without significant attenuation because of guidance by electric or magnetic fields—from too far away to be detected in terms of neutral gas which undergoes inverse square law attenuation in dispersing from its source. The SIDE observations were of 49.6 eV ions of mass-to-charge ratio in the range 18–23.3 (Freeman *et al.*, 1971) and of nearly monoenergetic ions near 70 eV/unit charge. The absence of lower energy ions is important as this indicates that the source cannot be photoionization of contaminant gases arising from the landing site.

The observed gas concentration at the landing site at the time of the observations in question was about 5×10^6 cm^{-3}, as can be seen from Fig. 6. The concentration of contaminant gases arising from the landing site must fall off according to an

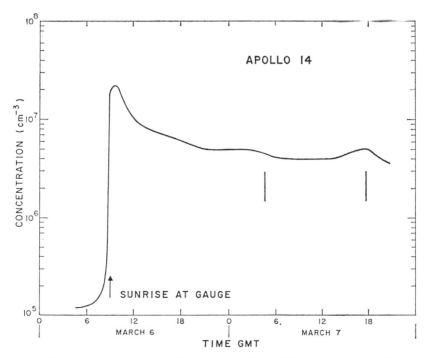

Fig. 6. Gas concentration at Apollo 14 site at time of first lunar sunrise. The two vertical lines indicate the time interval over which the observation of water vapor ions has been claimed by Freeman (1971).

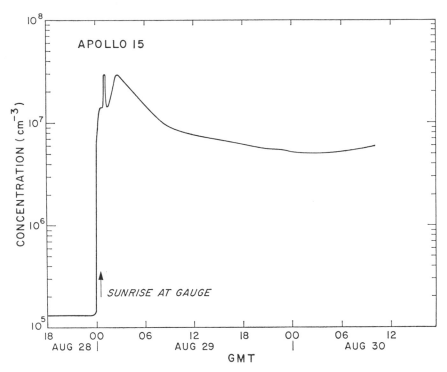

Fig. 7. Gas concentration at Apollo 15 site at time of first lunar sunrise.

inverse square law beyond some distance. Assuming an inverse square law of distance from the LM to apply from the ALSEP outward, the gas in a unit column would be about 5×10^{10} cm^{-2}. The photoionization rate for most gases is about 10^{-7} sec^{-1}. Thus a source strength of about 5×10^3 ions cm^{-2} sec^{-1} can be expected from the contaminant gas. Once ionized, the particles will be accelerated in the direction of the electric field and they should be seen by the SIDE only if the field direction happens to coincide with the field of view, which is nearly vertical and about 6° square (Freeman *et al.*, 1971). Since the SIDE field of view is near 10^{-2} ster, this would be reported as a flux of 5×10^5 ions cm^{-2} sec^{-1} ster^{-1}. Since most of the gas in a unit column is nearby (90% within the first km and 99% within the first 10 km), the spectrum should fall off very rapidly with energy. Changing the distance beyond which the distribution of gas particles varies according to an inverse square law affects the arguments only slightly. Neither the total observed flux nor the observed monoenergetic spectrum near 70 eV could be produced by such a cloud of contaminant gas.

Another possibility is that the observed gas might not be water vapor, but rather neon. The ions were observed in channel 5, which passed ions in the mass range 18–23.3 (Freeman *et al.*, 1971), and hence neon would seem to be as acceptable an identification as water vapor. Assuming the principal gas measured at night by the cold cathode gauge to be neon, the concentration early in the day should be near 10^4 cm^{-3}. The scale height is near 100 km, so the total gas in a unit vertical column is about 10^{11} atoms. The ionization rate is about 10^{-7} sec^{-1}, giving a total production in a unit column of 10^4 ions sec^{-1}. If the electric field orientation were such as to direct these into the SIDE, the observed flux would be 10^6 ions cm^{-2} sec^{-1} ster^{-1}. This is an order of magnitude below the observed flux and the lack of low energy ions also argues against this interpretation. However, with a scale height of 100 km, a sizable fraction of the ions would have energies in the observed range, as the electric field associated with the solar wind is normally of the order of 1 V km^{-1}. Further, these ions should be seen whenever conditions are right to observe ions from any other source within the field of view of the instrument.

The absence of low energy ions argues that the source must be localized and at an altitude in the vicinity of 100 km. Such a cloud could be produced by a waste water dump, a possibility suggested by El Baz (1971). The water would quickly freeze and the ice could survive in orbit for a relatively long time, gradually releasing vapor which could serve as a source of the ions seen by the SIDE.

Acknowledgment—This work was supported by NASA under contract number NAS 9-5964. We specially acknowledge the cooperation and assistance of Mr. Frank Torney at Norton Research Corporation in connection with the gauge, Dr. John Freeman at Rice University in connection with the gauge electronics, Dr. George Chang at Bellcomm, Inc. in connection with the lunar environment. Mr. James Sanders at NASA Manned Spacecraft Center and Mr. Derek Perkins at Bendix Aerospace Systems Division in connection with the incorporation of the gauge into the ALSEP, Mr. Charles Gosselin at Midwest Research Institute in connection with the gauge calibration, and many others too numerous to mention throughout the Apollo project.

REFERENCES

Bame S. J., Asbridge J. R., Hundhausen A. J., and Montgomery M. D. (1970) Solar wind ions: ^{56}Fe^{+8} to ^{56}Fe^{+12}, ^{28}Si^{+7}, ^{28}Si^{+8}, ^{28}Si^{+9}, and ^{16}O^{+6}. *J. Geophys. Res.* **75**, 6360–6325.

Buehler F., Cerutti H., Eberhardt P., and Geiss J. (1972) Results of the Apollo 14 and 15 solar wind composition experiment (abstract). In *Lunar Science—III* (editor C. Watkins), pp. 102–104, Lunar Science Institute Contr. No. 88.

Cameron A. G. W. (1971) A new table of abundances in the solar system. In *Origin and Distribution of the Elements* (editor L. H. Ahrens), pp. 124–145, Pergamon.

Eberhardt P., Geiss J., Graf H., Grogler N., Krahenbuhl U., Schwaller H., Schwarzmuller J., and Stettler A. (1970) Trapped solar wind noble gases, exposure age, and K/Ar age in Apollo 11 lunar fine material. *Proc. Apollo 11 Lunar Sci. Conf., Geochim. Cosmochim. Acta* Suppl. 1, Vol. 2, pp. 1037–1070. Pergamon

El Baz F. (1971) Private communication.

Freeman J. W. (1971) Various newspaper accounts, October 16.

Freeman J. W. Jr., Fenner M. A., Hills H. K., Lindeman R. A., Medrano R., and Meister J. (1971) Suprathermal ions near the moon. Presentation at XVth General Assembly, International Union of Geodesy and Geophysics, Moscow, August 1971.

Freeman J. W. Jr., Hills H. K., and Vondrok R. R. (1972) Water vapor, whence comest thou? (abstract). In *Lunar Science—III* (editor C. Watkins), p. 283, Lunar Science Institute Contr. No. 88.

Geiss J., Buehler F., Cerutti H., Eberhardt P., and Meister J. (1971) The solar wind composition experiment. In *Apollo 14 Preliminary Science Report*, NASA SP 272, pp. 221–226.

Hundhausen A. J. (1970) Composition and dynamics of the solar wind plasma. *Rev. Geophys. Space Phys.* **8**, 729–811.

Hundhausen A. J., Bame S. J., Ashridge J. R., and Sydoriak S. J. (1970) Solar wind proton properties: Vela 3 observations from July 1965 to June 1967. *J. Geophys. Res.* **75**, 4643–4657.

Hodges R. R. Jr. and Johnson F. S. (1968) Lateral transport in planetary exospheres. *J. Geophys. Res.* **73,** 7307–7317.

Johnson F. S. (1971) Lunar atmosphere. *Rev. Geophys. Space Phys.* **9,** 913–823.

Jokipii J. R. (1971) Propagation of cosmic rays in the solar wind. *Rev. Geophys. Space Phys.* **9,** 27–87.

Manka R. H. and Michel F. C. (1971) Lunar atmosphere as a source of lunar surface elements. *Proc. Second Lunar Sci. Conf., Geochim. Cosmochim. Acta* Suppl. 2, Vol. 2, pp. 1717–1728. MIT Press.

Manka R. H., Michel F. S., and Freeman J. W. (1972) Evidence for acceleration of lunar ions (abstract). In *Lunar Science—III* (editor C. Watkins), pp. 504–506, Lunar Science Institute Contr. No. 88.

Ness N. F., Behannon K. W., Scearce C. S., and Cantarano S. C. (1967) Early results from the magnetic field experiment on lunar Explorer 35. *J. Geophys. Res.* **72,** 5769–5778.

Sonett C. P. and Colburn D. S. (1967) Establishment of a lunar unipolar generator and associated shock and wake by the solar wind. *Nature* **216,** 340–343.

Proceedings of the Third Lunar Science Conference
(Supplement 3, *Geochimica et Cosmochimica Acta*)
Vol. 3, pp. 2243–2257
The M.I.T. Press, 1972

Some surface characteristics and gas interactions of Apollo 14 fines and rock fragments

D. A. CADENHEAD, N. J. WAGNER,* B. R. JONES, and J. R. STETTER

Department of Chemistry, State University of New York at Buffalo,
Buffalo, N.Y. 14214

Abstract—Over the past year we have carried out a comprehensive survey of the physical surface characteristics of Apollo 14 fines, two fragments of a breccia (14321) and of a crystalline rock (14310). The survey was carried out with optical and both scanning and transmission electron microscopy and by studying the adsorption of a variety of gases including nitrogen, hydrogen, and water vapor.

Our objective in the optical microscope study was to relate the visible geological and petrological features to the surface properties. Electron microscopy particularly helped relate surface roughness and particle fusion to gas adsorption and pore structure. The fine sample (14163,111) had a surface area of 0.210 m²/g and a helium density of 2.9 g/cc. Similar values have been observed with breccia fragments. Other observations include physical adsorption of molecular hydrogen at low temperatures and of water vapor at ambient temperatures.

Our present conclusions are that these particular lunar materials, while capable of adsorbing water vapor, do not retain it for any significant time at low pressures, nor, under lunar conditions, is there any indication of absorption or penetration. In an earth-like atmosphere, weathering would be expected to be a relatively slow process, and brief exposure to water vapor should produce little or no damage to these or similar samples.

Finally, although molecular hydrogen will not chemically affect such lunar sample surfaces, the possibility that the solar wind may bring about some surface reduction of exposed materials remains a very real one.

INTRODUCTION

AN UNDERSTANDING OF gas interactions with lunar materials can clearly provide at least partial answers to many important questions. Thus, it would be interesting to know what effect the solar wind might have on the surface chemical composition of lunar soil and rocks. Also bearing in mind the apparent inevitability of at least some contamination of collected samples by astronaut, lunar module, or earth atmospheres, we should investigate what physical and chemical effects these might have on lunar sample surfaces and to what extent such exposures could lead to erroneous interpretations of lunar history. Finally, would brief exposure to an earth-like atmosphere adversely affect such samples by producing a rapid initial "weathering?"

In addition to answering such questions, it is important to correlate gas adsorption behavior with the differing species of samples returned so that "typical" behavior may be predicted. Needless to say, any such prediction will improve as the range of samples examined is broadened and at this initial stage must be regarded as somewhat tentative.

Bearing this in mind, we have accompanied our gas adsorption work with an

---

* Present address: Department of Chemistry, Alliance College, Cambridge Springs, Pa.

extensive optical and electron microscope study and the results, which we are reporting here, will be listed under the two main headings of microscopy and gas-interaction studies.

I. Microscopy

Optical microscopy

Optical microscopy studies were carried out primarily using two microscopes: an American Optical Stereoscopic and a Reichert Photo-Plan. Studies of the bulk fines (14163,111) with crossed nicols and a typical magnification of $50\times$, revealed a wide variety of minerals in both microcrystalline and glassy forms including spheres. An evaluation of macroscopic external shape and size distribution (≤ 0.1 mm) of randomly selected articles indicated that a minimum value of a surface area should be ≥ 0.1 m^2/g.

Many of the features described for larger rocks were evident in the surface fragments supplied (Mason, 1971). Typically in the breccias the fragmental nature of the samples was self-evident (Fig. 1a; 14321,74). More relevant to surface studies were the highly fragmented surface regions and the considerable defect structure as well as an amorphous, porous surface structure seen in Fig. 1b.

Fig. 1a. Breccia 14321, clast inclusions, $50\times$.

Fig. 1b. Breccia 14321, defect structure and amorphous surface, $50\times$.

By contrast, optical studies of crystalline rock fragments, even those possessing external exposed surfaces, indicated relatively low surface areas. Thus, while microcrystalline mineral irreversible differentiation and a certain degree of surface roughness are observable, there is no evidence of either extensive internal pore or defect structure except for some geode formation (sample 14310,98; Fig. 2). Micrometeorite impact on the crystalline samples resulted in relatively shallow indentations with the formation of glass-lined pits and, in this case, appeared to decrease rather than increase the surface area (Fig. 3; 14310,192).

Scanning electron microscopy

The studies described here were carried out on a Hitachi Scma Model I, scanning electron microsccpy (SEM). The SEM permitted a search for possible microbreccia and/or pore structure in bulk fines (14163,111). Figures 4a and 4b, respectively, represent 325× and 2200× magnification of the central clump of material in Fig. 4a. It is clear that a considerable amount of surface roughness still exists even at the submicron level with at least some possible internal pore structure.

Fig. 2. Crystalline fragment 14310, geode formation, 100×.

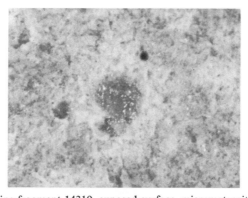

Fig. 3. Crystalline fragment 14310, exposed surface, micrometeorite impact, 50×.

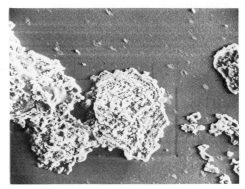

Fig. 4a. Fines 14163,111, 325× (SEM).

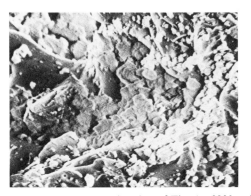

Fig. 4b. Fines 14163,111 central region of Fig. 4a, 2200× (SEM).

Fig. 5. Breccia fragment, seetion of 14321, 80× (SEM).

Minute fragments of breccia material were also used for the SEM studies. As with the fines, indications of both surface roughness and pore structure at the submicron level were found. A few of the fragments, however, were clearly glassy and, while showing considerable surface roughness, gave no indication of internal pore structure (14321,126; 80×, Fig. 5). Clearly, the precise nature of the overall breccia surface reflects the surfaces of the fragments from which the breccia is formed, as well as their individual exposure history.

Transmission electron microscopy

These studies were carried out with a J.E.M. Model 6 Transmission Electron Microscope (TEM). The major objective in this portion of the work was to obtain an evaluation of maximum possible surface roughness and to observe internal pore structure by beam penetration through randomly scattered fine particles. A typical fine (14163,111) TEM picture is shown in Fig. 6a (14163,111; 20,000×), and it can be seen that pores arising through the compaction of smaller particles are present in the size range 100 Å (i.e., "intermediate pores" (Dubinin, 1955). Studies of microbreccia fragments reveal essentially similar features (Fig. 6b), i.e., pore formation apparently arising through pressure fusion of smaller particles (rather than by gas evolution through solidifying molten rock).

II. Gas-Solid Adsorption Studies

Surface areas

The conventional method of measuring surface areas involves measuring the amount of nitrogen adsorbed at monolayer coverage and liquid nitrogen temperatures usually with a volumetric system (Faeth and Willingham, 1955), and, from a knowledge of the area occupied by an individual nitrogen molecule (McClellan and Harnsberger, 1967), calculating the surface area from a BET plot. Anticipating somewhat low surface areas we chose to use a gravimetric system rather than switch to the use of krypton or xenon as adsorbates. The system (Cadenhead and Wagner, 1968) and the specialized adsorption techniques (Cadenhead and Wagner, 1971) have been described elsewhere. However, for all adsorption measurements one particular problem merited special attention: that of sample density measurement.

When measuring surface areas or studying gas adsorption in a gravimetric microbalance system, it is important to know the density of the sample under study with some precision so that accurate taring of the sample may be achieved. If this is not done microbalance corrections are magnified and the precision of the gas adsorption study greatly reduced. Where samples having inclusions, defects, or pore structure are involved, the density measuring fluid should penetrate all such regions, while maintaining a uniform bulk density in contact with the sample (i.e., the fluid should not be adsorbed and should, in no way, contaminate the sample surface). Clearly helium constitutes an ideal density measuring fluid and we have been measuring sample densities in helium using a classical BET system based on that of Faeth and Willingham, with a specially designed pycnometer for breccia fragments. The density of the

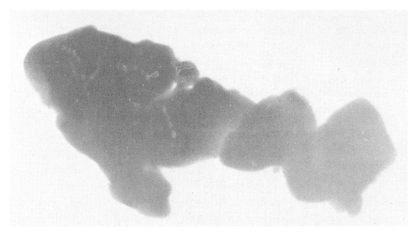

Fig. 6a. Fines 14163,111, 20,000× (TEM).

LUNAR AND
PLANETARY UNIT.

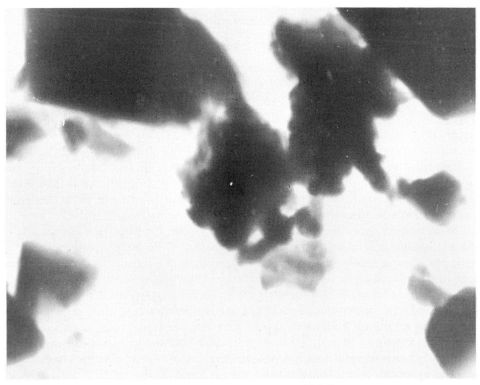

Fig. 6b. Breccia 14321, 20,000× (TEM).

bulk fines (14163,111) in helium was observed to be 2.9 g/cc while those of breccia fragments 14321,74 and 14321,156 were both 3.2 g/cc, all values ±0.1 g/cc. Even a 0.1 g/cc density error can produce a significant buoyancy error. Thus, for a sample having a density of 3.2 g/cc and about 0.3 m²/g surface area, the buoyancy error resulting from 0.1 g/cc density measurement error would be 8% at 100 Torr, but only 4% at 50 Torr. Since a monolayer is usually adsorbed well below 50 Torr our buoyancy error is less than 4%. A failure, however, to carry out such a density measurement could clearly result in a much larger error.

Corrections due to electrostatic and thermomolecular effects have been minimized or allowed for in the calibration curve. One final precaution was taken to ensure that the sample was in thermal equilibrium at −196°C before commencing the measurement of the nitrogen adsorption isotherm. A partial pressure of helium was admitted for an hour or so until equilibrium was perceptibly achieved. After this the sample was again evacuated and nitrogen admitted.

Based on the microscopy studies the obvious decision was to concentrate gas-adsorption studies on the lunar fines and breccia fragments. Samples were outgassed at 140°C and then immersed in a liquid nitrogen bath stirred by bubbling nitrogen gas through the bath. The temperature was read on a vapor pressure thermometer. The bulk fines surface area, determined by nitrogen gas adsorption, was 0.21 m²/g. The BET plot for this system is illustrated in Fig. 7. A similar measurement was carried out on selected samples from breccia fragments 14321,156 giving an area of 0.34 m²/g. While the precision of the measurement is similar for both fines and breccia fragments, the value for the former is more representative, because the fines by their very nature may be considered as already constituting a randomized sample, whereas the latter value is specific to those particular fragments studied.

Hydrogen adsorption

Hydrogen adsorption at this time has been carried out only on the fines 14163,111. The hydrogen gas used was purified by passage through a palladium-silver thimble. Adsorption studies so far have been carried out at three temperatures: −196°, −78°C, and 25°C. At −196° and −78°, the measurements were made gravimetrically, while at 25°C a classical volumetric system was used. Hydrogen adsorption, if it took place at −78°C was negligible (less than 1 μg adsorbed on a 0.6492 g sample). Similarly, no adsorption was observed at 25°C. Adsorption was, however, observed at −196°C, with about 14μg being adsorbed at 16 Torr. The isotherm was a relatively featureless Type I isotherm, i.e., a typical Langmuir Type isotherm, yielding a surface area some 2% higher than that obtained with nitrogen. While the error in the hydrogen adsorption values appears of the order of 2%, it is possible that the difference is a real one, and that hydrogen molecules can penetrate regions inaccessible to nitrogen.

The disappearance of an adsorption present at lower temperatures when the temperature is raised from −196° to −78°C and then to 25°C is typical of a weak form of adsorption, presumably physical adsorption. The fact that it is possible to completely remove all adsorbed hydrogen and return to the original outgassed weight of the sample (within 0.5 μg) is further confirmation. It would, therefore, seem that

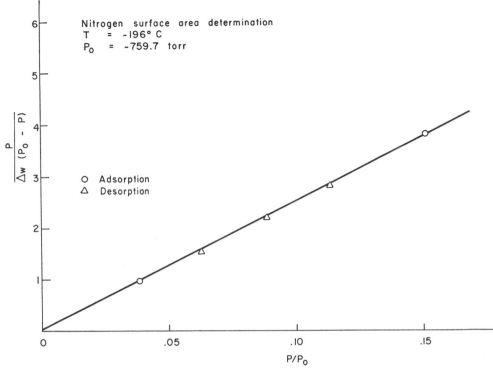

Fig. 7. BET nitrogen surface area of 14163,111. *Abscissa:* Relative pressure P/P_0. *Ordinate:* $P/\Delta W (P_0 - P)$ where ΔW is the weight of material absorbed.

some molecular hydrogen will physically adsorb at the lowest temperatures observed on the moon ($-133°C$) (Reed *et al.*, 1971). It may also be that at higher temperatures (i.e., up to a lunar high-noon temperature of 150°C), it might be possible to chemisorb molecular hydrogen; however, such temperatures are probably still too low to dissociatively adsorb hydrogen on such a surface.

Water adsorption

Water adsorption has, so far, been carried out only on the lunar fines 14163,111. The water was triply distilled from an all-glass apparatus (Cadenhead, 1969), and thoroughly degassed before use. All water adsorption was measured on the gravimetric (Cahn RG) system previously described (Cadenhead and Wagner, 1968 and 1971), the pressures being measured on a calibrated quartz Bourdon spiral gauge. Isotherms were measured at 15°, 20° and 37.5°C. In all cases, a water bath stirred with bubbling nitrogen gas was used to thermostat the system, the vibration due to the stirring limiting the balance weigh change detectability to 1.0 μg. At 37.5° the bath was maintained at about 36° with a non-regulated heater, the final temperature being achieved to within 0.02°C with a second heater controlled by a proportionating controller. The remainder of the balance chamber was maintained several degrees above 37.5°C by use of regulated heating tapes. Even so, problems arose on approaching saturation vapor pressure, on this particular isotherm, and it was terminated at about

0.81 relative pressure. At 15° and 20°C the temperature of the water bath was maintained at about one degree below the desired temperature, using a Lauda T-30 com-
pressor refrigerator, the final temperature again being achieved to within $\pm 0.02°$
using a heater controlled with a proportionating controller. Temperatures were read
using a calibrated Chromel-Alumel thermocouple. Sample outgassing, as before, was
carried out at 140°C; however, a second set of isotherms at 15°C were determined
after outgassing at 350°C.

In addition to the many problems already outlined involving microbalance use for
gravimetric adsorption studies, it would seem that even more serious difficulties exist
specifically for water adsorption studies involving the Cahn microbalance. Our attention was drawn to this when on changing sample buckets we observed that the calibration curve had also changed, i.e., that the calibration curve was in fact (for water
adsorption, as measured on a Cahn balance) a function of the symmetrical weight
loading. That it is due to the Cahn RG balance and not to some other feature of our
adsorption system is shown by the similar observation of Evans and White (1967).
In Fig. 8 calibration curves are illustrated for two symmetrical weight loadings of the

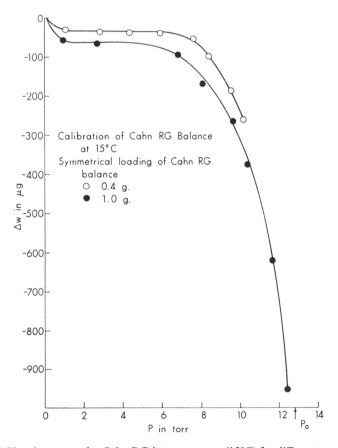

Fig. 8. Calibration curves for Cahn RG in water vapor (15°C) for different symmetrical
loading.

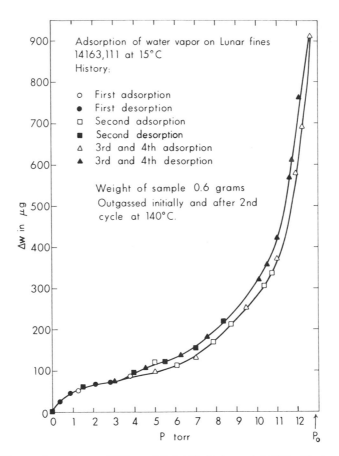

Fig. 9. Water adsorption at 15°C on 14163,111. Outgassed 140°C. *Abscissa:* Vapor pressure in Torr. *Ordinate:* Amount absorbed in μg.

Cahn RG at 15°C. (The calibration curves are also temperature sensitive.) We also observed, as did Evans and White, that the shift in the calibration curve was approximately proportional to the symmetrical weight loading. Thus, when adjusted to a common symmetrical weight, the curves shown in Fig. 8 become one and the same. All subsequent calibration curves were obtained at 0.948 g symmetrical weight loading.

In determining the curves obtained in Fig. 8, it was noted that equilibrium was rapidly attained at pressures up to 7 and pressures above 10 Torr. Between 7 and 10 Torr, however, equilibrium was attained very slowly, taking up to six hours or more. Typically these problems arose near 0.9 relative pressure on adsorption and 0.8 relative pressure on desorption and, noting the large corrections necessary as S.V.P. is approached, it is clear that unless considerable care is taken in determining these calibration curves, particularly at high relative pressures, spurious weight changes will be observed. Finally, it should be noted that while the specific effects may vary from one system and/or Cahn balance to the next, there is little doubt that similar effects will be seen when observing water adsorption with *any* Cahn RG gravimetric system.

The adsorption isotherm at 37.5°C has been previously published (Cadenhead, *et al.*, 1972). The adsorption isotherm at 15°C is shown in Fig. 9. Sample 14163,111 was outgassed at 140°C after re-equilibrium at 15°C and the data listed as Run 1 and 2 were obtained. The sample was then again outgassed at 140°C and the data in Runs 3 and 4 obtained. The reproducibility of the data indicated sample stability under these pretreatment conditions. The sample was then outgassed at 350°, the results subsequently obtained being illustrated in Fig. 10. In Run 1 adsorption was terminated at essentially monolayer coverage, desorption being completely reversible. Run 2 was taken to S.V.P. and held overnight at a pressure just below S.V.P. On desorbing the isotherm exhibited extensive hysteresis as well as a 40 μg water retention after an overnight outgassing at 15°C. On repeating the adsorption desorption cycle without high temperature outgassing and allowing for the observed water retention, the isotherm showed considerable resemblance to that exhibited after an initial 140°C outgassing was used. (Runs 3 and 4, Fig. 10.)

Using only data obtained after 140° outgassing, and including a 20°C isotherm,

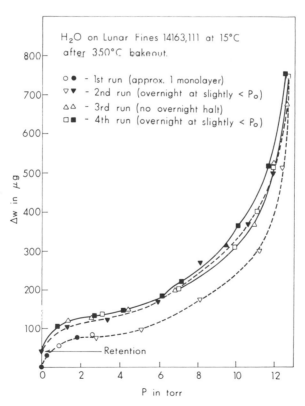

Fig. 10. Water adsorption at 15°C on 14163,111. Outgassed at 350°C. *Abscissa:* Vapor pressure in Torr. *Ordinate:* Amount adsorbed in μg.

an isosteric heat was calculated for $\frac{1}{2}$ monolayer coverage. The value obtained was 3.6 kcal relative to the heat of liquification of water, a value clearly reasonable for this type of surface and of the correct order of magnitude for a metal oxide surface (Adamson, 1966). Finally, monolayer coverage for water and nitrogen (and hydrogen) can be made to coincide if the area of the adsorbed water molecule is taken to be 8.6 Å², a not unreasonable though possibly slightly low value for a silicate (McClellan and Harnsberger, 1967). Alternatively, it may be that (as could be the case for hydrogen) water can "see" slightly greater surface than nitrogen.

Discussion

Microscopy

Our microscopy studies have clearly shown that there would be little point in carrying out gas-interaction studies with crystalline rock fragments, even heavily bombarded surface fragments. Simple calculations of surface area, bearing in mind observed surface roughness and internal pore structure, suggest areas of 0.01 m²/g magnitude. Such surface areas would adsorb amounts close to our limit of detectability and we would expect that only in exceptional cases would such crystalline samples justify the extensive investigation that would be necessary to establish their adsorption characteristics. This is particularly the case when both fines and breccia surface fragments available for study easily fall within the scope of gas adsorption investigational techniques.

The relatively high surface areas seen for fines and breccia surface fragments (0.1 to 1.0 m²/g) can readily be explained through microscopy. For the fines, surface area enhancement arises through small particle size, surface roughness (for some, but not all particles) and, to a lesser degree, internal pore structure. Electron microscopy reveals that this pore structure is a result of the pressure fusion of even small particles. With breccia surface fragments, considerable surface roughness, coupled with fracturing, leads to amorphous surface regions of apparently moderately high surface porosity. Again, pressure fusion seems to result in the formation of significant internal pore structure.

While the above conclusions are strictly limited to the Apollo 14 samples studied (bulk fines 14163, surface fragments of 14310 and 14321) there is considerable evidence to suggest that they may have considerably wider applicability. Thus, the mechanism of creation of internal pore structure through pressure fusion would seem to be the correct one for many of the samples returned so far. However, there are instances where high surface area, through internal pore structure, appear to result from gas evolution accompanying rock solidification (e.g. 15015).

Nitrogen adsorption and surface areas

The surface areas as determined by nitrogen adsorption at liquid nitrogen temperatures are consistent with microscopy observations and are of the same order of magnitude as that observed for sample 10087,5 by Fuller et al. (1971). The shape of their adsorption isotherm was also similar to those observed by us, i.e., a Type II BET classification, as would be expected for the physical adsorption of nitrogen on a complex silicate surface. The only other nitrogen isotherms published are those reported by Grossman et al. (1972), and since these show considerable differences from both our results and those of Fuller et al. (1971) they clearly merit further consideration.

We have always assumed that similar samples selected from different regions would exhibit qualitative similarities and quantitative differences. One sample which exhibited significantly different behavior (12001,118) was referred to as an "ultra-high vacuum" (UHV) sample ($<10^{-8}$ Torr) and its different behavior attributed at least in part to this special handling. However, the following points should be borne in mind:

(1) While this sample might have suffered less contamination than 14163,111 (found to be at 1 atm pressure when opened at the Lunar Receiving Laboratory) exposure to astronaut handling, to a 10^{-2} Torr pressure, and to a high vacuum ($>10^{-7}$ Torr) adsorption system all indicate at least partial contamination since at 10^{-6} Torr a typical gas will achieve monolayer coverage in a matter of seconds.

(2) The prediction of differing behavior for "clean" lunar samples is not unreasonable; however, the evidence presented by Grossman et al. appears somewhat contradictory. Thus the adsorption of nitrogen and argon some 174° and 149°C above their respective triple points is difficult to accept as physical adsorption, yet the ease of removal is not consistent with a chemisorbed gas.

(3) The nitrogen isotherms reported by Grossman et al. (1972), for an "exposed" sample (14259,93), should not be compared with those of Fuller et al. (1971) or with ours since the former are clearly nonequilibrium measurements.

Hydrogen adsorption

Clearly the molecular hydrogen adsorption we have observed is physically adsorption, the disappearance of adsorption at −78°C being consistent with this view. Raising the temperature could lead to dissociative chemisorption with the energetics of the surface dictating the precise temperature at which this would occur. We know that it does not happen at 25°C and suspect that an increase in temperature to 150°C (the Lunar mid-day temperature), would be insufficient. While we do intend to check this point we presently plan to proceed with an attempt to adsorb atomic (i.e., pre-dissociated) hydrogen in order to evaluate the possibility that the solar wind may reduce exposed surfaces of lunar materials.

Water adsorption

The water isotherms and the heat of adsorption reported for sample 14163,111 are in line with values obtained on other silica surfaces (Gaines, 1962). We were, therefore, concerned to observe considerable differences between our results and those obtained by Fuller et al. (1971) on sample 10087,5. As has already been stated, we would have expected these two samples to behave in a qualitatively similar way. Because of this, we turned our attention to the prime difference in our pretreatment: that of the outgassing temperature. Sample 14163,111 was outgassed at 140°C while 10087,5 was outgassed at 300°C. Our selection of 140°C was based on the idea that most of the physically adsorbed material would be removed at this temperature and that it was comparable with the maximum temperature seen under lunar conditions.

In order to evaluate this difference in pretreatment, we outgassed a portion of 14163,111 at 350° with the results already described (Fig. 10). Clearly the features of the isotherm initially obtained (Run 2) bear considerable resemblance to isotherm 1 reported by Fuller *et al.* (1971). In particular, we now observed extensive hysteresis and water retention. Moreover, their isotherm No. 8 resembles our isotherms No. 3 and No. 4 in that hysteresis is essentially a multilayer-capillary condensation phenomenon. Other features they reported, which we did *not* observe, included the desorption of water vapor with increasing water vapor pressure and the abrupt breaks in the adsorption and desorption isotherms at 0.9 and 0.8 relative pressure, respectively. These latter features we regard as artifacts and refer the reader to our discussion of specific calibration problems of the Cahn RG microbalance (used by Fuller *et al.*, 1971) when studying water vapor adsorption.

From the above, we may conclude that outgassing at higher temperatures (300–350°C) not only removes physically adsorbed but also chemisorbed material. In other words, the chemical nature of the surface is altered. Clearly this change weakens the bonding between fused particles such that the stress of water adsorption is now capable of rupturing such links enabling large amounts of near liquid water to be sorbed close to saturation vapor pressure and permitting the chemisorption and retention of water at low pressures on the newly created surface (Bailey *et al.*, 1970). Provided the sample is not again subjected to high temperature outgassing (Runs 3 and 4, Fig. 10) the isotherm returns to the form of that found with a stable silicate. (cf. Fig. 9) since no further rupturing of particle linkages occurs.

Since the recent reports of water vapor on the moon (Freeman *et al.*, 1972) (irrespective of its source) it behoves us to consider possible interactions of water vapor with lunar soil under lunar-like conditions. It would seem that the behavior to be expected, *once any highly active sites were occupied*, would be that of the low pressure region of the isotherm shown in Fig. 9, i.e., that water vapor would be reversibly adsorbed and desorbed as long as it persisted in the lunar atmosphere either as a function of pressure or temperature. Further irreversible adsorption would only take place if some newly created surface became available. Such an explanation is not inconsistent with the experimental observations of gas pulses in the lunar atmosphere in the vicinity of the Apollo 14 and 15 landing sites (Johnson, 1972). Finally, if we ever do succeed in bringing back a true ultrahigh vacuum sample to earth (and it is our opinion that this has not so far been achieved) then we would anticipate on exposure to water vapor, a possible small irreversible adsorption followed by reversible physical adsorption (Runs 3 and 4, Fig. 10).

Acknowledgments—We gratefully acknowledge the assistance of Dr. Carl H. Bates in utilization of the scanning electron microscope. We are pleased to acknowledge the financial assistance of the National Aeronautics and Space Administration through Grant NGR-33-183-004, in the completion of this work.

References

Adamson A. W. (1966) Adsorption of gases and vapors on solids. In *Physical Chemistry of Surfaces*, 2nd edition, Chap. XIII, pp. 565–648, Wiley-Interscience.

Bailey A., Cadenhead D. A., Davies D. H., Everett D. H., and Miles A. J. (1970) Low pressure hysteresis in the adsorption of organic vapours by porous carbons. *Trans, Faraday Soc.* **67,** 231–243.

Cadenhead D. A. (1969) Monomolecular films at the air-water interface. *Ind. and Eng. Chem.* **61,** 22–28.

Cadenhead D. A. and Wagner N. J. (1968) Low temperature hydrogen adsorption on copper-nickel alloys. *J. Phys. Chem.* **72,** 2775–2781.

Cadenhead D. A. and Wagner N. J. (1971) Gravimetric adsorption studies of hydrogen on granular metal surfaces using a vacuum microbalance. *Vacuum Microbalance Techniques*, Vol. 8, pp. 97–109. Plenum Press.

Cadenhead D. A., Wagner N. J., Jones B. R., and Stetter J. R. (1972) Some surface characteristics and gas interactions of Apollo 14 fines and rock fragments (abstract). In *Lunar Science—III* (editor C. Watkins), pp. 110–112 Lunar Science Institute Contr. No. 88.

Dubinin M. M. (1955) A study of the porous structure of active carbons using a variety of methods. *Quart. Rev.* **9,** 101–114.

Evans B. and White T. E. (1967) A gravimetric study of the influence of water on the adsorptive properties of low-surface-area glass fibers. *Vacuum Microbalance Techniques*, Vol. 6, pp. 157–172. Plenum Press.

Faeth P. A. and Willingham C. B. (1955) In "The Assembly, Calibration and Operation of a Gas Adsorption Apparatus for the Measurement of Surface Area, Pore Volume Distribution, and Density of Finely Divided Solids." *Mellon Institute Bulletin.*

Freeman J. W. Jr., Hills H. K., and Vondrak R. R. (1972) Water vapor, whence comest thou? (abstract). In *Lunar Science—III* (editor C. Watkins), pp. 283–285, Lunar Science Institute Contr. No. 88.

Fuller E. L., Holmes H. F., Gammage R. B., and Becker K. (1971) Interaction of gases with lunar materials: Preliminary results. *Proc. Second Lunar Sci. Conf.*, Geochim. Cosmochim. Acta Suppl. 2, Vol. 3, pp. 2009–2019. MIT Press.

Gaines G. L. Jr. (1962) Sorption of gases by active charcoal and silicates (including glasses) and cellulose. In *Scientific Foundation of Vacuum Technique*, 2nd edition, Chap. 7 (editors S. Dushman and J. F. Lafferty), pp. 435–515, Wiley.

Grossman J. J., Mukherjee N. R., and Ryan J. A. (1972) Microphysical, microchemical, and adhesive properties of lunar material, III: Gas interaction with lunar material. *Proc. Third Lunar Sci. Conf.*, Geochim. Cosmochim. Acta Suppl. 3, Vol. 3, MIT Press.

Johnson F. S. (1972) Lunar atmosphere (abstract). In *Lunar Science—III* (editor C. Watkins), pp. 436–438, Lunar Science Institute Contr. No. 88.

Mason B. (1971) The lunar rocks. *Scientific American* **225,** 49–58.

McClellan A. I. and Harnsberger H. F. (1967) Cross-sectional areas of molecules adsorbed on solid surfaces. *J. Colloid Interface Sci.* **23,** 577–599.

Reed G. W., Goleb J. A., and Jovanovic S. (1971) Surface-related Hg in lunar samples. *Science* **172,** 258–261.

Proceedings of the Third Lunar Science Conference
(Supplement 3, *Geochimica et Cosmochimica Acta*)
Vol. 3, pp. 2259–2269
The M.I.T. Press, 1972

Microphysical, microchemical, and adhesive properties of lunar material III: Gas interaction with lunar material

J. J. GROSSMAN, N. R. MUKHERJEE, and J. A. RYAN

Space Sciences Department, McDonnell Douglas Astronautics Company,
West Huntington Beach, California 92647

Abstract—Gas adsorption measurements on an Apollo 12 ultrahigh vacuum-stored sample and Apollo 14 and 15 N_2-stored samples, show that the cosmic ray track and solar wind damaged surface of lunar soil is very reactive. Room temperature monolayer adsorption of N_2 by the Apollo 12 sample at 10^{-4} atm was observed. Gas evolution of Apollo 14 lunar soil at liquid nitrogen temperature during adsorption/desorption cycling is probably due to cosmic ray track stored energy release accompanied by solar gas release from depths of 100–200 nm.

INTRODUCTION

THE PURPOSE OF THIS STUDY is to determine the manner in which various gases interact with lunar materials and the interaction differences, if any, between lunar and terrestrial materials. Such studies are important because the lunar and terrestrial materials have had differing histories, the former being directly exposed to the various space radiations and the latter being exposed to the earth's atmosphere, in both instances for a considerable period of time. Information provided by such studies could provide knowledge toward the understanding of the interaction of lunar surface material with gases of various origin. The work reported here is concerned with surface area measurements of Apollo 14 and 15 samples plus gas adsorption/desorption studies of UHV-stored Apollo 12 and N_2-stored Apollo 14 samples.

This study is an outgrowth of our previous studies on Apollo 11 and 12 lunar samples, particularly gas exposure effects on the samples (Grossman *et al.*, 1970a, 1971). Upon exposure to O_2, H_2O vapor, their mixtures, organic and inorganic acids and bases, disruptions of some of the bonded particles were observed. Exposure to dry N_2 did not show any noticeable change, although N_2-exposed samples gained weight during storage in N_2 for about one year. The lunar soil under vacuum and upon exposure to ambient atmosphere exhibited noticeable electrostatic charging. The bonding between gas-disrupted particles could well be of the same origin, but other bonding mechanisms cannot be excluded at present. For instance, prior to gas exposure, the bonded particles were purposely subjected to considerable mechanical disturbance, such as lateral motions, roll, and impact by other particles. Some of the particles that appeared bonded separated (or changed in orientation) as a result, but others disrupted only upon exposure to gases. A detailed examination of O_2 and H_2O vapor disrupted surfaces by petrographic microscope ($\times 500$) and scanning electron microscope ($\times 20,000$) revealed that each disrupted area was sufficiently porous to permit interparticle diffusion of a gas, and that the geometrical area was much smaller than the real area due to fractures, voids, craters, and particle agglomerates. The shape of individual small particles and protrusions on larger plate-like crystals suggested some deposition and agglomeration might have occurred, such as from a liquid spray(s).

Samples examined by us and others (Crozaz et al., 1970 and 1971; Fleischer et al., 1970 and 1971; Comstock et al., 1971; Arrhenius et al., 1971) showed high cosmic ray track densities ($\sim 10^8$ to $\sim 10^{10}$ cm^{-2}) within 100 nm of the surface. Most particles were also found to be coated with an amorphous or ultramicrocrystalline layer of thickness 20 to 100 nm (Gold et al., 1970; Hapke et al., 1970; Borg et al., 1971). If these coatings were composed of purely radiation-damaged material, they would contain high stored energy. Hoyt et al. (1970) found calorimetrically that stored energy is large in thermodynamically unstable lunar glassy spherules and grains, but small due to cosmic ray damage in the interior of minerals. However, they calculated that the relatively large amount of stored energy present in the highly damaged particle surfaces could not be detected by calorimetry. This strongly suggests a need for the determination of surface reactivity directly in experiments which are insensitive to the bulk volume effects of lunar material to improve our understanding of solid phenomena of space-environment exposed materials.

An important technique for the studies of surface reactivity is the measurement of adsorption/desorption behavior for selected gases. It is highly desirable to use ultrahigh vacuum lunar samples for this because some of the surface reactivity is likely to be destroyed by prior gas exposure.

Fuller et al. (1971) determined surface areas of Apollo 11 fines (sample 10087,5) and Apollo 12 lunar fines (sample 12033,46), and studied the adsorption/desorption on the former sample. Their samples had previously been exposed to laboratory atmosphere for a week or so at $25 \pm 2°C$ and relative humidity of $60 \pm 5\%$. Prior to their studies they degassed the samples at 300°C and 10^{-5} Torr for 24 hours. The specific surface area measured by using N_2 and CO at $-196°C$ gave virtually identical results of 1.1_5 m^2 gm^{-1} for the Apollo 11 sample and less than 0.05 m^2 gm^{-1} for the Apollo 12 sample. Adsorption/desorption of A, N_2, O_2, and CO on the Apollo 11 sample showed the material surfaces to be quite nonpolar. Repeated and prolonged H_2O adsorption/desorption modifies the water adsorption isotherm from a hydrophobic to a hydrophilic character, although the capacity of N_2 adsorption remained unaltered. Such behaviour they interpreted to be due to a unique micropore structure in the amorphous, radiation-damaged surfaces of the lunar particles. Fanale et al. (1971) determined the surface areas of two fractions of sieved size ranges (74–147 μm and 147–256 μm) of Apollo 11 fines (sample 10084) by using Kr at liquid nitrogen temperature (LN$_2$), and found them to have effective surface areas similar to ground terrestrial mafic rock powders of about the same size distribution. They suggest on the basis of their finding of the BET parameter C that apparently low heats of adsorption for Kr on lunar fines are consistent with the presence of glassy or glass-coated particles.

EXPERIMENTAL TECHNIQUES

Experimental techniques utilized were dependent on the parameters to be measured as well as lunar sample type. Surface area measurements were made at LN$_2$ ($-196°C$) using krypton as the measuring gas in a LN$_2$ trapped oil diffusion pumped system (volumetric). The adsorption/desorption studies were performed by the same volumetric system at LN_2 ($-196°C$) and by a gravimetric system using a Cahn RH balance at room temperature ($25 \pm 0.3°C$).

Sample type received and handling

We received two types of samples: a vacuum sample from the Apollo 12 mission and N_2-stored samples from the Apollo 14 and 15 missions. Since it was received by us the Apollo 12 sample was stored under ultrahigh vacuum (UHV) condition ($\lesssim 10^{-8}$ Torr) at room temperature for about one year. This sample, however, was exposed to a pressure of about 10^{-2} Torr during transit from the moon to the earth which undoubtedly filled all reactive, chemisorbing sites. Since the sample was kept under the UHV condition for about one year, most of the nonchemisorbed contaminant gases from the 10^{-2} Torr exposure were probably removed from the sample.

The techniques for handling and transfer of the UHV-stored sample are similar to that given by Grossman *et al.* (1971), the sample never having been exposed to any gas at or subsequent to receipt from the LRL prior to the adsorption runs. The N_2-stored sample portion used in the gravimetric system was handled and loaded in the system in dry N_2 atmosphere. The sample portion used in the volumetric system was exposed to air for about 20 min during insertion into the system.

Apparatus and procedure

The volumetric apparatus used for studies at LN_2 temperature is of standard design and further details are not given.

For room temperature studies, gravimetric systems were used. The one used for N_2-stored samples consists of two main parts: (a) the gas adsorption subsystem made of stainless steel and (b) the reference subsystem for measuring pressure in the adsorption chamber containing a Cahn RG microbalance. Ultrahigh vacuum $\sim 10^{-8}$ Torr was achieved by a system of two LN_2 trapped diffusion pumps and two ion pumps (capacity 500 liter sec^{-1} and 50 liter sec^{-1} pumping speed). The pressure measurement up to 10^{-5} Torr were made with Bayard-Alpert ion gages directly connected to (a).

Measurements from 10^{-4} to 760 Torr were made using the reference system and capacitance manometers. The microbalance chamber was vibration isolated from the floor, and decoupled as well as possible from the rest of the system by use of stainless steel flexible tubing. The microbalance zero was checked repeatedly throughout the runs utilizing a manipulation system which allowed removal and replacement of the sample and tare while at vacuum.

The UHV-stored Apollo 12 sample (12001,118) was transferred under ultrahigh vacuum conditions into a specially designed glass cell through a copper tube which was then pinch-off sealed (Fig. 1). After transferring the cell to the Cahn RH gravimetric microbalance in a metal bell jar system and evacuating to 10^{-7} Torr, the sample cell was opened by a differential thermal expansion device at the breakoff constriction and adsorbate gas introduced within one hour. For each gas, corrections were made for buoyancy, thermomolecular effects, microbalance temperature, and zero shift effects at the different temperatures and pressures.

The UHV-stored sample was not baked, but the N_2-stored samples were baked before each run at 110°C for about 2 hours (the 110°C temperature being chosen so as not to exceed the maximum sample exposure temperature at the lunar surface).

RESULTS AND DISCUSSION

Surface areas

At LN_2 temperature our Kr measured surface areas were 0.61 m^2 gm^{-1} for the Apollo 14 sample and 0.40 m^2 gm^{-1} for Apollo 15. The results are in the same order of magnitude range for samples returned by other Apollo missions, plus terrestrial fines of about the same grain size distribution (Table 1). In the table we have shown surface area, P_m/P_s, $E_1 - E_l$ and C obtained by us and others for lunar and terrestrial samples. These symbols are defined in the table. The C value of the BET equation is

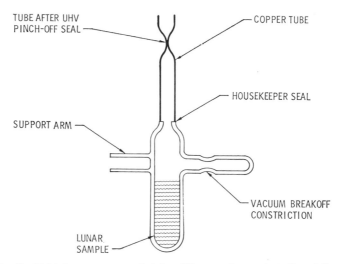

Fig. 1. Apollo 12 high vacuum sample tube. (The sample was transferred through the copper tube that was pinch-off sealed. It was suspended with the vacuum breakoff constriction side arm vertical, held by a support arm until opened at high vacuum and then lowered onto the microbalance.)

Table 1. Surface areas and the BET parameters of lunar and terrestrial materials (P_m is the pressure at which the monolayer formation is complete, P_s is the saturation pressure, E_1 is the energy of adsorption which is assumed to be always greater than the liquefaction energy E_l, and C is the BET constant which is a measure of the adsorption energy).

Apollo mission and sample no.	Measuring gas	Surface area (m²gm⁻¹)	P_m/P_s	$E_1 - E_l$	C	Authors
Apollo 11						
10087,5	N_2	1.1	0.19	450	40	Fuller *et al.* (1971)
10087,5	CO	1.1	0.08	750	130	Fuller *et al.* (1971)
10048 sieved fraction						
74–147 μm	Kr	0.26	0.20	—	16	Fanale *et al.* (1971)
147–256 μm	Kr	0.31	0.16	—	29	Fanale *et al.* (1971)
Apollo 12						
12033,46	N_2	0.05		—	—	Fuller *et al.* (1971)
12001,118*	N_2	0.08	(10^{-4} atm)	—	—	Grossman *et al.*
Apollo 14						
14163,111	N_2	0.21	—	—	—	Cadenhead *et al.* (1972)
14321,156	N_2	0.34	—	—	—	Cadenhead *et al.* (1972)
14259,93	Kr	0.61	0.087	720	109	Grossman *et al.*
14259,93	O_2	0.37	0.048	916	386	Grossman *et al.*
Apollo 15						
15401,48	Kr	0.40	0.108	650	68	Grossman *et al.*
Crushed terrestrial						
Gabbro	Kr	0.19	0.16	—	28	Fanale *et al.* (1971)
Basalt cinder	Kr	0.26	0.13	—	44	Fanale *et al.* (1971)
Vacaville Bas.	Kr	3.98	0.10	—	74	Fanale *et al.* (1971)

* Note: Room temperature measurement of UHV-stored sample.

related to the surface energy of an adsorbent, the lower the C value the smaller is the surface energy. For active inorganic oxides the C values for N_2 adsorption may be as high as 1000. All values of C shown in Table 1 are less than 140 except for O_2 on sample 14259,93. This is the sample for which we observed hysteresis with O_2 at LN_2 temperature.

Calculations for the C value in the BET equation is not possible for N_2 adsorption at room temperature because the saturation pressure P_s cannot be defined. However, the maximum heat of adsorption for the UHV-stored Apollo 12 sample's Langmuir Type I isotherm (N_2 first run, Fig. 2) was approximately 9 kcal, a value which is still considered physical adsorption (Hobson, 1967). The surface area inferred from monolayer adsorption that occurred at the unusually low pressure of 10^{-4} atm is 0.08 m^2 gm^{-1}. This surface area value of ours is in reasonable agreement with the measured value (<0.05 m^2 gm^{-1}) of Fuller et al. (1971) for Apollo 12 sample 12033,46.

Apparent geometric surface area alone is a relatively poor measure of sample properties. Petrographic, scanning electron micrographic, and transmission and replica electron micrographic examination of lunar samples readily provides convincing evidence that inhomogeneous, fractured samples have surface areas far in excess of the apparent geometrical area (Fanale et al., 1971).

Adsorption/desorption at room temperature

Figure 2 shows the research grade N_2 and A adsorbate gas runs at room temperature for the UHV-stored Apollo 12 sample (weight used 2.315 gm). The N_2 run

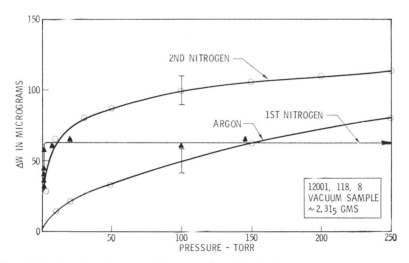

Fig. 2. Gravimetric nitrogen and argon adsorption isotherms at room temperature (25 ± 0.3°C) for UHV-stored Apollo 12 vacuum sample. (There was no permanent retention of N_2 within the experimental error, and results were reproducible in two subsequent runs (not shown). The reproducible Langmuir knee for the first nitrogen adsorption measurements occurs at $P = 10^{-4}$ atmospheres. The isotherm is modified after air exposure (2nd N_2). The Argon isotherm obtained after N_2 runs is the last of the series).

was done first and this run showed high adsorption initially resulting in the entire weight gain at pressures less than 0.5 Torr. The sample weight increased by about 60 μgm, corresponding to approximately 1 μmole gm^{-1}, which is reasonably close to the 0.6 μmole gm^{-1} amount required for monolayer formation (Table 1; also Fuller *et al.*, 1971). The sample weight then remained essentially constant up to 760 Torr. On desorption there was no net N_2 retention within the experimental error (see Fig. 2 for the absolute error bar). These results were reproducible in two subsequent runs (not shown) indicating that this lunar sample retained high reactivity, although there appears to be no permanent retention of N_2 within the experimental error. For clarity, the data points with no permanent retention are shown in the figure.

In order to ascertain the change in the surface reactivity, the sample was exposed to air for a short period (~ 5 Torr sec) at a pressure less than 2 Torr and then evacuated to a pressure less than 10^{-7} Torr. The subsequent N_2 run showed a change in the desorption isotherm and the knee moved to about 25 Torr (second N_2, Fig. 2) with larger total weight gain. Argon also exhibited a similar (Type I) adsorption isotherm. These results strongly suggest that the surface characteristics of this UHV-stored samples changed even upon a very brief exposure to air. Unfortunately, there was no more UHV-stored sample left to conduct another series of these studies. However, in our previous studies, Apollo 11 and 12 samples also exhibited high reactivity when exposed to oxygen, water vapor and their mixtures.

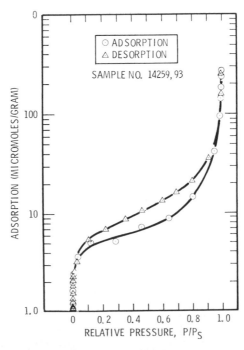

Fig. 3. Oxygen adsorption and desorption isotherms on an Apollo 14 sample at liquid N_2 temperature ($-196°C$) using a volumetric system. (The cumulative error for a 16 point adsorption/desorption isotherm is 1.6 μmole gm^{-1}.)

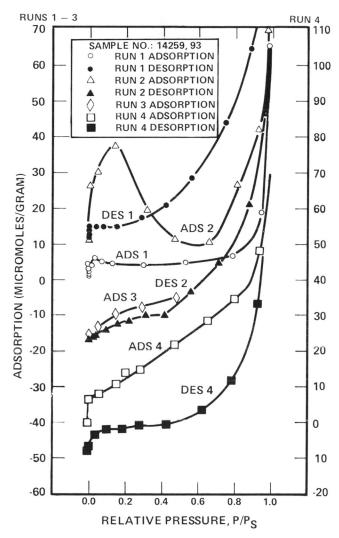

Fig. 4. Nitrogen adsorption and desorption isotherm series at liquid N_2 temperature on the Apollo 14 sample after O_2 runs in Fig. 3. (The left-hand ordinate is for the upper curves and the right-hand ordinate is for the fourth adsorption/desorption curves (lower two). The shape and the order indicate the unusual gas evolution properties of lunar samples. The cumulative error for a 16 point adsorption/desorption isotherm is 1.6 μmole gm^{-1}.)

Adsorption/desorption at liquid nitrogen temperature ($-196°C$)

The adsorption isotherms of research grade O_2 first and then N_2 were determined at LN_2 temperature by using the volumetric system for Apollo 14 sample 14259,93 (see Figs. 3 and 4). A definitive hysteresis (see the error range in the figure title) that was not found by prior studies on an Apollo 11 sample (Fuller *et al.*, 1971) was observed. Although the sample was baked at 110°C for about two hours, this hysteresis was unexpected, because of the following: (1) this N_2-stored sample was exposed to air for about 20 min during weighing and insertion into the volumetric system, (2) no hysteresis observed for the UHV-stored Apollo 12 sample with N_2, and (3) Fuller *et al.* (1971) did not observe any hysteresis with O_2 on Apollo 11 sample which remained exposed to air for a week but baked at 300°C and 10^{-5} Torr for 24 hours prior to their adsorption experiment. For the O_2 experiment the BET surface area of 0.37 m^2 gm^{-1} at $P/P_s = 0.37$ is comparable to our Kr measured surface area (see Table 1). The sigmoidal behavior of our data should be compared with the previously observed slight undulation in the O_2 data of Fuller *et al.* (1971).

Four N_2 runs at LN_2 temperature, which followed the O_2 measurements, are shown sequentially in Fig. 4. The first N_2 run shows a slight decrease in adsorption with increasing pressure above $P/P_s = 0.05$, no additional effects up to 0.8 and then multilayer condensation. The 4 μmole gm^{-1} between $P/P_s = 0.05$ to 0.08 is consistent with one monolayer calculated from the O_2 isotherm or 0.6 monolayer from the Kr surface area determination. The first desorption isotherm starts from 400 μmoles gm^{-1} showing hysteresis down to zero pressure and a net retention of 10.4 μmoles gm^{-1}. The second adsorption rises to a maximum of five monolayers (O_2 surface area) at $P/P_s = 0.15$ the amount decreasing from there to $P/P_s = 0.66$ and then increasing again to 300 μmoles gm^{-1} at $P/P_s = 1.0$. The second desorption shows negative hysteresis (i.e., the desorption curve lies below the adsorption curve at every pressure P), and at $P = 0$, 18 μmoles gm^{-1} more gas was lost than put into the system. The third adsorption isotherm appears to be more normal, though the desorption end point was not determined. Starting then with a new zero, the fourth adsorption isotherm (right-hand ordinate) has the more classical sigmoidal shape. However, the desorption again shows negative hysteresis over the entire pressure range. This behavior should be compared with results of Fuller *et al.* (1971) for H_2O vapor on Apollo 11 sample 10087,5. They observed a dip and two plateaus in their adsorption curve of Sorption Cycle #1; however, repreated cycling (adsorption/desorption) led to more conventional isotherms with a more and more pronounced sigmoidal shape.

The adsorption decrease is indicative of gas release from the adsorbent. As mentioned previously, Fuller *et al.* (1971) also observed such a decrease for their H_2O adsorption run at 25°C. We ascribe this behavior to be due to a group of processes which could be irreversible and cyclical opening and closing of micropores in lunar material grains. As a result, some of the gases, which could have been solar, lunar, terrestrial or otherwise trapped, might have been released. All low temperature runs were indicative of a nonterrestrial type behavior for the Apollo 14 sample, probably as a result of radiation damage.

Funkhouser *et al.* (1970); Epstein *et al.* (1970); Eberhardt *et al.* (1970); Friedman *et al.* (1970); Kirsten *et al.* (1971); Moore *et al.* (1971); Funkhouser *et al.* (1971); Gibson *et al.* (1971), and others have reported the composition, distribution, and thermal release of solar wind trapped gases in lunar soil. Up to 90 μmoles gm^{-1} have been reported, H_2 and He, making up the largest fraction. Well over 90% lies between 100 and 200 nm from the surface (Eberhardt *et al.*, 1970). An appreciable amount can be driven off by heating at 300°C but not at 100°C.

An hypothesis for the observed lunar material adsorption behavior, based essentially on the irreversible behavior of cosmic ray damaged surfaces, is as follows: As adsorption reaches saturation pressures, the liquid penetrates the radiation damaged pores releasing both stored energy and some solar wind implanted gases, an unusual example of the well known Rebinder effect. With each adsorption more of the surface damaged stored energy is released, and then the lunar sample starts behaving more like terrestrial silicates.

In our experiments the O_2 run might have sensitized the sample as suggested by the change in the adsorption isotherm of the UHV-stored sample after a short air exposure so that even liquid nitrogen could penetrate the sample and be retained (first run, Fig. 4). After annealing at 100°C the second adsorption releases the trapped gas. If only pore trapped N_2 is released the penetration of the damaged layer is less than 100 nm since H_2 and He lie deeper in the sample (Eberhardt *et al.*, 1970). Continued adsorption deepens the damaged layer penetration and the negative hysteresis on the second desorption would then be merely a release of the solar wind H_2 and He at that depth. Further annealing and adsorption deepens the penetration further with continued release of trapped gas. Ultimately the sample should become thermodynamically stable and further adsorption/desorption behavior should be like terrestrial materials.

The water adsorption work reported by Fuller *et al.* (1971) led them to postulate 5 nm pores which we believe to be cosmic ray damage produced. The longer they cycled the Apollo 11 sample the more the adsorption isotherm approached terrestrial silicates. They reported the 100°C bakeout weight loss was 300 μgm gm^{-1} and the 300°C, 24-hour weight loss was 940 μgm gm^{-1}, the latter probably being all solar wind trapped gases. Their air preconditioned sample had already reacted with O_2 (and H_2O); thus the sample had changed from its pristine condition. Such a change was observed by us, as discussed previously, for the UHV-stored Apollo 12 sample.

Conclusions

(1) An UHV-stored lunar sample was observed to have unique room temperature gas adsorption properties for N_2 with Langmuir monolayer uptake at 10^{-4} atm. This behavior may be typical of pristine lunar material.

(2) Short exposure to air modifies the behavior but the material appears to be relatively stable in dry nitrogen.

(3) Reactive gas may sensitize the surface so that further reaction with liquid N_2 continues release of stored energy, and solar wind trapped gases.

It is evident that further studies are required for the following reasons: First, none of the studies to date are really representative of "pristine" lunar material. Second, differing adsorption behavior has been found by different investigators of lunar samples (Fuller *et al.*, 1971; Cadenhead *et al.*, 1972; present authors). The exact reasons for these differences need to be determined. Finally, the observed non-terrestrial type behavior of lunar materials needs further confirmation and specific reasons for such behavior need to be determined.

The findings indicate that the degree of adsorption is highly dependent on the nature of the surface. In those cases where the adsorbent surface area might be sensitized by a reactive gas, the sequential use of an inert gas, a reactive gas, and followed by the same inert gas may prove to be a valuable technique for probing the nature of the adsorbent surface, and the effects caused by the reactive gas. Additionally, means should be provided to measure the gas composition in order to definitively determine whether or not gases are indeed released from the samples and if so, the types.

Acknowledgments—The authors thank A. D. Pinkul especially for his work on the Apollo 12 High Vacuum Samples, C. L. Wilkensen for Apollo 12 microbalance measurements, L. Atkins for the volumetric analyses and W. M. Hansen for his help on all phases of the program. This work was supported by NASA Contract NAS 9-11680.

REFERENCES

Arrhenius G., Liang S., Macdougall D., Wilkennig L., Bhandari N., Bhat S., Lal D., Rajagopalan G., Tamhane A. S., and Venkatavaradan V. S. (1971) The exposure history of the Apollo 12 regolith. *Proc. Second Lunar Sci. Conf., Geochim. Cosmochim. Acta* Suppl. 2, Vol. 3, pp. 2583–2598. MIT Press.

Borg J., Maurette M., Durrieu L., and Jouret C. (1971) Ultramicroscopic features in micron-sized lunar dust grains and cosmophysics. *Proc. Second Lunar Sci. Conf., Geochim. Cosmochim. Acta* Suppl. 2, Vol. 3, pp. 2027–2040. MIT Press.

Cadenhead D. A., Wagner N. J., Jones B. R., and Stetter J. R. (1972) Some surface characteristics and gas interactions of Apollo 14 fines and rock fragments. In *Lunar Science—III* (editor C. Watkins), pp. 110–112, Lunar Science Institute Contr. No. 88.

Comstock G. M., Evwaraye A. O., Fleischer R. L., and Hart H. R. (1971) The particle track record of lunar soil. *Proc. Second Lunar Sci. Conf., Geochim. Cosmochim. Acta* Suppl. 2, Vol. 3, pp. 2569–2582. MIT Press.

Crozaz G., Haack U., Hair M., Maurette M., Walker R., and Woolum D. (1970) Nuclear track studies of ancient solar radiations and dynamic lunar processes. *Proc. Apollo 11 Lunar Sci. Conf., Geochim. Cosmochim. Acta* Suppl. 1, Vol. 3, pp. 2051–2080. Pergamon.

Crozaz G., Walker R., and Woolum D. (1971) Nuclear tract studies of dynamic surface processes on the moon and the constancy of solar activity. *Proc. Second Lunar Sci. Conf., Geochim. Cosmochim. Acta* Suppl. 2, Vol. 3, pp. 2543–2558. MIT Press.

Eberhardt P., Geiss J., Graff H., Grögler N., Krähenbühl U., Schwaller H., Schwarzmuller J., and Steitler A. (1970) Trapped solar wind noble gases, Kr^{81}/Kr exposure ages and K/Ar ages in Apollo 11 lunar material. *Science* **167**, 558–560.

Epstein S. and Taylor H. P. Jr. (1970) $^{18}O/^{16}O$, $^{30}Si/^{28}Si$, D/H, and $^{13}C/^{12}C$ studies of lunar rocks and minerals. *Science* **167**, 533–535.

Fanale F. P., Nash D. B., and Cannon W. A. (1971) Lunar fines and terrestrial rock powders: Relative surface areas and heats of adsorption. *J. Geophys. Res.* **76**, 6459–6461.

Fleischer R. L., Haines E. L., Hart H. R., Woods R. T., and Comstock G. M. (1970) The particle track record of the Sea of Tranquility. *Proc. Apollo 11 Lunar Sci. Conf., Geochim. Cosmochim. Acta* Suppl. 1, Vol. 3, pp. 2103–2120. Pergamon.

Fleischer R. L., Hart H. R., Comstock G. M., and Evwaraye A. O. (1971) The particle track record of the Ocean of Storms. *Proc. Second Lunar Sci. Conf., Geochim. Cosmochim. Acta* Suppl. 2, Vol. 3, pp. 2550–2568. MIT Press.

Friedman I., O'Neil J. R., Adami L. H., Gleason J. D., and Hardcastle K. (1970) Water, hydrogen, deuterium, carbon, carbon-13, oxygen 18 content of selected lunar material. *Science* **167**, 538–541.

Fuller E. L., Holmes H. F., Gammage R. B., and Becker K. (1971) Interaction of gases with lunar materials: Preliminary results. *Proc. Second Lunar Sci. Conf., Geochim. Cosmochim. Acta* Suppl. 2, Vol. 3, pp. 2009–2019. MIT Press.

Funkhouser J. G., Schaeffer O. A., Bogard D. D., and Zähringer J. (1970) Gas analysis of the lunar surface. *Science* **167,** 561–563.

Funkhouser J., Jessberger E., Muller O., and Zähringer J. (1971) Active and inert gases in Apollo 12 and Apollo 11 samples released by crushing at room temperature and by heating at low temperatures. *Proc. Second Lunar Sci. Conf., Geochim. Cosmochim. Acta* Suppl. 2, Vol. 2, pp. 1381–1396. MIT Press.

Gibson E. K. Jr. and Johnson S. M. (1971) Thermal analysis-inorganic gas release studies of lunar samples. *Proc. Second Lunar Sci. Conf., Geochim. Cosmochim. Acta* Suppl. 2, Vol. 2, pp. 1351–1366. MIT Press.

Gold T., Campbell M. J., and O'Leary B. T. (1970) Optical and high-frequency electrical properties of lunar samples. *Proc. Apollo 11 Lunar Sci. Conf., Geochim. Cosmochim. Acta* Suppl. 1, Vol. 3, pp. 2149–2154. Pergamon.

Grossman J. J., Ryan J. A., Mukherjee N. R., and Wegner M. W. (1970a) Surface properties of lunar samples. *Science* **167,** 743–745.

Grossman J. J., Ryan J. A., Mukherjee N. R., and Wegner M. W. (1970b) Microchemical, microphysical, and adhesive properties of lunar material. *Proc. Apollo 11 Lunar Sci. Conf., Geochim. Cosmochim. Acta* Suppl. 1, Vol. 3, pp. 2171–2181. Pergamon.

Grossman J. J., Ryan J. A., Mukherjee N. R., and Wegner M. W. (1971) Microchemical, microphysical, and adhesive properties of lunar material, II. *Proc. Second Lunar Sci. Conf., Geochim. Cosmochim. Acta* Suppl. 2, Vol. 3, pp. 2153–2164. MIT Press.

Hapke B. W., Cohen A. J., Cassidy W. A., and Wells E. N. (1970) Solar radiation effects on the optical properties of Apollo 11 samples. *Proc. Apollo 11 Lunar Sci. Conf., Geochim. Cosmochim. Acta* Suppl. 1, Vol. 3, pp. 2199–2212. Pergamon.

Hobson V. P. (1967) Physical adsorption at extremely low pressures. In *The Solid-Gas Interface* (editor E. A Flood), Vol 1, pp. 447–489. Marcel Debber.

Hoyt H., Kardos J. L., Miyajima M., Seitz M. G., Sun S. S., Walker R. M., and Wittels M. C. (1970) Thermoluminescence, x-ray, and stored energy measurements of Apollo 11 samples. *Proc. Apollo 11 Lunar Sci. Conf., Geochim. Cosmochim. Acta* Suppl. 1, Vol. 3, pp. 2269–2287. Pergamon.

Kirsten T., Steinbrunn F., and Zähringer J. (1971) Location and variation of trapped rare gases in Apollo 12 lunar samples. *Proc. Second Lunar Sci. Conf., Geochim. Cosmochim. Acta* Suppl. 2, Vol. 2, pp. 1651–1669. MIT Press.

Moore C. B., Lewis C. F., Larimer J. W., Delles F. M., Gooley R. C., and Nichiporuk W. (1971) Total carbon and nitrogen abundances in Apollo 12 lunar samples. *Proc. Second Lunar Sci. Conf., Geochim. Cosmochim. Acta* Suppl. 2, Vol. 2, pp. 1343–1350. MIT Press.

Proceedings of the Third Lunar Science Conference
(Supplement 3, *Geochimica et Cosmochimica Acta*)
Vol. 3, pp. 2271–2286
The M.I.T. Press, 1972

Magnetic fields near the moon

PAUL J. COLEMAN, JR.

Department of Planetary and Space Science
and
Institute of Geophysics and Planetary Physics,
University of California, Los Angeles, California 90024

B. R. LICHTENSTEIN

Department of Planetary and Space Science
and
Institute of Geophysics and Planetary Physics,
University of California, Los Angeles, California 90024

C. T. RUSSELL

Institute of Geophysics and Planetary Physics,
University of California, Los Angeles, California 90024

L. R. SHARP

Department of Planetary and Space Science
and
Institute of Geophysics and Planetary Physics,
University of California, Los Angeles, California 90024

G. SCHUBERT

Department of Planetary and Space Science,
University of California, Los Angeles, California 90024

Abstract—Magnetic field measurements made with the Apollo 15 subsatellite have established the existence of remanent magnetization over much of the moon's surface. The major features of the structure of the measured remanent field are apparently associated with large craters, the most apparent detected to date being that associated with the Van de Graaff crater. A preliminary contour map of relative magnetic intensity has been produced for a limited region of the moon. Magnetic field measurements taken while the moon was in the solar wind have revealed the frequent presence of regions just downstream from the limbs characterized by substantially increased field strengths. Similar temporary regions have been detected previously at greater distances with the lunar satellite Explorer 35. They have been attributed to limb shocks produced by the grazing interaction of the solar wind with regions of locally high remanent magnetization at the limbs. Our studies of the occurrence of these shock effects at the subsatellite orbit indicate that one of the regions of the lunar surface that is apparently highly effective in producing limb shocks is also a region of relatively strong remanent fields, thus supporting this explanation as to the general cause of the limb shocks.

INTRODUCTION

ON AUGUST 4, 1971, shortly before they left lunar orbit to start their trip home, the crew of Apollo 15 launched an 80 lb satellite from their spacecraft. The scientific instruments on board the subsatellite include a biaxial fluxgate magnetometer, two

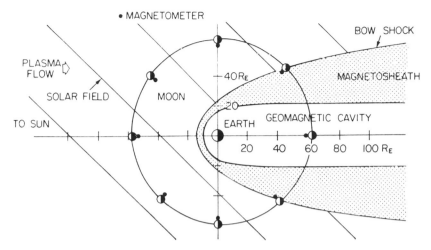

Fig. 1. Schematic diagram showing the three regions of near-earth space traversed by the moon. For most of each lunation the moon is in the solar wind. The other two regions are the magnetosheath and the tail of the geomagnetic cavity. The plane of the figure is essentially the ecliptic plane. The large dot adjacent to the lunar surface marks the approximate location of the Apollo 12 lunar surface magnetometer. (Courtesy NASA Ames.) Earth and moon are not to scale.

solid state telescopes, and several electrostatic analyzers. An *S*-band transponder also on board is used to measure gravitational accelerations of the spacecraft.

This paper is the second report on our preliminary analysis of the magnetic field measurements obtained with the subsatellite magnetometer. It is essentially an expanded version of our initial report (Coleman *et al.*, 1972).

The objectives of this experiment are to measure the moon's magnetic field, both the permanent and the induced components, and to measure the magnetic effects of the moon's interaction with the electromagnetic fields and charged particles of its environment. The multiplicity of our objectives is made possible by the geometry of the moon's orbit, which, as shown in Fig. 1, passes through three fundamentally different regions in near-earth space. These are the "solar wind" region upstream from the earth in which the solar wind flow is essentially undisturbed by the presence of the earth; the magnetosheath in which the solar wind flow is drastically modified by its interaction with the earth's magnetic field; and the geomagnetic cavity that consists of the space threaded by the magnetic field from the earth.

ORBITAL PARAMETERS AND SPACECRAFT ORIENTATION

The initial orbit of the Apollo 15 subsatellite was somewhat elliptical with perilune altitude at 104 km and apolune altitude at 135 km. The orbital plane is inclined to the moon's spin axis at roughly 28°. The orbital period is 1 h 59 min 40 sec.

The irregular gravitational field in the vicinity of the moon causes the orbit of the subsatellite to vary in a rather complicated fashion. Among the important effects are variations in the altitudes of perilune and apolune and variations in the longitudes of these points. Thus, for example, by the first week in September, the perilune altitude had dropped to 80 km and the apolune altitude had increased correspondingly to 159 km.

The subsatellite is spin stabilized at a spin rate of approximately 12 rpm. The spin axis is essentially normal to the ecliptic plane and therefore within 5° of the moon's spin axis.

INSTRUMENTATION

The magnetometer system carried on board the Apollo 15 subsatellite consists of two orthogonal fluxgate sensors mounted at the end of a 6-foot boom and an electronics unit housed in the spacecraft. A block diagram of the instrument is shown in Fig. 2. The specifications of the instrument are listed in Table 1. Three sampling rates are listed in the table, one for each of the three modes of operation of the instrument.

An important item of subsatellite equipment is the data storage unit. This unit records field measurements on the far side of the moon for playback when the subsatellite is in view from the earth.

The two magnetometer sensors are mounted orthogonally with one parallel and the other transverse to the spin axis of the spacecraft. The measured quantities used to define the vector field are: B_P, the vector component parallel to the spin axis and positive northward; $|\vec{B}_T|$, the absolute value of the vector component transverse to the spin axis; and ϕ the angle from the satellite-sun line to $\vec{B}_T$ measured in the right-handed sense relative to the northward B_P. Thus, the system may be described as one that provides measurements of the vector magnetic field in a right-handed, orthogonal, spacecraft centered coordinate system with the X axis toward the sun, the Z axis normal to the ecliptic and positive northward (parallel to the spacecraft spin axis), and the Y axis completing the system.

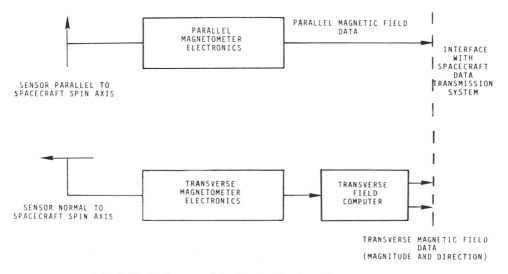

Fig. 2. Block diagram of the Apollo 15 subsatellite magnetometer.

Table 1. Apollo subsatellite magnetometer specifications.

Type:	Second-harmonic, saturable core fluxgate.
Sensor configuration:	Two sensors, one sensor parallel to the satellite spin axis (B_P) and one perpendicular to this axis.
Mounting:	Sensor unit at end of 6-foot boom. Electronics unit in spacecraft body.
Dynamic range:	Two ranges, automatically selected. 0 to $\pm 50\ \gamma$ at higher sensitivity. 0 to $\pm 200\ \gamma$ at lower sensitivity.
Resolution:	0.4 γ and 1.6 γ depending on range.
Sampling rates:	
Real time:	B_P every 2 seconds, B_T every second.
High-rate storage:	B_P, B_T magnitude and B_T phase once every 12 seconds.
Low-rate storage:	B_P, B_T magnitude and B_T phase once every 24 seconds.
Power:	0.70 watts.
Weight:	Electronics unit: 1.8 lb. Sensor unit: 0.5 lb.
Size:	Electronics unit: 11″ × 6.25″ × 1.5″. Sensor unit: 0.6″ diam. × 3″.
Operating temperature range:	$+160°F$ to $-60°F$.

When the satellite is in sunlight, the angle ϕ is measured directly, but when the satellite is in solar eclipse, ϕ must be computed from interpolations of spin rate measurements taken before and after each particular eclipse period. This calculation is usually performed with a computer. However, only "quick-look" data in printed form were available for the work described here. Consequently, the results are based on studies of B_P and $|\vec{B}_T|$ only.

Remanent Magnetism

The exploration of the moon under the Apollo program has revealed the existence of significant remanent magnetism on the moon. Rocks brought from the moon by the crews of Apollos 11, 12, and 14 exhibit measurable remanent magnetization (Runcorn *et al.*, 1970; Strangway *et al.*, 1970). Magnetometers landed on the moon by the Apollo 12 and 14 astronauts recorded measurable magnetic fields at both landing sites (Dyal *et al.*, 1970, 1971).

During intervals of relatively low levels of geomagnetic activity, the magnetic field in the geomagnetic tail is quite constant. Thus, although this field is, on the average, stronger than that in either of the other two regions through which the moon passes, remanent fields are most easily detectable when the moon is in the geomagnetic tail during such intervals. However, at the orbit of the moon, the geomagnetic tail consists of three regions: the northern lobe in which the magnetic field points toward the earth; the southern lobe in which the magnetic field points away from the earth; and the plasma sheet, with the neutral sheet at its center, that lies between the two lobes and across which the magnetic field reverses. The electrical currents associated with this field reversal flow in this last region and its position and magnetic field

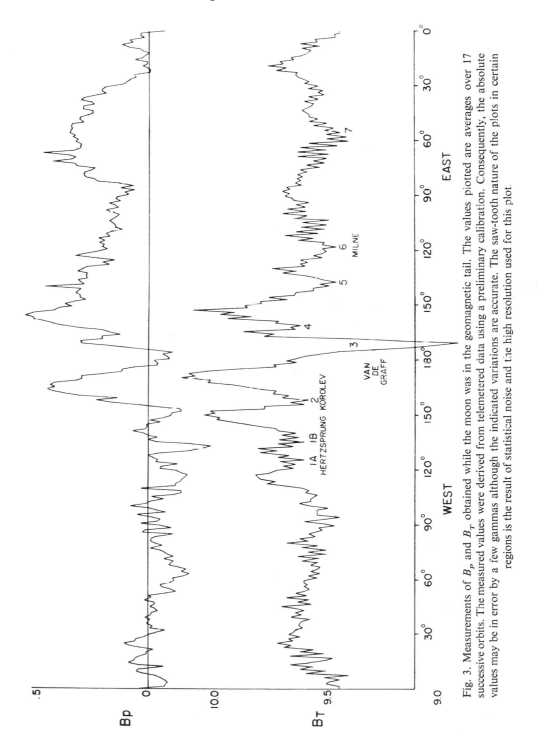

Fig. 3. Measurements of B_P and B_T obtained while the moon was in the geomagnetic tail. The values plotted are averages over 17 successive orbits. The measured values were derived from telemetered data using a preliminary calibration. Consequently, the absolute values may be in error by a few gammas although the indicated variations are accurate. The saw-tooth nature of the plots in certain regions is the result of statistical noise and the high resolution used for this plot.

strength are rather variable. Thus, with the relatively simple statistical analysis used in our study of the quick-look data, we were able to employ only data obtained in the northern and southern lobes during geomagnetically quiet times. However, even with so limited a sample, it is apparent that remanent magnetization producing measurable fields at 50–150 km above the surface is a characteristic feature of the moon.

The quick-look data recorded during the first traversals of the geomagnetic tail revealed the existence of measurable levels of remanent magnetism over much of the subsatellite orbit. Figure 3 shows average values of B_P and $|\vec{B}_T|$ for 17 orbits during which the moon was in the geomagnetic tail.

The major features of the structure in the traces, particularly in that of $|\vec{B}_T|$, appear to be associated with large craters lying within 10° of the band defined by the ground tracks of the 17 orbits. We have numbered the seven more obvious local minima in $|\vec{B}_T|$ and named four of them with the associated craters. The most obvious feature is that apparently associated with the crater Van de Graaff, which produces a 1 γ variation in the field at 130 km as the satellite sweeps past it. Van de Graaff is approximately 9° across and its center is located about 8° from the satellite ground track. Other prominent features of the data are associated with the craters Hertzsprung, Korolev, and Milne. The location of the average satellite ground track is shown in Fig. 4. The numbered points along the track correspond to the numbered points in Fig. 3.

Subsequent passes through earth's tail covering different ground tracks expanded the coverage and enabled us to construct a preliminary contour map. Figures 5 and 6 display contours of relative magnetic intensity of $|\vec{B}_T|$ for the near and far sides of the moon, respectively. Due to the limited data available in constructing these maps, the resulting contours must be treated as rough approximations.

The values shown on the map are the values measured in tenths of gamma in $|\vec{B}_T|$ normalized to 93 km altitude. The normalization process involved normalizing the values of $|\vec{B}_T|$ measured at other altitudes to the reference altitude of 93 km using the scaling law $B \propto (h_r/h)^{2.5}$ where h is the altitude of a measurement and $h_r = 93$ km. The exponent 2.5 gives the best fits to the observed mean altitude dependence and indicates that the satellite is within a few scale sizes of the sources of interest.

Data were utilized from both lobes of the earth's magnetotail with an appropriate sign change to be consistent with the field direction. For example, the Van de Graaff signature is a minimum in the southern lobe and a maximum in the northern lobe. Thus, the field has a substantial radial component ($\simeq 3$ γ at 93 km) directed *into* the lunar surface. The zero level has been arbitrarily chosen at the minimum of the Van de Graaff crater which is the most prominent feature in the magnetic field observed thus far.

Comparing the two figures, the most striking difference is the great variability of magnetic field on the far side and the local minima which seem to be strongly correlated with the crater positions. These same features are apparent in Fig. 3. Even on the near side where the magnetic field is fairly constant, the local minima seem to be associated with craters, as in the Balmer, Langrenus, and Copernicus areas. Local

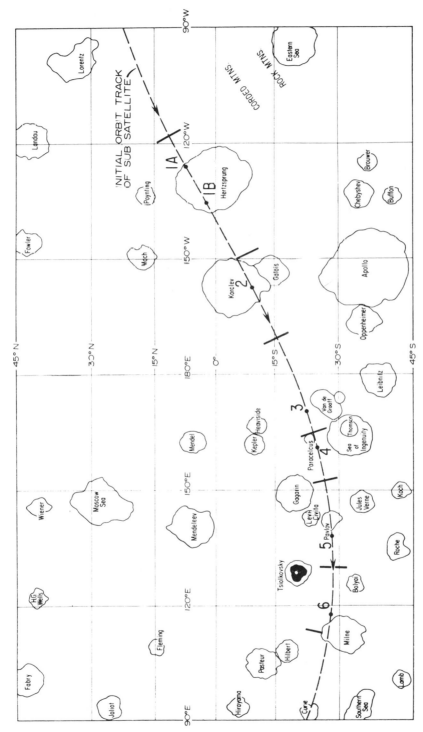

Fig. 4. Mercator projection of the far side of the moon showing the center ground track of the satellite for the 17 orbit sequence used in Fig. 3. The numbered points along the tracks correspond to the numbered points in Fig. 3.

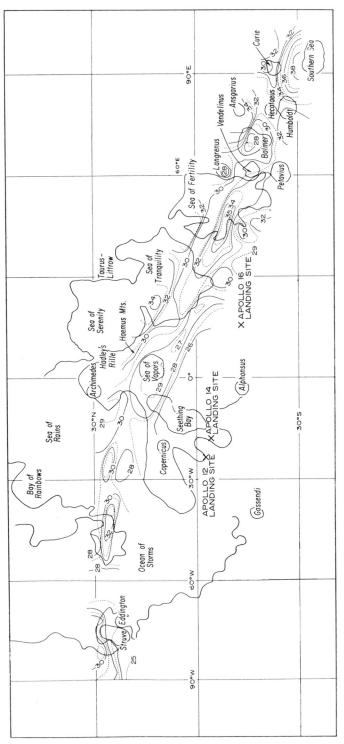

Fig. 5. Contours of the relative magnetic field intensity of B_T at 93 km altitude for the near side of the moon.

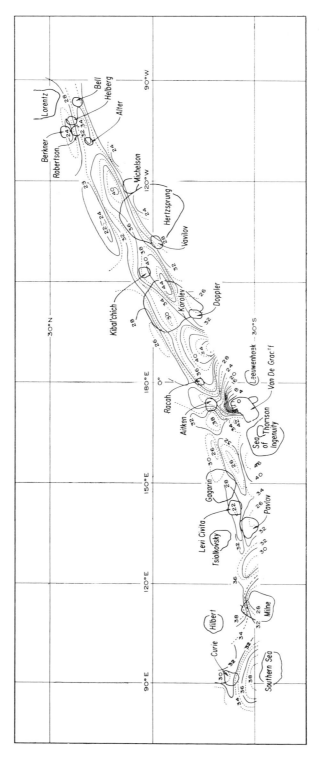

Fig. 6. Contours of relative magnetic field intensity of B_T at 93 km altitude for the far side of the moon.

maxima in the field on the near side appear to exist in the Southern Sea and Sea of Fertility.

We would emphasize that the two contour maps shown are not plots of expected ground level values. Also, the values used are relative values since we have arbitrarily assigned the value zero to the minimum at Van de Graaff.

As discussed in our initial report, the indications that the field is smoother and possibly weaker on the near side than on the far side leads us to the speculation that the measured variations are due to irregularities in a magnetized crust. Thus, it seems likely that on the near side the crust has been disturbed over a broad region, possibly by the same processes that created the ringed maria, while, on the far side, localized crater formation has produced the variable field observed, possibly by shock de-magnetization of the uniformly magnetized crust at impact.

MAGNETIC DIPOLE MOMENT OF THE MOON

Instruments on board the lunar orbiting satellite, Explorer 35, have already yielded a great deal of information on the properties of the moon and its interaction with its environment. Orbits of this satellite for times separated by six months are shown in Fig. 7. The area traversed by Explorer 35 during a 12-month period is indicated by the bounding circles marking the loci of its perilune and apolune. The orbit of the Apollo 15 subsatellite is also shown.

Explorer 35 data recorded when the moon is in the geomagnetic tail have been used to determine upper limits on the dipole moment, M_M, of the moon's magnetic field. The first of these estimates (Sonett *et al.*, 1967) was $M_M \leq 6.10^{20}$ cgs units or about 7.10^{-6} that of the earth. Subsequent refinements led to a lower value of $M_M \leq 4.10^{20}$ cgs units (Ness *et al.*, 1967). Working with a more extensive data set from Explorer 35, Behannon (1968) later reduced the upper limit to $M_M \leq 1.10^{20}$ cgs units.

A preliminary estimate of this upper limit based on the quick-look data from the Apollo 15 subsatellite is $M_M \leq 8.10^{19}$ cgs units, a value essentially consistent with Behannon's. A centered dipole with M_M equal to the upper limit would produce surface field strengths ranging from 1.5 to 3.0 γ.

BOUNDARY LAYER STUDIES

One of the objectives of the subsatellite magnetometer experiment is to study the interaction of the solar wind plasma with the moon in and near the boundary layer of the flow. When the moon is in the solar wind, the region directly behind the moon, i.e., the region directly downstream from the moon, is essentially devoid of solar wind plasma and the boundary between this downstream diamagnetic cavity and the adjacent solar wind flow is defined by a characteristic rarefaction wave in the flow. This situation is shown schematically in Fig. 7. The plasma void region behind the moon was discovered by Lyon *et al.* (1967) with the Explorer 35 plasma detector. The diamagnetic cavity in the plasma void and the magnetic effects later attributed to the rarefaction wave were discovered by Colburn *et al.* (1967) with the Explorer 35 magnetometer. The magnetic field structure associated with this moon-solar wind interaction is readily observable at the Apollo 15 subsatellite during intervals in which the interplanetary magnetic field is relatively steady. The characteristic variation over an orbit is apparent, for example, in the plots of $|\vec{B}_T|$ shown in Fig. 8 and is qualitatively similar to that at the greater altitudes of Explorer 35. The major features of the magnetic field pattern are (1) a typical solar wind field in the region upstream from the moon, (2) an increased field in the region directly downstream, and (3) a decreased field at the boundary between the two. The field increase on the downstream side is associated with the diamagnetic downstream cavity as discussed by Colburn *et al.* (1967) in their initial paper on this phenomenon. The decreases at the boundary of the cavity in the near lunar wake are associated with a rarefaction wave as shown by Siscoe *et al.* (1969) in their study of simultaneous measurements of the plasma and the

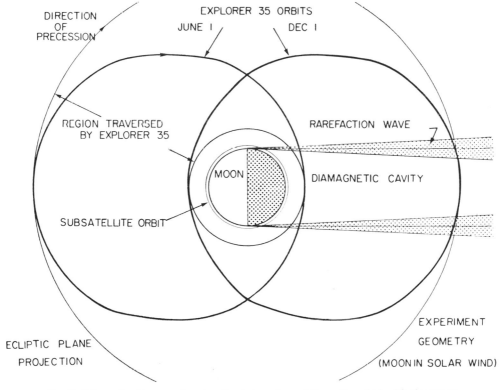

Fig. 7. Schematic diagram showing the interaction of the solar wind with the moon. (Courtesy NASA Ames.) Also shown are the orbits of the lunar subsatellite and Explorer 35. The sketch is drawn as though the orbital planes of both are parallel to the ecliptic plane. Actually, the former is inclined to the ecliptic by about 25–30° while the latter is inclined by about 15°. Orbits of Explorer 35 are shown for times separated by 6 months. The region swept out by the Explorer 35 orbit over a 12-month interval is indicated by bounding circles.

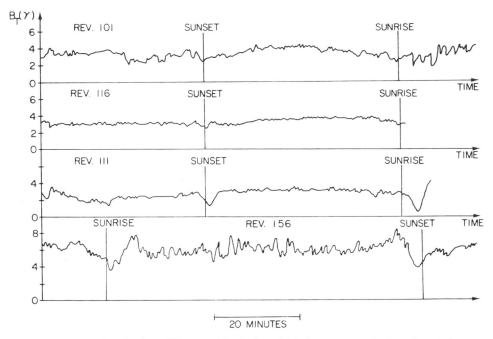

Fig. 8. Plot of B_T for four different orbits during which the moon was in the solar wind.

magnetic field obtained with the instruments on board Explorer 35. The Mach cone of the rarefaction wave front has been observed by Whang and Ness (1970) with the GSFC magnetometer on board Explorer 35.

Often the interaction just described is modified by the addition of a region of increased field at the outer boundary of the region associated with the rarefaction wave. Data recorded during revolution 156 and shown in Fig. 8 exhibit the phenomenon. This effect also is sometimes detected at the greater distances of Explorer 35. It was described by Mihalov *et al.* (1971) as the magnetic effect of a "limb interaction." It has since been interpreted by Barnes *et al.* (1971) and by Sonett and Mihalov (1972) in terms of a limb shock. Because of the significant correlation of the occurrence of this phenomenon with the presence at the limb of certain surface regions, these authors suggest, among other possibilities, that limb shocks are produced by the interaction of the grazing solar wind with localized regions of relatively high remanent magnetization.

In this preliminary study we have mapped the areas of the moon associated with large increases in $|\vec{B}_T|$ as observed at the lower altitude of the subsatellite. The details of this interaction between the solar wind and remanent fields at the lunar limb have not been worked out. Consequently, there is an uncertainty in identifying the selenographic position of the region responsible for a particular observed limb shock. We have used the simplified geometry illustrated in Fig. 9 to calculate disturbance source locations for limb shocks observed on a particular subsatellite revolution. This posi-

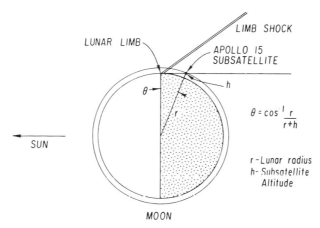

Fig. 9. Geometry used to define the lunar limb.

tion is defined as being θ degrees east of the subsatellite selenographic position at sunset and θ degrees west of the subsatellite selenographic position at sunrise. A limb region was associated with a limb shock if there was a well defined increase in $|\vec{B_T}|$ of at least 2 γ exterior to the rarefaction decrease. Because of the low altitude of the Apollo 15 subsatellite, the regions associated with large increases in $|\vec{B_T}|$ should be near enough to the actual source to identify any large regions of the moon which cause limb shocks.

Data from the first month were tabulated and the results are shown in Fig. 10. Orbital revolutions during which $|\vec{B_T}|$ was unusually variable were excluded. During the first month the subsatellite always encountered sunset near latitude 20° N and sunrise near latitude 28° S. Each box in the lower Mercator projection represents at least four observations. The 20° latitude extent of the boxes is only a crude estimate of the possible latitude variation of the source of an observed disturbance. Figure 11 shows some examples of large disturbances in $|\vec{B_T}|$ associated with regions A and B. Region B produces the largest perturbations of any of the regions identified.

We tentatively conclude that there are regions of the moon of at least 30° longitude which cause shock disturbances in $|\vec{B_T}|$ of the type under discussion when they are exposed to the grazing solar wind and that there are other regions of at least equal size that have little or no shock effect at the subsatellite altitude. The results shown in Fig. 10 agree very well with the Explorer 35 results of Sonett and Mihalov (1972) if one takes into account the uncertainties in the latitudes that are inherent in this preliminary determination of the disturbing regions. A comparison of these results with the direct measurements of the remanent fields discussed in the previous section indicates that Region B is probably the region of relatively high field values, which is shown in Fig. 6, centered about 170° W, 17° S, thus supporting the earlier suggestion that regions of greater remanent magnetization are responsible for the limb shocks.

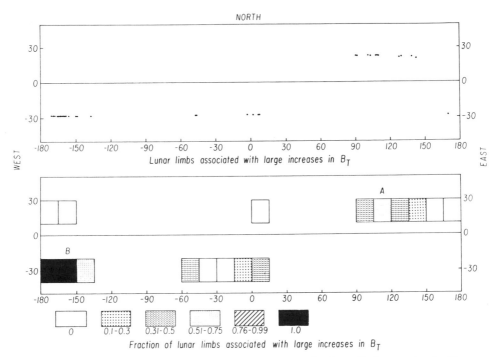

Fig. 10. Mercator projections of the lunar surface showing lunar limbs associated with large increases in B_T at the outer boundary of the rarefaction wave and the fraction of such limbs in areas 15° longitude versus 20° latitude.

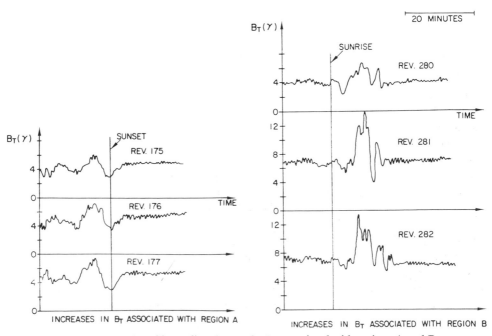

Fig. 11. Examples of large disturbances in B_T associated with regions A and B.

SUMMARY

Our preliminary analysis of the data from the UCLA magnetometer on board the Apollo 15 subsatellite indicates that remanent magnetization is a characteristic property of the moon, that its distribution is such as to produce a rather complex pattern or fine structure at 100 km altitude, and that the regions of more variable topography tend to have more variable and probably stronger fields than the mare regions. Further, the craters, at least those with diameters of 100 km and greater, have qualitatively similar magnetic signatures. The analysis also indicates that the plasma void or diamagnetic cavity that forms behind the moon when the moon is in the solar wind is detectable at the satellite's orbit and that the flow of the solar wind across the limbs is usually rather strongly disturbed. Some specific characteristics of the limb flow, in particular the presence of field effects tentatively associated with temporary limb shocks, depend partly on which surface regions are at the limbs. Regions that produce the more variable remanent fields at the satellite orbit are found to be more effective in producing limb shocks.

Acknowledgments—We are indebted to G. Takahashi and his staff at Time Zero, Inc., for their efforts in the design and fabrication of the subsatellite magnetometer; and to T. Pederson, R. Brown, and their staff at TRW Systems, Inc., for their work in the design, fabrication and testing of the subsatellite and the integration of the magnetometer. We are particularly grateful to C. Thorpe who was charged with the difficult task of controlling the magnetic fields of the subsatellite.

The UCLA engineering team was led by R. C. Snare. Preliminary circuit designs were done by the late R. F. Klein. The testing and calibration of the magnetometer were supervised by F. R. George.

Extensive assistance in the data reduction was provided by J. Wilkinson and J. Tomassian.

We are also grateful to the many people at the Manned Spacecraft Center who contributed to the success of this project, especially to J. Johnson, the program manager, and P. Lafferty, the technical monitor for our project.

No list of acknowledgments for this experiment can be complete without an expression of appreciation to astronauts Scott, Irwin, and Worden. Through their efforts Apollo 15 opened a new era of space exploration.

REFERENCES

Barnes A., Cassen P., Mihalov J. D., and Eviatar A. (1971) Permanent lunar surface magnetism and its deflection of the solar wind. *Science* **172**, 716–718.

Behannon K. W. (1968) Intrinsic magnetic properties of the lunar body. *J. Geophys. Res.* **73**, 7257–7268.

Colburn D. S., Currie R. G., Mihalov J. D., and Sonett C. P. (1967) Diamagnetic solar wind cavity discovered behind the moon. *Science* **158**, 1040–1042.

Coleman P. J. Jr., Schubert G., Russell C. T., Sharp L. R. (1972) Satellite measurements of the moon's magnetic field. *The Moon,* in press.

Dyal P., Parkin C. W., and Sonett C. P. (1970) Apollo 12 magnetometer: Measurements of a steady magnetic field on the surface of the moon. *Science* **196,** 762–765.

Dyal P., Parkin C. W., Sonett C. P., DuBois R. L., and Simmons G. (1971) Lunar portable magnetometer experiment. Preprint, NASA Ames Research Center.

Lyon E. F., Bridge H. S., and Binsack J. H. (1967) Explorer 35 plasma measurements in the vicinity of the moon. *J. Geophys. Res.* **72**, 6113–6117.

Mihalov J. D., Sonett C. P., Binsack J. H., and Moutsoulous M. D. (1971) Possible fossil lunar magnetism inferred from satellite data. *Science* **171**, 892–895.

Ness N. F., Behannon K. W., Scearce C. S., and Cantarano S. C. (1967) Early results from the magnetic field experiment on lunar Explorer 35. *J. Geophys. Res.* **72**, 5769–5778.

Runcorn S. K., Collinson D. W., O'Reilly W., Battey M. H., Stephenson A. A., Jones J. M., Manson A. J., and Readman P. W. (1970) Magnetic properties of Apollo 11 lunar samples. *Proc. Apollo 11 Lunar Sci. Conf., Geochim. Cosmochim. Acta* Suppl. 1, Vol. 3, pp. 2369–2374.

Siscoe G. L., Lyon E. F., Binsack J. H., and Bridge H. S. (1969) Experimental evidence for a detached lunar compression wave. *J. Geophys. Res.* **74,** 59–69.

Sonett C. P. and Mihalov J. D. (1972) Lunar fossil magnetism and perturbations of the solar wind. *J. Geophys. Res.* **77,** 588–603.

Sonett C. P., Colburn D. S., and Currie R. G. (1967) The intrinsic magnetic field of the moon. *J. Geophys. Res.* **72,** 5503–5507.

Strangway D. W., Larson E. E., and Pearce C. W. (1970) Magnetic studies of lunar samples— breccia and fines. *Proc. Apollo 11 Lunar Sci. Conf., Geochim. Cosmochim. Acta* Suppl. 1, Vol. 3, pp. 2435–2451.

Whang Y. C. and Ness N. F. (1970) Observations and interpretation of the lunar mach cone. *J. Geophys. Res.* **75,** 6002–6110.

Proceedings of the Third Lunar Science Conference
(Supplement 3, *Geochimica et Cosmochimica Acta*)
Vol. 3, pp. 2287–2307
The M.I.T. Press, 1972

Surface magnetometer experiments: Internal lunar properties and lunar field interactions with the solar plasma

PALMER DYAL, CURTIS W. PARKIN, and PATRICK CASSEN

NASA-Ames Research Center, Moffett Field, California 94035

Abstract—The remanent magnetic fields measured to date on the moon are $38 \pm 3 \, \gamma$ at Apollo 12 in Oceanus Procellarum; 103 ± 5 and $43 \pm 6 \, \gamma$ at two Apollo 14 sites separated by 1.1 km in Fra Mauro; and $6 \pm 4 \, \gamma$ at the Apollo 15 Hadley-Apennines site. Measurements show that the 38-γ remanent field at Apollo 12 is compressed to $54 \, \gamma$ by a solar wind pressure increase of 7×10^{-8} dyne/cm^2. The change in magnetic pressure is proportional to the change in plasma pressure and the field is compressed primarily in the z (northerly) component. The whole-moon relative magnetic permeability has been calculated to be $\mu/\mu_0 = 1.01 \pm 0.06$ from measurements taken while the moon was immersed (magnetized) in the geomagnetic tail. The electrical conductivity of the lunar interior has been determined from magnetic step transient measurements made on the lunar dark side. A range of monotonic conductivity profiles is calculated that provides a fit to the normalized data curve within error limits. Deeper than 90 km into the moon, the conductivity rises from 3×10^{-4} mhos/m to 10^{-2} mhos/m at 1000 km depth. These conductivities, when converted to temperatures for an assumed lunar material of peridotite, suggest the existence of a thin outer layer (perhaps 90 km thick) in which the temperature rises sharply to 850–1050°K, then increases gradually to 1200–1500°K at a depth of about 1000 km.

INTRODUCTION

IN THIS PAPER we report lunar magnetic field measurements and calculations of magnetic permeability, electrical conductivity, and temperature of the lunar interior. We also describe the interaction between the fossil remanent field and the solar wind plasma.

Magnetometers have been deployed on the lunar surface at the Apollo 12 site in Oceanus Procellarum, at the Apollo 14 site in Fra Mauro, and at the Apollo 15 Hadley-Apennines site. This network of instruments has been used to measure the intrinsic lunar magnetic field and the global response of the moon to large scale solar and terrestrial magnetic fields. The input (driving) magnetic field is measured by the lunar orbiting Explorer 35 magnetometer and the output (response) fields are measured by the Apollo surface instruments.

The fossil remanent field, first discovered by the Apollo 12 experiment (Dyal *et al.*, 1970a) provides a record of the magnetic field environment that existed at the moon 3 to 4 b.y. ago at the time the crustal material cooled below its Curie temperature. This fossil record points to the possible existence of an ancient lunar dynamo or to a solar or terrestrial field much stronger than exists at present.

The remanent magnetic field is found to be compressed in direct proportion to solar wind pressure. Properties of the compression are used to study the scale size of the remanent field. The magnetic field interaction causes the solar wind to be deflected and is therefore also important for the study of solar gas accretion in the lunar regolith.

Relative magnetic permeability is calculated from measurements obtained when the moon is immersed in a steady geomagnetic field. The permeability is related to the amount of permeable material, such as iron, that exists in the outer layers of the moon.

A continuous electrical conductivity profile of the lunar interior is calculated by analysis of the decay of eddy-current magnetic fields induced in the moon by solar-field step transients. This is an extension of the three-layer conductivity model reported earlier (Dyal and Parkin, 1971a). The conductivity is in turn related to internal temperature, which is calculated for assumed lunar material compositions.

Magnetic Fields at the Lunar Surface

The external driving magnetic field in the lunar environment can vary considerably with the lunar orbital position (see Fig. 1). Average magnetic field conditions vary from relatively steady fields of magnitude $\sim 9\ \gamma$ ($1\ \gamma = 10^{-5}$ Gauss) in the geomagnetic tail to mildly turbulent fields averaging $\sim 5\ \gamma$ in the free-streaming solar plasma region to turbulent fields averaging $\sim 8\ \gamma$ in the magnetosheath. Average solar wind velocity is ~ 400 km/sec in a direction approximately along the sun-earth line.

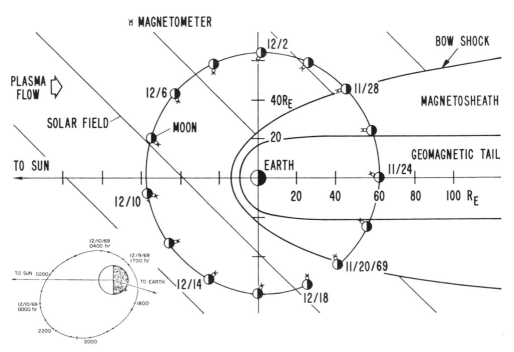

Fig. 1. Lunar orbit projection onto the solar ecliptic plane, showing the Apollo 12 magnetometer during the first post deployment lunation, 1969. During a complete revolution around the earth, the magnetometer passes through the earth's bow shock, the magnetosheath, the geomagnetic tail, and the interplanetary region dominated by solar plasma fields. The insert shows the Explorer 35 orbit around the moon, also projected onto the solar ecliptic plane. The Explorer 35 period of revolution is 11.5 hours.

Various induced lunar and plasma-interaction fields are assumed to exist at the lunar surface; for reference we write the sum of these fields as

$$\vec{B}_A = \vec{B}_E + \vec{B}_S + \vec{B}_\mu + \vec{B}_P + \vec{B}_T + \vec{B}_D + \vec{B}_F \qquad (1)$$

Here $\vec{B}_A$ is the total magnetic field measured on the surface by an Apollo lunar surface magnetometer (LSM); $\vec{B}_E$ is the total external (solar or terrestrial) driving magnetic field measured by the Explorer 35 and Apollo 15 subsatellite lunar orbiting magnetometers while outside the antisolar lunar cavity; $\vec{B}_S$ is the steady remanent field at the surface site; $\vec{B}_\mu$ is the magnetization field induced in permeable lunar material; $\vec{B}_P$ is the poloidal field caused by eddy currents induced in the lunar interior by changing external fields; $\vec{B}_T$ is the toroidal field corresponding to unipolar electrical currents driven through the moon by the $\vec{V} \times \vec{B}_E$ electric fields; $\vec{B}_D$ is the field associated with the diamagnetic lunar cavity; and $\vec{B}_F$ is the field associated with the hydromagnetic solar wind flow past the moon.

The relative importance of these different fields varies with orbital position, and therefore different magnetic fields and lunar properties can be investigated during selected times of each lunation. When the moon is passing through a quiet region of the geomagnetic tail, solar wind interaction fields ($\vec{B}_T$, $\vec{B}_D$, and $\vec{B}_F$) and the induced poloidal lunar field $\vec{B}_p$ are negligible, and the total field at the lunar surface is

$$\vec{B}_A = \vec{B}_E + \vec{B}_\mu + \vec{B}_S \qquad (2)$$

The magnetic moment of the field $\vec{B}_\mu$ is proportional to $\vec{B}_E$, i.e., $\vec{m}_\mu = K\vec{B}_E$ (the proportionality constant K in turn depends on the permeability and the dimensions of the permeable region of the moon). Therefore when $\vec{B}_E = 0$, $\vec{B}_A = \vec{B}_S$, and the Apollo magnetometer measures the steady remanent field alone. Once $\vec{B}_S$ is known, K can be determined and the relative magnetic permeability μ/μ_0 can be calculated.

A different set of field terms in equation (1) is dominant when the moon is immersed in free-streaming solar wind and the magnetometer is on the lunar sunlit side. $\vec{B}_D \rightarrow 0$ outside the cavity, and the global fields $\vec{B}_\mu$ and $\vec{B}_T$ can be neglected in comparison to $\vec{B}_P$ (Dyal and Parkin, 1971b). The interaction field $\vec{B}_F$ has been found to be important during times of high solar wind particle density (Dyal et al., 1972); therefore the interaction term $\vec{B}_F$ is not to be assumed negligible in general, and equation (1) becomes

$$\vec{B}_A = \vec{B}_P + \vec{B}_E + \vec{B}_S + \vec{B}_F \qquad (3)$$

At low frequencies ($\leq 3 \times 10^{-4}$ Hz), $\vec{B}_P \rightarrow 0$ and the interaction field $\vec{B}_F$ can be investigated.

When the magnetometer is located on the dark (antisolar) side of the moon, it is generally isolated from solar plasma flow and $\vec{B}_F \rightarrow 0$. Then for dark side data, equation (3) reduces to

$$\vec{B}_A = \vec{B}_P + \vec{B}_E + \vec{B}_S \qquad (4)$$

where cavity effects ($\vec{B}_D$) are neglected to a first approximation for measurements made near lunar midnight (Dyal and Parkin, 1971b). After $\vec{B}_S$ has been calculated from geomagnetic tail data, only the poloidal field $\vec{B}_P$ is unknown. Equation (4) can then be solved for certain assumed lunar models, and curve fits of data to the solution determine the model dependent conductivity profile $\sigma(R)$. Furthermore, electrical conductivity is related to temperature, and the lunar interior temperature can be calculated for assumed lunar material compositions.

Experimental Technique

Magnetic fields at the lunar surface have been measured by magnetometers emplaced by astronauts on the Apollo 12, 14, and 15 missions. The three orthogonal vector components are measured as a function of time and position and telemetered to earth. Simultaneously a magnetometer in the lunar orbiting Explorer 35 spacecraft measures the ambient solar and terrestrial field and transmits this information to earth. Three different types of lunar magnetometers are described in the following paragraphs. Both a portable and stationary surface magnetometer are planned for the Apollo 16 mission to Descartes.

Stationary lunar surface magnetometer (LSM)

The stationary magnetometer deployed at the Apollo 15 site in the Hadley-Apennines region of the moon is shown in Fig. 2. Characteristics of this and a similar instrument deployed at the Apollo 12 site are given in Table 1. A more detailed description of the stationary magnetometer is reported by Dyal et al. (1970b).

The three orthogonal vector components of the lunar surface magnetic field are measured by three fluxgate sensors (Gordon et al., 1965) located at the ends of three 100-cm-long orthogonal booms. The sensors are separated from each other by 150 cm and are 75 cm above the ground.

The instrument can also be used as a gradiometer by sending commands to operate three motors in the instrument which rotate the sensors such that all simultaneously align parallel first to one of

Table 1. Apollo magnetometer characteristics.

Parameter	Apollo 12 LSM	Apollo 15 LSM	Apollo 14 LPM
Lunar locations	3.2°S, 23.4°W	26.4°N, 3.5°E	3.7°S, 17.5°W
Ranges (γ)	0 to ± 400 0 to ± 200 0 to ± 100	0 to ± 200 0 to ± 100 0 to ± 50	0 to ± 100 0 to ± 50 —
Resolution (γ)	0.2	0.1	0.5
Frequency response	dc to 3 Hz	dc to 3 Hz	dc to 0.01 Hz
Angular response	Proportional to cosine of angle between magnetic field vector and sensor axis.		
Sensor geometry	Three orthogonal sensors at ends of 100 cm booms	Three orthogonal sensors at ends of 100 cm booms	Three orthogonal sensors in 6 cm cube
Azimuthal orientation	$\pm 0.5°$	$\pm 0.5°$	$\pm 3°$
Tilt orientation	$\pm 0.2°$	$\pm 0.2°$	$\pm 1°$
Analog zero determination	180° flip of sensor	180° flip of sensor	180° flip of sensor
Power (watts)	3.5	3.5	1.5
Weight (kg)	8.9	8.9	4.6
Size (cm)	25 × 28 × 63	25 × 28 × 63	56 × 15 × 14
Operating temperature (°C)	−50 to +85	−50 to +85	−30 to +60

Fig. 2. The Apollo 15 lunar surface magnetometer (LSM) deployed on the moon in the Hadley-Apennines region. Sensors are at the top ends of the booms and approximately 75 cm above the lunar surface, separated by 150 cm. Properties of the instrument are listed in Table 1.

the boom axes, then to each of the other two boom axes in turn. This rotating alignment permits the vector gradient to be calculated in the plane of the sensors and also permits an independent measurement of the magnetic field vector at each sensor position.

Lunar portable magnetometer (LPM)

The portable magnetometer developed for the Apollo 14 mission to Fra Mauro is shown in Fig. 3, and the instrument characteristics are given in Table 1. A more detailed description of the instrument is reported by Dyal *et al.* (1971).

The instrument was designed to be a totally self-contained, portable experiment package. Three orthogonally oriented fluxgate sensors are mounted on the top of a tripod, positioned 75 cm above the lunar surface. These sensors are connected by a 15-meter-long cable to an electronics box which contains a battery, electronics and three milliammeters used to read the field output.

Lunar orbiting Explorer 35 magnetometer

The ambient steady-state and time-dependent magnetic fields in the lunar environment are measured by the Explorer 35 satellite magnetometer. The satellite, launched in July 1967, has an orbital period of 11.5 hours, aposelene of 9390 km, and periselene of 2570 km (see Fig. 1 insert). The Explorer 35 magnetometer measures three magnetic field vector components every 6.14 sec at 0.4 γ resolution; the instrument has an alias filter with 18 dB attenuation at the Nyquist frequency (0.08 Hz) of the spacecraft data sampling system. A more detailed description of the instrument is reported by Sonett *et al*, (1967).

The Apollo 15 subsatellite magnetometer, orbiting approximately 100 km above the lunar surface, has also measured fields intrinsic to the moon. This instrument is described by Coleman *et al*. (1972).

Remanent Magnetic Fields at the Apollo Sites

Local permanent fields $\vec{B}_S$ at four Apollo surface sites have been measured and are listed in Table 2. The field components are expressed in their respective local

Fig. 3. The Apollo 14 lunar portable magnetometer (LPM), a self-contained, battery powered instrument, shown on the lunar surface in the Fra Mauro region. The electronics box and field indicating meters are located on the two-wheeled cart (left), and the sensor-tripod assembly (right) is shown during measurement of the 103-gamma remanent magnetic field at Fra Mauro.

Table 2. Remanent magnetic field measurements at Apollo 15, 14, and 12 sites.

Site	Coordinates (deg)	Field magnitude (γ)	Magnetic field components (γ)		
			Up	East	North
Apollo 15	26.4°N, 3.5°E	6 ± 4	+4 ± 4	+1 ± 3	+4 ± 3
Apollo 14	3.7°S, 17.5°W				
Site A		103 ± 5	−93 ± 4	+38 ± 5	−24 ± 5
Site C′		43 ± 6	−15 ± 4	−36 ± 5	−19 ± 8
Apollo 12	3.2°S, 23.4°W	38 ± 3	−24.4 ± 2.0	+13.0 ± 1.8	−25.6 ± 0.8

surface coordinate systems. All fields are attributed to remanent magnetization in nearby subsurface material.

The gradients of the fields at the landing sites were also measured and found to be relatively low: less than 0.13 γ/m at the Apollo 12 and 15 sites (i.e., below the instrument resolution) and 0.06 γ/m at the Apollo 14 site. In addition, the scale size of the Apollo 12 field has been found to be greater than 2 km from solar wind plasma interaction measurements (Dyal et al., 1972).

The three high field measurements (at the Apollo 12 and 14 sites) were taken at separations of no more than 185 km, whereas the Apollo 15 measurement, taken 1200 km distant, showed a comparatively small field. The similarities between the Apollo 12 and 14 field measurements (viz., all vectors are pointed down and toward the south and have magnitudes that correspond to within a factor of 3) suggest that the two Apollo 14 sites and possibly the Apollo 12 site are located above a near-surface layer of material that was uniformly magnetized at one time. Subsequently, the magnetization in the layer was altered by local processes, such as tectonic activity or fracturing and shock demagnetization from meteorite impacts; indeed, preliminary analysis of Apollo 15 subsatellite data (Coleman et al., 1972) shows correlation of surface magnetic field with some large impact craters such as Van de Graaff on the lunar far side.

No obvious mechanism for such large-scale magnetization of near-surface layers exist at present. Magnetization of Apollo samples would have required an external field greater than 10^3 γ (Pearce et al., 1972; Helsley, 1970, 1971; Runcorn et al., 1970). Ambient fields of this magnitude have not been measured in space near the moon; the largest measured so far have been transient fields of magnitude of approximately 10^2 γ (Dyal et al., 1970c); these transient fields last only a few minutes. It is evident, therefore, that at one time in the lunar past, an ambient field much stronger than at present existed over much or all of the lunar surface. Possible origins of the ancient ambient field could have been external to the moon (sun or earth) or intrinsic to the moon (e.g., a lunar dynamo).

REMANENT MAGNETIC FIELD INTERACTION WITH THE SOLAR WIND

Compression of the remanent lunar magnetic field by the solar wind has been measured at the Apollo 12 site. Simultaneous surface magnetic field and plasma data show, to first order, a compression of the remanent field in direct proportion to the solar wind pressure.

In order to study the compression of the remanent field, it is advantageous to define a field $\overrightarrow{\Delta B}$ as

$$\overrightarrow{\Delta B} = \overrightarrow{B_A} - (\overrightarrow{B_E} + \overrightarrow{B_S}), \tag{5}$$

where $\overrightarrow{B_A}$ is the total surface magnetic field, measured by the Apollo 12 lunar surface magnetometer; $\overrightarrow{B_E}$ is the extralunar (solar or terrestrial) driving magnetic field, measured by the lunar orbiting Explorer 35 magnetometer; and $\overrightarrow{B_S}$ is the steady remanent field at the site due to magnetized material. For low frequencies, i.e., 1-hr averages of magnetic and solar wind data, the eddy current poloidal field can be neglected and equation (5) contains all the vector fields which are dominant at the lunar surface.

We shall show that to first order $\overrightarrow{\Delta B}$ is the vector change in the steady remanent field due to the solar wind pressure. Figure 4 shows simultaneous 1-hr average plots

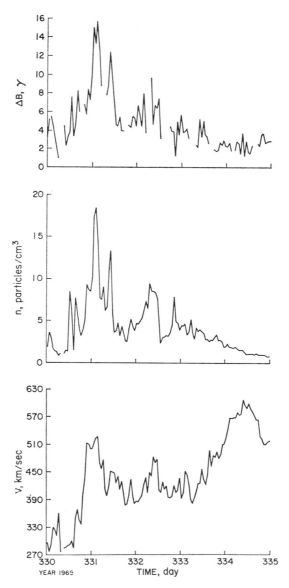

Fig. 4. One-hour-average graphs of magnitude of the compressed field vector $\overrightarrow{\Delta B} = \overrightarrow{B_A} - (\overrightarrow{B_E} + \overrightarrow{B_S})$, plasma proton density n and plasma velocity V. $\overrightarrow{B_A}$ is the total surface field measured by the Apollo 12 lunar surface magnetometer, $\overrightarrow{B_E}$ is the extralunar field measured by lunar orbiting Explorer 35, and $\overrightarrow{B_S}$ is the 38 γ remanent field at the Apollo 12 site.

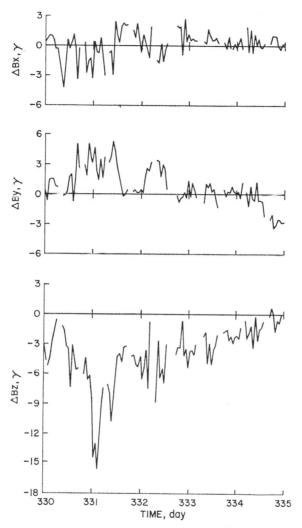

Fig. 5. Components of the magnetic field difference vector $\vec{\Delta B}$. Data are expressed in the surface coordinate system ($\hat{x}$, $\hat{y}$, $\hat{z}$) which has its origin at the Apollo 12 magnetometer site; $\hat{x}$ is directed radially outward from the surface while $\hat{y}$ and $\hat{z}$ are tangent to the surface, directed eastward and northward, respectively. The uncompressed steady remanent field at the Apollo 12 site is $\vec{B_s} = (-24.4\,\gamma,\ +13.0\,\gamma,\ -25.6\,\gamma)$.

of the magnitude of the vector field difference $\vec{\Delta B}$, solar wind proton density n, and solar wind velocity magnitude V measured at the Apollo 12 site. Figure 1 shows the orbital position of the moon during the time interval of Fig. 4, which includes measurements made both in the magnetosheath and free-streaming solar wind. All data are expressed in the surface coordinate system ($\hat{x}$, $\hat{y}$, $\hat{z}$) which has its origin at the Apollo 12 magnetometer site; $\hat{x}$ is directed radially outward from the surface while $\hat{y}$ and $\hat{z}$ are tangent to the surface, directed eastward and northward, respectively.

Components of the steady remanent field at the Apollo 12 site have been determined (Dyal and Parkin, 1971b) to be $B_{Sx} = -24.4\,\gamma$, $B_{Sy} = +13.0\,\gamma$, and $B_{Sz} = -25.6\,\gamma$. By inspection we see a strong relationship between the magnitude ΔB and plasma proton density (n); no such strong correlation, however, is apparent between ΔB and velocity V alone.

Figure 5 shows the x (vertical), y (easterly), and z (northerly) components of $\overrightarrow{\Delta B}$. Correlation with density is seen in all three components, but it is stronger in the horizontal components y and z and strongest in the z direction. The correlations between the field difference components ΔB_i and plasma proton density n suggest that the field change $\overrightarrow{\Delta B}$ is due to a compression of the local remanent field $\overrightarrow{B_s}$ by the solar wind, with the largest compression occurring in the z-component. This preferential compression of the z-component may be due to the geometric configuration of the remanent field.

The nature of the correlation between the magnetic field and plasma pressure is illustrated in Fig. 6. The pressures are related throughout the measurement range and the plot contains data from two sequential lunations. A fit to the data averages by a least-squares straight line permits a slope $K \cong 0.01$ to be calculated which is approximately the ratio of magnetic pressure change to plasma pressure.

The ratio of plasma dynamic pressure to *total* magnetic pressure is expressed

$$\beta = \frac{nmV^2}{B_{ST}^2/8\pi} \tag{6}$$

where $\overrightarrow{B_{ST}} = \overrightarrow{B_s} + \overrightarrow{\Delta B}$ is the total surface compressed field. During times of maximum plasma pressure shown in Fig. 6, we calculate $\beta = 5.9$; $\beta \le 1$ would imply

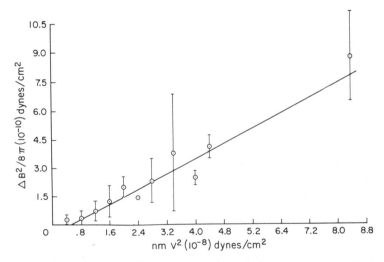

Fig. 6. Magnetic field pressure difference versus solar wind plasma pressure at the Apollo 12 site. The ratio of magnetic pressure change to plasma pressure is given by the slope to be $\cong 0.01$. The maximum ratio of dynamic plasma pressure to *total* magnetic pressure is calculated to be $\beta = 5.9$.

that the field had been compressed to the stagnation magnitude required to stand off the solar wind and possibly form a local shock. Compression of the remanent field alone, therefore, does not cause the stagnation condition to be reached during the time period of these data.

The experimental measurement of this solar-wind remanent field interaction allows us to estimate the size (Barnes et al., 1971) and orientation (Siscoe and Goldstein, 1971) of the magnetic field region produced by these magnetic concentrations ("magcons"). If the Apollo 12 field were assumed to be due to a single dipolar source, then Explorer 35 measurements would constrain its size to $L < 200$ km (Dyal et al., 1972). Since the field does interact with the solar wind, its scale length $L \gtrsim 2$ km; thus a single dipolar Apollo 12 magcon would likely be in the size range 2 km $\lesssim L < 200$ km. Finally, we note that since there is a measured interaction between the remanent magnetic field (Dyal et al., 1972; Neugebauer et al., 1972) and the solar wind particles, it follows that the geometry of accretion of solar wind particles into the lunar regolith will be partially determined by the topology of the magnetic field.

RELATIVE MAGNETIC PERMEABILITY OF THE MOON

The bulk relative magnetic permeability of the moon is calculated to be $\mu/\mu_0 = 1.01 \pm 0.06$. This value has been determined from simultaneous measurements of the Apollo 12 surface magnetometer and Explorer 35 orbiting magnetometer obtained when the moon was "permed" by a steady geomagnetic tail field. Equation (2), which described the sum of magnetic fields measured at the lunar surface, is solved for the case of a spherically symmetric lunar permeable shell following the method of Jackson (1962). Components of the field solutions, in local surface coordinates ($\hat{x}$ is directed radially outward from the lunar surface; $\hat{y}$ and $\hat{z}$ are tangential with respect to the surface, directed eastward and northward, respectively), are expressed

$$B_{Ax} = (1 + 2F)B_{Ex} + B_{Sx} \qquad (7)$$

$$B_{Ay,z} = (1 - F)B_{Ey,z} + B_{Sy,z} \qquad (8)$$

where

$$F = \frac{(2k_m + 1)(k_m - 1)\left(1 - \left(\dfrac{R}{R_m}\right)^3\right)}{(2k_m + 1)(k_m + 2) - 2\left(\dfrac{R}{R_m}\right)^3 (k_m - 1)^2} \qquad (9)$$

In the latter equation, k_m is the relative permeability μ/μ_0; R_m is the lunar radius; and R is the radius of the boundary which encloses lunar material with temperature above the Curie point.

Equations (7) and (8) are linear equations, and their slopes are proportional to the factor F. Figure 7 shows a plot of radial components of Apollo 12 surface field (B_{Ax}) versus the geomagnetic tail field (B_{Ex}) measured by Explorer 35. A least-squares fit and slope calculations determine the factor $F = 0.0030$, which is used to determine the relative magnetic permeability for an assumed inner radius R, as shown in Fig.

P. Dyal, C. W. Parkin, and P. Cassen

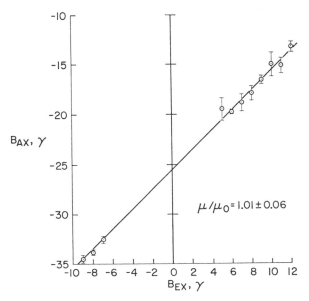

Fig. 7. Graphical representations of equation (7), used to calculate relative magnetic permeability of the moon. Radial component of Apollo 12 total surface magnetic field B_{Ax} as a function of the radial component of external driving field B_{Ex}. Data points consist of averages of measurements in quiet regions of the geomagnetic tail taken during the first five postdeployment lunations. The B_{Ax} intercept of the least-squares best-fit solid line gives the radial component of the Apollo 12 remanent field; the best-fit slope corresponds to a value of 1.01 ± 0.06 for the bulk relative permeability μ/μ_0 of the moon.

8a. For the bulk permeability of the moon (the case $R = 0$), $\mu/\mu_0 = 1.01 \pm 0.06$. For a thinner permeable shell inside the moon the permeability is higher, as illustrated in Fig. 8b.

A more accurate calculation of lunar permeability will be determined in the future from network measurements obtained at three locations on the lunar surface. The increased accuracy will make possible a meaningful calculation of the percentage of permeable iron in the outer layer of the lunar sphere.

Transient Magnetic Field Response:
Internal Electrical Conductivity and Temperature

1. Discussion of the analytical method

The experimental method of determining the radial electrical conductivity profile of the moon by analysis of the response to a step function input has been described by Dyal and Parkin (1971a). A discontinuity in the interplanetary magnetic field which is swept by the moon induces eddy currents within the moon. These currents induce a field which tends to oppose the external field, retarding its penetration of the moon. The currents diffuse and decay in a manner governed by the lunar electrical conductivity distribution. The transient behavior of the induced magnetic field

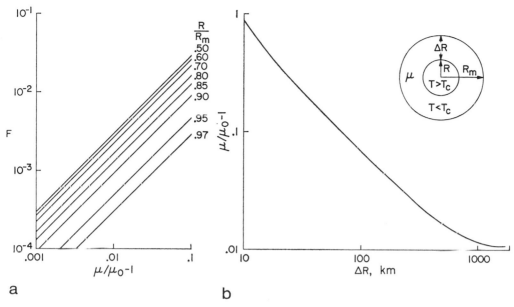

Fig. 8. (a) The function F (in a graphical representation of equation (9)), related to relative magnetic permeability $k_m = \mu/\mu_0$ for various values of R/R_m. R and R_m are internal and external radii, respectively, of a global permeable shell. (b) Lunar relative magnetic permeability as a function of lunar shell thickness $\Delta R = R_m - R$, for homogeneous permeable lunar material. ΔR is the region wherein permeable material is below its Curie temperature (T_c) and can be magnetized ("permed") by an external magnetic field.

(recorded at the surface by the Apollo 12 and 15 magnetometers) is related to the conductivity function through the diffusion equation for the induced fields. Hence the experimental method consists of scanning time-series data from the lunar orbiting Explorer 35 magnetometer (which monitors the solar wind field undisturbed by eddy currents within the moon) to find interplanetary field discontinuities in the time record of the data from Explorer 35 that have smooth fields before and after the discontinuity, then examining the corresponding induced field response recorded at the lunar surface. A conductivity distribution is then sought which yields this response as a solution of the magnetic field diffusion equation.

Before we proceed to a more general treatment, it is useful to consider the idealized case in which the moon is represented by a uniformly conducting sphere in a vacuum. This case is solved analytically by Smythe (1968). Suppose that initially there is no magnetic field, but at $t = 0$ an external magnetic field $\overrightarrow{\Delta B_E}$ is switched on which is uniform far from the sphere. At this time a surface current is induced on the sphere which excludes the applied field completely from the interior (Fig. 9). The current then diffuses through the sphere, and eventually decays to zero strength. Correspondingly, the external field, which was originally excluded, diffuses into the sphere so that finally a uniform field occupies all space. If the conductivity of the sphere is

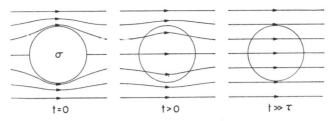

Fig. 9. Transient magnetic response of a conducting sphere in a magnetic field. At the time of the step transient ($t = 0$) eddy currents are generated which exclude field lines to the outer edge of the sphere. At later times the eddy currents diffuse inward, permitting the field lines to permeate the entire sphere at a time which is much greater than the time response of the conductor.

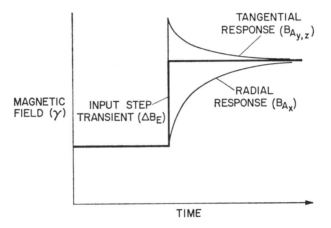

Fig. 10. Theoretical solutions for the lunar night vacuum poloidal magnetic field response of a homogeneous conducting sphere to a step function transient in the driving solar wind magnetic field. For a step-function change $\Delta \vec{B}_E$ in the external driving field (measured by Explorer 35), the total magnetic field at the surface of the moon $\vec{B}_A$ (measured by the Apollo 12 and 15 magnetometers) will be damped in the radial (B_{Ax}) component and will overshoot in the tangential (B_{Ay} and B_{Az}) components.

large, the eddy currents will be sustained for a long time, and the external field will only slowly diffuse to the interior. At the surface, the tangential component of the field is initially greater than the corresponding external field component, and the radial surface component is initially zero (Fig. 10). Both components asymptotically approach the values of external field components as the induced currents decay.

The response of the moon, as measured by the Apollo 12 magnetometer on the dark side, is qualitatively similar to the curves shown in Fig. 10. (It deviates from these insofar as the moon's electrical conductivity is not uniform.) In order to determine the conductivity profile from the shape of the transient response curves, we assume that the dark side response is that of a sphere in a vacuum. This is justified in view of the low plasma density on the dark side surface reported by Snyder et al. (1970).

Furthermore, the lunar response has been found not to be a function of position if the magnetometer is more than 400 km inside the optical shadow. Therefore, the effect of currents at the boundary of the lunar cavity (and on the front surface of the moon) on the shape of the response curves is neglected. It has been shown experimentally that the currents in the moon driven by the solar wind $\vec{V} \times \vec{B}_E$ field can be neglected (Dyal and Parkin, 1971b). The driving field is taken to be spatially uniform, which requires that its scale be much greater than the diameter of the moon, and that the discontinuity be swept past the moon in a time short compared to the lunar response time. Both of these conditions are generally fulfilled. These assumptions allow us to model the dark side transient response by that of an inhomogeneous conducting sphere in a vacuum.

2. Theory for the response time

To describe the response of the sphere to an arbitrary input, we define the magnetic vector potential $\vec{A}$ such that $\vec{\nabla} \times \vec{A} = \vec{B}$ and $\vec{\nabla} \cdot \vec{A} = 0$. We seek the response to an input $\vec{B} \Delta_E b(t)$, where $b(t) = 0$ for $t < 0$ and $b(t)$ approaches unity as $t \to \infty$. (Since the governing equations are linear, the response to a more general input is readily found by superposition.) The direction of $\Delta \vec{B}_E$ is taken to be the axis of a spherical coordinate system (R, θ, ϕ). If the conductivity is spherically symmetric, the transient magnetic field response has no ϕ component, and hence $\vec{A} = A \hat{e}_\phi$ and all $\partial/\partial\phi = 0$. Under these conditions (and neglecting displacement currents) the laws of Faraday, Ampère, and Ohm combine to yield the diffusion equation for the magnetic potential (in MKS units):

$$\nabla^2 \vec{A}(R, \theta; t) = \mu\sigma(R) \frac{\partial \vec{A}}{\partial t}(R, \theta; t). \tag{10}$$

We have shown in the preceding section on magnetic permeability that we may take $\mu \cong \mu_0$ everywhere. Then, for $t > 0$, the magnetic field must be continuous at the surface, so that $\vec{A}$ and $\partial\vec{A}/\partial R$ must always be continuous at $R = R_m$. We also have the boundary condition $\vec{A}(0, t) = 0$ and the initial condition $\vec{A}(R, \theta; \phi) = 0$ inside the moon. Outside of the moon, where $\sigma = 0$,

$$A = \Delta B_E \left(\frac{R}{2}\right) b(t) \sin\theta + \frac{\Delta B_E f(t)}{R^2} \sin\theta. \tag{11}$$

The first term on the right is a uniform magnetic field modulated by $b(t)$; the second term is the (as yet unknown) external transient response, which must vanish as $R \to \infty$ and $t \to \infty$. Note that at $R = R_m$,

$$A = \Delta B_E \sin\theta \left(\frac{b(t)}{2} + f(t)\right) \tag{12}$$

and

$$\frac{\partial A}{\partial R} = \Delta B_E \sin\theta \left(\frac{b(t)}{2} - 2f(t)\right). \tag{13}$$

Therefore, at $R = R_m$,

$$\frac{\partial A}{\partial R} = -2A + \tfrac{3}{2}(\Delta B_E \sin \theta \, b(t)). \tag{14}$$

Since the magnetic field is continuous at $R = R_m$, this is a boundary condition for the interior problem. Letting $G(R, t) = A/\Delta B_E \sin \theta$ and $\bar{G}(R, s)$ be the Laplace transform of G, equation (10) becomes

$$\frac{1}{R}\left(\frac{\partial^2}{\partial R^2}(r\bar{G}) - \frac{2}{R}\bar{G}\right) = s\mu_0 \sigma \bar{G} \tag{15}$$

for the interior. The boundary conditions are

$$\frac{\partial \bar{G}}{\partial R} = -2\bar{G} + \tfrac{3}{2}\bar{b}(s) \qquad \text{at } R = R_m \tag{16}$$

and

$$\bar{G} = 0 \qquad \text{at } R = 0. \tag{17}$$

For a given $\sigma(R)$ and $b(t)$, this system is numerically integrated to obtain $\bar{G}(R, s)$ in the range $0 \le R \le R_m$. The function $\bar{G}(R_m, s)$ is then numerically inverse Laplace transformed to find the characteristic transient response function $f(t)$ for the system.

For comparison with the data from the Apollo 12 magnetometer, cases in which the input (as monitored by Explorer 35) was a step function were sought. However, following Schubert and Colburn (1971), we have in our analysis chosen $b(t)$ to be a ramp function with a time constant of 15 sec, a time characterizing the passage of a discontinuity past the moon. (For a 400 km/sec solar wind, this time is 10–20 sec, depending on the thickness of the discontinuity and the inclination of its normal to the solar wind velocity.) Although the driving field is still approximated by a spatially uniform field, this procedure provides a better model for the very short time response.

3. Comparison with the data; electrical conductivity profile

The normalized data from Apollo 12, giving the step function response obtained from the radial surface field component (for 11 events) is shown in Fig. 11. All of these events occurred when the magnetometer was more than 900 km inside the optical shadow, so that plasma effects are assumed to be absent. The error bars are standard deviations of the measurements.

The theoretical response curves corresponding to a large number of lunar conductivity profiles were compared with the data of Fig. 11. It was found that a range of monotonic conductivity profiles, defining the shaded region in Fig. 12, provided fits to the data curve that fall within the error bars. As pointed out by Schubert and Colburn (1971), the early response ($t < 20$ sec) is dominated by the finite rise time of the driving function, and hence detailed information on the conductivity at shallow depths is limited. On the other hand, a perfectly conducting lunar core with a radius of about 300 km would be undetectable even with large magnitude ($\sim 40 \gamma$) inputs, so there is an inherent limitation on the conductivity information to be gained at large

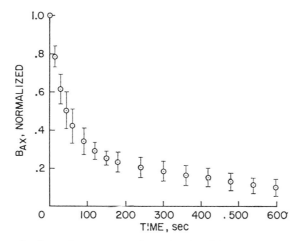

Fig. 11. Normalized transient response data, showing $f(t)$ decay characteristics of the radial component of the total surface field B_{Ax} after arrival of a step transient which changes the external magnetic field radial component by an amount ΔB_{Ex}, here normalized to one.

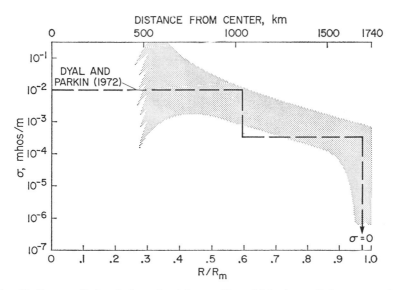

Fig. 12. Range of electrical conductivity profiles which give radial response time-dependent curves $f(t)$ that fall within the error bars of Fig. 11. A step-function input magnetic field, modified by an initial ramp input of 15-sec rise time, is used in the analysis. The information at shallow depths is limited by the uncertainty in the rise time of the interplanetary magnetic field; that at large depths is limited by the sensitivity of the surface magnetometer. The superimposed dashed line shows the three-layer approximation of an earlier work.

depths. Nevertheless, the data allow only a rather restricted range of conductivities at intermediate depths, if the conductivity function is monotonic. The curvature of the response curve is greatest for times around 100 sec after the step input. The curvature of the theoretical curve at these times is sensitive to the conductivity at depths of 300 to 700 km. Hence the shape of the data curve puts fairly restrictive limits on the conductivity in this region. Deeper than 90 km, the conductivity is seen to rise from about $10^{-3.5}$ mhos/m to 10^{-2} mhos/m at $R/R_m = 0.45$. This is in general agreement with the three layer model of Dyal and Parkin (1972). However, we note that conductivities greater than $10^{-1.5}$ mhos/m for $R/R_m < 0.4$ are compatible with the transient data. Although the conductivity in the outermost layers of the moon must be very low, less than 10^{-9} mhos/m (Dyal and Parkin, 1971a), this region is apparently very thin.

Schubert and Colburn (1971) have pointed out that a reasonable fit to the transient data can be made for a case in which $\sigma(R)$ is sharply peaked in a very narrow region, corresponding to the suggested conductivity profile of Sonett et al. (1971). In this case $\mu\sigma d R_p$ (d is the thickness of the peak, R_p is the radius at which it occurs) is comparable to $\mu\sigma R_m^2$ of a monotonic function giving roughly the same response curve. Our calculations indicate that the response to a peaked conductivity profile ($\sigma_{max} \sim 10^{-2}$ mhos/m, $d \sim 17$ km at $R_p = 1480$ km), superimposed on a suitable smooth function, falls within the error bars of the data. However, if the peak is broadened to occupy a region greater than about 45 km, the response curve cannot be made compatible with the data. Hence, at this point, the data does not exclude the possibility of a sharply peaked profile, but it does seem to rule out a broad peak.

In order to define further the conductivity of the moon, the error bars in Fig. 11 must be reduced by the analysis of more data and the theoretical treatment of more sophisticated models. Probably the most serious approximations are the neglect of the asymmetries associated with the passage of the interplanetary field discontinuity, and those associated with the boundary currents of the lunar cavity. The former affects only the very short time response, and can be studied by examining events when the solar wind speed is high. The main effect of the boundary currents is to confine the induced field to a region roughly defined by the interior of the moon and its wake. To the order of approximation used so far, this would change the normalization of the response curve, but not its shape, and hence it does not change $\sigma(R)$. Higher order time-dependent distortions of induced fields, due to asymmetric boundary conditions (Reisz et al., 1972) have been neglected.

4. Implications concerning the lunar internal temperature

By assuming the material composition of the lunar interior and using a known conductivity-temperature relation of that material, we can calculate an internal temperature distribution of the moon from its conductivity profile. Figure 13 was obtained by using the expressions for the electrical conductivity as a function of temperature given by England et al. (1968), for olivine and peridotite:

$$\sigma_{\text{olivine}} = 55 \exp\left(-0.92/kT\right) + 4 \times 10^7 \exp\left(-2.7/kT\right) \tag{18}$$

$$\sigma_{\text{peridotite}} = 3.8 \exp\left(-0.81/kT\right) + 10^7 \exp\left(-2.3/kT\right), \text{ in mhos/m} \tag{19}$$

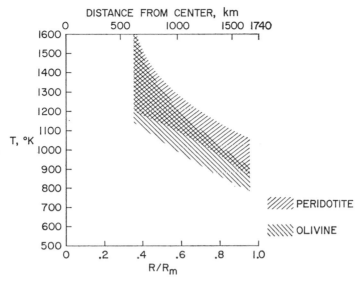

Fig. 13. Temperature estimates for assumed lunar compositions of peridotite and olivine, calculated from the electrical conductivity profile of Fig. 12.

together with the results shown in Fig. 12 for the lunar conductivity. (In the above expressions, kT is in electron volts.) For the example of a peridotite moon, a temperature profile that rises sharply to $850°-1050°K$ at $R/R_m \sim 0.95$ and then gradually to $1200°-1500°K$ at $R/R_m = 0.4$ is suggested by the data. At depths greater than $R/R_m = 0.4$ the temperature could be higher than $1500°K$.

CONCLUSIONS

1. Remanent magnetic fields at the Apollo sites

The remanent magnetic fields measured to date on the moon are $38 \pm 3\ \gamma$ at Apollo 12 in Oceanus Procellarum; 103 ± 5 and $43 \pm 6\ \gamma$ at two Apollo 14 sites separated by 1.1 km in Fra Mauro; and $6 \pm 4\ \gamma$ at the Apollo 15 Hadley-Apennines site.

Simultaneous measurements by the lunar orbiting Explorer 35 satellite have shown that the sources of these remanent fields are not global in extent. The Apollo 12 remanent field, if a single dipolar source, must be confined to a region within 200 km of the landing site. Gradient measurements and the extent of compression of this remanent field place a lower bound of 2 km for the scale size of the Apollo 12 field source. Possible origins of the large ancient ambient field required to magnetize the remanent sources could have been external to the moon (sun or earth) or intrinsic to the moon (e.g., a lunar dynamo).

2. Remanent magnetic field interaction with the solar wind

Measurements show that the remanent field at the Apollo 12 site is compressed by the solar wind. The 38-γ remanent field is compressed to 54 γ by a solar wind pressure increase of 7×10^{-8} dyne/cm^2. The ratio of plasma dynamic pressure to total magnetic pressure (β) is 5.9 during the time of maximum field compression. The change in magnetic pressure is proportional to the change in plasma dynamic pressure and the field is compressed primarily in the z (northerly) component. This compression asymmetry is perhaps dependent upon the remanent field geometry.

3. Relative magnetic permeability of the moon

The whole-moon relative permeability has been calculated to be $\mu/\mu_0 = 1.01 \pm 0.06$ from measurements taken while the moon was immersed (magnetized) in the geomagnetic tail. If an inner core of radius $R \sim 0.6\ R_m$ is above the Curie temperature ($\approx 800°C$), then $\mu/\mu_0 \cong 1.012$ is calculated for the outer layer.

4. Transient magnetic field response, internal electrical conductivity, and temperature implications

The electrical conductivity of the lunar interior has been determined from magnetic step transient measurements made on the lunar dark side. The general aspects of the data fit the classical theory of a conducting sphere in a magnetic field. Radial and tangential magnetic field component measurements indicate a global rather than a local response to these step transients. A continuous conductivity profile, with error limits, has been determined from normalized radial step-transient response data. The conductivities, when converted to temperatures for an assumed lunar material of peridotite, suggest the existence of a thin outer layer (perhaps 90 km thick) in which the temperature rises sharply to 850°–1050°K, then increases gradually to 1200°–1500°K at a depth of about 1000 km. In the deep interior higher temperatures are compatible with the data.

Acknowledgments—We thank Drs. C. P. Sonett and D. S. Colburn of NASA-Ames Research Center for use of Explorer 35 data. We have especially appreciated working with Drs. C. W. Snyder and D. R. Clay of the Jet Propulsion Laboratory in a joint study of solar wind spectrometer and magnetic field measurements. We wish to thank Messrs. J. Keeler and C. Privette for experiment fabrication and testing; K. Levis, R. Marraccini, M. Legg, and team members for help in computer programming and data reduction; and Messrs. J. Szabo and L. McCulley for numerical analysis and computer programming of electrical conductivity solutions.

REFERENCES

Barnes A., Cassen P., Mihalov J. D., and Eviatar A. (1971) Permanent lunar surface magnetism and its deflection of the solar wind. *Science* **171**, 716–718.

Coleman P. J., Schubert G., Russell C. T., and Sharp L. R. (1972) The particles and field subsatellite magnetometer experiment. In *Apollo 15 Preliminary Science Report*, NASA SP-289, in press.

Dyal P. and Parkin C. W. (1971a) Electrical conductivity and temperature of the lunar interior from magnetic transient-response measurements. *J. Geophys. Res.* **76**, 5947–5969.

Dyal P. and Parkin C. W. (1971b) The Apollo 12 magnetometer experiment: Internal lunar properties from transient and steady magnetic field measurements. *Proc. Second Lunar Sci. Conf., Geochim, Cosmochim. Acta* Suppl. 2, Vol. 3, pp. 2391–2413. MIT Press.

Dyal P. and Parkin C. W. (1972) Lunar properties from transient and steady magnetic field measurements. Proceedings of Conference on Lunar Geophysics, *The Moon*, in press.

Dyal P., Parkin C. W., and Sonett C. P. (1970a) Apollo 12 magnetometer: Measurement of a steady magnetic field on the surface of the moon. *Science* **196,** 762.

Dyal P., Parkin C. W., and Sonett C. P. (1970b) Lunar surface magnetometer. *IEEE Trans. on Geoscience Electronics GE-8* **4,** 203–215.

Dyal P., Parkin C. W., and Sonett C. P. (1970c) Lunar surface magnetometer experiment. In *Apollo 12 Preliminary Science Report*, NASA SP-235, 55–73.

Dyal P., Parkin C. W., Sonett C. P., Dubois R. L., and Simmons G. (1971) Lunar portable magnetometer experiment. In *Apollo 14 Preliminary Science Report*, NASA SP-272, 227–237.

Dyal P., Parkin C. W., Snyder C. W., and Clay D. R. (1972) Measurements of lunar magnetic field interaction with the solar wind. *Nature*, in press.

England A. W., Simmons G., and Strangway D. (1968) Electrical conductivity of the moon. *J. Geophys. Res.* **73,** 3219.

Gordon D. I., Lundsten R. H., and Chiarodo R. A. (1965) Factors affecting the sensitivity of gamma-level ring-core magnetometers. *IEEE Trans. on Magnetics, MAG-1* **4,** 330.

Helsley C. E. (1970) *Proc. Apollo 11 Lunar Sci. Conf., Geochim. Cosmochim. Acta* Suppl. 1, Vol. 3, pp. 2213–2219. Pergamon.

Helsley C. E. (1971) Evidence for an ancient lunar magnetic field. *Proc. Second Lunar Sci. Conf., Geochim. Cosmochim. Acta* Suppl 2, Vol. 3, pp. 2485–2490. MIT Press.

Jackson J. D. (1962) *Classical Electrodynamics*. John Wiley.

Neugebauer M., Snyder C. W., Clay D. R., and Goldstein B. E. (1972) Solar wind observations on the lunar surface with the Apollo 12 ALSEP. *Planetary and Space Science* (submitted).

Pearce G. W., Strangway D. W., and Gose W. A. (1972) Remanent magnetization of lunar samples (abstract). In *Lunar Science—III* (editor C. Watkins), pp. 599–601, Lunar Science Institute Contr. No. 88.

Reisz A. C., Paul D. L., and Madden T. R. (1972) The effects of boundary condition asymmetries on the interplanetary magnetic field-moon interaction. Proceedings of Conference on Lunar Geophysics, *The Moon*, in press.

Runcorn S. K., Collinson D. W., O'Reilly W., Battey M. H., Stephenson A. A., Jones J. M., Manson A. J., and Readman P. W. (1970) Magnetic properties of Apollo 11 lunar samples. *Proc. Apollo 11 Lunar Sci. Conf., Geochim. Cosmochim. Acta* Suppl. 1, Vol. 3, pp. 2369–2387. Pergamon.

Schubert G. and Colburn D. S. (1971) Thin highly conducting layer in the moon: Consistent interpretation of dayside and nightside electromagnetic responses. *J. Geophys. Res.* **76,** 8174–8180.

Siscoe G. L. and Goldstein B. (1971) Solar wind interaction with lunar magnetic fields. *J. Geophys. Res.*, submitted.

Smythe W. R. (1968) *Static and Dynamic Electricity* (3rd edition), pp. 378–380. McGraw-Hill.

Snyder C. W., Clay D. R., and Neugebauer M. (1970) The solar-wind spectrometer experiment. In *Apollo 12 Preliminary Science Report*, NASA SP-235, pp. 55–81.

Sonett C. P., Colburn D. S., Currie R. G., and Mihalov J. D. (1967) The geomagnetic tail; topology, reconnection, and interaction with the moon. In *Physics of the Magnetosphere* (editors R. L. Carovillano, J. F. McClay, and H. R. Radoski), D. Reidel.

Sonett C. P., Schubert G., Smith B. F., Schwartz K., and Colburn D. S. (1971) Lunar electrical conductivity from Apollo 12 magnetometer measurements: Compositional and thermal inferences. *Proc. Second Lunar Sci. Conf., Geochim. Cosmochim. Acta* Suppl. 2, Vol. 3, pp. 2415–2431. MIT Press.

Proceedings of the Third Lunar Science Conference
(Supplement 3, *Geochimica et Cosmochimica Acta*)
Vol. 3, pp. 2309–2336
The M.I.T. Press, 1972

The induced magnetic field of the moon: Conductivity profiles and inferred temperature

C. P. Sonett, B. F. Smith, and D. S. Colburn

NASA Ames Research Center,
Moffett Field, California 94035

G. Schubert

Department of Planetary and Space Science,
University of California, Los Angeles, California 90024

and

K. Schwartz

American Nucleonics Corporation,
Woodland Hills, California 91364

Abstract—Electromagnetic induction in the moon driven by fluctuations of the interplanetary magnetic field is used to determine the lunar bulk electrical conductivity. The earlier data are now augmented by an order of magnitude. The present data clearly show the North-South and East-West transfer function difference as well as the high-frequency rollover suggested earlier. The difference is shown to be compatible over the midfrequency range (10^{-3} to 10^{-2} Hz) with a noise source associated with the compression of the local remanent field by solar wind dynamic pressure fluctuations. The rollover of the transfer functions is shown to result from higher order magnetic multipole radiation; electric multipoles appear suppressed, though a vestigial TM interaction may still be present. Models for two, three, and four layer; current layer, double current layer, and core plus current layer moons are generated by inversion of the data, using a theory that incorporates higher-order multipoles. Resolution, limited by signal/noise ratio and frequency range, restricts present models to 3 or 4 layers. Core radii conductivities generally are in the range $1200 \leq R < 1300$ km and $10^{-3} \leq \sigma \leq 3 \times 10^{-3}$ mhos/m; and for the conducting shell (of 3 layer models) $1500 \lesssim R \lesssim 1700$ with $10^{-4} \lesssim \sigma \lesssim 7 \times 10^{-4}$ mhos/m with an outer layer taken as nonconducting. The conductivity model reported earlier, with a local maximum at a depth ~ 250 km remains a possible configuration but is not unique. Uncertainties in the conductivity from noise effects introduce uncertainties in thermal estimates small compared to those introduced by conductivity temperature relations. Core temperature based on available olivine data is $700°C \lesssim T \leq 1000°C$, well below estimated convection thresholds. If early convection and outward transport of radioactives did not occur, the primordial radionuclide distribution in the moon was either sharply stratified and concentrated near the surface or substantially below chondritic levels. This model cannot accommodate a lunar dynamo. Even if the bulk of the moon were formed "cold," its outer part could have been sufficiently hot (e.g., through accretional heating) to account for the near-surface melting evidenced by the existence of the maria.

INTRODUCTION

THIS PAPER REPORTS PROGRESS in the determination of the electrical conductivity profile of the lunar interior, using electromagnetic induction in the moon caused by the solar wind. Earlier reports of this work which use the data on the sunlit lunar hemisphere have shown that a strong global response takes place (Sonett *et al.*,

1971a, b, c). The interpretation of this response showed that the deep layers of the moon have substantially greater electrical conductivity than the near-surface region; a conductivity $\sigma \approx 10^{-3}$ mhos/m at a depth of about 800 km has been inferred together with a rapidly decreasing conductivity as the surface was approached. Also a "spike" in the conductivity was found at a depth of about 250 km with $\sigma \approx 10^{-2}$ mhos/m. Profiles of this sort place rather stringent limits upon lunar models. The relatively low conductivity at depth implies that the deep interior is well below the melting point at the present time, a view supported by the existence of mascons (Muller and Sjogren, 1968) and the low seismicity of the moon (Latham *et al.*, 1971). The "spike" has been criticized on the grounds of uniqueness (Kuckes, 1971); a major point of this paper is to consider this further. The alternate analysis of lunar induction, which uses the response of the moon to interplanetary field discontinuities as observed on the lunar darkside (Dyal and Parkin, 1971a, b) provides a consistent picture of the deep temperature, though the two analyses still differ in several important details. A later section considers possible sources of the remaining differences and how they may be resolved.

Studies of electromagnetic induction in the earth have a long history; following Gauss, Schuster (1889) demonstrated that fluctuations in the geomagnetic field could be separated into fields of internal and external origin from which a profile of the interior conductivity could be found. Chapman and Price (1930) and Lahiri and Price (1939), among others, have investigated this problem in detail. In the earth a steep rise of the conductivity with depth is found.

In the case of the earth the field is analyzed into spherical harmonic components using surface data alone. Of the two modes, transverse magnetic (TM) and transverse electric (TE), the former is usually ignored because of the insulating property of the atmosphere, though polarization currents must still flow. For the moon the problem is posed differently because the dynamic pressure of the solar wind tends to force the induced fields back into the moon (Sonett *et al.*, 1971a). This effect is modeled by introducing a surface current layer on the sunward hemisphere of the moon that is taken to confine the fields more or less perfectly (Sonett and Colburn, 1968; Johnson and Midgley, 1968; Blank and Sill, 1969; Schubert and Schwartz, 1969). Since the fields are compressed into the less conducting outer shell of the moon, a strong amplification of the induction signal takes place which aids in establishing a large signal-to-noise ratio. As for the earth, the TM mode appears suppressed or vanishingly small, though there remains a possibility that some TM fields are contributing to the lunar response. The TM mode has not entered into any calculations designed to invert the lunar response function into a conductivity profile (Sonett *et al.*, 1971b, c).

Induction in the moon has a formal similarity to scattering of radiation from a radially inhomogeneous sphere. Because of the effects of the solar wind, the scattering takes places in a supermagnetosonic stream which compounds the theoretical difficulties significantly. The results reported here assume complete confinement of the induced fields in the moon, even on the dark side. This is an inexact representation of the problem, since confinement on the dark side is incomplete. The errors introduced by the assumption of symmetry are not thought to be crucial (Blank and Sill, 1969).

For the earth the spectrum is available over a range of some nine decades, whereas we are presently limited by various data gap generating features of the spacecraft systems and orbits to only about two decades. In spite of this limitation, we can place important restrictions upon acceptable lunar conductivity profiles. The small frequency interval together with certain noise-generating phenomena that appear on the sunward hemisphere in the Apollo 12 data create a further restriction upon the signal-to-noise ratio. Darkside data are presumably free of this source of interference, but other factors can contaminate them and influence their interpretation. These include noise from the diamagnetic rarefaction wave, the possibility of contamination from volume currents in the cavity, and currents on the boundaries as well as the time-dependent sweeping back of lines of force into the cavity. In addition, errors exist in the application of present theory based on a symmetric vacuum response, which ignores the solar wind confinement.

The present work treats a significantly larger set of time series than previously available, consisting of more than 120 hours of data. Some swaths have a time duration of 10 hours extending the low frequency limit down to $f = 5 \times 10^{-4}$ Hz. Although the earlier work suggested a "spike" at a radius of 1500 km, this conductivity function should be recognized as only one member of a larger set of possible profiles. We shall give quantitative fits to within one SDM (standard deviation of the means) for two layer (2L), three layer (3L), four layer (4L), current layer (CL), dual current layer (DCL), and core plus current layer (CCL) models. A choice between these models would be aided by more data and possibly by improvement in identification of effects seated in the solar wind. However, it is important to recognize that the different models discussed in this paper share certain properties that characterize the lunar conductivity at depth in an average sense.

We shall show that the frequencies associated with the dominant response of the moon cover the interval where higher order multipoles become significant (Schubert and Schwartz, 1972). The inclusion of multipoles $1 \leq l \leq 5$ ($l = 1$ for dipole) provides a satisfactory fit of the empirical data with models at the higher frequencies. Previously the high-frequency behavior of model transfer functions was inconsistent with the data. The high-frequency rollover in the empirical transfer function suggested in the early data is confirmed by the increased data which reduce the error estimates. The inclusion of higher orders means that the phase velocity of the incoming wave field v_p and the central angle θ between the position vector of the Lunar Surface Magnetometer (LSM) and the wave vector $\mathbf{k}$ must be taken into account as variables in model fitting.

The difference between the North-South (A_z) and East-West (A_y) transfer functions in the earlier data remains. At low frequencies this difference is probably caused by the modulation of the remanent field at the Apollo 12 site by fluctuations of the solar wind dynamic pressure. At higher frequencies the higher order multipoles, which are basically asymmetric, complicate the interpretation of the A_y, A_z difference.

EMPIRICAL TRANSFER FUNCTIONS

Figure 1 shows the amplitudes of the three transfer functions A_x, A_y, and A_z based upon data from the first three lunations of the Apollo 12 LSM and from the Ames

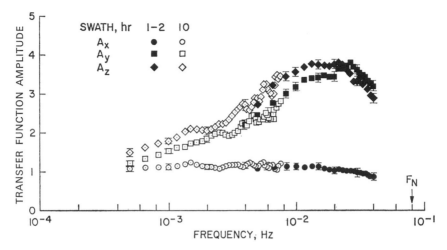

Fig. 1. Experimental transfer function amplitudes for the frequency interval $5 \times 10^{-4} < f < 4 \times 10^{-2}$ Hz. Coordinates x, y, and z are upward, eastward, and northward, respectively, at the Apollo 12 site. The lower frequency bound is determined by the time series swath of maximum length and the upper limit by the Nyquist frequency F_N of the Explorer 35 magnetometer and by noise considerations discussed in the text. The solid points are from one and two hour swaths and the open points from ten hour swaths of combined Explorer 35 and Lunar Surface Magnetometer (LSM) data. The error bars are the one standard deviation limits of the means. Error bars shown are representative of error bars for neighboring frequencies. Both A_y and A_z show the characteristic increase in response with frequency up to $f = 2 \times 10^{-2}$ Hz, beyond which a roll-over begins that can be ascribed to orders higher than dipole in interaction of the solar wind with the moon. A_x generally is about unity with a slight departure over the curve and a small roll-off at higher frequency. These data come from a much wider data base than those reported by Sonett *et al.* (1971a,b).

magnetometer on the Explorer 35 lunar orbiter. The time series used in determining these transfer functions include 47 one-hour swaths and 20 nonoverlapping two-hour swaths from lunations 1, 2, and 3. Seven 10-hour swaths which partially overlap the above data are included to extend the frequency downward. The definition of the coordinate system follows that used before, where positive x is in the direction of the outward pointing unit normal to the lunar surface and positive y and z are eastward and northward, respectively, at the magnetometer site (Sonett *et al.*, 1971b, c). The transfer functions are defined by

$$A_i(f)h_{1i}(f) = h_{1i}(f) + h_{2i}(f), \qquad (1)$$

where $h_{1i}(f)$ and $h_{2i}(f)$ are the Fourier transformed time series of the free stream interplanetary magnetic field and the magnetic field induced in the moon, respectively. The total field transform $h_{1i} + h_{2i}$ is measured by the LSM and the incident field by the Explorer 35 Ames magnetometer. The parameter f is the frequency and the subscript i is x, y, or z. To obtain these transfer functions a large number of data time series are employed in generating an equal number of Fourier transforms. For each pair of transforms obtained from LSM and Explorer 35 we compute the amplitudes of the transfer functions. An average of these at each frequency then yields composite transfer functions of improved accuracy.

The standard deviation of the mean is determined by the spread of the values from the ensemble of individual transfer functions according to

$$(SDM)_i^2 = \frac{1}{n(n-1)} \sum_{k=1}^{n} \left\{ |A_i(k)| - \frac{1}{n} \sum_{j=1}^{n} |A_i(j)| \right\}^2, \qquad (2)$$

where j and k are the indices labeling particular time swaths and n is the total number of time swaths used. Sonett *et al.* (1971c) showed that when the individual differences between values of $A(f)$ and the mean were normalized at each value of f by the standard deviation, the final distribution was approximately Gaussian (normal).

The data shown in Fig. 1 correspond to the lunar response versus frequency for the sunlit hemisphere of the moon. The reduced scatter in the data from that reported earlier reflects the significantly increased volume of Apollo 12 data now available and the elimination of data with faulty or otherwise noisy properties. Although the Explorer 35 Nyquist frequency F_N is 0.08 Hz, the data shown are restricted to 0.04 Hz to eliminate data possibly subject to digitization noise and filter recoloring uncertainty which are more important at high frequencies. The low-frequency limit is determined by the length T of the longest records available. We use a low-frequency limit of $15/T$.

As before, A_x is nearly unity over the frequency interval, and the scatter is substantially reduced. This lends confidence to a model where confinement is nearly perfect. However, at intermediate and low frequencies, A_x appears to be slightly elevated over unity while at high frequency, A_x approaches 0.8. The former suggests the creation of noise (see section on Corrections for Plasma Noise) while at the high frequencies the simplest model suggests imperfect confinement, though more complex possibilities exist.

Examination of the record for A_y and A_z shows the same general features as before though with considerably less noise. The distinctive feature of the difference in these transfer functions stands out clearly at all frequencies; the rolloff in response at the high-frequency end of the data suggested in our earlier reports is reproduced here with great clarity. Although $A_z > A_y$ over most of the frequency span, at $f = 0.02$ Hz a crossover occurs beyond which $A_y > A_z$.

Figure 1 also shows the frequency ranges identified with swaths of 1–2- and 10-hour lengths. The low frequencies up to about $f = 0.0075$ Hz are determined from 10-hour swaths, while the upper frequency range uses the 1- and 2-hour lengths, with a midfrequency range where both are used. In this range data taken from the 1–2- and 10-hour swaths are in agreement. We attribute the high-frequency rollover in the response to the effect of higher order multipoles, a belief confirmed by the crossover phenomenon and model calculations, both discussed in detail later. Some of the residual noise we believe due to plasma sources associated with the permanent field at the Apollo 12 site.

Since at the lowest frequency A_y approaches unity, TM magnetic field fluctuations in the east-west direction lie below the detectability threshold. For the low-frequency A_y, A_z difference to be ascribed solely to TM interaction, the TM magnetic fluctuation field must be preferentially oriented in the north-south direction. Belcher and Davis

(1971) have shown such a preferential orientation of the microscale wave field in the free stream solar wind. This is a necessary but not sufficient condition for explaining the A_y, A_z difference by TM interaction. It should also be noted that a measurable TM interaction requires what seems like an unacceptably high crustal bulk electrical conductivity, i.e., $\gtrsim 10^{-7}$ mhos/m for a uniform composition and reasonable thermal gradient. The next section discusses the problem of the A_y, A_z difference in connection with anisotropic plasma noise.

CORRECTIONS FOR PLASMA NOISE

In this section we present evidence that suggests that a plasma noise source associated with the modulation of the local remanent magnetic field by fluctuations in the solar wind dynamic pressure contributes to the A_y, A_z difference. The A_y, A_z difference shows that the lunar response in the tangential magnetic field components is anisotropic (at least at the Apollo 12 site) even at low frequencies where the induction theory predicts no such effect (Schubert and Schwartz, 1972). To understand this anisotropy, the directional properties of the driving and response functions have been investigated by considering the variations of each in the plane tangential to the lunar surface. The basic coordinate system was rotated about the x axis by an angle α measured counterclockwise from the y axis (east).

The power spectral densities P in the rotated coordinate system (denoted by primes) were obtained by using

$$P_{y'}(\alpha) = \frac{(P_y + P_z)}{2} + \left(\frac{P_y - P_z}{2}\right) \cos 2\alpha + Q \sin 2\alpha$$

$$= P_{z'}(\alpha \pm 90°) \tag{3}$$

where Q is the real part of the cross power spectral density. The power spectral density $P_{y'}$ is a periodic function of α with period 180°.

Figure 2 shows the effect of this rotation on the power spectral densities of both the Explorer 35 and the LSM data at $f = 0.005$ Hz. The power spectral densities are averages over the combined data for the first three lunations comprising 67 one- and two-hour data swaths. The average Explorer 35 power varies sinusoidally about a steady offset showing that the incident radiation has an apparent elliptical polarization as seen by LSM. The LSM data reflect this polarization, but the maximum in the LSM power is shifted in angle with respect to that of the Explorer power. The maximum in the LSM power at this frequency is in a direction approximately along that of the local remanent field at this site, $\alpha \approx -63°$. This shift in the maximum is responsible for the observed difference in the A_y and A_z transfer functions.

The transfer functions for each swath were computed at 5° intervals for $-90° \leq \alpha \leq 90°$ at all frequencies and then averaged over all swaths at each frequency. This average transfer function is shown as curve a in Fig. 2. An alternative method of computing the average transfer function is to take the square root of the ratio of the average powers at a given angle; this is shown as curve b in Fig. 2. The two methods give essentially the same mean transfer function in this example.

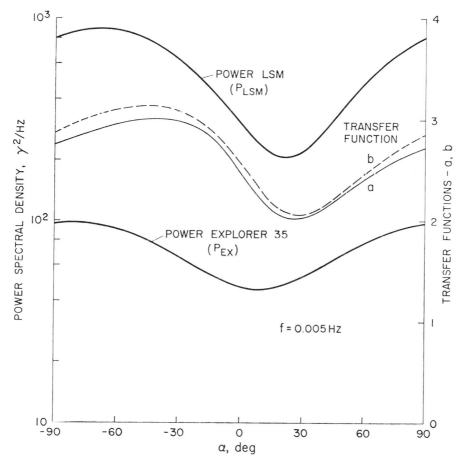

Fig. 2. The power spectral densities of Explorer 35 and LSM shown as a function of the "compass" heading angle α on the lunar surface. A_y corresponds to $\alpha = 0°$ while A_z corresponds to $\alpha = 90°$. Both show quasi-sinusoidal variations but the angles for the minima do not coincide. The data shown are for the frequency $f = 0.005$ Hz. The lighter traces labeled a and b correspond to the transfer functions determined from the power spectral densities according to the two algorithms given in the text. The variation of P_{EX} and P_{LSM} with angle is indicative of elliptically polarized radiation.

If the fluctuations in the noise field are predominantly aligned in one direction, then the LSM power normal to this direction is noise free and represents the global lunar response to induction. For both a noise field that is incoherent with the induction field and one that is coherent and in phase with the induction field, the transfer function in a direction perpendicular to the noise field is a minimum. For the case in which the noise field is coherent and out of phase with the induction field, the transfer function in a direction normal to the noise field is a maximum. There is no *a priori* reason to expect that the solar wind dynamic pressure fluctuations (assumed to be driving the noise field) would be coherent with the forcing field for the induction, which consists mainly of Alfvén waves traveling along or near the mean field direction. Note from Fig. 2 that the minimum in the transfer function occurs in a direction approximately normal to that of the remanent magnetic field.

Thus we construct a transfer function by taking the minimum at each frequency of the average transfer function computed according to the method used for curve a in Fig. 2. This function A_{min} is shown in Fig. 3 along with the original $A_y(\alpha = 0)$. Figure 3 also shows the frequency dependence of the direction α_{min} along which $A = A_{min}$. Curve $\alpha_{min}(f)$ rises smoothly from $\alpha = -15°$ at $f = 5 \times 10^{-4}$ Hz to a plateau in the midfrequency range, where $\alpha \sim 20°$–$30°$. Above $f = 10^{-2}$ Hz, α again rises monotonically reaching $55°$ at $f = 4 \times 10^{-2}$ Hz. The midfrequency values of α_{min} agree with the idea that the noise source is aligned with the permanent magnetic field.

The monotonic increase in α_{min} for $f \gtrsim 10^{-2}$ Hz can be attributed to the effects of high-frequency induction. At high frequencies the amplification of tangential magnetic fields is anisotropic (see next section) and can produce a shift in the direction of minimum power in the response field relative to the incident field. This phenomenon

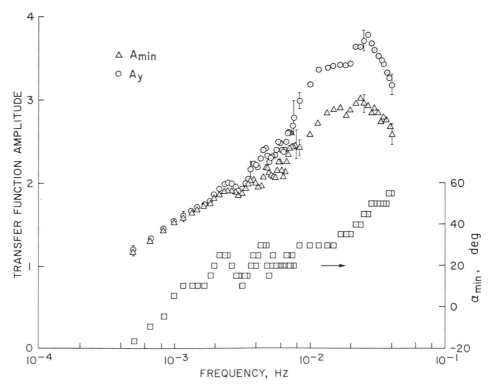

Fig. 3. Transfer function (A_y and A_{min}) amplitudes versus frequency together with the direction α_{min} corresponding to the minimum value of A_y. The effect of rotation is most pronounced at the higher frequencies where the higher order multipoles supplement the dipole interaction. The plateau in α_{min} is consistent with a noise source attributable to the interaction of the dynamic pressure fluctuations of the solar wind with the permanent field at the Apollo 12 site. The low-frequency behavior of α_{min} is unexplained at the present time.

is illustrated by the following computation. Consider a spectrum of circularly polarized incident waves at several different frequencies. The power in the incident wave as it appears in the plane tangent to the moon's surface at LSM is shown in Fig. 4 as a function of α. LSM is assumed to be at an angle $\theta = 130°$ from the incident wave vector direction. The power in the tangential components of the total field at LSM is also shown as a function of α. The induced field was computed for a model moon

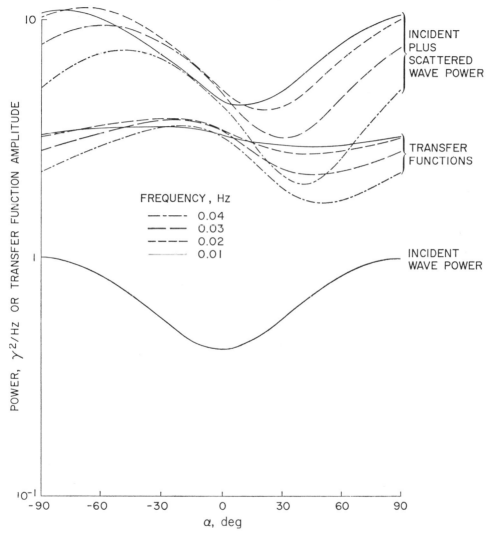

Fig. 4. Illustrative calculation of the high-frequency transfer function behavior versus angle α for 4 frequencies. The hypothesized Explorer power (incident wave), the LSM response power (incident + scattered wave), and the transfer function are shown for the different frequencies. The calculation is carried out for a single current layer moon with $\theta = 130°$ and $v_p = 200$ km/sec. The current layer is at $R = 1500$ km with a $\sigma\delta$ value of 150 mhos. The increase in angle of minimum response with frequency is similar to the increase seen in the data.

which typifies the models obtained from our inversions. At low frequency the LSM power is a simple multiple of the incident power. As f increases the position of the minimum in LSM power increases to larger values of α. From the curves of LSM and incident power, the transfer function can be calculated for any value of α. The results are also shown in Fig. 4. For $f = 0.01$ Hz the transfer function is almost independent of angle even though both the incident and LSM power depend strongly on α. As the frequency increases there is an increasing shift toward higher α in the minimum of the transfer function. At $f = 0.04$ Hz the minimum in the transfer function occurs at $\alpha \approx 50°$.

The behaviour of α_{min} at the lowest frequencies suggests that there is some contribution, in addition to TE induction and the compressive noise source previously discussed, to the magnetic field fluctuations in the north-south direction. This contribution might be associated with TM induction (see the preceding section); however, this suggestion is a tentative one and requires further investigation.

The preference for a preferred alignment of the noise source along the permanent magnetic field direction is further illustrated by the distribution of angles at which individual transfer functions are minimized, as shown in Fig. 5 for $f = 0.01$ Hz and

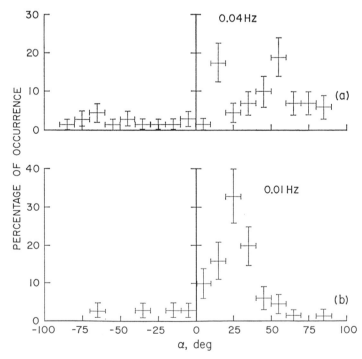

Fig. 5. Histograms of α_{min} at $f = 0.04$ Hz and 0.01 Hz showing the statistical favoring of the direction perpendicular to the permanent field B_p in the lower frequency data and a bimodal modification at the upper frequency limit. The latter shows the peak perpendicular to B_p, i.e., $\alpha_{min} = 27°$, and an additional peak suggesting a two-component source of noise. The wide low-level background, i.e., the distribution of values of α_{min}, is indicative of possible additional components that contribute to the noise vector.

$f = 0.04$ Hz. At $f = 0.01$ Hz there is one peak in the distribution at an angle of about 25° that typifies other low-frequency distributions. This is consistent with an interpretation based on a noise source due to an incoherent compressive effect on the steady field caused by fluctuations in the dynamic pressure of the solar wind. At $f = 0.04$ Hz the distribution displays two peaks; this suggests that the compressive effect is augmented by another significant source at about 50°.

The second peak at $\alpha = 50°$ follows the anisotropic behavior of lunar induction at high frequency that rotates the direction of minimum response as discussed previously. A noise source associated with solar wind modulation of the global induction seems unlikely because the peak power in the solar wind driving field is in the north-south direction, $\alpha = \pm 90°$.

The evidence presented in this section for the existence of a noise source associated with the modulation of the permanent magnetic field by solar wind dynamic pressure fluctuations is further substantiated by several independent investigations. Neugebauer *et al.* (1972) have shown evidence that the solar wind ion component is decelerated by roughly 50 km/sec near the lunar surface. It is difficult to explain this deceleration by means other than the interaction of the solar wind plasma with the steady magnetic field at the site. Also Dyal *et al.* (1972) have shown that the dynamic pressure of the solar wind and the surface magnetic field intensity are related. This result was obtained by considering hourly averages ($f \cong 3 \times 10^{-4}$ Hz) so that the effects of internal induction were assumed to be small.

THE THEORETICAL LUNAR TRANSFER FUNCTION

A theory for obtaining lunar magnetic field transfer functions for arbitrary moon models is a prerequisite to inverting the experimental data presented in the previous sections. Schubert and Schwartz (1972) have shown that there are two distinct transfer functions corresponding to the two orthogonal components of the tangential surface magnetic field. Their formulae are

$$
\begin{Bmatrix} A_\theta \\ A_\phi \end{Bmatrix} = \left| \sum_{l=1}^{\infty} \beta_l \frac{\lambda}{2\pi i} j_l \left(\frac{2\pi a}{\lambda} \right) \left(\frac{dG_l}{dr} \right)_{r=a} \begin{Bmatrix} \dfrac{dP_l^1(\cos\theta)}{\cos\theta \, d\theta} \\[2ex] \dfrac{P_l^1(\cos\theta)}{\sin\theta} \end{Bmatrix} \right|,
\tag{4}
$$

where j_l are the spherical Bessel functions,

$$
\beta_l = \frac{i^l(2l+1)}{l(l+1)},
\tag{5}
$$

$P_l^1(\cos\theta)$ are the Legendre functions and $G_l(r)$ are determined from the radial differential equations

$$
\frac{d^2G_l}{dr^2} + \left\{ k^2 - \frac{l(l+1)}{r^2} \right\} G_l(r) = 0,
\tag{6}
$$

with

$$
k^2 = \omega^2 \mu\varepsilon + i\omega\mu\sigma, \qquad \omega = 2\pi f, \qquad \lambda = f/v_p,
\tag{7}
$$

and $G_l(r = 0)$ finite, $G_l(r = a) = 1$. The parameters μ and ε are the magnetic permeability and electric permittivity, respectively, and σ, the electrical conductivity, is an arbitrary function of radius. In the present work we assume free space values for μ and ε, and MKS units are used. The geometry is shown in Fig. 6. A spherical coordinate system with the polar axis in the direction of the incident wave vector and origin at the moon's center has been employed.

Because of the difference in the θ dependence, A_θ and A_ϕ exhibit marked differences in their behavior as a function of the colatitude angle θ. This is shown in the upper part of Fig. 7 where the results for transfer function versus frequency are presented for a three-layer moon model for different values of θ but for a fixed value of the wave velocity. From equation (4) we can easily show that $A_\theta = A_\phi$ at $\theta = 180°$. Numerical calculations show that A_ϕ changes only slightly as θ is varied. As one can see in Fig. 7, A_θ changes markedly as θ goes from $180°$ to $120°$ for those frequencies where $2\pi a/\lambda \gtrsim 1$. For low frequencies, only the dipole ($l = 1$) term makes any significant contribution to the transfer function, which is then independent of θ.

For a given frequency, the velocity determines how many multipoles are required in the sum of equation (4). This is strongly evidenced in the second half of Fig. 7 where θ is held fixed but the velocity is varied. As the velocity is decreased, the turnover frequency in the transfer function also decreases. For a given lunar model, both the maximum in the transfer function and the frequency at which it occurs decrease with decreasing wave propagation velocity. In the calculations of this paper up to five multipoles are used for the higher frequencies. The use of the transfer function formulas of equation (4) overcomes the difficulties found in our earlier work where we were unable to match theoretically the slope and roll-off of the experimental transfer functions at high frequencies.

INVERSION OF THE TRANSFER FUNCTION INTO CONDUCTIVITY PROFILES

In our earlier work numerical integration of the radial induction equation for the TE mode was used to generate theoretical transfer functions for a given conductivity

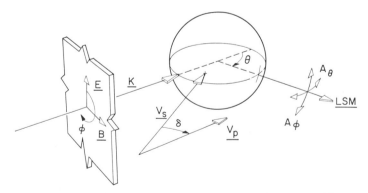

Fig. 6. Geometry of the generalized induction problem. A plane wave with wave vector **k** is shown incident upon the moon. In general the solar wind bulk velocity vector, $\mathbf{v}_s$, will not be collinear with **k**. A_θ and A_ϕ are the transfer functions at the site of LSM resolved into a spherical polar system whose polar angle is given by θ and azimuth by ϕ. Generally $A_\theta \geq A_\phi$.

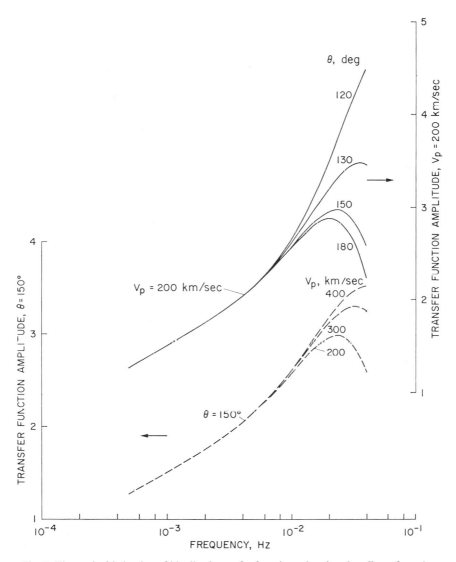

Fig. 7. Theoretical behavior of idealized transfer functions showing the effect of varying θ and v_p. These effects are most pronounced at the higher frequencies where the geometry dependence becomes important. The calculations are carried out for the three layer (3L) moon conductivity model discussed in the text. The importance of the assigned values of θ and v_p are amply demonstrated. The wavelength varies with v_p modifying the response in the intermediate scattering regime where most of the lunar response lies.

profile. An iterative procedure (Newton-Raphson) adjusted the conductivity model to provide a best fit of the theoretical to the empirical transfer functions. The calculation used a modal transfer function that is independent of θ but has the disadvantage that the forcing function cannot be experimentally resolved into modes. Here the higher order multipoles, up to and including $l = 5$, are included in the inversions based on equation (4). It can be seen from equation (4) that the inversion requires specification of the parameters v_p and θ. The data being inverted are averages over a spectrum of these parameters, the detailed nature of which is not presently known. In lieu of a superposition of single plane wave induction solutions over this spectrum we are limited to carrying out inversions using plane waves characterized by one set of values of v_p and θ. We have investigated a range of velocities and directions to determine the sensitivity of the inversions to these parameters. Further, we have used A_θ in the inversions for the following reasons: First, the function A_ϕ is insensitive to θ variation; so, to explore the sensitivity of the inversions to θ variations, A_θ is preferable. Second, over most of the frequency range, the direction corresponding to $A_{\min}$ is close to east-west. Since the wave vectors lie on the average in the ecliptic plane, A_θ is in fact the east-west transfer function.

In an attempt to minimize the possible contamination due to noise, the data fitted in the inversions are the $A_{\min}$ previously described. We will also present results obtained by fitting A_y to demonstrate the relative unimportance of the noise to the resultant models.

The models considered are 2, 3, and 4 layer (2, 3, 4L) single and double current layers (CL and DCL) and core plus current layer (CCL). The parameters in each of these cases have been determined by iteration to minimize the difference between the amplitudes of the theoretical and the empirical transfer functions in the least squares sense. The calculation is carried out for 63 frequencies in the range $5 \times 10^{-4} < f < 4 \times 10^{-2}$ Hz. In the 2-, 3-, and 4-layer models the outer shell is assumed to be nonconducting, and the iterative process yields the conductivities and radii for the inner layer(s). Thus for each conducting layer, two parameters σ_i, the conductivity, and R_i, the outer radius, are determined.

Current layers are characterized by their radial position and $\sigma\delta$, the productivity and thickness δ. The current layer concept assumes $\sigma \to \infty$ as $\delta \to 0$, while maintaining $\sigma\delta$ invariant. In the pure current layer models the conductivity outside the current layers is assumed to be zero. Both a double current layer model and a core plus current layer model are specified by four iteratively determined parameters. All of the above models are calculated for a number of choices of v_p and θ.

In the present work the measure of goodness of fit is defined by ε^2, the sum of the squares of the differences between the calculated and the empirical values of $A(f)$.

Figure 8 shows a representation of the different model conductivities computed for $v_p = 200$ km/sec and $\theta = 150°$ in a fit to $A_{\min}$. The values of ε^2 are to be compared with the sum of the squares of the SDM for all the 63 frequencies (0.65 for $A_{\min}$). The hyperbolas associated with the current layers are the loci of $\sigma\delta =$ constant centered at the calculated R values. However, it should be noted that as δ increases beyond about 20 km, both the location of the current layer and the $\sigma\delta$ product will change significantly.

The dual current layer model (DCL) is shown with the outer current layer having electrical admittance 59 mhos and the inner current layer 424 mhos. The admittance of the outer current layer in the CCL model is 46 mhos and that of the CL model is 100 mhos. The admittances and positions of the outer current layers in the DCL and CCL models are similar. The cores of the 3L, 4L, and CCL models have essentially the same conductivities and nearly the same size. The 3L model improves upon the

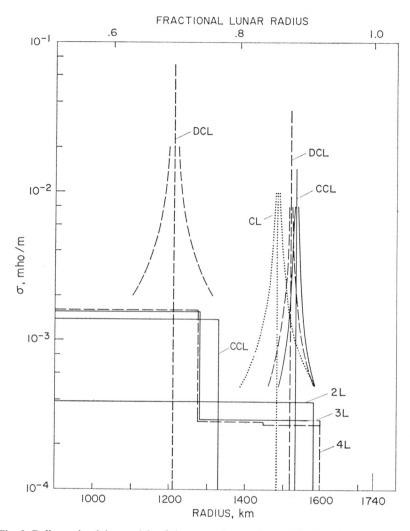

Fig. 8. Bulk conductivity models of the moon for two layer (2L), three layer (3L), four layer (4L), current layer (CL), double current layer (DCL), and core plus current layer (CCL) models. All models here are iterative best fits to A_{min} data using $v_p = 200$ km/sec and $\theta = 150°$. These models are not the best fits when other values of v_p and θ are permitted; for those models the reader is referred to the tabulations. The hyperbolas define current layers for which $\sigma\delta =$ constant, approximating the effect of the very thin layers used in the current layer calculations.

2L one by the addition of a core while retaining an outer conducting layer having nearly the same radius and conductivity as that of the 2L model. There is a negligible change in the fit and in the character of the conductivity profile between the 4L and 3L models.

The theoretical transfer functions corresponding to the models of Fig. 8 are shown in Fig. 9 together with the empirical data for A_{min}. The transfer functions of the 2L

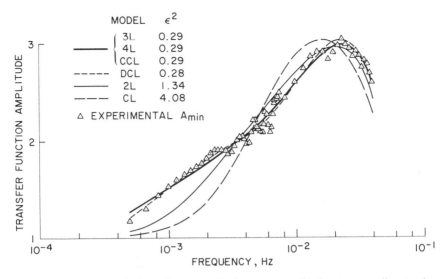

Fig. 9. Forward calculation of the transfer function amplitude corresponding to the conductivity models shown in Fig. 8. The 3L, 4L, and CCL models all yield transfer functions having values too close to plot separately on this scale. The experimental A_{min} values are shown for reference. The legend lists the values of ε^2 belonging to each model fitted. The experimental value of ε^2 (the sum of the squares of the standard deviations of the means) for A_{min} is 0.65; thus all models shown here are reasonable fits with the exception of the 2L and CL cases.

and CL models are poor fits to the data particularly *at the low frequencies*, emphasizing the importance of the inner core in the 3L, 4L, and CCL models. The locations of the current layers in the DCL model are consistent with the 3L model according to the following argument: The transfer function for a DCL model is expected to be only moderately changed as the layer thicknesses increase. Following the hyperbolas to a model in which the conductivities of the two current layers match those of the two conducting layers of the 3L model, it is seen that the outer edges of the current layers coincide with the corners of the 3L model profile. In both models damping tends to occur at two regions in the lunar interior, at approximately $R = 1600$ km for the higher frequencies and $R = 1300$ km for the lower frequencies.

The dependence of 3L models on the parameter θ is illustrated in Fig. 10 for $v_p = 200$ km/sec. From the ε^2 values it is clear that the best fits are obtained for $\theta = 150°$ and $180°$. This can also be seen by comparing the theoretical transfer functions of the models to the empirical data A_{min} (also shown in Fig. 10). For the larger angles the radii of the conducting layers decrease and their conductivities increase with decreasing θ. At $\theta = 120°$ the iteration routine converged to nearly a CL model. As judged by ε^2 the fit was relatively poor.

The way in which the 3L models depend on v_p is shown in Fig. 11 for $\theta = 150°$. The outer layer radius decreases and its conductivity increases with increasing v_p. The behavior of the inner layer is less predictable. Best fits are obtained for phase velocities of 200 or 300 km/sec. The preference of these better fits for the smaller phase

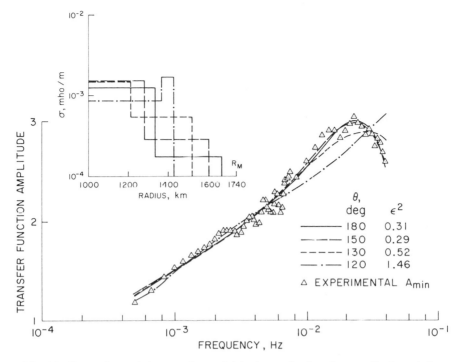

Fig. 10. Comparison of the experimental (A_{min}) transfer function amplitude and the transfer function amplitudes corresponding to three layer (3L) models using a variable value of θ. Each model is an iterative best fit under the conditions $v_p = 200$ km/sec and θ the value associated with the model. All models are shown in the insert for reference. The fit $\theta = 120°$ fit yields a "spike" in the conductivity.

velocities (compared to the solar wind speed of about 400 km/sec) suggests that wave normals oblique to the solar wind are present, since the phase velocity of incident waves is the sum of the solar wind speed resolved along **k** and the wave velocity in a reference frame comoving with the plasma.

The parameters of all our models best fit to the data A_{min} are summarized in Tables 1 to 5. The 3L and 4L monotonic models for which $\varepsilon^2 \lesssim 0.7$ have outer conducting layers whose conductivities lie between about 1.1 and 6.6×10^{-4} mhos/m and whose outer radii lie between about 1500 and 1710 km. Even the two layer models wherein $\varepsilon^2 \lesssim 1$, though they are poor fits to the low-frequency data, have conductivities and radii that fall within these ranges. The cores of the 3L and 4L models have conductivities varying between 1.3 and 2.2×10^{-3} mhos/m and radii varying between 1170 and 1330 km. The cores of the CCL models for which $\varepsilon^2 \lesssim 0.7$ vary in conductivity between 1.2 and 2.4×10^{-3} mhos/m and vary in radius between 1200 and 1360 km. Although no unique model can be inferred from inversion of the data, the 3L and 4L monotonic models are all in substantial agreement and characterize the average properties of monotonic conductivity profiles. The cores of the 3L, 4L, and C + CL models are in particularly close agreement among themselves. Since the

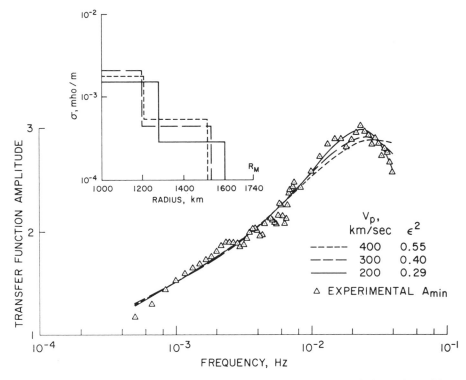

Fig. 11. Models and transfer function amplitudes similar to those of Fig. 10 but with v_p as variable and θ fixed at 150°.

Table 1. Two layer and current layer model fits to A_{min}.

v_p (km/sec)	θ(deg)	2L				CL		
		R(km)	σ(mhos/m)	ε^2		R(km)	$\sigma\delta$(mhos)	ε^2
400	180	1520	5.45 (−4)	0.78		1442	127.0	2.55
	150	1513	5.80	0.81		1438	131.3	2.43
	130	1493	6.87	0.95		1422	146.9	2.17
	120	1469	8.56	1.30		1404	169.2	2.03
300	180	1544	4.56	0.86		1459	114.0	3.05
	150	1530	5.06	0.80		1449	122.6	2.74
	130	1496	6.68	0.90		1425	143.3	2.19
	120	1456	9.59	1.47		1394	183.0	1.97
200	180	1606	3.03	1.90		1504	88.5	5.21
	150	1578	3.65	1.34		1484	99.6	4.08
	130	1509	6.00	0.74		1435	135.1	2.32
	120	1435	1.17 (−3)	1.61		1376	209.9	1.74

electrical conductivity of geologic material is an exponential function of the inverse temperature, the differences in conductivity among the various 3L and 4L models are relatively unimportant in assessing the lunar temperature at depth.

Certain 4L fits resemble more closely the CCL model rather than a monotonic fit. For example, at $v_p = 400$ km/sec, $\theta = 180°$, the outer layer is 14 km thick centered

Table 2. Three layer model fits to $A_{\min}$.

v_p (km/sec)	θ(deg)	R_1(km)	R_2(km)	σ_1(mhos/m)	σ_2(mhos/m)	ε^2
400	180	1206	1527	1.90 (-3)	5.13 (-4)	0.47
	150	1208	1519	1.80	5.49	0.55
	130	1181	1497	1.56	6.71	0.83
	120	1201	1471	1.06	8.33	1.28
300	180	1224	1553	1.88	4.18	0.33
	150	1194	1537	2.10	4.85	0.40
	130	1168	1500	1.83	6.56	0.76
	120	1405	1438	7.12 (-4)	1.72 (-3)	1.35
200	180	1330	1661	1.28 (-3)	1.79 (-4)	0.31
	150	1278	1600	1.55	2.92	0.29
	130	1209	1515	1.54	5.76	0.52
	120	1362	1422	8.72 (-4)	1.71 (-3)	1.45

Table 3. Four layer model fits to $A_{\min}$.

v_p (km/sec)	θ(deg)	R_1(km)	R_2(km)	R_3(km)	σ_1(mhos/m)	σ_2(mhos/m)	σ_3(mhos/m)	ε^2
400	180	1290	1465	1479	1.58 (-3)	2.76 (-5)	2.42 (-3)	0.29
	150	1198	1513	1520	2.17	5.66 (-4)	4.06 (-4)	0.48
	130	1280	1487	1519	1.08	6.46	3.61	1.00
	120	1286	1488	1496	9.34 (-4)	7.75	8.12 (-5)	1.70
300	180	1244	1457	1548	1.73 (-3)	3.59	4.82 (-4)	0.31
	150	1281	1440	1508	1.63	1.16	1.05 (-3)	0.35
	130	1256	1499	1540	1.16	6.14	1.12 (-4)	0.88
200	180	1263	1456	1713	1.56	4.45	1.09	0.29
	150	1277	1449	1600	1.56	2.91	2.89	0.29
	130	1249	1442	1483	1.76	2.62	1.50 (-3)	0.37

Table 4. Core plus current layer model fits to $A_{\min}$.

v_p (km/sec)	θ(deg)	R_1(km)	R_2(km)	σ_1(mhos/m)	$\sigma\delta$(mhos)	ε^2
400	180	1254	1467	1.91 (-3)	84.0	0.39
	150	1236	1459	2.13	91.9	0.44
	130	1207	1440	2.38	109.7	0.64
	120	1170	1418	2.58	134.8	0.98
300	180	1282	1491	1.71	68.1	0.30
	150	1260	1476	1.91	78.5	0.34
	130	1212	1443	2.32	106.5	0.59
	120	1185	1408	2.11	141.6	1.12
200	180	1364	1587	1.20	31.8	0.32
	150	1329	1537	1.40	46.4	0.29
	130	1241	1457	2.03	92.6	0.42
	120	1155	1388	2.08	170.8	1.22

around $R = 1472$ km with a conductivity giving $\sigma\delta = 34$ mhos. This is separated from the core by a region of low conductivity. The CCL model gives $\sigma\delta = 84$ mhos at $R = 1467$ km with a comparable core. Similarly at $v_p = 300$ km/sec, $\theta = 150°$, the relatively highly conducting outermost layer can be described by $R = 1474$ km, $\sigma\delta = 68$ mhos. The CCL model has $R = 1476$ km, $\sigma\delta = 78$ mhos, the cores again being similar. The inclusion of a conductivity peak superimposed on a monotonic profile as proposed earlier is seen by the low ε^2 values of the CCL fits to be among the possible fits to the present data.

Table 5. Double current layer fits to A_{min}.

v_p (km/sec)	θ(deg)	R_1(km)	R_2(km)	$\sigma_1\delta_1$(mhos)	$\sigma_2\delta_2$(mhos)	ε^2
400	180	1154	1463	505.1	93.3	0.31
	150	1147	1457	512.1	98.6	0.36
	130	1129	1440	528.1	113.5	0.55
	120	1108	1419	531.4	134.5	0.90
300	180	1176	1485	470.7	78.7	0.24
	150	1165	1473	485.8	86.6	0.27
	130	1132	1443	527.7	111.1	0.50
	120	1108	1409	455.4	144.6	1.07
200	180	1237	1559	380.0	45.1	0.39
	150	1209	1522	423.9	59.5	0.28
	130	1147	1454	509.8	100.7	0.34
	120	1100	1389	443.2	171.2	1.18

The CCL and DCL models for which $\varepsilon^2 \lesssim 0.7$ have outer current layers whose $\sigma\delta$ products vary between about 32 and 115 mhos and whose radii vary between about 1440 and 1590 km. The poorer fitting CL models also tend to group in this region. As was the case with the monotonic models, the models with current layers are in substantial agreement among themselves. Furthermore, as the thickness of current layers is allowed to increase, the models yield conductivities in agreement with those of the 3L and 4L monotonic models.

A scanning of the ε^2 values shows a preference for certain velocity and angle ranges. The best fits are obtained for $\theta = 180°$ or $150°$, with the fits becoming substantially poorer for $\theta = 120°$. The ε^2 values for $v_p = 200$ or 300 km/sec are generally smaller than those for $v_p = 400$ km/sec, although acceptable fits are found for $v_p = 400$ km/sec.

Generally it appears that the electromagnetic transfer function is reasonably sensitive to changes in the conductivity model, but differences in the determination of the lunar temperature will be attributed more to lack of knowledge of the proper conductivity-temperature function than to differences in the various conductivity profiles.

In Tables 6–10 we present the results of the different model fits to the A_y data,

Table 6. Two layer and current layer model fits to A_y.

v_p (km/sec)	θ(deg)	R(km)	σ(mhos/m)	ε^2	R(km)	$\sigma\delta$(mhos)	ε^2
400	180	1593	4.3 (−4)	1.40	1509	109	3.29
	150	1586	4.5	1.51	1505	112	3.21
	130	1568	5.2	1.92	1493	121	3.10
	120	1547	6.2	2.53	1478	133	3.16
300	180	1613	3.8	1.24	1523	101	3.69
	150	1602	4.1	1.31	1515	106	3.43
	130	1572	5.1	1.79	1495	119	3.09
	120	1537	6.7	2.73	1470	139	3.17
200	180	1668	2.7	1.85	1560	84	6.00
	150	1644	3.1	1.45	1545	91	4.76
	130	1586	4.6	1.38	1504	114	3.13
	120	1520	7.6	2.72	1456	152	2.94

Table 7. Three layer model fits to A_y.

v_p (km/sec)	θ(deg)	R_1(km)	R_2(km)	σ_1(mhos/m)	σ_2(mhos/m)	ε^2
400	180	1198	1594	2.1 (-3)	4.4 (-4)	0.95
	150	1185	1587	2.3	4.7	1.12
	130	1141	1568	2.9	5.4	1.64
	120	1127	1547	3.0	6.4	2.34
300	180	1199	1615	2.1	3.8	0.62
	150	1181	1603	2.4	4.2	0.78
	130	1156	1572	2.7	5.2	1.48
	120	1123	1538	2.7	6.8	2.59
200	180	1297	1681	1.3	2.4	0.52
	150	1244	1648	1.7	3.1	0.50
	130	1189	1587	2.2	4.7	1.00
	120	1289	1520	0.8	7.7	2.72

Table 8. Four layer model fits to A_y.

v_p (km/sec)	θ(deg)	R_1(km)	R_2(km)	R_3(km)	σ_1(mhos/m)	σ_2(mhos/m)	σ_3(mhos/m)	ε^2
300	180	1253	1389	1605	1.9 (-3)	0.13 (-4)	4.5 (-4)	0.55
	150	1267	1465	1579	1.8	0.26	6.9	0.74
200	180	1240	1433	1694	1.6	4.3	2.1	0.50
	150	1231	1414	1650	1.8	3.6	3.0	0.51

Table 9. Core plus current layer model fits to A_y.

v_p (km/sec)	θ(deg)	R_1(km)	R_2(km)	σ_1(mhos/m)	$\sigma\delta$(mhos)	ε^2
400	180	1235	1523	2.2 (-3)	87	0.71
	150	1230	1518	2.3	90	0.79
	130	1205	1504	2.6	101	1.09
	120	1180	1487	2.7	114	1.59
300	180	1270	1541	1.8	75	0.54
	150	1246	1531	2.1	83	0.61
	130	1211	1507	2.5	98	1.00
	120	1169	1479	3.0	121	1.78
200	180	1371	1607	1.1	45	0.55
	150	1326	1575	1.4	58	0.49
	130	1232	1517	2.2	90	0.72
	120	1153	1464	3.0	133	1.82

Table 10. Double current layer model fits to A_y.

v_p (km/sec)	θ(deg)	R_1(km)	R_2(km)	$(\sigma\delta)_1$ (mhos)	$(\sigma\delta)_2$ (mhos)	ε^2
400	180	1149	1521	529	92	0.63
	150	1141	1516	546	95	0.71
300	180	1171	1538	487	82	0.48
	150	1158	1529	512	87	0.54
200	180	1241	1593	355	55	0.67
	150	1210	1567	410	66	0.51
	130	1145	1516	532	95	0.64

for which $\varepsilon^2 = 1.1$. For the most part the models which best fit the A_y data have parameter values that lie within the ranges of variability defined by the model fits to A_{min}.

DISCUSSION OF CONDUCTIVITY MODELS

The model inversions have been based upon two versions of the transfer function. Also two new variables θ and v_p have been introduced to account for the more complex geometry encountered when the higher order multipoles are included and the inversions have been carried out using A_θ. The possibility that A_ϕ is sometimes important must be considered. Although A_θ is a strong function of θ, A_ϕ varies only slowly, and for all values of θ investigated here $A_\theta \geq A_\phi$, the equality holding at $\theta = 180°$. Using the model of the calculation whose results are shown in Fig. 4, the value of A_ϕ changes from 2.1 to 2.3 when θ is changed from 180° to 130° at $f = 0.04$ Hz. Thus A_ϕ ($\theta = 180°$) is a reasonable approximation to A_ϕ over a wide range of θ. For cases where **k** is out of the ecliptic, a mixture of A_θ, A_ϕ responses is expected. This will modify the net response somewhat, decreasing the sensitivity of the transfer function to changes in θ. A similar effect is expected when a superposition over a spectrum of plane waves is used in determining transfer functions. Thus the values for θ and v_p are tentative and subject to improvement.

We have shown a wide range of model conductivities based upon both A_y (taken to be identical to A_θ in the average sense) and a noise corrected version of the transfer function, A_{min}. Both may be subject to some mixing between A_θ and A_ϕ. These effects cannot be important at mid and low frequencies where the induction is nearly purely dipolar. The models all show a conductivity at about $R = 1200$ to 1350 km in the range of about $1–2.5 \times 10^{-3}$ mhos/m. Conductivity below about 1100 km radius cannot yet be measured because of the strong damping of the low frequencies at that level. An extension to lower frequency and possibly improvement in the signal-to-noise ratio will be required. The alternate model of a current layer at that depth as part of the DCL model also cannot be ruled out and would likely be due to material of anomalous conductivity at that depth; there appears no reason to pursue this line of reasoning at the present time.

The close fit of the DCL and CCL models show that the high-frequency behavior can also be modeled by a near surface equivalent current layer. Although there is a basis in lunar evolutionary theories for introducing a model with a conductivity "spike" at about $R = 1500$ km, and the earlier calculations yielded such a model, it is only one of a class of possible models. Although the low-frequency behavior is relatively independent of details of the solar wind geometry, the latter has an important influence upon the radius of the outer shell and less so upon the average conductivity of the shell. An extension of our model calculations to include the superposition of plane waves is required before accurate values for the conductivity and radius of the outer conducting shell are determined. The model fits indicate that these parameters vary between about $1–6 \times 10^{-4}$ mhos/m and 1500–1700 km. None of the variations discussed are large, and therefore the existence of a noise field of the magnitude suggested by the rotation investigation does not pose a fundamental difficulty at the level of resolution available.

The lunar conductivity profiles based upon darkside data (Dyal and Parkin, 1971a, b) differ somewhat from the ones presented here, but the temperatures which the models imply are never different by more than about 200°C. The deep conductivity reported by Dyal and Parkin (1971a, b) is $\approx 10^{-2}$ mhos/m based upon five observations of the extended "tail" of the transient response. It is not possible to reconstruct their argument completely but the principal issue is that the response is thought to last in excess of 15 min. If their detectability implies that the transient decays to 15% of the final value, this equals two e-fold times; if 5%, three e-folds. Thus the "time constant" is 450 sec for the first case and 300 sec for the second. "Time constants" determined from the models of this paper lie in the range of about 300 sec to 200 sec; thus if the Dyal-Parking 15 min represents a detection limit, the value is in accord with our "time constant" and their internal conductivity should be altered downward. However, it is difficult to determine their time bound since at least in one case it is stated to last longer than 15 min. An additional potential complication arises from the possibility of contamination of the transient response by a time-variable forcing function. Inspection of Dyal and Parkin's Fig. 13 (1971a) shows a large scatter of cases for transients lasting longer than 4 min; indeed anywhere from 28 to 38% of the cases examined in x, y, or z show evidence for complete relaxation by 4 min, suggesting the possibility of a nonsteady forcing function.

A similar problem associated with the assumption of step forcing functions has been discussed by Schubert and Colburn (1971). In this case account must be taken of the finite time for the whole moon to become immersed in a convected field change, requiring as a first approximation a ramp input for the forcing function. Errors in the interpretation of the high-frequency response are connected with this effect and have an important influence upon the final values found for the shell radius and conductivity.

Although an accurate high-frequency comparison of the dark side and sunlit side data cannot yet be made because of complications with the solar wind geometry, it is instructive to show a forward calculation for the comparative models. In Fig. 12 we show the three layer best fit models for both A_y and A_{min} compared to a similar calculation for the Dyal-Parkin three layer models using the same solar wind parameters, i.e., $v_p = 200$ km/sec and $\theta = 150°$. The curves labeled DP_1 and DP_2, respectively, are based on the models of Dyal and Parkin (1971a, b) and Dyal and Parkin (private communication). The departures at low frequency are associated with the core (their core is more highly conducting but deeper), while at high frequencies the DP_2 model fits the A_y data quite well. The DP_1 model is a poor fit everywhere to both the A_{min} and A_y transfer functions. It is worth noting that an incoherent noise source for the sunlit side models should mean that A_{min} is a more realistic transfer function; however, the DP_2 model does not fit A_{min}. Clearly, further work is required on both approaches in order to effect a reconciliation between the results.

LUNAR TEMPERATURE

Transformation of the conductivity profile into a thermal profile rests upon use of conductivity-temperature functions that are poorly known. Also some assumption must be made regarding the composition of the interior at the relevant depths. The

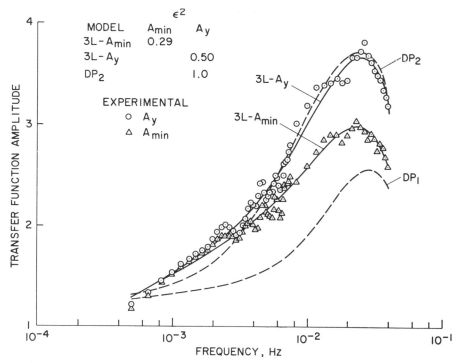

Fig. 12. Comparison of the 3L best fits to A_y and A_{min} with two versions of the Dyal and Parkin (1971a,b; private communication) models based upon the transient response measured on the dark side of the moon. These latter correspond to three layer lunar models: for DP_1 the shell (between the core and crust) conductivity is 10^{-4} mhos/m and the radius is 0.95 Rm for DP_2 the conductivity of the shell is 3×10^{-4} mhos/m and the radius is 0.95 Rm. Both models have a core of conductivity 10^{-2} mhos/m and a radius of 1044 km. All fits are for $v_p = 200$ km/sec and $\theta = 150°$. Rm is the lunar radius.

more resistive the matter, the higher the estimate of the temperature. Following our earlier estimate we use olivine as the material. Much laboratory work has been carried out on this substance, though the values for activation energy E and mobility σ_0 vary widely. The early values determined by Hughes (1955) yield conductivities generally lower than later determinations, but all values are somewhat suspect. England *et al.* (1968) show a conductivity function which is widely quoted; it uses an electronic term from Hughes (1955) and an ionic term from Bradley *et al.* (1964). Thus this is a composite once removed from direct determination. However, the activation energies in the various determinations are not too far apart, and the greatest variation is in the mobility, where differences of an order or two can occur.

Kobayashi and Maruyama (1971) have studied single olivine crystals. The principal cause of variation in activation energy and mobility is the iron content expressed as the fayalite fraction (Fa). The variation is mostly in σ_0 with $\sigma_0 = 8 \times 10^{-2}$ mhos/m for Fa = 0 (Shankland, 1969) and $\sigma_0 = 20$ mhos/m for Fa = 0.20. Shankland (1969) found that the activation energy changed by an order of magnitude as Fe was added to pure forsterite, but this might be a saturation effect at a very small concentration of Fe, since the results of Kobayashi and Maruyama (1971) for

0.074 < Fa < 0.126 show constant single-crystal activation energies. In summary, from this work we can expect about a 2.5 order difference in σ_0 as the Fa content varies from zero up to 20%. Recently Housley and Morin (1972) have shown that spurious effects associated with thermionic emission can raise the conductivity esti- mates of olivine and introduce an error of several hundred degrees at a temperature in the neighborhood of 1000°C.

In spite of these limitations, we are substantially aided by the logarithmic de- pendence of temperature upon conductivity. Thus an order error means approximately a 100–200° error in temperature. The England et al. (1968) formula yields a deep temperature of about 800°C and a shell temperature of about 600°C. Although the constants differ, the Noritomi (1961) olivines yield approximately the same tempera- ture, i.e., about 700°C. It should also be noted that the single crystal determinations of Kobayashi and Maruyama (1971) yield temperatures ranging from about 750 to 1000°C, while the Fa mole fraction decreases from 12.6 to 7.4%. The final conclusion arrived at is that a central temperature in the neighborhood of 800–1000°C is probable, based upon current data for olivines. More conducting matter would depress the computed temperature while an increase in resistivity at a given temperature would require an upward revision. For the outer conducting shell further work is required to refine the temperature estimate. Values cluster about our present estimate of 600°C based upon the England et al. (1968) formula and a uniform olivine moon.

A "cold" moon places serious constraints upon any theory of its origin and evolution (Reynolds et al., 1972; Hays, 1972; Papanastassiou et al., 1970; Urey and MacDonald, 1971). A chondritic moon would by today have melted unless perhaps convection would preclude this (Runcorn, 1967; Turcotte et al., 1972). Other possible sources of thermal energy, such as fossil nuclides (Fish et al., 1960), accretional energy (Wood, 1972; Hanks and Anderson, 1969; Sonett and Colburn, 1970), electrical heating during formation (Sonett et al., 1970), tidal friction during a hypothetical capture (Kaula, 1966), and lastly a "hot" start by formation within a dusty neighbor- hood at an elevated temperature would require the long-lived radionuclide budget to be reduced in order to meet the boundary condition of a low central temperature.

A potentially serious additional constraint upon the thermal history is the pre- sence of ubiquitous magnetism in lunar rocks. Possible origins for the magnetism have been discussed elsewhere (Sonett and Runcorn, 1971). The simplest source for the background field required for rock magnetization lies in a lunar dynamo which would have had to operate as early as 0.6 aeons after formation, implying a molten convect- ing core (presumably iron) at that time. This would be difficult to achieve without a "hot" start and additional heating, both incompatible with a cold moon unless the heat were subsequently convected out.

The comments made assume a moon in which the radioactivity was laid down uniformly; a gross inhomogeneity in the radial concentration of nuclides could alter the picture and possibly yield a viable model, though much more complex.

CONCLUSIONS

We have shown a number of conductivity profiles for the moon which differ at most in detail and mostly in regions near to the surface. All models yield a deep con- ductivity ($R = 1100–1300$ km) of about $1–3 \times 10^{-3}$ mhos/m. Conversion of this value to a temperature using reasonable olivine conductivity functions yields a temperature at that depth of about 800°C which could be low by as much as 200°. Thus a major conclusion of this paper that substantiates earlier reports (Sonett *et al.*, 1971b, c; Dyal and Parkin, 1971a, b) is that the central temperature of the moon is well below the melting point. Large increases in temperature below the level where the deepest sounding takes place are ruled out for a moon that is reasonably well behaved thermally because of the excessive thermal gradients required at depth.

The conclusion that the lunar interior is well below the melting point at the present time is in accord, with the low seismicity (Latham *et al.*, 1971), and the existence of mascons (Muller and Sjogren, 1968). The recent heat flow measurement (Langseth, 1972) is in disagreement if this measurement is assumed to represent a global value, but that viewpoint in turn requires a global concentration of radioactivities about twice chondritic, so it appears likely that local influences are important in the interpretation of the heat flow.

The presence of thin, highly conducting current layers, such as are assumed in the CL, DCL, and CCL models, are consistent with certain evolutionary models of the moon. The low surface abundance of iron has led Urey *et al.* (1971) to propose the existence of a sunken layer enriched in Fe. Also Rama Murthy *et al.* (1971) have suggested the formation of subsurface Fe–FeS mixtures as a way of depleting the lunar surface Fe abundance. A nonunique test for a conducting layer, consistent with our CCL, DCL, and CL models implies a threshold for the layer thickness based upon pure, consolidated Fe of 10 μ. Of course, realistically an Fe layer would likely consist of an unconsolidated layer whose mean conductivity $\bar{\sigma}$ was significantly lessened. Nevertheless even a reduction of $\bar{\sigma}$ by many orders would result in an electromagnetic response very sensitive to such a layer.

Lastly, the deep temperature surmised for the moon appears too low to permit solid state convection to operate at the present time. However it does not rule out such convection in the past. For example, if the conductivity function is in error requiring an upward revision of temperature toward an extremum of ~1000°C, it is possible that early convection was followed by conduction, the latter causing a drop of 100 to 200°C at 0.6–0.7 Rm over, say, 2 aeons consistent with the lunar thermal "time constant." Thus an early temperature in the neighborhood of 1200°C could have existed, consistent with current views on the low temperature limit on solid state convection. Alternatively it is entirely possible that the rheology of lunar material is sufficiently different from that of the earth that 1200°C is not an absolute lower limit for the occurrence of solid state convection.

NOTE ADDED IN REVIEW

In work so far reported, including this paper, we have dealt only with the amplitude of the transfer function. Additional information is contained in the phase of the complex transfer function but extraction of phase information is made difficult because of several effects that are believed to occur.

The Explorer 35 and Apollo 12 magnetometers are in relative motion; the effect is to introduce a Doppler shift that affects the coherence between Explorer and LSM signals. This effect is by no means trivial, since there are complications introduced by the nature of the solar wind itself. Tests of the data and theoretical analysis show that the relative motion introduces a randomization that can severely limit or even destroy the coherence between the signals observed by the two instruments. Changes in the properties of the solar wind between the two points of observation will also distort the coherence. For example, the solar wind can vary between conditions wherein turbulence is locally being generated or destroyed, or a steady state level of turbulence can exist. Also changes in the magnitude or direction of the bulk velocity of the solar wind will temporarily modify the Doppler shift, providing an additional source of randomization.

In view of these as yet unresolved problems, we have so far discussed only the transfer function amplitude, understanding that some phase information which would likely increase resolution within the moon is thereby lost. In later work, provided that suitable information is available, we shall attempt to reconstruct the entire transfer function and thus be in a position to comment upon the effects of the introduction of phase.

REFERENCES

Belcher J. W. and Davis L. (1971) Large amplitude Alfvén waves in the interplanetary medium, 2. *J. Geophys. Res.* **76**, 3534–3563.

Blank J. and Sill W. R. (1969) Response of the moon to the time varying interplanetary magnetic field. *J. Geophys. Res.* **74**, 736–743.

Bradley R. S., Jamil A. K., and Munro D. C. (1964) The electrical conductivity of olivine at high temperatures and pressures. *Geochim. Cosmochim. Acta* **28**, 1669–1678.

Chapman S. and Price A. T. (1930) The electrical and magnetic state of the interior of the earth as inferred from terrestrial magnetic variations. *Proc. Royal Soc.*, ser. A, **229**, 427–460.

Dyal P. and Parkin C. W. (1971a) Electrical conductivity and temperature of the lunar interior from magnetic transient-response measurements. *J. Geophys. Res.* **76**, 5947–5969.

Dyal P. and Parkin C. W. (1971b) The Apollo 12 magnetometer experiment: Internal lunar properties from transient and steady magnetic field measurements. *Proc. Second Lunar Sci. Conf.*, *Geochim. Cosmochim. Acta* Suppl. 2, Vol. 3, pp. 2391–2413. MIT Press.

Dyal P., Parkin C. W., Snyder C. W., and Clay D. R. (1972) Measurements of lunar magnetic field interactions with the solar wind. *Nature* **236**, 381–385.

England A. E., Simmons G., and Strangway D. (1968) Electrical conductivity of the moon. *J. Geophys. Res.* **73**, 3219–3226.

Fish R. A., Goles G. G., and Anders E. (1960) The record in the meteorites. III—On the development of meteorites and asteroidal bodies. *Ap. J.* **132**, 243–258.

Hanks T. C. and Anderson D. L. (1969) The early thermal history of the earth. *Phys. Earth Planet. Int.* **2**, 19–29.

Hays J. F. (1972) Radioactive heat sources in the lunar interior. *Phys. Earth Plan. Int.* **5**, 77–84.

Housley R. M. and Morin F. J. (1972) Electrical conductivity of olivine and the lunar temperature profile. *The Moon* **4**, 35–38.

Hughes H. (1955) The pressure effect on the electrical conductivity of peridot. *J. Geophys. Res.* **60**, 187–191.

Johnson F. S. and Midgley J. E. (1968) Notes on the lunar magnetosphere. *J. Geophys. Res.* **73**, 1523–1532.

Kaula W. M. (1966) Thermal effects of tidal friction. In *The Earth–Moon System* (editors B. G. Marsden and A. G. W. Cameron), pp. 46–52. Plenum, New York.

Kobayashi Y. and Maruyama H. (1971) Electrical conductivity of olivine single crystals at high temperature. *Earth Planet. Sci. Lett.* **11**, 415–419.

Kuckes A. F. (1971) Lunar electrical conductivity profile. *Nature* **232**, 249–251.

Lahiri N. and Price A. T. (1939) Electromagnetic induction in nonuniform conductors and the determination of the conductivity of the Earth from terrestrial magnetic variations. *Proc. Royal Soc.*, ser. A, **784**, 509–540.

Langseth M. G. Jr., Clark S. P. Jr., Chute J. Jr., and Keihm S. (1972) The Apollo 15 lunar heat flow measurement (abstract). In *Lunar Science—III* (editor C. Watkins), pp. 475–477, Lunar Science Institute Contr. No. 88.

Latham G., Ewing M., Dorman J., Press F., Toksoz N., Sutton G., Meissner R., Duennenbier F., Nakamura Y., Kovach R., and Yates M. (1971) Moonquakes. *Science* 170, 620–626.

Muller P. M. and Sjogren W. L. (1968) Mascons: Lunar mass concentrations. *Science* 161, 680–684.

Neugebauer M., Snyder C. W., Clay D. R., and Goldstein B. E. (1972) Solar wind observations on the lunar surface with the Apollo 12 ALSEP. *Planetary and Space Sci.*, in press.

Noritomi K. (1961) The electrical conductivity of rock and the determination of the electrical conductivity of the earth's interior. *J. Mining Coll.*, Akita Univ., ser. A, **7**, 27–59.

Papanastassiou D. A., Wasserburg G. J., and Burnett D. S. (1970) Rb–Sr ages of lunar rocks from the Sea of Tranquility. *Earth Planet. Sci. Lett.* **8**, 1–19.

Rama Murthy V., Evenson N. M., and Hall H. T. (1971) Model of early lunar differentiation. *Nature* **234**, 267–290.

Reynolds R. T., Fricker P., and Summers A. L. (1972) Thermal history of the moon. In *Lunar Thermal Characteristics* (editor J. W. Lucas), AIAA progress series, in press.

Runcorn S. K. (1967) Convection in the moon and the existence of a lunar core. *Proc. Roy. Soc.*. ser. A, **296**, 270–284.

Schubert G. and Schwartz K. (1969) A theory for the interpretation of lunar surface magnetometer data. *The Moon* **1**, 106–117.

Schubert G. and Colburn D. S. (1971) Thin highly conducting layer in the moon: Consistency of transient and harmonic response. *J. Geophys. Res.* **76**, 8174–8180.

Schubert G. and Schwartz K. (1972) High frequency electromagnetic response of the moon. *J. Geophys. Res.* **77**, 76–83.

Schuster A. (1889) The diurnal variation of terrestrial magnetism. *Phil. Trans. Roy. Soc.*, ser. A, **180**, 467–512.

Shankland T. J. (1969) Transport properties of olivine. In *The Application of Modern Physics to the Earth and Planetary Interiors* (editor S. K. Runcorn). John Wiley, London.

Sonett C. P. and Colburn D. S. (1968) The principle of solar wind induced planetary dynamos. *Phys. Earth Planet. Interiors* **1**, 326–346.

Sonett C. P. and Colburn D. S. (1970) The role of accretionally and electrically inverted thermal profiles in lunar evolution. *The Moon* **1**, 483–484.

Sonett C. P., Colburn D. S., Schwartz K., and Keil K. (1970) The melting of asteroidal-sized parent bodies by unipolar dynamo induction from a primordial T Tauri sun. *Astrophys. Space Sci.* **7**, 446–488.

Sonett C. P. and Runcorn S. K. (1971) How to use magnetic fields for fun and profit. *Comments on Astrophys. and Space Phys.* **3**, 149–154.

Sonett C. P., Dyal P., Parkin C. W., Colburn D. S., Mihalov J. D., and Smith B. F. (1971a) Whole body response of the moon to electromagnetic induction by the solar wind. *Science* 172, 256–258.

Sonett C. P., Colburn D. S., Dyal P., Parkin C. W., Smith B. F., Schubert G., and Schwartz K. (1971b) Lunar electrical conductivity profile. *Nature* 230, 359–362.

Sonett C. P., Schubert G., Smith B. F., Schwartz K., and Colburn D. S. (1971c) Lunar electrical conductivity from Apollo 12 magnetometer measurements: Compositional and thermal inferences. *Proc. Second Lunar Sci. Conf.*, *Geochim. Cosmochim. Acta* Suppl. 2, Vol. 3, pp. 2415–2431. MIT Press.

Turcotte D. L., Hsui A., Torrance K. E., and Oxburgh E. R. (1972) Thermal structure of the moon. Preprint.

Urey H. C. and MacDonald G. J. F. (1971) Origin and history of the moon. In *Physics and Astronomy of the Moon* (editor Z. Kopal), Academic Press.

Urey H. C., Marti K., Hankins J. W., and Liu M. K. (1971) Model history of the lunar surface. *Proc. Second Lunar Sci. Conf.*, *Geochim. Cosmochim. Acta* Suppl. 2, Vol. 3, pp. 987–998. MIT Press.

Wood J. A. (1972) History and early magmatism in the moon. *Icarus* 16, 229–241.

Proceedings of the Third Lunar Science Conference
(Supplement 3, *Geochimica et Cosmochimica Acta*)
Vol. 3, pp. 2337–2342
The M.I.T. Press, 1972

Iron–titanium–chromite, a possible new carrier of remanent magnetization in lunar rocks

S. K. Banerjee

Department of Geology and Geophysics,
University of Minnesota,
Minneapolis, MN 55455

Abstract—Magnetic measurements of synthetic titanchromites reveal a characteristic transition at low temperature enabling one to identify the presence of titanchromites in lunar rocks 12063, 14321 (dark clasts), and 14310. It is shown that titanchromites with a high chromium content could be the carriers of intermediate stability NRM observed in lunar rocks. Such NRM is acquired by TTRM at low temperature and cannot be duplicated by laboratory TRM imparted above room temperature. The latter process will not yield the correct lunar paleointensity.

Introduction

A major reward of the magnetic study of lunar rocks is a knowledge of the time variation of lunar paleointensity, assuming that the carriers and origin of natural remanent magnetization (NRM) in such rocks are well-defined. Although fine-grained iron is a prime candidate as carrier of the stable component of NRM in lunar rocks, to possess stable NRM such grains should be single domain and hence, have average diameters of the order of 150 Å. Gose *et al.* (1972) have presented viscous magnetization data for a lunar rock that support an iron grainsize distribution model with an upper cut-off of 150 Å, which supports iron as the carrier of stable NRM in this sample. Even then, it is still necessary to convince ourselves that the same model could indeed be used for most of the rocks and that no other magnetic mineral is an alternative carrier. Second, there is still a need to identify the carrier of that part of observed NRM in lunar rocks that is intermediate in stability between high-stability (Helsley, 1971; Hargraves and Dorety, 1972) single-domain iron and low-stability multidomain iron. These intermediate stability carriers seem to lose their NRM at apparent blocking temperatures between 100°C and 300°C. Such low apparent blocking temperatures in terms of TRM in suitably sized iron grains are difficult to understand in view of the thermal cycling on lunar surface that would have washed out such NRM, if acquired by a thermoremanent process due to cooling from above the blocking temperature. Our present study of synthetic titanchromites provides an answer to this dilemma and further, we postulate a process for the acquisition of NRM by titanchromites that is different from TRM. Conventional paleointensity determination methods do not reproduce this process.

Experimental Results and Discussion

Although titanchromites and chromian ulvöspinels have been discovered by mineralogists and petrologists studying lunar samples, no magnetic study of these had been undertaken before. Schmidtbauer (1971) has studied the magnetic properties of chromium substituted titanomagnetite that contains both Fe^{2+} and Fe^{3+} ions and thus is not representative of lunar titanchromites which have no Fe^{3+} ions. Our samples represent solid solution members between ulvöspinel (Fe_2TiO_4) and pure iron chromite ($FeCr_2O_4$). These can be represented by the general chemical formula, $Fe_{1+n}^{2+}Cr_{2-2n}^{3+}Ti_n^{4+}O_4$, and were synthesized by Prof. A Muan by equilibriating a charge for several days at 1300°C at an oxygen pressure of 10^{-10} atm. Although these compounds do not contain additional Al^{3+} as seen in the lunar samples (Haggerty, 1972), so far as magnetic properties go, lack of Al^{3+} does not make a big difference. Both Al^{3+} and Ti^{4+} are diamagnetic ions strongly preferring the octahedral sublattice in a spinel lattice. Thus, synthetic iron-titanium chromite of a certain titanium substitution behaves magnetically like a natural lunar titanchromite of a lower titanium content. The three synthetic samples reported here have n values of 0.14, 0.29, and 0.44 in the chemical formula, $Fe_{1+n}Cr_{2-2n}Ti_nO_4$.

Our first discovery was that although pure $FeCr_2O_4$ is antiferromagnetic and in the absence of lattice defects do not have a net moment and cannot carry NRM; iron-titanium chromites studied by us do have a net moment and can carry NRM. The explanation is diagrammatically shown in Fig. 1. Because of strong intra-sublattice interaction (Jacobs, 1960) between paramagnetic Cr^{3+} ions there is nearly zero moment (M) leading to the observed antiferromagnetism ($M = 0$) in $FeCr_2O_4$. A small substitution of Ti^{4+} (or Al^{3+}) in octahedral sublattice (B) decreases the intrasublattice interaction, thus allowing the magnetism of paramagnetic Cr^{3+} ions to be expressed. A net positive (using the convention that net M is along B sublattice) moment results. We observed this to be the case for $n = 0.14$. With more Ti^{4+}, the intrasublattice canting of Cr^{3+} is almost gone. Our sample of $n = 0.29$ nearly showed this behavior because M at 8°K and about 50 k Oe was positive but only 10% of the previous sample of $n = 0.14$. Finally, with even more Ti^{4+} substitution at $n = 0.44$, a negative M was experimentally observed. Thus we show the importance of iron-titanium chromite as a magnetic (ferrimagnetic) mineral with a substantial moment although it could be regarded as a solid solution between one antiferromagnetic

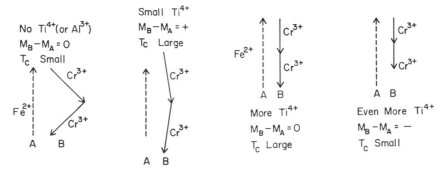

Fig. 1. Orientation of magnetic spin moments M of Fe^{2+} and Cr^{3+} ions occupying tetrahedral A and octahedral B sublattices of titanchromite lattice. Net magnetic moment for the lattice is given by $M_B - M_A$ while T_c denotes Curie point. On Ti^{4+} (or Al^{3+})substitution net moment first increases, then decreases to zero and then becomes negative.

mineral, $FeCr_2O_4$, and another imperfectly antiferromagnetic mineral, Fe_2TiO_4, neither carrying any substantial magnetic moment.

The second discovery was that all the three samples showed a very sharp decrease in moment on warming through characteristic transition temperatures (T_t). Figure 2, curve A through the hollow circles, represents M (in 116 Oe) versus T data obtained for $n = 0.14$. A similar decrease is found whether the applied field is zero or nearly 50,000 Oe, thus distinguishing it from magnetite (Fe_3O_4) where a similar sharp decrease is obtained on application of fields greater than 10,000 Oe. We consider curve A to be a characteristic magnetic signature of titanchromites, and in this light we can now look at data points B (crosses) and C (solid circles) in Fig. 2 that were obtained for a sample of lunar rock 12063 obtained through the courtesy of Prof. R. B. Hargraves. These data were obtained by warming (crosses) in 116 Oe and cooling (solid circles) in 41 Oe to delineate clearly the observed transition temperature T_t. There are three features to note. (1) A shoulder at 60°K, representing the Néel point of limenite, (2) a sharp decay of magnetization following Curie law ($M \alpha T^{-1}$) above 20°K due to free Fe^{2+} and/or ultrafine Fe and most important, (3) an additional moment between 60°K and 100°K whose temperature dependence is paralleled by our synthetic titanchromite sample of $n = 0.14$. The observed sharp rise in magnetization

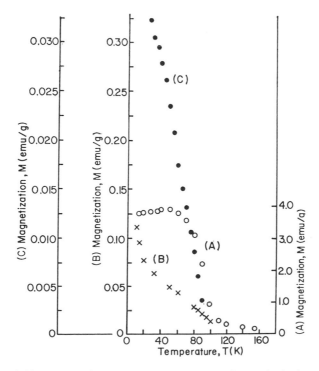

Fig. 2. Low field magnetization M versus temperature T for synthetic titanchromite of $n = 0.14$ (curve A through hollow circles) and rock 12063 (crosses denote warming in 116 Oe, solid circles denote cooling in 41 Oe). Note the presence of a characteristic magnetic transition temperature ($T_t = 100°K$) in both synthetic and natural samples.

at 11°K above the base level of 0.013 emu/g is mainly due to contributions from ilme-nite, paramagnetic (or superparamagnetic) iron and iron-titanium chromite. A conservative way to test for the presence of iron-titanium chromite in sample 12063 would be to assume that all the magnetization at 11°K is only due to paramagnetic (or superparamagnetic) iron whose magnetization decays inversely with temperature. Then any observed magnetization, say at 80°K, that is significantly above the baselevel (due to bulk iron) and inexplicable by inverse temperature-dependence of the low temperature magnetization must be due to iron-titanium chromite, the contribution from ilmenite at 80°K being very small. This is indeed what we have observed. An inverse temperature-dependence of the 11°K-value of magnetization leads to 0.015 emu/g at 80°K while the observed value is 0.030 emu/g. Subtracting the baselevel contribution of 0.013 emu/g due to bulk iron, this leads to a paramagnetic (or superparamagnetic) iron contribution to magnetization at 80°K of 0.002 emu/g and the remainder, 0.015 emu/g remains unexplained unless we conclude that magnetic iron-titanium chromite of $n = 0.14$ (which has the required Curie point) is indeed present in lunar rock 12063. Optical microscopy also revealed the presence of titanchromite in this sample. Using the same method, we have found the presence of titanchromite in a sample of rock 14321 (in the dark clasts) and 14310. We note that part of the magnetic evidence on the basis of which Runcorn *et al.*, (1971) based their claim of discovering magnetite in lunar rock 12020 was a similar T_t at 140°K in a warming experiment. It would seem that titanchromite, not studied magnetically before, would be a plausible and perhaps, preferable alternative. According to Table 1 where we have listed variation of T_t with n, a sample of $n = 0.44$ will have a T_t similar to the one seen by Runcorn *et al.*

Associated with the sharp decrease of M at T_t, we also observed a sharp decrease in coercivity H_c of magnetization. In Fig. 3 we have shown hysteresis loops obtained at 10°K for sample FCT-1 ($n = 0.14$) with and without field cooling (TRM). If the sample is cooled in zero applied field (no TRM), the loop is conventionally symmetrical although the observed lack of magnetic saturation at 40 k Oe attests to the intrinsic high value of H_c. The lower inset provides a physical picture of the lack of saturation. However, if the same sample is cooled in 40 k Oe through T_t the asymmetric and sheared loop, marked "After TRM," results. The shearing proves (Smit and Wijn, 1959) that there are two magnetic (not chemical) components in the sample. One has a lower coercivity and a higher susceptibility leading to the apparent constriction near $H = 0$. The other component has a higher coercivity and a lower susceptibility and is responsible for the observed lack of saturation even at $H = 45$ k Oe. Next, consider the $-H$-axis intercept of the two loops marked "After TRM" and

Table 1. Variation of transition temperature of titan-chromites with Ti^{4+} substitution.

Ti^{4+} Content (atoms per molecule)	Transition (temperature (°K))
0	< 80
0.14	100
0.29	120
0.44	140

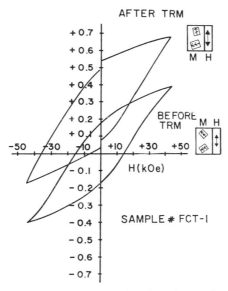

Fig. 3. Hysteresis loops at 10°K for synthetic titanchromite sample no. FCT-1 ($n = 0.14$) without field-cooling through T_t ("Before TRM") and with field-cooling ("After TRM"). The insets depict the respective physical situations in the samples in terms of magnetic (not chemical) components of high and low microscopic coercivities.

"Before TRM." The difference is equal to about 20 k Oe and is a measure (Jacobs and Bean, 1963) of the large intrinsic value of coercivity below T_t in the abovementioned second and "hard" magnetic phase. At a temperature greater than T_t the second phase no longer has a high coercivity and the loop was found to be no longer sheared. By thermal cycling across T_t, it is therefore possible to alter the coercivity of the "hard" phase by a factor of 10^3 to 10^4.

CONCLUSION

Such unusual H_c versus T dependence leads us to suggest a novel mechanism for acquisition of NRM in titanchromites and by implication, in some lunar rocks. The physical process is outlined in Fig. 4 and has been called "inverse thermoremanent magnetization (ITRM)" by Ozima et al. (1963). A more accurate name is transition thermoremanent magnetization (TTRM). Thermal cycling across T_t in zero field results in a realignment of the individual magnetic moments in a rock. A large enough number of cycling in zero field will demagnetize the sample. But when cycled in a finite field, there will be a net positive NRM ($M_R^+ > M_R^-$). If the sample has a Curie point that is greater than room temperature a part of the above NRM will be present and can be mistaken for a TRM acquired at a higher temperature by the usual process of magnetization blocking. The magnitude of blocked NRM by our postulated TTRM process depends, in the first order, on the ratio of H_c, above and below T_t. In magnetite where TTRM has been observed, this ratio of coercivities is only about 5 or 6, as compared to 10^3 or 10^4 for titanchromites. Therefore, TTRM in titanchromite should be a very efficient process. Second, we observe a Curie point of about 150°C for our

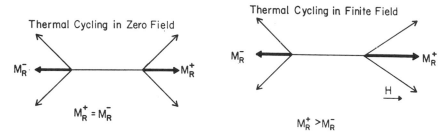

Fig. 4. Physical picture of transition thermoremanent magnetization (TTRM). In zero field the result is demagnetization, in finite field TTRM parallel to field direction results.

sample FCT-1 ($n = 0.14$). This would allow lunar titanchromite to (1) have a TTRM due to thermal cycling between lunar day and lunar night and (2) provide an intermediate stability NRM which will *appear* to have a blocking temperature of 150°C. For those lunar rocks mostly carrying stable TTRM due to titanchromites, the conventional paleointensity experiment (Helsley, 1971) will not yield true lunar paleointensity. Second, even those lunar rocks where titanchromite contribution to NRM is overshadowed by that of the more common iron grains, it may be the former that is responsible for the observed intermediate-stability NRM. We need not postulate an unusual grainsize of iron for that.

Acknowledgments—This work was supported by NASA Grant NGR 24-005-248 to the author. Grateful thanks are due to Professors A. Muan and R. B. Hargraves for providing samples and to Mrs. Naoma Dorety for the Curie point measurement.

REFERENCES

Gose W. A., Pearce G. W., Strangway D. W., and Larson E. E. (1972) On the magnetic properties of lunar breccias (abstract). In *Lunar Science—III* (editor C. Watkins), p. 332, Lunar Science Institute Contr. No. 88.

Haggerty S. E. (1972) Subsolidus reduction and compositional variations of lunar samples (abstract). In *Lunar Science—III* (editor C. Watkins), p. 347, Lunar Science Institute Contr. No. 88.

Hargraves R. B. and Dorety N. (1972) Magnetic property measurements on several Apollo 14 rock samples. *Proc. Third Lunar Sci. Conf., Geochim. Cosmochim. Acta* Suppl. 3, Vol. 3. MIT Press.

Helsley C. E. (1971) Evidence for an ancient lunar magnetic field. *Proc. Second Lunar Sci. Conf., Geochim. Cosmochim. Acta* Suppl. 2, Vol. 3, pp. 2485–2490. MIT Press.

Jacobs I. S. (1960) High field magnetization study of ferrimagnetic arrangements in chromite spinels. *J. Phys. Chem. Solids* 15, 54–65.

Jacobs I. S. and Bean C. P. (1963) Fine particles, thin films, and exchange anisotropy. In *Magnetism—III* (editors G. T. Rado and H. Suhl), Academic Press, New York.

Ozima M., Ozima M., and Nagata T. (1963) Role of crystalline anisotropy energy on the acquisition of stable remanent magnetization: Inverse type of thermoremanent magnetization. *Geofis. Pura Appl.* 55, 77–90.

Runcorn S. K., Collinson D. W., O'Reilly W., Stephenson A., Battey M. H., Manson A. J., and Readman P. W. (1971) Magnetic properties of Apollo 12 lunar samples. *Proc. Roy. Soc. London* 325, Ser. A, 157–174.

Schmidtbauer E. (1971) Magnetization and lattice parameters of Ti substituted Fe–Cr spinels. *J. Phys. Chem. Solids* 32, 71–76.

Smit J. and Wijn H. P. J. (1959) *Ferrites*. John Wiley.

Proceedings of the Third Lunar Science Conference
(Supplement 3, *Geochimica et Cosmochimica Acta*)
Vol. 3, pp. 2343–2361
The M.I.T. Press, 1972

Magnetic properties of Apollo 14 rocks and fines

D. W. Collinson, S. K. Runcorn, A. Stephenson, and A. J. Manson

Department of Geophysics and Planetary Physics, School of Physics,
The University, Newcastle upon Tyne, England

Abstract—The magnetic properties of rocks and fines returned by the Apollo 14 mission have been investigated. Following previous work on Apollo 11 and 12 samples, native iron in a range of grain sizes is again the most important magnetic mineral present. Ilmenite, troilite, ulvöspinel, and iron in paramagnetic form are of minor importance.

The rocks possess two well-defined components of permanent magnetization, a comparatively strong but unstable component and a weaker, hard component of high stability. The origin of the former is not clear; the characteristics of the stable magnetization suggest it is a primary one acquired at the time of formation of the rocks in a weak magnetic field. Evidence is accumulating that this field was of internal lunar origin.

Introduction

This paper describes the continuing investigations into the magnetic properties of returned lunar samples, and reports on the results obtained from Apollo 14 rocks and fines. The results obtained from Apollo 11 and 12 rocks by the present authors (Runcorn *et al.*, 1970, 1971) and other investigators (see, for instance, Nagata *et al.*, 1970, 1971, for magnetic mineralogy, and Strangway *et al.*, 1971, in which remanent magnetic properties are summarized) have proved to be of considerable interest, both as regards the magnetic minerals present and the presence and characteristics of the remanent magnetism that the rocks are found to possess.

The object of the investigations is to determine the nature and quantity of the magnetic minerals present, and to investigate the weak permanent magnetization (natural remanent magnetization, NRM). The latter property may be of importance in the history and structure of the moon, in that the NRM requires a magnetic field for its origin. If the character of the NRM suggests the existence of a field of internal origin on the moon when the rocks were formed, the most likely origin of this, as for all planetary magnetic fields, is some form of dynamo action in a molten, electrically-conducting lunar core: this feature would clearly be of importance in any model of the moon's history and structure.

The samples investigated in this paper are a small quantity of fines (14259,54) and two chips each from rocks 14053 (crystalline) and 14066, 14305, and 14318 (all fragmental).

Magnetic mineralogy

Magnetic mineralogy is similar to the Apollo 11 and 12 material. The total FeO content is in the range 10%–20%, with the basalt sample 14053 showing the highest content of 17% (Bence and Papike, 1972; Compston *et al.*, 1972). As has been previously found, most of the iron is in the paramagnetic form, including some ilmenite, amounting to about 1% by weight. Native iron or iron-nickel occurs in all samples; up to 0.5% is reported in soil samples (Wlotzka et al., 1972), and Gose et al. (1972) report 0.2%–0.6% in fragmental rocks and 0.1% in the basalt sample 14310. These values are probably typical of other Apollo 14 samples. Other minor magnetic minerals present are troilite (FeS), iron titanium chromite, and ulvöspinel (Haggerty, 1972). It is interesting to note that ferric iron has been detected in some Apollo 14 samples; Agrell et al. (1972) report the presence of goethite (FeOOH) in the fragmental sample 14301, and Weeks et al. (1972), using electron magnetic resonance, have demonstrated the presence of Fe^{3+} in some fragmental rocks, with the majority of it in paramagnetic form. The present authors (Runcorn et al., 1971) have presented evidence from magnetic measurements for small amounts ($\sim 0.05\%$) of magnetite (Fe_3O_4) in some Apollo 12 samples; although no other investigators have reported its presence and none has been detected in Apollo 14 rocks, there is now some independent evidence that free Fe^{3+} does occur in some lunar rocks.

Rock Magnetism

Fines

Temperature dependence of initial susceptibility. The initial susceptibility of Apollo 14 fines sample 14259,54 was measured over the range $-196°$ to $800°C$ in a field of 2.5 Oe rms at a frequency of 1.5 kHz, the sample being sealed in an evacuated ampoule. The results were very similar to those obtained from Apollo 11 and 12 fines, although the room temperature value (3×10^{-3} emu g^{-1} Oe^{-1}) was higher by 25% and 15% respectively. The Curie point was $775° \pm 3°C$ and the room temperature susceptibility decreased to 1×10^{-3} emu g^{-1} Oe^{-1} after heating to $800°C$. This is a decrease similar to that shown by Apollo 11 and 12 fines.

The high value of the initial susceptibility and the similarity of behavior to that of the fines returned from previous missions is indicative of the presence of superparamagnetic iron grains with a size distribution comparable to that of the Apollo 11 fines, which have been studied in detail by Stephenson (1971a, b), who showed that the number of iron grains within a given volume range is inversely proportional to the square of the volume, for grain sizes in the range 30–120 Å.

Induced and remanent magnetization in high fields. The induced magnetization was measured as a function of applied field up to 8 kOe. The curve is identical in shape to that measured for Apollo 11 fines (see Fig. 4, Runcorn et al., 1970), the maximum value of induced magnetization being 2.2 emu g^{-1}, 23% higher (c.f. 25% higher susceptibility in Apollo 11). The acquisition of IRM is also very similar, the respective values of saturation remanence being 0.20 and 0.13 emu g^{-1}. The coercivity of remanence, 400 Oe, was also the same as in the Apollo 11 material.

The variation with temperature of the induced magnetization in a field of 11 kOe shows an iron Curie point only; the decrease in induced magnetization after heating to $800°C$ is probably due to oxidation and occurs predominantly at temperatures above $600°C$, since on holding the sample at this temperature for 5 min a decrease of only 0.6% was observed. The heating was carried out with the sample sealed in an evacuated ampoule.

Crystalline and fragmental rock

Induced and isothermal remanent magnetization. Sample 14053,35 (crystalline) gave a $J_i - H$ curve almost identical to that of the Apollo 11 fines, with a maximum value of J_i about 10% lower (1.6 emu g^{-1}). This implies a metallic iron content of not less than 0.5% by weight. The actual concentration of iron is difficult to estimate from this curve, since there is almost certainly a significant contribution from superparamagnetic iron grains of diameter smaller than about 20 Å, which is the smallest size that will saturate in the maximum applied field of 8 kOe. Since the measurements on Apollo 15 fines indicate a size distribution of iron grains very similar to that of Apollo 11 fines, there is the possibility of up to a further 0.5 wt.% of iron grains of diameter less than 20 Å (Stephenson, 1971b).

Further evidence for the presence of superparamagnetic grains comes from the measurement of initial susceptibility of 14053,35; the value of 0.73×10^{-3} emu g^{-1} Oe^{-1} is about one-quarter that of the fines. Like the fines, the susceptibility remains constant over the range $-196°C$ to $20°C$. This suggests an iron grain size distribution similar to that of the fines. This low temperature behavior of the susceptibility is unlike that of the Apollo 12 crystalline samples 12020,34 and 12018,47 (Runcorn *et al.*, 1971).

The $J_i - T$ curve for 14053,35 in a field of 11 kOe is shown in Fig. 1. An iron

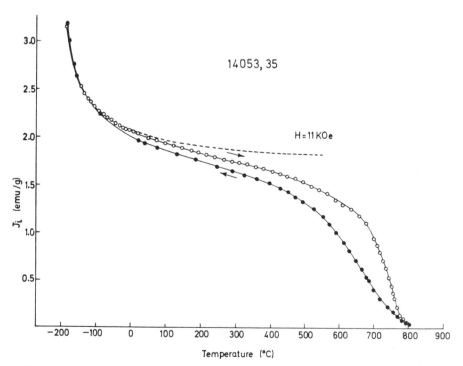

Fig. 1. The dependence of induced magnetization (J_i) on temperature, in a field of 11 kOe, for sample 14053,35. The presence of a component varying as $1/T$ is shown by the dashed line.

Curie point is observed, but, unlike the fines, very little decrease is observed in the room temperature value of J_i after heating to 800°C. Thus, oxidation in this rock is apparently not such a problem. The sharp rise in J_i at low temperatures, which exhibits a $1/T$ variation, is indicative of a paramagnetic component or of superparamagnetic grains of small volume. The rock powder was sealed in an evacuated capsule during heating.

For the fragmental rock 14305,56, $J_i - H$ and $J_i - T$ curves were also obtained. The former was similar to that of 14053,35 in shape, with about a quarter of the intensity. The temperature dependence of J_i over the range 20°C to 800°C in a field of 4.2 kOe showed an iron Curie point and was almost identical in shape to that obtained from Apollo 11 fines (Runcorn *et al.*, 1970). After heating to 800°C, J_i at room temperature had decreased to 41% of its initial value.

The acquisition of isothermal remanent magnetization, J_r, was measured in 14053,35, 14305,56, 14318,31, and 14066,24. It was found during the course of this and other measurements that the saturation IRM of different chips from the same rock varied greatly; hence considerable magnetic inhomogeneity must be present. An extreme case was that of 14053,35, in which $J_{r\,(sat)}$ differed by a factor of 10 in two chips. The coercive forces of remanence of the four samples were 70, 370, 400, and 270 Oe respectively, and $J_{r\,(sat)}$ was in the range $3-30 \times 10^{-3}$ emu g^{-1}.

Low temperature experiments. Samples 14053,35 and 14318,38 were cooled to -196°C in a spinner magnetometer (Molyneux, 1971) to observe the changes in NRM and saturated IRM. All measurements showed that irreversible decreases were taking place on cooling, but there were no discontinuities attributable to magnetite as found previously (Runcorn *et al.*, 1971). The decreases are tentatively ascribed to irreversible domain wall movements in multidomain grains.

Further experiments involving the thermal demagnetization of a low temperature TRM produced by cooling the samples to -196°C in a field of 1.2 Oe were carried out. On warming in zero field, 14053,35 showed a sudden decrease of 90% at about -150°C. This is almost certainly due to the Néel point of ulvöspinel (-153°C), which is estimated by comparison with the similar curves obtained for synthetic ulvöspinel (both in powder form and as a single crystal) to have a concentration of about 2% by weight. This is a similar result to Apollo 12 samples 12018,47 and 12020,34 (Runcorn *et al.*, 1971). Since no transition TRM was produced on allowing a demagnetized sample to warm from -196°C to 20°C in 1.2 Oe, the presence of magnetite can be ruled out. The fragmental sample 14318,38 did not show any evidence for the presence of ulvöspinel or magnetite.

Rotational hysteresis (W_R) as a function of applied field. These measurements were made on the fines at room temperature and at -196°C before and after heating to 830°C in an evacuated capsule. The W_R peak (Fig. 2) of the unheated sample again appears to arise from the iron-troilite system and is two to three times stronger than that of Apollo 11 fines 10084,13 and Apollo 12 fines 12070,130 (Runcorn *et al.*, 1970, 1971). The persistence of W_R at high fields is possibly due to superexchange anisotropy in the intergrown iron-troilite (Meiklejohn and Bean, 1957). After heating, the main peak broadens and the overall losses decrease by a factor of two, as was observed in Apollo 11 and 12 fines (Runcorn *et al.*, 1970, 1971). In each case an increase in W_R

occurs when the measurements are made at low temperature. The $W_R - H$ characteristic of a sample of crushed rock (14053) at room temperature (Fig. 2) is similar to that for 5 μ iron powder (Runcorn *et al.*, 1970), not to that of the iron-troilite system.

Microprobe analyses. To establish possible correlations between composition and magnetic properties, microprobe analyses of multidomain iron grains greater than about 10 μ across, which presumably contribute to the NRM of the rock, have been carried out on probe mounts supplied by NASA. Sample 14306,63 contained nine such grains, which were homogeneous and predominantly iron, with nickel contents ranging from 1–18%. In addition to the Ni, a small amount of Co is also present in concentrations which are very strongly correlated with the Ni content (Fig. 3). The ratio Co : Ni is 1 : 11.2 $\pm$ 0.2. Although 14306 is a fragmental rock, the constancy of the Co : Ni ratio implies that all the iron grains originate from the same source. A second fragmental rock 14318,10 did not show a strong correlation between Co and

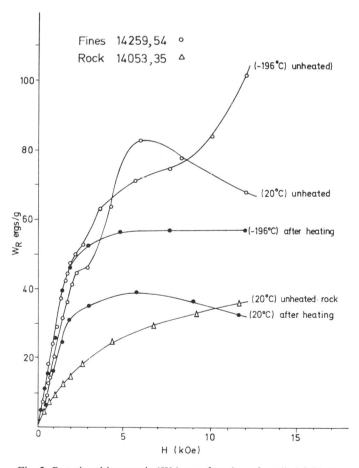

Fig. 2. Rotational hysteresis (W_R) as a function of applied field H.

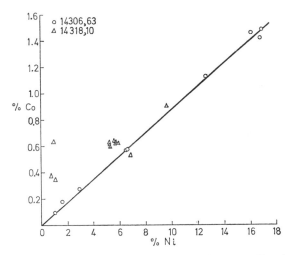

Fig. 3. Co–Ni content of iron grains in fragmental samples 14306,63 and 14310,10 (thin sections). The line shown is fitted to the values for 14306,63.

Table 1. Remanent magnetism of lunar samples.

Rock no.	Type	NRM (emu g^{-1} × 10^6)	Hard* component (emu g^{-1} × 10^6)	IRM (emu g^{-1} × 10^3)
14053,35	crystalline	218	3	2.86
14053,49	crystalline	153	—	—
14066,24	fragmental	8.7	1	2.19
14066,28	fragmental	1.8	—	—
14305,56	fragmental	8.3	0.5	2.80
14305,83	fragmental	9.4	—	3.30
14306,33	fragmental	22.4	—	3.08
14306,40	fragmental	67.5	1	—
14318,31	fragmental	95.8	—	—
14318,38	fragmental	41.2	2	5.65
10020,4F	crystalline	16.8	3	0.75

* Approximate NRM remaining after demagnetization in 300 Oe.
A dash indicates no measurement made.

Ni content for 12 grains examined (Fig. 3). Sample 14063,28 had only one grain of the required size present with an Ni content of 24% and a Co : Ni ratio of 1 : 51. With the exception of the last result, the observed nickel contents are consistent with there being no significant decrease in the Curie points of the rocks (Brailsford, 1951).

Remanent Magnetism

Characteristics of the NRM

All the Apollo 14 rocks measured by us possess a measurable remanent magnetization, as did the Apollo 11 and 12 samples, but the former show a wider range of intensity (Table 1). There is also a marked variation in the intensity of NRM of

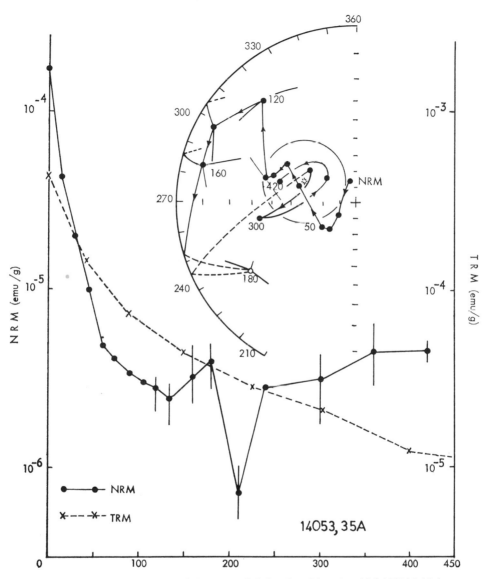

Fig. 4. A.F. demagnetization of the NRM (left-hand scale) and artificial TRM (right-hand scale) of 14053,35. The vertical lines on the intensity curve and lines from the points on the stereogram show the range of intensities and directions observed after repeated demagnetization at the same field. The mean direction and 95% circle of confidence based on all the measurements between 240 Oe and 420 Oe is also shown, to give an indication of the significance of the final direction.

chips from the same rock sample; this is particularly noticeable in the fragmental rocks and is doubtless due to the inhomogeneity of the structure of these rocks.

Alternating field demagnetization reveals similar behavior in all the samples. A relatively soft component of magnetization, generally the dominant contributor to the NRM, is removed in a peak alternating field of 100 Oe or less after which a weak, harder component is revealed which persists up to 400 Oe or more. Because of the

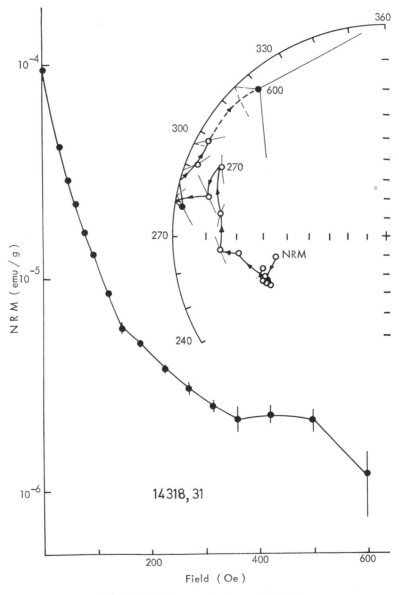

Fig. 5. A.F. demagnetization of 14318,31.

weakness of this hard component (about $1-5 \times 10^{-6}$ emu g^{-1}) and the facility with which some of the rocks pick up anomalous magnetizations during the demagnetization process, it was not possible to demagnetize the rocks completely. However, the shape of the demagnetization curves and the behavior of the directions of magnetization (Figs. 4, 5, and 6) leaves little doubt that there is a hard component present in these rocks. Table 1 shows estimates of the intensity of the hard NRM remaining after treatment at 300 Oe in the samples investigated. The similarity of the saturation IRM values suggests that there are similar quantities of material able to carry an NRM in these rocks.

Two further features may be noted. The change in the direction of NRM during demagnetization is not large, generally staying within an octant on the stereogram, and the anomalous maxima and minima of intensity, particularly in 14053,35. The repeatability of these readings is tested during demagnetization. The significance of these observations is discussed later.

Figure 7 shows the thermal demagnetization of the NRM of 14053,35 and 14306,40. There is a steady decay of the NRM in each sample, 14053 showing a stable

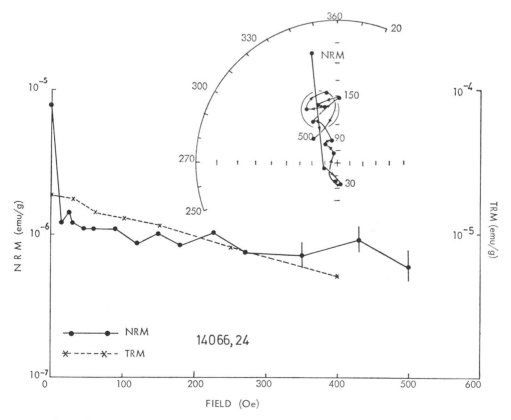

Fig. 6. A.F. demagnetization of the NRM (left-hand scale) and artificial TRM (right-hand scale) of 14066,24. The mean direction and 95% circle of confidence based on the measurements between 180 Oe and 500 Oe are shown.

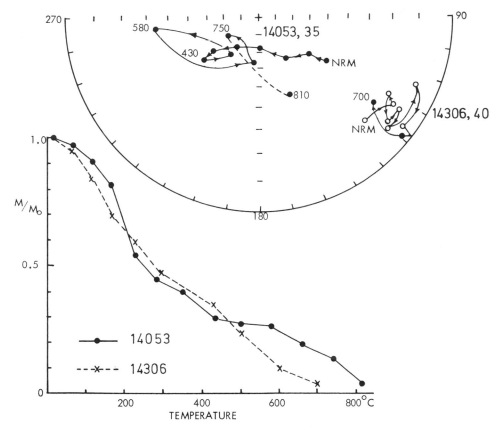

Fig. 7. Thermal demagnetization of the NRM of 14053,35 and 14306,40.

direction about 350°C; although the direction of the NRM during A.F. demagnetization of 14053,35 changes in a complicated fashion, the angular separation of the initial and "demagnetized" NRM in the two chips from 14053 is approximately the same. Unfortunately, because of lack of orientation data, it is not possible to compare the absolute changes of direction in each case.

The thermal demagnetization was carried out in a vacuum of about 10^{-3} Torr (continuously pumped), and there is good evidence that iron is the carrier of the hard magnetization. There was evidence of chemical changes occurring in 14306,40 at 700°C, and the measurements were discontinued, but 14053 appeared to remain essentially unaltered. The lunar rocks from the different missions now show the presence of hard and soft NRM in differing proportions. Figure 8 shows A.F. demagnetization of a recently-measured sample of Apollo 11 crystalline rock (10020,4F), which possesses only a hard component of NRM. Another sample (10049,1) gave a similar result. Most of the Apollo 14 samples possess a hard and soft component (see also Hargraves and Dorety, 1972; Pearce *et al.*, 1972; Nagata *et al.*, 1972), while in the Apollo 12 rocks the hard component was very weak with the soft component dominant (Strangway *et al.*, 1971).

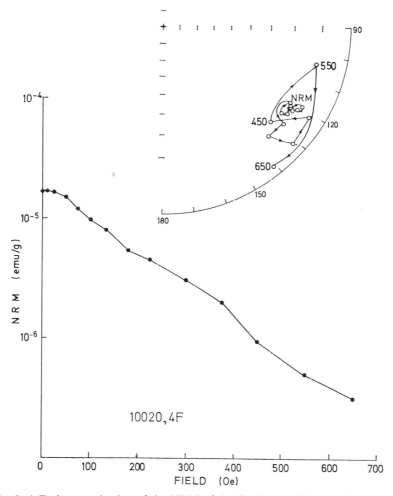

Fig. 8. A.F. demagnetization of the NRM of Apollo 11 crystalline sample 10020,4F.

Origin of the NRM

The hard component. Some tests were carried out to determine whether the hard NRM has properties consistent with its being of thermoremanent origin in a weak lunar field. Two samples, 14053,35 and 14066,28, were given an artificial TRM by heating to 800°C in a vacuum (10^{-3} Torr) and allowing them to cool in a magnetic field of 0.32 Oe (cooling time about 10 min). They were then demagnetized in an alternating field, the results of which are shown in Figs. 4 and 6. The 14066 curve is similar to the hard portion of the NRM curve; the same is approximately true of 14053, if the NRM curve is smoothed through its irregularities, although the fit is less satisfactory. The decay of the TRM of 14053 suggests that some part of the soft NRM may be of thermoremanent origin. The initial susceptibility and low field IRM of the rocks showed no significant change after heating (although 14066 was starting to change), evidence that a negligible degree of oxidation has occurred. The approximate values of the ancient lunar field intensity from the TRM and NRM data lie between 1000 γ and 10,000 γ; in the absence of more systematic and thorough ancient intensity investigations, even this wide range of values must be regarded cautiously.

The soft component. Several experiments were carried out to investigate the likelihood of different processes of magnetization contributing the soft component.

There was significant acquisition of viscous remanent magnetization (VRM) in samples 14053,35, 14318,38, and 14305,56, with viscosity index values (slope of the VRM $-\log t$ graph) of 3.3, 1.8, and 0.3 $\times 10^{-6}$ emu g^{-1} Oe^{-1} respectively. Assuming that the VRM intensity is proportional to the applied field and to the logarithm of time, it does not appear to be a likely source of the NRM in any reasonable lunar field or time interval.

A feature of the lunar surface is meteoritic bombardment and the associated shock waves in surrounding rocks. Piezo-remanent or shock magnetization has long been known as a possible source of remanence in rocks, both terrestrial (e.g., Hargraves and Perkins, 1969) and lunar (Nagata *et al.*, 1971); simple experiments show its ready acquisition in 14053,35 and 14318,31, the former sample acquiring an intensity of about 10^{-5} emu g^{-1} through light tapping in an ambient field of 1 Oe. In the presence of a field of suitable magnitude, this process is a possible source of the soft component of the NRM.

Comparatively weak fields are required, in the range 5–20 Oe, to give the samples an IRM of equivalent intensity to their initial NRM. However, the stability of this equivalent IRM against both A.F. and thermal demagnetization is significantly less than is observed in the NRM (Fig. 9).

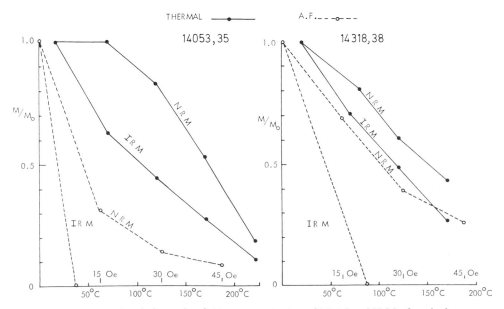

Fig. 9. Thermal and alternating field demagnetization of NRM and IRM of equivalent intensity in samples 14053,35 and 14318,38.

Finally, the acquisition of partial thermoremanent magnetization (PTRM) up to a temperature of 220°C was investigated in samples 14053,35 and 14306,33. Although significant PTRM is acquired by both rocks, comparison with the decay of the NRM of the rocks during thermal demagnetization in similar temperature intervals shows considerable inconsistencies (Fig. 10), suggesting that this is not the origin of the softer magnetization.

The carrier of the NRM. Because of the importance of iron in contributing to the magnetic properties of lunar rocks, it is the most likely carrier of the NRM. There is evidence for this from the thermal demagnetization of 14053, and some further experiments have been carried out, using a new small-scale magnetometer developed by one of the authors (D.W.C.) for the measurement of individual magnetic grains. Several such particles were extracted from a crushed chip of 14053 and their NRM and saturated IRM intensity and initial susceptibility measured. The average values from a group of six particles were 0.09 emu g^{-1}, and 0.03 emu g^{-1} Oe^{-1}, respectively. The saturating field was of the order of 600 Oe, and the NRM was removed completely in less than 100 Oe alternating field. The grains were not homogeneous, but all of them consisted partly of a greyish, matte-surfaced material which rusted in a moist atmosphere. This feature, combined with the strong magnetic properties, suggests that the material is native iron. It is very difficult to estimate the mass of iron in a grain. The above values are based on the mass of the whole grain, and should, therefore, be regarded as minimum values.

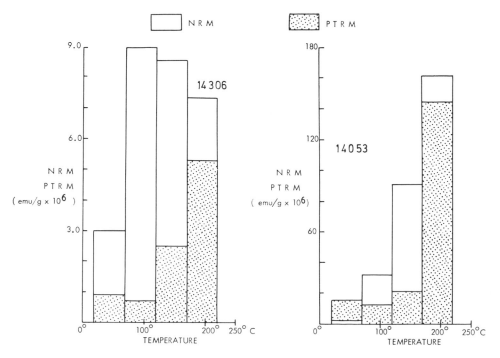

Fig. 10. Decay of NRM and acquisition of PTRM up to 220°C in samples 14306,33 and 14053,49. PTRM acquired in 1.0 Oe.

No grains possessing a hard component of remanence have been found. Other particles show a range of weaker magnetic properties; some of these particles have been identified by reflection microscopy as ilmenite, although microprobe analysis shows they are not pure. Also, some of them are capable of carrying an IRM, which pure ilmenite would not. A possible explanation is the existence of small inclusions of iron in low concentration below the resolution limit of the microprobe. These might be expected to be present if the ilmenite is formed by the breakdown of ulvöspinel (El Goresy et al., 1972). An alternative explanation might be the presence of Ti^{3+}, which has been found in the ilmenite from 14053 by Pavicevic et al. (1972). Although this possibility is somewhat conjectural, the presence of Ti^{3+} which carries a moment of 1 μ_B would almost certainly render the ilmenite ferrimagnetic and also increase the Néel point.

Discussion

The Apollo 14 rocks again demonstrate the importance of iron in contributing to the magnetic properties, which can satisfactorily be explained by the existence of a range of grain sizes from superparamagnetic through to multidomain. It is worth noting the similarity between the magnetic properties and their magnitudes of 14053 and the fines; this may in part be due to the exceptionally reducing conditions in which 14053 is believed to have been formed, and the preservation of much of the finer-grained iron (El Goresy et al., 1971).

The existence of hard and soft components of magnetization is now well established in lunar rocks. More evidence has also accumulated that iron, or dilute nickel-iron is the most important magnetic constituent, both in contributing to the bulk magnetic properties and as the carrier of the NRM. A feature of the Apollo 14 rocks is the variability of their properties within a hand sample. As an example, 14053,35 and 14053,49, studied in this paper, have an NRM of 218 and 153×10^{-6} emu g^{-1} respectively, and the former chip has a viscosity coefficient of 3.3×10^{-6} emu g^{-1} Oe^{-1}; Nagata et al. (1972) report an NRM of 2030×10^{-6} emu g^{-1} in another chip from 14053 and a negligible growth of viscous magnetization. This variability of properties is presumably associated with the lithological inhomogeneity of the rocks, particularly the breccias.

The soft component of NRM

The origin of the soft NRM presents several problems. In some rocks, e.g., 14066,24 (Fig. 6) it appears to be distinct from the hard component and to represent a separate magnetization event. In 14053,35 comparison of the A.F. demagnetization curves of NRM and artificial TRM suggests that some of the soft magnetization is a TRM in grains of appropriate size. This latter case goes some way towards explaining the observation that the difference in direction between initial NRM and the remaining hard component is not great. If part of the soft NRM is in the same direction as the hard, then the addition of another soft component in a random direction (as would be expected from a process associated with the solar wind or some temporary lunar magnetic field) would have a smaller effect on the direction of the total magnetization of the rock.

None of the processes investigated appears to be entirely satisfactory as a source of a distinct secondary component, and there is a further puzzling feature associated with it.

It can be seen in Figs. 7 and 9 that there is a significant decay of NRM during thermal demagnetization up to a temperature of 120°C of samples 14053,35, 14318,38, and 14306,40. This observation is difficult to explain, since the maximum temperature of a surface rock almost certainly exceeds this value. Unless this component of NRM has been acquired by the rocks since removal from the lunar surface, there should be no decrease in NRM on heating to temperatures higher than that of the lunar surface. On the basis of single domain theory (Néel, 1949), no change in a laboratory heating of duration 10 min should occur below 460°C for a sample which has spent, say, 10^6 yr at 120°C in zero field. However, this calculation is not strictly valid, since the major part of the NRM of this sample (14053) is not carried by single domain grains. This can be shown by relating the demagnetizing effects of heating to those of applying an alternating field of peak intensity h. For single domain grains, a field applied for a time of, say, 1 min will demagnetize all grains of volume less than v_1, where

$$v_1 = v_0/(1 - 2h/H_c).$$

In the equation, v_0 is the volume of the grains with a relaxation time of 10 min in the absence of an applied field and H_c is their coercive force in the absence of thermal fluctuations (Stephenson, 1971a). The temperature T required to demagnetize grains of volume v_1 is, however, given by the equation $v_1 = v_0 T_1/T_0$, and so the effect of heating is related to the peak field of A.F. demagnetization through the equation

$$T = T_0/(1 - 2h/H_c).$$

Figure 11 shows this relation between T and h for single-domain grains for fields up to about 30 Oe. In Fig. 11, H_c has been taken to be 1700 Oe, which is the value calculated for the single-domain iron grains in the Apollo 11 fines (Stephenson, 1971b). The diagram also shows the experimental relationship derived from Figs. 4 and 7. The result clearly shows that the major part of the soft NRM of this particular rock does not reside in single-domain grains, even though there is evidence for their

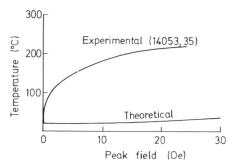

Fig. 11. The relationship between thermal and alternating field demagnetization. The lower line is a theoretical one for single-domain grains with $H_c = 1700$ Oe, and the upper one was obtained from the demagnetization of the NRM of 14053,35.

presence from the high susceptibility and from the behavior of the $J_i - T$ curve in high fields at low temperatures.

An explanation of this is that the remanence of the single domain material is much less than the total NRM of the rock. The low field susceptibility of 14053,35 is about one-quarter that of the fines; hence, assuming that the same distribution of single-domain grains is present in the rock and fines, and that the NRM of the single-domain grains is a TRM, then the magnitude of this NRM should be about one-quarter of that acquired by the fines. For the Apollo 11 fines this was about 3×10^{-3} emu g^{-1} in 0.42 Oe (Runcorn et al., 1970). Assuming an ancient lunar field intensity of about 1000 γ, the TRM of the single-domain grains in the rock would be about 20×10^{-6} emu g^{-1}, very much less than the observed NRM of about 220×10^{-6} emu g^{-1}, and hence the absence of a significant contribution from the single-domain grains is accounted for.

The origin of the soft component still presents some problems, and the introduction of all or part of it into the samples after removal from the lunar surface cannot yet be entirely ruled out.

The hard component of NRM

The only mechanisms which seem capable of producing an NRM of the observed hardness are a thermoremanent or piezo-remanent magnetization in a weak field, or an IRM in a strong field of the order of 1000 Oe. The latter does not seem likely. Even allowing for some demagnetization due to prolonged thermal cycling, the intensity of NRM would be expected to be at least a hundred times stronger than is observed, and there is the difficulty of proposing the source of such a field. Whether such a field can be produced as a result of a meteoritic impact is not known. Terrestrial evidence from the NRM of rocks sampled in Meteor Crater, Arizona (Hargraves and Perkins, 1969) indicates no abnormal field, but this is, of course, only a moderately sized crater by lunar standards.

Strangway et al. (1971) have summarized the remanent magnetic properties of Apollo 11 and 12 samples, to which are now added data from Apollo 14 reported in this paper and by Hargraves and Dorety (1972), Nagata et al. (1972), and Pearce et al. (1972). The picture that emerges is of the existence of a hard component of NRM persisting at a level $0.5–5 \times 10^{-6}$ emu g^{-1} in demagnetizing fields of the order of 400 Oe; the time span represented by the samples is more than 800 m.y. (Wasserburg et al., 1972). Investigations of the origin of the hard NRM (thermal demagnetization, artificial TRM and its stability) are consistent with the rocks having acquired a TRM in a weak field on the moon. The persistence of this field over long periods of time at a seemingly fairly steady intensity strongly suggests that it is of internal origin, and is against any "catastrophic" theory of the origin of the hard remanence.

Although many of the measurements lack the thorough tests considered necessary in ancient field intensity determinations in terrestrial rocks, the lunar samples generally give an ancient field in the range 1000 γ to 10,000 γ, based on the acquisition of TRM in known fields (this paper, Helsley 1970, 1971; Dunn and Fuller, 1972).

The behavior of the NRM during A.F. demagnetization is complex. In Fig. 6 the rock, a glass-poor fragmental rock (class III of Engelhardt *et al.*, 1972), shows the presence of a soft component removed in about 30 Oe; the 10% of the initial NRM remaining at that point consists of two components. In Fig. 5 the successive directions fall approximately on a great circle, showing the presence of two components of moderately high stability. In Fig. 4 three components must be present, the soft component and the hardest having approximately the same direction, and an intermediate one that is in the opposite quadrant and accounts for the sharp dip in the A.F. demagnetization curve.

In the case of the fragmental rocks, the explanation may be in a stable "secondary" magnetization combined with residual magnetization(s) of intermediate hardness present in the clasts in the rocks prior to the impact event in which the rock was formed. The fragmental rocks contain quite large clasts (up to 0.5 cm in our samples), each of which may have a primary magnetization before the event during which they were incorporated into their present form. During ejection after impact, the rock temperature may have risen as high as 900°C (Anderson *et al.*, 1972). If the breccia was formed at higher than about 800°C it would acquire a new TRM in any ambient weak field, but some of the clasts might retain a remanant of their original magnetization, according to the temperature that they reached in the rock. These prior magnetizations would be randomly directed and qualitatively it can be seen that the demagnetization of such a rock might give rise to the observed variations in intensity and direction. This explanation is less satisfactory for the crystalline samples.

CONCLUSIONS

The work reported in this paper and on the magnetic studies of the Apollo 11 and 12 rocks shows that permanent magnetization is widespread in rocks of the lunar crust. This conclusion is in accord with the fact that a mean magnetic field of about 36 γ is recorded by the magnetometer at the Apollo 12 site on Oceanus Procellarum, and up to about 100 γ at the Apollo 14 site. Further, as has previously been shown (Runcorn *et al.*, 1971) a field of this magnitude would be expected, given suitable geometry, from the NRM of 10^{-4}–10^{-5} emu g^{-1} measured.

A.F. demagnetization has shown that in many instances NRMs of at least two stabilities are present in the lunar rocks, the intensity of the most stable component being only of the order of 10^{-6} emu g^{-1}. As this is inadequate to produce the observed fields of greater than 10 γ near the lunar surface, the conclusion is reached that the less stable component, requiring about 100 Oe for demagnetization, was not only present when the samples were lying on the lunar regolith but was also present in the parent body, i.e., in the case of the crystalline samples, the maria lava flows. Thus in many rocks there appear to be two groups of magnetic grains carrying NRM components of different stability but of early origin. To explain the data otherwise would seem to require the hypothesis that the less stable grains could pick up from the local field of the hard NRM in the lava a viscous magnetization parallel to it. The size of the field around the grains, if they are randomly distributed, would only be of the order of 1 γ and the results of the VRM experiments appear to exclude this mechanism.

The extensive unbroken lavas of the maria basins, magnetized to the intensities observed in the returned samples, would not give anomalies as much as 10 γ except over length scales of the order of the thickness of a few flows, or at the edges of the basins. Thus the absence of "magcons" on the maria is consistent with the lavas being magnetized. Magnetometers in orbit round the moon have also detected magnetization of the highlands.

Thus, magnetizations in rocks of widely different age on the moon exclude the most plausible of the external sources of the magnetizing field—that of the earth. The law of tidal friction would cause the moon to retreat very rapidly from a position close to the earth, say of the order of a few million years.

An internal lunar magnetic field which was present some 300 m.y. earlier than that inferred from previous investigations is concluded from the present studies. A molten core at this early stage in lunar evolution challenges the basic assumptions of current discussions of the moon's thermal history.

Acknowledgments—Helpful discussions with members of the Department of Geophysics and Planetary Physics are gratefully acknowledged. Dr. L. Molyneux gave valuable assistance with the low temperature measurements of remanent magnetism, and Dr. J. M. Jones of the Department of Geology, polished and carried out reflectivity measurements on extracted grains. We are also grateful to Dr. W. O'Reilly for discussion of rock magnetic aspects of the work, to Mr. G. Hoye for assistance with measurement of the fines and to Mr. W. Davison for carrying out the measurements for the electron microprobe analyses.

One of the authors (A.S.) is in receipt of a N.E.R.C. Senior Research Associateship. We are grateful to N.A.S.A. for providing the samples under the Lunar Sample Analysis Program, British participation in which was facilitated by the Science Research Council.

REFERENCES

Agrell S. O., Scoon J. H., Long J. V. P., and Coles J. N. (1972) The occurrence of geothite in a microbreccia from the Fra Mauro formation (abstract). In *Lunar Science—III* (editor C. Watkins), pp. 7–9, Lunar Science Institute Contr. No. 88.

Anderson A. T. Jr., Braziunas T. F., Jacoby J., and Smith J. V. (1972) Breccia populations and thermal history: Nature of pre-Imbrian crust and impacting body (abstract). In *Lunar Science—III* (editor C. Watkins), pp. 24–26, Lunar Science Institute Contr. No. 88.

Bence A. E. and Papike J. J. (1972) Crystallization histories of pyroxenes from lunar basalts. In *Lunar Science—III* (editor C. Watkins), pp. 59–61, Lunar Science Institute Contr. No. 88.

Brailsford F. (1951) *Magnetic Materials*. Methuen.

Compston W., Vernon M. J., Berry H., Rudowski R., Gray C. M., and Ware N. (1972) Age and petrogenesis of Apollo 14 basalts (abstract). In *Lunar Science—III* (editor C. Watkins), pp. 151–153, Lunar Science Institute Contr. No. 88.

Dunn J. R. and Fuller M. (1972) Thermoremanent magnetization (TRM) of lunar samples (abstract). In *Lunar Science—III* (editor C. Watkins), pp. 195–197, Lunar Science Institute Contr. No. 88.

El Goresy A., Ramdohr P., and Taylor A. (1971) The geochemistry of the opaque minerals in Apollo 14 crystalline rocks. *Earth Planet. Sci. Lett.* **13**, 121–129.

El Goresy A., Ramdohr P., and Taylor L. A. (1972) Fra Mauro crystalline rocks: Petrology, geochemistry and subsolidus reduction of the opaque minerals (abstract). In *Lunar Science—III* (editor C. Watkins), pp. 224–226, Lunar Science Institute Contr. No. 88.

Engelhardt W. von, Arndt J., and Stöffler D. (1972) Apollo 14 soils and breccias, their compositions and origin by impact (abstract). In *Lunar Science—III* (editor C. Watkins), pp. 233–235, Lunar Science Institute Contr. No. 88.

Gose W. A., Pearce G. W., and Strangway D. W. (1972) On the magnetic properties of lunar breccias. In *Lunar Science—III* (editor C. Watkins), pp. 332–334, Lunar Science Institute Contr. No. 88.

Haggerty S. E. (1972) Subsolids reduction and compositional variations of lunar spinels (abstract). In *Lunar Science—III* (editor C. Watkins), pp. 347–349, Lunar Science Institute Contr. No. 88.

Hargraves R. B. and Perkins W. E. (1969) Investigations of the effect of shock on natural remanent magnetism. *J. Geophys. Res.* **74**, 2576–2589.

Hargraves R. B. and Dorety N. (1972) Magnetic property measurements on several Apollo 14 rock samples (abstract). In *Lunar Science—III* (editor C. Watkins), pp. 357–359, Lunar Science Institute Contr. No. 88.

Helsley C. E. (1970) Magnetic properties of lunar 10022, 10069, 10084, and 10085 samples. *Proc. Apollo 11 Lunar Sci. Conf., Geochim. Cosmochim. Acta* Suppl. 1, Vol. 3, pp. 2213–2219. Pergamon.

Helsley C. E. (1971) Evidence for an ancient lunar magnetic field. *Proc. Second Lunar Sci. Conf., Geochim. Cosmochim. Acta* Suppl. 2, Vol. 3, pp. 2485–2490. MIT Press.

Meiklejohn W. H. and Bean C. P. (1957) New magnetic anisotropy. *Phys. Rev.* **105**, 904 913.

Molyneux L. (1971) A complete result magnetometer for measuring the remanent magnetization of rocks. *Geophys. J. R. Astr. Soc.* **24**, 429–433.

Nagata T., Ishikawa Y., Kinoshita H. Kono M., Syono Y., and Fisher R. M. (1970) Magnetic properties and natural remanent magnetization of lunar materials. *Proc. Apollo 11 Lunar Sci. Conf., Geochem. Cosmochim. Acta* Suppl. 1, Vol. 3, pp. 2325–2340. Pergamon.

Nagata T., Fisher R. M., Schwerer F. C., Fuller M. D., and Dunn J. R. (1971) Magnetic properties and remanent magnetization of Apollo 12 lunar materials and Apollo 11 lunar microbreccia. *Proc. Second Lunar Sci. Conf., Geochim. Cosmochim. Acta* Suppl. 2, Vol. 3, pp. 2461–2476. MIT Press.

Nagata T., Fisher R. M., Schwerer F. C., Fuller M. D., and Dunn J. R. (1972) Magnetism of Apollo 14 lunar materials (abstract). In *Lunar Science—III* (editor C. Watkins), pp. 573–575, Lunar Science Institute Contr. No. 88.

Néel L. (1949) Theory of magnetic viscosity of fine grained ferromagnetics with application to baked clays. *Ann. Geophys.* **5**, 99.

Pavioevic M., Ramdohr P., and El Goresy A. (1972) Microprobe investigations of the oxidation state of Fe and Ti in ilmenite in Apollo 11, Apollo 12, and Apollo 14 crystalline rocks (abstract). In *Lunar Science—III* (editor C. Watkins), pp. 596–598, Lunar Science Institute Contr. No. 88.

Pearce G. W., Strangway D. W., and Gose W. A. (1972) Remanent magnetization of lunar samples (abstract). In *Lunar Science—III* (editor C. Watkins), pp. 599–601, Lunar Science Institute Contr. No. 88.

Runcorn S. K., Collinson D. W., O'Reilly W., Battey M. H., Stephenson A., Jones J. M., Manson A. J., and Readman P. W. (1970) Magnetic properties of Apollo 11 lunar samples. *Proc. Apollo 11 Lunar Sci. Conf., Geochim. Cosmochim. Acta* Suppl. 1, Vol. 3, pp. 2369–2387. Pergamon.

Runcorn S. K., Collinson D. W., O'Reilly W., Stephenson A., Battey M. H., Manson A. J., and Readman P. W. (1971) Magnetic properties of Apollo 12 lunar samples. *Proc. Roy. Soc. London,* Ser. A, **325**, 157–174.

Stephenson A. (1971a) Single domain grain distributions I. A method for the determination of single domain grain distributions. *Phys. Earth Planet. Interiors* **4**, 353–360.

Stephenson A. (1971b) Single domain grain distributions II. The distribution of single domain iron grains in Apollo 11 lunar dust. *Phys. Earth Planet. Interiors* **4**, 361–369.

Strangway D. W., Pearce G. W., Gose W. A., and Timme R. W. (1971) Remanent magnetization of lunar samples. *Earth Planet. Sci. Lett.* **13**, 43–52.

Wasserburg G. J., Turner G., Tera F., Podosek F. A., Papanastassiou D. A., and Huneke J. C. (1972) Comparison of Rb–Sr, K–Ar, and U–Th–Pb ages; lunar chronology and evolution (abstract), In *Lunar Science—III* (editor C. Watkins), pp. 788–790, Lunar Science Institute Contr. No. 88.

Weeks R. A., Kolopus J. L., and Kline D. (1972) Magnetic phases in lunar material and their electron magnetic resonance spectra (abstract). In *Lunar Science—III* (editor C. Watkins), pp. 791–793, Lunar Science Institute Contr. No. 88.

Wlotzka F., Jagoutz E., Spettel B., Baddenhausen H., Balacescu A., and Wänke H. (1972) On lunar metallic particles and their contribution to the trace element content of the Apollo 14 and 15 soils (abstract). In *Lunar Science—III* (editor C. Watkins), pp. 806–808, Lunar Science Institute Contr. No. 88.

Proceedings of the Third Lunar Science Conference
(Supplement 3, *Geochimica et Cosmochimica Acta*)
Vol. 3, pp. 2363–2386
The M.I.T. Press, 1972

On the remanent magnetism of lunar samples with special reference to 10048,55 and 14053,48

J. R. DUNN and M. FULLER

Department of Earth and Planetary Sciences,
University of Pittsburgh, 506 Langley Hall,
Pittsburgh, Pennsylvania 15213

Abstract—The magnetic behavior of 10048,55 and 14053,48 is remarkably dissimilar. Remanence is carried by multidomain iron in 14053,48 and is consequently predominantly very soft. TRM is a strongly nonlinear function of the applied field. The sample is not magnetically viscous. In contrast, 10048,55 is viscous and contains predominantly superparamagnetic and single domain iron. TRM is acquired linearly with field and is extremely stable against AF demagnetization. In contrast,

Attempts have been made to simulate the NRM of lunar samples by processes which may have given rise to the original NRM of the rocks. The AF and thermal demagnetization characteristics of the various remanent magnetizations have been determined and are used as criteria of the success of the simulation. The best simulation achieved so far with either rock is a two-component magnetization, composed of a weak field thermoremanent magnetization (TRM) and an isothermal remanent magnetization (IRM) acquired in a field of tens of oersted. There is little doubt that the TRM like stable component of NRM was acquired on the moon. The fit between the soft IRM like component in the simulation of NRM and the NRM demagnetization curves is improved by thermal cycling of the simulated NRM, in a manner similar to that experienced by the rocks on the lunar surface. Hence, although dangers of magnetic contamination must not be discounted, it seems that at least part of the soft NRM was also acquired on the moon.

The acquisition of remanence by the soil sample 14259,79 as a result of thermal cycling in the presence of weak fields has been demonstrated.

INTRODUCTION

THIS PAPER DESCRIBES our continuing studies of the natural remanent magnetization (NRM) of lunar samples. The remanence of 14053,48 and 10048,55 are discussed in some detail. In addition, preliminary results for 14301,65 and 15555,132 are reported.

With the demonstration of the AF and thermal demagnetization characteristics of numerous lunar samples (Doell *et al.*, 1970; Grommé and Doell, 1971; Gose *et al.*, 1972a; Hargraves and Dorety, 1971, 1972; Helsley, 1970, 1971, 1972; Nagata *et al.*, 1970, 1971, 1972; Nagata and Carleton, 1970; Pearce *et al.*, 1971, 1972; Runcorn *et al.*, 1970, 1971; Strangway *et al.*, 1970, 1971), it is now appropriate to compare these characteristics with those of remanent magnetization generated in the laboratory by processes which may be relevant.

The NRM of lunar samples may consist of several components with a variety of origins. Igneous rocks will acquire a thermoremanent magnetization if they cool in the presence of a magnetic field. Similarly, if a breccia is heated above its Curie point during its formation and subsequently cools in a magnetic field, it too will acquire thermoremanent magnetization. If either type of rock is later subjected to shock or is reheated in the presence of a field, additional remanence may be acquired. Since all rocks available for study have come from the lunar surface, they have all been subjected to the lunar diurnal temperature cycle. They will also have experienced varying degrees of irradiation. In very weak fields, each of these processes of heating, shocking, thermal cycling, and irradiating can act as a magnetizing process if the sample is essentially demagnetized, or as a demagnetizing process if the sample already carries substantial remanence. If the processes take place in strong fields, they will generate remanence. Finally, the samples are subjected to magnetic contamination of a somewhat unassessed nature during the return to earth and later on earth during preparation. With our simulations of the NRM of lunar rocks, we hope to distinguish mechanisms of magnetization that can account for the NRM, using the AF and thermal demagnetization characteristics as the criteria of simulation of the original NRM. As part of this effort samples 14053,48 and 10048,55 have already been subjected to treatments such as thermal cycling and shocking in addition to more standard magnetization processes such as thermoremanent magnetization. We have also undertaken a study of the remanent magnetization of a soil sample. In particular we investigated the acquisition of remanence during thermal cycling comparable to that experienced on the lunar surface.

SAMPLE 14053,48

Sample description and history

Sample 14053 was collected by the side of a large boulder at Station C2 which is on the flank of Cone Crater. It is thought to be a clast weathered from the boulder (Swann *et al.*, 1971). The rock is a fine-grain basalt containing olivine, two pyroxenes, and about 60% felspar. It is very inhomogeneous with the concentration of olivine and pyroxene varying on the scale of a few centimeters. The iron occurs in part as a breakdown product of fayalite (El Goresy and Ramdohr, 1972). After initial cooling the rock was reheated to about 1000°C and cooled rapidly (Clayton *et al.*, 1972; Bence and Papike, 1972; Finger *et al.*, 1972; Ghose *et al.*, 1972). The age of the rock by the Ar^{40}/Ar^{39} method is 3.9 b.y. (Schaeffer *et al.*, 1972), which suggests that the recorded age is the Imbrian event. Burnett *et al.* (1972) report an exposure age of 24 m.y., which is thought to be the age of Cone Crater. Thus, 14053 cooled initially as a lava. It was then excavated and ejected by the Imbrian event at 3.9 b.y. In this process it was heated to about 1000°C, so that any earlier remanent magnetization would have been erased at this time. The rock then formed a part of the Fra Mauro blanket until it was excavated again by the Cone Crater impact. It was not heated sufficiently at that time to reset the Ar^{40}/Ar^{39} age, so that the maximum temperature is not likely to have approached the Curie point of iron.

Magnetic properties (*Nagata et al., 1972*)

Sample 14053,48 contains an unusually large amount of metallic iron for a lunar igneous rock (1.02 wt.%) and the saturation magnetization is correspondingly higher than that of other samples. The ratio of the remanent coercivity to the coercive force (H_{RC}/H_C) is four, indicating the presence of hard and soft components of magnetization. Since neither the coercive force nor the saturation remanent magnetization (IRM_S) changes substantially on cooling to 5°K, the prominent soft component of magnetization at room temperature is likely to be multidomain and not superparamagnetic. The lack of magnetic viscosity exhibited by 14053,48 is consistent with this interpretation and makes the sample particularly convenient for analysis.

Natural remanent magnetization (NRM)

Both Nagata *et al.* (1972) and Runcorn *et al.* (1972) have reported the remarkably strong NRM of this sample ($\approx 10^{-3}$ Gauss cm^3 g^{-1}). The NRM is variable in intensity. For example, on cutting 14053,48,2 into two pieces, we found that although the direction of NRM remained effectively constant the magnitudes of NRM per unit mass of 2a and 2b were 8×10^{-4} and 4×10^{-3} respectively. The sum of these individual moments is approximately equal to the initial moment of the whole sample. It appears that part of the variability is due to inhomogeneity of the rock, since the bulk magnetic properties differ somewhat. However, not all the variation in NRM can be accounted for in this way, because neither the artificially induced IRM nor the TRM of 2a and 2b differ by the factor of 5 noted in the NRM. Hence, the NRM is itself intrinsically variable on the scale of a few mm.

AF demagnetization of NRM has revealed a predominantly soft magnetization with a more stable component which survives to fields of about 100 Oe. Thermal demagnetization (Fig. 1) demonstrated that much of the NRM is blocked below

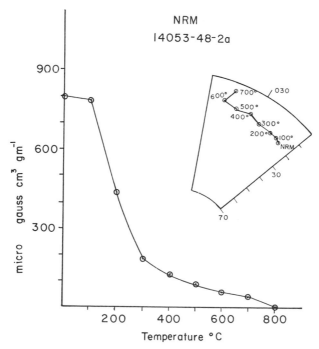

Fig. 1. Thermal demagnetization of the NRM of 14053,48,2a. Inset shows change of direction of NRM during demagnetization.

400°C. The thermal demagnetization curve exhibits a curious plateau between room temperature and 100°C indicating that little remanence is blocked in this temperature range. The direction of NRM does not change by more than about 20° during either AF or thermal demagnetization.

Thermoremanent magnetization (TRM)

A preliminary description of the TRM of 14053 has been given by Dunn and Fuller (1972). The TRM was interpreted to be carried by multidomain iron because of its extreme softness and its AF demagnetization characteristics (Lowrie and Fuller, 1971). New results have extended the range of fields for which the TRM has been determined (Fig. 2). The nonlinearity in high fields is reminiscent of the approach to saturation observed in the TRM of large multidomain samples of magnetite and other ferrites in much higher fields (Syono *et al.*, 1962). The low field departure from linearity has not been reported previously, although some indications that such phenomena might exist have been observed (Steele, private communication). This pronounced nonlinearity in the very low field range could be of some importance in lunar paleomagnetism. The TRM acquired in a field of 200 γ is only approximately a factor of 2 smaller than the TRM acquired in a field of 1000 γ, so that estimates of ancient field intensity made on the assumption of linearity could be critically misleading.

The stability against AF demagnetization of the artificially induced TRM was found to be low, although a small moment survived fields of 100 Oe (Dunn and Fuller, 1972). The stability of TRM increased with increasing strength of the inducing field. In comparing the stability of NRM and TRM, we found that the NRM was appreciably softer than TRM. Thermal demagnetization of TRM revealed a dependence of blocking temperature upon the inducing field (Fig. 3). Thus, low field TRM acquired in fields of 10^3 γ is blocked predominantly at high temperature, but a 1 Oe TRM was blocked more evenly across the temperature range. It is evident from

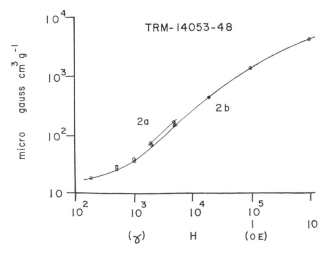

Fig. 2. Field dependence of thermoremanent magnetization (TRM).

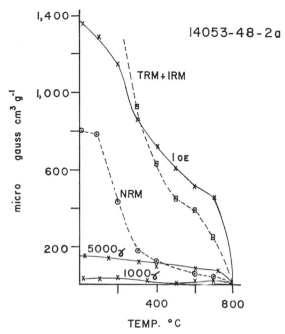

Fig. 3. Thermal demagnetization of (a) NRM, (b) TRM acquired in fields of 1 Oe, 5000γ and 1000γ, (c) composite remanence– 2000γ TRM with 50 Oe IRM superposed.

Fig. 3 that the thermal demagnetization of NRM is significantly different from that of TRM; if the NRM were essentially thermal a higher proportion of remanence should be blocked at higher temperatures. During thermal demagnetization of the 1000 γ TRM, the moment decreased to a minimum at 500°C and at higher temperature recovered magnitude.

Partial thermoremanent magnetization (pTRM)

Partial thermoremanent magnetization (pTRM) is acquired when a magnetic material is field cooled through part of the temperature interval between its Curie point and the observation temperature. Figure 4 gives AF demagnetization curves for several pTRM. The 0.5 Oe pTRM were field cooled from 400°C and 250°C to room temperature. They are too small to account for the NRM but have somewhat similar demagnetization curves to NRM. The 10 Oe pTRM which was field cooled from 400°C is of approximately the appropriate magnitude but is too stable, having AF demagnetization characteristics similar to a 0.2 Oe TRM. It therefore appears that like TRM, pTRM does not simulate the NRM adequately in this rock.

Isothermal remanent magnetization (IRM)

The AF stability of IRM of the same magnitude as the NRM has been investigated. The necessary inducing field is approximately 50 Oe and the IRM is much softer than is NRM.

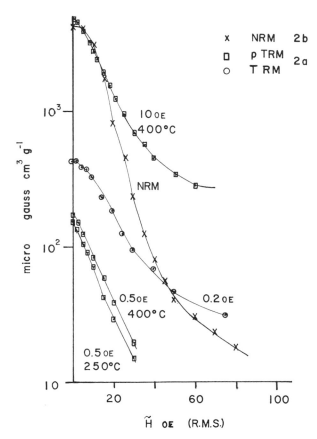

Fig. 4. AF demagnetization of (a) NRM (14053,28,2b), (b) TRM and various pTRM
(14053,48,2a).

TRM with superposed IRM

The occurrence of NRM of a hardness intermediate between IRM and TRM
suggests that NRM may be a TRM with superposed IRM. We have therefore investigated the AF demagnetization characteristics of such two component magnetizations.
Figure 5 illustrates the comparison between the AF demagnetization of NRM and of
composite remanence consisting of a 5000 γ TRM and IRM acquired for tens of
oersted. In the low field demagnetization which is dominated by the IRM the curves are
very similar. However, in higher fields the NRM is softer than either of the two combinations of TRM and IRM which were tried. This is probably an indication that the
TRM contribution is too large. We therefore carried out a thermal demagnetization
of 50 Oe IRM superposed on a 2000 γ TRM. The sample used was 2a, which has a
smaller NRM than 2b, so the comparison is not good in the low temperature blocking
range. However, by normalizing at 200°C, we can conveniently compare the thermal
demagnetization of the NRM and the composite magnetization. As is seen in Fig. 6,

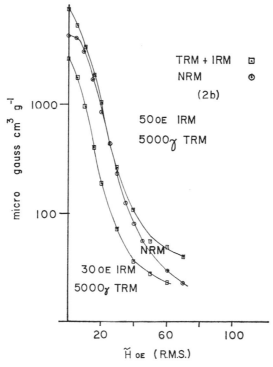

Fig. 5. AF demagnetization of (a) NRM and (b) composite TRM and IRM (14053,48,2b).

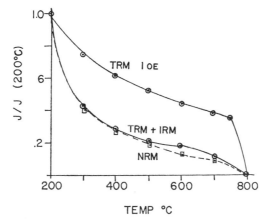

Fig. 6. Normalized thermal demagnetization curves of (a) NRM (14053,48,2a), (b) composite remanence (14053,48,2b).

the normalized curves are very similar. The absolute values are plotted in Fig. 3. It is evident that in the range from 500° to 700°C thermal demagnetization is less effective for the composite remanence than for the NRM. Nevertheless, the two demagnetizations are remarkably similar between 200° and 500°C. By comparison, a 1 Oe TRM is much less like the NRM.

Shock remanent magnetization (SRM)

A sample of 14053 was exposed to known mechanical impulses in the presence of a magnetic field and the shock or dynamic remanence determined using equipment built by Nagata and described in Nagata *et al.* (1971). The remanence which is acquired under these circumstances is significantly different from that acquired under static uniaxial loading in the presence of a field (Shapiro and Ivanov, 1967; Nagata 1971). Since it was impracticable to expose the samples to severe shock, we have studied the acquisition of SRM in high fields of the order of tens of oersted under very weak impulses of momentum 10^4 g cm sec^{-1}. The SRM is slightly harder than IRM but it is a poor simulation of the NRM being much too soft (Fig. 7).

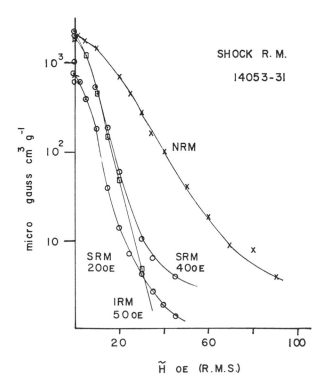

Fig. 7. AF demagnetization of (a) shock remanent magnetization (SRM), (b) IRM, and (c) NRM.

Thermal cycling

Samples have been thermally cycled from 100°C to −196°C as a convenient simulation of the lunar surface diurnal cycle. Two magnetic effects have been studied: first the acquisition of pTRM when the sample is cooled in a field, and second the demagnetization brought about by cycling samples carrying remanence in null field conditions.

In a field of 0.5 Oe, 14053,47,2a acquired a moment of order 10^{-4} Gauss cm^3 g^{-1} on being cooled from 100°C to −196°C. This moment decreased to 10^{-5} on warming to room temperature in null field. It is therefore quite clear that thermal cycling cannot generate a moment comparable to NRM in the weak fields encountered on the moon.

Typical results of demagnetization brought about by thermal cycling are shown in Fig. 8. The effect of cycling can be seen by comparing the curves of AF demagnetization of the 30 and 50 Oe IRM that are shown with and without thermal cycling. The hatched area represents the difference due to the cycling. The effect upon IRM is substantial, but the curves remain softer than NRM. On the other hand, by cycling a composite 5000 γ TRM and a 50 Oe IRM a good fit between the demagnetization curves of the NRM and the artificial remanence can be achieved in the low field range. The effect of cycling to 100°C after the low temperature cycle is small. Thermal

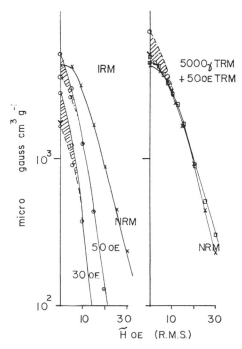

Fig. 8. The effect of thermal cycling upon remanence and its AF demagnetization characteristics (a) IRM and (b) composite remanence. The hatched area represents the difference between the AF demagnetization curves brought about by cycling.

demagnetization has already demonstrated that thermal demagnetization up to 100°C is small for low field TRM but is greater for high field TRM or IRM.

Interpretation of NRM

The major aspects of NRM that must be accounted for by models of its origin are (1) its large and variable intensity and (2) its AF and thermal demagnetization characteristics. The experimental results demonstrate that (1) IRM is too soft to account for the demagnetization characteristics of NRM, while (2) TRM and pTRM are too hard to be consistent with the demagnetization behavior, (3) TRM with superposed IRM gives similar AF and thermal demagnetization curves to NRM, (4) shock remanent magnetization (SRM) induced by light shocking in high fields is harder than IRM but softer than NRM, (5) thermal cycling in fields as large as 0.5 Oe fails to generate remanence comparable to NRM, (6) in null-field thermal cycling brings about demagnetization of artificially induced remanence and improves the fit between the demagnetization curves of NRM and soft IRM-like remanence acquired in fields of tens of oersted.

The NRM clearly includes a substantial soft component of magnetization, which is like IRM in its AF and thermal demagnetization characteristics. Since both AF and thermal characteristics of the artificial soft remanence are made more like those of the NRM by thermal cycling such as the rock experiences on the lunar surface, a conservative interpretation of the magnetic data is that this soft component was indeed acquired on the moon. Nevertheless, it is unwise to rule out the possibility that some of the soft component is due to contamination in the space vehicles or on earth.

In turning to the question of the existence of a hard TRM-like component of magnetization, it must be recognized that 14053,48 is fundamentally soft because it is multidomain so that separation of hard components is difficult. However, a small moment does survive demagnetization to about 100 Oe. Thermal demagnetization of NRM demonstrates that any weak field TRM like component is very small because so little NRM is blocked at high temperature (Fig. 3). It is then possible that a moment of thermal origin acquired in a field of $10^2 \gamma$ or $10^3 \gamma$ exists, but if so, it is small and appears to have been acquired in a very similar direction to the soft component of NRM.

If it is assumed that the similarity in direction between hard and soft components of NRM is coincidental, a possible two-component model of NRM is a hard TRM-like component with a soft moment acquired later as contamination on earth. However, the field required to generate the soft moment is approximately 50 Oe, which is rather high for a stray field. Moreover, it would not then be clear why the NRM appears to have been exposed to cycling, such as it would experience upon the lunar surface. An alternative two-component model, which takes into account the thermal cycling result, assumes that the IRM-like soft component is due to shock on the moon in weak fields. However, the similarity of direction of the hard and soft components makes this model unsatisfying. We are at present investigating the possibility of generating NRM by a single process such as severe shock, which might give rise to distributed microscopic coercivities and blocking temperatures as are seen in the NRM.

SAMPLE 10048,55

Sample description

Sample 10048 is a relatively well-indurated breccia in which metallic iron occurs as fine particles in association with troilite. The fragments that comprise the breccia show evidence of substantial shocking, e.g., displacement of translational twins lamellac and partial recrystallization of ilmenite, deformed and partially devitrified glass sphere, and mosaic structure in glass fragments (Haggerty *et al.*, 1970). Following Warner (1972) sample 10048 would be classified as a surface breccia on the basis that the rock matrix has a detrital or accumulative origin.

Magnetic properties

Sample 10048,55 has a saturation magnetization of 1.8 Gauss cm^3 g^{-1} at room temperature. The ratio of remanent coercivity to coercive force is 10 ($H_{RC}/H_C = 10$) which indicates the presence of a large soft component of magnetization in addition to the hard moment. The coercive force increases by a factor of 4 from 50 to 190 Oe, when the sample is cooled to 5°K. At the low temperature, the saturation remanence (IRM$_s$) also increases (from 1.3 × 10^{-2} to 54 × 10^{-2} Gauss cm^3 g^{-1}). Hence, the large soft contribution to magnetization at room temperature is clearly due to a superparamagnetic fraction which exhibits stable single domain behavior at low temperature. There is also an extremely hard magnetic phase whose influence will be seen in the following experiments. However, the superparamagnetic fraction is prominent at room temperature and makes the sample particularly viscous in its pristine state. The sample is degraded by heating, even in a vacuum of the order of 10^{-6} Torr. The effect of heating is indicated by the decrease in saturation isothermal remanent magnetization after the heat treatment. We found that the heating destroyed carriers throughout the whole range of coercivities, but that low microscopic coercivity particles were preferentially destroyed. After heating, the magnetic viscosity was much reduced, indicating that the fine particles were no longer present in abundance.

Natural remanent magnetization (NRM)

Two AF demagnetization curves of NRM are available for this sample, one by Larochelle and Schwarz (1970) and one by Nagata *et al.* (1972). Unfortunately, the extreme viscosity of the sample makes AF demagnetization difficult. However, it is clear that the NRM is reduced from 5 × 10^{-4} Gauss cm^3 g^{-1} to about one-half of this value by 100 Oe demagnetization (Fig. 9). In very low fields the NRM increases slightly indicating that a moment opposed in direction to the main NRM is being preferentially demagnetized. The small increase is truncated by the onset of the demagnetization of the major part of NRM. It is not evident whether the curve in the higher fields represents the demagnetization of a single component or of a two-component magnetization. Moreover, since no stable high field direction is established, some caution is advisable in interpreting the curve in this range. No thermal demagnetization of the NRM is available at present.

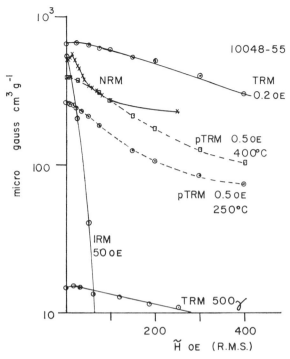

Fig. 9. AF demagnetization of 10048,55, (a) NRM, (b) TRM, (c) pTRM, and (d) IRM.

Thermoremanent magnetization (TRM)

The acquisition of TRM by 10048,55 has been described in part previously (Dunn and Fuller, 1972). Figure 10 indicates multiple determinations of TRM. The numbers refer to the heating during which the TRM was acquired. Using three as a base value, it is easy to see that the departure from linearity increases with successive heatings. Hence, it is due to the progressive destruction of carriers of remanence. To the accuracy of our measurements the magnetic carriers intrinsically acquire TRM linearly in the field range examined. AF demagnetization of TRM (Fig. 9) revealed an extremely hard remanence. Thermal demagnetization of TRM showed that the magnetization was blocked rather continuously across the range of temperatures between room temperature and the Curie point but with a maximum in the high temperature range.

Partial thermoremanent magnetization (pTRM)

The AF demagnetization of pTRM like that of TRM reveals a stable magnetization acquired in cooling in a 0.5 Oe field from 400° or from 250°C. The curve approaches the NRM curve, but for the combinations of temperature and fields which we looked at the curves are too hard in the low field demagnetization range and too soft in the higher fields (Fig. 9).

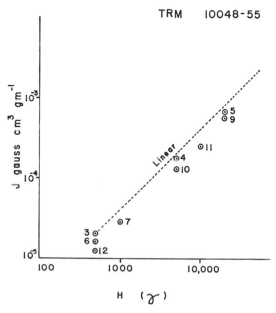

Fig. 10. Field dependence of TRM. The numbers record the order of the heatings during which TRM was acquired.

Isothermal remanent magnetization (IRM)

The stability of IRM of the appropriate magnitude to account for NRM is much too soft to give the AF demagnetization characteristics (Fig. 9).

Thermoremanent magnetization with superposed isothermal remanent magnetization (TRM and IRM)

Figure 11 shows the AF demagnetization of a 2000 γ TRM with a superposed 50 Oe IRM. In comparison with the NRM curve the two-component magnetization is too soft in the low field but there is some similarity in the overall form of the two curves. Thus, both contain hard and soft components of magnetization. However, the two-component magnetization reveals a more obvious separation of the individual components than does the NRM. In the low field range, the demagnetization of the composite TRM and IRM is identical with a 50 Oe IRM and in the high field range we see the demagnetization of essentially TRM.

Shock remanent magnetization (SRM)

SRM was investigated in this sample using the technique described in connection with the experiments on 14053,31. Like that sample, 10048,55 readily acquires SRM by light shocking (10^4 g cm sec^{-1}) in strong fields (50 Oe). The AF demagnetization characteristics are similar to those of IRM but some slight hardening is evident (Fig. 12).

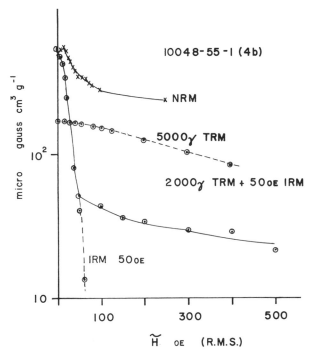

Fig. 11. AF demagnetization of (a) NRM, (b) TRM, (c) IRM, and (d) composite TRM and IRM.

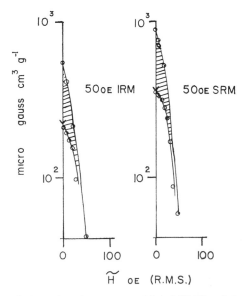

Fig. 12. The effect of thermal cycling upon IRM and SRM and their demagnetization characteristics. The hatched area represents the difference between the demagnetization curves brought about by cycling.

Thermal cycling

Demagnetized samples of 10048,55 were exposed to thermal cycling in fields of 0.5 Oe and acquired a remanence of 2.8×10^{-3} at $-196°C$ that reduced to 1.3×10^{-3} at room temperature. IRM and SRM were demagnetized by the thermal cycling. In the demagnetization experiments the low temperature cycle has little effect, but cycling to 100°C gave rise to considerable demagnetization. Thermal demagnetization experiments presented earlier (Dunn and Fuller, 1972) also demonstrate the effect of cycling to this temperature.

Interpretation of NRM

The characterization of the NRM of 10048 is less complete than for 14053; we only have an AF demagnetization curve for this rock. The curve is distinctive, although some caution in interpreting the high field results is advisable. The slight increases in remanence in very low fields is due to the removal of a small soft component which is in a direction opposed to the main NRM. Such a component may be terrestrial contamination or acquired on the moon. From about 20 to 100 Oe NRM of intermediate hardness is clearly demagnetized. It is still not clear whether a separate stable component is being demagnetized after 100 Oe demagnetization. It appears from the experimental analysis of the two-component magnetization that if the NRM were of the two component type, a clearer separation of hard and soft remanence should be seen.

Before any conclusion as to the origin of the NRM of this sample can be reached, it is essential to carry out a thermal demagnetization. At present we note that 10048,55 has a large NRM and that like 14053 this remanence is harder than IRM but softer than TRM or pTRM. It may be due to a superposition of a soft IRM-like component upon a stable thermal moment or due to a SRM with well distributed microscopic coercivities.

SAMPLE 14301: PRELIMINARY RESULTS

Sample 14301 is a fragmental rock of which the matrix comprises 80%. Clasts larger than 1 mm make up the remainder of the rock. These clasts are predominantly light colored. In the classification of Warner (1972) this rock is a surface breccia (group 2 of the low facies). In that of Jackson and Wilshire (1972) it is F2 which corresponds to shock compressed breccias of Chao *et al.* (1971). It was collected at Station G1—the north side of the north crater of Triplet (Swann *et al.*, 1972). Iron is present as kamacite with up to 6% Ni and 0.2% Co, but there is also some metallic iron with less than 0.2% Ni. The Curie point is 750°C. The rock is magnetically viscous. H_{RC}/H_C is approximately 20. On heating metallic iron is readily eliminated even under a vacuum of 10^{-6} Torr. Hence, this rock is more similar to 10048 than 14053 in this respect.

The intensity of NRM in the two samples studied is quite dissimilar (Fig. 13). Sample 14301,60 is almost an order of magnitude greater than that of 14301,65,1. Yet, the 17 Oe IRM of the two samples is very similar. During AF demagnetization

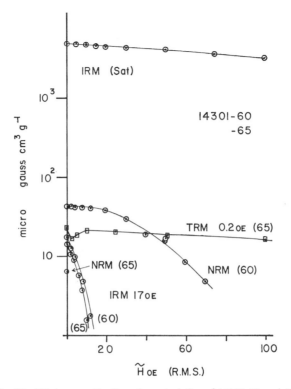

Fig. 13. AF demagnetization characteristics of 14301,60 and 65.

up to 70 Oe, little change in the direction of NRM was recorded and the magnitude of remanence decreased to about 10% of the initial value (Nagata, private communication). Sample 14301,65,1 was thermally demagnetized. Particular care was taken throughout the whole process to shield the sample, so that it would not acquire viscous magnetization. Nevertheless, large directional changes were observed during the thermal demagnetization. Thus, orthogonal components of magnetization were observed to switch sign between successive heatings and to recover at higher temperatures. Finally, after heating to above 800°C a substantial moment was observed on cooling to room temperature in a null field. A 0.2 Oe TRM was given to the sample and AF demagnetization carried out. During this experiment a remarkable decay and recovery of magnetization was observed between 5 and 10 Oe (Fig. 13). Although the rock is viscous, it is hard to explain these observations on the basis of viscosity alone since the sample was never exposed to more than 10^2 γ and moreover, viscosity in the earth's field could not reproduce the switching phenomena.

The magnetic behavior of this rock is hard to interpret and until more explicable thermal and AF demagnetization curves can be obtained, it should not be used to infer lunar magnetic fields. It is interesting that NRM is variable whereas the similarity of IRM suggest that as in 14053 the variability cannot be entirely attributed to sample inhomogeneity.

SAMPLE 15555,132: PRELIMINARY RESULTS

Sample 15555 was collected at Station 9a about 12 m north of the rim of Hadley Rille. Its age is 3.28 ± 0.6 b.y. which is similar to that of other Apollo 15 basalts (e.g., Schaeffer *et al.*, 1972). Its exposure age is 90 m.y. (Burnett *et al.*, 1972). It seems possible that this rock has been derived from local bedrock. The rock is a porphyritic vuggy basalt with two pyroxenes, olivine, and 50% plagioclase. The metallic iron occurs as rare discrete grains and blebs in troilite. Its magnetic properties have not yet been comprehensively studied, but the saturation magnetization is 0.127 Gauss cm^3 g^{-1} and the coercive force 7 Oe. In these respects it is very similar to other 15 basalts (Schwerer, private communication), which strongly suggests that these basalts are from a single unit and hence are of local origin. The NRM carried by the sample is weak, 3.3×10^{-6} Gauss cm^3 g^{-1} and soft; it is demagnetized to one-third of its initial value in a field of approximately 10 Oe (Fig. 14). The NRM is comparably stable to a 50 Oe IRM. A 0.2 Oe TRM and saturation IRM were extremely stable being almost unaffected by AF demagnetization to 100 Oe. Thus, from this very preliminary study, it appears that 15555,132 is similar to other lunar crystalline rocks in that it has a small NRM, which is more stable than the equivalent IRM but much less stable than TRM. The determination of the presence or absence of a stable TRM-like component must await analyses of larger samples than are presently available to us.

REMANENT MAGNETISM OF LUNAR SOIL

The possibility that the lunar soil carries NRM does not appear to have been considered. However, the soil contains plentiful iron whose transition from stable single domain to the superparamagnetic state takes place at temperatures within the

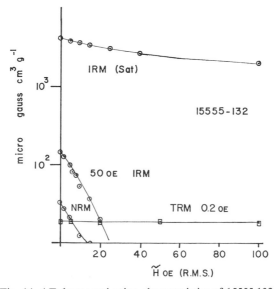

Fig. 14. AF demagnetization characteristics of 15555,132.

lunar surface diurnal temperature cycle. Hence, this iron could acquire pTRM during the cooling half of the cycle, if an ambient field were present. Moreover, on burial, the soil comes to an equilibrium temperature below the maximum of the diurnal cycle, until under deep burial the thermal gradient of the moon determines its temperature. It therefore appears that relatively shallowly buried soil, down to depths of some tens of meters, could carry NRM.

To test the suggestion of acquisition of remanence by soil 1 g of fines (14259,79) was subjected to cycling in a field of 0.5 Oe, in the manner described in the treatment of the rocks. The soil acquired a moment of 1.5×10^{-3} Gauss cm^3 g^{-1} at liquid nitrogen temperature. This moment decreased to 0.34×10^{-3} at room temperature. A second experiment in a 5000 γ field gave an initial moment at liquid nitrogen temperature of 2.5×10^{-4} which reduced at room temperature to 1.2×10^{-4}. During the warming to room temperature the magnetization was seen to change erratically. However, at room temperature a component in the direction of the inducing field was observed. The AF demagnetization curves of these two remanent magnetizations revealed a relatively hard remanence.

It is too early to assess the importance of this artificial remanence which can be acquired by the soil. Nevertheless, the similarity of its AF demagnetization to that of certain lunar NRM is intriguing. The values of remanence suggest that on shallow burial the soil could acquire a NRM of approximately 10^{-4} Gauss cm^3 g^{-1} in fields of 10^3 γ. If the soil does in fact carry NRM, it could have important implications for the study of soil dynamics and play a role in surface anomalies and the NRM of very lightly metamorphosed fragmental rocks.

Review of NRM of Lunar Material

A wide variety of interpretations of the NRM of lunar samples and of the fields in which it was acquired have been given. Runcorn et al. (1971) and Strangway et al. (1971) strongly favor an ancient lunar field generated by a core dynamo that has subsequently shut off. They interpret the stable component of NRM as a thermal moment acquired in the planetary wide ancient field of the moon. They cite the results of the Apollo 15 subsatellite magnetometer experiment (Coleman et al., 1972) as evidence that large regions of the moon were homogeneously magnetized in the ancient lunar field.

Among the other numerous explanations of the NRM which have been considered are: (1) it is a TRM acquired in (a) a solar wind field enhanced by a unipolar generator effect, or (b) in an intrinsically higher ancient solar wind field (Nagata et al., 1971), or (c) in the terrestrial field at a time when the moon was much closer to the earth (Helsley, 1971); (2) it is an IRM which was acquired in an extremely high solar flare field and was subsequently hardened by radiation demagnetization (Butler and Cox, 1971); (3) it is a weak field remanence enhanced by the introduction of defects due to thermal cycling or irradiation (Helsley, 1971); (4) it is a shock remanence acquired in the ancient field or in a field generated by the impact itself (Helsley, 1971; Nagata et al., 1972; Hide, 1972). Unfortunately it is impossible to distinguish between these various models at the present time because the critical information is not available. It should also be remembered that a number of mechanisms are liable to be involved in the explanation of the NRM of lunar samples.

With the exception of the strongly magnetized sample 14053, the lunar crystalline rocks exhibit NRM in the range of 6×10^{-4} to 2×10^{-6} Gauss cm^3 g^{-1}. Frequently, the NRM consists of a soft moment, which can be demagnetized in fields of a few tens of oersted and a stable moment of order of magnitude 10^{-6}. Samples from the same rock frequently exhibit substantially different intensities of NRM. This is readily explained, if the soft component is contamination acquired in varying amounts by each subsample. However, it is less clear why stable moments should also differ, if they are thermal in origin and acquired in a planetary wide field (for example, 12063, Strangway et al., 1971). It is not known, in general, how much of the variability is due to sample inhomogeneity and how much to an intrinsic variability of NRM. AF demagnetization curves of the NRM of crystalline rocks frequently exhibit a distinctive minimum at low fields with subsequent recovery in higher fields, followed by little further change in direction or magnitude. Curiously, we observed a similar curve in the AF demagnetization of the TRM given to certain weakly metamorphosed fragmental rocks (e.g., 14301, Fig. 13, and 14047). In some crystalline rocks the direction of NRM changes substantially during demagnetization, so that the magnetization appears to be multicomponent. In others such as 12063,98 (Strangway et al., 1971), AF demagnetization brings about a decrease of approximately two orders of magnitude in intensity but the direction remains unchanged. Hence, in this sample the large soft component and the stable moment were in the same direction. Sample 14053 is clearly different from other igneous rocks, having much more iron than crystalline rocks usually have. Moreover, the iron is multidomain unlike that in other rocks. It seems therefore likely that the reheating of this sample, which may be a shock heating, has fundamentally changed the rock.

The NRM of the fragmental rocks varies from 10^{-3} to 10^{-6} Gauss cm^3 g^{-1}. The fragmental rocks exhibit the same curious variability in intensity between different samples from the same rock as do the igneous rocks. Although much of the variability in the fragmental rocks is clearly due to sample inhomogeneity, in 14301 the NRM is also intrinsically variable. The NRM and the demagnetization characteristics of fragmental rocks correlates in a rough manner with the classification schemes of Warner (1972) and of Jackson and Wilshire (1972). Thus the Apollo 11 and 12 samples and the low-grade fragmental rocks from Apollo 14 tend to have a larger NRM than do the medium and high grade 14 rocks. As pointed out by Gose et al. (1972a, b) a decrease of magnetic viscosity accompanies increasing metamorphic grade. Thus 10021, 10048, 14047, and 14301 are all of low metamorphic grade and are high viscosity (Type II rocks in Nagata's classification). However 14303 and 14311 are of higher metamorphic grade and are Type I low viscosity rocks according to Nagata et al. (1972). These trends suggest that in the low metamorphic grade rocks there is magnetic viscosity due to very fine iron similar to that found in the soil, but in the higher grade rocks this iron is less abundant. Such an interpretation is consistent with the few reported ratios of saturation IRM at room temperature and 5°K.

This ratio serves as an indication of the amount of superparamagnetic iron in the rock which becomes stable single domain at low temperature. For 10048, 14047, and the fines 14259,79, the values of room temperature and low temperature IRM_S are in the ratio $1:4$, $1:5$, $1:7$, respectively. In complete contrast, the ratio for the higher grade rock 14303 is $1:1.6$. Thus, in the higher grade rock there is much less difference between the room temperature and low temperature IRM_S which indicates that the rock contains little superparamagnetic iron. The value of the saturation IRM suggests that the total amount of iron present in the various samples is similar. If the soil does in fact turn out to carry NRM, it is likely that this NRM may account for some part of the NRM of the most lightly metamorphosed fragmental rocks and that the destruction of iron may give demagnetization accompanying increasing thermal metamorphism.

The interpretation of the NRM of lunar samples remains unsatisfying. It is possible that both igneous and fragmental rocks carry a stable TRM acquired at their time of formation and subsequently acquired a soft component of magnetization as a result of secondary processes either on the moon (Runcorn et al., 1972) or as contamination on earth or in the space vehicles (Strangway et al., 1971; Gose et al., 1972a, b; Pearce et al., 1972). Nevertheless there remain a number of aspects of the NRM which need to be clarified before this interpretation is unequivocably accepted.

The interpretation assumes that the hard and soft components of NRM are independent of each other and should therefore be oriented at random to each other. In a number of rocks the hard and soft components of NRM are in a similar direction. It remains to be seen whether this is coincidental or whether some special explanation is necessary.

Lunar samples are subjected to a number of processes not generally encountered by terrestrial samples, e.g., shocking, thermal cycling, and irradiation. The magnetic effects of these processes are sufficiently poorly known that the possibility that they may play some role in the NRM cannot yet be discounted.

It is quite evident that thermal cycling in the present lunar fields is unable to generate a remanent moment comparable with the NRM of either 14053 or 10048. However, there is another aspect of the thermal cycling that may be of importance in certain rocks. When the sample is initially removed from the lunar surface, it may be at a higher temperature than that at which the remanence is eventually measured. The sample will acquire a pTRM, whenever it finally cools to room temperature. It is not immediately clear where this cooling takes place. However, at least in the case of 14053, it appears to have been in a very weak field environment since so little NRM is blocked below $100°C$.

The effect of radiation has already been invoked in one model of lunar NRM (Butler and Cox, 1971). Their suggestion is that an IRM was acquired by all samples sometime after the last lavas erupted due to an intense solar flare. This magnetization was then hardened by the preferential demagnetization of soft components during subsequent irradiation in the very weak fields such as are seen now on the lunar surface. This model depends more critically on the solar flare field than on the irradiation effect, since the thermal cycling could also harden the remanence. There are a number of descriptions of the effect of radiation upon magnetic characteristics such as coercive force and susceptibility of ferrites and ferromagnets (Brodskaya, 1967; Butler and Cox, 1971; Henry and Salkovitz, 1958; Smith, 1967). In Brodskaya's work and that of Butler and Cox evidence of demagnetization due to irradiation is presented. Although we need to determine the efficiency of the irradiation process as a magnetizing mechanism before we can be satisfied that it has not played a role in the NRM of the lunar samples, it seems extremely unlikely that it can generate large NRM such as is observed in 14053 or 10048.

The effects of shock upon magnetization are well known for relatively weak shock range (Shapiro and Ivanov, 1967, 1970; and Shapiro and Alova, 1970). The stability of shock remanence (SRM) induced by shocking comparable to, though larger than, that used in our experiments was demonstrated to be harder than IRM but considerably softer than TRM. It is blocked at low temperature. By carrying out shock experiments at elevated temperature Shapiro and Alova (1970) showed that both hardness and blocking temperatures were increased compared with the room temperature equivalent. By analogy with observations of static pressure remanent magnetization (PRM) (Carmichael, 1968), it appears that greater AF stability would ensue from stronger shocking in weaker fields. In turning to the effect of more extreme shock in the range of hundreds of kilobars little information is available. There are studies of changes in structure sensitive properties of magnetic materials as a result of shock (e.g., Crowe and Rose, 1971; Rose et al., 1969) and a change of remanence was observed in invar by Clator and Rose (1967), but no systematic study of the effects of shock upon remanence appears to be available. A study of natural material which has been shocked was carried out by Hargraves and Perkins (1967) and detectable effects were noted. Unlike irradiation and thermal cycling, shock is not readily eliminated as a source of NRM in cetrain lunar samples and further work in the shock magnetization acquired in the range of kilobars is warranted.

In conclusion, we note that the NRM of lunar samples may eventually prove to be of considerable importance in lunar studies but that until a satisfying understanding of the nature of this remanence is achieved it is unwise to base fundamental ideas of the moon such as the presence or absence of a conducting core upon NRM. In addition to further study of processes such as shock, we need stability tests of NRM which can give some idea of the time of acquisition of remanence. Tests which are used in terrestrial paleomagnetism are not immediately applicable. However, Gose et al., 1972a, have applied a modified version of Graham's conglomerate test to demonstrate that the NRM was acquired before the samples reached their present orientations on the lunar surface. A second test is to compare the NRM of very young glass spatter with that of the rock below. However, unfortunately we have not yet obtained suitable material to apply this test. Finally, the suggestion that the soil may carry NRM should be tested.

Acknowledgments—We are grateful to Professor Nagata of Tokyo University, Dr. R. Fisher, and Dr. F. Schwerer of the U.S. Steel Research Laboratory, Monroeville, Pa., for their comprehensive analyses of the magnetic properties of the samples whose remanence we studied and for many helpful discussions. We are also indebted to Dr. W. Cassidy and Mr. A. A. DeGasparis for their suggestions relating to the magnetization of the fragmental rocks and the soil. We thank Mrs. Anne Day for help with measurements of remanence and for drafting.

REFERENCES

Bence A. E. and Papike J. J. (1972) Crystallization histories of pyroxenes from lunar basalts (abstract). In *Lunar Science—III* (editor C. Watkins), pp. 59–61, Lunar Science Institute Contr. No. 88.

Brodskaya S. Y. (1968) Gamma-radiation induced changes in the magnetic properties of ferrimagnetic minerals. *Izv. Akad. Nauk SSSR* (*Physics of Solid Earth*), No. 10, 57–63, English trans., 614–617.

Burnett D. S., Huneke J. C., Podosek F. A., Russ G. P. III, Turner G., and Wasserburg G. J. (1972). The irradiation history of lunar samples (abstract). In *Lunar Science—III* (editor C. Watkins), pp. 105–107, Lunar Science Institute Contr. No. 88.

Butler R. F. and Cox A. V. (1971) A mechanism for producing magnetic remanence in meteorites and lunar samples by cosmic ray exposure. *Science* **172**, 939.

Carmichael R. (1968) Remanent and transitory effects of elastic deformation of magnetite crystals. *Phil. Mag.* **17**, No. 149, 911–927.

Chao E. C. T., Boreman J. A., and Desborough G. A. (1971) The petrology of unshocked shocked Apollo 11 and Apollo 12 microbreccia. *Proc. Second Lunar Sci. Conf., Geochim. Cosmochim. Acta* Suppl. 2, Vol. 1, pp. 797–816. MIT Press.

Clator I. G. and Rose M. F. (1967) Shock-induced second-order phase change in invar. *Brit. J. Appl. Phys.* **18**, 853.

Clayton R. N., Hurd J. M., and Mayeda T. K. (1972) Oxygen isotope abundances in Apollo 14 and 15 rocks and minerals (abstract). In *Lunar Science—III* (editor C. Watkins), pp. 141–143, Lunar Science Institute Contr. No. 88.

Coleman P. J. Jr., Russell C. T., Sharp L. R., and Schubert G. (1972) Magnetic fields near the moon (abstract). In *Lunar Science—III* (editor C. Watkins), pp. 148–150, Lunar Science Institute Contr. No. 88.

Crowe C. R. and Rose M. F. (1971) Isochronal recovery of the shape of the normal magnetization curve of nickel shocked at 400 k bar. *J. Appl. Phys.* **42**, No. 11, 4319.

Doell R. R., Gromme C. S., Thorpe A. N., and Senftle F. E. (1970) Magnetic studies of Apollo 11 lunar samples. *Proc. Apollo 11 Lunar Sci. Conf., Geochim. Cosmochim. Acta* Suppl. 2, Vol. 3 pp. 2097–2120. Pergamon.

Dunn J. R. and Fuller M. (1972) Thermoremanent magnetization (TRM) of lunar samples. Proceedings of Conference on Lunar Geophysics, *The Moon*, in press.

El Goresy A. and Ramdohr P. (1972) Fra Mauro crystalline rocks: Petrology, geochemistry, and subsolidus reduction of the opaque minerals (abstract). In *Lunar Science—III* (editor C. Watkins), pp. 224–226, Lunar Science Institute Contr. No. 88.

Finger L. W., Hafner S. S., Schurmann K., Virgo D., and Warburton D. (1972) Distinct cooling histories and reheating of Apollo 14 rocks (abstract). In *Lunar Science—III* (editor C. Watkins), pp. 259–260, Lunar Science Institute Contr. No. 88.

Ghose S., Ng G., and Walter L. S. (1972) Clinopyroxenes from Apollo 12 and 14: Exsolution, cation order and domain structure (abstract). In *Lunar Science—III* (editor C. Watkins), pp. 300–302, Lunar Science Institute Contr. No. 88.

Gose W. A., Pearce G. W., Strangway D. W., and Larson E. E. (1972a) On the applicability of lunar breccias for paleomagnetic interpretations. Proceedings of Conference on Lunar Geophysics, *The Moon*, in press.

Gose W. A., Pearce G. W., Strangway D. W., and Larson E. E. (1972b) On the magnetic properties of lunar breccias (abstract). In *Lunar Science—III* (editor C. Watkins), pp. 332–334, Lunar Science Institute Contr. No. 88.

Gromme C. S. and Doell R. R. (1971) Magnetic properties of Apollo 12 lunar samples 12052 and 12065. *Proc. Second Lunar Sci. Conf., Geochim. Cosmochim. Acta* Suppl. 2, Vol. 3, pp. 2491–2499. MIT Press.

Haggerty S. E., Boyd F. R., Bell P. M., Finger L. W., and Bryan W. B. (1970) Opaque minerals and olivine in lavas and breccias from Mare Tranquilitatis. *Proc. Apollo 11 Lunar Sci. Conf., Geochim. Cosmochim. Acta* Suppl. 1, Vol. 1, pp. 513–538. Pergamon.

Hargraves R. B. and Perkins W. E. (1970) Investigations of the effect of shock on remanent magnetism. *J. Geophys. Res.* **74**, 2576–2589.

Hargraves R. B. and Dorety N. (1971) Magnetic properties of some lunar crystalline rocks returned by Apollo 11 and Apollo 12. *Proc. Second Lunar Sci. Conf., Geochim. Cosmochim. Acta* Suppl. 2, Vol. 3, pp. 2477–2483. MIT Press.

Hargraves R. B. and Dorety N. (1972) Magnetic property measurements on several Apollo 14 rock samples (abstract). In *Lunar Science—III* (editor C. Watkins), pp. 357–359, Lunar Science Institute Contr. No. 88.

Helsley C. E. (1970) Magnetic properties of lunar 10022, 10069, 10084, and 10085 samples. *Proc. Apollo 11 Lunar Sci. Conf., Geochim. Cosmochim. Acta* Suppl. 1, Vol. 3, pp. 2213–2219. Pergamon.

Helsley C. E. (1971) Evidence for ancient lunar magnetic field. *Proc. Second Lunar Sci. Conf., Geochim. Cosmochim. Acta* Suppl. 2, Vol. 3, pp. 2485–2490. MIT Press.

Helsley C. E. (1972) Remanent magnetism of lunar samples. Proceedings of Conference on Lunar Geophysics. *The Moon*, in press.

Henry W. E. and Salkovitz E. I. (19) Reduction of saturation magnetization of Fe_2O_3 and Fe_3O_4 by pile irradiation. *J. Appl. Phys.*, ser. S, **30**, No. 4, 286.

Hide R. (1972) Comments on the moon's magnetism. Proceedings of Conference on Lunar Geophysics, *The Moon*, in press.

Jackson E. D. and Wilshire H. G. (1972) Classification of the samples returned from the Apollo 14 landing site (abstract). In *Lunar Science—III* (editor C. Watkins), pp. 418 420, Lunar Science Institute Contr. No. 88.

Larochelle A. and Schwarz E. J. (1970) Magnetic properties of lunar samples 10048,22. *Proc. Apollo 11 Lunar Sci. Conf., Geochim. Cosmochim. Acta* Suppl. 2, Vol. 3, pp. 2305–2308. Pergamon.

Lowrie W. and Fuller M. (1971) On the alternating field demagnetization characteristics of multi-domain thermoremanent magnetization in magnetite. *J. Geophys. Res.* **76**, 6339.

Nagata T. (1971) Introductory notes on shock remanent magnetization and shock demagnetization of igneous rocks. *Pure and Applied Geophysics* **89**, 159–177.

Nagata T. and Carleton B. (1970) Natural remanent magnetization and viscous magnetization of Apollo 11 lunar materials. *J. Geomag. Geoelec.* **22**, No. 4, 491.

Nagata T., Ishikawa Y., Kinoshita H., Kono M., Syono Y., and Fisher R. M. (1970) Magnetic properties and natural remanent magnetization of lunar materials. *Proc. Apollo 11 Lunar Sci. Conf., Geochim. Cosmochim. Acta* Suppl. 1, Vol. 3, pp. 2325–2340. Pergamon.

Nagata T., Fisher R. M., Schwerer F. C., Fuller M., and Dunn J. R. (1971) Magnetic properties and remanent magnetization of Apollo 12 lunar materials and Apollo 11 lunar microbreccia. *Proc. Apollo 11 Lunar Sci. Conf., Geochim. Cosmochim. Acta* Suppl. 1, Vol. 3, pp. 2461–2476. Pergamon.

Nagata T., Fisher R. M., Schwerer F. C., Fuller M. D., and Dunn J. R. (1972) Magnetism of Apollo 14 lunar materials (abstract). In *Lunar Science—III* (editor C. Watkins), pp. 573–575, Lunar Science Institute Contr. No. 88.

Pearce G. W., Strangway D. W., and Larson E. E. (1971) Magnetism of two Apollo 12 igneous rocks. *Proc. Second Lunar Sci. Conf., Geochim. Cosmochim. Acta* Suppl. 2, Vol. 3, pp. 2451–2460. MIT Press.

Pearce G. W., Strangway D. W., and Gose W. A. (1972) Remanent magnetization of lunar samples (abstract). In *Lunar Science—III* (editor C. Watkins), pp. 599–601, Lunar Science Institute Contr. No. 88.

Rose M. F., Villere M. P., and Berger T. L. (1969) Effect of shock waves on the residual magnetic properties of armco iron. *Phil. Mag.* **19**, 39.

Runcorn S. K., Collinson D. W., O'Reilly W., Battey M. H., Stephenson A., Jones J. M., Manson A. J., and Readman P. W. (1970) Magnetic properties of Apollo 11 lunar samples. *Proc. Apollo 11 Lunar Sci. Conf., Geochim. Cosmochim. Acta* Suppl. 1, Vol. 3, pp. 2369–2387. Pergamon.

Runcorn S. K., Collinson D. W., O'Reilly W., Stephenson A., Battey M. H., Manson A. J., and Readman P. W. (1971) Magnetic properties of Apollo 12 lunar samples. *Proc. Roy. Soc. London*, ser. A, **325**, 157–174.

Runcorn S. K., Collinson D. W., O'Reilly W., and Stephenson A. (1972) Magnetic properties of lunar rocks and fines (abstract). In *Lunar Science—III* (editor C. Watkins), pp. 669–671, Lunar Science Institute Contr. No. 88.

Schaeffer L., Husain L., Sutter J., and Funkhouser J. (1972) The ages of lunar material from Fra Mauro and the Hadley Rille-Apennine Front area (abstract). In *Lunar Science—III* (editor C. Watkins), pp. 675–677, Lunar Science Institute Contr. No. 88.

Shapiro S. and Ivanov N. A. (1967) Dynamic remanence and the effect of shocks on the remanence of strongly magnetic rock. *Dokl. Akad. Nauk SSSR* **173**, 1065–1068.

Shapiro S. and Ivanov N. A. (1969) Characteristics of dynamic impact generated magnetization in samples of natural ferromagnetic substances. *Izv. Akad. Nauk SSSR* (*Physics of Solid Earth*), No. 5, 298.

Shapiro S. and Alova N. I. (1970) Characteristics of dynamic remanent magnetization of ferromagnetic rocks at temperatures from 0 to 600°C. *Izv. Akad. Nauk SSSR* (*Physics of Solid Earth*), No. 11, 743.

Smith R. W. (1967) Thesis, University of Pittsburgh.

Strangway D. W., Larson E. E., and Pearce G. W. (1970) Magnetic studies of lunar samples—breccia and fines. *Proc. Apollo 11 Lunar Sci. Conf., Geochim. Cosmochim. Acta* Suppl. 1, Vol. 3, pp. 2435–2451. Pergamon.

Strangway D. W., Pearce G. W., Gose W. A., and Timme R. W. (1971) Remanent magnetization of lunar samples. *Earth Planet. Sci. Lett.* **13**, 43–52.

Syono Y., Akimoto S., and Nagata T. (1962) Remanent magnetization of ferromagnetic single crystal. *J. Geomag. Geoelec.* **14**, 113.

Swann G. A., Bailey N. G., Batson R. M., Eggleton R. E., Hait M. H., Holt H. E., Larson K. B., McEwen M. C., Mitchell E. D., Schaber G. G., Schafer J. B., Shepard A. B., Sutton R. L., Trask N. J., Ulrich G. E., Wilshire H. G., and Wolfe E. W. (1971) Preliminary geologic investigations of the Apollo 14 landing site. In *Apollo 14 Preliminary Science Report*, pp. 39–87, National Aeronautics and Space Administration SP-272.

Warner J. L. (1972) Apollo 14 breccias: Metamorphic origin and classification (abstract). In *Lunar Science—III* (editor C. Watkins), pp. 782–787. Lunar Science Institute Contr. No. 88.

Proceedings of the Third Lunar Science Conference
(Supplement 3, *Geochimica et Cosmochimica Acta*)
Vol. 3, pp. 2387–2395
The M.I.T. Press, 1972

Magnetic properties of Apollo 14 breccias and their correlation with metamorphism

W. A. GOSE

Lunar Science Institute, Houston, Texas 77058

G. W. PEARCE

Lunar Science Institute, Houston, Texas 77058 and
Department of Physics, University of Toronto

D. W. STRANGWAY

Geophysics Branch, NASA, Manned Spacecraft Center, Houston, Texas 77058

and

E. E. LARSON

Department of Geological Sciences, University of Colorado, Boulder, Colorado

Abstract—The magnetic properties of Apollo 14 breccias can be explained in terms of the grain size distribution of the interstitial iron which is directly related to the metamorphic grade of the sample. In samples 14049 and 14313 iron grains < 500 Å in diameter are dominant as evidenced by a Richter-type magnetic aftereffect and hysteresis measurements. Both samples are of lowest metamorphic grade. The medium metamorphic-grade sample 14321 and the high-grade sample 14312 both show a logarithmic time-dependence of the magnetization indicative of a wide range of relaxation times and thus grain sizes, but sample 14321 contains a stable remanent magnetization whereas sample 14312 does not. This suggests that small multidomain particles ($< 1\ \mu$) are most abundant in sample 14321 while sample 14312 is magnetically controlled by grains $> 1\ \mu$. The higher the metamorphic grade, the larger the grain size of the iron controlling the magnetic properties. Experiments on synthetic lunar glasses suggest that solid-state reduction during impact is an effective mechanism for producing the interstitial iron. It is therefore concluded that the remanent magnetization in the lunar breccias is of thermal origin and contemporaneous with the time of breccia formation.

INTRODUCTION

WITH THE RETURN OF the Apollo 14 rocks it has become necessary to examine closely the magnetization of breccias since most samples were of this type. As we have pointed out earlier (Gose *et al.*, 1972) the breccias offer the best hope of extending the record of the lunar magnetic field over a considerable portion of lunar history since igneous rocks seem to be restricted to the time between about 3 and 4 b.y. Before breccias can be used it must be demonstrated that they possess a stable magnetization which was acquired during the formation of the breccia.

Generally, the breccias show a pronounced viscous remanent magnetization (VRM), but it is often possible to isolate a component which is as stable as the magnetization which is typical of the igneous rocks (Doell *et al.*, 1970; Nagata and Carleton, 1970; Gose *et al.*, 1972). We have had the opportunity to examine four breccias from the Apollo 14 mission and it seems possible now to draw some general conclusions as to the origin and nature of the magnetic properties and remanent magnetization of the breccias.

Viscous remanent magnetization of breccias 14049, 14312, 14313, and 14321

The time dependence of the magnetization is of two basic types. The first type is illustrated in Fig. 1. Samples 14049,28 and 14313,25 were exposed to a magnetic field of 2.5 Oe for 8 min. At time $t = 0$ the field was removed and the decay of the magnetization in a field-free space was measured as a function of time. This behavior can be explained in terms of a Richter-type aftereffect (e.g. Becker and Döring, 1939). In the Richter model, the relaxation times of the magnetic particles are assumed to be restricted to a finite range between, say, τ_1 (lower limit) and τ_2 (upper limit). This model has been successfully applied to several lunar breccias by Nagata and Carleton (1970) and by Gose et al. (1972). From these curves it is not possible to obtain the lower limit of the relaxation times τ_1. But the shape of the curve allows one to estimate the upper limit, τ_2, with considerable accuracy. For both samples shown in Fig. 1 this value lies between 100 and 1000 min.

According to Néel's theory (Néel, 1949) the relaxation times are related to grain size. Using values for iron, the equation can be written in the form

$$\log \tau = 2.68 \times 10^{18} H_{\mathrm{RC}} \frac{v}{T} - 0.5 \log \frac{v}{T} - \log H_{\mathrm{RC}} - 16.8$$

where H_{RC} is the remanence coercive force, v the volume of the grain, and T the absolute temperature. The value of the coercive force is a rather critical quantity in this equation. For spherical single-domain particles H_{RC} is determined by the crystalline anisotropy, and for iron the value is about 170 Oe, assuming that there is no significant

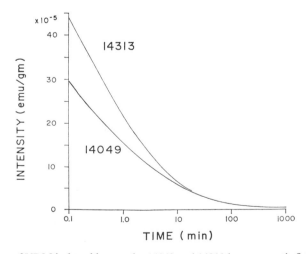

Fig. 1. Decay of VRM induced in samples 14049 and 14313 by a magnetic field of 2.5 Oe. Both samples were measured up to 1000 min.

strain. By adding a small amount of cobalt or nickel the value for H_{RC} decreases. If the particle deviates from sphericity, shape anisotropy becomes the dominant factor. For a prolate ellipsoid with an axial ratio larger than 1.1 the shape is the controlling factor for the coercive force. Figure 2 illustrates the dependence of the value of H_{RC} on the grain size for a given relaxation time and temperature. The vertical axis shows the diameter for a sphere with a volume equivalent to a prolate ellipsoid whose axial ratio is indicated along the curve. In this example the equivalent grain size can vary from about 120 Å to 270 Å depending on the possible range of H_{RC}. The precise value of this range is not of great importance considering that the lunar samples contain iron grains ranging in size up to 100 μ. It is, however, of theoretical interest in that the upper limit of the relaxation time coincides with the transition from superparamagnetic to single domain grains.

The experimentally determined value of the remanence coercive force is 460 Oe for sample 14313, which corresponds to a grain size of 185 Å for the transition from superparamagnetic to single-domain behavior. However, the value of 460 Oe has to be considered an average value for an assemblage of grains many of which will have considerably larger coercivities. This is evidenced by steady field demagnetization experiments such as shown in Fig. 3. The sample was given an isothermal remanent magnetization (IRM) in an 18 kilogauss field. Then a dc field was applied in the opposite direction. The figure shows the decrease of the IRM as a function of the applied dc field; a wide coercive force spectrum is clearly present, ranging from 0 to at least 2500 Oe. Thus it seems more realistic to use a larger value than 460 Oe for the remanence coercive force of the very small particles. Néel (1949) reports a value of about 1000 Oe for fine grains of iron, and the same value was observed by Bertaut (1949). Stephenson (1971) obtains a value of 1700 Oe for iron grains in the range 60–120 Å in lunar dust. If we adopt a value of 1000 Oe for the lunar breccias, a grain size of about 150 Å is obtained for the transition from superparamagnetic to single-domain behavior at room temperature.

In view of our later discussion the important observation is that breccia 14049

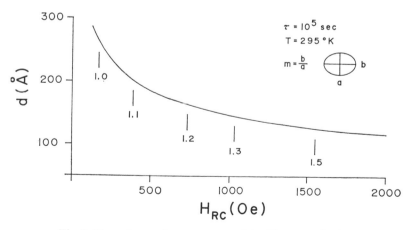

Fig. 2. Dependence of remanent coercivity H_{RC} on grain size.

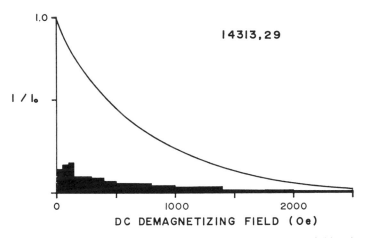

Fig. 3. D.C. demagnetization of IRM acquired in an 18-kilogauss field. Histogram indicates the portion of the intensity removed in a given interval.

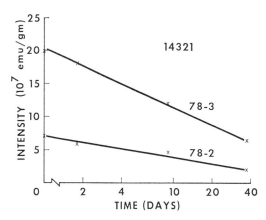

Fig. 4. Decay of natural remanent magnetization of samples 14321,78,2 and 14321,78,3 during storage in a field-free space.

and 14313 show a Richter-type aftereffect as the dominant VRM characteristic, and that this behavior is caused by superparamagnetic grains in the 100 Å range.

Two neighboring chips from breccia 14321 (78,1 and 78,3) show a different kind of VRM (Fig. 4). These samples were stored in a field-free space and measured over a 38-day period. The intensities decay linearly with the logarithm of time. One of these breccias was subjected to the same test as were the previous samples, and the log t relationship was found to be dominant down to 6 sec. Such behavior is observed when the relaxation times of the magnetic grains cover a time range which is very much larger than the duration of the experiment and is typical for multidomain grains (Néel, 1955). This sample has a stable magnetic remanence, however (Pearce *et al.*, this volume), which suggests that much of the magnetization is carried in grains less than a micron in size.

Sample 14312,07 exhibits a very weak response and a time dependence of magnetization which can be described by two log t functions after a 2.5 Oe field has been applied for 8 min (Fig. 5). But this sample differs from the other breccias in that no stable remanence could be isolated (Gose *et al.*, 1972; Pearce *et al.*, this volume). Upon AF demagnetization the directions of the natural remanence changed very erratically. Thus it is inferred that the magnetization of sample 14312 is controlled by rather large multidomain grains ($\gg 1 \mu$).

Hysteresis measurements

In an earlier paper (Gose *et al.*, 1972) we described the usefulness of room temperature magnetization curves in determining the amount of iron present and its grain size distribution. A sample which has a magnetization curve which is almost linear (ramp-shaped) to saturation at 6000–7000 Oe and which has a small saturation remanence ($J_r/J_s = 0.01$ or less where J_r is the saturation remanence and J_s is the saturation magnetization) contains only iron grains whose magnetization is controlled by the demagnetizing field ($4\pi/3 J_s$ for spherical grains). Equidimensional, multidomain particles have those properties. If the magnetization is measured on such a sample at higher temperatures (Fig. 6), the shape of the curve remains the same but saturation occurs at lower fields, since the demagnetizing field is proportional to the spontaneous magnetization of the grains. Superparamagnetic and stable single-domain particles, on the other hand, have quite different magnetization curves. Superparamagnetic grains have rounded curves which reach saturation in fields of less than 1000 Oe at room temperature if their radii are much above 20 Å. By definition they have no hysteresis, and the magnetization curves plotted against H/T can often be superimposed. An assemblage of stable single-domain grains, randomly oriented, shows rounded curves characterized by large hysteresis; for example, particles with uniaxial anisotropy have $J_r/J_s = 0.5$. The magnetization reaches approximate saturation in several thousand Oe for relatively equant grains (axial ratio $b/a = 1.5$ or less), but can approach 10,000 Oe for needlelike grains. (In the limit $H_c = 2\pi J_s = 10,800$ Oe.)

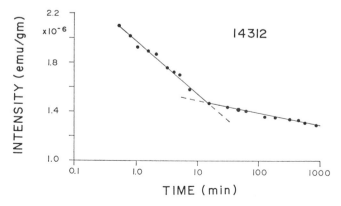

Fig. 5. Decay of VRM induced in sample 14312,7 by a magnetic field of 2.5 Oe.

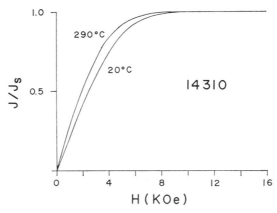

Fig. 6. Magnetization curves for igneous sample 14310 measured at 20°C and 290°C. These curves have been corrected for paramagnetic magnetization and then normalized to $J_s(T) = 1$.

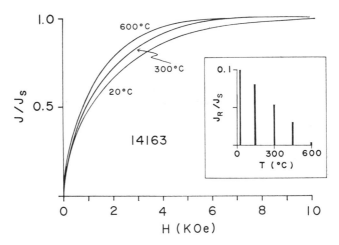

Fig. 7. Magnetization curves for soil 14163 measured at 20°C, 300°C, and 600°C. These curves have been corrected for paramagnetic magnetization and then normalized to $J_s(T) = 1$. (Inset) Histogram showing variation of ratio J_r/J_s with temperature for sample 14163.

Many igneous rocks and breccias which show log t-type after effect possess the simple ramp-shaped multidomain magnetization curve of Fig. 6, while the soils and those breccias which show a Richter-type aftereffect have complex curves, showing evidence of superparamagnetic and single-domain grains. For example, soil 14163 has a rounded curve (Fig. 7) with a J_r/J_s ratio of 0.1, suggesting that about 20 wt.% of the native iron grains in the sample are single-domain grains, assuming these to be uniaxial (that is, shape anisotropy predominates). The ratio decreases with increasing temperature, as the single-domain grains become superparamagnetic according to the theory of Néel (1949). The increase in the quantity of superparamagnetic grains at high temperatures also explains the lack of superposition between the curves when plotted as J versus H/T (Fig. 8).

Optical studies

The different iron grain size distributions inferred from the VRM and hysteresis experiments can be seen in the optical microscope as well, although the limit of resolution is only about 0.25 μ. The examination of polished thin sections shows that most of the interstitial iron in sample 14312 is larger than 1 μ in size, whereas in sample 14321 grains below 1 μ are most abundant. The dominant size range in samples 14313 and 14049 lies at the limit of optical resolution, i.e. <0.25 μ. Most interesting is the

Table 1.

Sample no.	Metamorphic grade*	Mean Fe grain size	VRM behavior	Shape of hysteresis loop	J_r/J_s	Metallic Fe wt. %‡	Initial susceptibility ($\times 10^{-4}$ emu/g Oe)
14313	1		Richter	Rounded	0.066	0.47	9.1
14049	1	< 500Å	Richter	Rounded	0.058	0.59	14.1
14047†	1		Richter				
14301†	2		Richter				
14063†	3		log t				
14321	4	< 1 μ	log t	Ramp	0.019	0.19	0.91
14311†	5		log t				
14303†	6		log t				
14312	7	> 1 μ	log t	Ramp	0.010	0.24	1.53
14310	Igneous			Ramp	0.019	0.10	0.57

* Warner, 1972.
† Nagata *et al.*, 1972.
‡ Determined from saturation magnetization curves.

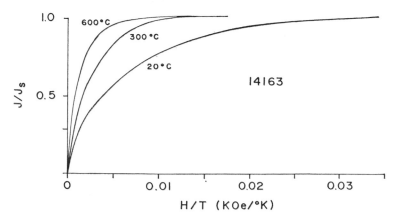

Fig. 8. Magnetization curve for soil with ratio H/T plotted on abscissa. The magnetization has been corrected for paramagnetic magnetization and then normalized to $J_s(T) = 1$.

fact that the latter two samples appear to contain less interstitial iron than samples 14321 and 14312, whereas the saturation magnetization clearly shows that the metallic iron concentration is considerably higher in 14049 and 14313 (Table 1). Since meteoritic iron makes up only a fraction of the metallic iron and is likely to be equally abundant in all the samples, it follows that most of the interstitial iron in 14049 and 14313 must be in the very fine grain size range ($\ll 0.25$ μ).

Discussion

Based on these experiments it is possible to classify the Apollo 14 breccias into three groups. The first group, which is represented by samples 14049 and 14313, is magnetically dominated by superparamagnetic and single-domain iron grains. This is evidenced by a Richter-type magnetic aftereffect and a rounded hysteresis curve. The second group shows a VRM which decays linearly with a logarithm of time, has a ramp-shaped hysteresis curve, and carries a stable remanence. This behavior is caused by small multidomain grains. Sample 14321 belongs to this group. Samples 14312 is an example of the third group, which is similar to the second group in its VRM and hysteresis characteristics but differs from group 1 as well as group 2 in not having a stable remanent magnetization. Iron grains above 1 μ in size are most abundant in this sample. The different magnetic behavior is related to the metamorphic grade of the breccias. Warner (1972) classified the Apollo 14 breccias according to their metamorphic grade based on the abundance of matrix glass, glass clasts, and matrix texture. The lowest metamorphic grade (class 1) has a detrital matrix with abundant glass whereas the highest grade (class 8) is a glass-free breccia with a totally recrystallized matrix.

Table 1 presents variation of magnetic properties with increasing metamorphic grade, and summarizes our data on the Apollo 14 breccias, and includes some results from Nagata et al. (1972). For comparison the data on the igneous sample 14310 are shown as well. The magnetic properties correlate well with the metamorphic classification. Magnetically, the fundamental change is the increase in grain size of the interstitial iron from the 100 Å range in the lowest metamorphic-grade samples to grains larger than 1 μ in the highest-grade samples. All the magnetic characteristics can be attributed to this variation.

The correlation between the magnetic properties and the degree of metamorphism has been tested experimentally by Pearce and Williams (1972) with synthetic lunar glass powders. Upon heating to about 900°C in a reducing environment not too unlike those expected to occur in a large ejecta blanket, native iron of predominantly single-domain size is precipitated from the silicate melt. Heating to about 1000°C produces essentially only multidomain grains. In addition, the experiment shows that solid-state reduction during impact melting is an effective mechanism for producing the interstitial iron observed in the lunar breccias.

Since the interstitial iron is the main carrier of the stable remanent magnetization, we can conclude that this remanence must be of thermal origin and that it is contemporaneous with the formation of the breccia. Such an origin of the magnetization implies that the direction of magnetization should be uniform within a sample. The three neighbouring chips of sample 14321 which we investigated yield similar directions (Pearce *et al.*, this volume). In addition, Hargraves and Dorety (1972) reported the same direction for another chip of the same sample which is separated from our chips by about 8 cm. It thus appears that the lunar breccias can indeed be used for reconstructing the history of the lunar magnetic field since they contain a stable magnetization (Gose *et al.*, 1972) which originated at the time of their formation.

Acknowledgments—Laboratory assistance in this work was ably provided by J. Love and E. Pahl of Lockheed Electronics Company. Financial support was provided to one of us (GWP) by the National Research Council of Canada and in part by the Lunar Science Institute. Financial support for WAG was provided by Contract No. NSR 09-051-001 between the National Aeronautics and Space Administration and the Universities Space Research Association. This paper constitutes the Lunar Science Institute contribution Number 93.

REFERENCES

Becker R. and Döring W. (1939) *Ferromagnetismus*. Springer Verlag, Berlin.

Bertaut F. (1949) Champ coercitif et dimension cristalline. *Comptes Rendue* **229**, 417–419.

Doell R. R., Grommé C. S., Thorpe A. N., and Senftle F. E. (1970) Magnetic studies of Apollo 11 lunar samples. *Proc. Apollo 11 Lunar Sci. Conf.*, Geochim. Cosmochim. *Acta* Suppl. 1, Vol. 3, pp. 2079–2102. Pergamon.

Gose W. A., Pearce G. W., Strangway D. W., and Larson E. E. (1972) On the applicability of lunar breccias for paleomagnetic interpretations. *The Moon*, in press.

Hargraves R. B. and Dorety N. (1972) Magnetic property measurements on several Apollo 14 rock samples (abstract). In *Lunar Science—III* (editor C. Watkins), pp. 357–359, Lunar Science Institute Contr. No. 88.

Nagata T. and Carleton B. J. (1970) Natural remanent magnetization and viscous magnetization of Apollo 11 lunar materials. *J. Geomag. Geoelectr.* **22**, 491–506.

Nagata T., Fisher R. M., Schwerer F. C., Fuller M. D., and Dunn J. R. (1972) Magnetism of Apollo 14 lunar materials (abstract). In *Lunar Science—III* (editor C. Watkins), pp. 573–575, Lunar Science Institute Contr. No. 88.

Néel L. (1949) Théorie du trainage magnétique des ferromagnétiques en grains fin avec application aux terres cuites. *Ann. Geophysique* **5**, 99–136.

Néel L. (1955) Some theoretical aspects of rock-magnetism. *Phil. Mag. Suppl.* **4**, 191–243.

Pearce G. W. and Williams R. J. (1972) Excess iron in lunar breccias and soils: A possible origin (abstract). EOS, *Trans. Amer. Geophys. Union* **53**, 360.

Stephenson A. (1971) Single domain grain distributions, II. The distribution of single domain iron grains in Apollo 11 lunar dust. *Phys. Earth Planet. Interiors* **4**, 361–369.

Warner J. L. (1972) Apollo 14 breccias: Metamorphic origin and classification (abstract). In *Lunar Science—III* (editor C. Watkins), pp. 782–784, Lunar Science Institute Contr. No. 88.

Proceedings of the Third Lunar Science Conference
(Supplement 3, *Geochimica et Cosmochimica Acta*)
Vol. 3, pp. 2397–2415
The M.I.T. Press, 1972

Evidence of lunar surface oxidation processes: Electron spin resonance spectra of lunar materials and simulated lunar materials

D. L. Griscom and C. L. Marquardt

Solid State Division, Naval Research Laboratory,
Washington, D.C. 20390

Abstract—Exposed lunar surface materials were examined by electron spin resonance (ESR) techniques at 9 and 35 GHz and temperatures from 77 to 573°K. The spectra of individual mm-sized glass and mineral fragments from 10085, 14230, 15302, and 15505 were compared with those of unsorted fine fines (12001, 14230, 15401, and 15302). Two principal resonance absorptions were discerned at 9 GHz in virtually every individual particle: Type-I, an asymmetric resonance centered near $g \approx 2.1$ with a line shape and temperature dependence effectively identical with those of the so-called "characteristic" resonance of lunar soils, and Type-II, an extremely broad resonance with hysteresis effects, having an absorption maximum in the range 200–900 G, which is observed only after cycling the sample to ≥ 5 kG. The "characteristic" resonance is found to display its greatest intensity in the finest soil fractions and its linewidth is noted to increase with increasing TiO_2 contents. The significance of these findings is inferred from an investigation of simulated lunar glass samples that were subjected to oxidizing and reducing treatments. Strong chemical reduction of a simulated lunar glass melt resulted in an intense Type-II resonance, indicating metallic iron precipitates as its probable source. Type-I resonances have been produced in powdered simulated lunar glasses by heating at 650°C in air pressures as low as ~ 0.5 Torr for periods of hours. It is concluded that this ESR signal results from oxidation of Fe^{2+} to Fe^{3+} followed or accompanied by crystallization of ferrite phases catalyzed by the presence of TiO_2. This observation suggests that exposed lunar rock surfaces and fines may have been oxidized while on the moon.

Introduction

Electron spin resonance (ESR) investigations of lunar material are undertaken with several objectives: (1) to determine the natures of any radiation-induced defects that may be paramagnetic, (2) to identify and determine the valence states of those transition-group ions whose resonances can be observed, (3) to make quantitative estimates of the abundances of these species, (4) to infer local atomic arrangements and chemical bonding on the basis of spectral structure, and (5) to apply these findings in forming a clearer picture of the lunar processes that shaped the returned samples.

Previous ESR studies (Geake *et al.*, 1970; Manatt *et al.*, 1970; Weeks *et al.*, 1970a, 1970b, 1971; Duchesne *et al.*, 1971; Hanneman and Miller, 1971; Kolopus *et al.*, 1971; Tsay *et al.*, 1971a, 1971b) have achieved some of these goals, while leaving certain important issues unresolved. In particular, all workers who have investigated lunar fines have reported an extremely intense, distinctively shaped ESR absorption with a width of ~ 800–1000 oersteds and a g value of ~ 2.1 ($g = h\nu/\beta H$, where h is Planck's constant, ν is the spectrometer frequency, β is the Bohr magneton, and H

is the value of the applied magnetic field at which resonance occurs). This signal has been called the "characteristic" resonance of the fines, since it has been reported for every sample of fines studied, with specific intensities ≥ 100 times those found in any rock fragments. Because the "characteristic" resonance absorption intensities were observed to be relatively independent of temperature and several orders of magnitude greater than those expected for any conceivable paramagnetic species, and because of the apparently high prevalence of metallic iron in the soils (see, for example Nagata et al., 1970), the "characteristic" resonance was ascribed to spherical particles of iron (Manatt et al., 1970; Tsay et al., 1971a, 1971b). However, various inconsistencies with this model have also been pointed out (Kolopus et al., 1971; Weeks et al., 1971).

One of the first objectives of the present investigation was to identify and characterize by ESR methods some of the transition-group elements known to be dissolved in lunar glasses, using the techniques and insights already developed in conjunction with studies of other glass systems (Griscom and Griscom, 1967). It became apparent, however, that the ESR spectra of individual lunar glass particles of all descriptions are dominated by just two or three resonances, one of which is essentially identical with the "characteristic" resonance of the fines. It was further found that a resonance resembling the "characteristic" resonance could be produced in synthetic glasses of lunar compositions by oxidizing them in air at temperatures $\sim 650°C$ (Griscom and Marquardt, 1972a). This represented the first time that the "characteristic" resonance had been successfully "modeled" in the laboratory and provided an additional strong clue that metallic iron may not be the source of this ubiquitous lunar ESR signal. In addition, an apparent correlation has now been noted between the line width of the "characteristic" resonance and the amount of TiO_2 in the sample. A positive correlation with such a highly refractory component in the local soil chemistry would clearly rule out the "metallic iron" hypothesis. In view of the observation that the "characteristic" resonance is more intense in the finer soil fractions and absent in the rocks, the evidence presently suggests the hypothesis that lunar soils have been oxidized subsequent to their formation. Studies of samples returned under high vacuum (Weeks et al., 1970a, 1970b), coupled with the high temperatures required to produce the resonance in simulated lunar materials, indicate that this oxidation could not have taken place on the earth. Thus, certain lunar-surface oxidation processes are inferred. The following text describes in detail the experimental techniques and the results leading up to this conclusion. Also provided are quantitative estimates of the amounts of Fe^{3+} in lunar soils and discussions of the probable chemical, physical, and selenological significance of these findings.

Experimental

On the basis of their morphologies and lusters, over a dozen mm-sized particles of various colors were selected from the Apollo 11 bulk sample 10085 as probably being mostly glass. A few additional particles, including some appearing to be crystalline, were selected from Apollo 14 core samples (14230) and from 0.25 g of coarse fines from 15302. A sample of coating glass from an Apollo 15 breccia (15505) was also separated and investigated. Most sorting and sample transfers were carried out in a Class-10,000 clean facility where the samples were sealed in predried quartz tubes evacuated to $\sim 10^{-1}$ Torr after flushing with dry nitrogen. Along with samples of unsorted fines, similarly packaged, these samples were examined at X-band (9.12–9.52 GHz) and at Ka-band (34.5–35.5 GHz) frequencies by means of a Varian E-9 spectrometer. The field was modulated at 100 kHz and the first derivatives of the absorption spectra were displayed on X-Y recording charts, accurate to better than 1% in the field variable. Measurements of g values were accomplished with reference to a pitch standard. Temperature variations between 93°K and 573°K were achieved at X-band by means of a Varian nitrogen flow insert. Ka-band spectra could be obtained at room temperature or ~ 100°K, using a Varian bridge and dewar system. Relative ESR intensities were determined by double numerical integration of the experimental derivative spectra (Ayscough, 1967; O'Reilly et al., 1971). Estimates of absolute spin density were made by comparison with calibrated standards (Hyde, 1961).

Mineral glasses were produced by briefly fusing terrestrial minerals (diopside, pidgeonite, augite, anorthite) with a standard hydrogen-oxygen welder's torch. These glasses could easily be made oxidized or reduced (as attested by their ESR spectra) by varying the oxygen content of the flame.

Two synthetic glass melts were delivered by Owens-Illinois. The first was a "simulated lunar glass," having a target composition (Table 1) close to the average for Apollo 12 red-brown-ropy and greenish-yellow streaky glasses (Wood et al., 1971); the second was a "control" glass to be described below. In the "simulated lunar glass," iron was entered as Fe^{2+} in synthetic fayalite (checked by x-ray diffraction) and melting was carried out at 1540°C for 6 hours under nominally reducing conditions (an atmosphere containing $\sim 3.5\%$ combustibles). For ease in identifying the major ESR species, all potentially paramagnetic elements present in the average lunar composition in amounts ≤ 1 wt% were deleted. The remaining elements that could assume magnetic states were iron (11.6% FeO) and titanium (2.4% TiO_2). The glass was quenched by allowing the 50 gm melt to cool in the Pt crucible in air; the billet was annealed in air at 620°C for one hour. In the "control" glass, all FeO was replaced by CaO + MgO, while a small amount of ferric iron was added as 0.1% Fe_2O_3 and melting was carried out under oxidizing conditions. Comparison of the ESR signals at $g = 4.3$ (cf., Castner et al., 1960) in the two glasses as delivered indicated nearly equal Fe^{3+} contents, implying that there was probably no more than 0.1% Fe_2O_3 in the "as delivered" simulated lunar glass. This conclusion was corroborated by Mössbauer measurements (Forester, 1971).

RESULTS

Lunar glasses

All of the 18 individual lunar particles investigated (most of them glassy, but a few definitely crystalline) displayed asymmetric ESR spectra centered near $g \approx 2.1$ with peak-to-peak first-derivative line widths of between 500 and 800 Oe at X-band frequencies and room temperature (see Fig. 1). Detailed line width-versus-temperature studies were carried out on some of these resonances, showing them to behave essentially identically to the "characteristic" resonance of fines from Apollos 14 and 15 (Fig. 2), Apollo 12 (Kolopus et al., 1971), and presumably Apollo 11. It will sometimes be convenient to refer to this "characteristic" lunar ESR response as the Type-I resonance (see Fig. 3a).

Most of the lunar glasses investigated also displayed what will be referred to as Type-II resonances. In Fig. 1, Type-II resonances are responsible for much of the broad "background" stretching from zero field to ~ 10 kOe. However, the "archetypical" Type-II signal as defined in this study is illustrated in Fig. 3b. The lunar sample referred to here is a single mm-sized dark brown chip, 10085,170,5a, which appeared under microscopic examination to be a cleanly fractured homogeneous glass with no inclusions or vesicles. The Type-II resonances of Fig. 3b are clearly distinguishable from the Type-I resonances of Fig. 3a. One noteworthy feature of the

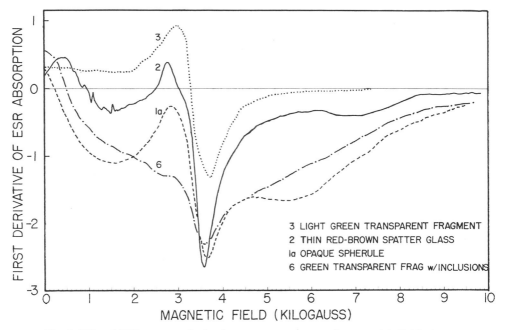

Fig. 1. X-band ESR spectra obtained at room temperature for several individual mm-sized fragments, selected for their glassy appearance from 10085,170. The first-derivative zero level was determined by subtraction of a "no-sample" baseline; otherwise, the scale of the y axis is arbitrary. Each spectrum may be regarded as a superposition of two components: Type I, the sharp features centered between ~2.5 and 4 kG, and Type II, the broad components characterized by negative first-derivative values from ~1 to 10 kG.

Type-II resonance is the presence of dramatic hysteresis effects: the large-amplitude feature centered near 900 Oe does not appear until the sample has been cycled to ≥ 5 kOe *in situ* (Griscom and Marquardt, 1972a, 1972b). The presence of one or more "shoulders" (relative derivative minima) between 6 and 9 kOe is also an integral feature of the Type-II resonance.

At Ka-band frequencies, the spectral distinction between Type-I and Type-II signals is less clear, since the major portions of resonance intensity for all samples investigated (including 10085,170,5a) occur in the vicinity of $g \approx 2$ ($H \approx 12.5$ kOe) when a spectrometer frequency of ~35 GHz is employed. On the basis of six mm-sized lunar glass fragments studied carefully, it appears that $2.20 \geq g_{\text{Type-II}}(35 \text{ GHz}) \geq g_{\text{Type-I}}(35 \text{ GHz}) \geq 2.00$. Further interpretation of the Ka-band spectra of lunar samples is facilitated by comparisons with the spectra of simulated lunar materials (see below).

Terrestrial minerals and mineral glasses

A search for terrestrial analogs of lunar ESR signals was carried out on a wide variety of terrestrial minerals and glasses prepared therefrom; the salient findings of this search are enumerated here.

(1) Resonances bearing some resemblance to Type-I lunar signals were observed in many terrestrial augites both before and after vitrification.

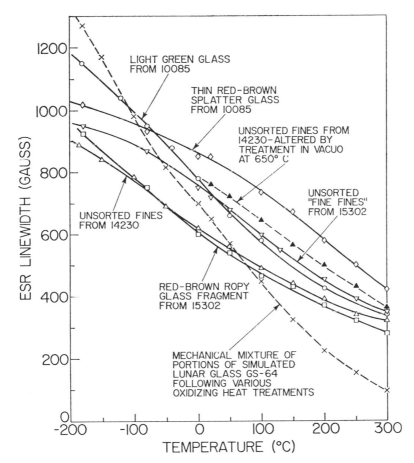

Fig. 2. Temperature dependence of X-band ESR line widths (measured between positive and negative first-derivative extrema) of Type-I resonances in a variety of individual mm-sized lunar particles, two lunar soils, and a mixture of oxidized simulated lunar glasses. Data pertaining to synthetic materials, or lunar materials altered by heat treatment, are distinguished by broken curves.

(2) An amber-colored crystal of diopside from Quebec, containing $\sim 7\%$ FeO but showing only the resonance of Mn^{2+} in its natural state, exhibited a strong signal with "Type-I" characteristics following prolonged treatment of 650°C in air. This resonance appeared with considerably less intensity when the heat treatment was carried out in vacuo.

(3) Glasses prepared from natural anorthite (Japan) displayed X-band ESR spectra having a sharp peak at $g \approx 4.3$ ($H \approx 1.5$ kOe), familiar as arising from Fe^{3+} (Castner et al., 1960). However, pyroxene glasses usually showed X-band spectra centered near $g \approx 2$ ($H \approx 3.4$ kOe) when similarly melted in an oxidizing flame. Various considerations suggest that the $g \approx 2$ signals are also due to Fe^{3+} (Griscom, 1971).

(4) Glasses prepared from high-iron pyroxenes such as Virginia pidgeonite exhibited resonances clearly identifiable as Type-II, when melted under highly reducing conditions.

(5) The resonance spectra of ilmenite did not resemble either type of lunar ESR signal at either X-band or Ka-band frequencies. This is in accord with the findings of Weeks et al. (1970b).

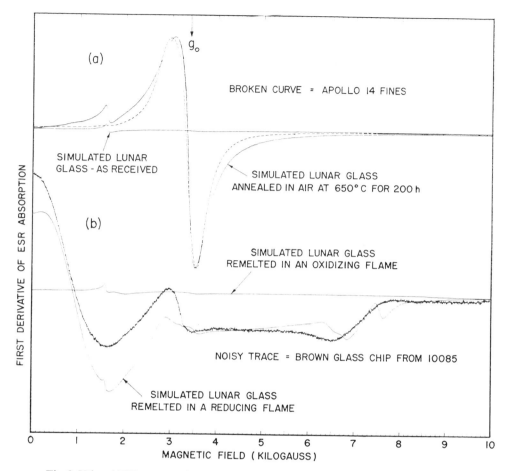

Fig. 3. X-band ESR spectra of a simulated lunar glass are compared with (a) the "characteristic" resonance of Apollo 14 fines and (b) the ESR spectrum of a single, mm-sized dark brown glass chip (10085,170,5a). The principal resonances in (a) are designated "Type I"; the resonances in (b) are denoted "Type II." All spectra were obtained at 9.52 GHz and room temperature at a microwave power level of 10 mW. An arrow locates the free electron g value, $g_0 = 2.0023$.

(6) Powdered pure Fe metal dispersed in SiO_2 displayed a broad featureless X-band signal which most closely resembled the spectrum of 10085,170,6 of Fig. 1 except for the absence of features near $g \approx 2$. Similar results have been reported by Weeks *et al.* (1971).

Simulated lunar glasses

The simulated lunar glass as delivered (Table 1) was dark brown in color and exhibited no ESR responses at X-band frequencies, save for a weak signal at $g \approx 4.3$ and a still-weaker signal at $g \approx 2$ (Fig. 3a). Based on the above results for pyroxene glasses, it was anticipated that remelting in an oxidizing flame might result in an enhancement of the $g \approx 2$ signal. However, only minimal enhancements were achieved by brief remelting in a hot hydrogen-oxygen flame ($\sim 1800°C$) followed by rapid quenching in liquid nitrogen. Only when the glass was first finely powdered in moist air did a $g \approx 2$ signal grow in on remelting in an oxidizing flame, but this was still weak by lunar standards. ^{60}Co γ-irradiations up to 4×10^7 R at room temperature also failed to alter the original resonances. Remelting in a reducing flame, however, resulted in an intense, well-defined resonance corresponding in all essential respects to the Type-II resonances of lunar glass particles (Fig. 3b).

In another series of experiments paralleling those of Kolopus et al. (1971) performed on actual lunar fines, powdered samples of simulated lunar glass were subjected to heat treatments in the temperature range 600–700°C under various atmospheres. Finely powdered samples, ground under dry N_2, showed no significant changes in their ESR spectra when sealed off in evacuated quartz tubes prior to annealing. By contrast, equivalent samples heated to 650°C in air or O_2 at 1 atm or in a continuously pumped soft vacuum (~ 0.5 Torr) showed enormous enhancements in the $g \approx 2$ spectral region. Samples heated at the lower pressure achieved "lunar intensities" in a span of ~ 200 hours, while samples heated at atmospheric pressure showed extremely rapid initial increase of resonance intensity followed by a prolonged period of inhibited growth (Fig. 4). After a period of 200 hours at 650°C in air at 1 atm, a sample of simulated lunar glass GS-64 exhibited an ESR response closely resembling the "characteristic" resonance of Apollo 14 fines. As seen in Fig. 3a, this resemblance extended to line shape (asymmetry), line width, and g value. In general, the line width appeared to depend inversely on the partial pressure of oxygen during the anneal. Thus, widths of ~ 750 Oe were achieved by heating at ~ 0.5 Torr, as compared with

Table 1. Chemical analyses of synthetic glasses.

	Simulated Lunar Glass GS-64		Control Glass GS-65[a]	
	Target	Analysis	Target	Analysis
SiO_2	50.0	49.88	50.0	49.99
Al_2O_3	15.6	16.16	15.6	15.55
CaO	10.4	10.65	18.1	17.91
MgO	8.4	8.80	14.6	14.66
TiO_2	2.4	2.43	—	—
Na_2O	0.6	0.71	0.6	0.67
K_2O	1.0	1.09	1.0	1.08
FeO	11.6	9.09	—	0.035
Fe_2O_3	—	1.13[b]	0.1	0.078
Total Fe as Fe_2O_3		(11.22)		(0.117)

Spectrographic analyses:[c]

Cr	0.0006–0.006
Cu	0.005 –0.05
Mn	0.01 –0.1
Sr	0.003 –0.03

[a] An additional control glass was also prepared by melting together 97.6% (GS-65) and 2.4% TiO_2.

[b] Mössbauer and ESR measurements indicate that the actual Fe_2O_3 content is closer to 0.1%. Possible sources of error in the wet chemical methods have not been evaluated.

[c] Performed on GS-65, but probably also pertain to GS-64.

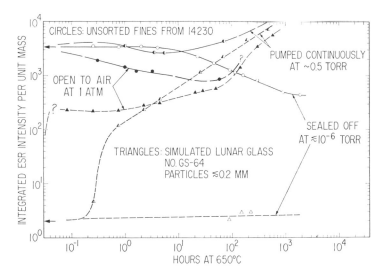

Fig. 4. The effects of isothermal anneals in various atmospheres upon the ESR intensities of roughly equivalent samples of Apollo 14 fines and simulated lunar glass No. GS-64. "ESR intensity per unit mass" is expressed in arbitrary units. As a calibration, a value of $\sim 2 \times 10^2$ would be equivalent to 100% of the iron ions in paramagnetic Fe^{3+} states; still larger numbers necessarily imply superparamagnetism (see text). Data for different lunar samples were normalized to an average initial value indicated by the upper arrow on the left-hand ordinate.

~ 500–600 Oe when the pressure was 1 atm. However, the relationships of ESR line width and intensity to these heat treatments are clearly complex and not yet fully elucidated. For example, samples initially heated in air and subsequently annealed at the same temperature in an evacuated sealed system show further increases in both intensity and line width that are dependent on the pressure and duration of the first treatment, as well as on the duration of the second. Also, certain subtleties in the "characteristic" resonance line shape have not yet been fully duplicated in the course of the experiments described here. In particular, the line width at $+300°C$ is narrower for the oxidized simulated glasses than for Type-I lunar signals (Fig. 2). Nevertheless, it has been adequately demonstrated that resonances very similar in all aspects to the "characteristic" resonance of lunar fines are ready consequences of mild oxidation of lunar-like materials, involving air pressures of $\sim 10^{-1}$–10^3 Torr, temperatures of 600–700°C, and durations of 10^1–10^3 hours.

Since these results must be placed in a selenological context, it is important to note that the rusty orange-brown color associated with oxidizing materials of lunar compositions has been observed to disappear after a few hours of subsequent annealing in a continuously pumped "hard" vacuum ($\leq 10^{-6}$ Torr). The physical appearance of an oxidized simulated lunar glass after the second treatment is very similar to that of the lusterous dark glassy components in many returned lunar soils, and, as mentioned, the oxidation-induced resonance is generally *enhanced* (at least initially) by the further heating in the absence of air.

The annealing experiments described above were duplicated on "control" glasses containing no FeO (Table 1). None of these treatments resulted in any increased resonance intensity in glasses with or without TiO_2, so long as Fe^{2+} was not present.

The resonances of simulated lunar glasses were also investigated at Ka-band frequencies and comparisons were made with the spectra of individual lunar glass particles. Both the oxidized and reduced simulated lunar glasses exhibited resonances with derivative zero-crossings near $g \approx 2$ ($H \approx 12.4$ kOe), with $2.20 > g$ (reduced glass) $> g$ (oxidized glass) > 2.00. Figure 5 suggests that the Ka-band spectrum of the red-brown spatter glass, 10085,170,2, may be a superposition of resonances due to oxidized and reduced magnetic species, such as have been produced in simulated glass GS-64. This is consistent with the observation that Type-I and Type-II resonances are both present with roughly equal intensities in the X-band spectrum of 10085,170,2 (Fig. 1). The "double-bump" structure of Fig. 5 was not apparent in the Ka-band spectra of most other lunar samples investigated. On the basis of the measured g values, it is inferred that Type-II resonances dominate the Ka-band spectra of most mm-sized fragments; while Type-I resonances are predominant in the finer fractions.

Lunar fines

Unsorted fines samples from Apollos 12, 14, and 15 were investigated over the same ranges of frequency and temperature as the actual and simulated lunar glasses. The resemblance at X-band and room temperature between the Type-I resonance in

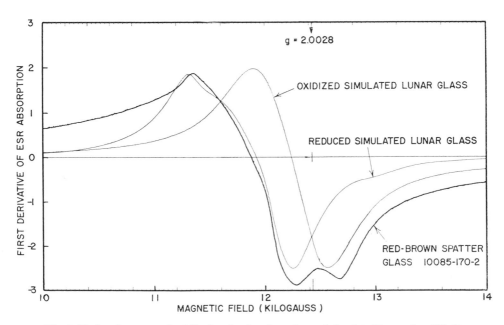

Fig. 5. Ka-band spectra of oxidized and reduced portions of simulated lunar glass GS-64 are compared with the spectrum of a red-brown spatter glass (10085,170,2) at room temperature.

an oxidized simulated lunar glass and the "characteristic" resonance of Apollo 14 fines has been shown in Fig. 3a. All three Apollo 14 fines samples studied (14230,88; 92; 96) exhibited "characteristic" resonance spectra with line widths significantly narrower than those reported for fines from Apollos 11 and 12 (~ 600 G versus 800–980 G at X-band and room temperature). The line width for Apollo 15 fines (15302,27) was found to be intermediate between these (~ 725 G).

The effects of various heat treatments upon the "characteristic" resonance of the Apollo 14 fines were explored in detail. One sample (14230,88,1) was subjected to isothermal anneals in air at 650°C; another (14230,96,1) was given a treatment in a continuously pumped "soft" vacuum (~ 0.5 Torr) at the same temperature. A third sample was handled in a way that should have precluded any significant chemical reactions with terrestrial contaminants; sample 14230,92,1 was baked out at ~ 130°C for 16 hours in a predried quartz tube continuously pumped to a pressure lower than 10^{-6} Torr. This tube was then sealed off *in situ* and the sample, so encapsulated, was subjected to the same isothermal anneals as were the previous two. The intensities of the "characteristic" resonance in these three samples are plotted versus time at 650°C in Fig. 4. The intensity and line width for the third sample are also plotted in Fig. 6, where two distinct time regimes may be noted. In the first 3 hours the line width increases monotonically with time at 650°C while the overall integrated intensity remains constant. In the second time regime (≥ 3 hours) the line width becomes virtually static while the intensity decreases monotonically. Gradual changes in line shape are also indicated in Fig. 6.

Samples of Apollo 12 and Apollo 15 fines were similarly baked out at 130°C and annealed at 650°C in evacuated sealed systems of small volume. In each case the line width was observed to increase during the initial 1–3 hours and then to stabilize near a maximum value which was never exceeded during the remainder of the experiments, lasting more than 1000 hours. Both the initial and ultimate line widths were found to

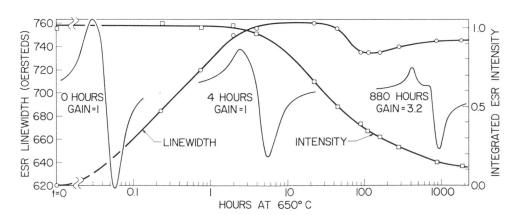

Fig. 6. Intensity, line width, and line shape of the "characteristic" resonance of an Apollo 14 fines sample are shown as functions of annealing time at 650°C in a sealed quartz tube having a volume of ~ 0.3 cm³, initially evacuated to $\leq 10^{-6}$ Torr at 130°C.

depend on the selenographic origin of the returned sample. Therefore, possible correlations with soil chemistry were explored. At best, only a weak correlation could be inferred with the FeO content of local soils. However, a plot of "characteristic" resonance line width versus wt% TiO_2 (Fig. 7) is highly suggestive of a causal relationship.

Intensities of the "characteristic" resonance in various lunar fines samples were determined by numerical integration techniques and the relative values, normalized to sample mass, are given in Table 2. The results clearly demonstrate an inverse correlation of "characteristic" resonance intensity with particle size and tentatively suggest another possible correlation with TiO_2 concentration. The integrated ESR intensity was found to be relatively independent of the temperature at which the spectra were obtained; samples from 12001, 14230, and 15302 all exhibited broad intensity maxima in the range -50 to $+50°C$ with drop-offs to $\sim 75\%$ at -200 and $+300°C$.

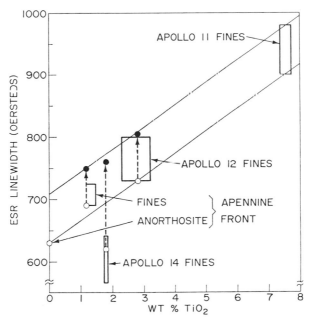

Fig. 7. Line widths of the "characteristic" resonance for fines from the first four Apollo landing sites are plotted versus the TiO_2 contents of the respective soils. ESR results for as-returned material are taken from Weeks *et al.* (1970a) (Apollo 11), Kolopus *et al.* (1971) (Apollo 12), and Weeks *et al.* (1972) (Apollo 14), as well as the present study (Apollos 12, 14, and 15; open circles). *Maximum* line widths displayed by samples heated to 650°C *in vacuo* are indicated by filled circles. Chemical data are derived from LSPET (1972), LSPET (1971), and other sources. Data points are placed on average compositions for Apollos 12 and 14, while the Spur Crater analysis (15301) was considered appropriate to 15302 fines. A single anorthosite fragment from 15302,27 was assumed to be similar in composition to 15415 (LSPET, 1972). ESR spectra were obtained at 300°K and 9 GHz.

Table 2. ESR absorption intensities[a] per unit sample mass for the "characteristic" resonance of various lunar soils and individual glass or mineral fragments.

Sample No.	Description	Mass (mg)	Specific Intensity[b]
10087,11	Mare Tranquillitatis fines	c	270 ± 27[c]
12003,24	Oceanus Procellarum fines	c	243 ± 27[c]
12001,15	Oceanus Procellarum fines	1.6	234
12070,11	Oceanus Procellarum fines	c	222 ± 27[c]
14230,88	Fra Mauro core tube fines (2 samples)	~2 (each)	224 ± 12
14230,96	Fra Mauro core tube fines (2 samples)	~2.5 (each)	190 ± 34
14230,92	Fra Mauro core tube fines (2 samples)	~2 (each)	161 ± 11
15302,27,11	Apennine Front (Spur Crater) fine fines	2.8	141
14230,88,7	Ten particles ~0.3 mm hand picked from 14230-88	0.8	46
15401,56,1	Fine fines, presumed to be disintegration products of a 3-m, friable breccia boulder[d]	1.8	46
10085,170,2	Thin red-brown spatter glass	1	20
10085,170,1b	Portion of opaque hollow spherule	3	12
15302,27,4	Fractured red-brown vesicular glass with flow textures	5	6
10085,170,5a	Dark brown (homogeneous?) glass with clean conchoidal surfaces and no vesicles	2	3.5
10085,170,3	Light green translucent angular glass fragment	2	0.9
15302,27,8	Anorthosite fragment, Spur Crater	2.8	0.6
15505,43,1	Grey-brown vesicular coating glass from rille-edge crater breccia[d]	2.3	0.5

[a] All intensities were determined by numerical integration of X-band spectra obtained near 25°C. Data for unsorted fine fines were obtained under identical experimental conditions and the integration extended from zero field to ~6 kOe. For individual particles, the region of integration was restricted to ~2–5 kOe in order to descriminate against Type-II resonances, whose relative contributions are more significant in the coarser soil fractions (see Fig. 3).

[b] Intensity divided by sample mass. Units are "wt % paramagnetic Fe_2O_3" (see text for explanation).

[c] Data taken from Tsay, et al. (1971a, 1971b). Sample masses are unknown.

[d] Sample descriptions are given by Swann et al. (1971).

DISCUSSION AND CONCLUSIONS

Origin of Type-II resonances

The Type-II resonances of Fig. 3b are attributed to metallic iron precipitates on the basis of the observed hysteresis effects and the method by which they are produced in the laboratory, i.e., by chemical reduction of a glass melt of lunar composition. Because of the short melting time required for this reduction (~1 min) and because of surface tension effects at the glass-metal interface, it is assumed that the iron particles are both small and nearly spherical. It has thus been shown that the ferromagnetic resonance spectrum of small, spheroidal iron particles in at least one class of lunar materials does not resemble the "characteristic" resonance of the fines (compare Fig. 3b with Fig. 3a) which has previously been ascribed to such particles (Manatt et al., 1970; Tsay et al., 1971a, 1971b).

Based on studies of metallic Fe powders and of highly reduced synthetic lunar materials supplied by R. M. Housley it is concluded that very broad, relatively shapeless Type-II resonances with only minor hysteresis effects (Fig. 1) are due to larger, more irregular particles of metallic iron. No reduced materials which have been investigated have exhibited clearly identifiable Type-I signals.

Origin of the "characteristic" resonance

In view of the apparent correlations of ESR line width with a non-siderophile element in the soil chemistry (Fig. 7), it seems very unlikely that the Type-I, or "characteristic" resonance, also arises from metallic iron phases. On the other hand, laboratory experiments show that resonances resembling the lunar Type-I signals are easily producible by relatively-low-temperature ($\sim 650°C$) oxidation of glasses or minerals of lunar-like compositions; the only crucial requirements appear to be the initial presence of Fe^{2+} and a relatively brief exposure to a tenuous oxidizing atmosphere. It is concluded that the Type-I resonances produced in simulated lunar materials are related to the conversion of Fe^{2+} to Fe^{3+} in the solid state, a process that has been corroborated by Mössbauer spectroscopic measurements (Forester, 1971). Inasmuch as the "characteristic" resonance of lunar fines is universally found in small, exposed glass and mineral fragments and it attains its greatest specific intensity in the finest soil fractions (Table 2), it seems possible that this ubiquitous lunar signal may have come about in the same way as its laboratory counterpart, i.e., by atmospheric oxidation.

Superparamagnetism of the "characteristic" resonance

It has been pointed out that the enormous, nearly-temperature-independent intensity of the "characteristic" resonance is a hallmark of ferromagnetic resonance (Tsay *et al.*, 1971a, 1971b). These properties will pertain to many systems with large net ferromagnetic moments. Superparamagnetism is the name commonly given to group paramagnetic behavior of small single-domain magnetic particles that are internally coupled ferromagnetically (all spins parallel), antiferromagnetically (all spins antiparallel), or ferrimagnetically (some spins antiparallel, but a net ferromagnetic moment remains). In general, the effect involves particles of 150 Å or less and is most easily seen when the strongly magnetic particles are embedded in, and separated by, an ideally nonmagnetic matrix. This ideal system is approximated by magnetic oxide particles precipitated within silicate glass compositions (Shaw and Heasley, 1967). Such magnetic oxide particles are most frequently ferrimagnets (often referred to as ferrites). As ferrimagnetic materials are poor conductors, no eddy current effects are expected, and none has been observed in the lunar samples (Tsay *et al.*, 1971a, 1971b). Moreover, many properties of ferrites can be analyzed in terms of their resultant magnetization; thus, it is essentially the ferromagnetic resonance of ferrimagnetic materials that is being studied (Morrish, 1965, p. 560). In particular, the functional expression (Standley and Stevens, 1956) for the magnetic fields at which ferromagnetic resonance occurs is expected to be applicable also to the case of *ferri*magnetic resonance.

Because of the broad line widths (Fig. 2) and enormous intensities (Fig. 4) of the Type-I resonances produced by oxidizing powders of simulated lunar glasses, it is concluded that the Fe^{3+} phases involved here are fine grained ferrimagnetic precipitates, possibly similar to those investigated by Shaw and Heasley (1967) or Collins and Mulay (1970). Thus, it could be entirely fortuitous that Tsay *et al.* (1971a, 1971b) were able to computer simulate the "characteristic" resonance using the expression of Standley and Stevens (1956) and the g value, anisotropy constant (K_1) and saturation magnetization (M_s) for metallic iron. Judging from the good superpositioning of the "characteristic" resonance and the resonance of the oxidized simulated lunar glass of Fig. 3a, it is inferred that the Fe^{3+} ferrites responsible for the latter are characterized by virtually the same values of g and $2K_1/M_s$ as iron metal.

Interpretation of line width variations

Variations of the "characteristic" resonance line width with soil chemistry have not been remarked upon by other workers; however, the data of Fig. 7 exhibit an unmistakable trend. It is consistent to suggest, therefore, that the "characteristic" resonance arises from the precipitation in lunar fines of ferrite phases involving Fe^{3+} and Ti^{4+}. Indeed, TiO_2 is a well known catalyzing agent for the nucleation of crystallites in glasses; Stookey (1960) has described over 100 glass systems which may be effectively crystallized by the presence of titania. Thus, the results of Fig. 7 may reflect a variation in average crystallite size as affected by the concentration of nucleating centers and/or a variation in crystallite stoichiometry.

The results for samples annealed *in vacuo* suggest that some of the variations observed for samples from the same selenographic site may be due to differences in thermal histories rather than cross-contamination with materials from distant localities. Thus, the anomalously low widths for the as-returned Apollo 14 samples may be indicative of lower ambient temperatures at Fra Mauro, as compared with the other regions sampled.

Lunar Fe^{3+} concentrations

It is appropriate to estimate the Fe^{3+} contents of lunar soils as predicted by the proposed "ferrite model" for the "characteristic" resonance and to demonstrate that these estimates do not conflict with the findings of other physical and chemical methods. Tsay *et al.* (1971a, 1971b) have explained the fact that an ensemble of electronic spins behaving superparamagnetically will exhibit resonance absorptions up to 1.3×10^3 times more intense than the absorption of an equivalent number of independent spins behaving paramagnetically, when observation is made at 9 GHz and room temperature. Following the method described by these authors the intensities of various Type-I lunar resonances were determined in units of the fictitious quantity "wt% paramagnetic Fe_2O_3." The numbers so obtained (Table 2) were typically of the order of 200 for unsorted fine fines, in good general agreement with the results of Tsay *et al.* (1971a, 1971b) and indicating, of course, that superparamagnetism rather than paramagnetism is involved. Assuming ferromagnetic coupling of Fe^{3+} states, then, a typical sample of <1 mm fines would contain the equivalent of 0.15 ± 0.05 wt% Fe_2O_3 and perhaps more if the actual phases should be ferrimagnetic.

Mössbauer studies of lunar materials have generally been interpreted as showing no Fe^{3+}. The only noteworthy exception has been the observation by Gay *et al.* (1970) of a hyperfine sextet that they attributed quite unambiguously to "iron spinel." Since this observation was made at room temperature, it is inferred that the spinel phases in this case were larger than single domain size ($\gtrsim 150$ Å). It is suggested here that the "characteristic" resonance may be due to chemically similar but finer grained Fe^{3+} ferrites. The nonobservation at room temperature of "iron spinel" in the Apollo 12 sample (Gay *et al.*, 1971) does not preclude this interpretation. Small, superparamagnetic ferrite particles are expected to contribute to the room-temperature Mössbauer spectrum in the vicinity of zero velocity and, indeed, such an "excess area" accounting for $\sim 5\%$ of the total iron has been reported by a number of workers (Hafner, 1971; Herzenberg *et al.*, 1971; Housley *et al.*, 1971).

Some of this "excess area" is apparently ascribable to single-domain metallic Fe (Housley *et al.*, 1972) but only $\sim \frac{1}{5}$ of it would be required to account for the "characteristic" resonance if the ferrite source is accepted.

The high initial susceptibilities along with the temperature dependence of the magnetizations have often been cited as evidence of superparamagnetic metallic iron in lunar fines (Nagata *et al.*, 1970; Runcorn *et al.*, 1970). However, large initial susceptibilities are also measured for ferrite phases precipitated in silicate glasses (Shaw and Heasley, 1967), and Néel temperatures in the range 700–1000°K are expected for many spinels (Morrish, 1965, p. 507). Thus, it may be difficult to disentangle the static magnetic effects of ferrite particles from those of metallic iron.

Wet chemical analyses of lunar materials generally show Fe_2O_3 as "nil." However, this result apparently comes from the difference between analyses for Fe^{2+} and total iron (Maxwell *et al.*, 1970), and the error in each of these numbers may be larger than the 1–2% of the total iron which is ascribed here to Fe^{3+} in a ferrite phase.

Explicit evidence of goethite has been detected in a weakly recrystallized Apollo 14 breccia 14301 (Agrell *et al.*, 1972). This finding along with indications of topotactic reactions in some lunar pyroxenes and olivines separated from a partially recrystallized fragmental rock 14321 (Gay and Bown, 1972) further support the notion that fine lunar surface materials may have been oxidized on the moon.

Finally, in a recent preprint Walker *et al.* (1972) have interpreted the longer interval of olivine crystallization observed for comprehensive fines 14259 (vis-à-vis igneous rock 14310) as probably being an oxidation effect.

Lunar oxidation processes

In light of the above discussion, the conclusion of the present paper that lunar soils generally contain from 0.1 to 0.3 wt% Fe_2O_3 does not seem to be in obvious conflict with the literature. This result contrasts, however, with the $Fe^{3+}/(Fe^{2+} + Fe^{3+})$ ratio of between 0.001 and 0.003 estimated for lunar clinopyroxenes (Hafner *et al.*, 1971). Thus, the inverse correlation of "characteristic" resonance intensity with particle size (Table 2) readily suggests a surface-dependent atmospheric oxidation process as being responsible for the higher Fe^{3+} content of the fines. Various exogenous sources of oxidants can be proposed, e.g., cometary impacts or water input from carbonaceous chondrites, as well as endogenous sources, such as venting of a subsurface "pockets" of steam (Freeman *et al.*, 1972) or disproportionation of the regolith into reduced melts and oxygen-enriched vapors under impact. Cometary or volcanic events are expected to be highly episodic, and hence there is hope that records of such occurrences may be preserved in core tube samples in terms of differing "characteristic" resonance intensities for different stratographic units. Strata laid down in the course of an oxidizing event should exhibit greater specific intensities than adjacent units emplaced, say, by the impact of an iron meteorite or by electrostatic transport (Gold, 1971).

However, "characteristic"-resonance intensity data must be treated with certain cautions. Clearly it will be necessary to study equivalent size fractions from the strata being compared because of the apparent particle-size dependence. Less obvious, but equally important, it must be established from geologic evidence that the soil being investigated has not been subjected to intense, postdepositional thermal transients; it has been shown that 30-minute vacuum anneals at temperatures as low as 800°C can reduce the "characteristic" resonance intensity by 99%, presumably due to re-solution or chemical reduction of the superparamagnetic phases (Kolopus *et al.*,

1971). Thus, specific resonance intensities of sintered soil breccias are expected to be lower than those of the fines from which they were derived. For example, the resonance of soil (15401) sampled near the top of the high fillet on the large breccia boulder at Sta. 6a on the Apennine Front is found to be only $\frac{1}{5}$ to $\frac{1}{3}$ as intense as that of (slightly coarser) fines from other sites (Table 2). This tends to support the conclusion of Swann *et al.* (1971) that this fillet material derives at least partly from the boulder itself rather than downslope movement of debris from the Front. Final interpretation of the relatively small intensity variations among the three unsorted Apollo 14 core samples (Table 2) must await detailed studies of sieved fractions, although the present results are consistent with the "fineness" estimates of Fryxell and Heiken (1972).

The findings of the present study of simulated lunar glasses have shown that heating at $\sim 650°C$ in an air pressure of ~ 0.5 Torr is a highly efficacious means of producing Type-I resonances of "lunar intensities" in the laboratory (see Fig. 4). It is presumed that both oxidation and crystallization proceed simultaneously under these conditions. Crystallization of glasses will also take place in lower temperature regimes, although the time scale may grow large (Gottardi and Bonetti, 1959); indeed, Type-I resonances have been acquired in simulated lunar materials by heating at 600°C, but the reaction proceeded too slowly for practical purposes. Inasmuch as the average surface exposure age of lunar soils ranges from 5 to 100 m.y. (Arrhenius *et al.*, 1971), it would not be surprising if some degree of oxidation and crystallization has taken place even at temperatures as low as that of the lunar daytime and O_2 partial pressures as low as the present lunar average. It is conceivable then that a continual oxidation/crystallization process is responsible for the precise ferrite phases which give rise to the "characteristic" resonance. Nevertheless, the close superpositioning of the spectra shown in Fig. 3a tempts one to consider the possibility that the "characteristic" resonance was produced on the moon in much the same way as similar resonances have been obtained in the laboratory. One means by which this could have occurred would be the lunar ash flow (Pai *et al.*, 1972). It has been pointed out (O'Keefe and Hsieh, 1972) that such flows may be sustained for periods of hours or days by steam emission from new ash heated to temperatures of the order of 850°C. Since overall pressures of ~ 1 atm may be attained (Pai *et al.*, 1972), partial pressures of $O_2 \approx 0.1$ Torr seem possible. Figure 4 suggests that such a flow, upon cooling through the 600–700°C temperature range, would inevitably produce a "characteristic" resonance. Preliminary laboratory modelling experiments seem to confirm this.

SUMMARY

It has been shown that the "characteristic" resonance of lunar fines probably arises from fine grained ferrites and not from fine grained metallic iron as previously supposed. The significance of this finding is several fold:

(1) It shows that lunar soils may have been oxidized *in situ*.

(2) It provides a means of measuring lunar Fe^{3+} concentrations and, hence, the degree of oxidation (or chemical disequilibrium).

(3) It necessitates the re-evaluation of previous measurements of metallic iron in the lunar soils by Mössbauer and susceptibility methods, since a fraction of the signal intensities ascribed to superparamagnetic Fe probably arises from lunar ferrites.

(4) These ferrites may be the carriers of remanent magnetism in some classes of lunar breccias.

Acknowledgments—The authors are indebted to G. H. Sigel, Jr. for pointing out several important references and for contributing in other ways to this work. R. D. Kirk is thanked for preparing a number of synthetic glasses and C. E. Schott and Owens-Illinois are gratefully acknowledged for melting and furnishing the simulated lunar glass. Thanks are also due D. W. Forester for examining the Mössbauer spectra of the simulated materials and granting permission to quote his results.

REFERENCES

Agrell S. O., Scoon J. H., Long J. V. P., and Coles J. N. (1972) The occurrence of goethite in a microbreccia from the Fra Mauro Formation (abstract). In *Lunar Science—III* (editor C. Watkins) pp. 7–9, Lunar Science Institute Contr. No. 88.

Arrhenius G., Liang S., Macdougall D., Wilkening L., Bhandari N., Bhat S., Lal D., Rajagopalan G., Tamhane A. S., and Venkatavaradan V. S. (1971) The exposure history of the Apollo 12 regolith. *Proc. Second Lunar Sci. Conf., Geochim. Cosmochim. Acta* Suppl. 2, Vol. 3, pp. 2583–2598. MIT Press.

Ayscough P. B. (1967) *Electron Spin Resonance in Chemistry*, p. 442. Methuen.

Castner T., Newell G. S., Holton W. C., and Slichter C. P. (1960) Note on the paramagnetic resonance of iron in glass. *J. Chem. Phys.* **32**, 668.

Collins D. W. and Mulay L. N. (1970) Crystallization of beta $NaFeO_2$ from a glass along the Na_2SiO_3–Fe_2O_3 join. *J. Am. Ceram. Soc.* **53**, 74.

Duchesne J., Depireux J., Gerard A., Grandjean F., and Read F. (1971) A study by electron paramagnetic resonance and Mössbauer spectroscopy of some lunar samples collected by Apollo 12, Proc. Second Lunar Sci. Conf. (unpublished proceedings).

Forester D. W. (1971), private communication of unpublished results.

Freeman J. W. Jr., Hills H. K., and Vondrak R. R. (1972) Water vapor, whence comest thou? (abstract). In *Lunar Science—III* (editor C. Watkins), pp. 283–285, Lunar Science Institute Contr. No. 88.

Fryxell R. and Heiken G. (1971) Description, dissection, and subsampling of Apollo 14 core sample 14230. NASA Technical Memorandum X-58070.

Gay P., Bancroft G. M., Bown M. G. (1970) Diffraction and Mössbauer studies of minerals from lunar soils and rocks. *Proc. Apollo 11 Lunar Sci. Conf., Geochim. Cosmochim. Acta*, Suppl. 1. Vol. 1, pp. 481–497. Pergamon.

Gay P., Bown M. G., Muir I. D., Bancroft G. M., and Williams P. G. L. (1971) Mineralogical and petrographic investigation of some Apollo 12 samples. *Proc. Second Lunar Sci. Conf., Geochim. Cosmochim. Acta* Supp. 2, Vol. 1, pp. 377–392. MIT Press.

Gay P. and Bown M. G. (1972) Topotactic reactions in some lunar pyroxenes and olivines (abstract). In *Lunar Science—III* (editor C. Watkins), pp. 291–293, Lunar Science Institute Contr. No. 88.

Geake J. E., Dollfus A., Garlick G. F. J., Lamb W., Walter C., Steigmann G. A., and Titulaer G. (1970) Luminescence, electron paramagnetic resonance and optical properties of lunar material from Apollo 11. *Proc. Apollo 11 Lunar Sci. Conf., Geochim. Cosmochim. Acta* Suppl. 1, Vol. 3, pp. 2127–2147. Pergamon.

Gold T. (1971) Evolution of mare surface. *Proc. Second Lunar Sci. Conf., Geochim. Cosmochim. Acta* Suppl. 2, Vol. 3, pp. 2675–2680. MIT Press.

Gottardi V. and Bonetti G. (1959) Evoluzione termica della structura di vetri fosfatici (in Italian). *Vetro e Silicati* **III**, pp. 31–38.

Griscom D. L. (1971) ESR Studies of Fe^{3+} in fused natural diopside, (unpublished work).

Griscom D. L. and Griscom R. E. (1967) Paramagnetic resonance of Mn^{2+} in glasses and compounds of the lithium borate system. *J. Chem. Phys.* **47,** 2711–2722.

Griscom D. L. and Marquardt C. L. (1972a) Electron spin resonance studies of iron phases in lunar glasses and simulated lunar glasses (abstract). In *Lunar Science—III* (editor C. Watkins), pp. 341–343, Lunar Science Institute Contr. No. 88.

Griscom D. L. and Marquardt C. L. (1972b) Electron spin resonance studies of fine grained metallic iron in reduced glasses of lunar compositions, (to be published).

Hafner S. S. (1971) State and location of iron in Apollo 11 samples. In *Mössbauer Effect Methodology* (edited by I. J. Gruverman), Vol. 6, pp. 193–207. Plenum.

Hafner S. S., Virgo D., and Warburton D. (1971) Cation distributions and cooling history of clino-pyroxenes from Oceanus Procellarum. *Proc. Second Lunar Sci. Conf., Geochim. Cosmochim. Acta* Suppl. 2, Vol. 1, pp. 91–108. MIT Press.

Haneman D. and Miller D. J. (1971) Clean lunar rock surfaces: unpaired electron density and adsorptive capacity for oxygen, *Proc. Second Lunar Sci. Conf., Geochim. Cosmochim. Acta* Suppl. 2, Vol. 3, pp. 2529–2541. MIT Press.

Herzenberg C. L., Moler R. B., and Riley D. L. (1971) Mössbauer instrumental analysis of Apollo 12 lunar rock and soil samples. *Proc. Second Lunar Sci. Conf., Geochim. Cosmochim. Acta* Suppl. 2, Vol. 3, pp. 2102–2123. MIT Press.

Housley R. M., Grant R. W., Muir A. H. Jr., Blander M., and Abdel-Gawad (1971) Mössbauer studies of Apollo 12 samples. *Proc. Second Lunar Sci. Conf., Geochim. Cosmochim. Acta* Suppl. 2, Vol. 3, pp. 2125–2136. MIT Press.

Housley R. M., Grant R. W., and Abdel-Gawad M. (1972) Study of excess Fe metal in the lunar fines by magnetic separation (abstract). In *Lunar Science—III* (editor C. Watkins), pp. 392–394. Lunar Science Inst. Contr. No. 88.

Hyde J. S. (1961) EPR standard sample data. (Available from Varian Associates, Palo Alto, Calif.)

Kolopus J. L., Kline D., Chatelain A., and Weeks R. A. (1971) Magnetic resonance properties of lunar samples: mostly Apollo 12. *Proc. Second Lunar Sci., Conf., Geochim. Cosmochim. Acta* Suppl. 2, Vol. 3, pp. 2501–2514. MIT Press.

LSPET (Lunar Sample Preliminary Examination Team) (1971) Preliminary examination of lunar samples from Apollo 14. *Science* **173,** 681–693.

LSPET (1972) (Lunar Sample Preliminary Examination Team) (1972) The Apollo 15 lunar samples: A preliminary description. *Science* **175,** 363–365.

Manatt S. L., Elleman D. D., Vaughan R. W., Chan S. I., Tsay F.-D., and Huntress W. T. Jr. (1970) Magnetic resonance studies of some lunar samples. *Science* **167,** 709–711; *Proc. Apollo 11 Lunar Sci. Conf., Geochim. Cosmochim. Acta* Suppl. 1, Vol. 3, pp. 2321–2323. Pergamon.

Maxwell J. A., Peck L. C., and Wiik H. B. (1970) Chemical composition of Apollo 11 lunar samples 10017, 10020, and 10084. *Proc. Apollo 11 Lunar Sci. Conf., Geochim. Cosmochim. Acta* Suppl. 1, Vol. 2, pp. 1369–1374. Pergamon.

Morrish A. (1965) *The Physical Properties of Magnetism.* John Wiley.

Nagata T., Ishikawa Y., Hinoshita H., Kano M., Syono Y., and Fisher R. M. (1970) Magnetic properties and natural remanent magnetization of lunar materials. *Proc. Apollo 11 Lunar Sci. Conf., Geochim. Cosmochim. Acta* Suppl. Vol. 3, pp. 2325–2340. Pergamon.

O'Keefe J. A. and Hsieh T. (1972) (private communication).

O'Reilly D. E., Salamony J. E., and Squires R. G. (1971) Electron paramagnetic resonance of chromia supported on silica. Enhancement of dispersed chromia by oxalic acid complexing. *J. Chem. Phys.* **55,** 4147–4148.

Pai S. I., Hsieh T., and O'Keefe J. A. (1972) Lunar ash flows: How they work (abstract). In *Lunar Science—III* (editor, C. Watkins), pp. 593–594, Lunar Science Institute Contr. No. 88.

Runcorn S. K., Collinson D. W., O'Reilly W., Battey M. H., Stephenson A., Jones J. M., Manson A. J., and Readman P. W. (1970) Magnetic Properties of Apollo 11 samples. *Proc. Apollo 11 Lunar Sci. Conf., Geochim. Cosmochim. Acta* Suppl. 1, Vol. 3, pp. 2369–2387. Pergamon.

Swann G. A., Hait M. H., Schaber G. G., Freeman V. L., Ulrich G. E., Wolfe E. W., Reed V. S., and Sutton R. L. (1971) Preliminary description of Apollo 15 sample environments. United States Geological Survey. Interagency Report: 36.

Shaw R. R. and Heasley J. H. (1967) Superparamagnetic behavior of $MnFe_2O_4$ and α-Fe_2O_3 precipated from silicate melts. *J. Am. Ceram. Soc.* **50,** 297–302.

Standley K. J. and Stevens K. W. H. (1956) The anisotropy correction in ferromagnetic resonance. *Proc. Phys. Soc. London* **69B,** 993–996.

Stookey S. D. (1960) Method of making ceramics and product thereof. U.S. Pat. 2,920,971; *Ceram. Abstr. 1960,* p. 142a.

Tsay F.-D., Chan S. I., and Manatt S. L. (1971a) Magnetic resonance studies of Apollo 11 and Apollo 12 samples. *Proc. Second Lunar Sci. Conf., Geochim. Cosmochim. Acta* Suppl. 2, Vol. 3, pp. 2515–2528. MIT Press.

Tsay F.-D., Chan. S. I., and Manatt S. L. (1971b) Ferromagnetic resonance of lunar samples. *Geochim. Cosmochim. Acta* **35,** 865–875.

Walker D., Longhi J., and Hays J. F. (1972) Experimental petrology and origin of Fra Mauro rocks and soil, (preprint supplied by authors).

Weeks R. A., Chatelain A., Kolopus J. L., Kline D., and Castle J. G. (1970a) Magnetic properties of some lunar material. *Science* **167,** 704–707.

Weeks R. A., Kolopus J. L., Kline D., and Chatelain A. (1970b) Apollo 11 lunar material: Nuclear magnetic resonance of [27]Al and electron resonance of Fe and Mn. *Proc. Apollo 11 Lunar Sci. Conf., Geochim. Cosmochim. Acta* Suppl. 1, Vol. 3, pp. 2467–2490. Pergamon.

Weeks R. A., Kolopus J. L., and Arafa S. (1971) Ferromagnetic and paramagnetic resonance spectra of lunar material: Apollo 12, (preprint of paper submitted for publication).

Weeks R. A. and Kolopus J. L. (1972) Magnetic phases in lunar material and their electron magnetic resonance spectra: Mostly Apollo 14 (abstract). In *Lunar Science—III* (editor C. Watkins), pp. 791–793. Lunar Science Institute Contr. No. 88.

Wood J. A., Marvin U. B., Reid J. B. Jr., Taylor G. J., Bower J. F., Powell B. N., and Dickey J. S. Jr. (1971) Mineralogy and petrology of the Apollo 12 sample. Smithsonian Astrophysical Observatory, Special Report 333.

Proceedings of the Third Lunar Science Conference
(Supplement 3, *Geochimica et Cosmochimica Acta*)
Vol. 3, pp. 2417–2421
The M.I.T. Press, 1972

Natural remanent magnetization in lunar breccia 14321

R. B. Hargraves and N. Dorety

Department of Geological and Geophysical Sciences,
Princeton University, Princeton, New Jersey 08540

Abstract—Studies of the natural remanent magnetization of breccia 14321,194 involving alternating field and thermal demagnetization of contiguous fragments suggests that it consists of a uniform partial thermoremanent overprint acquired on cooling after agglomeration at moderate temperatures, superimposed on a stable but diversely oriented remanence of individual clasts. Results of thermal demagnetization plus partial thermoremanence (pTRM) acquisition are consistent with the $n \times 10^3 \gamma$ ambient fields inferred by other workers. The mild heating applied (160°C maximum) however, is within the lunar diurnal heating cycle and these results are of uncertain significance (Helsley, 1971).

Introduction

Lunar sample 14321 is a polymict macrobreccia composed of three major lithologic components: basaltic fragments (up to 5 cm), microbreccia clasts (three classes or generations), and a light-colored friable matrix (Grieve *et al.*, 1972; Duncan *et al.*, 1972). Grieve *et al.* report little textural evidence of recrystallization of the light colored matrix material. Warner (1972) reports the matrix texture as being partly annealed, and classifies it in Group 4 of his eight-member progressive metamorphic series.

Cliff *et al.* (1972) and Compston *et al.* (1972) have dated individual clasts from 14321 at 4.0 to 4.1 b.y. Compston *et al.* (1972) found evidence to suggest a slight difference in the age of a troctolitic clast, versus two basaltic clasts.

Magnetic Measurements: Procedures and Results

The sample supplied to us (14321,194, 20.5 gm) was an elongate sawed piece from what was the base of the sample as it lay on the lunar surface. It is a polymict breccia, as described above, with numerous rock fragments and clasts, some up to 1 cm, many about 4 or 5 mm in size, in a finer-grained, lighter-colored matrix. It was broken in half by hand, the relative and absolute orientations recorded, and natural remanent magnetization (NRM) measured. The result of progressive AF demagnetization of these two pieces to 50 Oe is shown in Fig. 1 (change in orientation) and Fig. 2 (change in intensity). It can be seen that there is marked dissimilarity in orientation and specific intensity in these two contiguous fragments.

In quest of better understanding of this anomaly further investigation of the stronger sample (B) was undertaken as follows:

Upon opening the specimen holder after the initial AF demagnetization measurements, the wrapped sample was found to have cracked. From the original 9.1 gm, two large coherent pieces 5.3 gm and 2.7 gm were recovered (plus 1 gm powder). The

2417

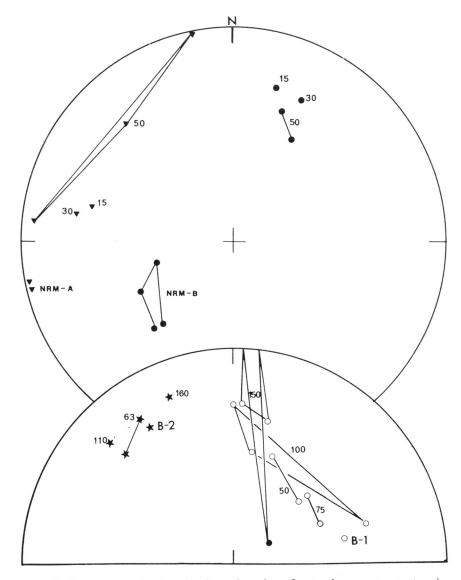

Fig. 1. Equal area projection showing orientation of natural remanence vectors in breccia 14321, 194 samples before and after various demagnetization treatments. Directions are with reference to lunar orientation of the sample; closed (open) symbols are lower (upper) hemisphere. Numbers pertain to AF demagnetizing field (Oe). Lines join vectors obtained on repeated demagnetization at a particular stage. Upper figure shows triangles and circles pertaining to contiguous halves (A and B) of original 20 gm sample, both stepwise AF demagnetized to 50 Oe. Lower figure shows individual vector orientations of the two pieces (B_1 and B_2) into which original B sample was broken. B_1 (circles, mostly on upper hemisphere) was stepwise AF demagnetized to 200 Oe (due to wide scatter, only vectors to 150 Oe are shown). Solid stars (all on lower hemisphere) show migration of B_2 vector on thermal demagnetization to 160°C. In each case, the B_2 vectors shown are those after 50 Oe AF demagnetization in addition to the thermal demagnetization.

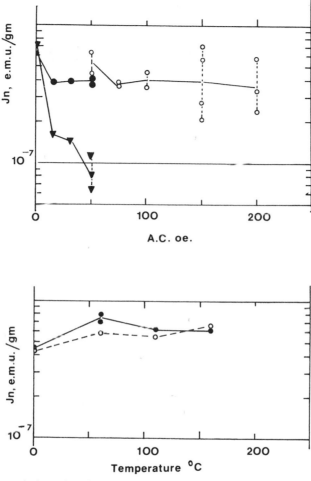

Fig. 2. Change in intensity of remanence (J_N) of 14321 samples as a result of various demagnetization treatments. Top: Solid figures; AF demagnetization of samples A (triangles) and B (circles) to 50 Oe; small circles show demagnetization of B_1 to 200 Oe. Bottom: Thermal demagnetization to 160°C; solid circles: natural; open circles: after 50 Oe AF demagnetization.

original orientation of each was preserved, their remanence remeasured, and measured again after repeat 50 Oe AF demagnetization (Figs. 1 and 2). The specific intensities of each fragment (Fig. 2) were both similar to the original, but the orientations were 130° apart, the vectors roughly bracketing the original (Fig. 1). In progressive AF demagnetization of the 5.3 gm fragment the vectors after repeated demagnetization were stable to 100 Oe, erratic at 150 Oe, and apparently lost at 200 Oe (Fig 1).

The smaller (2.7 gm) fragment was given a pTRM treatment to 160°C as follows: It was heated in field-free space ($< 20 \gamma$) in a continuously pumped vacuum chamber

($< 10^{-5}$ Torr). The sample was first thermally demagnetized to a particular temperature, the remanence measured after cooling, AF demagnetized in 50 Oe, and remeasured. The sample was then reheated to the same temperature, cooled in an applied field of 10,000 γ, and the remanence measurements repeated. Small heating cycles were employed: to 60°C, 110°C, and 160°C; the results are also shown in Figs. 1 and 2. In Table 1 are shown the orientation and magnitude of the cumulative vector removed by thermal demagnetization, and the magnitude of the TRM vector added by each pTRM run in a 10,000 γ field. Although the orientation of the vector erased on thermal demagnetization varies, the ratio of vector removed to vector added is moderately consistent (0.33 to 0.51) and the implied original field (3300 γ to 5100 γ) is similar to that obtained by other workers (Helsley, 1971; Nagata et al., 1971). In each case the vectors used are those remaining after 50 Oe AF demagnetization. The equivalent ratios without the 50 Oe demagnetization were much more varied (0.47, 0.69, and 1.47, respectively) and in view of the notorious ease with which lunar samples can acquire a viscous RM are considered less reliable.

Discussion

Our data suggest that on a small scale, the natural remanence of sample 14321 is extremely heterogeneous. On the other hand, the stable vector orientation of the stronger of our initial two fragments is similar to that reported for two samples of 14321 by Pearce et al. (1972). An overall, average, consistent orientation is suggested. That the two smaller parts of our stronger original sample should have divergent vectors reinforces the evidence of small-scale heterogeneity.

The moderate degree of annealing and metamorphism of the matrix of 14321 (Warner, 1972; Grieve et al., 1972; Duncan et al., 1972) indicates that the rock has not been subjected to very high temperature subsequent to its "agglomeration." It is likely, therefore, that individual clasts retain their own individual remanence, which would explain the small-scale heterogeneity. By the same token, if the clast vectors are random, a consistent "large-scale" average pTRM vector would result from terminal cooling in an ambient field. The paleointensity experiments were carried only to 160°C, but as this is within the range of diurnal thermal cycling on the lunar surface, the significance of the pTRM ratios is uncertain (Helsley, 1971). The cumulative vectors erased in the three thermal demagnetization steps vary (Table 1) which

Table 1. Vectors erased by progressive thermal demagnetization from natural remanence $\Delta J(0)$ compared with pTRM acquired on cooling in 10,000 γ field $\Delta J(10,000)$. In each case the vectors used are those remaining after 50 Oe.

		Vector erased	
		D	I
63°C	$\dfrac{\Delta J(0)}{\Delta J(10,000)} = \dfrac{0.103 \times 10^{-6}}{0.259 \times 10^{-6}} = 0.40$	59°C	$-18°C$
110°C	$\dfrac{\Delta J(0)}{\Delta J(10,000)} = \dfrac{0.176 \times 10^{-6}}{0.533 \times 10^{-6}} = 0.33$	129°C	$-3°C$
160°C	$\dfrac{\Delta J(0)}{\Delta J(10,000)} = \dfrac{0.188 \times 10^{-6}}{0.371 \times 10^{-6}} = 0.51$	180°C	33°C

adds to the uncertainty. It is possible that this inconsistency may reflect chemical changes during heating, whereby a thermally stable TCRM was acquired during the intervening pTRM runs. The apparent paleointensity value inferred from these data is in keeping with results obtained by other workers (Helsley, 1971; Nagata *et al.*, 1971) but their precise significance remains very uncertain.

CONCLUSIONS

The remanent magnetization of breccia sample 14321 is heterogeneous on a small scale and probably consists of a uniform partial thermal overprint on the diversely oriented, stable remanence of individual clasts. We attribute the anomalous apparent weakness of one of our original samples (A) to a fortuitous near cancellation of such stable clast vectors.

REFERENCES

Cliff R. A., Lee-Hu C., and Wetherill G. W. (1972) K, Rb, and Sr measurements in Apollo 14 and 15 material (abstract). In *Lunar Science—III* (editor C. Watkins), pp. 146–147, Lunar Science Institute Contr. No. 88.

Compston W., Vernon M. J., Berry H., Rudowski R., Gray C. M., Ware N., Chappell B. W., and Kaye M. (1972) Age and petrogenesis of Apollo 14 basalts (abstract). In *Lunar Science—III* (editor C. Watkins), pp. 151–153, Lunar Science Institute Contr. No. 88.

Duncan A. R., Lindstrom M. M., Lindstrom D. J., McKay S. M., Stoeser J. W., Goles G. G., und Fruchter J. S. (1972) Comments on the genesis of Breccia 14321 (abstract). In *Lunar Science—III* (editor C. Watkins), pp. 192–194, Lunar Science Institute Contr. No. 88.

Grieve R., McKay G., Smith H., and Weill D. (1972) Mineralogy and petrology of polymict breccia 14321 (abstract). In *Lunar Science—III* (editor C. Watkins), pp. 338–340, Lunar Science Institute Contr. No. 88.

Helsley C. E. (1971) Evidence for an ancient lunar magnetic field. *Proc. Second Lunar Sci. Conf., Geochim. Cosmochim. Acta* Suppl. 2, Vol. 3, pp. 2485–2490. MIT Press.

Nagata T., Fisher R. M., Schwerer F. C., Fuller M. D., and Dunn J. R. (1971) Magnetic properties and remanent magnetization of Apollo 12 lunar materials and Apollo 11 lunar microbreccia. *Proc. Second Lunar Sci. Conf., Geochim. Cosmochim. Acta* Suppl. 2, Vol. 3, pp. 2461–2476. MIT Press.

Pearce G. W., Strangway D. W., and Gose W. A. (1972) Remanent magnetization of lunar samples (abstract). In *Lunar Science—III* (editor C. Watkins), pp. 599–601, Lunar Science Institute Contr. No. 88.

Warner J. L. (1972) Apollo 14 breccias: Metamorphic origin and classification (abstract). In *Lunar Science—III* (editor C. Watkins), pp. 782–784, Lunar Science Institute Contr. No. 88.

Proceedings of the Third Lunar Science Conference
(Supplement 3, *Geochimica et Cosmochimica Acta*)
Vol. 3, pp. 2423–2447
The M.I.T. Press, 1972

Rock magnetism of Apollo 14 and 15 materials

T. Nagata

Geophysics Research Laboratory, University of Tokyo,
Tokyo, Japan

R. M. Fisher and F. C. Schwerer

U.S. Steel Corporation Research Laboratory,
Monroeville, Pennsylvania 15146

and

M. D. Fuller and J. R. Dunn

Department of Earth and Planetary Sciences, University of Pittsburgh,
Pittsburgh, Pennsylvania

Abstract—Intrinsic and structure-sensitive magnetic parameters of seven Apollo 14 materials (an igneous rock 14053,48, fines 14259,69 and five breccias, 14047,47, 14063,47, 14301,65, 14303,35, and 14311,45) and two Apollo 15 materials (an igneous rock 15058,55 and a breccia 15418,41) have been measured. The magnetic transition temperatures have suggested that the ferromagnetic constituent is either almost pure Fe or CoFe alloy of less than 2% of Co content in igneous rocks, while it is either nearly pure Fe or NiFe alloy of 3–7% of Ni content in breccias and fines.

The content of native iron in a highland igneous sample, 14053,48, is considerably larger than that in Apollo 11 and 12 igneous rocks which ranges from 0.02 to 0.1% in weight, while the Fe-contents in Apollo 14 breccias and fines are within the Fe-content range of Apollo 11 and 12 breccias and fines which is 0.2–1.0% by weight. The abundance of native irons in the lunar breccias and fines are discussed in terms of meteoritic origin.

The thermomagnetic curves of lunar materials for a low temperature range between 4.2°K and 300°K imply that some antiferromagnetic phases of 5–50°K in Néel temperature may be present in addition to the antiferromagnetic ilmenite and the paramagnetic pyroxene and olivine.

The viscous magnetization caused by very fine iron grains is dominant in the lunar fines and weakly impacted breccias. It has been experimentally confirmed that the superparamagnetic behavior of these fine single-domain particles at room temperature almost disappears at 5°K.

Seven grains selected at random from an assemblage of coarse fines, 14161,38, can be magnetically classified according to their saturation magnetization into strong, intermediate, and weak groups. The intensity of natural remanent magnetization (NRM) is not always proportional to the Fe-content in these coarse grains.

NRM's and their stability for eight Apollo 14 rocks and six Apollo 15 rocks have been examined, their intensity ranging from 2×10^{-6} to 1.3×10^{-4} emu/gm except for sample 14053,31 that has an unusually high value, 2.2×10^{-3} emu/gm. Statistically, the lunar breccias have stronger NRM than the lunar igneous rocks, and the thermal and AF demagnetization experiments have indicated that NRM of some lunar rocks consists of a soft component and a hard one, the latter being attributable to the thermoremanent magnetization acquired in a magnetic field of about $10^3 \gamma$ in intensity.

Introduction

General magnetic characteristics of the lunar materials have been revealed, at least qualitatively, by various studies on Apollo 11 and 12 lunar samples. The experimental results obtained independently by different investigators are in good agreement with one another in regard to several characteristic features of magnetism of the

lunar materials. From observed data of the Curie point temperature, it has been concluded that almost all ferromagnetic components in the lunar materials are metallic
irons. It seems however that the metallic irons are not always pure but may contain
some other metals such as Ni and Co to a certain small extent. This point should be
examined in more detail.

One of the striking features of the lunar materials is that a certain portion of the
native irons is in the form of very fine particles, which behave superparamagnetically
even at room temperature. In the lunar fines and weakly impacted clastic rocks in
particular, superparamagnetic and nearly superparamagnetic grains comprise a considerable fraction of the total metallic iron, resulting in a remarkable viscous magnetization in a weak magnetic field. Characteristics of the viscous magnetization caused
by these fine iron particles should be studied in more detail, because any magnetic
contamination that might have taken place in the lunar rocks after they were returned
to the earth must be carefully eliminated in examining their natural remanent
magnetization. An experimental method to approach this problem would be a
comparison of the magnetization curve at room temperature with that obtained at
sufficiently low temperatures which can block the fine iron particles.

Another characteristic feature of the magnetic properties of the lunar materials
is the paramagnetic magnetization. In this connexion, it seems that the antiferromagnetic minerals in the lunar materials must be looked for more carefully, because
a combination of the known antiferromagnetic minerals, the known paramagnetic
minerals, and the metallic iron cannot provide a full interpretation of the observed
magnetization versus temperature relationship of the lunar materials in the low
temperature range. It has been suggested that antiferromagnetic ferrosilite or similar
minerals of the pyroxene composition may be commonly present in the lunar materials
(Nagata et al., 1970).

The outstanding problem in lunar rock magnetism is the natural remanent magnetization and its acquisition mechanism. A number of investigators have suggested
that the most probable mechanism of the acquisition of natural remanent magnetization of the lunar rocks is the thermoremanent magnetization acquired in a magnetic
field of $1000 \sim 3000 \gamma$ in intensity. It does not seem, however, that this question has
been solved. Further experimental studies are needed to identify the physical processes responsible for origin of the natural remanent magnetization of lunar clastic
rocks as well as lunar igneous rocks. Results are presented in the following section of
special studies made on an unusually magnetic igneous rock, 14053, together with
other Apollo 14 clastic rocks.

Descriptions of Apollo 14 and 15 Lunar Materials

Various magnetic properties of six lunar materials returned by the Apollo 14
mission have been examined in detail. According to the preliminary descriptions
given in Apollo 14 Preliminary Science Report (NASA, 1971), the petrographic
characteristics of these six samples are as follows:

14047: A friable, fine-grained clastic rock with a small percentage of subangular
leucocratic clasts in a medium-gray matrix.

14053: An equigranular, fine-grained crystalline rock.

14301: A coherent, medium-gray clastic rock with sparse subangular leucocratic clasts and less abundant melanocratic clasts in a fine-grained matrix.

14303: A very friable, fine-grained clastic rock with less than 1% of subrounded leucocratic clasts in a medium-gray matrix.

14311: A coherent clastic rock with a small percentage of subangular clasts, mostly leucocratic, in a fine-grained crystalline groundmass.

14259: Fines whose grain size distribution is similar to those of Apollo 11 and 12 soils.

In addition to the six samples, the natural remanent magnetization and its stability of two other Apollo 14 samples were measured. The petrographic characters of the two samples are as follows:

14063: A friable breccia with approximately 40% of subangular to subrounded clasts in a very light-gray, fine-grained matrix. Melanocratic clasts are subordinate to leucocratic clasts.

14310: A fine-grained, medium-gray equigranular crystalline rock.

The chemical composition of four samples, 14053, 14301, 14310, and 14259, have also been reported (NASA, 1971). Results of the chemical analyses have shown that the chemical composition of Apollo 14 materials, with the exception of that of 14053, are clearly distinct from those of Apollo 11 and 12 samples; namely, the abundances of Fe, Ti, Mn, and Cr in Apollo 14 samples are considerably lower than those in Apollo 11 and 12 samples, while Si, Al, K, and Na are more abundant in Apollo 14 samples. Since Fe and Mn are the principle elemental carriers of lunar rock magnetism, the observed lower abundance of the two elements is an important factor in relation to various magnetic properties of Apollo 14 materials.

Three Apollo 15 samples also have been under magnetic examinations. The petrographic characteristics of the three samples are as follows (Lunar Sample Information Catalog, Apollo 15, NASA-MSC 03209):

15058: A porphyritic clinopyroxene basalt which contains discrete grains and inclusions in troilite of Fe-Ni.

15418: A partially shock melted, devitrified, and recrystallized rock, in which the total opaque mineral content is less than 0.01%.

15556: A very vesicular clinopyroxene basalt, in which the metallic phase of Fe-Ni are present as discrete grains, blebs in troilites, and inclusions in olivines.

The chemical compositions of the three samples have been reported (LSPET, 1972). FeO contents in the preliminary results are 19.97, 5.37, and 22.25% by weight in samples 15058, 15418, and 15556 respectively.

BASIC MAGNETIC PROPERTIES

The observed values of intrinsic and structure-sensitive magnetic properties of samples 14053,48 (igneous rock), 14259, 69 (fines), 14047,47, 14301,65, 14303,35, 14311,45 (clastic rocks), 15058,55 (igneous rock), and 15418,41 (clastic rock) are summarized in Table 1. The tabulated quantities are the three intrinsic magnetic parameters (the saturation magnetization I_s and the paramagnetic susceptibility χ_a

Table 1. Magnetic properties of some Apollo lunar materials.

Magnetic parameter	Igneous rocks		Clastic rocks							Fines	Unit
	14053,48	15058,55	10048,55	14047,47	14063,47	14301,65	14303,35	14311,45	15418,41	14259,69	
χ_0(300°K)	2.24	—	—	8.0	—	0.93	0.69	0.45	—	10.0	10^{-3} emu/gm
χ_a 300°	4.6	3.6	4.3	2.4	2.14	2.1	2.4	2.3	1.5	2.5	10^{-5} emu/gm
χ_a 5°	50.0	—	36.2	32.0	—	33.0	42.0	38.0	—	31.0	10^{-5} emu/gm
I_s 300°	2.20	0.13	1.8	1.40	0.034	0.69	1.27	0.74	0.15	1.5	emu/gm
I_s 5°	3.0	—	3.0	2.1	—	1.45	1.73	1.2	—	2.6	emu/gm
I_R 300°	3.2	≲0.10	1.3	6.1	—	0.64	2.10	0.43	≲0.13	6.0	10^{-2} emu/gm
I_R 5°	4.0	—	54.0	32.0	—	—	3.4	—	—	44.0	10^{-2} emu/gm
H_c 300°	20	<15	50	26	—	27	27	19	—	19	Oe
H_c 5°	21	—	190	175	—	—	33	—	—	140	Oe
H_{RC} 300°	80	—	500	350	—	450	180	140	—	300	Oe
H_{RC} 5°	—	—	500	450	—	—	—	—	—	350	Oe
Θ_c	765(±5)	790(±5)	—	760(±5)	—	750(±5)	735(±5)	750(±5)	760(±5)	750(±5)	C°
Θ_c^*	—	—	—	—	—	640(±5)	620(±5)	710(±5)	670(±5)	—	C°
χ_0/I_s, 300°K	1.02	—	—	5.71	—	1.35	0.54	0.62	—	6.67	$\times 10^{-3}$
I_R/I_s, 300°K	1.82	0.77	0.72	4.36	—	0.93	1.65	0.58	0.87	4.00	$\times 10^{-2}$
H_{RC}/H_c, 300°K	4.0	—	10.0	13.5	—	16.7	6.7	7.4	—	15.8	—
Metallic Fe	1.01	0.06	0.82	0.64	0.016	0.32	0.58	0.34	0.069	0.69	wt.%
Ni content Kamacite	~0	~0	—	~0	—	5.5 ± (0.5)	7.0 ± (1.0)	3.0(±0.5)	4.0(±0.5)	~0	wt.%
$W'(Fe)/W(Fe)^{**}$	0.97	—	0.64	0.67	—	0.86	0.97	0.97	—	0.67	—
Type of VRM	(I)	—	(II)	(II)	—	(II)	(I)	(I)	—	(II)	—

** Remarks: $W(Fe)$ and $W'(Fe)$ are content of metallic iron before and after heating sample to 820°C in 10^{-5} Torr.

at room temperature and the Curie temperature Θ_c), and four structure sensitive magnetic parameters (the initial reversible magnetic susceptibility χ_0, the saturation remanent magnetization I_R, the coercive force H_c, and the remanence coercive force H_{RC} which were all observed at room temperature).

Curie-point temperature Θ_c

In addition to the Θ_c value of six Apollo 14 samples and two Apollo 15 samples listed in Table 1, Curie temperatures of six of Apollo 11 and ten of Apollo 12 lunar materials have been reported (Doell and Grommé, 1970; Grommé and Doell, 1971; Helsley, 1970; Larochelle and Schwarz, 1970; Nagata *et al.*, 1970, 1971; Runcorn *et al.*, 1970; Schwarz, 1970; and Strangway *et al.*, 1970). The histogram of Θ_c values of all these Apollo 11, 12, 14, and 15 samples is shown at the bottom of Fig. 1, where two maxima, one around 770°C and the other around 750°C, of the occurrence frequency are observed. It may be generally concluded in Fig. 1, however, that all the observed values of Curie temperature of the lunar materials are reasonably close to that of pure metallic iron (770°C). The top and the middle in Fig. 1 show the histogram of Θ_c values of lunar breccias and that of lunar igneous rocks respectively. It may be tentatively pointed out in this histogram that Θ_c of lunar breccias (or clastic rocks) deviates considerably to the low temperature side, the median value being

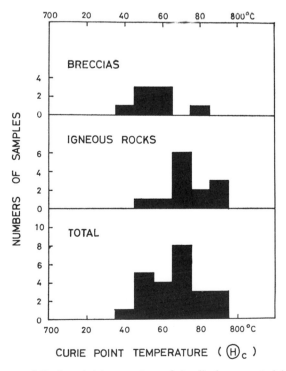

Fig. 1. Histogram of Curie point temperature of Apollo lunar materials observed in the initial heating process.

about 750°C, while the median value of Θ_c of lunar igneous rocks is in agreement with Curie temperature of pure metallic iron. A certain number of igneous rocks have their Curie temperatures appreciably higher than 770°C. As already discussed by Grommé and Doell (1971), it seems most likely that the observed increase of Θ_c of these lunar igneous rocks could be attributed to a small content of cobalt in iron as an FeCo alloy. It has been suggested (Strangway *et al.*, 1970, Schwarz, 1970) on the other hand, that the observed decrease of Θ_c of lunar materials may be caused by FeNi alloys, in which Ni content is several percent. In the present analyses of Apollo 14 and 15 lunar materials also, unseparated samples of clastic rocks 14310, 14303, 14311, and 15418 are distinctly associated with the same magnetization versus temperature hysteresis as illustrated in Fig. 2. The apparent low Curie temperatures (or the magnetic transition temperatures) Θ_c^* in the cooling process of the four samples also are summarized in Table 1. The apparently different Curie points in the heating and cooling curves could be attributed to the $\alpha \rightleftharpoons \gamma$ transitions of kamacites. The exact $\alpha \rightleftharpoons \gamma$ transition temperature is dependent on the heating or cooling rate of a sample. Approximate values of Ni content in kamacites, based on the discussions made by Lovering and Parry (1962), are given in Table 1. Electron probe analyses

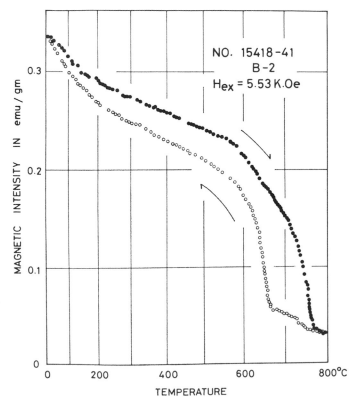

Fig. 2. Example of the α–γ transition of kamacite in the heating and cooling thermomagnetic curves. Full circles: heating curve. Hollow circles: cooling curve.

show that the percentage of Co in the Fe-Ni-Co alloys does not exceed 1% and that the content of Ni does not exceed 8% in these lunar breccias.

Saturation magnetization (I_s)

Another intrinsic magnetic parameter of materials is their saturation magnetization (I_s). In Table 1, the observed values of I_s of Apollo 14 fines and clastic rocks are comparable with those of the same kinds of Apollo 11 and 12 materials. However, an extremely large value of I_s of 14053 must be considered anomalously large in comparison with the I_s values of Apollo 11 and 12 igneous rocks. Because the chemical composition of native irons in the lunar materials is close to pure metallic iron, the saturation magnetization of the native irons may be approximately given as 218 emu/gm (or 1714 emu/cm³) at room temperature. Assuming this value for I_s of the native irons, the weight percentage of native iron in each sample can be estimated from its I_s values.

Figure 3 shows the histogram of weight percentages of native iron in the Apollo 11, 12, 14, and 15 lunar materials. As seen in the histogram, the weight percentage of native iron in lunar igneous rocks ranges from 0.02 to 0.1, the median value being about 0.06. Rock number 14053 is exceptional. The weight content of native iron in lunar fines and clastic rocks is much larger than that in igneous rocks and ranges from 0.2 to 1.0, the median value being about 0.5. (An exceptionally small value for a breccia is that for the gabbroic anorthosite, 15418, which contains only 4.17% by weight of Fe.) As already reported (*Proc. Apollo 11 Lunar Sci. Conf.*, Vol. 3, 1970; *Proc. Second Lunar Sci. Conf.*, Vol. 2, 1971; LSPET, 1971), the content of cobalt is 30 to 50 ppm throughout all lunar materials, while the content of nickel is 15 to 30 ppm in lunar igneous rocks and 150 to 300 ppm in lunar fines and clastic rocks. Adding the present data of the native iron content to the chemical data of Co and Ni, it may be concluded that the abundances of Fe and Ni are proportionally increased by one

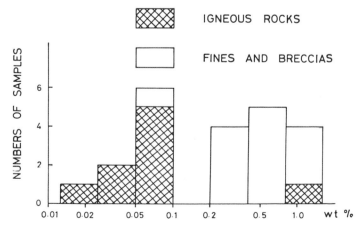

Fig. 3. Histogram of the weight content of metallic iron in Apollo lunar materials, deduced from their saturation magnetization values.

order of magnitude in the lunar fines and clastic rocks in comparison with the lunar igneous rocks. It is very likely then that part of the excess iron and nickel in the lunar fines and clastic rocks are due to the mixing of Fe and Ni which originates in meteorites, as already suggested by a number of investigators. This interpretation could be further justified by the observed fact that the Co content does not change much in all lunar materials, because the Co content is 10% or less of that of Ni in chondorites and iron meteorites. It is possible then that only about 10% of native iron in the lunar fines and clastic rocks have come from the original lunar rocks, while the remaining parts may have come from Fe with several percent of Ni contained in impacting meteorites. Alternate sources for excess iron have been discussed by Ghose *et al.* (1972) and Housley *et al.* (1971).

Paramagnetic susceptibility (χ_a)

The paramagnetic susceptibility (χ_a) is also one of the intrinsic magnetic parameters of the lunar materials. The paramagnetic contribution of Fe^{2+} in the materials controls χ_a. The observed values of χ_a of Apollo 14 and 15 materials at room temperature (Table 1) range, except for 15418,41, from 2.0×10^{-5} to 4.6×10^{-5} emu/gm. This range is not particularly different from those of Apollo 11 and 12 materials (Nagata *et al.*, 1970; Thorpe *et al.*, 1970; Nagata *et al.*, 1971; Sullivan *et al.*, 1971). The χ_a values observed at room temperature include not only a pure paramagnetic component of paramagnetic pyroxenes but also the paramagnetic phase of antiferromagnetic ilmenite and probably ferrosilite ($FeSiO_3$) and/or pyroxferroite.

Initial reversible magnetic susceptibility (χ_0)

In Table 1 χ_0 represents the initial reversible magnetic susceptibility measured at room temperature by a susceptibility bridge operated at 1000 Hz. It should be noted that the ratio χ_0/I_s is particularly large for fines (14259,69) and the friable, fine-grained clastic rock 14047,47. As pointed out by Runcorn *et al.* (1970) and Grommé and Doell (1971), the observed initial magnetic susceptibility of the lunar materials is almost always too high to be entirely attributable to the magnetic susceptibility of multidomain iron grains which is approximately 3×10^{-2} emu/gm. The corresponding value of χ_0/I_s amounts to approximately 1.4×10^{-4}. Runcorn *et al.* (1970) and Nagata and Carleton (1970) suggested that the high values of χ_0/I_s are very likely due to the superparamagnetic susceptibility of single-domain grains of metallic iron contained in the lunar materials. Actually, samples 14259,69 and 14047,47 have markedly large viscous components as described in a later section.

Saturation remanent magnetization (I_R), *coercive force* (H_c), *and remanence coercive force* (H_{RC})

The saturation remanent magnetization I_R evaluated from the hysteresis curve measured between -16 and $+16$ kOe at room temperature is a typical structure-sensitive magnetic parameter (refer to Table 1). The ratio I_R/I_s (Table 1) ranges from 0.6×10^{-2} to 4.4×10^{-2} for different samples, implying a considerably larger range for the blocking mechanism of isothermal remanent magnetization.

The coercive force H_c and the remanence coercive force H_{RC} in Table 1 are structure-sensitive magnetic parameters independent of I_s, provided that the ferromagnetic component is of a single phase. It is certain, however, that the ferromagnetic component of lunar materials is composed of at least two phases; for example, a very soft component I_F consisting of superparamagnetic or nearsuperparamagnetic particles which have practically no remanent magnetization, and soft multidomain particles which have small remanent magnetization and a magnetically hard component I_H that can carry a fairly stable remanent magnetization at room temperature. From these theoretical considerations, the remanence coercive force of an assemblage of fine grains of ferromagnetics has been estimated to be twice or less as large as its coercive force (Gaunt, 1960). However, the ratio H_{RC}/H_c of the lunar materials given in the table is far larger than two. As discussed generally by Wohlfarth (1963) and others, the abnormally large value of H_{RC}/H_c of an assemblage of single-domain ferromagnetic particles can be attributed only to a superposition of a magnetically very soft component and a hard one. Then the total magnetization $I(H)$ is expressed by

$$I(H) = (1 - \alpha)I_H(H) + \alpha I_F(H), \tag{1}$$

where $\alpha < 1$ denotes the weight content of the soft component and $I_F (H = 0) \approx 0$. Equation (1) leads to

$$I_R = I (H = 0) = (1 - \alpha)I_H (H = 0), \tag{2}$$

$$(1 - \alpha)I_H (H = -H_c) = \alpha I_F (H = -H_c). \tag{3}$$

In equation (3), $H_c = 0$ for $\alpha = 1$ and $H_c = H_c^0$ (i.e., the coercive force of the hard component which is about a half of H_{RC}) for $\alpha = 0$. Thus, the ratio H_{RC}/H_c can increase from about two to infinity with an increase of α from zero to unity. The extremely large values of H_{RC}/H_c of the lunar fines (14259) and the very friable clastic rocks (14047 and 14301) can be interpreted as due to a larger portion of the magnetically soft component in these materials.

MAGNETIC PROPERTIES AT LOW TEMPERATURES

Magnetic hysteresis curves of six Apollo 14 lunar materials were measured at 5°K also. The magnetic hysteresis of 14311,45 was measured at 12.5°K. The magnetic parameters at the low temperature were compared with those at room temperature. Figure 4 illustrates an example of a pair of magnetic hysteresis curves measured at 300°K and at 5°K. As summarized in Table 1 all ferromagnetic parameters, I_s, I_R, H_c, and H_{RC}, change considerably at the low temperature as compared with the corresponding values at room temperature; particularly the I_R and H_c values of samples 14259, 14047, and 10048 increase markedly with decreasing temperature from 300°K to 5°K.

As discussed in the foregoing section, the ratio H_{RC}/H_c becomes increasingly larger than 2 as the percentage of superparamagnetically soft component in the material increases. The I_R (300°K) value represents the saturation remanent magnetization of only the hard component at room temperature. The increase in I_R with

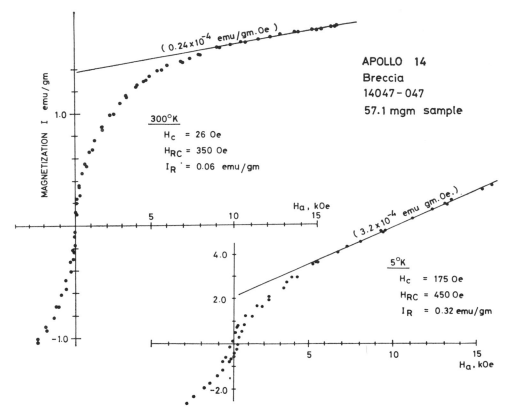

Fig. 4. Example of the magnetic hysteresis curve measured at 300°K and 5°K.

decreasing temperature can reasonably be interpreted as a transition of fine particle irons from the superparamagnetic state at 300°K to a normal ferromagnetic state at 5°K. Similarly, an increase of H_c with decreasing temperature also is attributable to the same transition of magnetic state of ferromagnetic grains caused by a decrease in temperature. It will be seen in Table 1 that the ratio H_{RC}/H_c which is over 10 at room temperature approaches 2 at the low temperature. The approach of H_{RC}/H_c toward 2 at the low temperature may represent an approach of α toward zero in equation (1), whence I_R approaches to I_H ($H = 0$), in equation (2), and H_c in equation (3) approaches the coercive force (H_c^0) of the hard component with the decrease of α. More quantitative discussion on this problem, however, meet some difficulties regarding the distribution spectrum of the grain size of superparamagnetic iron particles and the magnetic properties of individual particles. Further experimental examinations for individual samples is needed.

In Table 1 the I_s values at the low temperature are considerably larger than that at room temperature for all samples. The ratio I_s (5°K)/I_s (300°K) is particularly large (being nearly 2) in those materials whose H_c(5°K)/H_c (300°K) and I_R (5°K)/I_R (300°K) are abnormally large. If the observed I_s values are considered to represent the true magnitudes of the saturation magnetization at respective temperatures, the rate of increase of the saturation magnetization with a decrease in temperature becomes unreasonably large for metallic irons or similar alloys. This increase could be attributed to superparamagnetic particles only if a significant fraction of native iron is in extremely fine particles. In Fig. 5 the thermomagnetic curves (from 4.2°K to 70°K) of five Apollo 14 materials in a magnetic field of 13.5 kOe are illustrated. These magnetization versus temperature curves in the temperature range between 4.2°K and 20°K cannot be represented by a simple expression as

$$I(H, T) = \left(\frac{C}{T}\right) H + I_s(H), \tag{4}$$

where C represents the Curie constant and H represents a constant high magnetic field. In Fig. 6, an example of the $\partial \chi / \partial (1/T)$ versus temperature curve is shown together with the original I versus T curve, where $\chi = I/H$. If $I_s(H)$ in equation (4) is independent of temperature in the low temperature range concerned, then there must hold

$$\frac{\partial \chi}{\partial (1/T)} = \frac{1}{H} \frac{\partial I}{\partial (1/T)} = C. \tag{5}$$

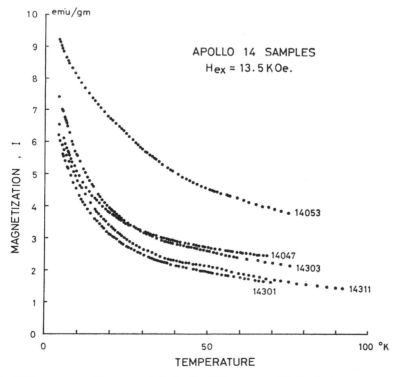

Fig. 5. Thermomagnetic curves of six Apollo 14 lunar materials for the low temperature range between 4.2°K and 70°K.

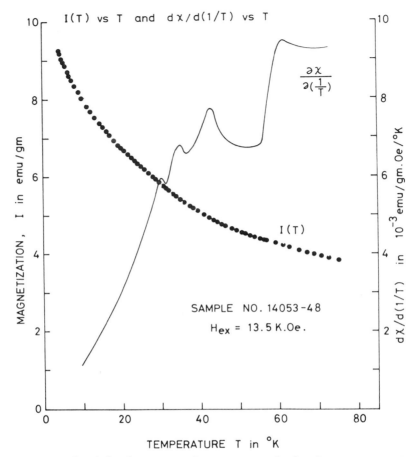

Fig. 6. Example of the thermomagnetic curve, magnetization I versus temperature T, and the differential thermomagnetic curve $(1/H)[\partial I/\partial(1/T)]$ versus T.

However, the $\partial \chi/\partial(1/T)$ of all observed data is not at all constant but decreases sharply with a decrease in temperature. This would mean that Curie constant C and consequently the paramagnetic susceptibility χ_a become relatively smaller at the lower temperature. In Table 1, the apparent value of Curie constant derived at $C = \chi_a T$

Table 2. Magnetic properties of individual coarse grains in sample no. 14161,38.

Grain no.	Weight (m gm)	I_s (emu/gm)	χ_a (at 300°K) (emu/gm/Oe)	H_c (Oe)	Metallic Fe (wt. %)	I_n (emu/gm)	I_n/I_s
1	25.0	1.13	0.21×10^{-4}	($\lesssim$30)	0.52	4.48×10^{-6}	4.0×10^{-5}
2	26.2	1.4	0.33×10^{-4}	28.5	0.64	127.0×10^{-6}	9.1×10^{-5}
3	33.6	0.02	0.24×10^{-4}	($<$70)	0.009	17.6×10^{-6}	8.8×10^{-4}
4	42.3	0.19	0.24×10^{-4}	($<$70)	0.09	8.3×10^{-6}	4.4×10^{-5}
5	34.2	~0	0.12×10^{-4}	—	~0	—	—
6	33.2	0.06	0.25×10^{-4}	—	0.027	—	—
7	44.0	1.6	0.32×10^{-4}	27	0.73	93.0×10^{-6}	5.8×10^{-5}

at the low temperature is considerably smaller than that at room temperature in all examined samples.

Obviously, the presence of antiferromagnetic components in the materials seriously affects the $\partial \chi / \partial(1/T)$ versus temperature relationship. If a material contains several antiferromagnetic components, then the I versus T relation of the material at temperatures higher than its highest Néel point temperature may be expressed, in place of equation (4), as

$$I(H, T) = \left(\frac{C}{T}\right) H + H \sum_i \frac{C_i}{T - \Theta_i} + I_s(H) \tag{6}$$

where the C_i are the Curie-Weiss constants and Θ_i the Curie-Weiss temperature. Then,

$$\frac{\partial \chi}{\partial(1/T)} = C + \sum_i \frac{C_i}{(1 - \Theta_i/T)^2}. \tag{7}$$

Below the Néel point of each antiferromagnetic component, the change of antiferromagnetic susceptibility $\chi_A(T)$ with temperature is considerably smaller than that above the Néel point and the average value of $\partial \chi_A / \partial(1/T)$ is negative. The $\partial \chi_A / \partial(1/T)$ just below Néel point is always negative. Hence, $\partial \chi / \partial(1/T)$ must discontinuously increase at each Néel point temperature. For example, the abrupt change of the $\partial \chi / \partial(1/T)$ curve between $55°K$ and $50°K$ in Fig. 6 corresponds to the Néel point ($56°K$) of ilmenite.

The second term in equation (7) must be replaced by a negative value below each Néel point temperature, giving rise to a peak in $\partial \chi / \partial(1/T)$ at each transition temperature. It seems likely in Fig. 6 that at least several antiferromagnetic phases are present in addition to ilmenite, and that their Néel points are below $50°K$. It is difficult at present to deal more quantitatively with this specific problem. It could be suggested however that the antiferromagnetic phases may be due to special compositions of pyroxene, including ferrosilite. In addition to these effects, a detailed quantitative analysis must consider contributions from fine iron grains.

VISCOUS MAGNETIZATION

The viscous magnetization is one of the remarkable characteristics of the lunar rock magnetism. All lunar materials examined so far are more or less subjected to the effect of viscous magnetization, which is mainly due to superparamagnetically fine grains of native iron contained in the lunar materials (Nagata et al., 1970; Nagata and Carleton, 1970; Nagata et al., 1971; Stephenson, 1971).

In laboratory demonstration, the viscous magnetization ΔI and the stable magnetization I_0 can be defined by a time-dependence of the decay in a nonmagnetic space of isothermal remanent magnetization (IRM) acquired after an extended time in a magnetic field (in practice the order of one day is sufficient). This is empirically expressed as

$$I(t, H) = I_0(H) + \Delta I(t, H), \tag{8}$$

where $\Delta I(t, H)$ decays from $\Delta I_0(H)$ at $t = 0$ to zero for $t = \infty$. In paleomagnetic studies using lunar rocks, the viscous magnetization acquired in the geomagnetic field may be comparable with the natural remanent magnetization (NRM : I_n). In Table 3, the ratios $\Delta I(H_0)/I_n$ and $\Delta I(H)/I_0(H)$ for $H = 10$ Oe of six Apollo 14 lunar rock samples are listed. No exact measurement was carried out for the fine sample 14259, but it was experimentally checked that $\Delta I_0(H)/I_0(H) > 10$ for this sample. It is clear in Table 3 that the lunar samples can be classified into a group whose $\Delta I_0(H)/I_0(H)$ is smaller than 0.1 (Type I) and the other group, in which the ratio is larger than unity (Type II). In Table 1 also, the Apollo 14 lunar materials are classified into Types I and II. Generally speaking, the ratios, I_R (5°K)/I_R (300°K), H_c (5°K)/ H_c (300°K), and H_{RC} (300°K)/H_c (300°K), are extremely large for Type II samples, while these ratios are much smaller for Type I. In addition, the saturation magnetization of Type I samples is not appreciably reduced by heating to 820°C in 10^{-5} Torr atmosphere. However, the saturation magnetization of Type II samples is considerably reduced by the same experimental process, as shown in Table 1 in terms of the ratio of metallic iron contents before and after the heat treatment. All these experimental results consistently indicate that Type II lunar materials contain a considerable portion of very fine particles of metallic iron which behave superparamagnetically at room temperature and are easily oxidized by the heat treatment, while Type I materials contain only a small fraction of such fine metallic irons.

As shown by Néel (1949), the relaxation time τ of a single domain particle of v in volume and H_{RC} in remanence coercive force is given by

$$\frac{1}{\tau} = f_0 \exp\left(-\frac{H_{RC}J_s^0 v}{2kT}\right) \tag{9}$$

where J_s^0 and f_0 denote, respectively, the saturation magnetization of magnetic particle per unit volume and a numerical constant of about 10^9 sec^{-1} depending slightly on temperature T and H_{RC}. Thus, a small increase in v or a small decrease in

Table 3. Natural Remanent magnetization and its stability of Apollo 14 and Apollo 15 lunar samples.

Sample no.	I_n (emu/gm)	I_n/I_s	$\tilde{H}_0$ (Oe rms)	$\tilde{H}^*$	h (Oe)	$\tilde{H}_0'$ (Oe rms)	$\Delta I_0(H)/ I_0(H)$	$\Delta I_0(H_0)/I_n$	S_0' (emu/gm)
14053,31	2.2×10^{-3}	1.0×10^{-3}	19	80	50	11	0.055	1.1×10^{-4}	1.9×10^{-7}
14310,159	2.4×10^{-6}	—	>25	>25	12	—	—	—	—
14047,47	8.2×10^{-6}	5.9×10^{-6}	45	30	7.5	8	16.3	4.3	—
14063,47	14.1×10^{-6}	4.1×10^{-4}	145	150	47	16	0.046	1.4×10^{-3}	2.1×10^{-8}
14301,65	41.2×10^{-6}	6.0×10^{-5}	48	40	23	8	2.64	7.3×10^{-4}	—
14303,35	131.2×10^{-6}	1.0×10^{-4}	18	40	29	9	0.034	1.9×10^{-4}	2.1×10^{-8}
14311,23	8.1×10^{-6}	1.1×10^{-5}	8	10	14	4	0.050	4.8×10^{-3}	4.2×10^{-8}
14259,69	—	—	—	—	—	—	>10	—	—
15058,55	10.6×10^{-6}	8.2×10^{-5}	17	20					$<1 \times 10^{-7}$
15058,70	17.9×10^{-6}	—	—	—					—
15418,41	7.7×10^{-6}	5.1×10^{-5}	50	60					$<7 \times 10^{-8}$
15418,46	2.5×10^{-6}	—	—	—					—
15556,37	1.7×10^{-6}	1.4×10^{-5}							
15556,38	20.6×10^{-6}								

T results in an exponentially large increase in τ. If the distribution spectrum of τ is expressed by $\psi(\tau)$ for a sample, then the viscous magnetization, $\Delta I(t, H)$, is given by

$$\Delta I(t, H) = \int_0^\infty \psi(\tau) e^{-t/\tau} d\tau, \qquad (10)$$

where $\psi(\tau)$ is dependent on both T and H and τ is dependent on T. For Type I samples,

$$\int_0^\infty \psi(\tau) d\tau = \Delta I_0(H)$$

is much smaller than I_0, and therefore the functional form of $\psi(\tau)$ is difficult to determine exactly from experimental data. Assuming, however, that $\psi(\tau) = S/\tau$ for $\tau_1 \leq \tau \leq \tau_2$ and $\psi(\tau) = 0$ for $\tau < \tau_1$ and $\tau > \tau_2$ for experimental data of $\tau_1 \ll t \ll \tau_2$, $\Delta I(t, H)$ is given by constant-$S' \log_{10} t$ (Néel, 1949; Nagata and Charleton, 1970).

As is well known (Nagata, 1961), S' is approximately proportional to H and T (in °K). The S' values for $H = 1$ Oe and $T = 300°K$ (S_0') are summarized for Type I samples in Table 3. A comparison of the intensity of natural remanent magnetization I_n with the coefficient S_0' for each of the Type I samples in Table 3 indicates that the acquisition of viscous remanent magnetization in a magnetic field of $H = 1$ Oe at room temperature up to the same intensity as the observed value of I_n does need at least 10^{200} seconds. Therefore, a hypothesis that the observed natural remanent magnetization of Type I lunar rocks may be due to VRM acquired in the geomagnetic field after these samples were returned to the earth, can safely be ruled out.

The dominant VRM of Type II lunar materials has already been discussed from various viewpoints (e.g., Doell et al., 1970; Larochelle and Schwarz, 1970; Nagata and Carleton, 1970; Nagata et al., 1971). It has been concluded in these studies that very fine grains of metallic iron of 100–200 Å in diameter are dominantly present in Type II lunar materials. In the present studies on Apollo 14 lunar materials of Type II, the spectral form of $\psi(\tau)$ has been determined from the observed curve of $\Delta I(t, H)$ versus t with the aid of equation (10) (Nagata et al., 1972). Results of the analyses have shown that $\psi(\tau)$ is a widely spread continuous spectrum with respect to τ; a peak of $\psi(\tau)$ is around 5 sec for sample 14047 and an extremely high peak of $\psi(\tau)$ is about 10–15 sec for sample 14301. With the aid of equation (9), the dominant grain size of metallic iron in these samples is estimated to be about 170 Å in mean diameter. Although the stable component of remanent magnetization I_0 can be apparently identified in these samples of Type II after storing them in a nonmagnetic space for a sufficiently long time, say about a week, extreme care must be taken in discussing the stable natural remanence and its interpretation for these samples.

MAGNETISM OF INDIVIDUAL COARSE GRAINS OF LUNAR FINES

Magnetic properties of seven comparatively large individual grains selected at random from an assemblage of Apollo 14 coarse fines (14161–38) have been examined. Figure 7 shows three examples of representative magnetization curves measured at room temperature. Grain #3, #4, and #7 contain metallic iron of 0.01, 0.09, and 0.73

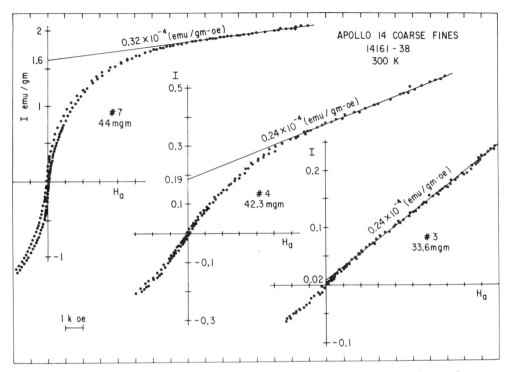

Fig. 7. Examples of the magnetic hysteresis curves at room temperature of strongly magnetic (#7), intermediately magnetic (#4) and weakly magnetic (#3) grains in coarse fines 14161,38.

wt.%, respectively, while their paramagnetic susceptibilities are approximately the same, being $(2-3) \times 10^{-5}$ emu/gm. The measured values of I_s and χ_a of seven grains are summarized in Table 2. It may be noted in the table that the magnetic characteristics of the coarse grains of lunar fines can be classified into three groups, namely (i) magnetically strong grains such as #1, #2, and #7 whose I_s values are larger than 1 emu/gm, (ii) magnetically intermediate grains such as #4, and (iii) magnetically weak grains such as #3, #5, and #6 whose I_s values are less than 0.06 emu/gm. This result may suggest that the distribution of metallic iron in individual grains of the lunar coarse fines ranges about two orders of magnitude. The distribution spectrum of the metallic iron content is in agreement with a similar spectrum for Apollo 11 and 12 glass spherules (Thorpe *et al.*, 1970; Sullivan *et al.*, 1971).

The natural remanent magnetization (NRM) of five coarse grains was also measured, the results are shown in Table 2. The group of grains which contain the largest amount of metallic iron (0.5–0.7 wt.%) have considerably larger NRM than the other two groups. It must be noted however that the intensity I_n of NRM is not simply proportional to I_s among five grains selected at random from a group of the coarse lunar fines. Particularly, #3 grain has an extremely large value of the ratio I_n/I_s compared with the other grains. It may be suggested that #3 grain was at least once subjected to some mechanism of aquisition of a comparatively strong NRM— such as thermoremanent magnetization or shock remanent magnetization.

NATURAL REMANENT MAGNETIZATION AND ITS STABILITY

The natural remanent magnetization (NRM) of seven rock samples (igneous and clastic) of Apollo 14 lunar materials and six samples of Apollo 15 lunar materials has been examined. The intensity of NRM and the stability against the AF demagnetization of nine samples are summarized in Table 3, where the effective AF-demagnetization field, $\tilde{H}_0$, is defined as the intensity of demagnetization field to reduce the intensity of NRM to $1/e$ of its initial value and the critical AF-demagnetization field $\tilde{H}^*$ represents the maximum demagnetization field below which the direction of NRM is approximately invariant, that is within $\pm 10°$ in most cases (Nagata *et al.*, 1971), and h represents the intensity of magnetic field which can produce IRM whose intensity is the same as that of NRM. The possibility was ruled out in the preceding section that VRM in the geomagnetic field produced the observed NRM of Type I lunar rocks. In Table 3, the possibility of IRM may also be ruled out as the origin of observed NRM of samples 14053,31, 14063,47, and 14303,35, because it can hardly be considered that these lunar rock samples were once exposed to a steady magnetic field larger than 30 Oe in intensity.

On the other hand the values of $\tilde{H}_0$ are smaller than 50 Oe and some of them are smaller than 20 Oe. Sample 14063,67 is an exception. This comparatively weak coercivity against the AF-demagnetization may imply that the major part of the observed NRM can hardly be attributable to the thermoremanent magnetization (TRM). Another method of testing the stability of NRM is a comparison of the AF-demagnetization curve of NRM with that of IRM whose intensity is the same as NRM. In Fig. 8, for example, the AF-demagnetization curve of the stable component of IRM

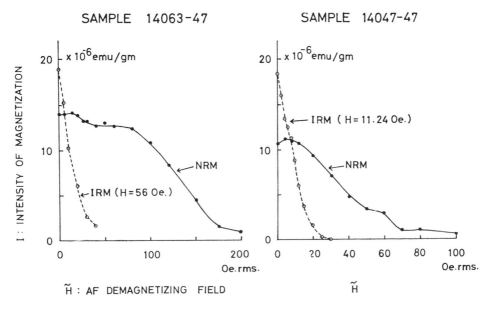

Fig. 8. Comparison of the AF-demagnetization curve of NRM with that of IRM for two Apollo 14 clastic rocks.

whose intensity is a little larger than that of NRM is compared with the AF-demagne-
tization curve of the NRM. For the two examples, NRM is definitely more stable than
IRM against the AF-demagnetization. For the sake of comparison, the $\tilde{H}_0'$ values in
Table 3 represent the effective AF-demagnetization field for the IRM whose intensity
is the same as NRM of the same samples. It may be concluded from these results that
NRM of sample 14063,47 is particularly stable and reliable.

As shown in Fig. 9 for example, the AF-demagnetization curve of sample 15418,41
seems to indicate that NRM of this sample consists of a comparatively soft component
and a hard one. Since the hard component of IRM cannot be destroyed by an AF-
demagnetization field of 250 Oe rms, this component could be considered a stable
and reliable NRM, which may be attributable to TRM or a similar stable remanent
magnetization.

Sample 14310,159 has been oriented in the laboratory based on weathering
characteristics. Based on this orientation, the inclination of NRM of this sample was
39.5° downward from the horizontal plane. This compares with observed values of
64° at Site A and 22° at Site C′ (Dyal *et al.*, 1971).

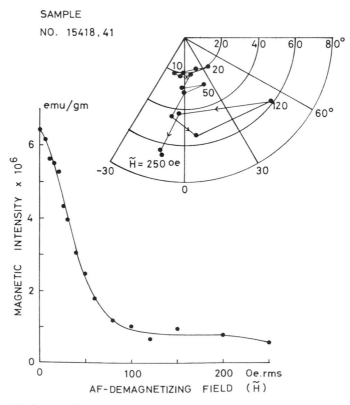

Fig. 9. AF-demagnetization curve of an Apollo 15 breccia. Top: changes in the
direction. Bottom: changes in the intensity.

Special Studies of an Unusually Magnetic Igneous Rock 14053

As shown in Tables 1 and 3, Apollo 14 igneous rock 14053 is unusually strong in every aspect of its magnetic properties. NRM of this sample also is unusually intense, amounting to about 2×10^{-3} emu/gm. Two specimens of this rock sample, 14053,31 (5.27 gm) and 14053,48 (0.97 gm), were examined magnetically: NRM of 14053,31 is 2.2×10^{-3} emu/gm while that of 14053,48 is 2.7×10^{-3} cmu/gm. A chip (0.304 gm in weight) taken from 14053,48 has only 0.80×10^{-3} emu/gm in its NRM intensity. Thus, the distribution of NRM intensity within a rock mass of 14053 seems to be appreciably heterogeneous, but still the intensity of NRM of this rock is unusually high, because the NRM intensity of all other lunar igneous rocks so far reported is smaller than 2×10^{-3} emu/gm.

Anhysteretic remanent magnetization (ARM)

It has been pointed out by several people that some lunar rocks shielded in a teflon bag such as sample 14053,31 might have acquired the anhysteretic remanent magnetization (ARM) under the effect of an alternating magnetic field in the presence of the steady geomagnetic field. To examine the ARM characteristics of sample 14053,31, the dependence of the intensity of ARM (I_{ARM}) on applied alternating magnetic fields $\tilde{H}$ in a steady constant magnetic field $h = 0.5$ Oe, and the dependence of I_{ARM} on h with a constant alternating magnetic field, $\tilde{H} = 30$ Oe rms, were experimentally studied. The experimental results have given the following dependence of I_{ARM} on $\tilde{H}$ (in Oe rms) and h (in Oe),

$$I_{ARM}(\tilde{H}, h) = (8.81 \, \tilde{H} - 0.086 \, \tilde{H}^2) \cdot h \times 10^{-6} \text{ emu/gm.} \tag{11}$$

This experimental result has indicated that I_{ARM} is only 1.08×10^{-4} emu/gm for $\tilde{H} = 40$ Oe rms in $h = 0.5$ Oe. There would be almost no possibility therefore that the observed NRM of sample 14053 is due to the ARM acquired in the geomagnetic field.

Pressure remanent magnetization (PRM) and shock remanent magnetization (SRM)

The static pressure remanent magnetization (PRM) of this sample has been experimentally demonstrated for a weak pressure range from zero to 50 bar. The dependence of the intensity of PRM (I_{PRM}) on a magnetic field H and a uniaxial compression P, when the direction of H is parallel to the axis of P, is represented by an empirical equation as

$$I_{PRM} = 2.28 \times 10^{-7} \text{ (emu/gm/Oe bar) } H \cdot P. \tag{12}$$

Experimentally, the above relationship can hold for $H \lesssim 10$ Oe and $P \lesssim 200$ bars. If we assume that the dependence of I_{PRM} on P can be extended linearly to much higher values of P, then the product $H \cdot P$ to produce $I_{PRM} = 2 \times 10^{-3}$ emu/gm amounts to about 10^4 Oe bar, which corresponds to $H = 10^4 \, \gamma$ even for $P = 10^2$ kbar.

The acquisition of shock remanent magnetization (SRM) in case of $H//S$ also has been experimentally demonstrated on this sample. The experimental data can be summarized by

$$I_{SRM} = 5.1 \times 10^{-4} \text{ (emu/gm/Oe bar sec) } H \cdot S. \tag{13}$$

The shock wave applied in the experiment is a single peak shock of about 0.45 millisec in time width (Nagata, 1971), whence equation (13) can be rewritten in terms of the maximum pressure peak (P_m) as

$$I_{SRM} = 2.3 \times 10^{-7} \text{ (emu/gm/Oe bar) } H \cdot P_m, \tag{13'}$$

which is practically identical to equation (12).

Thermoremanent magnetization (TRM)

Dunn and Fuller (1972) have experimentally demonstrated the acquisition of TRM of a chip of this sample, the NRM intensity of which is 0.80×10^{-3} emu/gm. The acquisition rate of TRM in this sample is not too high. The acquisition of TRM of 8×10^{-4} emu/gm in intensity requires about 0.62 Oe for an applied magnetic field in the course of TRM acquisition process. This result contrasts to previous results reported for other lunar igneous and clastic rocks (e.g., Nagata *et al.*, 1970; Helsley, 1970; Runcorn *et al.*, 1970; Doell *et al.*, 1970; Nagata *et al.*, 1971; Helsley, 1971; Pearce *et al.*, 1971), which have suggested that the required magnetic field for producing TRM of the same intensity as that of observed NRM is several thousands of gamma.

Dunn and Fuller have shown that the dependence of TRM intensity (I_{TRM}) on the applied magnetic field deviated considerably from the linear relationship, $I_{TRM} \sim H$ even for a weak magnetic field range of 10^3–10^5 γ. The dependence is better represented by

$$I_{TRM} = I_0 \tanh (aH) \approx 5 \times 10^{-4} \text{ (emu/gm) } \tanh (6H), \tag{14}$$

which is an experimentally derived expression (Nagata, 1943) for terrestrial basalts for high magnetic fields (i.e., $10 \sim 40$ Oe), and which was theoretically introduced by Néel (1949) for an assemblage of single-domain particles of ferromagnetics. For an assemblage of iron particles of mean diameter 300 Å or less, the coefficient of H in equation (14) would be much smaller than the empirical value given. This result suggests that this sample contains a considerable amount of multi-domain iron particles.

AF-demagnetization of NRM, IRM, and TRM

As shown in Table 3, the intensity of NRM of this sample is reduced to $(1/e)$ of the initial value by the AF-demagnetization only up to 19 Oe rms, but its direction is kept substantially invariant against the AF-demagnetization of less than 80 Oe rms in intensity. This result may indicate that NRM of this sample is magnetically soft, though the direction of the major parts of NRM is not at random. As given in Table 1, the H_{RC} value of this sample is extremely small, indicating that single-domain iron particles occupy only a small portion of native irons in this sample. Hence it may be reasonably presumed that the majority of NRM of this sample is borne by magnetically soft multi-domain particles of iron.

Dunn and Fuller (1972) have compared that AF-demagnetization curve of the total TRM of this sample acquired in $H = 5 \times 10^3\,\gamma$ and $H = 2 \times 10^4\,\gamma$ with the AF-demagnetization curve of the NRM.

The AF-demagnetization curve of the total TRM can be approximated for a range of $0 \sim 50$ Oe rms of $\tilde{H}$ by

$$I_{\text{TRM}}(\tilde{H}) = I_{\text{TRM}}(0) \exp(-\alpha\tilde{H}), \tag{15}$$

where the decay constant α amounts to about 5×10^{-2} (Oe rms)$^{-1}$. The observed value of α for TRM is approximately the same as the decay constant of the AF-demagnetization of NRM and the saturated IRM for the same range of $\tilde{H}$. It may thus be summarized that TRM of this sample also is unusually soft compared with the magnetic hardness of TRM of any other lunar rock. This result also may suggest that multi-domain iron particles are dominant and the major parts of TRM are borne by these iron particles in this sample.

Thermal demagnetization of NRM

Figure 10 illustrates a thermal demagnetization curve of NRM of sample 14053,48. The major parts of NRM can be thermally demagnetized by heating to about 350°C in a nonmagnetic vacuum space. However, a small fraction of the NRM remains above 350°C until it disappears at the Curie point of this sample, the dependence of the demagnetization of the remaining portion on temperature being reasonably similar to that of the stable TRM of other lunar rocks (e.g., Dunn and Fuller, 1972).

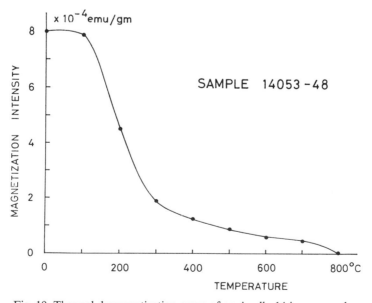

Fig. 10. Thermal demagnetization curve of an Apollo 14 igneous rock.

If we divide, in Fig. 10, the TRM component of this sample into the high temperature blocking group and the low temperature blocking group (which can be thermally demagnetized by heating to 350°C), the intensity of the high temperature blocking TRM component would be approximately 1×10^{-4} emu/gm at room temperature. A possible interpretation is that the high temperature blocking part represents the true TRM acquired by single-domain iron particles.

Discussions of the possible origin of the intense NRM

All experimental results have led to a conclusion that magnetically soft multi-domain particles comprise the major portion of metallic iron in sample 14053. There is little doubt that the major parts of NRM of this sample are borne by such multi-domain particles. However, no definite evidence has been derived from these experiments for identifying the origin of the observed intense NRM of this sample. It seems that the TRM mechanism is the most probably in the sense of having no serious contradiction between the NRM characteristics and the TRM ones, with the provision that a magnetic field as high as 0.6 Oe could be assumed to be present at the locality of this rock when it was cooled. A considerable hypothesis would be that there is a certain unknown mechanism which produces the remanent magnetization as a combined effect of a strong mechanical shock and a field-cooling in a weak magnetic field from an adiabatically raised high temperature.

CONCLUDING REMARKS

Basic magnetic properties of Apollo 14 and 15 lunar materials examined in the present work have revealed that their ferromagnetism is substantially due to the metallic iron grains contained in them. High temperature $I(T)$ data for four samples of lunar breccias have demonstrated that their native irons are dominantly kamacites, in which the Ni content is less than 7 wt.%. The magnetic viscosity characteristics of these lunar materials can be classified into two distinctly separated groups: (a) Type I which contains only a small portion of superparamagnetically soft component of fine iron particles and (b) Type II which contains the superparamagnetically soft component as its dominant magnetic constituent. Here, the superparamagnetically soft component is defined as an assemblage of fine single-domain magnetic particles whose relaxation time ranges between 0.1 and 10^5 sec at room temperature. At liquid helium temperature, the relaxation time of these particles is much greater and consequently the magnetization of these particles becomes completely nonviscous. It is evident that the viscous magnetization due to the nearly superparamagnetical soft component must be carefully eliminated from the remanent magnetization when one deals with NRM of returned lunar rocks for paleomagnetic purposes. The important point in this regard is that Type II materials also contain a certain amount of stable single-domain iron particles whose relaxation time at room temperature is extremely long, say $10^{20} \sim 10^{30}$ seconds, whereas the stable component of Type I materials does not always represent the stable single-domain particles but sometimes the multi-domain iron grains whose magnetic coercivity is much smaller than that of stable single-domain iron particles. The above-mentioned physical situation may be typically represented by relatively large values of H_{RC} of Type II materials and relatively small values of H_{RC} of Type I materials, particularly in case of sample 14053. With the aid of equation (9), the volume of an iron particle whose relaxation time is 10 sec at room temperature (the peak value of τ in sample 14047 for example) is estimated to be about 3×10^{-18} cm^3. On the other hand, the largest possible volume of a single-domain particle of iron is about 2×10^{-17} cm^3, the relaxation time of which is 7×10^{61} sec and 5×10^{12} sec, respectively, at room temperature and at 750°C. Since a continuous distribution of iron particle volume has been experimentally found for the lunar clastic rocks (Nagata et al., 1972), a certain portion of the iron particles will have the grain size which can acquire the stable TRM. For example, the relaxation time of an iron particle of 1×10^{-17} cm^3 in volume is 5.5×10^{25} sec at room temperature but it becomes only 26 sec at 750°C. As shown in Table 3, the stable component of NRM of Apollo 14 clastic rocks of Type II, 14047 and 14301, have relatively large values of H_0. According to Dunn and Fuller (1972) the total TRM of sample 14047,47 acquired in $H = 5 \times 10^3 \gamma$ and $H = 2 \times 10^4 \gamma$ are 4.3×10^5 and 2.1×10^{-4} emu/gm, respectively. Hence the measured intensity of the stable component of NRM of this sample could be attributed to TRM acquired in a magnetic field of about 800 γ. Because of the dominant viscous magnetization of this sample, however, a more exact identification of the origin of NRM of this sample may still be necessary.

Conversely, as previously discussed, full reliance cannot be given to an interpretation of the acquisition mechanism of NRM of Type I sample 14053 because the remanent magnetization of this sample is mostly attributable to the soft magnetization of multidomain iron particles and the magnetic field required for the TRM interpretation is unusually high. Among seven samples of Apollo 14 lunar rocks examined in the present studies, it seems that NRM of sample 14063,47 can be sufficiently reliable in every aspect of experimental data of the stability tests. Examinations in further detail of such a stable NRM would be worthwhile.

A lunar rock sample whose magnetic properties have been studied in the most comprehensive ways is the Apollo 11 clastic rock 10048,55 (Nagata et al., 1971; Dunn and Fuller, 1971). Approximately half the NRM of this sample is stable against the AF-demagnetization up to 350 Oe rms. The acquisition rate of the total TRM is 3.3×10^{-3} emu/gm/Oe, and TRM thus acquired is very stable against the AF-demagnetization up to 400 Oe rms ($\alpha = 2 \times 10^{-3}$ (Oe rms)$^{-1}$). If we assume that the stable component of NRM is due to the TRM mechanism, then the magnetic field required for the TRM acquisition amounts to about 800 γ. A similar statement could also be made for sample 15418,41. Only preliminary studies have been made on Apollo 15 samples; thus, discussions in detail of Apollo 15 samples must depend on future experiments.

It seems most likely at the present stage of knowledge that NRM of these lunar rocks were acquired by the TRM mechanism in the presence of a magnetic field of about $10^3 \gamma$ on the lunar surface.

Acknowledgments—The authors' thanks are due to S. V. Radcliffe of the Case-Western Reserve University who has allowed them to make measurements of NRM of sample Nos. 14161,38, 14063,47, and 14310,159. They are indebted also to Y. Aoki of University of Tokyo for his assistance during the whole course of thermomagnetic experiments.

REFERENCES

Doell R. R. and Grommé C. S. (1970) Magnetic studies of Apollo 11 lunar samples. *Proc. Apollo 11 Lunar Sci. Conf., Geochim. Cosmochim. Acta* Suppl. 1, Vol. 3, pp. 2097–2102. Pergamon.

Dunn J. R. and Fuller M. D. (1972) Thermoremanent magnetization (TRM) of lunar samples. Proceedings of Conference on Lunar Geophysics. *The Moon* (in press).

Dyal P., Parkin C. W., Sonett C. P., DuBois R. L., and Simmons G. (1971) Lunar portable magnetometer experiment. *Apollo 14 Prelim. Sci. Rep.*, NASA SP-272, pp. 227–237.

Gaunt P. (1960) A magnetic study of precipitation in a gold-cobalt alloy. *Phil. Mag.* **8**, No. 5, 1127–1148.

Gose W. A., Pearce G. W., Strangway D. W., and Larsen E. E. (1972) On the magnetic properties of lunar breccias (abstract). In *Lunar Science—III* (editor C. Watkins), pp. 332–334, Lunar Science Institute Contr. No. 88.

Grommé C. S. and Doell R. R. (1971) Magnetic properties of Apollo 12 lunar samples 12052 and 12065. *Proc. Second Lunar Sci. Conf., Geochim. Cosmochim. Acta* Suppl. 2, Vol. 3, pp. 2491–2499. MIT Press.

Helsley C. E. (1970) Magnetic properties of lunar 10022, 10069, 10084, and 10085 samples. *Proc. Apollo 11 Lunar Sci. Conf., Geochim. Cosmochim. Acta* Suppl. 1, Vol. 3, pp. 2213–2219. Pergamon.

Helsley C. E. (1971) Evidence for an ancient lunar magnetic field. *Proc. Second Lunar Sci. Conf., Geochim. Cosmochim. Acta* Suppl. 2, Vol. 3, pp. 2485–2490. MIT Press.

Housley R. M., Grant R. W., and Abdel-Gawad M. (1972) Study of excess Fe-metal in the lunar fines by magnetic separation (abstract). In *Lunar Science—III* (editor C. Watkins), pp. 392–394, Lunar Science Institute Contr. No. 88.

Larochelle A. and Schwarz E. J. (1970) Magnetic properties of lunar sample 10048,22. *Proc. Apollo 11 Lunar Sci. Conf., Geochim. Cosmochim. Acta* Suppl. 1, Vol. 3, pp. 2305–2308. Pergamon.

Lovering J. F. and Parry L. G. (1962) Thermomagnetics analysis of co-existing nickel-iron metal phases in iron meteorites and thermal histories of the meteorites. *Geochim. Cosmochim. Acta* **26**, 361–382.

LSPET (Lunar Sample Preliminary Examination Team) (1971) Preliminary examination of samples from Apollo 14. *Science* **173**, 681–693.

LSPET (Lunar Sample Preliminary Examination Team) (1972) Preliminary examination of samples from Apollo 15. *Science* **175**, 363–375.

Nagata T. (1943) The natural remanent magnetism of volcanic rocks and its relation to geomagnetic phenomena. *Bull. Earthquake Res. Inst.* **21**, 1–196.

Nagata T. (1961) *Rock Magnetism*, pp. 1–350, Maruzen, Tokyo.

Nagata T. (1971) Introductory notes on shock remanent magnetization and shock demagnetization of igneous rocks. *Pure Appl. Geophys.* **89**, 159–177.

Nagata T. and Carleton B. J. (1970) Natural remanent magnetization and viscous magnetization of Apollo 11 lunar materials. *J. Geomag. Geoele.* **22**, 491–506.

Nagata T., Ishikawa I., Kinoshita H., Kono M., Syono Y., and Fisher R. M. (1970) Magnetic properties and natural remanent magnetization of lunar materials. *Proc. Apollo 11 Lunar Sci. Conf., Geochim. Cosmochim. Acta* Suppl. 1, Vol. 3, pp. 2325–2340. Pergamon.

Nagata T., Fisher R. M., Schwerer F. C., Fuller M. D., and Dunn J. R. (1971) Magnetic properties and remanent magnetization of Apollo 12 lunar materials and Apollo 11 lunar microbreccia. *Proc. Second Lunar Sci. Conf., Geochim. Cosmochim. Acta* Suppl. 2, Vol. 3, pp. 2461–2476. MIT Press.

Nagata T., Fisher R. M., and Schwerer F. C. (1972) Lunar rock magnetism. Proceedings of Conference on Lunar Geophysics. *The Moon* (in press).

NASA (1971) *Apollo 14 Preliminary Science Report*, NASA SP-272, U.S. Government Printing Office.

Néel L. (1949) Theorie des trainage magnétiques des ferromagnétiques an grains fin avec applications aux terrestres ciutes. *Ann. Géophys.* **5**, 99–136.

Pearce G. W., Strangway D. W., and Larson E. E. (1971) Magnetism of two Apollo 12 igneous rocks. *Proc. Second Lunar Sci. Conf., Geochim. Cosmochim. Acta* Suppl. 2, Vol. 3, pp. 2451–2461. MIT Press.

Runcorn S. K., Collinson D. W., O'Reilly W., Battey M. H., Stephenson A., Jones J. M., Manson A. J., and Readman P. W. (1970) Magnetic properties of Apollo 11 lunar samples. *Proc. Apollo 11 Lunar Sci. Conf., Geochim. Cosmochim. Acta* Suppl. 1, Vol. 3, pp. 2369–2387. Pergamon.

Schwarz E. J. (1970) Thermomagnetics of lunar dust sample 10084,88. *Proc. Apollo 11 Lunar Sci. Conf., Geochim. Cosmochim. Acta* Suppl. 1, Vol. 3, pp. 2389–2397. Pergamon.

Stephenson A. (1971) Single domain grain distribution, II. The distribution of single domain iron grain in Apollo 11 lunar dust. *Phys. Earth Planet. Interior* **4,** 361–369.

Strangway D. W., Larson E. E., and Pearce G. W. (1970) Magnetic studies of lunar samples— breccia and fines. *Proc. Apollo 11 Lunar Sci. Conf., Geochim. Cosmochim. Acta* Suppl. 1, Vol. 3, pp. 2435–2451. Pergamon.

Sullivan S., Thorpe A. N., Alexander C. C., Senftle F. E., and Dwornik E. (1971) Magnetic properties of individual glass spherules, Apollo 11 and Apollo 12 lunar samples. *Proc. Second Lunar Sci. Conf., Geochim. Cosmochim. Acta* Suppl. 2, Vol. 3, pp. 2433–2449. MIT Press.

Thorpe A. N., Senftle F. E., Sullivan S., and Alexander C. C. (1970) Magnetic studies of individual glass spherules from the lunar sample 10084,86,2, Apollo 11. *Proc. Apollo 11 Lunar Sci. Conf., Geochim. Cosmochim. Acta* Suppl. 1, Vol. 3, pp. 2453–2462. Pergamon.

Wohlfarth E. P. (1963) Permanent magnetic materials. In *Magnetism* (editors G. T. Rado and H. Suhl, pp. 351–393, Academic Press, New York.

Proceedings of the Third Lunar Science Conference
(Supplement 3, *Geochimica et Cosmochimica Acta*)
Vol. 3, pp. 2449–2464
The M.I.T. Press, 1972

Remanent magnetization of the lunar surface

G. W. Pearce

Lunar Science Institute, Houston, Texas 77058
and
Department of Physics, University of Toronto

D. W. Strangway

Geophysics Branch, NASA, Manned Spacecraft Center,
Houston, Texas 77058

and

W. A. Gose

Lunar Science Institute, Houston, Texas 77058

Abstract—Two lines of evidence support each other in suggesting that a large volume of the rocks near the lunar surface possess a uniform remanent magnetization with an intensity of about 2×10^{-6} emu/g. The first line is the discovery by several groups of investigators of weak but stable remanent magnetizations in igneous samples returned from the first four Apollo missions. Although the mechanism of acquisition of this remanence has not been definitely established, several lines of evidence, including thermal demagnetization, suggest that it is a thermoremanent magnetization (TRM) carried by iron. Many of the breccias are similarly magnetized. The second line is the measurement of significant fields at the Apollo sites and the discovery of large-scale anomalies by the sub-satellite magnetometer experiment. It appears that magnetized rocks are the source for these fields, and simple model calculations suggest that the magnetization typical of mare basalt can account for these features. This implies that the anomalies which appear on the far side of the moon are due to basalt beneath major craters. These fields are not so clear on the front side, but this is undoubtedly due to the edge anomalies expected over large sheets of lava. The implication of this magnetization is that the moon possessed a magnetic field at least during the period from about 4.0 to 3.0 b.y. The most probable explanation for the origin of such a field is magnetohydrodynamic dynamo action associated either with a molten metallic core or with large pockets of molten metal at depth in the moon.

Natural Remanence and Alternating Field Demagnetization

We have summarized our results on the natural remanent magnetization of lunar samples in two previous papers (Strangway *et al.*, 1971; Gose *et al.*, 1972) so in this paper we will only discuss the results of measurements on samples from Apollo 14 and some preliminary results from Apollo 15. Typically most of the igneous samples have a soft component of magnetization which is random in direction and can be readily removed in alternating fields of 50 Oe or less. We believe that this soft remanence is not of lunar origin and is due to exposure to fields in the spacecraft or back on earth. The typical value for the stable part of the remanent magnetization is about 2×10^{-6} emu/g. At the same time we have examined the remanent magnetization of several breccias and find that they behave in a manner very similar to the igneous rocks

2449

(Gose *et al.*, 1972). In addition, however, the breccias show a pronounced time-dependent viscous remanent magnetization (VRM) which is related to the grain-size distribution of the interstitial iron (Gose *et al.*, 1972). Superparamagnetic grains show VRM characterized by a limited range of relaxation times, whereas multidomain grains show a seemingly unlimited range of relaxation times.

Apollo 14 breccias

The paleomagnetic results from the Apollo 14 breccias are included in Figs. 1–4 of this paper. These breccias range in their degree of metamorphism from almost unmetamorphosed to highly recrystallized on a scale from 1 to 8 (Warner, 1972).

In Fig. 1 the results from sample 14313,25, which is in Warner's lowest metamorphic grade, are shown. In this case alternating field (AF) demagnetization revealed a stable component (curve A) at a cleaning field of 100 Oe. On subsequent storage in the earth's field the sample reacquired a remanent magnetization, but upon AF demagnetization the original direction was recovered (curve B). Later the sample was exposed to a 10 Oe field, in a VRM experiment, and again on AF demagnetization it was possible to recover the original direction (curve C). It is interesting to note that the magnetization acquired from exposure to a field of only 10 Oe required demagnetization in a field of over 50 Oe to remove. Sample 14049,28 is in the same low metamorphic grade and shows very similar behavior (Fig. 2). On AF demagnetization to 100 Oe a definite stable component was found with a well-defined direction. We can therefore conclude that there is a stable remanence present in these lowest metamorphic grade samples, but care needs to be taken to remove VRM effects.

Sample 14321,78, which is of medium metamorphic grade (4), also shows viscous

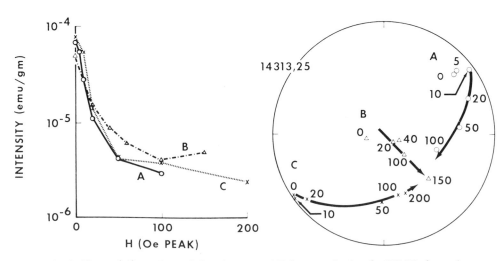

Fig. 1. Change in intensity and direction upon AF demagnetization for NRM of sample 14313,25. The numbers in the stereographic projection correspond to the applied alternating AF field. See text for further explanation of three curves. All directions are in lower hemisphere, and the inclination is true lunar inclination.

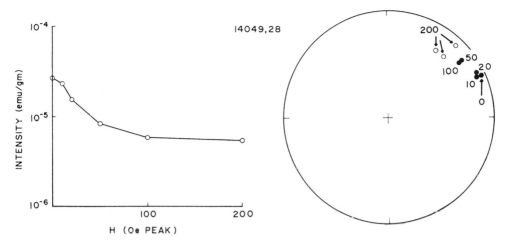

Fig. 2. AF demagnetization of NRM of sample 14049,28. The numbers in the stereographic projection correspond to the applied AF field. Open circles represent lower hemisphere; solid circles, upper hemisphere. Lunar orientation for this sample is unknown.

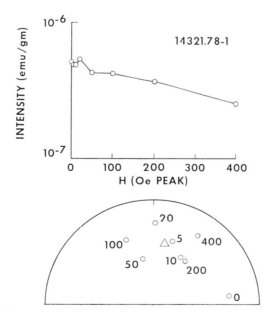

Fig. 3. AF demagnetization of NRM of sample 14321,78,1. The triangle in the stereographic projection represents the mean value of all demagnetized data. Directions correspond to the true lunar orientation.

effects which are characteristic of a large range of grain sizes and are probably due to small multidomain grains (Gose *et al.*, this volume). The net result, however, as shown in Fig. 3, is that there is a stable remanence present which can be readily detected and measured. There was no appreciable unstable component remaining in the sample after storage in a field-free chamber for several weeks. This is in general agreement with the findings of Hargraves and Dorety (1972), who also indicate the presence of a weak stable component in sample 14321,194.

The final breccia sample we have examined is 14312,07. This is in Warner's metamorphic grade 7. The remanent magnetization of this sample, as shown in Fig. 4, is quite different from that of the others just discussed. The intensity of the magnetization decreases to the low value of 2×10^{-7} emu/g on cleaning in alternating fields up to 100 Oe, and no stable direction is indicated since there is great scatter above 10 Oe. Moreover, on storage the sample acquired a new magnetization and it was not possible on subsequent AF demagnetization to reacquire anything close to the original directions. The scatter of directions was again quite large. We therefore conclude that there is not a useful stable remanence in this sample. This is consistent with the results discussed by Gose *et al.* (this volume) and suggests the presence of many large multidomain grains incapable of carrying a stable remanence.

We therefore conclude that samples which have large numbers of small particles either as single domain or as small multidomains ($<1\ \mu$) do carry a useful stable remanence and give a measure of an ancient lunar field. Since the ages of the Apollo 14 breccias all appear to cluster around values of 3.8 and 3.9 b.y., we feel that these samples record the presence of a magnetic field at that time.

Igneous rocks (14310, 15555, and 15415)

Since our earlier paper we have examined the properties of sample 14310,89 in some detail and find that it behaves much like other basalts from Apollo 11 and 12.

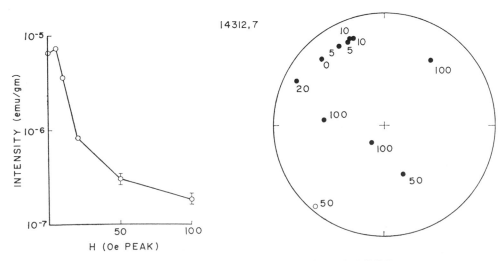

Fig. 4. AF demagnetization of NRM of sample 14312,7.

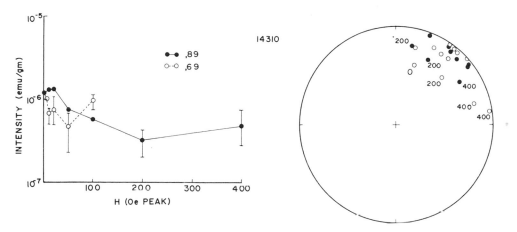

Fig. 5. AF demagnetization of NRM of samples 14310,69 and 14310,89. Only sample 89 is represented in the stereographic projection. Open circles represent the upper hemisphere and solid circles, the lower hemisphere. The inclination of the magnetization corresponds to true lunar orientation.

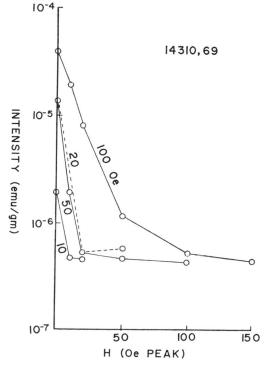

Fig. 6. AF demagnetization of IRMs acquired by sample 14310,89. Numbers on curves refer to magnetic fields in which IRMs were acquired.

The remanent magnetization data are shown in Fig. 5. In this case, there is no super-imposed soft component and the direction of the magnetization remains essentially constant up to 400 Oe. The stable component of remanent magnetization in this sample is measured in two chips and turns out to be a little less than 10^{-6} emu/g in both cases. One chip (89) is lunar oriented (inclination only), and the stereographic plot in Fig. 5 shows the actual lunar magnetic inclination. The other chip is not oriented, and it is not used in the stereonet; however, the scatter in its direction is similar. The direction of chip 89 is approximately the same as that of chip 57 as measured by Hargraves and Dorety (1972). It is dated at 3.78 b.y. by Schaeffer *et al.* (1972).

We have induced a series of isothermal remanent magnetizations (IRM) in 14310,89 in various fields up to 100 Oe and then subjected it to AF demagnetization (Fig. 6). The base-level intensity to which the magnetization returns in each case after elimination of these magnetizations is that of the demagnetized NRM. Even when exposed to dc fields of 100 Oe it is possible to recover the original direction and in-tensity of magnetization. The similarity of the 20-Oe and 50-Oe curves suggests that there is a gap in the coercive force spectrum. The intensity of the IRM induced at 10 Oe is such that this sample has probably not seen magnetic fields much greater than the earth's field in its journey from the moon to our laboratory.

We have examined the behavior of two samples from Apollo 15 in some detail. The first of these is sample 15555, a basalt from the vicinity of Hadley Rille. This sample has been dated at about 3.2–3.5 b.y. by several authors (Compston *et al.*, 1972, 3.53 b.y.; York *et al.*, 1972, 3.3 b.y.; Schaeffer *et al.*, 1972, 3.28 b.y.; Alexander *et al.*, 1972, 3.33. b.y.; Wasserburg *et al.*, 1972, 3.22 b.y.; Cliff *et al.*, 1972, 3.34 b.y.). The magnetic behavior of this sample is much like that of other lunar basalts. Figure 7

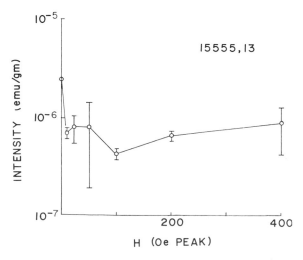

Fig. 7. Change in intensity of magnetization of NRM of sample 15555 upon AF demagnetization.

shows that there is a very small, soft component which is readily removed, leaving a stable component with an intensity just under 10^{-6} emu/g.

The final sample for which magnetic results have been obtained is 15415, which is an anorthosite. This sample is of particular interest since it may provide a clue to the remanent magnetization that is typical of the anorthositic parts of the moon believed by some to be quite widespread. These results are illustrated in Fig. 8. It can be seen that there is indeed a stable component of remanent magnetism present, but the value of this is extremely low at around 10^{-7} emu/g and therefore hard to measure on a 2-g sample. At this level it is one of the most weakly magnetic lunar samples, probably due to the very small amount of iron present in the sample as indicated by Hargraves and Hollister (1972) and by James (1972). We have no estimate of the metallic iron content at this time, but it seems likely that it is extremely small, since the total iron content is 0.18% (LSPET, 1971). The age of this sample, as given by several authors, is about 4 b.y. (Husain et al., 1972, 4.09 b.y.; Stettler et al., 1972, 3.92 b.y.), and so it is the oldest sample yet studied magnetically. Since it does contain a definite remanence this continues to confirm that a lunar field existed at least from 4 b.y. to about 3 b.y. So far all samples in the age range of 3–4 b.y. have been found to carry a definite remanent magnetization. This measurement indicates that the field that created this remanent magnetism was of widespread lunar extent and not just confined to the Apollo 11, 12, and 14 sites.

THERMAL DEMAGNETIZATION

One of the most nagging problems that remains to be solved is the mechanism by which the remanent magnetization in the lunar samples was acquired. The presence of a magnetic field is an almost certain requirement, but there have been no convincing tests done yet that demonstrate the mechanism of acquisition, although there is much

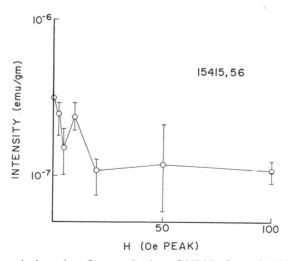

Fig. 8. Change in intensity of magnetization of NRM of sample 15415 upon AF demagnetization.

evidence pointing to thermoremanent magnetization (TRM). In our earlier work (Pearce *et al.*, 1971) we showed that the soft component of magnetization could well be due to exposure to a steady magnetic field and the acquisition of an isothermal remanent magnetism (IRM). We also showed that heating a sample and allowing it to cool from above 800°C in the presence of a weak magnetic field could lead to the acquisition of a very stable TRM much like that seen in the samples. Dunn and Fuller (1972) have examined TRM in several samples and in various low fields and found it to be of considerable stability except in igneous sample 14053, whose magnetic properties seem to be determined by an unusually large amount of multidomain iron.

Two authors (Helsley, 1971; Grommé and Doell, 1971) conducted thermal demagnetization tests on the natural remanence of lunar samples and found drastic changes occurring at around 300°C. Since their work was done on samples which had not been cleaned by AF demagnetization, they were looking at both the soft and the stable components of the natural remanent magnetization. For this reason we have conducted such tests on two samples in which only the stable component is present. The apparatus used, consisted of a vacuum furnace with a vacuum of less than 10^{-6} Torr and a field-free space controlled by a feedback network to a value of less than 1 γ.

We first examined a breccia sample, 14321, which, as discussed earlier, had essentially no soft component. Thermal demagnetization results for this sample are shown in Fig. 9. It is seen that the natural remanence disappears quite sharply at about 800°C. The second is an igneous rock, 12002, which we studied in some detail earlier (Strangway *et al.*, 1971). The sample 12002,86 initially had a soft component which was eliminated by AF demagnetization in about 20 Oe. A series of IRMs were induced in another sample of 12002 and on AF cleaning showed that this soft com-

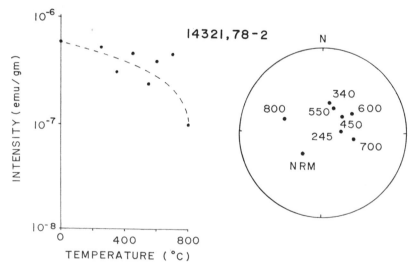

Fig. 9. Change of intensity and direction of NRM of sample 14321,78,2 upon thermal demagnetization. The numbers in the stereographic projection represent the temperatures to which the sample was heated. Directions correspond to true lunar orientation.

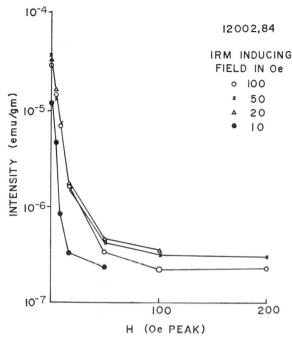

Fig. 10. A.F. demagnetization of IRMs acquired by sample 12002,84 in fields indicated.

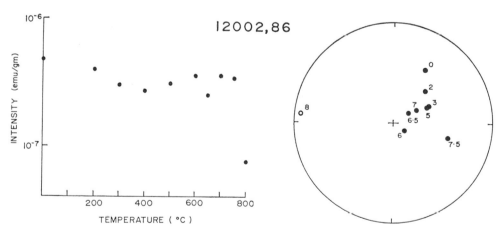

Fig. 11. Thermal demagnetization of 12002,86 after the soft component of magnetization had been removed by AF cleaning to 100 oersted. Numbers in stereographic projection refer to temperatures to which the sample was heated in hundreds of degrees C.

ponent might be an IRM acquired in a field of about 10 Oe (Fig. 10). The thermal demagnetization of the remaining stable component in 12002,86 is shown in Fig. 11. The stable component again disappears fairly sharply at about 800°C.

These two experiments suggest that the natural stable remanence in both breccias and igneous rocks is acquired when the samples cool through the Curie temperature of iron in the presence of a weak magnetic field. Although we have not yet determined the strength of this field with any accuracy, it appears that about 1000 γ is required to account for the observed TRM. Also in the case of the breccia 14321 a lower limit of about 800°C can be given for its temperature of formation.

Conglomerate Test

In an earlier paper (Strangway *et al.*, 1971) we described a conglomerate test based on the magnetic inclinations of oriented samples from Apollo 11 and 12. The regolith can be looked at as a conglomerate with an age determined by the average tumbling rate due to meteorite impacts. This is on the order of 10^6 yr. The returned rocks are samples of this conglomerate; if they show random magnetization directions the magnetization predates the regolith. Inclinations only are considered, since many of the rocks have been oriented by means giving only this information, for example, pit counts, gamma-ray spectroscopy, and track counting.

The combined Apollo 11 and 12 data suggested randomness of orientation and so a minimum age of a million years for the stable magnetization is implied. Samples 14310, 14313, and 14321 have been added to the histogram (Fig. 12). They support the previous data and strengthen the indication that the magnetization predates the regolith tumbling.

Magnetic Anomalies

Dyal and Parkin (1972) have reported on steady magnetic fields observed at the Apollo 12, 14, and 15 sites. Using the lunar surface magnetometer (LSM) they found a field of 38 ± 3 γ at the Apollo 12 site. The lunar portable magnetometer (LPM) showed fields of 43 ± 6 and 103 ± 5 γ at two stations at the Apollo 14 site about 1.1 km apart. The lower value was measured high up on the flank of Cone Crater. At the Apollo 15 site the LSM did not clearly detect the presence of a steady magnetic field since the value reported was 6 ± 4 γ. In view of the fact that the orbiting spacecraft Explorer 35 indicated that any lunar-wide field must be less than a few gammas at the lunar surface, the discovery of these fields is quite surprising. Subsequent analysis of the Explorer 35 data has suggested the presence of local surface

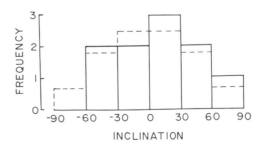

Fig. 12. Conglomerate test: Distribution of lunar magnetic inclinations in oriented samples from Apollo 11, 12, and 14. (Solid line is observed data; dashed line is theoretically predicted for random orientation.)

anomalies on the moon (Mihalov *et al.*, 1971), which seem to be associated with the highlands parts of the moon.

With the launching of the Apollo 15 subsatellite a powerful new tool for mapping lunar magnetic fields became available. The preliminary results of this experiment have been described by Coleman *et al.* (1972). They show that there are definite anomalies of lunar origin which are more distinctive on the far side of the moon and seem to be associated with the large craters. Measurements made in the magnetic tail are relatively free of time fluctuations, and anomalies of up to 1 γ at an orbital height of 110 km are definitely resolved. In Fig. 13 we have shown some of the data from Coleman *et al.* (1972) for the two components B_p and B_T. B_p is essentially the local horizontal N-S component since it is the component parallel to the subsatellite spin axis. This axis is normal to the plane of the ecliptic. B_T is the vector sum of the vertical component and the horizontal E-W component, since it is measured by a sensor normal to the spin axis.

In this same figure we have shown a simple calculation which can apply to either of two models. The first model is that of a uniformly magnetized crust in which a hole has been cut. This hole might represent a crater with a diameter of about 200 km and a depth of about 40 km. The magnetization chosen is 2×10^{-6} emu/g (6×10^{-6} emu/cc for an assumed density of 3 g/cc) and is similar to the values discussed earlier for lunar samples. This gives anomalies as shown with peak values of about 0.4 γ. It is particularly interesting to note in the theoretical models that if the magnetization is vertical the vertical component (Z) reaches its maximum value over the center of the crater, while the horizontal component (H) reaches a zero value at the center and

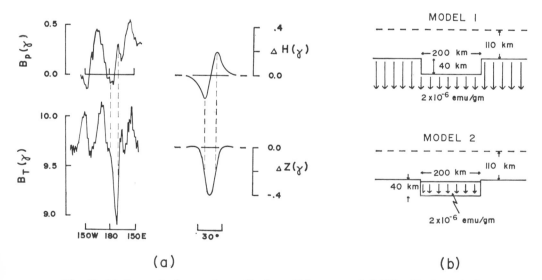

Fig. 13. (a) Segment of anomaly profile from Coleman *et al.* (1972) with theoretical calculation based on models shown in (b). The segment is near the crater Van de Graff. (b) Two possible models to explain observed anomaly profile. See text for further explanation.

is antisymmetrical about this axis. This behavior is very much like that found in the observed field where the peaks of one component correspond to the zero crossings of the other.

This model seems unrealistic for two reasons. First, the only anorthosite measured so far is very weakly magnetized with an intensity of about 2×10^{-7} emu/g or less. Many authors think that the highlands are composed of anorthositic gabbros or norites, as suggested by the x-ray data of Adler et al. (1972). We have not measured the magnetization of such material but assume that it is between that of mare basalts and anorthosite. If such rocks are representative of the farside highland rocks, the magnetization would be inadequate to account for the observed anomalies. Second, the crater geometry chosen seems quite unrealistic and the hole in the crust must involve less volume than our model. We therefore propose that the anomalies found are due to the presence of mare-type basalt associated with the major features on the back side of the moon. It is well known that Tsiolkovsky is flooded with basalt, and it is reasonable to suppose that basalt, with the magnetic properties of other mare basalts, is present beneath these craters. The sense of the magnetization is roughly vertical, suggesting that the source magnetic field was radial at these latitudes ($30°$ or less).

It is difficult to account for the large magnetic anomalies of 38, 43, and 100 γ observed at the lunar surface by the presence of materials with the low magnetization of about 2×10^{-6} emu/g, since in general, anomalies are rarely greater than about $4\pi I$, where I is the magnetization in emu/cc. Using the value of 6×10^{-6} emu/cc used for the calculation shown in Fig. 13 and which correlates well with the sub-satellite data, the maximum field expected at the surface is 7 or 8 γ. A few of the igneous samples and some of the breccias do show a stable component of magnetization up to about 10^{-5} emu/g (3×10^{-5} emu/cc), so that the higher fields measured in the surface experiments could be due to local regions with more magnetic rocks.

It is also possible that the soft component of magnetization observed in many of the rocks is of lunar origin and that the stable components we are considering might only be a part of the story. It does not seem necessary at this point to call upon this as a source of the anomalies, since we feel that these can be accounted for by the stable remanent magnetization of typical mare basalts, while some of the surface anomalies require that there be local regions where the magnetization of the basalt or breccias is up to 5 or 10 times that of the mean value.

The lack of clear-cut anomalies on the front side of the moon in association with the mare is probably due to the fact that they cover a very large area and therefore to a rough approximation give rise to anomalies only at the edges (Fig. 14). It is a well-known fact that sheets of magnetic material which are thin relative to their lateral dimensions only give anomalies at the edges, so that the front side does not have bull's eye anomalies that can be clearly correlated with obvious features. This magnetic picture, then, is in general agreement with the gravity data of Sjogren et al. (1972), suggesting that gravity and magnetic anomalies are related and associated with basalts. This conforms remarkably well with the models of Wood et al. (1970), in which a 25-km-thick crust or anorthosite overlies a more gabbroic interior. With minor modifications this model is compatible with the seismic velocity structure described by Toksöz et al. (1972). They suggest a 25-km-thick crust of basalt overlying

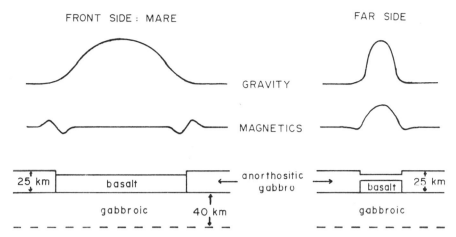

Fig. 14. Nature of anomaly patterns expected from front side mares and far side craters flooded with basalt. In each case the basalt is magnetized vertically and the surrounding material is unmagnetized.

a layer of anorthosite, gabbro, or norite which in turn is overlying a deeper layer of more mafic rock at 65 km in the mare areas.

DISCUSSION

The discovery of remanent magnetization in returned lunar samples and the detection of local magnetic fields on the moon must be regarded as one of the most interesting discoveries of the Apollo program. It is important to note that the findings of the magnetometer studies suggest that large volumes of the lunar crust have a remanent magnetization which is generally uniform in sense and of an intensity comparable to the level observed in the basaltic samples studied. The single anorthosite studied to date suggests that the levels of magnetization in this rock type (and probably in gabbroic anorthosites) is not adequate to account for any of the lunar magnetic anomalies. Therefore it seems likely that the moon has basaltic material underlying at least some of the major craters on the far side. On the front side the magnetic anomalies are not nearly so distinctly associated with the mare, but this is probably due to the large lateral extent and the consequent development of edge anomalies.

The origin of this remanent magnetization is of real interest, since it applies not to just a few surface samples but to large volumes of the crust. This implies that this deep-seated remanence cannot be due to thermal cycling effects as proposed by Banerjee (1972) for surface-collected samples, since thermal effects at the lunar surface can be expected to penetrate to only a few tens of centimeters. The origin of this remanence as shown by the thermal demagnetization experiments is most likely due to cooling in the presence of a field of a few hundred gammas. As discussed in a variety of other papers (e.g., Strangway et al., 1971), there are several possible causes of this field.

It seems to us most likely that this field is of lunar origin, the result of fluid motions in an ancient metallic lunar core or in other large masses of molten metal. There are many objections to a moon which was ever hot enough or differentiated enough to have a fluid metallic core (Gast, 1972; Toksöz et al., 1972). The main arguments are: (1) If the moon was hot at one time it could not be cold now, and there seems to be strong evidence that it is presently cold (see Dyal and Parkin, 1972, for example); (2) The presence of mascons and the disequilibrium shape requires that large parts of the moon are cold enough now and have been cold enough for the last 3.5 b.y. to maintain at least a rigid shell; and (3) The problem of differentiating the moon totally to produce both a metallic core and a radioactive crust is quite severe, since there is evidence of crustal radioactive material about 4.4 b.y. This means total differentiation in 200 or 300 m.y., which puts a serious constraint on the initial temperature distribution of the moon. Tozer (1972) has examined the problem in detail and finds that the moon is likely to undergo convective motion as soon as the interior temperature rises above about 1000°C. At the same time, since viscosity is strongly dependent on the temperature, it is quite possible to maintain the outer 200–300 km of the moon quite rigid. Following this line of reasoning, it is therefore possible to have had a hot moon in early history which has subsequently cooled by convective transfer after the internal temperature reaches about 1000°C. This takes care of the basic objection to an early hot moon and permits considerable differentiation of the moon early in its history. It also explains a cutoff in lunar volcanic activity. The convective model permits the moon to maintain mascons and its disequilibrium figure over long periods of time by keeping the outer parts of the moon cool.

This model has the appeal that it permits the formation of a liquid core in the early history. As shown by Runcorn (1967), Solomon and Toksöz (1968), and others, a heavy, metallic core of 0.2 times the radius of the moon in no way violates the restrictions on the mean density and the mean moment of inertia. From considerations of the self-generating dynamo we know that if $U\mu\sigma L \gg 1$ that a magnetic field can be generated where σ is the electrical conductivity, μ the magnetic permeability, U the velocity, and L the diameter. In the lunar case it is reasonable to take $\sigma = 3 \times 10^5$ mhos/m (typical for the earth's core), $\mu = 4\pi \times 10^{-7}$ henries/m, and $L = 3.5 \times 10^5$ m, so that $\sigma\mu L \sim 1.3 \times 10^5$. The velocity required then is only about 10^{-5} m/sec or about one revolution every 7000 yr. This is a remarkably small limiting velocity and such a body could readily maintain a magnetic field. Other models of the moon (Gast, 1972) suggest that only the outer few hundred kilometers were differentiated. One consequence of this model is the probable accumulation of pockets of molten iron several hundred kilometers deep in the moon. A substantial body of fluid metal located at some depth in the moon as predicted by the partial differentiation model would also be adequate to generate one or more self-sustaining dynamos.

Acknowledgments—The magnetic model calculations were performed by R. K. McConnell of Earth Science Research Inc. One of us (GWP) is supported by the National Research Council of Canada and in part by the Lunar Science Institute. WAG is supported by the National Aeronautics and Space Administration under Contract No. NSR 09-051-001 and the Universities Space Research Association. We thank J. A. Love and E. Pahl of Lockheed Electronics Company for assistance in the laboratory. This paper constitutes the Lunar Science Institute contribution Number 94.

REFERENCES

Adler I., Trombka J., Gerard J., Lowman P., Yin L., and Blodgett H. (1972) Preliminary results from the S-161 x-ray fluorescence experiment (abstract). In *Lunar Science—III* (editor C. Watkins), pp. 4–6, Lunar Science Institute Contr. No. 88.

Alexander E. C. Jr., Davis P. K., Lewis R. S., and Reynolds J. H. (1972) Rare gas analyses on neutron irradiated lunar samples (abstract). In *Lunar Science —III* (editor C. Watkins), pp. 12–14, Lunar Science Institute Contr. No. 88.

Banerjee S. K. (1972) Iron–titanium–chromite, a possible new carrier of remanent magnetization in lunar rocks (abstract). In *Lunar Science—III* (editor C. Watkins), pp. 38–40, Lunar Science Institute Contr. No. 88.

Cliff R. A., Lee-Hu C., and Wetherill G. W. (1972) K, Rb, and Sr measurements in Apollo 14 and 15 material (abstract). In *Lunar Science—III* (editor C. Watkins), pp. 146–147, Lunar Science Institute Contr. No. 88.

Coleman P. J. Jr., Schubert G., Russell C. T., and Sharp L. R. (1972) Satellite measurements of the moon's magnetic field: A preliminary report. *The Moon,* in press.

Compston W., Vernon M. J., Berry H., Rudowski R., Gray C. M., and Ware N. (1972) Age and petrogenesis of Apollo 14 basalts (abstract). In *Lunar Science—III* (editor C. Watkins), pp. 151–153, Lunar Science Institute Contr. No. 88.

Dunn J. R. and Fuller M. D. (1972) Thermoremanent magnetization (TRM) of lunar samples. *The Moon* **4,** 49–62.

Dyal P. and Parkin C. W. (1972) Lunar properties from transient and steady magnetic field measurements. *The Moon* **4,** 63–87.

Gast P. W. (1972) The chemical composition and structure of the moon. *The Moon,* in press.

Gose W. A., Pearce G. W., Strangway D. W., and Larson E. E. (1972) On the applicability of lunar breccias for paleomagnetic interpretations. *The Moon,* in press.

Grommé C. S. and Doell R. R. (1971) Magnetic properties of Apollo 12 lunar samples 12052 and 12065. *Proc. Second Lunar Sci. Conf., Geochim. Cosmochim. Acta* Suppl. 2, Vol. 3, pp. 2491–2499. MIT Press.

Hargraves R. B. and Dorety N. (1972) Magnetic property measurements on several Apollo 14 rock samples (abstract). In *Lunar Science—III* (editor C. Watkins), pp. 357–359, Lunar Science Institute Contr. No. 88.

Hargraves R. B. and Hollister L. S. (1972) Mineralogic and petrologic study of lunar anorthosite slide 15415,18. *Science* **175,** 430–432.

Helsley C. E. (1971) Evidence for an ancient lunar magnetic field. *Proc. Second Lunar Sci. Conf., Geochim. Cosmochim. Acta* Suppl. 2, Vol. 3, pp. 2485–2490. MIT Press.

Husain L., Schaefer O. A., and Sutter J. F. (1972) Age of a lunar anorthosite. *Science* **175,** 428–430.

James O. B. (1972) Lunar anorthosite 15415: Texture, mineralogy, and metamorphic history. *Science* **175,** 432–436.

LSPET (Lunar Science Preliminary Examination Team) (1972) The Apollo 15 lunar samples: A preliminary description. *Science* **175,** 363–375.

Mihalov J. D., Sonett C. D., Binsack J. H., and Moutsoulas M. D. (1971) Possible fossil lunar magnetism inferred from satellite data. *Science* **171,** 892–895.

Pearce G. W., Strangway D. W., and Larson E. E. (1971) Magnetism of two Apollo 12 igneous rocks. *Proc. Second Lunar Sci. Conf., Geochim. Cosmochim. Acta* Suppl. 2, Vol. 3, pp. 2451–2460. MIT Press.

Runcorn S. K. (1967) Convection in the moon and existence of a lunar core. *Proc. Roy. Soc. London* **A296,** 270–284.

Schaeffer O. A., Husain L., Sutter J., and Funkhouser J. (1972) The ages of lunar material from Fra Mauro and the Hadley–Rille–Apennine Front area (abstract). In *Lunar Science—III* (editor C. Watkins) pp. 675–677, Lunar Science Institute Contr. No. 88.

Sjogren W. L., Muller P. M., and Wollenhaupt W. R. (1972) Apollo 15 gravity analysis from the S-band transponder experiment. *The Moon,* in press.

Solomon S. C. and Toksöz M. N. (1968) On the density distribution in the moon. *Phys. Earth Planet. Interiors* **1**, 475–484.

Stettler A., Eberhardt P., Geiss J., and Grogler N. (1972) Ar^{39}/Ar^{40} ages of Apollo 11, 12, 14, and 15 rocks (abstract). In *Lunar Science—III* (editor C. Watkins), pp. 724–725, Lunar Science Institute Contr. No. 88.

Strangway D. W., Pearce G. W., Gose W. A., and Timme R. W. (1971) Remanent magnetization of lunar samples. *Earth Planet. Sci. Lett.* **13**, 43–52.

Toksöz M. N., Solomon S. C., Minear J. W., and Johnston D. H. (1972) Thermal evolution of the moon. *The Moon* **4**, 190–213.

Tozer D. C. (1972) The moon's thermal state and an interpretation of the lunar electrical conductivity distribution. *The Moon*, in press.

Warner J. L. (1972) Apollo 14 breccias: Metamorphic origin and classification (abstract). In *Lunar Science—III* (editor C. Watkins), pp. 782–784, Lunar Science Institute Contr. No. 88.

Wasserburg G. J., Turner G., Tera F., Podosek F. A., Papanastassiou D. A., and Huneke J. C. (1972) Comparison of Rb–Sr, K–Ar, and U–Th–Ob ages; lunar chronology and evolution (abstract). In *Lunar Science—III* (editor C. Watkins), pp. 788–790, Lunar Science Institute Contr. No. 88.

Wood J. A., Dickey J. S., Jr., Marvin U. B., and Powell B. N. (1970) Lunar anorthosites and a geophysical model of the moon. *Proc. Apollo 11 Lunar Sci. Conf., Geochim. Cosmochim. Acta* Suppl. 1, Vol. 1, pp. 965–998.

York D., Kenyon W. J., and Doyle R. J. (1972) ^{40}Ar–^{39}Ar ages of Apollo 14 and 15 samples (abstract). In *Lunar Science—III* (editor C. Watkins), pp. 822–824, Lunar Science Institute Contr. No. 88.

Proceedings of the Third Lunar Science Conference
(Supplement 3, *Geochimica et Cosmochimica Acta*)
Vol. 3, pp. 2465–2478
The M.I.T. Press, 1972

Temperature-dependent magnetic properties of individual glass spherules, Apollo 11, 12, and 14 lunar samples

A. N. Thorpe and S. Sullivan

Howard University, Washington, D.C. 20001

C. C. Alexander, F. E. Senftle, and E. J. Dwornik

U.S. Geological Survey, Washington, D.C. 20242

Abstract—Magnetic susceptibility of 11 glass spherules from the Apollo 14 lunar fines have been measured from room temperature to 4°K. Data taken at room temperature, 77°K, and 4.2°K, show that the soft saturation magnetization was temperature independent. In the temperature range 300 to 77°K the temperature-dependent component of the magnetic susceptibility obeys the Curie law. Susceptibility measurements on these same specimens and in addition 14 similar spherules from the Apollo 11 and 12 mission show a Curie-Weiss relation at temperatures less than 77°K with a Weiss temperature of 3–7 degrees in contrast to 2–3 degrees found for tektites and synthetic glasses of tektite composition. A proposed model and a theoretical expression closely predict the variation of the susceptibility of the glass spherules with temperature. The model infers the Weiss temperature is associated not only with the antiferromagnetic mineral inclusions within the glass but also with a distortion of the octrahedral ligand field of the Fe^{+2} ions in the glass phase.

The data show that in the glass spherules of the Apollo 14 fines the concentration of antiferromagnetic inclusions is greater and the ligand field distortion is less than for similar specimens in the Apollo 11 and Apollo 12 fines. The concentration of Fe^{+2} in the glassy phase is essentially constant from spherule to spherule in a given sample of lunar fines and there is evidence that the particulate minerals in the glass suffered little or no dissolution after injection into the glass. The later conclusions are a strong indication that the glass phase is saturated with respect to iron and that some or all the metallic iron spherules were formed by reduction. The saturation iron concentration on the glassy phase of the Apollo 14 spherules is less than in the Apollo 11 and 12 specimens indicating a lower temperature of formation for the Apollo 14 specimens.

Introduction

A silicate glass containing only dissolved iron, i.e., no particulate iron or iron minerals, will have a temperature-dependent paramagnetism, χ, which follows the Curie law ($\chi = C/T$; C = Curie constant; T = temperature) from 300 to 77°K. Similarly, it has been shown that the temperature-dependent paramagnetism in the lunar glass spherules follows the Curie law down to 77°K (Thorpe *et al.*, 1970). Thus, the particulate iron and iron minerals known to be present in the lunar glass spherules either follow the Curie law or contribute only a temperature independent Pauli paramagnetic component to the total paramagnetism in this temperature range. However, as suggested by Tsay *et al.* (1971) and determined independently by Sullivan *et al.* (1971), at lower temperatures the temperature-dependent magnetism in the same specimens does not follow the Curie law but rather can be approximately fitted to a Curie-Weiss relation ($\chi = C/(T + \theta)$; θ = Weiss temperature). The Weiss temperature was found to be 3 to 7° for the lunar glass in contrast to 2 to 3° for tektite glass.

These measurements were made on glass from Apollo 14 samples, selected Apollo 11 and 12 samples, and additional terrestrial samples in order to determine the cause of the deviation of the temperature-dependent paramagnetism from the Curie law.

The Weiss temperature, for the lunar glass spherules was found to be related to both (1) the presence of one or more antiferromagnetic phases in the glass and (2) a distorted octahedral ligand field about the Fe^{+2} ion in the glassy phase of the spherules. A model and a theoretical analytical expression is proposed that explains the variation of the paramagnetic susceptibility of the lunar glass spherules down to liquid helium temperatures.

Physical Measurements

The methods of measurement are the same as previously reported (Thorpe *et al.*, 1970; Sullivan *et al.*, 1971). An approximate diameter was first determined for each specimen. Several specimens deviated considerably from perfect spheres and only an estimate could be determined in these cases. The masses were determined on the same helical spring balance used for the magnetic susceptibility measurements. Susceptibility measurements were made as a function of magnetic field at room temperature, 77°K, and 4.2°K, to determine the temperature dependency, if any, of the soft saturation magnetization. Susceptibility measurements at constant magnetic field were also made as a function of temperature from 300 to 4.2°K on all the specimens to determine the temperature-dependent and the temperature-independent components of the susceptibility. Following the magnetic studies, electron microprobe analyses were made on polished sections of six of the glass spherules.

The estimated error in the magnetic susceptibility measurements is $\pm 2\%$, and the reported electron microprobe analyses are good to $\pm 10\%$.

Experimental Results

Of the eleven glass spherules from the Apollo 14 fines selected for detailed study, samples 14-2 and 14-12 were particularly inhomogeneous and contained relatively large inclusions. The electron microprobe analyses of six of the specimens are shown in Table 1. Repetitive analyses of specimens 14-10, 14-11, and 14-12 showed a wide range of iron values and indicated inhomogeneity of the glass. Voids, mineral inclusions, devitrified areas, and metallic inclusions of the order of 1–2 μg were observed (total masses given in Table 2). Specimens 14-15, 14-18, and 14-19 were a light yellowish green and were homogeneous with no obvious large inclusions.

Table 1. Electron microprobe analyses of selected glass spherules from lunar fines (No. 14,049,37). Composition in weight percent.*

	14–10	14–11	14–12	14–15	14–18	14–19
SiO_2	47.7	47.6	43.9	52.5	44.2	44.3
Al_2O_3	11.9	14.8	11.5	12.4	11.9	12.2
FeO	13.3	9.7	12.1	12.1	10.1	10.4
MgO	8.9	6.9	8.3	4.8	12.1	5.4
CaO	10.3	10.1	15.9	10.2	9.4	9.4
Na_2O	0.1	0.2	<0.1	0.1	<0.1	0.1
K_2O	0.4	0.8	—	0.8	<0.1	0.9
TiO_2	2.8	2.6	<0.1	1.7	2.3	2.7

* Nickel, if present, is <0.1%.

Table 2. Room-temperature and low-temperature measurements for glass spherules from Apollo 14 fines.

(1) Spherule	(2) Mass (mg)	(3) Diameter (μ)	(4) Calc. density (g/cm^3)	(5) χ_0 ($\times 10^{-4}$ emu/g)	(6) σ ($\times 10^{-4}$ emu/g)	(7) I_s (emu/g)	(8) H_s (kOe)	(9) Curie constant ($\times 10^{-3}$)	(10) χ_t ($\times 10^{-6}$ emu/g)
					Room temperature (300°K)			Low temperature (300°K–77°K)	
			Lunar fines sample no. 14,003,30						
14–1	0.388	700	2.2	29.4	61.9	—	—	6.25	8.83
14–2	0.205	540	2.5	499	375	2.2	6.8	3.26	488
			Lunar fines sample no. 14,049,37						
14–10	0.173	440	3.9	83.9	103	0.46	7.5	4.39	69.7
14–11	0.192	525	2.5	84.3	234	0.42	7.2	3.54	72.8
14–12	0.054	400	1.6	337	3578	1.9	6.2	9.03	307
14–13	0.111	510	1.6	15.2	76.3	—	—	3.28	4.3
14–14	0.181	520	2.5	42.2	20.1	0.23	8.0	3.81	29.8
14–15	0.135	500	2.1	58.9	406	0.34	7.5	4.61	43.9
14–16	0.114	450	2.4	41.7	92.6	0.24	7.0	3.35	30.9
14–18	0.035	300	2.5	85.7	169	0.44	7.5	4.09	72.4
14–19	0.025	275	2.3	68.9	83.6	0.42	7.5	3.26	58.4

Prior to making the very low temperature measurements ($< 77°$K), the Apollo 14 specimens were first analyzed from 300 to 77°K to obtain the basic parameters as shown in Table 2. These are the same parameters which were determined for the Apollo 11 and 12 specimens reported earlier (Sullivan et al., 1971). In the previous investigation the total measured magnetic susceptibility χ_a of the lunar glass spherules was found to be the sum of several identifiable components which could be expressed as follows:

$$\chi_a = C/T + \chi_I + \chi_g + \sigma/H_a + f(H_a)/H_a = C/T + \chi_t. \tag{1}$$

The first term is the paramagnetism of the Fe^{+2} ions in the glass where C is the Curie constant and T is the absolute temperature, χ_I is the temperature-independent paramagnetism of the metallic iron or nickel-iron spherules embedded in the glass, χ_g is the basic diamagnetism of the glass, σ/H_a is the susceptibility of the irregularly shaped iron particles in the glass where σ is the soft saturation magnetization and H_a is the applied magnetic field, and $f(H_a)/H_a$ is the susceptibility of rounded ellipsoidal or quasi-spherical particles. The latter term was found to be negligible and was ignored in this investigation. In general, the data are similar to those obtained from previous Apollo specimens. Specimen 14–12, however, had a very large magnetization σ, which is indicative of gross ferromagnetic inclusions.

Magnetic susceptibility measurements were subsequently made down to liquid helium temperatures not only on the Apollo 14 specimens but also on a number of specimens of Apollo 11 and 12 that were reported previously (Thorpe et al., 1970; Sullivan et al., 1971). In addition, some tektites and synthetic glasses of tektitic composition were also measured to low temperatures. In magnetic fields ranging from 1.41 to 4.45 k Oe no temperature dependency of σ was observed indicating no significant superparamagnetic component. The experimental susceptibilities were least square fitted to a Curie-Weiss relation

$$\chi = \frac{C}{T + \theta} + \chi_t \tag{2}$$

to determine C, θ, and χ_t where χ_t is the temperature-independent paramagnetic component of the susceptibility. The values of the Weiss temperature θ, which were so determined, are listed in Table 3, column 1. The corresponding value of χ_t obtained by this method was always larger and C was always smaller than the values determined from the linear part of the curve from room temperature to 77°K. As the true values of χ_t and C can be evaluated much more accurately from the high temperature

Table 3. Weiss temperatures (θ) of lunar glass spherules: (1) θ found from least squares fit of all data, (2) χ_t and C found from room temperature to liquid nitrogen temperature then θ found from least squares fit of low-temperature data, and (3) zero field splitting parameter from equation (8).

Sample no.	(1) θ (°K)	(2) θ (°K)	(3) D (cm^{-1})
10084,86,2			
3	6.2	6.9	10.5
7	3.2	4.9	7.49
10084,22			
102	3.9	3.6	6.99
104	3.6	5.4	7.70
105	2.6	2.6	6.79
10085,17			
201	3.2	5.4	7.35
202	2.7	3.7	6.72
203	2.5	2.8	6.63
102070,90			
301	3.4	4.6	7.51
302	3.1	6.4	7.23
303	3.7	4.9	7.45
304	3.6	4.6	7.83
307	2.4	2.2	6.53
12057,10			
GB-1	3.4	4.3	7.59
14,003,30			
14–1	3.0	4.7	7.14
14–2	6.0	7.9	10.12
14,049,37			
14–10	3.7	5.1	7.97
14–11	2.6	4.1	6.54
14–12	3.9	16.5	5.85
14–13	2.6	3.2	6.75
14–14	2.1	4.5	5.74
14–15	2.4	6.8	6.22
14–16	2.3	3.4	6.13
14–18	3.2	4.3	7.46
14–19	2.3	3.5	6.58
Synthetic glasses			
STG-2	2.2	2.4	6.25
STG-72	2.3	2.4	6.52
Tektite glasses			
If-1	3.2	2.5	7.44
Ed-1	1.4	2.3	4.44
Ps-65	2.0	2.6	5.78
Js-3	2.1	2.9	6.04
Js-11	1.5	3.2	4.38
B-87	1.9	2.9	5.40
B-6	1.4	2.3	4.69
Id-1	1.3	2.4	4.07
42-2	1.1	2.0	3.20
77611	1.1	1.1	2.01
1771-6	2.2	2.8	6.17

linear portion of the χ versus $1/T$ plot (Thorpe and Senftle, 1964), the experimental data were fitted to a Curie-Weiss relation using χ_t and C determined from the high-temperature data only. The values of θ were recomputed as shown in column 2, Table 3. The experimental points at the low and high temperatures fit reasonably well but the points in the middle of the temperature range tend to deviate from a linear relationship. A typical example is shown in Fig. 1. This treatment of the data was not entirely satisfactory, and it was clear that a simple Curie-Weiss plot equation (2) does not accurately

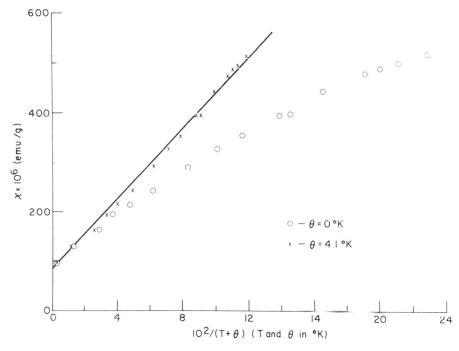

Fig. 1. Magnetic susceptibility of specimen 14-11 as a function of reciprocal temperature (o) and the same data (x) least squares fitted to $1/(T + \theta)$ plot where $\theta = 4.1$ degrees and where the intercept χ_t and the Curie constant C were determined from the susceptibility data above 77°K.

represent the data. It is therefore pertinent to examine the Curie-Weiss law and in particular the Weiss temperature in relation to the lunar samples.

THE WEISS TEMPERATURE

The Weiss temperature is not a true temperature but only a mathematical artifice to account for a magnetic interaction taking place within the specimen at low temperatures. The most likely source of the Weiss temperature in the lunar glasses is antiferromagnetic inclusions, such as ilmenite or ferrosilite (Nagata *et al.*, 1970; Pickart and Alpern, 1971; Nagata *et al.*, 1971), which will cause a deviation from the Curie law at low temperatures. Assuming this to be the cause of the Weiss temperature, then, the value of θ should be a function of the concentration of the antiferromagnetic particulate phases present in the glass. However, if this is the only source of the Weiss temperature, it is difficult to explain the Weiss temperature reported for tektites (Sullivan *et al.*, 1971) which contain no known particulate antiferromagnetic phases. Moreover, it has been observed (Thorpe *et al.*, 1972) that in the lunar glass the Curie constant, which is proportional to the total ironic iron generally increases with the Weiss temperature, but there is a large scatter in the points. At the same time there is little change in the Curie constant with the Weiss temperature in tektites. It appears, therefore, that the Weiss temperature is at least in part associated with the ionic iron in the glass phase of the spherules.

In our studies on tektites we observed (Thorpe *et al.*, 1963) that the magnetic moment of Fe^{+2} ion in tektite glass (4.92 Bohr magnetons) was lower than that due to Fe^{+2} ion in reduced synthetic silicate glasses (~ 5.4 Bohr magnetons). Further examination of this phenomenon (Thorpe and Senftle, 1964) led us to suggest that the low magnetic moment may be associated with a distortion of the ligand field produced by the six oxygen ions surrounding the Fe^{+2} ions in the glass. A calculation of the concentration of iron in the lunar glass of Apollo 11 and 12 samples based on 4.92 Bohr magnetons has yielded results closely paralleling those obtained by electron microprobe analysis (Thorpe *et al.*, 1970; Sullivan *et al.*, 1971). Therefore, an alternative source of the Weiss temperature in tektites and lunar glass might be attributed to an axial distortion of the octahedral crystal fields due to the oxygen ions surrounding the ferrous ions in the glass.

For axial distortion of the octahedral ligand fields of fluorine ions about Fe^{+2}, Tinkham (1956) has shown that there is a splitting of the lowest $S = 2$ spin quintuplet which will give rise to anisotropy of the magnetic susceptibility of the Fe^{+2} ion. Assuming a similar splitting exists in glass due to the oxygen ions around the Fe^{+2} and letting the direction of the major distortion define the z axis, i.e., assuming cylindrical symmetry, the ground state may be represented by the fine structure spin Hamiltonian $H = DS_z^2$ where D is the zero field splitting parameter, and S_z is the z component of the total spin S. Application of a magnetic field H gives rise to a Zeeman splitting term given by $H = g \beta \vec{H} \cdot \vec{S}$ where g is the spectroscopic splitting factor and β is the unit Bohr magneton. The total Hamiltonian thus becomes (Pake, 1962)

$$H = DS_z^2 + g \beta \vec{H} \cdot \vec{S}. \tag{3}$$

If there are N Fe^{+2} ions per mole and the magnetic field is parallel to the axis of distortion of the Fe^{+2} ion, then the paramagnetic susceptibility is given by

$$\chi_{\parallel} = \frac{2N\beta^2 g^2}{ZkT} [4 + \exp(-3\xi)]. \tag{4}$$

Perpendicular to the magnetic field it is

$$\chi_{\perp} = \frac{2N\beta^2 g^2}{ZkT} \left[\frac{2}{3\xi} \{1 - \exp(-3\xi)\} + \frac{3}{\xi} \{\exp(-3\xi) - \exp(-4\xi)\} \right] \tag{5}$$

where $Z = 2 + 2 \exp(-3\xi) + \exp(-4\xi)$, $g = 2$, and $\xi = D/kT$ (Van Vleck and Penney, 1934). As the axis of the Fe^{+2} ions are randomly distributed, one can substitute equations (4) and (5) into the powder susceptibility formula for anisotropic crystals,

$$\chi_p = \tfrac{1}{3}(\chi_{\parallel} + 2\chi_{\perp}) \tag{6}$$

in order to obtain an expression for the bulk susceptibility. A plot of the susceptibility given by equation (6) as a function of reciprocal temperature gives a linear relation at high temperatures but deviates from linearity at low temperatures and can be approximated by a Curie-Weiss relation, e.g. see Fig. 3, described below. Thus, the zero field splitting parameter, D, which is a measure of the distortion of the ligand field about the Fe^{+2} ions in the glass, can also be a source of the observed θ.

DATA ANALYSIS

If one assumes that the lunar glass does not contain any antiferromagnetic mineral inclusions, a hypothetical situation, an attempt can be made to least squares fit all the data to an equation of the form

$$\chi = \chi_p + \chi_t \tag{7}$$

where χ_p is determined from the proper substitutions in equation (6). When this is done, essentially the same fitting problems are encountered as were found when the data were fitted to a Curie-Weiss law. We therefore have added a correction term to equation (7) to account for the antiferromagnetic mineral inclusions. Above the Néel point where $T > \theta$, it can be shown for a glass containing several types of anti-ferromagnetic inclusions that to a first approximation the Curie constants and the Weiss temperatures of the various antiferromagnetic minerals are additive. Thus,

$$\text{for } T > 55°K; \chi = \chi_p + \frac{C'}{T + \theta_a} + \chi_t \tag{8a}$$

where θ_a and C' are the effective values of the Weiss temperature and the Curie constant respectively for all the antiferromagnetic species present. Below the effective Néel temperature the paramagnetic term continues to increase rapidly whereas the antiferromagnetic term in comparison changes relatively little. Therefore, the term $C'/T + \theta_a$ was assumed constant below 55°K. This assumption is valid if the susceptibility of the antiferromagnetic minerals is much less than the susceptibility of the paramagnetic ions in the glass above the Néel points, and, if the Néel points of the antiferromagnetic species are in the range 45–65°K. Thus,

$$\text{for } T < 55°K; \chi = \chi_p + \frac{C'}{55 + \theta_a} + \chi_t. \tag{8b}$$

The data were fitted by computer to equation (8) and the appropriate constants were determined when the standard deviation between the observed and calculated points were a minimum. The experimental susceptibilities and corresponding temperatures were found to fit remarkably well for $\theta_a = 20°$. A typical fit of the data is shown in Fig. 2. The value of 20° for θ_a is somewhat surprising in that the Weiss temperature for ilmenite, the most likely antiferromagnetic phase, is about $-17°$. Evidently other antiferromagnetic species are present with significantly higher Weiss temperatures.

Table 3 gives the zero field splitting parameters D of the antiferromagnetic inclusions as determined from the best fit of the data to equation (8). This treatment of the data suggests that the observed Weiss temperature of the lunar glass spherules is a function of both the antiferromagnetic mineral inclusions in the glass plus a component due to the octahedral ligand field distortion about the ferrous ions in the glass.

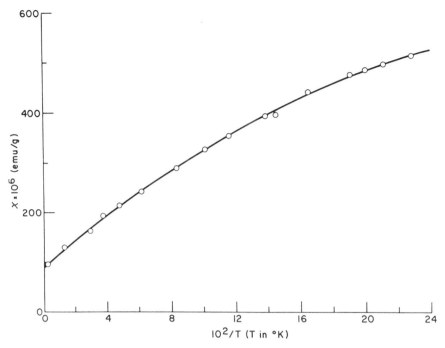

Fig. 2. Typical fit using equation (8) of the theoretical susceptibility (solid line) to the observed susceptibility (points) as a function of the reciprocal temperature for glass spherule 14-11.

The Weiss Temperature of the Lunar Glass

Using the proposed model it is possible to obtain an average value for that component of the Weiss temperature which is caused by the octahedral ligand field distortion about the Fe^{+2} ions in the glass. We assume that the conditions of formation of the glass spherules in any given sample of lunar fines is nearly the same for all the spherules in that sample, but not necessarily between samples from different locations on the moon. Thus, the component of the measured Weiss temperature θ_g, which is due to the glass should be approximately constant for all spherules, say from a given Apollo 14 sample of fines. For example, if we average the values of D for the nine specimens of sample 14,049,37 and substitute this value into equation (6), a curve of the susceptibility of the Fe^{+2} ions in the glass as a function of temperature can be generated. This curve will deviate from a straight line at temperatures of less than 77°K in accordance with the average octahedral field distortion in these lunar glass specimens. If one now fits these data to a Curie-Weiss relationship, a value of θ_g can be determined. The generated and fitted curves for this sample are shown in Fig. 3, and the Weiss temperature is calculated to be $\theta_g = 1.64°$. This is an average θ_g for the lunar glass in this particular Apollo 14 sample. Similar calculations for samples from other lunar sites also yield values of θ_g of the order of 1.5–2.5°.

An alternative method can be used to determine θ_g. In equation (8) C' is the Curie

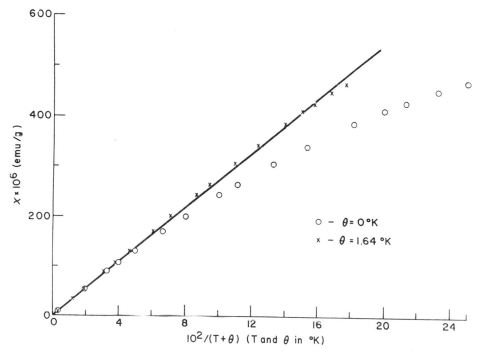

Fig. 3. Susceptibility values (o) generated from equation (6) versus reciprocal temperature and corrected values (x) using a Curie-Weiss relation.

constant associated with the amount of antiferromagnetic mineral inclusions actually present in the glass in contrast to the much larger Curie constant of the pure antiferromagnetic mineral. Thus, C' is proportional to the concentration of antiferromagnetic minerals present in a given specimen. Likewise, the measured total Weiss temperature of a specimen will also be a function of the concentration of antiferromagnetic inclusions, and, hence the intercept on the θ versus C' plot should be the component θ_g of the Weiss temperature which is due to the glass phase only. In determining the best fit for equation (8) the Curie constant C' of the antiferromagnetic minerals was determined for each specimen. Figure 4 shows how the total θ for the same specimens varies with C'. For $C' = 0$, i.e., for no antiferromagnetic inclusions, the intercept, θ_g, is 1.89°. This is in acceptable agreement with the value determined using the proposed model. The linearity (slope $= 2.17 \times 10^{-3}$) between θ and C' indicates that the increase in θ with C' (i.e., with the concentration of iron in the antiferromagnetic phases) is due to antiferromagnetic inclusions. Further, the positive intercept, θ_g, confirms the assumption used in the model, namely that the basic lunar glass has an effective Weiss temperature caused by the distorted octahedral ligand crystal field about the Fe^{+2} ions in the glass phase.

A plot of similar data for Apollo 11 and 12 specimens is shown in Fig. 5. Although there is a greater spread in the data, the linear relationship is evident. The value of θ_g is somewhat larger (2.23°) and the slope of the line is less (1.52×10^{-3}) than for

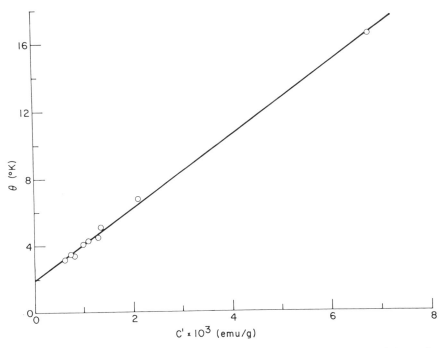

Fig. 4. The Weiss temperature θ as a function of the Curie constant C' of the anti-ferromagnetic mineral inclusions computed from "best fit" of equation (8) for Apollo 14 specimens.

the Apollo 14 specimens. It is interesting that both Apollo 11 and 12 spherules lie close to the same straight line whereas the Apollo 14 specimens lie close to a different straight line. We interpret the greater slope of the curve in Fig. 4, as being due to a higher concentration of antiferromagnetic inclusions in the Apollo 14 specimens. This is substantiated by visual observations of the greater number of particles in these specimens. A smaller ligand field distortion about the Fe^{+2} ions is indicated by the smaller value of θ_g in the Apollo 14 spherules and this probably reflects slower quenching from the molten state and/or lower temperatures, in agreement with Griscom and Marquardt (1972).

The Curie Constant of the Lunar Glass

If one considers the glassy phase as that part of the glass spherules in which the iron is in true solution, i.e., excluding all particulate matter even submicroscopic particles, and further, if one assumes that all the glass spherules in a given sample of lunar fines were formed from the same melt, then it is reasonable to assume that the dissolved iron is approximately constant in the glassy phase from spherule to spherule in a given sample of lunar fines. It is true that electron microprobe analyses of glass spherules from the same sample of fines does show a variation of the iron concentration from specimen to specimen, but the observed iron may be to a significant extent

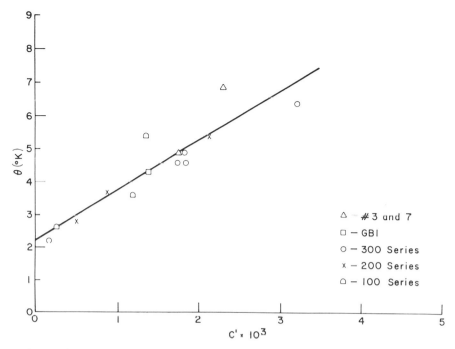

Fig. 5. The Weiss temperature θ as a function of the Curie constant C' of the anti-ferromagnetic mineral inclusions for Apollo 11 (No. 3, 7, 100 series, and 200 series) and Apollo 12 (GB1, and 300 series) glass spherules.

associated with submicroscopic inclusions. The assumption of the constancy of the concentration of iron dissolved in the glass as ionic iron can be tested by studying the Curie constant C_g of the glass phase.

It can be shown that the measured Curie constant C can be expressed as

$$C = C' \frac{T + \theta}{T + \theta_a} + C_g \frac{T + \theta}{T + \theta_g}. \tag{9}$$

In the previous discussion it was shown that $\theta_a = 20°$, θ_g is 1.5–2.5°, and θ is 3–7° for the lunar glass spherules. Thus, for $T > 55°K$, $(T + \theta)/(T + 20)$ will be 0.75 to 1.0 and $(T + \theta)/(T + \theta_g)$ will be close to one. As an approximation one can write.

$$C \simeq C' + C_g. \tag{10}$$

Figure 6 shows a plot of C versus C' for the specimens from sample 14,049,37. The nearly linear relation yields a value of 2.73×10^{-3} for the intercept, $C_g[(T + \theta)/(T + \theta_g)]$, and 0.93 for the slope $(T + \theta)/(T + 20)$. If $\beta = 4.92$ Bohr magnetons, the intercept of the curve corresponds to about 5.1% iron as Fe^{+2} in the glassy phase. A similar plot for the Apollo 11 and 12 specimens could not be made because of the small number of specimens we had from any given sample of fines. However, a plot of C versus C' for the few specimens we had in each suite appeared to be linear but

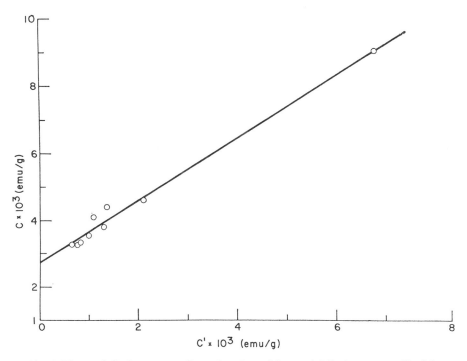

Fig. 6. The total Curie constant C as a function of the partial Curie constant C' of the antiferromagnetic mineral inclusions.

the curves yielded values of C_g that ranged from (1.5 to 4.5) $\times$ 10^{-3}, i.e., iron concentrations up to about 8.5%, a higher value than found in the Apollo 14 spherules.

The concentration of ionic iron in the glassy phase of individual glass spherules from a given sample of fines evidently is relatively constant from spherule to spherule. If the glass phase varied widely in iron concentration between spherules there would be a significant scatter of points in Fig. 6. The fact that C_g is constant and that the slope of the curve in Fig. 6 is nearly one indicates that the observed variation of the Curie constant from spherule to spherule in a given sample of lunar fines simply reflects the concentration of antiferromagnetic inclusions in the glass. The linearity of the data in the figure also points to the fact that if the particulate inclusions were injected or picked up by the glass while still in a molten state, there was essentially no dissolution of the particles. Partial dissolution of the particles would move the points above the curve.

The fact that (1) dissolved iron in the glassy phase in a given sample of fines is relatively constant from spherule to spherule, and (2) there is no evidence of solution of the injected particulate material in the glassy phase, suggests that the glass is essentially saturated with respect to iron for the particular conditions existing at the time of formation. Saturation of the iron in the glassy phase strongly suggests that some of the iron spherules found in the lunar glass spherules were produced by reduction of iron from the glass. This is in agreement with the conclusions of Housley *et al.* (1972). If one assumes that the glassy phase in those spherules with a large number of inclusions is saturated with respect to iron, then the above analysis indicates that the Apollo 14 glass spherules were formed at a lower temperature than their Apollo 11 and 12 counterparts. This is true because of the higher concentration of iron in the saturated glassy phase of the Apollo 11 and 12 specimens further confirming the conclusions in the previous section.

SUMMARY OF CONCLUSIONS

The magnetic parameters determined by susceptibility measurements of the glass spherules from the Apollo 14 fines are similar to those in the Apollo 11 and 12 samples. At temperatures less than 77°K the magnetic susceptibility of all the glass spherules approximately follows a Curie-Weiss relation with a Weiss temperature of 3–7°. A theoretical expression for the magnetic susceptibility based on the presence of antiferromagnetic inclusions and a distortion of the octahedral ligand field of the Fe^{+2} ions in the glass fits the experimental data. With the help of this expression the Weiss temperature of the glassy phase was determined.

By plotting the experimental and calculated parameters, evidence is obtained that shows:

(a) There are two components to the Weiss temperature, one due to ligand field distortion in the glassy phase and the other due to antiferromagnetic inclusions in the glass. From the model and theoretical expression the magnitude of the Weiss temperature associated with each component can be determined.

(b) The Apollo 14 glass spherules have a higher concentration of antiferromagnetic inclusions than those from the Apollo 11 and 12 samples. The distortion of ligand field about the Fe^{+2} ions in the glassy phase of the specimens from Apollo 14 is less than in those specimens from Apollo 11 and 12 lunar fines. We attribute the latter effect to a lower temperature of formation of the Apollo 14 glass and/or to a less rapid quenching of the glass.

(c) The Curie constant of the glassy phase is essentially constant for all the spherules in a given sample of fines. As the principal source of the Curie constant is Fe^{+2} ions in the glass, the iron in the glassy phase is also constant, and, although it will vary somewhat from place to place on the lunar surface, the concentration of iron is about 5% for the Apollo 14 specimens, and somewhat higher in the Apollo 11 and 12 specimens.

(d) The particulate iron mineral inclusions did not dissolve in the glass. This fact and the constancy of the iron in the glassy phase indicates that the glass is essentially saturated with respect to iron, and that some of the iron spherules in the glassy phase were formed by reduction of the iron in the glassy phase.

Saturation of the glassy phase of the Apollo 14 specimens and relatively low iron concentrations indicates lower temperatures of formation compared to the Apollo 11 and 12 samples.

Acknowledgments—We gratefully acknowledge the assistance of our colleague Richard R. Larson of the U.S. Geological Survey who made some of the electron microprobe analyses. The lunar samples were made available for this investigation by the National Aeronautics and Space Administration, which supported part of this work under contract NGL-09-11-006. Publication authorized by the Director, U.S. Geological Survey.

REFERENCES

Griscom D. L. and Marguardt C. L. (1972) Evidence of lunar surface oxidation processes: Electron spin resonance spectra of lunar materials and simulated lunar materials. *Proc. Third Lunar Sci. Conf., Geochim. Cosmochim. Acta* Suppl. 3, Vol. 3. MIT Press.

Housley R. M., Grant R. W., and Abdel-Gawad M. (1972) Study of excess Fe metal in the lunar fines by magnetic separation (abstract). In *Lunar Science—III* (editor C. Watkins), pp. 392–395, Lunar Science Institute Contr. No. 88.

Nagata T., Ishikawa Y., Kinoshita H., Kono M., Syono Y., and Fisher R. M. (1970) Magnetic properties and natural remanent magnetization of lunar materials. *Proc. Apollo 11 Lunar Sci. Conf., Geochim. Cosmochim. Acta* Suppl. 1, Vol. 3, pp. 2325–2340. Pergamon.

Nagata T., Fisher R. M., Schwerer F. C., Fuller M. D., and Dunn J. R. (1971) Magnetic properties and remanent magnetization of Apollo 12 lunar materials and Apollo 11 lunar microbreccia. *Proc. Second Lunar Sci. Conf., Geochim. Cosmochim. Acta* Suppl. 2, Vol. 3, pp. 2461–2476. MIT Press.

Pake G. E. (1962) *Paramagnetic Resonance*, Chap. 3, pp. 65–70, W. A. Benjamin.

Pickart S. J. and Alpern H. (1971) Neutron diffraction study of lunar materials. *Proc. Second Lunar Sci. Conf., Geochim. Cosmochim. Acta* Suppl. 2, Vol. 3, pp. 2079–2082. MIT Press.

Sullivan S., Thorpe A. N., Alexander C. C., Senftle F. E., and Dwornik E. (1971) Magnetic properties of individual glass spherules, Apollo 11 and Apollo 12 lunar samples. *Proc. Second Lunar Sci. Conf., Geochim. Cosmochim. Acta* Suppl. 2, Vol. 3, pp. 2433–2449. MIT Press.

Thorpe A. N., Senftle F. E., and Cuttitta F. (1963) Magnetic and chemical investigations of iron in tektites. *Nature* **197,** 836–840.

Thorpe A. N. and Senftle F. E. (1964) Submicroscopic spherules and color of tektites. *Geochim. Cosmochim. Acta* **28,** 981–994.

Thorpe A. N., Senftle F. E., Sullivan S., and Alexander C. C. (1970) Magnetic studies of individual glass spherules from the lunar sample 10084,86,2, Apollo 11. *Proc. Apollo 11 Lunar Sci. Conf., Geochim. Cosmochim. Acta* Suppl. 1, Vol. 3, pp. 2455–2462. Pergamon.

Thorpe A. N., Sullivan S., Alexander C. C., Senftle F. E., and Dwornik E. (1972) Temperature dependent magnetic properties, Apollo 11, 12, and 14 lunar samples (abstract). In *Lunar Science—III* (editor C. Watkins), pp. 752–754, Lunar Science Institute Contr. No. 88.

Tinkham M. (1956) Paramagnetic resonance in dilute iron group fluorides, I. Fluorine hyperfine structure. *Proc. Roy. Soc. London* **236,** Ser. A, 535–549.

Tsay F. E., Chan S. I., and Manatt S. L. (1971) Ferromagnetic resonance of lunar samples. *Geochim. Cosmochim. Acta* **35,** 865–875.

Van Vleck J. H. and Penney W. G. (1934) The theory of the paramagnetic rotation and susceptibility in manganous and ferric salts. *Phil. Mag.* **17,** Ser. 7, 961–987.

Proceedings of the Third Lunar Science Conference
(Supplement 3, *Geochimica et Cosmochimica Acta*)
Vol. 3, pp. 2479–2493
The M.I.T. Press, 1972

Mössbauer studies of Apollo 14 lunar samples

T. C. Gibb, R. Greatrex, and N. N. Greenwood,

Department of Inorganic and Structural Chemistry,
The University of Leeds, Leeds LS2 9JT, England

and

M. H. Battey

Department of Geology, The University of Newcastle upon Tyne,
Newcastle upon Tyne NE1 7RU, England

Abstract—The iron-bearing minerals in ten Apollo 14 lunar samples have been examined by Mössbauer spectroscopy. The results parallel earlier data on the Apollo 11 and 12 samples, but show significant differences in mineral content. Mineral separations on soil 14259 have shown that the metallic iron-nickel phase is largely associated with the glassy particles; microprobe analyses on 25 Fe–Ni granules gave an approximate average composition of 5.4 at. % Ni, 0.6 at. % Co. One particle had a kamacite-taenite intergrowth which had equilibrated at 740°K. The clinopyroxene in rock 14310 features a high degree of cation order, consistent with equilibration at a low temperature of $\sim 900°$K.

Introduction

The iron-bearing minerals in ten Apollo 14 lunar samples have been examined by Mössbauer spectroscopy. The allocation comprised four fines (14003,20, 14162,48, 14163,50, and 14259,17), and six rock chips (14301,15, 14303,36, 14310,66, 14311,32, 14318,35, and 14321,179). The spectra give information about the oxidation state, site symmetry, and magnetic state of iron in these samples. The results parallel earlier data on Apollo 11 (Gay *et al.*, 1970; Greenwood and Howe, 1970; Hafner and Virgo, 1970; Herzenberg and Riley, 1970; Housley *et al.*, 1970) and Apollo 12 samples (Gay *et al.*, 1971; Hafner *et al.*, 1971c; Herzenberg *et al.*, 1971; Housley *et al.*, 1971) but show significant differences in the mineral content. Sample 14259,17 was subjected to several physical separation procedures in an attempt to provide specimens of the component minerals. A strongly magnetic separate (7 mg) proved to be iron-nickel fragments.

Experimental

The Mössbauer spectrometer has been described previously (Gibb *et al.*, 1970). Two velocity ranges were scanned ± 10 mm s^{-1} to reveal magnetic hyperfine interactions, and ± 4 mm s^{-1} to provide better resolution in the principal region of silicate absorption, and the spectra were calibrated from an iron foil enriched in ^{57}Fe. In Figs. 1–7 the absorption is expressed as a percentage of the total baseline count.

The radioactive source was 50 mCi of ^{57}Co in a palladium matrix at room temperature. The absorbers were prepared by spreading the sample evenly between two polythene disks to a thickness of ~ 50 mg cm^{-2}. The fines were used as received and the rock samples were prepared by crushing in an agate mortar. Samples were examined at 295°K, and 78°K, and sometimes also at 4.2°K, the cryogenic temperatures being achieved using a Texas Instruments Inc. cryostat (CLF-3).

Details of the various fractionation procedures applied to sample 14259,17 are covered in the discussion sections.

Results and Discussion

Fines from Fra Mauro

The Fra Mauro region is believed to be an ejecta blanket of pre-mare crust from the Imbrium basin. Chemical analysis of the Apollo 14 soils (Schnetzler and Nava, 1971; Brunfelt et al., 1971) has shown them to be related to but distinct from those of other lunar areas. Fines 14163 and 14259 are similar and show lower Ti, Cr, Mn, and Fe, and higher Si, Al, and K analyses than the Apollo 11 (Compston et al., 1970; Rose et al., 1970, and refs therein), Apollo 12 (Maxwell and Wiik, 1971), and Luna 16 (Schnetzler and Nava, 1971) results. The major constituent of 14163 is glassy material (Brunfelt et al., 1971) and the same appears to be true of 14259. A microprobe analysis of 856 grains from 14259,26 included 45% of glassy types subdivided into seven classifications (Apollo Soil Survey, 1971). Of the remaining mineral fragments, 47% were various feldspars including plagioclase, 16.7% orthopyroxene, 8.7% pigeonite, 15.8% augite, and 7.3% olivine. Opaques were ilmenite and other oxides (3.1%), iron-nickel (1.4%), and only four grains of troilite. The large orthopyroxene content (Apollo Soil Survey, 1971; Fuchs, 1971) contrasts with the mere trace quantities in Apollo 11 and 12 samples. The composition ranged from about $En_{82}Fs_{15}Wo_3$ to $En_{63}Fs_{34}Wo_3$. The clinopyroxene showed a very wide scatter throughout the En–Fs–Hd–Di quadrilateral with the iron content ranging from Fs_{15} to Fs_{70}.

Mössbauer spectra at 78°K with a velocity scan of ± 10 mm s^{-1} are shown in Fig. 1 for samples 14003 and 14259. All four fines gave a central broad asymmetric doublet from the iron silicates, and a weak magnetic pattern with broad lines attributable to iron-nickel alloys (see below). No evidence was found for troilite.

Spectra of all four fines at 295°K and 78°K covering a velocity scan of ± 4 mm s^{-1} are given in Figs. 2 and 3. The ilmenite present may be seen in the spectra at 295°K as a pair of weak unresolved lines in the central absorption region. The proportion of the iron as ilmenite is much lower than that of Apollo 11 fines (23%), but is similar to that found for Apollo 12 (5%–8%). The iron silicate lines from 14003, 14163, and 14259 are very broad and unresolved, and 14162 is the only sample to show significant line structure at 295°K. The detailed analysis of these spectra by least-square curve-fitting to obtain the distribution of iron amongst each individual phase presents problems no less serious than those encountered in the treatment of the data from Apollo 11 and 12 samples.

The terrestrial orthopyroxenes have been studied in detail (Virgo and Hafner, 1970). For an Fe/(Fe + Mg) ratio of <0.4 one finds an almost complete ordering of the iron onto the M2 sites. However, the exact site population depends on the thermal history of the orthopyroxene. At room temperature the quadrupole splittings are ~2.5 mm s^{-1} for the M1 site and ~2.1 mm s^{-1} for the M2 site (Evans et al.,

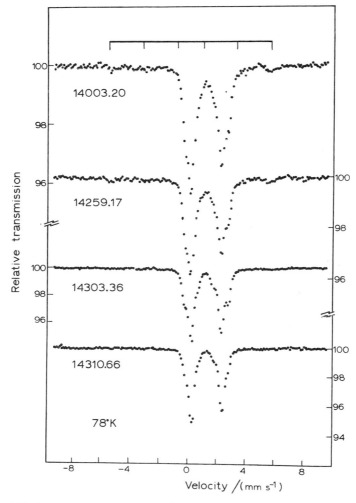

Fig. 1. Mössbauer spectra at 78°K of Apollo 14 fines and rocks. The bar diagram represents the spectrum of iron-nickel alloy.

1967). The two site symmetries are more clearly resolved at 77°K with splittings of ~3.0 and ~2.1 mm s^{-1} respectively (Virgo and Hafner, 1968). These parameters are not very sensitive either to the M1/M2 populations or to the degree of ordering, although the splitting of the M1 site does decrease slightly while that of the M2 increases with increasing magnesium content (Hafner and Virgo, 1970). A terrestrial clinopyroxene has been found to give similar parameters to the orthopyroxenes with splittings for the M1 and M2 sites of ~2.8 and ~2.1 mm s^{-1} at 77°K. Similarly, lunar clinopyroxenes from Apollo 11 rocks, though more inhomogeneous, have shown average splittings in the ranges 2.84–3.01 and 2.03–2.10 mm s^{-1} (Gay et al., 1970; Hafner and Virgo, 1970).

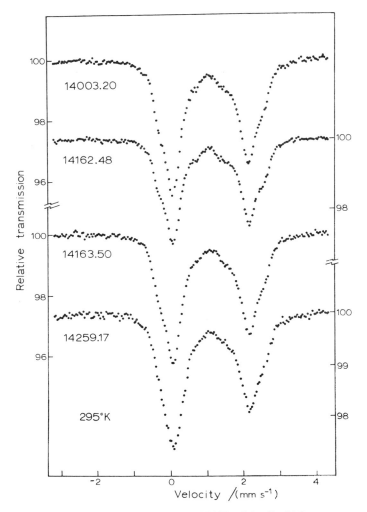

Fig. 2. Mössbauer spectra at 295°K of Apollo 14 fines.

On this basis, although both ortho- and clinopyroxenes are present in the Apollo 14 fines, they can not be differentiated easily in the Mössbauer spectra.

A hand-picked olivine separate from an Apollo 11 rock sample gave a quadrupole splitting of 2.86 mm s⁻¹ at 295°K (Gay et al., 1970). This value is greater than that for the pyroxenes at this temperature, but the distinction might be expected to be less at 78°K. The two site symmetries in olivines are not normally distinguishable.

A major problem is generated by the iron-bearing glasses. They have been associated in one report with the lines from the M2 pyroxene sites (Gay et al., 1970). Following the results described in the next section, we incline to the view that iron silicate glass will give a spectrum similar to that of crystalline pyroxenes, but with substantially broader lines because of greater variations in site symmetry. Similar

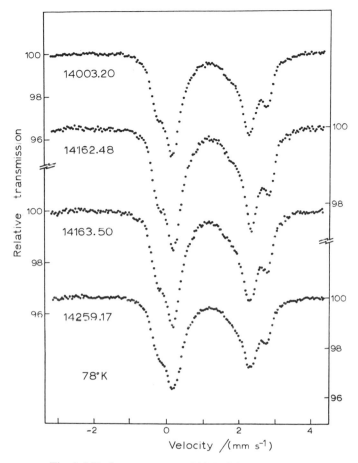

Fig. 3. Mössbauer spectra at 78°K of Apollo 14 fines.

conclusions have been drawn from data for a partial glass separate from an Apollo 11 soil (Hafner *et al.*, 1971a).

We have analyzed our data by least-squares fitting of three Lorentzian doublets. In increasing magnitude of quadrupole splitting, the doublets are assigned to: (1) ilmenite, (2) M2 pyroxenes + glass, and (3) olivine + M1 pyroxenes + glass. The poor resolution of these components has necessitated the estimation of the ilmenite content from the spectra at 295°K, and the percentage of iron as olivine + M1 pyroxenes + glass from the spectra at 78°K. The results are as follows:

Percentage of Fe present as	14003	14162	14163	14259
ilmenite	8	8	11	11
M2 pyroxene + glass	73	71	71	70
olivine + M1 pyroxene + glass	19	21	18	19

These figures were calculated under the assumption that the areas of the high-velocity components of the three quadrupole doublets are linearly proportional to the concentration of iron. The low-velocity components were ignored because of uncertainty as to the origins of the asymmetry in the spectra. Possible additional contributions to the lower velocity peaks are discussed in later sections. The line broadening of the glassy material may have resulted in a small overestimation of the ilmenite content.

Mineral separates from 14259,17

The sample of the 14259,17 fines contained a high proportion of glass. Many particles appeared to be crystalline but glass-coated, thus preventing optical recognition of the core grains. Olivine and pyroxene were identified. The quantity of this sample available to us was 4.6 g compared to <0.5 g for all other samples, and efforts were made to obtain mineral separations on this sample.

The fines were first sieved using B.S.S. sieve sizes to provide a particle size fractionation. The result was

>500 μ	0.192 g	4.2%
355–500 μ	0.147 g	3.2%
250–355 μ	0.207 g	4.5%
180–250 μ	0.288 g	6.2%
150–180 μ	0.172 g	3.7%
125–150 μ	0.291 g	6.1%
75–125 μ	1.482 g	32.1%
<75 μ	1.849 g	40.0%

This size fractionation produced no noticeable separation within the group of iron bearing minerals.

The two portions with grain size <125 μ were washed and reunited and then a small hand magnet was used to attract any strongly magnetic particles by moving it over the sample at a height of 1–3 mm. In this way, 7 mg of a highly magnetic fraction was obtained. Measurements on this sample are the subject of a later section.

After the magnetic extraction, the remaining mineral was separated into seven approximately equal fractions with different densities using flotation in Clerici solutions. Mössbauer spectra at 78°K for five of these fractions are illustrated in Fig. 4. The heaviest fraction shows the narrowest lines observed for any separate from 14259,17 and, although the degree of mineral separation is disappointing, the spectrum does indicate substantial enhancement in olivine + M1 pyroxenes + glass sites. Ilmenite was also clearly present. Grains of olivine, which are usually denser than pyroxene, were identified visually.

The poorly resolved resonance from the lightest fraction is undoubtedly associated with a high glass content in this sample, which is also noteworthy in containing a discernible residual quantity of metallic iron. The greater part of the latter is therefore associated with the glassy phases.

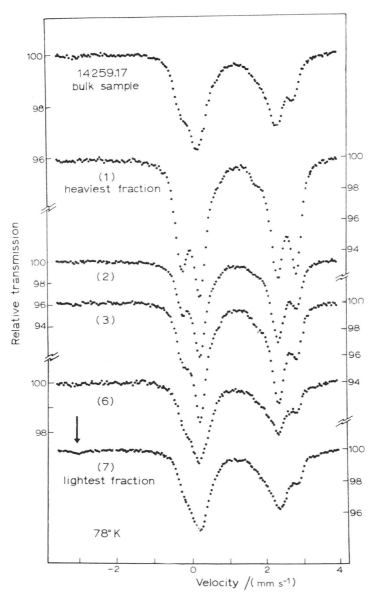

Fig. 4. Mössbauer spectra at 78°K of separations by density from sample 14259,17. The arrow indicates one of the iron-nickel resonance lines which is strongest in the lightest fraction.

Rocks from Fra Mauro

Mössbauer spectra have been obtained for partly crushed samples from six different Apollo 14 rocks. In all cases the metallic iron and troilite content was smaller than in the fines and could not be detected in the Mössbauer spectra. This is illustrated in Fig. 1 by spectra for samples 14303 and 14310. The spectra at 295°K and 78°K with a velocity scan of ± 4 mm s^{-1} are shown in Figs. 5 and 6.

The resonance lines are significantly narrower than in the spectra of the fines. This we attribute mainly to the much lower glass content in the rocks. The spectra

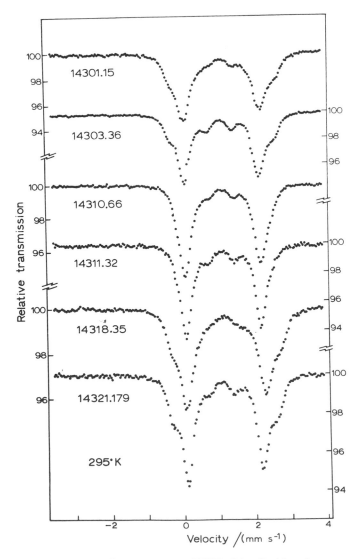

Fig. 5. Mössbauer spectra at 295°K of Apollo 14 rocks.

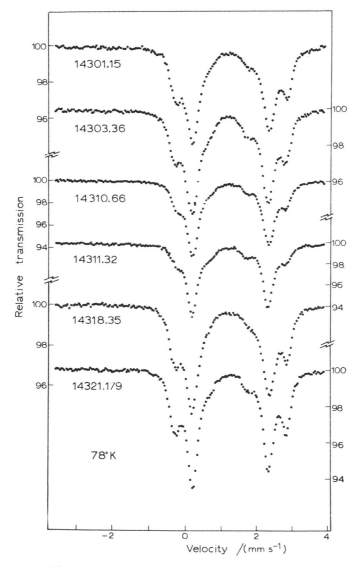

Fig. 6. Mössbauer spectra at 78°K of Apollo 14 rocks.

have been computer analyzed as three overlapping quadrupole doublets. In those cases where one or more components are not well resolved, it was found advantageous to introduce line-width constraints in the curve-fitting. The results are summarized in Table 1.

In some instances, notably rock 14318 and the room-temperature spectra of 14310 and 14311, it was not possible to obtain reasonable parameters for all six peaks, and the quadrupole splitting Δ and chemical isomer shift δ for the appropriate components are not given. The rock compositions were determined in the same manner as were those of the fines, and the results are given in Table 2.

Fragmental rock 14318 showed unusually broad lines and significantly greater pyroxene quadrupole splittings at room temperature than did 14301, 14303, and 14321, consistent with a higher glass content (Frederiksson et al., 1972). Fragmental rock 14311 and igneous rock 14310 show the lowest olivine/M1 pyroxene content ($\sim 16\%$) with a concomitant reduction in the quadrupole splittings for these sites.

The crystalline rock fragment 14310,66 was the only igneous rock examined. The ilmenite content is the lowest of all the six rocks studied. A report of microprobe analysis on 14310,6 gives the composition as plagioclase (50%), clinopyroxene (augite and pigeonite, 40%), ilmenite (3%), with minor amounts of Fe metal (containing 6–16 wt.% Ni), troilite, olivine and glass (Gancarz et al., 1971). Chromian ulvöspinel and orthopyroxene were absent. The pyroxenes feature an almost constant iron content with individual grains having a pigeonite core exemplified by $[Ca_{0.27}Mg_{1.10}Fe_{0.56}Mn_{0.01}Cr_{0.02}Ti_{0.03}]Si_{1.95}Al_{0.08}O_6$ ranging through to an augite rim $[Ca_{0.72}Mg_{0.66}Fe_{0.51}Mn_{0.01}Cr_{0.01}Ti_{0.07}]Si_{1.89}Al_{0.12}O_6$. However, independent work on this rock reports a wider range of composition (Brown and Peckett, 1971). For the purpose of our discussion we adopt an approximate average composition (neglecting minor elements) of $En_{46}Fs_{28}Wo_{26}$. Calcium is generally considered to prefer the M2 position in pyroxenes exclusively, and this assumption was used to

Table 1. Mössbauer parameters for rocks*.

	T (°K)	Ilmenite		M2 pyroxene		Olivine/M1 pyroxene	
		Δ	δ	Δ	δ	Δ	δ
14301,15	77	—	—	2.15	1.27	3.09	1.29
	295	0.70	1.08	2.09	1.15	2.87	1.16
14303,36	77	1.08	1.20	2.13	1.27	3.11	1.28
	295	0.68	1.08	2.08	1.14	2.94	1.14
14310,66	77	1.07	1.18	2.12	1.27	3.03	1.29
	295	0.65	1.11	2.07	1.14	—	—
14311,32	77	1.01	1.18	2.16	1.24	3.04	1.28
	295	0.72	1.07	2.06	1.15	—	—
14318,35	77	—	—	2.14	1.27	3.11	1.29
	295	—	—	2.24	1.19	3.00	1.20
14321,179	77	1.07	1.20	2.12	1.27	3.10	1.28
	295	0.71	1.06	2.09	1.15	2.88	1.15

* Quadrupole splitting, Δ mm s^{-1}; chemical isomer shift, δ mm s^{-1} relative to metallic iron at room temperature.

Table 2. Percentage distribution of iron amongst lunar minerals.

	Ilmenite	M2 pyroxene	Olivine/M1 pyroxene
14301,15	6	70	24
14303,36	8	69	23
14310,66	4	78	18
14311,32	8	77	15
14318,35	6	72	22
14321,179	6	66	28

derive site occupancies in pyroxenes from the Apollo 11 rocks (Hafner and Virgo, 1970). The quadrupole splitting of the M1 site of ~ 3.03 mm s^{-1} is less than the ~ 3.10 mm s^{-1} shown by four of the fragmental rocks, and is consistent with a higher magnesium content at these sites in 14310. Assuming that the four apparent silicate lines are due entirely to M1 and M2 sites in clinopyroxene, the ratio of M1/M2 occupations is 0.23. From this can be derived a site occupancy of $(Mg_{0.90}Fe_{0.10})_{M1} \cdot [Mg_{0.02}Fe_{0.46}Ca_{0.52}]_{M2}Si_2O_6$. On this formulation the magnesium is almost entirely ordered onto the M1 sites. The restricted compositional variation at this site is fully consistent with the sharper lines and smaller quadrupole splitting observed.

These results can be compared with those for a pigeonite sample from 12021 with a modal formula of $En_{60}Fs_{31}Wo_9$ (Hafner $et\ al.$, 1971b). A thermal disordering study proved that the naturally occurring order corresponds to an equilibrium temperature of about 840°K, implying either a slow subsolidus cooling after the initial rapid quench from the liquidus or a subsequent thermal annealing. Our observation of cation ordering in 14310, which has been confirmed independently by Finger $et\ al.$ (1972) and Ghose $et\ al.$ (1972), is an indication of equilibration at a low temperature of about 900°K.

Further interpretations await the availability of more data on compositional variations in the pyroxenes by microprobe analysis.

Iron-nickel fragments

The 7 mg of highly magnetic material separated from soil 14259 as described in the section on Mineral Separates were subjected to intensive study. Mössbauer spectra at 78°K and 4.2°K are shown in Fig. 7. The sample contained a comparatively small silicate residue, and the majority of the iron was ferromagnetic with a large magnetic

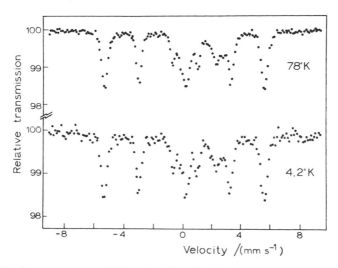

Fig. 7. Mössbauer spectra at 78°K and 4.2°K of a highly magnetic separate from soil 14259,17.

hyperfine field. The observed value of 333 ± 1 kG at 295°K was 3 kG higher than found in an iron foil, the resonance lines being considerably broader than found in pure iron, and the spectrum centroid was shifted by +0.02 mm s^{-1}. The figures for the hyperfine field at 77°K and 4.2°K were 345 kG and 346 kG as against 337 kG and 338 kG for pure iron. The majority of metals which alloy with iron reduce the average hyperfine field, the exceptions being cobalt and nickel. From published data on dilute nickel in iron alloys, our data are consistent with an average nickel content of 3 at.% (Johnson et al., 1963). It is noteworthy that the separate contained no troilite.

The presence of nickel in the iron fragments was confirmed by x-ray fluorescence measurements. The relative proportions of Ti, V, Cr, Mn, Fe, Co, and Ni were determined in the sample and compared with the analysis of the bulk soil. Considerable enrichment of nickel and cobalt was found and, after correction for the iron in the silicate residue, an approximate average alloy composition of 4.5 at.% Ni, 1.0% Co, and 94.5% Fe was deduced. The manganese and chromium were both depleted in proportion to the residual silicate and were not present in the alloy in significant quantity. Vanadium was not detected (<0.05%).

A detailed microprobe study was made of some of the iron-nickel granules. Analyses of Fe, Ni, and Co were made on a total of 25 fragments, and the results are illustrated in Fig. 8. None of the 25 particles was associated with silicates. A phosphorus analysis was made on 10 particles, but in all cases the amount detected was less than 0.1 at.% and usually not more than 0.01%. The average composition based on these 25 fragments was 5.4 at.% Ni and 0.6 at.% Co, in reasonable agreement with the Mössbauer and x-ray fluorescence measurements.

Five of the particles showed some degree of nickel concentration zoning. None of these minor zones is included in Fig. 8. Most interesting was a particle of 40 × 25 μ cross section. A sulphur analysis proved negative, and the nickel showed several small

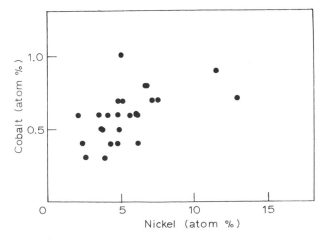

Fig. 8. Nickel and cobalt analyses for some of the iron-nickel alloy fragments in the highly magnetic separate from soil 14259,17.

zonings. Of particular note was a small area at one edge with 37 at.% Ni, 0.7% Co against a neighboring matrix of 5.6% Ni, 1.2% Co. The boundary appeared to be sharp and this is clearly a case of the γ-phase (taenite) precipitating from the α-phase (kamacite). This process does not usually occur by cooling of the γ-phase, but only by secondary thermal annealing for long periods. Above 1180°K there is a region of complete Fe-Ni solubility called the f.c.c. γ-phase (Goldstein and Ogilvie, 1965). Below this temperature an iron-rich b.c.c. α-phase exists which extends to about 8 at.% Ni at 750°K. However, cooling of the γ-phase alloy to within the $\alpha + \gamma$ field does not allow a $\gamma \rightarrow \alpha$ transformation, and on further cooling the α_2-phase is formed. In small particles of γ-phase, however, the α_2 transformation is suppressed for >18 at.% Ni at room temperature (Asano, 1969). Reheating of α_2 alloy to the two-phase region causes a very slow (i.e., years) precipitation of the γ-phase, generating kamacite-taenite intergrowths. In this instance the compositions of the α and γ phases indicate an equilibration at ~740°K (Goldstein et al., 1965).

A similar microprobe examination of about 15 typical particles from rock 14321 revealed trace quantities of Fe-Ni in only one granule, the largest inclusion being 16% Ni, 1.8% Co. Independent measurements on a crystalline fragment from this rock have given 14–17 wt.% Ni, and for igneous rock 14310, 6–16 wt.% Ni (Gancarz et al., 1971).

It is important to consider the effect of compositional variation of 0–40 wt.% Ni on the Mössbauer spectrum. It has been shown that small particle Fe-Ni with a composition of 24 at.% Ni gives no magnetic hyperfine splitting even at 4.2°K, and this is largely true for 18–30 at.% Ni (Asano et al., 1969, and refs. therein). Above 30% Ni, hyperfine splitting is recorded, as indeed it is in low Ni concentrations where the magnetic field is greater than that in iron (Johnson et al., 1963). In the latter case it is not clear which of the multiphase alloys is referred to, but it seems likely that it is the α_2.

It seems probable that our magnetic separate would concentrate the magnetic α_2 and α alloys to the exclusion of any γ phase with a lower magnetization. The iron-nickel Mössbauer spectrum in the bulk samples is likely to include a broad "paramagnetic" peak from both the γ-phases and any superparamagnetic microparticles of iron metal (Housley et al., 1971) at the center of the hyperfine pattern from the low-Ni α alloys. This component will coincide with the low velocity M2 pyroxene line, which is invariably seen to be more intense than the high-velocity line. This is particularly noticeable in the glass separate from fines 14259. (See Fig. 4.)

The foregoing is one of the reasons why we have calculated the silicate site occupancies from the high-velocity components only. However, one need not attribute all the area difference to Fe-Ni; a substantial Goldanskii-Karyagin asymmetry from the more distorted M2 sites may be anticipated, and this is also likely to contribute to the observed asymmetry.

A major problem attaches to deciding the origin of the iron particles. Many of the petrological studies on Apollo 11 and 12 samples have reported the presence of Fe-Ni alloys. The low nickel in the iron in Apollo 11 igneous rocks led to the belief that the Fe-Ni and FeS intergrowths in the soils and breccia were of extralunar origin. However, in Apollo 12 rocks evidence has been found for formation of high Ni content alloys (15–30 wt.% Ni) in early formed olivine and chromite, with only 3–15 wt.% Ni in the later phases in the crystallization sequence (Reid *et al.*, 1970). Further evidence in favor of indigenous Fe-Ni comes from an Apollo 11 low-Ti-anorthositic rock with 6–29% Ni for which a lunar Ni fractionation is also considered likely (Dickey, 1970). Iron meteorites are generally considered to have >5 wt.% Ni and <1.0% Co, and over 80% of metallic inclusions in an Apollo 12 soil were outside these limits (Goldstein and Yakowitz, 1971). However, even in igneous rocks it is not impossible that the iron results from meteoritic impact predating the crystallization of the rock in its present form.

From the present Fe-Ni analyses for both the soil 14259 and the fragmental rock 14321, it therefore is feasible that some of the Fe-Ni is not extralunar in origin.

Acknowledgment—We thank Dr. G. Hornung of the Department of Earth Sciences, University of Leeds, for assistance in obtaining the x-ray fluorescence and microprobe data, and the S.R.C. for financial support.

REFERENCES

Apollo Soil Survey (1971) Apollo 14: Nature and origin of rock types in soil from the Fra Mauro formation. *Earth Planet. Sci. Lett.* **12**, 49–54.

Asano H. (1969) Magnetism of γ Fe–Ni invar alloys with low nickel concentration. *J. Phys. Soc. Japan* **27**, 542–553.

Brown G. M. and Peckett A. (1971) Selective volatilization on the lunar surface; evidence from Apollo 14 feldspar-phyric basalts. *Nature* **234**, 262–266.

Brunfelt A. O., Heier K. S., Steinnes E., and Sundvoll B. (1971) Determination of 36 elements in Apollo 14 bulk fines 14163 by activation analysis. *Earth Planet. Sci. Lett.* **11**, 351–353.

Compston W., Chappell B. W., Arriens P. A., and Vernon M. J. (1970) The chemistry and age of Apollo 11 lunar material. *Proc. Apollo 11 Lunar Sci. Conf., Geochim. Cosmochim. Acta* Suppl. 1, Vol. 2, pp. 1007–1027. Pergamon.

Dickey J. S. (1970) Nickel-iron in lunar anorthosites. *Earth Planet. Sci. Lett.* **8**, 387–392.

Evans B. J., Ghose S., and Hafner S. S. (1967) Hyperfine splitting of ^{57}Fe and Mg–Fe order-disorder in orthopyroxenes ($MgSiO_3$–$FeSiO_3$ solid solution). *J. Geol.* **75**, 306–322.

Finger L. W., Hafner S. S., Schürmann K., Virgo D., and Warburton D. (1972) Distinct cooling histories and reheating of Apollo 14 rocks (abstract). In *Lunar Science—III* (editor C. Watkins), pp. 259–261, Lunar Science Institute Contr. No. 88.

Frederiksson K., Nelen J., and Noonan A. (1972) Apollo 14: Glasses, breccias, chondrules (abstract). In *Lunar Science—III* (editor C. Watkins), pp. 280–282, Lunar Science Institute Contr. No. 88.

Fuchs L. H. (1971) Orthopyroxene and orthopyroxene-bearing rock fragments rich in K, REE, and P in Apollo 14 soil sample 14163. *Earth Planet. Sci. Lett.* **12**, 170–174.

Gancarz A. J., Albee A. L., and Chodos A. A. (1971) Petrologic and mineralogic investigation of some crystalline rocks returned by the Apollo 14 mission. *Earth Planet. Sci. Lett.* **12**, 1–18.

Gay P., Bancroft G. M., and Brown M. G. (1970) Diffraction and Mössbauer studies of minerals from lunar soils and rocks. *Proc. Apollo 11 Lunar Sci. Conf., Geochim. Cosmochim. Acta* Suppl. 1, Vol. 1, pp. 481–497. Pergamon.

Gay P., Brown M. G., Muir I. D., Bancroft G. M., and Williams P. G. L. (1971) Mineralogical and petrographic investigations of some Apollo 12 samples. *Proc. Second Lunar Sci. Conf., Geochim. Cosmochim. Acta* Suppl. 2, Vol. 1, pp. 377–392. MIT Press.

Ghose S., Ng G., and Walter L. S. (1972) Clinopyroxenes from Apollo 12 and 14: Exsolution, cation order, and domain structure (abstract). In *Lunar Science—III* (editor C. Watkins), pp. 300–302, Lunar Science Institute Contr. No. 88.

Gibb T. C., Greatrex R., Greenwood N. N., and Sarma A. C. (1970) Mössbauer spectra of some tellurium complexes. *J. Chem. Soc.* (*A*), 212–217.

Goldstein J. I. and Ogilvie R. E. (1965) A reevaluation of the iron-rich portion of the Fe–Ni system. *Trans. Met. Soc. AIME* **233**, 2083–2087.

Goldstein J. and Yakowitz H. (1971) Metallic inclusions and metal particles in the Apollo 12 lunar soil. *Proc. Second Lunar Sci. Conf., Geochim. Cosmochim. Acta* Suppl. 2, Vol. 1, pp. 177–191. MIT Press.

Greenwood N. N. and Howe A. T. (1970) Mössbauer studies of Apollo 11 lunar samples. *Proc. Apollo 11 Lunar Sci. Conf., Geochim. Cosmochim. Acta* Suppl. 1, Vol. 3, pp. 2163–2169. Pergamon.

Hafner S. S., Janik B., and Virgo D. (1971a) State and location of iron in Apollo 11 samples. In *Mössbauer Effect Methodology* (editor I. J. Gruverman), Vol. 6, pp. 193–207. Plenum Press.

Hafner S. S. and Virgo D. (1970) Temperature-dependent cation distributions in lunar and terrestrial pyroxenes. *Proc. Apollo 11 Lunar Sci. Conf., Geochim. Cosmochim. Acta* Suppl. 1 Vol. 3, pp. 2183–2198. Pergamon.

Hafner S. S., Virgo D., Warburton D., Fernande H., Ohtsuki M., and Hibino A. (1971b) Subsolidus cooling history of coarse grained lunar basalt from Oceanus Procellarum. *Nature Phys. Sci*, **231**, 79.

Hafner S. S., Virgo D., and Warburton D. (1971c) Cation distributions and cooling history of clinopyroxenes from Oceanus Procellarum. *Proc. Second Lunar Sci. Conf., Geochim. Cosmochim. Acta* Suppl. 2, Vol. 1, pp. 91–108. MIT Press.

Herzenberg C. L., Moler R. B., and Riley D. L. (1971) Mössbauer instrumental analysis of Apollo 12 lunar rock and soil samples. *Proc. Second Lunar Sci. Conf., Geochim. Cosmochim. Acta* Suppl. 2, Vol. 3, pp. 2103–2123. MIT Press,

Herzenberg C. L. and Riley D. L. (1970) Analysis of the first returned lunar samples by Mössbauer spectrometry. *Proc. Apollo 11 Lunar Sci. Conf., Geochim. Cosmochim. Acta* Suppl. 1, Vol. 3, pp. 2221–2241. Pergamon.

Housley R. M., Blander M., Abdel-Gawad M., Grant R. W., and Muir A. H. (1970) Mössbauer spectroscopy of Apollo 11 samples. *Proc. Apollo 11 Lunar Sci. Conf., Geochim. Cosmochim. Acta* Suppl. 1, Vol. 3, pp. 2251–2268. Pergamon.

Housley R. M., Grant R. W., Muir A. H., Blander M., and Abdel-Gawad M, (1971) Mössbauer studies of Apollo 12 samples. *Proc. Second Lunar Sci. Conf., Geochim. Cosmochim. Acta* Suppl. 2, Vol. 3, pp. 2125–2136. MIT Press.

Johnson C. E., Ridout M. S., and Cranshaw T. E. (1963) The Mössbauer effect in iron alloys. *Proc. Phys. Soc.* **81**, 1079–1090.

Maxwell J. A. and Wiik H. B. (1971) Chemical composition of Apollo 12 lunar samples 12004, 12033, 12051, 12052, and 12065. *Earth Planet. Sci. Lett.* **10**, 285–288.

Reid A. M., Meyer C., Harmon R. S., Butler P., and Brett R. (1970) Metal in two Apollo 12 igneous rocks. *Trans. Amer. Geophys. Union* **51**, 584.

Rose H. J., Cuttitta F., Dwornick E. J., Carron M. K., Christian R. P., Lindsay J. R., Ligon D. T., and Larson R, R. (1970) Semimicro x-ray fluorescence analysis of lunar samples. *Proc. Apollo 11 Lunar Sci. Conf., Geochim Cosmochim. Acta* Suppl. 1, Vol. 2, pp. 1493–1497. Pergamon.

Schnetzler C. C. and Nava D. F. (1971) Chemical composition of Apollo 14 soils 14163 and 14259. *Earth Planet. Sci. Lett.* **11**, 345–350.

Virgo D. and Hafner S. S. (1968) Reevaluation of the cation distribution in orthopyroxenes by the Mössbauer effect. *Earth Planet. Sci. Lett.* **4**, 265–269.

Virgo D. and Hafner S. S. (1970) Mg order-disorder in natural orthopyroxenes. *Amer. Mineral.* **55**, 201–223.

Proceedings of the Third Lunar Science Conference
(Supplement 3, *Geochimica et Cosmochimica Acta*)
Vol. 3, pp. 2495–2501
The M.I.T. Press, 1972

Nuclear magnetic resonance properties of lunar samples*

D. KLINE

State University of New York at Albany

and

R. A. WEEKS

Solid State Division, Oak Ridge National Laboratory,
Oak Ridge, Tennessee

Abstract—Nuclear magnetic resonance spectra of ^{23}Na, ^{27}Al, and ^{31}P in fines samples 10084,60 and 14163,168 and in crystalline rock samples 12021,55 and 14321,166, have been recorded over a range of frequencies up to 20 MHz. A shift in the field at which maximum absorption occurs for all of the spectra relative to the field at which maximum absorption occurs for terrestrial analogues is attributed to a sample-dependent magnetic field at the Na, Al, and P sites opposing the laboratory field. The magnitude of these fields internal to the samples is sample dependent and varies from 5 to 10 G. These fields do not correlate with the iron content of the samples. However, the presence of single-domain particles of iron distributed throughout the plagioclase fraction that contains the principal fraction of Na and Al is inferred from electron magnetic resonance spectra shapes. Shapes and widths of the ^{23}Na and ^{27}Al spectra correlate with the magnitude of the glassy fraction of the samples.

INTRODUCTION

NUCLEI IN LUNAR ROCKS AND SOILS, whose nuclear magnetic resonance (NMR) spectra have been detected, are ^{23}Na, ^{27}Al, and ^{29}Si (Weeks *et al.*, 1970; Kolopus *et al.*, 1971). The elemental abundance of Na, Al, and Si in Apollo 12 and 14 samples is, on the average, 0.28% and 0.48%, 9% and 6%, and 21% and 23%, respectively (LSPET, 1970; LSPET, 1971). The isotopic abundance of ^{23}Na, ^{27}Al, and ^{29}Si is 100%, 100%, and 4.8%, respectively. The largest fractions of these isotopes are in the plagioclase minerals in the crystalline and fragmental rocks, and the plagioclases comprise from 40% to 70% of most of these rocks (LSPET, 1971; LSPET, 1972). In the soils, these nuclei are present not only in the plagioclase fraction, but also in the glasses which comprise from 5% to 50% of the material (Marvin *et al.*, 1971; Carr and Meyer, 1972; Fredriksson *et al.*, 1972). The amount of Al and Na in the glassy fraction of 14163, for example, is 8 to 32 wt.% Al as Al_2O_3 and 0.1 to 3.6 wt.% Na as Na_2O (Fredriksson *et al.*, 1972). Another nucleus that is of interest is ^{31}P, an isotope with 100% natural abundance. Although the elemental abundance of P in lunar rocks is low, ~ 0.03 wt.% P_2O_5 in 14321, for example (Compston *et al.*, 1972), its abundance is higher in the soils, ~ 0.52 wt.% P_2O_5 in 14163 (Compston *et al.*, 1972). This higher abundance in the soils may be related to the higher glass content and in particular to the KREEP glass, in which its abundance is 0.5 to 1 wt.% at P_2O_5 (Meyer *et al.*, 1971). Since the nuclear

* Research sponsored by the U.S. Atomic Energy Commission and supported by NASA Contract MSC-T-76458.

spin of ^{31}P is $I = \frac{1}{2}$, the resonance transition is not perturbed by quadrupolar effects. The anomalous shift in the absorption of ^{23}Na and ^{27}Al which has been observed (Kolopus *et al.*, 1971) should be particularly evident in the spectrum of ^{31}P.

EXPERIMENTAL RESULTS AND DISCUSSION

Nuclear magnetic resonance measurements have been made in the dispersion mode at room temperature on samples of the following fines: 10084 and 14163,168; and crystalline rocks: 12021,55 and 14321,166. Spectra from the following nuclei

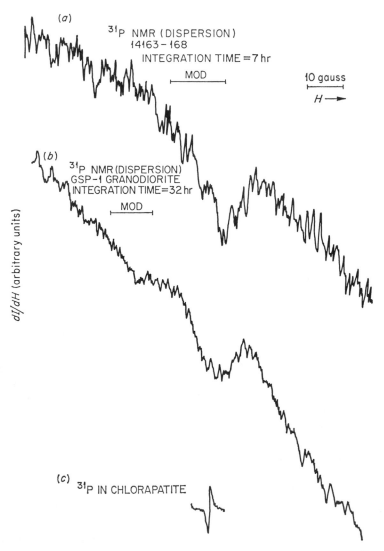

Fig. 1. Comparison of nuclear magnetic resonance dispersion mode spectra of ^{31}P at 30 MHz from (a) Apollo 14 fines 14163,168, (b) GSP-1 granodiorite, and (c) chlorapatite.

have been studied in one or more of the lunar samples and compared to those observed from selected terrestrial samples: $^{31}P(I = \frac{1}{2})$, $^{27}Al(I = \frac{5}{2})$, and $^{23}Na(I = \frac{3}{2})$. The experimental apparatus and procedures have been described elsewhere (Weeks *et al.*, 1970).

The dispersion derivative of ^{31}P from 14163,168 recorded at 30 MHz (Fig. 1) is very weak and rather broad and appears to be shifted by a very slight amount, possibly as much as 5 G, to the high-field side of the unperturbed resonant field, H_0. This shift was measured relative to a sharp, narrow absorption derivative of ^{31}P from a sample of chlorapatite, used to locate the position of H_0. For comparison, the ^{31}P spectrum from a terrestrial sample of GSP-1 granodiorite (0.28 wt.% P_2O_5) was also very weak and broad, with a shape similar to that of ^{31}P in 14163,168, but with apparently little or no shift measurable to within ± 2 G. The ^{31}P shift in 14163,168 may be associated with small internal magnetic fields similar in magnitude to those that oppose the laboratory field at the aluminum and sodium sites reported in earlier NMR studies of lunar material (Kolopus *et al.*, 1971).

The ^{23}Na and ^{27}Al nuclear magnetic resonance spectra from all lunar samples studied thus far, as well as those from several terrestrial samples such as BCR-1 basalt, exhibit an inverse dependence of their line widths on ν_0, the unperturbed resonant frequency, and a broad asymmetric frequency-dependent line shape characteristic of second-order nuclear electric quadrupole perturbations. Only their "central" ($m_I = \frac{1}{2} \leftrightarrow m_I = -\frac{1}{2}$) transitions, arising from slightly differing distributions of quadrupole interactions from sample to sample, appear in the lunar spectra. The satellite transitions, $m_I = \pm\frac{3}{2} \leftrightarrow m_I = \pm\frac{1}{2}$ and $m_I = \pm\frac{5}{2} \leftrightarrow m_I = \pm\frac{3}{2}$, have been spread out over such a broad frequency range that they are unobservable. The distributions of quadrupole interactions in the Apollo 14 samples resemble qualitatively those found previously for Apollo 11 and 12 specimens and are taken to indicate, principally, the degree of atomic disorder at the aluminum and sodium sites and the presence of mineral components which contain strains. The relation between this disorder and strain and the factors that produced it, such as the rate of quenching and the crystallization history of the sample, has not been clearly established.

Despite interference from the extraneous Knight-shifted aluminum response arising from the NMR probe, the shapes of the ^{27}Al high-field (low-frequency) peaks and their splittings from H_0 are easily resolved. The interference from the background probe line is less serious at the highest frequency used, 20 MHz, than at the lower frequencies used previously (Kolopus *et al.*, 1971), due to the relatively stronger sample intensities and the greater separation of the probe line from H_0 at high frequencies. In addition, use of an "aluminum-free" NMR probe which operates in the range 6–15 MHz and shows no ^{27}Al spectrum has permitted more precise studies of both ^{27}Al and ^{23}Na spectra at lower frequencies than were possible in earlier studies (Kolopus *et al.*, 1971; Weeks *et al.*, 1970).

For sample 14163,168 (Fig. 2), the width of the high-field side of the ^{27}Al spectrum, measured from H_0 at half-maximum intensity, was found to be $39(\pm 1)$, $46(\pm 2)$, $76(\pm 5)$ G at spectrometer frequencies of 20, 13, and 8 MHz, respectively. Accompanying this broadening were a decrease in intensity and an increase in the splitting of the ^{27}Al high-field peak from H_0, found to be $10(\pm 1)$, $12(\pm 1)$, and $17(\pm 2)$ G,

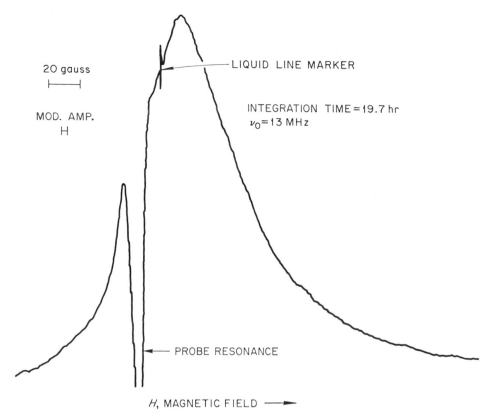

20 gauss

MOD. AMP.
H

LIQUID LINE MARKER

INTEGRATION TIME = 19.7 hr
$\nu_0 = 13$ MHz

PROBE RESONANCE

H, MAGNETIC FIELD

Fig. 2. Nuclear magnetic resonance of ^{27}Al in Apollo 14 fines, sample 14163,168.
The dispersion mode is recorded at $\nu_0 = 13$ MHz with an integration time of 19.7
hours. The very narrow "liquid line marker" is due to ^{27}Al resonance in a solution
of AlCl$_3$.

respectively. The inverse behavior with frequency indicates the dominant presence of
second-order quadrupole perturbations in the ^{27}Al spectrum of 14163,168. Similar
results have been found for the other lunar samples as well, with the fines samples in
general tending to show somewhat larger values of widths and peak splittings, and
thus atomic disorder relative to those found in samples of crystalline rocks. At a
spectrometer frequency of 20 MHz the high-field widths at half-maximum and the
peak splittings are shown in Table 1. In addition, the experimentally measured ratios
of the ^{27}Al high-field widths at half-maximum (Table 1) are possible indications that
small internal magnetic fields exist at the aluminum sites in the lunar samples, opposing the laboratory field. Such a conclusion may be made by comparing the above
ratios with those obtained from computer simulations of ^{27}Al lunar spectra using the
quadrupole parameters of the eight sites of anorthite (e^2qQ/h = 8.42, 7.25, 6.81,
5.54, 4.90, 4.30, and 2.66 MHz and η = 0.66, 0.76, 0.65, 0.88, 0.42, 0.42, 0.53, 0.66)
with varying amounts of isotropic magnetic field shifts (Kolopus *et al.*, 1971). These
simulations show that the computed ratio of high-field widths at half-maximum at

Table 1. Parameters of ^{27}Al NMR Spectra at 20 MHz and the ratio of widths at 13 and 20 MHz.

Sample	ΔH $H_0 - H$ ($\frac{1}{2}$ max. ampl.) (gauss)	$H_0 - H$ (max. ampl.) (gauss)	% Glass
14163,168	39 ± 1	10 ± 1	51–65[a, b, c]
14321,166	33 ± 1	7 ± 1	∼10[d]
12021,55	28 ± 1	6 ± 1	< 1[e]
BCR-1[f]	26 ± 1	8 ± 1	< 1[f]

	ΔH (13 MHz)/ΔH (20 MHz)	Fe wt. %[g]
14163,168	1.18 ± 0.08	0.58
14321,166	1.12 ± 0.06	0.19
12021,55	1.18 ± 0.07	0.06
BCR-1	1.23 ± 0.09	< 0.01[f]
Anorthite (calculated)	1.32	

[a] Masson et al., 1972.
[b] Marvin et al., 1972.
[c] Carr and Meyer, 1972.
[d] Warner, 1972. The amount of glass was estimated on the basis of Warner's description.
[e] Warner, 1971.
[f] Flannagan, 1967.
[g] Gose et al., 1972.

13 MHz and 20 MHz decrease somewhat when magnetic field shift parameters are included relative to the value of approximately 1.32 obtained from simulations when no magnetic field shifts are included. For comparison, the comparable ratio for BCR-1 basalt is 1.23(±0.09). Taken together, the experimental ratios quoted for the lunar samples are considered to deviate sufficiently from the calculated value of 1.32 to suggest, at least indirectly, the presence of internal magnetic fields of the order of 5–10 G. The magnitudes of these internal magnetic fields cannot be quoted precisely because of the relatively large experimental uncertainties involved and, also, because it is not clear if these internal fields remain constant or vary with varying external magnetic fields.

Previous NMR studies of ^{23}Na in lunar samples have been hampered by very weak signal intensity (Kolopus et al., 1971). More recently, improved signal-to-noise for ^{23}Na, particularly in two samples of fines, 10084 and 14163,168, has been obtained through use of larger samples than used previously, 7–8.5 g, and study at higher frequencies, up to 20 MHz. Examination of the ^{23}Na spectra from 14163,168 (Fig. 3) and 10084 over a range of frequencies from 20 MHz to 16 MHz showed that the spectral widths were somewhat larger in the Apollo 11 sample as compared to the Apollo 14 sample, indicating greater apparent site disorder in the former. Computer simulations indicate that the ^{23}Na spectra of 10084 and 14163,168 are not inconsistent with distributions of quadrupole interactions whose coupling constants lie mainly in the range 2–4 MHz and whose asymmetry parameters lie largely in the range 0.25–0.75. No special significance may be attached to such distributions. The criteria for choosing the limits of quadrupole parameters noted above are based mainly on the magnitude and ratio of the high-field peak splitting and width at half-maximum as calculated from a given distribution. In addition, a comparison of the

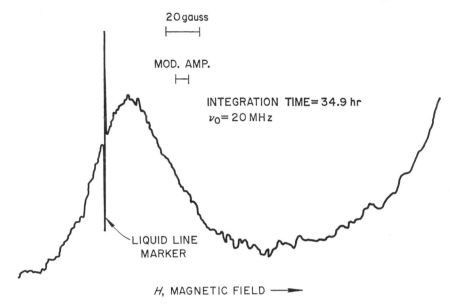

20 gauss

MOD. AMP.

INTEGRATION TIME = 34.9 hr
$\nu_0 = 20$ MHz

LIQUID LINE
MARKER

H, MAGNETIC FIELD ⟶

Fig. 3. Nuclear magnetic resonance of ^{23}Na in Apollo 14 fines, sample 14163,168. The dispersion mode is recorded at $\nu_0 = 20$ MHz with an integration time of 34.9 hours. The very narrow "liquid line marker" is due to ^{23}Na in a solution of NaCl.

^{23}Na high-field widths at half-maximum, at 16 MHz and 20 MHz, in a manner similar to that used for the case of the aluminum spectra, suggests the possibility of small internal magnetic fields at the sodium sites in the lunar samples.

CONCLUSIONS

It is evident from the data in Table 1 that large shifts in the peak position and increased widths, ΔH, correlate with glass content of the samples. The difference in ΔH between the two crystalline lunar rocks is larger than the error in the measurements and correlates with differences in formative processes (Warner, 1971; Warner, 1972) which have produced different levels of disorder and local strains at the Al sites. The shift in the field at which maximum absorption occurs correlates with glass content but is less sensitive to crystalline disorder. The ratio of ΔH measured at two frequencies, which is indicative of the assumed local magnetic field, does not correlate with the wt.% iron in the samples. However, such fields could be due to a distribution, throughout the Al-containing fraction, of iron particles with sizes in the 100 Å range which comprise only a relatively small fraction of the total iron. There is an indication that this is the case in the plagioclase fraction of 14053, whose ferromagnetic resonance spectrum contains a relatively intense C component (Weeks, 1972, this volume). In this case the wt.% iron is estimated to be <0.05%.

Acknowledgments—We thank H. S. Story (State University of New York at Albany) for his aid with the computer simulations and for his comments on the NMR results. J. L. Kolopus, whose death occurred on October 16, 1971, made many contributions to the research reported here, and we are deeply grateful for these.

REFERENCES

Carr M. H. and Meyer C. E. (1972) Petrologic and chemical characterization of soils from the Apollo 14 landing site (abstract). In *Lunar Science—III* (editor C. Watkins), pp. 116–118, Lunar Science Institute Contr. No. 88.

Compston W., Vernon M. J., Berry H., Rudowski R., Gray C. M., Ware N., Chappell B. W., and Kaye M. (1972) Age and petrogenesis of Apollo 14 basalts (abstract). In *Lunar Science—III* (editor C. Watkins), pp. 151–153, Lunar Science Institute Contr. No. 88.

Flannagan F. J. (1967) U.S. Geological survey silicate rock standards. *Geochim. Cosmochim. Acta* **31**, 289–308.

Fredriksson K., Nelen J., and Noonan H. (1972) Apollo 14: Glasses, breccias, chondrules (abstract). In *Lunar Science—III* (editor C. Watkins), pp. 280–282, Lunar Science Institute Contr. No. 88.

Gose W. A., Pearce G. W., Strangway D. W., and Larson E. E. (1972) Magnetic properties of lunar samples (abstract). In *Lunar Science—III* (editor C. Watkins), pp. 332–334, Lunar Science Institute Contr. No. 88.

Kolopus J. L., Kline D., Chatelain A., and Weeks R. A. (1971) Magnetic resonance properties of lunar samples: Mostly Apollo 12. *Proc. Second Lunar Sci. Conf., Geochim. Cosmochim. Acta* Suppl. 2, Vol. 3, pp. 2501–2514. MIT Press.

LSPET (Lunar Sample Preliminary Examination Team) (1970) Preliminary examination of lunar samples from Apollo 12. *Science* **167**, 1325–1339.

LSPET (Lunar Sample Preliminary Examination Team) (1971) Preliminary examination of lunar samples from Apollo 14. *Science* **173**, 681–693.

LSPET (Lunar Sample Preliminary Examination Team) (1972) Preliminary examination of lunar samples from Apollo 15. *Science* **175**, 681–693.

Marvin U. B., Wood J. A., Taylor G. J., Reid J. B. Jr., Powell B. N., Dickey J. S. Jr., and Bower J. F. (1971) Relative proportions and probable sources of rock fragments in the Apollo 12 soil samples. *Proc. Second Lunar Sci. Conf., Geochim. Cosmochim. Acta* Suppl. 2, Vol. 1, pp. 679–699. MIT Press.

Marvin U. B., Reid J. B. Jr., Taylor G. J., and Wood J. A. (1972) A survey of lithic and vitreous types in the Apollo 14 samples (abstract). In *Lunar Science—III* (editor C. Watkins), pp. 507–509, Lunar Science Institute Contr. No. 88.

Masson C. R., Smith I. B., Jamieson W. D., and McLachlan J. L. (1972) Chromatographic and mineralogical study of Apollo 14 fines (abstract). In *Lunar Science—III* (editor C. Watkins), pp. 515–517, Lunar Science Institute Contr. No. 88.

Meyer C. Jr., Brett R., Hubbard N. J., Morrison D. A., McKay D. S., Aitkin F. K., Takeda H., and Schorfield E. (1971) Mineralogy, chemistry, and origin of the KREEP component in soil samples from the Ocean of Storms. *Proc. Second Lunar Sci. Conf., Geochim. Cosmochim. Acta* Suppl. 2, Vol. 1, pp. 393–411. MIT Press.

Warner J. L. (1971) Lunar crystalline rocks: Petrology and geology. *Proc. Second Lunar Sci. Conf., Geochim. Cosmochim. Acta* Suppl. 2, Vol. 1, pp. 469–480. MIT Press.

Warner J. L. (1972) Apollo 14 breccias: Metamorphic origin and classification (abstract). In *Lunar Science—III* (editor C. Watkins), pp. 782–784, Lunar Science Institute Contr. No. 88.

Weeks R. A. (1972) Magnetic phases in lunar material and their electron magnetic resonance spectra: Apollo 14. *Proc. Third Lunar Sci. Conf., Geochim. Cosmochim. Acta* Suppl. 3, Vol. 3. MIT Press. (Next paper, this volume.)

Weeks R. A., Kolopus J. L., Kline D., and Chatelain A. (1970) Apollo 11 lunar material: Nuclear magnetic resonance of ^{27}Al and electron magnetic resonance of Fe and Mn. *Proc. Apollo 11 Lunar Sci. Conf., Geochim. Cosmochim. Acta* Suppl. 1, Vol. 3, pp. 2467–2490. Pergamon.

Proceedings of the Third Lunar Science Conference
(Supplement 3, *Geochimica et Cosmochimica Acta*)
Vol. 3, pp. 2503–2517
The M.I.T. Press, 1972

Magnetic phases in lunar material and their electron magnetic resonance spectra: Apollo 14

R. A. WEEKS

Solid State Division, Oak Ridge National Laboratory
Operated by Union Carbide Corporation for the U.S. Atomic Energy Commission
Oak Ridge, Tennessee 37831

Abstract—Electron magnetic resonance spectra of soil samples 14163,68, 14148,31, 14149, 47, 14156,31, and 14003,60, and of fragmental rocks 14301,66, 14303,42, 14310,68, 14311,36, 14318,36, and 14321,166 have been recorded at 9 and 35 GHz at 300°K and at 9 GHz at 130°K. One spectral component, the "characteristic" ferromagnetic resonance, of all the soil samples is 50 to 1000 times more intense than any other component in the soils or in the spectra of the rocks. The intensity of this component in Apollo 11, Apollo 12, and Apollo 14 soils varies only within one order of magnitude. It varies with depth below lunar surface but is not correlated with depth. The intensity does not have any correlation with the fraction of glassy particles nor with the fraction of anorthositic particles. This component is present in the spectra of at least two fragmental rocks, 14301,66 and 14318,36. Its presence and intensity appear to correlate inversely with degree of consolidation and of recrystallization of the fragmental rocks. At least two other ferromagnetic resonance components are resolved in the spectra of the fragmental rocks. One of these with a width at 9 GHz of ~ 7000 gauss is attributed to multi-domain particles of iron and the second with a width at 9 GHz of ~ 3500 gauss is tentatively attributed to single-domain particles of iron. Spectral components due to paramagnetic Fe^{3+}, Ti^{3+}, and Mn^{2+} are resolved in the spectra of most of the rocks. These components are present in plagioclase fractions separated from Apollo 11, Apollo 12, and Apollo 14 rocks. A hypothesis is proposed, consistent with all the data presently available, to explain the occurrence of the "characteristic" resonance in the soils, and in some of the fragmental rocks, but not all of them. In this hypothesis the characteristic resonance is attributed to a ferric oxide phase with an abundance < 0.1 wt.% in the soils.

INTRODUCTION

LUNAR ROCKS AND FINES contain minor amounts of antiferromagnetic, ferrimagnetic, and ferromagnetic mineral phases such as spinels, chromites, ilmenite, troilite, iron, and alloys of iron with nickel and cobalt (Simpson and Bowie, 1971; Cameron, 1970; El Goresy *et al.*, 1971; Masson *et al.*, 1971; Masson *et al.*, 1972; Haggerty, 1972), and goethite (Agrell *et al.*, 1972). Major minerals, pyroxene, and olivine, are iron-rich with most of the iron present as Fe^{2+} (LSPET, 1970; LSPET, 1971). Electron magnetic resonance (EMR) absorption spectra due to some of these minerals would be expected; e.g., spinels, chromites, iron and its alloys, nonstoichiometric troilite (if present), and goethite. The EMR spectra of these ferrimagnetic and ferromagnetic mineral components are determined by particle shapes and sizes, by the magnitude of the saturation magnetization, and by the relative magnitude and sign of the anisotropy constants. For example (Lin and Neaves, 1962), nickel ferrite particles in the size range 20 to 200 Å have an EMR absorption with one maximum, whose width decreases from ~ 1300 to 600 G with decreasing particle size and whose g_{eff} value is ≥ 2 ($g_{eff} = hv/\beta H$ where h, v, β, and H are Planck's constant, spectrometer frequency, Bohr magneton, and the magnetic field at which absorption maximum occurs, respectively). The line shape, which is symmetric for the smallest particle sizes, becomes increasingly asymmetric with increasing particle size. Similar effects have been observed in the spectra of γ–Fe_2O_3 particles (Morrish and Valstyn, 1962). Hence the spectra of samples containing a variety of ferrimagnetic and ferromagnetic minerals with a distribution of particle sizes (10^{-6} to 10^{-1} cm) and of shapes (acicular to spherical) is expected to be complex.

Samples of lunar fines contain the same mineral components as are contained in igneous and fragmental rocks, since a major fraction of the fines are comminuted rocks (Marvin, 1971). Glassy fragments, agglutinates, and spheroids comprise the remaining fraction (5–50%) (Marvin et al., 1971; Marvin et al., 1972; Carr and Meyer, 1972).

EMR spectra of igneous rocks (10047,49, 12021,55, and 12075,19) have at least two components which are due to ferrimagnetic and ferromagnetic phases, and spectral components due to three paramagnetic species (Weeks et al., 1970a; Weeks et al., 1970b; Kolopus et al., 1971; Weeks et al., 1972). The magnetic phases which are the source of the ferromagnetic components have not been unambiguously identified. (Hereafter, the term "ferromagnetic components" will refer to spectral components that are due to both ferrimagnetic and ferromagnetic mineral phases.) One of these components is certainly due to iron and its alloys, since their abundance ranges from $<0.1\%$ for most igneous rocks to $\sim 1\%$ for many samples of fines (Cameron, 1970; Carter and MacGregor, 1970; Herzenberg et al., 1971; Housley et al., 1971). The spectra of fines (<1 mm particle size), with one exception, have one component whose intensity is 100 to 1000 times more intense than any other spectral component (Weeks et al., 1970a; Weeks et al., 1970b; Kolopus et al., 1971; Tsay et al., 1971a). The width, shape, and g_{eff} of this component are almost independent of the site from which a sample was collected. The temperature dependence of the width and intensity are characteristic of a ferromagnetic resonance (Kolopus et al., 1971). It was observed in weakly consolidated breccias from the Apollo 11 collection (Weeks et al., 1970b), but was very weak, if present, in the spectra of igneous rocks (Weeks et al., 1972). Iron is approximately 10 times more abundant in fines than in igneous rocks (Herzenberg et al., 1971; Housley et al., 1970; Housley et al., 1971). Thus the ratio of Fe (igneous rocks) to Fe (soil) ranges from <0.01 to 1, while the ratio of the intensity of this spectral component in the spectra of igneous rocks (when it is present) to the intensity in the spectra of fines ranges from $<10^{-4}$ to 10^{-2}. Hence it is apparent that no correlation exists between intensity of this spectral component and the total iron content determined by various methods.

Paramagnetic species which have been identified are Fe^{3+}, Mn^{2+}, and tentatively Ti^{3+} (Weeks et al., 1970a; Weeks et al., 1972). These paramagnetic ions have been observed in the plagioclase fraction of two igneous rocks (10047,49, 12021,55).

Specimens from the Apollo 14 collection, upon which measurements have been made, are shown in Table 1. Most of the specimens weighed ~ 0.5 g, with the exception of 14321,166, which was larger and contained a plagioclase clast ~ 1 cm diameter. A portion of it was extracted, providing a plagioclase fraction (estimated to be $>90\%$ plagioclase on the basis of the fraction of gray, red-brown, yellow, and opaque par-

Table 1. EMR samples.

Number	Station	Comments
14003,60[1]		Contingency, fines < 1 mm.
14148,31[1]	G	Fines < 1 mm, top of trench.
14156,36[1]	G	Fines < 1 mm, middle of trench.
14149,47[1]	G	Fines < 1 mm, bottom of trench.
14163,68[1]		Fines < 1 mm, bulk sample.
14301,66[1]	G1	Grab sample, fragmental rock, coherent, F2.[2]
14303,42[1]		Comprehensive sample, fragmental, coherent, F4.[2]
14311,35[1]	Dg	Grab sample, fragmental, coherent, F4.[2]
14318,36[1]	H	Grab sample, fragmental, coherent, F2.[2]
14321,166	C1	Fragmental, coherent, F4.[2]
14310,68[1]	G	Igneous, basaltic.

[1] Specimen weights ~ 0.5 g.
[2] Jackson and Wilshire (1972).

ticles which were present when viewed at a magnification of 30 times). On one end of specimen 14318,36 a layer of gray-brown glass, ~1 mm thick × 0.5 cm diameter, was observed. It was broken away with only a small part of the host rock adhering. The contact with the host rock was well defined and planar. None of the other specimens exhibited such well-defined glassy particles. Lustrous metallic-appearing particles were observed in all the rock specimens; these were magnetic and are presumed to be iron.

Measurements were made at 9 and 36 GHz and at temperatures ranging from 130° to 470°K on homodyne spectrometers (Weeks *et al.*, 1970b). Samples of fines were prepared by encapsulating ≤3 mg of the smallest particle sizes (<0.05 mm) in quartz tubes (Spectrosil, Thermal American Fused Quartz Company). Whole rock samples were prepared by picking fragments from the specimens which had the same appearance as the bulk of the specimens and encapsulating them in quartz tubes. Samples for 9-GHz measurements usually weighed ~30 mg, and samples for 35-GHz measurements were ≤3 mg. These smaller samples had a much lower probability of being representative of the whole rock than did the 9-GHz samples.

EXPERIMENTAL RESULTS

Spectra of the fines samples are shown in Fig. 1. Shape, width, and g_{eff} of the "characteristic" absorption varied slightly from sample to sample. The relative intensity, $I = A \Delta H^2$, where A = amplitude of the dI/dH curve measured between the maximum and minimum values (magnetic field modulation was constant) and ΔH = width measured between the fields at which the maximum and minimum of dI/dH curves occurred, varies over an order of magnitude. Also shown is the low-field part of the derivative curve recorded at much higher signal gain. This portion of the spectra is similar for all the samples, with a shoulder at $H = 100$ gauss and another one at $H = 1500$ gauss. The relative amplitude of these two features is least in spectra of the samples from 14003,60 and 14148,31. Temperature dependence of ΔH and A of the "characteristic" absorption was the same as for the "characteristic" resonance in the spectra of Apollo 12 samples (Kolopus *et al.*, 1971). At 120°K the shoulders at 100 and 1500 gauss are less well resolved. There is no indication of a paramagnetic temperature dependence of the intensity (I proportional to $e^{-h\nu/kT}$) of the shoulder at 1500 gauss.

Some parameters of the "characteristic" absorption are given in Table 2 for both Apollo 14 and Apollo 12 samples. The range of g_{eff} is $2.066 \leq g_{eff} \leq 2.090$ (± 0.009) and the range of values for Apollo 14 samples overlaps the range for Apollo 12 samples. The width ΔH is different for the two

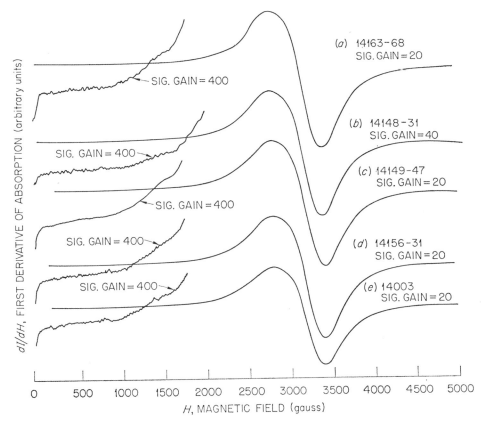

Fig. 1. Electron magnetic resonance spectra of Apollo 14 "fines" (< 0.05 mm) at 9.02 GHz and 300°K. All of the samples weighed ~2 mg. The curves marked Sig. Gain = 400 are the low-field part of the spectrum of each sample at higher amplifier gain. Magnetic field modulation amplitude and microwave power level (−20 db) were constant for all measurements.

sets of samples: for Apollo 14, $590 \leq \Delta H \leq 615$ gauss and for Apollo 12, $750 \leq \Delta H \leq 815$ gauss. Intensities show a much greater variation, ranging from 7 to 52 relative units, with the exception of 12033,50. The intensity of this component in the spectra of samples of 12033,50 are < 7 relative units with the shoulder at 100 gauss having an amplitude > 100 times that in the spectra of the other Apollo 12 samples and all the Apollo 14 samples. This feature has been discussed in some detail in Weeks *et al.* (1972). For the trench samples, I ranges from 7 to 25 to 52 for samples from the top, bottom, and middle, respectively, and hence is apparently indicative of the layer from which they were taken.

At 35 GHz ΔH is greater, and g_{eff} is invariant within ±0.01 for the characteristic resonance, a weak absorption peak with $I \leq 0.01\ I$ of the characteristic absorption is resolved at $g = 4.1$, and no shoulder is resolved in the region of $H = 100$ gauss. The fractional increase in ΔH of the "characteristic" resonance with increase in ν (9 to 35 GHz) is greater for Apollo 14 samples, 600 to 750 gauss, than for Apollo 12 samples, 800 to 950 gauss.

At 9 GHz and 130°K the spectra of samples of the fragmental and igneous rock specimens are shown in Fig. 2. All of these spectra contain paramagnetic and ferromagnetic components. The paramagnetic components are resolved from the ferromagnetic ones on the basis of their temperature dependence (Weeks *et al.*, 1970a; Weeks *et al.*, 1970b). The resolution of the ferromagnetic absorption

Table 2. Parameters of "characteristic" ferromagnetic resonance
of Apollo 12 and Apollo 14 fines.

Sample	$g \; (h\nu/\beta H)$* (ν = 9.020 GHz)	ΔH (Oe)	A Amp., dI/dH curve† (relative units)	$\dfrac{A \; \Delta H \ddagger}{mm^3}$
12001	2.090	780	0.50	40.0
12033	2.090	815	0.07	5.0
12030	2.090	750	0.35	20.0
12070	2.079	750	0.93	52.0
14003	2.073	615	0.40	15.0
14148,31,1	2.080 (surface of trench)	600	0.20	7.0
14149,47,1	2.066 (bottom of trench)	615	0.75	28.0
14156,31,1	2.080 (middle of trench)	610	1.24	52.0
14163,68,1	2.083	590	0.82	29.0

* The relative error between these values is ± 0.001, while the absolute error is ± 0.010.
† The amplitude of the dI/dH curve is measured between its extreme values.
‡ $A \; \Delta H^2$ is proportional to area beneath an absorption curve and hence to the concentration of the magnetic phase producing the curve when the dI/dH curve is symmetric. For asymmetric curves with the same shape, the proportionality of the values is correct. In this case the curve shapes do vary slightly from sample to sample and ΔH varies; hence the relative error is greater. Taking into account the error in measuring the amplitude ($\pm 20\%$), ΔH ($\pm 5\%$), and estimating the error in assuming that $A \; \Delta H^2$ is proportional to area ($\pm 25\%$), the total error in each of these values is $\pm 55\%$.

into distinct components is based upon a comparison of the spectra from all whole-rock samples and mineral separates upon which measurements have been made (10047,49, 10062,21, 12021,55, 12075,19, the Apollo 14 rocks listed in Table 1, plagioclase fractions from 12021,55, 14053,47, 14321,166, an olivine fraction from 12075,19, and a pyroxene fraction from 12021,55 and 10047,49 (Weeks et al., 1970a; Weeks et al., 1970b; Kolopus et al., 1971; Weeks et al., 1972). The resolved components are identified by the symbols A, B, and C at the bottom of Fig. 2. The A component is well resolved in the spectra of 14301,66, 14303,42, 14310,68, 14318,36, and 14321,166 (Fig. 2a, b, c, e, and f), the B component in the spectra of 14301,66 and 14318,36 (Fig. 2a and e), the C component in the spectra of 14301,66, 14310,68, 14311,36, and 14321,166 (Fig. 2a, c, d, and f). The C components may be present in the spectra of 14303,42 but the B component is not. The low-field maximum in the dI/dH curve for the A and the C components is in the region of $H = 0$.

Paramagnetic components in these spectra (Fig. 2) are indicated by the symbols Fe^{3+} and Ti^{3+}, and have been attributed to the 3+ valence state of iron and titanium (Weeks et al., 1970a; Kolopus et al., 1971; Weeks et al., 1972). The amplitude of the Fe^{3+} component is greatest in the spectrum of 14310,68 (Fig. 2c) and least intense in the spectrum of 14311,36 (Fig. 2d). The Ti^{3+} component is well resolved in the spectra of 14303,42, 14310,68, 14318,36, and 14321,166 (Fig. 2b, c, e, and f).

The D component, present only in the spectrum of 14321,166 (Fig. 2f) is observed in the spectrum at 300°K. Its intensity increases $\sim 20\%$ at 130°K but its width does not appear to change.

A considerable alteration in the spectra occurs when the spectrometer frequency is increased to 35 GHz (Fig. 3). The widths of the various ferromagnetic components change slightly, and most of the absorption occurs over a smaller fraction of the magnetic field compared to the absorption at 9 GHz. There is a greater overlap of the various components and hence their resolution is less certain. Also, because of the small sample size (~ 3 mg), the mineral phase responsible for one of the components may, by chance, be more abundant than the phases responsible for the other two components. The A and B components in the spectra of 14318,36 (Fig. 3c) are well resolved and there is very little contribution from the C component. The C component is strongest, and the B component is indicated by an inflection in the slope of the dI/dH curve at $H = 11,500$ gauss in the spectra of 14301,66 and 14311,35 (Fig. 3a and b). The spectrum of the sample from 14321,166 (Fig. 3d) is particularly interesting in that only one component is present. This component has a width corresponding to that of the C component, but the shape is more symmetrical. The width of the A component

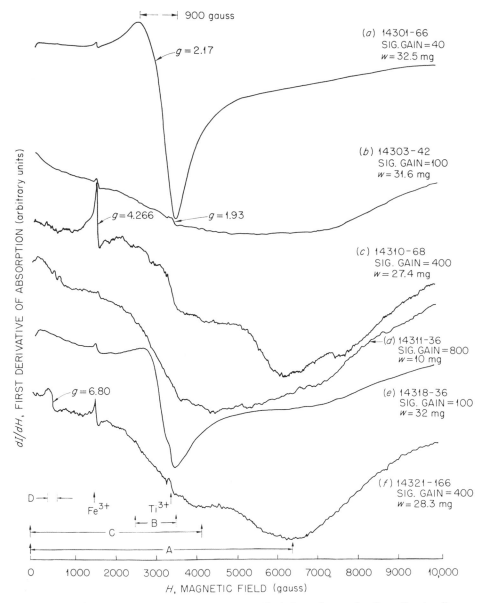

Fig. 2. Electron magnetic resonance spectra of whole rock samples from the Apollo 14 fragmental and igneous rock specimens. Sample weights and amplifier gain are shown on the right side for each curve. Magnetic field modulation amplitude and microwave power level (−10 db) were constant for all of the measurements.

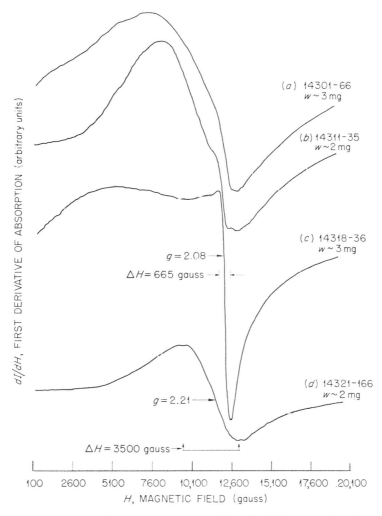

Fig. 3. Electron magnetic resonance spectra of Apollo 14 fragmental rocks at 35.2 GHz and 300°K. Microwave power level (−6 db) and magnetic field modulation amplitude were constant for all of the measurements.

in the spectrum of 14318,36 is $\Delta H \sim 7500$ gauss, and in the spectra of 14301,66 and 14311,35 the combined effects of the A and C components produce a width $\Delta H \sim 5000$ gauss.

A plagioclase fraction of 14053,47 (prepared by S. Hafner and colleagues) has a spectrum in which the C component is the only ferromagnetic absorption detected. Its width at 9 GHz, $\Delta H = 3500$ gauss and at 35 GHz, $\Delta H \sim 2000$ gauss. With the assumption that a similar decrease in width occurs in other samples, a resolution of the B and C components in the spectra of 14301,66 and 14311,35 (Fig. 3a and b) at 35 GHz is difficult, and the assignment made above is therefore tentative. The spectrum of 14321,166 at 35 GHz (Fig. 3d) may be due to a fourth component, which is much more intense than in the spectra of other samples if it is present in them. The temperature dependence of ΔH and A of the C component at 9 GHz are given in Fig. 4. Since the low-field maximum of the dI/dH curve was not resolved from $H = 0$, the width was determined from the field at which the

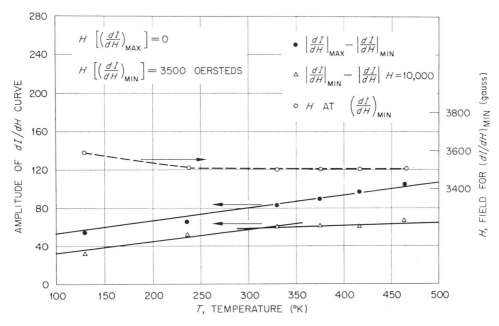

Fig. 4. Amplitude of the dI/dH curve and the field at which the minimum in the dI/dH curve occurs for the C component in the spectrum of a lunar plagioclase from 14053,47 are shown as a function of temperature. Magnetic field modulation amplitude, microwave power level, and the cavity Q were constant throughout the measurements. The curves are straight-line segments joining the data points, and the arrows indicate the appropriate ordinate.

minimum in the dI/dH curve occurred. It increased by only 100 gauss at the lowest temperature. The amplitude A, which was measured by two methods, decreased linearly and monotonically, and both methods gave the same rate of decrease at the lower temperatures.

Paramagnetic components due to Fe^{3+}, Ti^{3+}, and Mn^{2+} were resolved in the spectra of plagioclase fractions of 14321,166 and 14053,47. The data and discussion of the magnetic resonance properties of the plagioclase fraction of 14053,47 will be presented elsewhere (Weeks, to be published).

The glassy layer separated from one end of 14318,36 has a spectrum at 9 GHz in which the B component is more intense than in the whole rock sample, the C component is present, and the A component is not detected. Parameters of the B component are $\Delta H = 420$ gauss and $g_{eff} = 2.065$ at 300°K and $\Delta H = 800$ gauss and $g_{eff} = 2.065$ at 120°K. For the same decrease in temperature the amplitude of the C component decreased from 94 to 45 relative units and ΔH increased by $\sim 50\%$. Paramagnetic components Fe^{3+} and Ti^{3+} were not detected in this sample.

DISCUSSION

Compared to lunar igneous and fragmental rocks from which EMR spectra have been obtained, the EMR spectra of lunar soils (particle size <1 mm) are remarkably uniform, with the exception of 12033,50. The intensity of the "characteristic" resonance of the soils is a function of the collection site. This function is most clearly indicated by the order of magnitude variation in intensity of the three trench samples (Table 2) and hence may be a function of stratigraphy. The intensity, on the basis of the limited number of samples measured, is not correlated with glass content or mineral content. For example, the ratio of glass content in 12070 to 12033 is ~0.5 (Marvin *et al.*, 1971), while the ratio of intensities is ~10. Basalt fragments and basaltic microbreccias are twice as abundant in Apollo 14 soils as they are in Apollo 12 soils (Carr and Meyer, 1972; McKay *et al.*, 1971); yet the same range of intensities is observed in both sets of samples (Table 2).

It has been suggested that the "characteristic" resonance is due to spherical particles of iron with diameters $\leq 1\,\mu$ (Tsay *et al.*, 1971b). Iron abundance in soils is, relative to the rocks, approximately 10 times more abundant (Herzenberg *et al.*, 1971; Housley *et al.*, 1971; Gose *et al.*, 1972; Goldstein *et al.*, 1972; Goldstein and Yakowitz, 1971). Wlotzka *et al.* (1972) estimate that Fe (12001)/Fe (14163) ~ 0.3. The ratio of intensities is $I(12001)/I(14163) = 1.4$, and hence a correlation between wt.% iron and intensity does not exist in this case. It is possible that the EMR sample and Wlotska *et al.*'s samples have iron contents different from those on which EMR measurements were made. However, there is good agreement on the wt.% iron in different samples taken from the same specimen (Gose *et al.*, 1972; Runcorn *et al.*, 1970). Some fraction of the iron particles in soil samples have dimensions exceeding the penetration depth (~1 μ) of 9-GHz electromagnetic waves. Contributions to EMR from these particles will be limited to their surfaces to a depth of ~1 μ, and as a result contributions to resonance intensity will not be proportional to the weight fraction of such particles. The weight fraction of particles with dimensions $>1\,\mu$ has not been determined, although in one glass sample it has been found that particles with dimensions in the 0.1 to 1 μ range are more abundant than either smaller or larger particles (Wosinski *et al.*, 1972, and references). The magnetization and "acquisition of IRM" data of Runcorn *et al.* (1970) indicate that for one sample of fines (10084) there are 10 times as many iron particles with diameters $<160\,\text{Å}$ as there are in the 160 to 300 Å range. No data on the fraction of particles with diameters $>300\,\text{Å}$ are given. Particles with diameters $<160\,\text{Å}$ are superparamagnetic, in the 160 to 300 Å range they are single domain, and in the range $>300\,\text{Å}$ they are multidomain (Néel, 1949).

Contributions to EMR from such iron particles will be primarily a function of their shape, since crystalline anisotropy constants of iron are less than the saturation magnetization, i.e., $2\kappa_1/M_s < M_s$ (Morrish, 1965). Nonspherical particles which are noninteracting (i.e., the field at a given particle due to nearest-neighbor particles is small compared to the fields induced by shape anisotropy) will have peaks distributed over a wide range of laboratory fields, *Hr*. Noninteracting spheres will have resonant peaks which are determined by anisotropy constants, $2\kappa_1/M_s$ and $2\kappa_2/M_s$, and the angle between the easy magnetization direction and the applied field, *Hr* (Morrish, 1965). Considering the case of the sphere,

$$Hr\,(\text{max}) - Hr\,(\text{min}) = \left[\frac{10\kappa_1}{3M_s} + \frac{4\kappa_2}{9M_s}\right],$$

and for iron $\kappa_1/M_s \approx 300$ Oe and $\kappa_2/M_s \approx 100$ Oe (Morrish, 1965), which gives

$$\Delta Hr \approx 1000 \text{ gauss.}$$

The absorption for an ensemble of randomly oriented spheres will thus have a line width determined by an appropriate average over all possible orientations of the spheres (Schlömann, 1958) which gives $\Delta H \approx 800$ gauss. The width of the resonance of an individual sphere will in general be much less and will be determined by the anisotropy constants, strain, defects, and particle size (Morrish, 1965; Lin and Neaves, 1962). Tsay *et al.* (1971b) have shown that the absorption of such an ensemble does have a shape and width similar to the "characteristic" resonance observed in Apollo 11 and 12 fines. As can be seen from Table 2, line widths of the "characteristic" resonance of Apollo 14 fines are $\sim 30\%$ less than the line widths of Apollo 12 fines and hence less than the line width of iron spheres. The glassy fragment from specimen 14318,36 has a line width 50% smaller than that expected for iron spheres.

The resonance spectra of particles with other shapes will have Hr's determined by shape anisotropies. For example, a 10% deviation from sphericity, i.e., the ratio of the axes a and b of an ellipsoid of revolution varying between $m = b/a = 0.90$ to $m = 1.10$ will have values of ΔHr, when Hr is parallel to the axis of revolution, $\Delta Hr \approx 1500$ gauss, a 50% increase in the range of values of Hr over those for spheres with random orientations. The extreme cases, i.e., a disk and cylinder, will have $\Delta Hr \approx 26,000$ gauss (Morrish, 1965). It is thus evident that small deviations from $m = 1$ will increase the line width of iron particles to values greatly in excess of those observed for the "characteristic" resonance.

Within the framework of the preceding discussion the variations in line width of the "characteristic" resonance cannot be explained by the iron spheroid hypothesis. It should be noted that a basic assumption of this discussion is that the particles are noninteracting. If there is magnetic interaction between the particles greater than the fields produced by shape anisotropies, then the line width of the EMR absorption will change (Schlömann, 1969). Griscom and Marquardt (1972) have shown that the line width of the characteristic resonance is proportional to wt.% TiO_2. There does not appear to be any relation between iron content and TiO_2 content.

Weeks *et al.* (1972) have suggested that the "characteristic" resonance may be due to a ferric oxide phase formed on and in the smaller particles of the soil. The formation of this oxide could occur as the result of an oxidizing event produced by a cometary impact (Shoemaker, 1972). Another possible source is H_2O (Freeman *et al.*, 1972) and other oxidizing gases which may be emitted from the lunar interior. The identification of goethite in 14301 by Agrell *et al.* (1972) has shown that ferric oxide phases are present in lunar fragmental rocks. The intensity of the "characteristic" resonance in the spectrum of samples of this rock was greater than in any of the other rocks, but less than in the fines in which goethite has not been detected. It has been suggested on the basis of optical absorption spectra that Fe^{3+} ions are present in some glass particles (Rao *et al.*, 1972). If this resonance is due to a ferric oxide phase, then heating at very low oxygen pressures should reduce it to a ferrous phase or to iron. Kolopus *et al.* (1971) have shown that the intensity decreased by a factor of 50 when a sample of fines was heated at 800°C for one hour in a closed system in which the initial number of oxygen molecules was sufficient for 10% of the Fe atoms to be oxidized to Fe^{2+}. Synthetic lunar glasses fused at very low oxygen pressures have very weak resonance spectra that have no similarity to the "characteristic" resonance. But when powders of these glasses are heated at oxygen pressures of ~ 0.5 Torr and temperatures of 650°C, a resonance similar to the characteristic resonance is produced (Griscom and Marquardt, 1972). If such a phase is present, then its concentration must be less than that detectable in a Mössbauer experiment, which concentration is ~ 0.5 wt.% (S. Hafner, private communication). Intensities of the same magnitude as the "characteristic" resonance are found for < 0.1 wt.% Fe_3O_4 particles (< 43 μ diameter) dispersed in an insulating diamagnetic matrix (Weeks et al., 1972), although the characteristics of the absorption at 9 GHz differed from those of the "characteristic" resonance.

In breccias that are poorly consolidated and that show little recrystallization, the "characteristic" resonance is as intense as in soil samples (Weeks et al., 1970b); e.g., the Apollo 11 breccias (LSPET, 1970). Breccias from the Apollo 14 collection do exhibit a range of consolidation and recrystallization (Jackson and Wilshire, 1972). The B component, which is similar with respect to g_{eff} and ΔH, has an intensity in these breccias that correlates with the degree of consolidation and recrystallization according to the classification scheme of Jackson and Wilshire (1972). Assuming that the B component and the characteristic resonance are due to the same phase, we suggest that, since meteorite impacts have apparently been responsible for the consolidation and reheating which produced the breccias, and since these impacts do not produce an oxidizing event, their net effect in the low-ambient oxygen pressure of the moon is to reduce the ferric oxide phase produced by earlier oxidizing events. Rocks 14301 and 14318 are then representative of differing degrees of reduction of the ferric oxide phase, and in 14303 it has been completely reduced. The temperature and the times at temperatures which are required are not unreasonable if the laboratory heating experiments are indicative (Kolopus et al., 1971; Griscom and Marquardt, 1792) of the temperatures and times required for the two reactions.

In the spectra of all soil samples on which measurements at 9 GHz have been made, a shoulder in the dI/dH curve has been observed at $H \sim 100$ gauss. It is also present in the spectra of some rock samples; e.g., 14301,66 and 14318,36. In fines, heat-treated at low oxygen pressure ($< 10^{-10}$ Torr), in black basaltic-appearing pieces and in scoriaceous glasses from 12001,15 and 12033,50 (Weeks et al., 1972), this shoulder has an amplitude maximum at $H \approx 100$ gauss that exceeds the amplitude of the characteristic resonance (see Curve E, Fig. 2). The shoulder is not present in the spectrum of the plagioclase fraction of 14053,47, in which the C component was much more intense than either the A or B component. In the spectra of the fines there is no correlation between the amplitude of this feature and the amplitude of the characteristic resonance. Its intensity does appear to be associated with the abundance of the two kinds of material, scoriaceous glasses and black basaltic-appearing fragments (Weeks et al., 1972), in which it is particularly intense. The absorption of microwave energy by multidomain particles as a function of applied field will change rapidly between zero field and the field for which all of the particles have their magnetization saturated. In the case of single crystals of iron fields, between 50 and 300 gauss are required, depending upon orientation of a crystal with respect to the direction of the applied field (Morrish, 1965, p. 312). Thus the shoulder at ~100 gauss is attributed to the saturation magnetization of multidomain particles. The resonance condition is still determined by the shape and crystalline anisotropies discussed above.

The A component is attributed to multidomain iron particles for the following reasons: (1) Synthetic lunar glasses containing iron particles with a distribution of sizes and shapes have an absorption whose width is ~7000 gauss and a strong absorption at $H = 0$ (one glass was fused and iron particles were identified by R. Housley); and the shape of the minimum in the dI/dH curve in the region of $H = 7000$ gauss varies from the broad, shallow minimum present in the spectrum of 14303,42 (Fig. 2b) to the more sharply peaked minimum in the spectrum of 14321,166 (Fig. 2f). (2) Dilute mixtures (0.1 wt.%) of iron particles ($<43 \mu$) in a diamagnetic insulating matrix have an absorption spectrum (Weeks *et al.*, 1972) similar in shape and intensity to that of 14303,42 (Fig. 2b), which has a similar wt.% of iron (Gose *et al.*, 1972). (3) The ratio of intensity of the A component in the spectrum of 14321,166 to the intensity in the spectrum of 14310,68 is approximately equal to the ratio of wt.% iron in these two rocks (Gose *et al.*, 1972).

With decrease in particle size the ferromagnetic resonance width decreases (Lin and Neaves, 1962), and single-domain particles will have absorption widths less than multidomain particles. Hence, it is suggested that the C component may be due to single-domain particles of iron. Nagata *et al.* (1972) have shown that the wt.% in rock 14053 is greater (~1 wt.%) than in other rocks. The intense C component observed in the plagioclase fraction of 14053,47 may be due to this iron present as single-domain particles. Contributions of iron to the Mössbauer spectrum of this fraction have not been detected (S. Hafner, private communication). The concentration may be too low to be detected in the Mössbauer spectrum. Other possible sources for the C component are the spinels, but data on the ferromagnetic resonance absorption of lunar spinels is lacking, and thus this suggestion is tentative.

Measurements of silica activity by Nicholls *et al.* (1971) have shown that as much as 0.1 wt.% Fe_2O_3 is present in Apollo 11 basalts, and based on petrographic similarities (LSPET, 1970; LSPET, 1972), similar amounts would be expected in Apollo 14 rocks. The ferric ion may be present in the spinels, but it is also present as goethite (Agrell *et al.*, 1972). In either case, components in the EMR absorption from these forms of the ferric ion should be present. The component observed in the 35-GHz spectrum of 14321,166 (Fig. 3d) has a shape, width, and g_{eff} that correspond to those found for the spectrum of reagent grade Fe_2O_3 after heating in vacuum (Weeks *et al.*, 1970b). The component in the spectrum of 14321,166 is tentatively attributed to a reduced form of Fe_2O_3.

An extensive discussion of paramagnetic Fe^{3+} and Ti^{3+} and their presence in lunar plagioclases will be published elsewhere. The presence of these ions in lunar plagioclase was suggested in earlier papers of Weeks *et al.* (1970a) and Kolopus *et al.* (1971). It is sufficient to note here that all of the rocks whose spectra are shown in Fig. 2 contain paramagnetic Fe^{3+} and some of them contain Ti^{3+}. These rocks contain abundant plagioclase (40 to 70%) (LSPET, 1971). Since paramagnetic Fe^{3+} and Ti^{3+} have been shown to be present in samples of lunar plagioclase from both Apollo 12 and 14 rocks, it is reasonable to assume that the paramagnetic Fe^{3+} and Ti^{3+} are present in the plagioclase fraction of these rocks. Geake *et al.* (1972) have shown that luminescent emission from some lunar plagioclases is affected by the presence of Fe^{3+} ions. Finger *et al.* (1972) attributed a portion of the Mössbauer spectrum of lunar plagioclases to Fe^{3+} ions.

Conclusions

The magnetic phase which is the source of the "characteristic" resonance observed in the spectra of Apollo 11, 12, and 14 soils is of widespread occurrence in lunar soils. All the data available on the magnetic and thermal properties of the resonance and from laboratory modeling experiments can be consistently explained with the hypothesis that the resonance is due to a ferric oxide phase present in amounts ~ 0.1 wt.%. But a ferric oxide phase in lunar soil has not been reported, and hence the evidence for the existence of such a phase is circumstantial, while there is abundant evidence for iron particles. The limits on the parameters of the resonance of iron spheroids are exceeded by the parameters of the "characteristic" resonance, particularly the parameters of the spectra of Apollo 14 soils and glasses. Brecciation of the soil by shock and consequent heating in the lunar atmosphere reduce the phase which is the source of the characteristic resonance by amounts which correlate with the degree of consolidation and reheating. Spectral components due to metallic iron have been identified and, on the basis of available data, are distinct from the "characteristic" resonance of the soils. Paramagnetic Fe^{3+} and Ti^{3+} have been identified and are associated primarily with the plagioclase fraction.

Acknowledgments—Discussions with D. Griscom about the possible existence of a ferric oxide phase and with S. Hafner about Fe^{3+} in lunar plagioclases have been helpful. I thank S. Hafner for his generous loan of a plagioclase fraction and R. Housley for samples of a synthetic lunar glass. Comments of the referees have been useful in assessing some of the consequences of the iron spheroid hypothesis.

J. L. Kolopus was principal investigator for this program during 1971. He was most effective in directing it during that time. His death on October 16, 1971, has been a great loss to his colleagues. I acknowledge his many contributions, both scientific and personal, with deep gratitude.

This research was sponsored by the U.S. Atomic Energy Commission and supported by NASA Contract NSC-T-76458.

References

Agrell S. O., Scoon J. H., Long J. V. P., and Coles J. N. (1972) The occurrence of goethite in a microbreccia from the Fra Mauro formation (abstract). In *Lunar Science—III* (editor C. Watkins), pp. 7–9, Lunar Science Institute Contr. No. 88.

Cameron E. N. (1970) Opaque minerals in certain lunar rocks from Apollo 11. *Proc. Apollo 11 Lunar Sci. Conf., Geochim. Cosmochim. Acta* Suppl. 1, Vol. 1, pp. 221–245. Pergamon.

Carr M. H. and Meyer C. E. (1972) Petrologic and chemical characterization of soils from the Apollo 14 landing site (abstract). In *Lunar Science—III* (editor C. Watkins), pp. 116–118, Lunar Science Institute Contr. No. 88.

Carter J. L. and MacGregor I. D. (1970) Mineralogy, petrology, and surface features of some Apollo 11 samples. *Proc. Apollo 11 Lunar Sci. Conf., Geochim. Cosmochim. Acta* Suppl. 1, Vol. 1, pp. 247–265. Pergamon.

El Goresy A., Randohr P., and Taylor L. A. (1971) The opaque minerals in the lunar rocks from

Oceanus Procellarum. *Proc. Second Lunar Sci. Conf., Geochim. Cosmochim. Acta* Suppl. 2, Vol. 1, pp. 219–236. MIT Press.

Finger L. W., Hafner S. S., Schurmann K., Virgo D., and Warburton D. (1972) Distinct cooling histories and reheating of Apollo 14 rocks (abstract). In *Lunar Science—III* (editor C. Watkins), pp. 259–261, Lunar Science Institute Contr. No. 88.

Freeman J. W., Hills H. K., and Vandrak R. R. (1972) Water vapor, whence comest thou? (abstract). In *Lunar Science—III* (editor C. Watkins), pp. 283–285, Lunar Science Institute Contr. No. 88.

Geake J. E., Walker G., Mills A. A., and Garlick G. F. J. (1972) Luminescence of lunar material excited by protons or electrons (abstract). In *Lunar Science—III* (editor C. Watkins), pp. 294–296, Lunar Science Institute Contr. No. 88.

Goldstein J. I. and Yakowitz H. (1971) Metallic inclusions and metal particles in the Apollo 12 lunar soil. *Proc. Second Lunar Sci. Conf., Geochim. Cosmochim. Acta* Suppl. 2, Vol. 1, pp. 177–191. MIT Press.

Goldstein J. I., Yen F., and Axion H. J. (1972) Metallic particles in the Apollo 14 lunar soil (abstract). In *Lunar Science—III* (editor C. Watkins), pp. 323–325, Lunar Science Institute Contr. No. 88.

Gose W. A., Pearce G. W., Strangway D. W., and Larson E. E. (1972) On the magnetic properties of lunar breccias (abstract). In *Lunar Science—III* (editor C. Watkins), pp. 332–334, Lunar Science Institute Contr. No. 88.

Griscom D. L. and Marquardt C. L. (1972) Electron spin resonance studies of iron phases in lunar glasses and simulated lunar glasses (abstract). In *Lunar Science—III* (editor C. Watkins), pp. 341–343, Lunar Science Institute Contr. No. 88.

Haggerty S. E. (1972) Subsolidus reduction and compositional variations of lunar spinels (abstract). In *Lunar Science—III* (editor C. Watkins), pp. 347–349, Lunar Science Institute Contr. No. 88.

Herzenberg C. L., Moler R. B., and Riley D. L. (1971) Mössbauer instrumental analysis of Apollo 12 lunar rock and soil samples. *Proc. Second Lunar Sci. Conf., Geochim. Cosmochim. Acta* Suppl. 2, Vol. 3, pp. 2103–2123. MIT Press.

Housley R. M., Blander M., Abdel-Gawad M., Grant R. W., and Muir A. H. Jr. (1970) Mössbauer spectroscopy of Apollo 11 samples. *Proc. Apollo 11 Lunar Sci. Conf., Geochim. Cosmochim. Acta* Suppl. 1, Vol. 3, pp. 2251–2268. Pergamon.

Housley R. M., Grant R. W., Muir A. H. Jr., Blander M., and Abdel-Gawad M. (1971) Mössbauer studies of Apollo 12 samples. *Proc. Second Lunar Sci. Conf., Geochim. Cosmochim. Acta* Suppl. 2, Vol. 3, pp. 2125–2136. MIT Press.

Jackson E. D. and Wilshire H. G. (1972) Classifications of the samples returned from the Apollo 14 landing site (abstract). In *Lunar Science—III* (editor C. Watkins), pp. 418–420, Lunar Science Institute Contr. No. 88.

Kolopus J. L., Kline D., Chatelain A., and Weeks R. A. (1971) Magnetic resonance properties of lunar samples: Mostly Apollo 12. *Proc. Second Lunar Sci. Conf., Geochim. Cosmochim. Acta* Suppl. 2, Vol. 3, pp. 2501–2514. MIT Press.

Lin C. J. and Neaves O. (1962) Ferromagnetic resonance in very fine particle ferrites. *Proc. Int. Conf. on Mag. Cryst., J. Phys. Soc. Japan* 17, Suppl. B-1, pp. 389–392.

LSPET (Lunar Sample Preliminary Examination Team) (1970) Preliminary examination of the lunar samples from Apollo 12. *Science* 167, 1325–1339.

LSPET (Lunar Sample Preliminary Examination Team) (1971) Preliminary examination of lunar samples from Apollo 14. *Science* 173, 681–693.

LSPET (Lunar Sample Preliminary Examination Team) (1972) *Science* 175, 363–375.

McKay D. S., Morrison D. A., Clanton U. S., Ladle G. H., and Lindsay J. F. (1971) Apollo 12 soil and breccia. *Proc. Second Lunar Sci. Conf., Geochim. Cosmochim. Acta* Suppl. 2, Vol. 1, pp. 755–773. MIT Press.

Marvin U. B., Reid J. B. Jr., Taylor G. J., and Wood J. A. (1972) Lunar mafic green glasses, howardites, and the composition of undifferentiated lunar material (abstract). In *Lunar Science—III* (editor C. Watkins), pp. 507–509, Lunar Science Institute Contr. No. 88.

Marvin U. B., Wood J. A., Taylor J. G., Reid J. B. Jr., Powell B. N., Dickey J. S., and Bower J. F. (1971) Relative proportions and probable sources of rock fragments in the Apollo 12 soil samples. *Proc. Second Lunar Sci. Conf., Geochim. Cosmochim. Acta* Suppl. 2, Vol. 1, pp. 679–699. MIT Press.

Masson C. R., Gotz J., Jamieson W. D., and McLachlan J. L. (1971) Chromatographic and minera-logical study of lunar fines and glass. *Proc. Second Lunar Sci. Conf., Geochim. Cosmochim. Acta* Suppl. 2, Vol. 1, pp. 957–971. MIT Press.

Masson C. R., Smith I. B., Jamieson W. D., and McLachlan J. L. (1972) Chromatographic and mineralogical study of Apollo 14 fines (abstract). In *Lunar Science—III* (editor C. Watkins), pp. 515–516, Lunar Science Institute Contr. No. 88.

Morrish A. H. (1965) *The Physical Principles of Magnetism*, Chapter 10, John Wiley.

Morrish A. H. and Valstyn E. P. (1962) Ferrimagnetic resonance of iron-oxide micropowders. *Proc. Int. Conf. Mag. Cryst., J. Phys. Soc. Japan* **17**, Suppl. B-1, pp. 392–397.

Nagata T., Fisher R. M., and Schwerer F. C. (1972) Magnetism of Apollo 14 lunar materials (abstract). In *Lunar Science—III* (editor C. Watkins), pp. 573–575, Lunar Science Institute Contr. No. 88.

Néel L. (1949) Theorie du trainage magnetique des ferromagnetiques au grains fin avec applications aux terres cuites. *Ann. Geophys.* **5**, 99–136.

Nicholls J., Carmichael I. S. E., and Stormer J. C. (1971) Silica activity and P_{total} in igneous rocks. *Contrib. Mineral. Petrol.* **33**, 1–20.

Rao K. J., Sarkar S. K., Klein L., and Cooper A. R. (1972) Study of the optical properties of lunar glass spherules (abstract). In *Lunar Science—III* (editor C. Watkins), pp. 633–634, Lunar Science Institute Contr. No. 88.

Runcorn S. K., Collinson D. W., O'Reilly W., Battey M. H., Stephenson A., Jones J. M., Manson A. J., and Readman P. W. (1970) Magnetic properties of Apollo 11 lunar samples. *Proc. Apollo 11 Lunar Sci. Conf., Geochim. Cosmochim. Acta* Suppl. 1, Vol. 3, pp. 2369–2387. Pergamon.

Schlömann E. (1958) Ferromagnetic resonance in polycrystalline ferrites with large anisotropy—I. *J. Phys. Chem. Solids* **6**, 257–266.

Schlömann E. (1969) Inhomogenous broadening of ferromagnetic resonance lines. *Phys. Rev.* **182**, 632–645.

Shoemaker E. M. (1972) Cratering history and early evolution of the moon (abstract). In *Lunar Science—III* (editor C. Watkins), pp. 696–698, Lunar Science Institute Contr. No. 88.

Simpson P. R. and Bowie S. H. U. (1971) Opaque phases in Apollo 12 samples. *Proc. Second Lunar Sci. Conf., Geochim. Cosmochim. Acta* Suppl. 2, Vol. 1, pp. 207–218. MIT Press.

Tsay F. D., Chan S. I., and Manatt S. L. (1971a) Magnetic resonance studies of Apollo 11 and Apollo 12 samples. *Proc. Second Lunar Sci. Conf., Geochim. Cosmochim. Acta* Suppl. 2, Vol. 3, pp. 2515–2528. MIT Press.

Tsay F. D., Chan S. I., and Manatt S. L. (1971b) Ferromagnetic resonance of lunar samples. *Geochim. Cosmochim. Acta* **35**, 865–875.

Weeks R. A., Chatelain A., Kolopus J. L., Kline D., and Castle J. G. (1970a) Magnetic resonance properties of some lunar material. *Science* **167**, 704–707.

Weeks R. A., Kolopus J. L., Kline D., and Chatelain A. (1970b) Apollo 11 lunar material: Nuclear magnetic resonance of ^{27}Al and electron resonance of Fe and Mn. *Proc. Apollo 11 Lunar Sci. Conf., Geochim. Cosmochim. Acta* Suppl. 1, Vol. 3, pp. 2467–2490. Pergamon.

Weeks R. A., Kolopus J. L., and Arafa S. (1972) Ferromagnetic and paramagnetic resonance of lunar material: Apollo 12. *The Moon*, to appear.

Wlotzka F., Jaqoutz E., Spettel B., Baddenhausen H., Balacescu A., and Wänke H. (1972) On lunar metallic particles and their contribution to the trace element content of the Apollo 14 and 15 soils (abstract). In *Lunar Science—III* (editor C. Watkins), pp. 806–808, Lunar Science Institute Contr. No. 88.

Wosinski J. F., Williams J. P., Korda E. J., Kane W. T., Carrier G. B., and Schreurs J. W. H. (1972) Inclusions and interface relationships between glass and breccia in lunar sample 14306,50 (abstract). In *Lunar Science—III* (editor C. Watkins), pp. 811–813, Lunar Science Institute Contr. No. 88.

Proceedings of the Third Lunar Science Conference
(Supplement 3, *Geochimica et Cosmochimica Acta*)
Vol. 3, pp. 2519–2526
The M.I.T. Press, 1972

Moonquakes and lunar tectonism results from the Apollo passive seismic experiment

Gary Latham, Maurice Ewing, James Dorman and David Lammlein

Lamont-Doherty Geological Observatory

Frank Press and Nafi Toksöz

Massachusetts Institute of Technology

George Sutton and Frederick Duennebier

Hawaii Institute of Geophysics

and

Yosio Nakamura

Fort Worth Division of General Dynamics

Abstract—The natural seismicity of the moon appears to be very low relative to that of the earth. However, moonquakes do occur. They are detected by the stations of the Apollo seismic network at an average rate of 1800/yr at Station 14 and at lower rates at Stations 12 and 15. All of the moonquakes are small, and in the few cases for which the foci have been located, they occur at great depth (about 800 km). The frequency of occurrence of moonquakes is strongly correlated with lunar tides. The dynamic processes that generate quakes are clearly much less vigorous within the moon than they are within the earth.

Introduction

As shown by Latham *et al.* (1972a) and Toksöz *et al.* (1972), analysis of seismic signals from man-made impacts (LM ascent stages and Saturn S-IVB stages) has revealed that the moon possesses a crust, at least in the Fra Mauro region of Oceanus Procellarum. The crust in this region is about 65 km thick and probably consists of two layers: an upper layer about 25 km thick and a lower layer about 40 km thick. If these layers were formed by gravitational differentiation, the outer shell of the moon must once have been molten to depths of at least several hundred km. Alternatively, the layers may have formed during the original accumulation process, if the chemical composition of the reservoir of accumulating materials varied with time. Whether the deeper layer of this crust outcrops in the lunar highlands or not, we hope to learn from the results of Mission 16. This paper deals with the natural seismic activity of the moon—moonquakes.

We came to the moon with superstethascopes to listen to her heartbeat, as had been done for many years on earth, with the successful landing of Apollo 15, we now have three of the superstethascopes, called seismic stations, in operation on the moon, and we have found that her heartbeat is feeble indeed by comparison with the earth. The dynamic processes which generate quakes are clearly much less vigorous within

2519

the moon than they are within the earth. Yet, the moon is by no means completely inactive. Moonquakes do occur. Understanding the sources of energy for these moonquakes, and their focal mechanisms, is basic to our understanding of the dynamics of the lunar interior. Coupled with the magnetometer data, seismic measurements represent our most direct means for the eventual determination of the structure of the deep lunar interior.

The moonquakes have a rhythm. They occur at semi-monthly and monthly intervals in correlation with lunar tides, but we believe that the tides are only the trigger that sets off the moonquakes, not the fundamental source of strain energy.

Moonquakes have been recorded at an average rate of 600/yr at Station 12 and 1800/yr at Station 14. The larger number at Station 14 is a consequence of the thick regolith at the Frau Mauro site which makes this station more sensitive. Data from the Apollo 15 station are not yet sufficient to permit meaningful estimates of the detection rate there.

The numbers of detected moonquakes given above are higher than have been reported in the past. It was not until recently, when we were able to compare records from several stations and to apply improved data processing techniques, that we began to recognize numerous, very small moonquakes that we had not been able to identify before.

All of the moonquakes are small. The largest yet recorded has an equivalent Richter magnitude of between two and three. The total seismic energy release from the moon is estimated to be not more than 10^{15} erg/yr; nine orders of magnitude less than that of the earth.

Seismic signals are also recorded from meteoroid impacts. However, we defer the problem of deriving an estimate of meteoroid flux from these data until a better understanding is obtained of the long-range propagation characteristics of lunar seismic signals and the coupling of impact kinetic energy into seismic energy. Information on these factors will be derived principally from the study of the LM and S-IVB impact signals.

Results from the passive seismic experiment have been presented by the experiment team in 11 previous papers (Ewing, *et al.*, 1971; Latham *et al.*, 1969, 1970a, b, c, d, 1971a, b, 1972a, b; Toksöz *et al.*, 1972).

SIGNAL CHARACTERISTICS

A typical moonquake record is shown in Fig. 1. As shown in previous seismic team papers, the general characteristics of the lunar seismic signals; particularly, their emergent beginnings and long durations can be explained as resulting from intensive scattering of seismic waves in a dry, heterogeneous layer which probably blankets the entire lunar surface. The possible thickness of this layer, referred to as the "scattering zone," ranges between several km and 20 km. This is not a particularly surprising discovery. By nearly any theory of lunar evolution, there is to be expected a lunar surface layer formed as a consequence of a long and thorough process of fragmentation by cratering processes. But this layer must also be nearly completely free of volatiles to account for the very long duration, i.e., low damping, of the lunar seismic

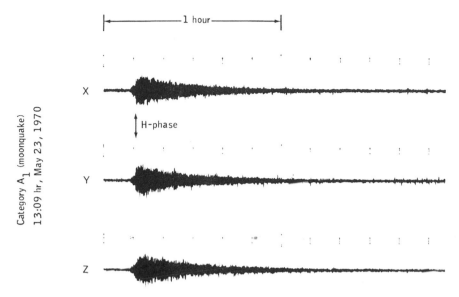

Fig. 1. Compressed time-scale records of a moonquake recorded at station 12. Z is the vertical component seismometer; X and Y are the horizontal component seismometers. The moonquake originated within the zone of greatest activity (A_1 zone). The H-phase is prominent on the seismograms from the horizontal component seismometers. This phase is tentatively identified as the direct shear wave arrival.

signals. The presence of complex sequences of lava flows may contribute to the general complexity of the scattering zone.

Moonquake signals are distinguished from meteoroid impact signals (and man-made impact signals) primarily by the presence of an abrupt change in signal amplitude, referred to as the *H*-phase. Ground motion associated with the *H*-phase is predominately horizontal. Thus, in the absence of any other likely phase, the *H*-phase is presumed to be the direct shear wave arrival. The prominence of this phase in moonquake signals, and its weak development in impact signals (weak shear waves can be identified in the LM and S-IVB impact signals), is consistent with terrestrial experience in which earthquakes usually generate much stronger shear waves than do explosions of equivalent energy. Other criteria used in distinguishing moonquake signals from meteoroid impact signals have been discussed in earlier papers (Ewing *et al.*, 1971; Latham *et al.*, 1971a, b, 1972a, b).

MOONQUAKE LOCATIONS AND FOCAL MECHANISMS

Among the moonquake signals, a remarkable similarity exists between some of the signals. Ten groups of matching signals, designated Type A events, have been identified. Long duration signals that are virtually identical must traverse precisely identical paths. This can occur in a complex medium only if the events have a common point of origin. Thus, there are at least ten points within the moon at which repeating moonquakes occur (designated A_1 through A_{10}).

   The nearly exact repetition of moonquake signals from a given focal zone over periods of many months requires that the focal zones must be small, 10 km in diameter or less, and fixed in location over periods approaching two years. If moonquake foci were separated by as much as one wavelength, larger differences would be observed among moonquake signals.

   Of the ten sources of repeating moonquakes, one source, the A_1 source, is by far the most active, accounting for nearly 80% of the total seismic energy detected. Using the observed arrival times of the seismic signals recorded from A_1 moonquakes at stations 12, 14, and 15, and a lunar model derived from the LM and S-IVB impact data, the epicenter (point on the surface directly above the focus) has been tentatively located at 21° S and 28° W, as shown in Fig. 2. The depth of the focus is approxi-

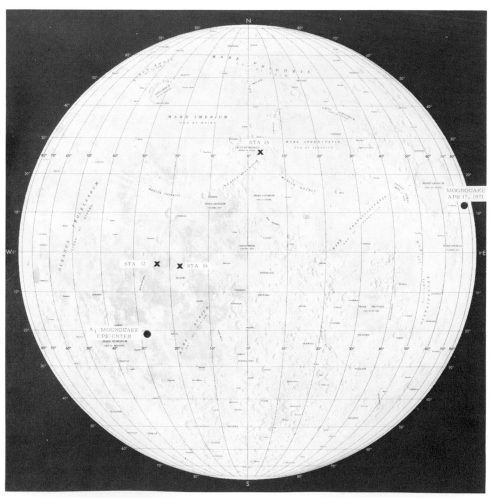

Fig. 2. Locations of the Apollo seismic network stations and the epicenter of the most active source of moonquakes (A_1 zone). The epicenter of the largest moonquake recorded thus far (April 17, 1971) is also shown. A depth of focus of 800 km was assumed in deriving the location of the April 17 event.

mately 800 km: slightly deeper than any known earthquake. Another source, the A_6 focus, is located within about 20 km of the A_1 focus. The A_7 focus is located somewhere along the southern edge of Mare Serenitatis, at unknown depth. The remaining moonquake foci have not been located because moonquakes originating at these foci of sufficient magnitude to be recorded at all three stations have not yet occurred.

LOCATION AND FOCAL MECHANISM OF MOONQUAKES

The source of strain energy released as Type A moonquakes is not known; but, if significant depth of focus for all moonquakes is verified by future data, it will have profound implications concerning the lunar interior. In general, this result would require that the shear strength of the lunar material at a depth of 800 km must be large enough to sustain appreciable stress.

As noted previously (Ewing *et al.*, 1971; Latham *et al.*, 1970d, 1971a, b, 1972a, b), the category A moonquakes occur in monthly cycles near times of apogee and perigee. This suggests that they are triggered by lunar tides. This hypothesis is strengthened by the observation that the total seismic energy release and the interval between the times of occurrence of the first moonquakes each month and perigee, both show seven-month periodicities which also appear in the long-term gravity variations. With a few possible exceptions, the polarities of signals belonging to a set of matching events are identical. This implies that the source mechanism is a progressive dislocation and not one that periodically reverses in direction. It is conceivable, of course, that detectable movements in one direction are compensated by many small, undetectable movements in the opposite direction. A progressive source mechanism suggests a secular accumulation of strain periodically triggered by lunar tides. Whether this strain is local, regional, or moonwide is an intriguing problem for further study. Several possible sources are (1) slight expansion of the moon by internal radiogenic heating or slight contraction on cooling; (2) a gradual settling of the lunar body from an ellipsoidal form to a more nearly spherical form as the moon gradually recedes from the earth; (3) localized strains due to uncompensated masses; (4) localized thermal stresses; or (5) weak convective motion.

What can be said about the majority of moonquake signals that do not match one another? These events, designated Type B moonquakes, occur principally during episodes of greatly increased activity. By analogy with similar patterns of earthquake activity, these episodes are called moonquake swarms. Each swarm is a distinctive sequence of moonquakes closely grouped in time, containing no conspicuous event. A moonquake swarm is characterized by an abrupt beginning and ending of activity. Events are recorded at a nearly constant rate of 8 to 12 per day during a swarm, as compared to an average of 1 to 2 per day between swarms.

The number of moonquakes versus time recorded over a 13-month period at Station 12 is shown in Fig. 3. This plot includes both types A and B moonquakes, but of the total of 650 events, only 99 events or 15% of the total are Type A moonquakes. The remainder are either Type B or are too small to be classified by present methods. Thus the plot is essentially the time history of the occurrence of Type B moonquakes. The pattern is periodic with principal harmonic components at 13.5 and 27 days, as shown in the spectra of Fig. 4. Types A and B moonquakes have been separated for

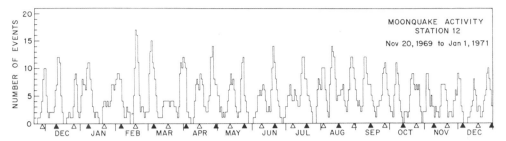

Fig. 3. Number of moonquakes of all types recorded at station 12. The plot has been smoothed by summing over 3-day intervals. Thus, to obtain the average daily rate of moonquake detection, divide ordinate values by 3.

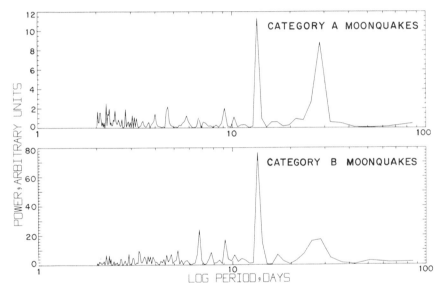

Fig. 4. Spectra of the frequency of occurrence of types A and B moonquakes recorded at station 12 over a period of 256 days. The raw data used in the harmonic analysis are given in Fig. 3.

purposes of the spectral analysis. The two spectra differ only in the relative amplitude of the monthly component which is much more prominent for Type A moonquakes than it is in the time series for Type B moonquakes. Thus, we conclude that Type B moonquakes are also induced by tidal stresses. However, in contrast with Type A moonquakes, the foci of Type B moonquakes must be widely distributed. Based upon terrestrial experience, we expect that they are concentrated in certain regions of the moon. The degree of success we eventually achieve in delineating the active zones will depend primarily upon the time interval over which multi-station data are acquired. In one case, however, we have a clue. In April 1971, the series of small

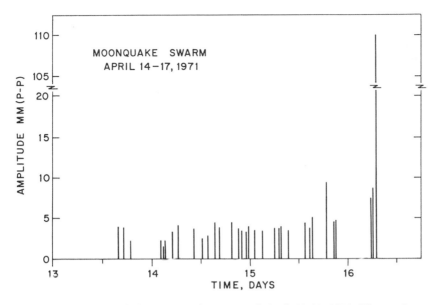

Fig. 5. Time history of the moonquake swarm of April 14–17, 1971. The maximum trace amplitude recorded at station 14, in millimeters peak-to-peak, is plotted at the time of occurrence of each moonquake.

moonquakes shown in Fig. 5 was detected at Stations 12 and 14. Station 15 had not yet been installed. The largest moonquake yet recorded occurred at the end of the swarm and activity ceased abruptly. This moonquake is distinguished not only by its magnitude, but also by the unusually high frequency of the signal generated. The moonquake epicenter can be located only if a depth-of-focus and a velocity model for the lunar interior are assumed. For a depth-of-focus of 800 km and the same velocity model used in locating Type A moonquakes, the epicenter is located near Mare Crisium, as shown in Fig. 2. The epicenter moves north-westward for decreasing focal depth. Thus, if the series of small moonquakes are foreshocks of the main event, they also must have originated in the vicinity of Mare Crisium.

On earth, swarms occur most frequently in association with volcanic activity. In the Lesser Antilles, for example, swarms preceed most of the major volcanic eruptions (Robson *et al.*, 1962). Thus, while the main episode of lunar volcanism appears to have ended about three billion years ago, the occurrence of the swarm of April, 1971, suggests that minor activity may be continuing today.

Acknowledgments—The contributions of Don Anderson, Robert Kovach, Christopher Scholtz, and Tosi Matumoto in critically reviewing the manuscript are gratefully acknowledged. This work was supported by the National Aeronautics and Space Administration under contracts and grants NAS9-5957 and NAS9-12334. Computer facilities were provided by the NASA Goddard Space Flight Center, Institute for Space Studies, New York. Lamont-Doherty Geological Observatory Contribution No. 1812, Hawaii Institute of Geophysics Contribution No. 473.

REFERENCES

Ewing M., Latham G., Press F., Sutton G., Dorman J., Nakamura Y., Meissner R., and Duennebier F. (1971) *Seismology of the Moon and Implications on Internal Structure, Origin, and Evolution.* Highlights of Astronomy, D. Reidel Pub. Co. (Dordrecht, Holland).

Latham G., Ewing M., Press F., Sutton G., Dorman J., Nakamura Y., Toksöz N., Wiggins R., Derr J., and Duennebier F. (1969) Passive seismic experiment. Sec. 6 of Apollo 11 Preliminary Science Report. NASA SP-214.

Latham G., Ewing M., Press F., Sutton G., Dorman J., Nakamura Y., Toksöz N., Wiggins R., Derr J., and Duennebier F. (1970a) Apollo 11 passive seismic experiment. *Science* **167**, 455–467.

Latham G., Ewing M., Press G., Sutton G., Dorman J., Nakamura Y., Toksöz N., Wiggins R., Derr J., Duennebier F. (1970b) Apollo 11 passive seismic experiment. *Proc. Apollo 11 Lunar Sci. Conf., Geochim. Cosmochim. Acta* Suppl. 1, Vol. 3, pp. 2309–2320.

Latham G., Ewing M., Press F., Sutton G., Dorman J., Nakamura Y., Toksöz N., Wiggins R., and Kovach R. (1970c) Passive seismic experiment. Sec. 3 of Apollo 12 Preliminary Science Report. NASA SP-235.

Latham G., Ewing M., Press F., Sutton G., Dorman J., Nakamura Y., Toksöz N., Meissner R., Duennebier F., and Kovach R. (1970d) Seismic data from man-made impacts on the moon. *Science* **170**, 620–626.

Latham G., Ewing M., Press F., Sutton G., Dorman J., Nakamura Y., Toksöz N., Duennebier F., and Lammlein D. (1971a) Passive seismic experiment. Sec. 6 of Apollo 14 Preliminary Science Report. NASA SP-272.

Latham G., Ewing M., Press F., Sutton G., Dorman J., Nakamura Y., Toksöz N., Lammlein D., and Duennebier F. (1971b) Moonquakes. *Science* **174**, 687–692.

Latham G., Ewing M., Press F., Sutton G., Dorman J., Nakamura Y., Toksöz N., Lammlein D., and Duennebier F. (1972a) Passive seismic experiment. Sec. 8 of Apollo 15 Preliminary Science Report. NASA SP-289.

Latham G., Ewing M., Press F., Sutton G., Dorman J., Nakamura Y., Toksöz N., Lammlein D., and Duennebier F. (1972b) Moonquakes and lunar tectonism. Proc. of the Lunar Science Institute Conference on Lunar Geophysics.

Robson G., Barr K., Smith G. (1962) Earthquake series in St. Kitts-Nevis, 1961–62. *Nature* **195**, 972–974.

Toksöz N., Press G., Anderson K., Dainty A., Latham G., Ewing M., Dorman J., Lammlein D., Sutton G., Duennebier F., Nakamura Y. (1972) Velocity structure and properties of the lunar crust. Proc. of the Lunar Science Institute Conference on Lunar Geophysics.

Proceedings of the Third Lunar Science Conference
(Supplement 3, *Geochimica et Cosmochimica Acta*)
Vol. 3, pp. 2527–2544
The M.I.T. Press, 1972

Structure, composition, and properties of lunar crust

M. N. Toksöz, F. Press, A. Dainty, and K. Anderson

Department of Earth and Planetary Sciences,
Massachusetts Institute of Technology,
Cambridge, Massachusetts 02139

G. Latham, M. Ewing, J. Dorman, and D. Lammlein

Lamond-Doherty Geological Observatory,
Palisades, New York 10964

and

G. Sutton and F. Duennebier

University of Hawaii, Honolulu, Hawaii 96822

Abstract—Lunar seismic data from three Apollo seismometers are interpreted to determine the structure of the moon's interior to a depth of about 100 km. The travel times and amplitudes of P and S arrivals from Saturn IV-B and LM impacts are interpreted in terms of a velocity profile. The most outstanding feature of the model is that, in the Fra Mauro region of Oceanus Procellarum, the moon has a 65 km-thick layered crust. Other features of the model are: (i) rapid increase of velocity near the surface due to pressure effects on dry rocks, (ii) a discontinuity at a depth of about 25 km, (iii) near-constant velocities between 25 and 65 km deep, (iv) a major discontinuity at 65 km marking the base of the lunar crust, and (v) very high apparent velocities(about 9 km/sec for P waves) in the lunar mantle below the crust. Velocities in the upper layer of the crust match those of lunar basalts while those in the lower layer fall in the range of terrestrial gabbroic and anorthositic rocks. The high apparent velocities in the mantle, if they persist, imply a differentiated lunar mantle whose composition varies with depth.

Introduction

With the successful recording of Lunar Module (LM) ascent stage and Saturn (S-IV B) rocket impacts by the Apollo 12, 14, and 15 seismometers, discrete seismic phases that can be interpreted in terms of a velocity structure inside the moon have become available. Travel times, amplitudes, and wave shapes of compressional (P) and shear (S) waves have been obtained for distances between $\Delta = 67$ km and $\Delta = 357$ km. These data are inverted to determine the seismic velocity structures in the outer 100 km of the lunar interior. In this paper we describe the data, the inversion techniques, velocity models, compositional implications, and the properties of the lunar interior in the light of laboratory measurements of velocities of lunar and terrestrial rocks. The data and the interpretation are more expansive than those of the two previous reports (Toksöz *et al.*, 1972a,b).

In the study of the earth's interior data from a large number of earthquakes of all magnitudes, numerous artificial sources (e.g., underground nuclear explosions) and more than a thousand seismic stations are being utilized. In the case of the moon the natural seismicity (both the number and energy of moonquakes) is many orders of magnitude lower than that of the earth (Latham et al., 1971, 1972). With only three stations it is not possible at this stage to specify the epicenter coordinates, focal depth, and origin time of these events with sufficient accuracy to use them in structural studies. It has not been possible to detect long-period surface waves or free oscillations of the moon from such very small moonquakes. Thus we must rely on seismic waves from six artificial impacts, as recorded by the Apollo 12, 14, and 15 seismometers, for the study of the lunar interior.

SEISMIC DATA AND VELOCITY MODELS

Compressional and shear velocity profiles were obtained by interpreting the travel times and amplitudes of P and S waves recorded by the Passive Seismic Experiment (PSE) instruments. The locations of the three operating stations, impact points and wave paths are shown in Fig. 1; the distances are listed in Table 1. The general characteristics of all impact signals are very similar to each other. They are extremely prolonged with gradual build-up of the signal and very long exponential decay of signal intensity. Typically, duration of the signals is about two hours. These characteristics, which are believed to be caused by intensive scattering, have been discussed previously (Latham et al., 1971; Nakamura et al., 1970; Dainty and Anderson, 1972)

Table 1.

A. Coodinates of Seismic Stations and Impact Points and Relevant Distances.

Stations and impacts	Coordinates (degrees)	Distances (in km) from Apollo 12 site	Apollo 14 site	Apollo 15 site
Apollo 12 Site	3.04S, 23.42W	—	—	—
Apollo 14 Site	3.65S, 17.48W	181	—	—
Apollo 15 Site	26.08N, 3.66E	1188	1095	—
Apollo 12 LM Impact	3.94S, 21.20W	73	—	—
Apollo 13 S-IV B Impact	2.75S, 27.86W	135	—	—
Apollo 14 S-IV B Impact	8.00S, 26.06W	170	—	—
Apollo 14 LM Impact	3.42S, 19.67W	114	67	—
Apollo 15 S-IV B Impact	1.36S, 11.77W	357	186	—
Apollo 15 LM Impact	26.36N, 0.25E	1130	1049	93

Table 1. (Cont.)

B. Impact Parameters.

Impact	Date (day–month–yr)	Time (hr:min:sec)	Velocity (km/sec)	Mass (kg)	Kinetic energy (ergs)	Angle from horizontal (degrees)
Apollo 12 LM	20–11–69	22:17:17.7	1.68	2383	3.36×10^{16}	3.7
Apollo 14 LM	7–02–71	00:45:25.7	1.68	2303	2.35×10^{16}	3.6
Apollo 15 LM	3–08–71	03:03:37.0	1.70	2385	3.44×10^{16}	3.2
Apollo 13 S-IV B	15–04–70	01:09:41.0	2.58	13925	4.63×10^{17}	76
Apollo 14 S-IV B	4–02–71	07:40:55.4	2.54	14016	4.52×10^{17}	69
Apollo 15 S-IV B	29–07–71	20:58:42.9	2.58	13852	4.61×10^{17}	62

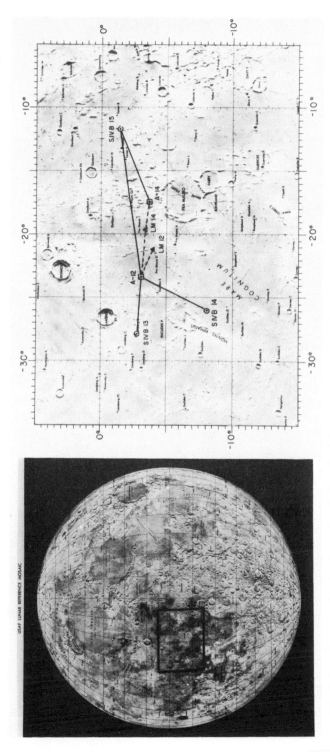

Fig. 1. Location map showing the region (in frame) of the moon where most of the seismic observations have been made. Three operating seismic stations are shown as black triangles. Enlarged picture on the right shows Apollo 12 and Apollo 14 seismic stations, impact points, and seismic wave paths.

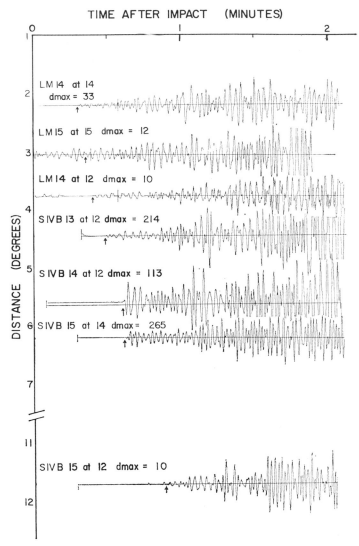

Fig. 2. A composite illustration of vertical components of seismograms recorded from all artificial impacts. Notation "LM 14 at 14" means LM-14 impact recorded at Apollo 14 seismic station. The amplitude normalization scale is "dmax"; to obtain true relative amplitudes, seismogram amplitudes must be multiplied by this factor. Distance is in degrees; 1 degree = 30.3 km. Arrows indicate the first arrival. Note the change of character of S-IV-B seismograms with distance.

and will not be treated here. The signals corresponding to the arrival of discrete seismic phases can be identified in the early parts of the records. These are of most interest for our study.

The initial portions of the impact seismograms are shown in Fig. 2. In order to understand this figure, it is important to keep in mind that the S-IV B impact precedes the lunar landing and thus is recorded by stations operating prior to the landing. The LM-ascent stage impact which occurs after the lunar landed mission is completed can be recorded with the instrument of the same mission. Furthermore, the kinetic energy of a typical S-IV B impact is about 13 times larger than that of a typical LM impact (see Table 1). The larger amplitudes of S-IV B impact signals relative to those of LM impacts are clear. The change in the signal characteristics from one record to another, at different distances such as $\Delta = 135$, 172, and 357 km, is also observed.

The travel times of seismic arrivals provide the most direct means of looking at the lunar interior. For P waves times are read from high-pass filtered or unfiltered traces. Positive identification of the signal is made by the linear polarization of the particle motion. Shear wave phases have been tentatively identified on the records; primarily low-pass filtered seismograms are used. As with P waves, an attempt was made to use particle motion to identify S phases, but this did not always succeed. Many of the arrivals were identified on the transverse (T) traces (Fig. 3). Theoretically one would expect the S waves to be generated less efficiently by the impacts than by the moon-quakes. This is observed in the case of the lunar data analyzed. The S-wave ampli-tudes are generally small in the S-IV B and LM impact records as well as the records of natural meteoroid impacts. The S-wave travel times are not as accurate as those of P waves. As a result of this, conclusions drawn concerning S-wave velocities should be treated with caution. The travel time-distance curves shown in Figs. 4 and 5 are composites of data for direct arrival P and S waves and later phases (surface-reflected PP, PPP, SS, SSS). Since both source and receiver are at the surface of the moon, for phases such as PP and SS the times can be plotted at equivalent distances for direct arrivals. The overall characteristics of the travel times indicate rapidly increasing velocities near the surface. An intermediate zone with nearly constant velocity and below this a high velocity zone are indicated by the two branches of the travel time curve between distances of 186 km and 357 km. In addition to these, later arrivals observed at 186 and 357 km indicate the presence of very high velocity gradients or discontinuities inside the moon.

The amplitudes (Figs. 4, 5) provide further evidence for velocity discontinuities. Because the geometric focusing, defocusing and reflection of seismic rays are con-trolled by the velocity gradients (see ray paths in Fig. 4), the amplitudes are very sensitive indicators of rapid velocity variations. The amplitude data come from both the LM and S-IV B impact records. Since the sources have different energies, the amplitudes need to be scaled. We chose an empirical approach for this purpose and required a smooth transition from LM signal amplitudes to S-IV B signal amplitudes as a function of distance. This required a correction (multiplication) factor of 20 for LM amplitudes, a value larger than the square root of energy ratios. However, since the angles of impact (about 3° from horizontal for LM versus 62° or greater for S-IV B) and sizes of the impacting objects were different, it is reasonable that S-IV B impacts would be more efficient generators of seismic waves at frequencies below 1 Hz. The agreement between theoretical curves and observed amplitude and travel-time data is very good both for P and S waves.

A more definitive approach to interpreting the observed lunar seismograms is to compute their theoretical equivalents (Helmberger, 1968; Helmberger and Wiggins,

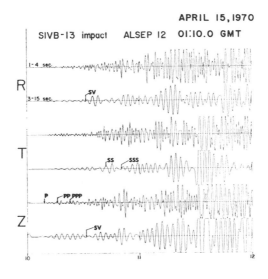

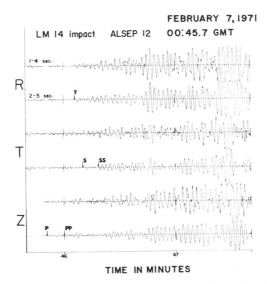

Fig. 3. Seismograms of the Apollo 13 S-IV-B impact recorded at the Apollo 12 station and the LM-14 impact recorded at the Apollo 12 station. *R*, *T*, and *Z* indicate radial, transverse, and vertical components of the motion, respectively. Each seismogram is band-pass filtered at two period intervals: 1–4 sec and 3–15 sec. *P*, *PP*, *PPP*, *S*, and *SS*, arrivals are shown.

1971). This requires, in addition to the velocity structure inside the moon, knowledge of the seismic source pulse due to the impact and the exact impulse response of the seismometer emplaced at the surface on a very low velocity regolith complex. To isolate the effect of the velocity structure, one must keep the other variables fixed. We chose the three S-IV B impact seismograms recorded by the Apollo 12 station. This had the advantage of keeping the instrument response fixed while the

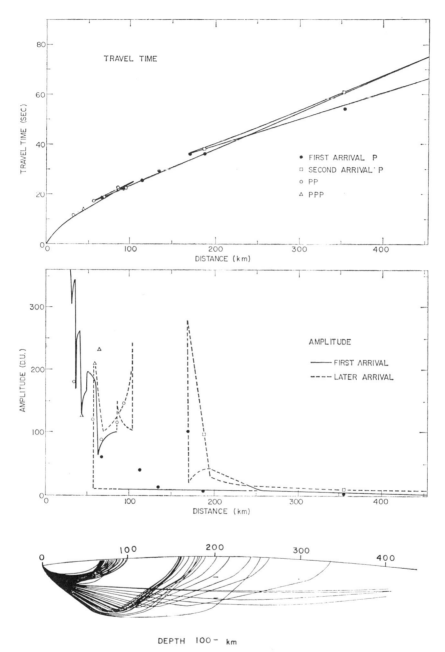

Fig. 4. Travel times, amplitudes, and ray paths for *P* wave pulses. "Second arrival" *P* denotes a relatively large amplitude pulse that arrives after *P* that is associated with a travel-time cusp. *PP* and *PPP* are surface reflected phases. Lines are theoretical curves for velocity model given in Fig. 7. "D.U." signifies digitization unit. Seismic ray paths inside the moon (at the bottom) show the effects of high velocity gradients. Ray crossings correspond to multiplication of travel times. High density of rays indicates focusing of energy and hence large amplitudes.

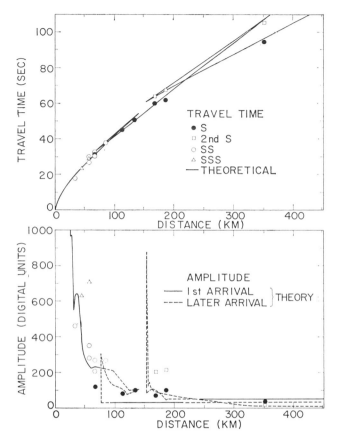

Fig. 5. Travel times and amplitudes of *S* waves. Theoretical curves are for the velocity model shown in Fig. 7. All other designations same as those in Fig. 4.

source-receiver distance increased for three nearly identical impacts. Because of the low signal-to-noise ratios, LM impact seismograms were excluded from matching.

The observed and computed seismograms at distances of $\Delta = 135$, 172, and 357 km are shown in Fig. 6. The source function was chosen such that at 172 km the seismograms were matched exactly for the first 8 seconds. The general characteristics of the observed seismograms change significantly from $\Delta = 135$ km to 172 km and again from $\Delta = 172$ km to 357 km. These are matched very closely by the theoretical seismograms. Only *P* arrivals are matched in Fig. 6.

The *S*-wave data were interpreted using travel-time analysis and ray theory amplitudes. Since scattering was obviously important (as evidenced by the presence of transverse motion), synthetic seismograms were not computed. Amplitudes were measured on the radial component, unfiltered, even if the arrival was originally identified on the low-pass filter transverse trace. Figure 5 shows the fit of travel times and measured amplitudes to the theoretical values for the *S* velocity model shown in

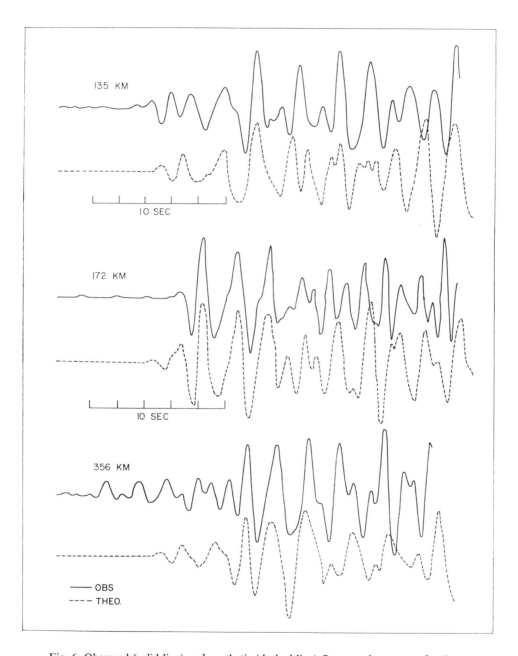

Fig. 6. Observed (solid line) and synthetic (dashed line) *P* wave seismograms for three S-IV-B impacts recorded at station 12. The change in seismic pulse shapes and relative amplitudes of first and later arrivals with increasing distance is obvious. At 357 km the first two peaks of observed seismograms are noise pulses and can be clearly identified as such in unfiltered seismograms.

Fig. 7. This model was constrained to follow the *P* velocity model, insofar as this was possible.

The velocity models finalized on the basis of the fit to the travel times, amplitudes, and synthetic seismograms are shown in Fig. 7. These models provide direct evidence for the presence of a layered lower crust. The main features of the velocity profiles are:

    i. Very rapid increase at shallow depths from the surface to 10 km depth. The details of the models and whether the velocities increase smoothly or stepwise in the upper 5 km cannot be resolved without additional data at distances closer than 30 km.

   ii. A sharp increase ("discontinuity") at a depth of about 25 km.

  iii. Near constant values between 25 and 65 km.

  iv. A significant and discontinuous increase at the base of the lunar crust (65 km).

   v. As can be determined from a single data point corresponding to the distance of 357 km, very high apparent velocities below the lunar crust.

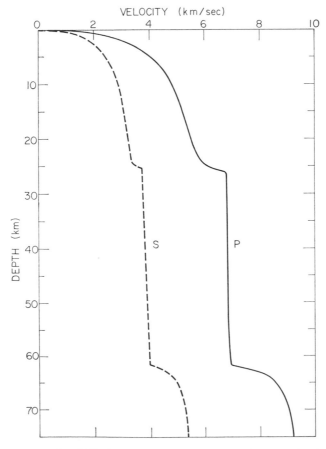

Fig. 7. Compressional (solid line) and shear (dashed line) velocity versus depth profiles for the moon in the vicinity of Fra Mauro.

From comparisons with velocities of the earth's crust and mantle, it is appropriate to define the base of the "lunar crust" at the discontinuity at 65 km. Although this is greater than the average thickness of the earth's crust, the jump of the P velocity at this interface from about 7.0 to 9.0 km/sec is very similar to the increase at the Mohorovic discontinuity. If hydrostatic pressure instead of depth is used as the variable, the base of the lunar crust occurs at a pressure of 3.5 kbar, which is reached at a depth of 10 km inside the earth.

COMPOSITION AND PROPERTIES

The compositional implications of the lunar velocity models can be explored with the aid of high-pressure laboratory measurements on lunar and terrestrial rocks. Velocity measurements have been made on lunar soils, breccias, and igneous rocks from four Apollo missions (Anderson et al., 1970; Kanamori et al., 1970, 1971; Wang et al., 1971; Warren et al., 1971, 1972; Mizutani et al., 1972; Tittman et al., 1972; Todd et al., 1972). Regardless of composition, these rocks are characterized by very low velocities at low pressures relative to terrestrial rocks. This can be attributed to the absence of water in the lunar rocks combined with the effects of porosity and microcracks. Laboratory measurements on terrestrial igneous rocks have demonstrated this effect. As shown in Fig. 8 taken from Nur and Simmons (1969), with only

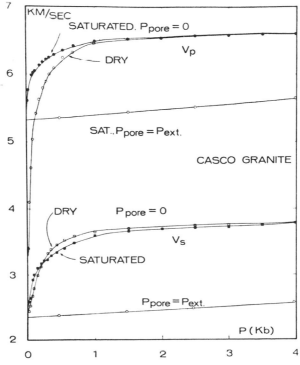

Fig. 8. Compressional and shear velocity versus pressure curves (laboratory data) for dry and water-saturated Casco granite (porosity = 0.7%). (After Nur and Simmons, 1971).

0.7% porosity, the velocity of compressional waves at pressures less than 0.5 kb is much lower in the dry state than in the saturated state. In dry-state conditions appropriate for the moon, the pressure gradient of compressional velocity is very high at low pressures, a behavior very similar to that of the lunar P velocity profile.

The measured velocities of lunar samples are plotted together with observed compressional and shear wave velocity profiles in Figs. 9 and 10, respectively. These figures show all available lunar rock velocities except the values for porphyritic basalts 12002 and 12022 (Wang *et al.*, 1971). These two rocks have higher velocities than those of all other lunar rocks (about 0.6 km/sec higher than those of Apollo 11 and 14 basalts at 3 kb pressure). At present, the significance of this deviation is being investigated. Compressional velocities of terrestrial rocks shown in Fig. 9 represent either an average value for one or more measurements, or specify general bounds between which most values fall (see Anderson and Lieberman, 1966; Press, 1966 for tabulations). The Poisson ratio σ $[\sigma = (V_p^2 - 2V_s^2)/2(V_p^2 - V_s^2)$, where V_p and V_s are velocities], shown in Fig. 10b, is a strong indication of the rock conditions in the lunar crust. To a depth of 25 km the σ calculated from observed velocities falls among the lunar sample values and close to terrestrial dry granite. At greater depths, both the model values and Poisson's ratios of lunar samples increase, indicating the closing of microcracks. All lunar basalts (except for sample 10020) have higher Poisson's ratios than the observed value below about 30 km depth.

From the comparison of the laboratory data and the lunar velocity profile, the following units can be identified:

(i) Near the surface the extremely low seismic velocities are very similar to those of lunar fines (soils) and broken rocks. The velocity increases very rapidly as a result of self-compaction under pressure. The exact depth to the bottom of the lunar regolith and brecciated and fractured layer cannot be determined without additional travel-time data in the distance range of 0.1 to 5.0 km.

(ii) Below a depth of a few kilometers, the measured velocities of lunar basaltic rocks fit both the compressional and shear velocity profiles to a depth of about 25 km. The rapid increase of velocity to a depth of about 10 km can be explained by the pressure effect on dry rocks having micro and macrocracks. The observed velocity models seem to fall between the measured values for lunar basaltic rocks, samples 10057 and 12065, paralleling these curves to a depth of 25 km. They coincide with laboratory velocities of lunar samples 14310 and 10020. Whether this layer consists of a series of flows or fairly thick intrusive basalt cannot be resolved with our data. Nor is it possible to rule out other compositions or rock types that may have similar velocities under a lunar environment.

There are other data indicating similar values for thicknesses of mare basalts. A model proposed by Wood (1970), on the basis of density contrast and relative elevations of basaltic mare and "anorthositic" highlands, requires 25 km of basalt in the mare for isostatic compensation to occur. The latest gravity data from Apollo 15 subsatellite observations fits a model of about 20 km of basalt filling in mascon basins (Sjogren *et al.*, 1972).

(iii) The second layer of the lunar crust (25 km to 65 km) appears to be made of competent rock. The increase in pressure affects velocities very little (velocity at the

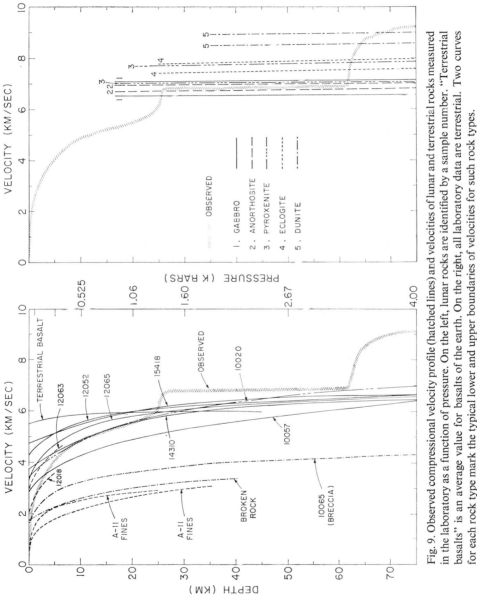

Fig. 9. Observed compressional velocity profile (hatched lines) and velocities of lunar and terrestrial rocks measured in the laboratory as a function of pressure. On the left, lunar rocks are identified by a sample number. "Terrestrial basalts" is an average value for basalts of the earth. On the right, all laboratory data are terrestrial. Two curves for each rock type mark the typical lower and upper boundaries of velocities for such rock types.

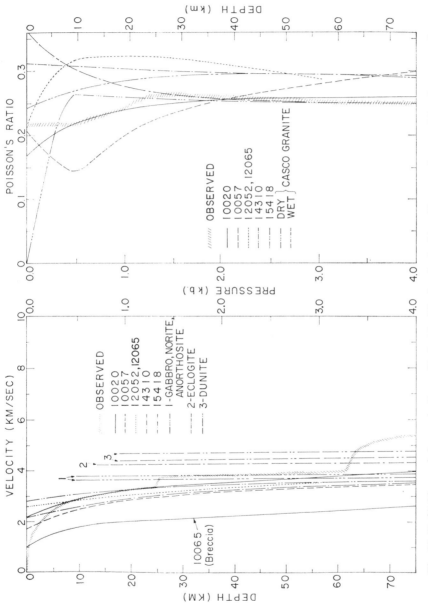

Fig. 10. (Left) Observed shear velocity profile (hatched line) and velocities of lunar and terrestrial rocks measured in the laboratory. (Right) Poisson ratio versus depth curve for the moon (computed from *P* and *S* velocity profiles) and those of lunar rocks and terrestrial dry and saturated granite.

bottom is about 7.0 km/sec). The petrological interpretation of the velocity curve is not very simple. It is clear from Figs. 9 and 10 that available lunar sample velocities are lower than the observed curve if we exclude the two anomalous rocks 12002 and 12022. A rock of special interest is 15418, a recrystallized or annealed gabbroic anorthosite (LSPET, 1972). Laboratory values of velocities of 15418 are higher than other lunar values plotted in Figs. 9 and 10. However, they are still lower than the observed values below a depth of 25 km. We compared the laboratory data on terrestrial rocks to observed lunar velocity curves. The examples plotted on Figs. 9 and 10 represent some possible candidates. The observed P and S velocity curves fall in the middle of the laboratory data for gabbros. It is very close to the anorthosite values. The lower bound for terrestrial pyroxenite is slightly higher than the observed P curve. Eclogite and dunite velocities are definitely higher than the observations, and these cannot be considered as serious contenders.

Petrological evidence as well as velocities favor a composition of anorthositic gabbro or gabbroic anorthosite (Wood et al., 1971; Gast and McConnell, Jr., 1972; Reid et al., 1972; Walker et al., 1972). However, the above evidence cannot rule out other interpretations.

(iv) The discontinuity at 65 km depth is required to satisfy the amplitudes, travel times, and most clearly the seismogram characteristics at $\Delta = 357$ km. The apparent velocities below this discontinuity increase to about 9 km/sec for P and 5 km/sec for S waves. Although these velocities are based on a single observation and are tentative until more seismic data become available from future impacts, the discontinuity is clearly a major structural boundary. It represents the lunar crust-mantle interface. If the high velocity in the outer portion of the lunar mantle persists, it cannot be matched with terrestrial laboratory values.

Let us consider four petrologically feasible models of the lunar mantle to compare with the observed velocities. With some simplification, these models can be classified as: (a) olivine (Smith et al., 1970), (b) pyroxenite [orthopyroxene, clinopyroxene, olivine, spinel, or garnet] (Ringwood and Essene, 1970; Green et al., 1970), (c) peridotite (Biggar et al., 1971; O'Hara et al., 1970), and (d) a high-pressure phase of anorthosite [grossularite, kyanite plus quartz] (Anderson and Kovach, 1972). It is clear from laboratory values plotted in Figs. 9 and 10 and estimates made for lunar rocks by Anderson and Kovach (1972) that the observed apparent velocities at the 70 km depth are greater than those of olivines, eclogites, pyroxenites (including garnet pyroxenites), and peridotites. Magnesium-rich olivines and eclogites with low FeO ratios are closest to the model. The high-pressure form of anorthosite could have compressional velocities as high as 9.3 km/sec at 3 kb, but the mineralogical constraints for such a phase transformation are very stringent (Boettcher, 1971).

Before the upper mantle composition can be narrowed down on the basis of seismic results, it is necessary to have additional data to verify the observed apparent velocities. If the interface between the crust and mantle is dipping 3° eastward, the observed apparent P velocities may be 0.4 km/sec higher than the true velocity, thus reducing the upper mantle velocities to about 8.6 km/sec. The Apollo 16 S-IV-B is planned to impact to the west of the Apollo 12 and 14 stations and, if successful, it should resolve this uncertainty.

Any of the above petrological models implies a differentiated lunar mantle whose composition must vary with depth to satisfy mean density and moment of inertia constraints.

CONCLUSIONS

1. Seismic evidence directly shows that in the Fra Mauro region of Oceanus Procellarum the moon has a layered crust and differentiated mantle.

2. The uppermost 25 km of the crust probably consists of basaltic material similar to that collected at Apollo 11, 12, 14, and 15 sites. A combination of very dry conditions and microcracks in rocks contribute to a rapid increase of velocity with depth, low attenuation, extensive scattering of seismic waves, and reverberating lunar seismograms.

3. At 25 km there is a change to what is believed to be a different composition. The most likely candidates on seismic and petrological grounds are gabbroic rocks such as anorthositic gabbro or gabbroic anorthosite.

4. At the investigated site, the lunar crust is about 65 km thick. The mantle below has very high apparent velocities based on a single impact recording. Any definitive interpretation of mantle mineralogy requires additional data.

5. All feasible mantle models compatible with petrological, chemical, and seismic data as well as mass and moment of inertia constraints require a differentiated, radially inhomogeneous lunar mantle.

Acknowledgments—We would like to thank Mr. E. Stolper for his assistance in the analysis of the data, Drs. G. Simmons and H. Wang, and Messrs. T. Todd and S. Baldridge for making their laboratory data available to us. Contributions of Drs. B. Julian, S. Solomon, and R. Wiggins are gratefully acknowledged.

This research was supported by NASA under contracts NAS9-12334 at M.I.T. and NAS9-5957 at Lamond-Doherty Gelogical Observatory.

REFERENCES

Anderson D. L. and Kovach R. L. (1972) The lunar interior. *Earth Planet. Sci. Lett.*, in press.

Anderson O. L. and Lieberman R. C. (1966) Sound velocities in rocks and minerals. *VESIAC State-of-the-Art Report*, No. 7885-4-X, Willow Run Laboratories, University of Michigan, 189 pp.

Anderson O. L., Scholz C., Soga N., Warren N., and Schreiber E. (1970) Elastic properties of a micro-breccia, igneous rock and lunar fines from Apollo 11 mission. *Proc. Apollo 11 Lunar Sci. Conf.*, *Geochim. Cosmochim. Acta*, Suppl. 1, Vol. 3, pp. 1959–1973. Pergamon.

Biggar G. M., O'Hara M. J., Peckett A., and Humphries D. J. (1971) Lunar lavas and the achondrites: petrogenesis of protohypersthene basalts in the maria lava lakes. *Proc. Second Lunar Sci. Conf.*, *Geochim. Cosmochim. Acta*, Suppl. 2, Vol. 1, pp. 617–643. MIT Press.

Boettcher A. L. (1971) The nature of the crust of the earth, with special emphasis on the role of plagioclase, in "The Structure and Properties of the Earth's Crust," J. C. Heacock, editor, *Amer. Geophys. Un. Mono.* **14**, pp. 264–278.

Dainty A. M. and Anderson K. R. (1972) Seismic scattering in the moon and the earth. Abstract submitted to the 53rd Annual Meeting, American Geophysical Union.

Gast P. W. and McConnell R. H. Jr. (1972) Evidence for initial chemical layering of the moon (abstract). In *Lunar Science—III* (editor, C. Watkins), p. 289. Lunar Science Institute Contr. no. 88.

Green D. H., Ringwood A. E., Ware N. G., Hibberson W. O., Major A., and Kiss E. (1971) Experimental petrology and petrogenesis of Apollo 12 basalts. *Proc. Second Lunar Sci. Conf.*, *Geochim. Cosmochim. Acta*, Suppl. 2, Vol. 1, pp. 601–615. MIT Press.

Helmberger D. V. (1968) The crust-mantle transition in the Bering Sea. *Bull. Seism. Soc. Amer.* **58**, 179–214.

Helmberger D. V. and Wiggins R. A. (1971) Upper mantle structure of midwestern United States. *J. Geophys. Res.* **76**, 3229–3245.

Kanamori H., Nur A., Chung D. H., Wones D., and Simmons G. (1970) Elastic wave velocities of lunar samples at high pressures and their geophysical implications. *Science* **167**, 726–728.

Kanamori H., Mizutani H., and Hamano Y. (1971) Elastic wave velocity of Apollo 12 rocks at high pressures. *Proc. Second Lunar Sci. Conf., Geochim. Cosmochim. Acta*, Suppl. 2, Vol. 3, pp. 2323–2326. MIT Press.

Latham G., Ewing M., Press F., Sutton G., Dorman J., Nakamura Y., Toksöz N., Duennebier F., and Lammlein D. (1971) Passive seismic experiment. *Apollo 14 Preliminary Science Report*, NASA SP-272, pp. 133–161.

Latham G., Ewing M., Press F., Sutton G., Dorman J., Nakamura Y., Toksöz N., Lammlein D., and Duennebier F. (1972) Moonquakes and lunar tectonism—results from the Apollo Passive Seismic Experiment (abstract). In *Lunar Science—III* (editor, C. Watkins), p. 478. Lunar Science Institute Contr. no. 88.

LSPET (Lunar Sample Preliminary Examination Team) The Apollo 15 lunar samples: a preliminary description. *Science* **175**, 363–375.

Mizutani H., Fujii N., Hamano Y., Osako M., and Kanamori H. (1972) Elastic wave velocities and thermal diffusivities of Apollo 14 rocks (abstract). In *Lunar Science—III* (editor, C. Watkins), p. 547. Lunar Science Institute Contr. no. 88.

Nakamura Y., Latham G. V., Ewing M., and Dorman J. (1970) Lunar seismic energy transmission. *EOS* **51**, 776.

Nur A. and Simmons G. (1969) The effect of saturation on velocity in low porosity rocks. *Earth Planet. Sci. Lett.* **7**, 183–193.

O'Hara M. J., Bigger G. M., Richardson S. W., Jamieson B. G., and Ford C. E. (1970) The nature of seas, mascons, and the lunar interior in the light of experimental studies. *Proc. Apollo 11 Lunar Sci. Conf., Geochim. Cosmochim. Acta*, Suppl. 1, Vol. 1, pp. 695–710. Pergamon.

Press F. (1966) Seismic velocities. *Handbook of Physical Constants*, S. P. Clark, editor, *Geol. Soc. Amer. Mem.* **97**, 195.

Reid A. M., Ridley W. I., Warner J., Harmon R. S., Brett R., Jakes P., and Brown R. W. (1972) Chemistry of Highland and Mare basalts as inferred from glasses in the lunar soils (abstract). In *Lunar Science—III* (editor, C. Watkins), p. 640. Lunar Science Institute Contr. no. 88.

Ringwood A. E. and Essene E. J. (1970) Petrogenesis of Apollo 11 basalts, internal constitution and origin of the moon. *Proc. Apollo 11 Lunar Sci. Conf., Geochim. Cosmochim. Acta*, Suppl. 1, Vol. 1, pp. 769–799. Pergamon.

Sjorgen W. L., Gottlieb P., Muller P. M., and Wollenhaupt W. (1972) Lunar gravity data via Apollo 14 Doppler radio tracking. *Science* **175**, 165–168.

Smith J. V., Anderson A. T., Newton R. C., Olsen E. J., Wyllie P. J., Crewe A. V., Isaacson M. S., and Johnson D. (1970) Petrologic history of the moon inferred from petrography, mineralogy, and petrogenesis of Apollo 11 rocks. *Proc. Apollo 11 Lunar Sci. Conf., Geochim. Cosmochim. Acta*, Suppl. 1, Vol. 1, pp. 897–925. Pergamon.

Tittmann B. R., Abdel-Gawad M., and Housley R. M. (1972) Rayleigh wave studies of lunar and synthetic rocks (abstract). In *Lunar Science—III* (editor, C. Watkins), p. 755. Lunar Science Institute Contr. no. 88.

Todd T., Wang H., Baldridge W. S., and Simmons G. (1972) Elastic properties of Apollo 14 and 15 rocks, this volume.

Toksöz M. N., Press F., Anderson K., Dainty A., Latham G., Ewing M., Dorman J., Lammlein D., Nakamura Y., Sutton G., and Duennebier F. (1972a) Velocity structure and properties of the lunar crust. *The Moon*, in press.

Toksöz M. N., Press F., Anderson K., Dainty A., Latham G., Ewing M., Dorman J., Lammlein D., Sutton G., Duennebier F., and Nakamura Y. (1972b) Lunar crust: structure and composition. Submitted to *Science*.

Walker D., Longhi J., and Hays J. F. (1972) Experimental petrology and origin of Fra Mauro rocks and soil (abstract). In *Lunar Science—III* (editor, C. Watkins), p. 770. Lunar Science Institute Contr. no. 88.

Wang H., Todd T., Weidner D., and Simmons G. (1971) Elastic properties of Apollo 12 rocks. *Proc. Second Lunar Sci. Conf., Geochim. Cosmochim. Acta*, Suppl. 2, Vol. 3, pp. 2327–2336. MIT Press.

Warren H., Schreiber E., Scholz C., Morrison J. A., Norton P. R., Kumazala M., and Anderson O. L. (1971) Elastic and thermal properties of Apollo 12 and Apollo 11 rocks. *Proc. Second Lunar Sci. Conf., Geochim. Cosmochim. Acta*, Suppl. 2, Vol. 3, pp. 2345–2360. MIT Press.

Wood J. A. (1970) Petrology of the lunar soil and geophysical implications. *J. Geophys. Res.* **75,** 6497–6513.

Wood J. A., Marvin U. B., Reid J. B., Jr., Taylor G. J., Bower J. F., Powell B. N., and Dickey J. S., Jr. (1971) Mineralogy and petrology of the Apollo 12 lunar sample. *Smithsonian Astrophysical Obs. Spec. Rep.* **333,** 272 pp.

Note Added in Proof

Seismic signals from the Apollo 16 S-IV B impact were recently recorded at Apollo Lunar Stations 12, 14, and 15. The time and location of this impact were uncertain due to a loss of tracking capability. These parameters can be estimated, however, from the arrival times of P and S waves at the Apollo 12 and 14 stations. The first detectable motion at the Apollo 15 seismometer (at a distance of 1095 kilometers) indicates an average lunar mantle velocity of about 8 km/sec near a depth of 130 kilometers. Whether the high velocity (9 km/sec) zone in the uppermost portion of the mantle, reported above, is a universal feature or not cannot be determined from the new data.

The high frequencies (0.5–3.0 Hz) of compressional and shear waves observed at station 15 from the latest impact and also from deep moonquakes indicate the absence of high attenuation, precluding widespread melting in the outer several hundred kilometers of the lunar mantle.

Proceedings of the Third Lunar Science Conference
(Supplement 3, *Geochimica et Cosmochimica Acta*)
Vol. 3, pp. 2545–2555
The M.I.T. Press, 1972

Measurements of the acoustical parameters of rock powders and the Gold-Soter lunar model

Barrie W. Jones

Center for Radiophysics and Space Research, Space Science Building,
Cornell University, Ithaca, N.Y. 14850

Abstract—We are measuring the low frequency velocities and damping factors of acoustic waves in rock powders in vacuo with a view to understanding the lunar seismic signals. Our results combined with those of others, when extrapolated to lunar conditions, lend support to the Gold-Soter model of the outer moon in which the lunar surface is covered with a fine rock powder to a depth of at least 4 km.

Introduction

We are determining the low frequency velocities and damping factors of acoustic waves in rock powders in vacuo with a view to understanding the lunar seismic signals. These signals differ markedly from those seen in the earth: They have first arrivals approximately three times later; they display no clear P-wave S-wave structure; and they have rise and fall times about 1000 times greater.

Many features of the lunar seismic signals can be understood in terms of the Gold-Soter model (Gold and Soter, 1970; Watkins, 1971) in which the outer layer of the moon to a depth of at least 4 km is largely composed of rock powder, the compaction of which increases in some manner with depth. The long rise and fall times of the lunar seismic signals would result from the diffusion of acoustical energy through the powder. The scattering mechanism needed to turn wave propagation into diffusion would arise from inhomogeneities such as undulations in the lunar surface and powder density fluctuations on a scale of a wavelength or more. The energy would be confined to such a powder layer by what is essentially total internal reflection, the increasing compaction with depth providing the necessary increase in velocity with depth.

This diffusion process results in long path lengths that can only result in the observed large signals at long delays if the acoustical damping factor η ($= 1/2Q$) is small. This factor thus emerges as the crucial rock powder parameter. The value of Q necessary to account for the observed lunar seismic signals depends, in diffusive models, on the velocities assumed at various depths. In many cases the lower the velocities the lower need be Q. Therefore, the low velocities typical of rock powders ease the requirement that Q be high. It would seem that to satisfactorily account for the lunar seismic signals Q need not exceed 2000, and it could be less.

Previous work

Measurements on coarse sand by Hunter *et al.* (1961) at frequencies above 7 kHz gave values of $Q \sim 5$. We did not consider this a discouraging result because most of the energy in the lunar seismic signals is near 1 Hz, at which frequency, for a given wave amplitude, the particle strain is about 10^4 times less than that at 10^4 Hz, which could result in much lower damping at the lower frequency. At a given particle strain we also expect lower damping in smaller grains, provided that the thickness of the deformable grain-grain contact is less than proportional to grain diameter. In fact Hunter *et al.* detected a rise in Q from 4 to 6 when the mean grain diameter was decreased from 0.86 mm to 0.26 mm, which was encouraging, because lunar surfaces fines have a mean volume corresponding to a grain diameter of only about 0.01 mm.

Another result on sand by Schmidt (1954) but at 20 Hz, also gave $Q \sim 5$ but in this case adequate precautions against container damping had not been taken. None of this earlier work had been done in vacuo and no convincing attempts to dry the powders had been made. Also, the importance of working at small particle strain does not seem to have been appreciated.

We are therefore measuring the Q of small signals in vacuo and as close to 1 Hz as is convenient, which at present is at about 10 Hz in well settled powder. The lowest limit is imposed upon us by the size of the laboratory and is about 6 Hz.

METHOD

We set up longitudinal standing waves in a horizontal trough of rock powder in vacuo and observe the free decay. Figure 1 illustrates the instrumentation system used. We chose to observe the free decay rather than measure resonant peak widths because such widths can be increased by spurious effects (Jeffreys, 1970). In order to reduce

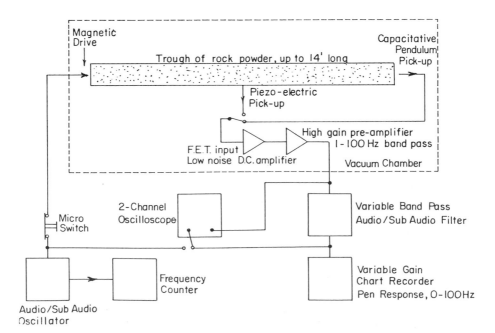

Fig. 1. Instrumentation system for observing Q of rock powders in vacuo.

support system effects, the trough was made very compliant horizontally, is of low mass, typically 1% of the rock powder, and is very lightly supported by fine wires on low loss supports. Any residual support system effects can only decrease Q and therefor any result we obtain may be a lower limit on the Q of the powder. From measurements performed on the system, we estimate that losses in the support system probably mask rock powder Q values in vacuo when these are much greater than 1000. To test for support system effects, we have worked with two different masses of powder in the trough and the results were essentially the same in each case. We are therefore not yet being limited by damping in the support system.

Material used

We are using powdered basalt in which 90% of the particle volume consists of particles less than 10 μ diameter, with a mean volume corresponding to a particle diameter of 5 μ. This is fairly close to the size distribution of the lunar fines recovered from the Apollo sites and as determined by Gold *et al.* (1972). The density of the solid alone after powdering was a little in excess of 2.90 gm/cm^3 which is close to the 2.99 gm/cm^3 estimated for the lunar fines by Anderson *et al.* (1970). The bulk physical properties of our powdered basalt correspond well enough to those of the lunar fines.

RESULTS

Stable Q values

The powder is settled by mechanical vibration and specifically by not applying external pressure, and some, but not all, of the fluids initially present are removed by exposing the powder to low ambient air pressures. At the end of such ageing processes, the density ρ, longitudinal wave velocity c_l, and Q have all risen from lower starting values to some fairly stable final values. In this steady condition the powder has Q values at various ambient air pressures as shown in Fig. 2. The Q rises with falling pressure until 0.1 Torr (0.1 mm Hg), below which Q becomes less sensitive to pressure and assymptotes to a value of about 120 which is roughly double the value at 760 Torr. Below 0.1 Torr it is likely that water and not air is the limiting fluid. Note that on Mars the surface atmospheric pressure is likely to significantly lower the Q of any surface rock powder layers that may be there.

Q: Insensitivity to amplitude because of moisture

Figure 3 shows a free decay in the aged powder at 0.07 Torr. The most interesting feature is the goodness of fit of a single exponential, which implies that the value of Q of 110 $\pm$ 5 holds over at least a factor of 5 in amplitude, from 0.05 μ down to at least 0.01 μ and possibly less. This is part of a more general result which indicates that between 100 μ and 10 μ amplitude there is a factor of 2 increase in Q but that at smaller amplitudes there is negligible further increase. The trough is 2.32 m long and therefore particles are strained by a 10 μ amplitude standing wave so that at the node $\Delta l/l \sim 10^{-5}$, and less elsewhere.

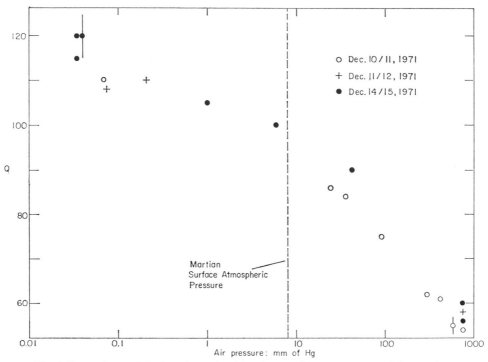

Fig. 2. Dependence of Q of aged powdered basalt on ambient air pressure. All the points were obtained with driven wave amplitudes near 0.05 μ and the fundamental mode frequency was in the range 10.0–10.4 Hz in all cases.

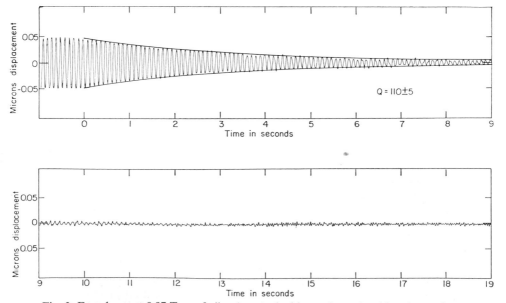

Fig. 3. Free decay at 0.07 Torr of vibrations excited in aged powdered basalt at a fundamental mode frequency of 10.3 Hz. The signal has been passed through a 5–20 Hz filter with Gaussian response at upper and lower band edges.

It is important to stress that this amplitude insensitivity tells us very little about dry grain-grain contacts because we certainly have present the remnants of lubricating films especially water, which, if they are limiting Q, as seems likely, and if they behave as simple viscous liquids, *will* result in the observed amplitude insensitivity of Q. However, a difficulty with a simple viscous model for loss is that such a model predicts a strong dependence of loss upon frequency, which, as we shall see, is not so. Nevertheless, complete removal of such films could substantially raise Q. (*Note:* A recent result of 230 for Q was obtained after heating the powder in vacuo. This rise in Q from the value of about 120 obtained for unheated powders suggests that trapped fluids were limiting Q to 120.)

Stable c_l values

The longitudinal wave propagation velocity c_l, in the aged powder is 47.3 ± 0.5 m/s and is insensitive to ambient air pressure and wave amplitude. This result is in excellent agreement with the 45 m/s obtained from Surveyor data (Latham, *et al.* 1970) for the top few *centimeters* of in situ lunar soil. The larger values of about 100 m/s from thumper experiments and other sources (Kovach and Watkins, 1972) are pertinent to the top few *meters* of soil, and because c_l for powders certainly increases with depth (Toksöz *et al.* 1972) these larger values are not in conflict with either our value or that from the Surveyor data.

Stable ρ values

The stable density of our settled powder is 1.34 ± 0.01 gm/cm^3. This value is significantly less than the 1.7–1.9 gm/cm^3 inferred for the lunar surface from Apollo 11 and 12 core samples (Carrier III *et al.*, 1971), but it is certain that these Apollo results seriously overestimate the density of the top few centimeters of the in situ lunar surface fines. The more reliable Apollo 15 value of 1.36 gm/cm^3 (Carrier III *et al.*, 1972) for the upper few cms is close to our laboratory powder density and it is not far from the value of 1.20 gm/cm^3 derived from Luna-16 data (Vinogradov, 1971) or the value of 1.5 g/cm^3 derived from Surveyor data (Carrier III *et al.*, 1972).

The porosity of our settled basalt powder is about 60%, which is at the upper end of the range derived for the moon from mechanical data (Houston *et al.*, 1972). Such a porosity would give a radio frequency dielectric constant in the range 2.4–2.8 and this is in good agreement with values of the lunar surface dielectric constant derived from radar measurements (Evans and Hagfors, 1968).

We conclude that our aged powder has a density and porosity that is at least representative of the lunar topsoil.

Q and c_l versus frequency

The aged powder showed no change in Q over the first 3 modes (10–30 Hz approx.) and no change in c_l over at least the first 4 modes (10–40 Hz approx.). We conclude that none of the values we have *yet* obtained for c_l or Q would have been much different at 1 Hz, which is the frequency at which most of the energy in the observed lunar seismic signal is concentrated. If anything, Q would be greater at 1 Hz.

Ageing processes without external pressure

In the absence of externally applied pressures, there are at least two distinct mechanisms that could alter ρ, c_l, and Q of a rock powder: (1) mechanical settling, and (2) removal of fluids.

Table 1 summarizes our present observations on ageing, some of which we think we can relate to these two mechanisms.

The settling of the powder by mechanical vibration displaces lubricating films and generates firmer grain-grain contacts. This will certainly result in stiffer grain-grain "springs" and this mechanism can therefore explain the initial rise of c_l with ρ. An appreciable increase in Q was not seen until the powder had been exposed to low ambient pressures. Therefore, the mere displacement of fluid films does not have much effect on Q, at least initially.

Thereafter the situation becomes confused: The general increase in Q is not matched by a general increase in c_l in which there are either small or no changes. We do not have a complete set of density measurements so we do not know if it is constant or if mechanical settling or if mechanical fluffing is occurring here. One thing *is* fairly clear, and that is that the longer the history of exposure to low ambient air pressures the greater is Q until values of ~ 60 at 760 Torr are reached. Moisture would be expected to attain an equilibrium layering of the dry grains in a minute or so at most and therefore an initial equilibrium layer would not give rise to the long-term improvement in Q that we observe over many cycles between atmospheric and low pressures.

In the *limiting* condition we reach, the reproducible Q versus pressure curve is undoubtedly associated with a combination of the removal of air and the partial removal of the 10–20 monolayers of adsorbed water that are known to accumulate

Table 1. Effect of ageing (settling and partial fluid film removal) on acoustic wave propagation in powdered basalt.

Date (1971)	Pressure (Torr)	Density (gm/cm^3)	Longitudinal wave velocity (m/s)	Q
Sept. 10	~ 760	0.88 ± 0.10	28.1 ± 0.6	10 ± 2
Sept. 22	~ 760	1.27 ± 0.05	35.2 ± 0.6	10 ± 2
Oct. 7	~ 760	—	39.1 ± 0.6	10 ± 2
Powder pumped down to 0.07 Torr and held for 1 day.				
Oct. 9	~ 760	—	43.7 ± 0.4	20 ± 1
Powder pumped down to 0.03 Torr and held for several days.				
Dec. 8	~ 760	—	42.0 ± 0.4	26 ± 1
Powder pumped down to 0.07 Torr and held for 2 days.				
Dec. 10 and later	~ 760	1.34 ± 0.01	47.3 ± 0.5	56 ± 2
First pump down.				
Oct. 10	0.09	—	42.3 ± 0.4	30 ± 2
Oct. 11	0.06	—	41.8 ± 0.4	40 ± 2
Second pump down.				
Nov. 12	0.03	—	41.6 ± 0.4	50 ± 2
Third and subsequent pumps down.				
Dec. 8 and later	0.03 −0.05	1.34 ± 0.01	47.3 ± 0.5	118 ± 5

on to silicate surfaces and which would be quickly readsorbed on exposure to normal damp air. Removal of such films from crevice corners in cracked rocks has been shown to significantly raise Q (Tittman et al., 1972). When such fluids are completely removed, we may expect that the rock powder will then be in the dry state it would be in were it on the moon, and that Q should be significantly higher than the value of 120 that we presently observe in vitro.

The value of $50 \lesssim Q \lesssim 100$ obtained by Kovach and Watkins (1972) for the in situ lunar surface is not necessarily applicable to low frequency low amplitude seismic vibrations. Even if their result were to prove to be true in such conditions then it does not embarrass the deep powder model which can tolerate a modestly lossy surface provided that Q increases with depth, which as we shall see it is expected to do.

We have been unable so far to unambiguously associate any change in c_l with fluid content rather than with mechanical settling. Any effect on c_l is therefore slight at the high moisture levels we are undoubtedly still working at.

The effect of external pressure

So far we have been concerned only with ρ, c_l, and Q in a mechanically settled powder. But if there is a deep powder layer on the moon, then most of it is subjected to considerable external pressure from the overburden, and in the past it would have been subjected to periods of higher pressures arising from impact events.

When external pressure is applied to a rock powder, then a rapid increase in c_l has been observed (Anderson et al., 1970). Thus at pressures that would be achieved by a lunar overburden at a depth of about 6 km, the velocity in lunar fines is 3000 m/s, and the density is about 2.5 gm/cm^3. When this pressure is released, the density only falls to 2.4 gm/cm^3 and the velocity to 1800 m/s. This hysteresis is indicative of the importance of the impact history and we might expect to find inhomegeneities of this sort in the moon that could scatter seismic waves.

Figure 4 illustrates an apparent inconsistency between laboratory results on c_l for Apollo fines, in which pressure has been converted into lunar overburden, and the velocity profile derived from lunar seismic data. Toksöz et al. (1972) believe that this apparent inconsistency rules out an outer rock powder layer on the moon greater than a kilometer or so deep. We believe that this inconsistency is only an apparent one and not a real one for three reasons.

First, and of least importance in this context, the laboratory work on pressure wave velocity in Apollo fines versus hydrostatic pressure was done at frequencies many decades higher than the 1 Hz appropriate to the lunar seismic signals. Second, and more important, we think that the results of Anderson et al. (1970) indicate that if, in addition to overburden pressure, a region of the moon has a history of impacts, then the memory of such impacts will be preserved in the form of higher wave velocities than would be expected from consideration of the overburden pressure alone. Third, and more important, we have many millions of years available on the moon in which time cold sintering could take place which would strongly bond the grains and thus raise c_l. This effect would be greater at greater depths and would be greater still if impact events had ever appreciably warmed the powder. Such cold sintering would

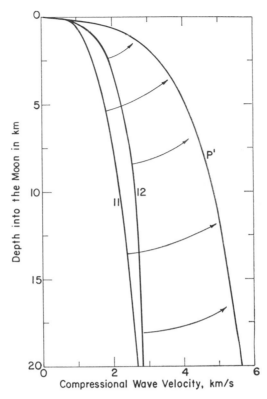

Fig. 4. A possible lunar seismic wave velocity profile P' versus depth compared with laboratory results on granular media. The arrows show the way in which the laboratory results will move when the effects of impacts, dryness, and cold sintering are added to overburden pressure.

11 = Apollo 11 fines, inferred from laboratory pressure effects.
12 = Apollo 12 fines, inferred from laboratory pressure effects.

still leave the rock powder essentially as a granular medium and not as a micro-fractured rock.

Cold sintering in nonmetals is not observed in the laboratory at temperatures much below one-third of the melting temperature (White, 1965) at least at the pressures normally expected within ten or so kilometers of the lunar surface, and the outer moon is probably nowhere always hotter than one-third of the melting temperature of silicates, though impact events will have caused local temporary heating. However, sintering is controlled by activation energies and therefore given sufficient time, bonding events between grains will take place at almost any temperature and especially deeper in the moon where external pressure forces the grains into large areas of contact. Some sintering mechanics have been discussed by Lenel and Ansell (1963), by Rossi and Fulrath (1965), and by White (1965).

It is clear that at this time insufficient data exist to enable us to speak quantitatively, but qualitatively we do expect the seismic wave velocities in subsurface in situ lunar rock powders to be considerably greater at a given depth than those values predicted from laboratory pressure measurements by simply translating such pressures into an equivalent lunar overburden. Therefore, there is not necessarily an inconsistency here between the acoustical properties of rock powders and the lunar seismic data.

Cold sintering can also raise Q for a given small wave amplitude because ultimately the grain-grain contact is so firm that the displacements are elastic. Until this elastic region is reached, then the Q of the rock powders should always decrease as amplitude falls unless these grain-grain bond losses are masked by losses in contaminating fluids. Ultimately we must be limited by the sort of mechanisms that operate in continuous solids, in which Q can attain values of several thousands.

SUMMARY

Our experimental results are summarized in Table 2.

The best value of Q obtained so far, 120 in as yet imperfect conditions, gives support to the lunar model in which the observed seismic signals result from propagation through variously compacted layers of rock powder. We think it possible that when fluids are completely removed from the powder then values of Q in excess of 1000 can occur, especially when the powder has been precompacted by external pressure. We should then observe a strong increase in Q with falling wave amplitude.

The value of c_l we observe is in good agreement with that derived from Surveyor data for the top few cm of lunar soil. The corresponding laboratory powder density is 1.34 ± 0.01 gm/cm^3 and the porosity is about 60%.

The strong pressure dependence of c_l in a rock powder, the low c_l values and the

Table 2. Summary table of results on acoustic wave propagation in powdered basalt

Parameter	Effect on Q	Effect on c
Powder support system	Negligible at intrinsic powder Q values below 1000	Negligible
Ageing: settling and *partial* removal of fluid films	6× increase, at least	50% increase at least
Frequency	<5% change over first 3 modes at least	<1% change over first 4 modes at least
Amplitude	2× increase between 100 μ and 10 μ, thereafter <5% down to at most 0.01 μ	Negligible
Ambient air pressure	2× increase between 760 torr and 0.1 torr, thereafter <5% increase down to 0.03 torr	Negligible

Result at low amplitudes on aged powder at <0.1 torr pressure:
       Density = 1.34 ± 0.01 gm/cm^3 (cf., 1.1–1.9 gm/cm^3 for the lunar surface).

Longitudinal wave velocity = 47.3 ± 0.5 m/s (cf., 45 m/s for the surface layer at Surveyor sites).
       $Q = 118 \pm 5$

likely occurrence of high Q and of surface and density inhomogeneities would result in containment of surface induced waves to the moon's upper few kilometers and the observed large signals at long delay.

The results and arguments presented here are considered to leave the deep dust lunar model both alive and well.

Acknowledgments—I would like to thank Professor T. Gold for introducing me to this problem and for subsequent practical suggestions. Dr. M. Campbell has provided useful discussion and Mr. J. Smith has provided extensive technical assistance for which I am very grateful. This work was supported by NASA grant number NGL-33-010-005.

References

Anderson O. L., Scholz C., Soga N., Warren N., and Schreiber E. (1970) Elastic properties of microbreccia, igneous rock, and lunar fines from Apollo 11 mission. *Proc. Apollo 11 Lunar Sci. Conf.*, *Geochim. Cosmochim. Acta* Suppl. 1, Vol. 3, pp. 1959–1973. Pergamon.

Carrier W. D. III, Johnson S. W., Werner R. A., and Schmidt R. (1971) Disturbances in samples recovered with the Apollo core tubes. *Proc. Second Lunar Sci. Conf.*, *Geochim. Cosmochim. Acta* Suppl. 2, Vol. 3, pp. 1959–1972. MIT Press.

Carrier W. D. III, Johnson S. W., Carrasco H. L., and Schmidt R. (1972) Core sample depth relationships: Apollo 14 and 15 (abstract). In *Lunar Science—III* (editor C. Watkins), pp. 122–124, Lunar Science Institute Contr. No. 88.

Evans J. T. and Hagfors T. (1968) *Radar Astronomy*, p. 264. McGraw-Hill.

Gold T. and Soter S. (1970) Apollo 12 seismic signal: Indications of a deep layer of powder. *Science* **169**, 1071–1075.

Gold T., Bilson E., and Yerbury M. (1972) Grain size analysis, optical reflectivity measurements, and determination of high frequency electrical properties for Apollo 14 lunar samples (abstract). In *Lunar Science—III* (editor C. Watkins), pp. 318–320, Lunar Science Institute Contr. No. 88.

Houston W. N., Hovland H. J., Mitchell J. K., and Namiq L. I. (1972) Lunar soil porosity and its variation as estimated from footprints and boulder tracks (abstract). In *Lunar Science—III* (editor C. Watkins), pp. 395–397, Lunar Science Institute Contr. No. 88.

Hunter A. N., Legge R., and Matsakawa E. (1961) Measurements of acoustic attentuation and velocity in sand. *Acustica* **11**, 26–31.

Jeffreys H. (1970) *The Earth*, p. 334. C.U.P.

Kovach R. L. and Watkins J. S. (1972) The near-surface velocity structure of the moon (abstract). In *Lunar Science—III* (editor C. Watkins), pp. 461–462, Lunar Science Institute Contr. No. 88.

Latham G., Ewing M., Dorman J., Press F., Toksöz N., Sutton G., Meissner R., Duennebier F., Nakamura Y., Kovach R., and Yates M. (1970) Seismic data from man-made impacts on the moon. *Science* **170**, 620–626.

Lenel F. V. and Ansell G. S. (1963) Metals, ceramics, polymers, interdisciplinary aspects of sintering, and plastic deformation. *Ind. Eng. Chem.* **55**, 46–50.

Rossi R. C. and Fulrath R. M. (1965) Final stage densification in vacuum hot pressing of alumina. *J. Amer. Cer. Soc.* **48**, 558–564.

Schmidt H. (1954) Die Schallausbreitung in Körnigen Substanzen. *Acustica* **4**, 639–652.

Tittman B., Abdel-Gawad M., and Housley R. M. (1972) Rayleigh wave studies of lunar and synthetic rocks (abstract). In *Lunar Science—III* (editor C. Watkins), pp. 755–757, Lunar Science Institute Contr. No. 88.

Toksöz M. N., Press F., Anderson K., Dainty A., Latham G., Ewing M., Dorman J., Lammlein D., Sutton G., Duennebier F., and Nakamura Y. (1972) Velocity structure and properties of the lunar crust (abstract). In *Lunar Science—III* (editor C. Watkins), pp. 758–760, Lunar Science Institute Contr. No. 88.

Vinogradov A. P. (1971) Preliminary data on lunar ground brought to earth by automatic probe Luna 16. *Proc. Second Lunar Sci. Conf., Geochim. Cosmochim. Acta* Suppl. 2, Vol. 1, pp. 1–16. MIT Press.

Watkins J. S. (1971) Seismic velocity models of the lunar near surface and their implications. *J.G.R.* **76,** 6246–6252.

White J. (1965) Sintering—An assessment. *Brit. Cer. Soc. Proc.* **3,** 155–176.

Proceedings of the Third Lunar Science Conference
(Supplement 3, *Geochimica et Cosmochimica Acta*)
Vol. 3, pp. 2557–2564
The M.I.T. Press, 1972

Elastic wave velocities and thermal diffusivities of Apollo 14 rocks*

Hitoshi Mizutani†, Naoyuki Fujii, Yozo Hamano, and Masahiro Osako

Geophysical Institute, The University of Tokyo,
Tokyo, Japan

Abstract—The compressional- and shear-wave velocities of Apollo 14 lunar rocks 14311,50 and 14313,27 as functions of pressure up to 10 kb and the thermal diffusivity of sample 14311,50 over the temperature range 100 to 550°K have been measured. Both samples 14311 and 14313 are polymict fragmental rocks. The overall elastic and anelastic behavior of the Apollo 14 samples are similar to those of Apollo 11 and 12 samples; low velocity and low Q at pressures below 1 kb and rapid increase of velocity and Q with pressure are also typical of the Apollo 14 rocks. The available data of P- and S-wave velocities of lunar rocks show that Birch's law ($v = a + b\rho$) holds for the lunar rocks. The thermal diffusivity of a lunar rock in vacuum is found to be significantly lower than that in air at one atmospheric pressure. The lower values in vacuum than those in one atmosphere air may be explained by the absence of thermal conduction of gases filled in pores and microcracks of the lunar samples in vacuum. The low thermal conductivity due to pores and microcracks may persist to depths up to 100 km and play an important role in the thermal evolution of the moon.

Introduction

This paper reports new experimental results of elastic properties and thermal properties of Apollo 14 rocks. First are described the compressional- and shear-wave velocity measurements on sample 14311,50 and 14313,27 under pressures up to 10 kb at room temperature. Second, we describe the thermal diffusivity measurement on sample 14311,50 over the temperature range 100 to 550°K in vacuum and at one atmosphere.

Both samples 14311 and 14313 are polymict fragmental rocks (LSPET, 1971a). The chemical composition of sample 14311 has been provided by Scoon (1972), from which the normative mineral composition, mean atomic weight, and theoretical density of sample 14311 are calculated (Table 1). Chemical analysis data on sample 14313 is not available at the present stage. According to the mineralogic description by LSPET (1971a), the chemical composition of sample 14313 may not be very different from that of sample 14311. Therefore the mean atomic weight and theoretical density of sample 14313 were tentatively assumed to be equal to those of sample 14311. The bulk densities were measured by Archimedes method; the densities of sample 14311 and 14313 are 2.86 g/cm³ and 2.39 g/cm³, respectively. Porosities are estimated at 6% for sample 14311 and 21% for sample 14313, using the bulk densities and the theoretical densities. The mean atomic weight and the theoretical density of the Apollo 14 rock are low compared with those of Apollo 11 and Apollo 12 rocks,

---

* Contribution No. 2139, Division of Geological and Planetary Sciences, California Institute of Technology, Pasadena, California 91109.

† Now at Seismological Laboratory, California Institute of Technology, Pasadena, California.

Table 1. Normative mineral composition, theoretical density, and mean atomic weight of sample 14311 (Scoon, 1972).

| | Norms | |
	(Weight %)	(Mole %)
Or	4.8	3.0
Ab	7.2	4.8
An	42.8	27.1
En	17.5	30.8
Fs	12.2	16.3
Wo	3.1	4.6
Fo	4.3	5.4
Fa	3.3	2.9
Ilm	3.4	4.0
Chr	1.4	1.1
Qtz	0.0	0.0

m (mean atomic weight) = 21.9 g
ρ_{ideal} = 3.05 g/cm³

reflecting the lower concentrations of Fe and Ti in Apollo 14 rocks than those in Apollo 11 and Apollo 12 rocks (LSPET, 1971b).

Ultrasonic Wave Velocities

Ultrasonic wave velocities were measured by the method described by Mizutani *et al.* (1970), using the high pressure system described by Kanamori and Mizutani (1965). Samples were jacketed by rubber tubing. 5 MHz lead zirconate transducers were used for both *P*- and *S*-wave measurements. The accuracy of the present method is better than 0.2% for both *P*- and *S*-waves at pressures above 1 kb if the dimensional change accompanied by pressure increase is properly corrected. At pressures below 200 b, however, the wave transmission efficiency is so poor (low *Q*) that the onset of the signal becomes blunt and the accuracy drops considerably, probably to a few percent.

The results are shown in Figs. 1 and 2 where the original readings are given. Because the samples have large compressibilities, the correction for the pressure shortening of the sample was made using the following equation by Cook (1975):

$$\frac{\alpha(P)}{\alpha'(P)} = \frac{\beta(P)}{\beta'(P)} = \left[1 + \frac{1}{3\rho_0} \int_0^P \frac{dP}{(\alpha'(P)^2 - \frac{4}{3}\beta'(P)^2)} \right]^{-1} \tag{1}$$

where $\alpha(P)$ and $\beta(P)$ are the corrected *P*- and *S*-wave velocities at pressure *P* and $\alpha'(P)$ and $\beta'(P)$ are the uncorrected *P*- and *S*-wave velocities. In Table 2 are listed the velocities and densities corrected by equation (1). The correction above, however, does not take account of the obvious discrepancy between static compressibility and ultrasonic compressibility. In general, static compressibility is larger for a porous sample than that obtained by ultrasonic measurement, because the former involves the volume change accompanied by closure of pores and microcracks. Since the present correction should be made using the static compressibility, the above method gives a minimum correction. If we use, for instance, the static compression data on

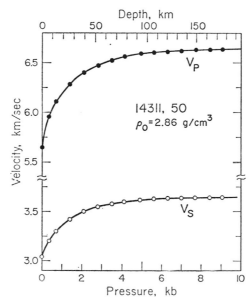

Fig. 1. The *P*- and *S*-wave velocities of sample 14311,50 as a function of pressure.

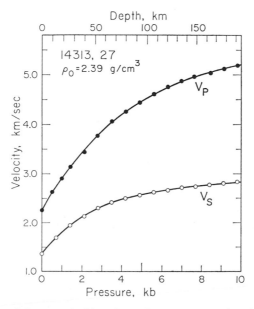

Fig. 2. The *P*- and *S*-wave velocities of sample 14313,27 as a function of pressure.

Table 2. Smoothed values of velocity and density corrected for dimensional change.

		Pressure (kb)							
		0.0	0.5	1.0	2.0	3.0	5.0	7.0	10.0
14311	$V_p{}^*$	5.65	6.02	6.18	6.36	6.47	6.56	6.59	6.62
	$V_s{}^*$	3.04	3.24	3.35	3.48	3.54	3.60	3.63	3.65
	ρ†	2.86	2.86	2.86	2.87	2.87	2.88	2.89	2.90
14313	V_p	2.25	2.61	2.88	3.39	3.82	4.41	4.81	5.16
	V_s	1.35	1.60	1.80	2.09	2.32	2.55	2.68	2.79
	ρ	2.39	2.41	2.42	2.44	2.46	2.49	2.50	2.52

$^*V_p$, V_s in km/sec; † ρ in g/cm^3.

an Apollo 11 crystalline rock (porosity $\sim 5\%$) by Stephens and Lilley (1970), corrections of about -1.7% and $+5.0\%$ at 10 kb should be made to the velocities and densities respectively, whereas, the corresponding corrections based on ultrasonic compressibility are -0.47% and $+1.4\%$ for sample 14311,50 and -1.8% and $+5.5\%$ for sample 14313,27.

The rapid increase of the velocity and the amplitude (Q) of the signal for the initial few kilobars increase found for Apollo 11 and 12 rocks (Kanamori et al., 1970; Anderson et al., 1970; Kanamori et al., 1971; Wang et al., 1971) is also typical of the Apollo 14 rocks. The velocity of sample 14311,50 begins to show intrinsic pressure dependence at 4 kb, while the velocity of sample 14313,27 continues to increase up to 10 kb at a larger rate than that expected from the intrinsic pressure dependence. This behavior of velocity versus pressure of the sample 14313,27 indicates that the sample contains numerous spherical pores which persist up to a fairly high pressure. The velocities of Apollo 14 rocks as well as Apollo 11 and 12 rocks are conformable with the seismic-velocity structure reported by Toksoz et al. (1972a).

In Fig. 3 are plotted the uncorrected velocities (at 5 kb) and densities (at 1 b) for all lunar samples, on which ultrasonic-wave velocity has been measured under pressures up to at least 5 kb (Kanamori et al., 1970; Kanamori et al., 1971; Wang et al., 1971). The linear dependence of both P- and S-wave velocities on density found for terrestrial rocks (Birch, 1961; Simmons, 1964b) is also valid for the lunar rocks. The velocity-density line for $m = 23.5$ given in Fig. 3, however, gives either slightly higher velocity or lower density than that expected from the relation for the terrestrial rocks with same mean atomic weights as those of lunar rocks. The deviation from the terrestrial relation is partly due to the high CaO content in the lunar rocks (for CaO effect on the sound velocity, see Simmons, 1964a) and partly due to the usage of the uncorrected velocity and density. Toksoz et al. (1972a) obtained very high P-velocity ~ 9 km/sec at a depth about 70 km. If chemical composition (mean atomic weight) at this depth is the same as that of the lunar surface material and the velocity-density systematics obtained here is applicable, the component at this depth must have density of 3.6 to 3.8 g/cm^3. The high density inferred here is in reasonably good agreement with the estimated density 3.75 g/cm^3 for a lunar eclogite by Ringwood and Essene (1970). Therefore it is possible that the high velocity layer $V_p \sim 9$ km/sec below 70 km depth is lunar eclogite, although Anderson and Kovach (1972) prefer a high pressure form of anorthosite (garnet + kyanite + quartz) to eclogite as a candidate

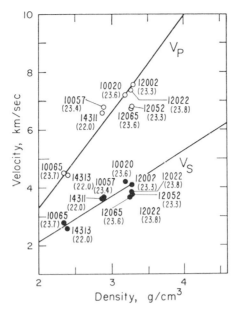

Fig. 3. *P*- and *S*-wave velocities at 5 kb versus density for lunar rocks. The numbers attached to points are sample numbers and the numbers in parentheses are mean atomic weights.

material of this layer. The high density, however, as pointed out by Ringwood and Essene and Anderson and Kovach, precludes such a high density and high velocity layer from being thicker than about 50 km. Thus if the high velocity layer obtained by Toksoz *et al.* (1972a) is interpreted as the eclogite layer, it indicates gravitationally unstable stratification in the moon.

THERMAL DIFFUSIVITY

The thermal diffusivity of the sample 14311 was measured over the temperature range 100 to 550°K by the modified Angstrom method (Kanamori *et al.*, 1969). The thermal diffusivity is determined from the phase lag and the amplitude decay with distance of a periodic temperature wave propagating through the sample. The accuracy of this method is estimated to be about $\pm 10\%$. Since heat transfer in porous media, including lunar rocks, is affected significantly by gases in the pores in the case where gas pressure is higher than 10^{-3} Torr (Wechsler and Glaser, 1965; Fujii and Osako, 1972), the measurement was made both in vacuum ranging from 10^{-3} to 10^{-5} Torr, and in air at one atmospheric pressure. The results are shown in Fig. 4 and in Table 3 which also lists the thermal conductivity calculated by using the data of heat capacity of an Apollo 14 sample 14321 (fragmental rock) by Hemingway and Robie (1972). The thermal diffusivity and the thermal conductivity in vacuum are substantially smaller than those measured in air at one atmosphere. The temperature dependence of the thermal diffusivity of the Apollo 14 sample in one atmospheric air

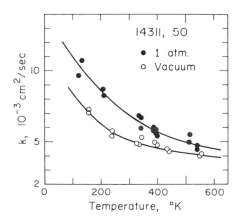

Fig. 4. Thermal diffusivity of sample 14311,50 as a function of temperature. Open circles represent the data in vacuum and solid circles the data in one atmospheric air.

is very similar to that of the Apollo 11 samples by Horai *et al.* (1970). The thermal conductivity measured in vacuum does not show such a large temperature dependence at temperatures 300 to 550°K as is observed for most terrestrial rocks, minerals, and for lunar rocks in air. Moreover, the thermal diffusivity and the thermal conductivity of the lunar rocks in vacuum are very low compared with those of terrestrial basalts and various rock-forming minerals over the temperature range covered in the present study. These characteristics of the thermal properties of the lunar rocks are probably due to the numerous pores and microcracks in the lunar samples and high content of plagioclase. The pore-free thermal conductivity of the sample 14311 is estimated to be $K = (5.7 \pm 0.1) \times 10^{-3}$ cal/cm sec °K at 300°K from the mineral composition (Table 1) and the thermal conductivity data of rock-forming minerals by Horai (1971), using Hashin and Strikman's (1962) formulas. Since pores and microcracks in the sample close as confining pressure increases, the thermal conductivity increases with pressure from the low value listed in Table 3 to the pore-free value estimated above as is the case with the ultrasonic wave velocity. If the pressure dependence of the thermal conductivity is similar to that of the ultrasonic-wave velocity, the pore-free intrinsic value will be attained at depth about 100 km. Therefore the thermal con-

Table 3. Thermal diffusivity and thermal conductivity of sample 14311,50.

	Temperature, °K									
	100	150	200	250	300	350	400	450	500	600
C_p^*	0.061	0.102	0.134	0.158	0.179	0.200	(0.211)	(0.220)	(0.228)	(0.239)
$k\dagger$ (air)	10.9	9.7	8.6	7.7	6.9	6.3	5.7	5.3	4.9	4.6
$K\ddagger$ (air)	1.90	2.83	3.30	3.48	3.53	3.60	3.43	3.33	3.20	3.14
k (vacuum)	8.4	7.2	6.3	5.6	5.2	5.0	4.6	4.4	4.2	4.0
K (vacuum)	1.47	2.10	2.53	2.61	2.75	2.77	2.77	2.73	2.73	2.73

* C_p in cal/g · °K; the numbers in parentheses are extrapolated values from the original data.
† The thermal diffusivities ($k = K/\rho C_p$) in 10^{-3} cm²/sec.
‡ The thermal conductivity in 10^{-3} cal/cm · sec · °K.

ductivity of the moon in the upper 100 km may be considerably lower than those used in previous thermal history calculations (MacDonald, 1959; Levin, 1962; McConnell *et al.*, 1967; Fricker *et al.*, 1967; Anderson and Phinney, 1967; Toksoz *et al.*, 1972b). The low thermal conductivity of the upper region of the moon may play an important role in the lunar thermal evolution.

Acknowledgments—We thank Professors Hitoshi Takeuchi and Syun-iti Akimoto for their constant encouragement. Professors Don L. Anderson, and Hiroo Kanamori, and Dr. Leon Thomsen kindly read the manuscript and offered valuable comments. This work was partially supported by National Aeronautics and Space Administration Contract NGL 05-002-069.

REFERENCES

Anderson D. L. and Phinney R. A. (1967) Early thermal history of the terrestrial planets. In *Mantle of the Earth and Terrestrial Planets* (editor S. K. Runcorn), pp. 113–126. Interscience.

Anderson D. L. and Kovach R. L. (1972) The lunar interior. Submitted to *Earth Planet. Sci. Letters.*

Anderson O. L., Scholz C., Soga N., Warren N., and Shreiber E. (1970) Elastic properties of a micro-breccia, igneous rock and lunar fines from Apollo 11 mission. *Proc. Apollo 11 Lunar Sci. Conf., Geochim. Cosmochim. Acta* Suppl. 1, Vol. 3, pp. 1959–1973. Pergamon.

Birch F. (1961) The velocity of compressional waves in rocks to 10 kilobars, Part 2. *J. Geophys. Res.* **66**, 2199–2244.

Cook, R. K. (1957) Variation of elastic and static strains with hydrostatic pressure; A method for calculation from ultrasonic measurements. *J. Acoust. Soc. Amer.* **29**, 445–449.

Fricker P. W., Reynolds R. T., and Summers A. L. (1967) On the thermal history of the moon. *J. Geophys. Res.* **72**, 2649–2663.

Fujii N. and Osako M. (1972) Thermal diffusivity of lunar rocks under atmospheric and vacuum condition (in preparation).

Hashin Z. and Shtrikman S. (1962) A variational approach to the theory of the effective magnetic permeability of multiple phase material. *J. Appl. Phys.* **33**, 3125–3131.

Hemingway B. S. and Robie R. A. (1972) The specific heats of Apollo 14 soil (14163) and breccia (14321) between 90 and 350°K. In *Lunar Science—III* (editor C. Watkins), p. 369, Lunar Science Institute Contri. No. 88.

Horai K., Simmons G., Kanamori H., and Wones D. (1970) Thermal diffusivity, conductivity and thermal inertia of Apollo 11 lunar material. *Proc. Apollo 11 Lunar Sci. Conf., Geochim. Cosmochim. Acta* Suppl. 1, Vol. 3, pp. 2243–2249. Pergamon.

Horai K. (1971) Thermal conductivity of rock-forming minerals. *J. Geophys. Res.* **76**, 1278–1308.

Kanamori H. and Mizutani H. (1965) Ultrasonic measurements of elastic constants of rocks under high pressures. *Bull. Earthquake Res. Inst. Tokyo Univ.* **43**, 173–194.

Kanamori H., Mizutani H., and Fujii N. (1969) Method of thermal diffusivity measurements. *J. Phys. Earth* **17**, 43–53.

Kanamori H., Nur A., Chung D., and Simmons G. (1970) Elastic wave velocities of lunar samples at high pressures and their geophysical implications. *Proc. Apollo 11 Lunar Sci. Conf., Geochim. Cosmochim. Acta* Suppl. 1, Vol. 3, pp. 2289–2293. Pergamon.

Kanamori H., Mizutani H., and Hamano Y. (1971) Elastic wave velocities of Apollo 12 rocks at high pressures. *Proc. Second Lunar Sci. Conf., Geochim. Cosmochim. Acta* Suppl. 2, Vol. 3, pp. 2323–2326. MIT Press.

Levin B. J. (1962) Thermal history of the moon. In *The Moon* (editors Z. Kopal and Z. K. Mikhailov), pp. 157–167. Academic Press.

LSPET (1971a) (Lunar Sample Preliminary Examination Team) Lunar sample information catalog—Apollo 14. LRL, NASA Manned Spacecraft Center.

LSPET (1971b) (Lunar Sample Preliminary Examination Team) Preliminary examination of the lunar samples from Apollo 14. *Science* **173**, 681–691.

MacDonald G. J. F. (1959) Calculations on the thermal history of the earth. *J. Geophys. Res.* **64,** 1967–2000.

McConnell R. K., McClaine L. A., Lee D. W., Aronson J. R., and Allen R. V. (1967) A model for planetary igneous differentiation. *Rev. Geophys.* **5,** 121–172.

Mizutani H., Hamano Y., and Akimoto S. (1970) Compressional wave velocities of fayalite, Fe_2SiO_4 spinel, and coesite. *J. Geophys. Res.* **75,** 2741–2747.

Ringwood A. E. and Essene E. (1970) Petrogenesis of Apollo 11 basalts, internal constitution, and origin of the moon. *Proc. Apollo 11 Lunar Sci. Conf., Geochim. Cosmochim. Acta* Suppl. 1, Vol. 1, pp. 769–799. Pergamon.

Scoon J. H. (1972) Chemical analyses of lunar samples 14003, 14311, and 14321. In *Lunar Science— III* (editor C. Watkins), pp. 690–691, Lunar Science Institute Contri. No. 88.

Simmons, G. (1964a) Velocity of compressional waves in various minerals at pressures to 10 kilobars. *J. Geophys. Res.* **69,** 1117–1121.

Simmons G. (1964b) Velocity of shear waves in rocks to 10 kilobars, 1. *J. Geophys. Res.* **69,** 1123–1130.

Stephens D. R. and Lilley E. M. (1970) Loading-unloading pressure-volume curves to 40 kbar for lunar crystalline rock, microbreccia and fine. *Proc. Apollo 11 Lunar Sci. Conf., Geochim. Cosmochim. Acta* Suppl. 1, Vol. 3, pp. 2427–2434. Pergamon.

Toksoz M. N., Press F., Anderson K., Dainty A., Latham G., Ewing M., Dorman J., Lammlein D., Sutton G., Duennebier F., and Nakamura Y. (1972a) Velocity structure and properties of the lunar crust. Submitted to *The Moon.*

Toksoz M. N., Solomon S. C., Minear J. W., and Johnston D. H. (1972b) Thermal evolution of the moon. Submitted to *The Moon.*

Wang H., Todd T., Weidner D., and Simmons G. (1971). Elastic properties of Apollo 12 rocks. *Proc. Second Lunar Sci. Conf., Geochim. Cosmochim. Acta* Suppl. 2, Vol. 3, pp. 2327–2336. MIT Press.

Wechsler A. E. and Glaser P. E. (1965). Pressure effects on postulated lunar materials. *Icarus* **4,** 335–352.

Proceedings of the Third Lunar Science Conference
(Supplement 3, *Geochimica et Cosmochimica Acta*)
Vol. 3, pp. 2565–2575
The M.I.T. Press, 1972

Elastic velocity and Q factor measurements on Apollo 12, 14, and 15 rocks

B. R. Tittmann, M. Abdel-Gawad, and R. M. Housley

North American Rockwell Science Center,
Thousand Oaks, California 91360

Abstract—The Rayleigh wave velocities (in one Apollo 12, one Apollo 15, and two Apollo 14 rocks) were measured by the impulse technique. For 14310 $v_R = 1.20$ km/sec; for 14321 $v_R \approx 0.9$ km/sec; for 12063, on which the orientation dependence was studied, $v_R - 1.16$–1.59 km/sec; for 15555 $v_R = 0.32$ km/sec; and for synthetic rock 10017 analogue $v_R = 2.26$ km/sec. This represents a larger spread by a factor of 3 than previously reported on lunar igneous rocks. Absolute Q factor measurements were performed on one Apollo 14 rock by the vibrating bar technique. For 14310 $Q \approx 70$ at STP, $Q \approx 10$ in water vapor, $Q \approx 150$ at 5×10^{-8} mm of Hg and $T = 25°C$, and $Q \approx 800$ at 5×10^{-8} mm of Hg and $T \approx -180°C$. Thus under exposure to high vacuum and low temperatures the Q factor is shown to increase towards values approaching the low end of the range of estimates from seismic data ($Q \approx 1000$).

Introduction

In this paper we present velocity and Q factor measurements that were made on Apollo 12, 14, and 15 returned lunar rocks 12063,96, 14321,211, 14310,86, and 15555,90.

Earlier elastic property measurements on returned lunar samples showed anomalous values with respect to corresponding values known for terrestrial rocks. Unexpectedly low velocities were observed for Apollo 11 rocks (Schreiber *et al.*, 1970; Kanamori *et al.*, 1970) and for Apollo 12 rocks (Kanamori *et al.*, 1971; Wang *et al.*, 1971; Tittmann and Housley, 1971; and Warren *et al.*, 1971). This same trend is again observed in recent measurements on Apollo 14 samples by us and by Wang *et al.* (1972) and an unusually low velocity has now been observed by us on Apollo 15 igneous rock 15555. The present investigation includes comparisons with terrestrial rocks and synthetic analogues of lunar rocks, studies of microfractures by the scanning electron microscope, correlations between microfracture orientation and Rayleigh wave anisotropy, and elastic measurements under conditions varying from rock saturation with a liquid to high vacuum.

The Q values deduced from lunar seismic data are known to be anomalously high ($Q \approx 1000$–3000) in comparison with the Q factors measured for the lunar return samples. Low Q factors were observed for Apollo 11 rocks (Kanamori *et al.*, 1970) and for Apollo 12 rocks (Warren *et al.*, 1971; Wang *et al.*, 1971). Similarly low Q factors were observed in our measurement on an Apollo 14 sample at standard temperature and pressure (STP). However, we show here that the Q factor can be made to rise strongly under exposure to high vacuum and low temperatures towards values approaching the low end of the range of seismic values.

VELOCITY MEASUREMENTS

Summary of experimental results

Rayleigh wave group velocity measurements were performed on portions of lunar rocks 12063, 14310, and 15555 as well as on a synthetic analogue of lunar rock 10017. The results are presented in Table 1 together with some additional data from the literature.

Discussion

Lunar igneous rocks 14310 and 12063 have velocities that are lower by a factor of about 3 from those typically measured in terrestrial basalts. Our measurement on a sample of augite-olivine basalt from northern California gave 2.97–3.15 km/sec for comparison. A similar difference was observed with bulk waves (Wang *et al.*, 1971; Warren *et al.*, 1971; Kanamori *et al.*, 1971) and has been explained as being due to the presence in lunar rocks of unfilled microfractures.

Rock 14321. Petrological studies of rock 14321 by Duncan *et al.* (1972), Grieve *et al.* (1972), Warner (1972), Swann *et al.* (1971), and LSPET (1971) indicate that rock 14321 has had a complex and multistage formational history, and is part of a group of moderately thermally recrystallized polymict breccias. Microscopic examination of our sample 14321,211 suggests that it is a breccia of low coherence. Because of the sample friability, measurements could not be made in the normal manner but estimates of the velocity were obtained by the time of flight method in which the total travel distance, and total travel time of the Rayleigh wave pulse are measured in an absolute manner and then divided to give the velocity directly. The value obtained for the velocity $v_R = 0.9$ km/sec has a rather uncertain meaning because of the extreme complexity of the sample. This value is similar to that calculated for another microbreccia, rock 10046, from bulk wave data of Anderson *et al.* (1970).

Table 1. Observed Rayleigh wave velocity data.

Sample	Velocity (km/sec)	Comments
12038	0.97 — 1.45	Granular basalt (Tittmann and Housley, 1971)
12063	0.94 — 1.59	Diabase
14310	1.20	Basalt
14321	0.90 ± 0.15	Complex breccia
15555	0.28 — 0.34	Olivine basalt, highly fractured
10046	0.7 (calc.)	Micro breccia (Anderson *et al.*, 1970)
Synthetic analogue of 10017	2.21 — 2.26	Basalt, analogue of lunar rock 10017
Terrestrial basalt	2.97 — 3.15	Augite olivine basalt with strong flow structure and mineralogy similar to lunar basalts.

Most of the measurements were performed by the impulse technique (Tittmann, 1971) which uses ceramic piezoelectric transducers as transmitter and receiver, both placed on one flat surface of a rock. The transducers are separated from the rock by thin soft aluminum foils which serve as acoustic bonds coupling the acoustic energy between transducer and rock. A 0.1 μsec wide voltage pulse applied to the transmitter produces an acoustic impulse which is detected some distance away by the receiver. In the measurements the distance between the two transducers is varied giving rise to changes in the signal arrival time of the Rayleigh wave pulse. Figure 1 is a plot showing sample data on rocks 14310, 15555, and synthetic analogue of rock 10017. The slope of the lines through the data points give the absolute group velocities independent of any inaccuracies in the absolute position or signal arrival time determinations. Spectrum analysis of the received signal gave a maximum at a frequency of about 2 MHz in the fine grained rocks. For rock 15555 the major frequency component dropped to about 700 kHz suggesting the presence of extensive scattering by the coarse grains in this rock.

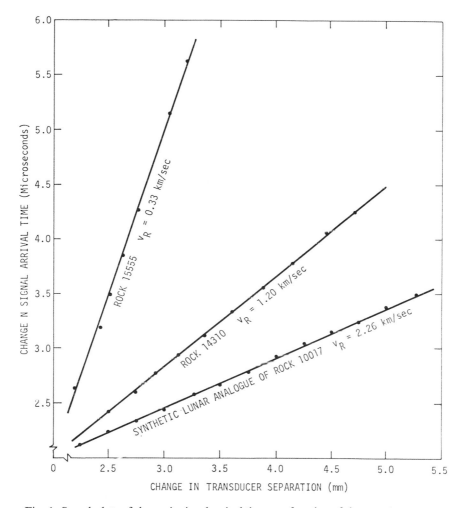

Fig. 1. Sample data of change in signal arrival time as a function of the transducer separation for rocks 14310, 15555, and a synthetic analogue of lunar rock 10017. The data were obtained by the impulse technique (Tittmann, 1971). The reciprocal of the slope gives the Rayleigh wave group velocity.

Rock 12063. Our previous measurements (Tittmann and Housley, 1971) of the Rayleigh wave group velocity on lunar rock 12038 and bulk wave data reported by Wang *et al.* (1971) and Warren *et al.* (1971) showed large changes of velocity with propagation direction. Since our sample 12063 was in the form of a cube about 2 cm on an edge, three of whose faces were smooth enough for Rayleigh wave measurements, we decided to make a fairly extensive study of the anisotropy and inhomogeneity. The rock indeed proved to be strongly inhomogeneous and anisotropic. Measured velocities ranged from about 0.90 to 1.6 km/sec. Part of the data is shown in Fig. 2 and some is given in Table 2.

Table 2. Rayleigh wave velocities in rock 12063

Direction*	Face*	Velocity (km/sec)	Comments
x	(1)	1.23	
y	(1)	1.15	Path contains pronounced fracture (~ 0.3 mm wide at surface)
z	(3)	0.91	
x'	(2)	1.00	
y'	(2)	1.23	

* Defined in Fig. 2.

Relative changes in signal arrival time as a function of angle at fixed transducer separation were systematically studied on faces 1 and 2. On face 1 as the angle θ between the propagation direction and one sample edge taken as reference was increased, a systematic rise in velocity was observed for low angles followed by a decrease and leveling out of the velocity at higher angles. This result is shown in the solid line in Fig. 2. Polished section optical microscopic examination of rock 12063 showed that microfractures, vesicles, and ilmenite blades tend to be preferentially elongated in the direction $10 \leq \theta \leq 25°$, which gave higher velocity values.

Measurements of the linear compressibility of Apollo 11 samples by Anderson et al. (1970) showed anisotropy at low pressure that they also suggested might arise from a preferential orientation of microfractures. We mapped the surface of 12063 by the scanning electron microscope at magnifications from about 50 to 500, and found that on a statistical basis the microfractures showed a preferred orientation in the direction in which the velocity was highest. Figures 3a, 3b, and 3c show sample micrographs of a small portion of the surface. The width of the microfractures observed ranged from 0.001 to 0.3 mm with most of them about 0.01 mm in width. A rough estimate of the surface area of microfractures ranged from 5 to 15% of the total surface area. Therefore it would appear that the dominant contribution to the velocity anisotropy may be a result of the material being more compliant for stresses perpendicular to the fractures. To shed more light on the effect of microfractures on the velocity, the sample was saturated with ethanol and then evacuated to 10^{-8} mm of Hg. During this treatment a net decrease of 25% was observed in the Rayleigh wave velocity.

Synthetic analogue of lunar rock 10017. In an effort to gain more insight into the source of microfractures and their influence on the Rayleigh wave velocity, we studied a synthetic rock having the chemical and phase composition of lunar basalt 10017, but differing considerably from it in texture. On the scanning electron microscope the synthetic rock showed considerably fewer microfractures (Fig. 3d) than either the lunar rock 10017 itself (Anderson et al., 1970) or our lunar rock 12063 (Figs. 3a, 3b, and 3c). The Rayleigh wave velocity values measured on the synthetic rock were found to be correspondingly high, i.e., $v_R = 2.21$–2.26 km/sec, a factor of more than two higher than the value $v_R \approx 0.96$ km/sec calculated from bulk wave velocity data (Anderson et al., 1970). This suggests that microfractures other than those introduced during the process of cooling from the melt may contribute significantly to the reduced velocities of lunar samples.

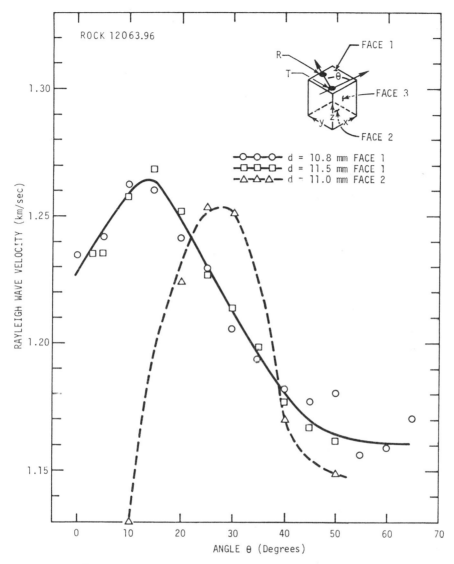

Fig. 2. Plot of Rayleigh wave velocity v_R as a function of angle θ between the direction of sound propagation and an edge of the sample chosen as arbitrary reference direction. The data were obtained for two opposite faces (solid line: face 1; dashed line: face 2) of rock 12063.96 (a cube of about 2 cm on edge) for several different separations d between transmitter T and receiver R.

Lunar rock 15555. Velocity data collected on 15555 were obtained on many locations on different sample faces, and fell into the range 0.28–0.34 km/sec (Tittmann *et al.*, 1972). These values are three to four times lower than those observed on lunar igneous rocks measured previously and less than half those obtained on lunar breccias, as shown in Table 1. Time of flight measurements of the bulk longitudinal wave

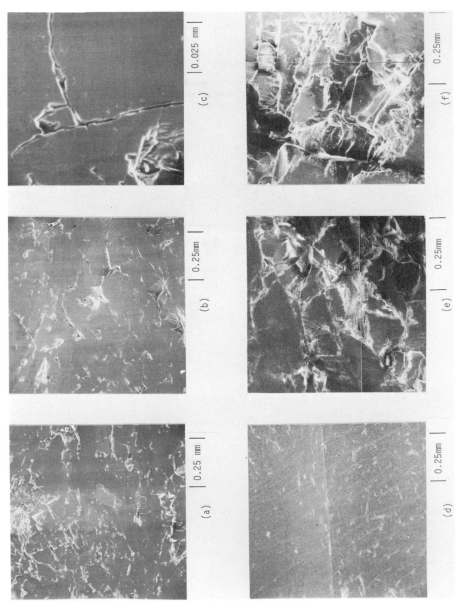

Fig. 3. Scanning electron beam micrographs of rock surfaces exhibiting microfractures. (a and b) Face 1 of rock 12063,96. (c) Same as in (a) but enlarged 10 times. (d) Synthetic rock lunar analogue 10017. (e and f) Rock 15555.

velocity also gave low values, i.e., $v_P \approx 0.70$–0.95 km/sec. Although these values are approximate, they appear to be somewhat higher than what would be expected from the observed Rayleigh wave velocity for an assumed density of $\rho = 3.1$ gm/cm^3 and Poisson's ratio of $\sigma \approx 0.23$. Microscopic examination showed that 15555 is a coarse grained olivine basalt with irregularly distributed vugs. Although the rock is highly fractured (see Figs. 3e and 3f) it shows little or no evidence of severe shock meta-morphism and appears coherent and competent in normal laboratory handling. Care was taken to make the Rayleigh wave velocity measurements on the more massive portions of the rock where the vugs were sparsely distributed and did not exceed about 0.1 mm in size.

The existence of a very low elastic wave velocity in this generally competent, igneous rock indicates that the existence of a thick, low velocity zone near the lunar surface does not necessarily require postulating the existence of a thick layer of fines on the moon's surface. Field observations (Swann et al., 1971) have concluded that the regolith in Station 9a at the Hadley Rille rim at which the sample was collected is very thin or absent and that most of the samples collected in this area are probably representative of the local bedrock exposed at the Rille rim. From the discussion by Swann et al. (1971) we are not sure of the exact source of rock 15555. However, this rock is classified (LSPET, 1972) as typical of the porphyritic olivine basalts and similar in this respect to 15535 which is judged most certainly representative of the bedrock. Thus the assumption that 15555 may have been derived from the upper part of vol-canic bedrocks exposed in the vicinity of the Hadley Rille seems justified. If this con-clusion is borne out by further studies, the unusually low velocities measured on this igneous rock should be taken into consideration in the interpretation of the low seis-mic velocities observed in the upper layer of the lunar surface.

Q-Factor Measurements

Introduction

The relatively high Q inferred for the lunar material may be in part a consequence of the nearly complete absence of fluids within the outer shell of the moon. Some experimental evidence supporting this possibility has been presented by Pandit and Tozer (1970) who state that evacuation of porous terrestrial rocks to pressures of 10^{-2} Torr typically increases the Q of the sample by a factor of 5 over the value measured at 1 atm. We used the impulse technique to measure relative changes in values for the Rayleigh wave attenuation as a function of environmental changes. In previous ex-periments (Tittmann and Housley, 1971), substantial relative changes in Rayleigh wave amplitude had been observed in rock 12038 when the absolute air pressure was changed from 1 atm to 6×10^{-7} mm of Hg. As the pressure was reduced the ampli-tude increased by a total of 25%, with most of the change occurring between 1 atm and 1 mm of Hg. In tests where the sample was previously outgassed, then pressurized with dry nitrogen gas, this change between atmospheric pressure and 1 mm of Hg was absent. This result is similar to the results obtained in bulk wave experiments by Warren et al. (1971). It is well known from adsorption studies (Brunauer, 1945) that an estimated 10 to 20 monolayers of H_2O can be adsorbed onto glass and silicate surfaces when exposed to the atmosphere. On the basis of these results and other considerations, we believe that, when the lunar return samples are allowed to come in contact with air, the trapping of water molecules in the microfractures becomes an important factor in decreasing the Q. The source of the attenuation is believed to arise from water molecules collected in the regions of the fracture tips. During the passage of a sound wave pulse, the relative displacements of opposing fracture faces can be expected to effectively shorten the crack depth. This motion of the small amount of liquid trapped in the fracture probably gives rise to losses due to the viscosity of the liquid. Experimental evidence for the attenuation of elastic waves due to the motion of liquids in porous media has been discussed by Wyllie *et al.* (1962).

Summary of experimental results

Absolute Q measurements were performed by the vibrating bar technique on a 2 cm long bar cut from rock 14310, a recrystallized basalt (Ridley *et al.*, 1972). The results are presented in Table 3.

Discussion

The vibrating bar technique was used to show the influence of water vapor on absolute Q of lunar rock. Sample 14310 was exposed to hot water vapors as the Q was being monitored continuously by electronic sweeping through the resonance. Upon exposing the rock to hot water vapors for only about 30 sec, the Q was lowered from a $Q \approx 90$ to $Q \approx 10$. Longer exposures rendered the Q so low it was unmeasurable. Repeated experiments confirmed the conclusion that the lunar rocks we studied are quite permeable, adsorb H_2O at a rapid rate, and as a result change their Q drastically. Figure 5 shows the effect of exposure to water vapor on the frequency response curve of 14310.

Table 3. Absolute vibrational attenuation data.

Environmental description	Absolute Q factor
Water vapor	10
Laboratory air (humid-dry day)	50–90
5×10^{-8} mm Hg	150
$-180°C$ and 5×10^{-8} mm Hg	800 (highest achieved)
Terrestrial augite basalt (see Table 1) in lab air	65

In the vibrating bar technique, the sample (typically 2 cm long, 2–3 mm thick) is held in the center by two needle point set screws and is vibrated in the longitudinal mode at its fundamental resonance by a magnetic drive. The transducers are small soft-iron buttons bonded on each end, one as transmitter the other as receiver. Because of the small separation between the coils of the magnetic drives, extensive shielding with μ-metal is necessary to reduce background due to direct coupling. The sample mount and magnetic drivers are mounted on a platform in a bell jar and can be cooled or heated as necessary, the temperature being monitored with a thermocouple in contact with one of the set screws holding the sample. In Fig. 4 we show a sample plot of data of the frequency response. The Q estimated from the half width of the resonance curve is $Q \approx 70$ at STP and $Q \approx 700$ at $T \approx -180°C$ and $P = 6 \times 10^{-8}$ mm of Hg.

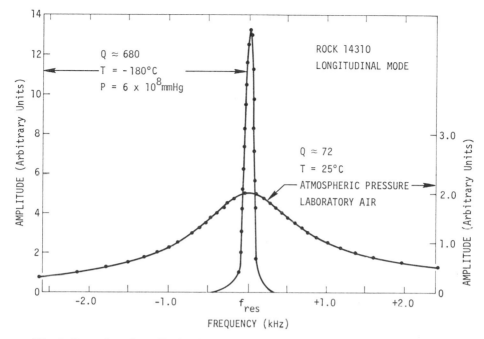

Fig. 4. Data plot of amplitude of vibration of a bar of rock 14310 as a function of frequency near resonance. The bar is driven in the free-free mode of compressional waves. The resonant frequency shifts with temperature.

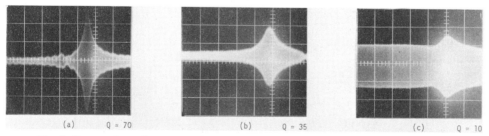

Fig. 5. Curves of amplitude of vibration as a function of frequency for rock 14310 in a series of oscilloscope traces showing from left to right how the Q diminishes with the length of time that the bar is exposed to hot water vapors.

In a similar manner the vibrating bar technique was used to monitor the absolute Q as the sample was exposed to a hard vacuum of 5×10^{-8} mm of Hg. At room temperature the Q was found to increase to values in the range $Q \approx 130$–150. Adsorption studies of H_2O on silicates suggest that this increase in Q is probably attributable to the removal of adsorbed water molecules from the microfractures.

The above experiment was extended by lowering the temperature of the sample while in the hard vacuum. This was accomplished by circulating cold N_2 gas through a Cu tube coiled around and soldered onto the sample holder platform. In this way sample temperatures down to $-185°C$ could be obtained. The Q was found to increase with decreasing temperature with the highest values $Q \approx 400$–800 found near $T \approx -180°C$ at a vacuum of 6×10^{-8} mm of Hg. The highest Q value achieved appears to depend somewhat on such factors as the lowest equilibrium temperature achieved concomitant with the hardness of the vacuum, the quality of the transducer bond, and the nearness of the sample holding set screws to the ideal nodal point. The sample itself was not ideal in shape having a rather nonuniform cross section. Therefore, the higher Q values are probably closer to the real Q and it is felt that optimization of all the parameters could achieve even higher values. The detailed mechanism for the increase in Q with decrease in temperature is not understood at this time. However, since the sample is allowed to be outgassed at 1×10^{-7} mm of Hg for a considerable amount of time the mechanism probably has little to do with the freezing of adsorbed water molecules. The well-known mechanism for damping by Coulomb friction between adjacent grains is not likely to give rise to any strong temperature dependent effect. A thermally activated relaxation mechanism seems to be the most probable source of damping. If so, detailed studies of the temperature and frequency dependence should reveal the source and nature of this mechanism.

SUMMARY DISCUSSION

The observations described above leave little doubt that the presence of microfractures influence both the velocity and Q factor values of lunar rocks. Velocity data but especially Q data are strongly influenced by contamination by the terrestrial atmosphere. This contamination can occur with very short exposure times because of the high permeability of the lunar rocks. Water vapor especially affects Q values drastically. Our highest observed Q value of $Q \approx 800$ still does not match those deduced from the seismic experiments but does approach the low end of the range of seismic Q values.

With the inclusion of the Rayleigh wave velocity data obtained on real lunar rocks and synthetic analogues we cover about one order of magnitude from $v_R \approx 0.3$ km/sec to $v_R \approx 2.2$ km/sec. These velocities have all been obtained at zero confining pressure and are therefore representative of the solid material in the upper few kilometers of the lunar surface. In view of this spread in velocities it is perhaps not surprising that seismic experiments show a high degree of scattering in this layer.

Acknowledgments—The authors are indebted to Prof. John A. Bastin for the loan of rock 12063.96, to Leroy H. Hacket for the work with the scanning electron microscope, and to G. V. Latham for suggesting the importance of measurements in high vacuum. The authors are grateful to G. A. Alers, O. Buck, L. J. Graham, T. C. Lim, E. A. Kraut, T. Smith, R. B. Thompson, T. Todd, and N. Warren for helpful discussion and suggestions. This work was supported in part by NASA contract NAS9-11542.

REFERENCES

Anderson O. L., Scholz C., Soga N., Warren N., and Schreiber E. (1970) Elastic properties of a micro-breccia, igneous rock, and lunar fines from Apollo 11 mission. *Proc. Apollo 11 Lunar Sci. Conf.*, *Geochim. Cosmochim. Acta* Suppl. 1, Vol. 3, pp. 1959–1973. Pergamon.

Brunauer S. (1945) The adsorption of gases and vapors. *Physical Adsorption*, Vol. I, Princeton University Press.

Duncan A. R., Lindstrom M. M., Lindstrom D. J., McKay S. M., Stoeser J. W., Goles G. G., and Fruchter J. S. (1972) Comments on the genesis of breccia 14321 (abstract). In *Lunar Science—III* (editor C. Watkins), pp. 192–194, Lunar Science Institute Contr. No. 88.

Grieve R., McKay G., Smith H., and Weill D. (1972) Mineralogy and petrology of polymict breccia 14321 (abstract). In *Lunar Science—III* (editor C. Watkins), pp. 338–340, Lunar Science Institute Contr. No. 88.

Kanamori H., Mizutani H., and Hamano Y. (1971) Elastic wave velocities of Apollo 12 rocks at high pressures. *Proc. Second Lunar Sci. Conf., Geochim. Cosmochim. Acta* Suppl. 2, Vol. 3, pp. 2323–2326. MIT Press.

Kanamori H., Nur A., Chung D., Wones D., and Simmons G. (1970) Elastic wave velocities of lunar samples at high pressures and their geophysical implications. *Science* **167**, 726–728.

LSPET (Lunar Sample Preliminary Examination Team) (1971) Preliminary examination of the lunar samples from Apollo 14. *Science* **173**, 681–693.

LSPET (Lunar Sample Preliminary Examination Team) (1972) Preliminary examination of the lunar samples from Apollo 15. *Science* **175**, 363–373.

Pandit B. I. and Tozer D. C. (1970) Anomalous propagation of elastic energy within the moon. *Nature* **226**, 335.

Ridley W. I., Williams R. J., Brett R., and Takeda H. (1972) Petrology of lunar basalt 14310 (abstract). In *Lunar Science—III* (editor C. Watkins), pp. 648–650, Lunar Science Institute Contr. No. 88.

Schreiber E., Anderson O. L., Soga N., Warren N., and Scholz C. (1970) Sound velocity and compressibility for lunar rocks 17 and 46 and for glass spheres from the lunar soil. *Science* **167**, 732–734.

Swann G. A., Bailey N. G., Batson R. M., Eggleton R. E., Hait M. H., Holt H. E., Larson K. B., McEwen M. C., Mitchell E. D., Schaber G. G., Schafer J. P., Shepard A. B., Sutton R. L., Trask N. J., Ulrich G. E., Wilshire H. G., and Wolfe E. W. (1971) Preliminary geologic investigations of the Apollo 14 landing site. *Apollo 14 Preliminary Sci. Report*, Sec. 3, pp. 39–85, NASA SP-272.

Swann G. A., Hait M. H., Schaber G. G., Freeman V. L., Ulrich G. E., Wolfe E. W., Reed V. S., and Sutton R. L. (1971) Preliminary descriptions of Apollo 15 sample environments. U.S. Geological Survey, Interagency Report 36, pp. 185–203.

Tittmann B. R. (1971) A technique for precision measurements of elastic surface wave properties on arbitrary materials. *Rev. Sci. Instr.* **42**, 1136–1142.

Tittmann B. R., Abdel-Gawad M., and Housley R. M. (1972) Rayleigh wave studies of lunar and synthetic rocks (abstract). In *Lunar Science—III* (editor C. Watkins), pp. 755–757, Lunar Science Institute Contr. No. 88.

Tittmann B. R. and Housley R. M. (1971) Surface elastic wave propagation studies in lunar rocks. *Proc. Second Lunar Sci. Conf., Geochim. Cosmochim. Acta* Suppl. 2, Vol. 3, pp. 2337–2343. MIT Press.

Wang W., Todd T., Simmons G., and Baldridge S. (1972) Elastic wave velocities and thermal expansion of lunar and earth rocks (abstract). In *Lunar Science—III* (editor C. Watkins), pp. 776–778, Lunar Science Institute Contr. No. 88.

Wang W., Todd T., Weidner D., and Simmons G. (1971). Elastic properties of Apollo 12 rocks. *Proc. Second Lunar Sci. Conf., Geochim. Cosmochim. Acta* Suppl. 2, Vol. 3, pp. 2327–2336. MIT Press.

Warner J. (1972) Apollo 14 breccias: Metamorphic origin and classification (abstract). In *Lunar Science—III* (editor C. Watkins), pp. 782–784, Lunar Science Institute Contr. No. 88.

Warren N., Schreiber E., Scholz C., Morrison J. A., Norton P. R., Kumazawa M., and Anderson O. L. (1971). Elastic and thermal properties of Apollo 11 and Apollo 12 rocks. *Proc. Second Lunar Sci. Conf., Geochim. Cosmochim. Acta* Suppl. 2, Vol. 3, pp. 2345–2360. MIT Press.

Wyllie M. R. J., Gardner G. H. F., and Gregory A. R. (1962). Studies of elastic wave attenuation in porous media. *Geophys.* **27**, 569–589.

Proceedings of the Third Lunar Science Conference
(Supplement 3, *Geochimica et Cosmochimica Acta*)
Vol. 3, pp. 2577–2586
The M.I.T. Press, 1972

Elastic properties of Apollo 14 and 15 rocks

Terrence Todd, Herbert Wang, W. Scott Baldridge, and Gene Simmons

Department of Earth and Planetary Sciences,
Massachusetts Institute of Technology,
Cambridge, Massachusetts 02139

Abstract—Ultrasonic *P*- and *S*-wave velocities of lunar samples 14310,72 and 15418,43 and *P*-wave velocities of sample 15015,18 were measured at room temperature to 5 kb confining pressure. The velocities of both igneous and breccia samples increased sharply over this pressure range. At low confining pressures the *shape* of velocity-pressure curves of rocks is determined by the distribution function of crack aspect ratios. We suggest that analogue studies on terrestrial rocks having a wide assortment of crack parameters may be used to infer the nature of cracks in lunar rocks. Velocity measurements on the more intensely cracked members of two sets of terrestrial rocks of increasing crack porosity yielded low V_p values similar to those found in lunar igneous rocks at low confining pressure. Velocity measurements on terrestrial analogues of lunar breccias gave velocity profiles similar to those of poorly compacted lunar breccias.

Introduction

The physical properties of lunar samples emphasize the roles of microcracks, absence of water, vacuum, texture, and prehistory. These factors are different for lunar and earth rocks. Previous investigations (Wang *et al.*, 1971; Kanamori *et al.*, 1970; Anderson *et al.*, 1970; and Latham *et al.*, 1970) indicate that the sharp increase in both *P*- and *S*-wave velocities of lunar samples over the first 2 kb, and the low *Q* measurements made in the laboratory on lunar samples, can be attributed to the presence of long, thin microcracks. We have attempted to isolate the role that microcracks play in determining velocity and *Q* values. We have thermally cycled several Fairfax diabase and Westerly granite cores (under vacuum) to temperatures between 100 and 1200°C, and produced two suites of rocks with increasing crack porosity. Warren and Latham (1970) showed that in the process of heating rocks, even to moderate temperatures, cracks were formed. Ide (1937) first demonstrated the effect of thermal cycling on velocity; he found that the velocity of thermally cycled samples, when rerun at room temperature, decreased systematically with each increase in the maximum temperature to which a sample was subjected. We report in this paper the preliminary results of the thermal cycling experiment, new data on an Apollo 14 igneous rock and two Apollo 15 breccias, and a comparison of the Apollo 15 breccias with two terrestrial rocks having structures, densities, and velocities analogous to those of poorly compacted lunar breccias.

Experimental Method

The velocity measurements were made with the pulse transmission method of Birch (1960, 1961). Sample preparation and jacketing techniques are similar to those previously described by Kanamori *et al.* (1970) and Wang *et al.* (1971). Two faces of each sample were *dry* surface ground parallel to

±0.001 inches and the sample was dried for 4 hours at 80°C in a vacuum oven. Coaxially plated, 1 MHz, barium titanate transducers were used for P-waves while AC-cut quartz transducers were used for S-waves (Simmons, 1964). The transducers were ⅜ inch in diameter and were bonded to the sample with Duco cement. All samples were jacketed in Sylgard, an electronic encapsulating material, to keep the pressure medium (petroleum ether) from penetrating the sample. Velocity measurements were made in a simple piston-cylinder high-pressure vessel. The estimated accuracy of the measurements is 2 to 3 % for P-waves and 5 % for S-waves; these values are larger than those normally obtained for terrestrial rocks because of the small sample size (about 1 cm) and the high attenuation at low confining pressure.

All Fairfax diabase and Westerly granite samples are cylinders, 2.5 inches in length and 1.0 inch in diameter, that were cored parallel to each other from the same large blocks of diabase and granite; the ends of each core were ground parallel to ±0.001 inches. A series of cores of each rock type were thermally cycled in vacuum to maximum temperatures (T_{max}) between 100 and 1200°C. One core of diabase ($T_{max} = 1200$°C) was partially melted ($\sim 5\%$) and the melted phase subsequently froze as a glass. The heating and cooling was gradual over a 24 hour period, with T_{max} held constant for about 12 hours. The P- and S-wave velocities of Fairfax diabase and the P-wave velocity of Westerly granite were then measured to 5 kb at room temperature. No change in the velocity values or in the shape of the velocity profiles was found for samples recycled to temperatures less than or equal to the initial T_{max}.

The Q measurements for Fairfax diabase and Westerly granite were made in a torsional pendulum apparatus specifically designed for attenuation measurements (Jackson, 1969) in the Hertz frequency range (as compared to MHz for the velocity measurements). All samples (⅜ inch in diameter and 6 inches long) were cored parallel to each other and the velocity cores. The Q cores and the velocity cores were from the same large blocks. The cores were vacuum dried prior to thermal cycling with the velocity cores. The estimated accuracy of each Q measurement is 10%.

Sample Description

New velocity measurements are reported for one igneous rock (14310,72) and two breccias (15418,43 and 15015,18). Sample 14310,72 is a fine grained, medium grey, crystalline basalt (Table 1). The dimensions of our sample are approximately $2 \times 2 \times 1$ cm and the density is 2.88 g/cm^3; velocities measured in the two long directions showed no anisotropy ($<1\%$). Sample 15418,43 is a medium dark, to dark grey, breccia of chemical composition similar to that of an anorthite-rich gabbroic anorthosite (Table 1). The dimensions of sample 15418,43 are approximately $2 \times 1 \times 1$ cm and the density is 2.80 g/cm^3; velocities are reported for the long direction. Sample 15015,18 is a somewhat more friable breccia. It is composed of a light grey matrix of fine grained glass and soil (the composition of the soil has not

Table 1. Modal analyses of samples 14310 and 15418 (in weight %) and Fairfax diabase (in volume %). The modal analysis for the lunar samples are from LSPET (1972); the modal analysis for the Fairfax diabase is from Fairbairn et al. (1952). A modal analysis for 15015 (density = 2.33 g/cm^3) was not available.

14310		15418		Fairfax Diabase	
Quartz	<1%	Orthoclase	<1%	Quartz	1.8%
Orthoclase	2.9%	Albite	2.6%	Potassium feldspar	3.0%
Albite	5.3%	Anorthite	71.5%	Plagioclase (Labradorite)	45.0%
Anorthite	50.8%	Olivine	5.3%	Pyroxene (Augite)	45.0%
Diopside	6.6%	Diopside	6.7%	Biotite	1.8%
Ferrohypersthene	29.9%	Ferrohypersthene	12.4%	Opaque	3.3%
Ilmenite	2.4%	Ilmenite	<1%	Other	0.2%
density = 2.88 g/cm^3		density = 2.80 g/cm^3		density = 3.00 g/cm^3	

yet been determined), containing white clasts ($\sim 5\%$) of sugary pyroxene and plagioclase (NASA, 1971). The dimensions of the sample are approximately $2 \times 1 \times 1$ cm and the density is 2.33 g/cm^3. The P-wave velocities were measured in the long direction; because of the poor coherence of the sample, S-wave velocity measurements were not possible.

Velocity Results

All velocity data are tabulated in Table 2. The P- and S-wave velocities for lunar sample 14310,72 and four thermally cycled Fairfax diabase samples are compared in Figs. 1 and 2. The P-wave velocities for lunar breccias 15418,43 and 15015,18 and two terrestrial analogues are given later in Fig. 5. All measurements were made at room temperature.

There are two obvious trends in the behavior of the thermally cycled Fairfax diabase samples shown in Fig. 1. First, the value of P-wave velocity at 5 kb for all Fairfax diabase samples is about 7.0 km/sec. The high pressure velocity values are relatively independent of T_{max} but do systematically decrease for the first several hundred degrees in T_{max}. Previously (Wang et al., 1971) we reported a 5 kb value of 7.3 km/sec for the compressional velocity of Fairfax diabase compared to present values around 7.0 km/sec. Of the difference 0.1 km/sec is attributed to the previous sample being vacuum heated to 80°C rather than 200°C or higher prior to the pressure run, another 0.1 km/sec is attributed to sample inhomogeneity and the remaining 0.1 km/sec to measurement error. Secondly, the low pressure values of P-wave

Table 2. Velocities (km/sec) of samples. T_{max} is the maximum temperature to which each sample was heated. All measurements were made at room temperature.

Samples		Confining Pressure (bars)										
		1	100	250	500	750	1000	1500	2000	3000	4000	5000
14310,72	P	3.93	4.21	4.55	4.96	5.23	5.51	5.86	6.08	6.33	6.50	6.68
(A direction)	S	2.08	2.23	2.40	2.63	2.80	2.95	3.15	3.30	3.46	3.57	3.66
14310,72	P	3.84	4.12	4.52	4.91	5.26	5.55	5.92	6.14	6.39	6.59	6.79
(B direction)	S	2.07	2.19	2.37	2.60	2.79	2.94	3.11	3.26	3.42	3.54	3.63
15418,43	P	4.85	5.00	5.20	5.50	5.77	6.02	6.33	6.50	6.64	6.69	6.75
	S	2.82	2.88	2.97	3.08	3.19	3.28	3.42	3.50	3.58	3.63	3.69
Fairfax diabase	P	5.95	6.15	6.33	6.54	6.66	6.71	6.78	6.82	6.88	6.93	6.97
($T_{max} = 200$°C)												
Fairfax diabase	S	3.53	3.61	3.69	3.77	3.82	3.84	3.85	3.86	3.87	3.87	3.88
($T_{max} = 300$°C)												
Fairfax diabase	P	4.20	4.75	5.35	6.01	6.30	6.48	6.66	6.74	6.83	6.89	6.94
($T_{max} = 600$°C)	S	2.74	3.04	3.29	3.55	3.65	3.71	3.76	3.79	3.82	3.84	3.85
Fairfax diabase	P	2.02	3.21	4.09	5.00	5.53	5.87	6.30	6.47	6.69	6.81	6.88
($T_{max} = 1000$°C)	S	1.59	2.28	2.65	3.06	3.31	3.46	3.61	3.71	3.77	3.81	3.85
Fairfax diabase	P	2.00	3.00	3.81	4.55	5.01	5.39	5.78	6.06	6.32	6.50	6.68
($T_{max} = 1200$°C)	S	1.33	1.73	2.19	2.68	2.99	3.19	3.43	3.56	3.67	3.74	3.82
Westerly granite	P	3.63	4.75	5.23	5.47	5.67	5.75	5.84	5.88	5.96	6.03	6.09
($T_{max} = 300$°C)												
Westerly granite	P	1.80	3.18	4.20	4.98	5.28	5.44	5.65	5.78	5.90	5.96	6.03
($T_{max} = 700$°C)												
Westerly granite	P	—	1.98	2.87	3.63	4.13	4.48	4.94	5.25	5.60	5.77	5.91
($T_{max} = 950$°C)												
15015,18	P	3.50	3.68	3.90	4.13	4.27	4.38	4.49	4.54	4.64	4.74	4.85
Rock 652	P	3.68	4.04	4.42	4.65	4.76	4.81	4.90	4.96	5.07	5.15	5.23
Rock 613	P	3.92	4.17	4.39	4.57	4.65	4.71	4.80	4.85	4.93	4.99	5.05

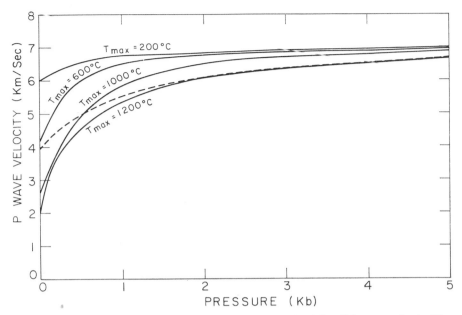

Fig. 1. Compressional wave velocities of thermally cycled Fairfax diabase samples (solid lines) and lunar sample 14310,72 (dashed line). The values of T_{max} in the figure are the *maximum* temperatures to which the samples were heated. The sample heated to 1200°C was partially melted. The velocities were measured at room temperature.

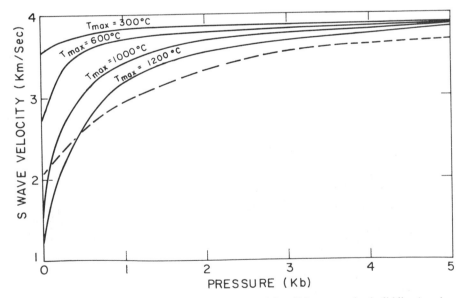

Fig. 2. Shear wave velocities of thermally cycled Fairfax diabase samples (solid lines) and lunar sample 14310,72 (dashed line). The values of T_{max} in the figure are the *maximum* temperatures to which the samples were heated. The sample heated to 1200°C was partially melted. The velocities were measured at room temperature.

velocity are not constant; they decrease systematically with T_{max}. For example, at a pressure of 1 bar, velocity values of 5.95, 4.20, and 2.02 km/sec correspond to T_{max} values of 200, 600, and 1000°C, respectively. The S-wave velocities of Fairfax diabase (Fig. 2) and the P-wave velocities of Westerly granite (Table 2) show similar behavior. With these observations in mind, we propose that: (1) thermal cycling does not alter

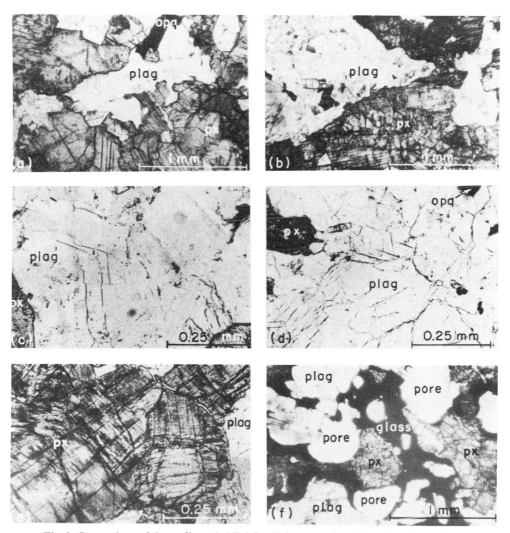

Fig. 3. Comparison of thermally cycled Fairfax diabase samples with unheated samples. (a) General view of unheated diabase, (b) general view of diabase heated to $T_{max} = 1000°C$, (c) expanded view of plagioclase grain in unheated diabase, (d) expanded view of plagioclase grain in diabase heated to $T_{max} = 1000°C$, (e) expanded view of pyroxene grain in diabase heated to $T_{max} = 1000°C$, and (f) general view of partially melted diabase ($T_{max} = 1200°C$).

significantly the mineralogy of our samples until a partial melt is formed, (2) thermal cycling cracks rocks, (3) crack density increases with T_{max}, (4) an increase in crack density decreases velocities at low confining pressures, and (5) velocity is independent of T_{max} at high confining pressures because cracks close. We will discuss each of these points in the following paragraphs. Because the mineralogy of Fairfax diabase is similar to that of the lunar igneous rocks (Table 1), we will primarily restrict our discussion to this rock type.

Two features suggest that little significant mineral alteration has occurred with heating in the Fairfax diabase cores: (1) observations of the thin sections (Fig. 3) and (2) the constancy of the 5 kb velocities for all T_{max}. In the pyroxene grains (Fig. 3e), this alteration (probably oxidation) is manifested in some grains by a slight darkening in color and by mottled irregular patches of alteration material concentrated along the cracks. The biotite has also darkened slightly, but it is still recognizable as biotite for $T_{max} = 1000°C$. The most striking feature in the thin sections of the heated cores is the high degree of fracturing in the plagioclase grains (compare Figs. 3c and 3d). This large increase in crack density with T_{max} is undoubtedly the cause of the dramatic increase of $(\partial V/\partial P)_T$ with T_{max} at low confining pressure. There is also evidence of cracking in the pyroxene and biotite grains at high T_{max}, but the amount is minor in comparison to that in the plagioclase. It was not possible to observe cracking along grain boundaries.

Most terrestrial rock samples, whether in situ or in the lab, contain some water. The process of thermal cycling under vacuum, in addition to forming new cracks, drives water from cracks originally in the rock. Dry samples have P-wave velocities up to 40% lower than saturated samples; S-wave velocities are unaffected by the presence of water (Nur and Simmons, 1969). Both P- and S-wave velocities of thermally cycled Fairfax diabase decrease with T_{max}; only P-wave values should decrease if desaturation is the dominant process. Apparently then, the effect of increased cracking on acoustic velocity (Nur and Simmons, 1970) dominates any effects on velocity due to the presence (or absence) of water in cracks of the Fairfax diabase.

Q Results

We also measured the quality factor Q in Fairfax diabase and Westerly granite. The value of Q should increase with desaturation, but decrease with cracking (Warren et al., 1971, and Wang et al., 1971). In Fig. 4 the values of Q for Fairfax diabase and Westerly granite are plotted as functions of T_{max}. For both rocks there is an initial increase in the value of Q, followed by a sharp decrease near 600°C. We interpret the initial rise in the value of Q as being due to the expulsion of water from the sample and the sharp decrease near 600°C as being due to the influence of cracking dominating that of water.

Discussion

Most of the cracks in all of our Fairfax diabase cores are closed by 1 kb confining pressure (i.e., the velocity at 1 kb for the thermally cycled samples is close to that of the unheated sample). Hence the thermal cycling process generates cracks with low

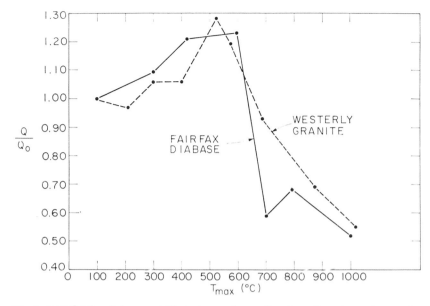

Fig. 4. Q of Fairfax diabase and Westerly granite as a function of maximum temperature (T_{max}) to which each sample was cycled. The values are normalized to Q_0 where Q_0 (Westerly granite) = 235 and Q_0 (Fairfax diabase) = 462. All measurements were made at room temperature.

aspect ratios. Some lunar samples have velocity profiles similar to those of the heated Fairfax diabase samples; for example, Apollo 12 samples (Wang *et al.*, 1971, and Kanamori *et al.*, 1971) have velocity profiles which can be fitted quite well by the $T_{max} = 600°C$ contour. We infer that the process of thermal cycling *may* be important as one cause of crack formation in lunar rocks. This inference is consistent with the interpretations of (1) the acquisition subsequent to initial cooling of part of the natural remanent magnetization of some Apollo 12 rocks (Grommé and Doell, 1971), (2) the recrystallization of pyroxene in sample 15415 (Stewart *et al.*, 1972), and (3) the compositional trends in the spinel group and related phases of several lunar samples being due to reduction caused by subsolidus heating (Haggerty, 1972). On the other hand, sample 14310,72, other Apollo 14 samples (Mizutani *et al.*, 1972), and Apollo 11 samples (Kanamori *et al.*, 1970) show a much wider distribution of aspect ratios in which cracks require 2 to 3 kb confining pressure to be closed. The discrepancy in the shape of velocity profiles implies that the cracks in these lunar rocks were probably not formed by a high temperature thermal cycling process. Repeated shock impacts, diurnal temperature variations, and small stress variations applied over geologic time (tidal stress) are other possible sources of crack production. The amount of cracking caused by long term temperature cycling and small stress variations is unknown, although larger stress variations of 1 to 2 kb (Hardy and Chugh, 1970) and shock events (Short, 1969) are known to produce cracks. It is also possible that cracks in lunar rocks were present with initial rapid crystallization and outgassing under lunar surface conditions (Warren *et al.*, p. 2350, 1971).

A wider distribution of aspect ratios has been simulated to some extent by partially melting ($\sim 5\%$) one of the Fairfax diabase cores. A portion of the melt flowed to the surface of the sample, leaving behind a wide assortment of irregularly shaped cracks and vesicules (Fig. 3f). The relatively lower velocity for this core throughout the 5 kb range (Figs. 1 and 2) shows the effect of this increase in porosity and wider distribution of aspect ratios.

In Fig. 5 the velocity profiles for lunar breccias 15015,18 and 15418,43 are compared with those of two terrestrial rocks. Rock 613 is a terrestrial volcanic rock of intermediate silica composition, containing phenocrysts of plagioclase, hornblende, biotite, and quartz ($\sim 40\%$) set in a very fine grained felty matrix ($\sim 60\%$). Its density is 2.42 g/cm^3. Rock 652 is a terrestrial vuggy volcanic rock of intermediate composition, composed of phenocrysts of olivine and biotite ($\sim 5\%$) in a fine grained matrix of plagioclase microlites, olivine, and opaque minerals ($\sim 95\%$). Its density is 2.52 g/cm^3. These rocks were chosen as analogue materials for lunar breccias because: (1) both samples are composed of crystal fragments set in a fine grained matrix, (2) both samples have low densities comparable to most terrestrial and lunar breccias (Kanamori *et al.*, 1970), and (3) both samples have low *P*-wave velocities (Fig. 5) similar to those of lunar breccias (Kanamori *et al.*, 1970). Sample 15015,18 closely resembles these terrestrial rocks in this respect. However, sample 15418,43, having a density of 2.80 g/cm^3 and a 5 kb velocity of 6.8 km/sec, is comparable with lunar igneous rocks. This similarity is apparently due to shock melting of the sample and subsequent recrystallization (NASA, 1971, and LSPET, 1972). On the other hand,

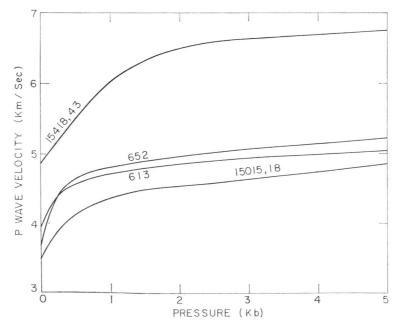

Fig. 5. Compressional wave velocities of lunar breccias 15418,43 and 15015,18 and two terrestrial analogues.

only the outer surface of 15015 has been melted; the inner section from which 15015,18 was cut has remained a compact, though somewhat friable, breccia.

Acknowledgments—We thank David Riach for preparing our samples and Bob Stevens for preparing our photographs. This work was supported by NASA contract NGR-22-009-540.

References

Anderson O. L., Scholz C., Soga N., Warren N., and Schreiber E. (1970) Elastic properties of a micro-breccia, igneous rock and lunar fines from Apollo 11 mission. *Proc. Apollo 11 Lunar Sci. Conf., Geochim. Cosmochim. Acta* Suppl. 1, Vol. 3, pp. 1959–1973. Pergamon.

Birch F. (1960) The velocity of compressional waves in rocks to 10 kilobars, Part 1. *J. Geophys. Res.* **65**, 1083–1102.

Birch F. (1961) The velocity of compressional waves in rocks to 10 kilobars, Part 2. *J. Geophys. Res.* **66**, 2199–2224.

Fairbairn H. W., Schlecht W. G., Stevens R. E., Dennen W. H., Ahrens L. H., and Chayes F. (1951) A cooperative investigation of precision and accuracy in chemical, spectrochemical and modal analysis of silicate rocks. *U.S. Geol. Survey Bull.* **980**, 1–71.

Grommé C. S. and Doell R. R. (1971) Magnetic properties of Apollo 12 lunar samples 12052 and 12065. *Proc. Second Lunar Sci. Conf., Geochim. Cosmochim. Acta*, Suppl. 2 Vol. 3, pp. 2491–2499. MIT Press.

Haggerty S. E. (1972) Subsolidus reduction and compositional variations of lunar spinels (abstract). In *Lunar Science—III* (editor C. Watkins) pp. 347–349, Lunar Science Institute Contri. No. 88.

Hardy H. R. Jr. and Chugh Y. P. (1970) Failure of geologic materials under low cycle fatigue. Paper presented at Sixth Canadian Symposium on Rock Mechanics, Montreal, Quebec.

Ide J. M. (1937) The velocity of sound in rocks and glasses as a function of temperature. *J. Geol.* **45**, 689–716.

Jackson D. D. (1969) Grain boundary relaxation and the attenuation of seismic waves. Sc.D. Thesis, Massachusetts Institute of Technology, Cambridge, Mass.

Kanamori H., Nur A., Chung D., and Simmons G. (1970) Elastic wave velocities of lunar samples at high pressures and their geophysical implications. *Proc. Apollo 11 Lunar Sci. Conf., Geochim. Cosmochim. Acta* Suppl. 1, Vol. 3, pp. 2289–2293. Pergamon.

Kanamori H., Mizutani H., and Hamano Y. (1971) Elastic wave velocities of Apollo 12 rocks at high pressures. *Proc. Second Lunar Sci. Conf., Geochim. Cosmochim. Acta*, Suppl. 1 Vol. 3, pp. 2323–2326. MIT Press.

Latham G., Ewing M., Dorman J., Press F., Toksoz N., Sutton G., Meissner R., Duennebier F., Nakamura Y., Kovach R. and Yates M. (1970) Seismic data from man-made impacts on the moon. *Science* **170**, 620–626.

LSPET (1972) (Lunar Sample Preliminary Examination Team) Preliminary examination of the lunar samples from Apollo 15. *Science* **175**, 363–373.

Mizutani H., Fujii N., Hamano Y., Osako M., and Kanamori H. (1972) Elastic wave velocities and thermal diffusivities of Apollo 14 rocks (abstract). In *Lunar Science—III* (editor C. Watkins) pp. 547–549, Lunar Science Institute Contri. No. 88.

NASA (1971) *Lunar Sample Information Catalogue*, Apollo 15. Manned Spacecraft Center, Houston.

Nur A. and Simmons G. (1969) The effect of saturation on velocity in low porosity rocks. *Earth Planet. Sci. Lett.* **7**, 99–108.

Nur A. and Simmons G. (1970) The origin of small cracks in igneous rocks. *Int. J. Rock Mech. Min. Sci.* **7**, 307–314.

Short N. M. (1969) Shock metamorphism of basalt. *Modern Geol.* **1**, 81–95.

Simmons G. (1964) Velocity of shear waves in rocks to 10 kilobars, 1. *J. Geophys. Res.* **69**, 1123–1130.

Stewart D. B., Ross M., Morgan B. A., Appleman D. E., Huebner J. S., and Commeau R. F. (1972) Mineralogy and petrology of lunar anorthosite 15415 (abstract). In *Lunar Science—III* (editor C. Watkins) pp. 726–728, Lunar Science Institute Contri. No. 88.

Wang H., Todd T., Weidner D., and Simmons G. (1971) Elastic properties of Apollo 12 rocks. *Proc. Second Lunar Sci. Conf., Geochim. Cosmochim. Acta* Vol. 3, pp. 2327–2336. MIT Press.

Warren N. W. and Latham G. V. (1970) An experimental study of thermally induced microfracturing and its relation to volcanic seismicity. *J. Geophys. Res.* **75,** 4455–4464.

Warren N., Schreiber E., Scholz C., Morrison J. A., Norton P. R., Kumazawa M., and Anderson O. L. (1971) Elastic and thermal properties of Apollo 11 and Apollo 12 rocks. *Proc. Second Lunar Sci. Conf., Geochim. Cosmochim. Acta* Vol. 3, pp. 2345–2360. MIT Press.

Proceedings of the Third Lunar Science Conference
(Supplement 3, *Geochimica et Cosmochimica Acta*)
Vol. 3, pp. 2587–2598
The M.I.T. Press, 1972

Applications to lunar geophysical models of the velocity-density properties of lunar rocks, glasses, and artificial lunar glasses*

NICK WARREN and ORSON L. ANDERSON

Institute of Geophysics and Planetary Physics,
University of California, Los Angeles 90024

and

NAOHIRO SOGA

Department of Industrial Chemistry, Kyoto University,
Sakyo-ku, Kyoto, Japan

Abstract—Theoretical and experimental results from studies on returned lunar rocks, lunar glasses, and artificial lunar glasses are used to discuss possible geophysical models consistent with the observed lunar velocity profiles (from Apollo 12, 14, and 15 PSE stations).

In the upper 25 kilometers of the mare, where the velocity profile is controlled much more by structure than by chemistry, rock type cannot be inferred from the velocity profile. However, assuming mare-type basalts, the observed seismic velocities and scattering of the seismic signal appear to be compatible with high *in situ* densities of about 3.2 g/cm³.

Velocity-density relations for lunar zero-porosity rocks are predicted using data from artificial lunar glasses and the fourth-power law. At zero porosity, mare-type (Apollo 12) basalts and feldspathic basalts (highland type) are predicted to have equal velocities at densities which differ by 0.3 to 0.4 g/cm³. A magmatic mineralogy may be able to account for the high-velocity layer below 65 kilometers.

INTRODUCTION

OBSERVATIONAL SEISMIC PROFILES of a lunar mare region to a depth of over 70 km are now available and can be compared with results from laboratory and theoretical studies on the physical properties of lunar and lunarlike materials.

The first section of this paper deals with our studies on artificial lunar glasses and uses the data to discuss properties of solid or zero-porosity rock at depths of 25 to 65 km. The problem of lunar structure at depths less than about 30 km is dealt with in the second section. Possible rock types in the high-velocity zone (depths >65 km) are discussed in the final section.

ARTIFICIAL LUNAR GLASSES AND LUNAR GLASSES

Elastic properties were determined for a number of artificial glasses whose compositions were similar to those of lunar rock samples from Apollo 12 and 14. Their chemical compositions are shown in Table 1. The comparison of major elements for these glasses and Apollo 12 and 14 rocks is shown in Fig. 1.

---

* Publication number 1034 of the Institute of Geophysics and Planetary Physics.

Table 1. Chemical composition and elastic properties of Apollo 12- and 14-type artificial glass.

| | Glass number and rock type | | | | |
	12009	14052	14049	14305	14065
Composition (wt.%)					
SiO_2	41	48	49	49	48
Al_2O_3	11	12	17	16	21
MgO	12.5	8.4	11	13	8.3
FeO	20	16	10	9.5	6.8
CaO	10	12	8.9	7.4	12
TiO_2	3.3	1.5	1.7	1.6	1.0
Na_2O	0.5	0.4	0.85	0.9	1.0
K_2O	—	—	0.53	1.2	—
Density (g/cm³)	3.066	2.907	2.802	2.798	2.748
Elastic properties					
Compressional velocity (km/sec)	6.693	6.494	6.519	6.633	6.553
Shear velocity (km/sec)	3.669	3.612	3.670	3.711	3.698
Poisson's ratio	0.285	0.276	0.268	0.272	0.266

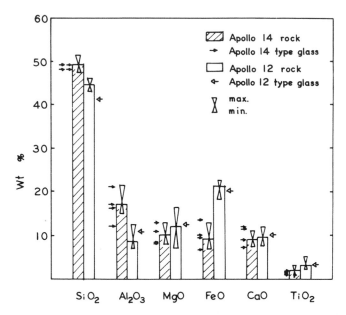

Fig. 1. The compositions of artificial glasses in comparison with those of Apollo 12 and Apollo 14 rocks. The data for these rocks are from the LSPET (1971) report.

The bulk chemistries represented range over mare (e.g., 12009) basalts (*m* basalts) to more feldspathic basalts (*f* basalts), e.g., 14065. Feldspathic basalts have been suggested by various authors as perhaps representative of the rock type making up the lunar highland or the lunar primitive crust (Gast and McConnell, 1972; Reid *et al.*, 1972; Smith *et al.*, 1970). Less feldspathic, more iron-rich basalt chemistries seem representative of lunar maria basin-fill material. This distinction in compositional variation will remain important throughout this paper.

Reagent-grade SiO_2, Al_2O_3, MgO, $FeC_2O_4 \cdot 2H_2O$, $CaCO_3$, TiO_2, Na_2CO_3, and K_2CO_3 were used for preparing the glasses. Iron oxalate in the glass batches is known to create reducing atmosphere during melting, but a small amount of metallic silicon (0.3%) was added as the reducing agent to the batches in order to prevent oxidation of ferrous ions as much as possible during melting. After mixing thoroughly, the batches were melted in sintered-alumina crucibles with a capacity of 50 cm^3 at 1450 to 1550°C for 1 to 2 hours in a closed electric furnace with silicon carbide heating elements. The glass was directly poured onto a steel plate and quenched in air. The glass samples obtained were prepared for sound-velocity measurements by grinding and polishing the two-basal surfaces flat and parallel (± 0.001 mm). Their bulk densities were obtained from the measured mass and measured dimensions.

The pulse-superposition method, developed and described in detail by McSkimin (1964), was employed to obtain the compressional- and shear-sound velocities of the specimens.

The electronic equipment used was the same as was used in the previous studies (Soga, 1969).

The measurement of sound velocity was made with 20 MHz X-cut or Y-cut 0.25-inch diameter quartz transducers bonded to the specimen with Dow-Corning resin 276-V9. The compressional- and shear-sound velocities were calculated from determined pulse repetition frequencies and from the length of the specimens. The results are listed in Table 1.

The densities of these artificial glasses are plotted as a function of FeO content in Fig. 2. The density appears to increase linearly with increasing FeO content for these glasses. This relation may serve as an indicator of the presence of FeO in the glasses whose compositions are similar to lunar rocks from Apollo 12 and 14.

The elastic properties are plotted as a function of density in Figs 3, 4, and 5. These figures include the data for lunar glasses reported previously from Apollo 11 (Anderson *et al.*, 1970). The correlations observed for lunar glasses from Apollo 11 also can

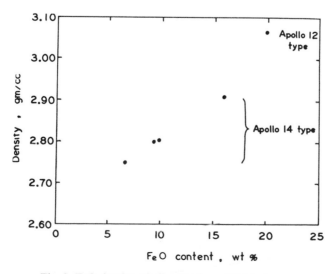

Fig. 2. FeO-density relationship for artificial glasses.

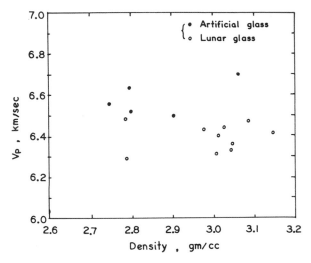

Fig. 3. Density-compressional velocity relationship for Apollo 12- and Apollo 14-type artificial glasses and for Apollo 11 lunar glasses.

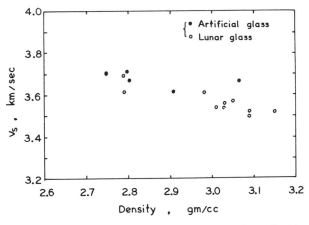

Fig. 4. Density-shear velocity relationship for Apollo 12- and Apollo 14-type artificial glasses and for Apollo 11 lunar glasses.

be seen for these artificial glasses. Thus, we believe that the data for lunar glasses from Apollo 12 and 14 would fit into this pattern, although we have not been able to determine them because of unavailability of glass spherules suitable for the velocity measurements. As indicated in the previous study, a useful relation between the Poisson's ratio and FeO content is apparent in Fig. 6. Also, the velocities remain about the same, in spite of the change in density. This again confirms the previous results for lunar glasses from Apollo 11, and is consistent with the fact that the density of rocks and minerals, but not their velocity, increases with the addition of Fe.

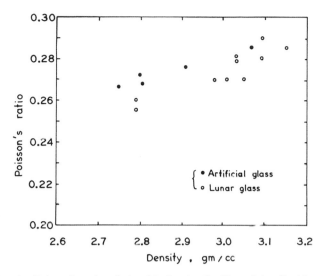

Fig. 5. Density-Poisson's ratio relationship for Apollo 12- and Apollo 14-type artificial glasses and for Apollo 11 lunar glasses.

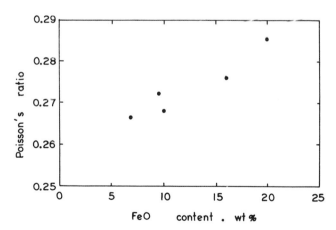

Fig. 6. FeO-Poisson's ratio relationship for artificial glasses.

We performed this study with a belief that the values of V_s and V_p for lunar glasses or pseudolunar glasses would yield typical values for lunar rocks with no porosity or cracks. A previous study indicated that the elastic moduli vary as the fourth power of the density, when the structure changes from a glassy to a crystalline state (Soga *et al.*, 1972). Fig. 7 is reproduced here to illustrate the point.

Although the density and sound velocities are affected by the manner in which the glass is melted, cooled, and annealed, these effects are much smaller than that of composition (usually less than 1%). Errors arising from the difference in these conditions are within the circle of data in Figs. 2 through 6. Applying this fourth-power relationship, one can estimate the velocities of pore-free rocks from those of glasses having

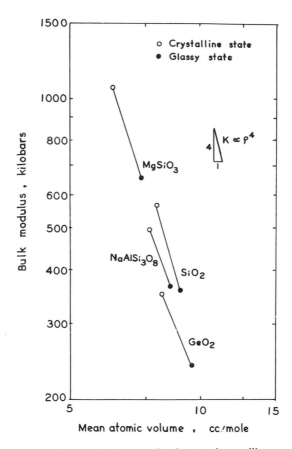

Fig. 7. Comparison in bulk modulus between the glassy and crystalline states as a function of volume (after Soga *et al.*, 1972).

identical compositions. The results are shown in Fig. 8, which will also be used to illustrate other aspects of the lunar model problem.

The hatched rectangles at P = 0, $V_p \approx 6.5$ km/sec represent V_p data from lunar glass spheroids (Schreiber *et al.*, 1970) (slant-hatched) and from the artificial glasses (vertical-hatched). Importantly, the *m*-basalt glass, 12009, has a velocity of 6.693 km/sec at $\rho = 3.066$ g/cm^3, while the *f*-basalt glass, 14065, has $V_p = 6.553$ km/sec at $\rho = 2.748$ g/cm^3. From the data on glass type 12009, if a density for the solid rock is assumed to be 3.2 to 3.3 g/cm^3, then intrinsic V_p is expected to be between 7.1 and 7.5 km/sec. For *f*-basalt glass, 14065, if solid rock densities of 2.9 to 3 are assumed, then intrinsic V_p is also expected to be between 7.1 and 7.5 km/sec. These results are in good agreement with density and 10-kb velocity data on real rock.

Densities of mare basalts are on the order of 3.2 to 3.3 g/cm^3 (Anderson *et al.*, 1970; Kanamori *et al.*, 1970; Wang *et al.*, 1971; Wood *et al.*, 1970). At 10 kb, measured velocities for these rocks are expected to be near the intrinsic value (Wang *et al.*, 1971;

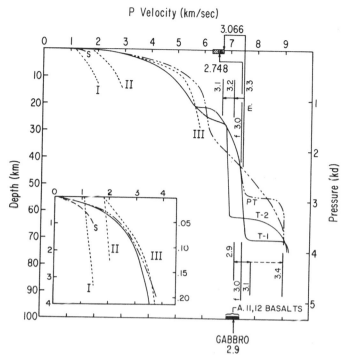

Fig. 8. Observed V_p versus depth (curves PT, T-1, and T-2) comparison with lunar glass, and rock data and models. Lunar and pseudolunar glass data at room pressure are indicated by the rectangles at the top of the graph. The lines with arrow tips which are labeled with density values 3.066 and 2.748 g/cm³ indicate the velocities for two glasses with chemistries similar to m basalt 12009 and f basalt 14065. The vertical lines joined to the arrows indicate predicted velocities for rocks with these chemistries, but at higher densities as indicated. Curves marked I, II, and III correspond to velocity models (Schreiber et al., 1970), and curve s corresponds to lunar soil under initial compaction (Warren et al., 1971). The black rectangle at the base of the graph indicates 10-kb velocities for Apollo 11 and 12 basalts (Wang et al., 1971; Kanamori et al., 1970, 1971) and a terrestrial gabbro (Birch, 1961).

Kanamori et al., 1970). The 10-kb results are indicated by the black rectangle labeled "A.11,12 basalts" at the base of Fig. 8.

In the case of f basalts, the predicted results are also consistent with observation. The bulk chemistry of 14310 is very close to that of the glass of rock type 14065. At room temperature and pressure, the bulk density of returned samples 14310,82 is 2.83 g/cm³. Ringwood and Green (1972) indicate a theoretical density of 2.95 g/cm³ for 14310 in its low-pressure, gabbroic phase. The compressional velocity at 10 kb for a terrestrial gabbro of very similar composition and $\rho = 2.93$ g/cm³ is indicated by the arrow at the base of Fig. 8 (Mellen Wis gabbro; Birch, 1961) and is in agreement with predicted V_p from the glass data.

Lunar seismic velocity profiles are indicated in Fig. 8 by dashed curve PT (two alternatives are given for depths between about 46–65 km) and solid curves T-1 and T-2 (curve PT from LSPET, 1972a; T-1 from Toksöz et al., 1971; T-2 from Toksöz et al., 1972). Curve PT is based on travel times. Curve T-1 is based on travel times plus amplitudes, and curve T-2 was derived by generating matching theoretical seismograms. The actual profile of V_p in this depth range is not yet uniquely determined. In the depth zones with approximately constant velocity, V_p may be as low as 6.9 to 7 km/sec, or may be closer to 7.3 to 7.5 km/sec. The higher estimates of V_p are consistent with either solid m or f basalts at their expected solid-state densities. The lower estimates of V_p (T-2) correspond more closely to expected values for true anorthosite (Toksöz et al., 1972), or bulk chemistries similar to 15415 or 15418 (Apollo 15, LSPET, 1972b).

If, as is generally assumed, the lunar crust underlying the maria basins is of f basalt to gabbroic anorthosite to anorthositic composition, the densities below the transition may be 2.9 to 3 g/cm^3, which is below the mean density of the moon, and very likely below the in situ density of the overlying mare basalts.

Structure of Mare Basin Fill—Depth Less than 25 Kilometers

The extensive scattering and low-velocity properties of the lunar mare leave much uncertainty about rock density and chemistries in this region.

In Fig. 8, three velocity models based on Apollo 11 data are shown (curves labeled I, II, III) from Schreiber et al. (1970) for comparison to observed velocity profiles in this region (expansion of the 0- to 4-km zone is shown in the inset).

The curves labeled I, II, and s refer to the very near surface and will be discussed briefly later. Model III corresponds to a simple, continuously welded, microfractured mare basalt "half space" in which cracks close as a function of pressure or depth. This model was derived from compressibility and compressional velocity data to 5 kb on rock 10017 (Anderson et al., 1970). The model has a corresponding density of 3.1 g/cm^3. A low-density model with similar velocity can also be generated assuming a more feldspathic basalt ($\rho = 2.9$ g/cm^3) in which velocity is also controlled by similar microfracture and fracture effects. The marked effect of microfractures on velocity has been recently experimentally demonstrated (Wang et al., 1972).

Scattering as well as velocity can be modeled for this layer assuming a joint or fracture distribution in the bedrock, in which the joints make up the large-scale structural end of a microfracture-fracture distribution compatible with velocity model III. On this bases, a simple and partially quantitative geological model is proposed.

Warren (1972) indicates that flat-crack porosities of as low as 10^{-3} to 10^{-4} are sufficient to explain the observed velocity pressure results, given a penny-shaped crack size distribution with a ratio of major to minor axis from about 10^4 to 10^3. Using such cracks as a first approximation to a joint, a simple, jointed, dry bedrock which satisfies the observed velocities may then also predict the observed scattering if major jointing dimensions are on the order of hundreds of meters.

Joints acting as scatterers are postulated to be open. Frictional attenuation is reduced if cracks and joints which are just closed under the local stresses become locked. Such locking under lunar conditions may be realistic. For joints or cracks in deep bedrock ($\gtrsim$ a few km) which have never been exposed to surface conditions, the characterization of fractured grains by Ryan (1966) may hold (see also NASA CR

1090). For example, if a simple joint opens up within a lava flow due to thermal contraction and is closed by an increased overburden of new lava flows, the initial fracture
surfaces may duplicate the clean fracture faces, hard vacuum conditions of Ryan's
experiments, and on closing the welding could be extremely strong even at low contact
pressure.

Such a low-porosity bedrock model is also consistent with both a lava flow model
of a mascon akin to that presented by Wood et al. (1970) and an interpretation of the
lunar mascons as flat disk features (Sjorgren, 1972). An uncompensated layer of m
basalt overlying f basalt could provide an in situ density contrast of about 0.3 to 0.4
g/cm^3. Importantly, if a superisostatic lava flow model of a mascon is correct, no
special effect on the lunar seismic profiles is predicted (no high-velocity or particular
seismic reflection or refraction interface need be observed.)

The above bedrock model cannot hold all the way to the mare surface. In the
uppermost 1 to 2 km, the observed velocity profile departs from that which is compatible with the simple, jointed bedrock model, and crosses the model curves for
heterogeneous (microbreccia) igneous medium (model II), a soil-microbreccia model
(model I), and grades into the velocity pressure (depth) profile for a lunar soil which
has not been precompacted (curve s) (Anderson et al., 1970; Warren et al., 1971).

A gradation with increasing depth from the surface may be expected from complete fragmentary material, through fractured loose block material, into welded bedrock (e.g., lava-flow structure with jointing). Clearly, interbedding at depths of welded
ash, and lava tube structures, are not discounted. However, extensive ash, lava-tube,
or meteor-impact damage to depths of 10 to 20 km do not seem to be required to
satisfy the observed velocity and the inferred possible density profiles of filled maria
basins. A schematic of a mare model as outlined previously (for the seismically pro-
filed eastern Procellarum region) is given in Fig. 9.

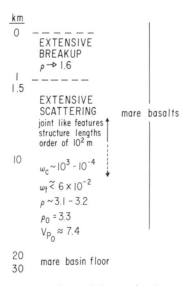

Fig. 9. Schematic mare model. ω_c is partial porosity due to joints (flat cracks) and
ω_t denotes total bulk or mean porosity. (See text for discussion.)

Velocity–Density for Depth over 65 Kilometers

The extremely high apparent velocity of over 9 km/sec at a depth over 65 km is based on a single data point, and the interpretation problem may be somewhat premature. However, this result raises the more general question of whether such a high velocity can be associated with low-density phase ($\rho = 3.3$ to 3.4 g/cm^3).

High velocity may be compatible with either high-pressure phases of rocks of lunar surface chemistry, or with magmatic residual material. The large step in V_p from 7 km/sec to 9 km/sec at only 3.4 kb (65 km) seems to make a high-pressure mineralogy interpretation for the high velocity layer difficult, since it is not apparent from the high-velocity data that intermediate-pressure mineral assemblages are present.

We suggest that a low-density, high-velocity zone may be modeled as a magmatic residual mantle of the alumina-rich crust. This proposal is compatible with the type of model proposed by Gast (1972).

Toksöz et al. (1972) have compared the lunar V_p to Mg-rich olivines. Gast (1972) has suggested olivine enrichment at these depths.

If low-pressure magmatic spinel can have been retained as a phase in association with the olivine (Turner and Verhoogen, 1960), then it provides an interesting model for the high-velocity zone. Spinel as a high-velocity phase ($V_p = 9.9$ km/sec, $\rho = 3.6$ g/cm^3) has been suggested by Schreiber (personal communication).

The occurrence of magnesium, slightly chromiferous, spinel on the moon has been reported by Christophe-Michel-Levy et al. (1972) and Mason et al. (1972). Walker et al. (1972) and Ford et al. (1972) have pointed out the apparent petrological relation of the high-aluminum basalts to the olivine-spinel-plagioclase cotectic.

Spinel can be retained as a phase due to gravity separation from the rest of the melt, or due to local silica undersaturation. The finding of rather pure spinel in some of the Apollo 15 samples would be of real interest. The exact shape of the possible spinel field has not been determined as a function of the presence of Fe, Mg, and Cr. However, unless it is considerably shrunk by an Fe/Mg ratio close to unity, the possibility of magmatic spinel formation seems reasonable.

An approximately 15% enrichment of spinel to a dunitic rock (assumed $\rho = 3.25$ g/cm^3, $V_p = 8.4$ km/sec) gives a density of 3.3 g/cm^3 and a Voigt-Reuss-Hill average velocity of $V_p = 8.6$ km/sec.

Although a magmatic origin to the high-velocity zone appears to be more reasonable than high-pressure phases, values of V_p for high-pressure phases of f basalts can be estimated using the fourth-power law. A rock of bulk chemistry 14065, in a phase with $\rho = 3.4$ g/cm^3, has a predicted $V_p = 9$ km/sec (Fig. 8).

Rocks of chemistry 14310 (and similar chemistries, e.g., 14065) have an eclogitic (Ca-rich garnet) phase with $\rho = 3.4$ g/cm^3 to 3.5 g/cm^3. However, the pressure of formation is 21 kb (Ringwood, 1972).

A gabbroic anorthosite or anorthosite (e.g., bulk chemistry of 15418, 15415) may have this high velocity associated with a phase of lower density. Again using the fourth-power law and assuming an anorthositic rock with $V_p = 7$ km/sec at $\rho = 2.8$ g/cm^3 (Clarke, 1966), then $V_p = 9.1$ km/sec is predicted for a high-pressure phase with $\rho = 3.34$ g/cm^3.

Acknowledgments—We wish to thank Dr. W. Gary Ernst, Dr. G. W. Wetherill, Dr. P. W. Gast, and Dr. N. M. Toksöz for important discussions on parts of this paper. Work was performed under NASA Grant NGL 05-007-330.

REFERENCES

Anderson O. L., Scholz C., Soga N., Warren N., and Schreiber E. (1970) Elastic properties of a microbreccia, igneous rock, and lunar fines from Apollo 11 mission. *Proc. Apollo 11 Lunar Sci. Conf.*, *Geochim. Cosmochim. Acta* Suppl. 1, Vol. 3, pp. 1959–1973. Pergamon.

Birch F. (1961) The velocity of compressional waves in rocks to 10 kbars, 2. *J. Geophys. Res.* **66**, 2199–2224.

Christophe-Michel-Lévy M., Lévy C., and Pierrot R. (1972) Mineralogical aspects of Apollo 14 samples: Lunar chondrules; pink spinel bearing rocks; ilmenites (abstract). In *Lunar Science—III* (editor C. Watkins), pp. 136–138, Lunar Science Institute Contr. No. 88.

Clarke S. P. Jr. (editor) (1966) *Handbook of Physical Constants*, revised edition, Geological Society of America, Memoir 97.

Ford C. E., Humphries D. J., Wilson G., Dixon D., Biggar G. M., and O'Hara M. J. (1972) Experimental petrology of high alumina basalt, 14310, and related compositions (abstract). In *Lunar Science—III* (editor C. Watkins), pp. 274–276, Lunar Science Institute Contr. No. 88.

Gast P. W. and McConnell R. K. Jr. (1972) Evidence for initial chemical layering of the moon (abstract). In *Lunar Science—III* (editor C. Watkins), pp. 289–290, Lunar Science Institute Contr. No. 88.

Kanamori H., Nur A., Chung D. H., and Simmons G. (1970) Elastic wave velocities of lunar samples at high pressures and their geophysical implications. *Proc. Apollo 11 Lunar Sci. Conf.*, *Geochim. Cosmochim. Acta* Suppl. 1, Vol. 3, pp. 2238–2293. Pergamon.

Kanamori H., Mizutani H., and Hamano Y. (1971) Elastic wave velocities of Apollo 12 rocks at high pressure. *Proc. Second Lunar Sci. Conf.*, *Geochim. Cosmochim. Acta* Suppl. 2, Vol. 3, pp. 2323–2326. MIT Press.

LSPET (Lunar Sample Preliminary Examination Team) (1971) *Apollo 14 Preliminary Science Report*. NASA SP-272.

LSPET (Lunar Sample Preliminary Examination Team) (1972a) Passive seismic experiment. In *Apollo 15 Preliminary Science Report*. NASA SP-289.

LSPET (Lunar Sample Preliminary Examination Team) (1972b) The Apollo 15 lunar samples: A preliminary description. *Science* **175**, 363–375.

Mason B., Melson W. G., and Nelen J. (1972) Spinel and hornblende in Apollo 14 fines (abstract). In *Lunar Science—III* (editor C. Watkins), pp. 512–514, Lunar Science Institute Contr. No. 88.

McSkimin H. J. (1964) Ultrasonic methods for measuring the mechanical properties of liquids and solids. Chapter 4. In *Physical Acoustics* (editor W. P. Mason), Vol. 1-A, Academic Press, New York.

NASA CR-1090 (1968) Ultrahigh vacuum adhesion related to the lunar surface. 89 pages.

Reid A. M., Ridley W. I., Warner J., Harmon R. S., Brett R., Jakes P., and Brown R. W. (1972) Chemistry of highland and mare basalts as inferred from glasses in the lunar soils (abstract). In *Lunar Science—III* (editor C. Watkins), pp. 640–642, Lunar Science Institute Contr. No. 88.

Ringwood A. E., Green D. H., and Ware N. G. (1972) Experimental petrology and petrogenesis of Apollo 14 basalts (abstract). In *Lunar Science—III* (editor C. Watkins), pp. 654–656, Lunar Science Institute Contr. No. 88.

Ryan J. A. (1966) Adhesion of silicates in ultrahigh vacuum. *J. Geophys. Res.* **71**, 4413–4425.

Schreiber E., Anderson O. L., Soga N., Warren N., and Scholz C. (1970) Sound velocity and compressibility for lunar rocks 17 and 46 and for glass spheres from the lunar soil. *Science* **167**, 732–734.

Sjogren W. L. (1972) Gravity measurements from Apollo 15 S-band transponder experiment (abstract). In *Lunar Science—III* (editor C. Watkins), pp. 707–709, Lunar Science Institute Contr. No. 88.

Smith J. V., Anderson A. T., Newton R. C., Olsen E. J., Wyllie P. J., Crewe A. V., Isaacson M. S., and Johnson D. (1970) Petrologic history of the moon inferred from petrography, mineralogy, and petrogenesis of Apollo 11 rocks. *Proc. Apollo 11 Lunar Sci. Conf.*, *Geochim. Cosmochim. Acta* Suppl. 1, Vol. 1, pp. 897–925. Pergamon.

Soga N. (1969) Pressure derivatives of the elastic constants of vitreous germania at 25, −78.5, and −195.8°C. *J. Appl. Phys.* **40**, 3382–3385.

Soga N., Ota R., and Kunugi M. (1972) Temperature and pressure effects on bulk moduli of inorganic glasses. *Proc. 1971 International Conf. Mechanical Behavior of Materials.* In press.

Toksöz M. N., Press F., Anderson K., Latham G., Ewing M., Dorman J., Lammlein D., Sutton G., Duennebier F., and Nakamura Y. (1971) Artificial impacts and internal structure of the moon (abstract). The Lunar Science Institute Conference on Lunar Geophysics, October 18–21.

Toksöz M. N., Press F., Anderson K., Dainty A., Latham G., Ewing M., Dorman J., Lammlein D., Sutton G., Duennebier F., and Nakamura Y. (1972) Velocity structure and properties of the lunar crust (abstract). In *Lunar Science—III* (editor C. Watkins), pp. 758–760, Lunar Science Institute Contr. No. 88.

Turner F. J. and Verhoogen J. (1960) *Igneous and Metamorphic Petrology*, second edition, McGraw-Hill.

Walker D., Longhi J., and Hays J. F. (1972) Experimental petrology and origin of Fra Mauro rocks and soil (abstract). In *Lunar Science—III* (editor C. Watkins), pp. 770–772, Lunar Science Institute Contr. No. 88.

Wang H., Todd T., Weidner D., and Simmons G. (1971) Elastic properties of Apollo 12 rocks. *Proc. Second Lunar Sci. Conf., Geochim. Cosmochim. Acta* Suppl. 2, Vol. 3, pp. 2327–2336. MIT Press.

Wang H., Todd T., Simmons G., and Baldridge S. (1972) Elastic wave velocities and thermal expansion of lunar and earth rocks (abstract). In *Lunar Science—III* (editor C. Watkins), pp. 776–778, Lunar Science Institute Contr. No. 88.

Warren N. (1972) Q and structure. Proceedings of Conference on Lunar Geophysics. *The Moon* (in press).

Warren N., Schreiber E., Scholz C., Morrison J. A., Norton P. R., Kumazawa M., and Anderson O. L. (1971) Elastic and thermal properties of Apollo 12 and Apollo 11 rocks. *Proc. Second Lunar Sci. Conf., Geochim. Cosmochim. Acta* Suppl. 2, Vol. 3, pp. 2345–2370. MIT Press.

Wood J. A., Dickey J. S. Jr., Marvin U. B., and Powell B. N. (1970) Lunar anorthosites and a geophysical model of the moon. *Proc. Apollo 11 Lunar Sci. Conf., Geochim. Cosmochim. Acta* Suppl. 1, Vol. 1, pp. 965–988. Pergamon.

Proceedings of the Third Lunar Science Conference
(Supplement 3, *Geochimica et Cosmochimica Acta*)
Vol. 3, pp. 2599–2609
The M.I.T. Press, 1972

Thermal expansion of Apollo lunar samples and Fairfax diabase

W. Scott Baldridge, Frank Miller, Herbert Wang, and Gene Simmons

Department of Earth and Planetary Sciences,
Massachusetts Institute of Technology,
Cambridge, Massachusetts 02139

Abstract—Measured values of the thermal expansion of seven Apollo samples over the temperature range −100°C to +200°C are reported and compared to terrestrial Fairfax diabase. For most of the lunar rocks the measured expansion is significantly lower (30%) than the intrinsic value calculated from Turner's equation for the thermal expansion of an aggregate. The measured expansion of the Fairfax diabase was equal to the intrinsic but was higher for samples of Fairfax which had been previously cycled to high temperatures. The effect of microfractures on the thermal expansion of rocks is discussed.

Introduction

The thermal expansion of rocks is a basic thermomechanical property of interest in computing density variation with depth. Recently several investigators have measured the thermal expansion of "lunar-analogue" materials in order to infer dilatational effects in the lunar environment (Thirumalai and Demou, 1970; Griffin and Demou, 1972). We have refined and extended our previous measurements made on lunar rocks (Baldridge and Simmons, 1971) and compared them to values obtained from terrestrial "analogues." We have also measured the thermal expansion of samples of Fairfax diabase which have been thermally cycled in order to understand better rock aggregate behavior.

Experimental Technique

The thermal expansion of seven Apollo rocks and of terrestrial Fairfax diabase was measured over the temperature interval −100°C to +200°C using a modified Brinkmann model TD IX dilatometer. The dilatometer, although designed to be mounted horizontally, was mounted vertically and the leaf spring push rod holder removed. This position reduced the friction contributed by movement of the transducer core against its enclosing quartz tube and permitted better contact between the sample and the quartz push rod. In addition we added a high-impedence ac amplifier near the displacement transducer which improved the signal-to-noise ratio and the sensitivity of the signal presented to the full wave synchronous demodulator. The temperature programming capabilities of the dilatometer were not used but rather all points were taken after both length and temperature had stabilized.

Four of the lunar samples were measured and the data reported previously (Baldridge and Simmons, 1971). Improved instrumental precision has made it desirable to remeasure these samples. All lunar samples were rectangular prisms with dimensions 1 × 1 × 2 cm except for rock 14318 which was an irregular chip 2 cm long. The Fairfax diabase samples were cylinders with dimensions approximately 1 cm diameter by 2.3 cm length. Opposite faces were ground flat and parallel to 0.003 cm. The (linear) thermal expansion was measured along the longest dimension only, and anisotropy was not investigated. The dilatometer was calibrated with a 2 cm long by 1 cm diameter rod of monocrystalline quartz

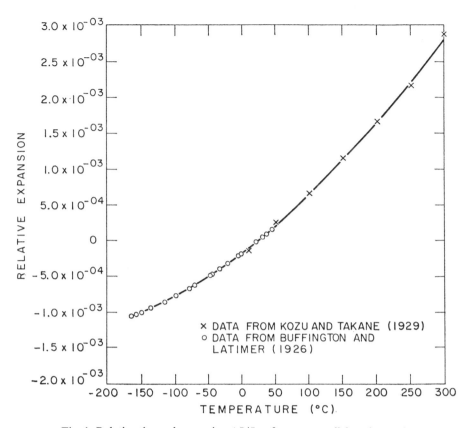

Fig. 1. Relative thermal expansion $\Delta L/L_0$ of quartz parallel to the c-axis.

cut parallel to the c-axis. We used the combined data of Buffington and Latimer (1926) and Kozu and Takane (1929) as the correct values for quartz over our experimental range of temperature. Each data set was recalculated to the relative expansion $\Delta L/L_0$, fitted to a second degree polynomial, and adjusted to a reference temperature of 25°C (Fig. 1). The expansion due to the quartz measuring system of the dilatometer was then subtracted from the total expansion of the rock samples and $\Delta L/L_0$ computed for the samples. The uncertainty in our measurements is 2% at 200°C.

Results

The relative expansions of the lunar breccias and igneous rocks are presented in Figs. 2 and 3, respectively, and of the Fairfax diabase for comparison in Fig. 4 (top). A second degree polynomial has been fitted to the data by the method of least squares. In Table 1 are tabulated the values of the mean volume coefficient α_v of thermal expansion

$$\alpha_v = 3\,\Delta L/[L_0 \cdot \Delta T(°C)]$$

derived from the data in Figs. 2, 3, and 4 as well as values for α_v calculated from Turner's equation for the thermal expansion of an aggregate (Kingery, 1960):

$$\alpha_r = \sum \alpha_i K_i V_i / \sum K_i V_i$$

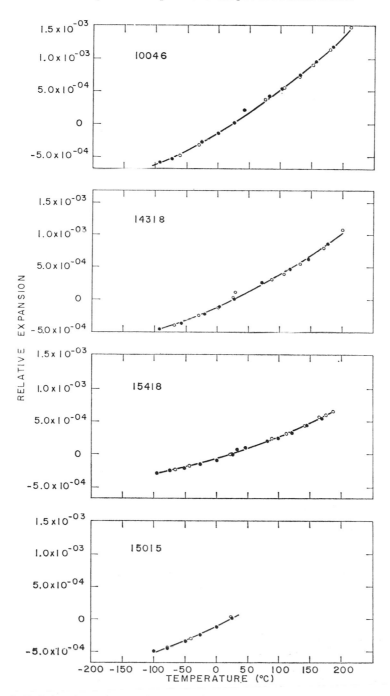

Fig. 2. Relative thermal expansion $\Delta L/L_0$ of lunar breccias. Open circles indicate measurements made at increasing temperature; solid circles indicate measurements made at decreasing temperature.

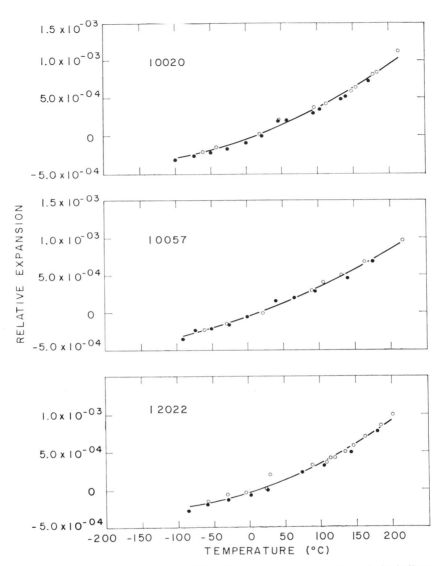

Fig. 3. Relative thermal expansion $\Delta L/L_0$ of lunar igneous rocks. Open circles indicate measurements made at increasing temperature; solid circles indicate measurements made at decreasing temperature.

where α_r and α_i are the volume coefficients of the aggregate and of the ith phase, respectively, and K_i and V_i are the bulk modulus and volume fraction of the ith phase. Several definitions of the thermal expansion coefficient α exist, but the mean volumetric definition has been chosen for convenience of calculation.

In applying Turner's equation to any material it is assumed that (1) the mixture is homogeneous, (2) each component of the mixture changes dimensions with temperature changes at the same rate as the aggregate, (3) no internal disruptions occur, and

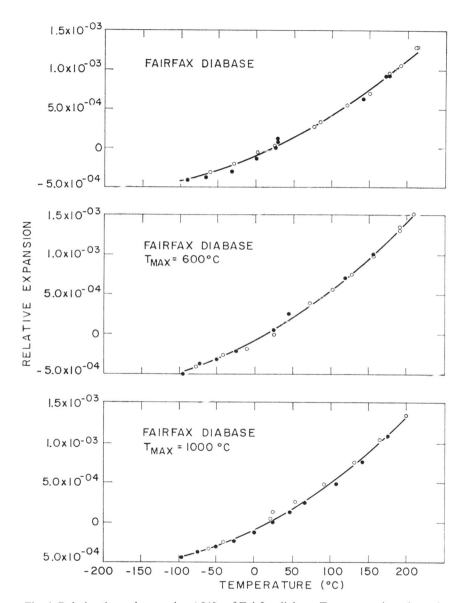

Fig. 4. Relative thermal expansion $\Delta L/L_0$ of Fairfax diabase. Top, no previous thermal cycling. Middle, cycled to 600°C ($T_{max} = 600°C$) prior to measurement. Bottom, cycled to 1000°C ($T_{max} = 1000°C$) prior to measurement. Open circles indicate measurements made at increasing temperature; solid circles indicate measurements made at decreasing temperature.

(4) bulk moduli of the various components are equal or nearly equal. A number of schemes exist for calculating the thermal expansion of an aggregate (Nielsen, 1967; Schapery, 1968). However the results obtained by these different schemes do not differ significantly from each other. For example, applying the rule of mixtures to samples 10020, 10046, and 10057 yielded values of α_v which were 6% lower, 1/2% lower, and 3% higher, respectively, than those derived from Turner's equation. Probably the best value for α_v lies between those obtained from Turner's equation and the rule of mixtures (Schapery, 1968). Hence Turner's equation is probably accurate within 2 or 3% and certainly adequate in view of the approximations which must be made for rocks.

Table 1. Measured and calculated values of $\Delta V/[\Delta T \cdot V_0][\times 10^6(^\circ C^{-1})]$.

		Measured −100 to 25°C	Measured 25 to 200°C	Calculated 25 to 200°C	Sample description reference
Igneous rocks	Fairfax diabase (uncycled)	10.2	19.5	18.4	Fairbairn et al. (1951)
	10020	7.2	16.2	22.5	Horai et al. (1970)
	10057	8.1	14.7	21.6	Haggerty et al. (1970)
	12022	5.7	15.9	23.7	NASA (1970)
Breccias	10046	15.0	22.2	22.1	Horai et al. (1970)
	14318	11.7	18.0	—	{Kurat et al. (1972) {Warner (1972)
	15418	7.5	12.3	—	NASA (1971)
	15015	12.6	—	—	NASA (1971)

Table 2. Values of thermal expansion and bulk modulus and sources of data used in calculating thermal expansion of Apollo samples and Fairfax diabase.

Mineral	Thermal expansion (at 200°C) $\alpha_v \times 10^6$ (°C^{-1})	Source of data
Quartz	43.3	Skinner (1966)
Augite	19.4	Skinner (1966)
Microcline	22.1	Skinner (1966)
Plagioclase (An$_{44}$)	13.3	Skinner (1966)
Plagioclase (An$_{95}$)	17.8	Skinner (1966)
Magnetite	28.5	Skinner (1966)
Olivine (Fo$_{85}$)	28.3	Skinner (1966)
Hematite	26.9	Skinner (1966)
Glass	29.1	Peters and Cragoe (1920)

Mineral	Bulk modulus (at 1 atm) K (Mb)	Source of data
Quartz	0.38	Simmons and Wang (1971)
Augite	0.96	Simmons and Wang (1971)
Microcline	0.55	Simmons and Wang (1971)
Plagioclase (An$_{53}$)	0.71	Simmons and Wang (1971)
Plagioclase (An$_{90}$)	0.82	Simmons and Wang (1971)
Magnetite	1.62	Simmons and Wang (1971)
Olivine (Fo$_{85}$)	1.30	Simmons and Wang (1971)
Hematite	2.07	Liebermann and Schreiber (1968)
Glass	0.63	Birch (1966)

The values of mineral thermal expansion and bulk modulus used in the calculation of α_v as well as the sources of these data are indicated in Table 2. All values for thermal expansion were obtained by dilatometer or interferometer. For those minerals whose compositions vary greatly as a result of solid solution, values of thermal expansion and bulk modulus of compositions appropriate to lunar rocks were chosen. Since the bulk modulus and thermal expansion coefficient of ilmenite have not been measured, data for hematite were used in these calculations. Because of the incompleteness of existing data for a number of minerals these calculated values can be considered as approximate only.

The improvement in instrument precision has resulted in an increase of the values of thermal expansion of samples 10020, 10046, 10057, and 12022 over those we previously reported (Baldridge and Simmons, 1971). However the major feature to be noted remains the same, *viz.*, that the measured values of the thermal expansion of lunar rocks (except for 10046) are 65 to 70% of the calculated (intrinsic) values where these comparisons can be made (Table 1). For three of the lunar breccias realistic modal analyses do not exist, hence calculation of intrinsic thermal expansion is impossible. In comparison the Fairfax diabase seems to have a measured thermal expansion approximately equal to the intrinsic (calculated) value. In Fig. 5 our data for the Fairfax diabase and for the lunar igneous rock 10057 are compared with typical values for "analogue" rocks studied by Griffin and Demou (1972).

We previously believed that the lowered values for the thermal expansion of lunar rocks were caused by the presence of numerous cracks in these rocks. To test this hypothesis we measured the expansion of two samples of Fairfax diabase that had been heated to 600°C and 1000°C, respectively. In both specimens large numbers of cracks were formed as a result of the thermal cycling (for a description of the cycling process and of resulting changes in the rocks see Todd *et al.*, 1972). In contrast to our expectations the measured value of thermal expansion was found to have increased for both specimens (Fig. 4).

DISCUSSION

Most of our measured thermal expansion values are lower than the intrinsic values. The several cases where the measured values equal or exceed the intrinsic values indicate that if the measured thermal expansion values reflect the presence of large numbers of cracks, then they do so in a way more complicated than if they were a function simply of crack density.

One might *a priori* expect that the value one measures for the thermal expansion of a porous elastic body would be the intrinsic value. Consider an isotropic body of volume V_0 containing pores of volume V_c. Let α be the thermal expansion of the solid. The volume change of the entire body is given by $\Delta V = \alpha V_0 \Delta T$, since the expansion of the pores ΔV_c is also determined by α as $\alpha V_c \Delta T$. The porosity V_c/V_0 remains constant with temperature.

Likewise in a porous multiphase material the presence of thermal stresses arising from anisotropic and mismatched thermal expansions does not change our *a priori* expectation that the measured thermal expansion will equal the intrinsic thermal

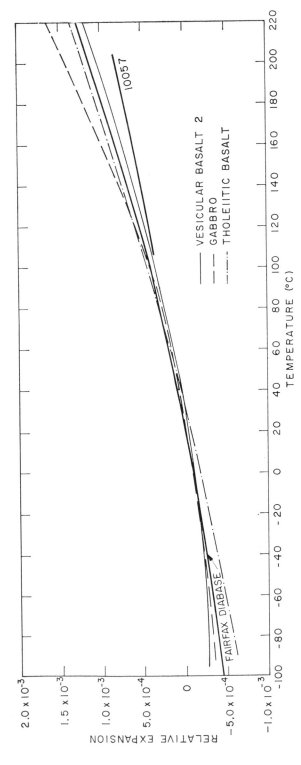

Fig. 5. Comparison of relative expansion $\Delta L/L_0$ of Fairfax diabase and lunar igneous rock 10057 with terrestrial "analogues" of Griffin and Demou (1972).

expansion as long as these stresses do not form cracks or cause sliding of grains relative to each other. The thermal stress system is in internal equilibrium since tensions and compressions average to zero (Goodier, 1958). Again the porosity remains constant with temperature.

Rocks however are known to depart from the ideal elastic behavior assumed in the *a priori* considerations above. For example, when rocks initially at room temperature are heated above a certain temperature (200 to 300°C) a permanent alteration in their structure occurs and they do not return to their original dimensions (Sosman, 1927; Thirumalai and Demou, 1970; Griffin and Demou, 1972). In addition we have observed two anelastic effects: (1) For many of the runs there is a tendency for the relative expansion to be lower for decreasing temperature than for increasing temperature (Figs. 2, 3, and 4). Although the magnitude of this effect lies on the margins of our experimental uncertainty, its consistency for many of the runs makes it appear real. (2) Rock 15418 was cycled twice between room temperature and −100°C, and each time the expansion-contraction path was slightly although systematically different (Fig. 6). This effect was not observed for any other rocks.

Since for most rocks the intrinsic thermal expansion is not measured we may conclude that the porosity changes with temperature. This porosity change depends in part upon the thermal history of the rocks. We will consider the cases of (1) repeated cycling, and (2) initial cycling (first two or three cycles only).

Repeated cycling

After being cycled a few times rocks reach a steady state expansion-contraction curve. Presumably when rocks have attained this condition, mineral grains are able to slide past each other and otherwise reorient themselves smoothly as their volumes change with temperature. The large number of cracks along grain boundaries (and within grains) permits many grains to expand or contract without contributing to the

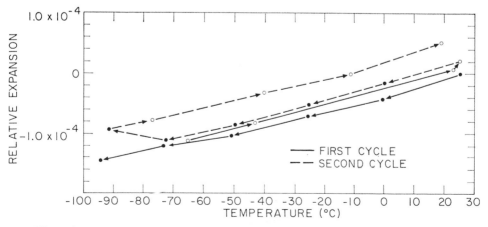

Fig. 6. Anelastic thermal expansion $\Delta L/L_0$ of lunar breccia 15418. Open circles indicate measurements made at increasing temperature; solid circles indicate measurements made at decreasing temperature.

overall expansion or contraction of the aggregate. The effect of a temperature change is then in part to change the percentage of porosity in the rocks. For such a model the upper limit for the thermal expansion would be the intrinsic value. This is the model we advance for the lunar rocks.

Rock 10046 apparently departs from this behavior. It has a measured value of thermal expansion approximately equal to its intrinsic value. This rock is a breccia, containing approximately 20% glass. The possible effect of glass is two-fold: (1) glass of an appropriate silica content may have a comparatively high thermal expansion, and (2) glass may tend to fill microfractures and otherwise "cement" grains together. Either or both of these explanations may apply to 10046.

Initial cycling

The measured expansion of the Fairfax diabase (Fig. 4, top) is approximately equal to its intrinsic expansion because it was not heated above the temperature at which significant irreversible structural alteration occurs (200 to 300°C). Based on the results of Thirumalai and Demou (1970) and Griffin and Demou (1972), we can expect that at higher temperatures the measured thermal expansion would have become significantly less than the intrinsic (when measured on the first cycle with increasing temperature). Apparently thermal stresses in the temperature range of 200 to 300°C form cracks that then act to reduce the expansion of the aggregate as described above. In view of this expectation the expansion of the samples of Fairfax diabase which were cycled (second cycle) to 200°C after being cracked initially at high temperature is anomalously high. No hysteresis was observed afterward at room temperature, hence it is unlikely that new cracks were formed. Possibly unbalanced stresses were set up in the rock by this high heating at room pressure which were relieved upon second heating.

Since the exact manner in which thermal expansion of rocks depends upon porosity cannot at present be determined, the extrapolation of our measurements over the pressure region where cracks are open (<5 kb) is uncertain. One can say that for rocks such as the lunar samples (with the exception of 10046) which have reached a steady-state expansion-contraction curve, thermal expansion probably increases rapidly with small pressure increases near 1 atm as cracks close and reaches intrinsic thermal expansion values in the range of 3 to 5 kb where cracks have closed.

Acknowledgments—We thank D. Riach for preparing our samples. This work was supported by NASA contract NGR-22-009-540.

REFERENCES

Baldridge W. S. and Simmons G. (1971) Thermal expansion of lunar rocks. *Proc. Second Lunar Sci. Conf., Geochim. Cosmochim. Acta* Suppl. 2, Vol. 3, pp. 2317–2321. MIT Press.
Birch F. (1966) Compressibility; elastic constants. In *Handbook of Physical Constants* (editor S. Clark), Section 7, pp. 97–173. Geol. Soc. Amer. Mem. 97.

Buffington R. M. and Latimer W. M. (1926) The measurement of coefficients of expansion at low temperatures, some thermodynamic applications of expansion data. *J. Amer. Chem. Soc.* **48,** 2305–2319.

Fairbairn H. W., Schlecht W. G., Stevens R. E., Dennen W. H., Ahrens L. H., and Chayes F. (1951) A cooperative investigation of precision and accuracy in chemical, spectrochemical, and modal analysis of silicate rocks. *U.S. Geol. Survey Bull.* **980,** 1–71.

Goodier J. N. (1958) Formulas for overall thermoelastic deformation. *Proc. of Third U.S. Natl. Congress of Appl. Mech.*, pp. 343–345. Am. Soc. Mech. Engineers.

Griffin R. E. and Demou S. G. (1972) Thermal expansion measurements of simulated lunar rocks. 1971 Symposium on Thermal Expansion, *Institute of Physics Conf. Proc. Series* (in press).

Haggerty S. E., Boyd F. R., Bell P. M., Finger L. W., and Bryan W. B. (1970) Opaque minerals and olivine in lavas and breccias from mare tranquilitatis. *Proc. Apollo 11 Lunar Sci. Conf., Geochim. Cosmochim. Acta* Suppl. 1, Vol. 1, pp. 513–538. Pergamon.

Horai K., Simmons G., Kanamori H., and Wones D. (1970) Thermal diffusivity, conductivity, and thermal inertia of Apollo 11 lunar material. *Proc. Apollo 11 Lunar Sci. Conf., Geochim. Cosmochim. Acta* Suppl. 1, Vol. 3, pp. 2243–2249. Pergamon.

Kingery W. D. (1960) *Introduction to Ceramics*, p. 478. Wiley.

Kozu S. and Takane K. (1929) Influence of temperature on the axial ratio, the interfacial angle, and the volume of quartz. *Sci. Rep. Tohoku Univ.* 3 (third series), 239–246.

Kurat G., Keil K., and Prinz M. (1972) A "chondrite" of lunar origin: Textures, lithic fragments, glasses, and chondrules (abstract). In *Lunar Science—III* (editor C. Watkins), pp. 463–465, Lunar Science Institute Contr. No. 88.

Liebermann R. C. and Schreiber E. (1968) Elastic constants of polycrystalline hematite as a function of pressure to 3 kilobars. *J. Geophys. Res.* **73,** 6585–6590.

NASA (1970) Lunar Sample Information Catalog, Apollo 12. Manned Spacecraft Center, Houston, Texas.

NASA (1971) Lunar Sample Information Catalog, Apollo 15. Manned Spacecraft Center, Houston, Texas.

Nielsen L. E. (1967) Mechanical properties of particulate-filled systems. *J. Comp. Mat.* **1,** 100–119.

Peters C. G. and Cragoe C. H. (1920) Measurements on the thermal dilatation of glass at high temperatures. *J. Optical Soc. Amer.* **4,** 105–144.

Schapery R. A. (1968) Thermal expansion coefficients of composite materials based on energy principles. *J. Comp. Mat.* **2,** 380–404.

Simmons G. and Wang H. (1971) *Single Crystal Elastic Constants and Calculated Aggregate Properties; A Handbook*, 2nd edition, 370 pp. MIT Press.

Skinner Brian J. (1966) Thermal expansion. In *Handbook of Physical Constants* (editor S. Clark), Section 6, pp. 75–96. Geol. Soc. Amer. Mem. 97.

Sosman R. B. (1927) *The Properties of Silica*, Amer. Chem. Soc. Monograph Series. Chem. Catalog. Co., Inc. (New York).

Thirumalai K. and Demou S. G. (1970) Effect of reduced pressure on thermal-expansion behavior of rocks and its significance to thermal fragmentation. *J. Applied Physics* **41,** 5147–5151.

Todd T., Wang H., Baldridge W. S., and Simmons G. (1972) Elastic properties of Apollo 14 and Apollo 15 rocks. *Proc. Third Lunar Science Conf., Geochim. Cosmochim. Acta* Vol. 3.

Warner J. L. (1972) Apollo 14 breccias: Metamorphic origin and classification (abstract). In *Lunar Science—III* (editor C. Watkins), pp. 782–784, Lunar Science Institute Contr. No. 88.

Proceedings of the Third Lunar Science Conference
(Supplement 3, *Geochimica et Cosmochimica Acta*)
Vol. 3, pp. 2611–2617
The M.I.T. Press, 1972

Thermal conductivity of Apollo 14 fines

C. J. Cremers

Department of Mechanical Engineering,
University of Kentucky, Lexington, Ky. 40506

Abstract—The thermal conductivity of the Apollo 14 fines, sample 14163,133, was measured under vacuum conditions as a function of temperature. Measurements were made for densities of 1100 and 1300 kg/m³. A least-squares curve of the form $k = A + BT^3$ is fitted to each data set in accordance with elementary theory. Comparisons are made with previously published data for terrestrial basalt and Apollo 11 and 12 samples.

Introduction

THE THERMAL CONDUCTIVITY is a transport property defined by the Fourier law of heat conduction. That is,

$$q_x = -k \, \partial T/\partial x. \tag{1}$$

Here q_x is the heat flux in the direction x caused by the temperature gradient acting in the negative x direction. The proportionality constant k is the thermal conductivity. It is obviously necessary to have known values of this property for heat flux calculations concerning systems with internal temperature gradients.

Heat-transfer calculations for either the lunar surface or engineering systems there involve, for the most part, the lunar fines. This is true at least for the regions so far visited by the Apollo manned flights and the Surveyor, Luna, and Lunakhod unmanned missions. Further evidence, as pointed out by Gold (1972), also suggests a lunar surface with a generally deep deposit of fairly compact but finely divided material. Photographs and returned samples show that rocks and boulders are present; however, they are infrequent and scattered and one might consider them to be more or less of a random perturbation on the fine particulate nature of the rest of the surface layer. Consequently, heat-transfer rates there should be determined by the properties of the fines and so the present measurements have been concentrated on these.

The porous nature of the lunar soil presents a problem. Any porous dielectric material with an internal temperature gradient will transfer heat through a complicated interaction between phonon conduction through the particles and their contact surfaces and thermal radiation which can be scattered in the voids and absorbed and reemitted by the solid material. There is no problem caused by gaseous conduction or convection because the 10^{-12} Torr pressure of the lunar atmosphere precludes these effects.

The internal heat-transfer problem then involves both conduction and radiation. If we wish to consider the heat flux as an entity, rather than considering separate radiative and conductive components, Fourier's law may be used provided one recognizes that the conductivity so defined is only an effective one rather than a basic property of the material. Elementary theory—e.g., Watson (1964) and Clegg, Bastin, and Gear (1966)—shows that such an effective conductivity of a particulate medium can be represented by the sum of a constant term, representing solid conduction, plus a term proportional to the temperature to the third power, representing radiation. That is,

$$k = A + BT^3 \tag{2}$$

where A and B are constants which can be obtained from experiment. The constraints on such an experiment are that the pressure must be kept low to suppress gaseous effects and also that a variable sample temperature be provided in order that the temperature dependence of the thermal conductivity be determined over the range of lunar temperatures. It is important that the "effective" nature of this conductivity be kept in mind in subsequent data interpretation.

It is apparent from the above arguments that the fractions of the heat flux which appear as phonon conduction and internal radiation depend on the porosity and geometry of the pores in the particulate matrix of the fines. The more dense a sample is, the more solid conduction one would expect to find. Consequently, the effective thermal conductivity at low pressures depends not only on the temperature but also on the sample density. To assess this effect, measurements are presented at two different sample densities, 1100 kg/m^3 and 1300 kg/m^3. Measurement at higher densities are in progress at this time.

One would expect further that external mechanical force exerted on the sample surface would also affect the conductivity as the integrity of the interparticle contact points would be enhanced. However, no attempt was made to assess the effect of the details of the particulate matrix in the present set of measurements.

The sample used for the experiments reported in this paper is fines sample 14163,133 as catalogued by the curator of Lunar Sample Analysis Program at the Manned Spacecraft Center, NASA, Houston.

THE EXPERIMENT

The size of the sample available for testing dictated the size of the experimental apparatus and, to a large degree, the approach to be used. It was decided to employ the line heat-source technique that has been used in the past for vacuum conductivity measurements of silicate materials. Briefly, the application of this method requires that a long (length to diameter ratio greater than 30) line heat source be imbedded in the material to be tested. For such a source in an infinite medium, it can be shown that after an initial period during which the probe heat-capacity is dominant, the temperature change at any point in the medium over a time period from t_1 to t_2 is given by

$$T_2 - T_1 = \frac{q}{4\pi k} \ln \frac{t_2}{t_1}. \tag{3}$$

Here q is the heat-source strength and k is the thermal conductivity of the medium. Note that k must be considered constant over the temperature range ($T_2 - T_1$) under consideration.

Equation (3) is the usual working equation for the line heat-source method. It is apparent that if the conditions for the model are met in the experiment, a plot of temperature versus the logarithm of time will result in a straight line, the slope of which is $q/4\pi k$. The measurement of q and the slope of the curve then yields the thermal conductivity k. The method is well suited for measurements of the thermal conductivity of small samples and was used for measurements on the Apollo 11 and 12 fines by Cremers et al. (1970, 1971). The basic mathematical treatment of the working equation is given by Carslaw and Jaeger (1959) and the experimental errors incurred in deviating from the mathematical model have been considered by Blackwell (1956, 1959). The present method is described in detail by Cremers (1971a).

The lunar samples which were made available for analysis were limited in volume so that it became imperative to measure the temperature as close to the source as possible to minimize deviations from the infinite medium assumption. For this reason the line source itself, a 36 AWG (0.127 mm diam.) Chromel-A wire, was calibrated as a resistance thermometer. Then with the wire in place in the sample the voltage change over about 22 mm of the wire was monitored during heating along with the voltage itself and the current flow. The current through the wire was controlled by a constant-current power supply which supplied a current constant to within four significant figures. The voltage change during a run was 0.25% at the maximum and so the heat generation was constant to within this value as well. Axial heat conduction loss was minimized by providing an extra 20 mm of heating wire beyond the voltage taps and outside of the sample.

The test cell was constructed of teflon and held about 5 g of the lunar fines at a density of 1300 kg/m^3. The size of the sample when in the cell was about 25 × 13 × 13 mm. The density of the fines when loosely poured was about 1100 kg/m^3 for the Apollo 14 fines. To achieve greater densities the cell was vibrated with a Vibrotool to cause uniform settling and packing.

The vacuum chamber is of stainless steel and is approximately 0.3 m diameter by 0.4 m high. The chamber was pumped with a Welch Turbomolecular pump to provide a pressure on the order of 10^{-6} Torr. A study by Cremers (1971b) showed that this is well below the pressure at which gaseous effects are apparent. Electrical feedthroughs were provided for power and for temperature control and sensing.

Table 1. Apollo 14 thermal conductivity data (1100 kg/m^3)

Temperature (K)	Thermal conductivity (W/m − K)
353	1.64 × 10^{-3}
352	1.67
351	1.75
350	1.69
344	1.63
337	1.70
329	1.60
328	1.60
314	1.72
313	1.64
387	1.31
285	1.36
284	1.33
283	1.36
274	1.27
271	1.23
260	1.12
256	1.09
252	1.02
245	1.16
219	0.98
210	0.94
185	1.13
181	1.02
179	0.99
178	0.75
156	0.73
142	1.04
139	0.79
126	0.89
126	1.09
122	0.90

An inner stainless steel chamber of double-walled construction was used for ambient temperature control. A heating tape wrapped about it provided higher than room temperatures and liquid nitrogen or expanding freon passed through the jacket provided low temperatures. A thermistor attached to the chamber actuated the heater for temperature control and a thermocouple immersed in the sample indicated the sample temperature.

RESULTS AND DISCUSSION

The measured thermal conductivity data for the Apollo 14 sample at densities of 1100 and 1300 kg/m^3, respectively, are given in Tables 1 and 2. It was mentioned above that elementary theory predicts a cubic dependence of the thermal conductivity on temperature. This was found to be a fair representation in the present case as it

Table 2. Apollo 14 thermal conductivity data (1300 kg/m^3)

Temperature (K)	Thermal conductivity (W/m − K)
404	2.16 × 10^{-3}
404	2.23
390	2.27
389	2.36
389	2.41
367	1.68
367	1.74
367	1.97
367	1.77
359	1.98
343	1.73
340	1.75
326	1.61
321	1.62
321	1.54
320	1.59
319	1.56
296	1.40
295	1.24
295	1.35
292	0.98
277	0.95
240	0.90
231	1.07
221	1.04
214	0.97
208	0.95
201	0.95
109	0.78
109	0.68

Table 3. Coefficients of equation (2)

Sample	Density (kg/m^3)	$A \times 10^3$ (W/m − K)	$B \times 10^{11}$ (W/m − K^4)
Apollo 14	1100	0.836	2.09
Apollo 14	1300	0.619	2.49
Apollo 12	1300	0.922	3.19
Apollo 11	1300	1.42	1.73
Basalt[a]	1130	0.887	1.90
Basalt[a]	1300	1.24	2.43

[a] Basalt values are from Fountain and West (1970).

was in the previous investigations of the Apollo 11 and 12 samples. Consequently, the numerical data are conveniently represented by equation (2) for which the constants were obtained by a least-squares analysis. Equation (2) is also plotted in Figs. 1 and 2 and the coefficients are given in Table 3. Consideration of the accuracy of representation of the data by an equation of the form of equation (2) suggests that a cubic dependence on temperature is probably the case. However, because of data scatter and present lack of data at lower temperatures, the evidence is far from conclusive.

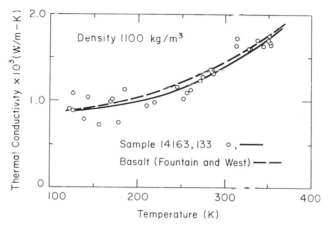

Fig. 1. Thermal conductivity of fines at a density of 1100 kg/m³.

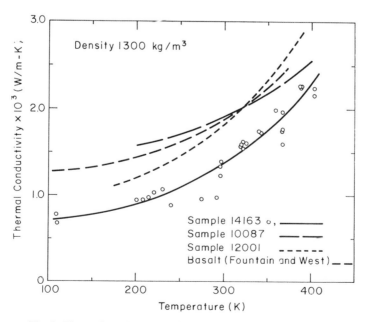

Fig. 2. Thermal conductivity of fines at a density of 1300 kg/m³.

The data given for the Apollo 11, 12, and 14 samples are not yet sufficiently complete for a critical analysis of the temperature dependence of the thermal conductivity. On the other hand, the Fountain and West data for terrestrial basalt, taken at a number of densities ranging from 790 to 1500 kg/m^3 are extensive. These were much larger samples and so the errors should be expected to be less significant. Analysis of these data shows that a cubic least-squares fit works well in some cases (one such case is the density of 1300 kg/m^3) but not so well for other densities. For the Apollo 12 and 14 samples the agreement is best at higher temperatures but in general is good. This suggests that the elementary theory is close to correct but that perhaps some vital elements are still missing.

A curve of the form $k = AT^{-1} + BT^3$ (suggested in a review by Kanamori of a previous paper by the author) reflecting an inverse temperature dependence for the lattice component of the conductivity was also tried for the Apollo 11 and 12 fines. The fit was acceptable at high temperatures but not at the low end of the range. This lack of success is probably due to the somewhat amorphous nature of the finely divided lunar samples. No attempt was made to fit this curve to the present data.

It is of interest to compare the thermal conductivity of the Apollo 14 fines with that of other samples. However, in the present case this is possible only with the higher of the two densities used as it was not possible to achieve such a comparably low density with the Apollo 11 and 12 fines. In both cases it was not possible to prepare a sample with a density less than about 1300 kg/m^3. This indicates that the Apollo 14 fines probably do not have as large a fraction of small particles as the previous Apollo samples and so although the voids are fewer in number they are much larger. This is borne out in particle size measurements by Gold *et al.* (1972). As a consequence it is possible to compare the 1100 kg/m^3 sample of Apollo 14 fines with only terrestrial basalt at a density of 1130 kg/m^3. These data are presented by Fountain and West (1970). They fitted a curve of the form of equation (2) to their data and the results are shown in Fig. 1. The coefficients for use with equation (2) are given in Table 3. The agreement between the basalt and Apollo 14 fines is outstanding in this case although there is no particular reason why it should be so. Figure 2 includes curves for the basalt as well as for the Apollo 11 and 12 data, all at the density of 1300 kg/m^3. The basalt data, again from Fountain and West (1970), are represented by curves of the form of equation (2) as are the Apollo 11 data of Cremers *et al.* (1970) and the Apollo 12 data of Cremers and Birkebak (1971). Because of its similarity in chemical composition to the lunar fines, basalt is probably the best terrestrial material for comparison. Agreement between the data for the several samples is not so good in this case. However, note that for the Apollo 14 sample, the conductivities at the two densities are nearly the same.

It is of interest to note that the magnitudes of all the conductivities shown in Fig. 2 are roughly the same. What is of perhaps more importance, however, is the apparent difference in temperature dependence between the different sets of data shown. The slope of each curve, at a given value of the temperature, is an indication of the magnitude of the radiative component which is expressed through the coefficient B. It appears that the radiative effects in the terrestrial basalt are slightly less important than in the Apollo 12 and 14 samples but of slightly greater magnitude than in the Apollo 11 sample. Most probably, the causes of these deviations are differences in particle size distribution and possibly shape as well as variations in the amounts of glassy material present in each of the lunar samples.

The data of Fountain and West (1970) are for particulate basalt 37–62 μ in diameter. That is a much narrower size range than was found for the Apollo lunar fines, e.g., Gold et al. (1970, 1971, 1972). The abundance of micron sized and smaller particles in the latter samples would tend to suppress the radiative transfer mode as the powder would more closely resemble a solid. The study of Watson (1964) indicated that the radiative component in powdered media depends strongly on particle size, being much more important for larger particles on the order of 100 μ diameter than for particles on the order of 10 μ diameter. The Apollo 12 data indicate that there is perhaps a greater abundance of larger particles present resulting in stronger radiative effects. That this is so is evidenced by the comparison of particle sizes from the fines samples of all three Apollo missions as presented by Gold et al. (1972).

Acknowledgment—The author wishes to acknowledge financial support from NASA under Grant NGR 18-001-060. The assistance of D. R. Talley in taking and processing the data is also gratefully acknowledged.

REFERENCES

Blackwell J. H. (1956) The axial flow error in the thermal conductivity probe. *Can. J. Phys.* **34**, 412–419.

Blackwell J. H. (1959) A transient method for determining the thermal constants of insulating materials in bulk. *J. Appl. Phys.* **25**, 137–144.

Carslaw H. S. and Jaeger J. C. (1959) *Conduction of Heat in Solids*, pp. 334–345. Oxford Press.

Clegg P. E., Bastin J. A., and Gear A. E. (1966) Heat transfer in lunar rock. *Mon. Not. R. Astr. Soc.* **133**, 63–66.

Cremers C. J. (1971a) Thermal conductivity cell for small powdered samples. *Rev. Sci. Inst.* **42**, 1694–1696.

Cremers C. J. (1971b) Density, pressure, and temperature effects on heat transfer in Apollo 11 fines. *Am. Inst. Aero. Astro. J.* **9**, 2180–2183.

Cremers C. J., Birkebak R. C., and Dawson J. P. (1970) Thermal conductivity of fines from Apollo 11. *Proc. Apollo 11 Lunar Sci. Conf.*, Geochim. Cosmochim. *Acta* Suppl. 1, Vol. 3, pp. 2045–2050. Pergamon.

Cremers C. J. and Birkebak R. C. (1971) Thermal conductivity of fines from Apollo 12. *Proc. Second Lunar Sci. Conf.*, Geochim. Cosmochim. *Acta* Suppl. 2, Vol. 3, pp. 2311–2314. MIT Press.

Fountain J. A. and West E. A. (1970) Thermal conductivity of particulate basalt as a function of density in simulated lunar and Martian environments. *J. Geophys. Res.* **75**, 4063–4069.

Gold T. (1972) The depth of the lunar dust layer (abstract). In *Lunar Science—III* (editor C. Watkins), pp. 321–322, Lunar Science Institute Contr. No. 88.

Gold T., Campbell M. J., and O Leary B. T. (1970) Optical and high-frequency electrical properties of the lunar sample. *Proc. Apollo 11 Lunar Sci. Conf.*, Geochim. Cosmochim. *Acta* Suppl. 1, Vol. 3, pp. 2144–2154. Pergamon.

Gold T., Campbell M. J., and O Leary B. T. (1971) Some physical properties of Apollo 12 lunar samples. *Proc. Second Lunar Sci. Conf.*, Geochim. Cosmochim. *Acta* Suppl. 2, Vol. 3, pp. 2173–2181. MIT Press.

Gold T., Bilson E., and Yerbury M. (1972) Grain size analysis (1972) optical reflectivity measurements and determination of high frequency electrical properties for Apollo 14 lunar samples (abstract). In *Lunar Science—III* (editor C. Watkins), pp. 318–320, Lunar Science Institute Contr. No. 88.

Watson K. (1964) The thermal conductivity measurements of selected silicate powders in vacuum from 150°–300°K. Part I of Ph.D. Dissertation, Cal. Inst. Tech., Pasadena.

Proceedings of the Third Lunar Science Conference
(Supplement 3, *Geochimica et Cosmochimica Acta*)
Vol. 3, pp. 2619–2625
The M.I.T. Press, 1972

Viscous flow behavior of lunar compositions 14259 and 14310

M. Cukierman, P. M. Tutts, and D. R. Uhlmann

Department of Metallurgy and Materials Science,
Center for Materials Science and Engineering,
Massachusetts Institute of Technology, Cambridge, Massachusetts

Abstract—The flow characteristics of lunar compositions 14310 and 14259 have been determined over a wide range of viscosity. The temperature ranges covered by the measurements are 1270° to 1440°C and 700° to 825°C for the 14310 composition and 1154° to 1431°C and 700° to 820°C for the 14259 composition. Reliable data could not be obtained over the intermediate ranges of temperature because of the occurrence of crystallization. The apparent activation energies for viscous flow in the high-temperature and low-temperature regions are respectively about 60 and 160 kcal (gm at)$^{-1}$ for the 14310 composition and 60 and 170 kcal (gm at)$^{-1}$ for the 14259 composition. The experimental data in the high temperature regions are compared with predictions of the semi-empirical model of Bottinga and Weill; and the agreement is found to be within a factor of $10^{0.2}$ in both cases. The implications of this agreement are discussed, as is the general flow behavior in both the high-temperature and low-temperature regions.

Introduction

THE VISCOUS flow behavior of lunar compositions is of importance in understanding a number of phenomena, including the rheology of lunar lavas, crystallization and melting behavior, the motion of crystallizing particles and other inclusions through molten phases, the occurrence and abundance of lunar glasses, and the phase morphologies expected for various thermal and mechanical histories.

Despite the utility of such information, only a few studies have been directed to the flow behavior of compositions of geophysical interest. None of these have covered a wide range of viscosity and temperature, and none have been concerned with conditions that simulate the state of reduction of the lunar compositions. For example, the few data which are available on Fe-containing silicate liquids are all in range of small Fe concentrations, correspond to poor control of the oxidation state, and do not permit any reliable estimate to be made of this constituent's effect on viscosity (see discussion in Bottinga and Weill, 1972).

A semi-empirical model has recently been developed (Bottinga and Weill, 1972) to describe the viscosity of magnetic silicate liquids. This model represents the logarithm of the viscosity as a linear function of composition:

$$ln\ \eta = \sum_i X_i D_i \tag{1}$$

where η is the viscosity, X_i is the mole fraction of the ith component and D_i is a constant associated with component i over a restricted range of composition. Different D_i constants are assigned for each temperature; and different constants are used for each of five ranges of SiO_2 concentrations. The model has provided a useful description of available flow data (generally within ± 0.20 in $ln\ \eta$) but detailed testing of its applicability to lunar compositions seems desirable. Further, almost no data are available on the flow behavior of these liquids in the high-viscosity region, where the semi-empirical model cannot be employed, and where important differences in flow behavior are usually observed for terrestrial liquids relative to their behavior in the low-viscosity region described by the model.

The present investigation was directed to providing direct information on the flow behavior of two selected lunar compositions: 14259B and 14310. The data to be reported cover both the molten range, to which the model of Bottinga and Weill may be applied, and the range of high viscosity as the glass transition is approached. In all cases, the initial melting as well as the viscosity measurements have been carried out under conditions of low oxygen activity to simulate the Fe^{2+}/Fe^{3+} conditions of the lunar material. In a companion paper (Scherer *et al.*, 1972), the present flow data will be used to help interpret the crystallization behavior of these compositions and to evaluate their glass-forming characteristics.

Experimental Procedure

The materials studied in this investigation, lunar compositions 14259B and 14310, were prepared from the reagent grade raw material powders shown in Table 1. Weighted powders of each component were combined and milled together for about 24 hours prior to melting. The melting was carried out in Mo crucibles under an atmosphere of 5% H_2 forming gas (0.95 N_2–0.05 H_2) at 1700°C for about 30 min. In some cases, the molten samples were poured onto preheated graphite blocks to form slabs of glass; in others, they were water quenched in the crucibles to form chunks of glass. An atmosphere of the forming gas was maintained during the pouring and quenching operations. The materials produced in this way were optically homogeneous glasses that were black in color for any thickness of the order of mm or larger.

Specimens for use in the beam-bending viscosimeter were cut from the glass slabs. Prior to the cutting operation, the slabs were annealed for 2 hours at 700°C. The chunks of glass were subsequently remelted in the rotating-cylinder viscosimeter. Samples of both the slab and chunk glass were also used in a concurrent study of crystallization kinetics (Scherer *et al.*, 1972).

Viscosities greater than about 10^8 poise were measured using the bending beam viscosimeter shown schematically in Fig. 1. This instrument is a modification of one described previously (Hagy, 1962); it permits the accurate determination of viscosities between 10^8 and 10^{15} poise in any desired atmosphere. The viscosity is determined from the rate of viscous deformation (bending) of a simple beam. The load is applied to the sample and the deformation rate measured with an Al_2O_3 hook-and-rod assembly coupled to a linear variable differential transformer. The ends of the specimen are supported on a SiC muffle located in the central portion of a resistance furnace 75 cm in length. The temperature uniformity over the sample region was well within ± 1°C; and with the proportional controller used, the maximum temperature variation in the sample region over the period of measure-

Table 1. Compositions of liquids investigated (wt. %)

Composition 14310		Composition 14259
50	$SiO_2(SiO_2)$	48
20	$Al_2O_3(Al_2O_3)$	18
8	$MgO(MgCO_3)$	9.2
7.7	$FeO(Fe_3O_4)$	10
11	$CaO(CaCO_3)$	11
1.3	$TiO_2(TiO_2)$	1.8
0.6	$Na_2O(Na_2CO_3)$	0.5
0.5	$K_2O(K_2CO_3)$	0.5

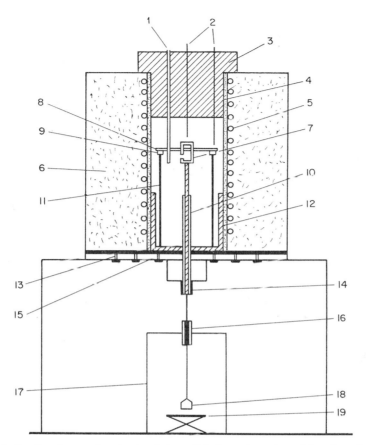

Fig. 1. Bending beam viscosimeter (schematic). 1 = atmosphere inlet; 2 = thermo-couples; 3 = insulating cover; 4 = Al_2O_3 furnace tube; 5 = Kanthal winding; 6 = insulation; 7 = hook for loading specimen; 8 = specimen; 9 = Al_2O_3 muffle stand; 10 = Al_2O_3 loading rod and tube; 11 = SiC stand; 12 = Al_2O_3 crucible; 13 = support screws; 14 = guide for loading device; 15 = table; 16 = linear variable differential transformer (LVDT); 17 = vertical adjustment device for LVDT; 18 = weights; 19 = micro-jack.

ment was also within this figure. For all the data to be reported here, an atmosphere of 5% H_2 forming gas was provided in the furnace.

Viscosities in the molten range were measured using the rotating-cylinder viscosimeter shown schematically in Fig. 2. The instrument is a modification of ones described previously (Bockris and Lowe, 1958; Napolitano et al., 1965). It provides accurate determinations of viscosities between 10^0 and 10^{10} poise in any desired atmosphere. The specimen is contained in an outer cylindrical crucible, 5 cm diameter × 10 cm high, and a cylindrical bob is suspended in the specimen. Con-nected to the bob is a hollow tube, through which a thermocouple extends to the center of the bob. The torque on the bob is measured using a large moving coil galvanometer system. The outer cylinder rests on a pedestal that passes out through seals at the bottom of the furnace and is attached to a turntable which can be driven over a range of speeds from 0.01 to 6 rpm. The data to be reported below were obtained by rotating the outer cylinder at a uniform angular velocity and measuring the torque on the inner cylinder (the bob).

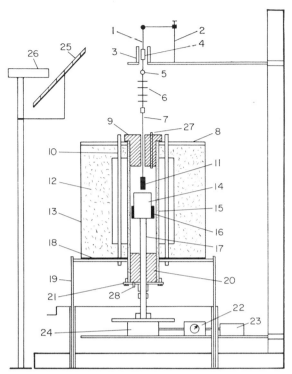

Fig. 2. Rotating cylinder viscosimeter (schematic). 1 = suspension; 2 = adjusting device for suspension; 3 = magnet; 4 = coil; 5 = mirror; 6 = radiation fins; 7 = hollow spindle with thermocouples; 8 = transite tap plate; 9 = insulating cover; 10 = SiC heating elements; 11 = bob; 12 = insulation; 13 = furnace shell; 14 = cylindrical cup to contain specimen; 15 = Al$_2$O$_3$ muffle tube; 16 = Al$_2$O$_3$ support for cylindrical cup; 17 = Al$_2$O$_3$ pedestal; 18 = transite bottom plate; 19 = lift table; 20 = insulation and seal; 21 = seal; 22 = speed reducer; 23 = motor; 24 = turntable; 25 = scale; 26 = cathetometer; 27 = atmosphere inlet; 28 = atmosphere outlet.

The specimen cylinder is located in the central region of a large volume SiC furnace, 75 cm in length, provided with a variety of heat shields and controlled by a proportional controller. The temperature uniformity over the sample volume at the highest operating temperature was within $\pm 2°C$, and was within closer limits at lower temperatures; the maximum temperature variation over the period of measurement was smaller than the spatial variation over the specimen region. For all results to be reported below, the outer cylinder, bob and tubing were fabricated of Mo, and an atmosphere of 5% H$_2$ forming gas was provided in the furnace.

Both viscosimeters were calibrated using NBS reference materials: Standard Glasses 710 and 711 for the range of high temperatures and Standard Oil S600 for low temperatures in the rotating cylinder viscosimeter. In both cases, the data on the reference materials were reproduced to better than $\pm 0.5\%$.

Results and Discussion

14310

The viscosity versus temperature relation for 14310 liquid is shown in Fig. 3. As shown there, reliable data could be obtained at temperatures above 1270°C and below

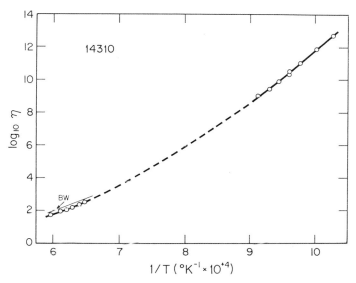

Fig. 3. Viscosity versus temperature relation for lunar composition 14310. Curve
labelled BW indicates predictions of Bottinga and Weill (1972).

825°C. At intermediate temperatures, the occurrence of crystallization during the
time required for the viscosity measurements prevented the determination of
reliable values for η. Data in the high temperature range below the liquidus (between
1310 and 1270°C) could be obtained only by superheating the specimen well above the
liquidus temperature prior to cooling (and presumably melting out crystalline em-
bryos retained in cavities in second-phase material). In the low temperature range, the
viscosity could be measured in times short relative to those required for significant
crystallization.

The glass transition temperature ($\eta = 10^{13}$ poise) for the 14310 composition is
about 710°C. The flow behavior appears to be Arrhenian as the glass transition is
approached; but at higher temperatures curvature is observed in the log η versus $1/T$
relation. The apparent activation energy for flow derived from the high temperature
data is about 60 kcal (gm at)$^{-1}$, while that derived from the low temperature data is
about 160 kcal (gm at)$^{-1}$. These values are in the range of those reported previously
for terrestrial compositions.

Also shown in Fig. 3 are predictions of the Bottinga and Weill model for the high
temperature flow behavior of composition 14310. As indicated there, the predicted
η-T relation is quite close to that observed experimentally. The model consistently
overestimates the viscosity by a small factor (of about $10^{0.2}$), and closely predicts
the variation of viscosity with temperature. The significance of these results will be
discussed after the data on lunar composition 14259 are presented.

14259

The viscosity versus temperature relation for composition 14259 is shown in
Fig. 4. As shown there, reliable data could be obtained at temperatures above 1154°C

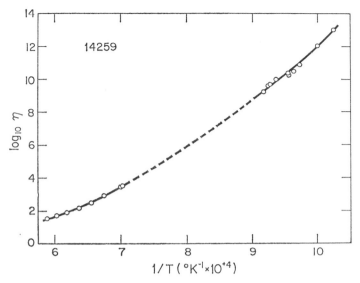

Fig. 4. Viscosity versus temperature relation for lunar composition 14259.

and below 820°C. The high temperature data could be extended to a range of lower temperature than with composition 14310, most likely because of the lower liquidus temperature (by about 70°C) for the 14259 composition. Again, prior superheating well above the liquidus was used to obtain data in the undercooled region.

The glass transition temperature of the 14259 composition is about 700°C. As with 14310, the low-temperature flow behavior appears to be Arrhenius, but curvature is observed in the log η versus $1/T$ relation at higher temperatures. The apparent activation energy for flow in the high temperature region is about 60 kcal (gm at)$^{-1}$, while that in the low temperature region is about 170 kcal (gm at)$^{-1}$. These values of the glass transition temperature and apparent activation energies are closely similar for both composition 14259 and composition 14310. Comparing the percentages of their constituents, it seems that the additional plagioclase material reflected in the 14310 composition has little effect on the viscosity.

The predictions of the Bottinga and Weill model for the high temperature flow behavior of composition 14259 lie within the width of the solid line in Fig. 4 of the experimental data. In this case, the viscosities as well as their variation with temperature are closely represented by the model.

The quality of agreement between predictions of the model and experimental data on both compositions was essentially within the range suggested by Bottinga and Weill. This very likely reflects the fact that many of the input data on Fe-containing liquids used in constructing the model were determined under reducing conditions. Extrapolation of the results to terrestrial compositions requires an evaluation of the role of the Fe oxidation state in affecting viscosity. Studies directed to this end are presently underway (Cukierman and Uhlmann, 1972), but the results are not yet available. It is expected, however, that Fe in the $3+$ state should result in increased viscosities relative to Fe in the $2+$ state.

There is no empirical model with which the present data on low temperature flow behavior can be compared. An Arrhenius temperature dependence in this region has been found for a number of other glass-forming liquids; but no available theoretical model can adequately represent the complexity of flow behavior observed when all such liquids are considered (Laughlin and Uhlmann, 1972). The magnitudes of the apparent activation energies for flow in the lunar compositions is in a range where transport over potential energy barriers seems reasonable for representing the flow process; but this suggestion must be regarded as a tentative one.

The curvature noted in the log η versus $1/T$ relations at more elevated temperatures for the lunar compositions is similar to that observed with other liquids. Such curvature is often well described by a free volume model of flow; but a detailed comparison in the present case is precluded by the occurrence of crystallization in the intermediate range of viscosity. When the glass transition temperature is used as a corresponding states parameter for the present liquids, and the log η versus Tg/T relation is constructed, results similar in form to those found with terrestrial silicate liquids are obtained. In detail, however, the viscosity seems to vary somewhat more rapidly with Tg/T for the lunar compositions.

The application of the present flow data to understanding kinetic processes on the moon will be presented in subsequent papers. The first such applications—to understanding the crystallization behavior of these compositions and evaluating their glass-forming characteristics—are described elsewhere in the present conference (Scherer *et al.*, 1972).

Acknowledgments —The authors are happy to acknowledge stimulating discussions with D. F. Weill of the University of Oregon and appreciate Dr. Weill's providing a preprint of his work in advance of its publication. Financial support for the present work was provided by the National Aeronautics and Space Administration under Grant NGR 22-009-646. This support is gratefully acknowledged.

REFERENCES

Bockris J. O'M. and Lowe D. C. (1953) An electromagnetic viscometer for molten silicates at temperatures up to 1800°C. *J. Sci. Inst.* **30**, 403–405.

Bottinga Y. and Weill D. F. (1972) The viscosity of magmatic silicate liquids: A model for calculation. *Amer. J. Sci.* (in press).

Cukierman M. and Uhlmann D. R. (1972) Effect of Fe oxidation state on the viscosity of silicate liquids. (To be published.)

Hagy H. E. (1963) Experimental evaluation of beam-bending method of determining glass viscosities in the range 10^9 to 10^{15} poises. *J. Am. Ceram. Soc.* **46**, 93–97.

Laughlin W. T. and Uhlmann D. R. (1972) Viscous flow in simple organic liquids. (To be published.)

Napolitano A., Macedo P. B., and Hawkins E. G. (1965) A wide-range (up to 10^{10} P) rotating cylinder viscometer. *J. Res. NBS* **69A**, 449–455.

Scherer G., Hopper R. W., and Uhlmann D. R. (1972) Crystallization behavior of glass formation of selected lunar compositions (abstract). In *Lunar Science—III* (editor C. Watkins), p. 678, Lunar Science Institute Contr. No. 88.

Proceedings of the Third Lunar Science Conference
(Supplement 3, *Geochimica et Cosmochimica Acta*)
Vol. 3, pp. 2627–2637
The M.I.T. Press, 1972

Crystallization behavior and glass formation of selected lunar compositions

G. Scherer, R. W. Hopper, and D. R. Uhlmann

Department of Metallurgy and Materials Science,
Center for Materials Science and Engineering, Massachusetts Institute of Technology,
Cambridge, Massachusetts

Abstract—The kinetics of crystal growth have been determined over a wide range of temperature, from 800 to 1219°C, for lunar compositions 14259 and 14310. At all temperatures for both compositions the extent of crystal growth is found to be a linear function of time. For both materials, the growth rate versus temperature relations exhibit the form generally found with glass-forming materials. At all temperatures measured the crystal growth rate of composition 14259 is smaller than that of composition 14310. The maximum growth rate for both compositions occurs at a temperature of about 1120°C, and is about 3.9×10^{-3} cm min^{-1} for the 14259 composition and about 6.6×10^{-3} cm min^{-1} for the 14310 composition.

The growth rate data are combined with viscosity data obtained on the same compositions to construct the reduced growth rate versus undercooling relations. The form of these relations, which exhibit positive curvature, is suggestive of growth by a surface nucleation mechanism; but a plot of the logarithm of (growth rate $\times$ viscosity) versus $1/T \, \Delta T$ does not display the simple form expected for this mechanism. The combined data are also used to construct time-temperature-transformation (TTT) curves corresponding to a just-detectable degree of crystallinity in the two materials. From these curves, the critical cooling rate required to form a glass of composition 14259 is estimated as about 1°C sec^{-1}, while that required for composition 14310 is about 40°C sec^{-1}. These values are consistent with experimental observations. Such TTT curves are considered to place meaningful limitations on the thermal histories of lunar specimens found in the glassy state.

Introduction

Studies of crystallization kinetics in glass-forming materials are interesting not only in their own right and in the information that they provide about an important kinetic phenomenon; they also can provide useful insight into the glass-forming characteristics of the materials and the critical conditions required for their occurrence as amorphous solids. The present study has been concerned with each of these aspects of the crystallization behavior of two lunar compositions: 14259 and 14310.

Crystal growth rates will be reported for each composition which cover a temperature range of more than 400°C. Also to be reported are observations of the interface morphology in each material. The crystallization results will be combined with data on the viscous flow behavior of the same compositions (Cukierman *et al.*, 1972) to develop an improved understanding of the crystal growth process, as well as to estimate the cooling rates required to form glasses of each material.

Kinetic Conditions for Glass Formation

In estimating the critical conditions for forming materials as glasses, use will be made of a recent analysis which will be described in detail elsewhere (Uhlmann,

1972a). This analysis is based upon the view that nearly any liquid can be formed as a glass if cooled sufficiently rapidly to a sufficiently low temperature without the occurrence of observable crystallization. In the present application, we shall take 10^{-6} as the volume fraction of crystals that can just be detected in a glass, and will be concerned with the minimum conditions for avoiding this fraction of crystals. The conditions are evaluated by constructing time-temperature-transformation (TTT) curves corresponding to the just-detectable fraction crystallized.

The TTT curves are constructed by applying the formal theory of transformation kinetics to the crystallization problem. According to this theory, the volume fraction, X, crystallized in time t at a given temperature may for small X be written:

$$X \approx I_v u^3 t^4. \tag{1}$$

Where I_v is the frequency of homogeneous nucleation per unit volume and u is the growth rate per unit area of the interface.

The nucleation frequency can be estimated from the relation:

$$I_v \approx N_v^0 v \exp\left[-\frac{1.024}{\Delta T_r^2 T_r^3}\right]. \tag{2}$$

Here N_v^0 is the number of atoms per unit volume; ΔT_r is the reduced undercooling, $\Delta T_r = (T_E - T)/T_E$, where T_E is the liquidus temperature; T_r is the reduced temperature, $T_r = T/T_E$; and v is the frequency factor for transport at the nucleus–liquid interface. This last factor is usually taken as inversely related to the viscosity as:

$$v = b/\eta. \tag{3}$$

The relation of equation (2) is in consonance with the results obtained in nucleation studies of a wide variety of materials, that the free energy of forming the critical nucleus is about $50kT$ at $\Delta T_r = 0.2$.

Taking $X = 10^{-6}$ and using measured values of the growth rate together with calculated values of I_v, the time required for the formation of a fraction crystallized of 10^{-6} can be estimated for each of a series of temperatures. When this is done, TTT curves similar in form to those shown in Fig. 1 are obtained. The nose in the curve results from a competition between the driving force for crystallization, which increases with decreasing temperature, and the atomic mobility, which decreases with decreasing temperature.

The cooling rate required to avoid a given fraction crystallized can be approximated by:

$$\left(\frac{dT}{dt}\right)_c \approx \frac{\Delta T_n}{t_n} \tag{4}$$

where T_n and t_n are, respectively, the temperature and time at the nose of the TTT curve and $\Delta T_n = T_E - T_n$. Correspondingly, the maximum thickness obtainable as a glass y_c can usually be estimated with about the same accuracy as:

$$y_c \approx (D_{TH} t_n)^{1/2} \tag{5}$$

where D_{TH} is the thermal diffusivity of the material.

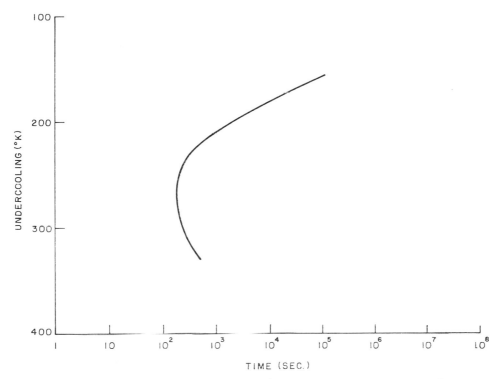

Fig. 1. Time-temperature-transformation curve for $Na_2O \cdot 2SiO_2$ for $X = 10^{-6}$.

Use of the frequency of homogeneous nucleation in the basic kinetic expression limits the above analysis to the minimum cooling rates required for glass formation. Nucleation on heterogeneities can only increase the overall extent of crystallinity in a given time at a given temperature, and hence increase the cooling rate needed to form a glass. Such heterogeneous nucleation is being treated in present extensions of the analysis; but for purposes of the present paper, the minimum conditions for glass formation will be sufficient.

EXPERIMENTAL PROCEDURE

Synthesis of lunar compositions 14259B and 14310 is detailed under "Experimental Procedures" and Table 1 of Cukierman *et al.* (preceding paper—this volume).

The growth rate measurements were carried out on small glass specimens prepared from the melted material which were 0.2 to 1 cm³ in size. For crystallization temperatures below 935°C, a horizontal resistance furnace with an internal Inconel sleeve and a 17 cm hot zone within ±0.5°C was used. A proportional controller provided temperature control within ±0.5°C in a 24 hour period. Specimens were wrapped in platinum foil and placed in a Vycor tube with a thermocouple. The tube was evacuated, sealed, and placed in the furnace. The time at which the specimens reached the crystallization temperature was determined using the thermocouple next to the specimens.

Some runs were carried out with an atmosphere of N_2 or a 5% H_2 forming gas maintained in the specimen tube.

For crystallization temperatures above 935°C, a second horizontal resistance furnace was employed. This furnace also was provided with a proportional temperature controller; but because of the elevated temperatures, the internal Inconel sleeve could not be used. With this furnace, the specimens were placed in a small molybdenum boat cemented to the end of a thermocouple. In some runs the furnace tube was flushed with dry nitrogen gas sufficient to prevent oxidation of the samples or the boat; in others a 5% H_2 forming gas atmosphere was provided.

In some cases with the 14259 composition it was found helpful to dust the glass surfaces with the crystallization products of previous runs prior to insertion in the heat treatment furnaces. This was done to effect a uniform nucleation of crystals on the external surfaces. It was not required on many runs with the 14259 composition and on none of the runs with the 14310 composition.

After heat treatment for desired periods of time at each crystallization temperature the specimens were quenched to room temperature, sectioned, and mounted in plasticine. The thickness of the external crystal layer on each specimen was determined using an optical microscope with a filar eyepiece. For each specimen, several measurements of the thickness were made and averaged. The morphologies of the crystal-liquid interfaces were determined using the optical microscope; and x-ray diffraction and optical microscopy were used to identify the crystallization products.

To determine the possible difference between the interface and furnace temperatures, a 3 mil Pt–Pt10Rh thermocouple was embedded in a 14310 sample. The sample was held at a temperature of fairly rapid growth (1100°C), and the temperature was recorded on a stripchart recorder at maximum sensitivity (full scale 2 mV) as the interface grew over the thermocouple.

Results and Discussion

Crystallization data

In all cases, crystals were observed to form on the external surfaces of the specimens and propagate into the interior regions. The thickness of this external crystalline layer after a given heat treatment time was found to be uniform within $\pm 15\%$. For both materials, the crystal–liquid interfaces were faceted under all conditions where the morphology could be examined in detail. In some runs with the 14310 composition, internally nucleated spherulitic crystals were observed at large undercoolings. These were not observed in runs with the 14259 composition. In all cases for which determinations were conducted the crystallization products were in consonance with the results of Walker et al. (1972).

For both materials at all temperatures the thickness of the crystal layer was found to increase linearly with time. A typical example of this behavior is shown in Fig. 2. The growth rates were determined from the slopes of such thickness versus time plots.

The variations of the growth rate with temperature for the 14310 composition run in atmospheres of N_2 and 5% H_2 forming gas are shown in Fig. 3. As seen there, the data cover an undercooling range of about 400°C. The forms of the growth rate

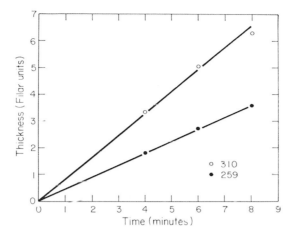

Fig. 2. Typical crystal thickness versus time relations for lunar compositions 14259 and 14310.

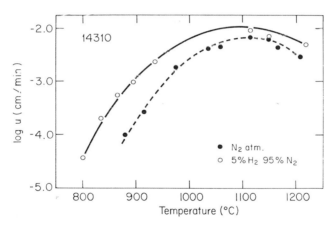

Fig. 3. Growth rate versus temperature relations for lunar composition 14310 in atmospheres of N_2 and 5% H_2 forming gas.

variations are similar to those observed with other glass-forming materials: a growth rate that increases with increasing undercooling below the liquidus temperature, passes through a maximum, and decreases with further increases in undercooling. The difference between the results obtained with the two atmospheres seems likely to reflect a difference in the state of reduction of the surface regions of the specimens (where the crystal growth was observed). This in turn could be reflected in a difference in the liquidus temperature or mobility at the interface or both. Since the effect of atmosphere seems most pronounced at low temperatures, however, the most important factor seems to be the effect on mobility.

The variation of the growth rate with temperature for the 14259 composition run in a N_2 atmosphere is shown in Fig. 4, where it is also compared with the results

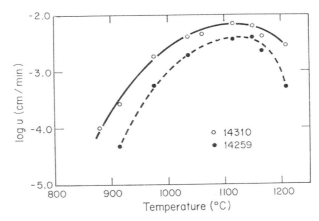

Fig. 4. Growth rate versus temperature relations for lunar compositions 14259 and 14310 in a N_2 atmosphere.

obtained under similar conditions for the 14310 composition. As shown there, the maximum growth rate for the 14259 composition, about 3.9×10^{-3} cm min^{-1}, is smaller by about a factor of 1.7 than the maximum growth rate for the 14310 composition, 6.6×10^{-3} cm min^{-1}. The maxima in both cases occur at temperatures in the range about 1120°C, corresponding to an undercooling of about 120°C for the 14259 composition and about 190°C for the 14310 composition. The viscosity of the 14259 composition at 1120°C is estimated to be smaller than that of the 14310 composition by a factor of about 1.4 (Cukierman *et al.*, 1972). The competing effects of driving force and molecular mobility are then reflected in the difference in maximum growth rates for the two materials.

Nature of the interface process

For the crystallization of the 14310 composition at 1100°C with a thermocouple embedded in the sample, the interface temperature was found to depart from the furnace temperature by less than 0.5°C, which is approximately the experimental accuracy of the measurement. This indicates that the growth rates are sufficiently small that the latent heat generated in the crystallization process can be removed from the interface region by the small gradients present in the system, and that the rate of interface advance is not limited by heat flow effects.

If it is assumed that the rate of interface advance is also not limited by mass transport—which seems reasonable for the extents of crystal growth and the growth rates measured in the present investigation—then the growth rate may be taken as controlled by interface attachment kinetics and useful information can be obtained about the nature of the interface process in these materials. It is recognized that significant changes in composition are involved in the crystallization of these liquids, and that the rate of advance of the crystal–liquid interfaces might possibly be limited by mass transport (diffusion) in the liquid phase: Arguing against this suggestion are: (1) the observed lack of a dependence of the growth rate on time over the range covered by the present investigation (see Fig. 2); and (2) for the extents of crystal growth covered here, the crystal size should not have been comparable with the extent of the diffusion fields in the liquid, and interface-controlled rather than diffusion-controlled growth should be expected. These points are discussed at greater length elsewhere and are used to interpret the crystallization data on other glass-forming systems (Uhlmann, 1972b).

In developing an improved understanding about the nature of the interface process in the lunar compositions, the data on crystallization kinetics will be combined with corresponding data on the viscosity versus temperature relations for the same compositions (Cukierman *et al.*, 1972). With such combined data, the approximate reduced growth rate, u_R, for each composition can be evaluated over a range of temperature. This quantity is defined:

$$u_R \approx \frac{u \cdot \eta}{\Delta T}. \tag{6}$$

This reduced growth rate is proportional to the interface site factor (the fraction of sites on the interface where growth preferentially takes place). The variation of u_R with undercooling thus provides information about the temperature dependence of the site factor and thereby serves to characterize the interface process.

In obtaining this information, it is generally assumed that a given phase is the crystallization product over the temperature range of interest. In the present case, this is true of plagioclase for the temperatures covered by the data; but at the inter-mediate- and low-temperature end of this range, other phases can be observed as devitrification products as well. Application of the various growth-rate expressions must then be predicated on the assumption that the primary crystallization product does not change over the range of interest and that this primary product is plagioclase.

Each of the standard models for crystal growth is based on a different assumption concerning the nature of the sites on the interface where atoms are added and removed. These models have been discussed at some length in a recent review (Uhlmann, 1972b), and may be briefly summarized here:

(1) *Normal growth:* According to this model, atoms can be added to or leave from any site on the crystal–liquid interface. The interface should be rough on an atomic scale and characterized by a large fraction of step sites where atoms can preferentially be added or removed. The reduced growth rate versus undercooling relation should be a horizontal line.

(2) *Screw dislocation growth:* According to this model, growth takes place at step sites provided by screw dislocations intersecting the interface. Such dislocations provide a self-perpetuating ledge as atoms are added to the interface. The interface should be generally smooth on an atomic scale and should be imperfect. The u_R versus ΔT relation should be a straight line of positive slope.

(3) *Surface nucleation growth:* According to this model, growth takes place at step sites provided by two-dimensional nuclei formed on the interface. The interface should be smooth on an atomic scale and should be perfect (free of intersecting screw dislocations). The u_R versus ΔT relation should be a curve with positive curvature which passes through the origin.

The reduced growth rate versus undercooling relations for compositions 14310 and 14259 are shown in Figs. 5a and 5b. For both materials, the relations exhibit positive curvature of the type expected for surface nucleation growth. The apparent maxima observed at the highest undercoolings undoubtedly reflect uncertainties in the viscosity interpolation employed. (Measurements of viscosity could be made at

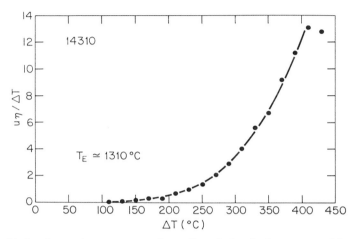

Fig. 5a. Reduced growth rate versus undercooling relation for lunar composition 14310.

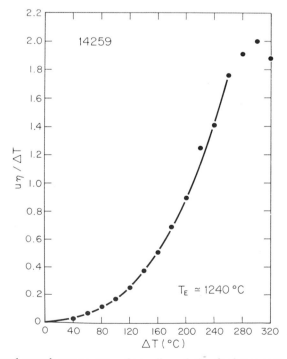

Fig. 5b. Reduced growth rate versus undercooling relation for lunar composition 14259.

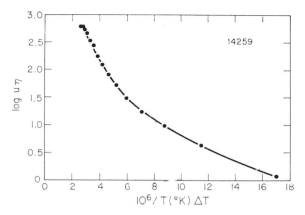

Fig. 6. Logarithm of (growth rate × viscosity) versus $1/T \, \Delta T$ for lunar composition 14259.

high temperatures and low temperatures, but could not be effected at intermediate temperatures because of crystallization.)

In testing further the applicability of the surface nucleation growth model, log ($u\eta$) versus $1/T \, \Delta T$ relations have been constructed for both materials. The results are shown in Fig. 6 for the 14259 composition; a relation of similar form was found for the 14310 composition. According to the simple models of surface nucleation growth, these relations should be straight lines of negative slope (assuming the edge surface energies of the surface nuclei are independent of temperature). Curved log ($u\eta$) versus $1/T \, \Delta T$ relations of the form shown in Fig. 6 have been found in nearly all studies of crystal growth where u_R versus ΔT relations of the surface nucleation growth form are observed. (See discussion in Uhlmann, 1972b; Scherer and Uhlmann, 1972.) The origin of the curvature in Fig. 6 cannot yet be established with any confidence. It likely reflects inadequacies in the standard models for surface nucleation growth but might also arise from an unanticipated increase in the edge surface energy with decreasing temperature.

Glass-forming characteristics

The kinetic model discussed in the second section can be used with the present data on crystallization kinetics and viscous flow behavior to estimate the critical conditions for forming glasses of the materials. Following the analytical procedure outlined above, the TTT curves shown in Fig. 7 were calculated. As noted there, the nose of the curve for the 14310 composition occurs at an undercooling of about 370°C and corresponds to a time of about 8.8 sec, while that for the 14259 composition occurs at an undercooling of about 300°C and corresponds to a time of about 260 sec.

From these values, the critical cooling rate required to form a glass of composition 14310 is estimated as about 40°C sec^{-1}, while for composition 14259 is somewhat lower, about 1°C sec^{-1}. The corresponding thicknesses obtainable as glasses are estimated as about 0.3 and 2 cm for the 14310 and 14259 compositions, respectively.

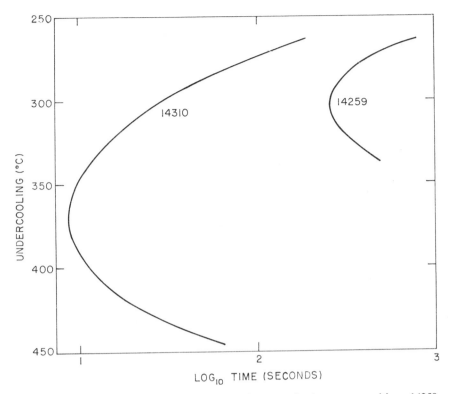

Fig. 7. Time-temperature-transformation (TTT) curves for lunar compositions 14259 and 14310. $X = 10^{-6}$.

Considering the approximations used in the original analysis, these estimates are not intended as valid to better than order-of-magnitude accuracy. The results then indicate that composition 14259 is a somewhat better glass-former than composition 14310; but the difference between the two is in the range that conceivable differences in the density of nucleating heterogeneities between the materials could outweigh the indicated difference in their expected behavior. The effects of such heterogeneities are being considered in a present extension of the analysis.

For comparison with other compositions, the critical cooling rate estimated for $Na_2O \cdot 2SiO_2$—using the same analysis with measured values of the growth rate and the viscosity—is about $1°C \, sec^{-1}$. The TTT curve for this material has been shown in Fig. 1 above. The glass-forming tendency of this material should then be comparable with that of lunar composition 14259.

In all cases, the calculated critical cooling rates and thicknesses obtainable as glasses are consistent with experimental observations. These observations include those carried out in the laboratory while preparing glass samples for the crystallization study as well as the sizes of the individual particles of the 14259 composition found on the lunar surface.

The occurrence of composition 14259 in the glassy state sets significant limitations on the thermal history of the individual particles, since the TTT curve shown in Fig. 7 and its extension to other ranges of temperature delineate the maximum times for which portions of the specimens could have remained in each temperature interval. The analysis outlined in the second section above is presently being extended to include detailed solutions to the heat flow problems, and these extensions should permit more definitive suggestions to be made concerning the origin and history of the lunar samples considered.

Acknowledgments—The authors are happy to acknowledge the experimental assistance of G. Wicks of M.I.T. and D. Walker of Harvard University, the computational assistance of M. Cukierman of M.I.T., and stimulating discussions with J. F. Hays of Harvard University as well as the above individuals. Appreciation is also due to the National Aeronautics and Space Administration who provided financial support for the present work under Grant NGR 22-009-646 and to Owens-Illinois, Inc., who provided one of the authors (G.S.) with the Owens-Illinois Fellowship in Materials Science.

REFERENCES

Cukierman M., Tutts P. M., and Uhlmann D. R. (1972) Viscous flow behavior of lunar compositions 14259 and 14310 (abstract). In *Lunar Science—III* (editor C. Watkins), p. 170, Lunar Science Institute Contr. No. 88.

Scherer G. and Uhlmann D. R. (1972) Crystallization behavior of α-phenyl o-Cresol. *J. Crystal Growth* (to be published).

Uhlmann D. R. (1972a) A kinetic treatment of glass formation. *J. Non-Cryst. Solids.* (to be published).

Uhlmann D. R. (1972b) Crystal growth in glass-forming systems. To be published in *Nucleation and Crystallization in Glasses and Melts*, American Ceramic Society, Columbus.

Walker D., Longhi J., and Hays J. F. (1972) Experimental petrology and origin of Fra Mauro rocks and soil (abstract). In *Lunar Science—III* (editor C. Watkins), pp. 770–772, Lunar Science Institute Contr. No. 88.

Proceedings of the Third Lunar Science Conference
(Supplement 3, *Geochimica et Cosmochimica Acta*)
Vol. 3, pp. 2639–2654
The M.I.T. Press, 1972

Direct observation of the lunar photoelectron layer

DAVID L. REASONER and WILLIAM J. BURKE

Department of Space Science, Rice University,
Houston, Texas 77001

Abstract—The Charged-Particle Lunar Environment Experiment (CPLEE), a part of the Apollo 14 ALSEP, is an ion-electron spectrometer capable of measuring ions and electrons with energies between 40 eV and 50 keV. Accordingly, the instrument, with apertures 26 cm above the surface, has detected a layer of photoelectrons, or photoelectron gas above the sunlit lunar surface with energies ranging up to 200 eV. The experimental data for periods when the moon was in the earth's magnetotail for electron energies between 40 and 200 eV follows a power-law spectrum of the form $j(E) = j_0(E/E_0)^{-\mu}$, with $j_0 = 2.5 \times 10^5$ electrons/cm^2 − sec, $E_0 = 40$ eV, and $\mu = 3.5$. The implications of this measurement are two-fold in that the lunar surface potential can be immediately determined to be at least 200 V, and a value of the photoelectron yield of the lunar surface material for photon energies above 40 eV may be computed.

A numerical solution for the variation of electron density and potential above the lunar surface was obtained. The method of solution was based on solving Poisson's equation, and computing the electron density at a height x by using the Liouville theorem along with the conservation of energy equation for the single-particle trajectories. Two parameters of the solution are the solar photon spectrum $I(h\nu)$ and the photoelectron yield function $Y(h\nu)$ of the surface materials. The solar photon spectrum was obtained from various experimental sources, and the solution for the energy spectrum at the height of the measurements (26 cm) was computed for various values of $Y(h\nu)$ until a fit to the experimental data was obtained. We used a functional form of $Y(h\nu)$ to be $Y(h\nu) = \alpha(h\nu - W)$ for 6 eV $\leq h\nu < 9$ eV, and $Y(h\nu) = Y_0$ for $h\nu \geq 9$ eV. The lunar surface work function W was chosen to be 6 eV. This procedure resulted in a value of Y_0 of 0.1 electrons/photon. The solution also showed that for the condition of low ambient plasma density in the magnetotail, the photoelectron density falls by 5 orders of magnitude within 10 m from the surface, but that the layer actually terminates several hundred meters above the surface.

The detailed temporal history of the photoelectron intensity during the total lunar eclipse on February 10, 1971 was studied in order to determine the source distribution of solar photons in the 40 eV to 200 eV range over the solar disc. It was found that the temporal behavior of the photoelectron intensity exhibited a classical penumbral-umbral behavior. Hence it is concluded that at the time of the eclipse the emission of these higher energy photons was uniform over the solar disc.

INTRODUCTION

THE GENERAL PROBLEMS of photoelectron emission by an isolated body in a vacuum and in a plasma have been the objects of several investigations. For example, Medved (1965) has treated electron sheath formations about bodies of typical satellite dimensions. Guernsey and Fu (1970) have considered the properties of an infinite, photo-emitting plate immersed in a dilute plasma. Grobman and Blank (1969) obtained expressions for the lunar surface potential due to photoelectron emission while the moon is in the solar wind. Walbridge (1970) developed a set of equations for obtaining the density of photoelectrons as well as the electrostatic potential as functions of height above the surface of the moon while the moon is in the solar wind. By assuming a simplified form of the solar photon emission spectrum he could provide analytic expressions for these quantities.

In this paper we report on observations of stable photoelectron fluxes, with energies between 40 and 200 eV by the Apollo 14 Charged Particle Lunar Environment Experiment (CPLEE). These observations, made in the magnetotail under near vacuum conditions, are compared with numerically calculated photoemission spectra to determine the approximate potential difference between ground and CPLEE's apertures (26 cm). Numerically calculated density and potential distributions, when compared with our measured values, help us estimate the photoelectron yield function of the dust layer covering the moon.

The Instrument

A complete description of the CPLEE instrument has been given by O'Brien and Reasoner (1971). The instrument contains two identical charged-particle analyzers, hereafter referred to as Analyzers A and B. Analyzer A looks toward the local lunar vertical, and Analyzer B looks 60° from vertical toward lunar west.

The particle analyzers contain a set of electrostatic deflection plates to separate particles according to energy and charge type, and an array of six channel electron multipliers for particle detection. For a fixed voltage on the deflection plates, a five-band measurement of the spectrum of particles of one charged sign and a single-band measurement of particles of the opposite charge sign are made. The deflection plate voltage is stepped through a sequence of three voltages at both polarities, plus background and calibration levels with zero voltage on the plates. A complete measurement of the spectrum of ions and electrons with energies between 40 eV and 50 keV is made every 19.2 sec. Of particular relevance to this study are the lowest electron energy passbands. With a deflection voltage of -35 V, the instrument measures electrons in five ranges centered at 40, 50, 65, 90, and 200 eV. With $+35$ V on the deflection plates, electrons in a single energy range between 50 and 150 eV are measured.

Observations

In this section we present data from the February 1971 passage of the moon through the magnetotail. Because these are so typical, the display of data from subsequent months would be redundant. At approximately 0300 UT on February 8 CPLEE passed from the dusk side magnetosheath into the tail. The 5 min averaged counting rates for Analyzer A, Channel 1, at -35 V measuring 40 eV electrons are plotted for this day in Fig. 1. Almost identical count rates are observed in Analyzers A and B during this period of observation. As CPLEE moves across the magnetopause the counting rate drops from ~ 200/cycle to the magnetotail photoelectron background of ~ 35/cycle (1 cycle = 1.2 sec). Enhancements at ~ 0530 hours and at ~ 0930 hours correspond to plasma events associated with substorms on earth (Burke and Reasoner, 1972). There is a data gap from 1000 to 1200 hours. With the exception of the short lived (≤ 1 hour) enhancements the detector shows a stable counting rate when the moon is in the magnetotail.

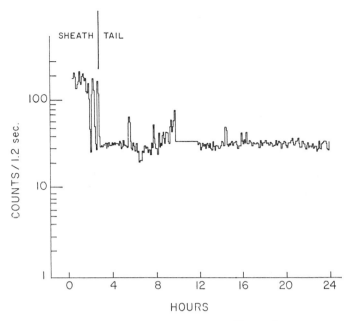

Fig. 1. Five minute average counting rates for CPLEE, Analyzer A, Channel 1 at −35 V, measuring 40 eV electrons on February 8, 1971. After 0300 U.T. counting rates fell from high magnetosheath to stable photoelectron levels.

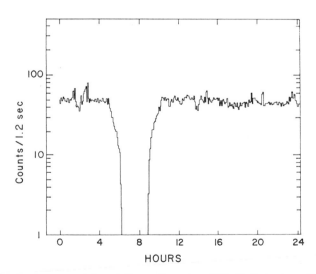

Fig. 2. Five minute averaged deep tail counting rates of 40 eV electrons on February 10, 1971. The lunar eclipse (0500–0900 U.T.) is marked by vanishing photoelectron counting rates.

Our contention is that these stable fluxes observed in the magnetotail during periods of low magnetic activity are photoelectrons generated by ultraviolet radiation from the sun striking the surface of the moon. In support of this thesis, we have reproduced the counting rates observed in the same detector on February 10 when the moon was near the center of the tail (Fig. 2). First, we note that the stable count level is the same at the center as it was when CPLEE first entered the tail. Secondly, from about 0500 to 1000 hours the moon was in eclipse. During this time we observe the counting rates go to zero. As the moon emerges from the earth's shadow, the counting rates return to their pre-eclipse levels. If the stable low energy electrons were part of an ambient plasma, rather than photoelectrons, the counting rates would not be so radically altered as the moon moved across the earth's shadow.

A detailed plot of the photoelectron flux on an expanded time scale for the period of lunar entry into the earth's penumbra and umbra is shown in Fig. 3. The times of penumbral and umbral entry are indicated, and these times were computed from ephemeris data appropriate to the lunar coordinates of CPLEE. In addition to con-

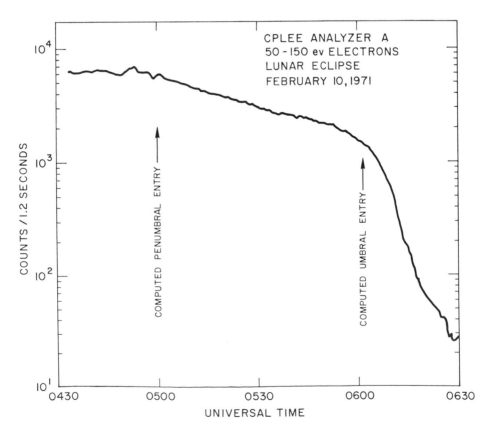

Fig. 3. Expanded time scale plot of photoelectron fluxes for the period of penumbral and umbra entry of the lunar surface region around CPLEE. The data is the counting rate of Channel 6 at +35 V, measuring electrons with 50 eV $<$ E $<$ 150 eV.

firming our earlier arguments, this plot also shows that the high energy (>40 eV) photons are radiated essentially uniformly over the solar disc. If these photons were emitted from a few isolated regions, then one would observe sharp transitions in the flux as the regions were progressively shadowed by the limb of the earth. As can be seen, however, the curve in the penumbral portion is smooth with no obvious discontinuities. Furthermore, one notices that the counting rate falls by more than two orders of magnitude as the moon traverses the umbral and penumbral regions (Fig. 3). The minimum counting rate of ~15/sec for this channel is very near the instrument background level, but even if these were all due to an ambient plasma our argument that the pre-eclipse counting rates were due entirely to photoelectrons is not significantly altered.

It could be argued that the observed counting rates were due to photons scattering within the detectors themselves and not due to external photoelectrons. This however is not the case. Preflight calibrations with a laboratory ultraviolet source showed enhanced counting rates only when the angle between the look direction of the detector and the source was less than 10°. Given the 60° separation between the look directions of Analyzers A and B, it would be impossible for the sun, essentially a point source, to produce identical counting rates in both analyzers simultaneously. There are times when we do observe ultraviolet contamination in one or the other channel. An example of such contamination is shown in Fig. 4 from February 11, 0600 hours, to February 12, 0900 hours. As the sun moves across the aperture of Analyzer A the counting rates increase a full order of magnitude. During this period Analyzer B continued to produce typical deep tail counting rates. Note that as the

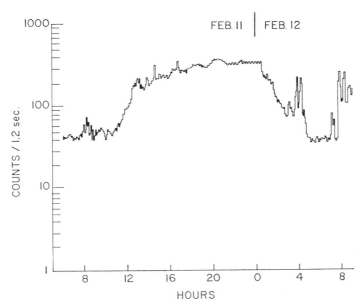

Fig. 4. An example of ultraviolet contamination of Analyzer A from ~1200 of February 11 to ~0300 of February 12, 1971.

detector came out of ultraviolet contamination it encountered typical magnetosheath plasma. At ~0345 it passed back into the magnetotail, then at ~0800 returned to the magnetosheath.

A typical spectrum of photoelectrons shown in Fig. 5 was observed by Analyzer A at ~0400 hours on February 10, shortly before the moon entered penumbral eclipse. The dark line marks the differential flux equivalent to a background cound of one per cycle in each channel. For all five channels, with the deflection plates at −35 V, the differential flux is well above this background level. During geomagnetically quiet times no statistically significant counts are observed when the deflection plates are at −350 or −3500 V corresponding to electrons with $E > 500$ eV (Burke and Reasoner, 1972).

With the exception of periods of ultraviolet contamination in Analyzer A, we always observe nearly the same counting rate due to photoelectrons in Analyzers A and B. For all purposes, we can say that the spectrum displayed in Fig. 5 is just as typical as for Analyzer B. We have found no case of anisotropy in the photoelectron fluxes. In all cases too, we found that the photoelectron spectra observed in both analyzers were close to a power law dependence on energy. If we write the differential flux in the form $j(E) = j_0(E/E_0)^{-K}$, K is between 3.5 and 4.0, $E_0 = 40$ eV and $j_0 \sim 2.5 \times 10^5$ electrons/cm^2−sec-sr-eV. In the following section the details of this spectrum are more carefully studied.

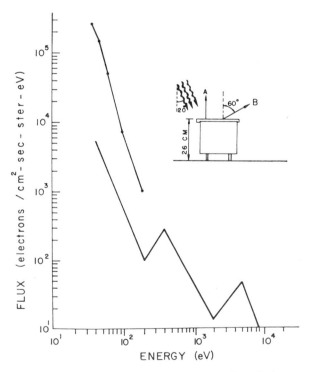

Fig. 5. Typical photoelectron spectrum observed by CPLEE at the lunar surface in the high latitude magnetotail.

Also in Fig. 5 we display a schematic cross section of our instrument as it is deployed on the surface of the moon. The apertures of both analyzers are elevated 26 cm from ground. Their geometry is such that they observe only electrons with a component of velocity in the downward direction. Since we continually observe photoelectrons with energies up to ~ 200 eV, we must assume that the lunar surface potential is on the order of 200 V during these times. This measurement will seem high to those familiar with the work of Walbridge (1970) and Grobman and Blank (1969), who calculate a surface potential that is at least an order of magnitude lower. The difference is that their models deal with photoemissions from the surface of the moon in the presence of the solar wind. Our measurements in the magnetotail are made under near-vacuum conditions. After further analysis of the problem we return to considerations of the surface potential.

To summarize: During geomagnetically quiet times, when the moon is in the magnetotail and not in eclipse, stable photoelectron fluxes with energies between 40 and 200 eV are observed. These fluxes are isotropic and obey a power law, $E^{-\mu}$, where μ is between 3.5 and 4. From the fact that CPLEE is observing downward moving electrons we conclude that in the magnetotail the lunar surface potential is on the order of 200 V.

NUMERICAL ANALYSIS

General theory

The variations of photoelectron density and electrostatic potential above the surface of the moon can be calculated numerically. Again we approximate the lunar surface by an infinite plane, with the x direction normal to the surface, and assume spatial variations of physical quantities only with the height.

At a height x above the surface the electron density is $\int f(\vec{v}, x) d^3 v$, where $f(\vec{v}, x)$ is the electron distribution function. If we assume an isotropic flux at the surface, the Liouville theorem can be used to show that the distribution function is independent of angles at all heights. Writing $d^3 v = \sqrt{2E/m^3}\, dE\, d\Omega$ and integrating over solid angles, the density is

$$n(x) = 4\pi \int_0^\infty \sqrt{\frac{2E}{m^3}} f(E, x)\, dE. \tag{1}$$

Since the distribution function is a constant along particle trajectories, $f(E, x) = f(E_0, x = 0)$, where $E = E_0 - q[\psi_0 - \psi(x)]$. By changing the variable of integration from E to E_0, equation (8) can be expressed

$$n(x) = 4\pi \int_{q[\psi_0 - \psi(x)]}^\infty \sqrt{2m(E_0 - q[\psi_0 - \psi(x)])}\, f(E_0, x = 0)\, dE_0. \tag{2}$$

To calculate the distribution function of photoelectrons at the surface consider the quantity

$$j(E_0)\, dE_0 = \left[\int_W^\infty I(h\nu)\, Y(h\nu)\rho(E_0, h\nu)\, dh\nu \right] dE_0 \tag{3}$$

the upward moving flux of photoelectrons emitted from the surface with energies between E_0 and $E_0 + dE_0$. $I(hv)\,d(hv)$ is the flux of photons reaching the lunar surface with energies between hv and $hv + d(hv)$. $Y(hv)$, the quantum yield function, gives the number of electrons emitted by the surface per incident photon with energy hv. $\rho(E_0, hv)\,dE_0$ is the probability that an electron emitted from the surface, due to a photon with energy hv, will have a kinetic energy between E_0 and $E_0 + dE_0$. $\rho(E_0, hv)$ is normalized so that

$$\int_0^\infty \rho(E_0, hv)\,dE_0 = 1.$$

W is the work function of the lunar surface material.

The total upward moving flux at the surface is

$$S_\uparrow(x = 0) = \int_0^\infty j(E_0)\,dE_0.$$

But

$$S_\uparrow(x = 0) = \int_0^{2\pi} \int_0^{\pi/2} \int_0^\infty \vec{v}_0 f(E_0, \theta, \phi, 0) v_0^2\,dv_0 \cdot \sin\theta\,d\theta\,d\phi.$$

Since $\vec{v}_0 = v_0[\vec{i}\cos\theta + \vec{j}\sin\theta\cos\phi + \vec{k}\sin\theta\sin\phi]$ and f is independent of angle,

$$S(x = 0) = \pi \int_0^\infty \frac{2E_0}{m^2} f(E_0, x = 0)\,dE_0. \tag{4}$$

Thus

$$f(E_0, x = 0) = \frac{m^2 j(E_0)}{2\pi E_0} \tag{5}$$

and

$$n(x) = 2 \int_{q[\psi_0 - \psi(x)]}^\infty \sqrt{2m(E_0 - q[\psi_0 - \psi(x)])} \frac{j(E_0)}{E_0}\,dE_0. \tag{6}$$

The potential as a function of height is evaluated by multiplying Poisson's equation, $\partial^2\psi/\partial x^2 = -4\pi q n(x)$, by $\partial\psi/\partial x$ and integrating in from $x = \infty$ to get

$$\left(\frac{\partial\psi}{\partial x}\right)^2 = -8\pi q \int_{\psi(x)}^0 n(\psi')\,d\psi' \tag{7}$$

where we have written

$$\int_x^\infty n(x') \frac{d\psi}{dx'}\,dx' = \int_{\psi(x)}^0 n(\psi')\,d\psi'.$$

A further integration out from the surface, gives us the potential at a point x.

Computational methods and results

To determine the upward moving differential flux at the surface, upon the knowledge of which the distribution function, number density, and potential depend, we must first solve the integral in equation (3). The solar photon differential flux at 1 A.U., $I(hv)$, is taken from Friedman (1963) for the range 2000 to 1800 Å and from Hinteregger (1965) for the range 1775–1 Å and is plotted in Fig. 6. Following the suggestion of Walbridge (1970) we have:

(1) Adopted a work function of lunar material of 6 eV.

(2) Assumed a photoelectron yield function of the form

$$Y(hv) = \begin{cases} u(hv - 6) \, Y_0 \, (hv - 6)/3 & 6 < hv < 9 \text{ eV} \\ Y_0 & hv > 9 \text{ eV} \end{cases} \tag{8}$$

where $u(hv)$ is a unit step function and Y_0 is a free parameter of our calculation.

(3) Chosen a probability function

$$\rho(E, hv) = \begin{cases} 6E(E_1 - E)/E_1 & 0 \leq E \leq E_1 \\ 0 & E > E_1 \end{cases} \tag{9}$$

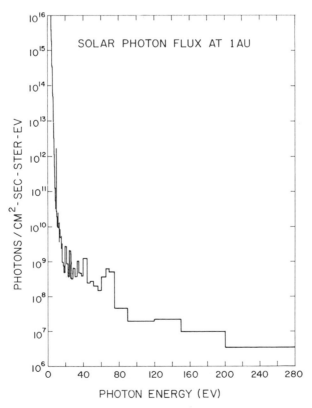

Fig. 6. Solar photon energy spectrum at 1 a.u. from 2000 to 1 Å.

where

$$E_1 = \begin{bmatrix} hv - W & & hv \geq W \\ 0 & & hv < W \end{bmatrix}$$

In general the probability function is a complicated function depending on the nature of the photoemission material. However, Grobman and Blank (1968) have shown that for the purpose of calculating equation (3) any broad function with zeros at $E = 0$ and $E = E_1$ and a width $\Delta E \sim hv$ will suffice. A plot of $\rho(E, hv)$ is shown in Fig. 7 for various values of E_1.

The upward directed differential flux in electrons/cm²-sec-sr-eV for the values $Y_0 = 1, 0.1, 0.01$ were numerically computed and have been plotted in Fig. 8. We have also inserted the photoelectron differential flux observed by CPLEE at 26 cm. The Liouville theorem allows us to set a lower bound on Y_0 of 0.1. That is if there were no potential difference between the ground and 26 cm the yield function would be 0.1 electrons/photon. After estimating the potential difference between 26 cm and ground we can also determine an upper bound on Y_0.

Solving the integro-differential equation (7) for $\psi(x)$ involves an integration from the surface outward, with an assumed value of ψ_0. However the expression of $\partial \psi / \partial x$ involves an integral from infinity in to x, or equivalently from $\psi = 0$ to $\psi(x)$. Integrals of this type are ordinarily impossible to evaluate numerically. By the expedient of dividing the integral into pieces in E_0 space and using an analytic approximation to the function $j(E_0)$ in each of these intervals, a solution was effected. In this way it was only necessary to know the values of $\psi = \psi(x)$ and $\psi = 0$ at the end points of the

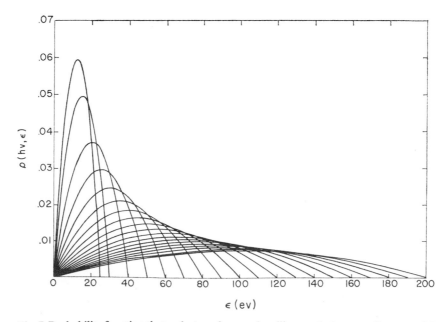

Fig. 7. Probability function that a photon of energy hv will cause the lunar surface material to emit a photoelectron of energy E, with different values of $E_1 = hv - W$.

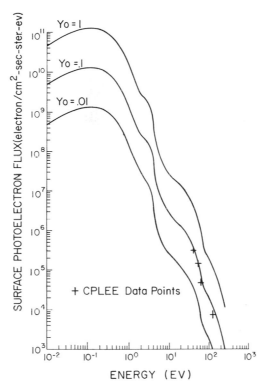

Fig. 8. Numerically computed photoelectron spectra emitted for the yield functions $Y_0 = 1$, 0.1 and 0.01 electrons per photon. The photoelectron spectrum measured by CPLEE is found to fall close to the $Y_0 = 0.1$ line.

interval, and the solution would proceed. In Fig. 9 we show families of solutions for $\psi(x)$ with several values of the parameter Y_0.

The value of Y_0 calculated by assuming no potential difference between the surface and $x = 26$ cm was 0.1. Figure 9 shows that for $Y_0 = 0.1$, the potential difference $\psi(x = 0) - \psi(x = 26 \text{ cm})$ is only 3 V. Obviously, we could now use an iterative procedure, modifying our spectral measurement at 26 cm to obtain the surface spectrum according to the equation $f(E, x) = f(E_0, 0)$ and hence obtain a new estimate of Y_0. However, the procedure is hardly justified considering the small potential difference (~ 3 V) and the energy range of the measured photoelectrons (40–200 eV). Hence we conclude from our numerical analysis and measured photoelectron fluxes a lunar surface potential on the order of 200 V and a value of the average photoelectron yield of $Y_0 = 0.1$ electrons/photon.

THE LUNAR SURFACE POTENTIAL ψ_0

The solar photon energy spectrum (Fig. 6) shows a marked decrease at $h\nu = 200$ eV. For the case of the moon in a vacuum the potential of the lunar surface would be equal to the highest energy photon present minus the lunar surface work function.

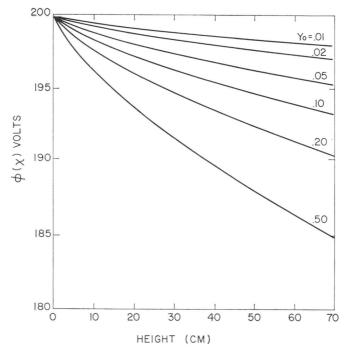

Fig. 9. Numerically computed potential distribution above the lunar surface for several values of the yield function Y_0. For $Y_0 = 0.1$ the potential difference between ground and 26 cm is about 3 V.

Hence we estimate the lunar surface potential ψ_0 to be 200 V. This is confirmed by the experimental measurements, as the photoelectron energy spectrum shows a measurable flux at 200 eV but no significant flux in the next highest energy channel at 500 eV.

The lunar surface potential can be decreased however by the presence of a hot ambient plasma which furnishes an electron return current which partially balances the emitted photoelectron current. In effect, the highest energy photoelectrons can escape from the potential well, since electrons from the ambient plasma furnish the return current to balance these escaping photoelectrons. Quantitatively, if F_s is the net negative flux to the lunar surface from the ambient plasma, and $j(E_0)$ is the emitted photoelectron energy spectrum in units of electrons/cm²-sec-eV, then:

$$F_s = \int_{q\psi_0}^{\infty} j(E_0)\,dE_0 \tag{10}$$

and this equation can be solved for ψ_0, the lunar surface potential.

Our measurements of photoelectrons were taken during periods in the magnetotail when all of the channels of the instrument except the lowest-energy electron channels were at background levels. Thus we can establish an upper limit to the electron flux from the ambient plasma for electrons with 40 eV $< E <$ 50 keV. Figure 5 shows the "background spectrum", calculated by converting the background counting rate of ~ 1 count/sec to equivalent flux in each of the energy channels. Integrating over this spectrum and converting to flux over the hemisphere gives $F_s \leq 3.4 \times 10^6$ electrons/cm^2-sec. We feel that this is a valid upper limit, as the range of measurement in energy includes both the peak energy of the plasma sheet spectrum (~ 1 keV) and of the magnetosheath spectrum (~ 40–60 eV).

We note that Vasyliunas (1968) obtained an upper limit to the electron concentration for locations outside of the plasma sheet based on OGO-3 data. The relation expressed was $NE_0^{\frac{1}{2}} < 10^{-2}$ cm^{-3} keV$^{\frac{1}{2}}$ where N is the electron density and E_0 is the energy at the peak of the spectrum. For an isotropic plasma where the bulk motion can be neglected relative to the thermal motion, the electron flux to a probe is given by $F_s = N\bar{v}/2\sqrt{\pi}$. Applying the appropriate conversion of factors, the expression of Vasyliunas results in an upper limit to the electron flux of $F_s < 5.6 \times 10^7$ electrons/cm^2-sec.

The emitted photoelectron energy spectrum $j(E_0)$ is shown in Fig. 8. The procedure involved in calculating the surface potential ψ_0 is to integrate the function $j(E)$ from the maximum energy of 200 eV backward until the total flux is equal to the upper limit of the return flux. The computation was done for $Y_0 = 1, 0.1,$ and 0.01, and for the two values of the upper limit of the return flux derived above. The results are shown in Table 1.

The lower half-height of the Channel 5 energy passband is 160 eV. Hence the surface potential could be as low as 160 V and still result in particle fluxes in Channel 5. This estimate of the potential is seen to be not inconsistent with a value of $Y_0 = 0.1$, $F_s \leq 3.4 \times 10^6$ resulting in a surface potential (Table 1) of 114 V.

Discussion

For the sake of displaying the complete numerical solution, Fig. 10 shows the numerically calculated density, pressure, and potential difference from $x = 0$ to $x = 200$ m. In this the pressure is defined by conservation of momentum consideration by the equation $p = (1/8\pi)\,[\partial\psi/\partial x]^2$. The electron density and potential difference curves are seen to follow approximately power law functions, and the marked deviations correspond to the irregularities in the emitted photoelectron spectrum (see Fig. 8) at $E \approx 10$ eV and $E \approx 100$ eV. Recalling that for the vacuum case studied here we require $n(x)$ and $\partial\psi/\partial x \to 0$ as $x \to \infty$, we see that indeed the potential difference curve begins to flatten at $x = 100$ m and consequently the absolute potential approaches zero asymptotically while the density continues to decrease

Table 1.

Electron flux	Y_0	Φ_0 (volts)
3.4×10^6	1.0	181
3.4×10^6	0.1	114
3.4×10^6	0.01	44
5.6×10^7	1.0	96
5.6×10^7	0.1	36
5.6×10^7	0.01	8

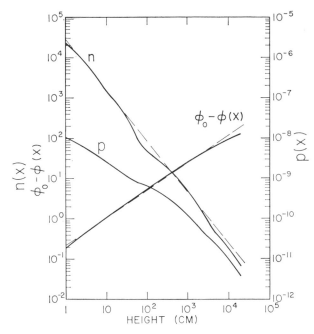

Fig. 10. Numerically computed values of electron density, potential difference and pressure as a function of distance from the lunar surface. The dashed lines represent power law curves.

approximately according to a power law. For all practical purposes, however, the photoelectron layer is seen to terminate 100 m above the lunar surface.

Feuerbacher *et al.* (1972) have measured the photoelectron yield of a lunar fine sample. In the photon energy range $5 < E < 20$ eV, they found a yield function that reaches a maximum value of ~ 0.08 at 15 eV then drops to 0.01 at 20 eV. B. Fitton (private communication, 1971) has suggested that our value of 0.1 is more a measurement of the CPLEE instrument case yield function than that of the lunar surface. We find it difficult to understand how this could be the case.

First, were CPLEE an electrically isolated package then the requirement that the net current to the instrument be zero would result in a measured yield function that is representative of the case material. Relative to the lunar surface the case would bear a positive potential in order to maintain an enhanced photoelectron density in its immediate vicinity.

The fact is, however, that CPLEE is not electrically isolated. CPLEE is connected to the central ALSEP station. Further, at any given time, only the top surfaces and one or two sides of ALSEP packages are illuminated by the sun, while the remaining area is shadowed. These unilluminated surfaces provide receptor areas for return current from the photoelectron gas. Thus no large potential difference can develop between CPLEE and the lunar surface, and photoelectrons emitted at the lunar surface and at the CPLEE case are undistinguishable. Geometrical considerations of electron trajectories would lead us to expect that the bulk of the photoelectrons measured by CPLEE were emitted at the lunar surface at least several meters from CPLEE. If the ALSEP instrument cases had a photoelectron yield much larger or much smaller than the yield of the lunar surface, then one would expect a pertubation of the photoelectron flux in the vicinity of ALSEP. This pertubation would depend not only on the ratio of photoelectron yield, but more importantly on the ratio of the area of the ALSEP instruments to the area encompassed by the trajectories of $E > 40$ eV photoelectrons from a point source. This last area is on the order of the square of the scale height of photoelectrons with $E > 40$ eV (~ 10 m, Fig. 10). Since the ALSEP area is ~ 3 m^2, the ratio of areas is on the order of 2%. Thus, even if the yield of the instrument cases was a factor of 10 greater or smaller than the surface yield, the flux pertubation would only be 20%.

Second, were CPLEE measuring its own photoelectrons one would expect to observe changes in the relative fluxes observed in the two analyzers with solar zenith angle as the electron cloud surrounding the instrument adjusts to changing illumination conditions. Specifically, the ratio of the flux observed by analyzer B (looking 60° west of vertical) to the flux observed in Analyzer A (looking to the vertical) should be larger after than before lunar noon. Our data shows isotropic photoelectron fluxes across the entire magnetotail.

SUMMARY AND CONCLUSIONS

In this paper we have reported the observation of stable, isotropic photoelectron fluxes 26 cm above the lunar surface. In the energy range $40 \leq E \leq 200$ the flux obeys a power law of the form $j(E) = j_0(E/E_0)^{-\mu}$ where μ is between 3.5 and 4, $j_0 = 2.5 \times 10^5$ electrons/cm^2-sr-sec-eV, and $E_0 = 40$ eV. Because these fluxes were moving down we conclude that in the near vacuum conditions of the high latitude magnetotail the lunar surface potential is at least 200 V. It was shown that these electrons can be explained in terms of the measured solar photon spectrum producing an isotropic flux of photoelectrons at the surface. A photoelectron yield function of $Y_0 = 0.1$ electron/photon was calculated.

Acknowledgment—This work was supported by National Aeronautics and Space Administration Contract No. NAS9-5884.

REFERENCES

Burke W. J. and Reasoner D. L. (1972) Absence of the plasma sheet at lunar distance during geomagnetically quiet times. *Planetary and Space Sci.* **20**, 429–436.

Feuerbacher B., Anderegg M., Fitton B., Laude L. D., and Willis R. F. (1972) Photoemission from lunar surface fines (abstract). In *Lunar Science—III* (editor C. Watkins), pp. 253–255, Lunar Science Institute Contr. No. 88.

Friedman H. (1963) Rocket spectroscopy. In *Space Science* (editor D. P. LeGallet), p. 549, John Wiley and Sons.

Geurnsey R. L. and Fu J. H. M. (1970) Potential distribution surrounding a photo-emitting plate in a dilute plasma. *J. Geophys. Res.* **75**, 3193–3199.

Grobman W. D. and Blank J. L. (1969) Electrostatic potential distribution above the sunlit lunar surface. *J. Geophys. Res.* **74**, 3943–3952.

Hinteregger H. E., Hall L. A., and Schmidtke G. (1965) Solar XUV radiation and neutral particle distribution in the July 1965 thermosphere. In *Space Research V* (editors D. K. King-Hele, P. Muller, and G. Righini), pp. 1175–1190, North Holland.

Medved D. B. (1968) On the formation of satellite electron sheaths resulting from secondary emission and photoeffects. In *Interaction of Space Vehicles with an Ionized Atmosphere* (editor S. F. Singer), p. 305, Pergamon.

O'Brien B. J. and Reasoner D. L. (1971) Charged-particle lunar environment experiment. In *Apollo 14 Preliminary Science Report*, pp. 193–213, NASA Manned Spacecraft Center SP-272.

Vasyliunas V. M. (1968) A survey of low-energy electrons in the evening sector of the magnetosphere with OGO 1 and OGO 3. *J. Geophys. Res.* **73**, 2839–2884.

Walbridge E. (1970) The lunar photoelectron layer, I. The steady state. Preprint, High Altitude Observatory, National Center for Atmospheric Research, Boulder, Colorado.

Proceedings of the Third Lunar Science Conference
(Supplement 3, *Geochimica et Cosmochimica Acta*)
Vol. 3, pp. 2655–2663
The M.I.T. Press, 1972

Photoemission from lunar surface fines
and the lunar photoelectron sheath

B. Feuerbacher, M. Anderegg, B. Fitton,
L. D. Laude, and R. F. Willis

Surface Physics Division

and

R. J. L. Grard

Ionospheric Physics Division,
European Space Research Organization, Noordwijk, Holland

Abstract—Measurements are reported on the photoelectric properties of lunar surface fines. The experimental results are used to calculate the photoelectric saturation current and its energy distribution under solar irradiation. A Maxwellian approximation for the energy distribution permitted a preliminary calculation of the photoelectron sheath properties in terms of electron density and electric field as a function of distance above the lunar surface. The sheath was found to be very tenuous: The surface electron density was 130 electrons/cm^3 and the surface shielding distance was 78 cm.

Introduction

A Number of authors have discussed theoretically the possible existence of a sheath of low energy electrons above the lunar surface, consisting of photoelectrons and secondary electrons, which are released from the lunar surface material by the incoming flux of particles and radiation (Öpik and Singer, 1960, 1962; Weil and Barasch, 1963; Kopal, 1969; Grobman and Blank, 1969; Walbridge, 1970). Quantitative estimations of the density and extension of this sheath were highly speculative due to the lack of data on the photoemission and secondary electron emission properties of the lunar material. Recently direct evidence for the photoelectron component was provided by the charged particle lunar environment experiment (CPLEE) during the Apollo 14 mission on the surface of the moon (O'Brien and Reasoner, 1971). The results of this experiment clearly show a flux of low-energy electrons in the energy range 40 to 200 eV, which was shown to be due to photoemission (Reasoner and Burke, 1972).

In this paper preliminary results are presented on the photoelectric characteristics of lunar surface fines number 14259,116. Secondary electron emission data are presented in the following paper (Anderegg *et al.*, 1972). Measurements on the work function, photoelectric yield, and energy distribution of the photoemitted electrons have been performed for photon energies ranging from 4 to 21 eV. The major contribution to the total photoemitted current is expected to occur in this range, as suggested both by the rapid increase of solar photon flux and the measured electron flux from the CPLEE data. The high energy limit of 21 eV for these measurements was

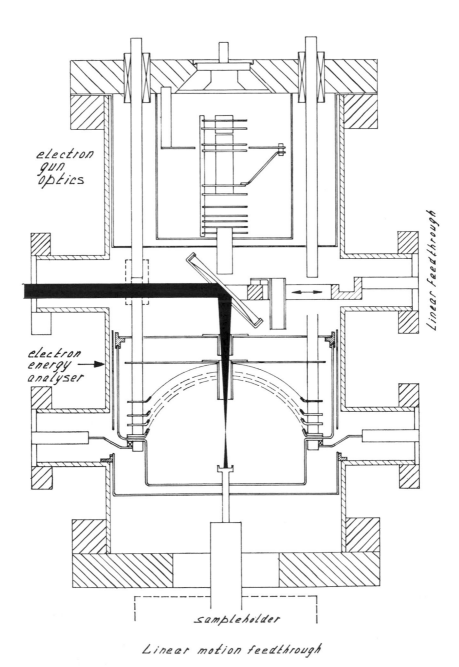

Fig. 1. Outline of the experimental chamber for photoemission and secondary electron emission experiments on lunar dust samples.

set by light source limitations. Future measurements using a synchrotron light source will extend this data up to about 100 eV and allow direct comparison with the CPLEE data in the region of overlap.

These photoemission data taken from the lunar fines have been used, together with the data on the incoming photon flux given by Tousey (1966) and Hinteregger *et al.* (1964), to calculate the emitted photoelectron flux and electron energy distribution under solar irradiation. The energy distribution curve was then approximated by a Maxwellian distribution to give an estimate of the electron density and the electric field distribution in the sheath formed by the photoelectrons as a function of distance above the lunar surface.

EXPERIMENTAL

The apparatus used for the experiments described here is shown in Fig. 1. It consists of a stainless steel vacuum container of 6 inch internal diameter. Light from a vacuum ultraviolet monochromator enters from the left as indicated by the shaded band. It is focused under 90° onto the sample, which is contained in a stainless steel cup of 1.5 cm diameter. The sample is introduced from a glove box below, in which handling takes place under a nitrogen atmosphere containing less than 10 ppm impurities. A commercial three-grid LEED optics is used to collect and analyze the photoemitted electrons. By applying a collecting voltage on the surrounding shield and grids, the total electron current leaving the sample is recorded to give the total photoelectric yield. Energy distribution profiles are measured by ramping a retarding voltage between the sample and the grids. A small ac voltage is superimposed on this ramp, allowing electronic differentiation of the current-voltage characteristic to obtain the energy distribution of the emitted electrons directly. The energy resolution of the instrument is about 300 meV. The electron gun in the upper part of the chamber is used for secondary electron emission measurements.

The absolute calibration of the light source was derived from a standard light source (Physikalisch Technische Bundesanstalt, Braunschweig, Germany). A thermopile was used to transfer the standard from the visible spectral region into the far ultraviolet and to calibrate the photoemission from the gold coated 90° ellipsoidal mirror in terms of photocurrent per photon reaching the sample. For the long wavelength region, where this secondary standard could not be used, a sodium salicylate fluorescent screen together with a photomultiplier could be inserted in the light path.

PHOTOEMISSION PROPERTIES

Work function

The photoelectric threshold or work function is the lowest photon energy for which a photoelectron may escape the material. Figure 2 shows how this property has been derived for the sample 14259,116 of lunar surface fines. The figure gives a plot of the square root of the photoelectric yield against the incoming photon energy. According to Fowler (1931), a straight line is obtained for energies near threshold, allowing extrapolation to zero yield. The measured work function of 5 eV is high compared to common materials, for which the threshold is usually about 4.5 eV. This fact will tend to lower the total photoelectron flux due to solar irradiation, since the solar flux decreases by roughly an order of magnitude between 4 and 5 eV.

Photoelectric yield

Figure 3 shows preliminary results on the photoelectric yield of the sample. The yield is given per incoming photon and plotted as a function of light wavelength in

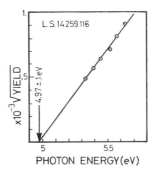

Fig. 2. Determination of the photoelectric work function of lunar sample 14259,116
by the Fowler method.

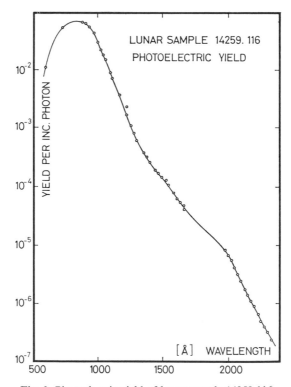

Fig. 3. Photoelectric yield of lunar sample 14259,116.

the range from 500 Å to 2500 Å. The photoelectric threshold of 5 eV corresponds to
a wavelength of 2480 Å.

A maximum yield value of about 7% is measured at 900 Å wavelength. This value
is remarkably low compared to materials commonly used in space research. A gold
coated satellite skin for example reaches a yield of 14% at the same energy, as shown
by Feuerbacher and Fitton (1972). For an insulator—and the lunar dust consists

mainly of insulating material—one would expect even higher yields in the order of 30% (Metzger, 1965; Samson, 1967). The low photoelectric yield measured with the lunar dust samples is attributed to the fact that the material is in the form of a fine grained powder. For a photoelectron released at the surface of a powder, the probability of reabsorption is much higher than on a flat surface. This reabsorption is due to the roughness of the powder surface and is similar to the effects leading to the low albedo of the moon. The reabsorption will tend to reduce the number of electrons escaping the surface, even though the number of electrons released per incident photon might be the same. The electrons will probably be emitted into a narrow angle centered to the direction of the incident light, similar to the light reflected by the moon (for a discussion of this, see Kopal, 1969a).

Another feature of the yield curve in Fig. 3 is the fact that the curve bends downward steeply after the peak, in particular for wavelengths shorter than 900 Å (14 eV). For an insulator one would expect the yield to increase further for photon energies between 12 and 20 eV. Instead a decrease is observed, dropping to 1.2% at 584 Å wavelength (21.2 eV). This decrease of the yield towards high energies will tend to lower the contribution from photoelectrons with high energies, say above 10 eV, under solar irradiation.

The contributions of the various regions of the solar spectrum to the total photoelectron flux can be seen on Fig. 4. Here the yield curve has been multiplied by the solar flux curve to obtain the differential photoelectron flux under solar irradiation. An asterisk marks the emission due to the hydrogen Lyman-α flux. This line is about an order of magnitude stronger than any other line in the spectrum. In the diagram the asterisk should be read as a column 1 eV wide if compared to the contributions of the histogram. Two spectral regions dominate the curve. One peaks around 6 eV, due to the rapid increase of the solar photon flux towards lower energies, together with the cutoff from the photoelectric threshold of the sample. The other region around

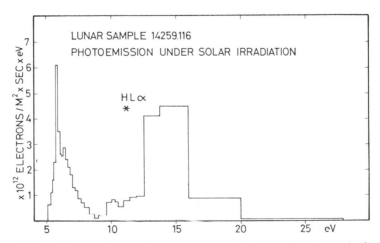

Fig. 4. Differential photoelectron flux from lunar sample 14259,116 due to solar irradiation. The asterisk gives the flux due to the hydrogen Lyman-α line in 10^{12} electrons · $m^{-2} \cdot sec^{-1}$.

14 eV is related to the peak in the yield curve. Integration of the curve in Fig. 4 with respect to the photon energy gives the saturation photoelectron flux under solar irradiation. Numerical evaluation gives a value of 4.5 μA/m^2, corresponding to $2.8 \cdot 10^{13}$ electrons $\cdot$ m^{-2} $\cdot$ sec^{-1}. This value may be compared to the value calculated for a gold surface under solar irradiation conditions, which gave 29 μA/m^2 or $1.8 \cdot 10^{14}$ electrons $\cdot$ m^{-2} $\cdot$ sec^{-1} (Feuerbacher and Fitton, 1972).

Energy distribution of photoelectrons

The photoelectron energy distribution curves, i.e., the velocity spectra of the electrons after escape from the surface, have been measured for a series of photon excitation energies. The result is shown in Fig. 5. The number of electrons emitted per incoming photon in an energy interval of 1 eV is plotted against electron energy. The third axis gives the incident photon energy, for which the curve has been measured. All curves are seen to peak at low energies, around 2 to 3 eV. A remarkable feature of this curve is the fact that a very small proportion of the electrons are emitted with energies close to the highest possible energy, namely the photon energy minus the work function. This fact is in contradiction to most estimates made previously on the energy distribution of lunar photoelectrons. Taken together with the observed low yield at high photon energies, it will result in a low total number of photoelectrons with high energies, that is, above 5 eV.

PHOTOELECTRON SHEATH

The data presented in Figs. 4 and 5 have been used to calculate the energy distribution of the electrons emitted from the lunar surface due to solar irradiation. Each curve for monochromatic light in Fig. 5 has been multiplied by the solar intensity at

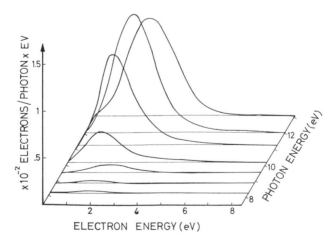

Fig. 5. Energy distribution of photoemitted electrons from lunar sample 14259,116. The curves are measured with monochromatic light. The photon energy is given on the right hand scale.

the respective wavelength. Then the curves have been integrated with respect to the photon energy. The result is plotted in Fig. 6, which shows the energy distribution of lunar photoelectrons due to solar irradiation. The curve is normalized such that the area under it is unity. The bulk of the electrons are seen to be emitted with energies between 1 and 4 eV, with a mean kinetic energy of 2.2 eV. The tail of electrons with energies higher than 6 eV is seen to be extremely weak, as expected from the measured energy distributions and the decreasing yield for higher photon energies.

The curve shown in Fig. 6 may be used, together with the known saturation photoelectron current, to calculate the properties of the photoelectron sheath. In the preliminary calculations presented here, however, an approximation was used, replacing the energy distribution curve by a Maxwellian with the same mean energy

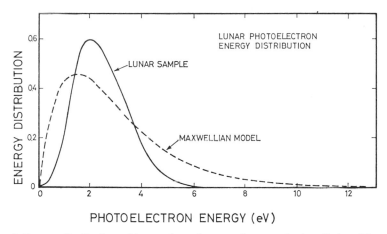

Fig. 6. Energy distribution of lunar photoelectrons due to solar irradiation. The area under the curve is normalized to unity. The dashed curve shows the Maxwellian approximation used to calculate the sheath profiles.

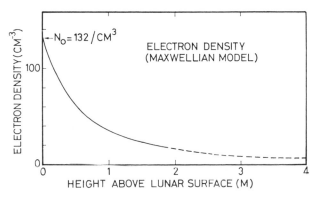

Fig. 7. Electron density in the lunar photoelectron sheath, calculated using a Maxwellian approximation.

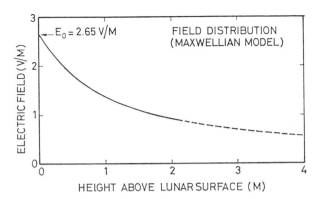

Fig. 8. Field distribution in the lunar photoelectron sheath calculated using a Maxwellian approximation.

and the same saturation current. This approximation is given in Fig. 6 as a dashed line. It is obviously a crude approximation, especially at high electron energies, where it merely gives an upper limit.

The electron density and electric field profile of the photoelectron sheath is readily derived for a Maxwellian distribution using a formalism derived by Grard and Tunaley (1971). The results are shown in Figs. 7 and 8, where electron density and field distribution are plotted as a function of distance above the lunar surface. The distance scales are characterized by a surface shielding length of 78 cm. The electron density of 130 electrons/cm^3 at the surface is an order of magnitude lower than that calculated by Walbridge (1970) and estimated by Reasoner and Burke (1972) from their CPLEE data.

DISCUSSION

The data have been taken from a single sample of lunar surface fines, and there is no proof as yet that this sample is representative. This problem will be resolved with measurements from different locations now being carried out. There remains still the most important problem concerning the cleanliness of the samples. Photo-emission is strictly a surface property, and even monolayers of contamination may change the properties drastically. Since no high-vacuum samples are available, these measurements had to be performed with a sample handled in an inert atmosphere of less than 30 ppm impurity. It is not possible to remove surface adsorbates by the usual heat treatment in high vacuum, since this heat treatment may well alter the optical properties considerably due to the annealing out of the radiation damage defects produced by solar irradiation and particles. The samples have therefore been measured in an oil free vacuum of 10^{-7} Torr. It is hoped that the low surface energy of the lunar material together with the reversibility of the adsorption curves (Fuller *et al.*, 1971) will lead to a contamination free surface under vacuum.

With these limitations in mind, the following conclusions may be drawn. The photoelectric yield of the lunar dust is small, reaching peak values of about 7%. This low value is attributed to the fact that the material is in the form of a powder, thus favoring reabsorption of photoemitted electrons within the porous surface. The work function of 5 eV is relatively high, cutting off the photoemission in the rapidly increasing part of the solar spectrum. Both these facts lead to a low total saturation photocurrent of 4.5 $\mu A/m^2$ due to solar irradiation. The energy distribution curves peak at low energy values of the order of 2 to 3 eV. Together with a decreasing yield for photon energies above 14 eV, this results in an electron energy distribution under solar irradiation with a mean energy of 2.2 eV and an extremely weak tail with energies above 6 eV. The photoelectron sheath calculated using a Maxwellian energy distribution is characterized by a surface shielding distance of 78 cm. The surface electron density is determined to be 130 electrons/cm^3.

The present work shows that the lunar photoelectron sheath is very tenuous. Photoemission from materials used in space research is roughly one order of magnitude larger (Feuerbacher and Fitton, 1972). Therefore care has to be taken in the interpretation of low energy electron measurements taken on the lunar surface, since a considerable amount of the measured electrons might be due to contamination of the local photoelectron sheath by the instrument itself.

Acknowledgments—The authors thank M. R. Adriaens, W. P. Fischer, and D. K. Skinner, for their skilled technical assistance. Many stimulating discussions with Dr. A. Pedersen are gratefully acknowledged.

REFERENCES

Anderegg M., Feuerbacher B., Fitton B., Laude L. D., and Willis R. F. (1972) Secondary electron emission characteristics of lunar science fines (abstract). In *Lunar Science—III* (editor C. Watkins), pp. 18–20, Lunar Science Contr. No. 88.
Feuerbacher B. and Fitton B. (1972) *J. Appl. Phys.* **43**, 1563.
Fowler R. H. (1931) *Phys. Rev.* **38**, 45.
Fuller E. L. Jr., Holmes R. B., Gammage R. B., and Becker K. (1971) *Proc. Second Lunar Sci. Conf.*, Geochim. Cosmochim. Acta Suppl. 2, Vol. 3, pp. 2009–2019, MIT Press.
Grard R. J. L. and Tunaley J. K. E. (1971) *J. Geophys. Res.* **76**, 2448.
Grobman W. D. and Blank J. L. (1969) *J. Geophys. Res.* **74**, 3943.
Hinteregger H. E., Hall L. A., and Schmidtke G. (1964) in *Space Research V*, p. 1175. North Holland.
Kopal Z. (1969) *The Moon*, p. 418, D. Reidel, Dordrecht.
Kopal Z. (1969a) *The Moon*, p. 357, D. Reidel, Dordrecht.
Metzger P. H. (1965), *J. Phys. Chem. Solids* **26**, 1879.
O'Brien B. J. and Reasoner D. L. (1971) Apollo 14 Preliminary Science Report, p. 193, NASA SP-272.
Öpik E. J. and Singer S. F. (1966) *J. Geophys. Res.* **65**, 3065.
Öpik E. J. and Singer S. F. (1962) *Planet. Space Sci.* **9**, 221.
Reasoner D. L. and Burke W. J. (1972) Direct measurement of the lunar photoelectron layer (abstract). In *Lunar Science—III* (editor C. Watkins), pp. 635–636, Lunar Science Institute Contr. No. 88.
Samson J. A. R. (1967) *Vacuum Ultraviolet Spectroscopy*, p. 224, John Wiley.
Tousey R. (1966) in *The Middle Ultraviolet*, p. 1, A. E. S. Green (ed.), John Wiley.
Weil H. and Barasch M. L. (1963) *Icarus* **1**, 346.

Proceedings of the Third Lunar Science Conference
(Supplement 3, *Geochimica et Cosmochimica Acta*)
Vol. 3, pp. 2665–2669
The M.I.T. Press, 1972

Secondary electron emission characteristics of lunar surface fines

M. Andfregg, B. Feuerbacher, B. Fitton,
L. D. Laude, and R. F. Willis

Surface Physics Division,
European Space Research Organization, Noordwijk, Holland

Abstract—The secondary electron yield and energy distribution of a lunar dust sample are reported over an incident electron energy range of 50 to 2000 eV. Preliminary results are discussed in terms of their pertinence to calculations of the electrostatic potential and charge distribution of the lunar surface.

Introduction

Calculations to date have shown that the steady-state electrostatic potential and charge distribution of the sunlit lunar surface is determined primarily by the photo-emissive properties of the lunar surface materials and the impacting solar wind flux. Only a small fraction of the emitted photoelectrons can escape as the sunlit lunar surface attains a small positive electrostatic potential, the magnitude of which is determined by the energy distribution of the photoemitted electrons and the incident solar wind particle velocity distribution and density (Öpik and Singer, 1960, 1962). Secondary electron emission due to the impact of solar wind particles has been assumed to be negligible in view of the low energies of solar wind electrons (< 10 eV) and protons (~ 1 keV) (W. D. Grobman and J. L. Blank, 1969; J. C. Brandt, 1970). However, recent measurements by the charged particle lunar environment experiment (CPLEE) during the Apollo 14 mission (O'Brien and Reasoner, 1971) have shown that, in addition to a stable low-energy photoelectron flux, observed for energies ranging from 200 eV down to 35 eV, rapidly varying fluxes of low energy electrons of magneto-spheric origin occur, with intensities greater than that of the photoelectron background. When the lunar surface is illuminated, the electron spectrum between 40 and 100 eV is dominated by the photoelectron continuum, but in the higher energy ranges, prominent peaks in the electron flux density were observed in the ranges 300 to 500 eV and at several keV respectively. These higher energy electrons will give rise to secondary electron emission at the lunar surface when the moon passes through the geomagnetic tail of the magnetosphere, which in turn may well effect the lunar surface charge and potential distribution significantly, depending on the secondary emission characteristics of the lunar surface materials. Also, secondary electron emission may play a significant role along the moon's terminator, where solar particles impact the surface in the shadow region due to their incident angle relative to the shadow line of the solar light.

Measurements have been made, therefore, of the secondary electron yield and energy distribution of lunar dust sample 14259,116 over a primary electron energy range of 50 to 2000 eV in order to obtain a more complete description of the low energy electron layer and the electrostatic potential of both the sunlit and dark-side lunar surface.

EXPERIMENTAL

No high vacuum samples were available for the present experiments. The measurements were performed therefore on samples handled in a dry nitrogen atmosphere. The secondary electron yield and the energy distribution of the secondary electrons were measured in the three grid hemispherical analyzer described by Feuerbacher *et al.* (1972) (see Fig. 1 of this reference). The mirror had been moved by a linear motion vacuum feedthrough such that electrons from the electron gun could reach the sample through the drift tube. The available electron energy range was 50 to 2000 eV. The samples were not baked under vacuum to remove adsorbed gases, in order to avoid changes in the optical properties of the lunar material as observed by Hapke *et al.* (1970). The measurements have been carried out in a vacuum of 10^{-7} Torr.

All elements of the electron collector, namely the three grids, the screen and the shielding can, were electrically connected for the yield measurements. A positive voltage was applied to the collector and the total collected electron current was recorded. The sample was grounded through an amperemeter to measure the current through the sample, which is the difference between the incident electron current and the secondary current leaving the sample. Variation of the collecting voltage and the observation of the collected current allowed an estimation of the positive voltage to which the sample was charged, due to it's rather low conductivity. For the beam currents used in this experiment, typically 1 μA, this charging was found to be less than 20 volts for incident beam energies in the range 150 to 1750 eV.

To measure the energy distribution of the secondary electrons, the sample, shielding can, and first grid where coupled together to obtain field free conditions around the sample. A retarding voltage together with a superimposed ac modulation, was ramped between the sample and the two coupled intermediate grids, providing electronic differentiation of the current-voltage characteristic in order to obtain the energy distribution spectra. No compensation was provided to avoid charging of the sample, so that some slight distortion resulted in the energy distribution spectra of the secondary electrons.

RESULTS

A preliminary curve for the secondary electron yield of sample 14259,116 of lunar surface fines is shown in Fig. 1. The curve shows a yield maximum of 1.4 at approximately 350 eV excitation energy. This yield value is somewhat lower than that observed for secondary electron emission from insulating materials (Gibbons, 1966) such as SiO_2, Al_2O_3, CaO, FeO, and MgO, the major constituents of the sample. The low yield is attributable to the particulate nature of the lunar dust sample which reduces the escape probability of the secondaries, i.e., due to scattering within microscopic cavities, as illustrated in Fig. 2. No elastically back-reflected electrons were detected in the energy distribution curves, even for primary beam energies of about 50 eV when the elastic reflection coefficient becomes quite large for most materials. This might be due to similar scattering and absorption in the microcavities of the dust surface. Furthermore it is also possible that the direction of the elastically back-reflected electrons may be confined to a small angle about the direction of the incident beam and, as such, would go undetected due to the large angular aperture of the drift tube of the electron gun. This would be an effect similar to that found for the reflection of light from lunar material.

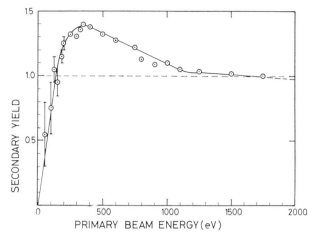

Fig. 1. Secondary electron yield from lunar sample 14259,116 for incident primary energies of 50 to 2000 eV.

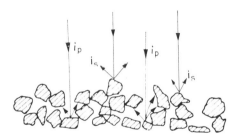

Fig. 2. Diagram showing the particulate nature of the lunar surface and its influence on the escape of secondary i_s and elastically reflected primary electrons i_p electrons.

The low energy crossover point at approximately 150 eV primary electron energy was difficult to locate in view of the rapidly changing yield with energy and associated dust charging effects. The effect of this charging can be seen in the distortion of the energy distribution spectra at primary energies for which the total secondary electron emission yield is different from unity, Fig. 3. The "true" secondary electron energy distribution is that shown for a primary energy of 1750 eV, for which the yield is unity and the net current passing through the sample vanishes. This curve is typical of that observed for insulating materials, the major part of secondaries occurring with energies less than 4 or 5 eV. In the energy range for which the yield is greater than unity the dust charges positively and reduces the number of electrons emitted at low energies as shown by the spectra for primary energies around 500 eV in Fig. 3.

It is significant perhaps that the increased electron flux density observed by the CPLEE at energies of 300 to 500 eV, during the moon's passage through the geomagnetic tail of the magnetosphere, is also the energy at which the lunar dust samples possess maximum secondary yield. This implies that such events will cause the lunar

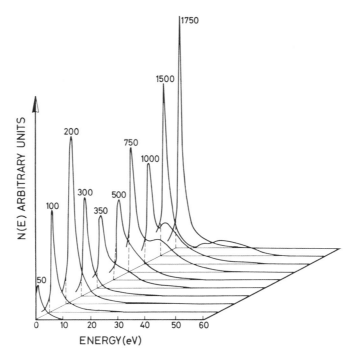

Fig. 3. Energy distribution curves of secondary electrons emitted from lunar sample 14259,116 for various primary energies. Note the reduced size of the low energy secondary peak at approximately 1–5 eV and the distorted spectra compared with the curve obtained at 1750 eV primary energy due to charging of the sample.

surface to charge even more positive and so modify both the photoelectron and secondary electron energy distributions, particularly close to the moon's terminator. These preliminary results suggest that incident electron energies below 100 eV, for which the secondary yield will be negligible, and above 2000 eV, at which energies increased electron flux density has also been observed (O'Brien and Reasoner, 1971) will cause the surface to charge negatively.

Concluding Remarks

Bearing in mind the limitations on the nature of these preliminary results pointed out in the previous paper (Feuerbacher et al., 1972), the secondary electron emission data indicates that the lunar surface charge and potential will depend not only on the solar wind flux and the photoemissive properties of the lunar material but also on the energy distribution of the magnetospheric electrons and secondary electron emission from the lunar surface. The surface charge and the potential of the sunlit lunar surface will be dominated by the denser photoelectron continuum and secondary electron emission will be negligible in comparison. However, any surface charging on the dark side of the moon and along the moon's terminator may be significantly affected by secondary electron emission, particularly during the moon's transit through the magnetosphere. Future calculations should take such considerations into account in view of their pertinence to related phenomena such as lunar soil charging effects (Gold, 1964).

Acknowledgments—The authors wish to thank M. R. Adriaens, W. P. Fischer, and D. K. Skinner for their expert technical assistance, and Dr. E. A. Trendelenburg for his encouragement of this project. We are also grateful to Dr. G. Eglinton of Bristol University and Dr. M. B. Duke the NASA Lunar Sample Curator for their advice and assistance concerning the lunar samples.

REFERENCES

Brandt J. C. (1970) in *Introduction to the Solar Wind*, W. H. Freeman.

Chang C. C. (1971) *Surface Sci.* **25,** 53 (1971), for a review of Auger electron spectroscopy.

Feuerbacher B., Anderegg M., Fitton B., Laude L. D., Willis R. F., and Grard R. J. L. (1972) preceding paper, *Proc. Third Lunar Sci. Conf., Geochim. Cosmochim. Acta* Suppl., Vol. 3. MIT Press.

Gibbons D. J. (1966) in *Handbook of Vacuum Physics*, ed. A. H. Beck, Vol. 2, Part 3. Pergamon Press.

Gold T. (1964) in *The Lunar Surface Layer* (ed. J. W. Salisbury, P. E. Glaser) p. 345–353. Academic Press.

Grobman W. D. and Blank J. L. (1969) *J. Geophys. Res.* **74,** 3943.

Hapke B. W., Cohen A. J., Cassidy W. A., and Wells E. N. (1970) *Proc. Apollo 11 Lunar Sci. Conf., Geochim. Cosmochim. Acta* Suppl. 1, Vol. 3, p. 2199. Pergamon.

O'Brien B. J. and Reasoner D. L. (1971) Apollo 14 Preliminary Science Report, p. 193, NASA SP-272 (1971).

Öpik E. J. and Singer S. F. (1960) *J. Geophys. Res.* **65,** 3065.

Öpik E. J. (1962) *Planet. Space Sci.* **9,** 221.

Proceedings of the Third Lunar Science Conference
(Supplement 3, *Geochimica et Cosmochimica Acta*)
Vol. 3, pp. 2671–2680
The M.I.T. Press, 1972

Lunar dust motion

David R. Criswell

The Lunar Science Institute,
3303 NASA Road 1, Houston, Texas 77058

Abstract —Surveyor 7 photographed a bright glow along the western lunar horizon one hour after local sunset. This horizon glow must result from the forward scattering of sunlight by a swarm of dust grains extending 3 to 30 centimeters above the local horizon with a column density of 5 grains/cm^2. The observable glow is evoked by grains approximately 6 microns in radius. An annual churning rate of 10^{-3} gr/cm^2 is implied. A model is presented for dust cloud production in terms of electrostatic levitation of lunar surface fines.

Horizon-Glow

Figure 1, a composite of one daytime (lower portion) and two postsunset (upper portion) photographs of the lunar horizon 200 m west of the Surveyor 7 spacecraft, clearly shows a bright strip of light referred to as horizon glow (HG). This example of HG extended 2° to 3° on each side of the sunset line, was much brighter than the solar corona, and persisted for 90 min after local sunset (Gault *et al.*, 1968a, b; Rennilson, 1968). This HG silhouetted the rocks and surface irregularities that constituted the horizon (Allen, 1968). The HG could be mapped to a scale of 3–4 cm, which was the Surveyor 7 television resolution at 200 m. Horizon-HG vertical separation was introduced in the preparation of Fig. 1. Polarization was not apparent in the analog photography (Shoemaker *et al.*, 1968). An extended analysis of the Surveyor images will be presented by Criswell and Rennilson (1972).

Figure 2 is a digitized representation of section (a) in Fig. 1. Each number (PV) is proportional to the image photometric brightness (B) at the corresponding location (pixel) on the television photocathode. Extensive preflight calibrations (Rennilson, 1971) established B (foot-lamberts) = 3.364 × (PV) for the normal operating mode (0.150 sec exposure) of the camera as digitized in this observation. In cgs units, B (candles/cm^2) = 1.14 × 10^{-3} × (PV).

HG has a clearly resolved vertical extent of 3 to 30 cm (approximately 1 to 7 pixels) which is not a systems artifact due to spread of a bright image. This is demonstrated by the insert which is a digitized segment of a Surveyor 7 photograph (time exposure of 3 sec) of an earth-based laser (Alley and Currey, 1968). The central numbers (63's) indicate saturation of the photocathode. However, the PV values decrease to at least 50% of maximum (<32) within one to two pixels of the saturated pixels. Conversely, the HG does not saturate the television system and is in many places 4 to 7 pixels in width with constant brightness (PV ≃ 10–19). Three to 30 cm vertical extensions rule out scattering by surface grains (Gault *et al.*, 1968a), residual gas molecules (Rozenberg, 1970), and secondary meteorite ejecta (Gault *et al.*, 1963) as sources.

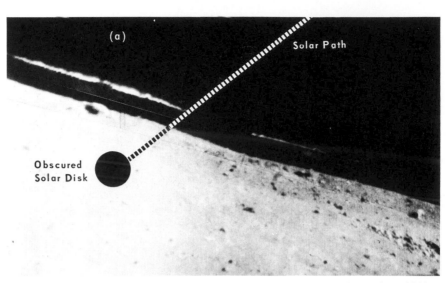

Fig. 1. Composite photograph of horizon west of Surveyor 7 (6.4° angular width). Slight horizontal scale differences are present in the composites. However, note that the horizon glow (upper bright lines) faithfully silhouettes the horizon features. The dashed line indicates the *approximate* apparent path of the sun. HG is occurring along two different ridges. The HG gap is the shadow zone of a background ridge (right) on the foreground ridge. The upper limb of the sun is approximately 1.3 solar diameters beneath the horizon as indicated by the black circle.

The proposed source mechanism is presented in Fig. 3. Electrically charged dust grains are levitated into the sunlight by an electrostatic field. The field is induced about partially illuminated rocks due to the ejection of photoelectrons by soft solar x-rays and the subsequent accretion of a portion of these photoelectrons in adjacent dark areas. The grains scatter sunlight to the shadowed Surveyor spacecraft. Large sphere diffraction applies producing a brightness described by (Van de Hulst, 1957)

$$\frac{B \ (\text{candles/cm}^2)}{I_0 \ (\text{lm/cm}^2)} = ND \ \frac{x^2 a^2}{4} \left(\frac{2J_1(x \sin \theta)}{x \sin \theta}\right)^2$$

where $I_0 = 13.7 \ \text{lm/cm}^2$ (solar intensity in cgs visual units); $a \sim$ grain radius (cm); $N \sim$ # scatters/cm³; $D \sim$ cloud depth along line of sight (cm); $\lambda \simeq 0.5 \times 10^{-4}$ cm (wavelength for peak response of Surveyor television system); $x = 2\pi a/\lambda$; $\theta \sim$ angle between sun-Surveyor and sun-HG lines; and $J_1 \sim$ first-order Bessel function with $J_1(3.832) = 0$. Assuming HG brightness approaches zero at $J_1(3.832)$ and $\theta = 3°$, we have $a = 6 \times 10^{-4}$ cm $= 6\mu$. Other particles are certainly present. However, $B \sim a^4$; thus smaller particles scatter much less light which Surveyor can detect. Conversely, larger particles have diffraction limits $\theta < 3°$ and will not contribute scattered light at $\theta = 3°$.

Using $\theta \simeq 1°$ and PV = 18 for the brightness at that angle away from the sun, we get

$$N \ (\#/\text{cm}^3) \times D \ (\text{cm}) = 4.7 \ \text{grains/cm}^2$$

```
11 12 17 23 18 23 13 19 20 15 23   3  3  3  3  4  5  9 16 18 19 19 15  5  0
12 13 12 18  9 23 26 18 23 18 22   3  3  4  4  4  6 12 18 19 19 19 12  1  0
16 13 12 13 17 24 24 15 20 16 16   3  4  3  3  7 13 18 18 18 18 19 14  2  0
18 23 15 15 20 33 26 21 20 18 19   3  3  3  3  5 12 17 17 17 18 17  8  1  0
17 15 15 16 30 59 47 11 11 16 19   3  4  3  5 13 17 17 18 17 10  1  1  2
17 26 17 14 38 63 63 30 13 17 11   3  3  3  3  4  7 14 16 17 17 12  2  1  1
20 18 25 14 20 25 23 12 11 18 12   3  3  3  3  5 10 15 16 17 13  3  0  1
11 15 12 19 13 10 20  2 21 21 15   3  3  3  5  8 14 17 17 16  9  1  1  2
14 15 13 14 11 12  8 17 13 11 11   3  3  3  3  5 10 16 17 16 15  8  0  0  2
11  1 11 10 18 13 12 13 15 14  8   3  3  3  5  7 12 16 16 16 15  7  0  1  2
 5 10  4  6 13 13 15  9 15 20  9   3  3  5  7 11 16 17 17 16 14  3  0  2  2
 3  3  3  3  3  3  3  3  3  3  3   3  3  5  9 15 17 16 17 16 11  2  0  2  3
 2  3  3  3  3  3  3  3  3  3  3   3  3  5  9 16 15 16 16 13  4  1  1  2  3
 3  2  2  3  3  3  2  3  3  3  3   3  4  7 11 15 15 16 15  9  1  1  2  2  3
 2  2  3  2  2  3  3  3  2  3  3   3  5  7 14 15 15 15 15  6  0  0  2  2  2
 3  3  2  2  3  2  2  2  2  3  3   4  8 14 15 15 15 14  4  0  1  1  1  3
 2  2  2  2  3  3  3  3  3  3  3   4  9 14 14 14 14  9  0  0  1  2  2  3
 3  2  2  2  3  3  3  3  2  3  3   3  6 13 15 14 14 12  1  0  1  2  2  3
 2  2  2  3  3  3  3  2  3  3  3   3  7 12 14 14 10  3  1  2  2  3  3
 2  2  2  3  3  3  3  3  3  3  3   4  5  7  9  9  6  3  3  3  3  3  3
 3  3  2  2  3  3  3  3  3  3  3   4  5  7  9 10  9  5  3  3  3  3  3  3
 3  2  3  2  3  3  3  3  3  3  3   5  7 11 14 14  9  4  1  3  3  3  3  3
 3  3  2  3  3  2  2  3  3  3  3   4  5 10 14 15 14 12  4  1  1  3  3  3  3
 3  3  3  2  2  3  3  3  3  3  3   5  9 14 15 14 14  8  1  1  2  2  3  3  3
 3  3  3  3  3  3  3  2  3  4  3   5  9 14 14 14 13  5  0  1  2  2  3  3  3
 2  2  3  2  2  3  3  2  3  3  7 13 14 14 14  8  1  0  1  2  2  2  3  3
 3  3  3  3  2  3  3  3  3  3  7 14 14 14 14  8  0  0  2  3  2  3  3  3
 2  2  2  2  3  2  2  2  3  4  5 10 14 14 14  6  0  1  2  2  3  2  3  3
 3  2  3  3  2  2  3  3  4  6  7  5  5 10 12  8  3  2  2  2  2  3  3  3  2
 3  3  3  2  3  2  3  3  3  8 12  8  2  3  4  3  3  3  3  3  3  3  3  3  3
 3  2  2  3  3  2  3  3  4  6 12 11  3  2  3  3  3  3  3  3  3  3  3  3  3
 3  3  2  3  2  3  3  3  3  4  5  5  4  3  3  3  3  3  3  3  3  3  3  3  3
 3  3  3  3  3  3  3  3  4  5  5  3  4  3  3  3  3  3  3  3  3  3  3  3  3
 3  2  3  3  3  3  5  6  7  4  3  3  3  3  3  3  3  3  3  3  3  3  3
 2  3  3  3  3  3  5  9  9  5  2  3  3  3  3  3  3  3  3  3  3  3  3  3
 3  3  3  3  3  3  5  9 12  5  1  2  3  3  3  3  3  2  3  3  3  3  3  3
 2  2  2  3  3  4  5  6  6  5  3  2  2  3  3  3  3  3  3  3  3  3  3  3
 2  3  3  3  5  6  8  7  5  3  3  3  2  3  2  3  3  3  3  3  3  3  3  3
 3  3  3  3  6 11 14 10  5  2  2  3  3  2  3  3  3  2  3  2  3  3  3  3
 2  3  3  3  4 11 14 13  7  0  1  3  3  2  3  3  3  3  3  3  3  3  3  3
 3  2  2  3  4 10 13 13  7  2  2  3  3  2  3  3  3  3  2  3  3  3  3  3
 3  3  2  3  3  4  5  6  5  4  3  3  3  3  3  3  3  3  3  3  3  3  3  2
 3  3  3  3  3  3  3  4  3  3  3  3  3  3  3  3  3  3  3  3  3  3  3  3
 2  3  3  3  3  4  3  3  3  3  3  3  3  3  3  3  3  3  3  3  3  3  3  3
 3  2  3  3  3  3  4  3  3  3  3  3  3  2  3  3  3  3  3  3  3  3  3  3
```

Fig. 2. Digitized representation of area (a) in Fig. 1 with the digital representation rotated approximately 60° counterclockwise from the photograph. Zero is black and sixty-three is white. The insert is the digitized segment of an earlier Surveyor 7 photograph of a laser signal which saturated (63 and 63) two pixels. Different calibration coefficients apply to the PV values of the HG and the insert.

Assuming $D \simeq d$ we have $N\ (\#/\mathrm{cm}^3) \simeq 0.16$ ($d = 30$ cm) to 1.6 ($d = 3$ cm). The corresponding column mass density is $M_c \simeq 1.3 \times 10^{-8}$ gr/cm^2 assuming $D = d$. An approximate columnar mass flow rate of $\dot{M}_c \simeq 2 \times 10^{-8}$ gr/cm^2 sec is obtained by assuming the grains accelerate at 10% lunar gravity (g_l), $d = 10$ cm, $\dot{M}_c \simeq M_c(2/t)$, and $t = (2d/.1g_l)^{\frac{1}{2}}$. HG persists the order of 5.4×10^3 sec, which implies a total churning per sunset of 10^{-4} gr/cm^2 and per yr of 10^{-3} gr/cm^2. An annual churning depth $d_c \simeq 6 \times 10^{-4}$ cm or 6 microns/yr is implied for a soil bulk density of 1.5 gr/cm^3, where 50% porosity is assumed.

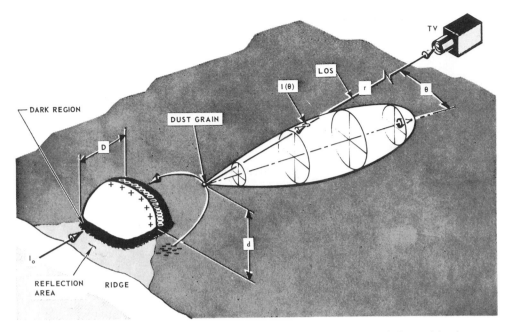

Fig. 3. Levitation of a dust grain about a partially illuminated rock located in the sunset terminator results in forward scattering sunlight to the completely shadowed Surveyor spacecraft. Large sphere diffraction describes the scattering. The distance D is the HG depth along the TV line of sight (LOS); d is the vertical height of the HG.

Surface bombardment by 10^{-5} gr micrometeoroids (Hartung *et al.*, 1972) at 6×10^{-17} gr/cm^2 sec (Dohnanyi, 1971) producing 6×10^{-12} primary impacts/cm^2 sec will eject approximately 2×10^{-5} secondaries/cm^2 sec in the 3–8 micron range (Braslau, 1970; Gold *et al.*, 1971). This is 10^5 times too low a secondary production rate to evoke HG. The production rate of tertiary ejecta is not known; however, tertiary particles should not be present in significant quantities due to the energy dissipation in primary and secondary impacts.

Shoemaker *et al.* (1970) predict 10^2 to 10^3 yr turn over times for the top 10μ of regolith due to micrometeorites and all their ejecta. This is 50 to 500 times slower than required for HG production.

LEVITATION CONDITION

Levitation will occur when the electrical force on loose, charged grains exceeds the gravitation force,

$$QE > \frac{4}{3} \pi a^3 \rho g_l$$

where the mass density of lunar grains $\rho \simeq 3$ gr/cm^3, $Q = (\pi a^2)[(5.5 \times 10^5$ electron/cm$^2) \times (E)]$ is the charge on a grain of radius a, and E is the surface electric field in volts/cm. It is assumed the surface electrical charge is uniformly distributed over small areas (<1 cm) and that a given grain simply acquires a net charge proportional to its circular cross-sectional area. Combining these equations we obtain the levitating field strength

$$E \text{ (volts/cm)} \gtrsim 270[a(\mu)]^{\frac{1}{2}}$$

A dust grain of $a = 5\,\mu$ will levitate in a field of 606 volts/cm. A charge imbalance of 262 electrons will be present on the grain. Levitation of a wide range of dust sizes will occur due to the square-root dependence of E on $a(\mu)$. The $5\,\mu$ grains will transport bound electrons at the rate $\dot{Q} \simeq 2100$ electrons/cm^2 sec between dark and light areas for the HG observed by Surveyor 7.

GENERATION OF THE ELECTRIC FIELD

Gold (1955), Heffner (1965), and Singer and Walker (1962) have suggested and examined various processes to evoke electrostatic transport of dust on the fully sunlit lunar surface. The suggested processes will not work because the pervasive dayside electric fields will be the order of a few volts/cm (Reasoner and Burke, 1972). The mechanism proposed in this paper operates only in the terminator zone. Reference to Fig. 3 will assist in following the explanation.

The rock is only partially illuminated due to shadowing by the western ridge. Soft solar x-rays of energy $\varepsilon_0 > 500$ eV evoke the ejection of photoelectrons ($\varepsilon \gtrsim 500$ eV) at the rate $j \simeq \Phi(\varepsilon_0 > 500 \text{ eV}) \times Y(>\varepsilon_0)$ where Φ is the integral solar flux (photons/cm^2 sec) for $\varepsilon_0 > 500$ eV or $\lambda \lesssim 25$ Å and $Y(>\varepsilon_0)$ is the yield of photoelectrons with energies near ε_0. Spicer (1971) estimates $Y(>500 \text{ eV}) \simeq 10^{-2}$ for the observed elemental abundances of lunar material (Wink and Ojanpera, 1970). The value $\Phi(>500 \text{ eV}) \simeq 5 \times 10^7$ photons/cm^2 sec, while the HG was detected, was derived from data and analyses of Drake *et al.* (1969), Teske (1970), and Wende (1972). Thus, $j(>500 \text{ eV}) \simeq 5 \times 10^5$ photoelectrons/cm^2 sec seems a reasonable ejection rate from the sunlit rock surface.

Photoelectrons will escape the rock until a sufficient positive charge (q) remains to retain newly ejected photoelectrons in the vicinity of the rock. A fraction (f) of j will accrete on the dark areas of the rock and remain fixed due to the rock's low electrical conductivity.

A computer model was developed to trace the motion of monoenergetic electrons (ε) sequentially ejected in random directions from randomly chosen ejection points on one hemisphere of an insulated sphere. The sphere was placed just above an infinite, nonconductive plane. Electrons moved under the influence of a smoothed, net positive charge on the ejection (sunlit) hemisphere and the isolated negative charges which had previously accreted on the opposite (dark) hemisphere and plane (lunar surface). The fraction f was calculated versus the equivalent electrical potential ($U = q/4\pi\varepsilon_0 d$) of a conductive sphere. Let $U_0 = \varepsilon/e$. It was found that U asymptotically approached $1.2\,U_0$ and that $f(U = U_0) \simeq 0.05$, $f(U = 1.1\,U_0) \simeq 0.01$, and $f(U \to 1.2\,U_0) \to 0$.

Applying these results to the previously calculated j we obtain an accretion current density $j_d \simeq 5\text{-}25 \times 10^3$ electrons/cm^2 sec. A potential difference $U \simeq 1.1\,U_0 > 550$ volts should be produced between the light and dark hemispheres. $E \simeq U/d$ will be greatest near the light-dark boundary and should exceed 550 volts/cm for $d \simeq 1$ cm due to charge concentration near both sides of the boundary.

The expected charging rate exceeds the discharging rate ($\simeq 2 \times 10^3$ electrons/cm^2 sec) due to the motion of charged dust grains between the light and dark areas.

Current densities in this process, the order of 10^{-15} amps/cm^2, are exceedingly small. The solar wind plasma, photoelectrons produced by scattered light, and conduction current through the rock must be considered as discharging agents. Solar wind does not directly impact the sunset terminator due to the ($1°–4°$) velocity abberation of the wind resulting from the earth-moon system's orbital motion about the sun. In addition, the accretion of solar wind electrons, which is driven by the 10 eV thermal energy of the electrons, will build up a negative charge layer in dark areas, thus preventing the entry of further solar wind electrons. The 1 keV solar wind protons cannot reach the dark negatively charged surfaces since these protons are traveling in essentially straight lines. The 10–15 eV plasma effects will not be significant because levitating electric fields are generated over scale lengths ($d \simeq$ few cm's) much smaller than the local Debye length ($> 10^3$ cm).

The monopole field will not occur if low energy photoelectrons can flow in from the dark region just below the (Fig. 3) directly illuminated area. Such an electron flow will replace escaping 500 to 1500 eV photoelectrons. An electron influx to the primary reflection area from the directly illuminated rock, westward sunlit areas, or the residual solar wind will complete the overall current loop in this neutralizing case. Neutralizing photoelectrons are evoked in greatest number by solar ultraviolet photons with energies of approximately ε_0 when the product $\Phi(>\varepsilon_0) \times Y(\varepsilon_0)$ is a maximum. Figure 3 helps one to visualize the light scattering pattern about the partially illuminated rock. Photons are scattered from the sunlit surface of the rock, to the primary reflection area in front of the rock, and then back to the dark region of the rock. Less than 2×10^{-8} of the directly incident photons reach the dark region of the rock. This "attenuation" is calculated on the assumptions of lambert scattering at both surfaces, the top quarter face of the rock being illuminated, and dust and rock albedos of 0.01 in the mid-ultraviolet. Lebedinsky, $et\ al.$ (1968) reported overall lunar backside albedos of 0.01–0.015 for 2200 Å.

$R < 1$ (equation (1)) must occur for the discharging photoelectron flux to be negligible.

$$R = (\text{Attenuation}) \left[\frac{Y(\varepsilon_0 \simeq w)\Phi(\varepsilon_0 \simeq w)}{Y(\varepsilon_0 \gtrsim 500\ \text{eV})\Phi(\varepsilon_0 \gtrsim 500\ \text{eV})} \right] \qquad (1)$$

Feuerbacher $et\ al.$ (1972) report $Y(w = 12\ \text{eV}) \simeq 5 \times 10^{-2}$ and $Y(w = 5\ \text{eV}) \simeq 10^{-7}$ for dust samples. Using $\Phi(>500\ \text{eV}) = 5 \times 10^{+7}$/cm^2 sec, $\Phi(>12\ \text{eV}) = 5.6 \times 10^{10}$, $\Phi(>5\ \text{eV}) = 3 \times 10^{14}$, and $Y(>500\ \text{eV}) = 10^{-2}$, we get $R(12\ \text{eV}) \simeq 10^{-4}$ and $R(5\ \text{eV}) \simeq 10^{-6}$. Even these approximate estimates indicate that secondarily evoked photoelectrons will not discharge the sunlit surface.

Conduction current is the final discharging mechanism. The effective electrical conductivity (η) of the rock must be $\eta < 10^{-18}$/ohm-cm. This is consistent with the stability over 10 days of surface charges induced triboelectrically on the fresh surface of lunar rocks cleaved in a vacuum as observed by Grossman $et\ al.$ (1970). Lunar rocks are heterogeneous assemblages of many small grains of differing mineralogies. The mineralogy indicates that water was not present when the rocks achieved their present state, and no water is present in the *in situ* rocks (Schmitt *et al.*, 1970). Review of Parkhomenko (1967) indicates that, for the lunar rocks, surface charge will be trapped along the many interfaces between fused grains and mineral phases. This results in the creation of an internal polarization electric field rather than conduction currents.

Schwerer *et al.* (1971) have experimentally determined $\eta(\simeq 300°K) \simeq 10^{-10}$ to 10^{-9}/ohm-cm for an igneous rock and a microbreccia. These results are probably not applicable for two reasons. The samples were very thin ($\simeq 2$ mm). High conductivity paths, such as an iron-rich phase, through the sample could produce a high apparent conductivity. These paths would not be connected over 1 to 20 cm lengths of the actual rock. More serious is permanent sample contamination and modification by trace levels of water during preparation of the thin sections and set-up of the experiment. *In situ* measurements may be necessary to determine η. It should be noted that Chung and Westphal (this volume) using lunar samples obtained $\eta \simeq 10^{-12}$/ohm-cm at 100 Hz for temperatures between 80°K and 250°K. Presumably the zero-frequency conductivity should be even lower.

DISCUSSION

The levitation mechanism will tend to preferentially remove dust from rocks and smooth the dust about rocks. Surface charge density, and therefore the levitation force, will be greater and more uniform on the sunlit than on surrounding dark areas because the negative charge can accrete over a larger dark area than exists on the sunlit rock surface. This is consistent with comments by Holt and Rennilson (1968) and Gold (1971b) on the lack of dust layers on rocks, even those which protrude only a few centimeters above the soil. It is also consistent with the lack of impact scars in the soil about secondary ejecta. However, careful study of the lifetimes of rocks in known surface orientations will be necessary to confirm this contention.

The mechanism will tend to selectively levitate the smallest grains (< 5 micron radii) from the topmost soil layer near crests and on illuminated slopes. Thus, these smallest grains will have enhanced exposure to the higher energy solar wind ions and solar cosmic rays. Levitation is enhanced when the sun is active. This is consistent with observations of enhanced particle track densities ($\gtrsim 10^{11}$/cm^2) in the outermost surface of micron-sized grains (Kreplin, 1970; Barber *et al.*, 1971).

Bibring *et al.* (1972) contend the lunar surface albedo decreases with increased lunar surface coverage by these intensely track-damaged grains. Conversely, Holt and Rennilson (1968) note the undisturbed lunar surface about Surveyor 7 is brighter than soil disturbed by Surveyor. In addition, sloping surfaces about Surveyor 7 have higher albedo than adjacent flat surfaces. The playas about the Surveyor 7 site are the lowest and darkest terrain. A speculative but plausible explanation is possible. The sloping areas are illuminated at either sunset or sunrise and are thus positively charged. The low areas are dark and negatively charged. The dark, micron-sized grains are preferentially extracted from the top 10 microns of the sloping surfaces and are slowly migrated toward the lower, negatively charged areas, where they gently come to rest on the surface. Thus, slopes are lightened while flat areas are darkened. The idea can be verified. The albedo of a highland playa of area A_p should be proportional to A_p/A_s where A_s is the slope area which is the source of the darkening grains. Micrometeorites should stir up both areas at the same rate and not selectively move one size range. Thus, the downslope electrostatic drifting of the smallest grains would be the differential, separating mechanism. This should apply for $A_p \lesssim 1 \text{ km}^2$.

Gold (1971a) suggested a different dust motion mechanism driven by the differential electrical charging of adjacent surface grains. Magnetotail electrons (0.5–1.5 keV) striking the earthward lunar surface evoke secondary electron emission at the rate r. The rate r varies with primary electron energy and grain chemistry (Anderegg et al., 1972). If $r < 1$ a negative grain charge results, while $r > 1$ evokes a positive grain charge. This mechanism will not directly produce the observed HG because there is a zero net surface charge over centimeter areas. Thus, the grains cannot be levitated the few centimeters necessary to produce HG. In addition, the Surveyor 7 longitude was 12°W, which makes it very unlikely that the moon was in the earth's magnetotail when the HG was observed.

Acknowledgments—I am especially grateful to Mr. J. J. Rennilson (California Institute of Technology) and Mr. J. N. Lindsley (Jet Propulsion Laboratory) for assistance in digitization of the Surveyor 7 pictures and critical discussions of the Surveyor camera systems. Credit must be extended to the scientific and engineering teams which developed the Surveyor television systems as precision photometric instruments. Miss Jo Ann Birchett (Manned Spacecraft Center—Computational and Analysis Division) programmed the computer model of electron accretion. This research was conducted at The Lunar Science Institute, which is operated by the Universities Space Research Association and supported by Contract NSR 09-051-001 with the National Aeronautics and Space Administration. This is Lunar Science Institute Contribution No. 95.

REFERENCES

Allen L. H. (1968) The lunar sunset phenomenon. *Surveyor Project Final Report, Part 2. Science Results*, JPL Tech. Rep. 32-1265, 459-465.

Alley C. O. and Currie D. G. (1968) Laser beam pointing tests, XI. *Surveyor Project Final Report, Part 2. Science Results*, JPL Tech. Rep. 32-1265, 441–448.

Anderegg M., Feuerbacher B., Fitton B., Laude L., and Willis R. F. (1972) Secondary electron emission characteristics of lunar surface fines (abstract). In *Lunar Science—III* (editor C. Watkins), pp. 18–20, Lunar Science Institute Contr. No. 88.

Barber D. J., Hutcheon I., and Price P. B. (1971) Extralunar dust in Apollo cores? *Science* **171**, 372–374.

Bibring J. P., Maurette M., Meunier R., Durieu L., Jouret C., and Eugster O. (1972) Solar wind implantation effects in the lunar regolith (abstract). In *Lunar Science—III* (editor C. Watkins), pp. 71–73, Lunar Science Institute Contr. No. 88.

Braslau D. (1970) Partioning of energy in hypervelocity impacts in loose sand targets. *J. Geophys. Res.* **75**, 3987–3999.

Criswell D. R. and Rennilson J. J. (1972) Surveyor observation of the lunar horizon glow (in preparation).

Dohnanyi J. S. (1971) Flux of micrometeoroids: Lunar sample analysis compared with flux model. *Science* **173**, 558.

Drake Jerry F. Sr., Gibson O. S. B. J., and Van Allen J. A. (1969) Iowa catalog of solar x-ray flux (2–12 Å). *Solar Physics* **10**, 433–459.

Feuerbacher B., Anderegg M., Fitton B., Laude L. D., Willis R. F., and Grard R. J. L. (1972) Photoemission from lunar surface fines (abstract). In *Lunar Science—III* (editor C. Watkins), pp. 253–255, Lunar Science Institute Contr. No. 88.

Gault D. E., Adams J. B., Collins R. J., Kuiper G. P., O'Keefe J. A., Phinney R. A., and Shoemaker E. M. (1968a) Post-sunset horizon glow. *Surveyor Project Final Report, Part 2. Science Results*, JPL Tech. Rep. 32-1265, 401–405.

Gault D. E., Adams J. B., Collins R. J., Kuiper G. P., Masursky H., O'Keefe J. A., Phinney R. A., and Shoemaker E. H. (1968b) Post-sunset horizon "afterglow," E. *Surveyor 7 Mission Report, Part 2. Science Results*, JPL Tech. Rep. 32-1264, 308–311.

Gault D. E., Shoemaker E. M., and Moore H. J. (1963) Spray ejected from the lunar surface by meteoroid impact. *NASA Tech. Note D-1767.*

Gold T. (1955) The lunar surface. *Monthly Notices Roy. Astron. Soc.* **115,** 585.

Gold T. (1971a) Evolution of the mare surface. *Proc. Second Lunar Sci. Conf., Geochim. Cosmochim. Acta* Suppl. 1, Vol. 3, pp. 2675–2680. MIT Press.

Gold T. (1971b) The nature of the lunar surface: Recent evidence. *Proc. of the American Philosophical Soc.* **115,** 74–82.

Gold T., O'Leary B. T., and Campbell M. (1971) Some physical properties of Apollo 12 lunar samples. *Proc. of the Second Lunar Sci. Conf., Geochim. Cosmochim. Acta* Suppl. 2, Vol. 2, pp. 2173–2181. MIT Press.

Grossman J. J., Ryan J. A., Mukhergee N. R., and Wegner M. W. (1970) Microchemical, microphysical and adhesive properties of lunar material. *Proc. Apollo 11 Lunar Sci. Conf., Geochim. Cosmochim. Acta* Suppl. 1, Vol. 3, pp. 2171–2181. MIT Press.

Hartung J. G., Hörz F., and Gault D. E. (1972) The origin and significance of lunar microcraters (abstract). In *Lunar Science—III* (editor C. Watkins), pp. 363–365, Lunar Science Institute Contr. No. 88.

Heffner H. (1965) Levitation of dust on the surface of the moon. N66-16171, *Minn. Univ. Report of August 1965 TYCHO meeting.*

Holt H. E. and Rennilson J. J. (1968) Photometry of the lunar regolith, as observed by Surveyor cameras. *Surveyor Project Final Report, Part 2. Science Report*, JPL Tech. Report 32-1265, 109–113.

Kreplin Robert W. (1970) The solar cycle variation of soft x-ray emission. *Ann. Geophys.* **26,** 567–574.

Lebedinsky A. I., Krasnopolsky V. A., and Aganina M. V. (1968) The spectral albedo of the moon's surface in the mid-ultraviolet according to data from the Zond III space probe. *Moon and Planets II* (editor A. Dollfus), pp. 47–54. North-Holland.

Parkhomenko E. I. (1967) *Electrical properties of rocks.* Plenum Press.

Reasoner D. L. and Burke W. J. (1972) Direct observation of the lunar photoelectron layer. *Proc. Third Lunar Sci. Conf., Geochim. Cosmochim. Acta* Suppl. 3, Vol. 3. MIT Press.

Rennilson J. J. (1968) Sunset observations. *Surveyor Project Final Report, Part 2. Science Results*, JPL Tech. Rep. 32-1265, 119–121.

Rennilson J. J. (1971) Private communication.

Rozenberg G. V. (1970) The atmospheric pressure on the moon according to the Surveyor 7 twilight photography. *Soviet Astronomy—AJ.* **14,** 361–363.

Schmitt H. H., Lofgren G., Swann G. A., and Simmons G. (1970) The Apollo 11 samples: Introduction. *Proc. Apollo 11 Lunar Sci. Conf., Geochim. Cosmochim. Acta* Suppl. 1, Vol. 1, pp. 1–54. Pergamon.

Schwerer F. C., Nagata T., and Fisher R. M. (1971) Electrical conductivity of lunar surface rocks and chondritic meteorites. *The Moon* **2,** 408–422.

Shoemaker E. M., Batson R. M., Holt H. E., Morris E. C., Rennilson J. J., and Whitaker E. A. (1968) Television observations from Surveyor 7, III. *Surveyor 7 Mission Report, Part 2. Science Results*, JPL Tech. Rep. 32-1264, 66.

Shoemaker E. M., Hait M. H., Swann G. A., Schleicher D. L., Schaber G. G., Sutton R. L., Dahlem D. H., Goddard E. N., and Waters A. C. (1970) Origin of the lunar regolith at Tranquility Base. *Proc. Apollo 11 Lunar Sci. Conf., Geochim. Cosmochim. Acta* Suppl. 1, Vol. 3, pp. 2399–2412. Pergamon.

Singer S. F. and Walker E. H. (1962) Electrostatic dust transport on the lunar surface. *Icarus* **1**, 112–120.

Spicer W. E. (1971) Private communication.

Teske Richard G. (1970) OSO satellite. *NASA Report N70-41817*.

Van De Hulst H. C. (1957) *Light Scattering by Small Particles*. John Wiley and Sons.

Wende Charles D. (1972) The normalization of solar x-ray data from many experiments. *Solar Physics* **22**, 492–502.

Wink H. B. and Ojanpera Pentti (1970) Chemical analyses of lunar samples 10017, 10072, and 10084. *Science* **167**, 531–532.

Proceedings of the Third Lunar Science Conference
(Supplement 3, *Geochimica et Cosmochimica Acta*)
Vol. 3, pp. 2681–2687
The M.I.T. Press, 1972

An explanation of transient lunar phenomena from studies of static and fluidized lunar dust layers

G. F. J. Garlick, G. A. Steigmann, and W. E. Lamb

Department of Physics, University of Hull, England

and

J. E. Geake

Department of Physics, University of Manchester Institute of Science and Technology,
Manchester, England

Abstract—Transient lunar phenomena (TLP's) such as brightening, color changes or obscuration of surface detail have been reported by many terrestrial observers. Explanations based on luminescence emission from the surface or from electrical discharges in gases venting through the surface fail for sunlit surfaces because the intensities would be many orders of magnitude lower than those needed for terrestrial observation. Experiments here reported show that when lunar dust layers lose their cohesion by suitable vibration, the albedo rises in sufficient magnitude to explain at least some of the TLP's observed within fully sunlit areas. The effect is very dependent on viewing angle but not on angle of incident sunlight. Dust flows associated with moonquakes etc. could produce such effects. Threshold conditions for dust fluidization and flow have been determined. Albedo changes are also accompanied by changes in the degree of polarization of reflected light from the layers. It is also shown that the albedo of dust layers is independent of temperature over a wide range including the lunar surface night-day extrema. In all experiments effects of postmission adsorption of gases can be found but can be eliminated by vacuum heat treatment.

Introduction

Over the past two centuries there have been many reports by observers of transient lunar phenomena (TLP's) in the form of brightenings, color changes, and temporary obscuration of surface details in the sunlit area of the moon and also of glows in the dark lunar areas. More than 700 such reports have been catalogued (Middlehurst *et al.*, 1968; Moore, 1971) and their degree of authenticity estimated. In the latest list by Moore (1971) we find that most of the hundred or so events for which durations were given lasted less than or about one hour. Some were of a few minutes' duration only. There is also some evidence of correlation of more recent events with lunar perigee. In a recent paper (Nash and Greer, 1970) explanations of lunar events based on luminescence from the sunlit surface have been shown to be invalid because the efficiency of luminescence in lunar soils as measured in the laboratory is much too low to give observable effects to be seen from the earth. These low efficiencies were confirmed by other workers (Blair and Edgington, 1970; Geake *et al.*, 1970). A similar objection can be made to Mills' (1970) suggestion that luminescence due to electrical discharges in gases venting through the lunar surface might be observed terrestrially at least in sunlit areas when these occur. However, his own experiments on gas venting and soil fluidization have a bearing on the work reported below. Any observation of lunar surface brightening in sunlit areas requires a contrast of about 10% or more for the TLP to be visible; this corresponds to an increase in light flux from the surface of about 10^{-2} w/cm^{-2}. It therefore occurred to us that the most obvious source of such an increment would be an increase in reflected sunlight due to a rise in albedo and that such an increase might be expected if dust flow occurred which involved temporary loss of cohesion between grains and removal of the strong optical shadowing effects of the normal "fairy castle" structure of the dust of the lunar regolith. Such flows would be most likely during moonquakes which cause gas venting, dust falling down slopes, and in electrostatic repulsion effects (Gold, 1972; Criswell, 1972) and should show some degree of correlation with lunar perigee. It is already well known that reflectance of light from the lunar surface is highly directional, being mainly back along the direction of incidence. This arises from the pileup and shadowing effects of the strong intergrain cohesion. Fluidization of the layers should remove the normal structure and in so doing raise the albedo toward that of terrestrial free-flowing dusts. We therefore constructed a system, described below, in which a dust layer can be vibrated so that cohesion is removed and the consequent change in albedo measured for various angles of incidence of light and of diffuse reflectance. The diffuse reflection spectra for static and fluid states were also measured.

In the recent Third Lunar Science Conference, Lloyd and Head (1972) reported anomalously higher albedos for the crater floor of Aristarchus under earthshine conditions and relative to the albedo of surrounding maria. In considering this report it seemed to us that the one parameter that would be very different for normal moonlight from that under earthshine conditions would be the surface temperature. We therefore included in our studies measurements of the albedo of various lunar dust samples over a temperature range from 77° to 473°K, this range including the lunar temperature extrema.

Experimental Equipment and Procedures

A moving coil transducer unit was fitted with a shallow, flat-bottomed cup in which a lunar dust sample of about 10 to 20 mg cm^{-2} was arranged as a horizontal layer and subjected to vertical, simple harmonic oscillations of selected amplitude. After finding that observed effects were not sensitive to vibration frequency over the usual audio range, a frequency of 208 Hz was selected for the experiments. Fluidization occurred for amplitudes of about 0.005 mm, and for all but threshold measurements an amplitude of 0.007 mm was used. A tungsten light source with collimator system and a photomultiplier with collimating apertures were mounted on separate circles concentric with the sample layer surface so that angles of incidence and reflectance could be separately varied. The output of the photomultiplier was fed to a pen recorder. For diffuse reflection spectra measurements the system was moved to a spectrometer which had a range from 0.4 to 1.0 μ. To measure variations of albedo with temperature, dust samples were mounted in a vacuum cryostat. All samples were preheated for several hours in vacuo at 140° to 160°C to remove laboratory-occluded gases. This eliminated spurious albedo changes due to the presence of adsorbed gases and also restored the high degree of cohesion characteristic of the uncontaminated dust samples. To simulate lunar regolith conditions, the dust layers were roughened by "teasing" with a needle point until they showed the right kind of appearance under microscopic observation (see, e.g., Geake et al., 1970). Thresholds for loss of intergrain cohesion were measured by visual and photoelectric observation as the vibration amplitude was slowly increased. The amplitude was precalibrated by observing the motion of the cup containing the sample through a microscope and recording the corresponding voltage across the transducer terminals with a digital voltmeter.

EXPERIMENTAL RESULTS

All lunar dust samples subjected to vertical vibrations showed a marked rise in albedo as soon as fluidization of the layers occurred, the effect being maintained until vibration was stopped. A full return to the previous low albedo values was not observed, but this was to be expected as the very random nature of the roughened surface is not restored under the very simple and highly directional disturbance in our system. The experiment is simply designed to show that dust fluidization can cause a marked rise in albedo sufficient to give an explanation of some TLP's. The albedo changes are very dependent on viewing angle, as shown by the selected data for an Apollo 14 sample in Fig. 1. A polar plot of reflectance before and during fluidization is presented for three different angles of incidence of light. No albedo change is observed in each case for reflection back along the incidence direction, but in each case the albedo change is a maximum at a viewing angle of about 20° to the surface. The variation of the reflection at 20° to the surface as a function of angle of incidence is given in Fig. 2 and is not very marked. The plots of Figs. 1 and 2 show that dust fluidization can yield rises in surface albedo of 50 to 100%.

Measurements of the diffuse reflection spectrum of a layer before and during fluidization show that the rise in albedo is the same at all wavelengths. This result may of course be peculiar to our experimental conditions, since dust flow and motion, e.g., due to gas venting on the lunar surface, will involve a wide dust grain-size distribution while in our experiments we had to use the sieved fines as provided.

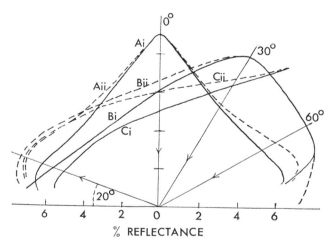

Fig. 1. Variation of lunar dust layer reflectance (sample 14259,56) with viewing angle for different angles of incident light and under static and fluidized conditions.

A(i)   Incidence angle 0° to normal: static condition.
A(ii) Incidence angle 0° to normal: fluidized condition.
B(i)   Incidence angle 30° to normal: static condition.
B(ii) Incidence angle 30° to normal: fluidized condition.
C(i)   Incidence angle 60° to normal: static condition.
C(ii) Incidence angle 60° to normal: fluidized condition.

(Line at 20° to surface shows direction of maximum albedo change.)

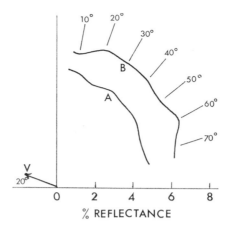

Fig. 2. Variation of lunar dust layer reflectance, observed at 20° to surface, with angle of incident light (sample 14259,56) A. Static condition, B. fluidized condition.

Table 1. Fluidization thresholds for lunar dust layers subjected to vertical vibrations.

Sample no.	Vibration amplitude ($\times 10^{-4}$ mm)		Max. acceleration of layer in msec^{-2}	
	Initial	After heat treatment	Initial	After heat treatment
10084,6	109 ± 10	119 ± 7	18.5 ± 1.7	20.2 ± 1.2
12032,39	110 ± 7	120 ± 4	18.7 ± 1.2	20.4 ± 0.7
12033,60	107 ± 5	114 ± 11	18.2 ± 0.8	19.4 ± 1.9
14163,51	102 ± 5	120 ± 7	17.3 ± 0.8	20.4 ± 1.2

Differential effects of different grain sizes on the spectral distribution would not be unexpected. Measurements of albedo over the temperature range 77° to 473°K showed that the albedo is independent of temperature over this range.

Finally, we have measured with the vibrator system the threshold amplitudes of vibration for breaking of cohesion between dust grains. The results are collected in Table 1, the maximum accelerations being calculated and inserted alongside the corresponding threshold amplitudes.

Discussion and Conclusions

It is evident that fluidization with consequent cohesion loss of lunar dust layers can cause a rise in albedo of a magnitude more than sufficient to satisfy the contrast requirements for terrestrial observation of such effects on sunlit lunar areas. The mass of dust needed to obtain the rise is of the order of 10 to 20 mg/cm^{-2} at the most. Only fluidization is needed without any resultant lateral flow of dust across the surface. Suggestions have been made by other workers (Pai, Hsieh, and O'Keefe, 1972) that processes such as gas venting may produce suspensions of dust which can persist for times comparable with those for many reported TLP's. In another case, Criswell (1972) suggests that dust suspensions might be created electrostatically by the excursion of photoelectrons from sunlit to dark areas across the terminator and that such suspensions could be an explanation of the "horizon" glow detected by the Surveyor 6 and 7 (soft-landing) spacecrafts just after lunar sunset. Intense light scattering relative to normal surface brightness is evident from the temporary dust clouds produced by the astronauts' boots as shown in several mission photographs. The removal of self shadowing of the dust-layer reflection by loss of intergrain cohesion has been shown to be critically dependent upon viewing angle; this places strong constraints on favorable viewing times for different sites on the lunar surface during the lunar phases, should dust disturbance be a cause of the TLP. To demonstrate this point we have used the data given in Fig. 1 to construct the polar diagrams of Fig. 3, which give the expected albedos, with and without fluidization of the dust layers for

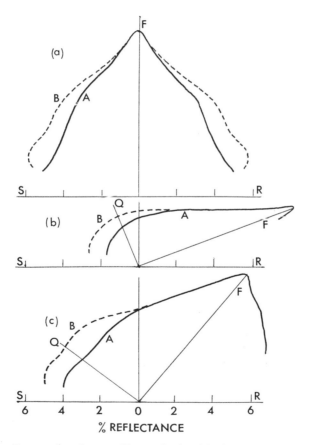

Fig. 3. Polar diagrams for change of lunar albedo with phase angle, with and without dust fluidization for different lunar surface sites. (a) Curves for sub-earth point. (b) Curves for sites of craters Hevelius (3°N, 67.5°W) and Grimaldi (5°S, 67.5°W). (c) Curves for site of crater Kepler (7.5°N, 37.5°W). A. static condition, B. fluidized condition, F. Full moon, R. Sunrise, S. Sunset, Q. Third quarter. (% reflectance contains the correction factor cos ε, where ε is the viewing angle relative to normal to lunar surface at site).

selected lunar sites, as functions of the phase angle (angle between incidence and viewing direction). For the subearth point, the best time for observation of TLP's due to soil disturbance is just after the first quarter or just before the third quarter. In other cases, e.g. for the sites of craters Hevelius, Grimaldi, and Kepler, the third quarter is the best time for observation.

Another optical effect that should change with fluidization is the degree of polarization of reflected light from the dust layer. We have designed a system to measure the polarization using the Lyot polarimeter from the Physics Department of the University of Manchester Institute of Science and Technology in combination with our vibrator system. Preliminary data show that the small negative polarization at small phase angles is slightly decreased on fluidization, but there is a large increase, sometimes of more than 100% in the positive degree of polarization at large phase angles. The general shape of the polarization curves follow those given by Dollfus et al. (1971), Geake et al. (1970), and others.

The albedo of lunar dust layers is independent of temperature over a range containing the lunar day and night extrema, which makes it unlikely that the differences in relative albedo for earthshine and moonlight conditions in the crater Aristarchus can be related to the large temperature difference. However, in making these experiments and those on thresholds for fluidization we have also shown that cohesion data and flow characteristics may be subject to terrestrial adsorption effects unless preheating and vacuum treatments are applied before measurements. The data also make it evident that soil mechanics studies should include dynamic conditions which may be very relevant to lunar surface slopes and stabilities involved in Apollo 16 and 17 missions.

Finally, we would re-emphasize that the reported effects of dust layer disturbance on albedo offer a particular contribution to explanations of lunar transients. They are relevant to TLP's occurring in the sunlit areas of the lunar surface and even within that group cannot account for the reported observations of spectral fine structure, suggestive of molecular species, by Kozyrev (1962, 1963) though, to be visible, the intensity of such emissions must have been associated with considerable local surface activity which would also engender albedo changes. Present data also do not give any explanation of the reddish coloration often reported by TLP observers. However, we have noted that our vibrator system and the sieved dust samples do not give a sufficiently close simulation of lunar surface conditions. It is worth noting that Greenacre (1965) refers to "sparkling" or "flowing" appearances of TLP's which might suggest possible surface motion. Such motion would also explain the temporary obscurations reported (Moore, 1971) if dust flow and even suspension were involved. Suffice it to say that the observational data on TLP's in sunlit lunar areas show a fair degree of correlation with the effects we have measured on lunar samples and their implications for surface motion.

References

Blair I. M. and Edington J. A. (1970) Luminescence and thermoluminescence under 159 MeV proton bombardment of lunar material returned by Apollo 11. *Proc. Apollo 11 Lunar Sci. Conf.*, *Geochim. Cosmochim. Acta* Suppl. 1, Vol. 3, pp. 2001–2012. Pergamon.

Criswell D. R. (1972) Horizon glow and motion of lunar dust (abstract). In *Lunar Science—III* (editor C. Watkins), p. 163, Lunar Science Institute Contr. No. 88.

Dollfus A., Geake J. E., and Titulaer C. (1971) Polarimetric properties of the lunar surface and its interpretation. Part III: Apollo 11 and Apollo 12 lunar samples. *Proc. Second Lunar Sci. Conf., Geochim. Cosmochim. Acta* Suppl. 2, Vol. 3, pp. 2285–2300. MIT Press.

Geake J. E., Dollfus A., Garlick G. F. J., Lamb W. E., Walker G., Steigmann G. A., and Titulaer C. (1970) Luminescence, electron paramagnetic resonance and optical properties of lunar material from Apollo 11. *Proc. Apollo 11 Lunar Sci. Conf., Geochim. Cosmochim. Acta* Suppl. 1, Vol. 3, pp. 2127–2147. Pergamon.

Gold T. (1972) The depth of the lunar dust layer (abstract). In *Lunar Science—III* (editor C. Watkins), pp. 321–322, Lunar Science Institute Contr. No. 88.

Greenacre J. C. (1965) The 1963 Aristarchus events. *Ann. N.Y. Acad. Sci.* **123**, 811.

Kozyrev N. A. (1962) Spectroscopic proofs for existence of volcanic processes on the moon. In *The Moon* (editors Z. Kopal and Z. K. Mikhailov), pp. 263–271, Academic Press.

Kozyrev N. A. (1963) Volcanic phenomena on the moon. *Nature* **198**, 979–980.

Lloyd D. D. and Head J. W. (1972) Earthshine and near terminator photography obtained on Apollo 15 (abstract). In *Lunar Science—III* (editor C. Watkins), pp. 489–490, Lunar Science Institute Contr. No. 88.

Middlehurst B., Burley J. M., Welther B. L., and Moore P. (1968) Chronological catalog of reported lunar events. NASA Tech. Report TR R-277.

Mills A. A. (1970) Transient lunar phenomena and electrical glow discharges. *Nature* **225**, 929–930.

Moore P. (1971) Extension of catalog of reported lunar events. *J. Brit. Astron. Assoc.* **81**, 365–390.

Nash D. B. and Greer R. T. (1970) Luminescence properties of Apollo 11 lunar samples and implication for solar excited luminescence. *Proc. Apollo 11 Lunar Sci. Conf., Geochim. Cosmochim. Acta* Suppl. 1, Vol. 3, pp. 2341–2450. Pergamon.

Pai S. I., Hsieh T., and O'Keefe J. A. (1972) Lunar ash flows: How they work (abstract). In *Lunar Science—III* (editor C. Watkins), pp. 593–595, Lunar Science Institute Contr. No. 88.

Proceedings of the Third Lunar Science Conference
(Supplement 3, *Geochimica et Cosmochimica Acta*)
Vol. 3, pp. 2689–2711
The M.I.T. Press, 1972

Lunar ash flow with heat transfer

S. I. Pai and T. Hsieh

Institute for Fluid Dynamics and Applied Mathematics,
University of Maryland, College Park, Maryland

and

J. A. O'Keefe

NASA, Goddard Space Flight Center, Greenbelt, Md.

Abstract—When the effects of heat transfer are taken into account, we find that a typical ash flow (where the rate of gas inflow at the bottom is not large) has a sharply defined, cold upper surface.

In a previous paper, we have studied the isothermal lunar ash flow. In this paper, we consider the lunar ash flow with heat transfer. The most important heat-transfer process in the ash flow under consideration is heat convection. Besides the four important nondimensional parameters of isothermal ash flow (Pai *et al.*, 1972), we have three additional important nondimensional parameters: the ratio of the specific heat of the gas γ, the ratio of the specific heat of the solid particles to that of gas δ, and the Prandtl number P_r. We reexamine the one dimensional steady ash flow discussed by Pai *et al.* (1972) by including the effects of heat transfer. Numerical results for the pressure, temperature, density of the gas, velocities of gas and solid particles, and volume fraction of solid particles as function of altitude for various values of the Jeffreys number J_e, initial velocity ratio $\bar{u}_0$, and two different gas species (steam and hydrogen, each with its density ratio G and ratio of specific heats δ) are presented. Our main results are as follows:

(i) The effects of heat transfer become important for the case of small initial velocity $\bar{u}_0$. For large initial velocities: $\bar{u}_0 \geq 8/G$, the isothermal approximation is a good one. For small initial velocities, especially when $\bar{u}_0 \to 0$, the effect of heat transfer is to give the ash flow a sort of outer boundary, with only a thin layer of dilute flow beyond it. The influence of heat transfer is strongest on the dilute phase of ash flow. Hence, heat transfer has a greater influence on the lunar ash flow than on the terrestrial ash flows.

(ii) The effect of heat transfer depends on the heat capacity of the gas species in the ash flow. The effect of heat transfer is larger if the gas is hydrogen than if it is steam.

The sharp boundary may explain some high-tide marks seen on Mt. Hadley and may help explain the lunar sinuous rilles.

I. Introduction

Strong support for the reality of the ash flow model (impact or volcanic in origin) for explaining the emplacement of the lunar soils and breccias has been furnished by the recent discovery by McKay *et al.* (1972) of crystals apparently deposited from a hot vapor phase. These crystals are reminiscent of crystals of similar minerals noted in terrestrial ash flow tuffs (Ross and Smith, 1961). With this evidence for the physical reality of the lunar ash flows, it becomes important to have satisfactory mathematical models. These must be constructed from theory, because we are not likely to observe a lunar ash flow.

Up to now, the mathematical models of lunar ash flows have been somewhat unsatisfactory, because they have involved extensive local atmospheres, reaching tens of kilometers above the lunar surface with substantial densities. These atmospheres were required in the models because without pressure from them, the gas within the denser layers tended to burst out, carrying the top layer of ash up into the vacuum. The notion of a dense local atmosphere was unsatisfactory, first because it was difficult to find sources for such a large volume of gas, and second because we can not explain the sinuous rilles in terms of ash flows unless the flows are supposed to hug the ground. In this paper, we show that these difficulties arose because thermal variation through the ash flow was neglected.

In a previous paper (Pai *et al.*, 1972), we have studied the isothermal lunar ash flow as one of the major processes responsible for certain features of the lunar soil. The justification for our assumption of uniform temperature is that it has been observed that the cooling process of an ash flow is a conspicuously slow process. Moreover, from our experience in the study of general problems of fluid dynamics, we have seen that the isothermal model will explain many important features of the flow field. Hence in our previous paper, we considered only the isothermal ash flow as a first approximation to the actual ash flow.

There is reason, however, to reexamine this assumption because the original temperature of the ash flow may be very high. This holds whether the origin of the ash flow is a great meteorite impact or is some sort of volcanism. As the ash flow expands into free space, the temperature of the ash flow will drop to a very low value. As a result, there may be large temperature variations in the ash flow. For a large temperature variation, we must examine whether the effects of heat transfer are negligible. In this paper, we study the effect of heat transfer on the ash flow process, particularly the lunar ash flow.

There are four main processes of heat transfer in an ash flow: (a) heat conduction, (b) heat convection, (c) radiative heat transfer, and (d) heat transfer between the gas species and the solid particles. For the kind of ash flow in which we are interested, it is easy to show that both heat conduction and thermal radiation are negligible. Furthermore, in Section IV, we shall show that the heat transfer between the gas species and solid particles in the cases we investigate is so rapid that the gas species and the solid particles are practically in local thermodynamic equilibrium throughout the flow field. We may assume that the temperature of the gas T_g is equal to the temperature of the solid particles T_p at all times. Hence the main heat transfer process in the ash flow is heat convection only.

We reexamine the one dimensional steady ash flow discussed by Pai *et al.* (1972), but now we include the effect of heat transfer. The flow field is sketched in Fig. 1 which is essentially the same as that described by Pai *et al.* (1972); but in the present paper the temperature of the ash flow T is not a constant but a function of the altitude so that the variation of gas density is not identical to the variation of pressure as in the isothermal case.

In general, for an ash flow with heat transfer, we have seven variables for the one dimensional steady flow, i.e., the temperature of the gas T_g, the temperature of the solid particles T_p, the gas density ρ_g, the pressure of the ash flow p, the velocity of the gas u_g, the velocity of the solid particles u_p, and the volume fraction of the solid

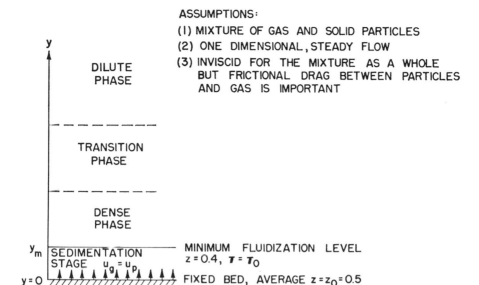

Fig. 1. Sketch of ash flow.

particles Z. These seven variables are governed by the following seven equations: the equation of state of the gas, the equation of continuity for the gas and that for the solid particles, the equation of motion of the mixture and that for the solid particles, the equation of energy of the mixture and the equation of energy for the gas species. Except for the two equations of energy, all of these equations are the same as those used in the isothermal ash flow, which has been discussed by Pai *et al.* (1972). Hence in Section II, we shall discuss only the two equations of energy, but we shall list all the seven nondimensional fundamental equations that we use in our numerical calculations. The boundary conditions for the velocities in the case with heat transfer are the same as those used for the isothermal case of Pai *et al.* (1972). Hence we do not repeat this discussion but list only the boundary values. However, we shall briefly discuss the boundary condition for the temperature.

In Section III, we discuss the important nondimensional parameters of the ash flow with heat transfer. Four important nondimensional parameters are carried over from the isothermal ash flow, namely: the Jeffreys number J_e, the initial velocity ratio $\bar{u}_0$, the density ratio G, and the dynamic pressure parameter H. These all remain in similar positions in equations for the ash flow with heat transfer. We shall not repeat these discussions here. We shall use approximations in this paper similar to those used by Pai *et al.* (1972). For instance, we shall continue to neglect the inertial terms by putting $H = 0$. We then have three new nondimensional parameters due to the heat transfer effects that will be discussed in Section III.

In Section IV, we solve the fundamental equations of one dimensional steady ash flow with heat transfer, particularly for the lunar ash flow, with proper approximations.

In Section V, the numerical results are presented, particularly for the lunar ash flow. These results are compared with those of isothermal ash flow of Pai et al. (1972). In some cases, the calculations for the terrestrial ash flow with heat transfer are also given.

II. Fundamental Equations of One Dimensional Steady Ash Flow with Heat Transfer and Associated Boundary Conditions

Since we do not consider the transition region near a solid boundary, we may assume that the viscous stresses and heat conduction of the gas and those of the pseudofluid of solid particles are negligible. Furthermore, since the temperature of the ash flow is not too high and the gas pressure is not too low, the effects of thermal radiation are also negligible (Pai, 1966) by comparison with those of conduction and convection. The equation of state of the gas, the two equations of continuity (for gas and for solid particles), and the two equations of motion of the ash flow are the same as those given by Pai et al. (1972). We must add the equation of energy of the ash flow to complete our system of fundamental equations for ash flow with heat transfer.

For the ash flow under consideration, the conservation of energy for the gas and for the pseudofluid of solid particles may be expressed respectively as follows (Pai, 1970):

$$\frac{d}{dy}\left[(1 - Z)\rho_g u_g(c_v T_g + \tfrac{1}{2}u_g^2 + gy) + u_g(1 - Z)p\right] = K_T(T_p - T_g) + \varepsilon_g \qquad (1)$$

$$\frac{d}{dy}\left[Z\rho_{sp}u_p(c_s T_p + \tfrac{1}{2}u_p^2 + gy) + u_p Z p\right] = K_T(T_g - T_p) + \varepsilon_p \qquad (2)$$

where c_v is the specific heat of the gas at constant volume, c_s is the specific heat of the solid particles, K_T is the thermal friction coefficient and ε_g, ε_p are, respectively, the energy source terms for the gas and solid particles, namely total energy associated with the transformation of the gas from the solid particles. Furthermore, in equation (1), $(1 - Z)\rho_g u_g$ represents the flux of mass of the gas species; $(c_v T_g + \tfrac{1}{2}u_g^2 + gy)$ represents the sum of internal energy, kinetic energy, and potential energy, respectively, per unit of mass for the gas; $u_g(1 - Z)p$ represents the work done by the gas; and $K_T(T_p - T_g)$ represents the heat transferred from the particles to the gas. The corresponding terms in equation (2) have the same meaning as explained for equation (1).

By definition of ε_g and ε_p we have

$$\varepsilon_g + \varepsilon_p = 0 \qquad (3a)$$

and

$$\varepsilon_g = bZ\rho_{sp}(c_v T_g + \tfrac{1}{2}u_g^2 + gy) \qquad (3b)$$

where b is the rate factor of emission of the gas from the solid particles. Therefore, we may add equations (1) and (2) and since the right-hand side of the resultant equation is zero, we may integrate it once to give

$$(1 - Z)\rho_g u_g(c_p T_g + \tfrac{1}{2}u_g^2 + gy) + Z\rho_{sp}u_p\left(c_s T_p + \tfrac{1}{2}u_p^2 + yg + \frac{p}{\rho_{sp}}\right) = c_1 \qquad (4)$$

where we have used the thermodynamic relation for the gas (considered as a perfect gas), i.e., $c_v T_g + p/\rho_g = c_p T_g$ with c_p defined as the specific heat of the gas at constant pressure. The c_1 in equation (4) is a constant that represents the total energy of the mixture and that may be determined by the boundary conditions as will be discussed later.

It is convenient to use equations (1) and (4) as the additional two energy equations in our system of fundamental equations for ash flow with heat transfer, because we may avoid going into the details of the process involved in the energy balance during the transformation of the gas from the solid particles. However, we still have some difficulties in writing the term K_T for finite values of Z, since no definite information is available up to the present time. One way to overcome this difficulty is to assume that the Reynold's analogy between heat transfer and drag force may be applied. Thus the thermal friction coefficient may be approximated by the formula (Pai, 1970 and Pai et al., 1972)

$$K_T = \frac{2}{3} \cdot \frac{\kappa_g}{\mu} \cdot K(Z, r_p, \mu) \tag{5}$$

$$K(Z, r_p, \mu) = \frac{1}{(1 - Z)^2} \cdot \frac{9}{2r_p^2} \left(1 + \frac{0.68}{h/r_p}\right) \tag{5a}$$

$$h/r_p = (1.35Z)^{-1/3} - 1 \tag{5b}$$

where κ_g is the coefficient of thermal conductivity of the gas, μ is the kinematic viscosity of the gas and r_p is the radius of solid particles.

The seven fundamental equations should be solved with appropriate boundary conditions. The boundary conditions for velocity are the same as those discussed by Pai et al. (1972). We assume that at $y = 0$, the temperature of gas and the solid particles are in equilibrium at a value T_0. Then the boundary conditions are as follows at $y = 0$:

$$u_{g0} = u_1, \qquad u_{p0} = u_0,$$

$$p = p_0, \qquad z = z_0 = 0.5 \text{ (average)}, \tag{6}$$

$$T_{p0} = T_{g0} = T_0.$$

The boundary value u_0 in the heat transfer case has the same restriction as in the isothermal case, namely that u_p should not be negative.

III. Important Nondimensional Parameters in an Ash Flow with Heat Transfer

As in the isothermal case, we introduce nondimensional quantities as follows:

$$\bar{p} = \frac{p}{p_0}, \qquad \bar{\rho}_g = \frac{\rho_g}{\rho_{g0}} = \frac{RT_0}{mp_0}\rho_g, \qquad \bar{y} = \frac{y}{L},$$

$$\bar{u}_g = \frac{u_g}{v_c}, \qquad \bar{u}_p = \frac{u_p}{v_c}, \qquad \bar{T}_g = \frac{T_g}{T_0}, \qquad \bar{T}_p = \frac{T_p}{T_0}, \tag{7}$$

where the bar refers to the nondimensional quantities. The characteristic length L is taken as follows:

$$L = \frac{p_0}{\rho_{sp}g} \tag{8}$$

so that L is a modified scale height; it is taken as the unit length in our analysis. The characteristic velocity v_c is defined by the following relation:

$$v_c = \frac{bRT_0}{mg} \tag{9}$$

where b is the rate of emission of the gas from the solid particles, m is the molecular weight of the gas, and g is the gravitational acceleration.

In terms of these nondimensional quantities, our fundamental equations of one-dimensional steady ash flow with heat transfer are as follows:

$$\bar{p} = \bar{\rho}_g \bar{T}_g \tag{10}$$

$$\frac{d}{d\bar{y}}(Z\bar{u}_p) = -\frac{Z}{G} \tag{11}$$

$$\frac{d}{d\bar{y}}[(1 - Z)\bar{\rho}_g\bar{u}_g] = Z \tag{12}$$

$$H\left(\frac{1 - Z}{G}\bar{\rho}_g\bar{u}_g\frac{d\bar{u}_g}{d\bar{y}} + Z\bar{u}_p\frac{d\bar{u}_p}{d\bar{y}}\right) = -\frac{d\bar{p}}{d\bar{y}} - \frac{(1 - Z)\bar{\rho}_g}{G} - Z \tag{13}$$

$$\frac{d\bar{p}}{d\bar{y}} = -\frac{Z}{(1 - Z)^2}\frac{9}{2J_e}\left(1 + \frac{0.68}{h/r_p}\right)(\bar{u}_g - \bar{u}_p) \tag{14}$$

$$(1 - Z)\bar{\rho}_g\bar{u}_g\left[\bar{T}_g + \frac{\gamma - 1}{\gamma G}(\tfrac{1}{2}H\bar{u}_g^2 + \bar{y})\right]$$
$$+ ZG\bar{u}_p\left[\delta\bar{T}_p + \frac{\gamma - 1}{\gamma G}(\tfrac{1}{2}H\bar{u}_p^2 + \bar{y} + \bar{p})\right] = c_1 \tag{15}$$

$$(1 - Z)\bar{\rho}_g\bar{u}_g\frac{d\bar{T}_g}{d\bar{y}} = \frac{27}{4}\frac{G}{P_rJ_eH}\frac{1}{(1 - Z)^2}\left(1 + \frac{0.68}{h/r_p}\right)(\bar{T}_g - \bar{T}_p) \tag{16}$$

and the boundary conditions are at $\bar{y} = 0$:

$$\bar{u}_{g0} = \bar{u}_{p0} = \bar{u}_0,$$
$$\bar{p} = 1, Z_0 = 0.5, \tag{17}$$
$$\bar{T}_{p0} = \bar{T}_{g0} = 1,$$

where equations (10) to (14) are the same as those for isothermal ash flow of Pai et al. (1972). In the boundary condition, we assume that $u_1 = u_0$ for simplicity.

There are seven nondimensional parameters in equations (10) to (17) that are defined as follows:

(i) The initial velocity ratio $\bar{u}_0$ is defined as

$$\bar{u}_0 = \frac{u_0}{v_c}. \tag{18}$$

(ii) The Jeffreys number J_e is defined as

$$J_e = \frac{\rho_{sp} r_p^2 v_c}{b \dfrac{R}{m} T_0}. \tag{19}$$

(iii) The density ratio G is defined as

$$G = \frac{\rho_{sp}}{\rho_{g0}}. \tag{20}$$

(iv) The dynamic pressure parameter H is defined as

$$H = \frac{\rho_{sp} v_c^2}{p_0}. \tag{21}$$

The significance of these four parameters: $\bar{u}_0$, J_e, G, and H has been discussed by Pai *et al.* and they play the same role in the ash flow equations with heat transfer as in the ash flow equations of the isothermal case.

(v) The ratio of specific heats of the gas γ is

$$\gamma = \frac{c_p}{c_v}. \tag{22}$$

This parameter γ depends on the complexity of the molecules of the gas species. In our analysis, we shall assume that γ is a constant for a given gas.

(vi) The ratio of specific heats δ is defined as

$$\delta = \frac{c_s}{c_p}. \tag{23}$$

This parameter indicates the importance of the heat capacity of the solid particles relative to that of gas; it is important in the determination of the temperature distribution in the mixture.

(vii) The Prandtl number of the gas P_r is defined as

$$P_r = \frac{\mu c_p}{\kappa_g}. \tag{24}$$

The Prandtl number is important in the study of heat transfer between the gas species and the solid particles. When the gas species and the solid particles are in local thermodynamic equilibrium, the Prandtl number drops out of the system of the fundamental equations as well as we shall show in Section IV.

As in the isothermal case, we shall put $H = 0$ in our fundamental equations because the initial terms are negligible. Furthermore, for the case $H = 0$, equation (16) shows that T_g must be equal to T_p. As a result, the Prandtl number will drop out of the system of fundamental equation. Hence for the ash flow with heat transfer, we need to consider five parameters only: $\bar{u}_0$, J_e, G, γ, and δ.

IV. Method of Solution

In equation (16), we find that the constant in the right-hand side is of the order of

$$\frac{27}{4} \cdot \frac{G}{P_r J_e H} \sim 10^6 \tag{25}$$

for the ash flow under consideration. Since the remaining terms in equation (16) are of the order of unity, for all practical purposes we may set

$$\bar{T}_p = \bar{T}_g = \bar{T}. \tag{26}$$

Hence for the ash flow with heat transfer, we need to solve six unknowns $\bar{T}$, $\bar{p}$, ρ_g, $\bar{u}_p$, $\bar{u}_g$, and Z from equations (10) to (15) with the boundary conditions (17).

The constant c_1 in equation (15) may be determined from the boundary conditions (17). However, since the solid particles are not yet fluidized in the sedimentation region $\bar{y} = 0$ to $\bar{y} = \bar{y}_1$, we do not know the thermal situation in the sedimentation region precisely, and it may be nonuniform at a given altitude. In order to get some definite idea of the thermal situation in the flow field $\bar{y} > \bar{y}_1$, we may replace the actual sedimentation by an isothermal layer with temperature $\bar{T}_0$ which is the average temperature of this thin layer. As a result, our boundary conditions become

$$\bar{y} = \bar{y}_1, \qquad Z = Z_1 = 0.4, \qquad \bar{u}_p = \bar{u}_{p1}, \bar{u}_g = \bar{u}_{g1},$$
$$\bar{p} = \bar{p}_1, \bar{T}_p = \bar{T}_g = 1. \tag{27}$$

With this thermal boundary condition (27), the total energy constant c_1 divided by $Z_0 G \bar{u}_0$ becomes

$$\bar{c}_1 = \frac{c_1}{Z_0 G \bar{u}_0} = 1 + (\delta - 1) \frac{0.4 \bar{u}_{p1}}{Z_0 \bar{u}_0} + \frac{\gamma - 1}{\gamma G} \left(\bar{y}_1 + \frac{0.4 \bar{u}_{p1}}{Z_0 \bar{u}_0} \bar{p}_1 \right). \tag{28}$$

With the help of an approximation (Pai, 1971) from equations (11) and (12)

$$\bar{u}_p = \frac{1}{ZG} \left[\tfrac{1}{2} \bar{u}_0 G - (1 - Z) \bar{u}_g \bar{p} \right] \tag{29}$$

and equation (26) and equation (15) give

$$\bar{T}_g = \bar{T} = \frac{\bar{c}_1 - \dfrac{\gamma - 1}{\gamma G} \left(\bar{y} + \dfrac{Z \bar{u}_p}{Z_0 \bar{u}_0} \bar{p} \right)}{1 + (\delta - 1) \dfrac{Z \bar{u}_p}{Z_0 \bar{u}_0}}. \tag{30}$$

We have used equations (10), (11), (12), (13), (14), and (30) with $H = 0$ to solve for the six unknowns $\bar{p}$, $\bar{T}$, $\bar{\rho}_g$, $\bar{u}_g$, $\bar{u}_p$, and Z in our one-dimensional steady ash flow with heat transfer subject to the boundary conditions (27) by numerical integration with the help of the high-speed computer UNIVAC-1180 of the Computer Science Center of the University of Maryland.

V. Numerical Results and Discussions

In our numerical calculations, we limit ourselves to the cases $\bar{u}_0 = O(1/G)$ and $H = 0$ for the reasons discussed by Pai *et al.* (1972). We examine the effects of $\bar{u}_0$, J_e, and G with two sets of γ and δ: one for the steam and the other for hydrogen gas. The following values have been used in our numerical calculations:

(i) General data:

$$g = 980 \text{ cm/sec}^2 \text{ for earth}; \quad g = 160 \text{ cm/sec}^2 \text{ for moon};$$

$$T_0 = 1130°\text{K}. \tag{31}$$

(ii) For solid particles:

$$\rho_{sp} = 2.4 \text{ gm/cm}^3; \quad r_p = 0.005 \text{ cm};$$

$$c_s = 0.3 \text{ erg/gm–°K}. \tag{32}$$

(iii) The properties of the gas species:

(a) Steam:

$$m = 18; \quad \gamma = 1.3; \quad c_p = 0.6 \text{ erg/cm–°K};$$

$$\mu = 4 \times 10^{-4} \text{ poise}; \quad b = 7.34 \times 10^{-7} \text{ cm/sec}. \tag{33}$$

(b) Hydrogen gas:

$$m = 2; \quad \gamma = 1.4; \quad c_p = 3.5 \text{ erg/gm–°K};$$

$$\mu = 2.6 \times 10^{-4}; \quad b = 1.25 \times 10^{-7} \text{ cm/sec}. \tag{34}$$

The choice of the values of b for different gas species will be discussed in Part B of this section. We present our numerical results in two parts: (A) Ash flow with heat transfer, and (B) the effects of different gas species as follows:

(A) *The effects of heat transfer on ash flow*

As we have mentioned before, the main heat transfer process in an ash flow is the convection of heat. Thus in studying the effects of heat transfer, we need to consider the temperature distribution $\bar{T}$ of the mixture of gas and solid particles through the heat convection process, equation (30), only. With heat transfer we should use equation (10) to determine $\bar{\rho}_g$ from the pressure distribution $\bar{p}$ and the temperature distribution $\bar{T}$. Numerical integrations of the fundamental equations have been carried out for various cases. Our numerical results are mainly for the lunar ash flows with heat transfer; but we have calculated a few cases for terrestrial ash flow in order to see the difference with respect to the effects of heat transfer between the moon and earth. For comparison with our isothermal ash flow calculations in Pai *et al.* (1972) that are based on steam as the fluidizing gas, we have made our first calculations of ash flows with heat transfer based on steam (Figs. 2 to 10).

Figure 2 shows the pressure distribution as a function of altitude for various values of $\bar{u}_0$ on the moon with heat transfer. Comparing with Fig. 5 of Pai *et al.* (1972), we see the effect of heat transfer on the pressure distribution. For $N > 1$ (note $\bar{u}_0 = N/G$), the pressure distributions with and without heat transfer are qualitatively the same; we find that as N decreases, the rate of pressure drop also decreases. But for $N < 1$, the effect of heat transfer has a great influence on the pressure distribution. For $N < 1$, the rate of decrease of pressure increases enormously as N decreases and the pressure drops very rapidly as N approaches zero. Hence for $N < 1$, the pressure distributions with heat transfer differ greatly from those in the isothermal case.

Figure 3 shows the distribution of gas density as a function of altitude. With heat transfer, the distribution of density $\bar{\rho}_g$ is different from that of pressure $\bar{p}$ while in the isothermal cases, we have $\bar{\rho}_g = \bar{p}$. It is interesting to notice that with heat transfer, for the case $\bar{u}_0 = 0$, the density of gas may increase to 20 times its original value. It then drops rapidly from this maximum value. This large variation of gas density makes the pressure distribution strikingly different from that of the isothermal case.

Figure 4 shows the temperature distribution for various $\bar{u}_0$ on the moon. For $N \geq 8$, the temperature variation is small in the greater part of the flow field, which suggests that the isothermal approximation is a good one for the lunar ash flow. For small values of N, particularly as N approaches zero, the variation of temperature in the flow field is large. Hence the isothermal approximation is not a good one for small N, particularly when N approaches zero.

In Fig. 5, the velocity distributions in the gas and in the solid particles are plotted

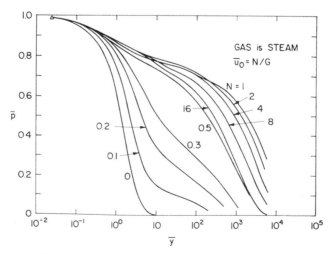

Fig. 2. Pressure distribution $\bar{p}$ as a function of altitude at various values of $\bar{u}_0$ on moon, $\bar{T} \neq 1$.

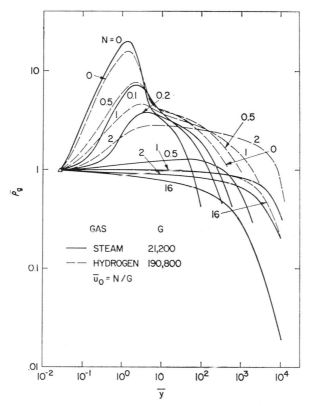

Fig. 3. Distribution of gas density $\bar{\rho}_g$ as a function of altitude at various values of $\bar{u}_0$ on moon, $\bar{T} \neq 1$.

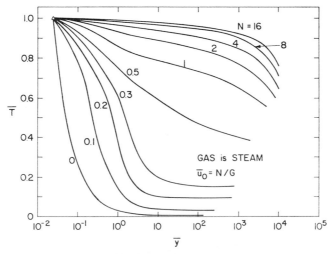

Fig. 4. Temperature distribution $\bar{T}$ as a function of altitude at various values of $\bar{u}_0$ on moon.

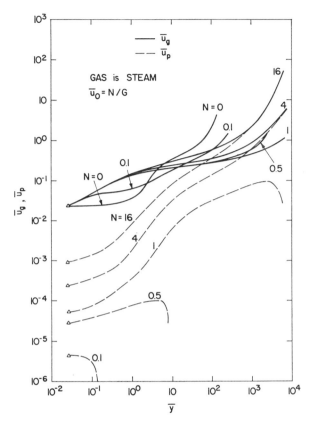

Fig. 5. Velocity distributions $\bar{u}_g$ and $\bar{u}_p$ as a function of altitude at various values of $\bar{u}_0$ on moon, $\bar{T} \neq 1$.

as function of altitude for various values of $\bar{u}_0$. We find that for cases where $N \leq 1$, the velocity of the solid particles becomes zero at certain level y_c, while in isothermal cases, this occurs at $N \leq 0.5$. Otherwise, the general trend of the velocity distributions of gas and solid particles is similar for the cases with and without heat transfer, i.e., heat transfer has little influence on velocity distribution.

In Fig. 6, we plot the distribution of the volume fraction of solid particles Z as a function of altitude for various values of $\bar{u}_0$. It is seen that for $N < 0.5$, the range of the dense phase ($Z \geq 0.2$) increases rapidly as N decreases. We do not have any similar situation in isothermal case. It is also interesting to notice the rapid drop in volume fraction of solid particles above the dense phase for small values of $\bar{u}_0$ in the case with heat transfer.

Figures 2 to 6 mainly give the results for lunar ash flows with heat transfer. We include, however, a few numerical calculations for the comparison between the flow field with and without heat transfer on the moon with corresponding cases on the earth for two typical values of initial velocities, i.e., $N = 0$ and 8 as follows:

Figure 7 shows the comparison of pressure distributions at $\bar{u}_0 = 0$ or $N = 0$.

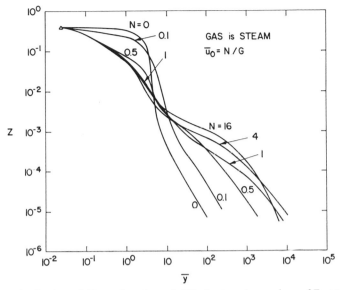

Fig. 6. Distribution of Z as a function of altitude at various values of $\bar{u}_0$ on moon.

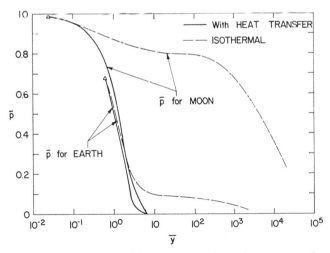

Fig. 7. Comparison of $\bar{p}$ with and without heat transfer and on moon and on earth at $\bar{u}_0 = 0$.

A tremendous change of pressure distribution is shown on the moon due to the influence of heat transfer; on earth, on the other hand, the effect of heat transfer is not so significant. It is interesting to notice that in this case ($N = 0$) with heat transfer, the pressure distributions as function of altitude on the moon and on the earth do not differ greatly from those in the isothermal cases.

Figure 8 shows the comparison of temperature and gas density distributions with

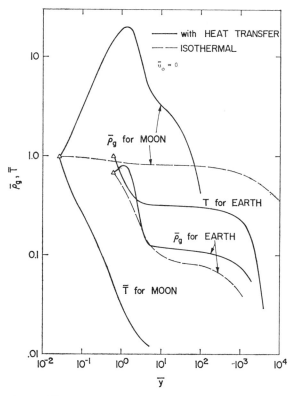

Fig. 8. Comparison of $\bar{T}$ and $\bar{p}_g$ with and without heat transfer and on moon and on earth.

and without heat transfer on the moon and on the earth for the case $\bar{u}_0 = 0$. The distributions of gas density on the moon with and without heat transfer bear no similarity to each other at all, while those on the earth do have roughly similar behavior. The temperature drop on the moon is much faster than that on the earth as shown in Fig. 8. As a result, for $\bar{u}_0 = 0$ the layer of hot ash is thicker on the earth than on the moon.

Figure 9 shows the comparison of gas velocity and volume fraction of solid particles at $\bar{u}_0 = 0$ on the moon with the same quantities on the earth. Again, the effects of heat transfer are much more significant on the moon than on the earth. The thickness of dense phase increases by 14 times due to the effect of heat transfer on the moon at $\bar{u}_0 = 0$, i.e., 3 m ($\bar{y} = 0.2$, $Z = 0.2$) in the isothermal case and 40 m ($\bar{y} = 2.7$, $Z = 0.2$) with heat transfer. On the other hand, the corresponding values for the thickness of dense phase on earth are 28.5 m for isothermal case (Pai *et al.*, 1972, gives 30 m) and 34 m for the case with heat transfer. Thus, we have only an increase of 1.2 times due to heat transfer.

In Fig. 10, comparisons of pressure, gas density, and temperature for $\bar{u}_0 = 8/G$,

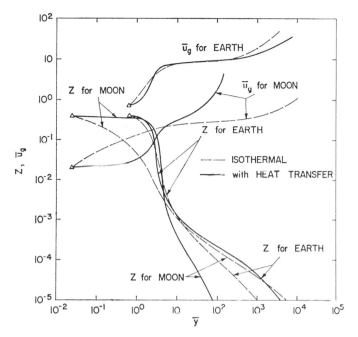

Fig. 9. Comparison of $\bar{u}_g$ and Z with and without heat transfer and on moon and on earth at $\bar{u}_0 = 0$.

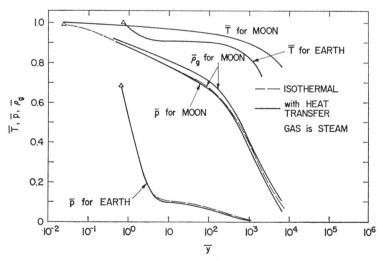

Fig. 10. Comparison of $\bar{p}$, $\bar{p}_g$, and $\bar{T}$ with and without heat transfer and on moon and on earth at $N = 8$ ($\bar{u}_0 = 8/G$).

i.e., $N = 8$, are shown. It is seen that for large N, the effect of heat transfer is small on the distributions of pressure and gas density, because the temperature variation for large N is small and we do not have such a large drop in temperature for $N = 8$ as in the case of $N = 0$. For $N = 8$, we also find that the effects of heat transfer on gas velocity and volume fraction of solid particles are insignificant. Hence we do not show these variations. In conclusion, for large initial velocity $\bar{u}_0 \geq 8/G$ or $N \geq 8$, the effects of heat transfer are small and isothermal approximation is good; for small initial velocity, however, when $N \leq 8$, the effects of heat transfer become important, particularly when $N \to 0$. On the moon, the effect of heat transfer at small initial velocity ($N \to 0$) is to make the whole ash flow layer more compact. The thickness of dense phase is increased considerably by the influence of heat transfer while that of the dilute phase is reduced. The direct consequences of the slow change of volume fraction of solid particles with height in the layer of dense phase on the moon for $N \to 0$ are that

(1) The pressure drop in the dense phase layer increases,

(2) The gas density also increases, and

(3) A steep vertical temperature gradient occurs.

(B) *The effects of different gas species on ash flow*

For small values of $\bar{u}_0$ or N, the flux of energy carried by the gas is relatively large in comparison with that carried by the pseudofluid of solid particles, while for large values of $\bar{u}_0$ or N, the flux of energy carried by the gas is relatively small in comparison with that carried by the pseudofluid of solid particles. Since gas is a compressible fluid, the effect of heat transfer on the gas is large. On the other hand, the pseudofluid of solid particles behaves as an incompressible fluid and thus the effect of heat transfer on the solid particles is small. As a result, we see in Part (A) of this section that the effect of heat transfer is large for small N while it is small for large N.

Since different gases have different specific heats or different heat capacities, the influence of heat transfer may be different for different gases. It is interesting to investigate how large would be the effects of heat transfer on the ash flow process for different kinds of gas species. In our previous numerical calculations, we considered only the case in which the gas is steam. Another possible fluidizing gas is hydrogen. Since the heat capacity of hydrogen is about six times larger than that of steam, we can investigate the principal effects of variation of gas species by carrying out some of our numerical calculations of ash flows with hydrogen as the gas species instead of steam.

Before we make the specific numerical calculations, it will be useful to consider the general effects on the ash flow that come from the physical properties of its gas species. The physical properties of the gas species which have some influence on the ash flow are: (i) the ratio of specific heat γ, (ii) the specific heat at constant pressure c_p, (iii) the mass of a molecule of the gas m, (iv) the coefficient of viscosity μ, and (v) the rate of emission of gas from the solid particles b. Among these five physical quantities γ, c_p, m, μ, and b, only the value of b is poorly known.

In order to compare the effects of physical properties of different gas species, we have to set up a criterion. Since in our study, we find that the Jeffreys number J_e, equation (19), is the most important parameter in the ash flow, we shall assume that the Jeffreys number J_e remains unchanged for different gas species when the other physical quantities remain unchanged. In other words, we may use the condition that $J_e = 37.5$ on earth to determine the value of b for different gas species. When the properties of solid particles, the gravitational acceleration and the initial temperature are the same, for a constant Jeffreys number, equation (19) gives

$$\frac{m}{\mu b} = \text{constant for any gas;} \qquad \text{or} \quad \frac{m_1}{\mu_1 b_1} = \frac{m_2}{\mu_2 b_2} \qquad (35)$$

where subscripts 1 and 2 denote the values for gases 1 and 2, respectively. Thus in equations (33) and (34), we have $b = 7.34 \times 10^{-7} \text{ sec}^{-1}$ for steam and $b = 1.25 \times 10^{-7} \text{ sec}^{-1}$ for hydrogen.

From equations (9), (20), (21), and (23), we have the following relations for the two different kinds of gas:

$$\frac{v_{c1}}{v_{c2}} = \frac{\mu_2}{\mu_1} ; \qquad \frac{G_1}{G_2} = \frac{m_2}{m_1} ; \qquad \frac{H_1}{H_2} = \frac{\mu_2^2}{\mu_1^2} ; \qquad \frac{\delta_1}{\delta_2} = \frac{c_{p2}}{c_{p1}} . \qquad (36)$$

Hence when we change the gas 1 in the ash flow to gas 2, we have to modify the reference velocity v_c, the density ratio G, the dynamic pressure parameter H, and the specific heat ratio δ in addition to the gas property parameter γ.

Since in our investigations, we consider only the cases of $H = 0$, then so long as we are considering the isothermal ash flow, we need to modify only the density ratio G. When the gas is hydrogen, from equation (36), we have

$$G_{\text{hydrogen}} = 9 G_{\text{steam}} \qquad (37)$$

because $m_2 = 18$ for steam and $m_1 = 2$ for hydrogen. From Pai et al. (1972), for the isothermal ash flow, the effect of G or the relative importance of the weight of gas becomes significant only for the dilute phase or for small values of Z (see Fig. 4 of Pai et al., 1972). Thus for hydrogen, we have a larger value of G than that for steam which implies that the effect of G comes at even smaller values for Z than those shown in Fig. 4 of Pai et al. (1972) for the case of steam. Since the isothermal approximation is poor for small values of $\bar{u}_0$ and an increase of G would influence the results in the region of small Z only, the change in the parameter G alone would give few new or interesting results for the isothermal ash flow. Hence we feel that we need not study in detail the results for the isothermal ash flow with hydrogen.

On the other hand, since the value of c_p for hydrogen is about six times greater than that of steam, we expect that some significant effects will occur due to the change from steam to hydrogen in the ash flow with heat transfer. Hence we recalculate the ash flow variables with hydrogen as the gas species for the case with heat transfer; the results are presented as follows:

Figure 11 shows the pressure distribution in the lunar ash flow with heat transfer as a function of altitude at various values of initial velocity ratio $\bar{u}_0$ (i.e., of N) with hydrogen as the gas species. We compare Fig. 11 with Fig. 2 which is for the case of steam. We find that there are significant differences between these results. For the case of steam, Fig. 2, the effects of heat transfer is not important for $N \geq 1$ and is important for $N \leq 1$. For the case of hydrogen, Fig. 11, the effects of heat transfer remain significant even if N is as high as 16. Qualitatively, the pressure curves for the case of hydrogen for all $N \leq 16$ behave similarly to those for the case of steam when $N \leq 1$. The reason is evidently that the larger the heat capacity of the gas is, the larger will be the relative influence of heat transfer in the mixture due to the contribution of the gas species. Thus the influence of heat transfer will extend to larger values of $\bar{u}_0$ of N for the gas with larger heat capacity.

Figure 12 shows the temperature distributions of the lunar ash flow with heat transfer as functions of altitude at various values of $\bar{u}_0$ or N with hydrogen as the gas species. It is evident that the isothermal approximation is not good even for $N = 16$, due to the large heat capacity of the hydrogen gas.

From the pressure and the temperature distributions, we calculate the density distribution of the gas which is given in Fig. 3 for both steam and hydrogen.

Figure 13 shows the corresponding distributions of velocities of the gas and the solid particles for the case of hydrogen that should be compared with the corresponding curves for the case of steam in Fig. 5.

Figure 14 shows the corresponding distributions of volume fraction for the case of hydrogen gas which should be compared with Fig. 6 for the case of steam. It is evident that as long as the heat transfer effects are important, the variations of the corresponding variables are similar. In other words, the curves of hydrogen for $N \leq 16$ are similar to those of steam for $N \leq 1$.

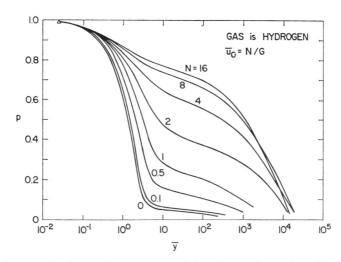

Fig. 11. Pressure distribution $\bar{p}$ as a function of altitude at various values of $\bar{u}_0$ on moon.

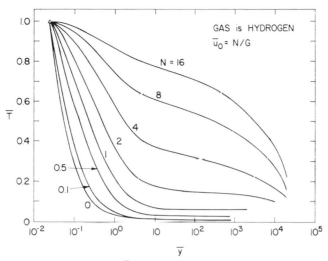

Fig. 12. Temperature distribution $\overline{T}$ as a function of altitude at various values of $\overline{u}_0$ on moon.

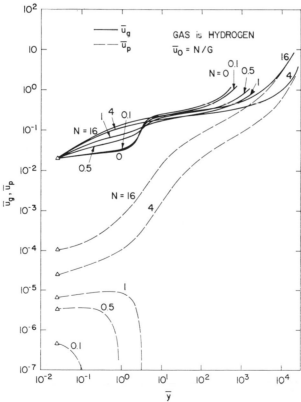

Fig. 13. Velocity distributions $\overline{u}_g$ and $\overline{u}_p$ as a function of altitude at various values of $\overline{u}_0$ on moon.

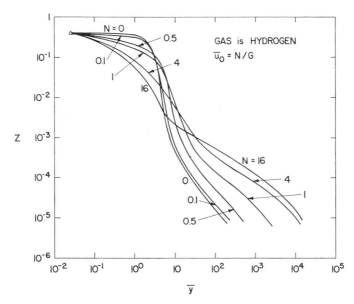

Fig. 14. Distribution of Z as a function of altitude at various values of $\bar{u}_0$ on moon.

VI. Summary and Conclusions

From our theoretical study and numerical results, the following conclusions may be drawn:

(1) A theoretical model of two phase flow of a mixture of gas and small solid particles is used in studying one dimensional steady ash flows for both the lunar and the terrestrial cases, with special emphasis on the lunar case. The fundamental equations and the boundary conditions are described. In general, there are seven unknowns: the temperature of the gas T_g, the temperature of the solid particles T_p, the velocity of the gas u_g, the velocity of the solid particles u_p, the pressure of the mixture p, the density of the gas ρ_g, and the volume fraction of the solid particles Z. For the ash flow under consideration, the temperature field is in local thermodynamic equilibrium so that $T_g = T_p = T$. We have to study six unknowns T, p, ρ_g, u_g, u_p, and Z only.

(2) From the fundamental equations and the boundary conditions of one dimensional steady ash flow, we find that there are seven nondimensional parameters that govern the ash flow; these are: (i) the Jeffreys number J_e, equation (19), which is one of the most important parameters in the ash flow process and which represents the ratio of gravitational force to viscous force; (ii) the density ratio G, equation (20), which represents the effect of the weight or gas in the ash flow; (iii) the dynamic pressure parameter H, equation (21), which represents the inertial effects in the ash flow; (iv) the initial velocity ratio $\bar{u}_0$, equation (18), which represents the effects of initial velocity u_0 of the ash relative to the characteristic velocity of emission v_c, equation (9), from the solid particles in the ash flow, (v) the gas property parameter γ, equation (22), which is the ratio of the specific heat at constant pressure to that at

constant volume; it is related to the complexity of the molecules of the gas, (vi) the specific heat ratio δ, equation (23), which is the ratio of the specific heat of the solid particles c to that of the gas at constant pressure c_p; it indicates the relation in heat capacity between the pseudofluid of solid particles and the gas; and (vii) the Prandtl number P_r, equation (24), which is important in the study of heat conduction in the mixture and heat transfer between species. In our study, we consider the cases where the inertial effects are negligible and the species in the mixture are in local thermodynamic equilibrium. Hence the parameters H and P_r will be dropped out in our analysis. We study mainly the influence of the five parameters J_e, G, $\bar{u}_0$, γ, and δ.

(3) Analytical solutions of the fundamental equations with and without heat transfer have been obtained for the dense phase region of the ash flow (Pai et al., 1972). Numerical solutions for the whole flow field of the ash flow have been obtained for both the isothermal ash flow and the ash flow with heat transfer. In the isothermal case, the effect of different gas species is small. For the ash flow with heat transfer, the heat capacity of the gas species has a large influence on the ash flow. Hence we calculate two cases of lunar ash flow: one for steam as the gas and the other for hydrogen as the gas and compare these two sets of results. The essential results for these two cases are given below:

(4) For the ash flow with heat transfer, we first reexamine our isothermal results for steam (Pai et al., 1972) with heat transfer. The most important heat transfer process in the ash flow is heat convection of the mixture. Our numerical results, given in Figs. 2 to 10, shows that the effect of heat transfer becomes important for the cases of small initial velocity $\bar{u}_0$. For the case of steam as the gas species, the effect of heat transfer is very small if $\bar{u}_0 \geq 8/G$, i.e., $N \geq 8$, but for $N < 1$, especially for $N \to 0$, the influence of heat transfer is very large. As $N \to 0$, the effect of heat transfer is to make the ash flow more like a single compact layer on the moon. The thickness of the layer of lunar ash flow, which is in the dense phase, is increased considerably by the influence of heat transfer while that of the layer which is in the dilute phase is reduced. The pressure drop in the dense phase of lunar ash increases. Most surprisingly, the gas density also increases until a maximum is reached and then decreases due to the effect of heat transfer.

(5) Since the effect of heat transfer depends greatly on the heat capacity of the gas species in the ash flow, we reexamine our numerical results for steam by using the hydrogen as the gas species for the lunar ash flow. Since the heat capacity of hydrogen is about 6 times that of steam, we do find that the effect of heat transfer with hydrogen is much larger than that with steam. Qualitatively, the effect of heat transfer still depends on the initial velocity $\bar{u}_0$. The larger the values of $\bar{u}_0$, the smaller the influence of heat transfer if other parameters remain unchanged. But for steam, we find that when $\bar{u}_0 \geq 8/G$ or $N \geq 8$, the effect of heat transfer is negligibly small. However, for hydrogen, the effect of heat transfer is not negligible even if $N = 16$.

VII. Applications

The most interesting result from this study is the existence of a sharply defined upper limit to the ash flow, even for a lunar ash flow, and even while the flow is still fluidized. A mark was observed on Mt. Hadley by the Apollo 15 investigators (Howard *et al.*, 1972), and interpreted by them as a sort of high-water mark. Similar high-water marks are observed above terrestrial ash flows.

A consequence of this sharp upper limit is increased credibility for the idea of Cameron (1964) and Schumm (1970) that lunar sinuous rilles such as the Hadley Rille are carved by ash flows. Earlier suggestions by various authors (Peale *et al.*, 1968, Urey, 1968, Burke *et al.*, 1970) to the effect that sinuous rilles were carved by running water became implausible when it became clear that the moon has been extremely dry since the formation of the maria rocks (LSAPT, 1970). Greeley (1971) suggested that the rilles were collapsed lava tubes, but this suggestion does not fit well with the observed thickness (only about 10 meters) of the rock layers in the walls of the Hadley Rille (Howard *et al.*, 1972). The isothermal calculation (Pai *et al.*, 1972) was not particularly favorable to an origin of the sinuous rilles by ash flows because it indicated very great pressure scale heights (tens of kilometers). This suggested that most of the gas, at least, would overflow the banks of a normal sinuous rille, although the actual solid particles were found to be somewhat concentrated toward the base. The new calculations with heat transfer show that in actuality, in a lunar ash flow, both gas and solid particles are largely confined to a layer only a few tens of meters in thickness. Since the Hadley Rille, for example, is about 400 meters deep, it is clear that a lunar ash flow could be confined within it.

Acknowledgment—This research is supported in part by the National Aeronautics and Space Administration under NASA Grant No. NGR-21-002-296 to the University of Maryland. The computer time was supported by NASA Grant NSG-398 to the Computer Science Center of the University of Maryland.

References

Burke J. D., Brenton R. G., and Muller P. M. (1970) Desert stream channels resembling lunar sinuous rilles. *Nature* **225**, 1234–1236.

Cameron W. S. (1964) An interpretation of Schröter's Valley and other lunar sinuous rilles. *J. Geophys. Res.* **69**, 2423–2430.

Greeley R. (1971) Lava tubes and channels in the Marius Hills. *The Moon, 3*, 284–314.

Green J. (1963) Some lunar resources, Proc. Lunar and Planetary Exploration Colloquium, 3 [3], 82–95, North American Aviation, Downey, California.

Howard K. A., Head J. W., and Swaun G. A. (1972) Geology of Hadley rille (abstract). In *Lunar Science—III*, pp. 353–354, Lunar Science Institute Contribution No. 88.

LSAPT (1970) Lunar Sample Analysis Planning Team, Summary of Apollo 11 Lunar Science Conference, *Science* **167**, 449–451.

McKay D. S., Clanton U. S., Heiken G. H., Morrison D. A., Taylor R. M., and Ladle, G. (1972) Vapor phase crystallization in Apollo 14 breccias, and size analysis of Apollo 14 soils (abstract). In *Lunar Science—III*, p. 95, Lunar Science Institute Contribution No. 88.

O'Keefe J. A. and Adams E. W. (1965) Tektite structure and lunar ash flow. *J. Geophys. Res.* **70**, No. 16, 3819–3829.

Pai S. I. (1966) Radiation Gasdynamics, Springer Verlag, Vienna and New York.

Pai S. I. (1971) A review of fundamental equations of the mixture of a gas with small particles, Tech. Note BN-668, Institute for Fluid Dynamics and Applied Mathematics, Univ. of Maryland, 1970, also *Zeit. f. Flugw.* **19**, Heft 8/9, pp. 353–360.

Pai S. I. and Hsieh T. (1971) One-dimensional lunar ash flow with and without heat transfer, Tech. Note BN-718, Institute for Fluid Dynamics and Applied Mathematics, Univ. of Maryland.

Pai S. I., Hsieh T., and O'Keefe J. A. (1972) Lunar ash flow: the isothermal approximation, to be published in *J. Geophys. Res.*

Peale S. J., Schubert G., and Lingenfeltes R. E. (1968) Distribution of sinuous rilles and water on the moon. *Nature* **220,** 1222–1225.

Ross C. S. and Smith R. L. (1961) Ash flow tuffs: their origin, geologie relation and identification. *Geological Survey Professional Paper 366*, Gov't Printing Office, Washington, D.C.

Schumm S. A. (1970) Experimental studies on the formation of lunar surface features by fluidization. *Bull. Geol. Soc. Am.* **81,** 2539–2552.

Urey, A. C. (1967) Water on the moon, *Nature* **216,** 1094–1095.

Proceedings of the Third Lunar Science Conference
(Supplement 3, *Geochimica et Cosmochimica Acta*)
Vol. 3, pp. 2713–2734
The M.I.T. Press, 1972

Effects of microcratering on the lunar surface

Donald E. Gault*

Max-Planck-Institut für Kernphysik,
Heidelberg, Germany

and

Friedrich Hörz and Jack B. Hartung

National Aeronautics and Space Administration, Manned Spacecraft Center,
Houston, Texas

Abstract—Based on new laboratory impact data and current "best" estimates of the lunar micrometeoroid flux, calculations have been made of (1) the survival times of rocks on the lunar surface before they are catastrophically ruptured by meteoritic impacts; and (2) the rate of mass wasting by single particle abrasion. The calculated results are in systematic disagreement with observations in a direction suggesting that the current micrometeoroid flux may be greater (factor of 10?) than the long-term average integrated over the past several tens (?) of millions of years. Most of the mass in the micrometeoroid flux is concentrated in particles with masses between 10^{-8} to 10^{-4} g, and it is shown to constitute a major geologic agent for erosion, mass transport of material across the lunar surface, and the source for fused and vaporized products in the regolith.

Introduction

Microcraters observed on lunar rocks and regolith fragments have provided a wealth of information and data that has opened new avenues for investigating lunar evolutionary processes and the meteoritic environment over extended periods of time. These ubiquitous features, which range in size from less than a micrometer to more than a centimeter, have been described in great detail by several groups of investigators (McKay *et al.*, 1970; Carter and McGregor, 1970; Neukum *et al.*, 1970; Bloch *et al.*, 1971; Hörz *et al.*, 1971; Morrison *et al.*, 1971; and others). Contrary to some suggestions (McKay *et al.*, 1970) we believe the craters were formed almost exclusively by primary events and, based on this judgment, explore in this paper some of the effects and consequences of microcratering on the lunar surface by the impact of micrometeoroids.

Because the effects of the micrometeoroid bombardment and their subsequent interpretation are strongly dependent on the assumed flux, we discuss first the development of a "best" estimate of the current lunar micrometeoroid flux. This flux, in combination with new data from laboratory impact studies, is then applied in subsequent sections to the quantitative evaluation of the effects of impacts against the lunar rocks and regolith, specifically (1) the erosion of rocks by catastrophic rupture

---

* Permanent address: Planetology Branch, NASA, Ames Research Center, Moffett Field, California 94035.

and small particle abrasion, and (2) the melting and vaporization produced by impacts into rocks and regolith in terms of both single events and the integrated effects of multiple impacts over long (geologic) periods of time.

METEOROID FLUX

Catastrophic rupture and abrasion of rocks on the lunar surface (and meteorites in space) and the "gardening" of the regolith are a direct function of the mass flux, velocity, and density distribution of the impacting bodies. Until recently when fluxes were derived from crater statistics on lunar rocks (Hartung et al., 1972; Neukum et al., 1972; Morrison et al., 1972) and the concentration of meteoritic components in the regolith (Laul et al., 1971; Dohnanyi, 1971), the available data on the meteoritic flux was confined to earth-based observation of meteors and zodiacal light, and from direct measurements on board spacecraft. Although fluxes derived from crater statistics holds great promise for future development, the most reliable information is still the accumulation of results from the terrestrial observations and spacecraft experiments. These latter results and their interpretation, however, are frequently contradictory, and one finds a wide assortment of meteoritic models and fluxes suggested by the various workers in the field. Three recent reviews of the meteoritic environment have appeared recently (Kerridge, 1970; Soberman, 1971; McDonnell, 1971) that, while agreeing in broad general terms, do not provide a quantitative model for a flux. These do provide a basis for selecting a model, and we have after a detailed review of the reviews and additional literature on the subject selected a "best" estimate for the mass-flux distribution and average values to use for the impact velocity and micrometeoroid mass density.

Data selected to establish the mass distribution and flux of the micrometeoroid environment for masses less than 10^{-6} g were taken from measurements returned from sensors flown on earth satellites and interplanetary probes; ground-based observations of the zodiacal light and gegenschein are omitted, because their interpretation is too model dependent. All data were required to fulfill three criteria that are similar but were applied in a more restrictive manner than those invoked by Kerridge (1970). First, the original data must have been from an experiment that gave a positive and unambiguous signal of a micrometeoroid event. Penetration of a thin sheet or membrane is considered to fulfill this requirement provided that reliable means of recording the penetration was provided. Simple microphone sensors were rejected because of their apparent contamination by spurious thermal, mechanical, and solar flare "noise" (Nilsson, 1966; Berg and Richardson, 1971). On the other hand, controlled microphone experiments that included a coincident signal of an impact event by a totally independent technique are accepted as a reliable source of data. Second, it was required that the experiment must have recorded sufficient micrometeoroid events to provide a statistically significant result. This requirement eliminates all results that are based on no events or, perhaps, a single event; such results furnish only questionable limiting values for the flux and are useless in defining a quantitative expression for mass distribution. Third, the response and sensitivity of the sensors must have been calibrated using hypervelocity impact facilities in order to provide, in so far as currently feasible with existing capabilities, a simulation of micrometeoroid impact events.

Review of the literature reveals that only 10 data points from six experiments meet these criteria. These results are presented in Table 1 together with explanations of their source and derivation for the convenience of other workers. Bulk density of the particles is assumed to be 1 g/cm^3 rather than 0.5 g/cm^3, which has been employed previously for evaluating masses for penetration sensors (Naumann et al., 1969) and has been adopted in the NASA model (Anon., 1969). The results are shown graphically in Fig. 1 and are compared with the NASA model, the power-law expressions adopted for calculations herein, and flux curves derived by Hartung et al. (1972) and Neukum et al. (1972) using the lunar rocks as micrometeoroid detectors. In Fig. 1 all near-earth fluxes from Table 1 have been decreased by a factor of 2 (Grew and Gurtler, 1971) as an approximate correction to the lunar environment. The conversion to a particle density of 1 g/cm^3 is based on a new analysis of photographic meteors by Givens (1972) and on recent results from the Prairie Network by McCrosky et al. (1971) and McCrosky (1967), indicating that meteor densities probably have been seriously underestimated in the past. These new results support earlier suggestions of

Table 1. Mass-flux data adopted from spacecraft measurements.

Spacecraft	Type of sensor	Log_{10} cumulative flux (N/cm²-sec-2π strd.)	Log_{10} mass (g)	Notes and references
Pegasus I, II, III	Penetration			
	0.038 mm Al	−9.171	−7.640	(a), (g), (h)
	0.2 mm Al	−10.184	−6.809	(1), (2), (3), (8)
	0.4 mm Al	−10.802	−6.189	
Explorer XVI	Penetration			
	0.025 mm Be–Cu	−9.135	−8.267	(b), (g), (i)
	0.05 mm Be–Cu	−9.489	−7.692	(2), (3), (8)
Explorer XXIII	Penetration			
	0.025 mm stainless steel	−9.129	−8.301	(c), (g), (h)
	0.05 mm stainless steel	−9.436	−7.726	(2), (3), (4), (8)
Lunar Orbiter I–V	Penetration			
	0.025 mm Be–Cu	−9.515	−8.301	(d), (g), (i)
				(2), (3), (5), (8)
Lunar explorer 35	Capacitor-microphone	−8.7	−10	(e), (6)
Pioneer 8, 9	Time-of-flight microphone	−7.62	−12	(f), (7)

(a) Flux from (1); mass from (2).
(b) Flux from (3); mass from (2).
(c) Flux from (4); mass from (2).
(d) Flux from (5); mass from (2).
(e) Flux and mass from (6).
(f) Flux and mass from (7).
(g) Flux corrected from penetration- to encounter-frequency as given in (3) and (8) for "fit II", except correction for 0.038 mm Al, ghich was taken to be a factor of 2.
(h) Masses converted to particles with density of 1.0 g/cm^3 with velocity of 20 km/sec using penetration equation in (2).
(i) Masses converted to particles with density of 1.0 g/cm^3 with velocity of 20 km/sec and corrected for material property differences using penetration equation in (2).

(1) Naumann (1972); (2) Naumann et al. (1969); (3) Naumann (1966); (4) O'Neal (1968); (5) Grew and Gurtler (1971); (6) Alexander et al. (1971); (7) Berg and Richardson (1971); (8) Anon., NASA TM X-53629 (1967).

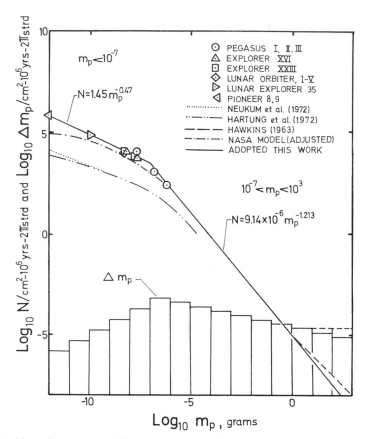

Fig. 1. The micrometeoroid flux measurements from spacecraft experiments which were selected to define the mass-flux distribution. Also shown is the incremental mass flux contained within each decade of m_p.

higher densities that were advanced by Allen and Baldwin (1967) and Ceplacha (1967).

The exclusion of certain data employed by others warrants some explanation. The Gemini S-10 experiment (Hemenway *et al.*, 1968) and the Gemini window data (Zook *et al.*, 1970) are not included because of the problems introduced by surface contamination obscuring craters and poor statistics. These Gemini data, however, agree well with the tabulated results and the derived power-law expression. All OGO data (Nilsson *et al.*, 1969; Alexander *et al.*, 1969; Arthur *et al.*, 1971) has been excluded due to both poor statistics and the possibility of spurious signals on the first detector (thin film) of the time-of-flight sensor system. The Russian Cosmos series of experiments were omitted because of the lack of hypervelocity-impact calibration and the lack of coincident means for detecting the events recorded by the microphone sensors. It is noteworthy, however, that unlike most of the microphone data the Cosmos 135 and 163 yielded very low fluxes in good agreement with the adopted distributions (Konstantinov *et al.*, 1968; Konstantinov *et al.*, 1969).

For masses between 10^{-6} and 10^0 g, recent analyses of photographic meteor data (Naumann, 1966; Lindblad, 1967; Erickson, 1968; Dalton, 1969; Dohnanyi, 1970; Dohnanyi, 1971) indicate with surprising agreement that the slope (in the log-log presentation) has a value of approximately -1.2 and that the terrestrial influx of 1 g objects is in the range from $10^{-17.7}$ to $10^{-18.3}$ per $cm^2/sec/2\pi$ strd. A good representation of these results is provided by the NASA meteoroid model (Anon., 1969) which has a slope of -1.213 and a 1 g influx rate of approximately $10^{-18.1}$ per $cm^2/sec/2\pi$ strd. *after correction from penetration to encounter frequency* (Naumann, 1966; Anon., 1967). The NASA model is adopted, therefore, and modified for the lunar environment for the present application; modifications are a factor of 2 decrease for the reduced gravitational field of the moon (Grew and Gurtler, 1971), an additional decrease of a factor of 1.3 for a change of particle density from 0.5 to 1 g/cm^3 (Naumann *et al.*, 1969), and a factor of 1.9 increase for the correction to encounter frequency. Extension of this modified NASA model to masses smaller than 10^{-6} g gives excellent agreement with the Pegasus penetration data.

For masses greater than 1 g, Hawkin's (1963) flux with a factor of 2 decrease is shown on Fig. 1 to indicate that the slope probably tends toward a value of -1. Although the Hawkin's flux provides an almost perfect extension of the NASA model from 1 g to larger masses, the lunar seismic data has been interpreted by Latham *et al.* (1971) to indicate that the flux of these larger bodies is a factor of 20 less than the Hawkin's values. On the other hand, results from the Prairie Network (McCrosky, 1968; McCrosky *et al.*, 1971) suggest that the flux is 10^2 and 10^3 greater than the seismic interpretations for, respectively, 10^3 and 10^6 g bodies. Both the seismic and Prairie Network results apparently suffer from inadequate "calibration" and the extension of the NASA model to 10^3 g as a compromise flux is merely a convenience.

Although the mass distributions from recent determinations from photographic meteors are in good agreement, interpretations of velocity distributions vary widely with average velocities from 16.5–22 km/sec and root-mean-square velocities, which are required because impact processes (ejected mass, fusion, etc.) are a function of projectile kinetic energy, from 17.5–25 km/sec. In addition, there are indications that the average velocities are dependent on the particle mass, and there are divergent interpretations of the correction to the lunar environment. Because of these uncertainties, a root-mean-square impact velocity $Vi = 20$ km/sec has been assumed in all calculations and, depending on the velocity distribution, corresponds to an average impact velocity of 18–19 km/sec.

To summarize, the mass distribution for the lunar meteoritic flux of N particles (per $cm^2/10^6$ yr/2π strd.) of mass m_p (g) is taken to be

$$N = 1.45 \, m_p^{\,0.47} \qquad \text{for} \qquad 10^{-13} \leq m_p \leq 10^{-7}$$
$$N = 9.14 \times 10^{-6} \, m_p^{\,1.213} \qquad \text{for} \qquad 10^{-7} \leq m_p \leq 10^3$$

with particles of density of 1 g/cm^3 impacting with a root-mean-square velocity $V_i = 20$ km/sec. The use of two exponential expressions with the resultant discontinuity is, of course, an artificial representation for the flux, but it is introduced for mathematical simplicity.

D. E. GAULT, F. HÖRZ, and J. B. HARTUNG

The incremental mass flux m_p in each decade of particle mass also is shown on Fig. 1 and illustrates that most of the mass impacting the lunar surface is concentrated in masses of the order of 10^{-6} g with relatively insignificant contributions from masses less than 10^{-9} g nor greater than a 10^{0} g. This predominance of a narrow range of the small particle masses has been pointed out previously (Zook *et al.*, 1970; Hartung *et al.*, 1972) although the values for the dominant mass range differ slightly because of the different mass distributions assumed. It should be noted, however, that, within the uncertainties of the mass distributions of large objects, the cumulative mass contributed by bodies larger than, say, 10^{10} g may be greater than for the micrometeoroids. Such large objects have obviously had a major effect upon the lunar surface, but their effect must be considered an intermittent mass pulse rather than the "steady" rain of the smaller more numerous particles. But it is important to note that the cumulative mass contained in this steady bombardment amounts to approximately 2×10^{-3} g/cm²-10^6 yr between the limits of 10^{-12} to 10^{0} g meteoroid masses. This influx rate onto the moon is remarkably similar to the 3.8×10^{-3} g/cm²-10^6 yr estimated from trace element studies by Laul *et al.* (1971) to be a primative meteoritic component of Type C1 carbonaceous chondrite in the regolith. The dominant contribution of the micrometeoroid masses must be considered the primary source of such material in the regolith, and because of the apparent generic relationship between meteors and comets, one is led to suggest that Type C1 meteorites probably have a cometary origin.

ROCK 14310

Fig. 2. Gradual mass wasting by single particle abrasion has rounded the upper portion of rock 14310, which was exposed to the meteoritic environment while the lower portion was buried in the regolith and retained its angular faces.

EROSION PROCESSES

Rocks and clods exposed on the lunar surface are subjected to two degrees of meteoritic degradation; (1) the violent and catastrophic disruption by particles which are capable of forming a crater with a diameter that is a significant fraction of the rock's dimensions; and (2) the slow abrasion by smaller particles that remove a minute mass per impact event. Examples of these two processes are shown in Figs. 2, 3, and 4. The crystalline rock 14310 is shown in Fig. 2 and illustrates the gradual mass wasting by single particle abrasion on an exposed surface. The lower half, buried in the regolith, was shielded from the meteoritic environment and retained its angular

2 cm

Fig. 3. A glass-lined pit, surrounded by its spall area, records the location of an impact on rock 14305 that almost ruptured the rock. Note deep fracture cutting across the spalled region.

Fig. 4. Three large boulders photographed on the lunar surface during the Apollo 14 mission that exhibit fractures which appear to have been produced subsequent to the emplacement of the rocks in their present position.

features that are undoubtedly the result of being broken or spalled off a larger rock. An example of a near catastrophic impact against the breccia 14305 is given as Fig. 3. Here the large, almost 1 cm, glass-lined pit is surrounded by a larger spall area and clearly marks the point of impact of a hypervelocity micrometeoroid. The impact was sufficiently energetic to fracture the rock in several locations, one of which may be seen cutting across the spalled area surrounding the glass pit. A slightly more energetic impact would have ruptured the rock into two or more smaller pieces. Possible examples of partial rupturing are shown in Fig. 4 where three boulders are visible with prominate cracks cutting through them. Although the cracks are not demonstrably attributable to impact, the overall roundness of the all three blocks in contrast to the sharp corners at the fracture surfaces strongly suggests that the splitting occurred *after* the rock was located in place on the surface. On this basis, impacts would appear to be the most direct explanation for the splitting.

The difference between slow abrasion and rupture were first considered by Gault (1969) and subsequently by other investigators (Shoemaker *et al.*, 1970; Hörz *et al.*, 1971; McDonnell and Ashworth, 1971). In the following sections we re-evaluate these two processes based on our assumed "best" estimate flux and, importantly, with the help of new laboratory impact data.

Catastrophic rupture

The mean exposure time on the lunar surface t_{rm} before a rock is shattered by the cumulative effects of impacting meteoritic particles has been calculated using the equation presented by Gault and Wedekind (1969). The original equation, however, has been modified to include the effects of rock strength as determined by recent laboratory results. This modification replaces the cumulative kinetic energy of the impacts that is required for rupture E_R (originally taken to be a constant with a value of 10^7 erg/g) with the expression

$$E_{RS} = 2.5 \times 10^5 \, S_c r^{-0.225}$$

where S_c is the unconfined compressive strength in kilobars (1 bar $= 10^6$ dynes/cm^2) and r is the radius in centimeters of an assumed spherical body. The expression is based on measurements varying S_c from 0.5 bars to 2.5 kbar. Implicit in this relationship is the assumption that the effective strength varies inversely with $r^{-0.5}$ as predicted theoretically (Griffith, 1925; Weibull, 1939) and demonstrated experimentally (Lundborg, 1967).

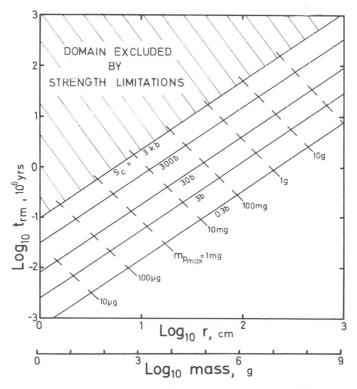

Fig. 5. Calculated mean residence time before destruction by catastrophic rupture t_{rm} for spherical rocks of radius r and with compressive strength S_c that are exposed on the lunar surface to the meteoritic environment. Masses of the largest particles $m_{p_{max}}$ that are considered to impact the rocks and contribute to the destruction process are indicated.

The calculated mean times of exposure on the lunar surface before rupture should occur are presented in Fig. 5; also indicated are the masses of the largest particles that are considered to have impacted the rocks before rupture occurred (the smallest masses considered effective are one-tenth of the maximum values).

With a limiting value of 3 kbar as an absolute upper limit for S_c for dense crystalline rocks, the maximum surface exposure times for rocks in the mass range 10^2 to 10^4 g, which are representative of the returned lunar samples, are from 7×10^5 to 6×10^6 yr. For each factor of 10 decrease in S_c the corresponding ages decrease by a factor of about 3.5.

Direct comparison is given in Table 2 between the calculated values and surface residence times determined from particle tracks for 16 rocks from Apollo 11 and 12. It can be seen that in general the calculated values are approximately an order of magnitude lower than the observed values. Some bias in the observed values toward higher values is to be expected because only those rocks that have survived could have been collected. Additionally, the rocks may be fragments broken off from larger rocks and contain a track record incurred as a surface piece from the larger rock. These effects should be small, less than a factor of 2, and excluding the unacceptable possibility that the track ages are grossly in error, the most likely sources for the disagreement are in the values used for E_{RS}, V_i, and the assumed meteoritic flux. Elimination of the difference being caused by E_{RS}, however, would require that the effective strength of lunar rocks must be of the order of 300 kbar, a value far in excess of high-strength steels. Similarly, the requisite change in V_i is unrealistic by requiring a reduction to about 6 or 7 km/sec in the root-mean-square impact velocity. The most acceptable change is the reduction of the meteoritic flux by a factor of about 10, and we delay discussion of this possibility until after abrasion by single particle impact has been considered.

Single particle abrasion

Mass wasting by single particle abrasion was calculated using new laboratory data to empirically formulate the mass of dense crystalline rock that is displaced per impact event. The functional form of the derived relationship is

$$M_e = k_1 E^\delta (\cos i)^2$$

Table 2. Calculated mean survival times for rocks exposed on the lunar surface compared to observed surface residence times determined from particle tracks.

Mass (g)	Number of rocks	Calculated* time, t_{rm} ($\times 10^6$ yr)	Average[†] residence time ($\times 10^6$ yr)	Range of[†] residence times ($\times 10^6$ yr)
53–350	6	0.4–1.2	8	1–16
350–10^3	5	1.2–2	14	1.3–32
10^3–2.5×10^3	5	2–3	11	1.5–24

* Calculations assume a strength of 3 kbar.
[†] Crozaz *et al.* (1970, 1971); Fleischer *et al.* (1970, 1971); Lal *et al.* (1970); Price and O'Sullivan (1970); Bhandari *et al.* (1971).

where M_e is the displaced mass in grams, E is the kinetic energy in ergs, δ is a constant, i is the angle of impact measured relative to the surface of the target, and k_1 is

$$k_1 = (\text{constant}) \times [\rho_p/\rho_t]^{1/2}$$

with ρ_p and ρ_t the mass densities in g/cm^3 of, respectively, the projectile and target. Because the differential probability of impact dP at an angle i is (Shoemaker, 1962)

$$dP = 2\,(\sin i)\,(\cos i)\,di$$

integration over all possible angles of impact leads to a correction factor F_0 for obliquity in the expression for M_e given above. In addition a correction factor F_s should be introduced to take into account the fact that only the top of a rock is exposed to 2π strd. while the sides, somewhat dependent on geometry, are limited to approximately π strd. On the average, therefore, a rock has a 1.5 π strd. exposure.

With these corrections the expression for M_e on the lunar surface can be rewritten

$$M_e = k_1 F_0 F_s E^\delta$$

When this relationship is combined with the flux equation expressed in the generalized form

$$N = k_2 m_p^\gamma$$

one obtains in a straightforward manner

$$h_m = \left(\frac{\gamma}{\gamma+1}\right)(\tfrac{1}{2}V_i^2)^\delta \left[\frac{k_1 k_2}{\rho_t}\right] F_0 F_s [m_1^{\gamma+\delta} - m_2^{\gamma+\delta}]$$

where h_m is the mean rate of abrasion in centimeters per 10^6 yr produced by impacts of meteoritic particles having masses between the limits of m_1 and m_2. Based on data from impacts of polythene, pyrex, aluminum, and iron into basalts and granites having densities from 2.6 to 2.86 g/cm^3, a least-squares fit to 64 data points covering an energy range from approximately 10 to 10^{12} erg gives values of $\delta = 1.13$ and the constant $k_1 = 8.63 \times 10^{-11}$ (Gault, 1972). F_0 and F_s were taken to be, respectively, 0.5 and 0.75; ρ_p and ρ_t were taken to be, respectively, 1 and 3.

The results of the calculation are shown in Fig. 6. Because on the average the largest micrometeoroid to impact a surface in a given time is a direct function of the exposed area of the object, the mean abrasion rate is a function of the size of the rock. There is, however, a maximum size particle that can impact without rupture. This limiting condition, in fact, limits the abrasion rates that can be realized.

Abrasion rates from 1 to 2 mm/10^6 yr (m.y.) are indicated by the calculations: In contrast, first results from studies of the particle track records in lunar rocks indicated that the erosion rate is between 0.1 and 1 mm/m.y. with the latter value a firm upper limit (Crozaz et al., 1970; Fleischer et al., 1970; Price and O'Sullivan, 1970). Subsequent studies, however, have shown that the upper limit for the mean erosion rate is probably much lower. Fleischer et al. (1971a, 1971b) suggest that the rate is between zero and 0.1 or 0.2 mm/m.y. Barber et al. (1971) give 0.3 mm/m.y. while Finkel et al. (1971) suggest 0.5 mm/m.y. On the other hand, Crozaz et al. (1972) still maintain the 1 mm/m.y. as an upper limit, but they also give 0.3 mm/m.y. as an

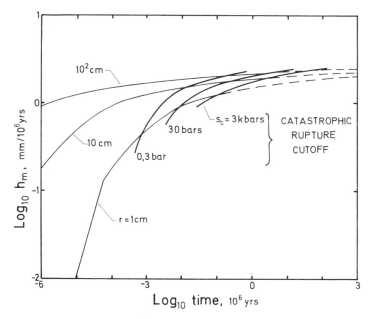

Fig. 6. Calculated temporal variation in the mean abrasion rates h_m by single particle impacts for spherical rocks of radius r exposed on the surface to the meteoritic environment. Limits due to destruction of rocks by catastrophic rupture are indicated for rocks of varying compressive strength S_c.

acceptable lower limit. Thus, as for the rupture calculations, the "concensus" of the observations approaches an order of magnitude difference which can be ameliorated by changes in the values of three major parameters k_1, k_2, or V_i. Although we cannot demonstrate that the empirical mass ejection expression for M_e is beyond question, it is awkward to try to rationalize how it could be in error by a factor of 10. An unacceptable reduction in the effective root-mean-square velocity V_i from 20 to 6 or 7 km/sec would also bring the abrasion rates into better agreement with the observations (as it would for the rupture times) so that we are left, again, with the most plausible choice for the difference being that the assumed flux is too high by a factor approaching an order of magnitude.

To first appearances it might seem that the results for rupture and abrasion are closely related, but it must be remembered that the data forming the basis of the two calculations are completely independent of each other. Although both processes are direct functions of the projectile kinetic energy, two totally different physical processes are involved. In the case of rupture the bodies are torn apart by tensile failures induced by the reflection of stress waves from free surfaces. In contrast for the cratering, crushing, and shearing play major roles in the comminution leading to the formation of the craters. Moreover, it is important to note that most of the abrasion is carried out by particles in the 10^{-6} range whereas rupture for typical lunar samples involves particles masses 3 to 5 orders of magnitude larger. The fluxes for these two ranges of mass have been determined by two totally different techniques, and it seems unlikely that the two techniques would both error in the same direction by the same amount. Although some of the differences may be attributable to weaknesses in the collective choice of values for the important parameters, we note that the fluxes derived from the lunar rocks by Hartung *et al.* (1972) and Neukum *et al.* (1972) yield abrasion rates from 0.1 to 0.3 mm/m.y. in excellent agreement with the observed values (similar calculations for rupture are precluded by the cut-off at 10^{-5} g in the flux derived from the lunar rocks). There are, of course, uncertainties in the fluxes derived from the lunar rocks, but the correlation between derived fluxes and the calculated and observed abrasion rate is impressive. We see no alternative at this time to the conclusion that the major reason for differences in abrasion rates and rupture times between observations and calculations based on the "best" estimate flux from satellite measurements is that the current meteoritic flux is significantly enhanced above the long-term average integrated over the past several tens of millions of years.

CRATERING IN THE REGOLITH

With the major fraction of the meteoritic kinetic energy deposited against the lunar surface by microgram masses, the mass transport, fusion, vaporization, etc. processes must be dominated by these microparticle events and play the major role in the geologic development of the upper surface. Until recently the effects of microparticle impacts into low density particulate targets representative of the outer few millimeters of the regolith were limited to uncertain extrapolations from macroscale experiments (Gault *et al.*, 1966). Vedder (1972) has now reported studies of craters formed in mineral dusts by hypervelocity microparticles that furnish valuable insight into this important aspect of the lunar environment. We will apply some of Vedder's results in the following sections to explore in a cursory manner the rates of mass transport, mixing rates, and the production of fused and vaporized material by micrometeoroid impact.

Mass transport and mixing

Vedder (1972) has studied craters formed in weakly cohesive basaltic powders of low density ($\rho = 0.65$ and 1.1 g/cm^3) by the impacts of polystyrene projectiles ($\rho = 1.06$ g/cm^3) with velocities up to 12.5 km/sec. Projectile diameter varied from 2 to 5 μ, comparable to the maximum grain size of the basaltic powders. The average crater-to-projectile diameter ratio is about 25 with a cratering efficiency (displaced mass/unit kinetic energy) of approximately 10^{-8} g/erg for normal incidence. It is significant that recent results from centimeter size projectiles impacting weakly cohesive basaltic fragments give very nearly the same results, differing only slightly in crater-projectile diameter ratio (30–35), and cratering efficiency (1 to 2 × 10^{-8} g/erg). Moreover, although the macroscale craters exhibit raised rims (Gault *et al.*, 1966) the basic shape of the inner craters are similar. Thus we now have strong evidence that cratering over the range of diameters of direct interest to the evolution of the upper few centimeters of the lunar surface is essentially the same over a diameter range from the order of 10 μ to several tens of centimeters.

Based on these macro- and microscale data we have calculated the incremental masses ΔM_e in each decade of m_p and the cumulative mass ΣM_e (summed up from the smallest particles) that will be produced by the meteoritic flux. In these calculations we have used a cratering efficiency of 10^{-8} g/erg (effectively assuming that $\delta = 1$) and an impact velocity of 20 km/sec. The results are shown as Fig. 7.

It will be seen that the total mass ejected from each cm^2 in 10^6 yr by particles between 10^{-12} and 10^3 g amounts to about 17 g. Expressed in different terms, 17 g are removed from each cm^2 of the lunar surface and redistributed across the lunar surface every 10^6 yr. We have no information on the trajectories of the ejected mass and cannot infer how widely dispersed the material will be scattered, but the quantities involved must constitute a powerful erosion and smoothing process. The major fraction of this mass, approximately 13 g, is contributed solely by the impacts of masses in the four decades of m_p from 10^{-8} to 10^{-4} g which will form craters whose diameters (after correction for oblique trajectories) will average 0.9 and 17.4 mm. With a diameter-depth ratio of approximately 6 (Vedder, 1972, and unpublished macroscale data) only the upper 0.15 to 3 mm of the regolith contributes to the 13 g for an average turnover or reprocessing rate of the order of 10^2–10^3 times per 10^6 yr assuming a bulk density of 1 g/cm^3 for this surficial layer. In contrast, between the 10^{-4} and 10^0 g decades, only 1.6 g/cm^2–10^6 yr is processed with craters having depths between approximately 3 and 60 mm. The mixing rate in this case is then of the order of 10^{-2} to 10^0 times per 10^6 yr, and it becomes clear that the rate of turnover decreases much more rapidly with increasing depth than previously estimated by Gault (1970). We will examine this subject in detail in a future paper.

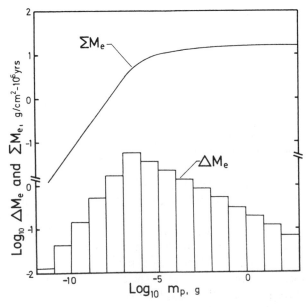

Fig. 7. Calculated incremental ΔM_e and cumulative ΣM_e masses that are displaced from the lunar regolith by impact of micrometeoroids with messes m_p.

Fusion and vaporization

Estimates of the masses of the fused and vaporized products of impact into lunar materials were carried out using the cratering model developed from the Charters-Summers theory (1959) by Gault and Heitowit (1963). Briefly, the crater is assumed to expand behind a hemispherical shock front while maintaining the total crater-forming energy in the system constant. Energy deposited irreversibly behind the shock front is considered trapped and stored as an increase in the internal energy in concentric shells of target material. The increase in internal energy, which is assumed to be manifested as heat energy for melting and vaporization, is evaluated by using the Hugoniot curve as an approximation for the release adiabat to ambient pressure from the local peak shock pressure. Initial conditions to start the calculations are determined from one-dimensional flow equations using the appropriate Hugoniot equation-of-state data for the target and projectile material. In practice, the model requires the simultaneous solution of two, linear, first-order differential equations. Hugoniot data for water ($\rho = 1$), a volcanic tuff ($\rho = 1.7$), and a basalt ($\rho = 2.86$) have been used to simulate, respectively, low density meteoroids, the regolith, and the crystalline rocks. The results are not sensitive to the choice of these specific materials provided that similar densities and porosities are employed. The Hugoniot data for plexiglass ($\rho = 1.18$), a quartz sand ($\rho = 1.65$), and the basalt give results that are virtually indistinguishable from those presented herein.

Figure 8 presents a representative result from these calculations showing the

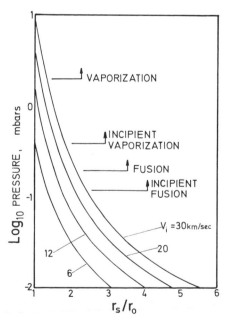

Fig. 8. Calculated variation of peak shock pressure with distance r_s from the point of impact for a projectile with density of 2.86 g/cm³ (basalt) against a target with density 1.7 g/cm³ (tuff).

variation of the peak shock pressure with radial distance r_s from the point of impact. The normalizing term r_0 is the radius of the initial shell of compressed target material and is a function of the initial energy partition between projectile and target. For cases presented herein r_0 is a factor of about 1.4 to 2.4 larger than the radii of the impacting masses. Attention is drawn to the rapid attenuation of the peak pressures with radial distance; only relatively small amounts of target material are subjected to conditions consistent with melting and vaporization. The pressures are indicated at which the various stages of fusion and vaporization will begin, without giving consideration to the differences in the requirements for melting and vaporization of specific mineral components in the rocks as have Ahrens and O'Keefe (1972). We have instead generalized the processes by assuming that fusion effectively will begin when approximately 1.2×10^{10} erg/g are irreversibly trapped in the material. Similarly, when 2, 4, and 12.5×10^{10} erg/g are deposited in the material it is assumed that fusion will be complete, vaporization will commence, and vaporization will be completed, respectively.

The variation with the impact velocity of these shock heated masses M_{eh} (normalized to the projectile mass) are shown in Figs. 9 and 10 for two cases which simulate the impact of (1) a low-density meteoroid against the lunar surface, and (2) a stoney meteoroid or rock or mineral fragment against a dense crystalline rock. The abrupt onset indicated for the various stages of shock heating is, of course, an artifact of the cratering model which considers shock heating as the only mechanism for

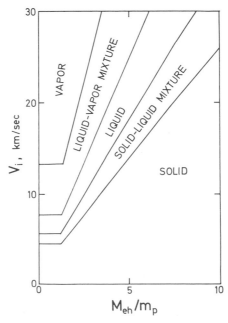

Fig. 9. Calculated variation with impact velocity V_i of the shock heated masses M_{eh} in a target having density of 1.7 g/cm³ (tuff) and normalized with the projectile mass m_p whose density is 1.0 g/cm³ (water).

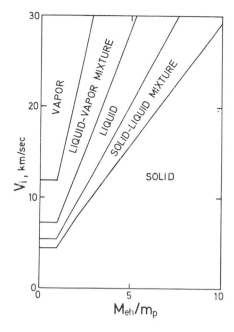

Fig. 10. Calculated variation with impact velocity V_i of the shock heated masses M_{eh} in a target having a density of 1.7 g/cm³ (tuff) normalized with the projectile mass m_p whose density is 2.86 g/cm³ (basalt).

energy dissipation during the cratering process. Inclusion of the lesser effects of deformational (shearing) losses, comminution, etc., should initiate the stages of melting and vaporization at slight lower impact velocities and also serve to make the onset of the heating effects more gradual than indicated.

Detailed numerical values for these calculations and two additional cases will be found in Table 3, which presents estimates of the material masses that are fused and vaporized in terms of the projectile masses; complete vaporization of the projectile material may be assumed for all impact velocities greater than 10 km/sec. For the root-mean-square impact velocity $V_i = 20$ km/sec, the results in Table 3 indicate that a mass equal to 7.5 projectile masses will be either totally or partially fused and/or vaporized by a low-density projectile ($\rho = 1$ g/cm³) into the simulated regolith material ($\rho = 1.7$ g/cm³). Use of the more dense material as a projectile ($\rho = 2.86$ g/cm³) results in a slight decrease in the mass that is fused and vaporized. This decrease is the cumulative effect of: (1) for a given mass (at constant velocity) the more dense projectile is physically smaller so that the transfer of energy and momentum to the target during the initial penetration of the projectile involves smaller masses; and (2) the peak shock pressure in these smaller masses is higher which, in turn, leads to greater losses across the shock front and a more rapid radial decay of the intensity of the shock wave. In effect, the higher pressure "overdrives" the system. Energy that otherwise could be expended in propagating the shock wave more deeply

Table 3. Estimated mass of lunar material fused and vaporized by meteoritic impact (normalized to the mass of the impacting body).

Impact velocity (km/sec)	Projectile density (g/cm³)	Target density (g/cm³)	Peak pressure (mbar)	Solid-fused mix	Fused-vapor		Vapor
					Fused	mix	
4.5	1.0*	1.7†	0.15		Incipient fusion		
7.8			0.36		Incipient vaporization		
10.0			0.56	0.9	0.7	1.2	0.6
20.0			1.9	1.9	1.6	1.7	2.3
30.0			4.2	3.0	2.6	2.6	3.5
3.2	2.86‡	1.7	0.15		Incipient fusion		
5.5			0.36		Incipient vaporization		
10.0			1.0	0.7	0.5	0.6	0.7
20.0			3.7	1.5	1.2	1.3	1.6
30.0			8.1	2.2	1.9	1.9	2.5
6.7	1.0	2.86	0.37		Incipient fusion		
10.0			0.73	0.9	1.0	—	—
10.9			0.85		Incipient vaporization		
20.0			2.5	1.9	2.0	1.8	2.3
30.0			5.3	3.4	3.0	3.0	3.7
4.5	2.86	2.86	0.37		Incipient fusion		
7.3			0.85		Incipient vaporization		
10.0			1.5	0.6	0.7	0.7	0.8
20.0			5.3	1.7	1.5	1.5	1.8
30.0			11.4	2.7	2.3	2.4	2.9

* Hugoniot for water (Van Thiel, 1966).
† Hugoniot for tuff (Shipman *et al.*, 1969).
‡ Hugoniot for Vacaville basalt (Jones *et al.*, 1968).

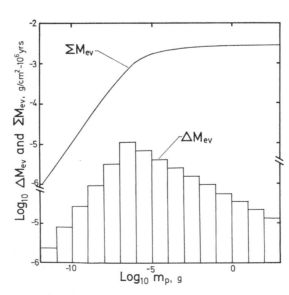

Fig. 11. Calculated incremental ΔM_{ev} and cumulative ΣM_{ev} masses from the regolith which are vaporized by the impact of micrometeoroids with mass m_p.

into the target is instead expended extravagantly to raise material to higher temperatures than required for fusion and vaporization.

If one assumes that each impact acts only on unfused material, the estimated yield of volatilized material from the lunar surface is given in Fig. 11, which shows the incremental mass ΔM_{ev} that is vaporized within each decade of m_p and the cumulative mass ΣM_{ev} (summed from the smallest masses) that is vaporized by the meteoritic flux from 10^{-12} to 10^0 g. The mass flux is, again, mirrored in these results. From a total of 2.7×10^{-3} g/cm^2 vaporized each 10^6 yr by the impact of particles with masses covering the range from 10^{-12} to 10^0 g, approximately 80% of the vaporized mass $(2.1 \times 10^{-3}$ g/cm$^2)$ is contributed by impacting particles with masses in the four decades from 10^{-8} to 10^{-4} g.

As indicated in Table 3, the corresponding values for the mass that is shock heated to the fusion point but not vaporized is a factor of about 2.3 greater; i.e., 6.1×10^{-3} g/cm^2 are raised to the fusion point with 4.7×10^{-3} g/cm^2 contributed in a million years by the middle four decades of m_p. For a material with a bulk density of 1 g/cm^3, the 4.7×10^{-3} g/cm^2 is equivalent to a layer 47 μ thick. If mixed uniformly through a thickness of material from 3 to 0.15 μ deep, as discussed in relation to ΣM_e in Fig. 7, the glass content would constitute from about 2 to 30%. This simple model of melting and mixing, however, assumes that each impact acts independently and this serves to maximize the derived concentration of the melt. A more realistic alternative approach would be to assume uniform mixing of the melt as it is formed so that some remelting is considered to occur. Such a model would reduce the glass content in comparison to the preceding estimate, but the effects would be small for concentrations being considered here. It must be noted, however, that glasses can be produced from the crystalline state by a direct solid-state transformation; preliminary considerations indicate that this source of glass could contribute significant amounts of glass as compared to the shock-melted fractions. We will pursue this point in detail in a future paper. It is sufficient to note here that the quantities of regolith that can be melted and vaporized by the current "best" estimate of the micrometeoroid flux are of the order of 30 g/cm^2 since the mare surfaces were formed more than 3×10^9 yr ago. The greatly enhanced fluxes before 3×10^9 yr that are inferred from the frequency of large craters also implies that still greater quantities of lunar material have been melted and vaporized early in lunar history. The enrichments of such elements as K, Rb, U, Th, and particularly Pb (Tatsumoto, 1970; Silver, 1970; Cliff *et al.*, 1971; Doe and Tatsumoto, 1972) in the regolith material is probably the end result of this process.

Acknowledgment—The senior author gratefully acknowledges the support of the John Simon Guggenheim Memorial Foundation.

REFERENCES

Ahrens T. J. and O'Keefe J. D. (1972) Shock melting and vaporization of lunar rocks and minerals. Contribution No. 2110, Division of Geological and Planetary Sciences, California Institute of Technology.

Alexander W. M., Arthur C. W., Corbin J. D., and Bohn J. L. (1969) Mariner 4 and OGO 2: Recent measurements of picogram dust particle flux in interplanetary and cislunar space. Preprint, 12th COSPAR meeting, Prague.

Alexander W. M., Arthur C. W., Bohn J. L., Johnson J. H., and Farmer B. J. (1971) Lunar Explorer 35: Dust particle data and analysis of shower related picogram lunar ejecta orbits in the earth-moon system. Abstract of paper presented at the 14th COSPAR meeting, June 18–July 2, Seattle, Washington, U.S.A.

Allen H. J. and Baldwin B. S. Jr. (1961) Frothing as an explanation of the acceleration anomalies of cometary meteors. *J. Geophys. Res.* **72**, 3483–3496.

Anon. (1967) Scientific Results of Project Pegasus. NASA TM X-53629.

Anon. (1969) Meteoroid Environment Model—1969 (Near Earth to Lunar Surface). NASA SP-8013.

Arthur C. W., Alexander W. M., Bohn J. L., Johnson J. H., and Farmer B. J. (1971) Results of a 1970 geminid dust particle rocket experiment and analysis of OGO 2 dust particle velocity measurements. Abstract of paper presented at the 14th COSPAR meeting, June 18–July 2, Seattle, Washington, U.S.A.

Berg O. E. and Richardson F. F. (1971) New and supplementary data from the pioneer cosmic dust experiments. Abstract of paper presented at the 14th COSPAR meeting, June 18–July 2.

Bhandari N., Bhat S., Lal D., Rajagopalan G., Tamhane A. H., and Venkatavaradan V. S. (1971) High resolution time averaged (millions of years) energy spectrum and chemical composition of iron-group cosmic ray nuclei at 1 A.U. based on fossil tracks in apollo samples. *Proc. Second Lunar Sci. Conf., Geochim. Cosmochim. Acta* Suppl. 2, Vo.l 3, pp. 2611–2619. MIT Press.

Bloch M. R., Fechtig H., Gentner W., Neukum G., and Schneider E. (1971) Meteorite impact craters, crater simulations, and the meteoroid flux in the early solar system. *Proc. Second Lunar Sci. Conf., Geochim. Cosmochim. Acta* Suppl. 2, Vol. 3, pp. 2639–2652. MIT Press.

Carter J. L. and MacGregor J. D. (1970) Mineralogy, petrology, and surface features of some Apollo 11 samples. *Proc. Apollo 11 Lunar Sci. Conf., Geochim. Cosmochim. Acta* Suppl. 1, Vol. 1, pp. 247–265. Pergamon.

Ceplecha Z. (1967) Classification of meteor orbits. In Proceedings of a Symposium on Meteor Orbits and Dust. NASA SP-135, pp. 33–65.

Charters A. C. and Summers J. L. (1959) Some comments on the phenomenon of high-speed impact. Decennial Symp. White Oaks U.S. Naval Ordnance Lab., Silver Springs, Md., pp. 1–21.

Cliff R. A., Lee-Hu C., and Wetherill G. W. (1971) Rb–Sr and U, Th–Pb measurements on Apollo material. *Proc. Second Lunar Sci. Conf., Geochim. Cosmochim. Acta* Suppl. 1, Vol. 2, pp. 1493–1502. Pergamon.

Crozaz G., Haack U., Hair M., Maurette M., Walker R., and Woolum D. (1970) Nuclear track studies of ancient solar radiations and dynamic lunar surface processes. *Proc. Apollo 11 Lunar Sci. Conf., Geochim. Cosmochim. Acta* Suppl. 1, Vol. 3, pp. 2051–2080. Pergamon.

Crozaz G., Walker R., and Woolum D. (1971) Nuclear track studies of dynamic surface processes on the moon and the constancy of solar activity. *Proc. Second Lunar Sci. Conf., Geochim. Cosmochim. Acta* Suppl. 2, Vol. 3, pp. 2543–2558. MIT Press.

Crozaz G., Drozd R., Hohenberg C. M., Hoyt H. P., Jr., Ragan D., Walker R. M., and Yuhas D. (1972) Solar flare and galactic cosmic ray studies of Apollo 14 samples (abstract). In *Lunar Science—III* (editor C. Watkins), p. 167, Lunar Science Institute Contr. No. 88.

Dalton C. D. (1969) Determination of meteoroid environments from photographic meteor data. NASA TR R-322.

Doe B. R. and Tatsumoto M. (1972) Volatilized lead from Apollo 12 and 14 soils (abstract). In *Lunar Science—III* (editor C. Watkins), p. 178, Lunar Science Institute Contr. No. 88.

Dohnanyi J. S. (1966) Model distribution of photographic meteors. Bellcomm Rep. TR-66-34-1.

Dohnanyi J. S. (1970) On the origin and distribution of meteoroids. *J. Geophys. Res.* **75**, 3468–3493.

Dohnanyi J. S. (1971) Flux of micrometeoroids: Lunar sample analyses compared with flux model. *Science* **173**, 558.

Erickson J. E. (1968) Velocity distribution of sporadic photographic meteors. *J. Geophys. Res.* **73**, 3721–3726.

Finkel R. C., Arnold J. R., Imamura M., Reedy R. C., Fruchter J. S., Loosli H. H., Evans J. C., and Delany A. C. (1971) Depth variation of cosmogenic nuclides in a lunar surface rock and lunar soil. *Proc. Second Lunar Sci. Conf., Geochim. Cosmochim. Acta* Suppl. 2, Vol. 2, pp. 1773–1789. MIT Press.

Fleischer R. L., Haines E. L., Hart H. R. Jr., Woods R. T., and Comstock G. M. (1970) The particle track record of the Sea of Tranquility. *Proc. Apollo 11 Lunar Sci. Conf., Geochim. Cosmochim. Acta* Suppl. 1, Vol. 3, pp. 2103–2120. Pergamon.

Fleischer R. L., Hart H. R. Jr., Comstock G. M., and Evwaraye A. O., (1971a) The particle track record of the Ocean of Storms. *Proc. Second Lunar Sci. Conf., Geochim. Cosmochim. Acta* Suppl. 2, Vol. 3, pp. 2559–2568.

Fleischer R. L., Hart H. R., Jr., Comstock G. M. (1971b) Very heavy solar cosmic rays: Energy spectrum and implications for lunar erosion. *Science* 171, 1240–1242.

Gault D. E. and Heitowit E. D. (1963) The partition of energy for hypervelocity impact craters formed in rock. Proc. 6th Hypervelocity Impact Sym., Vol. 2, pp. 419–456.

Gault D. E., Heitowit E. D., and Moore H. J. (1964) Some observations of hypervelocity impacts with porous media. In *The Lunar Surface Layer* (editors J. W. Salisbury and P. E. Glaser), pp. 151–178, Academic Press.

Gault D. E. and Wedekind J. A. (1969) The destruction of tektites by micrometeoroid impact. *J. Geophys. Res.* 74, pp. 6780–6794.

Gault D. E. (1969) Erosion and fragmentation of rocks on the lunar surface (abstract). *Trans. Amer. Geophys. Union* 50, No. 4, p. 219.

Gault D. E. (1972) Displaced mass, depth, diameter, and effects of oblique trajectories for impact craters formed in dense crystalline rock. *The Moon* (in press).

Givens J. J. (1972) Personal communication.

Grew G. W. and Gurtler C. A. (1971) The Lunar Orbiter meteoroid experiments. NASA TN D-6266. See also Gurtler C. A. and Grew G. W. (1968) Meteoroid hazard near the moon. *Science* 161, 462–464.

Griffith A. A. (1925) The theory of rupture. Proc. Int. Cong. Appl. Mech., Delft, Netherlands, 1924, pp. 69–75.

Hartung J. B., Hörz F., und Gault D. E. (1972) Lunar microcraters and interplanetary dust (abstract). In *Lunar Science—III* (editor C. Watkins), pp. 363–365, Lunar Science Institute Contr. No. 88.

Hawkins G. S. (1963) Impacts on the earth and moon. *Nature* 197, No. 4869, p. 781.

Hemenway C. L., Hallgren D. S., and Kerridge J. F. (1968) Results from the Gemini S-10 and S-12 micrometeorite experiments. *Space Res.* 8, 521–535.

Hörz F., Hartung J. B., and Gault D. E. (1971) Micrometeorite craters and lunar rock surfaces. *J. Geophys. Res.* 76, 5770–5798.

Jones A. H., Shipman F. H., and Isbell W. M. (1968) Material properties measurements for selected materials. Final Report, Contract NASA-3427, NASA, Ames Research Center, Moffett Field, California.

Kerridge J. F. (1970) Micrometeorite environment at the earth's orbit. *Nature* 228, No. 5272, 616–619.

Konstantinov B. P., Bredov M. M., Mazet E. P., Panov V. N., Aptekar' R. L., Golenetskiy S. V., Gur'Yan Yu. A., and Il'inskiy V. N. (1968) NASA TT-F-11753.

Konstantinov B. P., Predov M. M., Mazet E. P., Panov V. N., Aptekar' R. L., Golenetskiy S. V., Gur'Yan Yu. A., and Il'inskiy V. N. (1969) NASA TT-F-100751.

Lal D., MacDougall D., Wilkening L., and Arrhenius G. (1970) Mixing of the lunar regolith and cosmic ray spectra: Evidence from particle-track studies. *Proc. Apollo 11 Lunar Sci. Conf., Geochim Cosmochim. Acta* Suppl. 1, Vol. 3, pp. 2295–2303. Pergamon.

Latham G. V., Ewing M., Press F., Sutton G., Dorman J., Nakamura Y., Toksöz N., Duennebier F., and Lammlein D. (1971) Passive seismic experiment. NASA SP-272, Apollo 14 Preliminary Science Report, pp. 133–161.

Laul J. C., Morgan J. W., Ganapathy R., and Anders E. (1971) Meteoritic material in lunar samples: Characterization from trace elements. *Proc. Second Lunar Sci. Conf., Geochim. Cosmochim. Acta* Suppl. 2, Vol. 2, pp. 1139–1158.

Lindblad B. A. (1967) Luminosity function of sporadic meteors and extrapolation of the influx rate to micrometeorite region. In Proceedings of a Symposium on Meteor Orbits and Dust, NASA SP-135, pp. 171–180.

Lundborg N. (1967) The strength-size relation of granite. *Int. J. Rock Mech. Min. Sci.* 4, pp. 269–272.

McCrosky R. E. (1967) Orbits of photographic meteors. Smithsonian Astrophysical Observatory Special Report 252.

McCrosky R. E. (1968) Distribution of large meteoric bodies. Smithsonian Astrophysical Observatory Special Report 280.

McCrosky R. E., Posen A., Schwartz G., and Shao C. Y. (1971) Lost City meteorite—its recovery and a comparison with other fireballs. *J. Geophys. Res.* **76**, 4090–4108.

McDonnell J. A. M. (1971) Review of *in situ* measurements of cosmic dust particles in space. *Space Research XI*, pp. 415–435.

McDonnell J. A. M. and Ashworth D. G. (1971) Erosion phenomena on the lunar surface and meteorites. Preprint, 14th COSPAR, Seattle, Washington.

McKay D. S., Greenwood W. R., and Morrison D. A. (1970) Origin of small lunar particles and breccia from the Apollo 11 site. *Proc. Apollo 11 Lunar Sci. Conf., Geochim. Cosmochim. Acta* Suppl. 1, Vol. 1, pp. 673–694. Pergamon.

Morrison D. A., McKay D. S., Moore H., Bogard D., and Heikan G. (1972) Microcraters on lunar rocks (abstract). In *Lunar Science—III* (editor C. Watkins), pp. 558–560. Lunar Science Institute Contr. No. 88.

Naumann R. J. (1966) The near earth meteoroid environment. NASA TN D-3711.

Naumann R. J., Jex D. W., and Johnson C. L. (1969) Calibration of Pegasus and Explorer 23 detector panels. NASA TR R-321.

Naumann R. J. (1972) Personal communication.

Neukum G., Mehl A., Fechtig H., and Zähringer J. (1970) Impact phenomena of micrometeorites on lunar surface materials. *Earth Planet. Sci. Lett.* **8**, 31–35.

Neukum G., Schneider E., Mehl A., Storzer D., Wagner G. A., Fechtig H., and Bloch M. R. (1972) Lunar craters and exposure ages derived from crater statistics and solar flare tracks (abstract). In *Lunar Science—III* (editor C. Watkins), pp. 581–583, Lunar Science Institute Contr. No. 88.

Nilsson C. S. (1966) Some doubts about the earth's dust cloud. *Science* **153**, 1242–1246.

Nilsson G. S., Wright F. W., and Wilson D. (1969) Attempt to measure micrometeoroid flux on the OGO 2 and OGO 4 satellites. *J. Geophys. Res.* **74**, pp. 5268–5276.

O'Neal R. L. (1968) The Explorer 13 micrometeoroid satellite. NASA TN D-4284.

Price P. B. and O'Sullivan D. (1970) Lunar erosion rate and solar flare paleontology. *Proc. Apollo 11 Lunar Sci. Conf., Geochim. Cosmochim. Acta* Suppl. 1, Vol. 3, pp. 2351–2359.

Shipman F. H., Isbell W. M., and Jones A. H. (1969) High-pressure hugoniot measurements for several Nevada-test-site rocks. Materials and Structures Laboratory, Manufacturing Development, General Motors Corporation, Final report to NASA-2214, MSL-68-15.

Shoemaker E. M. (1962) Interpretation of lunar craters. *Phys. and Astron. Moon* (editor Z. Kopal), pp. 283–359, Academic Press.

Shoemaker E. M., Hait M. H., Swann G. A., Schleicher D. L., Schaber G. G., Sutton R. L., and Dahlem D. H. (1970) Origin of the lunar regolith at Tranquility Base. *Proc. Apollo 11 Lunar Sci. Conf., Geochim. Cosmochim. Acta* Suppl. 1, Vol. 3, pp. 2399–2412. Pergamon.

Silver L. T. (1970) Uranium-thorium-lead isotopes in some Tranquility Base samples and their implications for lunar history. *Proc. Apollo 11 Lunar Sci. Conf., Geochim. Cosmochim. Acta* Suppl. 1, Vol. 2, pp. 1533–1574. Pergamon.

Soberman R. K. (1971) The Terrestrial influx of small meteoritic particles. *Rev. Geophys. Space Phys.* **9**, No. 2, 239–258.

Tatsumoto M. (1970) Age of the moon: An isotopic study of U–Th–Pb systematics of Apollo 11 lunar samples. *Proc. Apollo 11 Lunar Sci. Conf., Geochim. Cosmochim. Acta* Suppl. 1, Vol. 2, pp. 1595–1612. Pergamon.

Van Thiel M. (editor) (1966) Compendium of shock wave data UCRL-50108, University of California, Livermore.

Vedder J. F. (1972) Craters formed in mineral dust by hypervelocity micrometer-size projectiles. *J. Geophys. Res.* (in press).

Weibull W. (1939) A statistical theory of the strength of materials. *Proc. Roy. Swedish Acad. Eng. Sci.* **151**.

Zook H. A., Flaherty R. E., and Kessler D. J. (1970) Meteoroid impacts on Gemini windows. *Planet. Space Sci.* **18**, 953–964.

Proceedings of the Third Lunar Science Conference
(Supplement 3, *Geochimica et Cosmochimica Acta*)
Vol. 3, pp. 2735–2753
The M.I.T. Press, 1972

Lunar microcraters and interplanetary dust

Jack B. Hartung and Friedrich Hörz

National Aeronautics and Space Administration, Manned Spacecraft Center,
Houston, Texas 77058

and

Donald E. Gault*

National Aeronautics and Space Administration, Ames Research Center,
Moffett Field, California 94035

Abstract—The ratio of spall to pit diameter of lunar microcraters decreases with decreasing crater size. The trend from micron-sized to cm-sized craters involves a transition of crater types from pit-only to pit-plus-spall craters. Some spall-only craters may have possessed a pit originally.

Most microcraters are formed by the impact of "primary" interplanetary dust particles because melting occurred during their formation. According to experimental and theoretical data such melting phenomena require impact velocities which are consistent only with the velocity distribution of extralunar particles.

Detailed crater statistics on two glass surfaces which are in a nonequilibrium state with respect to cratering yield a cumulative flux curve over the range of mass, 10^{-12} to 10^{-5} g, which agrees within an order of magnitude with independently derived flux data from satellite-borne experiments.

An average minimum flux at the moon of particles 3×10^{-5} g and larger based on lunar data alone is 2.5×10^{-6} particles/cm$^2 \times$ yr $\times 2\pi$ strd over less than the last 10^6 years. Particles less than 10^{-12} g in mass are the most abundant. Particles of masses 10^{-7} to 10^{-4} g are the dominant contributors to the cross-sectional areas of interplanetary dust. Particles in the 10^{-6} to 10^{-2} g range apparently deposit the most mass, and thereby energy, on the lunar surface and therefore are largely responsible for most impact phenomena observed in the lunar fines, such as fusion, vaporization, ionization, and the admixture of extralunar, meteoritic components.

Lunar Microcraters

Microcraters ranging in diameter from less than 1 micron to more than 1 cm have been observed on lunar materials. These features have been described in detail by Carter and MacGregor (1970), McKay *et al.* (1970), Neukum *et al.* (1970), Bloch *et al.* (1971a), Hörz *et al.* (1971), Morrison *et al.* (1972), and others. Similar features on the windows of manned spacecraft were described earlier by Zook *et al.* (1970) and Cour-Palais *et al.* (1972). The ubiquitous presence of microcraters on exposed lunar rock surfaces studied so far suggests that microcratering has affected literally every square centimeter of the lunar surface.

---

* Present address: Max-Planck-Institut für Kernphysik, Heidelberg, Germany.

Morphology

Lunar microcraters may consist of one or more of three important morphological elements (Fig. 1; see also Hörz *et al.*, 1971).

(1) Glass-lined pit: A cup-shaped depression is lined by smooth, rough, or vesiculated glass. The glass colors range from clear or colorless through honey-colored shades to black. Small pits in single crystals usually are lined with glass the same color as the host crystal, indicating that the glass is largely molten host rock caused by shock fusion.

(2) Halo zone: Surrounding and underlying the pit is a zone of highly microfractured host-rock material. The microfracturing intensity decreases radially outward from the pit. The albedo of most silicate materials is increased by microfracturing. Thus, glass-lined pits on crystalline rocks and, to some extent, on glass-coated and breccia surfaces are nearly always surrounded by a halo of brighter, microfractured material.

················· GLASSLINED PIT

— — — — — HALO ZONE

— · — · — · — SPALL ZONE

— — — — CONDENSATES

ROCK 12051

2 mm

Fig. 1. A large, apparently fresh, microcrater on crystalline rock 12051. The assymetry of the halo, spall, and condensation zones relative to the glass-lined pit is not typical. Such assymetries occur predominantly on crystalline rocks, as opposed to fine-grained breccias and glass surfaces. A smaller, 400-micron crater may be seen at the lower right of the photograph.

(3) Spall zone: A roughly concentric spall zone may be developed around the pit. The spall occurs along radial and concentric fractures in homogeneous noncrystalline materials and is controlled by grain boundaries or crystal lattice structure in non-homogeneous or crystalline materials. Occasionally the spalling action removes material surrounding the pit to such an extent that the remaining pit appears raised on a pedestal composed of underlying halo material. We expect that in some cases entire pits are removed in this way, thus leaving a depression or spall zone containing a central concentration of microfractured material.

Rarely, another concentric area of lower albedo outside the spall zone was observed. We interpret this darkening (Fig. 1) to be due to condensation of material vaporized during the impact event.

For millimeter-sized craters the ratios of the diameters of the pit, halo, spall, and condensation ring are roughly 1:2:4:10, respectively. Because the ratio of the spall diameter to the pit diameter is an important parameter in evaluating lunar rock crosion, a detailed study of this ratio was made using the crater population on the glass coating on rock 12054. The results indicate the spall-to-pit diameter ratio decreases with decreasing pit diameter. However, the ratios observed are highly variable.

Micron-sized craters on one glass spherule from the Luna 16 sample (Fig. 2) further illustrate the relationship between the pit and the spall zone. The largest micron-sized craters have a central pit and surrounding spall zone typical for millimeter-sized craters. Somewhat smaller craters have only partially developed spall zones or concentric fractures indicating incipient spallation. The smallest pits observed displayed no evidence of related spall zones. This gradual transition from pit-only to pit-plus-spall craters is qualitatively illustrated by comparing measured topographic profiles for craters of different sizes (Fig. 3). Statistical data for the above crater classes indicate considerable variability in the size-dependence of the transition from pit-only to pit-plus-spall craters. Other factors, such as impacting particle velocity or density, may account for the observed variability.

A transitional relationship may also exist between pit-plus-spall craters and pitless craters (McKay, 1970). The spall zones of the craters shown in Fig. 4 are of nearly the same size and character. The pit contained within the spall zone in Fig. 4a appears to be barely attached to the host material. Presumably, had slightly different conditions prevailed at the time of formation, the pit may have been removed during or after the crater-forming process, producing a crater similar to that in Fig. 4b. An alternate interpretation is that the crater shown in Fig. 4b originally was formed by an impact that produced no melting phenomena or glass-lined pit. Consequently, caution must be exercised when postulating vastly different impact velocities for craters with or without glass-lined pits, because many of the pitless craters could be craters with the glass-lined pit removed.

Origin

The existence of transitionally related crater types suggests that a single process is operating to produce the variety of craters described. Impact generally is accepted as the means by which craters are formed, but different interpretations have been

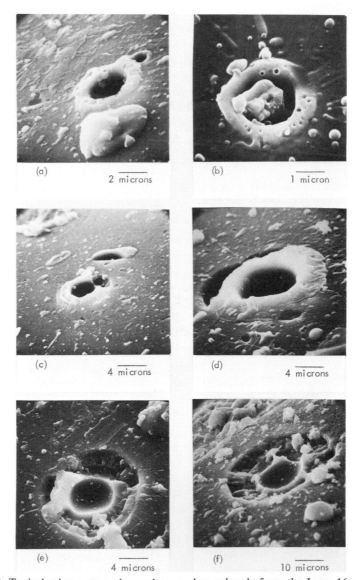

Fig. 2. Typical microcraters observed on a glass spherule from the Luna 16 sample. (a) and (b): Very small, pit-only craters showing a raised, presumably remelted, lip and no evidence of fracturing. We interpret the fragment in the crater in (b) to be not related to cratering event but deposited in the crater later. (c) and (d): Somewhat larger craters displaying concentric fractures indicating incipient spallation. (e): A crater similar in size to that in (d), but showing an almost complete spall zone and only a remnant of a raised lip. (f): A relatively large, pit-plus-spall crater showing complete development of the spall zone.

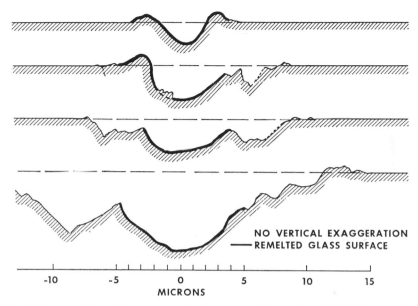

Fig. 3. Topographic profiles across typical craters at different sizes obtained by parallax measurements of stereoscopic scanning electron micrographs. Note the change in crater type, from pit-only to pit-plus-spall, and the increase in the spall to pit diameter ratio with increasing crater size. The assymetries of pit profiles suggest a rough direction of impact may be inferred for some craters.

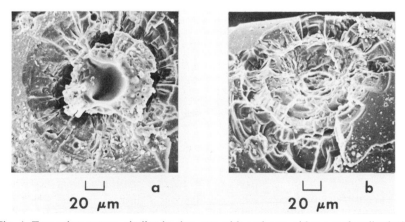

Fig. 4. Two microcraters, similar in size, one with and one without a glass-lined pit. Note that the glass-lined pit and its surroundings appear only loosely attached to the host spherule. Had slightly different conditions prevailed, the pit might have been removed from the crater in (a) to form a crater similar to that in (b). This possibility suggests that a transition from pit-plus-spall to spall-only craters may exist.

presented regarding the source of the impacting particles and the condition of the target rock or grain at the time of impact. The impacting particles may be "primary," coming to the moon from interplanetary space (Neukum *et al.*, 1970; Hörz *et al.*, 1971; Bloch *et al.*, 1971b; Fechtig, 1972; Hartung *et al.*, 1972a), or "secondary," originating on the lunar surface and put in motion by another larger impact event (Carter and MacGregor, 1970; Fredriksson *et al.*, 1970; Frondel *et al.*, 1970; Carter and McKay, 1971; Frondel *et al.*, 1971).

We argue in favor of a primary origin for the microcraters described based on one characteristic that distinguishes between the two sources of impacting particles. That distinguishing characteristic is the velocity distribution of the possible sources. Most microcraters were formed by high-velocity impacts, a fact which is only consistent with a primary origin.

The velocity distribution of interplanetary dust or meteoroids based on visual and radar meteors (e.g., NASA SP-8013, 1969) is bounded by the earth and the solar system escape velocities, and has a maximum at about 20 km/sec (Fig. 5), which con-

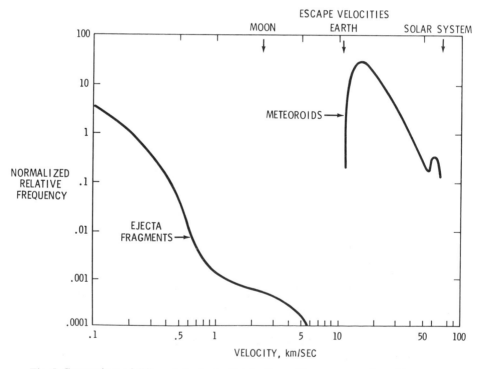

Fig. 5. Comparison of differential velocity distributions of impact crater ejecta fragments, i.e., "secondary" projectiles, and "primary" meteoroids. Data for the ejecta fragments is based on laboratory cratering experiments. Data for the meteoroids is based on visual and radar meteor observations at the earth. Even allowing for lower meteoroid velocities at the moon, the difference in the two distributions is so great that most microcraters, especially those containing impact-melted host material, are attributable to the impact of meteoroids.

ventionally is considered as a mean velocity for meteoroids. Particles impacting the moon may have a minimum velocity of 2.4 km/sec, the lunar escape velocity, but the mean is expected to remain near 20 km/sec because of the relatively lower effective cross section of the moon for slower particles. Inflight velocity measurements of micron-sized meteoroids by Berg and Gerloff (1970) and Berg and Richardson (1971) are generally consistent with this distribution. The velocity distribution of impact ejecta, i.e., "secondary" projectiles based on laboratory cratering experiments at an impact velocity of 6.25 km/sec (Gault *et al.*, 1963), indicates essentially an exponential increase in the number of particles moving at lower velocities (Fig. 5). The velocity of almost all impact cratering ejecta is below 1 km/sec.

Furthermore, there is a physical limit on the maximum velocity it is possible to accelerate solids or liquids by shock processes. If the pressure behind a given shock wave exceeds a critical value (typically 500–700 kilobars for silicates of zero porosity), the increase in internal energy of the ejected material will be sufficient to melt it. At somewhat higher pressures (700–1000 kilobars) the increase in internal energy is sufficient to vaporize the material. The mass velocities behind the shock fronts for these two regimes in material of zero porosity are typically 2.5–3.5 km/sec and 3.5–5 km/sec, respectively. When consideration is given to the geometry of the shock waves produced during an impact, these velocities represent good estimates for the maximum velocity the material will be ejected from the point of impact. Potential secondary ejecta in the solid or liquid state, therefore, are restricted to velocities less than about 5 km/sec. Pressures, mass velocities, and, hence, ejection velocities in low-density material are still lower. At higher impact velocities, the amount of material ionized, vaporized, and melted may increase, but the maximum ejection velocity for solid or melted material should remain about the same (Gault *et al.*, 1972).

Another mechanism for accelerating secondary particles is related to the rapid expansion of a gas cloud produced by a large impact event. Small ejecta particles may be carried by such a cloud at essentially the cloud expansion velocity. Rehfuss (1972) calculates that the gas cloud of a 150-km diameter cratering event caused by a Fe projectile of 3×10^{16} g mass impacting at 20 km/sec will only travel at about 4 km/sec.

Thus we have shown that the velocity distributions of primary and secondary projectiles differ drastically. There is virtually no overlap between their respective velocity regimes. The assessment of the absolute impact velocity of a projectile by only observing the resulting crater is not entirely straightforward. However, we suggest that if evidence of shock melting of the host target is present, then the projectile velocity is incompatible with the velocity regime for secondary particles because:

(1) In laboratory cratering experiments (Moore *et al.*, 1963; Gault *et al.*, 1968; Gault and Wedekind, 1970; Gault *et al.*, 1972) at room temperatures resulting in craters more than 5 mm in diameter, no glass-lined pits were produced using a variety of target materials (basalt, pumice, granite, glass, loose quartz sand) and projectile materials (iron, aluminum, lexan, pyrex). Carter and McKay (1971) succeeded in generating a glass-lined pit only after heating a soda lime glass target above 700°C and using a 7 km/sec aluminum projectile. Cour-Palais (1969) produced a similar pit by launching a tungsten carbide projectile into glass at 8 km/sec. However, the

above cases (heated target and projectile density of 15.3 gr/cm^2) are rather unlikely conditions for lunar rock surfaces. Thus we conclude that meteoroid impact velocities greater than 8 km/sec are required to produce mm-sized glass-lined pits.

(2) Cratering experiments using electrostatic microparticle accelerators resulting in pit diameters of 1 to 15 microns produce plastic flow phenomena and possibly melting at lower velocities (Bloch et al., 1971b; Vedder, 1971; Mandeville and Vedder, 1971). These phenomena were observed above 2 km/sec impact velocity on silica and soda-lime glasses, using Fe, Al, and polystyrene projectiles. However, for crystalline materials (norite, feldspar, olivine) velocities above 3 km/sec for micron-size craters are required.

(3) These cratering results are in fair agreement with considerations based on experimental equation of state data and thermodynamic considerations. According to Ahrens and O'Keefe (1972) and Gault et al. (1972), velocities of about 4, 5–6, and 6 km/sec are required for an iron projectile to shock melt quartz or plagioclase (An_{50}), olivine, and pyroxene, respectively. Corresponding velocities for a less dense projectile, e.g., aluminum, are about 6 km/sec (quartz), 5–6 km/sec (plagioclase), 8 km/sec (olivine), and 10 km/sec (pyroxene).

According to points 1, 2, and 3, a conservative limit of 2 km/sec can be postulated for all craters with glass-lined pits. For mm-sized craters this limit seems to be in excess of 8 km/sec. The vast majority of lunar microcraters possess such glass-liners, e.g., 95% of the total crater population on rock 12054. This striking preponderance of glass-lined pits is incompatible with the velocity distribution of impact ejecta, because for each "secondary" projectile potentially striking at above 2 km/sec, orders of magnitude more particles should generate craters without melting phenomena. Such low-velocity craters are, however, only rarely observed. Thus the exclusively "high-velocity" type of crater population is consistent only with the velocity distribution of extralunar particles.

Moreover, it has been observed that when differences in the frequency of craters are detected on various faces of one given rock specimen and the surface orientation of this specimen is reconstructed, the vertical sides or steeply inclined sides are less cratered than the top of the rocks (Hörz and Hartung, 1971). However, it is only the vertical sides which potentially could intercept secondary particles exceeding the lunar escape velocity of 2.4 km/sec. It is therefore concluded that the concentration of glass-lined pits on the topside of individual rock specimens is only compatible with an extralunar origin of high-velocity projectiles.

Hörz et al. (1971) have described abundant features on lunar materials attributable to secondary ejecta (ropy glass splashes, glass coatings, welded dust, thin film coatings). These features are distinct from high-velocity impact craters and correspond to the low-velocity distribution of lunar ejecta.

Furthermore, the size frequency of microcraters is compatible with the mass frequency distribution of interplanetary dust particles as demonstrated below. This agreement is additional evidence for a "primary" origin of lunar microcraters.

Consequently, the previous considerations lead us to the conclusion that the character and frequency of glass-lined pits are compatible only with the characteristics of primary micrometeoroids. Such a conclusion is less certain for craters which do not display glass-lined pits, though we have indicated before that at least some of them may be genuine hypervelocity craters with the glass-lined pits removed. Our conclusions also do not necessarily pertain to craters smaller than 1 micron.

Some evidence supporting a secondary origin for lunar microcraters has been presented and deserves consideration.

(1) Glass lining an impact pit found on an iron meteorite fragment had essentially lunar composition and was interpreted as the remnants of a secondary lunar projectile (Fredriksson *et al.*, 1970). The fact that the glass lining did not contain even traces of its target material possibly indicates that lunar material was analyzed which was deposited on the pit *after* pit formation. If the analyzed material were actually a remnant of the projectile, a significant amount of target elements should have been encountered. We therefore suggest that Fredriksson *et al.* may have analyzed one of the abundant thin film coatings or ropy splashes described by Hörz *et al.* (1971).

(2) A disproportionately high number of craters has been reported on glassy surfaces compared with crystalline materials in lunar fines (Frondel *et al.*, 1970; Carter and McKay, 1971). In the absence of a well-documented statistical analysis we suggest that this qualitative observation may be due to a much greater difficulty in recognizing craters in crystalline than in glass materials. We have experienced far greater difficulty in the quantitative observations of craters on crystalline rock surfaces as compared to those on glass surfaces. Also, the difference may be real because crystalline materials are fractured and shattered along crystal cleavage planes or grain boundaries much more readily than glass particles, thus producing statistically many more "young" crystalline surfaces with fewer craters as compared with their glassy counterparts.

Size distribution

Because of our interpretation that all craters with glass-lined pits are due to primary impact, a careful determination of the size distribution of these craters may lead to an improved mass distribution curve for interplanetary dust particles. Detailed crater size distribution data have been presented for a variety of lunar surface materials (Hörz *et al.*, 1971; Bloch *et al.*, 1971b; Fechtig, 1972; Hartung *et al.*, 1972a; Hartung *et al.*, 1972b). For this study we have selected only those data based on observations of surfaces not in equilibrium with respect to cratering; that is, surfaces displaying one crater for each impact event. Such a nonequilibrium condition is most easily verified for smooth glass surfaces. In addition, glass surfaces permit quantitative recognition of microcraters down to much smaller crater sizes. For craters with pit diameters greater than 50 microns, binocular microscope data from one glass-coated face of rock 12054 has been selected (Hartung *et al.*, 1972a). For craters with pit diameters less than 50 microns, scanning electron microscope (SEM) data from a single glass spherule from the Luna 16 sample was selected (Hartung *et al.*, 1972b). These data are presented in Fig. 6 as a curve showing the normalized number of craters, N, with pit diameters greater than some value versus pit diameter, D_p. The scale on the right of Fig. 6 shows the actual number of craters counted at the indicated magnifications. The data selected for different sized craters were taken at different

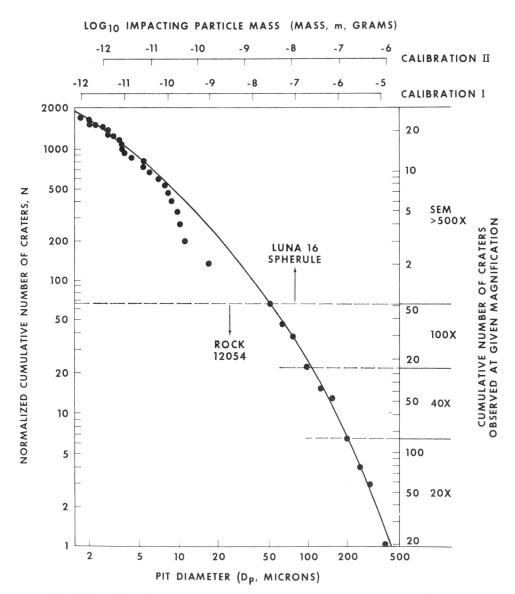

Fig. 6. Cumulative frequency of microcraters of different sizes observed on two carefully selected glass surfaces. The data are joined at a pit diameter of 50 microns, which corresponds to the largest crater observed on the Luna 16 glass spherule. Corresponding impacting particle masses are also shown according to two different calibration approaches. The statistical quality of the data is indicated by the actual number of craters counted, which is shown on the scale at the right for the different magnifications used.

magnifications to obtain reasonable statistics over as wide a range of crater sizes as possible. We estimate the uncertainty in the rock 12054 data to be less than $\pm 20\%$.

The statistical quality of the Luna 16 spherule data is significantly inferior to that of the rock 12054 data; however, the uncertainty is less than a factor of 2. Although the two sets of data are connected based on the occurrence of only one 50-micron crater, we believe the curve in Fig. 6 represents accurately the cumulative size distribution of microcraters for nonequilibrium surfaces.

INTERPLANETARY DUST

Cumulative mass distribution

In order to translate data on microcraters into data on interplanetary dust particles, a calibration must be made. Several approaches using the results of laboratory cratering experiments are possible. Micron-sized particles may be accelerated to extremely high velocities by electrostatic microparticle accelerators. Craters with characteristic glass-lined pits have been produced in this way (Bloch *et al.*, 1971b; Vedder, 1971; Mandeville and Vedder, 1971). Major problems with this technique are the limit on the size of particles accelerated, and therefore the size of craters produced, as well as the general negative correlation between the size of the particle accelerated and the velocity achieved. Nevertheless, we have chosen two approaches based on microparticle accelerator experiments to calibrate lunar microcrater pit diameters in terms of impacting particle mass.

(1) Both Bloch *et al.* (1971b) and Mandeville and Vedder (1971) have shown empirically, using iron and polystyrene projectiles, respectively, that the ratio of pit diameter, D_p, to projectile diameter, d, is independent of the projectile mass, m, over the range of masses studied, 10^{-14} to 10^{-9} g and 7×10^{-13} to 6.2×10^{-11} g, respectively. Thus, for any given projectile material the ratio, D_p/d, depends only on the impact velocity of the projectile. The ratio increases with increasing impact velocity. For simplicity, we assume a constant velocity for all meteoroids of 20 km/sec. Coincidentally, for this velocity, using either iron or polystyrene projectiles, a D_p/d ratio of about 2 results. Using $D_p/d = 2$ and assuming a spherical particle with a density of 3 g/cm^3 allows a relationship between pit diameter and particle mass to be calculated. The projectile masses obtained by this calculation are shown in Fig. 6 as "calibration I."

(2) Another calibration approach uses an experimentally determined relationship between the mass of material ejected from a microcrater pit, m_e, and the kinetic energy of the impacting particle, E_k. The relationship found by Mandeville and Vedder (1971) for glass targets is

$$m_e = 47 \, E_k^{1.1} \tag{1}$$

where m_e is in picograms and E_k is in microjoules. Again, for simplicity, we choose a constant impact velocity of 20 km/sec, and convert the unit of mass to grams to obtain a relationship between the mass ejected, m_e, and the impacting particle mass, m,

$$m_e = 124 \, m^{1.1} \tag{2}$$

where m_e and m are in grams.

The density of the ejected material is assumed to be 3 g/cm³, and the volume of the pit is assumed to be proportional to the cube of the pit diameter. The proportionality constant was determined empirically based on actual pit volumes and diameters determined by parallax measurements of stereoscopic scanning electron micrographs of pits less than 2 to more than 10 microns in diameter. The resulting relationship is

$$m_e = 0.5 \, D_p{}^3 \tag{3}$$

where m_e is in grams and D_p is in cm. The relationship of equations (2) and (3) is also shown in Fig. 6 as an impacting particle mass scale designated "calibration II."

In the picogram size range the two calibration approaches differ by about a factor of 3, while in the microgram size range the corresponding factor has increased to about 10. More detailed laboratory cratering experiments are needed to improve the interpretation of microcraters with respect to projectile properties. Apparently, calibration uncertainties are one of the most important limitations affecting the quality of the subsequent interpretations.

Differential distributions

The principle cumulative mass distribution curve is shown in Fig. 6. For subsequent presentations we will use the mass scale designated "calibration I." The cumulative number scale is arbitrarily normalized such that 10 particles of mass 10 micrograms or greater may be considered to have impacted a hypothetical surface.

Following Zook et al. (1970), three differential distributions of interest (Fig. 7) may be derived from the above cumulative mass distribution. The numbers on the vertical scales in Fig. 7 are based on the data given in Fig. 6 and a value for $d (\log m)$ of 0.2.

(1) A differential distribution of particles for equal logarithmic mass intervals is shown as the bottom curve in Fig. 7. The justification for indicating a maximum in this curve below 10^{-12} g is based on the observed trend of the cumulative distribution and a theoretically expected depletion of particles in the solar system less than one micron in diameter. The essential conclusion to be drawn from this curve is that for any constant logarithmic interval of mass a constant factor of 2, for example, the number of particles in each interval is greater than that in the interval for the next larger mass. This trend holds down to masses somewhere below 10^{-12} g, but probably reverses at still lower masses.

(2) By multiplying this differential distribution by the particle mass we obtain the total mass of all impacting particles in a given logarithmic interval as a function of particle mass. This distribution is shown in the middle curve in Fig. 7. The justification for the maximum at a mass higher than 10^{-5} g is based on the observed trend of the cumulative distribution combined with independent observations of radar and visual meteors corresponding to particle masses greater than 10^{-5} g (NASA SP-8013, 1969). Although greater numbers of particles occur in logarithmic mass intervals at masses below 10^{-12} g, the amount of mass contained in any given logarithmic interval is greatest in the 10^{-5} to 10^{-4} g range. A remarkably similar result was obtained by

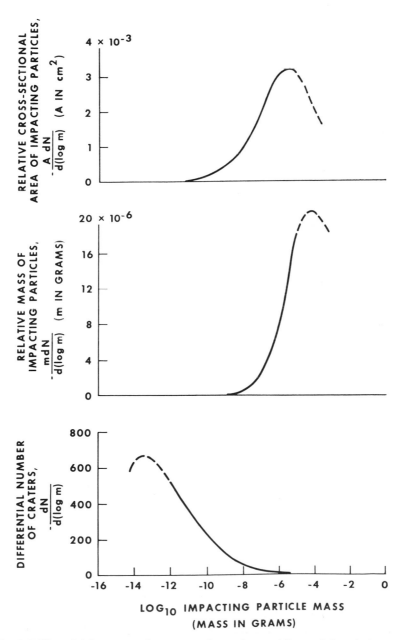

Fig. 7. Differential frequency of craters, or impacting particles, and the relative mass and cross-sectional area of impacting particles as a function of impacting particle mass. These curves are derived directly from the cumulative frequency curve using calibration I shown in Fig. 6.

Whipple (1967) based on an analysis of the characteristics of material in inter-planetary space.

If the velocity distribution does not depend on the masses of the particles, then we may conclude that greater amounts of impact energy are applied to the lunar surface in the form of 10^{-5} to 10^{-4} g particles than those in any other equivalent size class. And further, the overwhelming majority of all energy deposited at the surface of the moon by impact is delivered by particles in the 10^{-6} to 10^{-2} g range. Thus, the effects of all impacts, such as ionization, vaporization, fusion, ejection, and addition of extralunar material, are caused mainly by projectiles in this mass range, assuming a linear relationship between these effects and kinetic energy of impact (Gault *et al.*, 1972).

(3) By multiplying the differential distribution by cross-sectional area of the particle, A, we obtain the relative cross-sectional area of all meteoroids in a given logarithmic interval of particle mass shown at the top of Fig. 7. We assume spherical particles of density 3 g/cm^3. This curve emphasizes the dominance of particles in the 10^{-7} to 10^{-4} g range in producing zodiacal light and in causing exposure of fresh lunar rock surfaces by superposition of primary impact craters.

Flux

After determining the relative masses of micrometeoroids, their flux on the lunar surface may be obtained by correlating the areal density of craters on rock surfaces with surface exposure times for those sample rocks. Data of this type is presented in Fig. 8. On the vertical axis is the areal density of craters greater than 0.5 mm in diameter. Based on calibration I, a crater of this size corresponds to a particle mass of 2.5×10^{-5} g.

A 2 π-strd exposure geometry was assumed for all rock surfaces except 12054, which had a 1.5 π-strd exposure as determined from lunar surface documentation photography. The crater densities for all rock surfaces, except 12054, were judged to represent equilibrium crater populations based on the general character of the sur-faces and the relatively constant value of crater density for the different rock surfaces (Hörz *et al.*, 1971). These surfaces cannot be expected to yield accurate fluxes because previously existing craters, and therefore meteoroid records, may have been obliterated during the exposure lifetime of the rock surface. On the horizontal axis is the surface exposure time based on measurements made by other workers. One point is shown on the graph for each rock surface for which both crater density and exposure time data is available.

The slope of a straight line through the origin on this graph represents a flux. Such a straight line may be drawn through the origin, the rock 12054 point, and the rock 12017 data point. The 12054 exposure time is based on cosmogenic Al^{26} radioactivity measurements (Schonfeld, 1971), and the 12017 exposure time is based on cosmic ray track density measurements (Fleischer *et al.*, 1971). The slope of this line, which should be considered a minimum value, yields a flux of particles, 25 micrograms and larger, at the moon of 2.5×10^{-6} meteoroids/cm^2 $\times$ yr $\times$ 2 π strd. The large uncertainty in the exposure-time measurement for the nonequilibrium surface and the scarcity of

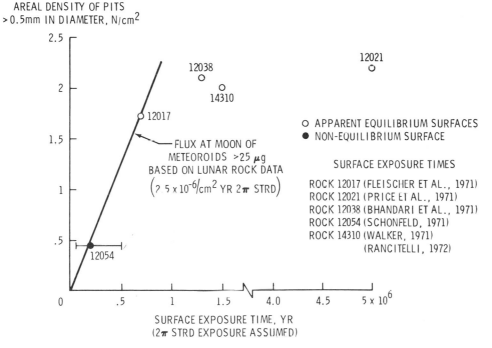

Fig. 8. Correlation between areal density of craters with pits 0.5 mm in diameter or greater (Hörz *et al.*, 1971) and surface exposure time for those rocks for which both kinds of data exist. The slope of the indicated line represents a flux of 2.5 × 10⁻⁶ meteoroids of mass 25 micrograms or greater per cm²-yr-2 π strd at the moon. This should be considered a minimum value for the flux averaged over a period of time as long as 10⁶ years. Note that all rocks having exposure times of more than 10⁶ years have essentially the same areal density of craters, thus confirming their equilibrium state with respect to cratering.

samples exposed less than 10^6 yr on the lunar surface indicate considerable improvement may be expected in the flux value obtained in this way.

Nevertheless, we may compare the meteoroid flux obtained based only on the study of lunar samples with similar results from satellite-borne particle-detection experiments. For the purposes of comparison, we take the cumulative flux at a single mass of 25 micrograms as shown in Fig. 8 together with the cumulative mass distribution data shown in Fig. 6. The marriage of these two results yields the cumulative flux curve over the range of masses from 10^{-12} to 10^{-4} g based on lunar data alone shown in Fig. 9. Results are shown for meteoroid densities of 3 and 1 gm/cm³. Also shown in Fig. 9 are the results of a variety of satellite-borne flux measuring experiments selected by Gault *et al.* (1972). Essentially, better than order-of-magnitude agreement exists between the two types of results, except at the smallest masses.

However, we still seek explanations for the apparent differences observed. The differences may be explained in terms of experimental uncertainties related to calibration of either the satellite-borne or lunar rock "detectors" or to the quantitative

J. B. HARTUNG, F. HÖRZ, and D. E. GAULT

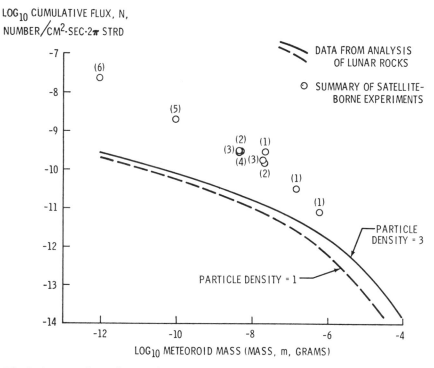

Fig. 9. A comparison of a cumulative flux versus particle mass curve based on data presented in Figs. 8 and 10 and flux data from satellite-borne experiments, (1) Pegasus (NASA TM X-53629, 1967 and Naumann et al., 1969); (2) Explorer 16 (Naumann, 1966); (3) Explorer 23 (O'Neal, 1968); (4) Lunar Orbiter 1 through 5 (Grew and Gurtler, 1971); (5) Lunar Explorer 35 (Alexander et al., 1971); (6) Pioneer 8 and 9 (Berg and Richardson, 1971). Satellite data are standardized for a particle impact velocity of 20 km/sec and density of 1.0 g/cm³. Data taken on earth orbit are corrected by a factor of ½ to get flux at the moon. See Gault et al. (1972).

detection of particles near the sensitivity threshold for the "instruments." Statistical uncertainties due to insufficient data may contribute to the differences observed, but this is not a major factor, in our judgment. Other possible explanations for the observed discrepancies are solar wind sputtering effects and variations in the flux over geologic time.

CONCLUSIONS

We have investigated many surfaces on crystalline rocks and breccias but report here only results based essentially on two glass surfaces. Reasons for this selection are: (a) These surfaces are in a nonequilibrium state with respect to cratering; the result of each impact event is observable; (b) these surfaces are extremely smooth, thus permitting quantitative observation of very small impact features; (c) projectile properties are more easily related to the size and shape of craters on glass because glass is mechanically more homogeneous than crystalline rocks and breccias, especially in a mm scale; and (d) microcrater calibration data obtained from small-scale cratering experiments on glass targets are available. These are not readily applied to crystalline rocks and especially breccias.

Based on binocular optical microscope and scanning electron microscope observations of these glass surfaces we offer the following conclusions:

(1) The ratio of spall to pit diameter of lunar microcraters decreases with decreasing crater size.

(2) A size-dependent, gradual transition from pit-only to pit-plus-spall craters exists.

(3) Some craters without a glass-lined pit are not necessarily low-velocity impacts.

(4) Microcraters with glass-lined pits are formed by the impact of primary interplanetary dust particles because melting occurred during their formation. According to experimental and theoretical data, such melting phenomena require impact velocities which are consistent only with the velocity distribution of extralunar particles.

(5) An average minimum flux of particles 25 micrograms and larger is 2.5×10^{-6} particles/cm^2 × yr × 2π strd on the lunar surface over the last 10^6 yr.

(6) A minimum cumulative flux curve over the range of mass, 10^{-12} to 10^{-4} g, based on lunar data alone is about an order of magnitude less than independently derived present day flux data from satellite-borne detector experiments.

(7) Particles of masses 10^{-7} to 10^{-4} g are the dominant contributors to the cross-sectional area of interplanetary dust particles. These particles are largely responsible for the zodiacal light and the exposure of fresh lunar rock surfaces by superposition of microcraters.

(8) The overwhelming majority of all energy deposited at the surface of the moon by impact is delivered by particles 10^{-6} to 10^{-2} g in mass. These projectiles are largely responsible for most impact phenomena observed in the lunar fines, such as fusion, vaporization, ionization, ejection, and admixture to extralunar meteoritic components.

REFERENCES

Ahrens T. J. and O'Keefe J. (1972) Shock melting and vaporization of lunar rocks and minerals. *The Moon* **4**, 214–249.

Alexander W. M., Arthur C. W., Bohn J. L., Johnson J. H., and Farmer B. J. (1971) Lunar Explorer 35: 1970 dust particle data and analysis of shower related picogram lunar ejecta orbits in the earth-moon system (abstract). 14th COSPAR Meeting, Seattle, Washington, June, 1971.

Berg O. E. and Gerloff U. (1970) Orbital elements of micrometeorites derived from Pioneer 8 measurements. *J. Geophys. Res.* **75**, 6932–6939.

Berg O. E. and Richardson F. F. (1971) New and supplementary data from the pioneer cosmic dust experiments (abstract). 14th COSPAR Meeting, Seattle, Washington, June 1971.

Bhandari N., Bhat S., Lal D., Rajagopalan G., Tamhane A. S., and Venkatavaradan V. S. (1971) High resolution time averaged (millions of years) energy spectrum and chemical composition of iron-group cosmic ray nuclei at 1 A.U. based on fossil tracks in Apollo samples. *Proc. Second Lunar Sci. Conf., Geochim. Cosmochim. Acta* Suppl. 2, Vol. 3, pp. 2611–2619. MIT Press.

Bloch M. R., Fechtig H., Gentner W., Neukum G., Schneider E., and Wirth H. (1971a) Natural and simulated impact phenomena: A photo documentation. Max-Planck-Institut fur Kernphysik, Heidelberg, Germany (private printing).

Bloch M. R., Fechtig H., Gentner W., Neukum G., and Schneider E. (1971b) Meteorite impact craters, crater simulations, and the meteoroid flux in the early solar system. *Proc. Second Lunar Sci. Conf., Geochim. Cosmochim. Acta* Suppl. 2, Vol. 3, pp. 2639–2652. MIT Press.

Carter J. L., and MacGregor J. D. (1970) Mineralogy, petrology, and surface features of some Apollo
11 samples. *Proc. Apollo 11 Lunar Sci. Conf., Geochim. Cosmochim. Acta* Suppl. 1, Vol. 1, pp.
247–265. Pergamon.

Carter J. L. and McKay D. S. (1971) Influence of target temperature on crater morphology and
implications on the origin of craters on lunar glass spheres. *Proc. Second Lunar Sci. Conf., Geochim.
Cosmochim. Acta* Suppl. 2, Vol. 3, pp. 2653–2670. MIT Press.

Cour-Palais B. G. (1969) The characteristics of very high speed small particle impacts in rocks and
glasses (abstract). *Meteoritics* **4**, 268–269.

Cour-Palais B. G., Flaherty R. E., Brown M. L., and McKay D. S. Apollo window meteoroid
experiment (abstract). In *Lunar Science—III* (editor C. Watkins), p. 157, Lunar Science Institute
Contr. No. 88.

Fechtig H. (1972) Cosmic dust in the atmosphere and in the interplanetary space at 1 AU today
and in the early solar system. *Proc. International Astron. Colloq. No. 13, The Evolutionary and
Physical Problems of Meteoroids* (in press).

Fleischer R. L., Hart H. R. Jr., Comstock G. M., and Evwaraye A. O. (1971) The particle track
record of the Ocean of Storms. *Proc. Second Lunar Sci. Conf., Geochim. Cosmochim. Acta* Suppl. 2,
Vol. 3, pp. 2559–2568. MIT Press.

Fredriksson K., Nelen J., and Melson W. G. (1970) Petrography and origin of lunar breccias and
glasses. *Proc. Apollo 11 Lunar Sci. Conf., Geochim. Cosmochim. Acta* Suppl. 1, Vol. 1, pp. 419–
432. Pergamon.

Frondel C., Klein C. Jr., Ito J., and Drake J. C. (1970) Mineralogical and chemical studies of Apollo
11 lunar fines and selected rocks. *Proc. Apollo 11 Lunar Sci. Conf., Geochim. Cosmochim. Acta*
Suppl. 1, Vol. 1, pp. 445–474. Pergamon.

Frondel C., Klein C. Jr., and Ito J. (1971) Mineralogical and chemical data on Apollo 12 lunar fines.
Proc. Second Lunar Sci. Conf., Geochim. Cosmochim. Acta Suppl. 2, Vol. 1, pp. 719–726. MIT
Press.

Gault D. E., Shoemaker E. M., and Moore H. J. (1963) Spray ejected from the lunar surface by
meteoroid impact. NASA TN D-1767.

Gault D. E., Quaide W. L., Oberbeck V. R. (1968) Impact cratering mechanics and structures. In
Shock Metamorphism of Natural Materials (editors B. M. French and N. M. Short), pp. 87–99.
Mono.

Gault D. E. and Wedekind J. A. (1969) The destruction of tektites by micrometeorite impact. *J.
Geophys. Res.* **74**, pp. 6780–6794.

Gault D. E., Hörz F., and Hartung J. B. (1972) Effects of microcratering on the lunar surface. *Proc.
Third Lunar Sci. Conf., Geochim. Cosmochim. Acta* Suppl. 3, Vol. 3. MIT Press.

Grew G. W. and Gurtler C. A. (1971) The lunar orbiter meteoroid experiments. NASA TN D-6266.

Hartung J. B., Hörz F., and Gault D. E. (1972a) Lunar rocks as meteoroid detectors. *Proc. International
Astron. Union Colloq. No. 13, The Evolutionary and Physical Problems of Meteoroids* (in press).

Hartung J. B., Hörz F., McKay D. S., and Baiamonte F. L. (1972b) Surface features on glass spherules
from the Luna 16 sample. *The Moon* (in press) .

Hörz F. and Hartung J. B. (1971) The lunar-surface orientation of some Apollo 12 rocks. *Proc.
Second Lunar Sci. Conf., Geochim. Cosmochim. Acta* Suppl. 2, Vol. 3, pp. 2629–2638. MIT Press.

Hörz F., Hartung J. B., and Gault D. E. (1971) Micrometeorite craters on lunar rock surfaces.
J. Geophys. Res. **76**, 5770–5798.

Mandeville J. C. and Vedder J. F. (1971) Microcraters formed in glass by low density projectiles.
Earth Planet. Sci. Lett. **11**, 297–306.

McKay D. S. (1970) Microcraters in lunar samples. *Proc. Twenty-Eighth Annual Meeting Electron
Microscopy Society of America* (editor C. J. Arceneaux), pp. 22–23. Claitors.

McKay D. S., Greenwood W. R., and Morrison D. A. (1970) Origin of small lunar particles and
breccia from the Apollo 11 site. *Proc. Apollo 11 Lunar Sci. Conf., Geochim. Cosmochim. Acta*
Suppl. 1, Vol. 1, pp. 673–694. Pergamon.

Moore H. J., Gault D. E., and Lugn R. V. (1963) Experimental impact craters in basalt. *Am. Inst.
Mining Metal. Engr. Trans.* **226**, 258–262.

Morrison D. A., McKay D. S., Moore H., Bogard D., and Heiken G. Microcraters on lunar rocks (abstracts). In *Lunar Science—III* (editor C. Watkins), pp. 558, Lunar Science Institute Contr. No. 88.

NASA SP-8013 (1969) Meteoroid environment model—1969 (near earth to lunar surface).

NASA TM X-53629 (1967) Scientific results of Project Pegasus.

Naumann R. J. (1966) The near earth meteoroid environment. NASA TN D-3711.

Naumann R. J., Jex D. W., and Johnson C. L. (1969) Calibration of Pegasus and Explorer 23 detector panels. NASA TR R-321.

Neukum G., Mchl A., Fechtig H., and Zähringer J. (1970) Impact phenomena of micrometeorites on lunar surface materials. *Earth Planet. Sci. Lett.* **8**, 31–35.

O'Neal R. L. (1968) The Explorer 23 micrometeoroid satellite. NASA TN D-4284.

Price P. B., Rajan R. S., and Shirk E. K. (1971) Ultra-heavy cosmic rays in the moon. *Proc. Second Lunar Sci. Conf., Geochim. Cosmochim. Acta* Suppl. 2, Vol. 3, pp. 2621–2627. MIT Press.

Rancitelli L. A. (1972) Personal communication.

Rehfuss D. E. (1962) Semiempirical model of lunar meteorite impact. *J. Geophys. Res.* (in press).

Schonfeld E. (1971) Personal communication. (See LSPET [Lunar Sample Preliminary Examination Team] Preliminary examination of the lunar samples from Apollo 12. *Science* **167**, 1325–1339.)

Vedder J. F. (1971) Microcraters in glass and minerals. *Earth Planet. Sci. Lett.* **11**, 291–296.

Walker R. M. (1971) Personal communication.

Whipple F. L. (1967) On maintaining the meteoritic complex. In *The Zodiacal Light and the Interplanetary Medium* (editor J. L. Weinberg), pp. 409–426. NASA SP-150.

Zook H. A., Flaherty R. E., and Kessler D. J. (1970) Meteoroid impacts on the Gemini windows. *Planet. Space Sci.* **18**, 953–964.

Proceedings of the Third Lunar Science Conference
(Supplement 3, *Geochimica et Cosmochimica Acta*)
Vol. 3, pp. 2755–2765
The M.I.T. Press, 1972

Simulated microscale erosion on the lunar surface by hypervelocity impact, solar wind sputtering, and thermal cycling

J. A. M. McDonnell, D. G. Ashworth, R. P. Flavill,
and R. C. Jennison

University of Kent, Canterbury, England

Abstract—Initial results from an experimental program investigating simulated erosion on a microscale by hypervelocity impact of iron spheres, thermal cycling and solar wind sputtering are presented. Hypervelocity impact craters show varying crater morphologies within differing velocity regimes. Fracture zones extend to typically 12 times the impacting particle diameter on Lunar rock for impacts of iron particles at velocities of several kilometers per second, showing radial and concentric fractures. Primary craters observed at impact velocities above several kilometers per second are typically 2 to 4 times the particle diameter. The value of the crater to particle ratio shows very little velocity dependence despite the differing morphology. Impact parameters are similar to those on quartz although the lunar rock exhibits greater tendency toward plastic deformation. At the highest velocities observed, of approximately 15 km sec^{-1}, a conical raised area of 2 to 4 times the primary crater diameter surrounds the primary crater; on quartz at this velocity, only one or two radial cracks typically remain, but on lunar rock distinct fractures are sometimes entirely absent. Penetration parameters applicable to the interpretation of natural lunar craters are presented. Thermal cycling has been performed by examination of a high density simulated impact area after thermal excursions of 210°K. No evidence for damage on a microscale was observed over the sample, including 90 hypervelocity impact areas and many grain boundaries and imperfections. It is concluded that thermal cycling is a weak erosion force on a microscale. Solar wind measurements have not yet been completed, but a review of recent satellite and lunar data leads to an estimated solar wind sputter rate of 0.04 Å yr^{-1} for the lunar surface. This would present an almost negligible contribution to erosion compared to microparticle impact, and also therefore present only a very weak erasure mechanism for submicron craters.

INTRODUCTION

The relative importance of the various parameters affecting lunar development depend on the scale of the field observed. In the widest sense the formation of the moon probably depends on accretion processes in the primordial meteor complex influenced primarily by gravitational forces and capture conditions. Modification of this accretion by heating leading to consolidation, melting, convection, and mineralogical differentiation has led to the formation of a solid crust exposed to the space environment. Major forces since then have been volcanism and meteor impact; crater distributions of kilometer-scale craters on maria show that impact saturation is not reached over this time scale but on the scale of hundreds of meters, since impact erosion accounts for the removal of some 10^{-7} cm yr^{-1} (McDonnell and Ashworth, 1972), impact saturation is reached in an exposure time of only 10^9 yr. Such impact erosion is probably the prime agent for the comminution and transport of lunar material and the subsequent generation of the lunar regolith. The magnitude

2755

of impact erosion from larger meteorites has been estimated from terrestrial observations prior to the Apollo program, but the opportunity for more detailed investigations, in particular the recovery of material has permitted an extension of the field of observation down to the submicron range. Particle impact studies promise to yield information not only on the moon but on the zodiacal meteor complex and its evolution in time.

Hypervelocity impact measurements

A 2 MeV electrostatic accelerator and associated dust particle injector is used to accelerate iron carbonyl microspheres to velocities of up to ~ 15 km sec^{-1}. Two types of impact studies are conducted: (1) high-density cratering studies, where a large number of particles ($\sim 10^3$) of varying diameters and velocities impact on an area of 3 mm^2, and (2) discrete impact studies, where craters formed by a small number of particles of measured diameter and mass are observed. In the high-density studies, correlation of crater profiles with impacting particle parameters may be estimated on a statistical basis from characteristics of the accelerator particle beam, but in the discrete impact tests a prediction of the impact position on the sample can be made for each particle; a positional accuracy of $< \cdot 2$ mm enables the use of several identifiable impacts per mm^2 of target leading to economy of material and a reduction in the microscope scanning time.

Prior to impact, the samples are cut and polished to optical finish. Crater examination is by optical microscopy and by stereoscan electron microscopy. For the latter purpose a coating of 50 Å to 100 Å of aluminum or gold is vacuum deposited under sample rotation.

Crater profiles observed on rock 14310,158, a tough medium grained basaltic rock (45% pyroxene and 45% feldspar), show a range of morphologies. At velocities of 2 km sec^{-1} to 4 km sec^{-1} extensive fracture zones show radial cracks that often terminate in concentric fractures or at crystal imperfections or interfaces as shown in the larger low velocity craters in Fig. 1a and in Fig. 3. Platelets formed in this fracture zone are raised from the surface, and this damage on high-density crater areas is even visible on a macroscale by a "blooming" of the otherwise specular surface reflectivity. Such platelets on the lunar rock are not at all easily detached, however, and even where hypervelocity impacts have subsequently occurred nearby little evidence of material detachment has been observed. Impacts on quartz show similar dimensions, Fig. 2, but by contrast the platelets formed on quartz are readily shed at low velocities (approximately 50% at 4 to 7 km sec^{-1}) when Conchoidal fracturing beneath these platelets is revealed. Crater depth is typically that of the primary crater diameter.

At higher velocities, in the region of 7 km sec^{-1} and above, the form of the damage surrounding the primary crater on the rock sample changes to a raised shallow cone, and evidence of plastic deformation without fracture is shown in Fig. 1 and Fig. 3. At the highest observed velocities of approximately 17 km sec^{-1}, the damage zone diameter is two to three times the primary crater diameter, but no shedding of platelets is observed even on quartz, and the incidence of radial cracks decreases. Due to the injector characteristics, particles impacting at higher velocities have much smaller

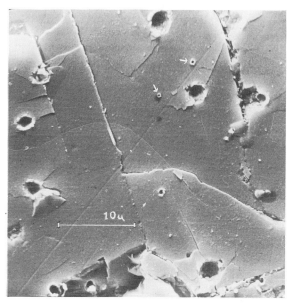

Fig. 1a. Hypervelocity impact craters of iron microspheres on polished section of lunar rock 14310,158: High density impact area: The larger craters correspond to slow but massive particles ($\sim 1\ \mu$ diameter at 3 km sec^{-1}). Smaller craters (arrowed) are typical of 10 km sec^{-1} impacts from particles of diameter $\sim 0.05\ \mu$.

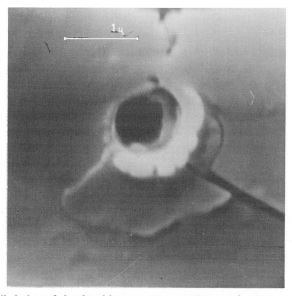

Fig. 1b. Detailed view of simulated impact crater at 6 km sec^{-1} velocity on lunar rock.

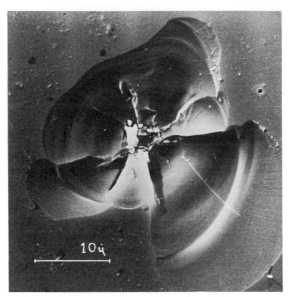

Fig. 2a. Hypervelocity impact craters of iron microspheres on quartz showing one
form of low velocity impact with conchoidal fracture.

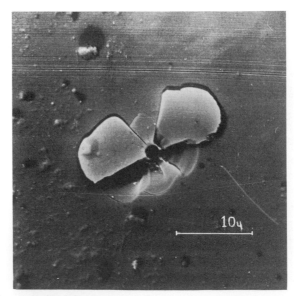

Fig. 2b. Hypervelocity impact craters of iron microspheres on quartz showing one
form of low velocity impact with conchoidal fracture.

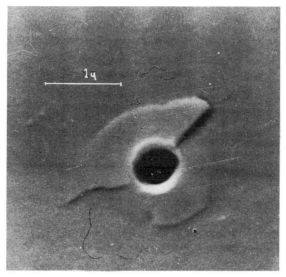

Fig. 2c. Impacts at higher velocities, ~ 7 km sec^{-1}, show only one or two radial fractures and no loss of fractured material.

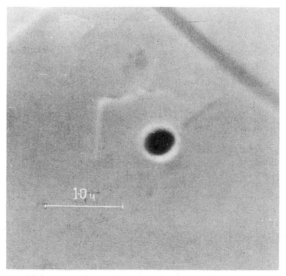

Fig. 2d. Impacts at higher velocities, 8 km sec^{-1}, show only one or two radial fractures and no loss of fractured material.

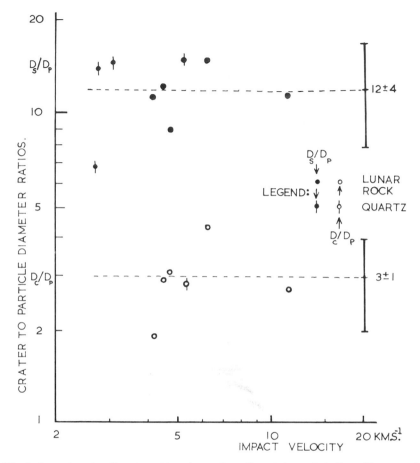

Fig. 3. Impact crater diameter ratios observed on discrete impact studies of lunar rock 14310,158 and on quartz. The data shows a high penetration resistance of lunar rock 14310 on a microscale and also the weak velocity dependence of the penetration ratios.

dimensions. Dimensional scaling of penetration parameters is not strong over the range of dimensions observed in this experiment.

The impact data on lunar rock and on quartz shows very high resistance of the lunar sample to hypervelocity impact, in agreement with other earlier data (Vedder, 1971) on similar materials and in direct contrast to impact on metallic samples (Auer *et al.*, 1968). A ratio of crater diameter to particle diameter of 3 ± 1 measured for this sample of lunar rock over a velocity range of 3 to 11 km sec^{-1}, and the very weak dependence of the impact velocity, indicates that this figure will not be grossly inaccurate at 20 km sec^{-1}. This velocity is typical of lunar primary impacts. Similarly a value of 12 ± 2 for the ratio of the diameter of the fractured zone (or the raised area at higher velocities) to the diameter of particle, measured for the lunar rock and for quartz from 3 to 11 km sec^{-1}, may be used tentatively at 20 km sec^{-1}. At 17 km sec^{-1} crater photographs show that the expected value of the damaged area is approximately two to three times the primary crater diameter although the absolute value of this figure has yet to be determined from the discrete impacts. Shown in Table 1 are several hypervelocity impact parameters which have been measured. Also shown are expected values at 20 km sec^{-1} impact velocity that may be used as a guideline for interpretation of lunar microcraters. At high velocities the late stage equivalence of the hypervelocity impact processes indicates that the impacting particle density is of minor importance except for extreme shape factors, and the value of the mass inferred from the dimensions of a crater is probably realistic for most particles with a density of 1 gm cm^{-3} upward.

THERMAL CYCLING

The possibility of the thermal erosion of lunar material resulting from hypervelocity stresses has been investigated by temperature cycling sample 14310,158 after hypervelocity impact but it has proved impossible to detect *any* surface changes at or near 90 impact areas after 10 cycles of 210°K excursion. The thermal cycles had a rise time of approximately 15 min and a periodicity of 24 hr. Since on a microscale the temperature differences across impact zones during heat flow are minute, it is probable that the temperature extremes are the major features of thermal cycling on the lunar surface, and it is likely that this preliminary test is fairly representative of the damage which could be induced on the lunar surface. However, it is impossible to say whether the results obtained with sample 14310 are representative of the surface as a whole as no previous experimental work is known on the thermal erosion of lunar material. The failure to detect induced displacement or fracture of even the spalled zones, in

Table 1. Hypervelocity penetration parameters reduced from impact craters of iron microspheres at 2 to 11 km sec^{-1} on lunar rock 14310 and on quartz.

Impact velocity (iron projectiles)	Target material	Spall diameter ÷ particle diameter (D_S/d_p)	Primary crater diameter ÷ particle diameter (D_C/d_p)	Damaged volume ÷ particle volume	Reduced formula for estimation of particle mass from observed crater diameters (microns) at typical lunar primary impact velocity
3 km sec^{-1} to 6 km sec^{-1} 6 km sec^{-1}	lunar rock 14310 quartz aluminum[1]	12 ± 2 no spall damage	3 ± 1 3.0	50 to 100^4 60	
11.3 km sec^{-1} 11.0 km sec^{-1} 11.0 km sec^{-1}	lunar rock 14310 norite[2] aluminum[1,3]	11.3 17.0 no spall damage	2.7 2.3 5.0	50 to 100^4 50 to 100^4 200	
			Tentative figures		
20 km sec^{-1}	lunar rock or glass	12 ± 4	3 ± 1	30 to 100	particle mass = 2.4 × 10^{-15} [D_S]^3 gm or = 1.5 × 10^{-13} [D_C]^3 gm
	aluminum	no spall damage	8.0^3	600	particle mass = 8.1 × 10^{-15} [D_C]^3 gm

[1] Auer *et al.*, 1968.
[2] Bloch *et al.*, 1971.
[3] Extrapolated from 8 km sec^{-1}.
[4] Includes material raised or deformed but not necessarily removed.

which platelets appear quite loosely bonded, however, indicates the strength of this type of sample on a microscale and suggests that thermal cycling is not a major force. Nevertheless the clearly crystalline and fractured nature of the surface of this sample (identified as top surface) defies explanation by hypervelocity impact alone. We are left without explanation of the generation of this surface condition unless (i) sample orientation deductions are false, or, less likely, (ii) the sample has been recently emplaced on the surface, or (iii) the sample has suffered recent spall damage from secondary impact. It appears from the results of our experimental observations to date that spall damage and thermal cycling are not responsible for the development of this crystalline surface. Figure 4a and b show a hypervelocity impact area before and after a thermal cycling run. The tilt of photograph 4(b) is 45° away from the observer, but 4a is photographed at normal incidence.

Solar Wind Sputtering

Sputtering of lunar rock samples, in conjunction with comparison surfaces, is being conducted to simulate the erosion effects of the solar wind at a velocity of 400 km sec^{-1}. The ion sources produce hydrogen ions at 0.82 keV and helium ions at 3.28 keV and simulate the effect of fluxes of 2.10^8 and 9.10^6 ions cm^{-2} sec^{-1} of hydrogen and helium, respectively. Measurements on a flat surface and near impact zones will be performed. It will be possible to collect sputtered lunar material, which might be of importance to long-term optical component degradation on the lunar surface. Recent measurements (Bühler, *et al.* 1969; Geiss *et al.*, 1970) have indicated

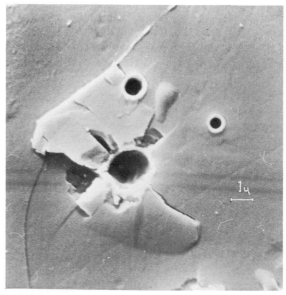

Fig. 4a. Specimen hypervelocity impact craters examined before thermal cycling. On 90 such impact sites examined, no platelets were observed to be dislodged by this cycling, nor were any fracture lines extended.

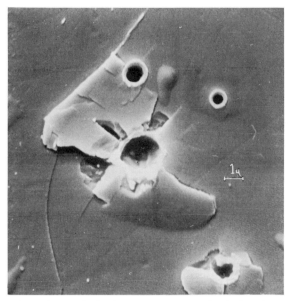

Fig. 4b. Specimen hypervelocity impact craters examined after thermal cycling.

Table 2. Estimated sputtering rates of lunar rocks and meteorites under solar wind bombardment, using recent solar wind flux measurements.

Solar wind component	Average solar wind flux $(cm^{-2} sec^{-1})$	Energy in keV (at 400 km $sec^{-1})$	Sputtering yield for lunar rocks (atoms $ion^{-1})$ [c]	Sputtering rate of lunar rocks (atoms $cm^{-2} yr^{-1})$	Sputtering rate of lunar rocks $(Å yr^{-1})$ [d]	Sputtering rate meteorites (space, sunlit face) $(Å yr^{-1})$
Hydrogen $^1H^+$	2.0×10^8 [a]	0.82	0.00214	1.35×10^{13}	0.016	0.064
Helium $^4H^{++}$	9.0×10^6 [b]	3.28	0.0188	5.34×10^{12}	0.006	0.025
		Total solar wind sputter rate		1.88×10^{13}	0.022	0.089

[a] Ness, 1965; Neugebauer, 1970; Clay et al., 1971.
[b] Buehler et al., 1972.
[c] McDonnell and Ashworth, 1972.
[d] A 50% reduction in the sputter efficiency is included due to contamination by constituents of the lunar atmosphere.

a much lower solar wind helium ion flux than used in earlier sputter estimates, and this had led to the suggestion that the average sputter rate on the lunar surface may be as low as 0.02–0.04 Å yr^{-1} (McDonnell and Ashworth, 1972) compared to previously accepted figures in the region of 1 Å yr^{-1} (Wehner et al., 1963). The low helium flux measurements have been confirmed (Buehler et al., 1972) by the Apollo 14 and 15 solar wind composition experiments, leading to an average flux of 9×10^6 ions $cm^{-2} sec^{-1}$. This average flux, together with the average solar wind proton flux (Ness, 1965; Neugebauer, 1970; Clay, 1971), leads to a value of 0.02–0.04 for the sputter rate on the lunar surface (Table 2). The implications of the low sputter rate have significance for pico-gram particle flux rates of interplanetary dust deduced from sub-micron surface crater counts (Neukum *et al.*, 1972) where equilibrium can be established between particle crater formation and erasure by solar wind sputtering.

Information about the interaction of adsorbed constituents from the lunar atmosphere and the solar wind could also be obtained by performing sputtering studies under a controlled atmosphere, maintaining a similar component ratio to that existing at the lunar surface. Implantation rates could be measured by gas analysis of subsequently heated samples and thereby deduce the effective interaction cross sections and implantation energies. Possible implantation effects have been investigated qualitatively by Ashworth and McDonnell (1972).

CONCLUSIONS

(1) Hypervelocity impact tests have established penetration profiles and criteria which permit the differentiation of primary lunar microcraters from secondary impacts. The spall diameter D_S is typically 12 times the particle diameter d_p and the primary crater diameter D_C is typically three times d_p for iron particles impacting at meteoric velocities. A very weak velocity dependence is indicated. No evidence for the spall diameter ratio of $D_S/d_p = 100$ as used to interpret impact craters on Apollo windows (Cour-Palais, 1972) was observed. Plastic deformation of lunar rock becomes evident at higher impact velocities, and at these velocities only one or two radial fractures are observed. The impacting particle for lunar microcraters may be estimated by the formula $m_p = 2.4 \cdot 10^{-15}(D_S)^3$ gm or $m_p = 1.5 \cdot 10^{-13}(D_C)^3$ gm where D_S and D_C are expressed in microns.

(2) Thermal cycling tests have failed to show any fractures of lunar material or any extensions of damage induced by hypervelocity impact, and thermal cycling seems unlikely, therefore, to contribute a major part of erosion at microscale dimensions.

(3) Solar wind sputter measurements have not been presented yet, but recent data on the solar wind combined with laboratory sputter yield measurements indicate that the lunar surface sputter rate may be as low as 0.02 Å yr^{-1} and 0.09 Å yr^{-1} for meteorites at 1 AU heliocentric distance.

Acknowledgments—We are much indebted to the Electron Physics Department at the University of Birmingham for the use of the 2 MeV accelerator (D. Smith, D. Bedford); the Mineralogy and Petrology Department of the University of Cambridge for specimen preparation (S. O. Agrell, Mr. Porter) and the Department of Crystalography, Birkbeck College, London, for Stereoscan Microscopy (M. Moore). We thank Miss D. Paine for outstanding effort in manuscript preparation.

REFERENCES

Ashworth D. G. and McDonnell J. A. M. (1972) Features of the lunar surface adsorptivity and atmospheric composition revealed by Apollo 14 pressure sensors. To appear.

Auer S., Grün E., Rauser P., and Rudolph V. (1968) Studies on simulated micrometeoroid impact. In *Space Research*, Vol. 8, pp. 606–616. North Holland.

Bloch R., Fechtig H., Gentner W., Neukum G., Schneider E., and Wirth H. (1971) Natural and simulated impact phenomena. Max Planck Institute fur Kernphysik, Research Report MPIH 6.

Bühler F., Cerutti H., Eberhardt P., and Geiss J. (1972) Results of the Apollo 14 and 15 solar wind composition experiments (abstract). In *Lunar Science—III* (editor C. Watkins), p. 102, Lunar Science Institute Contr. No. 88.

Bühler F., Eberhardt P., Geiss J., Meister J., and Signer P. (1969) Apollo 11 solar wind composition experiment: First results. *Science* **166**, 1502–1503.

Clay D. R., Neugebauer M., Snyder C. W., and Goldstein B. E. (1971) Solar wind observations on the lunar surface with the Apollo 12 ALSEP. Preprint.

Cour-Palais B. G., Flaherty R. E., Brown M. L., and McKay D. S. (1972) Apollo window meteoroid experiment, S-176 (abstract). In *Lunar Science—III* (editor C. Watkins), p. 157, Lunar Science Institute Contr. No. 88.

Geiss J., Eberhardt P., Bühler F., Meister J., and Signer P. (1970) Apollo 11 and 12 solar wind composition experiments: Fluxes of He and Ne isotopes. *J. Geophys. Res.* **75**, 5972–5979.

McDonnell J. A. M. and Ashworth D. G. (1972) Erosion phenomena on the lunar surface and meteorites. In *Space Research*, Vol. 12, Akademie-Verlag, Berlin.

Ness N. F. (1965) The interplanetary medium. In *Space Science* (editor W. N. Hess), p. 327, Blackie.

Neugebauer M. (1970) Initial deceleration of solar wind positive ions in the earth's bow shock *J. Geophys. Res.* **75**, 717–733.

Neukum G., Schneider E., Mehl A., Storzer D., Wagner G. A., Fechtig H., and Bloch M. R. (1972) Lunar craters and exposure ages derived from crater statistics and solar flare tracks (abstract). In *Lunar Science—III* (editor C. Watkins), p. 581, Lunar Science Institute Contr. No. 88.

Vedder J. F. (1971) Microcraters in glass and minerals. *Earth Planet. Sci. Lett.* **11**, 291–296.

Wehner G. K., KenKnight C., and Rosenberg D. L. (1963) Sputtering rates under solar wind bombardment. *Planet. Space Sci.* **11**, 885–895.

Proceedings of the Third Lunar Science Conference
(Supplement 3, *Geochimica et Cosmochimica Acta*)
Vol. 3, pp. 2767–2791
The M.I.T. Press, 1972

Microcraters on lunar rocks

D. A. MORRISON, D. S. MCKAY, and G. H. HEIKEN

NASA Manned Spacecraft Center,
Houston, Texas 77058

and

H. J. MOORE

U.S. Geological Survey,
Menlo Park, California 94025

Abstract—Microcrater frequency distributions have been obtained for nine Apollo 14 rocks and an exterior chip of an Apollo 12 rock. The frequency distributions indicate that five of the Apollo 14 rocks were tumbled more than once exposing different rock faces whereas four were not tumbled and represent a single exposure interval.

The cumulative frequency of craters per cm^2 was extended below optical resolution limits using a SEM scan of an exterior chip of breccia 12073. No craters with central pit diameters less than 15 microns were seen in a total area of 0.44 cm^2. A detailed SEM scan of crystal faces and glassy crater liners revealed no microcraters equal to or larger than the resolution limit of 5 microns. An upper limit of 170 craters per cm^2 with central pit diameters larger than 5 microns was set. The slope of the cumulative frequency curve for craters with central pit diameters less than about 75 microns is less than that obtained by other workers.

For distributions with the largest number of craters, the cumulative frequency of craters per cm^2 is inversely proportional to the square of their diameter. We interpret these distributions to represent a "steady-state" condition in which craters are destroyed as rapidly as they are produced. Using empirical data, we calculate coefficients between $10^{-0.398}$ to $10^{-0.096}$ for the steady-state equation. These values are about 4 to 8 times larger than that of the steady-state distribution of larger craters in the regolith.

Two current flux estimates predict a rate of crater production of $10^{-3.618}$ and $10^{-4.291}$ craters per cm^2 per year with *spall diameters* equal to or larger than 0.1 cm respectively. Applying these crater population rates to rock surfaces results in exposure ages which are generally less than ages determined by other means. Several factors may account for the discrepancy, including a lower flux in the past than the present rate. Using empirical data to account for these possibilities results in closer agreement between exposure ages determined by crater population studies and by particle track ages. Exposure age and crater population data for rock 14301 indicate a production rate of 10 craters per cm^2 per million years of 0.1 cm spall diameter or larger, but these data do not provide a quantitative flux model and do not resolve the question of variation of flux rates.

Metal particles of probable meteoritic origin are incorporated within glass filling a fracture in rock 14306, which has an exposure age of 24 m.y. The metal and glass may be products of the Cone Crater event and suggest that the meteorite producing Cone Crater was an iron or rich in iron.

INTRODUCTION

OUR OBJECTIVE IS TO present and interpret crater population data in terms of the flux and mass distributions of meteoroids in near-lunar space and to apply the results to the reconstruction of the history of rocks during their residence time near the lunar surface. To this end, we present data from an examination of the crater populations on the surfaces of nine rocks from the Apollo 14 site and one rock from the Apollo 12 site. Samples examined are listed in Table 1. The data are then combined with the results of other crater population studies and exposure ages.

Previous work (Horz *et al.*, 1971) establishes that glass-lined craters on lunar rock surfaces are produced by hypervelocity impacts of meteoroids, and that crater populations may indicate the mass distribution of the particles colliding with the lunar surface. We accept their work as a starting point.

Most of the rocks examined (see Table 1) are either documented or their approximate orientation on the lunar surface can be inferred with some confidence. Orientations of rocks 14305, 14306, 14313, 14318, and 14321 are well documented by lunar surface photography (Sutton *et al.*, 1971). The orientation of rock 14301 has been tentatively determined in the photography by Sutton *et al.* (1971), and lighting experiments with a model of 14301 indicate to us that the orientation determined by Sutton *et al.* (1971) is very plausible.

Rocks 14311, 14053, and 14073 were not documented by photography, but their lunar surface orientations have been determined by microcrater distribution studies (Horz *et al.*, 1972).

Rock 12073 formed part of the contingency sample at the Apollo 12 site, and the lunar surface orientation of this rock is imprecisely known (Shoemaker *et al.*, 1970). All surfaces of the rock are cratered and orientation is not determinable by crater distributions.

With the exception of 14053, 14073, and 12073 all the rocks are breccias of the Fra Mauro type. The breccias span the metamorphic classification of Warner (1972) from the low-grade facies (14313) to intermediate levels of the high-grade facies (14306, 14318, 14321). 14053 and 14073 are basalts. 12073 is a microbreccia of the Apollo 11 type and falls within the welded breccia classification of McKay *et al.* (1971), or the unshocked microbreccia class of Chao *et al.* (1971).

Rocks 14053, 14073, 14301, and 14311 have microcrater distributions suggestive of single exposure intervals. Surfaces of these rocks exposed at the time of collection are subangular to subrounded and cratered, whereas surfaces buried at the time of collection are angular and uncratered. Cratered and uncratered surfaces of rocks 14053 and 14311 are separated by a boundary. Buried and uncratered surfaces are light grey and cratered surfaces are light brown. The boundary between cratered and

Table 1.

Sample number	Rock type	Weight (grams)	Crater density
14301	Breccia	1361	Low
14305	Breccia	2498	High
14306	Breccia	584	High
14311	Breccia	3204	Moderate
14313	Breccia	144	Variable
14318	Breccia	600	High
14321	Breccia	9000	High to moderate
14053	Basalt	251	Moderate
14073	Basalt	10	Moderate
12073	Breccia-chip		High

uncratered surfaces of 14301 is not marked by a color change. The cratered surface of 14301 includes a glassy, flat, slickensided surface of approximately 20 cm² which has a low crater density.

The remaining samples (14321, 14305, 14313, 14318, and 14306) have large crater densities on all major surfaces. These rocks tend to be more rounded and have higher crater densities on all surfaces than the samples with simpler exposure histories.

Rocks 14318 and 14306 are rounded, coherent breccias of the same metamorphic grade (Warner, 1972). All surfaces on each rock have dense crater populations (Table 2). Both rocks have fractures filled with dark brown vesicular glass.

The fracture in rock 14306 is approximately 2 mm wide. Most of one wall and more than one-half of the glass have been stripped off. The glass filling the fracture contains three cylindrical metal particles 1 to 2 mm in length and a 200 μ diameter metal spherule. One of the cylindrical metal particles intersects a glass-walled vesicle. The section of the cylinder intersecting the vesicle is concave and conforms to the vesicle wall. Two other cylindrical fragments are separated by several millimeters but have ends which clearly could fit together. They appear to represent a single fragment which pulled apart. There is no obvious source of metal in the adjacent rock. Wosinski et al. (1972) have identified metallic spherules in the same glass ranging from 30 Å to 100 μ in diameter.

PROCEDURES

Whole rock surfaces were examined with a binocular microscope mounted on a cabinet in the curatorial facility at the Manned Spacecraft Center. The rocks were positioned under the microscope and fields of view of cratered surfaces were studied at magnifications ranging from 25× to 100×. Fields of view were divided into four quadrants and craters in each quadrant were counted separately.

Table 2. Cumulative crater size-frequency/cm².

Sample no.	Area counted cm	Lunar orientation	Ratio D_s/D_p	0.075	0.150	0.250	0.350	0.550	0.850	1.50	2.50
143061	0.80	Parallel to surface	3.4	124	113	81	43	18	8		
14306D1	4.9	90° to lunar horizontal	3.4	62	58	42	27	17	5	1.0	
14306C1	4.9	Buried	3.4	31	30	27	18	12	6	0.6	
14306B1	1.48	Parallel to surface	3.4	43	39	25	14	11	4	0.7	
14306C2	4.9	Buried	3.4	30	28	23	16	13	6	1.2	
14306A1	1.48	90° to lunar horizontal	3.4	36	33	24	15	12	6	2.0	
143211	4.9	25° to lunar horizontal	4.5		12	10	7.0	6	2.2	1.0	0.2
14321B9	4.9	50° to lunar horizontal	4.5		7.1	7	6	5	3	1.3	0.21
14321B10	4.9	50° to lunar horizontal	4.5		6	5.3	5	3	1.4	0.6	
14321C11	4.9	40° to lunar horizontal	4.5		7.2	6.5	5.3	4.3	2.0	1.4	0.4
14321E13	4.9	90° to lunar horizontal	4.5			5.2	4.5	2.9	1.2	1.0	0.2
14321E12	4.9	90° to lunar horizontal	4.5		6.7	6.3	4.4	3.1	0.6	0.4	
14311	4.9	Approx. parallel to surface	3.4	26.2	19.4	9.8	3.9	1.4	0.2		
14053D1	4.9	45° to lunar horizontal	3.6	12	11.7	9.8	6.5	5.3	2.2	1.2	
14053C1	4.9	70° to lunar horizontal	3.6		7.6	7.2	5.1	2.0	1.0	0.2	
14053A1	4.9	30° to lunar horizontal	3.6	10	9.4	7	4.5	4.1	2.2	0.4	
14053A2	4.9	30° to lunar horizontal	3.6	17.8	16.2	12.3	6.7	4.3	0.8	0.2	
14305D1	4.9	Approx. parallel to surface	3.5	29	28	25	16	11	5	1	
14305D3	4.9	Approx. parallel to surface	3.5	35	33	27	17	13	3.5	0.8	
14305E2	4.9	90° to lunar horizontal	3.5	26	25	22	13	6.5	3.4	0.6	
14318B2	4.9	Approx parallel to surface	3.3	70	57	38	23	13	3	1	
14301A2II	4.3	30° to lunar horizontal	3.0	25	14	5	2	0.9			
14301A2M	4.3	30° to lunar horizontal	3.0	30	21	8	2	0.9	0.23		
14301A1	4.8	30° to lunar horizontal	3.0	90	60	30	13	5			
14301A3	5.6	30° to lunar horizontal	3.0	16	10	5	1	0.7			
14301A5H	6.4	30° to lunar horizontal	3.0	19	14	8	4	3	1.4	0.6	0.3
14301A5M	6.4	30° to lunar horizontal	3.0	19	17	10	5	2	0.6		
14301AV	20.6	30° to lunar horizontal	3.0	33	23	11	3.5	2.4	0.74		

Two observers measured craters in the same field of view in several instances. Comparison of results between quadrants of the same field of view and between observers showed differences of up to 20%. All fields of view were located on rock models or photographs of each rock.

Each central pit observed was counted and its diameter measured. A spall zone where defined was also measured for each crater. Sufficient numbers of craters were measured on each distinct face of most rocks so that statistically significant differences in crater densities within and between distinct faces could be obtained. Crater population distributions were mapped on rock models and photographs. These data allowed reconstruction of the lunar surface orientation of some rocks. An exterior chip was scanned with a scanning electron microscope to extend crater statistics below practicable optical resolution limits. Approximately 4500 microcraters were measured with both techniques.

Results

As a result of our examination of the 10 lunar rocks and consideration of the data of others, we recognize the following: (1) the ratios of spall diameter (D_s) to central pit diameter (D_p) of craters on rock surfaces vary from sample to sample; (2) the form of the cumulative frequency distribution varies with crater diameter; (3) the cumulative frequency distributions on rock surfaces vary from rock to rock; (4) the cumulative frequency distributions on surfaces of a given rock vary from surface to surface; (5) the form of the cumulative frequency distribution for surfaces with low crater populations is different than those with the largest crater populations; (6) crater frequency distributions of surfaces with very large crater populations approach a limiting distribution; and (7) for surfaces which have not reached the limiting distribution, the cumulative frequency distributions of the cratered surfaces are correlative with exposure ages determined from analysis of particle tracks.

Spall-central pit diameters

Many craters examined are similar to those reported previously (Horz et al., 1971). They have a central glass-lined pit surrounded by a spall region. For smooth surfaces of glass and fine-grained rocks with few craters, spall surfaces are easily visible and nearly always present. Here, true spall diameters can frequently be measured. In contrast, spall diameters on surfaces with large crater populations are difficult to measure because erosion by subsequent micro-meteorite impacts reduces their topographic expression. However, evidence of a spall surface is normally found around each central pit. Thus, it is clear that central pit diameters do not represent the actual diameter of the original crater.

For each rock, however, ratios of spall diameter (D_s) to central pit diameter (D_p) can be measured for a number of craters. The ratios vary not only between rocks but also on a single surface. Ratios as high as 10 were measured whereas some craters had no discernible spall zone, but such extremes are rare, and for most surfaces we computed an average ratio. For the glassy slickensided surface of 14301, the average ratio is 3, the lowest average ratio measured. The average ratio on surfaces of 14321 is 4.5. The average ratio observed on 14053, a basalt, is 3.6 and somewhat less than the ratio of 4.5 reported for Apollo 12 basalts (Horz et al., 1971). Spall zones were not visible on the surface chip of 12073 we examined, but Horz et al. (1971) report a value of 4.5 for this rock.

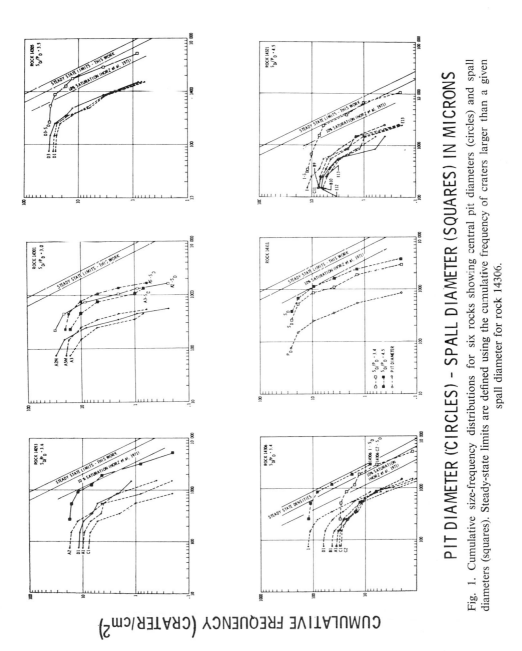

PIT DIAMETER (CIRCLES) - SPALL DIAMETER (SQUARES) IN MICRONS

Fig. 1. Cumulative size-frequency distributions for six rocks showing central pit diameters (circles) and spall diameters (squares). Steady-state limits are defined using the cumulative frequency of craters larger than a given spall diameter for rock 14306.

CUMULATIVE FREQUENCY (CRATER/cm²)

Other values are reported in Table 2. As seen in Table 2 there is no apparent correlation of spall diameter to pit diameter ratios and rock-type or metamorphic grade. For example, 14306, 14318, and 14321 fall within the same group (Warner, 1972) but have ratios of 3.4 and 4.5 respectively. Friable rocks appear to have higher ratios, but friability is a difficult quality to measure.

Form of distributions

The cumulative size-frequency distributions of craters obtained using optical methods exhibit the general characteristics discussed by Horz *et al.* (1971) and are related to crater diameter. Typically, slopes of the cumulative frequency distribution

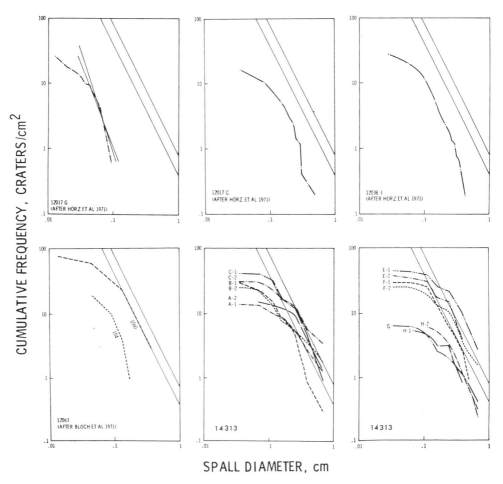

Fig. 2. Cumulative size-frequency distributions of craters for four rocks. Note variations between glass (12017G) and crystalline (12017C) surfaces of rock 12017, two different surfaces of 12063, and those of 14313. Two parallel lines represent steady state defined by 14306. Straight lines drawn through data for 12017G represent the form of distributions predicted using two "best estimates" flux (see text).

when plotted in log-log form are -2, -3, or more negative for craters with spall diameters near 850 μ and larger. For smaller craters, slopes change from -2 to zero with decreasing diameter. The general character of the distributions is shown in Figs. 1 and 2.

We extended the crater frequency distribution measurements to pit diameters of 15 μ using a Scanning Electron Microscope (SEM) and an exterior chip of breccia 12073 because studies of frequency distributions of large craters in the lunar regolith indicate that flattening of the distributions can be the result of incomplete counting at the limit of photographic resolution. The result of these measurements are shown in Fig. 3 along with data collected by Horz *et al.* (1971) for rock 12073. In Fig. 3, it may

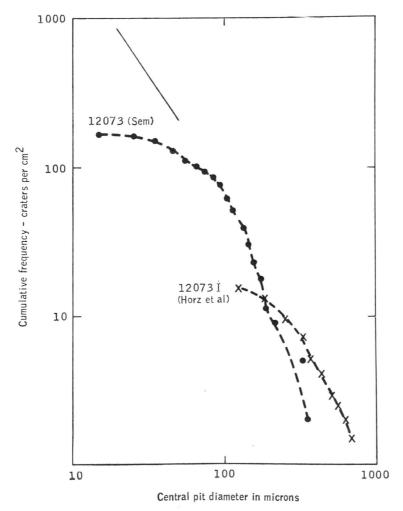

Fig. 3. Cumulative size-frequency distribution of craters for rock 12073. Data for curve labeled 12073 (SEM) obtained using scanning electron microscope. Data for curve labeled 120731 obtained using optical methods by Horz *et al.* (1971). Upper solid line refers to slope calculated from the data of Neukum *et al.* (1972).

be seen that crater counts are, in part, affected by the resolution of the instrument used to obtain the data. Instead of flattening of the cumulative size-frequency distribution implied by the optical microscope data for craters of 300 μ diameter or less, the SEM data show a flattening of the curve at 90 μ or less. The cumulative size-frequency distribution curve has a slope near -1.9 for craters with diameters between 90 and 300 μ. Below this, the curve flattens. This flattening must be real because the smallest crater measured is 15 μ and is larger than the resolution (1–2 μ) of the SEM, and the range of crater diameters is large enough to preclude recognition problems (Horz et al., 1971). For optical measurements, the flattening for smaller sizes is, at least in part, the result of resolution effects.

The 15 μ lower limit crater diameter measured on the host rock is considerably larger than the micron-sized craters reported on other surfaces (Bloch et al., 1971; Neukum et al., 1972). Therefore, we scanned the glassy liners of the larger craters on the host rock for superposed microcraters. No superposed microcraters were found which could be identified unambiguously as microcraters produced by hypervelocity impact. A further search was made of smooth crystal faces projecting above the rock surface. A total of 37 such faces with an aggregate area of 0.59 mm^2 were scanned at magnifications of 500 to 2000 ×. At these magnifications, we set a conservative resolution limit of 5 μ although microcraters of 1–2 μ could have been detected. No microcraters of 5 μ diameter or larger were seen. Although the 0.59 mm^2 is too small to represent a satisfactory statistical sampling, it sets an upper limit of one crater per 0.59 mm^2 or approximately 170 craters per cm^2 of 5 μ or larger central pit diameter.

The resulting distribution has a slope of -0.8 over the interval 50 μ to 95 μ and then flattens to nearly zero at smaller crater sizes (Fig. 3). The distribution for these crater diameters is different than that indicated by the data of Neukum et al. (1972) for which we calculate a slope of -1.3, and which is shown by the solid line in Fig. 3. Possible reasons for this discrepancy are discussed later.

Variations from rock to rock

Cumulative size-frequency distributions vary from rock to rock. Such variations can be seen in Fig. 1 where cumulative frequencies of craters for rock 14301 are substantially less than those of rock 14306. More striking differences are found between 12017G, which has a very low cumulative size-frequency distribution, and 12063,106A, which is higher by a factor of 20 to 30.

Variations on surfaces of one rock

Varying cumulative size-frequency distributions for different surfaces on the same rock are commonly found (see Figs. 1 and 2). For example, the cumulative size-frequency distributions for most of the surfaces on rock 14313 fall within the parallel straight lines at some point on the distribution curve. This is not so for surfaces B-2, G, H-1, and H-2 (Fig. 2). For these surfaces the cumulative size-frequency curves lie below those of other faces. Correlation of rock 14313 with surface documentation photographs (Swann et al., 1971, pp. 78–79) show that faces G, H-1, and H-2 were buried at the time of collection. A-1 and A-2 faced downward and were shielded but not buried.

In some cases, rocks have surfaces with different compositions, such as rock 12017, and the frequency distributions are different for the different materials. Figure 2 shows two distributions for 12017 from the data of Horz *et al.* (1971). The first is a glass surface (12017G in Fig. 2) coating the rock and the second (12017C in Fig. 2) is a crystalline surface that is older than the glass. The cumulative size-frequency of craters for the first is significantly lower than that of the second. In other cases, faces of a given rock may be similar, such as D-1, D 2, and D-3 of rock 14305 and faces A-1, B-1, C-1, and C-2 of rock 14306, although lunar surface orientations differ by as much as 90°.

Limiting frequency distribution

In addition to the varying slopes, the cumulative frequency curves exhibit a tendency to approach a limiting distribution which the cumulative frequency of craters per cm^2 may approach but not exceed. We define a range of limiting crater frequency distribution or crater density per cm^2, which is shown by the solid lines in Figs. 1 and 2, using rock 14306. The slopes of the cumulative frequency distributions for surfaces which fall within or near the limiting values are near -2 for craters larger than about 600 μ. This may be seen by juxtaposing the frequency curves for 14306,1 and 14321,1 in Fig. 1.

Cumulative frequency curves for rocks 14053, 14301, 14311 (Fig. 1) and the A-type surfaces of rock 14313 (Fig. 4, and Table 3) fall below the limits defined by rock 14306 crater densities, but they are near the 10% saturation level of Horz *et al.* (1971). Cumulative frequency curves for these rocks have a slope of -3 for intermediate crater sizes and steeper slopes for the larger crater sizes. This is distinctly different than the 14306 type of distribution which has a slope of -2 and a larger number of craters.

Table 3. Cumulative size-frequency-craters/cm^2—14313.

Sample	Area counted cm^2	Surface* type	0.338 / 0.075	0.675 / 0.150	1.125 / 0.250	1.575 / 0.350	2.475 / 0.550	3.825 / 0.850	6.75 / 1.5
14313H2	2.8	A	6.4	6.1	4.6	2.5	1.8	1.1	0.4
14313H2	1.3	A		5.4	4.6	3.1	2.1	0.8	
14313G	3.49	A			6.0	5.2	3.2	1.1	0.3
14313F2	1.3	C2	24.6	23.8	18.5	13.1	10.1	3.9	
14313F1	4.4	C2	29	28.6	25.2	13.2	5.9	2.5	0.23
14313E2	2.3	C2	37	33.5	30.5	17.5	14.5	3.5	1.5
14313E1	3.2	C2	44.1	42.5	38.1	26	21.6	10.2	2.5
14313C′	4.6	C3	40.9	40.4	32.4	17.2	12	4.8	0.9
14313C″	4.9	C	31	30	26.8	16.8	12.7	5.7	3.5
14313B2	0.55	C1	25	23.6	16.4	7.3	5.4	3.6	1.8
14313B1	2.8	C1	30	23	13.6	9.6	3.9	1.4	
14313A1	3.72	B	13.9	13.6	10.5	8.7	5	0.9	0.3
14313A2	3.23	B		15	13.2	7.5	5.6	2.7	1.3

The column headers above are: D_s (mm): 0.338, 0.675, 1.125, 1.575, 2.475, 3.825, 6.75; and D_p (mm): 0.075, 0.150, 0.250, 0.350, 0.550, 0.850, 1.5.

* As shown in Fig. 4.

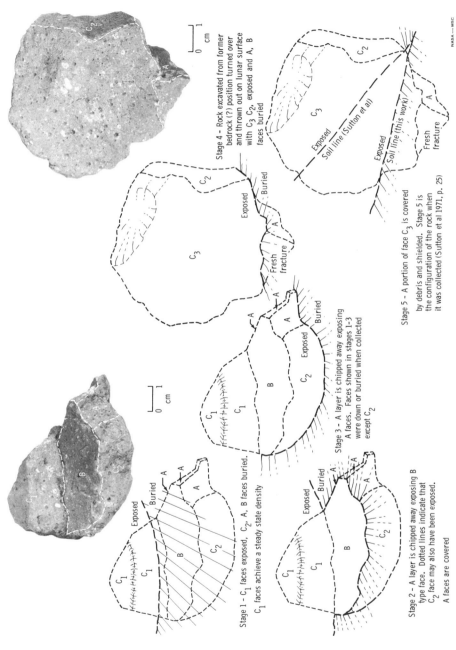

Fig. 4. Diagram illustrating tumbling history of rock 14313 as indicated by crater populations over its various surfaces. Estimates of the time span spent in each stage are listed in Table 5.

Frequency distributions and exposure ages

If the cumulative frequency distribution of a rock surface is less than the limiting value established by 14306, the cumulative frequency distribution may be a function of length of exposure to meteorite bombardment. Exposure ages for various rocks are listed in Table 4. For example, the particle track age of 12017G is 9×10^3 yr (Fleischer *et al.*, 1971) and that of 12038 is 1.3×10^6 yr (Bhandari *et al.*, 1971). A corresponding difference is seen in the cumulative frequency distributions. For the younger rock surface, 12017G, the cumulative frequency of craters of all crater sizes is less than that of the older rock 12038,1. A similar result is found for 14301 with an exposure age of 3.4×10^5 yr (Hart *et al.*, 1972), and 14311 with an exposure age of 1.1 to 3.4×10^6 yr (Hart *et al.*, 1972). For 14301, the cumulative number of craters with spall diameters larger than 0.1 cm is 3.5 (Fig. 1) whereas for 14311 it is 6.7 (Fig. 1). Such a correlation between cumulative frequency distributions and exposure ages is also found for different surfaces on one rock as seen for rock 12017 (Table 4 and Fig. 2). The glass surface (12017G) has a smaller exposure age and lower cumulative frequency distribution while that of crystalline surface (12017C) has a greater exposure age and a higher cumulative frequency distribution.

EXPOSURE AGE CALCULATIONS

The variation of the cumulative frequency distributions with exposure ages suggest that it should be possible to calculate the age of a rock surface by combining meteoroid flux estimates and crater population data. Therefore, in this section, we will compare particle track and spallation exposure ages of some of the rocks collected during the Apollo 12 and 14 missions with those calculated using "best estimates" of the meteoroid flux at the lunar surface and hypervelocity impact data. Then the results will be discussed.

Flux estimates

One current "best estimate" of the influx of meteoroids obtained using data collected from spacecraft (Soberman, 1971, p. 255) is expressed by two equations:

$$\phi_1 = 10^{-14.37} \, m_p^{-1.21} \tag{1}$$

Table 4. Exposure ages of Apollo 14 and 12 rocks.

Sample no.	Exposure age in yrs	Method	Source
14301	3.4×10^5	Tracks	Hart *et al.*, 1972
14311	3.4×10^6	Tracks	Hart *et al.*, 1972
	1.1×10^6		
14321	$24 \pm 2 \times 10^6$	Cosmic rays	Burnett *et al.*, 1972
14053	$24 \pm 2 \times 10^6$	Cosmic rays	Burnett *et al.*, 1972
14073	110×10^6	Cosmic rays	Burnett *et al.*, 1972
12063,106A top	$\leq 1.5 \pm 0.5 \times 10^6$	Tracks	Crozaz *et al.*, 1971
12063,104 bottom	$< 7 ^{+3}_{-7} \times 10^5$	Tracks	Crozaz *et al.*, 1971
12038	1.3×10^6	Tracks	Bhandari *et al.*, 1971
12021	Several m.y.	Tracks	Barber *et al.*, 1971
12017C	7×10^5	Tracks	Fleischer *et al.*, 1971
12017G	9×10^3	Tracks	Fleischer *et al.*, 1971

and

$$\phi_2 = 10^{-14.34} m_p^{(-1.58 - 0.063 \, \log_{10} m)} \tag{2}$$

where ϕ_1 and ϕ_2 are the cumulative number of impacts/m^2/sec/2π strd. of meteoroids of mass m_p and larger. ϕ_1 is valid for meteoroid masses between 10^{-6} and 10^0 g; ϕ_2 is valid for meteoroid masses between 10^{-12} and 10^{-6} g. The flux predicted by Whipple's curve B (1963) is expressed by:

$$\phi_3 = 10^{-13.80} m_p^{-1.0} \tag{3}$$

where ϕ_3 is the cumulative number of impacts/m^2/sec of meteoroids of mass m_p and larger. ϕ_3 is valid for meteoroid masses of 10^{-9} g and larger.

The flux equations may then be combined with an equation developed using hypervelocity impact data on rocks (Moore, Gault, and Heitowit, 1965):

$$M_e^{0.841} = (10^{-8.929}) \, (\rho_p/\rho_t)^{1/2} \, E_p \tag{4}$$

where M_e is the ejected mass in grams of a crater produced by hypervelocity projectile impact of energy E_p, ρ_p is the density of the projectile or meteoroid (g/cm^3), and ρ_t is the density of the target (g/cm^3). This equation was selected because it considers the effect of the projectile and target density and it agrees well with the data on micro-craters produced in glass by low density projectiles with masses between 0.7 and 62 picograms (Mandeville and Vedder, 1971).

Equation (4) can be recast in the form

$$m^{-1} = \frac{10^{-8.929}}{M_e^{0.841}} \left(\frac{\rho_p}{\rho_t}\right)^{1/2} \frac{V^2}{2} \tag{5}$$

where V is the projectile or meteoroid velocity. We then adopt: (1) a meteoroid density of 0.44 g/cm^3 which is consistent with Whipple's (1963) results, (2) a target density of 3 g/cm^3 which is consistent with the many lunar rocks, (3) a meteoroid velocity of 2×10^6 cm/sec, and (4) a spherical segment crater with a depth to spall diameter ratio of 1 to 5.

These adoptions yield:

$$m^{-1} = 10^{3.471} D^{-2.523} \tag{6}$$

where D is the crater spall diameter in centimeters. Substitution of equation (6) in equation (4) yields:

$$\phi_1 = 10^{-10.170} D^{-3.053} \text{ (impacts/m}^2\text{/sec/2}\pi \text{ strd.)} \tag{7}$$

or

$$\phi_1 = 10^{-6.66} D^{-3.053} \text{ (impact/cm}^2\text{/yr/2}\pi \text{ strd.);}$$

substitution of equation (6) in equation (2) yields

$$\phi_2 = 10^{-14.34} 10^{-3.471} D^{2.523(-1.58 - 0.063 \, \log_{10} 10^{3.471} D^{-2.523})} \cdot \text{(impacts/m}^2\text{/sec/2}\pi \text{ strd.)}$$

or

$$\phi_2 = 10^{-10.83} 10^{-3.471} D^{2.523(-1.58 - 0.063 \, \log_{10} 10^{3.471} D^{-2.523})}$$

$$\text{(impacts/cm}^2\text{/yr/2}\pi \text{ strd.);} \tag{8}$$

and substitution of equation (6) in equation (3) yields:

$$\phi_3 = 10^{-10.329}D^{-2.523} \text{ impacts/m}^2/\text{sec}$$

or

$$\phi_3 = 10^{-6.819}D^{-2.523} \text{ impacts/cm}^2/\text{yr} \qquad (9)$$

Equation (6) yields a spall diameter of 0.0995 cm (0.1 cm) if the micrometeoroid has a mass of 10^{-6} g and a spall diameter of 0.0399 (0.04 cm) if the micrometeoroid has a mass of 10^{-7} g. Thus, the cumulative numbers of craters produced with spall diameters larger than 0.1 cm per cm^2 per yr per 2π strd. are $10^{-3.601}$ for ϕ_1, $10^{-3.618}$ for ϕ_2, and $10^{-4.291}$ for ϕ_3. The predicted cumulative number of craters with spall diameters of 0.1 cm and larger have been plotted in Fig. 5 for the interval of 10^3 to about $10^{5.7}$ yr. For craters with spall diameters 0.04 cm and larger, the cumulative numbers of craters produced per cm^2 per yr per 2π strd. is $10^{-2.389}$ for ϕ_1, $10^{-2.848}$

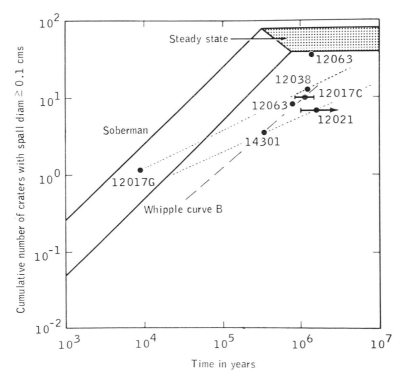

Fig. 5. Comparison between cumulative number of craters with spall diameters larger than 0.1 cm for seven lunar rocks and their exposure ages determined using particle tracks. Upper sloping solid line represents predictions made using Soberman's (1971) best estimate (ϕ_1) and lower sloping solid line represents predictions made using Whipple's (1963) curve B (ϕ_3). Short dashed line represents empirical expectations for meteoroid flux model with decreasing rates in the past. Long dashed line represents constant meteoroid flux model which is less than both ϕ_1 and ϕ_3. Shaded area represents steady-state condition defined by rock 14306.

for ϕ_2, and $10^{-3.289}$ for ϕ_3. If it is assumed that the terrestrial flux is higher than the lunar flux by a factor of 2 and such a correction is made, then the cumulative number of 0.04 mm craters produced per cm^2 per yr per 2π strd. is $10^{-2.692}$ for ϕ_1, $10^{-3.149}$ for ϕ_2, and $10^{-3.60}$ for ϕ_3.

Ages calculated for crater counts and crater production rates

With the combined knowledge of the estimates of the crater production rates and crater counts, it should be possible to estimate the age of a rock surface. We shall attempt this below and then place some limitations on the procedure later. In comparing calculated exposure ages to the published values listed in Table 4, we assume that the particle track ages indicate the amount of time the surfaces dated were directly exposed to the lunar environment.

Horz et al. (1971) have reported crater counts on a glass surface of an Apollo 12 rock (12017G). Their data is shown in Fig. 2, where it should be noted that the counts do not have the form of equation (7) or (9), and the intercept of the counts at 0.1 cm is near 0.34 craters/cm^2. We interpret this to be the result of an insufficient sample and fit curves of the form of equations (7) and (9) to the counts to yield intercepts at the 0.1 cm and larger. These curves are shown as solid lines in Fig. 2. Using the terrestrial flux estimates and ignoring solid angle effects, we calculate the age of the glass surface to be 4×10^3 yr using ϕ_1, and 20×10^3 using ϕ_3. About 10 craters/cm^2 are larger than 0.04 cm so that ϕ_2 yields 7×10^3 yr for the age of the surface. The ages calculated using ϕ_1 and ϕ_3 bracket the particle track age of 9×10^3 yr (Fleischer et al., 1971), whereas that for ϕ_2 is very close to it. If the intercept at the 0.1 cm size is take as 1 craters/cm^2 and the age of the surface is 9×10^3 years, then the production rate is $10^{-3.954}$ craters/cm^2 yr, greater than ϕ_3 but less than ϕ_2 for both lunar and terrestrial estimates. This has been plotted in Fig. 5. Thus, we find that the crater production rate computed for the glass surface of 12017G is more or less consistent with the predictions made from the "best estimate" fluxes. This agreement may, however, be fortuitous. Horz (personal communication) points out that the glass dated by Fleischer et al. (1971) is uncratered and was shielded at the time of collection; therefore, a track age based upon a solar particle flux is tenuously related at best to the cratered glass surface.

Frequency distributions of craters on crystalline surfaces of rock 12017C were also obtained by Horz et al. (1971). These are plotted in Fig. 5 using a pit to spall diameter ratio of 1/4. This distribution intersects the 0.1 cm diameter size at 8.5 craters/cm^2, yielding ages of 0.34×10^5 yr using ϕ_1 and 1.7×10^5 using ϕ_3. Extra polation of the counts to the 0.04 cm size gives 15 craters/cm^2 and an age near 0.1×10^5 yr using ϕ_2. All these calculated ages, which ignore the solid angle, are less than the particle track exposure age of the rock, which is 7×10^5 to 1.5×10^6 yr (Fleischer et al., 1971).

The crater frequency distributions and surface orientation for rock 14301 are known (Swann et al., 1971, p. 79) as previously discussed, and our distributions were obtained for the top surfaces. Thus, correction for the solid angle are small. Intercepts of the distribution at the 0.1 cm size range between 1.7 and 6 craters/cm^2, and the average value is near 3.5 craters per cm^2. From these intercepts ages calculated using ϕ_1 range between 0.68×10^4 to 2.4×10^4 and average 1.4×10^4 yr. For ϕ_3 ages are between 3.3×10^4 and 1.2×10^5 yr and average 6.8×10^4 yr. Using 0.04 cm craters and ϕ_2 we find ages between 0.85×10^4 to 1.6×10^4 yr. Thus, all the calculated ages for 14301 are less than the particle track age of 3.4×10^5 yr.

Calculation of exposure ages by identical means for rocks 12021, 12038 (Horz et al., 1971), and 12063 (Bloch et al., 1971) show similar results as indicated in Fig. 5.

In general, the ages calculated using crater frequency distributions fall below the expectations for current "best estimate" fluxes. There are a number of factors that could cause this. These factors are briefly discussed below.

Factors affecting age estimates

At least six factors can affect the ages calculated using crater frequency distributions: (1) orientation of the cratered surface and topography, (2) partial covering of the cratered surface by fine debris, (3) a complex history of the rock, (4) attainment of a "steady-state" crater distribution after a certain time span in which crater density is dependent of time, (5) a decreased crater production rate in the past, and (6) erosion of the cratered surface while being handled.

Orientation and topography at both large and small scales affect the rate of crater production per unit area of a rock surface. However, the flux equations used require ideal planar topography. This is rarely the case, and lack of such ideal surfaces would tend to yield apparent ages of rock surfaces that are smaller than their actual ages. Almost any rock collected can be shown to have surfaces oriented in a variety of directions. These directions may range from overhangs to nearly level. Thus, for cases where rock orientations are known, one may take the orientation of the surface into account by a suitable correction for the solid angle. Additionally, rock surfaces are irregular in detail. Here, corrections are more difficult. Other rocks lie near larger rocks that protect them from meteoroids. One such rock is 14314, which is shadowed by "Turtle Rock" (Swann et al., 1971, p. 55). For such a case, shadowing effects could be accounted for, at least approximately, using suitable topographic data. Shadowing due to larger features such as shallow depressions and surrounding hills is clearly present but more difficult to analyze.

Partial covering of rock surfaces by fine debris will also tend to give smaller ages calculated from crater frequency distributions than their real ages. This would be particularly true for the smaller craters on the rock surfaces when dust thicknesses are about the same as the crater depth.

A complex rock history may also reduce the apparent age of a rock surface. Rocks that have been exposed at the surface may be displaced and their orientations changed by subsequent events, such as larger impact events. Such complex histories are clearly the case, as shown by Fleischer et al. (1971) and the presence of microcraters on the buried surfaces of the rocks.

Rock surfaces may attain a "steady-state" frequency distribution of craters for which the area of crater population equals the rate of crater destruction. This distribution is independent of time, and ages of rock surfaces calculated using crater distributions that are in a "steady state" would, ideally, be constant. Such a "steady-state" condition has been demonstrated for larger lunar craters in the regolith theoretically (Moore, 1964; Marcus, 1970; Soderblom, 1970), experimentally (Gault, 1970; Moore, 1971), and in lunar studies (Shoemaker *et al.*, 1969; Horz *et al.*, 1971). For larger craters in the regolith a few centimeters across to several hundred meters across, this "steady-state" distribution may be approximately described by:

$$N = 10^{-1}D^{-2} \tag{10}$$

where N is the cumulative number of craters per unit area larger than spall diameter D.

It is clear that the "steady-state" distributions for small craters on rock surfaces are different than equation (10). Horz *et al.* (1971) find that the coefficient in equation (10) may exceed 10^{-1} reaching values higher than $10^{-0.7}$. Using the various models of Marcus (1970), which all predict an exponent of -2 for D, one can calculate coefficients as high as $10^{0.320}$. Using model 4 (Marcus, 1970) and $a = 1$, one calculates coefficients near $10^{-0.75}$ and $10^{-0.55}$ for values of the exponent in the crater production curve of 2.523 and 3.053, respectively. Additionally, the erosion effect of the solar wind and cosmic particles on the "steady-state" surface is unknown. Thus, the character of the "steady state" for rock surfaces is unclear. We have tentatively taken the highest crater size-frequency distributions of rock 14306 (see Fig. 1) as representing the "steady-state" condition for rock surfaces. For this "steady-state" condition we use an exponent of -2 for D, and a coefficient near $10^{-0.398}$ to $10^{-0.96}$. None of the rocks or surfaces discussed with the exception of 12063,106A (Fig. 5) approach this limit. Therefore, we postulate that those rocks which have crater size-frequency distributions less than the lower bound of our limiting frequency distribution are not in the "steady state."

It is entirely possible that the meteoroid flux in the past was either larger or smaller than the current flux. Although there is a strong argument for a larger rate of production of very large craters billions of years ago, this argument does not necessarily apply to small microcraters on rock surfaces. Indeed, some authors have postulated an increase of micrometer influx with increasing time.

Finally, some craters may be destroyed during handling. Such destruction would chiefly affect the smallest craters and be most extensive for friable rocks. This appears to be the case for rock 14321, the most friable and the largest of the rocks we examined. We postulate a distortion of the crater frequency distributions for 14321 because craters larger than about 0.5 to 0.6 cm fall within or near the assumed "steady-state" limits whereas smaller craters do not (Fig. 1). In addition, the smaller craters do not conform to the "steady-state" distribution or to the crater production distributions.

Empirical estimates

It does not appear possible to separate the various factors contributing to the apparent decrease of micrometeor flux with increasing time. Because of this and other possible unknown variables, an empirical-graphical approach to the problem of estimating exposure ages seems justified. Thus, in the empirical-graphical exposure age estimates discussed below, we assume that time-dependent factors and those due to topography, dust covering, and other factors are more or less the same for each rock and that the particle track ages are correct.

In order to empirically estimate exposure ages of some rocks, we have plotted the cumulative number of craters per cm^2 with spall diameters 0.1 cm and larger for rocks 12017G, 14301, 12063,104, 12063,106A, 12017C, 12038, and 12021 against their particle track exposure ages. For 12017G we have used the extrapolated intercept of 0.34 craters/cm^2 and for 14301 the average value of 3.5 craters/cm^2. Using these plotted points, we have drawn two parallel dotted lines (Fig. 4); one passes through 12017G and 12038 and the second is parallel to the first and passes through 14301. This results in a range of possible solutions for the cumulative number of craters larger than 0.1 cm and resulting exposure ages. Although the empirical graph described will be used below as a preliminary solution, three possibilities should be kept in mind. First, the empirical method assumes that the current flux estimates are valid and higher than those averaged over several hundred thousand years or more. Alternatively it may be assumed the current flux estimates are too high and that the flux indicated by 12017G is incorrect and high. The third possibility is that particle track ages are subject to the same or similar factors affecting the micrometer populations to the same or differing degrees and are incorrect. The first assumption was discussed previously. The second assumption is represented by a dashed line in Fig. 5 that passes through 14301, 12063, 12017C, and 12038 and is more or less parallel to the line designated Whipple's curve B.

Pits of craters on four surfaces on rock 14053 were counted and plotted using a spall to pit diameter ratio of 3.6 (Fig. 1). The orientation of this rock at the time of collection has been determined (Horz, Morrison, and Hartung, 1972), so that estimates of the effect of the solid angle can be made. Correlation of the faces we counted with the photographs indicates faces A-1 and A-2 (Table 2) were inclined about 30° to the lunar surface, face C-1 was inclined about 70° to the lunar surface, and face D-1 was about 45° to the lunar surface. The absence of pits on the buried surfaces suggest this rock had a simple history and that only one part has been exposed to the meteoroid flux. For face A-1, ϕ_1, and ϕ_3 indicate ages near 3×10^4 and 1.5×10^5 yr, while the empirical curve (dotted lines, Fig. 4) places the age near 3×10^5 to 1.9×10^6 yr. Those for A-2 are about 1.6 times larger than those for A-1. Face C-1 (Table 2), with fewest counts, is 1/1.7 times as old as A-1. Thus, we believe rock 14053 has been exposed to the surface for about 1.8×10^5 to 1.9×10^6 yr. Such a result is less than the spallation age of 26 to 30×10^6 yr for ejecta from Cone Crater. Because the spallation age may include time beneath the surface, rock 14053 may have been buried 20 or so million years and brought to the surface by an impact event some 1.8×10^5 to 1.9×10^6 yr ago, where it has remained without moving. Horz, Morrison, and Hartung (1972) suggest that the rock may have been shielded by enclosing breccia matrix.

Crater counts for rock 14313 (Figs. 2 and 4, Table 3) indicate a complex history and long exposure time at the surface. Correlation of the surfaces we counted with the orientation of the rock when it was collected (Swann et al., 1971, pp. 78–79) indicates faces A, B, C, G, and H (in Table 3) were facing downward, buried, or both as shown in Fig. 4, stage 5, face E was up and horizontal, while face F was perpendicular to the surface (Table 3). Most of the crater distributions fall within the "steady-state" region and thus are probably older than 20 to 40 $\times$ 10^6 yr. Only crater frequency distributions for faces G and H are clearly out of the "steady-state" region. Using the intercepts directly and the empirical curve yields ages of the surfaces between about 2 $\times$ 10^5 to 10^6 yr, while extrapolated intercepts are 4 to 5 times larger. Thus, this rock has apparently been exposed for more than 20 to 40 m.y., whereas faces G and H have been exposed 10^5 to 5 $\times$ 10^6 yr.

The positions occupied by the rock during this time span are shown diagrammatically in Fig. 4. The speculative estimates of the lengths of time spent in each position are shown in Table 5.

Our counts on the nearly horizontal surface of rock 14311 indicate it has a fairly low exposure age. Using a spall diameter to pit diameter ratio of 3.4, we obtain an intercept at the 0.1 cm spall diameter of 6.5 craters/cm^2 (Fig. 1). From the empirical graph the age of the surface is between 4 $\times$ 10^5 to 2 $\times$ 10^6 and probably near 10^6 yr. This is consistent with an independently determined exposure age of 1.1 $\times$ 10^6 yr based upon particle tracks (Hart *et al.*, 1972, Table 4). The second possibility (Fig. 5) indicated by the dashed line in Fig. 5 yields a similar result. Therefore, the data cannot distinguish between a lower and changing meteoroid flux in the past or a constant but lower meteoroid flux than current estimates.

Tumbling histories

Exposure age data and crater distributions allow some limits to be placed on the tumbling histories of some of the rocks we have examined. Five of the rocks have cratered surfaces which were buried at the time of collection and therefore were tumbled at least once. Four (14053, 14311, 14301, and 14073) have uncratered surfaces which were buried at the time of collection, and therefore their cratered surfaces represent a single exposure interval. The length of time of these intervals has been calculated for 14311, 14053, and 14313 and are listed in Table 5. Of this group, sample 14313 has the most complex tumbling history. A postulated sequence of events is

Table 5.

Sample no.	Calculated surface residence time in years	Other data
14313		
Stage 1	$\geq 10^7$ (?)	
Stage 2	2.5 $\times$ 10^6 (?)	
Stage 3	1 $\times$ 10^6 *	
Stage 4	$\geq 10^7$ (?)	
Stage 5	$\leq 10^6$	
Total	22.5 $\times$ 10^6 (to 40 $\times$ 10^6)	
14311	4.5 $\times$ 10^5–2 $\times$ 10^6	1.1 $\times$ 10^6 †
		3.1 $\times$ 10^6
14053	1.8 $\times$ 10^5–1.9 $\times$ 10^6	24 $\pm$ 2 $\times$ 10^6 ‡

* Stage 3 included in stage 2 span.
† Hart *et al.* (1972) particle track age.
‡ Burnett *et al.* (1972) cosmic-ray spallation age.

shown in Fig. 4. Data is listed in Table 3, which groups the crater populations into three sets based upon crater density and corresponding to the surfaces in Fig. 4. In stage 1, the rock apparently was part of a boulder on the lunar surface, and C_1 surfaces (Fig. 4, stage 1) were exposed. Stage 1 may have occupied 10^7 or more years. Stage 2 began when a catastrophic event chipped away part of the parent boulder, exposing the B surface for 10^6 yr or more. A third catastrophic event initiated stage 3 by removing a substantial fraction of the parent rock shielding the C_2 surface (Fig. 4, stage 3). The maximum exposure time in this stage is indicated by the low crater density of the A-type surfaces (Tables 3 and 5). A fourth catastrophic event removed sample 14313 from its parent and placed it on the lunar surface in about the position occupied when collected. The final event was a partial covering of face C_3 (Fig. 4, stage 5) by a shower of debris. Our data suggest that less of the rock was covered than suggested by the documentary photography (Sutton *et al.*, 1971). Because the C-type surfaces appear to be "steady-state" surfaces, the total amount of time represented by the crater populations in stages 1, 2, 4, and 5 can only be estimated as shown in Table 5. The probable minimum exposure suggested by the crater distribution is consistent with ejection of the parent boulder from Cone Crater.

DISCUSSION

Micrometeors and crater frequency distributions

In comparing the form and limits of our crater frequency distributions with the expectations of several estimates for the micrometeoroid mass distributions, we find that they do not agree for the smaller sizes. Whipple's curve B (ϕ_3) combined with equation (6) predicts that the cumulative frequency of craters should be inversely proportional to the 2.523 power of the diameter for craters from 10 cm or so to 0.0014 cm in diameter. The data collected for 12073 using the SEM indicates this is not the case (Fig. 6). Rather, the curve arches over and craters with spall diameters less than 0.005 cm are either rare or absent. Soberman's (1971) estimate arches for craters less than 0.1 m but not as markedly as the curve for 12073. The estimate of Cour-Palais *et al.* (1972) is:

$$\phi_4 = KM^{-0.56}, \tag{11}$$

which applies to very small craters, and predicts a slope near -1.4. Thus, this equation does not predict the observed bend in the cumulative frequency. None of these equations predicts the absence of craters with spall diameters less than 0.005 cm. Indeed, Cour-Palais' lower micrometeoroid mass of 10^{-13} g yields a crater diameter of 1.7×10^{-4} cm using equation (6), and we do not find such small craters. At this time, we cannot account for these differences because the factors that may affect flux estimates also may affect the form and limits of the size distribution. In the case of 12073, the combined curves for the SEM data and optical data (Fig. 6) yield a slope near -2. Thus, it is entirely possible that both 12073 and 14306 represent "steady-state" distributions. On the other hand, combining the optical and SEM data may not be justified because they represent different surfaces of the same rock and our data show variations from one face to the next face for any given rock. We do note, however, that our lower limit of crater size is in fair agreement with that of Whipple.

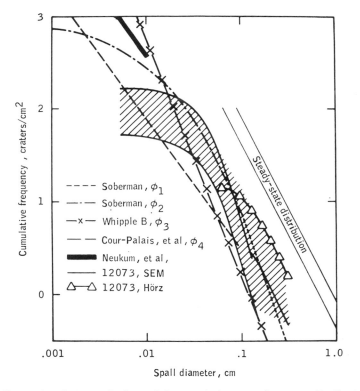

Fig. 6. Comparison between the form of the cumulative crater frequency distribution for rock 12073 and the expectations from the best estimates of Soberman (1971, ϕ_1 and ϕ_2), Whipple's curve B (1963, ϕ_3), and Cour-Palais (1972, ϕ_4). Ordinate is the log of the cumulative frequency.

Additionally, this result is in agreement with the results of Hartung *et al.* (1972), who depict a trivial frequency of craters produced by micrometeors with masses near 10^{-9} g and frequency distributions of craters which stop at central pit diameters near 0.0012 cm (spall diameters near 0.005 cm).

In any event, the relative abundance of meteorites which produce craters less than 100 μ in diameter is not well defined. Neukum *et al.* (1972) postulate a slope of approximately -1.3 for the cumulative crater frequency per cm^2 for craters with central pit diameters less than 100 to 75 μ (Fig. 3), whereas other data indicate a shallower slope. Several possibilities may account for this discrepancy: (1) the 2% of the area of 12073 we examined in detail for 5 μ or larger craters may be significantly younger than the host surface and the larger crater population, (2) the probability of detecting the predicted number of craters is low because the area examined is too small, (3) the slope of the distribution for crater diameters of 75 μ and less is less than -1.3, (4) the slope of the distribution for craters 75 μ is not constant, or (5) the surfaces examined by Neukum *et al.* (1972) are production surfaces in which small craters are less likely to have been removed by subsequent events. We believe the third possibility to be correct for the following reasons.

Hartung *et al.* (1972) have plotted cumulative frequency data for crater dimensions ranging from approximately 2μ (10^{-12} g) to 400μ (10^{-5} g), by juxtaposing cumulative frequency curves from rock 12054 and a Luna 16 spherule. The resulting slope is -0.3 on a log cumulative frequency versus log mass plot, in closer agreement with our results, and less than the value of Neukum *et al.* (1972).

Neukum *et al.* (1972) assume that a sputtering erosion rate of 1 Å per year results in a significant removal of micron-sized craters and correct their mass distribution curve upward to account for erosion. McDonnell *et al.* (1972), however, suggest that the helium concentration is overestimated and that the true sputtering rate is 0.02 Å per year. This lower rate would reduce the slope of the mass distribution curve of Neukum *et al.* (1972) because it requires a much smaller correction.

We conclude that the mass distribution of particles producing craters with central pit diameters of 75μ or less is not well defined, particularly for crater diameters of 15μ and less, and that the slope varies and is generally less than -0.8 when plotting log mass versus log cumulative frequency.

Micrometeor flux

Our results show fair agreement between micrometeor fluxes estimated using 12017G and a particle track age of 9×10^3 yr with Soberman's best estimates (ϕ_1, ϕ_2), Whipple's estimate (ϕ_3), and Cour-Palais' estimate (ϕ_4) for craters with diameters near 0.1 cm produced by micrometeoroids with masses near 10^{-6} g. This is not the case when rock 14301 and other rocks 3.4×10^5 yr and older are used. For these rocks, the estimated micrometeoroid flux falls below expectations. As noted previously, we are unable to assign a cause for this. If Cour-Palais' equation is modified to consider a lower density, then better agreement is obtained between our results for 14301 with a concomitant loss of agreement with 12017G (Fig. 7). Thus, this problem is unsolvable at this time. More data collected from rocks collected during the Apollo 16 mission may resolve these problems and permit a selection between the various possible causes. Further information could be obtained using existing rocks, for example, by determining the exposure age of the glassy surface of 12054. Crater population data for this surface have been carefully determined by Horz *et al.* (1972). The graphically determined age, assuming a lower flux rate (and a population of 1 crater of 333μ central pit diameter equivalent to 1 mm spall zone diameter), ranges from less than 10,000 to approximately 23,000 yr. The exposure age indicated by the dashed line in Fig. 5 is approximately 10^5 yr, assuming that 12017C and 14301 represent the flux. Therefore, exposure age determinations for rock 12054 would add a significant data point to flux estimates.

Steady-state surfaces

Our data indicate that steady-state surfaces have a slope of -2. All the cumulative frequency curves, however, show a steepening at larger crater sizes. This appears to be the result of restricted sample sizes relative to crater dimensions and the inevitable change of slope that occurs when a finite distribution is summed. Juxtaposition of the curves for 14306 (14306,1, Fig. 1) and 14321,1 (Fig. 1) indicates a slope of -2 over a

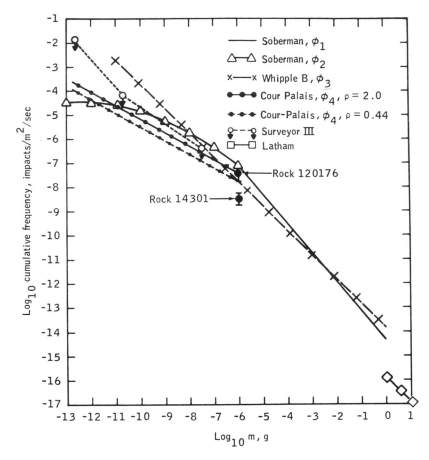

Fig. 7. Estimates of cumulative frequency distributions of meteoroid flux obtained by Soberman (1971), Whipple (curve B, 1963), Cour-Palais (1972), Latham (1971), and Brownlee (1971) compared with results from 12017G (shown as 120176 above) and 14301.

crater diameter range of 8 mm. In terms of absolute crater density on rock surfaces, the steady state remains poorly defined. If the highest crater frequency distributions for rock 14306 are taken as representative of the "steady state" then an exponent of -2 for D and a coefficient near $C = 10^{-0.398}$ to $10^{-0.096}$ can be calculated for an equation of the form $N = CD^{-2}$. The coefficient we calculate is larger than that calculated by Horz $et\ al.$ (1971). This difference may be the result of the larger spall diameter to pit diameter ratio observed by Horz $et\ al.$ (1971) on Apollo 12 basalts. Alternatively, both sets of coefficients could be correct. All these coefficients are higher than those theoretically derived for craters in the regolith (Moore, 1964; Soderblom, 1970) because the mechanism for crater obliteration is different, as pointed out by Marcus (1970) and Horz $et\ al.$ (1971).

The Cone Crater event

We have described metal particles in vesicular glass filling a penetrative fracture in rock 14306. The morphology of the particles, particularly the two cylindrical particles which appear to have pulled apart and the cylindrical particle which conforms to a glassy vesicle, suggest that the metal was injected with the glass into the fracture at a high temperature. There is no indication that metal fragments of the size and shape observed could have been derived from the host rock. Wosinski *et al.* (1972) have identified metallic spherules in the same glass as consisting of iron-nickel with segregations of troilite and schreibersite, and cite evidence pointing to rapid flow and cooling of the glass. These observations also suggest that the glass plus the metal particles were injected into the fracture at a high temperature and that the metallic particles are of meteoritic origin.

The cosmic-ray exposure age of 14306 is 24 m.y. (Crozaz *et al.*, 1972). Our data suggest an exposure age ranging from 20 to 40 m.y. Both these results are compatible with ejection of 14306 by the Cone Crater event. If the fracture was formed in 14306, and the glass and metallic particles injected by the Cone Crater event, then it is reasonable to postulate that the projectile which produced Cone Crater was an iron meteorite or rich in iron.

CONCLUSIONS

(1) Exposure age calculations based upon crater populations and the flux equations of Soberman (1971) and Whipple (1963) tend to be lower than exposure ages based upon particle tracks. This discrepancy could be the result of factors affecting crater populations such as rock orientation, partial shielding of cratered surfaces by debris, complex tumbling histories, achievement of steady-state crater densities, or decreased crater production rates in the past. If a decreased production rate with time is assumed, then an empirical-graphical method of calculating exposure ages from crater counts of non-steady-state surfaces yields reasonable results, particularly in the case of rock 14311. A decreased production rate with time, therefore, may be a valid hypothesis.

Alternatively, the true flux could be lower than present data would suggest. These two possibilities appear indistinguishable with current data. The single data point (Fig. 6) from our work pertinent to flux estimates is derived from crater population data and exposure age data (Hart *et al.*, 1972) for rock 14301. It indicates a production of 10 craters per million years of 0.1 cm spall diameter or larger, and is not sufficient to distinguish which point of view is correct.

(2) Rock surfaces in the steady state have a slope of -2 for craters with spall diameters larger than about 1 mm. We tentatively define the steady state with the crater density data of rock 14306. For this steady-state condition we use an exponent of -2 for D and a coefficient $C = 10^{-0.398}$ to $10^{-0.096}$ in an equation of the form $N = CD^{-2}$. Graphical interpretation of the data indicates that 10^7 or more years would be required to achieve this state.

(3) Crater population data for rocks 14053 and 14311 suggest exposure ages of

from 1.8×10^5 to 1.9×10^6 yr and 4×10^5 to 2×10^6 yr respectively. The higher exposure age for 14311 is in agreement with the track age of Hart *et al.* (1972).

(4) The tumbling history of rock 14313 is a complex five-stage sequence punctuated by catastrophic disruptions. Exposure ages determined empirically from crater densities are compatible with ejection of 14313 from Cone Crater, but also indicate that 14313 could have been exposed prior to the Cone Crater event.

(5) Data from rock 12073 suggest that the abundance of micrometeorites in the small mass ranges which produce craters of 75 μ or less in diameter is not well defined and may be less than that postulated by Neukum *et al.* (1972).

(6) Metal particles in glass filling a penetrative fracture in rock 14306 suggest that the projectile which excavated Cone Crater was an iron meteorite.

Acknowledgment—The work by H. J. Moore was performed under NASA Contract W13,130.

REFERENCES

Barber D. J., Cowsik R., Hutchen I. D., Price P. B., and Rajon R. S. (1971) Solar flares, the lunar surface, and gas-rich meteorites. *Proc. Second Lunar Sci. Conf., Geochim. Cosmochim. Acta* Suppl. 2, Vol. 3, pp. 2705–2714. MIT Press.

Bhandari N., Bhat S., Lal D., Rajagopalan G., Tamhane A. S., and Venkatavaradon V. S. (1971) High resolution time averaged (millions of years) energy spectrum and chemical composition of iron group cosmic ray nuclei at 1 A.U. based on fossil tracks in Apollo samples. *Proc. Second Lunar Sci. Conf., Geochim. Cosmochim. Acta* Suppl. 2, Vol. 3, pp. 2611–2619. MIT Press.

Bloch M. R., Fechtig H., Gentner W., Neukum G., and Schneider E. (1971) Meteorite impact craters, crater simulations, and the meteoroid flux in the early solar system. *Proc. Second Lunar Sci. Conf., Geochim. Cosmochim. Acta* Suppl. 2, Vol. 3, pp. 2639–2652. MIT Press.

Burnett D. S., Huneke J. C., Podosek F. A., Russ G. P. III, Turner G., and Wasserburg G. J. (1972) The irradiation history of lunar samples (abstract). In *Lunar Science—III* (editor C. Watkins), p. 105, Lunar Science Institute Contr. No. 88.

Chao E. C. T., Boreman J. A., and Desborough G. A. (1971) The petrology of unshocked and shocked Apollo 11 and Apollo 12 microbreccias. *Proc. Second Lunar Sci. Conf., Geochim. Cosmochim. Acta* Suppl. 2, Vol. 1, pp. 797–816. MIT Press.

Cour-Palais B. G., Flaherty R. E., Brown M. E., McKay D. S. (1972) Apollo window meteoroid experiment (abstract). In *Lunar Science—III* (editor C. Watkins), p. 0, Lunar Science Institute Contr. No. 88.

Cour-Palais B. G., Zook H. A., and Flaherty R. E. (1971) Meteoroid activity on the lunar surface from the Surveyor 3 sample examination. NASA TMX-58079.

Crozaz G., Droyd R., Graf H., Hohenberg C. M., Monnin M., Ragan D., Rolston C., Seitz M., Shirck J., Walker R. M., and Zimmermon J. (1972) Evidence for extinct Pu^{244} (abstract). In *Lunar Science—III* (editor C. Watkins), p. 164, Lunar Science Institute Contr. No. 88.

Crozaz G., Walker R., and Woolum D. (1971) Nuclear track studies of dynamic processes on the moon and the constancy of solar activity. *Proc. Second Lunar Sci. Conf., Geochim. Cosmochim. Acta* Suppl. 2, Vol. 3, pp. 2543–2558. MIT Press.

Fleischer R. L., Hart H. R. Jr., Comstock G. M., and Envoraye A. O. (1971) The particle track record of the Ocean of Storms. *Proc. Second Lunar Sci. Conf., Geochim. Cosmochim. Acta* Suppl. 2, Vol. 3, pp. 2559–2568. MIT Press.

Gault D. E. (1970) Saturation and equilibrium conditions for impact cratering on the lunar surface: Criteria and implications. *Radio Science* 5, 273–291.

Hart H. R. Jr., Comstock G. M., and Fleischer R. L. (1972) The particle track record of Fra Mauro (abstract). In *Lunar Science—III* (editor C. Watkins), p. 360, Lunar Science Institute Contr. No. 88.

Hartung J. B., Horz F., and Gault D. E. (1972) The origin and significance of microcraters (abstract). In *Lunar Science—III* (editor C. Watkins), p. 363, Lunar Science Institute Contr. No. 88.

Horz F., Hartung J. B., and Gault D. E. (1971) Micrometeorite craters on lunar rock surfaces. *J. Geophys. Res.* **76**, 5770–5798.

Horz F., Morrison D. A., and Hartung J. B. (1972) The surface orientation of some Apollo 14 rocks. *Modern Geology* **3**, 93–104.

Latham G. V., Ewing M., Press F., Sutton G., Dorman J., Nakamura Y., Toksoz N., Duennebier F., and Lammlein D. (1971) Passive seismic experiment. In *Apollo 14 Preliminary Science Report*, NASA SP-272, pp. 133–161.

Mandeville J. C. and Vedder J. F. (1971) Microcraters formed in glass by low density projectiles. *Earth Planet. Sci. Lett.* **11**, 297–306.

Marcus A. H. (1970) Comparison of equilibrium size distributions for lunar craters. *J. Geophys. Res.* **75**, 4977–4984.

McDonnell J. C., Ashworth D. G., Flavill R. P., and Jennison R. C. (1972) Micro-scale erosion on the lunar surface by hypervelocity impact, solar wind sputtering, and thermal cycling (abstract). In *Lunar Science—III* (editor C. Watkins), p. 526, Lunar Science Institute Contr. No. 88.

McKay D. S. and Morrison D. A. (1971) Lunar breccias. *J. Geophys. Res.* **76**, 5658–5669.

Moore H. J. (1971) Geologic interpretation of lunar data. *Earth Sci. Rev.* **7**, 5–33.

Moore H. J., Gault D. E., and Heitowit E. D. (1965) Change in effective target strength with increasing size of hypervelocity impact craters. *Proc. 7th Hypervelocity Impact Symposium*, Vol. 4, p. 341.

Morrison D. A., McKay D. S., Moore H. J., Bogard D., and Heiken G. (1972) Microcraters on lunar rocks (abstract). In *Lunar Science—III* (editor C. Watkins), p. 558, Lunar Science Institute Contr. No. 88.

Neukum G., Schneider E., Mehl A., Stoizer D., Wagner G. A., Fechtig H., and Bloch M. R. (1972) Lunar craters and exposure ages derived from crater statistics and solar flare tracks. In *Lunar Science— III* (editor C. Watkins), p. 581, Lunar Science Institute Contr. No. 88.

Shoemaker E. M., Morris E. C., Batson R. M., Holt H. E., Larson K. B., Montgomery D. R., Rennilson J. J., and Whitaker E. A. (1969) Television observations from Surveyor. TR-321265, JPL, Cal. Inst. Tech.

Soberman R. K. (1971) The terrestrial influx of small meteoric particles. *Rev. Geophys. Space Phys.* **9**, 239–258.

Soderblom L. A. (1970) A model for small impact erosion applied to the lunar surface. *J. Geophys. Res.* **75**, 2655–2661.

Sutton R. L., Batson R. M., Larson K. B., Schaffer J. P., Eggleton R. E., and Swann G. A. (1971) Documentation of the Apollo 14 samples. U.S. Geol. Survey Interagency Report No. 28.

Swann G. A. *et al.* (1971) Preliminary geologic investigations of the Apollo 14 landing site. In *Apollo 14 Preliminary Science Report*, pp. 39–85. Natl. Aeron. Space Admin. Special Publ., NASA SP-272.

Warner J. (1972) Apollo 14 breccias: Metamorphic origin and classification (abstract). In *Lunar Science—III* (editor C. Watkins), p. 782, Lunar Science Institute Contr. No. 88.

Whipple F. L. (1963) On meteoroids and penetration. *J. Geophys. Res.* **68**, 4929–4939.

Wosinski J. F., Williams J. P., Korda E. J., Kone W. T., Carrier G. B., and Schreurs J. W. H. (1972) Inclusions and interface relationships between glass and breccia in lunar sample 14306,50 (abstract). In *Lunar Science—III* (editor C. Watkins), p. 811, Lunar Science Institute Contr. No. 88.

Proceedings of the Third Lunar Science Conference
(Supplement 3, *Geochimica et Cosmochimica Acta*)
Vol. 3, pp. 2793–2810
The M.I.T. Press, 1972

Lunar craters and exposure ages derived from crater statistics and solar flare tracks

G. Neukum, E. Schneider, A. Mehl, D. Storzer,
G. A. Wagner, H. Fechtig, and M. R. Bloch

Max-Planck-Institut für Kernphysik, Heidelberg, Germany

Abstract—Measurements of microcraters performed on Apollo 14 and 12 rocks are discussed. The investigations have been performed optically and with the help of a scanning electron microscope. The crater densities found per cm² are

up to 20 craters ≥0.3 mm diameter;
up to 70 craters ≥0.1 mm diameter;
1000–3000 craters ≥ 1 μ diameter;

up to approximately 10,000 craters ≥0.25 μ diameter.

These confirm earlier results and extend the crater size frequency distribution to 0.25 μ diameter. The identification of craters and conversion of crater data to particle properties is based on new simulation experiments. With the assumption that micron- and submicron-size craters are mainly eroded by solar wind sputtering, a cumulative micrometeoroid flux of

$$\phi(m) = 6 \times 10^{-13} m^{-0.63} [\text{m}^{-2} \text{ sec}^{-1}], \qquad (10^{-15} \text{ g} < m < 10^{-9} \text{ g})$$

is calculated.

For sample 12024,8 an exposure age of 2500 years has been determined by solar flare track methods in addition to the crater size frequency. If sputter erosion is negligible a cumulative micrometeoroid flux of

$$\phi(m) = 10^{-7.54} m^{-0.283} [\text{m}^{-2} \text{ sec}^{-1}], \qquad (10^{-15} \text{ g} < m < 10^{-9} \text{ g})$$

results from the "production" crater size frequency.

From crater size frequency measurements of large craters with diameters >3 km the time development of the lunar surface can be studied. Using theoretical arguments of erosion of craters by impact superposition, mean formation ages of various parts of the lunar surface have been calculated.

Introduction

Most lunar rock surfaces show microcraters. They act as natural impact counters for interplanetary dust. In order to determine the meteoroid flux from crater size frequencies, three basic requirements have to be fulfilled: (a) crater morphology has to be studied in combination with impact simulation experiments; (b) exposure ages of the examined surface areas have to be measured; (c) erosion mechanisms altering the frequencies in different ways have to be taken into account.

In an earlier paper (Bloch *et al.*, 1971a) our investigations on microcraters on lunar rocks of Apollo 11 and 12 have been reported. Here, new measurements on microcraters performed at Apollo 14 and 12 rocks and on very large craters on the whole moon will be discussed in the frame of some theoretical considerations of erosion of craters (Neukum and Dietzel, 1971). Special attention was directed to submicron- and micron-size craters. Furthermore, new simulation experiments give us

2793

a stronger basis for identification of craters and conversion of crater data to particle properties.

MORPHOLOGY OF MICROCRATERS ON LUNAR SPECIMENS—COMPARISON WITH SIMULATED CRATERS

Microcraters with diameters larger than 10 μ found on lunar specimens usually show a central glass-lined pit surrounded by a zone of fractured material (spallation zone) which can partly or wholly be ejected. These observations were first reported by the Lunar Sample Preliminary Examination Team (LSPET, 1970). Comprehensive studies were published by Hörz *et al.* (1971a, b). Similar descriptions of microcraters on lunar material were given by McKay *et al.* (1970), Carter and McGregor (1970), Neukum *et al.* (1970), and Bloch *et al.* (1971a, b).

The identification of microcraters on lunar rock surfaces (crystalline rocks or breccias) is limited by the surface roughness of the specimens. By means of optical methods only craters larger than 100 μ in diameter can quantitatively be detected; the use of a Scanning Electron Microscope (SEM) lowers the quantitative detection limit to about 50 μ (although the resolution of the SEM is of the order of 300 Å). Below these sizes the statistics are incomplete.

Different conditions are found in the case of plane crystal surfaces or glass surfaces, such as glass spheres, dumbbells, glass splashes, or glass linings of larger craters. There, craters down to 0.1 μ in diameter can be identified with the help of the SEM.

The difficulty in this submicron- and micron-size range, however, is to discriminate genuine impact craters from bubbles or other crater-like depression features which often occur, especially on glass surfaces. Therefore additional simulation experiments have been performed to study the morphology of submicron-sized and micron-sized craters on various glass targets and a rock target (Norite). For the experiments spherical particles of iron, aluminum, carbon, and glass were fired in a 2 MV Van de Graaff accelerator.

Part of the results of these experiments was reported by Bloch *et al.* (1971a). Generally, the morphology of the artificial craters was found to be very similar to that of lunar micron-size craters. Now, the new results permit further conclusions on mass and velocity dependence of crater properties to be drawn.

The formation of the spallation zone around the central pit is dependant on target material. On quartz glass and window glass targets it occurs at lesser velocity than on Duran glass. On all targets used (window glass, Duran glass, quartz glass, Norite) we get the moon-like crater morphology at velocities of more than 5 km/sec, if some critical mass is exceeded. However, as reported by Bloch *et al.* (1970a) and Neukum (1971), the morphology of micron- and submicron-size craters also depends on projectile mass. It was shown that very small particles ($m < 10^{-13}$ g) shot on quartz glass at velocities higher than 5 km/sec form only a hemispherical pit without spallation or with only slight indication of fracturing. This effect was confirmed for iron particles at velocities up to 60 km/sec impacting on quartz glass and Duran glass targets.

In Fig. 1 micron- and submicron-sized craters on lunar crystalline material and

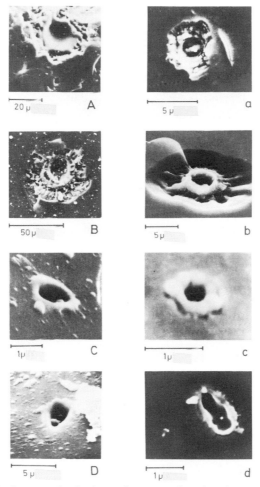

Fig. 1. Types of micron- and submicron-size craters found on lunar samples, in comparison to craters produced in the laboratory.

Lunar sample No.	Impact conditions
(A) 12063,106a Top (cryst. rock)	(a) Fe → Norite, $v = 7.6$ km/sec
(B) 14257 (coarse fines)	(b) Al → Quartz glass, $v = 5$ km/sec
(C) 14257 (coarse fines)	(c) Fe → Duran glass, $v = 50$ km/sec
(D) 12063,106a Top (cryst. rock)	(d) Fe → Duran glass, $v = 30$ km/sec, impact angle 45°

glass are compared to craters produced in the laboratory. The decreased definition of the spallation zones with decreasing crater size is apparent.

Sometimes the central pit of microcraters on lunar samples has a special form: It is situated upon a pedestal, and it is called a "stylus pit" by Hörz *et al.* (1971b). For the first time it could be simulated in our cratering experiments. It occurred if Al particles of 5-15 μ in diameter were shot on quartz glass targets at a velocity of about

3–5 km/sec. It is not necessary for its formation that the material is preshocked. Figure 2 shows a lunar stylus pit crater compared to an artificial one.

An important parameter for microcraters is the ratio of the central pit diameter D to the diameter d of the particle that formed the crater. It is a function of impact velocity v. Experiments in the mass range from 10^{-15} g to 10^{-10} g and velocity range from 5 km/sec to 60 km/sec showed a dependence of

$$D/d \sim v^\alpha,$$

with α approximately $\frac{2}{3}$. The relation permits the determination of particle properties from crater measurements. It is displayed in Fig. 3.

MICROCRATER STATISTICS

A basic question in the beginning of the microcrater investigations of lunar rocks was, are the microcraters of primary, i.e., extralunar origin, or are they secondary

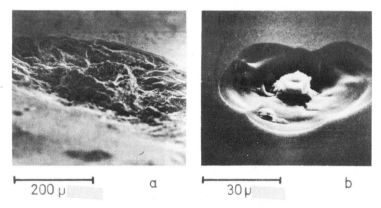

Fig. 2. "Stylus pit" craters: (a) Lunar Sample 12024,8,1; (b) Laboratory produced crater, 3 km/sec < v < 5 km/sec, Al → Quartz glass.

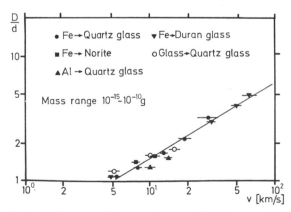

Fig. 3. Dependence of the crater diameter D to particle diameter d ratio on particle velocity v for silicate targets.

impacts of ejecta of larger craters? The investigations of Hörz *et al.* (1971b) and Hartung *et al.* (1972) on size frequency and morphology of the craters on various lunar rocks prove a primary origin at least for most of the craters in the diameter range $D \gtrsim 50\ \mu$. The velocity spectrum of impact ejecta in laboratory simulations (Gault and Heitowit, 1963) shows that most of the ejected material flies at velocities of some 100 m/sec. Very few particles have velocities exceeding a few km/sec. Therefore a secondary origin is very unlikely, because practically all microcraters with diameter $D \gtrsim 50\ \mu$ found on lunar samples show hypervelocity impact features and similar size distributions despite different orientations of the rocks on the lunar surface.

Carter and McKay (1971) concluded from the nonrandom distribution of the craters and from the complete absence of craters larger than about $10\ \mu$ in diameter on their samples that most of the micron- and submicron-size craters are due to secondary events. As those criteria do not apply to sample 12024,8,1—the distribution shown in Fig. 4 appears to be random and there are several craters with diameter $> 10\ \mu$—the craters reported here are believed to be of primary origin. This belief is strengthened by the fact that the size frequency distribution on various samples which had different locations on the moon are very similar.

In the course of our investigation of lunar material about 3000 craters have been counted. The statistical results are presented in Fig. 5. Up to 70 craters ≥ 0.1 mm diameter per cm^2 and 20 craters ≥ 0.3 mm diameter per cm^2 were found. The cumulative number of craters per unit area increases with decreasing crater diameter. On the average the slope of the distribution for the different specimens in the diagram is between -2 and -3 around 1 mm crater diameter and approaches

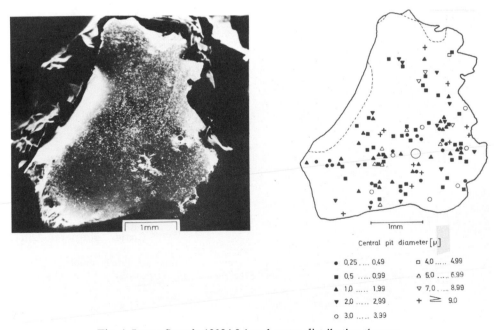

Central pit diameter [μ]	
● 0.25 .... 0.49	□ 4.0 .... 4.99
■ 0.5 ..... 0.99	∆ 5.0 .... 6.99
▲ 1.0 ..... 1.99	▽ 7.0 .... 8.99
▼ 2.0 ..... 2.99	+ ≧ 9.0
○ 3.0 ..... 3.99	

Fig. 4. Lunar Sample 12024,8,1 and crater distribution thereon.

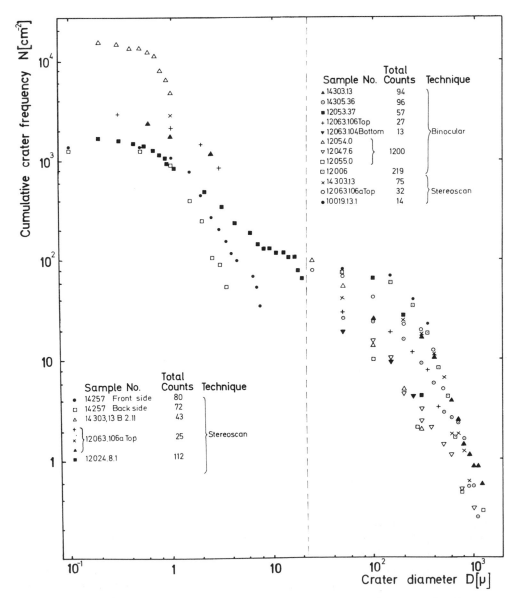

Fig. 5. Cumulative microcrater size–frequency distributions on lunar samples (the symbols and descriptions on the left and the right of the vertical dashed line belong together, respectively).

approximately -2 around 100 μ diameter. Below this size the slope is much smaller than -2, tending to zero. Although part of this decrease may be due to observational limitations as discussed above this behavior may indicate a depletion of interplanetary dust particles smaller than 100 μ in diameter.

For micron-size craters, smooth glass surfaces were scanned at magnifications up to 5000 fold. The following results could be found for the samples 12024,8 (glass splash), 12063,41 (crater glass), 12063,19,3 (crater glass), from coarse fines 14257 (dumbbell), 14303,13E2 (crater glass), and 14303,13B2,11 (crater glass): 1000–3000 craters ≥ 1 μ diameter and up to 10,000 craters ≥ 0.25 μ diameter per cm² surface area. These densities for micron- and submicron-size craters are plotted in Fig. 5. If one thinks the curves in Fig. 5 shifted as though normalized to age, they agree with and extend to smaller diameters the cumulative size frequency distributions determined for craters ≥ 50 μ diameter, although these were counted on three different types of glass surfaces. This also suggests that these craters have been produced predominantly by interplanetary micrometeoroids. Below 1 μ diameter again there seems to be indicated a depletion of interplanetary submicron-size dust, as the crater density tends to become constant. But part of the decrease of these very small craters may be due to incomplete counting caused by observational limitations, since we have to work near the resolution limit of the SEM. Another possible interpretation of the depletion below 1 μ will be given later.

A survey of crater frequencies from submicron size to kilometer size measured by different authors is given in Fig. 6. Our measurements (hatched area) are in good agreement with statistics published by Hörz et al. (1971a, b). If the crater frequency is plotted against the spallation diameter, which is important in the crater erosion process, the lower part of the hatched area (crater with diameter > 100 μ) will lie on the D^{-2}-steady state (equilibrium) crater frequency distribution measured by Shoemaker et al. (1970) in Mare Tranquilitatis and Oceanus Procellarum. This suggests that the crater distribution in this size range may be in equilibrium, too. But as indicated above and discussed later, this point of view is probably oversimplified.

CONVERSION OF CRATER FREQUENCIES TO METEOROID FLUXES

The reduction from crater diameter D to particle mass m in the micron-size range is performed with the help of the formula (after Gault and Moore, 1965)

$$D = km^{1/\beta} \qquad (k, \beta \text{ constants}),$$

where it is assumed that the interplanetary particles fall onto the moon with a mean constant velocity of about 20 km/sec. This assumption is supported by radar measurements of Miller (1970) and Cour-Palais (1969). The crater scaling exponent β can be set to about 3 in the micron-size range from about $D = 0.1$ μ to crater diameters of some tens of microns in accordance with our simulation experiments described earlier in this paper. For calculations ranging over more than 3 orders of magnitude in crater diameter (from micron-size range to meter sizes) it has to be replaced by a more adequate number of $\beta = 2.8$, as suggested by measurements of Rudolph (1969) and Kineke (1960). These experiments have been carried out with metal targets and therefore the absolute values of crater diameter (central pit) to particle diameter have to

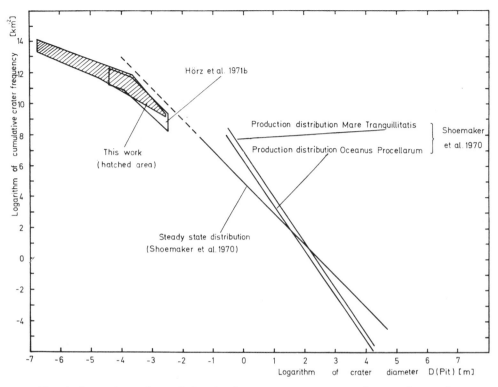

Fig. 6. Comparison of cumulative size–frequency measurements of craters from sub-micron size to kilometer size on the moon.

be determined from experiments performed with moon-like silicate targets (see Fig. 3). The assumption here is that the scaling law is similar. Then one takes into account that the D/d values slowly grow with increasing crater sizes (Gault and Moore, 1965). The constant k is about 3.3 (m in grams, D in centimeters) in the micron- to centimeter-size range (when $\beta = 2.8$).

If the exposure time of a certain lunar area is not too long, the randomly falling meteoroids will form craters far apart from each other. Practically no destruction of craters occurs. If it is assumed that there is no other erosion mechanism, the crater population on that area in this case represents the particle distribution in the inter-planetary space at 1 AU and is called "production distribution." The cumulative crater frequency $N_+(m, t_A)$ observed at time $t = 0$ (today) then is described by (in terms of mass m)

$$N_+(m, t_A) = \int_m^\infty \int_{t_A}^0 \varphi(m', t) \, dt \, dm' \qquad (t_A \text{ exposure age}).$$

The $\varphi(m, t)$ represents the differential number of particles falling per unit area and mass and time, i.e., the differential flux. It is defined as $\varphi(m, t) = am^{-\gamma}f(t)$ (a, γ are constants, m is the mass, $f(t)$ is a function of time t). In an earlier paper (Bloch *et al.*, 1971a) it was shown that $f(t) = e^{-Bt}$ (B constant).

If the exposure time is sufficiently long, craters may be eroded. The erosion of craters in the diameter range between centimeter-size to kilometer-size was treated by Moore (1964), Trask (1960), Ross (1968), and Soderblom (1970). They explained the destruction of craters by continuous infilling with ejecta of newly formed ones, or the gradual erosion of the rims with combined infilling of the crater interiors. Gault (1970) simulated the destruction of craters in the laboratory by bombarding a specially prepared target with centimeter-projectiles. These deliberations and experiments led to results that are in reasonable agreement with measurements performed by Shoemaker *et al.* (1970) in Mare Tranquilitatis and Oceanus Procellarum (Fig. 6 "equilibrium distribution"). However, these erosion processes do not play a great role in the case of crater populations on single lunar rocks. The main processes are believed to be (a) destruction of craters by direct superposition (spalling off of material; the rock size becomes smaller, the rocks become roundish), and (b) gradual erosion of micron- and submicron-size craters by solar wind sputtering (supposed the sputter erosion rate is of the order of 1 Å per annum). Both processes have been treated mathematically by Neukum and Dietzel (1971).

In order to convert the measured microcrater frequencies (shown in Fig. 5) to particle fluxes, the erosion state of the crater populations has to be determined. The micron-size frequencies measured on different samples of Apollo 12 have proved to be almost the same (Fig. 5). Therefore, we thought in the beginning that these populations could have reached sputter equilibrium state (as defined by Neukum and Dietzel, 1971) in the diameter range below 1 μ (where the curves are bending over). Application of the theory of Neukum and Dietzel (1971) leads to a cumulative micron-size particle flux of

$$\phi(m) = 6 \times 10^{-13} \, m^{-0.63} [\text{m}^{-2} \, \text{sec}^{-1}]$$

in the range of 10^{-15} g $< m < 10^{-9}$ g. It is displayed in Fig. 7 ("sputter model").

Sputter equilibrium for micron-size craters can be reached in reasonable times of 10^4–10^5 yr, if the erosion rate is of the order of 1 Å per annum. Such an erosion rate is reported by Wehner *et al.* (1963). In a recent publication of McDonnel and Ashworth (1971), however, sputter rates as low as 0.02 Å per annum have been reported. With this lower erosion rate, sputter erosion is not an important process for micron-size craters; the change in the crater frequency distribution around 1 μ diameter then is due to other effects, such as a real change in the distribution of interplanetary dust of that size or an observational lack.

In recent times we believe more and more that the low sputter rates are correct. We are strengthened in this idea, as the investigation of an Apollo 14 sample gave a micron-size crater frequency of about a factor of 10 higher than those of Apollo 12 specimens (cf. Fig. 5). The previously observed constancy of the Apollo 12 results can be explained by the scanned surface areas having the same exposure age. If there are no other effects influencing the statistics, such as very different sputter erosion rates of the investigated glass surfaces, we can conclude that the crater populations are not in sputter equilibrium state. We measure production populations. If we know the exposure ages of the examined surfaces, we can directly convert the measured crater frequencies to particle fluxes. Exposure ages can be determined by the particle track

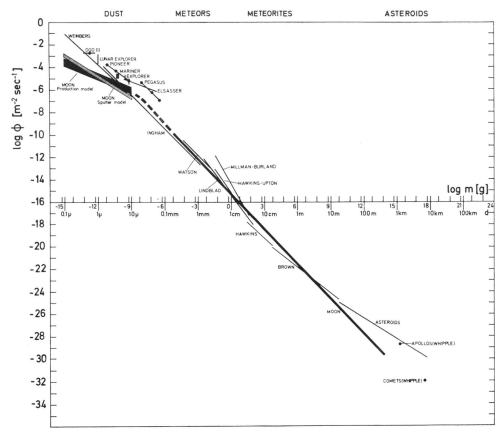

Fig. 7. The diagram shows the cumulative flux ϕ calculated from lunar crater distributions (denoted as MOON). It is compared to direct measurements with other techniques.

Table 1. Solar flare track (sft) density as a function of depth below the surface of the sample.

Depth (μ)	Track density in glass (sft/cm^2)	Track density in pyroxene (sft/cm^2)
10	3.5×10^8	—
20	9×10^7	—
30	3×10^7	2.7×10^7
40	1.45×10^7	1.4×10^7
50	7.5×10^6	7.6×10^6
60	—	4.7×10^6
70	3.2×10^6	3.4×10^6
80	—	2.3×10^6
90	—	1.8×10^6
100	1.25×10^6	1.4×10^6

method. For glasses and minerals which were resting in the lunar surface material, residence times at the very top of the surface can be calculated using tracks from Fe-group nuclei emitted in solar flares (Crozaz *et al.*, 1970; Fleischer *et al.*, 1970). These tracks are identified by their steep density gradient. Solar flare tracks were counted at different depths in the glass splash of sample 12024,8,1 and orthopyroxene crystals trapped within the glass during solidification (Table 1). The track density p decreases in the glass and in the pyroxenes between 10μ and 100μ depth R from the glass surface with $p = \text{const} \times R^{-\alpha}$, where $\alpha = 2.6$ and $\alpha = 2.4$, respectively. These values agree well with $\alpha = 2.5$ (Barber *et al.*, 1971; Crozaz and Walker, 1971; Fleischer *et al.*, 1971) which was found for the uneroded Surveyor 3 glass filter and indicate that the glass splash of sample 12024,8,1 was not affected by erosion.

Contrary to the pyroxenes, the solar flare tracks in the glass splash are strongly annealed. From our annealing experiments track fading must be expected in the glass splash at ambient temperatures on the lunar surface (Crecmers *et al.*, 1971). This contention is supported by the thermally affected appearance of the solar flare tracks. Their mean diameters are reduced to 35% of the diameters of thermally unaffected Fe-tracks in glass (Krätschmer, 1971). This diameter reduction corresponds to a track density reduction of about 10% of the original track density (Storzer and Wagner, 1969, 1972).

Assuming the track production rate versus depth derived from the uneroded Surveyor 3 glass filter (Barber *et al.*, 1971) and correcting for the observed track fading in the glass splash, we found concordant solar flare exposure ages of about 2.5×10^3 yr (with an uncertainty of probably a factor of 2) for both the glass splash and the trapped pyroxenes. The major uncertainty in applying the Surveyor III data to lunar material is that the Surveyor III spacecraft was on the moon for about 2.6 yr during the period of maximum solar activity and only during a fraction of the present solar cycle. In addition it is uncertain that the flux data relate to the average solar contribution over many solar cycles. With this age and from the crater size frequency measurements of sample 12024,8,1, a cumulative microparticle flux of

$$\phi(m) = 10^{-7.54} \times m^{-0.283} [\text{m}^{-2} \text{ sec}^{-1}]$$

can be calculated for $10^{-15} \text{ g} < m < 10^{-9} \text{ g}$. It is displayed in Fig. 7 ("production model").

Both methods, the sputter model and the production model, give flux results that are in agreement within a factor of 10 with satellite measurements of Pioneer 8 and 9 (Berg and Gerloff, 1971), Explorer 23 (Naumann, 1966), and Mariner 4 (Alexander *et al.*, 1970). Although we now prefer the production model, further investigations are required to substantiate the model.

STATISTICS OF LARGE CRATERS ON THE WHOLE MOON ($D > 3$ km)

If no erosion of craters takes place, a crater population on the moon (neglecting secondary craters) is in a "production state." Crater statistics of such populations on surface areas with different ages give an image of the mass distribution of the meteoroidal environment of the moon and the time development of the flux of meteoroids impacting the moon. Our first attempt to determine the time dependence of the flux was presented in Bloch *et al.* (1971a). An exponential function was found to be plausible and fitted best the data known at that time. The Apollo 14 and 15 samples of the Fra Mauro region and the Apennine front have given us two more radiometric ages. Various authors (Hartmann, 1972; Soderblom, 1971; Gault, 1972) have determined the crater frequency at the Fra Mauro site. Unfortunately there is no agreement. The absolute numbers for a given crater diameter vary by about a factor of 3. Furthermore, the statistics of the Apennine front, published by Hartmann (1972) are not secure enough. Because of the uncertainty underlying all these measurements, the time behavior of the meteoroid flux cannot be determined any better at the moment. Therefore, until we have more secure data, it is not necessary to assume a time development different from the exponential one with a half life of about 2.7×10^8 yr which was calculated by Bloch *et al.* (1971a).

Comprehensive counts of craters with diameters $D > 3$ km in Mare regions have been performed by Baldwin (1970). These crater frequencies are in production state, and one can be rather confident that no secondary craters vitiated the statistics. From these counts it can be deduced that the distribution of the masses that formed these craters did not change too much in the past between 3.7×10^9 and 3.3×10^9 yr. An impression of this fact can be obtained from the plots of the cumulative crater frequencies of Mare Tranquilitatis (MT) and Oceanus Procellarum (OP) in Fig. 8. In comparison, the statistics for craters with $D < 3$ km which have been reported by Shoemaker *et al.* (1970) are displayed. One recognizes that the slope in this logarithmic size–frequency plot changes from -2.9 to about -1.5 around $D = 3$ km. The absolute crater densities of MT and OP in the range $D > 3$ km show the same ratio as the densities for $D < 3$ km. This is a necessary result if the exponential decrease is correct and if the craters were formed by meteoroids. On the other hand, this indicates that the meteoroidal mass distribution in the past (not necessarily today) follows different laws in the size classes represented by $D > 3$ km and $D < 3$ km, respectively.

The crater counts of Baldwin (1970) give means to calculate formation ($=$ exposure ages) of the lunar maria. As one deals with production populations, the age can be directly determined from (exponential time development of the flux)

$$\int_{t_A(1)}^{0} e^{-2.6t}\, dt = R(1, 2) \int_{t_A(2)}^{0} e^{-2.6t}\, dt$$

(ages t_A to be inserted in aeons), where $R(1, 2)$ is the ratio of the crater densities of Mare 1 with age $t_A(1)$ and Mare 2 with age $t_A(2)$. For calculations, one can for instance take $t_A(2) = 3.65 \times 10^9$ yr which is the age of MT (Lunatic Asylum, 1970). In Table 2, the calculated mean formation ages of several mare regions are presented.

Statistics from the Southern Highlands have been published by Hartmann (1966). The crater frequency of the farside of the moon has been determined by Neukum (1971). The results are displayed in Fig. 8. One realizes a very similar size–frequency distribution for the lunar farside and for the Southern Highlands of the nearside. The discrepancy in the absolute numbers at smaller sizes may be explained by the different methods of measurement, since the density of craters with $D > 60$ km tends to become equal for both regions. Hartmann (1966) counted his craters from Lunar

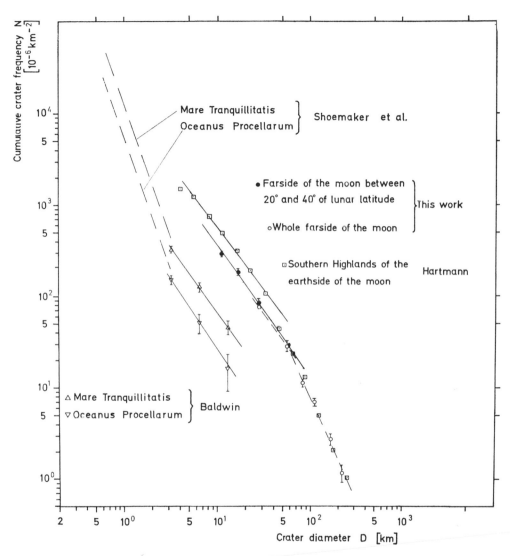

Fig. 8. Cumulative size–frequency distribution of craters with diameter $D \gtrsim 3$ km on the moon.

photographs, whereas the counts of Neukum (1971) were made from official lunar maps. Smaller craters may generally have been underestimated in the mapping process. The discrepancy can, however, be also due to a real difference in age since the statistics in the region $D > 60$ km are not as good as for $D < 60$ km.

The crater populations on the Southern Highlands and the lunar farside seem to be in a progressed state of erosion. Superposition of craters is believed to be the main erosion process; gradual erosion of crater rims by smaller impacts and ballistic

G. NEUKUM *et al.*

Table 2. Mean formation ages of special areas of the lunar surface.

Name	Radiometric ages in 10^9 yr	Calculated ages in 10^9 yr
Oceanus Procellarum	3.3*	—
Mare Serenitatis		3.30 ± 0.07
Mare Nectaris		3.37 ± 0.07
Palus Epidemiarum ⎫		
Mare Humorum ⎬		3.39 ± 0.07
Mare Nubium ⎭		
Mare Crisium		3.41 ± 0.07
Mare Imbrium		3.48 ± 0.06
Mare Fecunditatis	3.42 ± 0.17†	3.55 ± 0.05
Lacus Somniorum ⎫		
Mare Frigoris ⎬		3.61 ± 0.05
Mare Tranquilitatis	3.65*	—
Southern Highlands		≈ 4.6
Lunar Farside		≈ 4.4

* Lunatic Asylum, 1970; Papanastassiou and Wasserburg, 1970. (Ages here for calculations assumed to be without error).
† Rb–Sr age, Wasserburg *et al.* (1972).

sedimentation are not very effective in the crater size range $D > 3$ km. The development of crater populations on the moon under meteoroid bombardment with the effect of crater superposition has been treated by Neukum and Dietzel (1971), Marcus (1966), and Walker (1967). Marcus and Walker, however, did not consider in detail the case where the meteoroid mass distribution is such that the (cumulative) production population of craters reflecting this distribution follows the law $D^{-\alpha}$ (corresponding to a cumulative mass distribution of the meteoroid flux $\phi \sim m^{-\gamma+1}$), where $\alpha < 2$. Such a distribution can never reach the equilibrium distribution $\sim D^{-2}$, as predicted for $\alpha > 2$. On the contrary (Neukum and Dietzel, 1970), in this case the original distribution $\sim D^{-\alpha}$ is preserved in any state of erosion by superposition of craters. Such a case exists for $D > 3$ km, as is obvious from Fig. 8. Both populations, on the Southern Highlands and the lunar farside, show a size–frequency distribution in the range $D < 60$ km very similar to that in the maria for $D > 3$ km. Above $D > 60$ km the meteoroidal mass distribution changes again, and the slope of the crater size–frequency distribution therefore steepens. Presumably, the tail of this population shows us the original production distribution. In the range 3 km $< D < 60$ km the relations for crater superposition developed by Neukum and Dietzel (1971) can be applied. The cumulative number of craters per unit area in any state of superpositional erosion is given at time $t = 0$ (today) by

$$N(m, t = 0, t_A) = \frac{B}{\overline{m}^{2/\beta - \gamma + 1}} \, m^{-\gamma+1} \left[1 - \exp\left(\frac{a \overline{m}^{2/\beta - \gamma + 1}}{B} \left[F(t_A) - F(0) \right] \right) \right]$$

where t_A (in aeons) is the age of the surface, $\overline{m}$ is the largest mass impacted on the surface area under consideration, $F(t)$ is the antiderivative of $f(t) = e^{-2.6t}$ (t in aeons) and $B = 4(2/\beta - \gamma + 1)/k^2 \pi (\gamma - 1)$. The relation between meteoroid mass m and crater diameter D is given by $D = km^{1/\beta}$ (k, β constants). The γ is the mass exponent of the differential meteoroid flux $\varphi = am^{-\gamma}f(t)$, where a is a constant. As the production distributions for $D > 3$ km are known (MT and OP) (Fig. 8), one can calculate the ages t_A of the lunar farside and the Southern Highlands from the crater frequencies using the above formula. The results are presented in Table 2. On the basis of crater statistics, the Southern Highlands area and the highland like lunar farside belong to the oldest parts of the moon. The ages lie close to the supposed age of the moon (Wasserburg et al., 1972).

CONCLUDING REMARKS

The lunar rocks provide a collecting surface for interplanetary dust at 1 AU for extended time periods. The great number of craters which are formed allow determination of meteoroid fluxes at least within a factor of 10, provided the erosion of craters can quantitatively be understood. As was shown, this was possible in the micron-size range of craters.

Neukum and Dietzel (1971) studied the behavior of a crater population that depended on the exposure age of a certain surface and on a flux exponent γ at least constant over several orders of magnitude in mass. In the crater diameter range from 10 μ to about 1 mm the flux exponent apparently changes. The theory cannot be applied in this case. However, we know the submicron- to micron-size particle flux. The flux of particles that formed craters in the range 1 cm $\lesssim D < 3$ km was calculated by Neukum (1971) and Neukum and Dietzel (1971) from measurements of Shoemaker et al. (1970) as

$$\phi(m) = 10^{-15} \times m^{-1.04}[\mathrm{m}^{-2} \ \mathrm{sec}^{-1}].$$

It is in good agreement with the measurements of Watson (1956), Hawkins and Upton (1958), Millmann and Burland (1957), Hawkins (1959), Brown (1961), and Lindblad (1967) as shown in Fig. 7. We get a rough idea of the flux between this particle size and the micron size if we combine both regions and take as a basis the crater frequency behavior as measured in the average (see Fig. 5 and the related discussion). In Fig. 7 the result is displayed by the heavy dashed line in the mass range 10^{-9} g $< m < 10^{-4}$ g. It agrees within a factor of about 10 with data reported by Morrison et al. (1972).

As first indicated by the deep space satellites Mariner and Pioneer, our results show that below $m = 10^{-7}$ g there is a genuine depletion of dust particles. However, there is no cutoff of particles in the submicron-size range as reported by Berg and Gerloff (1971). The investigations on lunar microcraters show clearly the existence of submicron-sized particles in the solar system at 1 AU, as suggested by zodiacal light measurements of Weinberg (1964) and Ingham (1961). But the frequency appears to be much lower than expected.

Acknowledgments—We thank the National Aeronautics and Space Administration for providing the lunar samples. We acknowledge the hospitality of the Manned Spacecraft Center, Houston, during a short visit of G. Neukum, and thank for help and advice Drs. M. B. Duke, R. B. Laughton. F. Hörz, and J. B. Hartung. We also want to express our gratitude to Dr. N. Grögler, Bern, Switzerland, for providing the samples 12053,37 and 14305,36 for crater counting. We are indebted to Dr. D. E. Gault for discussions on interpretation of measurements and aid in linguistic problems.

References

Alexander W. M., Arthur C. W., and Bohn J. C. (1971) Lunar Explorer 35, and OGO 3: Dust particle measurement in selenocentric and cislunar space from 1967 to 1969. In *Space Research—XI.* pp. 279–285. Akademie-Verlag, Berlin.

Alexander W. M., Arthur C. W., Corbin J. D. (1970) Picogram dust particle flux: 1967–1968 measurements in selenocentric, cislunar and interplanetery space. In *Space Research X,* pp. 252–259. North-Holland.

Baldwin R. B. (1970) Absolute ages of the lunar maria and large craters, II. The viscosity of the moon's outer layers. *Icarus* **13,** 215–225.

Barber D. J., Cowsik R., Hutcheon I. D., Price P. B., and Rajan R. S. (1971) Solar flares, the lunar surface and gas-rich meteorites. *Proc. Second Lunar Sci. Conf., Geochim. Cosmochim. Acta* Suppl. 2, Vol. 3, pp. 2705–2714. MIT Press.

Berg O. E. and Gerloff U. (1971) More than two years of micrometeorite data from two Pioneer satellites. In *Space Research—XI,* pp. 225–235. Akademie-Verlag, Berlin.

Bloch M. R., Fechtig H., Gentner W., Neukum G., and Schneider E. (1971a) Meteorite impact craters, crater simulations, and the meteoroid flux in the early solar system. *Proc. Second Lunar Sci. Conf., Geochim. Cosmochim. Acta* Suppl. 2, Vol. 3, pp. 2639–2652. MIT Press.

Bloch M. R., Fechtig H., Gentner W., Neukum G., Schneider E., and Wirth H. (1971b) *Natural and Simulated Impact Phenomena—A Photo-Documentation.* Max-Planck-Institut für Kernphysik, Heidelberg, Germany (private printing).

Brown H. (1961) The density and mass distribution of meteoritic bodies in the neighborhood of the earth's orbit. *J. Geophys. Res.* **66,** 1316–1317.

Carter J. L. and MacGregor J. D. (1970) Mineralogy, petrology and surface features of some Apollo 11 samples. *Proc. Apollo 11 Lunar Sci. Conf., Geochim. Cosmochim. Acta* Suppl. 1, Vol. 1, pp. 247–275. Pergamon.

Carter J. L. and McKay D. S. (1971) Influence of target temperature on crater morphology and implications on the origin of craters on lunar glass spheres. *Proc. Second Lunar Sci. Conf., Geochim. Cosmochim. Acta* Suppl. 2, Vol. 3, pp. 2653–2670. MIT Press.

Cour-Palais B. G. (1969) Meteoroid environment model—1969. In *NASA Space Vehicle Design Criteria.* NASA SP-8013, pp. 1–31.

Creemers C. J., Birkebak R. C., and White E. J. (1971) Lunar surface temperatures from Apollo 12. *The Moon* **3,** 346–351.

Crozaz G., Haak U., Hair M., Maurette M., Walker R., and Woolum D. (1970) Nuclear track studies of ancient solar radiations and dynamic lunar surface processes. *Proc. Apollo 11 Lunar Sci. Conf., Geochim. Cosmochim. Acta* Suppl. 1, Vol. 3, pp. 2051–2070. Pergamon.

Crozaz G. and Walker R. M. (1971) Solar particle tracks in glass from Surveyor 3 spacecraft. *Science* **171,** 1237–1239.

Elsässer H. (1958) Interplanetare Materie. *Mitteilungen der Astronomischen Gesellschaft, 1957* **2,** 61–88.

Fleischer R. L., Haines E. L., Hanneman R. E., Hart H. R., Kasper J. S., Lifshin E., Woods R. T., and Price P. B. (1970) Particle track, x-ray, and mass spectrometry studies of lunar material from the Sea of Tranquility. *Science* **167,** 568–571.

Fleischer R. L., Hart H. R., and Comstock G. M. (1971) Very heavy solar cosmic rays: Energy spectrum and implications for lunar erosion. *Science* **171,** 1240–1242.

Gault D. E. (1970) Saturation and equilibrium conditions for impact cratering on the lunar surface: Criteria and implications. *Radio Science* **5,** 272–291.

Gault D. E. (1972) Private communication.

Gault D. E. and Heitowit E. D. (1963) The partition of energy for hypervelocity impact craters formed in rock. *Proc. 6th Symp. Hypervelocity Impact.,* Vol. 2, pp. 419–456.

Gault D. E. and Moore H. J. (1965) Scaling relationships for microscale to megascale impact craters. *7th Symp. Hypervelocity Impact,* Vol. 6, pp. 341–351.

Hartmann W. K. (1966) Martian cratering. In *Commun. Lunar and Planet. Lab. Univ. of Arizona 4,* Part 4, No. 65, pp. 121–131.

Hartmann W. K. (1972) Paleocratering of the moon: Review of post-Apollo data. *Astr. Space Sci.* **16**, in press.

Hartung J. B., Hörz F. and Gault D. E. (1972) The origin and significance of lunar microcraters (abstract). In *Lunar Science—III* (editor C. Watkins), pp. 363–365, Lunar Science Institute Contr. No. 88.

Hawkins G. S. (1959) The relation between asteroids, fireballs and meteorites. *Astron. J.* **64**, 450–454.

Hawkins G. S. and Upton E. K. L. (1958) The influx rate of meteors in the earth's atmosphere. *Astrophys. J.* **128**, 727–735.

Hörz F., Hartung J. B., and Gault D. E. (1971a) Micrometeorite craters and related features on lunar rock surfaces. *Earth Planet. Sci. Lett.* **10**, 381–386.

Hörz F., Hartung J. B., and Gault D. E. (1971b) Micrometeorite craters on lunar rock surfaces. *J. Geophys. Res.* **76**, 5770–5798.

Ingham M. F. (1961) Observations of the zodiacal light from a very high altitude station, IV. The nature and distribution of the interplanetary dust. *M.N. Royal Ast. Soc.* **122**, 157–176.

Kineke J. H. (1960) An experimental study of crater formation in metallic targets. *Proc. 4th Symp. Hypervelocity Impact.*, Vol. 1, pp. 1–36.

Krätschmer W. (1971) Die anätzbaren Spuren künstlich beschleunigter schwerer Ionen in Quarzglas. Doctor's thesis, Heidelberg.

Lindblad B. A. (1967) Luminosity function of sporadic meteors and extrapolation of the influx rate to the micrometeorite region. *Smithsonian contribution to astrophysics*, Vol. 11, pp. 171–180. NASA Sp.-135.

Lunatic asylum (1970) Ages, irradiation history, and chemical composition of lunar rocks from the Sea of Tranquility. *Science* **167**, 463–466.

LSPET (Lunar Sample Preliminary Examination Team) (1970) Preliminary examination of lunar samples from Apollo 12. *Science* **167**, 1325–1339.

Marcus A. H. (1966) A stochastic model of the formation and survival of lunar craters, II. Approximate distribution of diameter of all observable craters. *Icarus* **5**, 165–177.

McDonnell J. A. M. and Ashworth D. G. (1972) Erosion phenomena on the lunar surface and meteorites. In *Space Research—XII*, Akademie-Verlag, Berlin (in press).

McKay D. S., Greenwood W. R., and Morrison D. A. (1970) Origin of small particles and breccia from the Apollo 11 site. *Proc. Apollo 11 Lunar Sci. Conf.*, Geochim. Cosmochim. Acta Suppl. 1, Vol. 1, pp. 673–694. Pergamon.

Miller C. D. (1970) Empirical analysis of unaccelerated velocity and mass distributions of photographic meteors. NASA TN D-5710.

Millman P. M., and Burland M. S. (1957) Magnitude distribution of visual meteors. *Sky and Telescope* **16**, 22.

Moore H. J. (1964) Density of small craters on the lunar surface. In *U.S. Geol. Surv. Astrogeol. Stud. Annual Progr. Rep.*, Part D, p. 34, Government Printing Office, Washington D.C.

Morrison D. A., McKay D. S., Moore H., Bogard D., and Heiken G. Microcraters on lunar rocks (abstract). In *Lunar Science—III* (editor C. Watkins), pp. 558–560, Lunar Science Institute Contr. No. 88.

Naumann R. J. (1966) The near-earth meteoroid environment. NASA Tech. Note D-3117, Washington D.C.

Neukum G. (1971) Untersuchungen über Einschlagskrater auf dem Mond. Doctor's thesis, University of Heidelberg.

Neukum G. and Dietzel H. (1971) On the development of the crater population on the moon with time under meteoroid and solar wind bombardment. *Earth Planet. Sci. Lett.* **12**, 59–66.

Neukum G., Mehl A., Fechtig H., and Zähringer J. (1970) Impact phenomena of micrometeorites on lunar surface materials. *Earth Planet. Sci. Lett.* **8**, 31–35.

Papanastassiou D. A. and Wasserburg G. J. (1970) Rb–Sr ages from the Ocean of Storms. *Earth Planet. Sci. Lett.* **8**, 269–278.

Pegasus (1966) The meteoroid satellite Project Pegasus. First Summary Report, NASA Tech. Note TN D-3505.

Pegasus (1967) Scientific results of Project Pegasus. Interim Report, NASA TM X-53629.

Ross H. P. (1968) A simplified mathematical model for lunar crater erosion. *J. Geophys. Res.* **73**, 1343–1354.

Rudolph V. (1969) Untersuchungen an Kratern von Mikroprojektilen im Geschwindigkeitsbereich von 0,5 bis 10 km/sec. *Z. Naturforsch.* **24a**, 326–331.

Shoemaker E. M., Batson R. M., Bean A. L., Conrad C. Jr., Dahlem D. H., Goddard E. N., Hart M. H., Larson K. B., Schober G. G., Schleicher D. L., Sutton R. L., Swann G. A., and Waters A. C. (1970) Preliminary geologic investigation of Apollo 12 landing site, Part A: Geology of the Apollo 12 landing site. Apollo 12 Preliminary Science Report, NASA SP-235.

Soderblom L. A. (1970) A model for small-impact erosion applied to the lunar surface. *J. Geophys. Res.* **75**, 2655–2661.

Soderblom L. A. and Lebofsky L. A. (1971) Technique for rapid determination of relative ages of lunar areas from orbital photography. *J. Geophys. Res.* **77**, 279–296.

Storzer D. and Wagner G. A. (1969) Correction of thermally lowered fission track ages of tektites. *Earth Planet. Sci. Lett.* **5**, 463–468.

Storzer D. and Wagner G. A. (1972) Unpublished data, MPI, Heidelberg.

Trask N. J. (1966) Size and spatial distribution of craters estimated from Ranger photographs. Jet. Propul. Lab. Tech. Rep. 32-700, 252, Pasadena, California.

Walker E. H. (1967) Statistics of impact crater accumulation on the lunar surface exposed to a distribution of impacting bodies. *Icarus* **7**, 233–242.

Wasserburg G. J., Turner G., Tera F., Podosek F. A., Papanastassiou D. A., and Huneke J. C. (1972) Comparison of Rb–Sr, K–Ar, and U–Th–Pb ages: Lunar chronology and evolution (abstract). In *Lunar Science—III* (editor C. Watkins), pp. 788–790, Lunar Science Institute Contr. No. 88.

Watson F. G. (1956) *Between the Planets*. Harvard University Press.

Wehner G. K., Kenknight C. and Rosenberg D. L. (1963) Sputtering rates under solar wind bombardment. *Planet. Space Sci.* **11**, 885–895.

Weinberg J. (1964) The zodiacal light at 5300 Å. *Ann. d'Astrophys.* **27**, 718–738.

Whipple F. L. (1967) On maintaining the meteoritic complex. In *The Zodiacal Light and the Interplanetary Medium*, pp. 409–426. NASA SP-150.

Proceedings of the Third Lunar Science Conference
(Supplement 3, *Geochimica et Cosmochimica Acta*)
Vol. 3, pp. 2811–2829
The M.I.T. Press, 1972

Collision controlled radiation history of the lunar regolith

N. Bhandari, J. N. Goswami, S. K. Gupta, D. Lal,
A. S. Tamhane, and V. S. Venkatavaradan

Tata Institute of Fundamental Research,
Homi Bhabha Road, Bombay 5, India

Abstract—Several rock and soil samples from Apollo 12 and 14 sites have been studied for fossil records of cosmic ray tracks ($Z > 20$). The exposure history of samples derived from these data are analyzed in terms of erosion, fragmentation, and burial. It is shown that the accumulation of tracks up to depths of a few centimeters in case of rocks is definitely controlled by collisional processes; at depths of <0.1 cm, the track densities are found to be in equilibrium with production and erosion. The effective time of irradiation of samples at the lunar surface varies between 10^3–10^7 yr.

The mean total erosion rates applicable to near surface (<0.1 cm) regions of rocks are deduced to lie between 5×10^{-8} and 10^{-7} cm yr^{-1}.

Data on fossil tracks in a through slice of rock 14303 lead to an energy spectrum of iron group cosmic ray nuclei which is consistent with that derived earlier from two Apollo 12 rocks deduced to have simple exposure histories. Several other breccias show a complicated exposure history making it difficult to derive the primary composition or the energy spectrum of iron group nuclei. None except rock 14301, however, show any signs of pre-irradiation histories of individual fragments comprising the breccia. Adjacent feldspars from the interior of 14301 had track densities ranging between $(3–80) \times 10^6$ cm^{-2}.

The soil samples show a range of track densities similar to that found for Apollo 11 and 12 samples. These data confirm the general validity of the irradiation and the "throw-out" model presented earlier.

Based on burial depths of documented Apollo 14 rocks, it is deduced that most rocks on surface today represent fragments of larger rocks and that the average deposition rate of regolith at the Apollo 14 site is in the region of $(0.15–0.3)$ cm m.y.$^{-1}$.

Introduction

THE FOSSIL TRACK TECHNIQUE has been extensively applied to the study of the diverse components of the lunar regolith (Crozaz *et al.*, 1971; Lal *et al.*, 1970; Price and O'Sullivan, 1970; Arrhenius *et al.*, 1971; Fleischer *et al.*, 1971a, 1971b). The usefulness of this technique for study of regolith dynamics lies in the fact that the mean attenuation length in the rate of track formation is markedly depth sensitive and track production is appreciable up to depths of the order of 10 cm during a whole range of exposure times of interest. This allows an evaluation of the time a sample has spent at various depths within the top 10 cm of the lunar surface. In our earlier work (Bhandari *et al.*, 1971) based on examination of fossil tracks in Apollo 11 and 12 rocks as well as in meteorites, we were able to obtain an approximate relationship between the observed track density, $\rho(X)$ and the time of exposure, T, for rocks of different sizes. The observed track density profile, besides being a function of the geometry of the rock, depends on the exposure history and erosion/fragmentation during each of the exposure configurations. In view of this, only in the case of a single exposure history it is possible to obtain a relation between the track production spectrum $\dot{\rho}(X)$ and the observed $\rho(X)$. In the absence of erosion, the exposure age for the case of a single exposure history is given by:

$$T = \rho(X)/\dot{\rho}(X)$$

The modifications in the track profiles become important for depths, $X \leq \varepsilon T$, where ε is the total effective erosion rate due to different mechanisms.

It was discussed earlier (Bhandari *et al.*, 1971) that the relation of the type:

$$\dot{\rho}(X) = \left(\frac{d\rho}{dt}\right)_X = K(A + BX)^{-\alpha} \tag{1}$$

where K, A, B, and α are constants over the range of application of relation (1), holds between depths X_1 and X_2 measured radially from the surface. The normalised gradient in the track density $G(X)$ at any depth is then given by:

$$G(X) = \frac{1}{\rho}\left(\frac{d\rho}{dx}\right) = \frac{d(\ln \rho)}{dx} = -\frac{B\alpha}{A + BX}. \tag{2}$$

The *observed* values of the constants in equation (1) are summarized in Table 1, for the case of infinite size body of average composition corresponding to Apollo 11 and 12 rocks (specific gravity $= 3.4$ g cm^{-3}; in the case of regolith, the value of B is given by bulk density of soil/3.4).

The calculated values of $G(X)$ are also given in the Table 1. Alternatively, over short intervals of depth, if we express the track production as an exponential function:

$$\dot{\rho}(X) = K \, e^{-X/L} \tag{3}$$

then the mean free path for attenuation in track density by $1/e$ is given by $L = (A + BX)/\alpha B$. Corresponding values of $L(X)$ are also given in Table 1.

If the measured track density gradients in any given rock, expressed either in terms of $G(X)$ or $L(X)$ correspond to those given in Table 1, then it would imply that the rock had a simple exposure history, similar to those on which equation (1) is based. In the case of rocks which have had multiple exposure histories, the track profiles would be expected to differ markedly from equation (1). Combining the fossil track

Table 1. Values of various parameters in equation (1) based on Bhandari *et al.* (1971).

	"Observed" constants for a lunar rock of infinite size for depth intervals (units: tracks/cm² m.y.)		
Constants	$2 \times 10^{-3} < X < 0.1$ cm	$0.1 < X < 1.0$ cm	$1.0 < X < 25$ cm
K	1.2×10^6	1.4×10^6	6.36×10^{11}
α	0.75	0.7	6.15
A	0	0	7.5
B	1	1	1
$G(X)$	$-0.75/X$	$-0.7/X$	$-6.15/(7.5 + X)$
$L(X)$	$X/0.75$	$X/0.7$	$7.5 + X$
			6.15

data with other evidences regarding the exposure history of rocks based on cosmogenic effects (solar proton induced radioactivities, galactic cosmic ray produced Ne isotopes, etc.), it may become possible to make models of the exposure history of a given rock (c.f. Lal *et al.*, 1970; Fleischer *et al.*, 1971a).

It seems important to note here that $G(X)$ varies as X^{-1} for depths < 1.0 cm and this holds even for the production spectrum. As shown by Bhandari *et al.* (1971) and Crozaz *et al* (1971, 1972) the observed track spectrum for $X < 0.1$ cm is appreciably modified due to solar wind erosion and micrometeorite impacts. The corresponding value of α, α_p, at *production*, is deduced to be about 2.0–2.5. (Bhandari *et al.*, 1971.) Thus, for $2 \times 10^{-3} < X < 0.1$ cm, $G(X) = 2.25/X$, three times larger than the observed value and should be observable in samples exposed for a short duration.

In view of the obvious importance of knowledge of $\dot{\rho}(X)$, we have extended our earlier work of Apollo 12 and studied several Apollo 14 rocks as well as grains in soil. Based on these observations, we are now able to obtain fairly meaningful long-term average flux of iron group nuclei at 1 a.u. as well as a simple model of erosionary processes which includes continuous atomic sputtering due to solar wind and statistical "mass-wastage" (after R. M. Walker) due to micrometeorite impacts. The approximate rates of effective erosion due to these mechanisms are obtained. It is shown that the observed track densities in lunar grains and rocks can be fairly well understood in the framework of the simplified models presented here. Supplementary data on the rates of deposition of soil at the Apollo 14 site, based on fossil track records in fines and the documentary information on the distribution frequency of buried depths of centimeter size rocks is also presented.

Experimental Techniques

The experimental procedures for processing of rock and fines have been described in detail earlier (Lal *et al.*, 1968; Lal *et al.*, 1969; Bhandari *et al.*, 1971; Krishnaswami *et al.*, 1971; and Bhandari *et al*, 1972). Track densities up to 8×10^7 cm^{-2} were measured optically. For higher track densities,

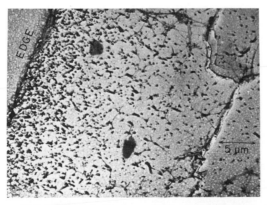

Fig. 1a. Electron micrographs of plastic replicas of etched lunar crystals: Olivine crystal from soil sample 12037,57, etched in WN solution (Krishnaswami *et al.*, 1971) for 40 min. The track density at the edge $\simeq 3 \times 10^8$/cm^2. Note the steep gradient in track density as a function of distance from edge.

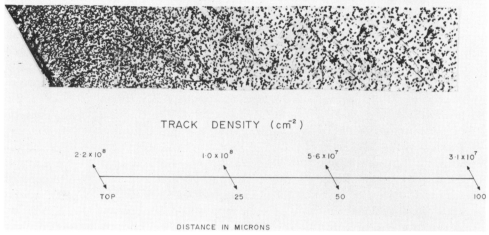

Fig. 1b. Electron micrographs of plastic replicas of etched lunar crystals: Tracks in a feldspar crystal from a thick section along *ZX* plane from the surface of rock 12038,17. The track density at surface is $2.2 \times 10^8/cm^2$. Note the steep gradient in track density as a function of distance from the edge.

we have employed the replication procedure (McDougall *et al.*, 1971). An intercomparison between optical and electron microscope track densities in the 10^7–10^8 cm^{-2} region shows that the track density is usually overestimated by up to a factor of 5 in the replicating technique. Furthermore there is a tendency of progressively overestimating the track densities at higher track densities. The present treatment refers to track data below 10^8 cm^{-2}, based on optical measurements; the electron microscope replica data are used only as a guidance towards evaluation of erosion of the samples,

It is known that various minerals have different registration characteristics. For an identical exposure, feldspars show slightly higher track densities as compared to pyroxenes but olivines show much lower track densities (Arrhenius *et al.*, 1971 and Bhandari *et al.*, 1972). In view of the small differences we have combined the observed track data in feldspars and pyroxenes but the observed track density in olivines is multiplied by a constant factor of 2 to take into account the differences in track registration characteristics.

Figures 1a and 1b show typical examples of track gradients observed in lunar rocks and soil grains.

RESULTS

The "best fit" steepest track profiles observed in the case of six rocks studied are shown in Fig. 2. The physical data on the rocks are given in Table 2. We have noticed that Apollo 14 samples distinguish themselves from the previous ones by the presence of a significant track contribution due to fission. Track clusters of different shapes and configuration are abundantly seen along cleavages and grain boundaries as well as in the central regions of crystals, in both rocks and fines. We have studied the two dimensional distribution patterns of tracks and selected regions of minimum fission contributions. In spite of this selection, the fission contribution becomes

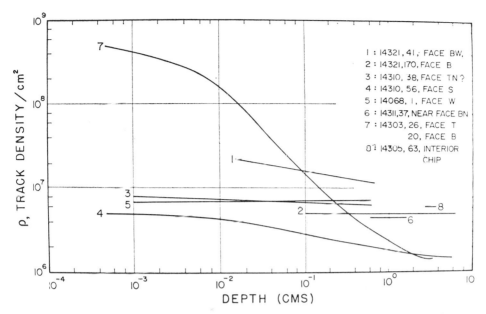

Fig. 2. Observed fossil track profiles in eight samples from six Apollo 14 rocks. Only smooth curves running through the experimental points corresponding to steepest gradient are shown. Location of sample 14311,37 is only approximately known.

Table 2. Exposure ages of Apollo 14 rocks.

Sample number	Mass (g)	Ellipsoid semi-axes used; $a \times b \times c$	Sample details	Range of observed track densities* (10^6 cm^{-2})	Sun-tan age (T_S), Sunny face	Sub-decimeter age (T_B)	Spallogenic neon exposure age (m.y.) Submeter age
14068,1	35.5	2.1 × 1.6 × 1.35	SC	6–7	?	15	20[d]
14301,54	1360	6.25 × 6 × 4	IC	3–80	—	8	102 ± 30[b]
14303,26	898	8 × 4.5 × 2.5	Slice I	2–300	2.5, T	0	17[c]
14303,20			Slice II	2			
14305,63	2496	7 × 7.5 × 3	IC	7	—	35	20[d]
14310,56	3439	8 × 7 × 6	Slice	1.5–5	1, S	[a]	262[e]
14310,38			SC	6–8	—	—	
14311,37	3202	7.5 × 3.5 × 2.5	IC	4–5	—	12	—
14321,41	8998	13 × 13 × 8	SC	12–20	4.2, BS	[a]	20[d]
14321,170			Slice	4–10	—	—	
14321,123			SC	10–100	2, B	—	

SC = Surface chip; IC = Interior chip.
* Track densities in near surface grains refer to a depth of ≈ 5 μ.
[a] In these cases a more detailed study has been made by the energy spectrum consortia (Yuhas *et al.*, 1972) and estimation of T_B will then become possible.
[b] Crozaz *et al.* (1972).   [c] Kirsten *et al.* (1972).   [d] Bogard and Nyquist (1972).   [e] Lugmair and Marti (1972).

important at track density values of $\leq 10^6$ cm^{-2}, as evidenced from the scatter in data at these values, e.g., in case of rock 14303 (Fig. 3).

The frequency distribution of track densities in fines is given in Figs. 4a, 4b, and 4c for different size fractions. These measurements were carried out in a manner similar to that discussed earlier (Arrhenius *et al.*, 1971): track counts, however, now include olivines which were etched using the WN etch (Krishnaswami *et al.*, 1971).

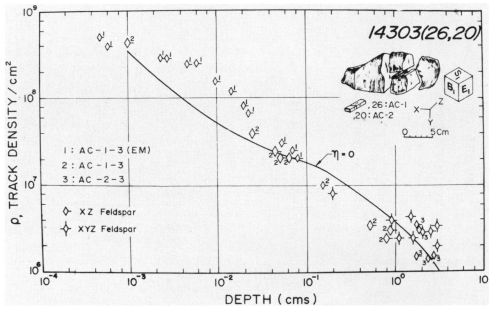

Fig. 3. Measured fossil track densities (for tracks of length > 1 μ) based on thick section studies of rock 14303(26,20). The position of slice in the rock is shown in the inset. Symbols refer to the orientation and position of the sample in the given section. $\diamond^1$ indicates that the track density data are for a feldspar crystal from a thick section in *XZ* plane, as measured by the electron microscope counting of replica. The solid curve is the expected profile for $\eta = 0$, in arbitrary units, based on the energy spectrum derived from a study of Apollo 12 rocks (Bhandari *et al.*, 1971).

DISCUSSION

Inference on energy spectrum of VH nuclei based on track profiles in rocks

As seen from Fig. 2, the steepest gradient in track densities is observed in the case of a through slice from rock 14303. In this case, the experimental data points are presented in Fig. 3 where we have also shown the expected track density profile (in arbitrary units) on the basis of the steepest track profiles observed in Apollo 12 rocks, 12038, 12018, and 12020 (Bhandari *et al.*, 1971).

We believe that there exists a fairly good agreement between the track profiles based on Apollo 12 rocks and that observed in 14303 indicating that this rock can be assumed to have a fairly simple exposure history. Furthermore, since we have not yet observed any case where the track profile is steeper, it supports our earlier contention that the data of some of the Apollo 12 rocks can be used to obtain the energy spectrum of iron group nuclei of cosmic rays up to energies of the order of 500 MeV/n. The deduced mean flux and energy spectrum for iron group nuclei averaged over irradiation exposure ages of Apollo 12 rocks has been given earlier for the energy interval, 0.5 MeV/n to 500 MeV/n (Bhandari *et al.*, 1971). We believe that these are still the best estimates, particularly for energies above 60 MeV/n. The following discussions,

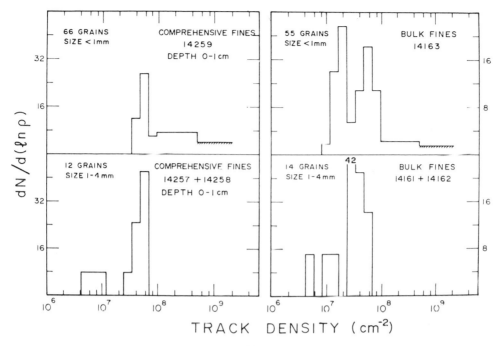

Fig. 4a. Histograms showing track density distribution in grains from Apollo 14 fines, normalized to 100 crystals. Number of grains examined is given on left top. As a convention, all grains having track density $> 5 \times 10^8/\text{cm}^2$ are uniformly spread over the interval 5×10^8 to $2 \times 10^9/\text{cm}^2$: Comprehensive and bulk fines for two grain size fractions.

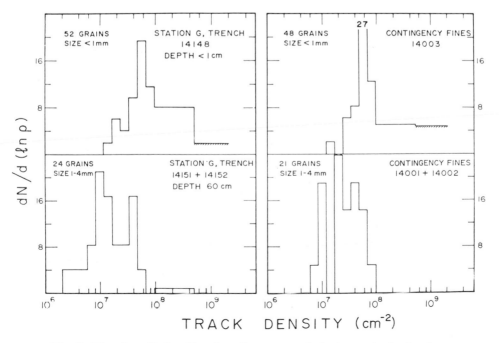

Fig. 4b. Fines from Station G and contingency sample for two grain size fractions.

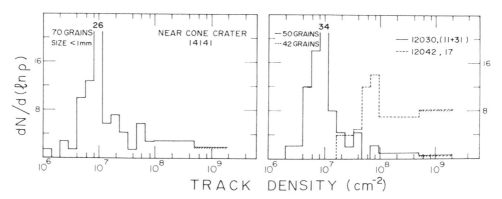

Fig. 4c. Fines (size fraction < 1 mm) collected from small crater south of Cone Crater, Station C, together with two extreme cases observed in Apollo 12 scoop samples.

however, suggest that we may have underestimated the magnitude of average erosion and, if so, this would imply that we have systematically underestimated the flux of iron group nuclei at lower energies, $E_k < 30$ MeV/n.

Exposure ages of rocks

In view of the fact that track profiles in rock 14303 confirm the inferred average prehistoric cosmic ray energy spectrum of VH nuclei based on Apollo 12 rocks, we have also deduced the cosmic ray exposure ages of Apollo 14 rocks using the model given by Bhandari *et al.* (1971). In this model, even a complex exposure history of a rock is described by a simplified two-stage time exposure. One of these, the "sun-tan" exposure age refers to the total time of "unshielded" irradiation of a *given face*. The second exposure time, "subdecimetre" exposure age, represents the integral exposure time of a rock shielded by 10 cm of regolith. Based on track data alone, it is not possible to ascertain the time sequence of "sun-tan" and "subdecimetre" exposures; it is only possible to deduce the total time integral for the unshielded and the shielded exposures.

In Table 2, we have summarized the calculated "sun-tan" and "subdecimeter" exposure ages for the various rocks. Following Bhandari *et al.* (1971), the sun-tan exposure ages are based on equation (1) with $X = 0.1$ cm. The values adopted for $\rho_P(X)$ are summarized in Table 3 for different orientations of the slice and the planes of measurement: the variables η and β characterizing the geometry of irradiation are the same as discussed earlier. In Table 3, we have also listed the average values of the "recordable" range of iron group nuclei used in the present calculations.

Subdecimeter ages are meaningfully deduced in cases where the track profiles for depths >0.1 cm differ appreciably from that at production (which is assumed to be given by rocks where the track profiles are steepest).

For obtaining subdecimeter ages, we have taken an average effective shielding depth of 3 cm rock equivalent corresponding to uniform exposure for different shieldings up to 10 cm of regolith. The calculated subdecimeter ages are given in Table 2.

Table 3. Track formation rates at 0.1 cm used for calculation of exposure ages.

Orientation		Track formation rate (10^6 per cm^2 m.y.) at 0.1 cm depth	
η	β	Pyroxene ($\Delta R^* = 8.5 \mu$)	Feldspar ($\Delta R^* = 10 \mu$)
0	0	6.8	8.0
0	$\pi/2$	5.2	6.2
0	$0 - \pi/2$	6.0	7.1
$\pi/2$	0	2.9	3.4
$\pi/2$	$\pi/2$	2.8	3.3
$\pi/2$	$0 - \pi/2$	2.8	3.3

* Values of ΔR have been obtained by averaging over recordable track lengths measured in a large number of lunar and meteoritic samples. However, these values may have been underestimated by about 15–20%.

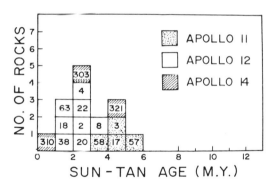

Fig. 5. Frequency distribution in *sun-tan* exposure ages for lunar rocks from various Apollo missions. These ages refer to a particular face of the rock. The rocks are identified by the last digits in the histogram. Whenever data for more than one face of a particular rock is available the higher age has been used.

The frequency distribution of "sun-tan" exposure ages of Apollo 11, 12, and 14 rocks is shown in Fig. 5. From these limited data, it is not possible to state unambiguously whether there exists a grouping in these ages, but such a tendency seems to be present for rocks collected from different sites. The subdecimeter exposure ages usually lie between 20–50 m.y. time bracket which is consistent with the cosmogenic submeter exposure ages (which refer to cumulative time spent by a rock up to depths of 1 m) being larger.

Surface irradiation ages of fines

The "surface" irradiation ages of several Apollo 14 fines have been deduced from the observed track density frequency distribution in grains of <1 and 1–4 mm size range (Fig. 4), using the earlier model developed and applied extensively for Apollo 12 scoop samples and double core (Arrhenius *et al.*, 1971). The number of grains

Table 4. Surface irradiation ages of Apollo 14 soil samples.*

Sample no.	Grain size (mm)	Location (depth)	No. of crystals measured	ρ_{av} ($\times 10^6$ cm^{-2}) for $\rho < 10^8$ cm^{-2}	$N\mu/N$, fraction of crystals exposed on the surface	Quartile track density (10^6 cm^{-2})	Surface irradiation age[†] (m.y.)
14161,40	2–4		2		0.07	20	60
14162,49	1–2	Bulk sample (0–6 cm)	12	28.0			
14163,121	< 1		55	37.2	0.27	20	60
14001,1	2–4		4		0.28	15	20
14002,1	1–2	Contingency sample (0–6 cm)	17	28.8			
14003,1	< 1		48	51.8	0.66	53	80
14257,4	2–4		2		0.25	7	3
14258,23	1–2	Comprehensive sample (0–1 cm)	10	34.0			
14259,81	< 1		66	52.4	0.35	55	20
14141,42	< 1	Station C (0–3 cm)	70	15.8	0.26	9	7
14148,37	< 1	Surface of trench (0–1 cm)	52	48.5	0.69	55	20
14151,14	1–2	Bottom of trench (36 cm)	13	18.3	0.25	10	15[‡]
14152,6	2–4		11				

* The ages are calculated on the basis of the model given by Arrhenius *et al.* (1971).
† The ages based on grains < 1 mm are statistically more reliable.
‡ This age is based on assumed layer thickness of 2 cm (cf., Arrhenius *et al.*, 1971).

analyzed in case of > 1 mm size is usually small and hence the ages calculated on the basis of < 1 mm size fractions are reliable. All relevant data are summarized in Table 4.

"Surface" exposure ages are found to range between 7–10 m.y. The Station C crater, south of Cone Crater dates younger than Cone Crater for which an exposure age of 20 m.y. has been deduced.

Evaluation of long-term average erosionary processes on the lunar surface

It is now well known (c.f., Bhandari *et al.*, 1971; Crozaz *et al.*, 1971, 1972; and Comstock, 1971) that track profiles at depths < 0.1 mm are usually in an erosion equilibrium in the case of most of the Apollo rocks studied. In view of the importance of this phenomenon in understanding both the irradiation history of rocks and lunar regolith grains as well as the long-term average flux of solar heavy nuclei we have reconsidered this problem, both theoretically and experimentally.

We characterize, as in equation (1), the track production spectrum as follows:

$$\dot{\rho}_p(X) = K_p(A_p + B_pX)^{-\alpha_p} \tag{5}$$

The subscript p is used to denote parameters applicable at production.

The exposed surface of a rock continually undergoes erosion. For the simple case of a single exposure of a rock (that is, in one position since its outcrop on the lunar surface), the rock surface, at a time t in the past, was, on an average, a distance $\Delta X = \varepsilon t$ higher than the present surface. In this case, the observed track profile $\rho(X)$ at $t = 0$, after a "sun-tan" exposure of T_s yr is given by the following relation:

$$\rho_{T_s}(X) = \int_0^{T_s} \dot{\rho}(X + \varepsilon t) \, dt$$

$$= \dot{\rho}(X) \cdot \frac{(A_p + B_pX)}{B_p\varepsilon(\alpha_p - 1)} \left[1 - \left(1 + \frac{B_p\varepsilon T_s}{A_p + B_pX} \right)^{-\alpha_p + 1} \right] \tag{6}$$

For the kinetic power law, spectrum, $J = dN/dE = \text{const } E^{-\gamma}$, the calculated production spectrum of tracks, $\dot{\rho}(X)$, can be fairly well approximated by relation (5) with $A_p = 0$ and $B_p = 1$:

$$\dot{\rho}_p(X) = K_p X^{-\alpha_p} \tag{7}$$

For values of γ lying between 1–5, we empirically calculate the following relations between α_p and γ:

$$5 \times 10^{-4} < X < 10^{-1} \text{ cm} \qquad \alpha_p = 1.17\gamma^{0.605} \tag{8a}$$

$$10^{-1} < X < 1 \text{ cm} \qquad \alpha_p = 1.02\gamma^{0.702} \tag{8b}$$

$$1 < X < 10 \text{ cm} \qquad \alpha_p = 1.63\gamma^{0.55} \tag{8c}$$

For the region of validity of equation (7), when $\varepsilon T_s \gg X$ (or if T_s is large), the track density at a depth X reaches an erosion equilibrium value, $\rho_{E.E.}(X)$, given by:

$$\rho_{E.E.}(X) = \dot{\rho}(X) \frac{X}{\varepsilon(\alpha_p - 1)} \tag{9a}$$

$$= K_p \frac{X^{-\alpha_p + 1}}{\varepsilon(\alpha_p - 1)} \tag{9b}$$

corresponding to a (saturation) effective bombardment time, $T_{E.E.}$:

$$T_{E.E.} = \frac{\rho_{E.E.}(X)}{\dot{\rho}(X)} = \frac{X}{\varepsilon(\alpha_p - 1)} \tag{10}$$

which is of course depth dependant.

Equations (6) and (9) show that erosion flattens the track profile in the region, $X \lesssim \varepsilon T_s$, and in the equilibrium region $\varepsilon T_s \gg X$ the slope in the track profile changes by unity, i.e., $\alpha_p = \alpha + 1$. The range of X up to which the slope changes will depend on the exposure time. Thus if a large number of rocks are analyzed which have been exposed over a wide range of T_s values, it should be possible to deduce the time-averaged values of the parameters, ε and α_p. Both track production and the effective rate of erosion depend on the orientation angle η (angle between the zenith and normal to the rock surface) on the moon. In the case of a spherical rock, assuming an isotropy in the flux of solar wind and micrometeorites, it can be shown that the average rate of erosion of the surface $\bar{\varepsilon}_\eta$, as a function of η, is given by

$$\bar{\varepsilon}_\eta = \varepsilon_\eta/\varepsilon_0 = \tfrac{1}{2}[1 \pm \sin^2(\eta - \pi/2)] \tag{11}$$

where ε_0 is the total erosion rate at the top of a spherical rock ($\eta = 0$). In equation (11) the sign within the parenthesis is $(+)$ or $(-)$ depending on whether $\eta < \pi/2$ or $> \pi/2$. Equation (11) simply describes the manner in which the solid angle changes as a function of η. The calculated relative erosion rates are plotted in Fig. 6. In the same figure, we have also shown the calculated variation in the track production rate, as a function of η relative to $\eta = 0$, for two depths. $X = 10^{-3}$ and 10^{-1} cm.

The results in Fig. 6 show that both erosion and track production rates vary in a

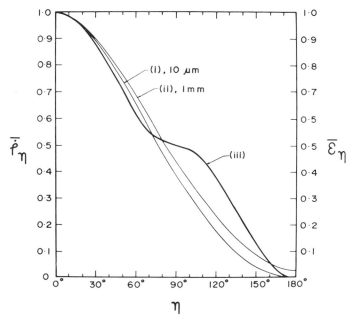

Fig. 6. The relative dependence of track production rates at (i) 10 μ, (ii) 0.1 cm on η for a lunar rock of radius 10 cm. Curve (iii) shows dependence of erosion on η. All the curves are normalized to values at $\eta = 0$.

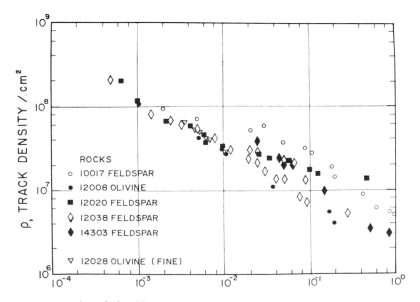

Fig. 7. Observed track densities in selected samples from various Apollo missions as a function of depth. The track profiles can be fitted to a spectrum of the type $\rho = KX^{-\alpha}$ with α between 0.8 to 1.0 up to a depth ≈ 1 mm.

similar manner except for values of η around 120°. From equation (9), we note that in the erosion controlled region the equilibrium track densities are proportional to the ratio $\dot{\rho}(X)/\varepsilon$ and therefore it follows that the extreme variations in the equilibrium track densities would be about a factor of 2.

Experimental results for $\rho(X)$ in the interval $10^{-3} - 1$ cm are plotted in Fig. 7. The observations are found to be in fair agreement with predictions. The point of inflexion in the track profile is different for different rocks and it occurs at greater depth, when the "sun-tan" age is higher. The absolute track densities at depths < 500 μ, which is the minimum erosion control depth for the four rocks under consideration, are confined within a factor of 2. There is also an indication that our earlier assumption (Bhandari et al., 1971) that $\rho(0.1$ cm$)$ is not effected to any appreciable extent by erosion may not be strictly valid in all cases. For example, in the case of rock 10017 we may have underestimated the "sun-tan" ages by a factor of ≈ 2.

Figure 8a shows the theoretically expected values of effective bombardment time $T_{\text{E.E.}}$ in the erosion controlled region for $\alpha = 2.25$ and 1.5 for $X < 0.1$ and > 1 cm depth intervals, as a function of X, for various assumed values of ε. The corresponding experimental values for several lunar and meteoritic samples, based on the 11 yr

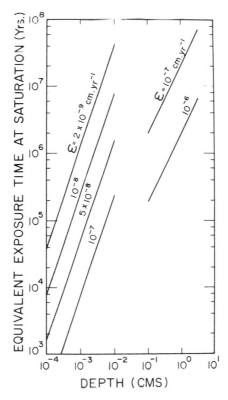

Fig. 8a. Effective irradiation times based on fossil tracks as a function of depth: Theoretical values of $T_{\text{E.E.}}$ for different values of ε.

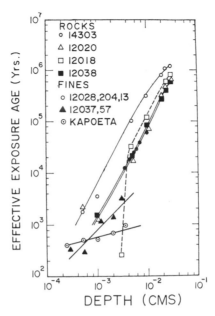

Fig. 8b. Experimentally determined values for different lunar samples and a track-rich grain from Kapoeta meteorite based on Surveyor glass production rates as given by Barber *et al.* (1971) for a mean solar cycle.

averaged solar flux deduced by Barber *et al.* (1971), Price *et al.* (1971) from observations in Surveyor III camera glass (see also Crozaz and Walker, 1971; Fleischer *et al.*, 1971b), for depths $< 10^{-1}$ cm are shown in Fig. 8b. Except for two samples (a lunar grain from 12037 and one Kapoeta grain), other correspond to erosion control with values of ε ranging between 3×10^{-8} and 10^{-7} cm yr^{-1}. It should be mentioned that the Surveyor data refer to observations over a short interval of time and moreover, since there does not exist a good agreement between various studies of the Surveyor's glass, the preceding discussion should be taken to indicate the importance of erosion in governing the cosmic ray track record in lunar rocks and a qualitative agreement with the model presented above.

Limits on erosion rates

The cumulative erosion rate, ε, is made up of two components $\varepsilon_{s.w.}$ and ε_m; $\varepsilon = \varepsilon_{s.w.} + \varepsilon_m$. The subscripts s.w. and m are used to denote contributions due to solar wind sputtering and micrometeorite impacts, respectively. In case $\varepsilon_{s.w.} > \varepsilon_m$, one would generally expect to find iso-track profiles parallel to the exposed surfaces of the rock (this may not hold in some cases, however, because erosion due to meteorite impacts is statistical in nature being the cumulative total of small and large discreet impacts, whereas one visualises solar wind sputtering as a continuous slow erosion on an atomic scale). We however found this not to be the case based on examination of several sections cut from surface chips and slices representing a perimeter of about 3 cm. The high track gradients are seen only in a small fraction of the rock surface indicating that micrometeorite impacts have denuded other regions due to recent impacts. It should be mentioned here that all the track profiles presented in this paper as well as those given earlier (Bhandari *et al.*, 1971) refer only to regions having the least impact erosion since the steepest observed track gradients are reported.

If we assume that a recent single meteorite impact results in a loss of surface of thickness ΔX_i after which no significant track accumulation occurred, then we can obtain values of ΔX_i at different points of the rock surface, using the highest observed track density profile, assumed to represent $\Delta X = 0$. In Fig. 9 we show the distribution of observed values of ΔX for rocks 12020 and 12038. It is estimated that the average area weighted values of ΔX for the rocks 12020 are 170 and 300 μ, respectively. Taking 10^6 yr as an upper limit for the average erosion equilibrium irradiation time for the depths interval $10^{-3} - 10^{-1}$ cm (see Fig. 8a, b), one obtains values of about $2 \times 10^{-8} - 5 \times 10^{-8}$ cm yr^{-1} – for ε_m as a lower limit.

The above value of ε_m is an order of magnitude higher than the corresponding estimate for $\varepsilon_{s.w.}$. The problem of solar wind sputtering of metals, oxides, and stone targets has been discussed by several authors (Whipple, 1959; Opik, 1962). The analysis by Eberhardt (1964) based on the Ar^{40} results of Heymann and Fluit (1963) taking into account the much smaller sputtering coefficients expected for solar wind hydrogen and helium as well as due to the lower solar wind energy yields for solar wind erosion rate a value of 3×10^{-9} cm yr^{-1}. This value is in excellent agreement with 4×10^{-9} cm yr^{-1} deduced by Wehner *et al.* (1963) primarily on the basis of their experimental data.

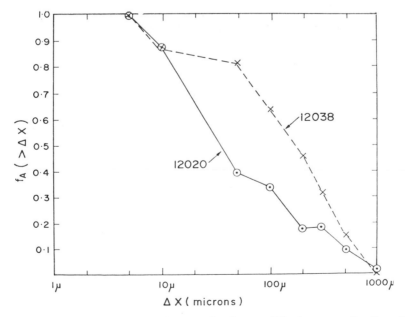

Fig. 9. The fractional area f_A, having a surface loss $> \Delta X$ is shown as a function of X in microns as deduced from solar flare maps in surface thick sections of rock 12020 and 12038.

An alternative approach for deducing total erosion rate, applicable to short intervals of depth, can be based on the facts that high track density grains in lunar regolith often reach an erosion equilibrium and that the duration of irradiation of these grains is approximately known. The model calculations presented below are based on a simplified picture of the depositional processes in the regolith, applicable for regions, where large craters have not obliterated records of a discrete deposition (regions having well-preserved stratification have been found on both Apollo 12 and 14 sites, LSPET, 1971). An extensive study of the frequency distribution of track densities in grains as well as the measurements of angular distributions of tracks in grains (McDougall *et al.*, 1972) having track density $< 10^7$ cm^{-2} lead us to postulate that before a layer is overlaid by a sudden deposition of another layer, the very upper regions get well mixed due to micrometeorite impacts, or by more complex physical processes. However, irrespective of the nature of the processes responsible for the mixing in the top layers, in view of the fact that a large fraction of grains have high densities characteristic of near surface irradiation, it must be concluded that a fraction f of the layer does undergo severe mixing so that individual grains come within depths of 1 mm from the surface. The fraction of grains having gradients or $\rho > 10^8$ cm^{-2}, N_H/N, has been experimentally determined for a large number of scoops and layers in the Apollo 12 double core (Arrhenius *et al.*, 1971). In the present model, the value of f is experimentally given by the following relation:

$$N_H/N = f \tag{12}$$

If we assume that the average grain size is $\langle d \rangle$ and the long-term average rate of deposition is S, then we obtain the following relation for the average sun-tan duration time T_s of a grain in the regolith:

$$T_s = \frac{\langle d \rangle}{fS} \tag{13}$$

Further, since our observations of track profiles in the regolith grains show that most grains with high track densities are in erosion equilibrium with production (track densities vary as X^{-1}), we must conclude that the cumulative erosion during irradiation must exceed the mean radius of the grain:

$$\varepsilon T_s \gg \frac{\langle d \rangle}{2} \tag{14}$$

The above two relations lead to the following inequality:

$$\varepsilon > \frac{fS}{2} \tag{15}$$

Taking for S an average value of 2×10^{-7} cm yr^{-1} which was observed for the Apollo 12 double core (Arrhenius *et al.*, 1971), for $\langle d \rangle$ a value of 10^{-2} cm (the size range most commonly applicable to track observations), and $f = 0.5$ we find a lower limit of 5×10^{-8} cm yr^{-1} for ε (for $X < 0.1$ cm) and an average time of irradiation of about 10^5 yr for individual grains which have high track densities.

We would like to stress here that the model considered above is an over-simplified one. However, it clearly brings out the mechanism that leads to the existence of an appreciable number of high track density grains in the lunar regolith. If we assume a typical value of 2 cm for the average thickness of the layer, then the mean mixing depth is of the order of $2f \approx 1$ cm. Furthermore, in the light of this model that the track densities up to 1 mm are governed by erosion, we can now understand the existence of a peak in track density observed in fines (typically 50–150 μ in size) around 40–80×10^6/cm^2 (Fig. 4a, b, c); similar peaks were observed earlier in several Apollo 12 samples (Arrhenius *et al.* 1971).

The present analysis differs from that given earlier (Bhandari *et al.*, 1971) to the extent that previously we may have underestimated the magnitude of erosion. The present data (Fig. 9) indicate that the product, εT_s, may be of the order of 10^{-1} – 5×10^{-1} cm in some rocks, in contrast to the value of $\leq 5 \times 10^{-2}$ cm considered by Bhandari *et al.* (1971).

An important consequence of the above deductions concerns the estimated sun-tan exposure ages. If the erosion equilibrium region exceeds up to 2–5 mm in individual rocks, then we have underestimated the exposure ages by factors of 5–10 in these cases, since we have assumed that the track densities at 0.1 cm depth are not affected by erosion. Thus in the case of the rock, 14303, in Table 2, the sun-tan age may be actually higher by a factor of up to 2.

In conclusion, the preceding analysis of the importance of erosion in governing the cosmic ray track records in the lunar regolith seems to be borne out by a variety of Apollo 11, 12, and 14 rocks and fines so far analyzed. It should be interesting to study samples representing a wider range of exposure times and irradiation histories.

Acknowledgments—We are extremely grateful to NASA for giving us the samples. Specially we are indebted to Dr. Michael B. Duke and Robert B. Laughon for a careful preparation of samples to meet our specific requirements.

We thank J. R. Arnold, G. O. S. Arrhenius, J. Hartung, D. S. McKay, D. McDougall, and C. Meyer for discussions and exchange of data on various samples. Study of some samples was carried out as a part of energy spectrum consortia. We are indebted to G. Crozaz, R. L. Fleischer, H. R. Hart, P. Pellas, P. B. Price, and R. M. Walker for several useful suggestions and discussions.

We would like to express our special appreciation to S. G. Bhat and S. Krishnaswami for their assistance in development of techniques as well as in study of samples. The laboratory assistance of Miss A. Padhye and N. Prabhu and technical help of P. B. Badle, P. K. Talekar and V. Noronha is gratefully acknowledged.

REFERENCES

Arrhenius G., Liang S., MacDougall D., Wilkening L., Bhandari N., Bhat S., Lal D., Rajagopalan G., Tamhane A. S., and Venkatavaradan V. S. (1971) The exposure history of Apollo 12 regolith. *Proc. Second Lunar Sci. Conf., Geochim. Cosmochim. Acta* Suppl. 2, Vol. 3, pp. 2883–2898. MIT Press.

Barber D., Cowsik R., Hutcheon I., Price P., and Rajan R. (1971) Solar flares, the lunar surface, and gas-rich meteorites. *Proc. Second Lunar Sci. Conf., Geochim. Cosmochim. Acta* Suppl. 2, Vol. 3, pp. 2705–2714. MIT Press.

Bhandari N., Bhat S., Lal D., Rajagopalan G., Tamahane A. S., and Venkatavaradan V. S. (1971) High resolution time averaged (millions of years) energy spectrum and chemical composition of iron group cosmic ray nuclei at 1 A.U. based on fossil tracks in Apollo samples. *Proc. Second Lunar Sci. Conf., Geochim. Cosmochim. Acta* Suppl. 2, Vol. 3, pp. 2611–2619. MIT Press.

Bhandari N., Lal D., Tamhane A. S., and MacDougall D. (1972) A study of the vestigial records of cosmic rays in lunar rocks. *Proc. Ind. Acad. Sci.* (to be submitted).

Bogard D. D. and Nyquist L. E. (1972) Noble gas studies on regolith materials from Apollo 14 and 15 (abstract). In *Lunar Science—III* (editor C. Watkins), pp. 89–91, Lunar Science Institute Contr. No. 88.

Comstock G. M. (1971) The particle track record of the lunar surface. General Electric Report No. 71-C-190.

Crozaz G. and Walker R. M. (1971) Solar particle tracks in glass from the Surveyor 3 spacecraft. *Science* **171,** 1237–1239.

Crozaz G., Walker R., and Woolum D. (1971) Nuclear track studies of dynamic surface processes on the moon and the constancy of solar activity. *Proc. Second Lunar Sci. Conf., Geochim. Cosmochim. Acta* Suppl. 2, Vol. 3, pp, 2543–2558, MIT Press.

Crozaz G., Drozd R., Hohenberg C. M., Hoyt H. P. Jr., Ragan D., Walker R. M., and Yuhas D. (1972) Solar flare and galactic cosmic ray studies of Apollo 14 samples (abstract). In *Lunar Science—III* (editor C. Watkins), pp. 167–169, Lunar Science Institute Contr. No. 88.

Eberhardt P. (1964) Rare gas measurements in meteorites and possible applications to the lunar surface. Ph.D. thesis, Physikalisches Institut, University of Berne, Switzerland.

Fleischer R. L, Hart H. R. Jr., Comstock G. M., and Evwaraye A. O. (1971a) The particle track record of the Ocean of Storms. *Proc. Second Lunar Sci. Conf., Geochim. Cosmochim. Acta* Suppl. 2, Vol. 3, pp. 2539–2568. MIT Press.

Fleischer R. L., Hart H. R. Jr., and Comstock G. M. (1971b) Very heavy solar cosmic rays: Energy spectrum and implications for lunar erosion. *Science* **171,** 1240–1242.

Heymann D. and Fluit J. M. (1963) Sputtering by 20 KeV Ar^{++} at normal incidence on meteorites. *J. Geophys. Res.* **67,** 2921.

Kirsten T., Deubner J., Ducati H., Gentner W., Horn P., Jessberger E., Kalbitzer S., Kaneoka I., Kiko J., Kratschmer W., Muller H. W., Plieninger T., and Thio S. K. (1972) Rare gases and ion tracks in individual components and bulk samples of Apollo 14 and 15 fines and fragmental rocks (abstract). In *Lunar Science—III* (editor C. Watkins), pp. 450–454, Lunar Science Institute Contr. No. 88.

Krishnaswamy S., Lal D., Prabhu N., and Tamhane A. S. (1971) Olivines; revelation of tracks of charged particles. *Science* **174,** 287–291.

Lal D., Murali A. V., Rajan R. S., Tamhane A. S., Lorin J. C., and Pellas P. (1968) Techniques for proper revelation and viewing of etch-tracks in meteoritic and terrestrial minerals. *Earth Planet. Sci. Lett.* **5,** 111–119.

Lal D., Rajan R. S., and Tamhane A. S. (1969) Chemical composition of nuclei of $Z > 22$ in cosmic rays using meteoritic minerals as detectors. *Nature* **221,** 33–37.

Lal D., MacDougall D., Wilkening L., and Arrhenius G. (1970) Mixing of the lunar regolith and cosmic ray spectra: New evidence from fossil particle-track studies. *Proc. Apollo 11 Lunar Sci. Conf., Geochim. Cosmochim. Acta* Suppl. 1, Vol. 3, pp. 2295–2303. Pergamon.

LSPET (Lunar Sample Preliminary Examination Team) (1971) Preliminary examination of the lunar samples from Apollo 14. *Science* **173,** 681–693.

Lugmair G. W. and Marti K. (1972) Neutron and spallation effect in Fra Mauro regolith (abstract). In *Lunar Science—III* (editor C. Watkins), pp. 495–497, Lunar Science Institute Contr. No. 88.

MacDougall D., Lal D., Wilkening L., Bhat S., Arrhenius G., and Tamhane A. S. (1971) Techniques for the study of fossil tracks in extraterrestrial and terrestrial samples—I: Methods of high contrast and high resolution study. *Geochem. J.* **5,** 95–112.

MacDougall D., Martinek B., and Arrhenius G. (1972) Regolith dynamics (abstract). In *Lunar Science—III* (editor C. Watkins), pp. 498–500, Lunar Science Institute Contr. No. 88.

Opik E. J. (1962) The lunar atmosphere. *Planet. Space Sci.* **9,** 211–244.

Price P. B. and O'Sullivan D. (1970) Lunar erosion rate and solar flare paleontology. *Geochim. Cosmochim. Acta* **3**, 2351–2359.

Price P. B., Hutcheon I. D., Cowsik R., and Barber D. J. (1971) Enhanced emission of iron nuclei in solar flares. *Phys. Rev. Lett.* **26**, 916–919.

Wehner G. K., Kenknight C., and Rosenberg D. L. (1963) Sputtering rates under solar-wind bombardment. *Planet. Space Sci.* **11**, 885–895.

Whipple F. L. (1959) Solid particles in the solar system. *J. Geophys. Res.* **64**, 1653–1654.

Yuhas D. E., Walker R. M., Price P. B., Hutcheon I. D., Lal D., Goswami J. N., Bhandari N., Reeves H., Poupeau G., Pellas P., Lorin J. C., Chetrit G. C., Berdut J. L., Hart H. R., Fleischer R. L., and Comstock G. M. (1972) Track consortium report on rock 14310. *Proc. Third Lunar Sci. Conf., Geochim. Cosmochim. Acta* Suppl. 3, Vol. 3. MIT Press.

Proceedings of the Third Lunar Science Conference
(Supplement 3, *Geochimica et Cosmochimica Acta*)
Vol. 3, pp. 2831–2844
The M.I.T. Press, 1972

The particle track record of Fra Mauro

H. R. Hart, Jr., G. M. Comstock,* and R. L. Fleischer

General Electric Research and Development Center,
Schenectady, New York 12301

Abstract—Apollo 14 breccias show a mixture of high and low track densities at most interior positions, indicating that the majority of the tracks have been inherited from the parent ingredients of the breccias. Using the lowest of these track densities as indicative of maximum postbrecciation surface residence times, we find a median 1.35 m.y., much younger than the less friable Apollo 11 and 12 igneous rocks. The igneous rock 14310 is studied as a part of a consortium, the results indicating a complex irradiation history. Soils are extremely variable, median track densities ranging over at least a factor of 200. Individual high density soil grains yield track density gradients having variable slopes, most of which are lower than expected from the Surveyor III filter glass results.

Introduction

THE COSMIC RAY particle tracks found in lunar samples yield a valuable record of the surface exposures of these samples. In earlier work (Arrhenius *et al.*, 1971; Barber *et al.*, 1971a, b; Bhandari *et al.*, 1971; Borg *et al.*, 1971; Comstock *et al.*, 1971; Crozaz *et al.*, 1970, 1971a; Fleischer *et al.*, 1970a, b, 1971a, b; Lal *et al.*, 1970; Price *et al.*, 1971; and Price and O'Sullivan, 1970) this record has been developed and presented in detail, with special emphasis on igneous rocks and soil core tube samples. In this paper the emphasis is somewhat different, for the Apollo 14 core tubes were not available for study and the rocks from the Apollo 14 site are overwhelmingly breccias or fragmental rocks. In interpreting the results obtained on breccias and random soil samples we have made use of the concept of deformation-induced track erasure, which is developed in separate papers (Fleischer *et al.*, 1972a, b).

After bringing up to date the description of our procedures, we shall present our results on igneous rocks, breccias, and soil samples, concentrating on the particle tracks formed by iron-group cosmic rays.

Procedures

The general techniques used in this work have been described in earlier papers (Fleischer *et al.*, 1970a, b, 1971a, b; Comstock *et al.*, 1971; Comstock, 1972). An improved etch for olivine has been used, that described by Krishnaswami *et al.* (1971): 2 cc H_3PO_4; 2 g oxalic acid; 80 g disodium salt of EDTA; 200 cc distilled water; NaOH pellets to pH of 8.0 (± 0.3) (~ 9 g NaOH). In the present work the olivines were etched for 6 hours at 98°C instead of the 2 or 3 hours at 125°C indicated by these authors. Individual crystals from both soils and breccias were most typically 50 to 200 μ in diameter. All optically counted track densities were inferred from tracks with lengths of 2 μ or greater.

In converting track densities to surface residence times the track production rate curves of Comstock (1972) and of Fleischer *et al.* (1967) were used. Pyroxene track densities are corrected for an etching efficiency of 0.7.

---

* Now at Centre de Spectrometrie de Masse du C.N.R.S., Orsay, France.

RESULTS

Igneous rocks

One of the few igneous rocks returned from Fra Mauro, 14310, was chosen for a detailed particle track study by a consortium. The consortium results are presented in another paper in this volume (Yuhas *et al.*, 1972). The track density profiles indicate a complicated surface history; the relatively flat interior profiles indicate an extended near surface exposure either in the interior of a larger rock or buried beneath a layer of soil. A plausible history is a relatively short surface exposure in its present form combined with an exposure of at least 400 m.y. in the interior of a rock having a radius of at least 20 cm.

The results of the detailed study of this rock emphasize the limitations of attempts to determine surface exposure histories of rocks for which only one or two small fragments are available. Extended track density profiles are needed; without these, misleadingly simplified interpretations are forced upon the investigation.

In rock 14310 the iron group cosmic ray track counts were made in feldspars in thin sections. In contrast with the igneous rocks of Apollos 11 and 12, fission fragment track concentrations were found in the feldspars. They were concentrated along the edges of crystals or along planar features in the crystals. In order to minimize the possibility of mistaking fission fragments for iron group cosmic rays, the counts were made away from crystal edges, planar features, and obvious stars. Errors are still possible, however, for the polishing operation may have just barely removed a uranium-rich layer adjacent to a feldspar crystal, leaving the fission fragment damage trails.

Three small igneous rocks from the 2–4 mm coarse fines (15233,5) yield surface ages of 3, 3, and 5 m.y. These ages were calculated on the assumption that the small rocks were just covered by level soil at the surface while being irradiated. The track densities and rock diameters were 6.3×10^6 cm^{-2}, $7.0 \times 10^{+6}$ cm^{-2}, and 7.6×10^6 cm^{-2}, and 0.27 cm, 0.29 cm, and 0.39 cm, respectively.

Breccias or fragmental rocks

The most abundant Apollo 14 rocks, the breccias, are found to have a mixture of high and low track densities at most positions in their interior. Figures 1–3 show detailed track density distributions in the interiors of three rocks; similar but less extensive results obtain for the other 8 breccias listed in Table 1. The wide distribution of track densities makes it clear that: (i) most of the tracks are inherited from the parent ingredients of the breccias, and (ii) track erasure has occurred during the breccia-forming event. Item (i) is supported by the work of Crozaz *et al.* (1970), who on the basis of high track densities in the interior of 10046 concluded that inherited tracks were present. Shock, the probable agent for producing the breccias, is shown in separate papers (Fleischer *et al.*, 1972a, b) to erase tracks in some crystals.

In order to establish upper limits for the surface residence times for these rocks we use the minimum track densities measured at each location, assuming that these crystals suffered either complete track erasure during the formation of the breccia or

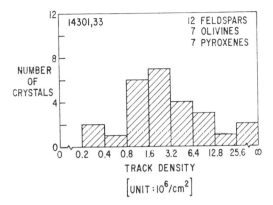

Fig. 1. Frequency distribution of cosmic ray track densities in the breccia 14301,33. The lowest track density was observed in both feldspar and pyroxene.

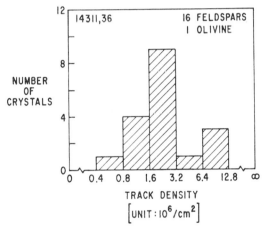

Fig. 2. Frequency distribution of cosmic ray track densities in the breccia 14311,36. The lowest track density was observed in a feldspar.

track fragmentation into segments of less than 2 μ, which would not have been counted according to our criteria. With this assumption, all tracks that were counted were created after the breccia-forming event. The limited extent of our sample of each breccia excludes the measurement of the track density profile necessary to derive the rock surface history. Rather, we have calculated the maximum direct exposure time for the rock surface which is closest to our sample. Our results are collected in Table 1. The median of these maximum surface ages for Apollo 14 and 15 breccias is a factor of ten lower than the indicated median for Apollo 11 and 12 igneous rocks. These lower surface ages appear to be the natural result of the friable nature of these samples, which yields a more rapid large scale erosion and a higher probability of catastrophic breakup under impact. Although the selection of only 11 breccias

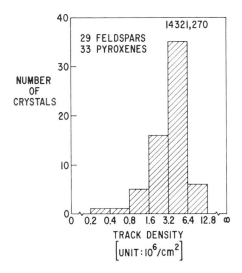

Fig. 3. Frequency distribution of cosmic ray track densities in the breccia 14321,270. The lowest track density was observed in a feldspar.

Table 1. Surface ages of breccias.

Rock Number	Min. Track Density ($\times 10^6$/cm^2)	Maximum Depth (cm)	Maximum Surface Age ($\times 10^6$ yr)
14047,42	1.1	0.5	3.4
14055,1	0.07	0.6	0.05
14066,22	0.47	0.7	0.49
14270,3	1.1	0.3	1.4
14301,33	0.27	0.5	0.34
14311,36*	0.73	⎰3.5	3.1
		⎱0.5	1.1
14321,270	0.39	10.8	8.2
15233,5,3	14	0.15	7.4
15233,5,14	1.6	0.08	0.3
15233,5,16	7.0	0.09	1.4
15233,5,17	3.5	0.13	1.3
		Median	1.35
	Median for Apollo 11 and 12		13 m.y.

* The two ages given result from an uncertainty in the location of our sample.

provides a somewhat limited sample, the order of magnitude difference between ages of these breccias and the igneous rocks appears to be adequate to indicate a genuine difference. The factor of ten in surface ages implies a factor of ten in large-scale erosion rates.

In choosing the single lowest observed track density we are, of course, forcing ourselves to work with poor statistics, and are therefore potentially able to be misled by a single crystal with a higher threshold for etched tracks, or by an effect of crystal orientation.* We do, however, infer from the fact that the minimum counts are not confined to any specific crystal type that such considerations are not dominant. In the future, a better understanding of the statistics of track erasure upon breccia formation may allow one to use more extensive track counts in determining the surface residence time. It is also worth noting that the width of the observed spread in track densities correlates with the degree of metamorphism as described by Warner (1972). For breccias with little metamorphism, such as 14047, 14055, and 14301, the spread in track densities exceeds two orders of magnitude; for those with greater metamorphism, such as 14066, 14311, and 14321, the spread reduces to a factor of 20 to 40. These results are consistent with effects expected from high temperatures and pressures. Indeed, Hutcheon *et al.* (1972) report a smooth variation of track density with depth in their sample, an observation that suggests that some portions of 14321 may not have retained tracks from prior irradiation.

Sample 14055,1 was covered with a glass splatter in which no tracks were found. Annealing studies indicated that fission fragment tracks could have annealed out at lunar surface temperatures.

Soils

The Apollo 14 fines (<1 mm) available for our study included: three Station G trench samples, top, 14148,32; middle, 14156,32; and bottom, 14149,48; a bulk sample, 14163,177; a comprehensive sample, 14259,70; and a sample, 14141,38,

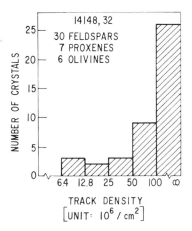

Fig. 4. Frequency distribution of cosmic ray track densities in a soil sample from the top of the trench at Station G, 14148,32.

---

* A lesser effect is that of near surface track gradients. For the observed (Crozaz *et al.* 1972) (depth)$^{-.7}$ behavior a 0.1 cm uncertainty in position at a depth of 0.5 cm gives rise to only a $\pm 14\%$ track density fluctuation, much less than the order of magnitude variations observed, and less than the $\pm 30\%$ etching efficiency variations in the pyroxenes, and than typical possible anisotropies due to the directionality of the incident cosmic ray particles.

from Station C′, at the edge of a morphologically very young 30 m crater near the edge of Cone Crater (Swann *et al.*, 1971).

The track density distributions for these samples (Figs. 4–8) are even broader than those of the breccias. The median track densities vary widely from sample to sample, by at least a factor of 60 within the previously mentioned samples. Within the trench alone the median densities vary by a factor of more than 30.

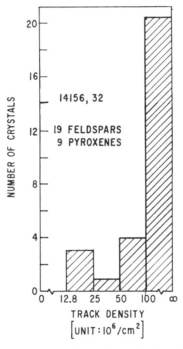

Fig. 5. Frequency distribution of cosmic ray track densities in a soil sample from the middle of the trench at Station G, 14156,32.

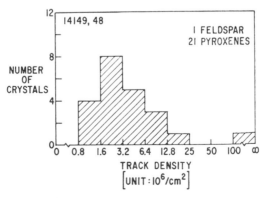

Fig. 6. Frequency distribution of cosmic ray track densities in a soil sample from the bottom of the trench at Station G, 14149,48.

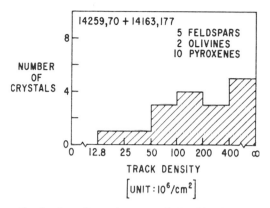

Fig. 7. Frequency distribution of cosmic ray track densities for a combination of two surface soil samples, the bulk sample 14163,177 and the comprehensive sample 14259,70.

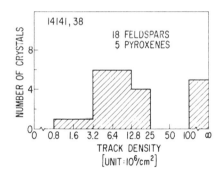

Fig. 8. Frequency distribution of cosmic ray track densities for a surface soil sample collected at Station C′ at the edge of a 30 m crater near Cone Crater.

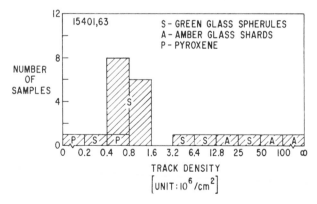

Fig. 9. Frequency distribution of cosmic ray track densities for a soil sample, 15401, 63, collected at the surface of a fillet of a large boulder on the Apennine Front. The glass track densities have been corrected for an etching efficiency of 0.1.

The lowest soil track densities we have measured have been in 15401,63, a sample of <1 mm fines collected from the top of a heavy fillet on a 3 m boulder lying on a steep slope at Station 6a on the Apennine Front. This sample is of special interest because of its high concentration of green glass spherules and amber glass shards. The track density distribution is shown in Fig. 9. For this figure the glass track densities have been corrected for the average glass etching efficiency of ~0.1. The rough agreement between the pyroxene track densities and those of the green glass spherules suggest that the glass retains tracks at lunar surface temperatures for extended periods of time. If laboratory annealing experiments confirm this suggestion, the plentiful and widely dispersed glass at the Apollo 15 site will be a useful dating medium.

The wide track density distributions observed in the soils can result from the mixing together of grains having different irradiation histories (Comstock *et al.*, 1971; Comstock, 1972). It can also result from variable track erasure during an impact event, e.g., track erasure caused by shock or mechanical deformation (Fleischer *et al.*, 1972a, b). In either case the minimum track density soil grains allow us to establish maximum irradiation times (Crozaz *et al.*, 1970; Fleischer *et al.*, 1972a, b). Within the several assumptions listed below, we can calculate a maximum age for a recent impact event. We assume: (i) the minimum track density soil grain lost all of its tracks during the impact event of interest; (ii) the soil grain resided throughout the subsequent irradiation at the deepest possible position in the collected soil sample; and (iii) no more recent event or transport process moved low track density grains into the sample or changed significantly the thickness of the soil over the grain.

Subject to these three assumptions an age can be rigorously calculated; therefore some discussion of their validity is appropriate.

(i) We have found that approximately half of the pyroxenes in the Apollo 12 deep core display evidence of fine scale deformation, so that track erasure is a common feature. We therefore have reasonable confidence that a statistical sampling of soil grains will provide cases of track erasure. If we fail to locate a crystal whose tracks were erased in the most recent soil event we can only infer too long a surface residence and therefore an upper limit on the surface age.

(ii) The original soil depths for grains in scoop samples are not known, but again a statistical selection of grains will insure that some were close to the maximum depth, and errors only increase the calculated age, giving once again a maximum age. For samples from a core, depths are known (at least in principle), and this approximation is not a problem.

(iii) Recent addition to a surface soil layer can occur. If yesterday a single freshly deformed grain or one excavated from depth were dropped on a soil layer, we would infer too young a soil age for the layer as a whole. Although this result is possible, it is highly unlikely. In an extreme case the young grain would show up as a displaced data point on graphs such as Figs. 4–9 rather than what we in fact observe—the lower edge of a distribution. A rough estimate of the relative probability of such an anomalously young grain can be made as follows. If we assume that each primary incoming particle mechanically erases tracks in a lunar grain of the same size, we can calculate from the flux of 100μ (3×10^{-6} g) particles (Soberman, 1971) the rate of generation of freshly reset $100\,\mu$ grains: $4 \times 10^{-5}/cm^2$ year. Thus in 10^6 years the spurious young grains would constitute only 0.003% of a 3 cm soil column. For a sampling of 100 soil grains the odds are thus at least 300 to 1 against a spurious result being obtained for a 1 m.y. old soil and 30 to 1 for a 10 m.y. old soil. With this possible but improbable uncertainty in mind we shall proceed to calculate what we call "maximum soil residence ages."

Using this approach we estimate the maximum ages of two young soil samples, 14141,38, from the edge of the 30 m crater near Cone Crater, and 15401,63, the fillet sample from the Apennine Front. In both cases we lack the exact dimensions of the sample collected; we shall take 3 cm as a reasonable estimate of the maximum depth sampled (Arrhenius *et al.*, 1971). Taking the production curve of Comstock (1972), a 3 cm depth, and minimum track densities of $1.0 \times 10^6\ cm^{-2}$ and $1.1 \times 10^5\ cm^{-2}$ for 14141,38 and 15401,63, respectively, we find maximum ages of 6 m.y. for 14141,38 and 0.7 m.y. for 15401,63, an exceptionally young soil layer.

A number of soil grains exhibit very steep track density gradients, reflecting the solar flare origin of these track-forming cosmic rays (Comstock, 1972; Fleischer *et al.*, 1970b). These surface-exposed grains potentially tell us about ancient solar flare energy spectra and fine scale erosion rates. Unfortunately the gradients as observed are determined by a combination of effects which make it difficult to extract separately the spectra and the erosion rates.

In Fig. 10 we show plots of track density versus depth for nine different grains from four different missions. Also shown as dashed lines are limits for near surface track densities in several rocks studied by Crozaz *et al.* (1972). For the mineral grains, the densities plotted are those measured from photographs taken of platinum shadowed replicas in a transmission electron microscope; no corrections for etching efficiency are made. The track densities in the glass, obtained from SEM photographs, have been corrected for an etching efficiency estimated to be 0.1.

There are large grain-to-grain variations in the track densities measured at each depth. The densities measured in the minerals vary by a factor of about 100 and extend beyond the limits shown for the rocks. The data for the gradients in all but two of the grains can be fit by a power law, the power varying from -0.7 to -1.5. For the other two samples, which have lower track densities, there is a definite break in the curve, with the steeper slopes occurring in the interior of the grains.

These results are not well understood. Until more such grains have been studied we shall be offering plausible, but probably not unique, interpretations of the data. As background let us note that:

(i) The Surveyor III filter glass, recently exposed to solar flare cosmic rays for 2.5 years, yielded (Price *et al.*, 1971; Crozaz *et al.*, 1971b; Fleischer *et al.*, 1971b) a track density as a function of depth D which can be fit for present purposes by a power law, D^m, with $m \approx -2.5$.

(ii) A uniform, fine-scale erosion during irradiation would alter the observed track density profile. Near the surface, the track production rate would be in equilibrium with the erosion rate, yielding a power-law profile, D^n, with $n = p + 1$, where p is the exponent characterizing the solar flare spectrum. The track densities in this equilibrium region would be proportional to the flux of solar flare cosmic rays,

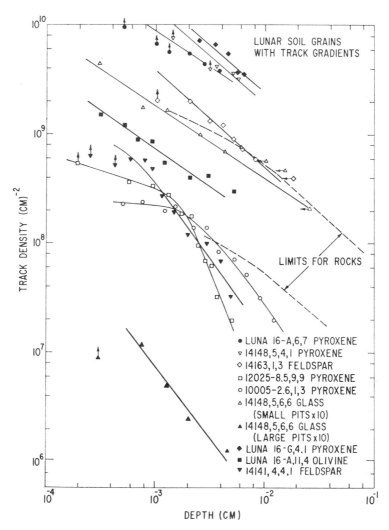

Fig. 10. Cosmic ray track gradients observed in several lunar soil grains. The dashed lines indicating limits for rocks are based on Crozaz *et al.* (1972). Corrections for etching efficiency have been made only for the glass sample, for which 0.1 is an average etching efficiency.

inversely proportional to the erosion rate, and independent of the exposure time. Deeper within the grain, the profile would change to the uneroded power law, D^p.

(iii) The irradiation of a soil or rock grain covered by a thin soil or dust layer would yield a profile having a definite break. Likewise, the chipping or abrading of a surface fragment subsequent to irradiation would yield a break in the profile. For a sufficiently long surface exposure, the grain might have been repeatedly covered (and later uncovered) by dust or soil layers of different thicknesses. The effect of such an exposure history would be to flatten the profile, much as does erosion, but to yield track densities which increase with total exposure time. The profile shape would depend on the cosmic ray spectrum, the distribution of covering thicknesses, and the rates of their removal.

(iv) In the laboratory, the random mounting and polishing of soil grains rarely yields the optimum etching plane, i.e., the plane having the steepest gradient. Thus, for depths small compared to the radius of the grain, the apparent depth is generally larger than the true depth by a constant trigonometric factor; for apparent depths approaching the radius of the grain, the true depth may saturate at a value much less than the apparent depth. The net effect is to yield a variable overestimate of the track density at each depth and an underestimate of the steepness of the gradient. (In the highly unlikely event that the grain was irradiated at the bottom of a pit of very small radius, it would, however, be possible to get an overestimate of the gradient steepness.)

In Fig. 10, the two lower density grains, from samples 10005 and 12025, show the previously mentioned breaks in their profiles. Concentrating on the 12025 grain, we note that an exposure to the Surveyor III spectrum for about 4,500 years while covered with about 1–2 mg/cm^2 of material would yield the observed profile. With the steep slope observed at depths greater than 15 μ, we have evidence that for 4,500 years at some time in the past there was a solar cosmic ray energy spectrum at least as steep as the recent Surveyor III spectrum. This conclusion is not changed by consideration of fine-scale erosion. For plausible erosion rates of 10^{-8} to 10^{-7} cm/year (Crozaz et al., 1971a; Fleischer et al., 1971b) and the short exposure time of this grain, the erosion equilibrium profile would not reach the region of steepest gradient. The 10005 grain profile is also consistent with an irradiation by a Surveyor III spectrum for 90,000 years with a 20 mg/cm^2 covering.

For most of the other grains, the gradients are not as steep; a typical power-law exponent would be -0.9. Comparing the mineral grain data with the rock surface data presented by Crozaz et al. (1972), we see that the power-law exponents are similar but the track densities in the grains are sometimes much higher. If the grains were in erosional equilibrium, their track densities would be independent of exposure time and would be determined by the ratio of the flux of solar cosmic rays to the rate of fine-scale erosion. With this interpretation, the data indicate that this ratio has varied by a factor of 50 over the times, locations, and materials sampled, and that the solar cosmic ray spectrum was harder at some time in the past than was recently found in the Surveyor III filter glass (e.g., $D^{-1.9}$ versus $D^{-2.6}$). The higher equilibrium track densities for some of the soil grains, relative to the rocks, could result if some modes of fine-scale erosion were effective against rocks, but not against loose soil grains.

If the profile flattening is instead caused predominantly by irradiation under a series of dust coverings of different thicknesses, the factor of 50 in track densities merely reflects different exposure times, and the power-law exponent does not necessarily imply a harder spectrum but perhaps only a particular dust thickness distribution. The dust covering interpretation is consistent with the observation that only the younger grains show the distinct profile break resulting from a single dust covering; the older grains show the roughly power-law profile which would result from a series of different dust coverings.

Some of the apparent spectral hardening and variability in track density could also be caused by problems in orienting the grains for mounting and polishing; the etch plane may not be the plane of steepest gradient.

Finally, the data from a single glass sample, 14148,32,5,6,6, are given as two curves in Fig. 10 for "small pits" and "large pits." In each case an etching efficiency of 0.1 was assumed. The levels of the two curves differ by a factor of >200, and the upper curve is flatter than the lower. We favor the interpretation that the smaller pits correspond to iron-group and somewhat lighter nuclei and the larger ones to charge >30 nuclei. One alternative explanation—that the numerous small pits are spallation recoil tracks—would not explain the steep gradient, nor would a subtraction of the possible residual spallation component alter the gradient significantly. Although range differences between charge 26 and charge $\gtrsim 30$ due to ionization loss differences will not explain the two slopes of the track density profiles, the damage densities that result may do so. Differences in damage could cause the etching efficiency for the small pits to be less than the average value of 0.1 used, so that the small pits would correspond to particles in a very limited solid angle range and with different and complicated geometrical relations to the exterior of the glass fragment. The observed pit density ratio of >200 to 1 could thus be consistent with the value of ~ 2000 to 1 observed in meteoritic crystals (Fleischer *et al.*, 1967; Price *et al.*, 1968).

Conclusions

We have found from the wide distribution of track densities in Apollo 14 breccias that the majority of the tracks were inherited from the parent materials. This finding puts a premium on the lowest track density grains, for they can be used to determine the maximum surface residence times of the breccias. We find that the median age of the breccias is one-tenth that of the Apollo 11 and 12 igneous rocks, a ratio that is consistent with the friable nature of the breccia samples that we have studied. The median soil track densities vary widely from sample to sample. Using minimum track density grains, we date a soil sample from the edge of a small crater near Cone Crater as being deposited no more than 6 m.y. ago and a soil sample from the Apennine Front at 0.7 m.y. ago. High gradient grains from the soil, clearly irradiated by solar cosmic rays, have proved difficult to interpret satisfactorily.

Acknowledgments—We are pleased to give thanks for experimental assistance to W. R. Giard, E. F. Koch, M. D. McConnell, G. E. Nichols, E. G. Stella, and G. N. Weltman. We are indebted to R. M. Walker for helpful comments, including the suggestion that metamorphism might correlate with track abundance distributions. This work was supported in part by NASA contract NAS 9-11583.

References

Arrhenius G., Liang S., Macdougall D., Wilkening L., Bhandari N., Bhat S., Lal D., Rajagopalan G., Tamhane A. S., and Venkatavaradan V. S. (1971) The exposure history of the Apollo 12 regolith. *Proc. Second Lunar Sci. Conf., Geochim. Cosmochim. Acta* Suppl. 2, Vol. 3, pp. 2583–2598. MIT Press.

Barber D. J., Cowsik R., Hutcheon I. D., Price P. B., and Rajan R. S. (1971a) Solar flares, the lunar surface, and gas-rich meteorites. *Proc. Second Lunar Sci. Conf., Geochim. Cosmochim. Acta* Suppl. 2, Vol. 3, pp. 2705–2714. MIT Press.

Barber D. J., Hutcheon I. D., and Price P. B. (1971b) Extralunar dust in Apollo cores? *Science* **171**, 372–374.

Bhandari N., Bhat S., Lal D., Rajagopalan G., Tamhane A. S., and Venkatavaradan V. S. (1971) High resolution time averaged (millions of years) energy spectrum and chemical composition of iron-group cosmic ray nuclei at 1 A.U. based on fossil tracks in Apollo samples. *Proc. Second Lunar Sci. Conf., Geochim. Cosmochim. Acta* Suppl. 2, Vol. 3, pp. 2611–2619. MIT Press.

Borg J., Maurette M., Durrieu L., and Jouret C. (1971) Ultramicroseopic features in micron-sized lunar dust grains and cosmophysics. *Proc. Second Lunar Sci. Conf., Geochim. Cosmochim. Acta* Suppl. 2, Vol. 3, pp. 2027–2040. MIT Press.

Comstock G. M. (1972) The particle track record of the lunar surface. Proceedings of Conference on Lunar Geophysics, *The Moon* (in press). This paper is presently available in preprint form as GE Report No. 71-C-190.

Comstock G. M., Evwaraye A. O., Fleischer R. L., and Hart H. R. Jr. (1971) The particle track record of lunar soil. *Proc. Second Lunar Sci. Conf., Geochim. Cosmochim. Acta* Suppl. 2, Vol. 3, pp. 2569–2582. MIT Press.

Crozaz G., Haack U., Hair M., Maurette M., Walker R., and Woolum D. (1970) Nuclear track studies of ancient solar radiations and dynamic lunar surface processes. *Proc. Apollo 11 Lunar Sci. Conf., Geochim. Cosmochim. Acta* Suppl. 1, Vol. 3, pp. 2051–2080. Pergamon.

Crozaz G., Walker R., and Woolum D. (1971a) Nuclear track studies of dynamic surface processes on the moon and the constancy of solar activity. *Proc. Second Lunar Sci. Conf., Geochim. Cosmochim. Acta* Suppl. 2, Vol. 3, pp. 2543–2558. MIT Press.

Crozaz G. and Walker R. (1971b) Solar particle tracks in glass from the Surveyor III spacecraft. *Science* **171**, 1237–1239.

Crozaz G., Drozd R., Hohenberg C. M., Hoyt H. P., Ragan D., Walker R. M., and Yuhas D. (1972) Solar flare and galactic cosmic ray studies of Apollo 14 samples (abstract). In *Lunar Science—III* (editor C. Watkins), pp. 167–169, Lunar Science Institute Contr. No. 88.

Fleischer R. L., Price P. B., Walker R. M., Maurette M., and Morgan G. (1967) Tracks of heavy primary cosmic rays in meteorites. *J. Geophys. Res.* **72**, 355–366.

Fleischer R. L., Haines E. L., Hanneman R. E., Hart H. R. Jr., Kasper J. S., Lifshin E., Woods R. T., and Price P. B. (1970a) Particle track, x-ray, and mass spectrometry studies of lunar material from the Sea of Tranquility. *Science* **167**, 568–571.

Fleischer R. L., Haines E. L., Hart H. R. Jr., Woods R. T., and Comstock G. M. (1970b) The particle track record of the Sea of Tranquility. *Proc. Apollo 11 Lunar Sci. Conf., Geochim. Cosmochim. Acta* Suppl. 1, Vol. 3, pp. 2103–2120. Pergamon.

Fleischer R. L., Hart H. R., Comstock G. M., and Evwaraye A. O. (1971a) Particle track record of the Ocean of Storms. *Proc. Second Lunar Sci. Conf., Geochim. Cosmochim. Acta* Suppl. 2, Vol. 3, pp. 2559–2568. MIT Press.

Fleischer R. L., Hart H. R., and Comstock G. M. (1971b) Very heavy solar cosmic rays: energy spectrum and implications for lunar erosion. *Science* **171**, 1240–1242.

Fleischer R. L., Comstock G. M., and Hart H. R. (1972a) Dating of mechanical events by deformation-induced erasure of particle tracks. *J. Geophys. Res.* (submitted).

Fleischer R. L., Comstock G. M., and Hart H. R. (1972b) Particle track dating of mechanical events (abstract). In *Lunar Science—III* (editor C. Watkins), pp. 265–267, Lunar Science Institute Contr. No. 88.

Hutcheon I. D., Phakey P. P., and Price P. B. (1972) Studies bearing on the history of the lunar breccias. *Proc. Third Lunar Sci. Conf., Geochim. Cosmochim. Acta* Suppl. 3, Vol. 3, pp. 2845–2866. MIT Press.

Krishnaswami S., Lal D., Prabhu N., and Tamhane A. S. (1971) Olivines: revelation of tracks of charged particles. *Science* **174**, 287–291.

Lal D., Macdougall D., Wilkening L., and Arrhenius G. (1970) Mixing of the lunar regolith and cosmic ray spectra: evidence from particle-track studies. *Proc. Apollo 11 Lunar Sci. Conf., Geochim. Cosmochim. Acta* Suppl. 1, Vol. 3, pp. 2295–2303. Pergamon.

Price P. B., Rajan R. S., and Tamhane A. S. (1968) The abundance of nuclei heavier than iron in the cosmic radiation in the geological past. *Astrophys. J.* **151,** L109–L116.

Price P. B. and O'Sullivan D. (1970) Lunar erosion rate and solar flare paleontology. *Proc. Apollo 11 Lunar Sci. Conf., Geochim. Cosmochim. Acta* Suppl. 1, Vol. 3, pp. 2351–2359. Pergamon.

Price P. B., Hutcheon I., Cowsik R., and Barber D. J. (1971) Enhanced emission of iron nuclei in solar flares. *Phys. Rev. Lett.* **26,** 916–919.

Soberman R. K. (1971) The terrestrial influx of small meteoric particles. *Rev. Geophys. and Space Phys.* **9,** 239–258.

Swann G. A., Trask N. J., Hait M. H., and Sutton R. L. (1971) Geologic setting of the Apollo 14 samples. *Science* **173,** 716–719.

Warner J. L. (1972) Apollo 14 breccias: metamorphic origin and classification (abstract). In *Lunar Science—III* (editor C. Watkins), pp. 782–784, Lunar Science Institute Contr. No. 88.

Yuhas D. E., Walker R. M., Reeves J. L., Price P. B., Poupeau G., Pellas P., Lorin J. C., Lal D., Hutcheon I. D., Hart H. R., Goswami J. N., Fleischer R. L., Comstock G. M. Chetrit G. C., Bhandari N., and Berdot J. L., (1972) Track consortium report on rock 14310. *Proc. Third Lunar Sci. Conf., Geochim. Cosmochim. Acta* Suppl. 3, Vol. 3, pp. 2941–2948. MIT Press.

Proceedings of the Third Lunar Science Conference
(Supplement 3, *Geochimica et Cosmochimica Acta*)
Vol. 3, pp. 2845–2865
The M.I.T. Press, 1972

Studies bearing on the history of lunar breccias

I. D. Hutcheon, P. P. Phakey,* and P. B. Price

Department of Physics, University of California
Berkeley, California 94720

Abstract—High-voltage electron microscopy of ion-thinned sections of breccias reveals details of substructure and radiation effects unobtainable by other techniques. Gas bubbles ~ 80 Å in diameter (originating in large part from solar wind implantation) and solar flare tracks inherited from a prior irradiation are observed in fine ($< 5\ \mu$) grains in 14301, 14315, 14318, 14311, and 14321, indicating that this material was once exposed on the surface as soil. The number density and visibility of the tracks, and the fraction of grains containing tracks and bubbles, decrease with the extent of metamorphism in the above order. Shock and recovery features in these rocks and in 15418 and 15086 are described. Large crystals have a different origin from the fine-grained material in matrix and clasts. Gradients of etched cosmic ray tracks studied optically in the large breccias 14311 and 14321 are similar to the gradient in igneous rock 12022 and the absence of large scatter indicates that no tracks inherited from irradiation prior to breccia formation show up by etching. Lexan maps reveal the heterogeneity of the uranium distribution in most breccias. Uranium is primarily concentrated in interstitial sites. The density of fission tracks in whitlockite in 14321 is consistent with a $\sim \frac{2}{3}$ contribution from ^{238}U and a $\sim \frac{1}{3}$ contribution from ^{244}Pu, as would be expected if its age were the same as that of an igneous clast in 14321 (~ 3.95 Gy).

Introduction

Breccias are an exceedingly complicated assemblage of individual fragments from a variety of locations and with a variety of histories. Part of their fascination is that with suitable techniques one may hope to find fragments essentially unaltered since before the period of mare formation began about 4 Gy ago. We have found that transmission electron microscopy is an essential tool in studying the complicated history of breccias. It has, for example, enabled us to see 80 Å solar wind gas bubbles and incompletely annealed fossil tracks in fine-grained material from matrix and clasts down to a micron in size, as well as intricate shock and recovery features on a submicron scale.

In this paper we discuss the substructure of several Apollo 14 and 15 breccias with emphasis on radiation and shock effects; we report optical microscope observations of solar flare and cosmic ray tracks; and we present a progress report on the fission track dating of breccia fragments.

Techniques

The nonbasaltic origin and visible compositional complexity of the Apollo 14 breccias raise many questions about the nature and extent of shock and thermal metamorphic features. We have examined selected areas of several breccias with the Berkeley 650 keV electron microscope, utilizing ion-thinned specimens in order to optimize the value of material available for observation. Good ion-thinned

---

* On leave from Department of Physics, Monash University, Victoria, Australia.

specimens are especially important in studying inter-relationships between features in adjacent grains and in permitting us to examine the problem of retention of solar flare tracks from a prior irradiation. Samples usually are viewed unetched so that submicroscopic shock features can be resolved, while high densities of tracks will be visible in dark field.

Optical microscopy of tracks in Apollo 14 breccias is done almost exclusively in feldspar crystals. We observe the grains in situ, in transparent rock sections ≤ 100 μ thick, mounted in epoxy. Track densities $\leq 5 \times 10^7/cm^2$ are best observed after etching in a boiling 6 g NaOH : 12 g H_2O solution for 45 min. We have not succeeded in etching *optically visible* tracks in the fine-grained matrix of any breccia. In the soil breccias the track density is simply too high to resolve optically. In other breccias, shock effects and thermal fading of the high track densities ($\leq 5 \times 10^8/cm^2$) in matrix crystals make electron microscopy a necessity.

The whole-rock uranium concentrations of Apollo 14 breccias are about an order of magnitude higher than in samples from Apollos 11 and 12 (LSPET, 1971). Individual crystals within a breccia tend to be fractured as a consequence of shock events and are best studied in situ to facilitate handling and the observation of fission tracks in uranium-rich crystals. Therefore, we prepared transparent polished sections, ~ 100 μ thick, mounted in epoxy, overlaid with sheets of Lexan plastic. Accurate quantitative maps of the uranium distributions in the rock sections were obtained by irradiating the sections and their plastic sheets with $\sim 1 \times 10^{16}$ thermal neutrons at a maximum temperature of 46°C. The flux was monitored by enclosing squares of calibrated glass containing a known ^{235}U and ^{238}U concentration (Schreurs *et al.*, 1971).

After etching for four hours at 40°C in 6.25 N NaOH solution, each plastic sheet contained a faint image (Kleeman and Lovering, 1967) that facilitated realigning it with the corresponding rock section, as well as fission track distributions quantitatively related to the uranium distributions in the rock sections. Observed uranium concentrations vary from ≤ 10 ppb to ≥ 20 ppm.

OBSERVATIONS ON SPECIFIC BRECCIAS

14301

This rock has attracted special interest because of its high trapped gas content (LSPET, 1971) and because of the compositional similarity of its fine-grained matrix to the Apollo 14 soil (Hubbard and Gast, 1972; Megrue and Steinbrunn, 1972). These observations suggest that if during the formation of the breccia, severe shock and thermal metamorphism did not occur, high track density regions might still exist in the interior of this rock.

Evidence that the brecciation process has not disturbed the trapped Xe content is reported by Drozd *et al.* (1972) and Kaiser (1972). Our electron microscope observations further illustrate the very low-grade metamorphism experienced by 14301.

1. Electron microscopy. The appearance of the fine-grained material, in which is embedded a small number of larger clasts, indicates that cohesion of this rock is caused by compression at very mild peak pressure without much temperature increase (cold shock compression). Virtually no porosity is present. The most common deformation features in this rock are fractures, twinning, and heterogeneous distribution of dislocations in plagioclose and pyroxene (Fig. 1). Shock-induced homogeneous glass of the type described below for 14315 and 14318 is not observed. The most striking substructural features are easily resolved "*b*" type antiphase domain boundaries, which are much larger than in other lunar plagioclase reported in the literature (Christie *et al.*, 1971; Wenk *et al.*, 1972; Fig. 2).

In contrast to the fine-grained material, the most striking feature of the larger clasts of 14301 is the presence of recrystallized nuclei where the grain structure is

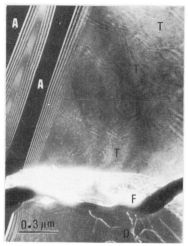

Fig. 1. Dark field (DF) micrograph showing characteristic features in plagioclase in breccia 14301. Microtwins (A), fractures (F), dislocations (D), and some weak antiphase domain boundaries (APB) of type b can be seen.

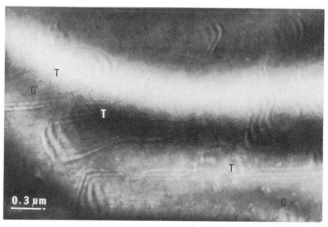

Fig. 2. Gas bubbles (G) and tracks (T) visible in diffraction contrast in plagioclase in 14301. The gas bubbles appear as dark or light dots reversed in contrast on either side of thickness fringes. Coarse scale APBs are marked by diffraction contrast fringes. DF micrograph, type b reflection.

more like that of a well-sintered ceramic (Fig. 3). We conclude that these clasts contain generations of breccia fragments which recovered and recrystallized prior to their consolidation in the fine-grained material.

A common feature of the micron-size grains in the matrix of 14301 is the presence of $\sim 10^{10}$ tracks/cm^2 independent of the location of the grains in the breccia (Figs. 1 and 2). As we have discussed elsewhere (Phakey *et al.*, 1972), the track production

Fig. 3. Bright field (BF) micrograph showing extensive development of recrystallized nuclei in a clast in 14301.

rate falls off so steeply with depth that such high track densities can only mean that the grains were once soil particles lying on top of the lunar surface. That the tracks have survived incorporation of the grains into the breccia means that the peak shock pressure during compaction probably did not greatly exceed ∼100 kbar (Ahrens *et al.*, 1970) and that the extent of subsequent metamorphism of the matrix was exceedingly small. These conclusions are consistent with the observations in the previous paragraphs.

We also observe gas bubbles ∼80 Å in diameter in the matrix grains (Fig. 2), which provide further strong evidence that they were once exposed to the sun as individual soil particles. The gas pressure in such a bubble is given by $p = 2\,\gamma/r$, where γ is the surface energy, ∼10^3 erg/cm^2. From the observed bubble densities, ∼10^{14} to 10^{15}/cm^3, and calculated pressure, ∼5000 atm, we can account for ∼0.1 to ∼1 cm^3 STP/g of gas in the form of bubbles, which is typical of the bulk concentrations measured in breccias by mass spectrometry. We see bubbles very similar in size and number/cm^3 to those seen in lunar fines (Phakey *et al.*, 1972) and in track-rich grains in gas-rich meteorites but not seen in track-poor grains in gas-rich meteorites nor in igneous lunar rocks.

Bubble densities up to ∼10^{14}/cm^3 could in principle result from the He gas released in decay of ∼20 ppm of U together with ∼80 ppm of Th over ∼4 Gy These concentrations are considerably higher than the average for Apollo 14 soil and breccias but are representative of U-rich interstitial phases. We would expect only slight gradients of bubble density in material whose grain-size is comparable to the range of alpha-particles (∼10 μ).

Bubble densities of ∼10^{15}/cm^3, commonly seen in grains of 14301, are too high to be attributed to U and Th decay except in rare phases with ∼200 ppm of U. They almost certainly must have resulted from solar wind implantation and thus act as tracers of soil that was once exposed to the solar wind while on the lunar surface.

Only a few hundred years of irradiation would suffice to supply the needed gas. The magnitude of the bubble density gradient depends on gas and bubble mobility but should be much steeper for solar wind gas than for gas formed from helium released in radioactive decay. Although hydrogen is the most abundant element in the solar wind, it diffuses easily (Merlivat *et al.*, 1972) and might not quantitatively survive after the compaction of grains into a breccia.

High densities of tracks also act as tracers of soil that was exposed to the solar wind and solar flares. Of all the breccias we have studied, the tracks in 14301, together with those in the soil breccia 15086, are most distinct and easily visible. In some of the more metamorphosed breccias, tracks present in high density in fine-grained material are extremely faint and cannot be adequately reproduced in photographs, even though they are visible on the negatives. Only under the most favorable diffracting conditions are those tracks in favorable orientations faintly visible. We attribute their faintness to a reduction of the strain around the tracks due to annealing.

We thus rank 14301 very low on the scale of thermal metamorphism shown in Table 1, which is consistent with the grade assigned by Warner (1972) based primarily on the abundance and type of glass.

2. Optical microscope observations of tracks. We have studied cosmic-ray track densities and uranium distributions in the two remaining constituents of 14301: chondrules and large feldspar crystals. Uranium is generally excluded from single crystals but when present is distributed heterogeneously, apparently associated with uranium-rich inclusive mineral phases such as zircon and baddelyite (Anderson and Hinthorne, 1972).

Feldspar crystals larger than $\sim 20\ \mu$, whether located randomly through the matrix or in a chondrule, exhibit two important features: (1) cosmic-ray track densities $\sim 1 \times 10^7/cm^2$, several orders of magnitude lower than in the fine-grained matrix material; and (2) an absence of any visible track density gradients within a crystal. The strong evidence that tracks from a prior irradiation have been retained in the fine-grained matrix and the lack of features associated with shock and/or thermal metamorphic processes lead us to conclude that the majority (if not all) of the feldspar crystals $\gtrsim 20\ \mu$ in size in 14301 had an origin and a history distinct from those of the micron-size matrix grains up to the time that the breccia was formed. Sample 14301 probably represents an assortment of track-rich soil particles and large unirradiated crystals, gently compacted into a breccia.

14311

According to Warner (1972), this is an annealed breccia characterized by recrystallization of its matrix constituents into a fine-grained agglomeration of pyroxene, plagioclase, and ilmenite (Dence *et al.*, 1972).

1. Uranium distribution. Dominating the Lexan map of section 14311,39 are the homogeneity of the level of uranium (3.1 ppm) and the striking absence of fission stars (representing "point sources" of uranium) and clusters of increased uranium concentration. This equilibrated distribution differs markedly from the heterogeneous uranium distributions observed by us in all the other Apollo 14 breccias and by Walker (1972) in his sample of 14311.

Table 1. Electron microscopic and radiation damage observations of breccias.

| Breccia | Thermal metamorphism (Warner, 1972) | Shock features; EM observations | Pre-existing solar flare tracks in small grains ($\leq 5\ \mu$) | | Gas bubbles in small ($\lesssim 3\ \mu$) grains | Total bulk ^{4}He concentration (cm^3 STP/g) (LSPET, 1971) | Uranium distribution |
			Density (cm^{-2})	Visibility; % of grains with tracks			
15086	—	Dislocations. No shock-induced glass. Very low peak pressure.	10^8 to $\gtrsim 10^{10}$	Strong; 70%	10^{14} to 10^{15}/cm^3	—	—
14301	Low	Fractures, dislocations, twinning of plagioclase and pyroxene. No shock-induced glass. Low peak pressure.	5×10^9 to 1×10^{10}	Strong; 70%	$\sim 10^{15}$/cm^3 common	0.012	Similar to 14318; uniform in some clasts.
14318	Moderate	Extensive subparallel fractures, twinning and bending in pyroxene. Abundant shock-induced glass. Highest peak pressure.	5×10^9	Strong; 10%	$\sim 10^{14}$/cm^3	—	Many large scale inhomogeneities; numerous clusters and groups of clusters.
14315	Moderate	Similar to 14318.	10^8 to $\sim 10^9$	Faint; 25%	$\sim 10^{14}$ in track-rich grains.	—	Exclusion of uranium from some clasts; concentration of uranium around crystal boundaries.
14321	High	Heterogeneously distributed re-crystallized zones. Some shock-induced glass. Mild peak pressure.	10^8 up to $\sim 10^{10}$	Very faint; 15%	10^{13} to 10^{14} in track-rich grains.	0.0013	Uniform in some clasts; relatively few fission stars.
14311	High	Fractures, twinning of plagioclase and pyroxene. No shock-induced glass. Low shock pressure.	10^8 to $\sim 10^{10}$	Faint; 25%	$\sim 10^{14}$ in track-rich grains.	—	Uniform; absence of fission stars.
15418	—	Highly deformed. Abundant re-crystallized regions. No shock-induced glass.	None	—	Large crystallographic bubbles.	—	—

2. Electron microscopy. Our observations are quite similar to those described for 14301: (i) the common deformation features are fractures, twinning and heterogeneous distribution of dislocations in plagioclase and pyroxene; (ii) porosity between grains is absent; (iii) shock-induced homogenous glass is *not* observed; and (iv) some grains contain bubbles and 1×10^8 to $>2 \times 10^9$ tracks/cm^2 that are clearly visible. The main difference between 14311 and 14301 is that the fraction of grains with tracks is considerably smaller in 14311, and these track-rich grains are distributed quite heterogeneously in the breccia. These observations suggest that the cohesion of both this rock and 14301 may be caused by cold shock compression. This raises the interesting question: Why do optical microscope observations (Warner, 1972) indicate that 14311 and 14301 are at opposite ends of the thermal metamorphism scale, whereas electron microscope observations show strong similarities in the substructure of small ($\sim 5\ \mu$) grains, which appear *not* to have suffered metamorphism?

There appears to be a good correlation between grain size and retained radiation damage that supports the differences in features observed optically and electron microscopically. Crystals larger than $\sim 5\ \mu$ generally have track densities of $\sim 10^7/$cm^2, consistent with densities observed optically and reported below. Crystals smaller than $\sim 5\ \mu$ frequently have track densities exceeding $10^9/$cm^2, and of course these small grains cannot be studied by optical petrographic techniques.

3. Optical microscope observations of tracks. Utilizing a preferential annealing technique (Price *et al.*, 1972) to erase the high background of solar and galactic cosmic-ray tracks, we have studied fission tracks in large feldspar crystals from 14311, 46. Crystals separated from bulk rock by gentle crushing are commonly surrounded by fine-grained material in the form of a microbreccia firmly shock-welded to the crystal (Fig. 4). Uranium is primarily located in the breccia portion, but the presence

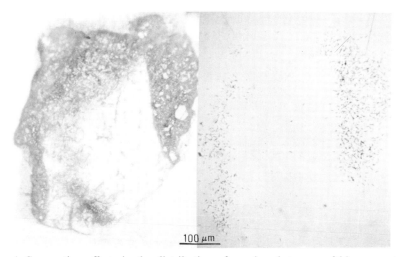

100 μm

Fig. 4. Segregation effects in the distribution of uranium between a feldspar crystal (commonly found in 14311 and 14321) and the surrounding fine-grained material as seen in the Lexan print. The microbreccia is shock-welded to the crystal and uranium has diffused across the interface giving rise to spontaneous fission tracks in the feldspar grain (not visible).

of spontaneous fission tracks in the closely adjacent feldspar portion suggests limited diffusion of uranium a distance of a few microns through the interface between the breccia and crystal. In order for the uranium to have diffused into the crystal and to have decayed sufficiently to produce the observed fission track density, the micro-breccia must have been welded to the feldspar grain billions of years ago, indicating that 14311 itself was possibly put together by impact early in the history of the moon.

Its large physical size (2.7 kg) makes it a good candidate in which to study the energy spectrum of high-energy galactic cosmic rays. Unfortunately, we were not allocated a vertical section of the rock, so that the data in Fig. 5 are taken from two chips, exterior and interior, and the depth is measured normal to two exposed surfaces. The cross-hatched area represents data from the interior chip whose location is not well documented. For purposes of comparison, also shown is a best-fit curve to the data obtained from 12022 (Barber *et al.*, 1971a), an igneous rock with a simple irradiation history and with regions of minimal erosion in between the impact pits on its top surface. At depths less than 2 mm, the slope of the density gradient in 14311 is appreciably shallower than that in 12022, indicative of the loss of a surface chip or the presence of a dust coating on the "exposed" surfaces. The friable texture of breccias suggests that an exterior chip ~1 mm thick might have been knocked off after the majority of the tracks had accumulated, perhaps even in the Lunar Receiving

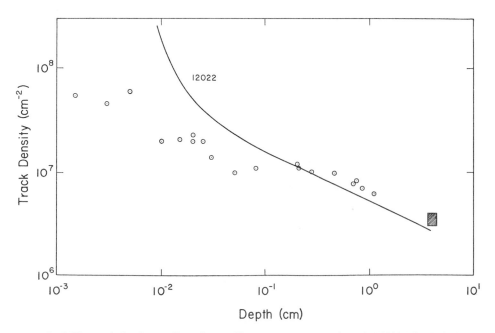

Fig. 5. The track density gradient observed in two chips from the breccia 14311. The circled points are densities at depths measured normal to two exposed surfaces; the cross-hatched area gives a measurement in the interior ship whose position is poorly documented. The smooth curve is the best fit to the density gradient measure in the basaltic rock 12022 (Barber *et al.*, (1971a).

Laboratory. The contribution from galactic cosmic rays dominates the solar flare flux at greater depths. There the slope of the density gradient is not sensitive to erosion or to surface loss on a mm scale and we find nearly equal slopes in the two rocks, implying a simple irradiation history for 14311, with no extended subsurface exposure.

14315 and 14318

1. Electron microscopy. The macroscopic similarity of breccias 14315 and 14318 extends to the submicroscopic features examined with the electron microscope. An impact origin for both rocks is inferred from the abundance of chondrule-like clasts and from features observed in glasses (Fredriksson *et al.*, 1972). Figure 6 illustrates the most common substructure found in these breccias. The intersecting zones of homogeneous glass separating crystalline remnants have also been observed in an electron microscope study (Phakey *et al.*, 1971) of naturally shocked quartz and plagioclase from the Ries crater; we therefore infer that the glass was produced by shock and was not derived from glassy fragments or splashes.

The crystalline fragments, varying in size from submicron to tens of microns, are distributed heterogeneously and are usually free of dislocations (Fig. 6). Pyroxene grains contain coarse exsolution lamellae identified by selected area electron diffraction as augite on (001) in host pigeonite, which typically shows antiphase boundary domains (APB) reported earlier by Christie *et al.* (1971) in Apollo 11 and 12 rocks. Indicative of the extent of deformation are observations of extensive sub-parallel fractures and twinning and bending in pyroxene and plagioclase (Fig. 7). Two sets of stacking faults were found in many pigeonites with or without showing the APB structure (Fig. 8).

We have found no evidence for vesiculation, melting, porosity, or large-scale

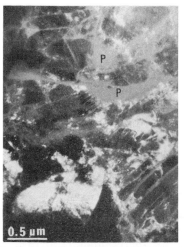

Fig. 6. BF micrograph of a typical area in breccia 14315 showing crystalline fragments separated by shock-induced zones of homogeneous glass (P). These fractures are similar to those found in naturally shocked rocks from Ries Crater.

Fig. 7. Simultaneous presence of microtwins and shock-induced zones of glass (P) in breccia 14318. DF micrograph.

Fig. 8. Intersecting sets of stacking faults in pigeonite in 14318. Fine scale APBs are present in region labelled A. DF micrograph.

vitrification in these rocks. This observation is consistent with the petrographic observations of Chao (1971) and indicates that on the whole the cohesion of rocks 14315 and 14318 is caused by shock compression at mild peak pressures. Other observations described above show that lithic fragments which have suffered moderately high shock peak pressures are common in 14315 and to a lesser extent in 14318. It is not clear whether these highly shocked fragments belong to previously consolidated breccia.

With electron microscopy, we have observed unetched tracks in several grains of 14315, but in only one grain of 14318. In 14315 the track density ranges from $\sim 8 \times 10^7/cm^2$ up to $\sim 10^9/cm^2$; in the one grain of 14318 the track density is $\sim 5 \times 10^9/cm^2$. In both rocks the tracks are extremely faint and are visible only under favorable diffracting conditions. Clearly they have been drastically modified during or after breccia formation, and the original track density in these grains was certainly higher than the numbers quoted above. No tracks were seen in the most highly shocked grains, and high track densities are not a common feature in these two breccias.

Gas bubbles at a density of $\sim 10^{14}/cm^3$ are common in small ($\lesssim 3\,\mu$) grains that are not too severely shocked, both in 14315 and in 14318. Decay of uranium plus thorium could contribute significantly to the total amount of gas present in these breccias. Until we study further the depth dependence of bubble density we will not be able to separate a radiogenic component from a solar wind component. Gas bubbles are observed in some grains unaccompanied by tracks. Apparently the bubbles are more resistant to thermal metamorphism and shock processes than are tracks, though this inference needs to be checked by controlled laboratory annealing experiments.

14321

The structural complexity of this polymict breccia is accurately reflected in the distributions of uranium found in several rock sections. Uranium concentrations varying over three orders of magnitude have been observed: from 20 ppm in a whitlockite crystal to 30 ppb in a polycrystalline igneous clast. Plagioclase and clinopyrozene crystals typically contain ~ 20 ppb of uranium except in small inclusions that are orders of magnitude richer in uranium. Uranium in microbreccia clasts comprises the majority of the whole-rock uranium concentration. Figure 4 shows such a microbreccia shock sintered to a feldspar crystal. The accompanying Lexan print illustrates the segregation of uranium in the fine-grained area.

1. Electron microscopy. Electron microscope studies of 14321 have been reported recently by Lally *et al.* (1972). Other workers (Duncan *et al.*, 1972; Grieve *et al.*, 1972) emphasize the necessity of a multiple impact origin and strong to moderate recrystallization at temperatures exceeding 1000°C observed in the several generations of microbreccia clasts.

With the electron microscope we find that this rock has an extremely heterogeneous distribution of recrystallized zones, moderately to heavily deformed areas showing twinning and bending in pyroxene and plagioclase (Fig. 9); crystalline fragments separated by zones of homogeneous glass similar to those in 14315 and 14318; and undeformed material (Fig. 10). Although the peak pressure suffered by this rock was less than that which deformed 14315 and 14318, on the basis of these observations it would be surprising to see tracks that predated breccia formation.

Nevertheless, in some grains 1–4 μ in size there are both gas bubbles and faint tracks that appear to have been incompletely annealed. The track densities range from $< 10^8/cm^2$ up to $\sim 10^{10}/cm^2$. Occasionally there is a very easily visible track that is so different in appearance from the more numerous faint ones that we suspect it was

Fig. 9. Microtwins and fractures in plagioclase in breccia 14321. Since some of the APB (D) are seen through the microtwin, they must have pre-existed in plagioclase prior to deformation twinning. DF micrograph, type b reflection.

Fig. 10. Simultaneous exsolution of augite on (001) major and (100) minor in host pigeonite in 14321. BF micrograph.

produced after the end of shock and metamorphic processes, presumably during the last 20 m.y. exposure to cosmic rays.

We have etched an ion-beam-thinned section of 14321 for 2 min in $2HF : 1H_2SO_4 : 280H_2O$. In contrast to our observations of gas-rich meteorites and of the soil breccia 15086 (discussed below), we found that the faint tracks could not be etched, indicating that the damage density along the track has been significantly reduced by shock and/or thermal metamorphism. Dran *et al.* (1972), using scanning electron microscopy on etched sections of 14321, also found no evidence for etchable tracks in the matrix that would have predated incorporation into the breccia.

As a preliminary to a future detailed depth-dependence study, we have crushed a 200 μ grain into fragments thin enough to observe by electron microscopy. The track density is no higher than would be seen by optical microscopy of a 200 μ etched grain, and no faint tracks were seen. However, we did see gas bubbles at a density somewhat less than $10^{14}/cm^3$. As we remarked earlier, we could account for such a density from decay of U and Th if the fragments we examined came from near the original surface of the 200 μ grain where there was a high concentration of U and Th in an inter-crystal phase.

At this time it is not possible to decide whether large ($\gtrsim 100$ μ) grains have a common origin with micron-size grains but have experienced greater thermal meta-morphism or if they originate from a separate source (such as ground up rock) and have never been irradiated by the solar wind.

2. Optical microscope observations of tracks. Further evidence for the formation of 14321 is provided by the overall lack of significant scatter in track densities at a given depth observed in the optical microscope. Figure 11 shows the track density as a function of depth measured in feldspar crystals in a 6 cm vertical section. Shown also are representative data from 12022 (Barber *et al.*, 1971a), a basaltic rock. The solid curve is the best fit to the data from 12022, while the squares illustrate the extent of variations in track density at the same depth found in igneous rocks. Variations

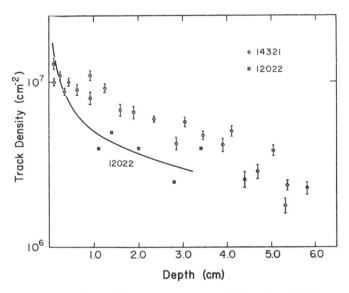

Fig. 11. The track density gradient observed in a vertical section of the breccia 14321 (circles). For comparison, data obtained from 12022 (Barber *et al.*, 1971) are shown in two forms: the smooth curve is a best fit to the density gradients, while the squares illustrate the variations in track density typical of those found in igneous rocks. The scatter of densities normal to the gradient in 14321 is generally not appreciably greater than in 12022.

about the mean track density in adjacent crystals at the same depth in 14321 are generally less than a factor of ~2, comparable to the scatter in 12022.

While we emphasize the overall smoothness of the density gradient, the presence of occasional fluctuations about the mean density of at least a factor of 3 cannot be overlooked. One of the difficulties inherent with studies of breccias is that adjacent grains may have experienced quite different shock or metamorphic histories. Alteration of their track-recording properties could account for the observed scatter in track density. We should also mention that Hart *et al.* (1972) find very large variations in track density within the same interior fragment of another portion of this breccia.

15418

A single large clast containing mainly plagioclase has been examined in the electron microscope. The optical structure of plagioclase in thin section showed heterogeneous extinction and composition changes. Electron microscope observations show that this clast is unlike those from the other breccias because the features indicative of deformation are distributed extremely heterogeneously: The most striking features are depicted in Fig. 12. There is extensive development of submicroscopic, partly recrystallized cells, some of which have a dislocation density in excess of 10^{12} cm^{-2}, deformation-induced polysynthetic twinning within such cells, and fine scale exsolution. There are also areas which are generally completely recrystallized, showing well-recovered structures exhibiting open networks and low dislocation density. The polygonal shaped and strain-free recrystallized grain in Fig. 13 contains stacking faults, dislocation loops, and bubbles. Some of these bubbles have a crystallographic shape and are quite different from the gas bubbles in matrix grains. It seems that in this breccia cohesion is caused by deformation, recrystallization, and perhaps by shock-welding.

Fig. 12. Heavily deformed region showing characteristic recrystallization features, fine scale twinning or exsolution in a plagioclase clast in breccia 15418. DF micrograph.

15086

This breccia is extremely friable (LSPET, 1971). Our sample consists of large (1–3 mm) agglomerates or clasts with weakly cohering, very fine ($<3\ \mu$) powder. Using the handling technique of Barber *et al.* (1971b), we etched the finest grains 2 min in $2HF : 1H_2SO_4 : 280H_2O$. Most of the grains (70%) examined in the electron microscope have track densities in the range of 10^8 to $\gtrsim 10^{10}/cm^2$ and were definitely etched, showing no evidence for partial track-annealing. This track density and the general appearance of the grains (Fig. 14) are characteristic of micron-size grains in

Fig. 13. Recrystallized region in 15418 showing polygonal shape and strain free character.

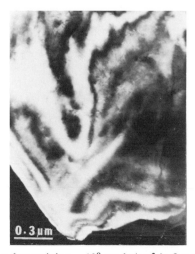

Fig. 14. Etched crystal containing $\sim 10^9$ tracks/cm^2 in fine powder from 15086.

lunar soil. That the tracks are not erased at the boundaries between small grains seems to suggest that this breccia was consolidated either by compaction by load of overburden or perhaps by a very mild, cold shock-compaction process. It represents the least metamorphosed breccia we have studied.

Table 1 briefly summarizes our observations of shock and recovery features, pre-existing tracks, gas bubbles, and uranium distribution in Apollo 14 and 15 breccias. The rocks are listed in order of increasing thermal metamorphism as inferred by Warner (1972) from petrographic observations of thin sections. With the exception of 14311, our electron microscope observations are compatible with this order. We have included in the table the available data on bulk ^{4}He concentrations, which correlate well with our observations of gas bubble density.

Fission-Track Dating: Evidence for ^{244}Pu

Whitlockite

The whole-rock uranium concentrations in the Apollo 14 breccia are high enough (2–4 ppm, LSPET, 1971) that in favorable circumstances fission tracks may dominate over solar and cosmic ray tracks, making it worthwhile to attempt fission track dating of individual crystals. Unfortunately, our Lexan mapping has shown that the majority of the uranium resides in interstitial areas unsuitable for observation of fission tracks. Uranium-rich mineral phases are, however, found (Lovering and Wark, 1972; Anderson and Hinthorne, 1972). We have recently studied a shocked 400 μ whitlockite crystal embedded in the fine-grained matrix of 14321. Here we summarize the evidence for ^{244}Pu fission tracks in that crystal; Hutcheon and Price (1972) have discussed the evidence in more detail elsewhere.

With the Lexan mapping technique we located the whitlockite and determined its uranium concentration, 19.4 $\pm$ 1.5 ppm. In order to measure the quite high track density and to discriminate against spallation recoil tracks by counting only long, well-formed tracks, we used a scanning electron microscope to examine a plastic replica of the whitlockite surface after it had been etched 30 sec in 0.25% HNO_3.

Contributions to the total observed track density, 9.36 $\pm$ 0.72 $\times$ 10^7/cm^2, arise from (1) spontaneous fission of U and possibly Pu; (2) Fe-group cosmic rays; (3) reactor neutron-induced fission; and (4) cosmic ray neutron-induced fission. The density of reactor-induced tracks is inferred from the Lexan map. The magnitude of the cosmic ray component depends, however, on the irradiation and metamorphic history of the crystal.

Our present optical and electron microscope observations, especially of crushed $\sim$100 μ grains, establish that there has been no retention in our portion of 14321 of etchable tracks by individual grains from an irradiation prior to breccia formation. We can therefore assign to the whitlockite an Fe-group track density equal to that measured in low-uranium adjacent grains. We can also conclude that the whitlockite has recorded tracks from cosmic ray neutron-induced fission only over the 25 m.y. exposure age determined for the whole rock (Lugmair and Marti, 1972).

The remainder of the tracks, 8.54 $\pm$ 0.73 $\times$ 10^7/cm^2, we attribute to spontaneous fission. We assume that the track retention age of the whitlockite crystal is equal to the Rb–Sr age of an igneous clast, 3.95 $\pm$ 0.04 Gy (Papanastassiou and Wasserburg, 1971). From the present U concentration we then calculate that in 3.95 Gy, the spontaneous fission contribution of ^{238}U would be $5.66 \pm 0.40 \times 10^7$ tracks/cm^2. The remaining tracks we ascribe to spontaneous fission of ^{244}Pu. The Pu contribution is about half that from U.

An alternative explanation which, though extremely implausible, is difficult to disprove, would require a complicated and unique exposure history for the whitlockite grain. An exposure to 10^{15} thermal neutrons/cm^2 at the onset of track retention 3.95 Gy ago, when the ^{235}U/^{238}U ratio was much higher, would produce a fission track density equivalent to that which we have ascribed to ^{244}Pu fission. Since this dosage is nearly twice the upper limit established by Lugmair and Marti for the whole rock, this model would require that the whitlockite was irradiated prior to incorporation and that it was compacted gently enough that the tracks were not erased.

Feldspar

Characteristic of many feldspar crystals are internal surfaces (cleavages) strongly enriched in U and other heavy elements such as Ba, Sr, K, and Zr (determined by ion microprobe). The uranium concentration in these cleavages is sufficiently high that we can study the tracks of fission fragments that recoil into the feldspar.

Although the uranium is heterogeneously distributed throughout a crystal as a whole, along a given cleavage the surface uranium concentration (measured in ppm-μ) is rather uniform. The background of solar and galactic cosmic ray tracks can be preferentially erased without disturbing fission tracks by annealing each mineral at a predetermined temperature (Kapuscik *et al.*, 1969; Maurette, 1970; Price *et al.*, 1972); in particular, these feldspars have been annealed at 625°C for 60 min (Fig. 15).

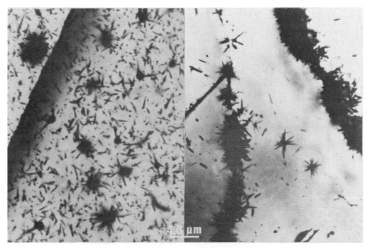

Fig. 15. Optical micrographs of feldspar crystals from 14311 illustrating the annealing technique developed by Price *et al.* (1972). The crystal on the left contains 1×10^7 cosmic ray tracks/cm^2 that effectively obscure any fission contribution. After annealing at 625°C for 60 min only fission tracks remain; they are shown in the crystal on the right.

The observed number of tracks, $N_{Cl}(\text{cm}^{-1})$, emanating from a cleavage (TINCLES) of width d is related to the concentration of fossil tracks (Bhandari *et al.*, 1971) by

$$N_{Cl} = \rho_s \Delta z f(\Theta) \tag{1}$$

where ρ_s is the "surface" track density (cm^{-2}), Δz is the depth of focus for observing TINCLES along a cleavage plane, and $f(\Theta)$ is a geometrical factor. For spontaneous fission the following equation relates N_{Cl} to the track retention time T and the concentration of uranium along the cleavage $C_u d$ (ppm-μ):

$$N_{Cl} = \frac{e^{\lambda_D T} - 1}{\lambda_D} \lambda_F N C_u \, d \Delta z f(\Theta),$$

where λ_D is the decay constant for ^{238}U, λ_F is the decay constant for spontaneous fission of ^{238}U, and N is the number of atoms per cubic centimeter of sample.

In large feldspars from 14311 we have determined uranium concentrations in cleavages from their Lexan prints (typically 5–30 ppm-μ). Figure 16 shows an example of spontaneous and induced fission track distributions in one crystal and in its Lexan print. Our quantitative studies of this and other feldspars are subject to large uncertainties resulting from the low reactor dose, the largely unknown exposure history of 14311, and the necessity to anneal out cosmic ray tracks without affecting the fission tracks. The following analysis is thus only semiquantitative.

The average of measurements along several cleavages in Fig. 16 yields $N_{Cl} \approx$ 42 ± 10 per 100 μ. Using an average value of $C_u d$ determined from the Lexan print and assuming a 660 m.y. exposure age of 14311 (Walker, 1972), we calculate that U alone would contribute $N_{Cl} \approx 27 \pm 9$ per 100 μ in ~3.95 g.y., whereas U + Pu in the same *ratio* as in the whitlockite in 14321 would contribute $N_{Cl} \approx 36 \pm 12$ per 100 μ. These results are not significant enough to distinguish between a zero contribu-

100 µm

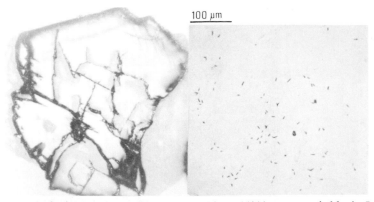

Fig. 16. Optical micrograph of a feldspar crystal from 14311 accompanied by its Lexan print. The distribution of tracks in the Lexan clearly outlines the cleavage planes in the crystal, illustrating the concentration of uranium along these internal surfaces. A higher magnification picture would show spontaneous fission tracks emanating from the cleavages (TINCLES).

tion of ^{244}Pu fission tracks and a contribution similar to that in 14321. They are, however, significant enough to rule out a large (factor of 5 or more) excess of fission tracks in cleavages that could be attributed to decay of superheavy transuranic elements (Bhandari *et al.*, 1971).

CONCLUSIONS

(1) Two features observable by high-voltage electron microscopy indicate that the fine-grained materials ($\lesssim 10\ \mu$) in Apollo 14 breccias are not fragments of rock crushed during breccia formation but were earlier exposed to solar radiation in their present size as loose grains of soil on the very surface: (a) Gas bubbles ~ 80 Å in diameter and similar in appearance to those in lunar fines are present in some of the matrix grains in each breccia at a density that appears to correlate with the bulk gas density determined mass spectrometrically by others. The ability to study the distribution of gases on a submicron scale provides complementary information to that obtained by mass spectrometry. (b) High track densities ranging from $\lesssim 10^8/\text{cm}^2$ to $\sim 10^{10}/\text{cm}^2$ occur intermittently in fine grains in matrix and some clasts. Their visibility, frequency of occurrence, and mean density generally correlate with a thermal metamorphism scale, being highest in 15086 and 14301 and much lower in 14315, 14318, and 14321. Surprisingly, they are present in the well-metamorphosed 14311. They are not observed in the recrystallized 15418. Their weak visibility, and the failure by Dran *et al.* (1972) to see them in etched sections of 14321 by SEM, suggest that they are vestigial tracks that have been strongly affected by metamorphism. They are not observed in strongly shocked regions.

(2) Large crystals ($\gtrsim 100\ \mu$) in which etched tracks are observed by optical microscopy do not generally show either a variability of track density at a given depth or a track density higher than can be accounted for by in situ production by penetrating heavy cosmic rays. The track gradients in 14311 and 14321 fall off smoothly with depth and are similar to that in igneous rock 12022 at depths below ~ 1 mm.

(3) Strong evidence that neither large nor small crystals contain etchable tracks from a prior irradiation supports our conclusion that excess tracks observed in a shocked whitlockite crystal in 14321 originate from spontaneous fission of ^{244}Pu. Measurements of tracks emanating from fractures in feldspar crystals in 14311 are compatible with spontaneous fission contributions from ^{238}U and ^{244}Pu similar to those in the whitlockite and with a 3.95 g.y. track retention age. No evidence was found for large excesses of tracks that could be attributed to fissions of hypothetical superheavy elements.

REFERENCES

Ahrens T. J., Fleischer R. L., Price P. B., and Woods R. T. (1970) Erasure of fission tracks in glasses and silicates by shock waves. *Earth Planet. Sci. Lett.* **8**, 420–426.

Andersen C. A. and Hinthorne J. R. (1972) U, Th, Pb, and REE abundances and Pb 206/207 ages of individual minerals in returned lunar material by ion microprobe mass analysis (abstract). In *Lunar Science—III* (editor C. Watkins). Lunar Science Institute Contr. 88, p. 21.

Barber D. J., Cowsik R., Hutcheon I. D., Price P. B., and Rajan R. S. (1971a) Solar flares, the lunar surface, and gas-rich meteorites. *Proc. Second Lunar Sci. Conf., Geochim. Cosmochim. Acta* Suppl. 2, Vol. 3, pp. 2705–2714. MIT Press.

Barber D. J., Hutcheon I. D., and Price P. B. (1971b) Extralunar dust in Apollo cores? *Science* **171**, 372–374.

Bhandari N., Bhat S., Lal D., Rajagopalan G., Tamhane A. S., and Venkatavaradan V. S. (1971) Spontaneous fission record of uranium and extinct transuranic elements in Apollo samples. *Proc. Second Lunar Sci. Conf., Geochim. Cosmochim. Acta* Suppl. 2, Vol. 3, pp. 2599–2609. MIT Press.

Chao E. C. T. (1971) Rock 14318 newsletter and private communication.

Christie J. M., Lally J. S., Heuer A. H., Fisher R. M., Griggs D. T., and Radcliffe S. V. (1971) Comparative electron petrography of Apollo 11, Apollo 12, and terrestrial rocks. *Proc. Second Lunar Sci. Conf., Geochim. Cosmochim. Acta* Suppl. 2, Vol. 1, pp. 69–89. MIT Press.

Dence M. R., Plant A. G., and Traill R. J. (1972) Impact-generated shock and thermal metamorphism in Fra Mauro lunar samples (abstract). In *Lunar Science—III* (editor C. Watkins). Lunar Science Institute Contr. 88, p. 174.

Dran J. C., Duraud J. P., Maurette M., Durrieu L., Jouret C., and Legressus C. (1972) The high resolution track and texture record of lunar breccias and gas-rich meteorites (abstract). In *Lunar Science—III* (editor C. Watkins). Lunar Science Institute Contr. 88, p. 183.

Drozd R., Hohenberg C. M., and Ragan D. (1972) Fission xenon from extinct ^{244}Pu in 14301: Age of the pre-Imbrium crust. To appear.

Duncan A. R., Lindstrom M. M., Lindstrom D. J., McKay S. M., Stoesser J. W., Goles G. G., and Fruchter J. S. (1972) Comments on the genesis of breccia 14321 (abstract). In *Lunar Science—III* (editor C. Watkins). Lunar Science Institute Contr. 88, p. 192.

Fredriksson K., Nelen J., and Noonan A. (1972) Apollo 14: Glasses, breccias, chondrules (abstract). In *Lunar Science—III* (editor C. Watkins). Lunar Science Institute Contr. 88, p. 280.

Grieve R., McKay G., Smith H., and Weill D. (1972) Mineralogy and petrology of polymict breccia 14321 (abstract). In *Lunar Science—III* (editor C. Watkins). Lunar Science Institute Contr. 88, p. 333.

Hart H. R., Comstock G. M., and Fleischer R. L. (1972) The particle track record of Fra Mauro (abstract). In *Lunar Science—III* (editor C. Watkins). Lunar Science Institute Contr. 88, p. 360.

Hubbard N. J. and Gast P. W. (1972) Chemical composition of Apollo 14 materials and evidence for alkali volatilization (abstract). In *Lunar Science—III* (editor C. Watkins). Lunar Science Institute Contr. 88, p. 407.

Hutcheon I. D. and Price P. B. (1972) Plutonium-244 fission tracks: evidence in a lunar rock 3.95 billion years old. *Science* **176**, 909–911.

Kaiser W. (1972) Kr and Xe in three Apollo 14 samples by stepwise heating technique. To appear.

Kapuscik A. Perelygin V. P., Tretiakova S. P., and Shadieva N. H. (1966) Search for ternary fission in the interaction of argon with heavy elements using mica detectors. *Proc. 6th Inter. Conf. on Corpuscular Photography*, p. 458. Florence.

Kleeman J. D. and Lovering J. F. (1967) Uranium distribution studies by fission track registration in Lexan plastic prints. *Atomic Energy in Australia* **10**, 3–8.

Lally J. S., Fisher R. M., Christie J. M., Griggs D. T., Heuer A. H., Nord G. L. Jr., and Radcliffe S. V. (1972) Electron petrography of Apollo 14 and 15 samples (abstract). In *Lunar Science—III* (editor C. Watkins). Lunar Science Institute Contr. 88, p. 469.

Lovering J. F., Wark D. A., Sewell D., and Frick C. (1972) Uranium geochemistry and late-stage (mesostasis) mineralogy of Apollo 14 lunar rocks (abstract). In *Lunar Science—III* (editor C. Watkins). Lunar Science Institute Contr. 88, p. 493.

LSPET (Lunar Science Preliminary Examination Team) (1971) Preliminary examination of lunar samples from Apollo 14. *Science* **173**, 681–693.

Lugmair G. W. and Marti K. (1972) Neutron and spallation effects in Fra Mauro regolith (abstract). In *Lunar Science—III* (editor C. Watkins). Lunar Science Institute Contr. 88, p. 495.

Maurette M. (1970) On some annealing characteristics of heavy ion tracks in silicate minerals. *Rad. Effects* **5**, 15–19.

Megrue G. H. and Steinbrunn F. (1972) Classification and source of lunar soils; clastic rocks; and individual mineral rock, and glass fragments from Apollo 12 and 14 samples as determined by the concentration gradients of the He, Ne, and Ar isotopes (abstract). In *Lunar Science—III* (editor C. Watkins). Lunar Science Institute Contr. 88, p. 533.

Merlivat L., Nief G., and Roth E. (1972) Deuterium analysis of hydrogen extracted from lunar material (abstract). In *Lunar Science—III* (editor C. Watkins). Lunar Science Institute Contr. 88, p. 537.

Papanastassiou D. A. and Wasserburg G. J. (1971) Rb–Sr ages of igneous rock from the Apollo 14 mission and the age of the Fra Mauro formation. *Earth and Planet. Sci. Lett.* **12**, 36–48.

Phakey P. P., Christie J. M., and Chao E. C. T. (1971) Unpublished results.

Phakey P. P., Hutcheon I. D., Price P. B., and Rajan R. S. (1972) Radiation damage in soils from five lunar missions (abstract). In *Lunar Science—III* (editor C. Watkins). Lunar Science Institute Contr. 88, p. 608.

Price P. B., Hutcheon I. D., Perelygin V. P., and Lal D. (1972) Lunar crystals as detectors of very rare nuclear particles (abstract). In *Lunar Science—III* (editor C. Watkins). Lunar Science Institute Contr. 88, p. 619.

Schreurs J. W. H., Friedman A. M., Rokop D. J., Hair M. W., and Walker R. M. (1971) *Radiation Effects* **7**, 231–233.

Walker R. M. (1972) Private communication of information in talk at Third Lunar Science Conference (not in revised abstracts).

Warner J. C. (1972) Apollo 14 breccias: Metamorphic origin and classification (abstract). In *Lunar Science—III* (editor C. Watkins). Lunar Science Institute Contr. 88, p. 782.

Wenk H. R., Ulbrich M., and Muller W. (1972) Lunar plagioclase (a mineralogical study) (abstract). In *Lunar Science—III* (editor C. Watkins). Lunar Science Institute Contr. 88, p. 797.

Proceedings of the Third Lunar Science Conference
(Supplement 3, *Geochimica et Cosmochimica Acta*)
Vol. 3, pp. 2867–2881
The M.I.T. Press, 1972

Track studies of Apollo 14 rocks, and Apollo 14, Apollo 15, and Luna 16 soils

J. L. Berdot, G. C. Chetrit, J. C. Lorin, P. Pellas, and G. Poupeau

Laboratoire de Minéralogie du Muséum National d'Histoire Naturelle and C.N.R.S.,
Paris 5éme, France

Abstract—Results from track analysis are reported for rocks 14301, 14307, 14310, 14311, and 14321. Maximum surface residence times for rocks 14310 and 14321 are inferred to be $23 \pm 7 \times 10^6$ yr and $23 \pm 5 \times 10^6$ yr, respectively. The value for 14321 is similar to the rare gas exposure age. Distributions of track-densities are given for the soils 14163, 14259, 14141, 14148, 14149, 14156, 15071, 15081, and Luna 16 soil C-118, as well as percentages of track-rich fragments in the grain-size 100–600 μ. A rough correlation is observed between the proportion of the track-rich fragments and the abundance of ^{36}Ar in bulk soils from four lunar sites, reflecting the fact that both high track-densities in the centers of the grains ($\geq 10^8/cm^2$) and ^{36}Ar are acquired by the fines in the uppermost part of the lunar regolith.

Introduction

The samples from the fragmental rocks (14307 and 14321) and rock 14310 consisted of slabs. In some cases the position and orientation of the slabs were known approximately. In other cases (14301 and 14311) the samples consisted of less documented chips or fragments. Track-density profiles were obtained from the slabs. Part of our work on lunar rocks is a contribution to a joint study on the very heavy cosmic ray energy spectrum (Yuhas *et al.*, this volume). The lunar material utilized in these studies was delivered to us in May 1971.

The samples of lunar soils from the Apollo 14, 15, and Luna 16 sites were investigated by track analysis for the grain-size range of 100 to 600 μ. Crystals were chosen during the course of this study that were suitable for examination by transmission light microscopy. Track density distributions are discussed in terms of current models of physical processes acting on the lunar surface.

We find that the proportion of track-rich fragments (TRF) in lunar soils roughly correlates with the bulk ^{36}Ar content. High-track densities and ^{36}Ar content are acquired in the uppermost part of the lunar regolith and are clearly solar particle irradiation effects. It is of interest to note that the correlation between TRF and ^{36}Ar contents is somewhat different from that found in gas-rich achondrites (Poupeau and Berdot, 1972).

Apollo 14 Rocks

Experimental

During our study of heavy cosmic-ray ion irradiation of Apollo 14 rocks, we relied chiefly on feldspar as the main track detector, following the suggestions of previous investigators (Crozaz *et al.*, 1971a, and references therein). The large abundance of

2867

feldspar in lunar rocks and its low fission track background make it suitable for this purpose. Furthermore, its track revelation properties for this work are superior to those of clinopyroxene according to Fleischer *et al.* (1970). Finally, its thermal track retentivity is superior to that of olivine (Price *et al.*, 1972). This characteristic is especially interesting for the study of crystals located on the surface of rocks subjected to the temperature of lunar day.

Track counts were performed by optical microscopy for track-densities smaller than $10^7/cm^2$. That was the case for all the rocks studied except rock 14310. Scanning electron microscopy (SEM) was used to deal with high track densities on exposed surfaces of rock 14310. In the range where both techniques could be applied track counts by SEM were approximately twice as high as track counts by optical microscopy for rock 14310. Thus, a systematic normalization to optical microscopy of the SEM data was used by lowering the latter by a factor of 2. The same procedure has previously been used (Crozaz *et al.*, 1970). Track counts were obtained in areas free from fission clusters and far from edges and cracks (frequently enriched in uranium).

Results

14301,55 (Station G1). In 20 large feldspar crystals removed from an internal 1.3 g chip, we found track densities ranging from 3 to 5.4 × $10^6/cm^2$. This variation can be entirely accounted for by random orientation of the detectors. In other locations of this rock, other groups have found both higher (10^7 t/cm², Hutcheon *et al.*, 1972a) and lower (0.27 t × $10^6/cm^2$, Hart *et al.*, 1972) track densities. These differences could be due either to the existence within the stone of a track density gradient or to the record of an irradiation predating the formation of the breccia. In this respect in our sample we have not found any evidence of such preirradiation effects. However, Hutcheon *et al.* (1972b, these proceedings) have been able to observe by high-voltage electron microscopy solar flare tracks inherited in fine ($<10 \mu$) grains of the matrix, while Crozaz *et al.* (1972a) have found that 14301 is enriched in both solar and fission gases.

14307,30 (Station G). This rock is a breccia in which both light fragments and dark matrix are enriched in solar type noble gases. However, the dark portion is enriched in ^4He by a factor of about 30 over the light ones (D. D. Bogard, personal communication, October 1971; L. Schultz and P. Signer personal communication, February 1972). Relics of the irradiation stages preceding the formation of the breccia have been sought in the dark matrix crystals. Unfortunately all the crystals were dissolved before track revelation, probably because of strong shock effects. Work is in progress to improve the revelation technique.

14310,50 (Station G?). Our track data were obtained on a slab about 9 cm long extending horizontally inwards from the western external pitted surface of this basaltic rock (LRL coordinates). These data are extensively reported elsewhere (Yuhas *et al.*, this volume). It should be noted that some of the feldspar sections were distinctly overetched under etching conditions that ensured an over-all correct revelation of the tracks. These differences in behavior, observable also among twinned crystals, are possibly due to variations in chemical composition (Ridley *et al.* 1972;

Wenk *et al.*, 1972) that are known to affect track revelation (Lal *et al.*, 1968) and/or to differences in crystallographic orientation. Figure 1 shows the track density gradient observed along our sample in a plane roughly perpendicular to the western face. The steepest near surface gradients were observed for orientations slightly tilted with respect to this one. For instance, a track density of about $3 \times 10^8/\text{cm}^2$ has been measured at $80\ \mu$ from the western external surface. This track density gradient clearly indicates that the western surface of the rock was exposed to solar flare VH particles for a substantial fraction of the surface residence time. We note indeed, that the track density varies by an order of magnitude between a depth of 2 mm and a depth of 5 cm in our slab.

14311,38 (Station Dg). Two small samples have been studied. In the first sample, 20 crystals from three different locations gave essentially the same track density dispersion from 1.6 to $3 \times 10^6/\text{cm}^2$. Two locations in the second chip showed a track density distribution between 0.9 and $2.2 \times 10^6/\text{cm}^2$. We note that a track density as low as $0.73 \times 10^6/\text{cm}^2$ has been reported in another fragment of the same rock (Hart *et al.*, 1972). In this breccia also we have not observed any evidence of irradiation predating the breccia formation.

14321,160; 14321,40 (308 and 310) (Station C). Besides a slab of 3.3 cm, we were

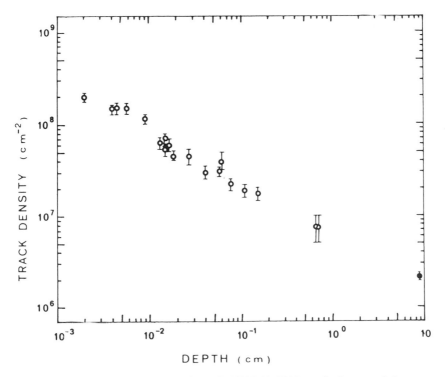

Fig. 1. Track density versus depth in rock 14310,50. Feldspar is the crystal detector. Open circle represents scanning electron microscopy counting normalized, by a factor of 0.5, to optical microscopy measurements (full circles).

provided with two spot samples. Therefore, we investigated crystals located along a line extending inward from the western external surface of the rock, approximately midway between top and bottom (LRL coordinates). Track densities were measured in crystals of random orientation with respect to the rock geometry and are extensively reported elsewhere (Yuhas *et al.*, this volume). They range between 2.1 ± 0.3 and $8.3 \pm 0.8 \times 10^6 t/cm^2$ in feldspars taken either in the slab or in the spot samples. A millimeter-sized olivine crystal broken into eight fragments inside the slab shows a track density variation from $0.7 \pm 0.1 \times 10^6/cm^2$ to $1.8 \pm 0.3 \times 10^6/cm^2$ using the etching method of Krishnawami *et al.* (1971). A millimeter-sized feldspar sampled near the preceding olivine shows the same dispersion of track densities (2.1 ± 0.3 to $5.1 \pm 0.8 \times 10^6/cm^2$). Part of this spread is due either to the poor conditions of the crystals as a consequence of shock effects or to relics of irradiation predating the rock formation (Hutcheon *et al.*, 1972a). Whatever the case it prevents an accurate determination of a track density gradient.

Discussion

Due to the rapid absorption of energetic iron group nuclei in matter, cosmic ray tracks are mainly induced in rocks when they are located in the uppermost part of the lunar regolith (within a depth of approximately 20 cm). Thus, by considering the track density gradient inside a lunar rock, it is in principle possible to infer its surface residence time. Such a study would first require, as has been done for the Saint Séverin meteorite which had a simple 4π irradiation history (Cantelaube *et al.* 1967; Lal *et al.*, 1969; Cantelaube *et al.*, 1969), a careful investigation of track densities on all surfaces of the rock in order to locate areas with the highest track density. Then a subsequent drilling can be made in an attempt to get a core with the steepest track density gradient. The rather good agreement between experimental results (Lal *et al.*, 1969) and theoretical expectations (Fleischer *et al.*, 1967) shows that such cosmic ray track analysis can indeed be used to derive "exposure ages."

For lunar rocks the situation is different because in most cases the irradiation geometries, break-up histories, erosion effects, and variable amounts of shielding are rather difficult to evaluate. Furthermore in fragmental rocks, such as those recovered by the Apollo 14 mission, the possible occurrence of crystal defects due to shock events, partial track erasure due to annealing, or even relics of irradiation predating the breccia formation constitute intrinsic complications that are difficult to disentangle. The loss of surface material from the sample is also possible during sample handling. Additionally, sample sections are sawn for track analysis with no knowledge of the overall surface track densities. Thus one can only expect to obtain semiquantitative information on the surface exposures. In this respect it is important to note that a flattening of the track density profiles will always result from erosion, variable shielding conditions, and erratic cuttings. The difficulty of cosmic-ray track analysis in lunar rocks is well illustrated by the following two examples:

(1) For rock 14301 the maximum surface residence time deduced by Hart *et al.* (1972) is 0.34×10^6 yr whereas Hutcheon *et al.* (1972a) estimate an exposure age $\gg 10^7$ yr.

(2) For rock 14310, we evaluate a maximum surface age of $23 \pm 7 \times 10^6$ yr (see below) whereas Crozaz *et al.* (1972b) from their data alone predict a true surface exposure age less than 3×10^6 yr.

We shall, however, keeping the above difficulties in mind, venture to estimate surface residence time for Apollo 14 rocks 14321 and 14310. Two methods can be applied:

First: A *maximum* surface exposure age can be derived from the smallest track density value measured in the center of a rock, provided the distances to the exposed surfaces are known.

Second: If one is lucky enough to observe a track density gradient (even if it is not the largest one inside the rock) it is then possible to bracket an "exposure age."

14321. The first method was used for rock 14321 in which the scatter in track densities prevents recognition of a well-defined track density gradient in our slab and chips. From the minimum track density value of 2.1×10^6 tracks per cm^2 measured (feldspar crystals) in the approximate center of the rock, we estimate a maximum surface residence time of $23 \pm 5 \times 10^6$ yr. We assume the stone has been tumbled about on the lunar surface. This is justified by the fact that high track densities (Yuhas *et al.*, this volume) have been found on the buried bottom of the rock (Swann *et al.*, 1971). The track production rates used in this evaluation are those given by Fleischer *et al.* (1971), corrected for 2π geometry. The track density gradient that would be observed, without assumptions, between a location at a depth of 1 cm and the center of the rock is not beyond the limits fixed by our data (i.e., a factor of 5 over a distance of about 8 cm, taking into account the error bars).

The maximum surface exposure age so deduced agrees with that derived by Crozaz *et al.* (1972b). Note, however, that Hart *et al.* (1972) found in 14321 a minimum track density of 0.39×10^6/cm^2 from which they deduced a maximum age of 8.2×10^6 yr. The maximum surface residence time of $23 \pm 5 \times 10^6$ yr happens to be similar to that found by means of different cosmogenic rare gas isotopes (Lugmair and Marti, 1972; Burnett *et al.*, 1972; Bogard and Nyquist, 1972). This age is most probably related to the impact event which formed Cone Crater. The sample was collected from the rim of Cone Crater.

14310. The steep track density gradient shown in Fig. 1 has enabled us to use both the above methods to bracket a surface residence time. The observed gradient is similar to that reported for a smaller stone (12022) by Barber *et al.* (1971). These authors inferred a surface exposure age of 10×10^6 yr for the stone in a fixed position on the lunar surface. It is probable that rock 14310 was split from a larger boulder and that the present western face of 14310 was exposed prior to splitting. Such a condition must be noted in obtaining an exposure age for 14310. This possibility is substantiated by the fact that two faces of the rock are fairly densely pitted whereas three others show no pitting (Anderson *et al.*, 1971). On the one hand, from the preceding geometrical considerations, we deduce that 10×10^6 yr would be a lower limit to the rock surface residence time. On the other hand, by considering the minimum track density (about 1.6×10^6/cm^2) and assuming a long term 2π irradiation geometry, an upper limit of $23 \pm 7 \times 10^6$ yr is set for the surface residence time. The values reported above do not exclude that rock 14310 comes from Cone Crater, as do the majority of rocks sampled in the Apollo 14 site (LSPET, 1971). The ejecta from the crater extend (Sutton *et al.*, 1972) beyond the LM landing site, i.e., *a fortiori* beyond Station G where this rock is presumed to have been collected (Swann *et al.*, 1971).

Our surface residence time estimate sharply disagrees with that obtained by Crozaz *et al.* (1972b). From their data alone, they "predict a true surface exposure age of $<3 \times 10^6$ yr leading to the prediction that ^{53}Mn should be undersaturated at the surface." Hart *et al.* (1972), after having proposed on the basis of their data a 21×10^6 yr surface residence time, now think (R. L. Fleischer, personal communication, April 1972) that the overall data would be consistent with 400×10^6 yr at a depth of 20 cm in a larger rock followed by a short surface exposure such as Crozaz *et al.* (1972a) calculate.

In this respect it is perhaps of interest to note that Rancitelli *et al.* (1972), with their data on ^{26}Al, consider that rock 14310 was at the surface for at least the last two or three million years.

Facing such great divergences in interpretations, albeit there is agreement on experimental data among the Energy Spectrum Consortium members (Yuhas *et al.*, this volume), it would be of great importance to try to settle the issue by a study of ^{53}Mn activity ($T_{\frac{1}{2}} = 3.7 \times 10^6$ yr) on carefully chosen samples. Of course, it remains the remote possibility that the main part of the surface irradiation occurred a long time ago, the rock being afterwards shielded from VH ions and then recently exhumed. Whatever the surface residence time, it is much shorter than the krypton age of about 260×10^6 yr (Lugmair and Marti, 1972; Kaiser, 1972a). That the main part of the irradiation took place at great depth (≥ 100 g/cm^2) in the lunar regolith is evidenced by the neutron fluence determined both by Burnett *et al.* (1972) and Lugmair and Marti (1972). At these depths, however, VH tracks are essentially not registered ($<10^5$ t/cm^2).

Lunar Soils

The following lunar soils (grain-size <1 mm) were studied: 14259 (comprehensive), 14163 (bulk), 14148 (top), 14156 (middle), and 14149 (bottom of the trench), 14141 (Cone Crater), 15071 and 15081 (Elbow Crater), as well as Luna 16, C-118 (20–22 cm depth in the core).

Experimental

Our survey has been centered on handpicked crystals (feldspars) and lithic fragments (igneous rocks and noritic breccias) having grain-sizes in the range of 100–600 μ. Work in progress is devoted to larger grain-sizes up to 4 mm.

Feldspar grains, as for the rocks, was used as the main track detectors. These were mounted in epoxy and etched under the same conditions previously described (Lal *et al.*, 1968). Shorter etching times were used for SEM observations. About 100–150 fragments were chosen, under a stereomicroscope equipped with polarizing system, for their apparently good cristallinity. In spite of our choice, 10% to 50% were lost during etching treatments depending on the soils studied. The remaining fragments (55 to 130) from each soil were counted. Only 30 fragments of 14141,39 were mounted on a polished section kindly provided by Maurette were counted. Both optical microscopy and SEM were utilized on soils 14259, 14163, and Luna 16 C-119. We frequently observed track densities two times higher with the SEM than with optical microscopy. Track densities lower than $10^8/cm^2$ were always counted by means of optical microscopy (immersion objective) because of the better definition of track criteria.

Results

Our results for Apollo 14, 15, and Luna 16 soils are presented in Table 1 and Fig. 2.

Table 1 reports the percentages of crystals and lithic fragments having track density higher than $10^8/cm^2$ *in their center* (track rich fragments, TRF). This cutoff in track density has been chosen according to Arrhenius *et al.* (1971), as giving a rather good partition between crystals mostly irradiated by solar flare VH nuclei ($\rho \geq 10^8/cm^2$) from those irradiated mostly by galactic cosmic rays ($\rho < 10^8/cm^2$). Experimental justification for this conventional partition will be given in the following discussion concerning the correlation between TRF percentages and ^{36}Ar contents of bulk soils. Frequency distributions of track densities in grains of different soils are shown in Fig. 2. On the whole, our data are in good agreement with those reported by other track groups (Crozaz *et al.*, 1972b; Comstock *et al.*, 1972; MacDougall *et al.*, 1972; Phakey *et al.*, 1972a; Walker and Zimmermann, 1972).

It appears from Fig. 2 that four soils, 14259 (comprehensive), 14148 (top), 14156 (middle of the trench), and Luna 16, C-118 (20–22 cm depth) have been extensively reworked as indicated by their large abundance of TRF (>90%). Three other soils, 14163 (bulk), 15071, and 15081 (Elbow Crater), are characterized by a lower percentage of the TRF (about 60 to 75%) whereas the other fragments peak at track density values of about 1 to 4 × $10^7/cm^2$. They are presumably less reworked materials. The latter two soils, 14149 (bottom of the trench) and 14141 (Cone Crater) contain only 30 to 40% of TRF, with other grains peaking at about 1 to 5 × 10^6 t/cm^2. In this

Table 1. Abundances of track-rich fragments (TRF) in the 100–600 μ grain size fraction of some lunar soils.

Sample number and description	Number of crystals measured	Percentage of crystals with $\rho^a > 10^8$ cm^{-2}
14148,34 (top)	80	92
14156,34 (middle)	71	94
14149,50 (bottom of the trench)	55	33
14141,39 (Cone Crater)	30	40
14259,57 (comprehensive)	128	98
14163,124 (bulk)	102	61
15071,34 (Station 1)	88	67
15081,29 (Station 1)	89	73
Luna 16-C118 (20–22 cm depth)[b]	114	100

[a] ρ: track density.
[b] 100–500 μ.

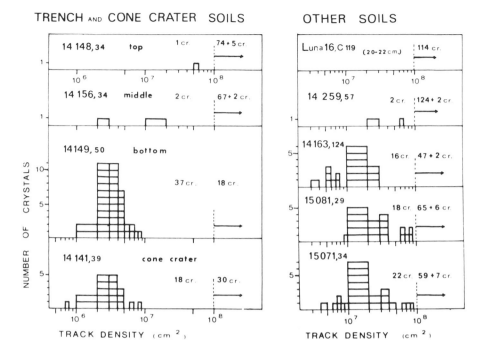

Fig. 2. Histograms of track densities observed in feldspars and rock fragments from soils of three lunar missions (Apollo 14, Apollo 15, and Luna 16). The number of crystals (and fragments) is reported both for grains with track densities less than $10^8/cm^2$ and higher than $10^8/cm^2$. In the last case the two number on the top of the arrows indicate: (a) the first number corresponds to the number of grains having track density *in their center* $> 10^8/cm^2$ (TRF); (b) the second number to grains with track density $> 10^8/cm^2$ on their edges only. (Note that the number of TRF reported for soil 14141 is 12 and not 30 as erroneously drawn).

respect they group with soils such as 12033, 12032, and layer VI from Apollo 12 double-core tube (Arrhenius *et al.*, 1971).

Only a few non-TRF grains ($< 10^8/cm^2$) possessed discernible track density gradients. In different fines samples, as observed by optical microscopy, there were $\sim 5\%$ for Apollo 14 and $\sim 8\%$ for Apollo 15 soils. Track density gradients in TRF-grains studied by SEM are often smooth and difficult to identify. They are recognizable only in a small percentage of TRF. For instance, we have been able to identify out of 114 TRF in Luna 16 soil only three which show a solar flare gradient (i.e., track density variations of more than a factor of 3). That differs markedly from track rich grains in gas-rich meteorites in which the gradient is easily observed.

Concerning the irradiation geometry of grains with discernible gradient, only 1 out of 24 non-TRF could present an isotropic irradiation geometry. SEM observations of TRF confirm that the anisotropic (2π) irradiation is much more frequent, as already observed by other authors (Arrhenius *et al.*, 1971).

Discussion

Soils 14149 (bottom of the trench) and 14141 (Cone Crater). These two soils are remarkable by the very low value ($\sim 1 - 5 \times 10^6$ t/cm^2) of track density in their non-track-rich grains. Soil 12033 (Arrhenius *et al.*, 1971) is the most similar on these grounds of all the lunar soils recovered up to now and also contains a very low amount of solar type gases (Funkhouser *et al.*, 1971a). In fact, our data reported in Table 1 and Fig. 2 misrepresent the true nature of soil 14149. As reported by Keith *et al.* (1972), it appears especially from ^{56}Co (and ^{54}Mn) activities, attributed to the large proton flare event of 25 January 1971, that the bottom of the trench has been contaminated by 40 to 50 wt.% surface materials. The upper levels of the trench (14148, top, and 14156, middle) show a very high percentage of TRF ($>92\%$). It appears likely that the 33% TRF present at the bottom of the trench constitute a very predominant contamination from the upper levels. This suggestion seems to be confirmed by our detailed results on 14149 shown in Table 2. From the data (Table 2), it appears clearly that *all* the sub-ophitic basalt (LSPET, 1971) fragments are non-track-rich grains. Therefore, we are led to consider that about half of the feldspar crystals and noritic breccia (the TRF ones) should correspond to the predominant contamination from the upper levels (a small lateral transport process having been possibly acting *before* the deposition of the upper levels of the trench).

In short, from the above analysis of the data, it appears rather convincing that the genuine soil 14149 contains not only an important fraction of sub-ophitic basalts but also feldspar crystals and noritic breccia fragments having essentially low track-density values in the range 1 to 5×10^6/cm^2.

Soil 14141 has been sampled at Station C' near Cone Crater. The impact event has been dated at about 25×10^6 yr (Burnett *et al.*, 1972). The preceding authors have shown that sized fractions of this soil give model ^{126}Xe exposure ages of 50 ($>300\ \mu$) and 99 ($<150\ \mu$) m.y. They suggest that the "excess exposure age" over 25×10^6 yr may be explained by various admixtures of more irradiated material with unexposed material ejected by the impact. Crozaz *et al.* (1972b) estimate from thermoluminescence the average depth of sample 14141 to be 2.1 ± 0.4 cm. These authors taking twice this depth value as the maximum depth and assigning the lowest track density crystal to this position calculate an exposure age of $\sim 18 \times 10^6$ yr for an undisturbed layer. However, these authors think that the distribution of track densities are consistent also with a 25×10^6 yr age, assuming a maximum stirring (or sampling) depth of 5 cm. These calculations seem rather unsatisfactory with regard to the number of assumptions made even though the age obtained in such a way beautifully fits

Table 2. Abundances of track-rich fragments (TRF) in the 100–600 μ grain size fraction of soil 14149 (bottom of the trench).

Sample description	Number of fragments measured	Number of fragments with $\rho > 10^8$ cm^{-2}
Single feldspar crystals	21	11
Basalts	22	0
Noritic breccias	12	7

that inferred by cosmogenic rare gases in rocks close to the crater (Burnett *et al.*, 1972). Bhandari *et al.* (1972) from their track data calculate an age of 7×10^6 yr for 14141. We think it most probable that the frequency distribution corresponds to a 25-m.y. soil. However, no convincing treatment has so far been proposed to derive this age from track data alone.

From the close similarity of the track-density frequency distributions between 14149 (bottom of the trench) and Cone Crater 14141 soils, along with the fact that unreworked layers are indeed not frequent in lunar soils, Crozaz *et al.* (1972a) and Poupeau *et al.* (1972a) have suggested that these strata might result from the same impact event. The fact that the locations of these two soils are 1.2 km distant do not rule out this possibility inasmuch as the Cone Crater ejecta extends far beyond Station G where 14149 was collected (Sutton *et al.*, 1972). However, the very different petrology of the two soils is a valid argument against a common origin. Rare gas analysis of basaltic fragments from sample 14149 would be very valuable in checking this point.

Other soils. Apollo 15 soils (15071 and 15081) were sampled at distances of 25 and 75 m from the rim of Elbow Crater. Their track density frequency distribution for non-track-rich grains show peaks at track densities an order of magnitude higher than those presumably related to Cone Crater (14141; 14149?). This is consistent with the fact that Elbow Crater, by its morphology, is considered to be one of the oldest impact events in Apollo 15 site (Swann *et al.*, 1971). Therefore it would be tempting to try to establish a relative time scale among craters having approximately the same dimensions. However, the close track density frequency distribution (Fig. 2) of soil 14163 (bulk) with the two Apollo 15 Elbow rim soils must be a warning against this type of extrapolation.

In this respect, it is also of interest to note the difference in track density frequency distributions between 14163 and 14259 (comprehensive) soils (Fig. 2). From our track data which agree with those of Crozaz *et al.* (1972b), we could be led to state that 14163 is a younger soil than 14259. However, the ^{126}Xe model "exposure age" would indicate the reverse is true: 550 and 710 m.y. for 14259 and 14163, respectively (Burnett *et al.*, 1972). These apparently conflicting results could be in fact accounted for by inhomogeneities in sampling. On the one hand, as demonstrated by Bogard and Nyquist (1972) a soil enriched in the smaller grain sizes will exhibit a higher ^{126}Xe model "exposure age"; on the other hand, performing track work on "good" crystal detectors carries along the risk of dealing with a material that has an exposure history somewhat different from that reflected by rare gas analysis or neutron fluence of bulk samples.

From track results alone the four soils in Fig. 2 (14148, 14156, 14259, and Luna 16, C-118) are not consistently different. When other solar or galactic irradiation monitors are used, differences come out.

As an example Luna 16 soil is interesting to consider. In this soil we lost the greatest amount of material during the chemical etching for track revelation. In the remaining 114 fragments (feldspars, olivines, pyroxenes, and basalts having grain sizes between 100 and 500 μ) we have been unable to find even one non-track-rich fragment.

[*Note added in proof:* further work on 5 different levels of Luna 16 core (including C-118) has shown that 7 out of 278 fragments have minimum track densities $<10^8/$ cm^2 (Poupeau *et al.*, to be published).] Two other levels (Comstock *et al.*, 1972; Walker and Zimmerman, 1972) give only one non-track-rich grain over 53 counted. By pooling all the data obtained by SEM or replica techniques, we observe the highest percentage (99.4%) of TRF compared to all other lunar soils. Higher resolution techniques using high-voltage electron microscopy (Borg *et al.*, 1971; Phakey and Price, 1972b) have shown that radiation damage in micron-size grains induced by heavy solar wind particles is strikingly greater in this soil than in any other soils recovered from all the Apollo missions. Concerning the ^{126}Xe model "exposure age," Kaiser (1972b) has found that it is higher by a factor of 2 (900 $\pm$ 300 $\times$ 10^6 yr) than the one obtained for 10084 (450 $\pm$ 150 $\times$ 10^6 yr). This model "exposure age" is also significantly higher than those obtained by Burnett *et al.* (1972) for 14259 and 14163 (see above). The neutron fluence (neutron/cm^2) of this peculiar soil, measured by Gd and Sm isotopic variations, is the highest of any soil (Burnett *et al.*, 1972; Lugmair and Marti, 1971; Russ, 1972). Furthermore the ^{36}Ar and "excess" ^{40}Ar contents are the largest at the different levels (Vinogradov, 1971; Heymann *et al.*, 1971; Kaiser, 1972b) among all the lunar soils recovered to date from the five missions.

Therefore Luna 16 appears to be the soil monitoring the largest solar and galactic irradiations of all the returned soils. However, we would point out that this conclusion does not come out convincingly enough from our track data reported in Fig. 2.

Correlation between TRF and ^{36}*Ar content.* In Fig. 3 we have plotted the percentage of TRF versus the total ^{36}Ar content as measured by different groups in 11 bulk lunar soils. These two parameters show a rough correlation. We note, however, that a large scatter exists in the track as well as in the rare gas data (see, for instance, the spread between 11 measurements of ^{36}Ar concentrations in sample 10084 and that in TRF for sample 14163). This type of scatter is likely to be due to sample inhomogeneities and to the use of different selection criteria for the choice of grains by different track groups (various mineral species having different track-registration and track-retentivity characteristics). In spite of its imperfection, this correlation appears to us significant. As it is well known that track production rate is a very sensitive function of depth, very high track densities must on the average correspond to exposure of the grains at the uppermost layer (<1 cm) of the lunar regolith. Same is true for the enrichment in ^{36}Ar content.

Conversely, the existence of this correlation indicates that the great majority of TRF must be solar flare irradiated grains and thus constitutes an *a posteriori* justification of the partition between TRF (track density $\geq 10^8$/cm^2 in the center) and non-track-rich grains. This procedure is valid for one range of grain-size studied (0.1–1 mm), and of course should not be applied without modifications to larger or smaller grain sizes.

From the above correlation one can predict that soils 14149 and 14141 are not very enriched in trapped ^{36}Ar and must have contents smaller than 2×10^{-4} cc STP/g.

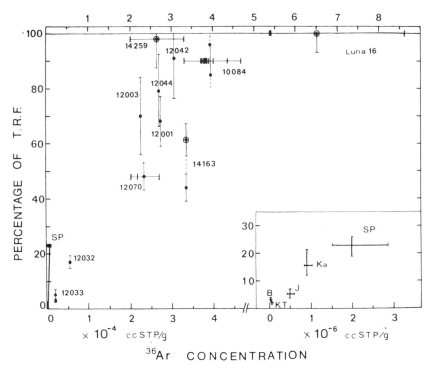

Fig. 3. Correlation between the percentages of track-rich-fragments (TRF) and ^{36}Ar contents in lunar soils. Full circles: TRF from references (Lal *et al.*, 1970), (Arrhenius *et al.*, 1971), (Crozaz *et al.*, 1971 and 1972b), (MacDougall *et al.*, 1972), (Phakey *et al.*, 1972), and from this work (dotted circles). Error bars in percentages represent standard deviations. Contents in ^{36}Ar (small vertical bars) of bulk soils are given (neglecting the errors) from references: (Heymann and Yaniv, 1970), (Hintenberger *et al.*, 1970, 1971), (Hohenberg *et al.*, 1970), (LSPET 1970, 1971), (Marti *et al.*, 1970), (Pepin *et al.*, 1970), (Funkhouser *et al.*, 1971a, 1971b), (Vinogradov, 1971), (Kaiser, 1972). The straight line drawn from the origin to point SP (Pesyanoe) represents similar correlation for gas-rich achondrites. An enlargement of this last type of correlation is shown in the rectangle (bottom right of the figure) from data reported by Poupeau and Berdot (1972): B (Bustee), KT (Khor Temiki), J (Jodzie), Ka (Kapoeta), SP (Pesyanoe).

Plotted also in Fig. 3 is the percentage of crystals with solar flare gradient in gas-rich achondrites (in the same range of grain-sizes as that of lunar TRF) versus ^{36}Ar contents as reported recently by Poupeau and Berdot (1972). We must specify, however, that the track density in the center of the meteoritic grains is an order of magnitude lower ($\sim 10^7$ to $< 10^8$ t/cm^2) than in lunar TRF. In this case also a distinct correlation is observed between the percentage of track-rich grains and the trapped ^{36}Ar. This latter correlation strengthens the preceding reasoning about the solar flare signature in TRF. It is worth noticing, however, that the two correlations are some-what different: The ^{36}Ar content in gas-rich achondrites is at least 2 orders of magnitude lower than that of the majority of lunar soils. This contrast is most prob-ably related to the drastically smaller solar particle irradiation to which gas-rich meteoritic material was subjected compared to lunar regolith material. This difference could have also important implications with regard to the relative effects of solar wind and solar flare particles in lunar soils (Poupeau *et al.*, 1972b).

Acknowledgments—We thank The National Aeronautics and Space Administration and the Soviet Academy of Sciences for providing us with lunar samples. We are indebted to Drs. D. D. Bogard, L. Schultz and P. Signer for communicating results before publications, and to Dr. M. Maurette who supplied a Cone Crater sample. We gratefully acknowledge Prof. J. Fabries and Prof. R. Lafitte for all the facilities they gave us during this study. Finally we thank Drs. D. S. Burnett, R. L. Fleischer, and R. M. Walker for their helpful criticisms and Mrs. D. Michel for technical assistance. The work has been made possible through the financial support of the Centre National de la Recherche Scientifique and of the Muséum National d'Histoire Naturelle.

REFERENCES

Anderson D. H., Lindsay J., and Ridley W. I. (1971) Apollo 14 Curator office preprint.

Arrhenius G., Liang S., MacDougall D., Wilkening L., Bhandari N., Bhat S., Lal D., Rajagopalan G., Tamhane A. S., and Venkatavaradan V. S. (1971) The exposure history of the Apollo 12 regolith. *Proc. Second Lunar Sci. Conf., Geochim. Cosmochim. Acta* Suppl. 2, Vol. 3, pp. 2583–2598. MIT Press.

Barber D. J., Cowsik R., Hutcheon I. D., Price P. B., and Rajan R. S. (1971) Solar flares, the lunar surface, and gas-rich meteorites. *Proc. Second Lunar Sci. Conf., Geochim. Cosmochim. Acta* Suppl. 2, Vol. 3, pp. 2705–2714. MIT Press.

Bhandari N., Bhat S. G., Goswami I. N., Gupta S. K., Krishnas-Wami S., Lal D., Tamhane A. S., and Venkatavaradan V. S. (1972) Collision controlled radiation history of lunar regolith (abstract). In *Lunar Science —III* (editor C. Watkins), p. 68, Lunar Science Institute Contr. No. 88.

Bogard D. D. and Nyquist L. E. (1972) Noble gas studies on regolith materials from Apollo 14 and 15 (abstract). In *Lunar Science—III* (editor C. Watkins), p. 89, Lunar Science Institute Contr. No. 88.

Borg J., Durrieu L., Jouret C., Kashkarov L., Lacaze J. C., and Maurette M. (1971) The high resolution radiation damage and texture record of the Luna 16-118-C dust grains. Progress report presented at the Intercosmos–C.N.E.S. Meeting, Nice, October 12, to appear.

Burnett D. S., Huneke J. C., Podosek F. A., Russ G. P. III, Turner G., and Wasserburg G. J. (1972) The irradiation history of lunar samples (abstract). In *Lunar Science—III* (editor C. Watkins), p. 105, Lunar Science Institute Contr. No. 88.

Cantelaube Y., Maurette M., and Pellas P. (1967) Traces d'ions lourds dans les minéraux de la chondrite de Saint Séverin, Radioactive dating and methods of low levels counting. *Meteorite Res.*, pp. 215–229, IAEA, Vienna.

Cantelaube Y., Pellas P., Nordemann D., and Tobailem J. (1969) Reconstitution de la météorite Saint Séverin dans l'espace. *Meteorite Res.* pp. 705–713, IAEA, Vienna.

Comstock G. M., Fleischer R. L., and Hart H. R. Jr. (1972) The particle track record of the Sea of Plenty. *Earth Planet. Sci. Lett.* **13**, 407–409.

Crozaz G., Haack U., Hair M., Maurette M., Walker R., and Woolum D. (1970) Nuclear track studies of ancient solar radiations and dynamic lunar surface processes. *Proc. Apollo 11 Lunar Sci. Conf., Geochim. Cosmochim. Acta* Suppl. 1, Vol. 3, pp. 2051–2080. Pergamon.

Crozaz G., Walker R., and Woolum D. (1971) Nuclear track studies of dynamic surface processes on the moon and constancy of solar activity. *Proc. Second Lunar Sci. Conf., Geochim. Cosmochim. Acta* Suppl. 2, Vol. 3, pp. 2543–2558. MIT Press.

Crozaz G., Drodz R., Graf H., Hohenberg C. M., Monnin M., Ragan D., Ralston C., Seitz M., Shirck J., Walker R. M., and Zimmerman J. (1972a) Evidence for extinct Pu^{244}: Implications for the age of the pre-Imbrium crust (abstract). In *Lunar Science—III* (editor C. Watkins), p. 164, Lunar Science Institute Contr. No. 88.

Crozaz G., Drozd R., Hohenberg C. M., Hoy H. P. Jr., Ragan D., Walker R. M., and Yuhas D. (1972b) Solar flare and galactic cosmic ray studies of Apollo 14 samples (abstract). In *Lunar Science—III* (editor C. Watkins), p. 167, Lunar Science Institute Contr. No. 88.

Fleischer R. L., Price P. B., Walker R. M., and Maurette M. (1967) Origins of fossil charged-particle tracks in meteorites. *J. Geophys.* **72,** 331–353.

Fleischer R. L., Haines E. L., Hart H. R., Woods R. T. and Comstock G. M. (1970) The particle track record of the Sea of Tranquility, *Proc. Apollo 11 Lunar Sci. Conf., Geochim. Cosmochim. Acta* Suppl. 1, Vol. 3, pp. 2103–2120. Pergamon.

Funkhouser J., Schaeffer O. A., Bogard D. D., and Zahringer J. (1971a) Noble gas abundances in lunar material, I. Solar wind implanted gases in Mare Tranquilitatis and Oceanus Procellarum. Preprint.

Funkhouser J., Bogard D., and Schaeffer D. (1971b) Noble gas analysis of core tube samples from Mare Tranquilitatis and Oceanus Procellarum. Paper presented at the Second Lunar Sci. Conf., Houston, January.

Hart H. R., Comstock G. M., and Fleischer R. L. (1972) The particle track record of Fra Mauro (abstract). In *Lunar Science—III* (editor C. Watkins), p. 360, Lunar Science Institute Contr. No. 88.

Heymann D. and Yaniv A. (1970) Inert gases in the fines from the Sea of Tranquility. *Proc. Apollo 11 Lunar Sci. Conf., Geochim. Cosmochim. Acta* Suppl. 1, Vol. 2, pp. 1247–1259. Pergamon.

Heymann D., Yaniv A., and Lakatos S. (1972) Inert gases in twelve particles and one "dust" samples from Luna 16. *Earth Planet. Sci. Lett.* **13,** 400–406.

Hintenberger H., Weber H. W., Voshage H., Wänke H., Begemann F., and Wlotzka F. (1970) Concentration and isotopic abundances of the rare gases, hydrogen and nitrogen in Apollo 11 lunar matter. *Proc. Apollo 11 Lunar Sci. Conf., Geochim. Cosmochim. Acta* Suppl. 1, Vol. 2, pp. 1269–1282. Pergamon.

Hintenberger H., Weber H. W., and Takaoka N. (1971) Concentrations and isotopic abundances of the rare gases in lunar matter. *Proc. Second Lunar Sci. Conf., Geochim. Cosmochim. Acta* Suppl. 2, Vol. 2, pp. 1607–1625. MIT Press.

Hohenberg C. M., Davis P. K., Kaiser W. A., Lewis R. S., and Reynolds J. H. (1970) Trapped cosmogenic rare gases from stepwise heating of Apollo 11 samples. *Proc. Apollo 11 Lunar Sci. Conf., Geochim. Cosmochim. Acta* Suppl. 1, Vol. 2, pp. 1283–1309. Pergamon.

Hutcheon I. D., Phakey P. P., Price P. B., and Rajan R. S. (1972a) History of lunar breccias (abstract). In *Lunar Science—III* (editor C. Watkins), p. 415, Lunar Science Institute Contr. No. 88.

Hutcheon I. D., Phakey P. P., and Price P. B. (1972b) Studies bearing on the history of lunar breccias. *Proc. Third Lunar Sci. Conf., Geochim. Cosmochim. Acta* Suppl. 3, Vol. 3. MIT Press.

Kaiser W. A. (1972a) Rare gas measurements in three Apollo 14 samples (abstract). In *Lunar Science—III* (editor C. Watkins), p. 442, Lunar Science Institute Contr. No. 88.

Kaiser W. A. (1972b) Rare gas studies in Lunar 16-G-7 fines by stepwise heating technique. A low fission solar wind Xe. *Earth Planet. Sci. Lett.* **13,** 387–399.

Keith J. E., Clark R. S., and Richardson K. A. (1972) Gamma ray measurements of Apollo 12, 14, and 15 lunar samples (abstract). In *Lunar Science—III* (editor C. Watkins), p. 446, Lunar Science Institute Contr. No. 88.

Krishnaswami S., Lal D., Prabhu N., and Tamhane A. S. (1971) Olivines: Revelation of tracks of charged particles. *Science* **174,** 287–291.

Lal D., Murali A. V., Rajan R. S., Tamhane A. S., Lorin J. C., and Pellas P. (1968) Techniques for proper revelation and viewing of etch-tracks in meteorite and terrestrial minerals. *Earth Planet. Sci. Lett.* **5,** 111–119.

Lal D., Lorin J. C., Pellas P., Rajan R. S., and Tamhane A. S. (1969) On the energy spectrum of iron-groups nuclei as deduced from fossil-track studies in meteoritic crystals. *Meteorite Res.* pp. 275–285, IAEA, Vienna.

Lal D., MacDougall D., Wilkening L., and Arrhenius G. (1970) Mixing of the lunar regolith and cosmic-ray spectra: New evidence from fossil particle-track studies. *Proc. Apollo 11 Lunar Sci. Conf., Geochim. Cosmochim. Acta* Suppl. 1, Vol. 3, pp. 2295–2303. Pergamon.

LSPET (Lunar Sample Preliminary Examination Team) (1969) Preliminary examination of lunar samples from Apollo 11. *Science* **165,** 1211–1227.

LSPET (Lunar Sample Preliminary Examination Team) (1970) Preliminary examination of lunar samples from Apollo 12. *Science* **167,** 1325.

LSPET (Lunar Sample Preliminary Examination Team) (1971) Preliminary examination of lunar samples from Apollo 14. *Science* **173**, 681–693.

Lugmair G. W. and Marti K. (1971) Neutron capture effects in lunar gadolinium and the irradiation histories of some lunar rocks. *Earth Planet. Sci. Lett.* **13**, 32–42.

Lugmair G. W. and Marti K. (1972) Exposure ages and neutron capture record in lunar samples from Fra Mauro. Preprint.

MacDougall D., Martinek B., and Arrhenius G. (1972) Regolith dynamics (abstract). In *Lunar Science—III* (editor C. Watkins), p. 498, Lunar Science Institute Contr. No. 88.

Marti K., Lugmair G. W., and Urey H. C. (1970) Solar wind gases, cosmic rays spallation products, and the irradiation history of Apollo 11 samples. *Proc. Apollo 11 Lunar Sci. Conf., Geochim. Cosmochim. Acta* Suppl. 1, Vol. 2, pp. 1357–1367. Pergamon.

Pepin R. O., Nyquist L. E., Phinney D., and Black D. C. (1970) Rare gases in Apollo 11 lunar material. *Proc. Apollo 11 Lunar Sci. Conf., Geochim. Cosmochim. Acta* Suppl. 1, Vol. 2, pp. 1435–1454. Pergamon.

Phakey P. P., Hutcheon I. D., Rajan R. S., and Price P. B. (1972a) Radiation damage in soils from five lunar missions (abstract). In *Lunar Science—III* (editor C. Watkins), p. 608, Lunar Science Institute Contr. No. 88.

Phakey P. P. and Price P. B. (1972b) Extreme radiation damage in soil from Mare Fecunditatis. *Earth Planet. Sci. Lett.* **13**, 410–418.

Poupeau G. and Berdot J. L. (1972) Irradiations ancienne et récente des aubrites. *Earth Planet. Sci. Lett.*, **14**, 381–396.

Poupeau G., Berdot J. C., Chetrit G. C., and Pellas P. (1972a) Lunar regoliths tribulations since September 19, 1969. Preliminary Abstracts of the Third Lunar Science Conference, p. 539, January.

Poupeau G., Berdot J. L., Chetrit G. C., and Pellas P. (1972b) Predominant trapping of solar-flare gases in lunar soils (abstract). In *Lunar Science—III* (editor C. Watkins), p. 613, Lunar Science Institute Contr. No. 88.

Price P. B., Hutcheon I. D., Lal D., and Perelygin V. P. (1972) Lunar crystals as detectors of very rare nuclear particles (abstract). In *Lunar Science—III* (editor C. Watkins), p. 619, Lunar Science Institute Contr. No. 88.

Rancitelli L. A., Perkins R. W., Felix W. D., and Wogman N. A. (1972) Cosmic ray flux and lunar surface processes characterized from radio nuclide measurements in Apollo 14 and 15 lunar samples (abstract). In *Lunar Science—III* (editor C. Watkins), p. 630, Lunar Science Institute Contr. No. 88.

Ridley W. I., Willimas J., Brett R., and Takeda H. (1972) Petrology of lunar basalt 14310 (abstract). In *Lunar Science—III* (editor C. Watkins), p. 648, Lunar Science Institute Contr. No. 88.

Russ G. P. (1972) Neutron capture on Gd and Sm in the Luna 16, G-2 soil. *Earth Planet. Sci. Lett.* **13**, 384–386.

Sutton R. L., Hatt M. H., and Swann G. A. (1972) Geology of the Apollo 14 landing site (abstract). In *Lunar Science—III* (editor C. Watkins), p. 732, Lunar Science Institute Contr. No. 88.

Swann G. A., Bailey N. G., Batson R. M., Eggleton R. E., Hait M. H., Holt H. E., Larson K. B., McEwen M. C., Mitchell E. D., Schaber G. G., Schafer J. B., Shepard A. B., Sutton R. L., Trask N. J., Ulrich G. E., Wilshire H. G., and Wolfe E. W. (1971) Preliminary geologic investigations of the Apollo 14 landing site. In *Apollo 14 Preliminary Science Report*, pp. 39–86, NASA Report SP-272.

Vinogradov A. P. (1971) Preliminary data on lunar grounds brought to earth by automatic probe Luna 16. *Proc. Second Lunar Sci. Conf., Geochim. Cosmochim. Acta* Suppl. 2, Vol. 1, pp. 1–16. MIT Press.

Walker R. M. and Zimmerman D. (1972) Fossil track and thermo-luminescence studies of Luna 16 material. *Earth Planet. Sci. Lett.* **13**, 419–422.

Wenk E., Glauser A., Schwander H., and Tommsdorff V. (1972) Optical orientation, composition and twin-laws of plagioclases from rocks 12051, 14053, and 14310 (abstract). In *Lunar Science—III* (editor C. Watkins), p. 794, Lunar Science Institute Contr. No. 88.

Proceedings of the Third Lunar Science Conference
(Supplement 3, *Geochimica et Cosmochimica Acta*)
Vol. 3, pp. 2883–2903
The M.I.T. Press, 1972

Track metamorphism in extraterrestrial breccias

J. C. Dran, J. P. Duraud, M. Maurette

Centre de Spectrométrie de Masse du C.N.R.S.,
91-Orsay, France

L. Durrieu, C. Jouret

Institut d'Optique Electronique du C.N.R.S.,
31-Toulouse, France

and

C. Legressus

Département de Physicochimie du C.E.A.,
91-Gif-sur-Yvette, France

Abstract—Nuclear particle tracks and various textural features stored in the constituent grains of the fine grained matrix of lunar and meteoritic breccias have been first studied by combined high voltage and scanning electron microscopics and then correlated to several characteristics of the breccias including their petrographic type, their scratch hardness, their rare gas concentration, their albedo, and the density of microfractures in their constituent grains. The main purpose of this work was both to scale in temperature the heat metamorphism evolved during brecciation, and to evaluate the degree of "dust sintering" in the breccias.

These investigations reveal the following features: (1) There are two very distinct groups of lunar breccias characterized by low ($\sim 10^7$ track/cm^2) and high ($\sim 10^9$ tracks/cm^2) densities of etched tracks respectively; (2) in the "high density" group, the grains still show the tracks they registered as dust particles, before their compaction into breccias and several characteristics in the track distribution can be used to scale in temperature the very mild heat metamorphism evolved during brecciation; the breccias in this group have also the lowest albedo and the smallest scratch hardnesses; (3) the majority of the breccias in the "low density" group have probably suffered a marked heat or shock metamorphism; (4) the scratch hardness of the breccias is correlated to their petrographic type; (5) the Luna 16 dust grains seem to have undergone a very complex history in the regolith, which is perhaps characterized by a temporary incorporation of the grains in very mildly metamorphosed breccias, which get subsequently eroded by micrometeorite chipping of the surface or "turn-over" grinding; (6) the various landing sites on the moon show distinct differences when they are classified as a function of the radiation damage annealing recorded in the breccias; (7) the track and texture records in grains from solar type gas-rich meteorites are quite different from those observed in lunar breccias, but their significance in terms of the irradiation and brecciation history of the "regolith" of the meteorite parent body is not clearly understood at the present time; (8) the difference in the albedo of the "light" and "dark" parts of gas-rich meteorites is still unexplained and remains a stumbling block for any albedo theory.

Introduction

The formation of breccias in the "regolith" of atmosphereless and magnetic field free planets depends on complex factors such as the flux of solid matter in interplanetary space, the planetary surface features, and the residence time of the dust grains in the top layers of the "regolith." The main objective of this investigation was to get a better insight into these extraterrestrial brecciation processes, as well as to study their possible variation in nature and in intensity both as a function of the distance to the sun and of the "age" of formation of the breccias, by examining simultaneously the *fine grained matrix* of lunar and meteoritic breccias.

This work is based on the discovery by Fleischer *et al.* (1965) of nuclear particle track annealing, in minerals which have been shocked or heated. This characteristic of tracks has already been used by various authors to trace back the thermal history of metamorphic contacts on the earth (Naeser and Faul, 1969), and that of tektites (Storzer and Wagner, 1969), as well as to change the sensitivity for track registration in terrestrial (Perelygin *et al.*, 1968), meteoritic (Maurette, 1970), and lunar materials (Price *et al.*, 1972). We first studied some features of the ultramicroscopic irradiation and texture record in lunar dust grains, which are exposed on the surface of the moon to low energy solar nuclear particles, by using high voltage and scanning electron microscopes. Then possible changes in this record as the dust grains are either heated up to temperatures of 1000°C or "sintered" into the fine grained matrix of lunar breccias were analyzed. Furthermore these changes were tentatively correlated to other potential indicators of the degree of metamorphism suffered by the breccias, such as their petrographic type, their scratch hardness, their albedo, their rare-gas concentration, and the extent of microfracturing in their constituent grains. Finally, with the view of changing both the time of brecciation and the distance to the sun, we studied the same features and correlations in various types of gas-rich meteorites, which are possibly breccias formed in the "regolith" of the meteorite parent body (Suess *et al.*, 1964).

At the present time we are not prepared to describe a geological model for brecciation, such as the "base surge" process proposed by McKay *et al.* (1970). Our limited contribution was to define new "solid state" indexes which can be used to scale the heat metamorphism and the degree of "dust sintering" produced during brecciation, as well as to study their variation as a function of several parameters. Preliminary results concerning our comparisons of lunar and meteoritic breccias have already been described (Dran *et al.*, 1970; Durrieu *et al.*, 1971; Dran and Duraud, 1971; Dran *et al.*, 1972), and a more extensive version of our work including combined optical and electron microscope petrographic observations, as well as electron and ion microprobe analysis is in preparation (Christophe-Michel-Levy *et al.*, 1972). Track "metamorphic" studies have also been reported at the Third Lunar Science Conference by Hutcheon *et al.* (1972) and Fleischer *et al.* (1972).

EXPERIMENTAL PROCEDURES

The following samples were examined for the present work: lunar fines: 10084, 14141, 14148, 14149, 14156, 14162, 14163, 14230,70, 14230,87, 14230,91, and Lunar 16-19; lunar breccias: 10046, 10059, 14267, 14305, 14083, 14063, 14321, and 14006; lunar dust "clod": 14049; lunar microbreccia: 14161,41; gas-rich meteorites: Kapoyeta, Weston.

Our electron microscope observation techniques have been described in detail elsewhere (Borg *et al.*, 1970, 1971), and will only be summarized below: (1) The finest lunar dust grains as well as micron-sized fragments extracted from the fine grained matrix of the breccias were directly deposited on the electron microscope substrates and then examined in dark field conditions with a 1 MeV electron microscope; (2) 200 mesh lunar dust grains as well as whole chunks of breccias were polished in epoxy mounts, slightly etched in boiling NaOH solution or in the "olivine" acid mixture developed by Krishnaswami *et al.* (1971), and the etched track distributions were observed with a high resolution scanning electron microscope, allowing the measurement of track densities up to 10^{10} tracks/cm^2.

In this investigation we have essentially used feldspar and olivine crystals, because it has been previously shown that there are serious problems in interpreting the track distribution in pyroxene crystals, both from lunar and meteoritic origin (Crozaz *et al.*, 1971; Dran *et al.*, 1972). Furthermore we have applied a new technique for obtaining the electron diffraction patterns of micron-sized grains, by monitoring the electron beam intensity with a Faraday cup, and by using an electronic shutter for exposing the plate for constant illumination.

The scratch hardnesses (SH) were measured as a function of the load (10, 15, 10, 25, 30 g), with the use of a Leitz microhardness apparatus giving scratches similar to those reported in Fig. 1. The SH values corresponding to a load of 30 g have been reported in Table 1 (column 6). The albedo data were collected by A. Dollfus. For thermal annealing experiments, the samples were heated 2 hours at various temperatures up to 1000°C, *under vacuum*; furthermore the *same* grains were examined both before and after heating (Fig. 2). The density of microfractures was estimated in grains larger than about 50 μ, by using the micromappings of polished sections obtained with the scanning electron microscope.

RESULTS

1. Scanning electron microscopy

In Fig. 3, we show the track densities ρ measured at the center of 200 mesh grains extracted from various soil samples from the Apollo 11, 14, and Luna 16 missions. These track distributions, which indicate the degree of irradiation of the soil samples in the solar flare cosmic rays (Crozaz *et al.*, 1970), give an estimate of the track distributions that should be expected in the fine grained matrix of nonmetamorphosed lunar breccias; they have the following characteristics: (1) With the exception of the Cone Crater fines (sample 14141) the values of ρ are generally spread over 2 orders of magnitude and clustered in the range $5.10^7 \lesssim \rho \lesssim 5.10^9$ tracks/cm^2; (2) the tracks are partially annealed at 600°C and they completely disappear above 700°C.

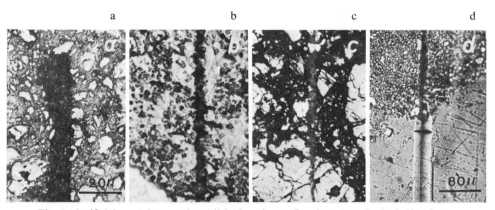

Fig. 1. Artificial scratches on the polished surfaces of various samples including lunar breccias 14049 and 14083 (Figs. 1a and 1b, respectively), the Weston chondrite (Fig. 1c), and a pellet of sintered lunar dust (Fig. 1d). The scratch hardness, which is inversely proportional to the width of the scratch markedly decreases in the following order: Weston > 14083 > 14049.

Table 1. Metamorphism in extraterrestrial breccias.

Samples	Type*	Rare gases† He⁴(ccSTP/g)	He⁴/Ne²⁰	Feldspar (tracks/cm²)	Olivine (tracks/cm²)	Fractures (nb/cm²)	Arbitrary unit, SH	OA** (%)	Major elements†† Al/Si	Fe/Si
14049	Group 2	4.7×10^{-2}	53.5	$(5–6) \times 10^9$	$(3–4) \times 10^8$	3×10^4	9,1	15	0.39	0.33
10046		2.6×10^{-1}	65.4	$(2–3) \times 10^9$		8×10^4	10,3	7.2	0.30	0.64
10059				$(1–2) \times 10^9$		5×10^4	12,5	9.0	0.34	0.68
14161		4.7×10^{-2}		$(8–9) \times 10^8$						
14063	Group 3	2.0×10^{-3}	110	$(1–2) \times 10^7$		3×10^4	12	50.3	0.50	0.20
14083	Group 3			$(1–2) \times 10^7$		6×10^4	13,2	52.5	0.47	0.22
14321	Group 4			$(4–5) \times 10^6$		5×10^4	15,3	27.1	0.40	0.30
14305	Group 6	1.3×10^{-3}	650	$(1–2) \times 10^7$	$(6–7) \times 10^6$	3×10^5	16,4	30.2	0.37	0.32
14006	Group 6	2.5×10^{-3}	5800	$(4–5) \times 10^7$	$(1–2) \times 10^7$	2×10^4	16,5	8.0		
14267				$(1–2) \times 10^7$		5×10^4	19,5	11.8		
15299						5×10^5	20,0	8.6		
Kapoyeta (D)		2.0×10^{-3}	56–96		$\{(1–2) \times 10^8\}‡$	6×10^5	18	28.0		
Weston (D)		$(2.2–17) \times 10^{-3}$	400–450		$(3–4) \times 10^8$	2×10^5	17,5			

* Petrologic type according to Warner (1972.)
† From Hinterberger et al. (1971) and LSPET (1971).
‡ These values have been measured in pyroxene crystals.
** These values have been measured by A. Dollfus.
†† From LSPET (1971).

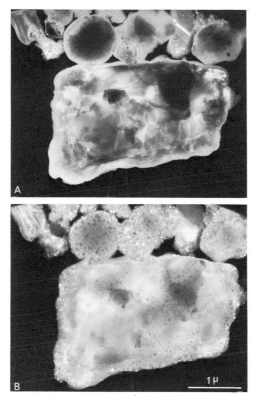

Fig. 2. 1 MeV dark field micrographs of micron-sized dust grains from sample 10084. The same grains have been observed both before (Fig. 2a) and after (Fig. 2b) heating. These micrographs illustrate the thermally activated transformation of the tracks and amorphous coatings into microcrystallites at a relatively low temperature (2 hours at 800°C, under vacuum). That the crystallites are not rare-gas bubbles is simply demonstrated by the observation that some of them are clearly bursting from the surface of the grains.

With the polished sections of various breccias the results of our investigations (Fig. 4) show: (1) The existence of two very distinct groups of lunar breccias characterized by low ($\sim 10^7$ tracks/cm^2) and high ($\sim 10^9$ tracks/cm^2) densities of etched tracks respectively; (2) a peculiar track distribution in the olivine grains from Weston, with ρ values that are "intermediate" between those of the two groups of lunar breccias.

During this scanning electron microscope survey of olivine and feldspar grains, both in lunar dust samples and in extraterrestrial breccias, we noted the following marked differences between the track distributions registered in the olivines and in the feldspars: (1) in olivine crystals from the Luna 16 soil sample and from the Weston chondrite, the diameter of the fossil tracks were about five times smaller than those measured for fission tracks (Fig. 5a); however this differential etching behavior was not observed for the olivines from breccia 10046; (2) the average track densities in

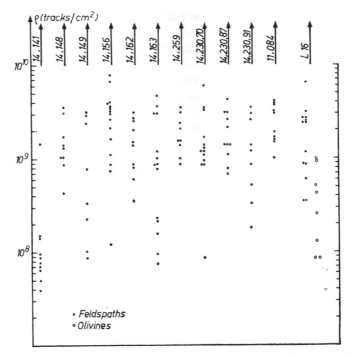

Fig. 3. Etched track distributions measured in 200 mesh feldspar and olivine grains from various lunar soil samples, from the Apollo 11, 14, and Luna 16 missions. These track distributions are also used for other purposes concerning for example the motion of the dust particles in the lunar regolith.

olivine grains from the Luna 16 sample and the 10046 breccia are about five times smaller than in the feldspars, and this characteristic has already been reported by the Washington University group, for dust grains extracted from sample 10084 (Crozaz *et al.*, 1971); (3) the proportion of grains showing an edge zoned track distribution— generally considered as a "marker" for a solar flare irradiation of the grains (Walker, 1971)—is much higher in the olivines ($\sim 20\%$) than in the feldspars or in the pyroxenes ($\lesssim 5\%$), and this feature was observed both for the Luna 16 grains and the Weston chondrite.

In a previous paper (Dran *et al.*, 1972), we pointed out that the high production rate of spallation recoils in lunar and meteoritic pyroxenes, as well as the inhomogeneous registration of heavier nuclei in these crystals, preclude their use for measuring the characteristics of solar flare cosmic rays, if the tracks are only observed with a scanning electron microscope. Therefore we decided to work with feldspar and olivine crystals, which are very insensitive for registering spallation recoils and which also show a homogeneous sensitivity for the registration of heavier nuclei. But the results presented in this paper still indicate that there are problems connected with the utilization of the olivine detectors for studying the VH nuclei of the solar flare and galactic cosmic rays, and this point has also been emphasized independently by the

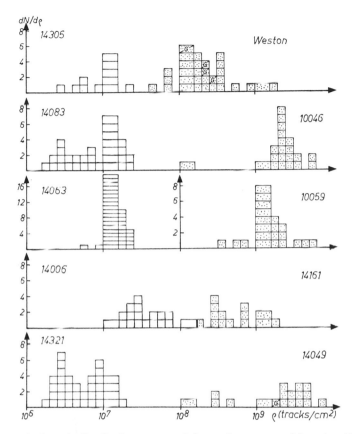

Fig. 4. Etched track distribution measured for various types of breccias. Feldspar grains have been used for samples 14006, 14049, 14063, 14083, 14161, 14305, 14321, 10046, and 10059; olivine grains have been examined for the Weston chondrite. The letter G identifies those of the grains showing a solar flare type track gradient.

Berkeley group (Price *et al.*, 1972). However such problems are probably not so serious for the interpretation of the track "metamorphic" studies reported in this paper, because we only need to start with tracks homogeneously distributed in the grains, without worrying too much about their nature. Furthermore the peculiar characteristics of the olivine should help in tracing back their heat metamorphic history, if the differences in the track records observed in the feldspars and olivines are due to an easier thermal annealing of the tracks in the olivines. This effect could possibly trigger a decrease in the etching rate of the tracks (Price *et al.*, 1972), as well as artificial edge zonings in the track distribution similar to those reported for annealed epidote crystals by Naeser *et al.* (1970).

2. Transmission electron microscopy

The finest and most irradiated lunar dust grains show the following ultramicroscopic features which have been extensively described elsewhere (Bibring *et al.*,

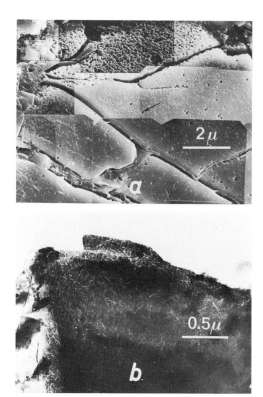

Fig. 5. Electron microscope study of etched tracks in extraterrestrial breccias. In Fig. 5a the olivine grain from the Weston chondrite has been observed by scanning electron microscopy; the diameters of the fission fragment tracks (single arrow) produced by irradiating the polished section with a Cf^{252} source, are much larger than those measured for the fossil tracks; the distribution of the fossil tracks is also inhomogeneous. The micron sized grain in Fig. 5b has been extracted from lunar breccia 10059 and then slightly etched in a dilute solution of HF; the very high density of etched tracks ($\rho \gtrsim 10^{11}$ tracks/cm^2) which appear as tiny etch canals in the 100 keV electron micrograph precludes a strong heating of the grain during brecciation.

1972a; Borg *et al.*, 1970, 1971, 1972a, 1972b; Dran *et al.*, 1970, 1972; Durrieu *et al.*, 1971): (1) they are generally markedly rounded and this feature has been attributed to solar wind sputtering; (2) they contain various forms of radiation damage features, such as ultrathin amorphous coatings, produced by an "ancient" solar wind implantation, and very high densities of nuclear particle tracks; (3) the tracks and the coatings disappear almost simultaneously at about 800°C, and they are then transformed into microcrystallites, the size of which increase markedly as a function of the annealing temperature (Fig. 2); (4) the Luna 16 dust soil contains a high proportion of dust particles loaded with similar crystallites (Fig. 6a); (5) the electron diffraction patterns of the grains can be classified as a function of the total number Σ_s of diffraction spots (Fig. 7) appearing in the patterns. The amount of lattice disorder in a given dust sample increases as the values of Σ_s get smaller. As the values of Σ_s markedly

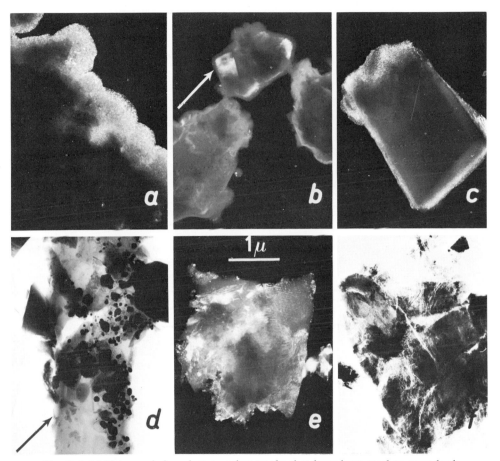

Fig. 6. 1 MeV transmission electron micrographs that have been used to trace back the "metamorphic" history of various types of extraterrestrial breccias. The fragment in Fig. 6a results from the crushing of a 200 mesh "microbreccia" from the Luna 16–19 soil sample, which appeared as "amorphous" in polarized light. The grain in Fig. 6b has been extracted from lunar dust clod 14049 and shows an amorphous coating (white arrow). The grain in Fig. 6c has been obtained from lunar breccia 10046 and it contains micro-crystallites similar to those observed in dust grains that have been artificially heated. The arrow in Fig. 6d identifies a type of glass lamella that was frequently observed in breccia 14006. Deformation bands are visible in the pyroxene reported in Fig. 6e, and which was extracted from breccia 15015. The grain in Fig. 6f has been extracted from the dark part of the Kapoyeta howardite and then very slightly etched in a dilute solution of HF, with the view to reveal the extensive microfracturing in the grain.

increased in Luna 16 dust particles heated at 900°C, we suggested a radiation damage origin for the lattice disorder stored in the grains (Borg *et al.*, 1972a). If this hypothesis is true, then the possible variation in the values of Σ_s estimated for the breccias, both before and after a thermal treatment, could be used to evaluate the degree of *natural* radiation damage annealing in the grains.

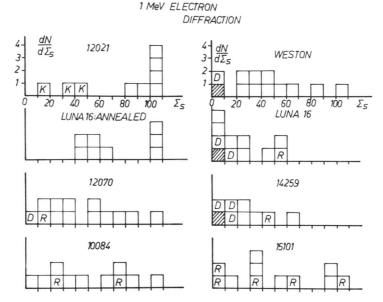

Fig. 7. Distribution of the total number of diffraction spots Σ_s obtained from the electron diffraction patterns of micron-sized grains from different origins: (1) The grains from lunar igneous rock 12021 have never been exposed to a heavy dose of solar ions because they have been extracted from an internal chunk of the rock; (2) the Luna 16–19 and 14259 grains show the smallest values for Σ_s and they are the most disordered samples in this series; as the values of Σ_s increase after a thermal annealing, this disorder is considered as being due to radiation damage; (3) some of the fragments from the fine grained matrix of Weston are highly disordered, but the disorder does not seem to anneal upon heating. The letters *D, K, R* refer to electron diffraction patterns showing distortion in the diffraction spots, Kikuchi lines, and Debye–Sherer rings, respectively.

For the micron-sized fragments from the breccias the following results have been obtained: (1) In lunar dust clods (sample 14049) the grains show nonmetamorphosed amorphous coatings and high track densities (Fig. 6b); (2) the breccias in the high "density" group (samples 10046 and 10059) contain a high proportion of grains that are loaded with very high densities of tracks (Fig. 5b); furthermore some of the grains show microcrystallites similar to those observed in the artificially heated dust grains (Fig. 6c); (3) in the low track density group of breccias the fragments were quite similar to those obtained by crushing lunar igneous rocks, and they did not show any heavily irradiated grains containing high track densities, amorphous coatings, microcrystallites, or low values of Σ_s; however peculiar textural features such as glass lamellae (Fig. 6d) and deformation bands in the pyroxenes (Fig. 6e) were frequently observed; (4) in Kapoyeta an extensive microfracturing, never found in lunar breccias, was discovered in grains extracted from several dark parts of this meteorite (Fig. 6f); however no heavily irradiated grains were observed during the survey of about 150 individual fragments; (5) in Weston the microfracturing was less extensive than in Kapoyeta but it was still present. Only 4 grains showing textural features which could be *optimistically* interpreted as a high density of microcrystallites, in three of the fragments and as an amorphous coating in the other, were found by examining about 300 fragments in this meteorite; (6) the lattice disorder, appearing as low values of Σ_s, was annealed upon heating in breccia 10046 but not in Kapoyeta.

The Berkeley group has used similar high-voltage electron microscope techniques to examine both lunar dust particles and small grains in lunar and meteoritic breccias (Barber *et al.*, 1971a, 1971b; Hutcheon *et al.*, 1972; Phakey and Price, 1972; Phakey *et al.*, 1972). For lunar dust grains their observations agree quite well with ours. However we still have not found any heavily irradiated grains in the fine grained matrix of gas-rich meteorites, whereas the Berkeley group reports a high proportion ($\sim 10\%$) of such grains in Kapoyeta and in Fayetteville (Barber *et al.*, 1971b). At the present time we have no simple explanation for such discrepancies. It could be that the fine grained matrices of solar type gas-rich meteorites are extremely inhomogeneous in their content of heavily irradiated grains, and that the ion thinning technique that was applied by the Berkeley group to prepare ultrathin sections of breccias allows a more efficient search for the irradiated grains.

3. Scratch hardness and albedo of the breccias, and artificial sintering of lunar dust grains

The scratch hardness (SH) and the orange light albedo (OA) of various breccias have been reported in Table 1 (column 6 and 7, respectively), as well as other parameters including the type of breccia (column 2), the rare gas concentrations (column 3), the "average" track densities as extrapolated from Fig. 4 (column 4), and the density of microfractures in the grains (column 5). Most of these values are very difficult to define, as the breccias are generally very inhomogeneous. To minimize this problem we only measured the SH and OA values as well as the density of microfractures and that of tracks in the same area of the fine grained matrix of the breccias.

We also heated up to temperatures of 900°C the 400 mesh residues extracted from various lunar fines including samples 10084 and 14141 that contain 90% and 7% of grains with an amorphous coating, respectively, and sample 10084 that was slightly etched to remove the amorphous coating on the grains. In Fig. 8 we show the state of aggregation of the dust samples when they have been heated at 900°C, in quartz tubes: the nonetched 10084 sample (Fig. 8a) was the only one to give a "pellet" of sintered dust, which was sufficiently hard to withstand the gentle unloading from the

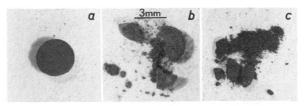

Fig. 8. Lunar dust sintering activated by thermal heating: A hard pellet of sintered dust was observed for the natural 10084 dust sample (Fig. 8a); a weakly bounded pellet resulted from the heating of the Cone Crater sample (Fig. 8b) and an extremely friable pellet was obtained from the etched 10084 dust sample (Fig. 8c).

quartz tube. Fragments from the pellets were then polished in epoxy mounts and their scratch hardnesses were determined; the nonetched 10084 pellet was the only one to yield a SH value that was distinctly higher than that of the epoxy. This characteristic is illustrated in Fig. 1d that shows that the width of the scratch is about three times smaller for the 10084 pellet than for the epoxy.

HEAT METAMORPHISM DURING BRECCIATION

In the following we estimate the intensity of the heat metamorphism evolved during brecciation. The breccias will be classified on a temperature scale T_B defined by comparing the degree of radiation damage annealing appearing in their constituent grains to that produced in dust particles from sample 10084, which have been heated two hours, under vacuum, at a temperature T_B. The main conclusions of these investigations are:

(1) The formation of lunar dust clods (sample 14049) involves a heat metamorphism that is insufficient to produce a measurable degree of radiation damage annealing, and that corresponds to $T_B < 600°C$.

(2) The Apollo 11 breccias have been processed by a very mild heat metamorphism, that has only partially annealed the "dust type" track record in a few grains; from the sizes of the microcrystallites in the grains we deduce that $600°C \lesssim T_B \lesssim 700°C$.

(3) At the present time we have some difficulty in characterizing the heat metamorphism responsible for the formation of the Apollo 14 breccias showing *low track density* because we have not detected large crystallites in their constituent grains. We might simply deduce that such breccias have been extensively metamorphosed, and that the corresponding values of T_B—as estimated from the lack of crystallites— exceed 900°C. This conclusion would be based on the working assumption that the grains in the breccias have been heavily irradiated before compaction, like the dust particles reported in Fig. 3. Therefore the observation of low track densities should reflect the occurrence of a severe track annealing during brecciation and the present track distribution could be due to the irradiation of the breccias in the galactic cosmic rays. This assumption certainly holds for regolith type breccias characterized by low albedo and high scratch hardnesses, because a random sampling of the Apollo 14 regolith yields a high proportion ($\sim 90\%$) of dark soil samples heavily irradiated in the solar flare cosmic rays (see Fig. 3). However, we cannot completely exclude the possibility that some of the breccias in the low track density group (particularly those showing both very high albedo and low scratch hardnesses and such as samples 14063 and 14083) have been formed during the deep excavation of nonirradiated soil samples that occurred during the formation of the Imbrium basin. Therefore the existence of low densities of tracks in such breccias would not be indicative of track annealing. A detailed study of the envelope of the track distribution reported in Fig. 4 should help in sorting the nonmetamorphosed breccias from the low "density" group. Indeed it is expected that the VH nuclei of the galactic cosmic rays should give a narrowly peaked track distribution, similar to that reported for breccia 14063, and characterized by a spread of about a factor of 2 in the ρ values (Fleischer *et al.*, 1967). However, before

concluding that the spread in the distributions measured for breccias 14006 and 14321 indicate that the tracks were inherited from the prebrecciation solar irradiation of the grains, the statistics in the ρ determination has to be markedly increased. Furthermore one should verify that the spallation recoils formed during the nuclear interactions of the galactic cosmic rays with the medium heavy elements in the grains do not broaden the VH distribution, by adding high ρ values in the track distributions.

(4). The Luna 16 dust grains give evidence for a more extensive formation of very mildly metamorphosed breccias at Fecunditatis base with T_B values of about 700°C.

(5) On the T_B metamorphic scale, it is quite apparent that the various landing sites on the moon are not "processing" the same type of breccias; in particular the constancy in the degree of radiation damage annealing of the Apollo 11 breccias has to be contrasted with the marked variations of the same parameter at the Apollo 14 landing site.

It is well known that shock can erase etched tracks in minerals (Fleischer *et al.*, 1965, 1972). However in the discussion presented here we have not attempted to relate the track distribution in the "metamorphosed" Apollo 14 breccias to shock features in the grains. Indeed no serious attempt has been made to clearly define and classify such features by transmission electron microscopy, although it could be argued that the glass lamella and the deformation bands reported in Figs. 6d and 6e, respectively, do in fact represent shock evidence. We are now developing a new method to identify *shock* metamorphosed breccias containing heavily irradiated grains. In such grains it can be expected that shock produces "dotted" latent tracks that would not be visible in the complex contrast structure due to the deformation bands. But a thermal annealing can certainly transform such invisible track "remnants" into large microcrystallites that would be much easier to observe with the view to identify shock metamorphism.

Dust "Sintering" during Brecciation

Now we want to discuss the degree of sintering of *the fine-grained matrix of the breccias*, in terms of the scratch hardness measurements reported in Fig. 9, and in Table 1 (column 6). The hardness of single crystals seems to be intimately connected with their volumetric lattice energy, as shown by Plendl and Gielesse (1962). However the physical formulation of hardness for a sintered powder is very difficult to establish, and we have simply considered that the fine grained matrices of lunar and meteoritic breccias are made of grains with identical hardnesses and with similar size distributions. Therefore during the scratching of this matrix with a diamond needle, the SH values should essentially depend on the strength of intergrain cementing and the least sintered matrix should give the deepest and broadest scratch, corresponding to the lowest SH value (Fig. 1)—this hypothesis is certainly not valid for the glass-welded type breccias, and should be corrected for coarse grained matrices.

In Table 1 it can be verified that the SH values greatly vary in the breccias investigated for the present work. However they are well correlated to the petrographic classification proposed by Warner (1972); they steadily increase from "group 2" to "group 6" of breccias. Figure 9 clearly shows that there are no simple relationships

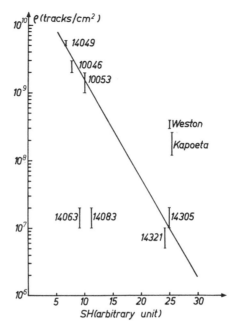

Fig. 9. Correlation plot between the track densities and the scratch hardnesses of lunar and meteoritic breccias.

between the track densities and the SH values, except for a subgroup of breccias characterized by the *lowest* albedo, and that includes samples 14049, 10046, 10059, 14006, and 14267. Finally, with the exception of breccia 14267, the gas-rich meteorites contain the grains that have been the most strongly cemented to each other.

If the breccias result from the "compaction" of soil particles, then the cementing of the grains should be made easier when the initial soil sample shows a high index of "maturity." Such an index has been first proposed by Adams and McCord (1971) to explain the optical properties of the lunar soil: a soil that is more mature contains a greater concentration of "exotic" materials such as opaque glasses, etc. We will use a slightly modified definition of "maturity": a soil that is more mature has resided a longer *integrated* time at the top surface of the regolith; therefore it will certainly be more contaminated with "exotic" components, but its constituent grains will also show a greater concentration of radiation damage, that has been induced by low energy solar particles, with a short penetration depth in the regolith. Therefore we will estimate the value of the index of maturity in a given soil sample by measuring its proportion of micron-sized grains showing an amorphous coating, which is due to solar wind radiation damage (Bibring *et al.*, 1972a).

Thus the easier cementing of the dust grains in a more mature soil could be attributed either to the release of the energy that has been stored as radiation damage in the amorphous coatings (Borg *et al.*, 1971), or to the redistribution of an "exotic" component with a low melting point which can help in cementing the grains to each other. The formation of pellets of lunar dust that was activated by a thermal treatment supports this hypothesis. Indeed the formation of a hard pellet giving SH values that were three times greater than those measured for the epoxy was only observed in the sample that contained the highest proportion of coated grains but not in the others. Furthermore the etching used to dissolve the coatings on the grains was too slight to remove a large fraction of the glassy objects in sample 10084. Therefore we think that the "solar wind" stored energy is one of the dominating factors in the preferential sintering of mature soil samples. From this result, it can be expected that very mildly metamorphosed breccias are preferentially formed in those of the regolith areas which have been filled up with a mature soil, containing a high proportion of grains heavily irradiated in the solar wind.

REDISTRIBUTION OF RARE GASES DURING BRECCIATION

Our attempts to correlate the rare-gas distributions to the SH, OA, and ρ values have been severely limited due to the lack of rare gas data and they can be summarized as follows:

(1) He^4 content (Table 1, column 3): the lunar breccias in the high and low track density groups show the highest and the lowest rare gas content respectively, and this supports the heat sintering hypothesis for the breccias.

(2) He^4/Ne^{20} ratios (Table 1, column 3): in the high track density group these ratios are slightly smaller than those (~ 90) measured for various soil samples; therefore the heat metamorphism involved during the formation of these breccias is insufficient to preferentially drive away a *large* fraction of the *lightest* rare gases, and this observation still agrees with our radiation damage annealing observations. However this ratio markedly increases in the low track density group, and this is contrary to expectation based on the thermally activated diffusion of rare gas atoms "pre-implanted" in the dust grains. Therefore it could be argued that the formation of breccias 14305 and 14321 has certainly involved complex processes, producing both a sharp decrease in the He^4 content and an unexplained increase in the He^4/Ne^{20} ratios, as well as dust sintering and track annealing.

(3) These remarks can also be applied to the solar type gas-rich meteorites: Weston is generally considered as nonmetamorphosed on the basis that the He^4/Ne^{20} ratio in this meteorite (~ 400) is undepleted with respect to those measured in aluminum catcher foils deployed on the lunar surface by the astronauts (Geiss et al., 1970). However it can be verified in Table 1 that a similar ratio has been measured for lunar breccia 14321, which has certainly been extensively metamorphosed, and which further show about the same He^4 content and the same scratch hardness as Kapoyeta and Weston; therefore it could be argued that the gas-rich meteorites examined in the present work are "metamorphosed" breccias.

This discussion, as well as those reported in the previous sections, show that great care should be exercised when deciphering the past history of extraterrestrial breccias. It is necessary to work with a combination of various metamorphic "probes" in the same area of the breccias. In particular the use of rare gas distribution or etched track densities alone is certainly not satisfactory for such a purpose.

THE "DIFFERENT" REGOLITH HISTORY OF SOLAR TYPE GAS-RICH METEORITES

If the correlations described in the previous section are verified by further work, then they will be helpful to trace back both the irradiation and metamorphic histories of the constituent grains in the breccias, but they will also show that the solar type gas-rich meteorites have suffered a very different "regolith" history. These differences in the nature or intensity of meteoritic and lunar brecciation processes have already been emphasized upon several occasions (Arrhenius *et al.*, 1971; Borg *et al.*, 1971; Dran *et al.*, 1970, 1972; Wilkening *et al.*, 1971). Our present results indicate that in these meteorites the track densities are intermediate between those of the two groups of lunar breccias and that there are no microcrystallites in the grains. Therefore we might deduce that the heat metamorphism suffered by Weston and Kapoyeta has been very mild and this conclusion would agree with that of Mason and Melson (1970), as well as with that of the Berkeley group which reported the observation of very high track densities in Kapoyeta (Barber *et al.*, 1971b). On the other hand the SH values reported in Table 1 for the same meteorites are very high. Then meteoritic brecciation processes should have the peculiar properties of producing a stronger intergrain cementing, without involving a measurable degree of radiation damage annealing. A more intensive shock history could be invoked, thus explaining the striking micro-fracturing of the grains, but then the lack of impact glass in the fine grained matrix of Kapoyeta and Weston is unexplainable.

We want to point out that these conclusions are based on the hypothesis that the constituent grains of the gas-rich meteorites still show the solar flare track record that was produced before the compaction of the grains into the breccias. This hypothesis is firmly advocated by a number of groups (Lal and Rajan, 1969; Pellas *et al.*, 1969; Wilkening *et al.*, 1971; Barber *et al.*, 1971a). However we still think that further work is needed to definitively ascertain the validity of this hypothesis. In particular one should understand the marked differences in the track distributions registered in the olivines and in the other minerals as well as the high degree of dust sintering in the fine grained matrix of these meteorites.

THE COMPLEX REGOLITH HISTORY OF THE LUNA 16 DUST GRAINS

By using an analytical expression similar to that proposed to describe the thermal annealing of etchable tracks (Fleischer *et al.*, 1965), we deduce from our studies of microcrystallites in artificially heated dust grains that the activation energy for the growth of such crystallites is about 2 eV. Then by applying the same expression, it can be shown that the Luna 16–19 dust grains should have been exposed for about 10^{10} years in the lunar thermal cycle to develop crystallites with a size of about 500 Å. Therefore we conclude that the natural crystallites in the grains are not due to solar heating. It is also unlikely that they result from an interaction between the electron beam and the grains in the microscope, because their size is independent of the electron dose.

An alternative way to form crystallites in the Luna 16 grains would be to incorporate temporarily the grains in dust clods or in Apollo 11 type breccias, which could subsequently get eroded away by micrometeorite chipping of the surfaces or turn-over "grinding." This would imply either that the grains are "older" than those returned to the earth by the Apollo missions, or that the rate of formation of relatively nonmetamorphosed "regolith" breccias at Fecunditatis base was higher than at the Apollo landing sites. In both cases the Luna 16 grains should have on the average longer integrated residence times on the top surface of the regolith, and this feature could explain the higher concentration of radiation damage observed in micron-sized grains (Borg *et al.*, 1972b; Phakey and Price, 1972), as well as the very high track densities ($\rho \gtrsim 10^{10}$ tracks/cm^2) that we reported for 200 mesh aggregates of the same soil sample, which look "amorphous" in polarized light (Borg *et al.*, 1972b).

Another explanation could be that the Luna 16 dust soil results from the mixing of two components (Borg *et al.*, 1972b); (1) A "maria" component, representing about 50% of the grains and looking very similar to the most irradiated and eroded soil samples so far returned to the earth by the Apollo missions, and (2) an "exotic" component, constituted of grains showing both a higher concentration of radiation damage and evidence for a more extensive thermal metamorphic history, and that is generally much less abundant in the Apollo samples. Such a component might in fact represent a contamination of Fecunditatis base by dust blankets ejected from the bordering highlands. Therefore the Luna 20 highland soil samples should be examined for higher concentrations of radiation damage and microcrystallites, before definitively choosing a "brecciation" history for the Luna 16 dust particles.

THE MYSTERIOUS ALBEDO OF EXTRATERRESTRIAL BRECCIAS

The albedo of the lunar soil has been discussed by various authors (Adams and McCord, 1971, 1972; Borg *et al.*, 1971; Bibring *et al.*, 1972b; Dollfus *et al.*, 1971; Hapke *et al.*, 1970; Nash and Conel, 1971). This quantity can be considered as a multiparameter function of several variables including: the type and abundance of glassy objects; the mineralogical composition; the concentration of strongly absorbing ions like Ti and Fe; the existence of various types of coatings on the grains. Therefore the albedo of the fine grained matrix of "regolith type" meteoritic and lunar breccias, should also depend on the same parameters *possibly corrected for additional effects produced during the metamorphism of brecciation*, and these effects in turn could help in tracing back the history of brecciation.

In Table 1 it can be verified that the albedo of lunar breccias is grossly correlated to the two metamorphic groups, as defined from etched track studies, with the "nonmetamorphosed" breccias being the darkest; this result contradicts the observations of Quaide (1972). This correlation could then lend support to the hypothesis that both glass and radiation damage are the dominating factors in the albedo of lunar breccias, as these "soil" parameters are at first glance, the only ones that can be severely modified by heating. However, there are serious problems associated with this naive conclusion; in particular the albedos of the "metamorphosed" breccias are much greater than the highest soil albedo so far measured, for sample 14141 (OA = 24%). Therefore it could be argued that the optical properties of the lunar soil are not correlated to those of the breccias, and this conclusion would be further supported by the observation that there is no marked difference between the mineralogical composition of the dark breccia, 14049, and those of the light breccias, 14305 and 14321, as indicated by their similar Al/Si and Fe/Si ratios (Table 1, columns 8 and 9).

The problem of understanding the albedo of gas-rich meteorites is even more severe. These meteorites, such as the Kapoyeta howardite which will be discussed in this section, have a *dark-light* structure with the dark parts showing both higher concentrations of rare gases and higher track densities. The various albedo parameters of the lunar soil can hardly explain the albedo differences between the light and the dark parts of Kapoyeta because: (1) There are no glassy objects or grains with ultra-thin coatings in the fine grained matrix of this meteorite (Durrieu et al., 1971); (2) the mineralogical composition of the light and dark parts are certainly similar for the major minerals (Fredriksson and Keil, 1963); (3) there is no variation in the chemical composition of the two structures at levels greater than about 1%—however marked variations in the concentrations of trace elements have been reported (Reid, 1963); (4) there is no serious evidence for a marked difference in the grain size distribution between the two parts (Suess et al., 1964).

The only remaining parameters which could be potentially responsible for the dark-light structures in Kapoyeta are: (1) The concentrations of rare gases which seem, correlated to the darkening, as Fayetteville has the lowest albedo but also the highest concentration of rare gas (Suess et al., 1964); (2) however it could be argued that several meteorites have dark-light structures similar to those observed for Kapoyeta and Fayetteville but no rare gas enrichment has been observed in their dark parts. These two features could be associated with a shock implantation of rare gases in the grains, because such an implantation can explain both the rare gas content and the microfracturing in the grains as well as the darkening of the rocks (Fredriksson et al., 1964; Zahringer, 1966; Mazor and Anders, 1967); however there are serious difficulties with this model (Suess et al., 1964). Therefore the albedo of dark-light structures in meteorites remains a stumbling block for any albedo theory.

CONCLUSIONS

The breccias investigated in this work were generally very inhomogeneous. To minimize this severe difficulty, we attempted to make combined track, hardness, and albedo studies in the same area of a given breccia, but we are not certain that the properties and the correlations that we reported for small chunks can be generalized for the whole breccias. Furthermore the possibly nonmetamorphosed breccias have to be sorted out from the low track density group and shock features have to be clearly identified in the grains. Therefore some of our conclusions will certainly be subject to considerable revision in the future, as more results concerning these complex "breccia within breccias" will be available. However we believe that both the radiation damage "thermometry" and the index of "dust sintering" that we have defined in this report will help in understanding the important problems related to the formation of extraterrestrial breccias, when they are combined with petrographic observations and ion microprobe determination of the rare gas concentrations in the same working area of the breccias.

Acknowledgments—This work has been made possible by the generous cooperation of Professors G. Dupouy and F. Perrier from the Institut d'Optique Electronique du C.N.R.S., at Toulouse. The efficient help we received from R. Sirvin and A. Saur from the same Institution is gratefully acknowledged. One of us (M. Maurette) is deeply indebted to Drs. R. Klapisch and R. Walker for their enthusiastic support and interest, and also wishes to thank Dr. J. B. Adams for fruitful discussions concerning the albedo of lunar soil samples. The manuscript profited from the critical reviews of Drs. H. A. Hart and R. L. Fleischer. We are very grateful to Drs. Courtel and A. Dollfus for their generous help in the gathering of the scratch hardness and albedo data respectively. One of us (J. P. Duraud) is supported by a Ph.D. fellowship from the French Atomic Energy Commission. It is a pleasure to acknowledge the clever assistance of B. Vassent and C. Thibault. We also thank the National Aeronautics and Space Administration and the USSR Academy of Sciences for lending us lunar samples and the Centre National de la Recherche Scientifique for funding this research.

References

Adams J. B. and McCord T. B. (1971) Optical properties of mineral separates, glass, and anorthositic fragments from Apollo mare samples. *Proc. Second Lunar Sci. Conf., Geochim. Cosmochim. Acta* Suppl. 2, Vol. 3, pp. 2183–2195. MIT Press.

Adams J. B. and McCord T. B. (1972) Optical evidence for regional cross-contamination of highland and mare soils (abstract). In *Lunar Science—III* (editor C. Watkins), pp. 1–3, Lunar Science Institute Contr. No. 88.

Arrhenius G., Liang S., McDougall D., Wilkening L., Bhandari N., Bhat S., Lal D., Rajagopalan G., Tamhane A. S., and Venkatavaradan V. S. (1971) The exposure history of the Apollo 12 regolith. *Proc. Second Lunar Sci. Conf., Geochim. Cosmochim. Acta* Suppl. 2, Vol. 3, pp. 2583–2598. MIT Press.

Barber D. J., Hutcheon I., and Price P. B. (1971a) Extralunar dust in Apollo cores? *Science* **171**, 372–374.

Barber D. J., Cowsik R., Hutcheon I. D., Price P. B., and Rajan R. S. (1971b) Solar flares, the lunar surface, and gas-rich meteorites. *Proc. Second Lunar Sci. Conf., Geochim. Cosmochim. Acta* Suppl. 2, Vol. 3, pp. 2705–2714. MIT Press.

Bibring J. P., Duraud J. P., Durrieu L., Jouret C., Maurette M., and Meunier R. (1972a) Ultrathin amorphous coatings on lunar dust grains. *Science* **175**, 753–755.

Bibring J. P., Durrieu L., Jouret C., Eugster O., Maurette M., and Meunier R. (1972b) Solar wind implantation effects in the lunar regolith (abstract). In *Lunar Science—III* (editor C. Watkins), pp. 71–73, Lunar Science Institute Contr. No. 88.

Borg J., Dran J. C., Durrieu L., Jouret C., and Maurette M. (1970) High voltage electron microscope studies of fossil nuclear particle tracks in extraterrestrial matter. *Earth Planet. Sci. Lett.* **8**, 379–386.

Borg J., Durrieu L., Jouret C., and Maurette M. (1971) Ultramicroscopic features in micron-sized lunar dust grains and cosmophysics. *Proc. Second Lunar Sci. Conf., Geochim. Cosmochim. Acta* Suppl. 2, Vol. 3, pp. 2027–2040. MIT Press.

Borg J., Durrieu L., Jouret C., Lacaze J. C., Maurette M., and Peter J. (1972a) Search for low energy $(10 \lesssim E \lesssim 300 \text{ keV/amu})$ nuclei in space: Evidence from track and electron diffraction studies in lunar dust grains and in Surveyor III material (abstract). In *Lunar Science—III* (editor C. Watkins), pp. 92–94, Luna Science Institute Contr. No. 88.

Borg J., Durrieu L., Jouret C., Kashkarov L., Lacaze J., and Maurette M. (1972b) The high resolution radiation damage and texture record of the Luna 16–19 dust grains. Special report of the USSR Academy of Sciences on Lunar 16 studies (in press).

Cristophe-Michel-Levy M., Dran J. C., Duraud J. P., Legressus C., Maurette M., and Slodzian G. (1972) Extraterrestrial brecciation processes. In preparation.

Crozaz G., Haack U., Hair M., Maurette M., Walker R. M., and Woolum D. (1970) Nuclear track studies of ancient solar radiations and dynamic lunar surface processes. *Proc. Apollo 11 Lunar Sci. Conf., Geochim. Cosmochim. Acta* Suppl. 1, Vol. 3, pp. 2051–2080. Pergamon.

Crozaz G., Walker R. M., and Woolum D. (1971) Nuclear track studies of dynamic surface processes on the moon and the constancy of solar activity. *Proc. Second Lunar Sci. Conf., Geochim. Cosmochim. Acta* Suppl. 2, Vol. 3, pp. 2543–2558. MIT Press.

Dollfus A., Geake J. E., and Titulaer C. (1971) Polarimetric properties of the lunar surface and its interpretation. *Proc. Second Lunar Sci. Conf., Geochim. Cosmochim. Acta* Suppl. 2, Vol. 3, 2285–2300. MIT Press.

Dran J. C., Durrieu L., Jouret C., and Maurette M. (1970) Habit and texture studies of lunar and meteoritic materials with a 1 MeV electron microscope. *Earth Planet. Sci. Lett.* **9**, 391–400.

Dran J. C. and Duraud J. P. (1971) Brecciation processes in extraterrestrial matter. 34th annual meeting of The Meteoritical Society, Tubingen, Germany. Unpublished proceedings.

Dran J. C., Duraud J. P., and Maurette M. (1972) Low energy solar nuclear particle irradiation of lunar and meteoritic breccias. In *The Moon* (editors H. A. Urey and K. Runcorn), p. 227–241.

Durrieu L., Jouret C., Leroulley J. C., and Maurette M. (1971) Applications of high voltage electron microscopy to cosmophysics. *Jernkont. Ann.* **155**, 535–540.

Fleischer R. L., Price P. B., and Walker R. M. (1965) Effect of temperature, pressure and ionization on the formation and stability of fission tracks in minerals and glasses. *J. Geophys. Res.* **70**, 1497–1507.

Fleischer R. L., Maurette M., Price P. B., and Walker R. M. (1967) Origins of fossil charged-particle tracks in meteorites. *J. Geophys. Res.* **72**, 1–16.

Fleischer R. L., Hart H. R. Jr., and Comstock G. M. (1972) Particle track dating of mechanical events (abstract). In *Lunar Science—III* (editor C. Watkins), pp. 265–267, Lunar Science Institute Contr. No. 88.

Fredriksson K. and Keil K. (1963) The light-dark structure in the Pantar and Kapoyeta stone meteorites. *Geochim. Cosmochim. Acta* **27**, 717–739.

Fredriksson K., DeCarli P., Pepin R. O., Reynold J. H., and Turner K. (1964) Shock emplaced argon in a stony meteorite. *J. Geophys. Res.* **69**, 1403–1411.

Geiss J., Eberhardt P., Buhler F., Meister J., and Signer P. (1970) Apollo 11 and 12 solar wind composition experiments: Fluxes of He and Ne isotopes. *J. Geophys. Res.* **75**, 5972–5979.

Hapke B. W., Cohen A. J., Cassidy W. A., and Wells E. N. (1970) Solar radiation effects on the optical properties of Apollo 11 samples. *Proc. Apollo 11 Lunar Sci. Conf., Geochim. Cosmochim. Acta* Suppl. 1, Vol. 3, pp. 2199–2212. Pergamon.

Hintenberger H., Weber H. W., and Takaoka A. (1971) Concentrations and isotopic abundances of the rare gases in lunar matter. *Proc. Second Lunar Sci. Conf., Geochim. Cosmochim. Acta* Suppl. 2, Vol. 2, pp. 1607–1625. MIT Press.

Hutcheon I. D., Phakey P. P., Price P. B., and Rajan R. S. (1972) History of lunar breccias (abstract). In *Lunar Science—III* (editor C. Watkins), pp. 415–417, Lunar Science Institute Contr. No. 88.

Krishnaswami S., Lal D., Prabhu N., and Tamhane A. S. (1971) Olivines: Revelation of tracks of charged particles. *Science* **174**, 287–291.

Lal D. and Rajan R. S. (1969) Observations relating to space irradiation of individual crystals of gas-rich meteorites. *Nature* **223**, 269–271.

LSPET (Lunar Sample Preliminary Examination Team) (1971) Preliminary examination of lunar samples from Apollo 14. *Science* **173**, 681–693.

Mason B. and Melson W. (1970) *The Lunar Rocks*, p. 111. Wiley-Interscience.

Maurette M. (1970) On some annealing characteristics of heavy ion tracks in silicate minerals. *Rad. Effects* **5**, 15–19.

Mazor E. and Anders E. (1967) Primordial gases in the Jodzie howardite and the origin of gas-rich meteorites. *Geochim. Cosmochim. Acta* **31**, 1441–1456.

McKay D. S., Greenwood W. R., and Morrison D. A. (1970) Origin of small lunar particles and breccia from the Apollo 11 site. *Proc. Apollo 11 Lunar Sci. Conf., Geochim. Cosmochim. Acta* Suppl. 1, Vol. 1, pp. 673–694. Pergamon.

Naeser C. W. and Faul H. (1969) Fission track annealing in apatite and sphene. *J. Geophys. Res.* **74**, 705–710.

Naeser C. W., Engels J. C., and Dodge F. C. W. (1970) Fission track annealing and age determination of epidote minerals. *J. Geophys. Res.* **75**, 1579–1584.

Nash D. B. and Conel J. E. (1971) Luminescence and reflectance of Apollo 12 samples. *Proc. Second Lunar Sci. Conf., Geochim. Cosmochim. Acta* Suppl. 2, Vol. 3, pp. 2235–2244. MIT Press.

Pellas P., Poupeau G., Lorin J. C., Reeves H., and Audouze J. (1969) Primitive low-energy particle irradiation of meteoritic crystals. *Nature* **223**, 272–274.

Perelygin V. P., Tretiakova S. P., Shadieva N. H., Boos A. H., and Brandt R. (1968) On ternary fission produced in gold, bismuth, thorium, and uranium with argon ions. Dubna Joint Institute for Nuclear Research, Report No. 102.

Phakey P. P. and Price P. B. (1972) Extreme radiation damage in soil from Mare Fecunditatis. *Earth Planet. Sci. Lett.* **13**, 410–418.

Phakey P. P., Hutcheon I. D., Rajan R. S., and Price P. B. (1972) Radiation damage in soils from five lunar missions (abstract). In *Lunar Science—III* (editor C. Watkins), pp. 608–610, Lunar Science Institute Contr. No. 88.

Plendl J. N. and Gielisse P. J. (1962) Hardness of nonmetallic solids on an atomic basis. *Phys. Rev.* **125**, 828–832.

Price P. B., Hutcheon I. D., Lal D., and Perelygin V. P. (1972) Lunar crystals as detectors of very rare nuclear particles (abstract). In *Lunar Science—III* (editor C. Watkins), pp. 619–621, Lunar Science Institute Contr. No. 88.

Quaide W. (1972) Mineralogy and origin of Fra Mauro fines and breccias (abstract). In *Lunar Science—III* (editor C. Watkins), pp. 627–629, Lunar Science Institute Contr. No. 88.

Reed G. W. (1963) Heavy elements in the Pantar meteorite. *J. Geophys. Res.* **68**, 3531–3535.

Storzer D. and Wagner G. A. (1969) Correction for thermally lowered fission track ages of tektites. *Earth Planet. Sci. Lett.* **5**, 463–468.

Suess H. E., Wänke H., and Wlotzka F. (1964) On the origin of gas-rich meteorites. *Geochim. Cosmochim. Acta* **28**, 595–607.

Walker R. M. (1971) Fossil track studies in extra-terrestrial materials. *Rad. Effects* **3**, 239–247.

Warner J. L. (1972) Apollo 14 breccias: Metamorphic origin and classification (abstract). In *Lunar Science—III* (editor C. Watkins), pp. 782–784, Lunar Science Institute Contr. No. 88.

Wilkening L., Lal D., and Reid A. M. (1971) The evolution of the Kapoyeta howardite based on fossil track studies. *Earth Planet. Sci. Lett.* **10**, 334–340.

Zähringer J. (1966) Primordial helium detection by microprobe technique. *Earth Planet. Sci. Lett.* **1**, 20–22.

Proceedings of the Third Lunar Science Conference
(Supplement 3, *Geochimica et Cosmochimica Acta*)
Vol. 3, pp. 2905–2915
The M.I.T. Press, 1972

Radiation effects in soils from five lunar missions

P. P. Phakey,* I. D. Hutcheon, R. S. Rajan, and P. B. Price

Department of Physics, University of California,
Berkeley, California 94720

Abstract—Optical and high-voltage electron microscopy were used to study radiation-produced defects. There is no systematic decrease in solar flare track distributions in lunar fines even to a depth of 2.5 m. The soil gradually accumulates in thickness; stirring by impacts is usually on a small scale. The accumulation rate at Luna 16 site is far lower than at the Apollo sites; the mean track density in micron-size grains there exceeds $\sim 10^{12}/\mathrm{cm}^2$; more than 90% of those grains are metamict; the Luna 16 site is at high longitude where electrostatic charging and transport of soil by geomagnetic tail electrons are minimal. A large fraction of soil grains on sloping ground have been isotropically irradiated, suggesting a tumbling downhill motion. Radiation exposure times of soils increase in the order: Fra Mauro trench bottom and Cone Crater soil; Apennine Front; Scarplet; Spur Crater; Mare Tranquillitatis and Oceanus Procellarum; Fra Mauro; Luna 16. Solar wind gases quantitatively precipitate as ~ 80 Å gas bubbles at ~ 5000 atm pressure in soil grains.

Introduction

The distribution of radiation effects in lunar soil is governed both by dynamical processes such as the production, transportation, erosion, and impact-welding of soil and by the spectrum and time-variation of the sources of radiation. Radioisotopes, inert gas atoms, tracks, amorphous coatings, and thermoluminescent centers have been extensively discussed in previous lunar science conference proceedings.

We wish to discuss here those solid state defects whose distribution and interactions in soil grains generally occur on a scale so fine that they can be most profitably studied by transmission electron microscopy. We shall emphasize some of the ways in which these observations, coupled with optical microscope observations, extend our understanding of the moon.

Rate of Accumulation of Tracks and Soil at the Lunar Surface

In this section we try to understand the distribution of soil grains less than a few microns in diameter that contain remarkably high densities of tracks ($\gtrsim 10^{11}/\mathrm{cm}^2$) left by low energy ($\lesssim 1$ MeV/nucleon) heavy ions emitted by the sun. Studies of the gradient of Fe tracks in glass from the Surveyor III camera (Crozaz and Walker, 1971; Fleischer *et al.*, 1971; Price *et al.*, 1971) gave a 2.6 yr snapshot of the recent average energy spectrum of solar heavy ions in virtually noneroded material. On an energy/nucleon basis, Price *et al.* (1971) found that the Fe/He ratio was enhanced more than a factor of 10 relative to the values in the solar atmosphere. Subsequent studies with detectors on satellites (Mogro-Campero and Simpson, 1972; Lanzerotti

---

* On leave from Physics Department, Monash University, Victoria, Australia.

et al., 1972) have confirmed that heavy ion enhancements are a common feature of low-energy solar flare particles. We assume that the Fe/He ratio during the 2.6 yr period sampled by the Surveyor glass is equal to the value averaged over millions of years.

All rocks studied until very recently have been found to contain Fe-track density gradients about one power shallower than the gradient measured in the Surveyor glass (data summarized by Crozaz *et al.*, 1972). Assuming similar energy spectra for all solar flares and a constant rate of rock erosion, the track density gradients in rocks can be made compatible with the Surveyor spectrum if the erosion rate is ~ 3 to 10 Å/yr. The maximum track density previously observed in any rock was $7 \times 10^9/\text{cm}^2$ in rock 12022 (Barber *et al.*, 1971), a perfectly consistent saturation value governed by a surface eroding at a rate of ~ 3 Å/yr.

It is then pertinent to ask why micron-size grains in most soils contain more than 10^{11} tracks/cm^2 (Borg *et al.*, 1970; Barber *et al.*, 1971) and why grains in Luna-16 soil can contain up to an order of magnitude more tracks (Phakey and Price, 1972). Apparently soil grains escape the normal erosion processes to which rocks are subjected.

This question was resolved when Hutcheon *et al.* (1972a) discovered a fresh surface of an Apollo 15 rock, 15499, with a maximum track density of $\gtrsim 5 \times 10^{10}/\text{cm}^2$ and a gradient almost identical to the Surveyor gradient. The data were obtained in a mm-sized crystal at the bottom of a vug that had been covered with rock material until a chip was removed $\sim 2 \times 10^4$ yr ago. This time was inferred by matching the spectra in 15499 and in Surveyor glass and correcting for differences in solid angle and recording efficiency. There were no micrometeorite impact pits on the crystal in 15499. Hutcheon *et al.* concluded that the rate of erosion by other processes such as solar wind sputtering did not exceed ~ 1 Å/yr and that most soil grains, like this fresh rock surface, contain very high track densities because they have escaped micrometeorite impacts while on the top of the soil. In other words, we study tracks only in the survivors.

The data of Hutcheon *et al.* on 15499 indicate that the track accumulation rate in a surface soil grain of size up to ~ 4 μ was $\sim 5 \times 10^6$ (cm^{2}yr 2π sr)$^{-1}$ during at least the last $\sim 2 \times 10^4$ yr. In the top ~ 4 μ the spectrum is fairly flat. At depths greater than ~ 4 μ, the track accumulation rate falls off so rapidly that an observed depth-independent distribution of grains with extremely high track density, ρ, could only be achieved if each such grain were irradiated for a time $\tau \approx 2 \times 10^{-7}\rho$ yr within ~ 4 μ of the surface. Irradiations at greater depths are much less effective.

To what depth might all the micron-size soil grains be found to contain at least ρ tracks/cm^2, assuming that the soil is deposited in an optimum way such that each small grain has resided at the surface for the same time? The answer depends strongly on the steepness of the soil size distribution, which we take for simplicity to have the form $dn/dr = Br^{-\alpha}$. As a further optimizing assumption, based on the observation that big grains are usually coated with small grains, we allow the topmost layer always to be made up of grains with $r \leq 4$ μ. Then if the soil has been building up for a time T, the number of such layers that are filled with ρ tracks/cm^2 is $T/\tau \approx 5 \times 10^6 T/\rho$.

The maximum depth H so irradiated is given approximately by

$$H = \frac{(5 \times 10^6 T/\rho)(4 \times 10^{-4} \text{ cm/layer})}{\text{fraction of grains with } r < 4 \times 10^{-4} \text{ cm}}$$

$$= (2 \times 10^3 T/\rho) \left(\frac{r_{max}}{4 \times 10^{-4}}\right)^{4-\alpha}$$

with the power law grain size distribution cut off at $r_{max} \approx 0.1$ cm.

Gold et al. (1971) have found that for Apollo 11 and 12 soils, α varies from ~ 2.75 for grains smaller than $\sim 10 \mu$ to ~ 4 for grains larger than $\sim 10 \mu$. The Luna-16 soil has a mean value of $\alpha \sim 3.2$. T is ~ 3 to 4 G.y.

For $\rho \approx 10^{11}/\text{cm}^2$ (typical of Apollo 11, 12, and 15 soils) the depth H ranges from ~ 6 m if $\alpha = 3.5$ to ~ 100 m if $\alpha = 3$. For $\rho \gtrsim 10^{12}/\text{cm}^2$ (typical of Luna-16 soil), H would range from $\lesssim 0.6$ m if $\alpha = 3.5$ to $\lesssim 10$ m if $\alpha \doteq 3$.

On this model there is an inverse relationship between mean track density ρ and rate of soil accumulation H/T which can be compared with observations.

ELECTRON MICROSCOPE OBSERVATIONS OF RADIATION DAMAGE IN SOILS

Recently two of us (Phakey and Price, 1972) made a comparative study of radiation damage in soils from five lunar missions, using the Berkeley 650 keV electron microscope to obtain both micrographs and diffraction patterns of a large, random sample of micron-size grains from each soil. Since then we have studied samples at six depths in the Apollo 15 deep core, and all the results are summarized in Table 1. The grains are divided into three categories on the basis of relative radiation damage: (1) *Those with good single-crystal electron diffraction patterns.* Track densities are invariably high, ranging from $\sim 10^9/\text{cm}^2$ to unresolvably high ($> 3 \times 10^{11}/\text{cm}^2$). The latter show a mottled pattern of dark and light blobs with strong diffraction contrast, but the spots in the diffraction pattern are fragmented. (2) *Those with faint diffraction spots on a diffuse background and low contrast in the micrographs.* Both Phakey and Price (1972) and Borg et al. (1972) have attributed the diffuse scattering and the reduction in total number of diffraction spots to intense radiation damage. Expressed in terms of Fe tracks/cm^2, the lower limit for inclusion in this category would be $\sim 5 \times 10^{11}/\text{cm}^2$. (3) *Those with a diffraction pattern containing no spots but only a diffuse background and having an amorphous structure with no diffraction contrast.* In this category we exclude glass balls and other products of melting and rapid cooling. From the presence of crystallographic angles on some of the grains and the absence of shock features which would indicate that they are shock-produced, we believe we are correct in attributing the amorphous structure and diffuse diffraction pattern to extreme radiation damage. Until heavy ion calibration experiments are done, we cannot determine the minimum Fe-track density for inclusion in this category. It is likely to be several times $10^{12}/\text{cm}^2$.

Phakey and Price (1972) have already emphasized that the Luna-16 soil had a strikingly higher radiation exposure than did the others, as can be seen from the entries in Table 1. Very few grains from either the 7 cm or 30 cm layer of Luna-16

Table 1. Percentage of micron-size grains in the three radiation-damage categories.

	1	2	3
Diffraction pattern	Strong spots	Weak spots; diffuse scattering	No spots; diffuse scattering
Contrast features in micrograph	Tracks or blobs	Low contrast	Amorphous
Estimated track density (cm^{-2})	up to $\sim 5 \times 10^{11}$	5×10^{11} to few $\times 10^{12}$	$>$ few $\times 10^{12}$
10084	55%	29%	16%
12028	43%	39%	18%
14259	29%	45%	26%
15501	65%	24%	11%
15000			
41 cm	60%	27%	12%
83 cm	65%	15%	20%
124 cm	64%	12%	24%
165 cm	67%	21%	12%
207 cm	75%	12%	12%
246 cm	75%	16%	9%
Luna 16			
7 cm	7%	33%	60%
30 cm	9%	23%	68%

showed good, crystalline diffraction patterns and individually resolvable tracks. Most had been converted into nearly or completely amorphous grains. In category 1, the mean track density in Luna-16 grains was far higher than in grains from other soils.

A possible explanation of the high radiation exposure of the Luna-16 grains is that, being at high longitude near the east limb, they were not directly exposed periodically to the energetic electrons in the earth's magnetospheric tail. Laboratory experiments (T. Gold, private communication) have shown that electrons with similar energy (~ 1 keV) may stir and transport the soil by electrostatic charging. In this view, little stirring and transportation took place at Mare Fecunditatis and therefore the radiation damage should be concentrated in a thin layer. Near the front of the moon the radiation damage would be distributed over a greater thickness. This is consistent with the smaller thickness of the regolith at Mare Fecunditatis (Vinogradov, 1971). If we take the regolith thickness to be ~ 0.5 m there, we would expect on the basis of the discussion in the previous section to be able in 3 G.y. to populate the micron-size grains with $\sim 10^{12}$ to 10^{13} tracks/cm^2, depending on the value of α. There thus appears to be no "source" problem.

As a control experiment we have examined micron-size grains from a very young soil, 14141, which had a very mild radiation exposure as measured by the low quartile track density at the center of large (~ 100 μ) grains studied in the optical microscope and reported in Table 2. Practically all of the micron-size grains of 14141 fell into category 1, but track densities were observed up to 10^{11}/cm^2, which is consistent with our model in the previous section if we take the mean depth of the grains to be 2 cm (Crozaz et al., 1972). In other words, if we are to distribute micron-size grains with $\sim 10^{11}$ tracks/cm^2 to a depth of tens of meters in ~ 3 G.y., we must in ~ 10 m.y. irradiate the micron-size grains in several cm with up to $\sim 10^{11}$ tracks/cm^2.

Table 2. Summary of optical microscope observations of tracks in soils

	Sample	Site	Quartile track density at center of grains (cm^{-2})	Fraction of grains with > 10^8/cm^2 at edge of grain	Fraction of grains with optical gradients indicating isotropic irradiation
"active areas"	14141	Conelet	8 × 10^6	0.32	0.41
	15301	Spur	~10^8	0.7	?
	15271	Apennine	3 × 10^7	0.78	0.58
	15501	Scarplet	5 × 10^7	0.9	0.86
Mare surfaces	10084	Tranquil.	> 10^8	0.96	?
	12025	Procell.	> 10^8	0.88	?
	14259	Fra Mauro	> 10^8	high	?
	14148	Trench top	> 10^8	0.9	?
	14156	Trench mid.	> 10^8	0.9	?
	14149	Trench bot.	8 × 10^6	0	< 0.03
	15006	Hadley	> 10^8	0.95	?
	Luna 16	Fecund.	> 10^8	1.0	?

DEPTH DEPENDENCE OF RADIATION DAMAGE IN THE APOLLO 15 DEEP CORE

This core material may prove to be the single most valuable item yet returned from the moon. Although we showed in the second section that micron-size grains throughout the top 6 to 100 m of lunar regolith could contain ~10^{11} tracks/cm^2 if the deposition were uniform, we must admit that it was thrilling to discover high track densities distributed throughout the entire 250 cm drill core from the mare at Hadley.

Figure 1 shows the depth dependence of radiation damage determined from samples at six depths. The micron-size grains monitor the solar wind and low-energy ($\lesssim 1$ MeV/nucleon) solar flare particles. We define as metamict the grains in categories 2 and 3 from Table 1, which are those with track densities greater than ~5 × 10^{11}/cm^2. The grains with diameter greater than 50 μ monitor mainly the multi-MeV/nucleon solar flare particles. We etched large grains, both feldspars and pyroxenes, and used the optical microscope to observe the fraction with track densities greater than ~10^8/cm^2. Our observations indicate that any gradient is slight in the upper 250 cm of the mare at Hadley. Even at the greatest depth the mean track density in micron-size grains in category 1 is at least a few times 10^{10}/cm^2.

Both sets of data in Fig. 1 are consistent with a roughly uniform rate of soil deposition, with small-scale stirring required to insure that all samples contain their share of grains that were once on the very surface. We do not believe that a statistical mixing model (Comstock *et al.*, 1971), even with a very rapid cratering rate in the past, can reproduce both the shallowness of the observed depth gradient and the high mean track densities. The presence of 58 strata within the core also attests to the rarity of deep stirring (LSPET, 1972).

OPTICAL MICROSCOPIC COMPARISON OF RADIATION DAMAGE IN
"YOUNG" OR "ACTIVE" AREAS AND IN MARIA

Table 2 summarizes our optical microscopic data on track densities in etched feldspar grains with diameter $\gtrsim 50$ μ and on track density gradients at the rims of the grains. Because the optical microscope can only resolve track densities $\lesssim 10^8$/cm^2, it

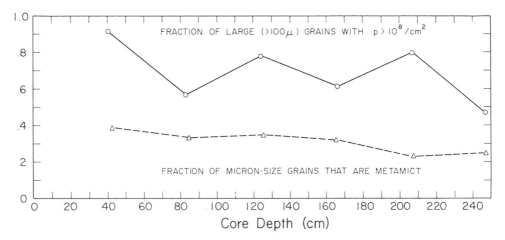

Fig. 1. Depth dependence of radiation damage in core 15001–15006. The circles refer to optical microscope observations on etched feldspars. The triangles refer to electron microscope observations of grains that have virtually lost their crystal structure (sum of entries in categories 2 and 3 in Table 1).

was possible with this technique only to search for gradients in grains with $\lesssim 10^8/\text{cm}^2$ at their center.

We see from the table that track densities in soils from rims of fresh craters and on sloping ground are considerably lower than in soils on level mare ground. Sample 14141 ("Conelet") was scooped from the rim of a fresh 10 m crater near the rim of Cone Crater and 15501 came from the rim of a 10 m crater near Scarp, on sloping ground near the edge of Hadley Rille. Sample 15271 came from sloping ground on the Apennine Front.

If the soil was thrown onto a crater rim from far enough below the surface that most of the tracks were formed after excavation, and if the event was recent enough that little vertical mixing has occurred, then we may estimate the age of the crater on the basis of the observed track density distribution. Arrhenius *et al.* (1971) have derived a relation between age T and lower quartile track density $\rho_{0.25}$ for a scoop of depth X, assuming no gardening before collection but a thorough mixing after collection:

$$T(\text{m.y.}) = \rho_{0.25}/[6.36 \times 10^{11}(7.5 + 0.3X)^{-6.15}] \qquad (1)$$

where X is the depth of the scooped layer. Thermoluminescence measurements have shown that $X \approx 4$ cm for scoop sample 14141 (Crozaz *et al.*, 1972), and if we use that value, equation (1) becomes

$$T(\text{m.y.}) = 10^{-6}\rho_{0.25}. \qquad (2)$$

We infer an age of ~ 8 m.y. for the 10 m crater on the rim of Cone, ~ 50 m.y. for the 10 m crater at Scarp and ~ 100 m.y. for Spur. Because of gradual soil mixing the latter two ages are quite uncertain, but at least their order is the same as that assigned by the field geology team on the basis of crater morphology.

The age of ~ 8 m.y. for the Cone sample indicates that the fresh 10 m crater was formed long after the Cone Crater event, which has been dated at ~ 25 m.y. ago on the basis of cosmic ray exposure ages of numerous breccias picked up from the vicinity of Cone Crater (Burnett et al., 1972). Our age of 14141 is consistent with that (~ 7 m.y.) reported by Bhandari et al. (1972) but much lower than that (~ 18 m.y.) assigned by Crozaz et al. (1972).

A large fraction of grains from Conelet, the Apennine front and Scarplet, including those several 100 μ in diameter, have strong track density gradients on all surfaces showing that they were irradiated isotropically (last column of table), whereas grains from the bottom of the trench at Fra Mauro site show no gradients. The track densities in the other soils are too high to be studied optically. Other workers, using a scanning electron microscope, have previously found that no more than $\sim 20\%$ of Apollo 11 and 12 soil grains with high track densities ($\gtrsim 4 \times 10^8/\text{cm}^2$) had track gradients and that isotropically irradiated grains were quite rare (Crozaz et al., 1971; Lal et al., 1970). We therefore must explain why only grains from certain soils were irradiated uniformly on all sides.

We believe it is significant that the soils from Conelet, the Apennine Front and Scarplet were collected from surfaces with greater than $10°$ slopes. Gradual downhill transportation, driven by gravity and perhaps assisted by electrostatic charging, could insure that each grain was barbecued on all sides while at the very surface of the soil. Grains on level mare surfaces would tend not to rotate as readily. For example, the sample from the trench bottom, which was on level ground, contained no grains with gradients.

The case for a parent-body origin of the gas-rich brecciated meteorites should now perhaps be reconsidered. The weight of current opinion is that this class of meteorite was formed by accretion of grains in space. The strongest evidence is the existence of isotropically irradiated grains inside the gas-rich meteorites. Our observations show that isotropic irradiation can also occur on a parent body, given a suitable driving force for rotation of soil grains.

We have tacitly assumed throughout this paper that the majority of tracks in soil grains were produced by heavy ions, mainly Fe, in solar flares. High local concentrations of uranium can contribute more than 10^8 fission tracks/cm^2 at boundaries and in inclusions within feldspar and pyroxene grains, but we have never seen uniform distributions of fission tracks at a level high enough in common minerals to be confused with solar flare tracks.

ELECTRON MICROSCOPE OBSERVATIONS OF SOLAR WIND GAS BUBBLES

The concentration of gases, mainly solar wind H and He, in lunar fines and breccias is typically ~ 0.1 to ~ 1 cm^3 STP/g or $\sim 10^{19}$ to $\sim 10^{20}$ atoms/cm^3 of rock. This far exceeds the solubility limit of gases in typical minerals and one might expect them to precipitate as bubbles if the temperature is high enough for limited diffusion to occur. Either as point defects or as bubbles their presence should be taken into account in considering the ways in which radiation can affect the lunar surface.

Some time ago Barber and Price (1971) reported the observation of ~ 100 Å gas

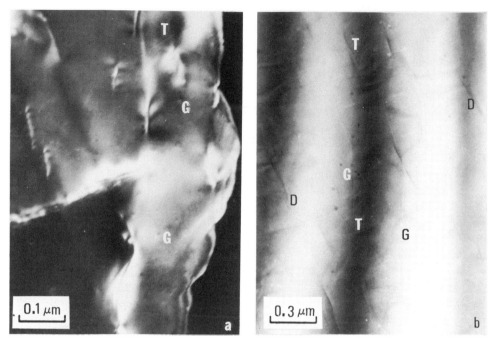

Fig. 2. High-voltage (650 keV) electron micrographs showing solar wind gas bubbles in (a) a micron-size grain from Cone Crater soil 14141; and (b) in a larger grain from bulk fines 10084 that had been ion-beam-thinned. Because of their small size (diameter ≲ 100 Å), the bubbles are difficult to see in the print except near thickness fringes. The grains contain ~10^{15} bubbles/cm³. Examples of tracks, gas bubbles and antiphase domain boundaries are labeled T, G, and D.

bubbles concentrated in the outer few microns of Apollo 11 and 12 soil grains that had been embedded in epoxy and ion-beam-thinned for high-voltage electron microscopy study.

We have observed these bubbles in micron-size grains of all fines in which the track density was low enough ($\lesssim 10^{11}$/cm²) not to obscure the bubbles. In the present work the grains were examined without any thinning, so the bubbles cannot be attributed to artifacts introduced by an ion beam. Figure 2 shows examples of gas bubbles in Cone Crater soil and in Apollo 11 bulk soil. We have also observed them in several breccias (Hutcheon *et al.*, 1972b) and in the track-rich grains within gas-rich meteorites. We have verified by the change in their contrast across a diffraction fringe that they are indeed *voids* and not *solid* precipitates.

The size and number of bubbles per unit volume that we see in a typical soil grain lead us to conclude that the majority of the gas in most small grains may be in the form of bubbles. If the only stresses on a bubble are the gas pressure p inside it and the surface energy γ of the wall, the pressure is related to the bubble radius by $p = 2\gamma/r$. For a typical bubble radius of ~40 Å and for $\gamma \approx 1000$ erg/cm², the pressure is ~5000 atm and the density of gas atoms in the bubble is $n = 2\gamma/rkT \approx 10^{23}$/cm³ or ~20,000 atoms/bubble. In small grains the number of bubbles per unit volume is typically $\sim 10^{15}/cm^3$, so that the total amount of gas in the grains in the form of bubbles is $\sim 2 \times 10^{19}/cm^3$, consistent with the densities measured by mass spectrometry on bulk samples and attributed to implanted solar wind gas atoms. This estimate is crude because it depends on the cube of the bubble radius, which is rather difficult to measure, and on the foil thickness, which is hardly known to within 50%.

In silicate grains up to ~ 2 microns in diameter we have not noticed any gradient in bubble density, whereas in grains larger than ~ 10 microns there is a noticeable decrease with depth. When subjected to a driving force F, bubbles should migrate with a velocity

$$v = \frac{D_s}{8\pi kT}\left(\frac{a_0}{r}\right)^4 F \tag{3}$$

where a_0 is a lattice parameter and D_s is the coefficient of surface diffusion of atoms of the solid around the walls of the bubble (Barnes and Mazey, 1963). To see whether bubble migration might account for a smoothing out of the gas concentration on a micron scale, we substitute the values $v = 1\ \mu/10^6$ yr, $T \approx 400°K$, $r \approx 40$ Å, $a_0 \approx 2$ Å, and $D_s \approx 10^{-13}\ cm^2/sec$ and calculate the required force. In the absence of any data for D_s in minerals, we have chosen the value at $400°K$ of the coefficient for volume diffusion of deuterium in quartz (Kats, 1962). We find that forces as small as $\sim 6 \times 10^{-12}$ dyne are sufficient to move bubbles a micron per million years. Such a force could result, for example, from a temperature gradient of $\sim 1°/cm$.

In solids with a very low value of D_s at lunar surface temperature, the distribution of bubbles may follow closely the original distribution of gas, which would normally be highest within the outer $\sim 0.1\ \mu$ of a grain. If the coefficient of volume diffusion of gas atoms is also extremely small, the bubbles may not even nucleate unless the temperature of the solid is temporarily raised by a nearby impact. These considerations may apply to ilmenite, for which the gas is known to be strongly concentrated in the outer few tenths of a micron (Eberhardt et al., 1970). The level of concentration at which as much gas diffuses out of a grain as is introduced may be determined not by volume diffusion but by migration of bubbles due to their mutual repulsion.

Bubbles and tracks may affect each other. We have found that the temperature for track annealing appears to depend on the extent of radiation damage of the crystal. For example, heavy ion tracks can be annealed out of terrestrial bytownite at $< 550°C$, out of large bytownite crystals from breccia 14311 at $\sim 625°C$, and out of bytownite from a heavily irradiated soil only at temperatures above $\sim 675°C$. Gas bubbles or interstitial gas atoms in a soil crystal may deposit along a track and help to stabilize it against annealing.

Tracks that intersect a grain surface may allow gas in bubbles to escape readily at low temperature in stepwise heating studies for mass spectrometry. In particular, Kaiser (1972) found that much of the gas in Luna 16 soil was released at unusually low temperature. One might also expect that the kinetics of gas release would depend on the total gas concentration. If low, the gas might all be present as interstitial atoms whose escape rate depends on *volume* diffusion; if high, it might have precipitated into bubbles whose migration rate depends on *surface* diffusion, i.e., diffusion by motion of material around the internal surface of a bubble.

If a grain is heated to a fairly high temperature, bubbles can coalesce into larger bubbles by migration. In some soils we have seen bubbles up to several 100 Å in diameter. Heymann and Yaniv (1971) have found that considerable gas can be released at room temperature in samples that are pulverized within the vacuum system and have suggested that the gas was present in voids.

Conclusions

(1) In the Apollo 15 deep core, the fraction of large ($>100\ \mu$) grains with high track densities and the fraction of micron-size grains that are metamict show little dependence on depth. Because of the steep dependence of track production rate on depth, grains with high track density *must* acquire them during the short interval that they reside on top of the soil. A mixing model, even a "fast" one, fails to account for the independence of *high* track densities on depth. The soil must gradually accumulate at a rate proportional to (mean track density)$^{-1}$.

(2) About half of all the finest soil grains ($\lesssim 4\ \mu$) from the Apollo missions are metamict as a result of irradiation by solar heavy ions. For Luna 16 the metamict fraction exceeds 90%. The accumulation rate of Luna 16 soil is inferred to be much lower than for the Apollo soils, perhaps because Mare Fecunditatis is less exposed to geomagnetic tail electrons that cause stirring and assist in transportation of soil.

(3) The fraction of large soil grains that have been isotropically irradiated is much greater in soil from sloping ground (rims of craters; Apennine Front; edge of Hadley Rille) than in soil from mare surfaces, suggesting that downhill motion is responsible for grain rotation.

(4) The age of the Cone Crater soil (14141) is ~ 8 m.y.; the Scarplet (15501) and Spur Crater (15301) soil are much older. The soil at the bottom of the Fra Mauro trench could not have been exposed on the lunar surface for more than a few million years before it was buried by more soil.

(5) Gas bubbles ~ 80 Å in diameter and at ~ 5000 atm pressure are present in soil grains in a concentration sufficient to account for much, perhaps most, of the gas detected by mass spectrometry. These bubbles may have detectable effects in mass spectrometric and track experiments.

Acknowledgment—This research was supported by NASA Grant NGL 05-003-410.

References

Arrhenius G. *et al.* (1971) The exposure history of the Apollo 12 regolith. *Proc. Second Lunar Conf.*, *Geochim. Cosmochim. Acta* Vol. 3, 2583–2598. MIT Press.

Barber D. J., Cowsik R., Hutcheon I. D., Price P. B., and Rajan R. S. (1971) Solar flares, the lunar surface, and gas-rich meteorites. *Proc. Second Lunar Sci. Conf.*, *Geochim. Cosmochim. Acta* Vol. 3, pp. 2705–2714. MIT Press.

Barber D. J. and Price P. B. (1971) Solar flare particle tracks in lunar and meteoritic minerals. *Proc. 25th Anniversary Meeting of EMAG*. Institute of Physics (U.K.).

Barnes R. S. and Mazey D. J. (1963) The migration and coalescence of inert gas bubbles in metals. *Proc. Roy. Soc.* (*London*) A **275**, 47–57.

Bhandari N. *et al.* (1972) Collision controlled radiation history of lunar regolith (abstract). In *Lunar Science—III* (editor C. Watkins), pp. 68–70, Lunar Science Institute Contr. No. 88.

Borg J., Dran J. C., Durrieu L., Jouret C., and Maurette M. (1970) High voltage electron microscope studies of fossil nuclear particle tracks in extra-terrestrial matter. *Earth Planet. Sci. Lett.* **8**, 379–386.

Borg J., Maurette M., Durrieu L., Jouret C., and Peter P. (1972) Search for low energy nuclei in space: Evidence from track and electron diffraction studies in lunar dust grains and in Surveyor III material (abstract). In *Lunar Science—III* (editor C. Watkins), pp. 92 94, Lunar Science Institute Contr. No. 88.

Burnett D. S., Huneke J. C., Podosek F. A., Russ G. P., Turner G., and Wasserburg G. J. (1972) The irradiation history of lunar samples (abstract). In *Lunar Science—III* (editor C. Watkins), pp. 105–107, Lunar Science Institute Contr. No. 88.

Comstock G. M., Evwaraye A. O., Fleischer R. L., and Hart H. R. (1971) The particle track record of lunar soil. *Proc. Second Lunar Sci. Conf., Geochim. Cosmochim. Acta* Vol. 3, pp. 2569–2582. MIT Press.

Crozaz G. and Walker R. M. (1971) Solar particle tracks in glass from the Surveyor 3 spacecraft. *Science* **171**, 1237–1239.

Crozaz G., Walker R. M., and Woolum D. (1971) Nuclear track studies of dynamic surface processes on the moon and the constancy of solar activity. *Proc. Second Lunar Sci. Conf., Geochim. Cosmochim. Acta* Vol. 3, pp. 2543–2558. MIT Press.

Crozaz G., Drozd R., Hohenberg C. M., Hoyt H. P., Ragan D., Walker R. M., and Yuhas D. (1972) Solar flare and galactic cosmic ray studies of Apollo 14 samples (abstract). In *Lunar Science—III* (editor C. Watkins), pp. 167–169, Lunar Science Institute Contr. No. 88.

Eberhardt P., Geiss J., Graf H., Grögler N., Krähenbühl U., Schwaller H., Schwarzmüller J., and Stettler A. (1970) Trapped solar wind noble gases, exposure age and K/Ar age in Apollo 11 lunar fine material. *Proc. Apollo 11 Lunar Sci. Conf., Geochim. Cosmochim. Acta* Suppl. 1, Vol. 2, pp. 1037–1070. Pergamon.

Fleischer R. L., Hart H. R., and Comstock G. M. (1971) Very heavy solar cosmic rays. Energy spectrum and implications for lunar erosion. *Science* **171**, 1240–1242.

Gold T., O'Leary B. T., and Campbell M. (1971) Some physical properties of Apollo 12 lunar samples. *Proc. Second Lunar Sci. Conf., Geochim. Cosmochim. Acta* Vol. 3, pp. 2173–2181. MIT Press.

Heymann D. and Yaniv A. (1971) Breccia 10065: Release of inert gases by vacuum crushing at room temperature. *Proc. Second Lunar Sci. Conf., Geochim. Cosmochim. Acta* Vol. 2, pp. 1681–1692. MIT Press.

Hutcheon I. D., Phakey P. P., and Price P. B. (1972a) To be published.

Hutcheon I. D., Phakey P. P., and Price P. B. (1972b) History of lunar breccias. *Proc. Third Lunar Sci. Conf.* (this volume).

Kaiser W. A. (1972) Rare gas studies in Luna-16-G-7 fines by stepwise heating technique. A low fission solar wind Xe. *Earth Planet. Sci. Lett.* **13**, 387–399.

Kats A. (1962) Hydrogen in alpha-quartz. Philips Research Reports No. 17, pp. 201–279.

Lal D., MacDougall D., Wilkening L., and Arrhenius G. (1970) Mixing of the lunar regolith and cosmic ray spectra: Evidence from particle-track studies. *Proc. Apollo 11 Lunar Sci. Conf., Geochim. Cosmochim. Acta* Suppl. 1, Vol. 3, pp. 2295–2303. Pergamon.

Lanzerotti L. J., Maclennan C. G., and Graedel T. E. (1972) Enhanced abundances of low energy heavy elements in solar cosmic rays. *Ap. J. Lett.* **173**, L39.

LSPET (Lunar Sample Preliminary Examination Team) (1972) Preliminary examination of the lunar samples from Apollo 15. *Science* **175**, 363–375.

Mogro-Campero A. and Simpson J. A. (1972) Enrichment of very heavy nuclei in the composition of solar accelerated particles. *Ap. J. Lett.* **171**, L5.

Phakey P. P. and Price P. B. (1972) Extreme radiation damage in soil from Mare Fecunditatis. *Earth Planet. Sci. Lett.* **13**, 410–418.

Price P. B., Hutcheon I. D., Cowsik R., and Barber D. J. (1971) Enhanced emission of iron nuclei in solar flares. *Phys. Rev. Lett.* **26**, 916–919.

Vinogradov A. P. (1971) Preliminary data on lunar ground brought to earth by automatic probe "Luna-16." *Proc. Second Lunar Sci. Conf., Geochim. Cosmochim. Acta* Vol. 1, pp. 1–16. MIT Press.

Proceedings of the Third Lunar Science Conference
(Supplement 3, *Geochimica et Cosmochimica Acta*)
Vol. 3, pp. 2917–2931
The M.I.T. Press, 1972

Solar flare and galactic cosmic ray studies of Apollo 14 and 15 samples

G. Crozaz,* R. Drozd, C. M. Hohenberg, H. P. Hoyt, Jr.,
D. Ragan, R. M. Walker, and D. Yuhas

Washington University, Laboratory for Space Physics,
St. Louis, Missouri 63130

Abstract—Thermoluminescence (TL) measurements in rock 14310 show a strong depth dependence consistent with that expected from solar flares. This effect should prove useful in studying solar flare fluctuations in the time interval 10^2 to 10^5 years. Solar flare track data in 14310 are similar to those previously measured and support the interpretation that the long term solar flare spectrum for heavy nuclei can be represented by $dN/dE = KE^{-2.6}$. Rare gas spallation ages for rock 14301, 14306, and 14311 are respectively 102 ± 30, 25 ± 2, and 661 ± 72 m.y. The 14306 value supports the idea that Cone Crater was formed 25 million years ago. Groupings of exposure ages suggest the dates of other major cratering events. Galactic track data in 14310 show little depth dependence. If the present cosmic ray spectrum is accepted, this implies that solar flare produced Mn^{53} ($T_{1/2} = 3.7 \times 10^6$ yr) should be undersaturated at the surface. Another possible interpretation is that galactic cosmic rays were more heavily modulated in the past at 1 a.u. Different soil samples show very different irradiation histories. Track data in the coarse fraction of the Cone Crater sample suggest that this material has been relatively undisturbed since its formation. Track densities increase with decreasing grain size in a way that suggests that the small crystals were derived by micrometeorite bombardment of initially coarser grained material. The bottom of the trench appears lightly irradiated but this sample is something of a puzzle. *All* positions of the Apollo 15 deep drill are heavily irradiated. Combined with neutron data (Burnett *et al.*, 1972), this shows that solar flares were active 5×10^8 years ago.

Introduction

The lunar rocks and soil preserve a record of their irradiation by energetic particles from the sun and the galaxy. The record can be used to learn about the past history of these radiations or, conversely, about the history of different lunar samples.

In this paper we present data from three different experimental approaches: track studies, rare gas studies, and thermoluminescence (TL). Tracks are produced directly by slowing down heavy particles present in the original beams striking the samples. Both rare gases and TL on the other hand are generated by secondary processes—the former by nuclear interactions and the latter by ionization. Because of these differences each of the techniques gives unique information about the radiation history of the moon.

Solar Flare Effects in Lunar Rocks

Low energy solar flare particles dominate the radiation history of the near surface regions of lunar rocks. Measurements of the depth dependence of different radio-isotopes (Finkel *et al.*, 1971; Rancitelli *et al.*, 1971) have previously been used to show

---

* Permanent Address: Université Libre de Bruxelles, Bruxelles, Belgium.

that the energy spectrum and absolute flux of solar flare particles averaged over the last 10^6 yr (Al^{26}) is about the same as that averaged over the last 2.6 yr (Na^{22}).

We present here a new method for studying solar flare effects in a time span (10^2 to 10^5 yr) previously inaccessible to radioactivity measurements. The method will be described in more detail in a separate publication (Hoyt *et al.*, 1972a) and only a brief summary will be given here.

The ionization produced by charged particles in crystals liberates electrons and holes, some of which are trapped in metastable states associated with such lattice defects as impurity sites. Heating of a sample frees the trapped charges, and in crystals with appropriate activation centers these charges can recombine with the emission of visible light. Previous work on this phenomenon of thermoluminescence (TL) has shown that at glow curve temperatures above $\sim 550°C$ the TL in lunar samples is near saturation. Below this temperature the stored TL is nonsaturated, and the level is determined by a dynamic equilibrium between the rate of ionization and the rate of thermal decay due to the high ambient lunar temperature (Dalrymple and Doell, 1970; Doell and Dalrymple, 1971; Hoyt *et al.*, 1970; Hoyt *et al.*, 1971).

The amplitude of the diurnal temperature fluctuations is attenuated exponentially with depth and in solid rock decreases by $1/e$ in ~ 70 cm (Hoyt *et al.*, 1971). Consequently, one would not expect significant gradients in rock fragments $\ll 70$ cm in size imbedded in soil of low thermal conductivity. Our measurements are made in rocks with characteristic dimensions < 10 cm and thermal gradients can be ignored (Hoyt *et al.*, 1972a). Thus TL measurements in rocks are expected to indicate the dose rate and near the surface will be dominated by solar flares. In Fig. 1 we show the normalized TL output as a function of depth in rock 14310. Each point represents the average of several determinations. A similar depth variation of TL exists for 12063.

Fitting the observed data to a solar flare differential energy spectrum requires a knowledge of the kinetics of recovery which at this time is not complete. The data in Fig. 1 can be fit assuming first order kinetics and a differential energy spectrum of the form $dN/dE = KE^{-2}$ (Hoyt *et al.*, 1972a). If second-order kinetics had been assumed, the exponent in the energy spectrum would have doubled. These spectral parameters are within the expected limits, and we conclude that the TL as a function of depth is what one would expect from solar flares.

Much work remains to be done in order to extract the maximum physical information from TL studies in rocks. In a companion paper (Hoyt *et al.*, 1972b) we show that the rather broad glow curve of bulk material can be separated into component glow peaks by measuring the TL of individual grains. Careful measurements of both stored TL and the kinetics of recovery in separated mineral fractions from different rocks should permit investigation of solar flare fluctuations in a time interval of 10^2 to 10^5 yr (as estimated from previous data, Hoyt *et al.*, 1971).

We and others (Crozaz *et al.*, 1971; Fleischer *et al.*, 1971a, 1971b; Barber *et al.*, 1971; Bhandari *et al.*, 1971; Crozaz *et al.*, 1970; Fleischer *et al.*, 1970; Lal *et al.*, 1970; Price and O'Sullivan, 1970) have previously reported studies of the variation of track density of VH nuclei in the surface regions of rocks. In Fig. 2 we show a log-log plot of selected data including our new work on 14310. The data for this rock were obtained by measuring from the highest hillock on a rock section which included a total of ~ 0.2 cm^2 of surface area.

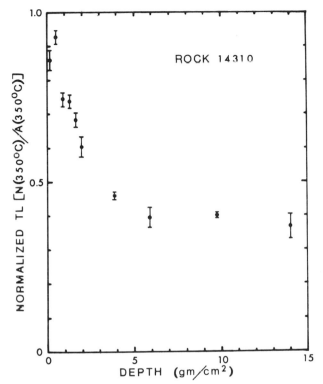

Fig. 1. Normalized TL as a function of depth in rock 14310. The points were obtained by dividing the natural TL intensity at 350°C by the intensity found in a second heating of the sample after an irradiation of 38 krad.

We have previously shown that the track density versus depth is well represented by a power law with an exponent of ~1 in the region 100 μ to 1000 μ. As can be seen in Fig. 2 the data for both 14310 and 12063 drop below the power law at depths <100 μ. If we assume that the drop off is due to a thin dust covering at the surface, and arbitrarily require that the thickness be such as to restore the simple power law dependence, we obtain the resultant slopes shown in Table 1.

Table 1. Summary of depth dependence of track density in selected lunar rocks
(100 μ to 1000 μ)

Rock	α_1	α_2	Absolute ρ at 300 μ
10057	1.04	1.14	1.1×10^8
12022*	1.10	1.21	4.4×10^7
12063	0.85	(0.91) 1.21†	3.3×10^7
14310	0.86	(1.10) 1.28†	2.5×10^7

α is the coefficient in the power law spectrum with depth, $\rho = \rho_0 D^{-\alpha}$
α_1 is calculated from raw data.
α_2 is corrected for galactic contribution.
* From Barber et al., 1971.
† 50 μ dust layer assumed on surface, values without dust layer are shown in parentheses.

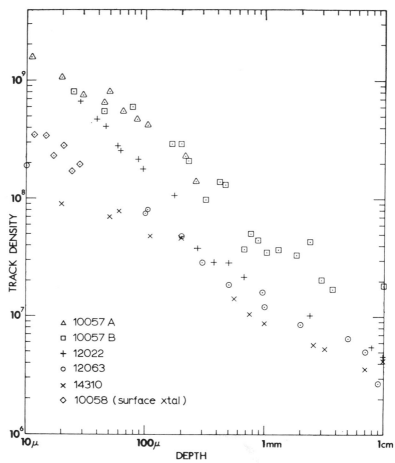

Fig. 2. Variation of track density (tracks/cm²) with depth in selected lunar rocks. The tracks in this depth range are produced predominantly by solar flare particles. The data for 12022 are taken from Barber *et al.*, 1971.

Table 2. Comparison of long and short term energy spectra.

	α_e	α_c	γ
Rocks (10 μ to 1000 μ)	1.20 ± 0.05	2.20 ± 0.05	2.6 ± 0.1
Surveyor (10 μ to 300 μ)	2.5 ± 0.2	2.5 ± 0.2	2.9 ± 0.2

Rocks 10057, 12022, 12063 and 14310 corrected for galactic contribution and for assumed 50 μ dust covering on 12063 and 14310.

α is the coefficient in the power law spectrum with depth, $\rho = \rho_0 D^{-\alpha}$, α_e is the experimental value, α_c has been corrected for rock erosion.

γ is the coefficient in the power law spectrum with energy, $dN/dE = AE^{-\gamma}$.

Although there is general agreement for different rocks, the average slope is very different from that found in a sample of Surveyor glass. As previously reported, we attribute the difference to lunar erosion which has the effect of changing the slope by one unit (Crozaz *et al.*, 1971; Fleischer *et al.*, 1971a). We also previously showed that a power law dependence of track density versus depth implies a power law in the energy spectrum of the slowing down particles. As shown in Table 2, the spectral index, γ, for the long term average spectrum $dN/dE = KE^{-\gamma}$ may be slightly harder than that previously found for the Surveyor glass.

In view of the various possible uncertainties, we consider the agreement in the experimental values for different rocks, measured in different laboratories, to be rather remarkable. Since 14310 agrees quite well with other rocks previously discussed (Barber *et al.*, 1971; Crozaz *et al.*, 1970; Crozaz *et al.*, 1971; Crozaz *et al.*, 1972b; Fleischer *et al.*, 1971a, 1971b), the same micro-erosion rate of 3×10^{-8} cm/yr to 10^{-7} cm/yr is implied by these data.

GALACTIC COSMIC RAYS IN ROCKS

In Table 3 we give data on the spallation ages of rocks 14301, 14306, and 14311, as measured by the Kr^{81}–Kr method. As discussed in a companion paper (Crozaz *et al.*, 1972a), rock 14301 has both a large solar component and a large fission excess due to Pu^{244} decay. Rocks 14306 and 14311 are dominated by spallation gases with little or no solar gas and no striking evidence for excess fission xenon. The Kr^{81}–Kr ages can therefore be computed with higher confidence for these rocks. Rock 14311 has an exposure age of 660 m.y. which is among the longest observed for any lunar sample.

The close agreement between the low exposure age of $25 \pm 2 \times 10^6$ yr for rock 14306 with similar ages for 14083, 14167, 14053, and 14321, from the flank of Cone Crater (Turner *et al.*, 1971; Lugmair and Marti, 1972), is further evidence that these rocks were simultaneously ejected by the Cone Crater event ~ 25 m.y. ago. It will be interesting to see if other groupings of exposure ages develop which may date different cratering events. At the present, three such groups seem to be suggested: rocks 14311 and 14001 (fragment) and a 14160 peanut have exposure ages in the 600–700 m.y. range; three 14160 peanuts, 14310, and a fragment from 14001 have ages in the 250–350 m.y. range; 14073 and 14301 have similar ages of about 100 m.y. (Burnett *et al.*, 1972; Lugmair and Marti, 1972).

Detailed three-dimensional measurements of the track profiles in rock 14310 are given elsewhere as part of a consortium report (Yuhas *et al.*, 1972). In Fig. 3 we show a comparison of our measurements, which were made on a vertical section running from the top to the center, with different exposure models. If we simply take the track density at the center of the rock and assume that the rock has accumulated all of its tracks while in its present position, we would calculate a surface exposure age of 15.7×10^6 yr.

Table 3. Cosmic ray exposure ages for Apollo 14 breccias (14306 has a pronounced light-dark structure; the two types of structure were individually studied).

Sample	$Kr^{81} - Kr^{83}$ Exposure age
14301	102 ± 30 m.y.
14306 (light)	25.4 ± 2.9
14306 (dark)	23.4 ± 1.4
14311	661 ± 72 m.y.

G. Crozaz *et al.*

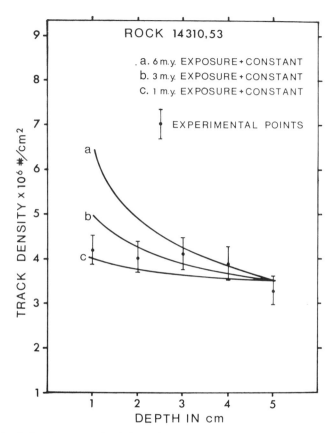

Fig. 3. Track density versus depth in rock 14310. The solid lines are theoretical curves representing a two stage irradiation history; the most recent being at the surface giving rise to a steep gradient, and the earlier an irradiation at deep depths which produces no gradient.

However, our track production model predicts a depth dependence that is totally at variance with what is observed. We must conclude either that the rock has had a complex irradiation history or that the cosmic ray spectrum that we have used to fit the data is wrong. There also exists the outside possibility that the lack of a depth dependence is due to tracks produced by an exotic fission component in the U-poor (<30 ppb) feldspar. The results are closely similar to those that we previously reported for 12063 (Crozaz *et al.*, 1971).

A complex irradiation history is not unreasonable for 14310 in view of its high spallation age and high neutron dose (Burnett *et al.*, 1972; Lugmair and Marti, 1972). In Fig. 3 we show fits to the data assuming that the tracks were accumulated in two epochs of exposure. One component is assumed to have been produced while the rock was either buried or was part of a much larger object (and thus shows no depth dependence). The other component is assumed to have been accumulated while 14310 was exposed in its present position. The lack of a large track gradient leads one to conclude that the rock could not have been in its present position for more than 3 m.y. The best fit is obtained with a "true surface exposure" age of 1.5–2.0×10^6 yr. This is consistent with the measurements of Rancitelli *et al.* (1972) who conclude that Al^{26} is saturated by solar flares at the surface.

This modest exposure leads to the prediction that Mn^{53} ($T_{1/2} = 3.7 \times 10^6$ yr) should be undersaturated at the surface. Although the result may well be confused by the effects of rock erosion (Finkel *et al.*, 1971), we believe that the measurement of the Mn^{53} concentration in a surface sample is extremely important.

The cosmic ray spectrum that we have used to calculate the absolute track densities and their variation with depth is the following:

$$\left.\begin{aligned}\frac{dN}{dE} &= 2.4 \times 10^6 & E &< 400 \text{ MeV/nuc} \\[2mm] &= \frac{5.0 \times 10^6}{(0.94 + E/1000)^{2.5}} & E &> 400 \text{ MeV/nuc}\end{aligned}\right\} \text{Part./cm}^2\text{-sr-m.y.-MeV/nuc} \quad (1)$$

A graph of the resultant track density for a 15.7 m.y. exposure is shown in curve a, Fig. 4. This spectrum is based both on contemporary measurements and comparison

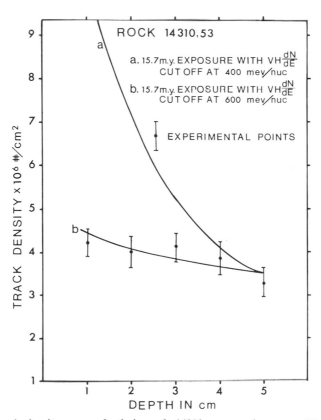

Fig. 4. Track density versus depth in rock 14310 compared to a modified galactic cosmic ray spectrum as described in the text.

with a drill core from the St. Severin meteorite (Cantelaube *et al.*, 1967; Maurette *et al.*, 1968). Conversion of this spectrum to track density also requires the assumption of a maximum etchable track length which we have taken to be 10 μ. Although there is some difference in detail in the spectra used by different track groups, conversion of track density at a given depth to exposure ages gives values that generally agree to within 30%, with our values consistently higher than those of other track groups. Our comparison includes data presented in the following papers: Barber *et al.*, 1971; Bhandari *et al.*, 1971; Crozaz *et al.*, 1971; Fleischer *et al.*, 1971a, 1971b; Crozaz *et al.*, 1970; Fleischer *et al.*, 1970; Lal *et al.*, 1970; and Price and O'Sullivan, 1970.

Recently Comstock (1971) has suggested that the absolute value of the track production rate should be lowered by a factor of two (with a concomitant increase in exposure ages) because of the abundance of elements which produce short tracks. However, a review of the most recent measurements of the abundances of elements in the VH group ($Z = 20$ to 26) (Webber *et al.*, 1971; Mewaldt, 1971), as well as the time variation due to solar modulation (Freier *et al.*, 1971), coupled with measurements of the etchable track length of Ca ions (Kirsten *et al.*, 1972) lead us to believe that equation (1) is probably correct within $\pm 20\%$.

The possibility remains, however, that the long term average cosmic ray spectrum at 1 AU may not be well represented by equation (1). In Fig. 4 (curve b) we show that the data on 14310 can be fit, without assuming a two stage irradiation history, by a spectrum of the following form:

$$
\left.
\begin{aligned}
\frac{dN}{dE} &= (2200)(E) + 3.3 \times 10^5 \qquad E < 600 \text{ MeV/nuc} \\
&= \frac{5 \times 10^6}{(0.94 + E/1000)^{2.5}} \qquad E > 600 \text{ MeV/nuc}
\end{aligned}
\right\} \text{Part./cm}^2\text{-sr-m.y.-MeV/nuc}
\tag{2}
$$

As far as we are aware, there exists no data of sufficient precision in a rock with a simple enough exposure history to rule out the possibility that modulation of low energy galactic cosmic rays may have been more severe in the past than at present. Until good agreement between theory and experiment can be obtained in a rock with a simple exposure history we regard this as an open question.

The lack of a depth dependence could also be due to a large fission track component. However, the measured U concentration is $\lesssim 30$ ppb and we would require a Pu^{244} excess of more than a factor of 50. This seems unlikely in view of the measured age of $\sim 3.9 \times 10^9$ yr (Papanastassiou and Wasserburg, 1971) for this rock.

Measurements of 8 feldspar crystals from the center of rock 14321 give an average track density of $1.8 \pm 0.2 \times 10^6$ corresponding to a surface exposure age of $25 \pm 3 \times 10^6$ yr. The agreement with the spallation age of 27×10^6 yr (Lugmair and Marti, 1972) is evidence that equation (1) is indeed correct, but this agreement may be fortuitous since other workers (Hart *et al.*, 1972) have reported highly variable track densities in this rock indicating that at least some crystals retain a memory of prior irradiation.

Although we received a large vertical section of rock 14306, track measurements turned out to be impossible because of the extreme attack of the crystals by the etching solution, presumably due to a high state of shock.

LIGHTLY AND HEAVILY IRRADIATED SOILS

Although most soil samples are characterized by very high track densities, some are lightly irradiated and clearly represent relatively recent additions to the regolith. In Apollo 12 this was true of the coarse grained layer of the double core and samples 12030, 12032, and 12033 (Crozaz *et al.*, 1971; Arrhenius *et al.*, 1971). In Fig. 5 we show the distribution of track densities for Apollo 14 and samples of the deep drill stem from Apollo 15. It can be seen that the bottom of the trench and the Cone Crater sample both have anomalously low track densities.

Most of the data represent SEM pit counts on the >100 mesh fraction although some crystals in the 100–200 mesh size range are also included. In some cases, notably Luna 16 (Walker and Zimmerman, 1972) and the Cone Crater sample 14141, the track counts were made with replicas counting only tracks $\geq 0.2 L_{max}$. As previously explained (Walker and Zimmerman, 1972) this gives good agreement with optical data.

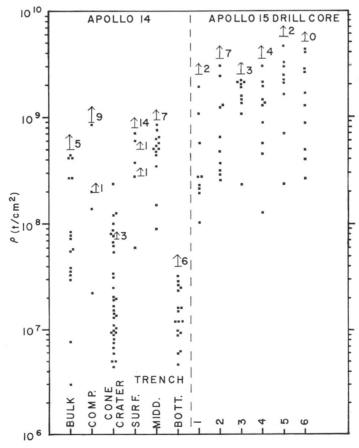

Fig. 5. Distribution of track densities in Apollo 14 and 15 soil samples. Most crystals were >100 mesh. The densities were generally measured by total pit counts in the SEM although some replica data are also included.

In Tables 4 and 5 we give a comparison of the track data with other measurable parameters in the soil. Positive correlations can be seen with grain size (all coarse grained samples are lightly irradiated, however, not all lightly irradiated samples are coarse grained), the fraction of glassy aggregates (McKay *et al.*, 1971; Quaide *et al.*, 1971), the amount of etchable CH_4 (Cadogan *et al.*, 1972), and the amount of solar rare gases (Cadogan *et al.*, 1972; Kaiser, 1972). Other positive correlations, such as the fraction of grains with amorphous coatings (Borg *et al.*, 1971), the fraction of shocked plagioclase and the fraction of spherules (Quaide *et al.*, 1971) also exist. These strong correlations are hardly surprising—the older a soil is, the more it is exposed to both irradiation and to micrometeorite impact and all the effects of both exposures will be positively correlated.

The very high track densities seen at all levels of the Apollo 15 long drill are extremely interesting. We have previously argued (Crozaz *et al.*, 1970; Walker and Zimmerman, 1972) that such high densities can be explained only by exposure of the grains to solar flare particles at or very near the surface of the moon. The neutron fluence (n/cm^2) results of Burnett *et al.*, 1972, indicate that the deep material has remained in its present position for $\gtrsim 5 \times 10^8$ yr. We may thus conclude that the sun had appreciable solar flare activity back to this time.

If we consider only the > 100 mesh fraction of the Cone Crater sample we conclude that the soil has lain relatively undisturbed since its formation. From TL measure-

Table 4. Correlation of track densities with other soil parameters: heavily irradiated samples.

Sample	Description	$>10^8/cm^2$ (%)	Average ρ^*	Median grain size (mm)	% Aggregates[e] 0.25 to 90 μ– 1 mm 150 μ		CH_4 $\mu g/g$[f]	Ar^{36} ccSTP $\times 10^{-8}$
12042	20 m N.W. of Halo	100	1.2×10^9	0.094[a]	—		—	—
12044	S. Rim Surveyor	100	1.2×10^9	—	26		—	—
12025 and 12028 (except unit VI)	Double core	78	8.6×10^8	0.050–0.070[b]	2.4–35		0.9–2.5	2.2 to 2.3[f]
14148	Surface of trench	95	1.5×10^9	0.087[a,c]	32	50	3.0	5.5×10^5[f]
14156	Middle of trench	90	1.0×10^9	0.068[c]	25	44	4.0	—
14163	Bulk, near LEM	48	3×10^8	0.065[c]	20	—	2.6	3.0×10^5[f]
14259	Comp., near Doublet	85	1.5×10^9	0.050[c]	30	52	—	—
Apollo 15	Deep drill	99	2.1×10^9	—	—	—	—	—
15231	Sta. 2, bott. of boulder	100	4.7×10^8	—	—	—	5.1	—
15471	Sta. 4, near Dune Crater	100	9.7×10^8	—	—	—	3.2	—
L16A14	6–8 cm Luna 16	100	1.71×10^9	0.070[d]	67[b]		—	—
L16G14	29–31 cm	100	5.1×10^8	0.090			—	8.6×10^4[g]

* Calculated assuming all crystals with unresolvably high track densities have $\rho = 2 \times 10^9$ t/cm².
[a] King *et al.*, 1971.
[b] Quaide *et al.*, 1971
[c] LSPET (1971).
[d] A. P. Vinogradov, 1971.
[e] McKay *et al.*, 1971; Apollo 14 (private communication).
[f] Cadogan *et al.*, 1972.
[g] W. A. Kaiser, 1972.

Table 5. Correlation of track densities with other soil parameters: light to medium irradiated samples.

Sample	Description	$>10^8/cm^2$ (%)	Average ρ	Median grain size (mm)	% Aggregates[d] 0.25 to 90 μ– 1 mm 150 μ		CH_4 $\mu g/g$[e]	Ar^{36}[e] ccSTP $\times 10^{-8}$
12028,61,67,69	Coarse layer double core	0	4.2×10^7	0.570[a]	0	—	—	—
12030	Between LEM and Head Crater	0	4.4×10^7	0.105[b]	0.5	—	—	—
12033	Trench sample near Head Crater	0	1.35×10^7	0.097[a]	2	—	0.1	2.4×10^3
14141	Cone Crater	11	5.2×10^7	0.735[c]	1	5–12	0.5	—
14149	Bottom of trench	26	6.4×10^7	0.410[c]	16	26.4	3.0	2.6×10^4

[a] Quaide *et al.*, 1971.
[b] King *et al.*, 1971.
[c] LSPET (1971).
[d] McKay *et al.*, 1971.
[e] Cadogan *et al.*, 1972.

ments (Hoyt *et al.*, 1972b) we estimate the average depth of sample 14141 to be 2 cm. Taking twice this as the maximum depth and assigning the lowest density crystals to this position, we calculate an exposure age of 15×10^6 yr for an undisturbed layer. Other track workers (Bhandari *et al.*, 1972; Phakey *et al.*, 1972) have given much lower exposure ages. We believe the difference lies in their having studied crystals with much lower densities. In this connection we wish to point out that the *distribution* of track densities in a *set* of crystals, rather than a rare crystal with very low density, may be more revealing. The curve of track density versus depth falls very rapidly near the surface and then tends to flatten out at deep depths. As a consequence, a scoop of undisturbed material will contain many more crystals with low track densities than those with high densities. In Fig. 6 we show a comparison between the calculated distribution function and the experimental results. It can be seen that a burial depth of 5 cm (9 gm/cm²) and 25×10^6 yr is a reasonable fit to the very limited statistics.

We do not maintain that 14141 is completely undisturbed. The Xe^{126} results of Burnett *et al.* (1972) show an admixture of well-irradiated material, presumably transported from a distant site. The fact that occasional very low track density crystals are found by others also indicates that some mixing has occurred.

Still another interesting manifestation of the fact that 14141 has been subjected to dynamic processes is shown in Fig. 7 which gives track density data for different grain size fractions. It is clear that the average track density increases steadily with decreasing grain size. At first glance it might appear as if the finer samples simply were more contaminated with well-irradiated material transported in from distant locations, as suggested by the data of Burnett *et al.* (1972). However, if this were the case

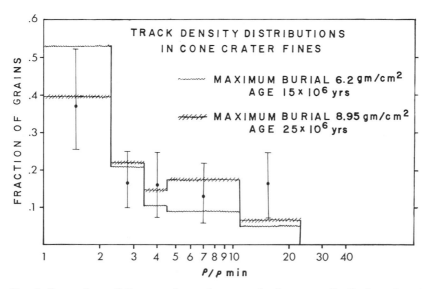

Fig. 6. Comparison of theory and experiment on the frequency distribution of track densities in an undisturbed soil sample. The abscissa is the track density divided by the lowest track density possible for the deepest buried crystal. Because of the form of the track density versus depth curve, there is a pronounced peaking at low track densities.

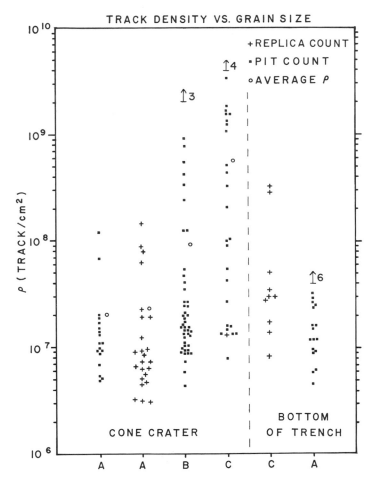

Fig. 7. Track density distribution as a function of grain size in Cone Crater material and the bottom of the Apollo 14 trench. A > 149 μ; B 74 − 149 μ; C 37 − 74 μ.

the minimum track densities should still stay clustered at the same values as for the coarse grained layer, the difference being an increase in the number of very high density crystals. Instead the minimum track density also increases with decreasing grain size. This result would be expected if the Cone Crater sample were originally coarser grained than at present with the fine grained material being derived by micrometeorite bombardment. If this were so, the fine grained crystals must have once been close to surface in order to have been exposed to the micrometeorites. Since the rate of track production also increases towards the surface, the correlation of grain size with track density would be expected.

Interpretation of data for the Apollo 14 trench sample 14149 (bottom), obtained by a variety of techniques, is still unresolved. Samples from the top, bottom, and middle all have low TL characteristic of surface samples (Hoyt *et al.*, 1972b). These results are consistent with those of Eldridge *et al.* (1972) which show uniform Na^{22}

and Al^{26} with depth and from which they conclude that extensive mixing has occurred and that the bottom sample is not representative of its nominal depth. In Tables 4 and 5 it can be seen that the percentage of glassy aggregates is rather high, suggesting a rather heavy irradiation, but the median grain size is large, suggesting just the opposite. The track data in Figs. 5 and 7 show the bottom to be a lightly irradiated sample that looks very similar to 14141 (Cone Crater) and is very different from the top and middle of the trench which are highly irradiated. Most probably the feldspar single crystals used for track work are biased towards material that was originally at the bottom, while both TL and radioactivity measure a mixture of original bottom material and material that was spilled in from the top and sides during sample collection. If this explanation is correct, the trench bottom could be original Cone Crater ejecta that has been recently covered over with older, more irradiated material.

Acknowledgments—G. Crozaz is deeply indebted to Professor J. Jedwab for the use of the scanning electron microscope in the University of Brussels, Belgium.

We acknowledge the assistance of P. Swan and S. Sutton in preparing and analyzing the samples. We thank B. Drozd for the drafting and S. Hoyt and M. Daggett for manuscript preparation. This work was supported by NASA Contract NAS 9-8165.

References

Arrhenius G., Liang S., MacDougall D., Wilkening L., Bhandari N., Bhat S., Lal D., Rajagopalan G., Tamhane A. S., and Venkatavaradan V. S. (1971) The exposure history of the Apollo 12 regolith. *Proc. Second Lunar Sci. Conf., Geochim. Cosmochim. Acta* Suppl. 2, Vol. 3, pp. 5283–5298. MIT Press.

Barber D. J., Cowsik R., Hutcheon I. D., Price P. B., and Rajan R. S. (1971) Solar flares, the lunar surface, and gas-rich meteorites. *Proc. Second Lunar Sci. Conf., Geochim. Cosmochim. Acta* Suppl. 2, Vol. 3, pp. 2705–2714. MIT Press.

Bhandari N., Bhat S., Lal D., Rajagopalan G., Tamhane A. S., and Venkatavaradan V. S. (1971) High resolution time averaged (millions of years) energy spectrum and chemical composition of iron-group cosmic ray nuclei at 1 A.U. based on fossil tracks in Apollo samples. *Proc. Second Lunar Sci. Conf., Geochim. Cosmochim. Acta* Suppl. 2, Vol. 3, pp. 2611–2619. MIT Press.

Bhandari N., Bhat S. G., Goswami J. N., Lal D., Tamhane A. S., and Venkatavaradan V. S. (1972) Study of heavy cosmic rays in lunar silicates (abstract). In *Lunar Science—III* (editor C. Watkins), pp. 65–67, Lunar Science Institute Contr. No. 88.

Borg J., Maurette M., Durrieu L., and Jouret C. (1971) Ultramicroscopic features in micron-sized lunar dust grains and cosmophysics. *Proc. Second Lunar Sci. Conf., Geochim. Cosmochim. Acta* Suppl. 2, Vol. 3, pp. 2027–2040. MIT Press.

Burnett D. S., Huneke J. C., Podosek F. A., Russ G. P., Turner G., and Wasserburg G. J. (1972) The irradiation history of lunar samples (abstract). In *Lunar Science—III* (editor C. Watkins), pp. 105–107, Lunar Science Institute Contr. No. 88.

Cadogan P. H., Eglinton G., Firth J. N. M., Maxwell R. Jr., Mays B. J., and Pillinger C. T. (1972) Survey of lunar carbon compounds, II: The carbon chemistry of Apollo 11, 12, 14, and 15 samples (abstract). In *Lunar Science—III* (editor C. Watkins), pp. 113–115, Lunar Science Institute Contr. No. 88.

Cantelaube Y., Maurette M., and Pellas P. (1967) Traces d'ions lourds dans les mineraux de la chondrite de Saint Severin. In *Radioactive Dating and Methods of Low-Level Counting*, pp. 215–229. International Atomic Energy Agency.

Comstock G. M. (1971) The particle track record of the lunar surface. General Electric Technical Information Series, No. 71-C-190.

Crozaz G., Haack U., Hair M., Maurette M., Walker R., and Woolum D. (1970) Nuclear track studies of ancient solar radiations and dynamic lunar surface processes. *Proc. Apollo 11 Lunar Sci. Conf.*, *Geochim. Cosmochim. Acta* Suppl. 1, Vol. 3, pp. 2051–2080. Pergamon.

Crozaz G., Walker R., and Woolum D. (1971) Nuclear track studies of dynamic surface processes on the moon and the constancy of solar activity. *Proc. Second Lunar Sci. Conf.*, *Geochim. Cosmochim. Acta* Suppl. 2, Vol. 3, pp. 2543–2558. MIT Press.

Crozaz G., Drozd R., Graf H., Hohenberg C. M., Monnin M., Ragan D., Ralston C., Seitz M., Shirck J., Walker R. M., and Zimmerman J. (1972a) Uranium and extinct Pu^{244} effects in Apollo 14 materials. Vol. 2, these proceedings.

Crozaz G., Drozd R., Hohenberg C. M., Hoyt H. P., Ragan D., Walker R. M., and Yuhas D. (1972b) Solar flare and galactic cosmic ray studies of Apollo 14 samples (abstract). In *Lunar Science—III* (editor C. Watkins), pp. 167–169, Lunar Science Institute Contr. No. 88.

Dalrymple G. B. and Doell R. R. (1970) Thermoluminescence of lunar samples from Apollo 11. *Proc. Apollo 11 Lunar Sci. Conf.*, *Geochim. Cosmochim. Acta* Suppl. 1, Vol. 3, pp. 2081–2092. Pergamon.

Doell R. R. and Dalrymple G. B. (1971) Thermoluminescence of Apollo 12 lunar samples. *Earth Planet. Sci. Lett.* **10**, 357–360.

Eldridge J. S., O'Kelly G. D., and Northcutt K. J. (1972) Abundance of primordial and cosmogenic radionuclides in Apollo 14 rocks and fines (abstract). In *Lunar Science—III* (editor C. Watkins), pp. 221–223, Lunar Science Institute Contr. No. 88.

Finkel R. C., Arnold J. R., Imamura M., Reedy R. C., Fruchter J. S., Loosli H. H., Evans J. C., DeLany A. C., and Shedlovsky J. P. (1971) Depth variation of cosmogenic nuclides in a lunar surface rock and lunar soil. *Proc. Second Lunar Sci. Conf.*, *Geochim. Cosmochim. Acta* Suppl. 2, Vol. 2, pp. 1773–1789. MIT Press.

Fleischer R. L., Haines E. L., Hart H. R. Jr., Woods R. T., and Comstock G. M. (1970) The particle track record of the Sea of Tranquillity. *Proc. Apollo 11 Lunar Sci. Conf.*, *Geochim. Cosmochim. Acta* Suppl. 1, Vol. 3, pp. 2103–2120. Pergamon.

Fleischer R. L., Hart H. R. Jr., and Comstock G. M. (1971a) Very heavy solar cosmic rays: Energy spectrum and implications for lunar erosion. *Science* **171**, 1240–1242.

Fleischer R. L., Hart H. R. Jr., Comstock G. M., and Evwaraye A. O. (1971b) The particle track record of the Ocean of Storms. *Proc. Second Lunar Sci. Conf.*, *Geochim. Cosmochim. Acta* Suppl. 2, Vol. 3, pp. 2559–2568. MIT Press.

Freier P. S., Long C. E., Cleghorn T. F., and Waddington C. J. (1971) The charge and energy spectra of heavy cosmic ray nuclei. Conference papers, Twelfth Int. Conf. on Cosmic Rays, Vol. 1, pp. 252–257.

Hart H. R. Jr., Comstock G. M., and Fleischer R. L. (1972) The particle track record of Fra Mauro (abstract). In *Lunar Science—III* (editor C. Watkins), pp. 360–362, Lunar Science Institute Contr. No. 88.

Hoyt H. P., Kardos J. L., Miyajima M., Seitz M. G., Sun S. S., Walker R. M., and Wittels M. C. (1970) Thermoluminescence, x-ray and stored energy measurements of Apollo 11 samples. *Proc. Apollo 11 Lunar Sci. Conf.*, *Geochim. Cosmochim. Acta* Suppl. 1, Vol. 3, pp. 2269–2287. Pergamon.

Hoyt H. P., Miyajima M., Walker R. M., Zimmerman D. W., Zimmerman J., Britton D., and Kardos J. L. (1971) Radiation dose rates and thermal gradients in the lunar regolith: Thermoluminescence and DTA of Apollo 12 samples. *Proc. Second Lunar Sci. Conf.*, *Geochim. Cosmochim. Acta* Suppl. 2, Vol. 3, pp. 2254–2263. MIT Press.

Hoyt H. P., Walker R. M., and Zimmerman D. W. (1972a) Solar flare induced thermoluminescence of lunar rocks. In preparation.

Hoyt H. P., Walker R. M., Zimmerman D. W., and Zimmerman J. (1972b) Thermoluminescence of individual grains from lunar fines. Vol. 3, these proceedings.

Kaiser W. A. (1972) Rare gas studies in Luna-16-G-7 fines by stepwise heating technique. A low fission solar wind Xe. *Earth Planet. Sci. Lett.* **13**, 387–399.

King E. A., Butler J. C., and Corman M. F. (1971) The lunar regolith as sampled by Apollo 11 and 12: Grain size analyses, and origins of particles. *Proc. Second Lunar Sci. Conf.*, *Geochim. Cosmochim. Acta* Suppl. 2, Vol. 1, pp. 737–746. MIT Press.

Kirsten T., Deubner J., Ducati H., Gentner W., Horn P., Jessberger E., Kalbitzer S., Kaneoka I.,

Kiko J., Kratschmer W., Muller H. W., Plieninger T., and Thio S. K. (1972) Rare gases and ion tracks in individual components and bulk samples of Apollo 14 and 15 fines and fragmental rocks (abstract). In *Lunar Science—III* (editor C. Watkins), pp. 452–454, Lunar Science Institute Contr. No. 88.

Lal D., Macdougall D., Wilkening L., and Arrhenius G. (1970) Mixing of the lunar regolith and cosmic ray spectra: Evidence from particle-track studies. *Proc. Apollo 11 Lunar Sci. Conf., Geochim. Cosmochim. Acta* Suppl. 1, Vol. 3, pp. 2295–2303. Pergamon.

LSPET (Lunar Sample Preliminary Examination Team) (1971) Preliminary examination of lunar samples from Apollo 14. *Science* **173**, 681–693.

Lugmair G. W. and Marti K. (1972) Neutron and spallation effects in Fra Mauro regolith (abstract). In *Lunar Science—III* (editor C. Watkins), pp. 495–497, Lunar Science Institute Contr. No. 88.

McKay D. S., Morrison D. A., Clanton U. S., Ladle G. H., and Lindsay J. F. (1971) Apollo 12 soil and breccia. *Proc. Second Lunar Sci. Conf., Geochim. Cosmochim. Acta* Suppl. 2, Vol. 1, pp. 755–773. MIT Press.

Maurette M., Thro P., Walker R., and Webbink R. (1968) Fossil tracks in meteorites and the chemical abundance and energy spectrum of extremely heavy cosmic rays. In *Meteorite Research* (editor P. Millman), pp. 286–315. D. Reidel.

Mewaldt R. A. (1971) The composition of relativistic VH and VVH cosmic rays. Ph.D. Thesis, Washington University.

Papanastassiou D. A. and Wasserburg G. J. (1971) Rb–Sr ages of igneous rocks from the Apollo 14 mission and the age of the Fra Mauro formation. *Earth Planet. Sci. Lett.* **12**, 36–48.

Phakey P. P., Hutcheon I. D., Rajan R. S., and Price P. B. (1972) Radiation damage in soils from five lunar missions (abstract). In *Lunar Science—III* (editor C. Watkins), pp. 608–610, Lunar Science Institute Contr. No. 88.

Price P. B. and O'Sullivan D. (1970) Lunar erosion rate and solar flare paleontology. *Proc. Apollo 11 Lunar Sc.. Conf., Geochim. Cosmochim. Acta* Suppl. 1, Vol. 3, pp. 2351–2359. Pergamon.

Quaide W., Oberbeck V., Bunch T., and Polkowski G. (1971) Investigations of the natural history of the regolith at the Apollo 12 site. *Proc. Second Lunar Sci. Conf., Geochim. Cosmochim. Acta* Suppl. 2, Vol. 1, pp. 701–718. MIT Press.

Rancitelli L. A., Perkins R. W., Felix W. D., and Wogman N. A. (1971) Erosion and mixing of the lunar surface from cosmogenic and primordial radionuclide measurements in Apollo 12 lunar samples. *Proc. Second Lunar Sci. Conf., Geochim. Cosmochim. Acta* Suppl. 2, Vol. 2, pp. 1757–1772. MIT Press.

Rancitelli L. A., Perkins R. W., Felix W. D., and Wogman N. A. (1972) Cosmic ray flux and lunar surface processes characterized from radionuclide measurements in Apollo 14 and 15 lunar samples (abstract). In *Lunar Science—III* (editor C. Watkins), pp. 630–632, Lunar Science Institute Contr. No. 88.

Turner G., Huneke J. C., Podosek F. A., and Wasserburg G. J. (1971) ^{40}Ar–^{39}Ar ages and cosmic ray exposure ages of Apollo 14 samples. *Earth Planet. Sci. Lett.* **12**, 19–35.

Vinogradov A. P. (1971) Preliminary data on lunar ground brought to Earth by automatic probe "Luna-16." *Proc. Second Lunar Sci. Conf., Geochim. Cosmochim. Acta* Suppl. 2, Vol. 1, pp. 1–16. MIT Press.

Walker R. and Zimmerman D. (1972) Fossil track and thermoluminescence studies of Luna 16 material. *Earth Planet. Sci. Lett.* **13**, 419–422.

Webber W. R., Damle S. U., and Kish J. M. (1971) The chemical composition of cosmic rays with $Z = 3$–30 at high and low energies. Conference papers, Twelfth Int. Conf. on Cosmic Rays, Vol. 1, pp. 229–234.

Yuhas D. E., Walker R. M., Reeves H., Price P. B., Poupeau G., Pellas P., Lorin J. C., Lal D., Hutcheon I. D., Hart H. R. Jr., Goswami J. N., Fleischer R. L., Comstock G. M., Chetrit G. C., Bhandari N., and Berdot J. L. (1972) Track consortium report on rock 14310. Vol. 3, these proceedings.

Proceedings of the Third Lunar Science Conference
(Supplement 3, *Geochimica et Cosmochimica Acta*)
Vol. 3, pp. 2933–2939
The M.I.T. Press, 1972

Charge assignment to cosmic ray heavy ion tracks in lunar pyroxenes

T. Plieninger, W. Krätschmer, and W. Gentner

Max-Planck-Institut für Kernphysik, Heidelberg, Germany

Abstract—Calcium is the first ion of which the total etchable range in pyroxenes has been measured by accelerator irradiations. It was found that energies of 2 MeV/amu are sufficient to produce total track lengths of about 4.5 μ in the interior of lunar pyroxene crystals. Direct comparison between the length distributions of cosmic ray and Ca tracks in the same crystals shows coincidence at the 4–5 μ peak. Investigations on the stability of 2 MeV/amu Ca and 1 MeV/amu Fe tracks in pyroxenes indicate that the shortening of tracks under lunar conditions is negligible, hence this peak may be ascribed to Ca.

Length distributions, measured in pyroxenes from Apollo 11 igneous rock 10047,13 and Apollo 11 fines 10084,32 show four distinguishable peaks which are ascribed to even atomic numbers. This leads to an Fe peak situated at about 16 μ. The peak at about 12 μ is ascribed to unresolved Cr and Mn. If one takes into account the width of the peaks, an average abundance ratio (Cr + Mn)/Fe of 0.8 ± 0.4 results.

Introduction

Fossil tracks stored in lunar and meteoritic minerals provide valuable information about the very heavy component of the cosmic radiation. By utilizing the track technique, flux, and energy of the cosmic radiation can be determined from measured track densities and from density gradients (see for example Barber *et al.*, 1971; Fleischer *et al.*, 1971). Moreover, the atomic numbers of the track forming ions can be deduced from their etchable ranges in the detector minerals (Price and Fleischer, 1971). The interpretation of tracks in terms of elemental abundances is interesting, because possible longtime variations in the chemical composition of the cosmic radiation, the fragmentation of highly energetic heavy ions, and shielding of the samples during exposure can be determined.

Much work has been done in measuring total track length distributions (Lal, 1969; Bhandari *et al.*, 1971), but problems in assigning atomic numbers to the observed peaks in the distributions still remain. These problems arise mainly from the unknown registration and track retaining properties of the detector materials. Nevertheless, four peaks are observed in most of the track length distributions measured in lunar and meteoritic pyroxene crystals (Lal, 1969; Bhandari *et al.*, 1971). The highest peaks are found at 12 μ and 16 μ. It is plausible to ascribe one of these peaks to iron ions. To make proper charge assignments, calibrations with artificially accelerated heavy ions are required (Price *et al.*, 1968; Plieninger and Krätschmer, 1971). Recent irradiations performed by Price *et al.* (1972) with Zn ions at Dubna yielded an etchable range of ≥ 33 μ in pyroxenes.*

---

* The value of 45 μ given by Price *et al.* (1972) had to be revised according to a private communication by D. Lal.

From theory (Fleischer *et al.*, 1967) one consequently would expect a total length of $\geq 16\ \mu$ for Fe tracks. This disagrees with the assumption of $12\ \mu$ for the length of fossil Fe tracks. However, the fossil tracks could have been shortened by annealing.

In this paper we present results about charge calibration measurements in lunar pyroxenes performed with 2 MeV/amu Ca ions and their implications on charge assignment to cosmic ray tracks. In order to ensure that our charge assignments are valid we have estimated the stability of the length of cosmic ray tracks in pyroxene crystals under lunar conditions from annealing experiments on tracks of artificial accelerated ions.

Experimental Procedures

The Heidelberg Tandem van de Graaff accelerators were used to irradiate lunar and terrestrial pyroxenes with 2 MeV/amu Ca, and 1 MeV/amu Fe ions. Polished surfaces of crystallographically unorientated samples were exposed perpendicular to the beam. After irradiation with Ca ions, pyroxene crystals from fines 10084,32 were polished parallel to the beam direction. Tracks in the interior of the crystals were revealed by application of the track in track (TINT) technique developed by Lal (1969). After decoration of the tracks with silver, an optical microscope was used to measure both Ca and cosmic ray track lengths (all TINTs) in the same crystals. The accuracy of the length measurements was about $0.5\ \mu$.

Crystals showing track densities of about $10^7\ cm^{-2}$ were chosen for the measurements. To increase the TINTs revelation probability especially for short tracks, the samples from rock 10047,13 and fines 10084,32 were bombarded with Fe ions perpendicular to the polished surface. By measuring TINT on induced and on preexisting tracks, length distributions were determined. Most of the measured TINTs were parallel to the surface within about 20° and corrections have been applied for the dip angles. All measurable TINTs down to $10\ \mu$ under the surface were used for the length distributions. The distributions were not corrected with respect to detection and revelation probabilities.

Detailed thermal annealing measurements were performed on terrestrial bronzite crystals which had been irradiated perpendicular to the surface with 1 MeV/amu Fe ions, and on lunar pyroxenes irradiated with 2 MeV/amu Ca ions. The samples were heated at temperatures between 400 and 550°C for times between 1 and 100 hours. The annealing of tracks due to ionizing radiation was studied with the electron beam of a microprobe. The electron energy was 40 keV and the current density $10^{-2}\ A/cm^2$. Length measurements in this case were performed on track replicas with a scanning electron microscope.

The samples were etched stepwise in a boiling NaOH solution (6g NaOH in 4g H_2O) until further etching did not change the positions of the maxima in the length distributions. This usually was achieved after 1 to 2 hours of etching.

Results

2 MeV/amu Ca ions produce tracks only in the interior of the detector crystals. Figure 1 shows a photomicrograph of such tracks in a pyroxene crystal from lunar fines 10084,32. The tracks start to appear at depths between 3.5 and $5.5\ \mu$ below the bombarded surface. They disappear at a depth of $9\ \mu$ which is constant within $0.5\ \mu$. This results in a mean track length of $4.5\ \mu$ with intrinsic fluctuations of $\pm 1\ \mu$. The mean total etchable range of Ca ions varied from 4 to $5\ \mu$ determined for a number of pyroxene crystals. This variation probably is due to different crystallographic orientations or differing chemical compositions of the crystals.

From range formulas given by Henke and Benton (1967) it can be deduced that the lower threshold energy for Ca is about 10 MeV corresponding to an undetectable

Ca – BEAM

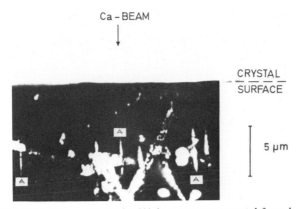

CRYSTAL
SURFACE

5 μm

Fig. 1. Artificial 2 MeV/amu Ca tracks (A) in a pyroxene crystal from lunar soil. The total etchable range lies within the crystal. Therefore only tracks intersecting cosmic ray tracks or cracks in the crystal are developed.

part of the range of 2.5 μ and an upper threshold energy of about 30–40 MeV. This confirms our previous deductions (Plieninger and Krätschmer, 1971).

A comparison between the length distribution of 2 MeV/amu Ca tracks and cosmic ray induced tracks in the same crystals is shown in Fig. 2a. Within statistical errors, the Ca lengths coincide with the 4–5 μ peak of the cosmic ray tracks. Therefore, this peak may be ascribed to Ca ion tracks from the cosmic radiation. Some of the measured length distributions obtained in pyroxenes from rock 10047,13 are shown in Figs. 2b and 2c. From the position of the Ca peak one can ascribe the 12 μ peak to unresolved Cr and Mn and the 16 μ peak to Fe. Assuming Gaussian distributions with 6 μ width for the 16 μ peak and 3 μ for the 12 μ peak, an average (Cr + Mn)/Fe cosmic ray abundance ratio of 0.8 $\pm$ 0.4 results. This ratio was found to be higher in the pyroxenes of the rock than in those of the soil. The (Cr + Mn)/Fe ratio deduced from our measurements is in agreement with the available data of cosmic ray element abundances (Shapiro and Silberberg, 1970). Its variation from rock to soil may be due to different degrees of fragmentation of the cosmic ray ions according to different shielding of these samples, or may reflect a possible energy dependence of the primary abundance ratio (Price and Fleischer, 1971).

Figure 3 shows the decrease of 1 MeV/amu Fe track lengths in terrestrial bronzite as a function of annealing time for different temperatures. At a given temperature, the length decreases linearly with the logarithm of time (Krätschmer, 1971):

$$L_0 - L = F(T) \, log \, (t/t_0)$$

where L_0 is the length at a given time t_0, and T the absolute temperature. For the slope $F(T)$ one expects an exponential dependency on T of the form

$$F(T) \propto e^{-E/kT},$$

where E is the activation energy of this process. A plot $F(T)$ versus $1/T$ is shown in Fig. 4. From this plot one can extrapolate the length decrease of the 1 MeV/amu Fe

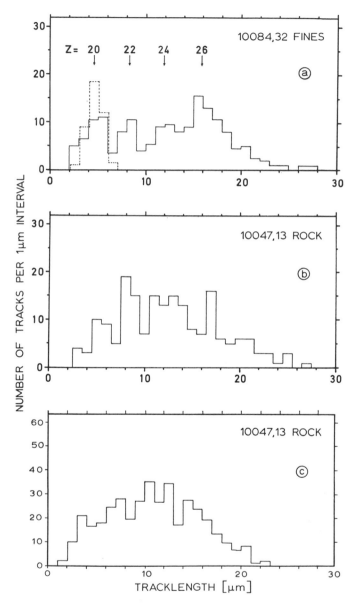

Fig. 2. Length distributions measured by the TINT method in lunar pyroxenes: (a) shows lengths of 2 MeV/amu Ca ions (dashed line) together with lengths of cosmic ray tracks measured in the same crystals from the lunar fines. The proposed charge assignment is indicated. (b) Shows a length distribution in a single grain, while in (c) the distributions from different crystals are superimposed.

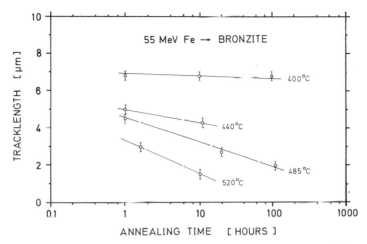

Fig. 3. The decrease in track length of 1 McV/amu Fe ions in terrestrial bronzite at different temperatures as a function of time. The data can be fitted by a linear decrease of length with the logarithm of time.

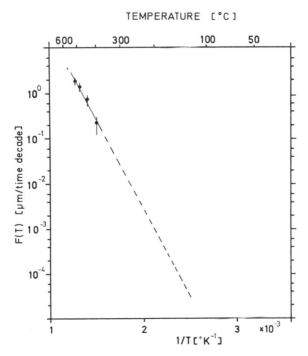

Fig. 4. Length decrease as function of temperature. From this plot, the thermal shortening rate of 1 MeV/amu Fe tracks at lunar temperature can be estimated.

tracks at lunar temperatures. Even a very conservative extrapolation yields less than 0.5 μ of shrinking in 10^9 yr. Similar to the behavior of 1 MeV/amu Fe tracks, Ca tracks in lunar pyroxenes were not measurably affected during 10 hours at 400°C. That indicates a stability comparable to Fe tracks. Above 400°C measurements were difficult because the track length became too small.

Ionizing radiation is another possible source of track length reduction. Its effects on 1 MeV/amu Fe tracks were studied using various doses of 40 keV electrons. In enstatite a dose dependent annealing effect was observed only for doses $> 10^{15}$ erg/g, while the tracks in hypersthene were stable up to doses of 10^{16} erg/g.

From the cosmic ray proton flux one expects a maximum dose of 10^{12} erg/g deposited in 10^9 yr. If annealing due to ionizing radiation is only dose dependent, then tracklength fading for lunar and meteoritic samples is negligible.

Discussion

The annealing experiments performed, indicate that lunar surface conditions do not affect the track lengths. This is supported by the small variations in the positions of the peaks in all length distributions measured in either meteoritic or lunar samples (Lal, 1969; Bhandari et al., 1971, 1972), even though these two kinds of samples probably have had very different histories. If the track lengths were affected by environmental conditions the peaks would have changed at least their positions.

All measured length distributions have prominent peaks at 4, 8, 12, and 16 μ (see also Bhandari et al., 1971). Since shortening of the Ca tracks is negligible, the 4 μ peak has to be due to Ca ions of the cosmic radiation. It seems reasonable to assign the other peaks to heavier elements. One unit of charge difference between subsequent peaks however would lead to a length region for Fe, where tracks are very rare. This assumption therefore is in disagreement with the observed large abundance of Fe in the cosmic radiation. An assignment of even charges to the peaks which moreover is consistent with the observed overabundances of even Z nuclei (Webber, 1971) leads unambiguously to the assignment of the 16μ peak to cosmic ray Fe ions. This would then contradict the assignment of the 12 μ peak to Fe which is only based on the argument of this peak being the most prominent one (Price et al., 1968; Lal, 1969; Bhandari et al., 1971). It must be noticed that the (Cr + Mn)/Fe ratio can reach values as high as 0.7 (Price and Fleischer, 1971). Therefore, one must be careful in using abundance arguments in this case. In addition, fragmentation will be another factor which influences this ratio.

Bhandari et al. (1971) measured large variations in the relative 16 μ track abundances between lunar soil and rocks. They assume an U^{238} fission track contribution to the 16 μ peak for pyroxenes of the lunar soil (Bhandari et al., 1972). But a significant fissiogenic contribution to the 16 μ peak seems rather unlikely because of the very low (< 5 ppb) uranium concentration in lunar soil and rock pyroxenes (Lovering, 1972), even if one takes neutron induced fission into account (Burnett et al., 1971). For a uniformly distributed Uranium concentration of 5 ppb in the lunar pyroxenes, the contribution of spontaneous and n-induced fission tracks to the 16 μ peak would be $< 10^{-2}$ for cosmic ray track densities of about 10^7 cm^{-2}. This indicates that nearly all measured tracks are cosmic ray tracks. The observed variations in the relative 16 μ track abundances between lunar rock and soil may be due to different burial histories of these samples.

Our calibration experiments with Ca ions are a first step to base charge assignments on experimental evidence alone. Further accelerator irradiations with heavier ions will improve this attempt. These experiments will yield more detailed knowledge about the tracklength-charge relation, length distribution widths and charge resolution of pyroxene track detectors.

Acknowledgments—We are grateful to NASA for providing the lunar samples. We appreciate the helpful discussions with Drs. A. El Goresy, T. Kirsten, J. F. Lovering, O. Müller as well as N. Bhandari and P. B. Price. We further acknowledge the help of the Accelerator crew of our Institute.

REFERENCES

Barber D. J., Coswik R., Hutcheon I. D., Price P. B., and Rajan R. S. (1971) Solar flares, the lunar surface, and gas-rich meteorites. *Proc. Second Lunar Sci. Conf., Geochim. Cosmochim. Acta* Suppl. 2, Vol. 3, pp. 2705–2714. MIT Press.

Bhandari N., Bhat S., Lal D., Rajagopalan G., Tamhane A. S., and Venkatavaradan V. S. (1971) Spontaneous fission record of uranium and extinct trasuranic elements in Apollo samples. *Proc. Second Lunar Sci. Conf., Geochim. Cosmochim. Acta* Suppl. 2, Vol. 3, pp. 2599–2609. MIT Press.

Bhandari N., Bhat S. G., Goswami I. N., Lal D., Tamhane A. S., and Venkataravaradan V. S. (1972) Study of heavy cosmic rays in lunar silicates (abstract). In *Lunar Science—III* (editor C. Watkins), pp. 65–67, Lunar Science Institute Contr. No. 88.

Burnett D. S., Huneke J. C., Podosek F. A., Price Russ G. III, and Wasserburg G. J. (1971) The irradiation history of lunar samples. *Proc. Second Lunar Sci. Conf., Geochim. Cosmochim. Acta* Suppl. 2, Vol. 2, pp. 1671–1679. MIT Press.

Fleischer R. L., Price P. B., Walker R. M., and Hubbard E. L. (1967) Criterion for registration in dielectric track detectors. *Phys. Rev.* **156**, 353–355.

Fleischer R. L., Hart H. R., and Comstock G. M. (1971) Very heavy solar cosmic rays: Energy spectrum and implications for lunar erosion. *Science* **171**, 1240–1242.

Henke R. P. and Benton E. V. (1967) A computer code for the computation of heavy-ion range energy relationship in any stopping material. Report USNRDL-TR-67-122.

Krätschmer W. (1971) Die anätzbaren Spuren künstlich beschleunigter schwerer Ionen in Quarzglas. Thesis, Max-Planck-Institut für Kernphysik, Heidelberg (unpublished).

Lal D. (1969) Recent advances in the study of fossil tracks in meteorites due to heavy nuclei of the cosmic radiation. *Space Sci. Rev.* **9**, 623–650.

Lovering J. F. (1972) Private communication.

Price P. B., Fleischer R. L., and Moack C. D. (1968) Identification of very heavy cosmic ray tracks in meteorites. *Phys. Rev.* **167**, 277–282.

Price P. B. and Fleischer R. L. (1971) Identification of energetic heavy nuclei with solid dielectric track detectors: Applications to astrophysical and planetary studies. *Ann. Rev. Nucl. Sci.* **21**, 295–333.

Price P. B., Hutcheon J. D., Lal D., and Perelygin V. P. (1972) Lunar crystals as detectors of very rare nuclear particles (abstract). In *Lunar Science—III* (editor C. Watkins), pp. 619–621, Lunar Science Institute Contr. No. 88.

Plieninger T. and Krätschmer W. (1971) (abstract). *Trans. Amer. Geophys. Union* **52**, No. 4, 268.

Shapiro M. M. and Silberberg R. (1970) Heavy cosmic ray nuclei. *Ann. Rev. Nucl. Sci.* **20**, 323–329.

Webber R. W. (1971) Recent measurements of charge and isotopic composition using counters. Published in "Isotopic Composition of the Primary Cosmic Radiation," proceedings of a symposium held in Lyngby, Denmark, March 1971, edited by Philip M. Dauber, Danish Space Research Institute, pp. 12–30.

Proceedings of the Third Lunar Science Conference
(Supplement 3, *Geochimica et Cosmochimica Acta*)
Vol. 3, pp. 2941–2947
The M.I.T. Press, 1972

Track consortium report on rock 14310

D. E. Yuhas, R. M. Walker

Laboratory for Space Physics, Washington University,
St. Louis, Missouri 63130

H. Reeves, G. Poupeau, P. Pellas, J. C. Lorin, G. C. Chetrit,
and J. L. Berdot

C.N.R.S. and Laboratoire Minéralogie du Museum,
Paris, France

P. B. Price, I. D. Hutcheon

Department of Physics, University of California,
Berkeley, California 94720

H. R. Hart, Jr., R. L. Fleischer, and G. M. Comstock

General Electric Research and Development Center,
Schenectady, New York 12301

and

D. Lal, J. N. Goswami, and N. Bhandari

Tata Institute of Fundamental Research,
Homi Bhabha Road, Colaba, Bombay, India

Abstract—Five different track laboratories each received a section running from the outside to the center of rock 14310. Data on the feldspars in individual sections, as well as sections exchanged between laboratories, are given here. In spite of the complications due to fission tracks, different laboratories generally obtain results that agree within the counting statistics. Solar flare tracks confirm the PET orientation of the rock. The depth dependence of track density outside the solar flare region is very flat, and is quite different than would be expected from prior studies of galactic cosmic rays. The tracks were apparently accumulated during the course of a complex irradiation history.

INTRODUCTION

THE IDEAL TRACK ANALYSIS of a lunar rock involves measurements taken along three mutually perpendicular axes with the origin at the center. In the case of a rock with a simple irradiation history this detailed analysis should permit measurement of the galactic cosmic ray energy spectrum at 1 a.u. averaged over the surface exposure age of the rock.

The ideal was approached for the first time with the allocation of samples from rock 14310 to each of five different track laboratories, each group receiving a section that ran from the exterior to the center of the rock. This paper summarizes the experimental results of the individual examination by each group of its section, as well as comparisons on samples exchanged between groups. Interpretations of these data can be found in the separate presentations of each laboratory.

The track data do not give the expected energy spectrum, and it is likely that the rock has had a complex irradiation history. Whatever the final significance of the data, it is clear that this consortium study has proved to be a useful exercise in interlaboratory comparison. Over the years, each laboratory has evolved its own sample preparation techniques (thick sections, thin sections, grain mounts), its own etching recipes, its own observational techniques (Scanning Electron Microscope—SEM, optical micro-scopy, replicas, silvered tracks), and finally, its own track recognition criteria. Under these conditions differences in absolute track counts might be expected, although normalized curves of depth dependence would be expected to agree much better. Differences do indeed occur, but in general the agreement is quite good.

A similar attack was planned for rock 14321, but due to sample cutting problems not all groups obtained rock sections. The five laboratories involved in this study are labeled 1–5 as follows: (1) Washington University, St. Louis, Missouri, (2) Musee d'Histoire Naturelle, Paris, France, (3) University of California, Berkeley, California, (4) General Electric Laboratory, Schenectady, New York, and (5) Tata Institute, Bombay, India.

ORIENTATION, CUTTING, AND SAMPLE COORDINATE SYSTEM

In Fig. 1 we show the cutting diagram of rock 14310 and the approximate location of the sections allocated to each group. Although the rock was not photographically documented on the lunar surface, the surface orientation was established during the preliminary examination both by observations of impact pits (F. Horz, private communication) and cosmogenic radioactivity (E. Schonfeld, private communica-tion). The surface orientation was also confirmed independently by thermolumines-

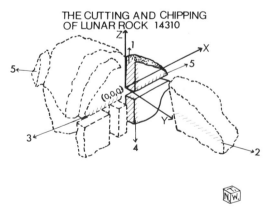

Fig. 1. This figure illustrates the relative orientation of the consortium pieces. Each slice, indicated by the hatched area, is numbered, and each number corresponds to a different laboratory. Although piece 2 is orthogonal to the other pieces as shown, its spatial location is not strictly correct. The compass directions are LRL notations and are not selenographic directions.

cense measurements of Laboratory 1 (Crozaz *et al.*, 1972) and by the track measurements reported here.

An (x, y, z) coordinate system was adopted by setting the $(0, 0, 0)$ point at one corner of the Washington University slab as shown in the diagram and assigning $+z$ the direction towards the top, $+y$ to the west and $+x$ to the south. The track density measured in a plane at a given point (x, y, z) depends on the orientation of the plane in space; for this reason we specify both the orientation and the coordinates in the data reported here.

Sample Preparation Techniques and Track Counting Criteria

Laboratory 1 used primarily oriented thick sections and measured tracks by counting replicas in an SEM. In independent studies of feldspar grain mounts from the same rock, it was established that agreement with optical counts was obtained within $\pm 5\%$ when only replicas of tracks $\geq (1/6)L_{max}$ were counted, where L_{max} is the maximum track length observed. This is equivalent to counting all tracks $\geq 2\ \mu$ in length.

Laboratory 2 used thin sections with optical counting ($\geq 1\ \mu$ track length) up to 7.5×10^6 t/cm²; higher densities were measured by total pit counts in the SEM. Laboratories 3, 4, and 5 used optical counts in polished thin sections. The recognition criterion required a definite cone-shaped "tail" which is characteristic of etched tracks and probably corresponded to a cut-off between 1 and 2 μ.

All groups used aqueous solutions of NaOH as the standard feldspar etchant, although the concentration varied from 1:1 to 1:2 (NaOH to H_2O by weight). Typical etch times were 12 min using the 1:1 concentration and 40–60 min using the 1:2 concentration.

Rock 14310 has one of the highest average uranium concentrations of any lunar rock studied, and all groups had difficulty in counting cosmic ray tracks due to contamination by fission tracks from inclusions or grain boundaries. Only areas far from grain boundaries and with an apparent uniform density were counted.

Experimental Data on Depth Dependences

The experimental data on the depth dependence are presented in logarithmic form in Figs. 2a–2c and in linear form in Figs. 3a–3c. The statistical uncertainties for the data are typically $\pm 10\%$, but may be as high as 30% in the solar flare region (0 to 1 mm).

From Fig. 2a it can be seen that the south face shows no strong evidence of being directly exposed to solar flares. The north face may have been exposed to flares but, if so, must have been partially shielded. The average track density observed by Laboratory 5 is close to that of Laboratory 3, but the scatter in the Laboratory 5 measurements is outside the statistical error. The reason for this is now known, but may represent a fluctuating fission track contribution.

In Fig. 2b the data indicate that the west face has been exposed to solar flares. The large difference in track counts between the optical and SEM data is typical when total pit densities are counted in the SEM. As shown in Figs. 2c and 3c, the top shows

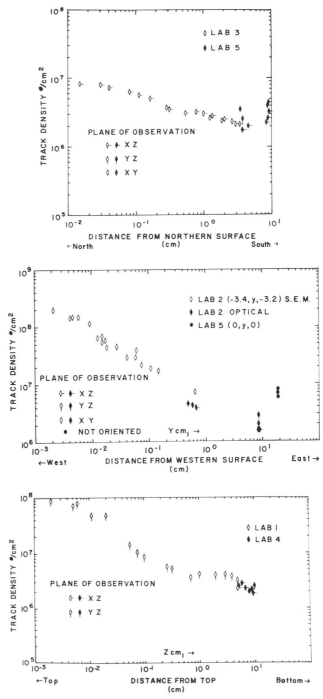

Fig. 2. Log-log plots of track density versus depth are along (a) a north-south, (b) an east-west, (c) a top-bottom line running through the center of rock 14310. Data is given only for feldspars. The orientation of the counting planes is noted in the figure.

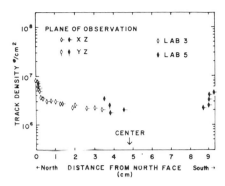

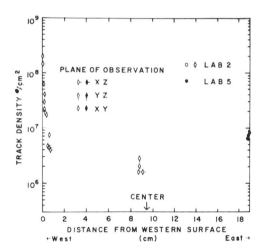

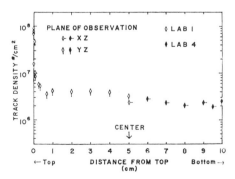

Fig. 3. Log-linear plot of track-density versus depth along (a) a north-south, (b) an east-west, (c) a top-bottom line running through the center of rock 14310. Data is given only for feldspars. The orientation of the counting planes is noted in the figure.

a clear solar flare effect while the bottom appears to be completely shielded. It should be realized, of course, that the track measurements refer to only a small portion of the area of any one face and are sensitive to the position of the soil line.

COMPARISONS OF ABSOLUTE TRACK DENSITIES BETWEEN LABORATORIES

Two types of comparisons are possible. First, where one section joins another, it is possible to compare independent measurements on equivalent planes at nearly equal points in space. Such a comparison is shown in Table 1. Of the 14 comparisons that can be made from Table 1, all but four measurements agree to within $\pm 16\%$ of the weighted average. The discrepancies greater than 16% occur in measurements on the xz plane. We are not able to account for this anomaly at this time.

A second comparison was made possible by an exchange of samples. As can be seen in Table 2 the agreement is generally within 20%, although occasional larger deviations occur.

Table 1. Interlaboratory comparison of track densities in feldspars near the center* of rock 14310.

Plane of observation	Track density ($\times 10^6$ per cm^2)					
	1 $x = 0$ $y = 0$ $z = 0$	2 $x = 2$ cm $y = 0$ $z = -2$ cm	3 $x = 0$ $y = 0$ $z = 0$	4 $x = 0$ $y = 0$ $z = 0$	5 $x = 0$ $y = +1$ $z = 0$	Weighted average†
xy	2.7 ± 0.1	2.1 ± 0.1	2.5 ± 0.2	2.4 ± 0.2	2.6 ± 0.2	2.4 ± 0.2
yz	3.3 ± 0.1	2.9 ± 0.1	2.6 ± 0.2	—	3.5 ± 0.3	3.1 ± 0.4
xz	2.4 ± 0.1	1.6 ± 0.1	2.2 ± 0.2	2.6 ± 0.2	1.5 ± 0.1	1.9 ± 0.5

1. Washington University, St. Louis, Missouri 63130.
2. Paris Museum, Paris, France.
3. University of California, Berkeley, California 94720.
4. General Electric Co., Schenectady, New York 12301.
5. Tata Institute, Bombay, India.
* Center is defined as $x = 0$, $y = 0$, $z = 0$.
† Error here is root mean square deviation from the average.

Table 2. Interlaboratory comparison of track densities in feldspars: track counts by different groups on the same sample.

Sample label	Laboratory number	ρ_1 density $\times 10^6$/cm^2	Laboratory number	ρ_2 density $\times 10^6$/cm^2	% difference†
AB-24	1	$3.5 \pm 0.2^*$	5	$3.0 \pm 0.3^*$	15
AB-24	1	3.9 ± 0.2	5	4.0 ± 0.5	2.5
AB-24	1	3.4 ± 0.2	5	2.8 ± 0.4	19
AB-24	1	1.9 ± 0.1	5	1.7 ± 0.3	11
AB-19	2	1.6 ± 0.2	5	2.0 ± 0.2	22
AB-19	2	1.6 ± 0.2	5	1.3 ± 0.2	22
P-End	3	2.6 ± 0.3	4	2.4 ± 0.2	8
E	1	4.1 ± 0.1	4	2.7 ± 0.1	40
P-Top	1	2.3 ± 0.1	3	2.5 ± 0.2	8.5
P-Top	1	2.3 ± 0.1	2	2.5 ± 0.1	8.5
P-Top	2	2.5 ± 0.1	3	2.5 ± 0.2	0
A-3	1	1.5 ± 0.1	2	1.6 ± 0.1	6.5

* Errors here are only statistical.
† % difference defined as $\dfrac{2|\rho_1 - \rho_2|}{\rho_1 + \rho_2} \times (100)$.

DISCUSSION

This paper reports only experimental results. However, it may be noted that the depth dependence is quite flat in the galactic region. This result is very different than that expected from the cosmic ray spectrum derived from the St. Severin meteorite (Cantelaube *et al.*, 1967; Lal *et al.*, 1968; Maurette *et al.*, 1968). The rock has a very high spallation age and a large neutron dose (Lugmair and Marti, 1972) and therefore very likely has had a complex irradiation history. The flatness could also be due to an exotic fission component in the feldspars, but only future work will be able to decide on a final interpretation.

REFERENCES

Cantelaube Y., Maurette M., and Pellas P. (1967) Traces d'ions lourds dans les mineraux de la chondrite de Saint Severin. *Symposium on Radioactive Dating and Methods of Low-Level Counting*, pp. 215–229, International Atomic Energy Agency, Vienna.

Crozaz G., Drozd R., Hohenberg C. M., Hoyt H. P., Ragan D., Walker R. M., and Yuhas D. (1972) Solar flare and galactic cosmic ray studies of Apollo 14 samples (abstract). In *Lunar Science—III* (editor C. Watkins), pp. 167–169, Lunar Science Institute Contr. No. 88.

Lal D., Lorin J. C., Pellas P., Rajan R. S., and Tamhane A. S. (1969) On the energy spectrum of iron group nuclei as deduced from fossil track studies in meteoritic minerals. In *Meteorite Research* (editor P. M. Millman), pp. 275–285, Springer-Verlag.

Lugmair G. W. and Marti K. (1972) Neutron and spallation effects in Fra Mauro regolith (abstract). In *Lunar Science—III* (editor C. Watkins), pp. 495 497, Lunar Science Institute Contr. No. 88.

Maurette M., Thro P., Walker R., and Webbink R. (1968) Fossil tracks in meteorites and the chemical abundance and energy spectrum of extremely heavy cosmic rays. In *Meteorite Research* (editor P. M. Millman), pp. 286–315, D. Reidel.

Proceedings of the Third Lunar Science Conference
(Supplement 3, *Geochimica et Cosmochimica Acta*)
Vol. 3, pp. 2949–2953
The M.I.T. Press, 1972

Thermoluminescence of Apollo 14 lunar samples following irradiation at −196°C

I. M. Blair

Nuclear Physics Division,
Atomic Energy Research Establishment, Harwell, U.K.

J. A. Edgington and R. Chen*

Department of Physics,
Queen Mary College, London, U.K.

and

R. A. Jahn

Department of Applied Physics,
Lanchester Polytechnic, Coventry, U.K.

Abstract—The nonthermal leakage and saturation properties of the traps responsible for proton induced thermoluminescence in Apollo 14 lunar samples have been studied. Our data on leakage are consistent with a short term component of about an hour, together with a longer term, possibly constant, component. Saturation sets in at a dose of about 1.5 Mrad. The consequences of our findings for a possible explanation of the transient lunar phenomena are discussed.

Introduction

The prime motivation for studying the thermoluminescence (TL) properties of lunar samples following irradiation at a temperature roughly equivalent to the lunar night temperature is to see if they provide a credible energy storage mechanism to explain the transient lunar phenomena (TLP) according to an hypothesis discussed by us previously (Blair and Edgington, 1968). As well as the absolute magnitude of the induced TL, its saturation and leakage characteristics are also relevant. It is with these latter two properties of the induced TL in Apollo 14 samples that this paper is chiefly concerned.

Experimental Details

Our equipment has been fully described in a previous publication (Blair and Edgington, 1970). The samples studied in this experiment are listed in Table 1. The irradiations were performed at −196°C in the 160 MeV proton beam from the Harwell synchrocyclotron using a dose rate of 140 rad per second. The heating rate was 0.5°C per second, and the TL was observed in the 435–485 nm wave band. The general feature exhibited by all samples studied was a broad hump extending from

---

* On attachment from the Department of Physics and Astronomy, Tel-Aviv University, Tel-Aviv, Israel.

Table 1.

Harwell No.	NASA No.	Type
M 409a	14321, 147	Dark fraction of crushed rock
M 410a	14321, 147	Light fraction of crushed rock
M 411	14163, 113	< 1 mm bulk fines

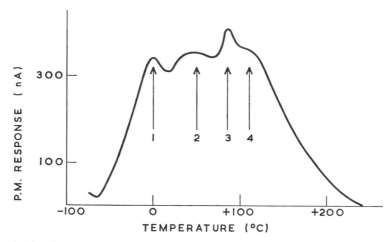

Fig. 1. The glow curve observed from the dark fraction of rock 14321,147 in the 435–485 nm waveband after a dose of 1 Mrad. Note the structure in the curve, four peaks being resolved.

−50°C to +150°C in which structure could be resolved. This is illustrated in Fig. 1 which shows a typical TL curve obtained from one of our samples. This behavior was similar in nature to that observed for Apollo 12 samples (Blair and Edgington, 1972). But as the magnitude of the effect was about twice as large we were able conveniently to study the leakage and saturation properties of the component peaks of the hump.

Leakage

If an irradiated sample is held at some temperature not too far below a TL peak, thermal leakage of energy from the traps will occur. Although the details of this process are sometimes obscure, in principle it is well understood. However, it has been pointed out recently by Garlick and Robinson (1971) that even if such a sample is held at a much lower temperature, at which thermal excitation effects would be negligible, nonthermal leakage can occur, possibly by a "tunnelling" process. We have studied this effect by starting the heating of the samples at different times after repeated irradiations of about 400 krad at −196°C. Figure 2 shows the variation of the response with delay time for one of the peaks in the curve shown in Fig. 1. Our experimental accuracy, and the number of points we have been able to measure in the time available, are insufficient to determine precisely the functional relationship between the response and the delay time. Depending on the mechanism involved this relationship can be of considerable complexity (Curie, 1963), but for the purpose of the present paper we shall assume that the decay is exponential. Thus neglecting retrapping effects the

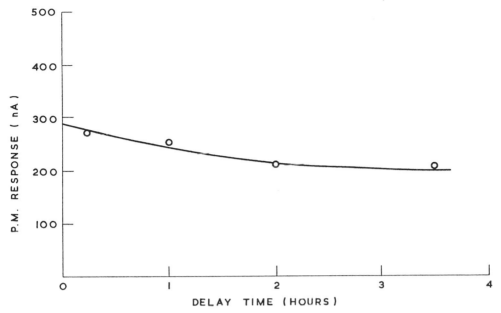

Fig. 2. Variation of the response observed at the temperature corresponding to peak 3 of the curve shown in Fig. 1 as a function of delay time after equal irradiations of 400 krad. The data have been fitted to a single exponential, but note that the two points at longest delay times have equal response suggesting a long term, possibly constant, component.

response (R) is related to the response for zero delay (R_0) and the delay time (t) by the following expression:

$$R = R_0 \exp(-\alpha t)$$

where α is the trap leakage rate.

In Table 2 we list, for each sample, the value of α for each of the peaks we were able to resolve from the glow curves. It will be noted that they are all of the same order of magnitude and correspond to mean leakage times of the order of an hour. Note, however, for the case shown in Fig. 2 that the two data points at longest delay times have equal response and thus our data are consistent with a short term decay plus a longer term component, possibly a constant of magnitude $\frac{2}{3}R_0$. We require further data to determine the form of the decay curve.

Table 2.

Sample No.	Peak temps. °C	α sec^{-1} ($\times 10^{-4}$)	γ sec^{-1} ($\times 10^{-3}$)	β sec^{-1} ($\times 10^{-3}$)	β' rad^{-1} ($\times 10^{-5}$)
M 409a	0	0.43	0.36	0.32	0.23
	+50	0.29	0.36	0.33	0.24
	+85	0.33	0.36	0.33	0.24
	+110	0.30	0.46	0.43	0.31
M 410a	−10	0.40	0.47	0.43	0.31
	+75	0.20	0.37	0.35	0.25
M 411	−15	1.20	0.50	0.38	0.27
	+90	0.67	0.35	0.28	0.20

Saturation

For small doses, i.e., those involving irradiation times small compared to the leakage time, and total doses small compared with the saturation dose, one might expect the magnitude of the peaks in the TL glow curves to vary linearly with dose. For larger doses these two conditions no longer hold, and one would expect the response to depart from linearity and approach an equilibrium value. We have obtained glow curves from our samples for a variety of irradiation times. In Fig. 3 we show the variation of the magnitude of the response for one of the peaks shown in Fig. 1 as a function of irradiation time. The approach to an equilibrium value is clearly seen. Making the same assumptions as in the previous section and assuming also that no new traps are produced during irradiation the response (R) is related to its equilibrium value (R_E) and the irradiation time (t) by the following expression:

$$R = R_E[1 - \exp(-\gamma t)]$$

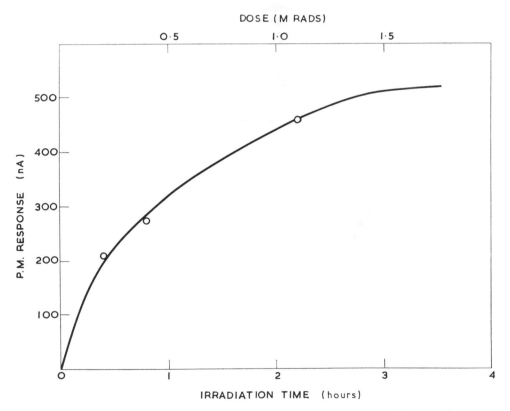

Fig. 3. Variation of the response observed at the temperature corresponding to peak 3 of the curve shown in Fig. 1 as a function of irradiation time at a constant dose rate of 140 rad per second. The corresponding doses are shown on the top scale. Note the onset of saturation at a dose of about 1.5 Mrad.

where $\gamma = \alpha + \beta$, and β is the trap excitation rate. From our data and using the values of α determined in the previous section we derive values of β for each of the component peaks observed. These are also listed in Table 2. The values of β in units of rad^{-1} (β') are also given. The saturation value (R_s) is related to the equilibrium value by the following expression:

$$R_E = \frac{\beta \cdot R_s}{\alpha + \beta}$$

Thus for our experimental conditions, which determine β, we note that the equilibrium value observed is only 10–20% below the saturation value for all the samples studied. As indicated by the similarity of the values of β obtained for each case, the onset of saturation occurred at similar doses, a typical value being ~ 1.5 Mrad.

DISCUSSION

Although the observed TL response from the Apollo 14 samples studied is too small by many orders of magnitude to provide an energy storage mechanism required to explain the TLP, it is of interest to discuss the consequences of the decay and saturation properties that we have observed. Certainly the mean leakage times of the order of an hour would indicate that the traps involved are incapable of storing energy for the required time of the order of 10^5 years. But the existence of a long term, possibly constant, component of the decay points the way to further study of this problem. Evidence of saturation at a dose of about 1.5 Mrad, which is small compared with the dose of about 100 Mrad that a sample buried at a depth of 10 gm/cm^2 would receive in 10^5 years on the moon, would indicate that much higher TL efficiencies would be required than hitherto measured to give a light output detectable against the background of reflected sunlight. The question of the energy conservation limit then arises, but a rough calculation indicates that a detectable effect is possible without requiring an efficiency in excess of unity.

REFERENCES

Blair I. M. and Edgington J. A. (1968) *Nature* **217**, 157.
Blair I. M. and Edgington J. A. (1970) *Proc. Apollo 11 Lunar Sci. Conf., Geochim. Cosmochim. Acta* Suppl. 1, Vol. 3, p. 2001. Pergamon.
Blair I. M., Edgington J. A., and Jahn R. A. (1971) *Earth Planet. Sci. Lett.* **13**, 116.
Garlick G. F. J. and Robinson I. (1971) Symposium No. 47 of the International Astronomical Union, Newcastle, U.K.
Curie D. (1963) *Luminescence in Crystals*, Methuen, p. 150.

Proceedings of the Third Lunar Science Conference
(Supplement 3, *Geochimica et Cosmochimica Acta*)
Vol. 3, pp. 2955–2970
The M.I.T. Press, 1972

Thermoluminescence of Apollo 12 samples: Implications for lunar temperature and radiation histories

S. A. Durrani, W. Prachyabrued, C. Christodoulides,*
and J. H. Fremlin

Department of Physics, University of Birmingham,
Birmingham B15 2TT, England

J. A. Edgington and R. Chen†

Department of Physics, Queen Mary College,
London, England

and

I. M. Blair

Nuclear Physics Division, Atomic Energy Research Establishment,
Harwell, England

Abstract—Thermoluminescence (TL), both natural and γ-ray induced, has been studied in contingency fines 12070,112 and in powdered samples from the interior rock chip 12051,15, in the temperature range 20°–600°C. In the natural samples glow peaks are observed at temperatures above ~ 300°C, those below having been thermally drained. Artificial irradiation introduces a prominent new peak at ~ 170°C, and enhances those in the region 300°–550°C. The growth of the high-temperature peaks with γ-ray irradiation is studied in order to deduce both the "natural (or equilibrium) dose" and the "half-dose" (needed to fill half the available "traps") for the samples. The TL parameters of the traps concerned are determined by the "initial rise" method. These data allow the "effective temperature" of lunar storage (358 $\pm$ 10°K for the fines and 346 $\pm$ 5°K for the rock chip) and the radiation dose rates received (~ 30 rad/y within a factor of 3, and ~ 60 rad/y within a factor of 2, respectively) to be calculated for the samples. By considering the attenuation of the diurnal heat wave on the moon, the mean subsurface depth of these samples can be estimated (~ 0.5 cm for the fines, and ~ 2.5 cm for the rock chip). Other topics discussed include spectral analysis of TL glow, inhomogeneities in the distribution of TL phosphors, and temperature sensitization of samples.

Introduction

The main purpose of studying the thermoluminescence (TL) of lunar material is to throw light on the thermal and radiation history of the moon. The radiation received by the moon (galactic cosmic rays and solar particles, in addition to internal radio-activity) leads to electrons being trapped in energy levels in the "forbidden" band gap of the solid, while the ambient temperature on the moon causes some of them to be released from these traps. By studying the properties of these traps, such as the growth

---

* Now at the Department of Applied Physical Sciences, University of Reading, England.

† On attachment from the Department of Physics and Astronomy, Tel-Aviv University, Tel-Aviv, Israel.

in the number of trapped electrons with radiation, and their release as a result of a systematic heating of the samples, it should, in principle, be possible to draw inferences regarding the two opposing processes: the radiation-filling and the thermal-drainage of traps (Christodoulides *et al.*, 1971). Several groups (see Hoyt *et al.*, 1970, 1971; Dalrymple and Doell, 1970; Doell and Dalrymple, 1971; Geake *et al.*, 1970; Garlick and Robinson, 1971) have studied lunar TL with these and related aims in mind. Two of us have reported work on Apollo 11 and 12 samples (Blair and Edgington, 1970; Blair *et al.*, 1971) with particular reference to a possible relation between TL and transient lunar phenomena. In the present investigation we have laid emphasis on (i) the determination of trap parameters corresponding to the various TL glow peaks observed (both natural and γ-ray induced), (ii) deducing the γ-ray equivalent dose in natural samples and investigating the saturation properties of various glow peaks, and (iii) attempting to interpret the history of lunar radiation and temperature environment from the above information.

Experimental Procedure

The samples investigated by us were (i) unsorted contingency fines 12070,112, and (ii) an interior rock chip 12051,15 (NASA type-code AB, fine to medium grained crystalline igneous rock (olivine basalt); part of a documented sample). The fines were in their natural (as received) state, while the chip had previously been irradiated at AERE, Harwell, with 160 MeV protons to an estimated dose of 200 krad (Blair *et al.* 1971).

Our experiments were carried out on two different sets of TL apparatus. On the first (called hereinafter set A), the amount of sample used was normally ≈ 2 mg, and the linear rate of heating β was $5°C \sec^{-1}$. On the second set (called B), the amount of sample used was normally ≈ 10 mg, and $\beta = 20°C \sec^{-1}$. The samples used in all cases were in powder form, and generally sieved to obtain various grain-size ranges; the rock samples were first crushed lightly in a mortar. The samples were spread uniformly over a tantalum strip and heated in an oxygen-free nitrogen atmosphere. The rate of heating was kept constant at a predetermined level with the help of a programed temperature controller operating via a thermocouple welded to the underside of the heating strip directly below the powder. A reflecting cone with an aperture at its apex (positioned over the sample) was used to limit the amount of black-body radiation received by the photomultiplier (PM). The photomultipliers used were specially selected tubes with low dark current. In set A, a quartz photomultiplier (EMI: 6256 SA) was used in a cooled housing (EMI assembly TE 102 TS, to cool the PM to $-20°C$), thus further reducing the dark current. Set B used a glass PM tube (EMI: 9502 SA). The PM output, measured by a sensitive electrometer, was recorded on an $X–t$, Y chart recorder. Set A had an extra pen which could record the integrated glow current simultaneously. Between the limiting cone and the PM a heat-shield glass window, and the filter in use, were interposed. For normal work, a broad-band filter was always employed, mainly to reduce black-body interference (on set A: Ilford "Bright Spectrum Blue" filter No. 622, transmission band 375 nm to 530 nm; on set B: Ilford "Bright Spectrum Violet" filter No. 621, transmission band 340 nm to 515 nm).

Results

Natural and artificially induced glow curves

Both the natural TL and that following γ-ray irradiation (from a 200 Ci cobalt-60 source) were studied in the unsorted virgin fines. Similar studies were carried out on the powdered rock chip sample (except that it had previously been subjected to a dose of 200 krad of 160 MeV protons). In what follows, TL apparatus A (with a heating rate $\beta = 5°C \sec^{-1}$) was always used, unless otherwise specified. Carefully weighed samples (usually of a preselected grain-size range but not otherwise separated) of approximately 2 mg were used for each run. Fresh samples were used for successive runs in order to avoid "temperature-sensitization" effects.

Figure 1 shows typical glow curves from the fines samples, both in their natural state and when subjected to artificial γ-ray doses ranging from 35 to 2700 krad (known to an accuracy of ~ ±10%). Each of the curves B–E was taken within a fairly short time (~15 min to 1 hour) after the end of the corresponding irradiation to minimize effects of thermal drainage at room temperature. It will be noticed that there is virtually no natural TL below 250°C; above this temperature there is a broad hump extending to ~550°C, containing more than one unresolved peak (note that the black-body radiation has been subtracted in all cases, though beyond ~500°C the subtraction becomes uncertain). As increasing amounts of γ-ray dose are added, a prominent new peak (termed peak I) is introduced around 175°C; it slowly moves towards lower temperatures as the artificial dose is increased (thereby enhancing the relative "weight" of peak I). The TL in the temperature region 300°–550°C (seen later by the initial rise method to contain peaks II and III) grows gradually with the γ-ray dose imparted, thus indicating that these peaks have not attained saturation in nature.

Figure 2 shows the TL glow curves from powdered samples of the rock chip 12051,15. Curve A represents TL from a natural sample (containing ~200 krad of

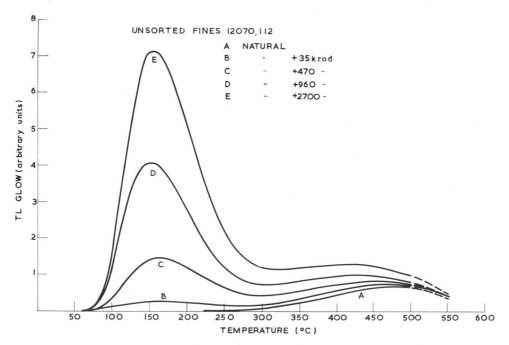

Fig. 1. Typical TL glow curves from the fines sample (grain diameter $d \leq 106\ \mu$; rate of heating $\beta = 5°C\ sec^{-1}$; a new 2.0 ± 0.1 mg sample used for each irradiation). The black-body contribution has been subtracted (though with increasing uncertainty beyond ~500°C); Ilford broad-band filter 622 used.

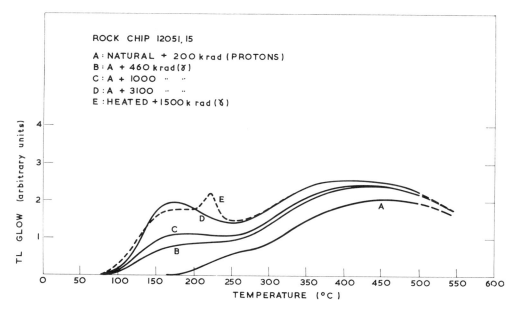

Fig. 2. Typical glow curves from (powdered) rock samples. Experimental conditions, and the scale of the ordinate, are as in Fig. 1 (but $d < 125\ \mu$ for curves A–D). The rock sample had, in all cases, an initial dose of ~ 200 krad of 160 MeV protons. The "pip" in curve E is discussed in the text. Notice the higher TL response of the heated sample ($d \sim 106$-$125\ \mu$). The near-equality of the areas under curves D and E results from the fact that the sample approaches saturation at ~ 1500 krad (see Fig. 5, 20-300°C).

160 MeV protons, i.e., a total dose N', say); while further doses of γ-rays (ranging from 460 krad to 3100 krad) have been superimposed upon N' in the case of curves B to D. The glow structure and its growth are broadly similar to those of fines (the hump in the region of $\sim 200°$–300°C is compatible with being the room-temperature remnant of the proton irradiation, which had been carried out some three months earlier).

Two points in Fig. 2 are worth noting. First, the rate of growth of peak I is much lower relative to the same peak in Fig. 1 (both drawn to the same scale in the ordinate): thus the relative heights of peak I at ~ 1 Mrad imparted γ dose are $\sim 1:4$ for rock to fines samples. Secondly, a sharp hump (at $\sim 225°$C) is observed from rock samples if powdered samples from a previous TL read-out are re-irradiated and then used for TL measurements. This was a persistent phenomenon and a number of experiments were carried out to investigate possible causes. The feature is further discussed below.

Spectral response

A series of Ilford filters (nos. 600, 607, and 621–626; typical band width at half-height ~ 50 nm) were used on apparatus B to determine the glow spectrum from various traps in the rock sample 12051,15 after irradiation with 270 krad of γ-rays. The photomultiplier current generated at any given temperature of read-out is a function of the emitted TL intensity $I(\lambda)$, the transmission characteristics $F(\lambda)$ of the filter in use, and the quantum efficiency $Q(\lambda)$ of the photomultiplier, λ being the wavelength. It is possible, by assuming as a first approximation that the intensity I at a given temperature is a slowly varying function of λ over the waveband of the filter, to deduce the (relative) value of $I(\lambda)$ from the observed current and from the known values of $Q(\lambda)$ and of $F(\lambda)$ for the various filters employed. This is done by carrying out the numerical integration $\int F(\lambda)\, Q(\lambda)\lambda\, d\lambda$ over the transmission band of each filter in turn (and by deriving a self-consistent solution by reiteration if necessary, though in our case it was not found to make a significant difference).

Figure 3 shows the emitted TL spectrum at three different temperatures (450°, 350°, and 200°C). The main conclusions derived from these and other similar curves (not shown) are that (i) very little TL intensity is observed at wavelengths $\gtrsim 550$ nm; and (ii) although at all temperatures the short wave length component (~ 380–490 nm, violet to blue) predominates, yet the long wavelength part (~ 500–570 nm, green to yellow) constitutes a more significant fraction of the total output at lower temperatures of emission ($\lesssim 300°C$).

Dose response and saturation effects

In Figs. 4 and 5 are shown the growth of the TL glow (integrated over the temperature intervals 400°–500°C, 300°–400°C, and 20°–300°C) as increasing amounts of γ-ray dose are imparted to the fines and the rock chip samples, respectively. Both types of sample show saturation in their 300°–500°C TL output at natural + ~ 2.5 Mrad γ-ray dose. By extrapolating backward the initial trend of the TL growth curve (see equation (6) below) it is possible to estimate the natural TL in each case. For

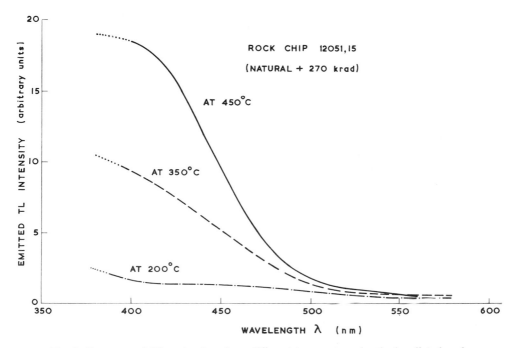

Fig. 3. Spectrum of TL emitted at three different temperatures by the irradiated rock sample (grain size, 125 $\mu < d \le 355\ \mu$). See text for method of analysis.

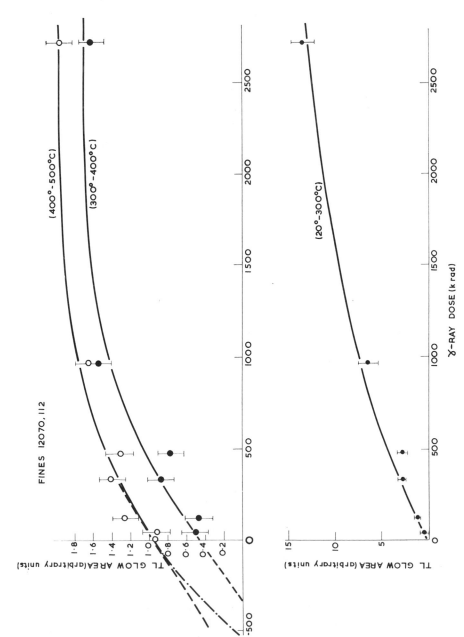

Fig. 4. Dose response of the fines (grain size $d < 106 \mu$). By extrapolation of the initial trend of the TL growth curve (shown as a series of dots and dashes (–·–) for the top curve (400–500°C)), the natural (or equilibrium) dose in each sample may be estimated (see equation (9)). The dashed line (– – –) is merely the tangent to the curve at the origin. The error bars result from repeated measurements, intrinsic inhomogeneities of samples, weighing uncertainties, etc.

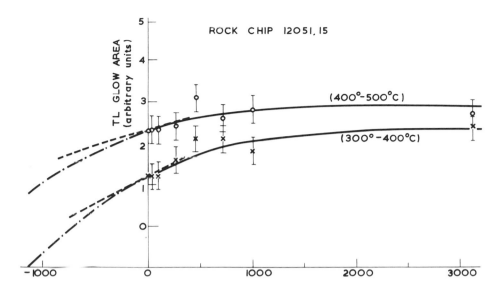

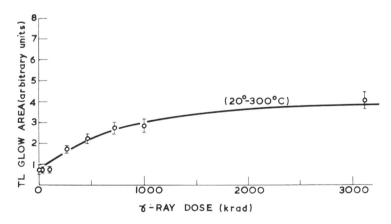

Fig. 5. Dose response of the rock sample ($d < 125\ \mu$) drawn to the same scale in the ordinate as Fig. 4. By extrapolation of the initial trend of the TL growth curve, the natural (or equilibrium) dose in each sample may be estimated (see discussion in text). For other information see caption to Fig. 4.

the fines, the natural dose in the temperature interval 400°–500°C is estimated to be ~550 krad, and in the rock chip ~1700 krad; while the estimates for the temperature interval 300°–400°C are: ~350 krad (fines), and ~1100 krad (rock chip).

The fact that the natural dose corresponding to the higher temperature region is larger than that for the lower region is reasonable, as the traps in the former case are deeper and are therefore better able to retain their dose. It must, however, be pointed out that there is a considerable scatter in the TL output from individual 2 mg samples. This effect, which is attributable to inhomogeneities in the distribution of the TL phosphors, is discussed below.

TL parameters

In discussing the thermal and radiation history of a thermoluminescent material, it is essential to know the values of the basic TL parameters E (trap depth) and s

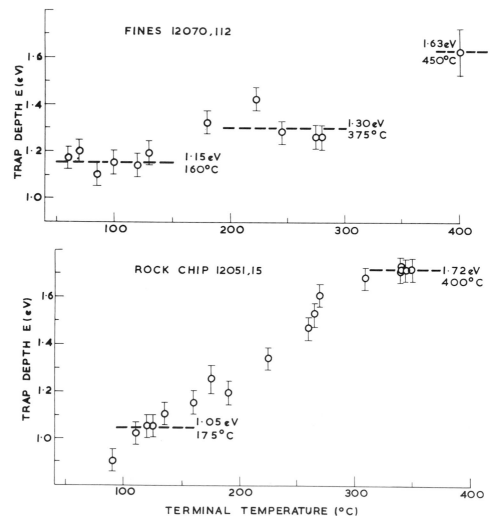

Fig. 6. Determination of the trap depth E for the various peaks in the fines and the rock samples by the initial rise method. The samples are heated to successively increasing terminal temperatures (followed by cooling). The horizontal clusters represent the peaks whose E and T^* values are marked. The values of T^* were obtained by a separate series of experiments involving the "thermal cleaning" of peaks.

(frequency factor) for it. In the first-order kinetics, these parameters are related as follows (Randall and Wilkins, 1945)

$$n(t, T) = n_0 \exp\left(-s \cdot t e^{-E/kT}\right) \tag{1}$$

where n is the number of filled electron traps at time t in a body kept at an absolute temperature T, n_0 is the number at $t = 0$, and k is Boltzmann's constant. This leads to a mean life τ ($= \tau_{\frac{1}{2}}/\ln 2$) for the trapped electrons at temperature T, given by

$$\tau(T) = \tau_0 \cdot e^{E/kT} \tag{2}$$

where $\tau_0 = s^{-1}$ (and is characteristic of the trap).

There are several ways of determining E and s experimentally in favorable circumstances. The one we found satisfactory for the Apollo 12 samples is the so-called initial rise method (Garlick and Gibson, 1948). This method is based on the fact that, at temperatures sufficiently below the peak temperature T^* for a given trap, the TL intensity $I(T)$ rises with temperature as $e^{-E/kT}$, irrespective of the order of kinetics and the rate of heating β involved. On plotting $\ln I$ against $1/T$ the initial part of the glow curve thus yields a straight line with slope $-E/k$. A number of heatings (interrupted at successively increasing temperatures, and followed by cooling) yield a series of values of E, until deviations from linearity become significant when the temperature of a glow peak T^* is approached. Figure 6 shows the resulting E values, plotted against the temperature of interruption, for both types of sample. The horizontal clusters correspond to the trap depths for the respective peaks. The s factor is then calculated from the (first-order kinetics) relation

$$\beta \cdot E/kT^{*2} = s \cdot e^{-E/kT^*} \tag{3}$$

Table 1 records the trap parameters for the peaks found in the fines and the rock chip samples by the initial rise method. Also shown in the table are the half-lives $\tau_{\frac{1}{2}}$ of trapped electrons corresponding to these peaks, computed from equation (2), at $-30°C$ (approximately the mean temperature of the moon unaffected by the diurnal heat wave), $20°C$ (room temperature), and $120°C$ (approximately the maximum lunar surface temperature).

Table 1. Trap parameters* and half-lives of peaks at different storage temperatures

Sample	Peak No.	Temp. (°C)	Trap depth E (eV)	Frequency factor s (sec^{-1})	Half-life $\tau_{\frac{1}{2}}$ at $-30°C$	$+20°C$	$+120°C$
12070,112	I	160	1.15 ± 0.05	$\sim 7 \times 10^{12}$	2×10^3 yr	73 days	55 sec
(Fines)	II	375	1.30 ± 0.05	$\sim 2 \times 10^9$	1×10^{10} yr	2.5×10^5 yr	182 days
	III	450	1.63 ± 0.10	$\sim 4 \times 10^{10}$	4×10^{15} yr	7×10^9 yr	5×10^2 yr
12051,15	I	175	1.05 ± 0.05	$\sim 2 \times 10^{11}$	6.5×10^2 yr	47 days	100 sec
(Rock chip)	II	400	1.72 ± 0.05	$\sim 1 \times 10^{12}$	1×10^{16} yr	7×10^9 yr	3×10^2 yr

* Determined by the initial rise method (see text); dose given: 3.9 mrad of γ-rays; rate of heating: $5°C \ sec^{-1}$.

DISCUSSION

Determination of lunar temperature and radiation dose rate

A good deal of information on the thermal and radiation history of samples may be gleaned from studying their natural and artificially induced TL curves. Thus it is seen (Fig 1) that peak I in the natural samples of lunar fines has been completely drained. This is readily understood since, from Table 1, its half-life at room temperature is only 73 days, whereas the experiments were performed some 18 months after retrieval of the samples from the moon. This underlines the importance of placing lunar samples, destined for TL study, in deep freeze (say at liquid nitrogen temperature, though even $-20°C$ storage would help), as well as of their prompt distribution for such work. Vital information, particularly from core tube samples which had previously remained at low ambient temperatures on the moon, may otherwise be irretrievably lost.

Glow peaks which have not reached saturation in nature (e.g., the peaks above $\sim 350°C$ in Figs. 1 and 2) but have attained a dynamic equilibrium in the prevailing radio-thermal environment (as a result of competition between the thermal emptying and radiative filling of the corresponding traps) can be used to estimate the environmental temperature if a certain dose rate is assumed, and vice versa. For details of the general method, reference may be made to Christodoulides *et al.* (1971) or Hoyt *et al.* (1971).

Briefly, consider a sample with N traps of depth E and frequency factor s irradiated with a dose rate r while being held at a temperature T. Then, if p is the probability per unit dose of filling one such trap, the number n of traps filled at time t is given by

$$n = \frac{N}{1 + \frac{s}{rp} \cdot e^{-E/kT}} \cdot \left\{ 1 - \exp\left[-rp\left(1 + \frac{s}{rp} \cdot e^{-E/kT} \right) t \right] \right\} \tag{4}$$

The number n_{eq} of traps filled at equilibrium, for large t, is thus

$$n_{eq} = \frac{N}{1 + \frac{s}{rp} \cdot e^{-E/kT}} = \frac{N}{1 + 1/r\tau p} \tag{5}$$

where, in the second equality in equation (5), we have used $\tau = \dfrac{e^{E/kT}}{s}$ given by equation (2).

If the sample is irradiated artificially at such high dose rates (or at such low temperatures) that decay (or drainage) of filled traps may be ignored, then equation (4) reduces to

$$n = N(1 - e^{-rt \cdot p}) \tag{6}$$

The probability p may then be expressed in terms of observable quantities by considering the total artificial dose rt for which the emitted TL under the glow peak is just half of the saturation TL (i.e., $N/2$ traps are filled). Such a dose may be termed $R_{\frac{1}{2}}$, or "half-dose" on the analogy of half-life; then

$$p = 0.693/R_{\frac{1}{2}} \tag{7}$$

Substituting p in equation (5), we obtain

$$n_{eq} = \frac{N}{1 + \{(R_{\frac{1}{2}}/r) \cdot se^{-E/kT}/0.693\}} = \frac{N}{1 + (R_{\frac{1}{2}}/r\tau_{\frac{1}{2}})} \quad (8)$$

where $\tau_{\frac{1}{2}} = 0.693\tau$.

If we wish to determine the artificial dose R_{eq} that would produce the same amount of TL in a sample (starting with all traps initially empty) as it had when at equilibrium in nature (usually called natural dose), we solve equation (6) (with $rt = R_{eq}$ and $n = n_{eq}$) simultaneously with the second equality in equation (8) to obtain the useful relation

$$R_{eq} = \frac{R_{\frac{1}{2}}}{0.693} \cdot \ln\left(1 + \frac{r\tau_{\frac{1}{2}}}{R_{\frac{1}{2}}}\right) \quad (9)$$

This formula is applicable to curves such as those in Figs. 4 and 5.

The ratio of the natural TL under the glow peak emitted by a sample at equilibrium to the TL emitted by an artificially saturated sample gives us the value of n_{eq}/N. Then, if the dose rate r in nature is known, the first of the equalities in equation (8) allows us to calculate T. Thus

$$T = \frac{E/k}{\ln\left[\dfrac{sR_{\frac{1}{2}}}{0.693r\left(\dfrac{N}{n_{eq}} - 1\right)}\right]} \quad (10)$$

The value of T is only weakly dependent on the values of r and N/n_{eq} used, since they occur in the logarithmic term—indeed, a factor of 2 in the product $r \cdot [(N/n_{eq}) - 1]$ changes the calculated value of T, in the cases of interest to us, by only $\sim 1\%$ in equation (10).

We have used the characteristics of peaks III and II in lunar fines and the rock chip, respectively, to calculate from equation (10) the effective lunar storage temperature in each case (see Table 2). The dose rate from galactic cosmic rays (in "2π geometry") in the vicinity of the earth is known (Haffner, 1967) to be ~ 9 rad/y, with a comparable contribution from solar flares. The efficiency of these radiations, relative to 1 MeV β and γ, in producing TL in materials is unknown; but preliminary experiments carried out by two of us (S.A.D. and W.P.) on meteorites bombarded with 7 GeV protons suggests the relative efficiency to be close to 1.

If the lunar temperature is assumed, for simplicity, to vary sinusoidally about a mean temperature T_0 with an amplitude A such that

$$T(\phi) = T_0 + A \sin \phi \quad (11)$$

where ϕ is the phase angle, the temperature T_{eff} (which will have the same effect on TL as the actual cycle) may be defined by the relation.

$$e^{-E/kT_{eff}} = \frac{1}{2\pi} \int_0^{2\pi} e^{-E/kT(\phi)} \, d\phi \quad (12)$$

The temperature T_0 of a body in space at 1 AU from the sun may be taken as 240°K, and the value of A at the surface of the moon, say A_s, as 150°K (Kopal, 1969). The amplitude of temperature modulation is sharply attenuated with depth d

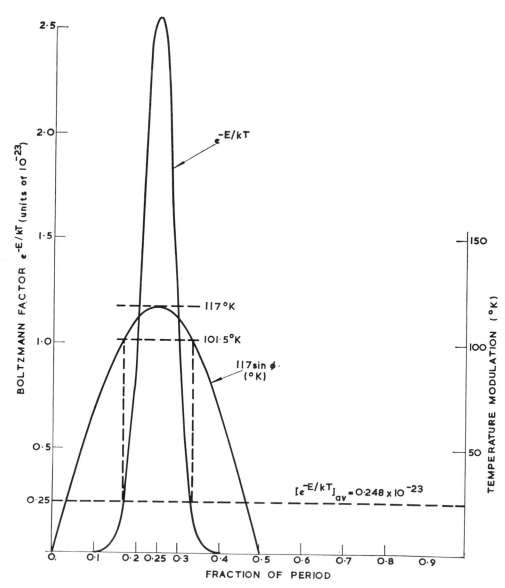

Fig. 7. Determination of the "effective temperature" T_{eff} that has the same effect on draining the TL as a sinusoidally varying temperature cycle on the moon. The average value $[e^{-E/kT}]_{\text{av}}$ is such that, if maintained over a lunation, it will have the same area under it as $\exp(-E/kT(\phi))$. By projecting the points of intersection of the average line and the Boltzmann factor onto the modulation curve, T_{eff} can be found graphically. (Alternatively, T_{eff} may be found by numerical integration.)

below the lunar surface. Following Hoyt et al. (1971), we assume a value of 24 cm for λ, the thermal wavelength for Apollo 12 fines, in the relation

$$A = A_s e^{-2\pi d/\lambda} \tag{13}$$

Figure 7 shows how the integral in equation (12) may be determined graphically. With $E = 1.6$ cV and $A = 117°K$ (this being the amplitude at $d \approx 1$ cm), the value of T_{eff} is found to be 341.5°K, which may be compared with $T_{\text{max}} = 240°K + 117°K = 357°K$ (at 1 cm depth).

It is possible to use the effective storage temperature found for the fines sample (Table 2) to estimate the mean depth below the lunar surface from which it came. Thus $T_{\text{eff}} = 358°K$ is found to correspond to $A = 136°K$ (i.e., $T_{\text{max}} = 376°K$), which yields a value of mean depth $d \approx 0.4$ cm.

As for the solid rock sample, if the thermal conductivity of lunar regolith in the top 1 m, reported from the Apollo 15 heat flow measurements (Langseth et al., 1972) to be ~ 10 times that for Apollo 12 fines, is used, this results in a $\lambda \sim 70$ cm. The temperature of the rock chip ($T_{\text{eff}} = 346°K$, i.e., $T_{\text{max}} = 362°K$ or $A = 122°K$) then yields a value of mean subsurface depth $d \approx 2.5$ cm. This depth, however, must only be treated as approximate in view of all the uncertainties involved.

Finally, if one takes the storage temperature as known, it is possible to estimate the dose rate experienced by the sample by using equation (9)—though not very accurately because of the logarithmic term involved. The value of $T_{\text{eff}} = 358 \pm 10°K$ for the fines leads to a half-life $\tau_{\frac{1}{2}} \sim 2 \times 10^4$ yr (varying by a factor of ~ 3 either way owing to the $\pm 10°K$ uncertainty in temperature). Hence the natural dose $R_{\text{eq}} \sim 550$ krad (corresponding to the temperature interval 400°–500°C for peak III in the fines (Table 2) yields a value of ~ 30 rad/y for the dose rate, accurate to a factor of ~ 3 either way. The value estimated for the rock chip using peak II (Table 2) is found, in a similar way, to be ~ 60 rad/y within a factor of ~ 2 either way.

It is clear from the foregoing that useful quantitative inferences can be drawn regarding the thermal and radiation histories of lunar samples by studying their TL. This will be increasingly true in future when data from the temperature sensing probes to be installed during the forthcoming Apollo missions (Langseth et al., 1972) become available so that some of the uncertainties regarding the temperature gradients in core tube samples can be eliminated (no such samples were available for use in the present investigation).

Table 2. Effective lunar storage temperature T_{eff} calculated from equation (10), assuming a dose rate of 10 rad/year*

Sample	Peak No.	Temp. (°K)	Trap depth E (eV)	Frequency factor s (sec^{-1})†	Half-dose $R_{\frac{1}{2}}$ (krad)	N/n_{eq}	T_{eff} (°K)
12070,112 (Fines)	III	723	1.63 ± 0.10	3.8×10^{10}	500	2.1	358 ± 10
12051,15 (Rock chip)	II	673	1.72 ± 0.05	1.4×10^{12}	500	1.3	346 ± 5

* For explanation of symbols, see text. Values of $R_{\frac{1}{2}}$ and N/n_{eq} are for the temperature interval 400–500°C. A change in the dose rate r by a factor of 2 changes the value of T_{eff} by $\sim 1.3\%$.
† The value of s changes inversely by a factor of ~ 2 for each 0.05 eV change in E.

One interesting use of the above calculations of the lunar temperatures is to specu-late on the temperature history of the earth's surface over the corresponding period of $\sim 10^4$–10^5 yr during which the equilibrium doses have been accumulated by the lunar samples. If the constancy of lunar temperatures over this period is accepted as a guide, then the glaciation cycles of the earth during the last $\sim 100{,}000$ yr cannot be attributed to sustained increases in solar heat output leading to enhanced cloud formation and precipitation of snow near the poles. Other causes for the glaciation pattern will then have to be invoked. Of course the existence of periods of low solar heat output in the past cannot be excluded on the above evidence.

Inhomogeneities and temperature effects in samples

In interpreting thermoluminescence data, a constant source of uncertainty is the variation in the TL output per unit mass from sample to sample. This is attributable to inhomogeneities in the distribution of phosphors responsible for TL, as first pointed out by Fremlin and Srirath (1964) in the case of pottery sherds, and also noted by Durrani and Christodoulides (1969) in meteorites. These nonuniformities are liable to be accentuated when small amounts of (independent) samples have to be used, as in the present investigation. The effect was particularly noticeable in the case of lunar fines, as can be seen from the scatter of experimental points on the dose response curve (Fig. 4). The ultimate solution of the problem may well be to work with indi-vidual grains of separated minerals as attempted by Hoyt *et al.* (1972).

It may be argued that better uniformity in TL response might be achieved by using physically the same sample over and over again after giving it varying artificial doses. But here one comes up against the problems of temperature sensitization and predose effects (Zimmerman *et al.*, 1965); for instance some tektite and meteorite samples have been shown (Durrani *et al.*, 1970; Christodoulides *et al.*, 1970) to yield six to eight times as much TL per unit dose after being annealed as prior to it. In the case of Apollo 12 samples, it was established that, for a test dose of 920 krad for fines and 1500 krad for the rock chip, the sensitivity of a given sample preheated to 500°C was enhanced by ~ 30–40% for fines and by ~ 10–20% for the rock.

Another effect of temperature which we noticed in the present investigation is the sharp hump (at ~ 225°C) visible in heated (and then irradiated) rock samples (Fig. 2). A number of subsidiary experiments were carried out to exclude possible causes (chemiluminescence; vacuum grease used in the previous proton-irradiation experi-ments; TaO_2 from the heating strip; oxidation of samples). It was established that the hump was observed only if the rock powder had been previously heated to at least 500°C; it became much more marked as a result of pre-heating up to 600°C. The area under the hump grew with dose and roughly in proportion with the body of the main curve. The only explanation that can be offered at this stage (in the absence of a definitive crystallographic analysis) is that a phase change brought about by heating the sample to above 500°C is apparently responsible for the phenomenon. No such "pip effect" was observed in the case of fines, or of any Apollo 14 samples sub-sequently examined.

In an effort to minimize the intrinsic inhomogeneities in the samples used, grain-size control was implemented in each series of related TL observations, as mentioned earlier. The advantage of such a procedure was established by a subsidiary investigation in which grains from unirradiated fines, subdivided into four groups by size (with diameters d ranging from 20 $\mu < d \leq$ 53 μ up to $d >$ 355 μ), were examined for TL output. It was found that the specific TL (output per unit mass) went down monotonically as the grain size increased (and the surface to volume ratio fell). The 2 mg samples were always spread out uniformly and thinly enough on the heating strip that loss of glow received by the photomultiplier through layering and intergrain scattering of light was minimal. Since each 2 mg sample consisted of several thousand grains, it was hoped that the phosphors chiefly responsible for TL output would be reasonably randomly distributed between samples (short of actual mineralogical separation) and that at least specific TL would be kept within narrow limits by means of grain-size control.

Acknowledgments—S. A. Durrani wishes to thank the Royal Society for a grant-in-aid of scientific investigations. W. Prachyabrued is grateful to the Colombo Plan authorities and the Thai Government for financial support.

REFERENCES

Blair I. M. and Edgington J. A. (1970) Luminescence and thermoluminescence. *Proc. Apollo 11 Lunar Sci. Conf., Geochim. Cosmochim. Acta* Suppl. 1, Vol. 3, pp. 2001–2012. Pergamon.

Blair I. M., Edgington J. A., and Jahn R. A. (1971) The luminescent and thermoluminescent properties of Apollo 12 lunar samples. *Earth Planet. Sci. Lett.* **13**, 116–120.

Christodoulides C., Durrani S. A., and Ettinger K. V. (1970) Study of thermoluminescence in some stony meteorites. *Modern Geology* **1**, 247–259.

Christodoulides C., Ettinger K. V., and Fremlin J. H. (1971) The use of TL glow peaks at equilibrium in the examination of the thermal and radiation history of materials. *Modern Geology* **2**, 275–280.

Dalrymple G. B. and Doell R. R. (1970) Thermoluminescence of lunar samples from Apollo 11. *Proc. Apollo 11 Lunar Sci. Conf., Geochim. Cosmochim. Acta* Suppl. 1, Vol. 3, pp. 2081–2092. Pergamon.

Doell R. R., and Dalrymple G. B. (1971) Thermoluminescence of Apollo 12 lunar samples. *Earth Planet. Sci. Lett.* **10**, 357–360.

Durrani S. A. and Christodoulides C. (1969) Allende meteorite: Age determination by thermoluminescence. *Nature* **223**, 1219–1221.

Durrani S. A., Christodoulides C., and Ettinger K. V. (1970) Thermoluminescence in tektites. *J. Geophys. Res.* **75**, 983–995.

Fremlin J. H. and Srirath S. (1964) Thermoluminescent dating: Examples of non-uniformity of luminescence. *Archaeometry* **7**, 58–62.

Garlick G. F. J. and Gibson A. F. (1948) The electron trap mechanism of luminescence in sulphide and silicate phosphors. *Proc. Phys. Soc., Lond.* **60**, 574–590.

Garlick G. F. J. and Robinson I. (1971) The thermoluminescence of lunar samples. Paper read at the 47th Meeting of the International Astronomical Union at Newcastle, U.K.

Geake J. E., Dollfus A., Garlick G. F. J., Lamb W., Walker G., Steigmann G. A., and Titulaer C. (1970) Luminescence, electron paramagnetic resonance and optical properties of lunar material from Apollo 11. *Proc. Apollo 11 Lunar Sci. Conf., Geochim. Cosmochim. Acta* Suppl. 1, Vol. 3, pp. 2127–2147. Pergamon.

Haffner J. W. (1967) *Radiation and Shielding in Space*, p. 279. Academic Press.

Hoyt H. P., Kardos J. L., Miyajima M., Seitz M. G., Sun S. S., Walker R. M., and Wittels M. C. (1970) Thermoluminescence, x-ray, and stored energy measurements of Apollo 11 samples. *Proc. Apollo 11 Lunar Sci. Conf., Geochim. Cosmochim. Acta* Suppl. 1, Vol. 3, pp. 2269–2287. Pergamon.

Hoyt H. P., Miyajima M., Walker R. M., Zimmerman D. W., Zimmerman J., Britton D., and Kardos J. L. (1971) Radiation dose rates and thermal gradients in lunar regolith: TL and DTA of Apollo 12 samples. *Proc. Second Lunar Sci. Conf.*, *Geochim. Cosmochim. Acta* Suppl. 2, Vol. 3, pp. 2245–2263. MIT Press.

Hoyt H. P., Walker R. M., Zimmerman D. W., and Zimmerman J. (1972) Thermoluminescence studies of lunar samples (abstract). In *Lunar Science—III* (editor C. Watkins), pp. 401–403, Lunar Science Institute Contr. No. 88.

Kopal Z. (1969) *The Moon*, p. 373. Reidel.

Langseth M. G., Clark S. P., Chute J., and Keihm S. (1972) The Apollo 15 lunar heat flow measurement (abstract). In *Lunar Science—III* (editor C. Watkins), pp. 475–477, Lunar Science Institute Contr. No. 88.

Randall J. T. and Wilkins M. H. F. (1945) Phosphorescence and electron traps. *Proc. Roy. Soc. London* **A184,** 366–407.

Zimmerman D. W., Rhyner C. R., and Cameron J. R. (1965) Thermal annealing effects on the thermoluminescence of LiF. *USAEC Rep. COO–1105–101.* Also in *Health Phys.* **12,** 525 (1966).

Proceedings of the Third Lunar Science Conference
(Supplement 3, *Geochimica et Cosmochimica Acta*)
Vol. 3, pp. 2971–2979
The M.I.T. Press, 1972

Luminescence of lunar material excited by electrons

J. E. Geake and G. Walker

U.M.I.S.T., Manchester, England

A. A. Mills

University of Leicester, England

and

G. F. J. Garlick

University of Hull, England

Abstract—The Manchester luminescence spectrophotometer, which previously used proton excitation has now been converted to use 25 keV electrons at a current density of about 50 μA/cm^2. The beam is pulsed at 2.5 to 1000 Hz, and a phase-sensitive detector with variable phase enables luminescence decay times to be measured. A spectral range of 4000–8500 Å may be scanned without a break, using an extended tri-alkali photomultiplier.

Luminescence emission spectra are shown for Apollo 14 fines and chips and for Apollo 15 fines, in comparison with the Apollo 11 and 12 material previously investigated. In all these materials the main luminescent constituent is plagioclase, which shows three main emission peaks, in the blue, green, and near IR, at about 4500, 5600, and 7800 A, respectively. The blue peak is probably caused by lattice defects. We previously ascribed the green peak to Mn^{2+} in Ca^{2+} sites; we now have some evidence that the IR peak may be due to Fe^{3+} in Ca^{2+} sites. Decay times of about 0.1 ms for the blue peak, 5 ms for the green peak and 1.5 ms for the IR peak are consistent with these explanations. The IR peak is much weaker in lunar than in terrestrial plagioclase, possibly because most of the lunar iron is in the Fe^{2+} form.

Luminescence emission photographs have been taken, in color, of two Apollo 14 chips; these both show only the characteristic plagioclase emission observed for the Apollo 11 and 12 chips.

Introduction

THIS PAPER describes the continuation of the luminescence investigation of lunar materials described in our earlier papers (Geake *et al.*, 1970, 1971, and 1972). Progress has been made in the explanation of the luminescence found, and data collection has continued using the following Apollo 14 and 15 samples:

14163,51   fines less than 1 mm
14259,56   fines less than 1 mm
14310,206 chip, many fragments and dust
14312,3   chip
15071,46   fines less than 1 mm
15091,53   fines less than 1 mm
15231,69   fines less than 1 mm
15601,103 fines less than 1 mm
15251,56   fines less than 1 mm

Apollo 15 chips 15086,16 and 15557,21 arrived too late for results to be included in this paper.

Major changes have now been made in our equipment for luminescence spectro-photometry: we have now gone over from proton to electron excitation, and we have also replaced our dc amplification system by a pulsed ac system that enables lumines-cence decay times to be measured, in addition to emission spectra. Also, the introduc-tion of the extended tri-alkali photomultiplier now enables us to scan the full spectral range from 4000 to 8500 Å with a single photomultiplier, and without needing to cool it for the IR region.

Electron Excitation Equipment

Our previous work on the luminescence of meteorites and of lunar samples was carried out with proton excitation, because we were also interested in proton damage effects. We are now concerned mainly with luminescence emission spectra, therefore we have converted our equipment to use electron excitation. As electrons cause less damage than protons for a given light output, we are now able to use a beam current that gives an order of magnitude more initial signal while resulting in an order of magnitude less efficiency-loss due to damage. We can revert to proton excitation if the need ever arises. There is no evidence of any changes in the emission spectra, caused by using electrons instead of protons.

A diagram of the modified equipment is shown in Fig. 1; the rf-excited ion bottle and accelerator tube have been replaced by a commercial electron gun (Twentieth Century Electronics ED221), which contains the accelerator electrodes and a beam-control grid. The filament is fed by an adjustable 0–6 V stabilized dc supply, which in turn is fed by a 12 V output 25 kV-insulated mains transformer. The filament supply unit and the grid bias batteries are in an insulated box up at the accelerator potential, which is supplied from a Brandenburg 0–30 kV unit. The gun anode is earthed.

Figure 2 shows a block diagram of the electronic system. There are two channels, one using an extended tri-alkali photomultiplier (EMI 9659A) to scan the spectrum, and the other using a tri-alkali photomultiplier (EMI 9558B) to monitor the total light emitted in order to correct the output for fluctuations and sample-efficiency reduction. The two channels each consist of an amplifier and a phase-sensitive detector, and they feed a two-channel pen recorder. A pulse generator feeds the grid of the electron gun, to modulate the beam, and also the phase-sensitive detectors, to supply the reference waveform. The grid is fed via a high-voltage capacitor, and another high-voltage capacitor is used for decoupling, from the filament to earth. Two alternative pulse types are used: for spectral scans a square wave is used because this gives the maximum signal-to-noise ratio. Frequencies of 2.5–1000 Hz are used. For decay-time measurements the time resolution depends on the narrowness of the pulses, and typically 1 ms pulses 100 ms apart are used; the grid is fed directly, and the phase-sensitive detector is fed via a variable delay, so that the emission is sampled a known time after excitation. A plot of signal against delay time gives the lumines-cence decay time. However, this very unequal mark-to-space ratio gives a poor

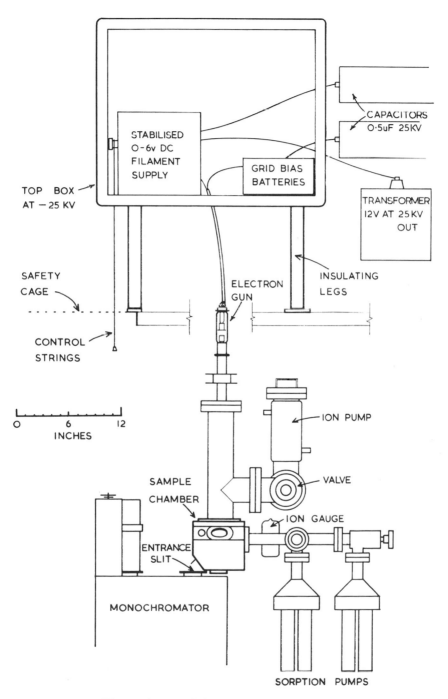

Fig. 1. Diagram of electron excitation equipment.

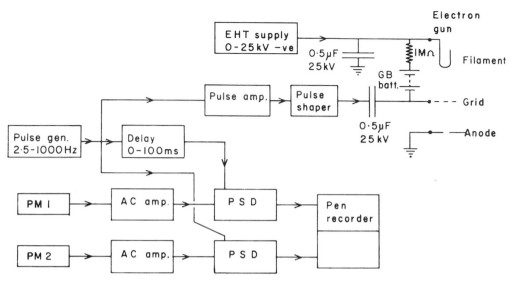

Fig. 2. Block diagram of electronic system.

signal-to-noise ratio; a semi-quantitative indication of decay time, with a much better signal-to-noise ratio, may be obtained simply by scanning the emission spectrum using a square wave at a wide range of frequencies, as shown later.

The rest of the equipment is as described earlier (Geake *et al.*, 1972).

Luminescence Emission Spectra

Luminescence emission spectra have now been investigated for fines samples from Apollo missions 11, 12, 14, and 15; Fig. 3 shows results for one sample from each mission. The other Apollo 14 and 15 samples investigated but not shown gave spectra very similar to those shown, in profile and intensity. Only a light-colored Apollo 12 sample is shown (12033); another light sample (12032) gave an identical profile but a slightly lower intensity, and a dark sample (12070) gave an identical profile but a much lower intensity. The Apollo 11 sample shown has a more prominent green peak than shown previously—usually they are like the Apollo 12 sample shown; this we believe to be sample variation. The Apollo 11 and 12 samples shown differ from our earlier results (Geake *et al.*, 1971) in that the weak IR emission of the fines that we then showed tentatively, is now shown, with our improved sensitivity, to be even weaker. The spectra for all these fines samples are rather similar in profile, with small variations in the prominence of the green peak, but differing mainly in overall intensity, which covers a range of a factor of 10, as shown in Table 1. The spectra are characteristic of plagioclase [(Na, Ca)(Al, Si)$_4$O$_8$], and a sample of separated lunar plagioclase investigated earlier (Geake *et al.*, 1971) showed the same spectral features as typical terrestrial plagioclases, with three main peaks at about 4500, 5600, and 7800 Å in the blue, green, and near IR, respectively. The blue peak is generally ascribed to lattice strain defects, and we were able to ascribe the green peak, which is dominant for lunar material, to the presence of a trace of Mn. We

Table 1. Relative emission intensites of fines.

Sample number	Relative Intensity at Green Peak
12033,60	10
12032,39	7
14163,51 ⎱ 14259.56 ⎰	4
12070.113	1.4
15071,46 ⎱ 15091,53 15601,103 ⎰	1
10084,6	1

conclude that the range of emission intensities found for the different fines samples are due to the presence of different proportions of plagioclase, superimposed on any overall reduction of efficiency due to radiation damage effects; also, that the small differences in spectral profile (mainly small differences in the blue/green peak height ratio) are due to the presence of different amounts of Mn in the relevant sites. Of the fines samples investigated, the Apollo 15 ones may be just significantly different from the others, in that although the emission intensity is low the green peak is rather prominent; this we ascribe to a low plagioclase content but a high Mn content in the plagioclase. We earlier found the energy conversion efficiency of the separated lunar plagioclase to be about 10^{-3}; this corresponds to about 100 on the scale of Table 1, giving lunar fines efficiencies in the range 10^{-4} to 10^{-5}, so these are the efficiencies to be expected for regions of the moon as observed from the earth. These efficiencies are far too low to give any surface luminescence capable of being observed from the earth, as shown by Nash and Greer (1970); our interest in luminescence now concerns the information it can give about the materials themselves, and their origin.

In addition to the fines samples, two Apollo 14 rock chips (14310,206 and 14312,3) have now been investigated. Color photographs have been taken of their luminescence emission under electron excitation, using the equipment at Leicester as described earlier (Geake et al., 1972). Both chips show only a fine-grained distribution of plagioclase crystals. Emission spectra, under electron excitation in the Manchester equipment, show plagioclase emission in both cases, but with distinct differences, as shown in Fig. 4; 14310,206 shows a spectrum very similar to that for the separated lunar plagioclase mentioned above, as did two Apollo 11 rocks investigated earlier (Geake et al., 1970); rock chips 14312,3 shows similar emission in the blue, but greatly reduced green emission, suggesting a very low Mn content. It is the first lunar sample we have found with the green peak less intense than the blue one.

When trying to find the cause of the different peaks in a luminescence emission spectrum, their luminescence decay times are useful evidence. As mentioned above, these may conveniently be found semi-quantitatively by scanning the spectrum at a range of different modulation frequencies. This is illustrated in Fig. 5, which shows the effect of scanning the same spectrum (that of an Apollo 14 fines sample) at high and low modulation frequencies. The green peak, with a decay time of a few milliseconds is missing from the high-frequency scan, whereas the blue peak, with a decay time of about a tenth of a millisecond, is present on both. This also illustrates the hazard of using modulated-beam ac systems for luminescence investigations, in that

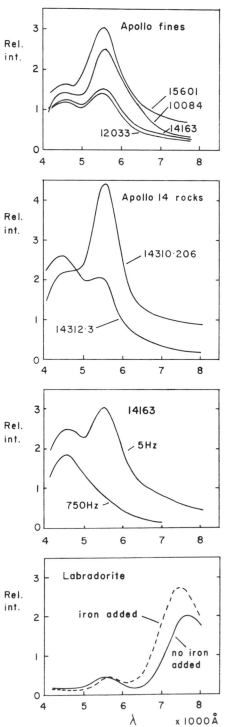

Fig. 3. Luminescence emission spectra for lunar fines.

Fig. 4. Luminescence emission spectra for lunar rock chips.

Fig. 5. Effect of modulation frequency on an emission spectrum.

Fig. 6. Effect of iron doping on the emission spectrum of labradorite.

features may be lost if too high a modulation frequency is used. For this reason, we normally use the very low frequency of 2.5 Hz.

Interest in the emission spectrum of plagioclase now centers on an attempt to explain the IR peak at 7800 Å, which is strong, and sometimes dominant, for terrestrial plagioclases, but weak or absent for lunar materials. Our sample of separated lunar plagioclase shows this peak; rock chip 14310,206 in Fig. 4 shows some emission there, but no peak. For the fines shown in Fig. 3 emission in this region is very weak. This difference between lunar and terrestrial plagioclases probably contains information about their different conditions of formation.

CAUSE OF IR EMISSION BY PLAGIOCLASE

We are now investigating the cause of the IR emission, using lunar samples and doped samples of terrestrial and synthesized plagioclases, and there is now some evidence that the cause may be ferric iron, Fe^{3+}, in Ca^{2+} sites.

Van Doorn and Schipper (1971) have recently found that sodalite ($Na_4Al_3Si_3O_{12}Cl$) doped with Mn^{2+} gave a green emission band at about 5400 Å; this is similar to the one we found for plagioclase doped with Mn^{2+}. However, they also found that sodalite doped with Fe^{3+} gave a broader emission band with its peak at about 7000 Å; this peak bears some resemblance in position and width to the IR peak which we find to be strong in terrestrial plagioclase, but weak or absent in lunar plagioclase, and may provide a clue as to its cause. We have therefore doped some terrestrial plagioclase (labradorite) with iron by heating it to 1200°C in air with ferrous sulphate, in the expectation that at this temperature, in air, most of the iron would go in as Fe^{3+}. Figure 6 shows emission spectra for doped and undoped samples: The IR peak has evidently been enhanced in relation to the green and blue peaks by adding iron. However, the "doped" spectrum has been scaled up by a factor of 2 so as to equalize the green peaks for comparison purposes, therefore the actual effect of doping has been a reduction everywhere, but less of a reduction for the IR peak than for the others. There are at least two possible explanations for this, but both are consistent with Fe^{3+} as the cause of the IR peak. One is that the iron actually went in partly as Fe^{2+} and partly as Fe^{3+}; Fe^{2+} is a powerful quencher of luminescence, so there might be an overall reduction in intensity, superimposed on some enhancement of the IR peak alone by the Fe^{3+} present. However, attempts to produce IR emission by iron doping of synthesized plagioclase, and also of CaO and wollastonite, have not so far been successful.

Another possible explanation is that the spectral changes shown in Fig. 6 are simply due to differential quenching by iron. Our decay-time measurements tend to support this explanation, but even these results, together with other features of the IR peak, seem consistent with theoretical predictions as to the properties of Fe^{3+} as a luminescence center. If we compare the probable effects of Mn^{2+} and Fe^{3+} as impurity ions in the same Ca^{2+} cation sites the expected luminescence transitions are very similar, in that they are both spin-forbidden and both have the same d^5 electron configuration. They therefore have the same energy level diagram, as shown in Fig. 7, and differ only in the value of the crystal field splitting parameter Δ; here Δ

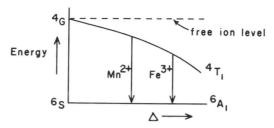

Fig. 7. Schematic energy level diagram for transition metals of d^5 electron configuration in a cubic crystal field. The $^4T_1$ level is actually triple: only the emitting state is shown.

is greater for the trivalent ion than for the divalent one, so a peak due to Fe^{3+} would be expected to occur at a longer wavelength than for the green Mn^{2+} peak, because the energy gap decreases as Δ increases. The shape of the $^4T_1$ level is not known, but it is probably either straight, or curved in the direction shown, giving an IR peak of width (in terms of frequency) either equal to or greater than that of the green peak; for example, if the level is curved as shown, the slope is greater at the Fe^{3+} position, so the energy spread is greater for a given Δ spread due to thermal vibrations, giving an Fe^{3+} peak broader than the Mn^{2+} one. The IR peak is in fact observed to be equal in width to the green one for some terrestrial plagioclases, and greater for others. The luminescence decay times for the two peaks should be similar, and fairly long (of the order of milliseconds), as the same spin-forbidden transition is involved, but nevertheless slightly shorter for Fe^{3+} as the energy jump is less, giving a greater chance of non-radiative deactivation. Our measurements show that the green peak for the iron-doped labradorite (attributed to the Mn^{2+} already there) has a decay time of about 5 ms, whereas for the IR peak it is about 1.5 ms, which is consistent with it being due to Fe^{3+}. The blue peak has a decay time of about 0.1 ms, which is consistent with it being caused by the lattice defects common in oxygen-dominated phosphors. As the IR peak has a slightly shorter decay time than the green one it should be somewhat less affected by a quenching agent such as Fe^{2+}, and this alone could qualitatively explain the spectral change produced by iron doping.

There is thus a good deal of circumstantial evidence that the IR peak for plagioclase is caused by Fe^{3+} in Ca^{2+} sites, which is consistent both with theoretical predictions and with such experimental observations as we have. Other transition metal ions could have rather similar properties, except that the most likely ones (e.g., Va or Ti) would probably produce emission much further into the IR than that observed. Furthermore, if the ion went into Si or Al sites rather than Ca ones there would be less room for them, and the resulting distortion would give a large value of Δ and therefore a small energy gap, giving emission that would again be too far into the IR.

CONCLUSIONS

We conclude that all the lunar material we have investigated shows some luminescence and that this is always characteristic of plagioclase. The green peak, which usually dominates the emission spectrum of lunar material, we had already ascribed to Mn^{2+}; the samples we have differ from each other mainly in the brightness of their emission, which probably depends on their plagioclase content, and in the relative intensity of the green peak, which probably depends on the Mn content of their plagioclase. We now have some circumstantial evidence that the IR emission may be due to Fe^{3+}; our tentative explanation of the much weaker IR emission from lunar plagioclase, as compared with terrestrial plagioclase, is that the lunar material probably has most of its iron in the form of Fe^{2+}, owing to a shortage of oxygen during its formation, whereas abundant oxygen would produce dominant Fe^{3+} in the terrestrial material. This further indication of oxygen shortage when the lunar surface material was formed is consistent with geological evidence, and also with our earlier work on the decomposition of heated lunar ilmenite (Geake *et al.*, 1970). We also reported earlier a similar absence of IR emission from another sample of extra-terrestrial plagioclase—the meteorite Juvinas.

Acknowledgments—We are grateful to M. L. Gould and D. Telfer for assistance with the construction and operation of the electron excitation equipment; to the Science Research Council for financial support, and to NASA for providing the lunar samples.

REFERENCES

Geake J. E., Dollfus A., Garlick G. F. J., Lamb W., Walker G., Steigmann G. A., and Titulaer C. (1970) Luminescence, electron paramagnetic resonance and optical properties of lunar material from Apollo 11. *Proc. Apollo 11 Lunar Sci. Conf., Geochim. Cosmochim. Acta* Suppl. 1, Vol. 3, pp. 2127–2147. Pergamon.

Geake J. E., Walker G., Mills A. A., and Garlick G. F. J. (1971) Luminescence of Apollo lunar samples. *Proc. Second Lunar Sci. Conf., Geochim. Cosmochim. Acta* Suppl. 2, Vol. 3, pp. 2265–2275. MIT Press.

Geake J. E., Walker G., and Mills A. A. (1972) Luminescence excitation by protons and electrons, applied to Apollo lunar samples. Proc. IAU Symp. 47, March 1971, *The Moon*, Urey and Runcorn, Eds., Reidel. (In press.)

Nash D. B. and Greer R. T. (1970) Luminescence properties of Apollo 11 lunar samples and implications for solar-excited lunar luminescence. *Proc. Apollo 11 Lunar Sci. Conf., Geochim. Cosmochim. Acta* Suppl. 1, Vol. 3, pp. 2341–2350. Pergamon.

Van Doorn C. Z. and Schipper D. J. (1971) Luminescence of O_2, Mn^{2+}, and Fe^{3+} in sodalite. *Phys. Letters* **34A**, 139–140.

Proceedings of the Third Lunar Science Conference
(Supplement 3, *Geochimica et Cosmochimica Acta*)
Vol. 3, pp. 2981–2995
The M.I.T. Press, 1972

Luminescence of Apollo 14 and Apollo 15 lunar samples

Norman N. Greenman and H. Gerald Gross

Space Sciences Department, McDonnell Douglas Astronautics Company—West,
Huntington Beach, California 92647

Abstract—Luminescence measurements have been made of Apollo 14 lunar samples with far uv, x-ray, and proton irradiation and of Apollo 15 lunar samples with x-ray irradiation. Preliminary efficiencies with the far uv are in the range 10^{-3} to 10^{-2}; efficiencies with x-rays and protons are in the range 10^{-8} to 10^{-6}. The crystalline igneous rocks show higher efficiencies, in general, than the breccias and glasses, and the ratio of intensity of the green to the blue luminescence peak tends to be higher for the crystalline igneous rocks than for the breccias and glasses. Therefore, both the efficiency and the spectral character appear to have a systematic relationship to lithologic type (granitic versus gabbroic versus fragmental) and to geologic history and processes on the moon (shocked versus unshocked or only mildly shocked material).

Introduction

We have been studying the luminescence of the Apollo lunar samples for the purpose of (1) understanding how the luminescence behavior reflects the origin, history, and environment of the lunar rocks; (2) discovering luminescence characteristics that might aid in geologic mapping and other lunar exploration activities; and (3) evaluating reports of luminescence on the moon based on astronomical observations. We have already reported the results of our studies of the Apollo 11 and Apollo 12 samples (Greenman and Gross, 1970a, 1970b, 1971) and some results from the Apollo 14 sample studies (Greenman and Gross, 1972). In this paper we present additional Apollo 14 and some Apollo 15 sample results.

Experimental Procedure

The experimental arrangements for the various irradiations were described in detail in our previous paper (Greenman and Gross, 1971). In this series of measurements the only major modification was the use of a McPherson 2 m spectrograph as a dispersing instrument in our far uv irradiations.

The samples used in the measurements are as follows: Apollo 14: 14003,19 (fines, contingency sample), 14163,30 (fines, bulk sample), 14259,37 (fines, comprehensive sample), 14301,50 (breccia, exterior), 14321,264 (breccia, interior), and 14310,155 (basalt, exterior); Apollo 15: 15015,22 (dark, vesicular, shiny glass from bottom exterior coating), 15015,22,1 (breccia, immediately adjacent to 15015,22), 15015,27 (dark, frothy, dull glass from top exterior coating), and 15085,28 and 15085,30 (both coarse-crystalline gabbro, exterior); terrestrial: granite (California), gabbro (California), and willemite (New Jersey).

Data Analysis and Results

Far uv irradiation (1100–2200 Å)

The terrestrial and lunar samples were first investigated with two vacuum monochromators, one to provide monochromatic excitation from the light source and the second to scan the sample luminescence, but no measurable results were obtained because of the low intensity of the hydrogen discharge light source ($\sim 10^8$ photon/sec). The procedure was then changed to measure the excitation spectrum; that is, the second monochromator was removed, and the variation in total sample luminescence was recorded as the spectrum of the light source was scanned (only luminescence above 3000 Å could be sensed because of the detector cutoff). A series of Corning filters was used to study the luminescence spectrally. The most intense radiation in the spectrum of the light source consisted of the group of lines in the band 1200–1400 Å and a continuum from about 1800 Å into the near uv.

The excitation spectrum showed (1) two luminescence maxima at excitation wavelengths of about 1370 Å and 1700 Å in all samples; (2) approximately five smaller bands produced in the interval 1400–1800 Å; and (3) the possibility of very narrow bands at about 1230, 1375, 1522, 1569, 1800, 1870, and 1879 Å in lunar sample 14310,155, with similar sets for each of the terrestrial samples (Figs. 1–4). Efficiencies determined on the basis of the first maximum at 1370 Å are given in Table 1. They are given as ranges because of the uncertainty as to the spectral character of the luminescence.

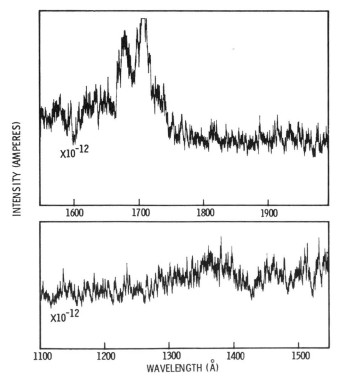

Fig. 1. Luminescence of Apollo 14 sample 14310,155 with scanned far uv (1100–2200 Å) irradiation.

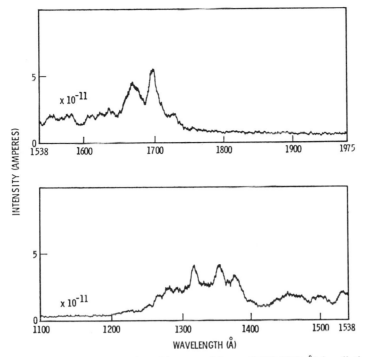

Fig. 2. Luminescence of granite with scanned far uv (1100–2200 Å) irradiation.

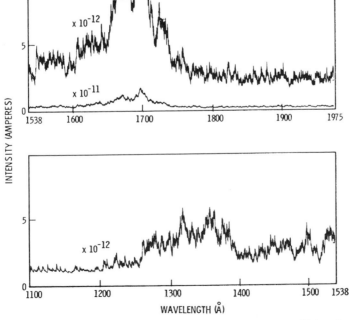

Fig. 3. Luminescence of gabbro with scanned far uv (1100–2200 Å) irradiation.

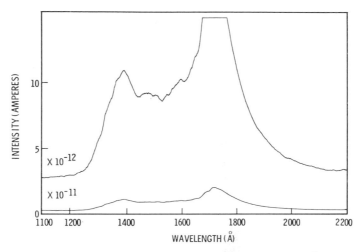

Fig. 4. Luminescence of willemite with scanned far uv (1100–2200 Å) irradiation.

The major excitation peak at around 1700 Å proved to be the most intense in all cases. The light source does not show any significant irradiation intensity in the band 1400–1800 Å. Therefore, this peak is produced either by a lower intensity band around 1600 Å or by higher orders of the extreme uv.

X-ray irradiation

The results of irradiation with x-rays from a tungsten target at 70 kV, 45 mA, are shown in Figs. 5–8. Distinct to prominent blue (4100–4500 Å range) and green (5300–5800 Å range) luminescence peaks are present in all samples. A red peak (6400–7200 Å range) is distinct in granite but barely discernible, though present, in gabbro and the Apollo 14 samples; it is not evident in the Apollo 15 samples. A faint to distinct near uv peak (3300–3600 Å range) is present in terrestrial granite and gabbro, in the Apollo 14 samples, and in the Apollo 15 breccia and glass but is absent in the Apollo 15 gabbro. A small middle uv peak (2800–3000 Å range) is present in terrestrial granite and gabbro but is absent in all lunar samples with the possible exception of a questionable peak in the 2600–2800 Å range in the Apollo 15 samples.

A significant feature of these spectra is that the intensity ratio of the green to the

Table 1. Luminescence efficiencies of lunar and terrestrial samples for the excitation band around 1370 Å.

Sample	Total Band Efficiency Range (ergs/erg)
Lunar:	
14310,155	3×10^{-2} to 5×10^{-3}
Terrestrial:	
Granite	3×10^{-2} to 5×10^{-3}
Gabbro	5×10^{-3} to 8×10^{-4}
Willemite	4×10^{-1} to 6×10^{-2}

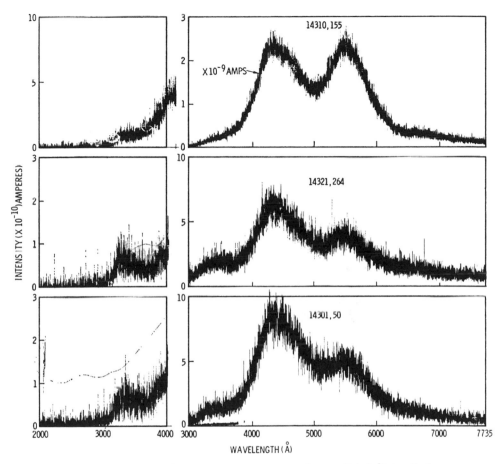

Fig. 5. Luminescence of Apollo 14 samples with soft x-ray (0.2–8 Å) irradiation.

blue peak, with one exception, appears to be related to lithologic type. The ratio is 1 or greater for the crystalline igneous rocks, both terrestrial and lunar, and less than 1 for the breccias and glasses. The exception is one of the two Apollo 15 gabbro samples, 15085,28, the ratio of which is in the breccia and glass range, although its companion sample from the same rock (15085,30) has a ratio in the igneous range, in conformity with the pattern. Because these two samples are small and the crystals of which they are composed large, this disparity in ratios may be the result of greatly different proportions of luminescent plagioclase to poorly or nonluminescent pyroxene in the two samples. According to Geake $et\ al.$ (1972), the green peak is caused by Mn^{2+} in Ca^{2+} sites, whereas the blue peak is attributed to strain defects common in silicates. The blue, associated with the silicate nature of plagioclase and pyroxene, might then be expected to vary less in intensity than the green, associated with the more abundant Ca^{2+} sites in plagioclase than in pyroxene. This seems to be the case here because the difference in the intensity of the green peak accounts for the entire difference in

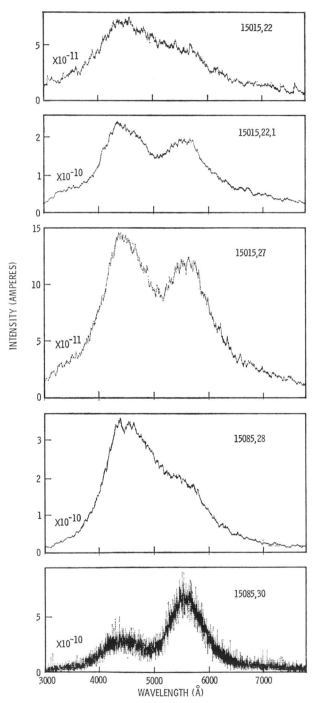

Fig. 6. Visible-wavelength luminescence of Apollo 15 samples with soft x-ray (0.2–8 Å) irradiation.

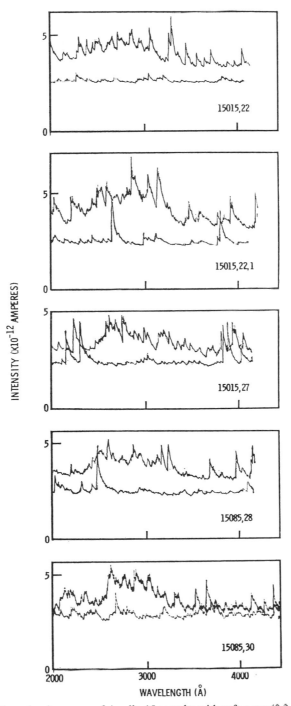

Fig. 7. Possible uv luminescence of Apollo 15 samples with soft x-ray (0.2–8 Å) irradiation. Upper curve, signal from sample; lower curve, dark current.

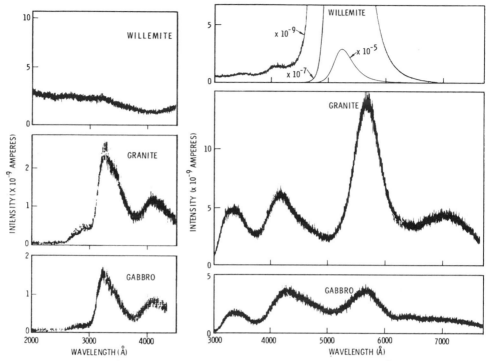

Fig. 8. Luminescence of terrestrial mineral and rocks with soft x-ray (0.2–8 Å) irradiation.

ratios between the two samples; the intensity of the blue peak is the same in both. The low ratio for sample 15085,28, therefore, may well be anomalous.

The green-to-blue peak relationships also appear to hold in the luminescence spectra obtained with high energy (100 keV) proton irradiation but not in those obtained with low energy (5 keV) proton irradiation (see below).

Data on the x-ray irradiation are given in Table 2.

Table 2. Luminescence data for lunar and terrestrial samples with x-ray irradiation.

Sample	Peak Wavelength (Å)	Bandwidth* (Å)	Total Band Efficiency (ergs/erg)
Lunar			
14301,50	3330, 4420, 5500, 6600	500, 880, 880, 480	7×10^{-7}
14321,264	3340, 4340, 5490, 6620	470, 880, 810, 440	9×10^{-7}
14310,155	3330, 4350, 5530, 6680	400, 890, 780, 580	3×10^{-6}
15015,22	($\sim$3500?), 4480, 5550	1140, 1100	3×10^{-8}
15015,22,1	($\sim$3300), 4340, 5590	690, 1100	2×10^{-7}
15015,27	($\sim$3300), 4380, 5640	790, 970	7×10^{-8}
15085,28	4450, 5380	830, 1280	6×10^{-7}
15085,30	4450, 5570	900, 780	1×10^{-6}
Terrestrial			
Granite	3410, 4230, 5760, 7180	580, 810, 610, 1220	2×10^{-5}
Gabbro	3370, 4260, 5690, 6470	480, 850, 820, 740	3×10^{-6}
Willemite	3540, 4160, 5330	410, 460, 420	2×10^{-3}

* Full width at half maximum.

Proton irradiation

Two series of proton irradiation measurements were made. One was with protons of 5 keV energy, flux density of 9.3×10^{13} protons/cm² sec, and energy flux density of 7.5×10^5 erg/cm² sec; the second was with protons of 100 keV energy, flux density of 7×10^{12} protons/cm² sec and energy flux density of 1.1×10^6 erg/cm² sec. The curves for the low energy proton irradiation are shown in Figs. 9–12; those for the high energy in Fig. 13. Both sets are generally similar to the x-ray irradiation results; the chief difference is the tendency for the blue and green peaks in the proton-excited luminescence to broaden and merge into the continuum level. This tendency is more marked with the low than with the high energy protons and makes for some blurring in the latter case and eradication in the former case of the green-to-blue

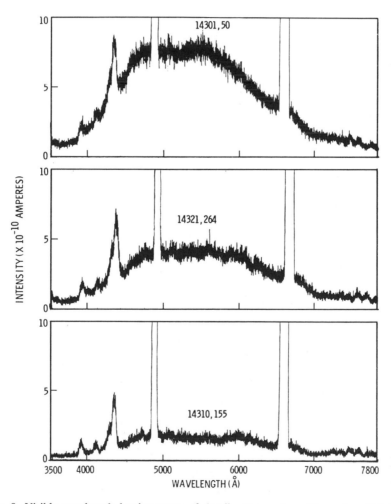

Fig. 9. Visible-wavelength luminescence of Apollo 14 samples with proton (5 kcV) irradiation.

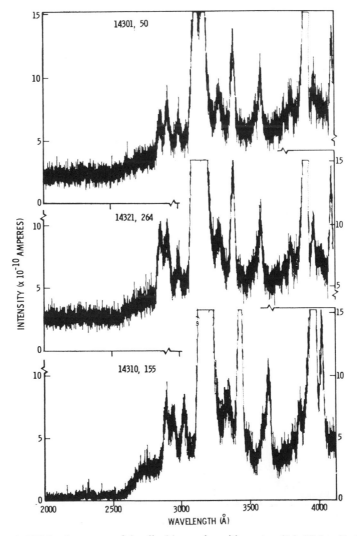

Fig. 10. UV luminescence of Apollo 14 samples with proton (5 keV) irradiation.

peak ratio relationships mentioned in the preceding section. Table 3 summarizes the luminescence data for the proton irradiation.

In the irradiation with protons of 5 keV energy the samples displayed a luminescence intensity decline over a time interval of the order of minutes. In our earlier studies we found that, in the 1000–4000 Å luminescence band with 100 keV proton irradiation, an Apollo 11 breccia (10048,36) showed a lower initial intensity and a lower rate of intensity decline than an Apollo 11 (10044,53) and two Apollo 12 (12002,114; 12020,55) crystalline igneous rocks (Fig. 14). All three Apollo 14 rocks, however, had lower decline rates than the Apollo 11 and 12 samples, and the crystalline rock rate was somewhat lower than the rates for the fragmental rocks. In part, this

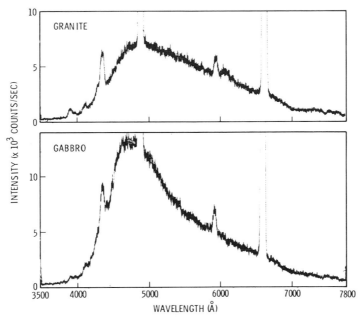

Fig. 11. Visible-wavelength luminescence of terrestrial rocks with proton (5 keV) irradiation.

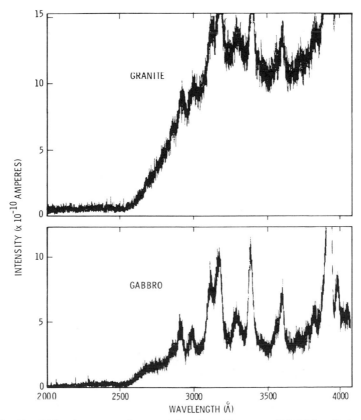

Fig. 12. UV luminescence of terrestrial rocks with proton (5 keV) irradiation.

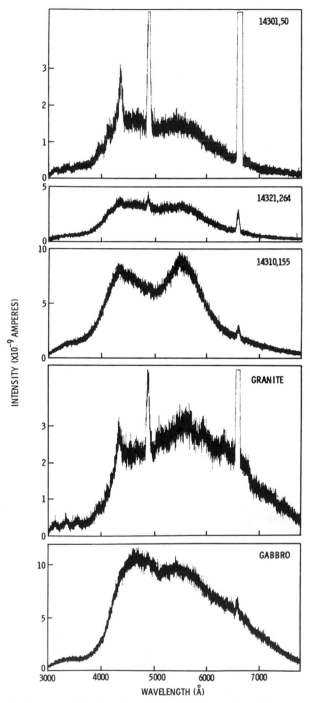

Fig. 13. Luminescence of Apollo 14 and terrestrial samples with proton (100 keV) irradiation.

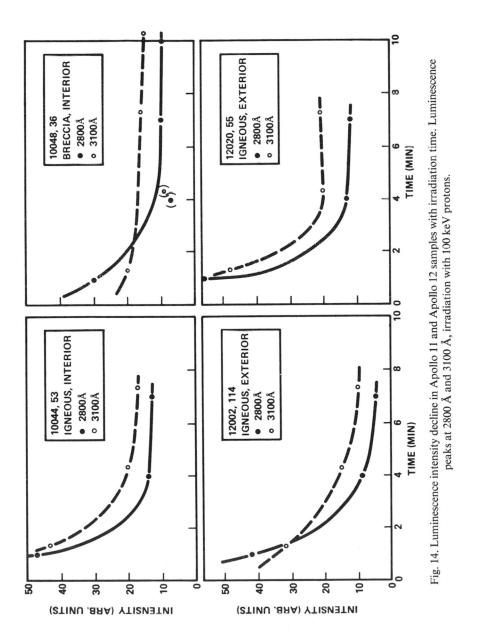

Fig. 14. Luminescence intensity decline in Apollo 11 and Apollo 12 samples with irradiation time. Luminescence peaks at 2800 Å and 3100 Å, irradiation with 100 keV protons.

Table 3. Luminescence data for lunar and terrestrial samples with proton irradiation.

Sample	Peak Wavelength (Å)	Bandwidth* (Å)	Total Band Efficiency (ergs/erg)
Low Energy (5 keV)			
Lunar:			
14301,50	4730, 5510 declining to broad band	2250	1×10^{-6}
14321,264	5300 (broad)	2270	1×10^{-6}
14310,155	5230	1990	1×10^{-6}
Terrestrial:			
Gabbro	4720	1620	5×10^{-6}
High Energy (100 keV)			
Lunar:			
14301,50	4590, 5450	1210, 1620	6×10^{-7}
14321,264	4380, 5520	760, 1280	2×10^{-6}
14310,155	4340, 5480	760, 1070	3×10^{-6}
Terrestrial:			
Granite	5620	2600	1×10^{-6}
Gabbro	4660, 5380	970, 2520	5×10^{-6}

* Full width at half maximum.

may be due to the fact that the two sets of data are not directly comparable. The rates for the Apollo 11 and Apollo 12 samples were calculated from 100 keV proton data on peaks in the middle and near uv whereas those of the Apollo 14 samples were calculated from 5 keV proton data on peaks and portions of the band in the visible wavelengths. Also, because of the time durations of the respective runs, the middle and near uv data points could be taken at about three minute intervals, whereas the data points in the visible band had to be taken at about seven minute intervals. It is also possible that the decline characteristics of the uv luminescence may yield better diagnostic information than those of the visible. Nash and Greer (1970) have reported luminescence decline characteristics associated with lunar rock type and exposure history, and Sippel and Spencer (1970) have reported what are probably related features, intensity and spectral differences in luminescence between shocked and unshocked feldspars. These decline characteristics with proton excitation, therefore, appear to contain important information on the geologic history of lunar rocks.

DISCUSSION

The luminescence characteristics of the Apollo samples and of the terrestrial comparison samples appear to contain information both as to lithologic type (granitic versus gabbroic) and as to geologic history and processes (unshocked or mildly shocked lunar igneous versus more strongly shocked lunar breccia). Despite some overlap of Apollo 14 and Apollo 15 samples, the efficiencies within each group tend to be higher for the igneous rocks than for the breccias and glasses; moreover, all, in general, tend to be comparable to that of terrestrial gabbro and less than that of terrestrial granite. This is in accord with previous studies (Greenman et al., 1965; Greenman and Milton, 1968; Nash, 1966), which have shown granitic rocks to have higher efficiencies than gabbroic ones, so that lunar rocks of more sialic character can be expected to show corresponding relationships. The lower efficiency of the lunar breccia as compared to that of the lunar igneous rock of comparable chemical composition is probably a reflection of shock degeneration of the luminescence (Sippel and Spencer, 1970) and probably also of an admixture of poorly luminescent glass.

The ratio of the green to the blue luminescence peak also appears to contain both lithologic and geologic process information. The higher ratio in granite than in gabbro and the relationships discussed above between the two Apollo 15 igneous samples 15085,28 and 15085,30 suggest that this ratio reflects lithologic differences. The higher values of the lunar igneous rocks as compared with those of the breccias and glasses, as in the case of the efficiency comparison, may result from greater shock degeneration of the green than of the blue peak and possibly also, in the breccias, an admixture of glass.

All the lunar sample efficiencies we have measured to date, with one exception, are low and, therefore, cannot account for the astronomical observations of luminescence on the moon. The exception, the efficiencies reported here for far uv irradiation, are high, but these are preliminary values and require further verification. The solar energy in these far uv wavelengths, however, is low enough so that even if these efficiencies are confirmed they still could not account for the reported observations.

Acknowledgments—We wish to thank W. M. Hansen, T. H. Mills, and R. R. Carlen for their valuable assistance. This work was supported by NASA Contract NAS9-11679.

REFERENCES

Geake J. E., Walker G., Mills A. A., and Garlick G. F. J. (1972) Luminescence of lunar material excited by protons or electrons (abstract). In *Lunar Science—III* (editor C. Watkins), pp. 294–296, Lunar Science Institute Contr. No. 88.

Greenman N. N., Burkig V. W., Gross H. G., and Young J. F. (1965) Feasibility study of the ultraviolet spectral analysis of the lunar surface. Douglas Aircraft Co. Rep. SM-48529.

Greenman N. N. and Gross H. G. (1970a) Luminescence of Apollo 11 lunar samples. *Science* **167**, 720–721.

Greenman N. N. and Gross H. G. (1970b) Luminescence studies of Apollo 11 lunar samples. *Proc. Apollo 11 Lunar Sci. Conf.*, *Geochim. Cosmochim. Acta* Suppl. 1, Vol. 3, pp. 2155–2161. Pergamon.

Greenman N. N. and Gross H. G. (1971) Luminescence of Apollo 11 and Apollo 12 lunar samples. *Proc. Second Lunar Sci. Conf.*, *Geochim. Cosmochim. Acta* Suppl. 2, Vol. 3, pp. 2223–2233. MIT Press.

Greenman N. N. and Gross H. G. (1972) Luminescence of Apollo 14 lunar samples (abstract). In *Lunar Science—III* (editor C. Watkins), pp. 335–337, Lunar Science Institute Contr. No. 88.

Greenman N. N. and Milton W. B. (1968) Silicate luminescence and remote compositional mapping. Proceedings of the Sixth Annual Meeting of the Working Group on Extraterrestrial Resources, NASA SP-177, pp. 55–63.

Nash D. B. (1966) Proton-excited silicate luminescence: experimental results and lunar implications. *J. Geophys. Res.* **71**, 2517–2534.

Nash D. B. and Greer R. T. (1970) Luminescence properties of Apollo 11 lunar samples and implications for solar-excited lunar luminescence. *Proc. Apollo 11 Lunar Science Conf.*, *Geochim. Cosmochim. Acta* Suppl. 1, Vol. 3, pp. 2341–2350. Pergamon.

Sippel R. F. and Spencer A. B. (1970) Luminescence petrography and properties of lunar crystalline rocks and breccias. *Proc. Apollo 11 Lunar Sci. Conf.*, *Geochim. Cosmochim. Acta* Suppl. 1, Vol. 3, pp. 2413–2426. Pergamon.

Proceedings of the Third Lunar Science Conference
(Supplement 3, *Geochimica et Cosmochimica Acta*)
Vol. 3, pp. 2997–3007
The M.I.T. Press, 1972

Thermoluminescence of individual grains and bulk samples of lunar fines

H. P. Hoyt, Jr., R. M. Walker, D. W. Zimmerman, and J. Zimmerman

Washington University, Laboratory for Space Physics,
St. Louis, Missouri 63130

Abstract—The thermoluminescence (TL) response of a typical bulk lunar fines is quite complex. The natural TL is produced by minor mineral constituents (typically less than 1 grain in 15). The low-temperature natural TL (around 200°C) is produced by a small fraction of the plagioclase grains, while the higher temperature TL is from grains rich in potassium and sometimes phosphorus. The dose rate on the moon varies from grain to grain of a sample, because although all grains receive the same cosmic ray dose rate, the internal dose rate from uranium and thorium varies significantly. Although the natural TL responds to the ambient temperature on the moon, it appears that data of sufficient accuracy to measure, for example, the temperature gradient in deep core material cannot be obtained from bulk measurements but require separation of the samples into fractions with more homogeneous TL characteristics. The TL of bulk samples is however useful in estimating the average depth of lunar fines from near the surface. The stored TL is affected by white room light but is relatively unaffected by red light.

INTRODUCTION

IN THIS PAPER we report the analysis of the thermoluminescence (TL) of several hundred individual grains, taken one at a time, from several samples of Apollo 12 and Apollo 14 fines. This work was undertaken to identify the minerals responsible for the TL and to divide the rather broad glow curve of bulk material into its component parts. Although fines samples from all the lunar missions to date have generally similar glow curves, the TL response of different grains in a given sample is highly variable. Resolving the detailed nature of the TL is a key problem if the full promise of lunar TL studies, e.g., measurement of lunar heat flux (Hoyt *et al.*, 1971), is to be realized.

In previous reports where measurements were made on bulk samples of lunar fines (Dalrymple and Doell, 1970; Doell and Dalrymple, 1971; Hoyt *et al.*, 1970; Hoyt *et al.*, 1971), the following results and interpretations have been presented. At glow-curve temperatures above ~500°C the TL is saturated for all samples except those within ~1 cm of the surface. At lower glow temperatures the TL is non-saturated and is in thermal equilibrium, the rate of acquisition of trapped charges being balanced by the rate of loss by thermal drainage. This interpretation of thermal equilibrium is supported by the depth variation of low temperature (≲200°C) TL in both the Apollo 11 and 12 cores. The TL increases rapidly to a depth of ~10 cm as a result of the attenuation of the diurnal heat wave and then decreases slowly with increasing depth (Apollo 12 core). The decrease at deep depths is interpreted as being partly due to the attenuation of galactic cosmic rays and partly due to the increasing temperature with depth due to the outward heat flow. Unraveling these

2997

two effects is important, because it would allow independent measurements of the lunar heat flow from long lunar cores (Hoyt *et al.*, 1971).

Two results have called this interpretation into question. First, Nash and Conel (1971) report that they have not seen the systematic variation of TL with depth in the Apollo 12 double core. However, they measure the natural TL as the glow-curve area from 300–400°C and our data likewise shows very little variation with depth in this temperature region compared to the 175–250°C range we have reported. The variations become smaller with increasing glow-curve temperature and of course there is no variation with depth above 550°C where the TL is saturated. Second, our more detailed measurements of the TL in bulk samples from the Apollo 12 double core reported here show certain anomalies in the monotonic variation of TL with depth. However, as we shall show, at least some of these anomalies are removed when measurements are made on individual grains.

In this paper we also give a survey of TL measurements made on bulk samples of several Apollo 14 and Apollo 15 fines; the average depths of the samples are estimated from these data. In addition, TL measured in lunar rocks decreases rapidly in the first few millimeters below the top surface. This effect is interpreted as being caused by the decrease in dose rate due to the attenuation of solar flare protons. These results can be used to obtain an estimate of the solar flare spectrum averaged over the TL equilibrium time as discussed in a companion paper (Crozaz *et al.*, 1972).

Finally we present data on the effects of light on lunar TL which show that red-light handling of TL samples is important.

EXPERIMENTAL TECHNIQUE

The experimental procedures and apparatus used for the TL measurements of bulk material are the same as those described previously (Hoyt *et al.*, 1970; Hoyt *et al.*, 1971). For the measurement of individual grains several changes are made to improve the signal-to-noise ratio: the TL is collected by a small diameter (6 mm) light guide and transmitted through a Corning 7-59 filter of the same diameter to the PM tube; the outer part of the photo-cathode is defocused with a magnetic lens to reduce the dark count rate; the signal from the PM tube is measured with fast photon counting electronics instead of (as previously) with a picoammeter. All handling of the samples is done in red light (in a 4 × 8 m room lit with eight General Electric F40R fluorescent tubes).

TL MEASUREMENTS OF SINGLE GRAINS

It is known from previous work that most of the natural TL from lunar fines is from the low-density fraction with specific gravity <2.89 (Dalrymple and Doell, 1970; Hoyt *et al.*, 1970). Therefore the low-density (sp gr <2.89) grains were separated from the 37–74 μm sieved fraction of samples 12033, 12025,57, and 14230,141 using bromoform; single grains were chosen at random, and their natural TL measured. Figures 1–3 show the glow curves of the brightest grains, as well as the glow curves of bulk samples.

The most complete measurements were made on sample 12033 and we will discuss them in detail; the conclusions are believed to be valid qualitatively for the other samples as well. Fifty-four transparent grains were measured. Most of them gave very little natural TL (33 gave <100 cps max). The sum of the glow curves of the 14

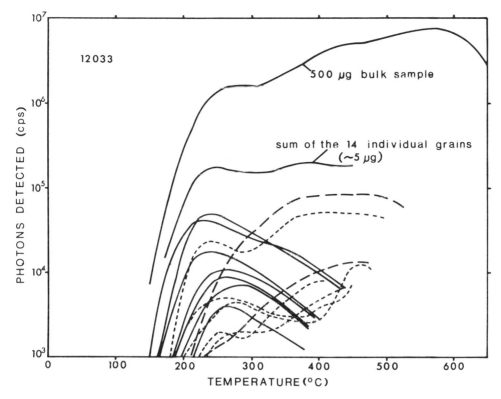

Fig. 1. Natural TL glow curves of fourteen grains of 12033; grain size 200–400 mesh. Blackbody has been subtracted. Also shown are the light sum of the grains and the TL emitted by a bulk sample.

brightest grains is similar in shape to the bulk-sample glow curve; therefore we conclude that the grains shown are representative of the bright TL grains in the bulk sample. It can be estimated from Fig. 1 that the majority of the natural TL from the 0.5 mg bulk sample is produced by about 100 grains. In other words, most of the natural TL of sample 12033 is produced by about 1 grain in 15. In the other samples shown in Figs. 2 and 3 the bright TL grains are even smaller fractions of the total.

The glow curves of the individual grains (Figs. 1–3) are divided roughly into three types. The most common type, shown with solid lines, has a low-temperature peak around 225–275°C. A microprobe analysis of four of these grains has shown just an AlSiCa phase (we are unable to detect Na), and they are probably some form of plagioclase. Other grains, whose glow curves are shown by long dashed lines, have higher temperature peaks and six of these grains were analyzed; all six contained an AlSiK phase, three contained an AlSiCa phase and two contained a phosphate phase. Still other grains (short dashed lines) have both the low and high temperature peaks and are presumably "composite" grains.

We reported elsewhere (Walker *et al.*, 1971) the results of a uranium map of 800

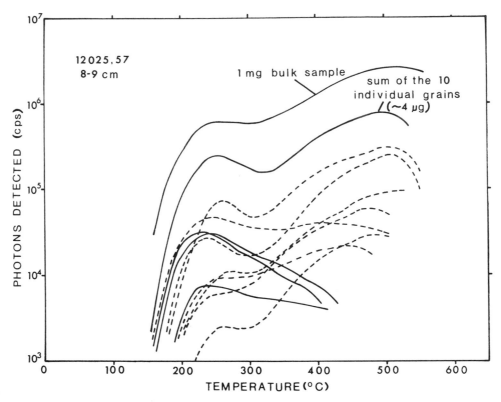

Fig. 2. Natural TL glow curves of ten grains of 12025,57; grain size 200–400 mesh. Blackbody has been subtracted. Also shown are the light sum of the grains and the TL emitted by a bulk sample.

grains of 14259, which showed that about one-third of the uranium-rich grains contained an AlSiK phase and sometimes a phosphate phase in addition. Therefore some of the bright TL grains are high in radioactivity which indicates that the TL is induced partly by internal radiation as well as by cosmic rays. At low glow curve temperatures the TL comes partly from plagioclase grains, in which the radiation should be mainly from cosmic rays, and partly from the "composite" grains in which the TL internal radiation dose rate could be important. Thus the radiation dose rates appropriate to the TL emitted by a bulk sample are very complex and could change from one sample to another with variation in radioactivity content and its distribution.

Measurements on bulk samples are further complicated by the presence of other grains that give very little natural TL but that contribute substantially to the artificially induced TL. This can be seen in Fig. 4 that shows the natural and artificial TL from a typical low-temperature "plagioclase" grain and from a bulk sample of 12033. For the single grain the natural peak height is about ten times the artificial induced by a 20 krad dose. For the bulk sample, on the other hand, the natural TL at 250°C (which we know from Fig. 1 is from grains of the low-temperature "plagioclase"

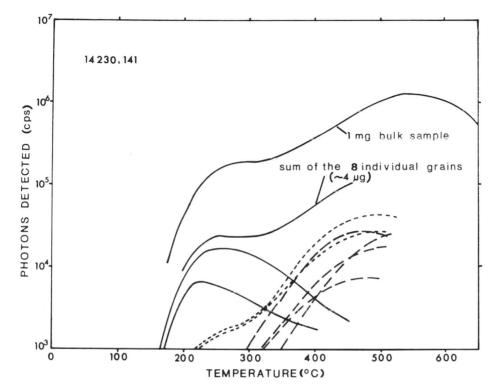

Fig. 3. Natural TL glow curves of eight grains of 14230,141; grain size 200–400 mesh. Blackbody has been subtracted. Also shown are the light sum of the grains and the TL emitted by a bulk sample.

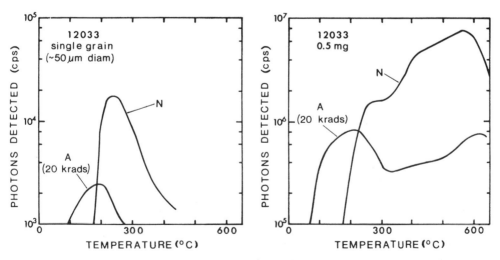

Fig. 4. Natural and artificial glow curves of sample 12033 for a single grain of the low-temperature "plagioclase" type and for a bulk sample.

type) is only twice the artificial peak height at 200°C. It follows that the grains giving most of the artificial TL at 200°C are different from those giving the natural TL at low temperatures. Although we have not yet identified these grains, it is clear that the use of the artificial TL to normalize the natural TL for the sensitivity of the bulk sample can be misleading.

The grains that give significant artificial TL but that are very low in natural TL may be significant in the decay of the artificial TL, which Geake *et al.* (1970), Garlick *et al.* (1971), and Blair *et al.* (1972) considered to be abnormally fast and suggested is "nonthermal" in nature. Of course the grains responsible for the natural TL do not decay abnormally fast: They retain a light level which is equivalent to a large dose ($\sim 10^5$ rad) and samples of Apollo 12 fines stored for 9 months at room temperature have shown no decay within experimental error ($\pm 15\%$ in this experiment) even at glow curve temperatures of $<250°C$.

Because of the complications involved in the bulk samples, quantitative measurement of the lunar TL must be made under conditions where the various components can be separated. For measurements of the depth dependence in core tubes the low-temperature "plagioclase" grains seem particularly promising. Because these grains are relatively free of uranium, the ionization rate will depend only on the cosmic ray dose rate. Shown in Fig. 5 is a comparison between bulk and single grain measure-

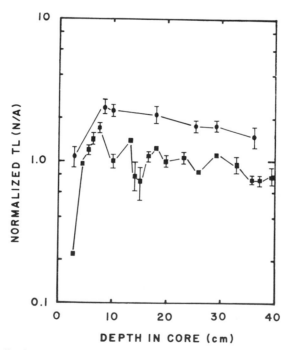

Fig. 5. Normalized TL (N/A) as a function of depth in the Apollo 12 double core 12025/ 12028 for single grains of the low-temperature "plagioclase" type (●) and for ~ 1 mg bulk samples (■). For the single grains N/A $\equiv$ N(peak)/A(peak); for the bulk samples N/A $\equiv$ N(250°C)/A(peak), see Fig. 4. The error bars are one S.D. based on duplicate measurements; typically 3–4 grains or 2–3 bulk samples.

ments of the depth dependence of TL in the Apollo 12 double core. Both sets of the data show the rapid decrease near the surface from the sun's heating and the more gradual decrease beyond 10 cm from the attenuation of galactic cosmic rays and the increasing temperature resulting from outward heat flow. However, the bulk TL data also show anomalies, particularly near 10 cm and the 14 cm coarse-grain layer. We have not identified the cause of these anomalies, but they presumably result from one of the difficulties with bulk measurements discussed earlier. The data for the low-temperature "plagioclase" grains on the other hand, although containing considerable statistical error, do not seem to show the anomaly at the 10 cm depth. Unfortunately we have no more coarse-grain layer material left to check the anomaly at 14 cm.

Although the low-temperature "plagioclase" grains look promising, greater accuracy must be obtained for precise measurements of the temperature gradient in the deep cores. The method of picking individual grains is too tedious, and we are now attempting mineral separations to concentrate the desired grains. Separation of the TL signals from different components may also be achieved by using filters to isolate different emission bands.

Although the bulk glow curves of many different lunar fines are qualitatively similar in shape, the sensitivities to ionizing radiation vary by two orders of magnitude, with 12033 being the brightest and Luna 16 the dullest of the samples.

The variation of sensitivity is probably influenced by a number of factors. Since even in the brightest samples only a small fraction of the plagioclase grains give detectable signals, it is likely that the sensitivity in part reflects differences in the trace element activator content of different grains. We have previously noted that the sensitivity is roughly correlated with uranium concentration (Walker and Zimmerman, 1972) and this element may be geochemically linked to the TL activators.

In this connection it is interesting to note that the TL sensitivity of rock 15415 is extremely low (see Fig. 6). This rock is 98% pure anorthite and has a very low U concentration. The glow curve shape is also very different from the bulk TL of the fines. If this rock is characteristic of the anorthositic component of the lunar regolith, it is clear that this exotic fraction contributes very little to the TL of lunar fines.

The variation of sensitivity is *not* obviously correlated with radiation damage. For example, sample 12028,69 is much less irradiated than the other double core samples (Crozaz *et al.*, 1971) but is very similar in TL sensitivity. Similarly, samples 14156 and L16 which are both typical heavily irradiated samples (Crozaz *et al.*, 1972; Walker and Zimmerman, 1972), have TL sensitivities different by a factor of 30. This lack of correlation is consistent with our observation that a sample bombarded with $\sim 10^{10}$ fission fragments per cm^2 from a ^{252}Cf source showed no alteration in its TL response to a standard β dose.

SURVEY OF APOLLO 14 AND 15 FINES

A survey has been made of the TL of a number of Apollo 14 and 15 fines using 1 mg bulk samples. Although the results of bulk measurements are subject to some uncertainty as discussed earlier, useful information can nevertheless be obtained from

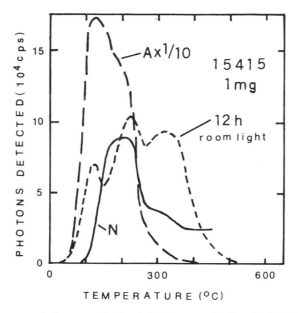

Fig. 6. Glow curves of a 1 mg sample of rock 15415; natural (N), artificial (A) produced by a beta dose of 20 krad, and the TL induced by a 12 h exposure to white fluorescent bulb roomlight. Curve A has been decreased by × 10. Blackbody has been subtracted.

such measurements. In particular, because of the rapid variation in TL within the first few centimeters (see Fig. 5), an estimate can be obtained of the average depth of fines from near the lunar surface (Hoyt *et al.*, 1970; Dalrymple and Doell, 1970). This is the depth averaged over the time required to reach the TL equilibrium level, which is of the order of 10^4 years. Table 1 summarizes the average depth estimates. We consider these results preliminary on two accounts: (1) the measurements are made on bulk samples; we hope to obtain more accurate and reliable results by

Table 1. Average depth in the regolith of various lunar fines estimated from TL measurements.

Sample	Average Depth (cm) Determined from TL Measurements of < 200 Mesh Fraction
14141 Cone Crater	1–3
14163 Bulk EVA 1	1–3
14259 Comprehensive	< 0.5
14148 Top of trench	< 0.5
14156 "Middle" of trench	< 0.5
14149 "Bottom" of trench	< 0.5
14230 "Disturbed core" (8 positions in 9 cm)	all > 5
15231 Sta. 2, Apennine Front, from underneath large rock	Equivalent to 3–4
15471 Sta. 4, Dune Crater	0.5–1
15531 Sta. 9a, Rille	< 0.5

measurements on separated mineral fractions, (2) the depths are estimated by comparing the normalized TL (N/A) of the various samples to the TL from various depths of the Apollo 12 core (Fig. 5). We have to assume the diurnal heat wave penetrates the same at the different sites. Eventually we should be able to compare directly to core samples from the Apollo 14 and 15 sites.

Most of the surface fines have average depths of less than ~ 3 cm. Sample 15231, which was collected from underneath a large boulder, reflects the shielding from the sun, having a temperature equivalent to a depth of about 3–4 cm. Samples from all depths of the "disturbed core" have natural glow curve shapes which are similar to the Apollo 12 core from about 5 cm downward; none look like surface material. This implies that at least 5 cm of material from the top of the core were lost when the follower fell out.

One surprising result in these data is that the samples from the top, middle, and bottom of the Apollo 14 trench do not show any depth dependence; all have low TL like the surface samples. This is consistent with the results of Eldridge et al. (1972) that the concentration of ^{26}Al and ^{22}Na are nearly uniform with depth; from this they concluded that extensive mixing had occurred and that the bottom sample was not representative of the 36 cm depth in the regolith. The TL results further imply that the middle as well as the bottom samples have been appreciably contaminated with surface (<0.5 cm) material. On the other hand, the track densities in grains from the bottom of the trench are much less than those from the middle and top samples (Crozaz et al., 1972), so the trench samples remain an enigma.

The glow-curve shapes induced by artificial radiation of the Apollo 14 fines are very similar to those of Apollo 11 and 12 fines, indicating similar TL minerals. Relatively more low-temperature TL ($<300°$C) is induced in the Apollo 15 fines. Our Apollo 14 and 15 samples are about as sensitive as the Apollo 12 core fines except for 14141 which is about four times brighter.

EFFECTS OF LIGHT ON STORED TL

Figure 7 shows the effects of white roomlight and red light on the stored TL of a bulk sample of Apollo 12 core material. Such a bleaching of the TL in the glow curve temperature range 200–250°C and an induction of TL at 150°C have been found in other lunar fines and in several lunar rocks. The TL of the low-temperature "plagioclase" grains (solid lines, Figs. 1–3) in particular is bleached by white light but unaffected by red light. The anorthosite rock 15415 has TL induced in it by exposure to white light (Fig. 6). It is clear that to obtain good accuracy, exposure to white light must be avoided. All our operations are carried out in red light and arrangements have been made with the Lunar Sample Curator to have TL samples taken in red light from core tubes.

Acknowledgments—We thank S. Sutton for much help with the laboratory measurements. This work was supported by NASA Contract NAS 9-8165.

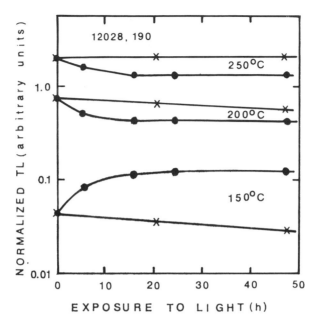

Fig. 7. Dependence of stored natural TL upon total time of exposure to white light (●) and red light (x), for 23-cm material from the Apollo 12 double core. For each sample, the natural TL measured after exposure to light has been normalized by the subsequent TL output (measured at 325°C) induced by a beta dose of 1.5 × 10⁵ rad.

REFERENCES

Blair I. M., Edgington J. A., Chen R., and Jahn R. A. (1972) Low Temperature Thermoluminescence of Apollo 14 Lunar Samples (abstract), in *Lunar Science—III* (editor C. Watkins), p. 86, Lunar Science Institute Contr. No. 88.

Crozaz G., Walker R., and Woolum D. (1971) Nuclear track studies of dynamic surface processes on the moon and the constancy of solar activity. *Proc. Second Lunar Sci. Conf., Geochim. Cosmochim. Acta* Suppl. 2, Vol. 3, pp. 2543–2558. MIT Press.

Crozaz G., Drozd R., Hohenberg C. M., Hoyt H. P., Ragan D., Walker R. M., and Yuhas D. (1972) Solar flare and galactic cosmic ray studies of Apollo 14 samples. Vol. 3, these proceedings.

Dalrymple G. B. and Doell R. R. (1970) Thermoluminescence of lunar samples from Apollo 11. *Proc. Apollo 11 Lunar Sci. Conf., Geochim. Cosmochim. Acta* Suppl. 1, Vol. 3, pp. 2081–2092. Pergamon.

Doell R. R. and Dalrymple G. B. (1971) Thermoluminescence of Apollo 12 lunar samples. *Earth Planet. Sci. Lett.* **10**, 357–360.

Eldridge J. S., O'Kelly G. D., and Northcutt K. J. (1972) Abundance of Primordial and Cosmogenic Radionuclides in Apollo 14 Rocks and Fines (abstract), in *Lunar Science—III* (editor C. Watkins) p. 221, Lunar Science Institute Contr. No. 88.

Garlick G. F. J., Lamb W. E., Steigmann G. A., and Geake J. E. (1971) Thermoluminescence of lunar samples and terrestrial plagioclases. *Proc. Second Lunar Sci. Conf., Geochim. Cosmochim. Acta* Suppl. 2, Vol. 3, pp. 2277–2283. MIT Press.

Geake J. E., Dollfus A., Garlick G. F. J., Lamb W., Walker G., Steigmann G. A., and Titulaer C. (1970) Luminescence, electron paramagnetic resonance and optical properties of lunar material from Apollo 11. *Proc. Apollo 11 Lunar Sci. Conf., Geochim. Cosmochim. Acta* Suppl. 1, Vol. 3, pp. 2127–2147. Pergamon.

Hoyt H. P. Jr., Kardos J. L., Miyajima M., Seitz M. G., Sun S. S., Walker R. M., and Wittels M. C. (1970) Thermoluminescence, x-ray and stored energy measurements of Apollo 11 samples. *Proc. Apollo 11 Lunar Sci. Conf.*, *Geochim. Cosmochim. Acta* Suppl. 1, Vol. 3, pp. 2269–2287. Pergamon.

Hoyt H. P. Jr., Miyajima M., Walker R. M., Zimmerman D. W., Zimmerman J., Britton D., and Kardos J. L. (1971) Radiation dose rates and thermal gradients in the lunar regolith: thermoluminescence and DTA of Apollo 12 samples. *Proc. Second Lunar Sci. Conf.*, *Geochim. Cosmochim. Acta* Suppl. 2, Vol. 3, pp. 2245–2263. MIT Press.

Nash D. B. and Conel J. E. (1971) Luminescence and reflectance of Apollo 12 samples. *Proc. Second Lunar Sci. Conf.*, *Geochim. Cosmochim. Acta* Suppl. 2, Vol. 3, pp. 2235–2244. MIT Press.

Walker R. M., Zimmerman D. W., and Zimmerman J. (1971) Thermoluminescence of lunar samples: measurement of temperature gradients in core material. Proc. Conf. of Lunar Geophysics, *The Moon*, in press.

Walker R. and Zimmerman D. (1972) Fossil track and thermoluminescence studies of Luna 16 material. *Earth Planet. Sci. Lett.* **13**, 419–422.

Proceedings of the Third Lunar Science Conference
(Supplement 3, *Geochimica et Cosmochimica Acta*)
Vol. 3, pp. 3009–3020
The M.I.T. Press, 1972

Spectral emission of natural and artificially induced thermoluminescence in Apollo 14 lunar sample 14163,147

C. Lalou, G. Valladas, U. Brito, and A. Henni

Centre des Faibles Radioactivités du CNRS,
91 Gif sur Yvette, France

and

T. Ceva and R. Visocekas

Laboratoire de Luminescence II,
Université de Paris VI, France

Abstract—Using either colored glass filters or interference filters, the spectral emission of natural thermoluminescence of lunar fines and of ultra-violet and γ induced thermoluminescence has been studied.

This analysis will probably be useful for further studies of lunar thermoluminescence as it may allow the differentiation of two types of thermoluminescence.

Introduction

THE OBJECT of this experiment was to study the natural and artificial thermoluminescence (TL) spectral emission of Apollo 14 lunar sample 14163,147.

This sample consisted of 0.989 g of fines from a bulk sample collected during EVA 1 of the Apollo 14 mission near the LEM at the Fra Mauro formation (3° 40′ 19″ S. Lat.; 10° 27′ 46″ W. Long.) on February 5 and 6, 1971.

The sample reached our laboratory at the beginning of October 1971. We had no information as to whether it had already been exposed to laboratory light.

This sample, with a granulometry less than 1 mm and a median size of 0.065 mm contains 40–75% glass (LSPET, 1971) which is not favorable for TL measurements.

Apparatus

The apparatus has been previously described (Valladas, 1972). It allows the use of small quantities of sample and numerous runs within a short period of time. The sample holder used in this study is a standard alumina cupel, guilded to reduce the blackbody emission. The heating system consists of a titanium plate on which the cupel is placed and heated from below by exposure to the radiation of a cavity maintained at a high temperature (blackbody). After measuring, the plate is rapidly cooled by replacing the heating system by a cooling plate.

The light, emitted in a large solid angle, is transmitted to the photomultiplier through a spherical diopter. This spherical diopter is transparent to ultraviolet light and reduces the blackbody radiation. Filters can be interposed in the beam of the light to analyze the wavelength of the light emitted and to cut down the thermal emission of the cupel. In this study, we used two sets of filters. One is a set of glass color filters which have a high transmission but a large bandwidth that allows a rough exploration of the light spectrum. The second is a set of interference filters. Those filters may be used as the spherical diopter produces a parallel beam of light.

The photomultiplier had a high quantum efficiency and a low dark current (56 DUVP, Radio-technique). Its photocathode can be cooled to lower the dark current.

Natural TL

All measurements were done on bulk samples without any mineral sorting.

The natural TL of this sample was very low. Figure 1 gives the glow curve for about 1 mg, without any filter. After subtraction of the blackbody radiation, a peak is found at about 480°C. The beginning of light output occurs at about 220°C. Above 480°C blackbody radiation is too high and TL cannot be seen in these conditions.

As the TL is very low, we first used glass color filters. Figure 2 gives the curves so obtained without correction for filter efficiency. It seems that TL may be divided into two parts: the TL below 400°C, which contains a predominance of green and yellow light over a background of blue and violet light, and the TL above 400°C, which contains only blue and violet light.

It has been shown by Hoyt *et al.* (1972) that TL in fines is due to two types of grains: grains having a TL emission around 225–275°C due essentially to the action of cosmic particles, and the other grains having a TL at higher temperature due to the irradiation by radionuclides and cosmic particles. As our measurements were done on bulk samples, it is possible that these two types of minerals were, respectively, responsible for the two lights emitted.

When observed with interference filters, the spectral distribution of light was not clearly resolved because the sample was only weakly thermoluminescent. New measurements are in progress on a more thermoluminescent sample (12033).

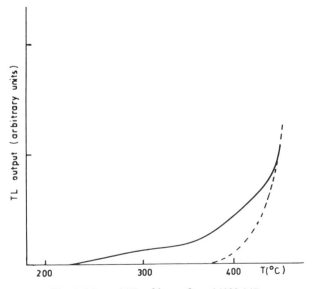

Fig. 1. Natural TL of lunar fines 14133,147.

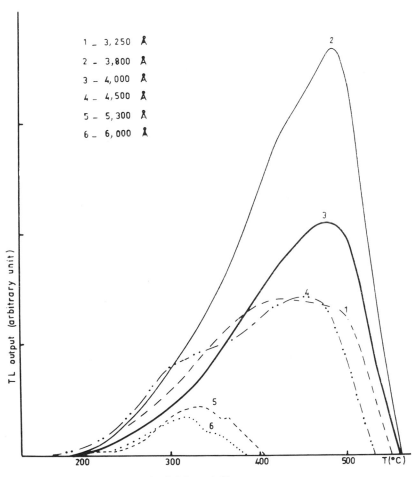

Fig. 2. Natural Glow curves.

ARTIFICIALLY INDUCED THERMOLUMINESCENCE

Short ultraviolet irradiations

A 1 mg sample of lunar fines was subjected, after drainage of its natural TL, to the action of a commercial uv lamp "Mineral Light U.V.S.12" (principal wavelength: 2537 Å) for diverse doses. Figure 3 shows the curves of fines subjected to uv energy fluxes from 170 to 40,000 erg/cm². The analysis of these curves enables one to resolve them into four elementary curves with maximums at about 120°C, 190°C, 260°C, and 310°C (Fig. 4). Each of these is related with a given group of traps. The relative populations of these traps depends on the irradiation time, the deeper ones filling up later.

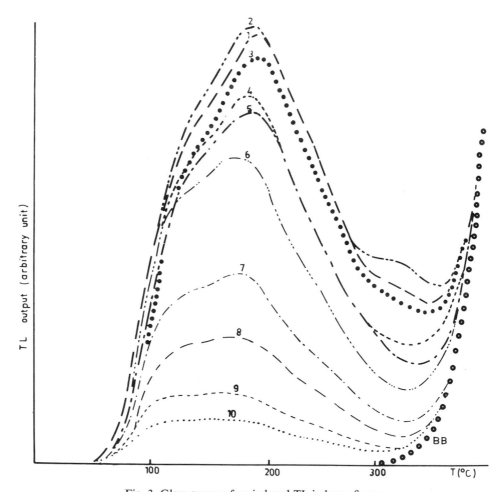

Fig. 3. Glow curves of uv induced TL in lunar fines:

(10)	170 erg/cm²	(5)	6800 erg/cm²
(9)	340 erg/cm²	(4)	10170 erg/cm²
(8)	680 erg/cm²	(3)	20340 erg/cm²
(7)	1360 erg/cm²	(2)	30500 erg/cm²
(6)	3390 erg/cm²	(1)	40700 erg/cm²

The color spectrum of this TL was studied with interference filters. The light was analyzed from 3565 Å to 6000 Å using 15 different filters. From 3565 to 3835 Å, the $\Delta\lambda$ of these filters was approximately 65 Å, and from 3835 to 6000 Å, it was approximately 200 Å. Thus analyzed, the TL presented two peaks, one near 190°C which emitted essentially blue to ultraviolet light, and the other near 310°C which was marked by yellow and green emission. Figures 5 and 6 give the intensities of TL as a function of wavelength for the 190°C peak and the 310°C peak, respectively.

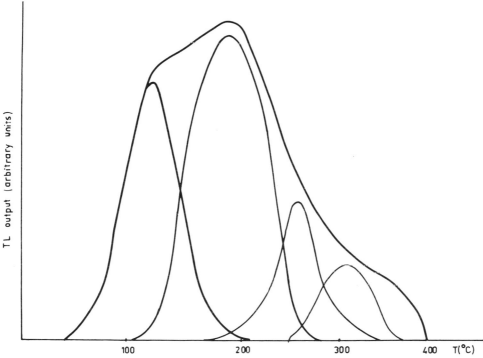

Fig. 4. Analysis of the glow curve of uv induced TL.

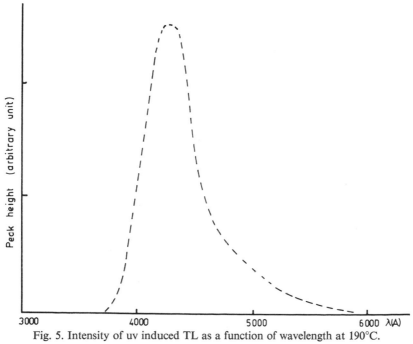

Fig. 5. Intensity of uv induced TL as a function of wavelength at 190°C.

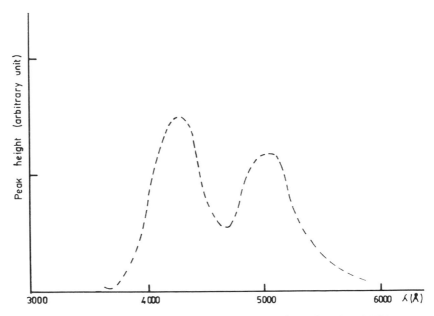

Fig. 6. Intensity of uv induced TL as a function of wavelength at 310°C.

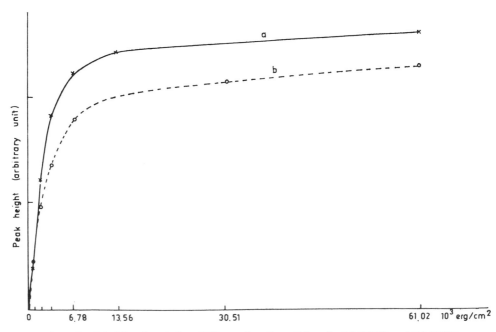

Fig. 7. Peak height of uv induced TL as a function of dose for (a) 190°C and (b) 310°C.

Next, calibration curves were obtained for these two peaks. Colored glass filters having a larger $\Delta\lambda$ and a greater yield were used. For the 190°C peak, the filter which transmitted 35% of the light between 3500 and 4500 Å was used, and at 310°C, the filter which transmitted 60% of the light between 5000 and 6000 Å. In Fig. 7, the curves of the peaks heights are shown as a function of uv doses (a) at 190°C and (b) at 310°C.

^{60}Co irradiations

The irradiations of the samples were done at the ^{60}Co source CAPRI, Centre d'Etudes Nucléaires de Saclay. At a distance of 30 cm, this source yields $2 \cdot 10^5$ rad/h.

Irradiations were performed on samples without the drainage of their natural TL.

Figure 8 shows the glow curve of ^{60}Co induced TL. Two distinct peaks are seen, one at 200°C and the other at 400°C.

The emission spectra of a sample irradiated with 0.8 Mrad were measured. The spectral distributions for the 200°C and 400°C peaks are shown in Figs. 9 and 10. The greatest intensity of emission for the two peaks was about 4000 Å, the first peak is shifted slightly toward the blue, while the second is shifted toward the ultraviolet.

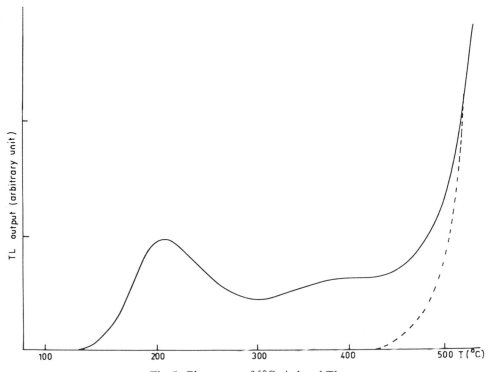

Fig. 8. Glow curve of ^{60}Co induced TL.

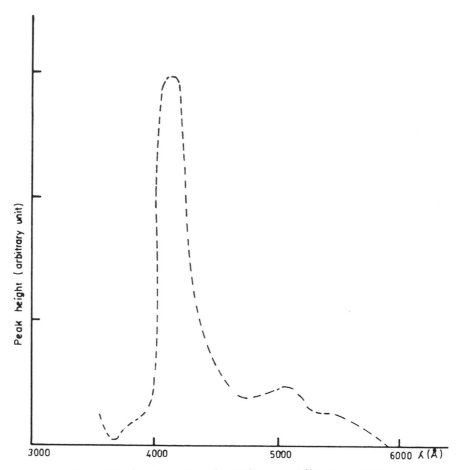

Fig. 9. Wavelength spectrum for 200°C peak of ^{60}Co induced TL.

Comparison with the Saint Severin Meteorite

As a comparison, the same study was performed on the Saint Severin meteorite whose TL is due to plagioclases (Lalou *et al.*, 1970).

In Fig. 11, the natural TL of the meteorite is shown. The glow curve presents two well-defined peaks at 200°C and 325°C. Figures 12 and 13 plot the wavelength spectrums for the 200°C and 325°C peaks, respectively. The light emitted in the 325°C peak is very similar to the light emitted by the lunar fines at high temperatures.

Ultraviolet irradiations create only one peak at about 280°C. The wavelength spectrum of this TL (Fig. 14) shows a bimodal distribution. The maximum is near 5000 Å, similar to the green light emission of the lunar fines at 300°C. The second maximum is farther in the ultraviolet than that of the lunar fines, 4000 Å instead of 4300 Å. The Saint Severin meteorite was also irradiated with ^{60}Co. The glow curve of ^{60}Co induced TL was similar to the one of natural TL, but with an enhancement of the two peaks. The color spectrum for the two peaks can be seen in Figs. 15 and 16, they are very similar to the spectrums of natural TL.

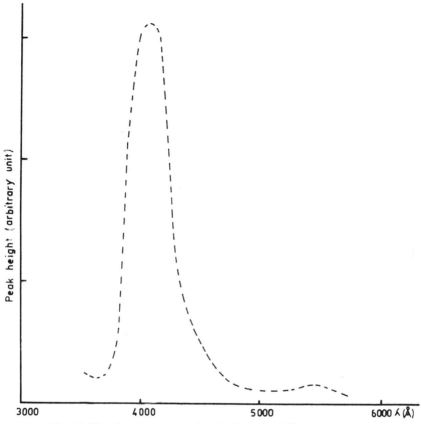

Fig. 10. Wavelength spectrum for 400°C peak of ^{60}Co induced TL.

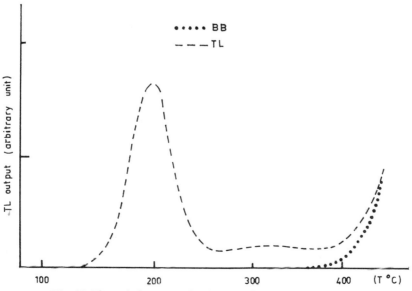

•••• BB

– – – TL

Fig. 11. Natural glow curve for 1 mg of Saint Severin meteorite.

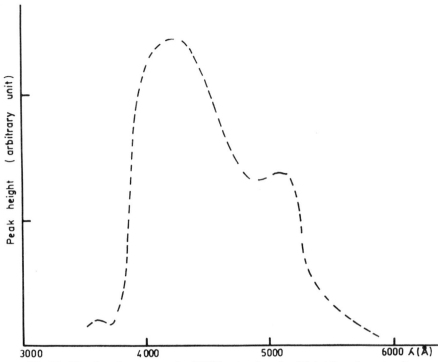

Fig. 12. Wavelength spectrum for 200°C natural peak of Saint Severin meteorite.

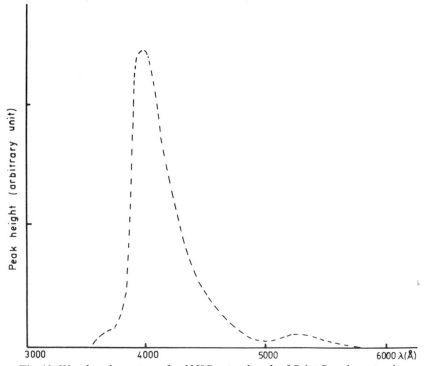

Fig. 13. Wavelength spectrum for 325°C natural peak of Saint Severin meteorite.

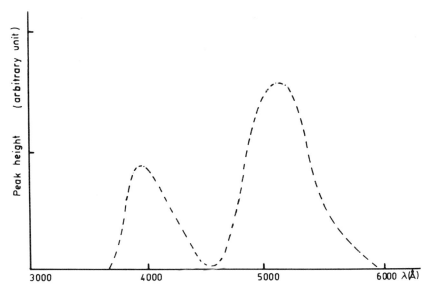

Fig. 14. Wavelength spectrum of uv induced TL in Saint Severin meteorite.

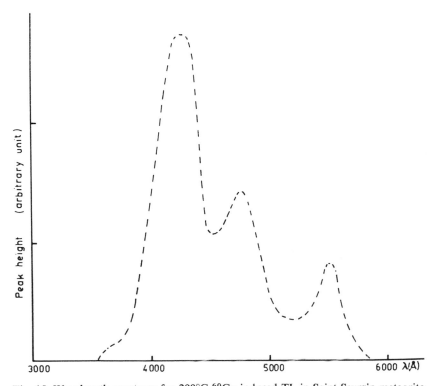

Fig. 15. Wavelength spectrum for 200°C ⁶⁰Co induced TL in Saint Severin meteorite.

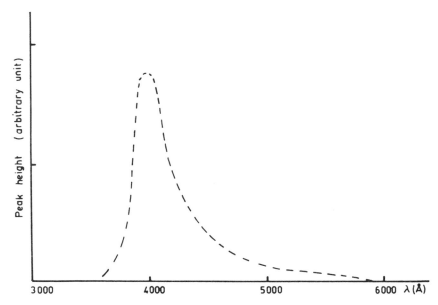

Fig. 16. Wavelength spectrum of 325°C ^{60}Co induced TL in Saint Severin meteorite.

CONCLUSION

As Hoyt *et al.* (1972) have shown that natural TL of bulk lunar fines is the result of two contributions and as the present study seems to enable us, using filters, to study separately these two contributions directly on the whole sample, this may be a useful tool for further studies of lunar TL.

REFERENCES

Hoyt H. P. Jr., Walker R. M., Zimmerman D. W., and Zimmerman J. (1972) Thermoluminescence studies of Lunar samples (abstract), in *Lunar Science—III* (editor C. Watkins), pp. 401–403, Lunar Science Institute Contr. No. 88.

Lalou C., Nordemann D., and Labeyrie J. (1970) Etude préliminaire de la thermoluminescence de la météorite Saint Séverin. C. R. Acad. Sc. Paris, t. 270 série D pp. 2401–2404 (20 mai 1970).

LSPET (1971) Lunar Sample Preliminary Examination Team, Preliminary Examination of Lunar Samples from Apollo 14, *Science* **173**, 681–693.

Valladas G. (1972) A sensitive apparatus to measure the thermoluminescence of small samples. (To be published in Journal of Physics E Scientific Instruments.)

Proceedings of the Third Lunar Science Conference
(Supplement 3, *Geochimica et Cosmochimica Acta*)
Vol. 3, pp. 3021–3034
The M.I.T. Press, 1972

Electronic spectra of pyroxenes and interpretation of telescopic spectral reflectivity curves of the moon

JOHN B. ADAMS*

Caribbean Research Institute, College of the Virgin Islands,
St. Croix, Virgin Islands 00820

and

THOMAS B. McCORD

Planetary Astronomy Laboratory, Department of Earth and Planetary Sciences,
Massachusetts Institute of Technology, Cambridge, Massachusetts 02139

Abstract—Data are presented that relate the wavelength positions of the major Fe^{2+} optical absorption bands in pyroxenes to overall pyroxene composition. The bands appear in reflectivity spectra of rock and soil samples from Apollos 11, 12, 14, and 15, and can be used to determine average pyroxene composition in the multiphase assemblages. Differences in average pyroxene content between rocks and soils at the four Apollo sites imply that the mare and highland soils have been cross-contaminated by one another. Spectral curves of lunar soil samples agree very closely with telescopic measurements of 8–18 km-diameter areas which include the landing sites. The characteristic telescopic spectral curve types for (a) background mare, (b) mare bright craters, (c) background uplands, and (d) upland bright craters are reproduced in the laboratory with the mare and upland samples. Based on lunar sample data it is now possible to make semiquantitative estimates for the telescopically observable areas of the moon of (a) the crystal : glass ratio, (b) the Ti content of the glass in the soil, and (c) the average pyroxene composition.

INTRODUCTION

IN PREVIOUS PAPERS on the optical properties of Apollo samples (Adams and Jones, 1970; Adams and McCord, 1970, 1971a, 1971b) we pointed out that the principal absorption bands in visible and near-infrared reflectance spectra of lunar rocks and soils arise from the mineral pyroxene. The wavelength positions of the bands, furthermore, are related to the pyroxene composition. In this paper we present new data on the optical spectra of pyroxenes that relate band positions to composition. We then discuss the classification of lunar rocks and soils in terms of average pyroxene composition. Finally, the laboratory data are compared with spectra of the moon obtained using earth-based telescopes.

PYROXENE SPECTRA

The electronic spectra of pyroxenes have been studied by several workers during the last seven years. For recent reviews see Burns *et al.*, 1971, and Lewis and White, 1972. Polarized absorption measurements of single crystals have yielded spectra of

---

* Present address: West Indies Laboratory, Fairleigh Dickinson University, St. Croix, Virgin Islands 00820.

high resolution, and at present there is reasonable agreement about the assignments of the major bands.

In pyroxenes the strongest absorptions arise from electronic transitions in Fe^{2+}. Only one spin-allowed transition is possible for the Fe^{2+} ion in octahedral co-ordination and this results in the band near 1 μ that is common to several minerals. In the pyroxene structure, however, the oxygen polyhedra around the cations are strongly distorted from octahedral symmetry so that additional energy levels are resolved and additional spin-allowed transitions arise. This splitting of the crystal field leads to two intense bands, one near 1 μ and another near 2 μ. These bands are polarization-dependent in terms of intensity and wavelength position. Figure 1 shows the spectra of a single grain of zoned augite-pigeonite from sample 12063. The measurements are through the courtesy of Dr. Peter Bell of the Geophysical Laboratory.

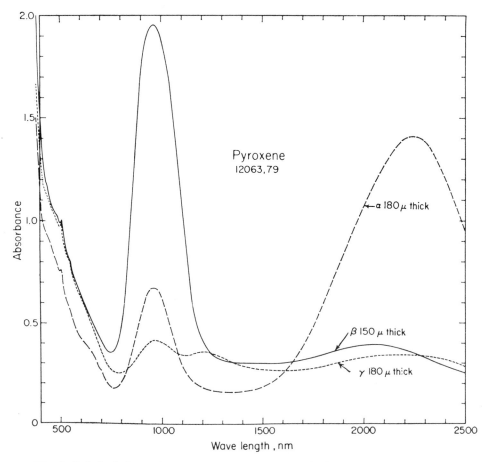

Fig. 1. Polarized absorption spectrum of a single crystal of clinopyroxene from rock 12063,79. Bulk of crystal is pigeonite ($Wo_{14} En_{48} Fs_{38}$) but rim is augite ($Wo_{36} En_{45} Fs_{19}$). Optical and microprobe analyses by Dr. Peter Bell, Geophysical Laboratory.

In diffuse reflected light the two main Fe^{2+} bands are easily resolved. Powdered samples (representing all possible crystallographic orientations) yield bands that are some average of the different absorptions seen in the single-crystal polarized spectra. The band positions in the reflectance spectra, however, still preserve evidence of the crystal structure and composition of the pyroxene. For example, as several authors have noted, the magnesian orthopyroxenes have intense bands near 0.9 μ and 1.8 μ whereas high-calcium pyroxenes exhibit bands near 1.0 μ and 2.3 μ. Figure 2 illustrates the reflectance curves for an enstatite and for the augite-pigeonite 12063 (also shown in Fig. 1).

The relationship between the wavelength positions of the two main bands and the pyroxene composition is shown in Fig. 3 (lower curve). The figure is a plot of the position of the short wavelength band (vertical axis) against the position of the longer wavelength band (horizontal axis). The points in Fig. 3 (lower curve) are derived from diffuse reflectance spectra of essentially pure pyroxene phases, although some of the pigeonites are zoned or intergrown with augite. Chemical analyses are completed for most of the pyroxenes shown. A more detailed discussion of these and other data is being prepared for separate publication. We are concerned here with the major compositional groupings.

Figure 3 (lower curve) shows a rather well-defined curving trend of points extending from the shorter to the longer wavelengths. All pyroxenes fall along the trend. The orthopyroxenes (open circles) occupy the short wavelength positions. In general,

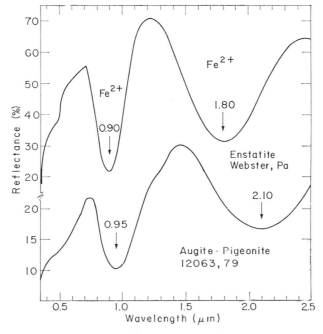

Fig. 2. Diffuse reflection spectra of pyroxene powders illustrating change in band positions with composition. Pyroxene 12063,79 is the same one shown in Fig. 1.

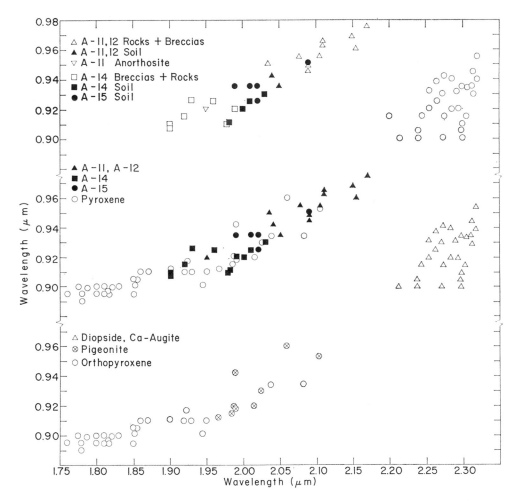

Fig. 3. Diagrams of absorption band positions in reflectance spectra for terrestrial pyroxenes and lunar samples. Pyroxene absorption band near 1 μ is shown on the vertical axis. Band near 2 μ is on the horizontal axis. Lower curve shows pure pyroxenes of different compositions. Middle curve has lunar samples superposed on pyroxene points. Upper curve compares lunar soil bands with bands in spectra of rocks and breccias.

the orthopyroxene bands shift to longer wavelengths as the Fe:Mg ratio increases. The two orthopyroxene points near 0.935 μ and 2.05 μ have Fe:Mg > 80%. The filled circles are pigeonites and subcalcic augites. In general, the bands in this group shift to longer wavelengths with increasing calcium content of the pyroxene. The open triangles represent the calcic augites and the members of the diopside-hedenbergite series. These pyroxenes cluster at the long wavelength end of the diagram.

Figure 3 (lower curve) is useful for determining the approximate composition of an unknown pyroxene from its diffuse reflection spectrum, providing that the two main bands are clearly developed. The figure cannot be used to identify unknowns when the two bands are absent or indistinct, as in some pyroxenes containing important amounts of trivalent ions (Al^{3+}, Ti^{3+}, Fe^{3+}). Mixtures of pyroxene and other minerals having bands within the given wavelength region fall off the trend line. For example the addition of olivine to a low-calcium pyroxene has the effect of moving the pyroxene point vertically on the diagram. The addition of mafic glass (bands near 1.0 μ and 1.9 μ) would move the point of the mixture above and/or to the left of the curve.

LUNAR SAMPLE SPECTRA

The lunar pyroxenes have two intense and clearly developed absorption bands (Figs. 1 and 2) and lie on the trend in Fig. 3. The low abundance of trivalent ions, especially of Fe^{3+}, allows the Fe^{2+} bands to be well resolved. Single-crystal polarized spectra of lunar pyroxenes have been measured by Burns (1971, 1972) and Bell and Mao (1972). Adams and McCord (1971b) discussed the diffuse reflection spectrum of a pyroxene separate from rock 12063. They showed, furthermore, that the main Fe^{2+} bands in the pyroxene account for the two strong bands in the spectrum of the whole rock. The presence of plagioclase, ilmenite, glass, and minor amounts of other phases does not affect the wavelength positions of the pyroxene bands, which persist in the spectra of rocks, breccias, and soils.

If the two strong pyroxene bands survive unaltered in wavelength in the spectra of lunar bulk rocks and soils, then the bulk materials should plot on the pyroxene compositional trend of Fig. 3. This is indeed the case as is shown in the middle curve of Fig. 3.

It is also evident from Fig. 3 (middle curve) that the Apollo samples spread along the pyroxene compositional trend, implying that there are significant differences in average pyroxene composition among the lunar materials. Apollo 14 samples (filled squares) cluster toward the low-calcium end of the compositional trend. This is in good agreement with published analyses, which indicate a preponderance of pigeonite and orthopyroxene in the Apollo 14 materials (LSPET, 1971). The Apollo 11 and 12 mare samples, on the other hand, have more calcic pyroxenes ranging from pigeonite to subcalcic augite and augite. These materials (solid triangles) have longer wavelength bands in the reflectance spectra and plot separately from the Apollo 14 rocks and soils. The one exception is a sample of anorthosite that was separated from the Apollo 11 bulk soil. The anorthosite has a low-calcium pyroxene (Wood et al., 1970) that accounts for the lone solid triangle at 0.92 μ and 1.95 μ.

Five samples of Apollo 15 soil are shown on Fig. 3. In general, the Apollo 15 samples are intermediate between the Apollo 14 materials and those from the Apollo 11 and 12 mare sites. As end members of the Apollo 15 sample suite (Apennine front and mare materials) are approached, the points in Fig. 3 plot with the appropriate upland or mare group. The Apollo 15 point at 0.95 μ and 2.10 μ, for example, is a pyroxene-rich soil from the mare site 9a. Only partial results of our Apollo 15 measurements are presented here. A more detailed account will be published at a later date.

It is of interest to note that the Apollo 11 and 12 samples extend into an unoccupied part of the pyroxene compositional trend that lies between the pigeonites and the calcic augites and diopsides. The gap apparently arises because the two main pyroxene bands are not clearly developed in the common terrestrial augites that would otherwise be expected to occupy this part of the diagram. Augites typically contain Fe^{3+} and other trivalent ions (Ti^{3+}, Al^{3+}). Their reflectivity curves show strong absorption throughout the short wavelength end of the spectrum, the result mainly of Fe^{3+}–Fe^{2+} charge transfers. The spin-allowed Fe^{2+} bands are only weakly developed and are superposed on a steeply sloping continuum, making it difficult to assign band positions from the diffuse reflection spectra.

In contrast, the lunar augites contain little or no Fe^{3+} (Hafner *et al.*, 1971). The strong absorption at short wavelengths is absent, and the spin-allowed Fe^{2+} bands are clearly developed. It appears, therefore, that pyroxenes with well developed bands near 0.97 μ and 2.15 μ are strongly reduced. Natural pyroxenes of this type may effectively be restricted to extraterrestrial sources.

SOILS

From the data in Fig. 3 (middle curve) it is apparent that the band positions in spectra of soils from a given site do not necessarily match those for the associated rocks and breccias. It thus appears that the average pyroxene compositions of most soils differ slightly from the average pyroxene compositions of the bedrock from which the soils were at least in part derived.

The top curve in Fig. 3 shows the Apollo sample points labeled according to (a) soils and (b) rocks and breccias. The Apollo 12 surface soils are at the short-wavelength end of the group of points that represent the mare rocks and breccias. The exact position of the Apollo 11 soil is uncertain, however, owing to the indistinct band near 2 μ. The average pyroxene of these mare soils appears to be less calcic than for the mare basalts. We interpret this to mean that the Apollo 12 soils (and possibly the Apollo 11 soils) contain a component of low-calcium pyroxene that moves the bands of the averaged pyroxene spectrum to shorter wavelengths. This foreign pyroxene component probably is accounted for by the presence of anorthositic and KREEP rock fragments, both of which contain low-calcium pyroxenes. (For reference, the bands for the anorthositic separate from the Apollo 11 soil are represented by the inverted triangular symbol on the same figure. KREEP materials are represented by several of the Apollo 14 breccia points.)

Several investigators have identified contaminant phases in the lunar soils. In the Apollo 11 soil (1–5 mm range) Wood *et al.* (1970) found 5.9% crystalline anorthositic material which typically contains low-calcium clinopyroxene. Although pyroxene is only a minor mineral ($<10\%$) in the anorthositic fragments, it contributes strong optical absorption bands (Adams and McCord, 1971b; Nash and Conel, 1972). Using analyses of glass in the Apollo 11 fines Reid *et al.* (1972) also report the presence of 6% anorthositic material.

The most abundant contaminant in the Apollo 12 soil is KREEP (Meyer *et al.*, 1971) which, including crystalline and glassy material, comprises approximately 30% to 50% of the soils and a larger proportion of some breccias. KREEP rock is dominated by plagioclase and orthopyroxene. Marvin *et al.* (1971) report that crystalline norite-anorthosite comprises from 8% to 19% of the 0.6–3 mm size fraction of the various Apollo 12 soils. Reid *et al.* (1972) report 29% Fra Mauro basalt (=KREEP =norite) and 3% anorthositic material in the Apollo 12 fines, based on calculations from glass analyses. Using the Marvin *et al.* (1971) figure of 16% norite-anorthosite (crystalline) for soil 12070 and assuming that orthopyroxene makes up 50% of the rock, we estimate that approximately 8% of the Apollo 12 soil shown on Fig. 3 may consist of orthopyroxene.

The Apollo 14 soils shown in Fig. 3 (top curve) as filled squares fall at the high-calcium end of the cluster of Apollo 14 breccias and rocks. The three filled squares that lie farthest to the right represent surface and near-surface bulk soils. The lower left solid square represents soil from the edge of Cone Crater (sample 14141). The Cone Crater soil shows a close affinity to the Apollo 14 breccias, from which it has been largely derived as evidenced by the abundance of breccia fragments (40–60%) even in the small-size fractions (62.5 μ) (LSPET, 1971).

On the basis of the inferred average pyroxene composition from Fig. 3, the Apollo 14 bulk soils cannot have been derived entirely from the Apollo 14 breccias. It is suggested instead that the Apollo 14 soils are contaminated by a component of high-calcium pyroxene. The only known source at present for high-Ca pyroxene is the mare materials.

Reid *et al.* (1972) estimate that there is approximately 11% mare-derived glass in the Apollo 14 soils. Chao *et al.* (1972) found titanium-rich ($TiO_2 = 7.8\%$) glasses, which they suggest are derived from a mare area. Glass (1972) also defines a category of mare-derived glasses in the Apollo 14 soils. The presence of mare-derived glasses implies that some amount of crystalline material (including high-calcium pyroxene) should be present also. Steele and Smith (1972) categorized igneous lithic fragments from the 1–2 mm fines. Their Type I high-alumina basalt (14310 type) has the highest-Ca pyroxenes (augite and pigeonite); however in Fig. 3, 14310 plots just to the left of the Apollo 12 anorthosite, indicating that the optical properties of the pigeonite component are dominant in this rock. Steele and Smith also comment that there is little correspondence between the 1–2 mm lithic fragments and the glass composition types of Reid *et al.* (1972). It thus appears that any high-Ca pyroxene contaminant must occur mainly in the <1 mm fraction of the soil. Carr and Meyer (1972) note that there is a maximum of 6% igneous (basalt) fragments (origin unspecified) in the <1 mm fines, whereas light and dark hornfels breccias are the dominant types of fragments. The task of finding mare contaminants is more difficult in the mature Fra Mauro soils, where the host rocks and the contaminants may have been homogenized through several generations of breccia formation. If our conclusions based on the pyroxene bands in Fig. 3 are correct, there remain to be identified a few percent of mare pyroxene in the Apollo 14 soils.

Apollo 15 soil samples from the area of the LM and from the Apennine front have similar band positions. The points on Fig. 3 lie between the Apollo 11 and 12 points and those for Apollo 14. The notable exception is the Apollo 15 soil sample from site 9a which is rich in calcic pyroxene (LSPET, 1971b) and which expectably plots in the cluster of mare basalts. The main group of Apollo 15 soils has an average pyroxene composition intermediate between that of highland and mare materials. We conclude that the Apollo 15 surface soils are cross-contaminated, and that both the Apennine front (upland) materials and the mare soils have been partially blended over the area sampled.

APPLICATIONS OF PYROXENE SPECTRA

A principal objective of our investigation of the optical spectra of the pyroxenes is to strengthen interpretation of telescopic spectral reflectivity data. It is significant, for example, that the main bands in the spectra of lunar rocks fall along the pyroxene trend diagram (Fig. 3). Remote spectra of rocky areas, such as fresh craters, therefore, should give information on the average pyroxene composition. Few telescopic data are available, however, in the 2 μ wavelength region owing to instrument-sensitivity limitations, although improved techniques are now being tested at the telescope. We presently rely on the wavelength position of the band near 1 μ for interpretation of the pyroxene composition. It is possible to measure the telescopic band near 1 μ (McCord et al., 1972) to within about 0.02 μ which would allow a distinction to be made, for example, between the Apollo 14 rocks (0.92 μ) and the Apollo 11 and 12 rocks (0.96 μ). As we have discussed already the absorption bands in the soils converge to similar values probably owing to contamination effects. The soils are not readily distinguished using the single band (see Fig. 3); and because telescopes sense mainly soil material, similar band positions are seen for highland and mare areas.

On the other hand the convergence of band positions for the soils becomes an indicator for contamination; and since contamination is time dependent, it should be possible to separate fresh from old craters in a given material by their pyroxene band positions. This can already be done on the basis of band depths and the overall shapes of the spectral curves, which are controlled by the crystal:glass ratio (Adams and McCord, 1971a, 1971b, 1972).

Cone Crater provides an example. The spectral reflectivity curve of the Cone Crater soil (14141) shows deep (12%) pyroxene band structure which correlates with the very low (<10%) glass content of the soil. In contrast, the soil near the LM (14421, 14259) has 40–75% glass (LSPET, 1971b) and weak (6%) pyroxene bands. If an earth-based or satellite-borne telescope with adequate spatial resolution (better than 1 km) were available, it would be possible to identify Cone Crater as having a high crystal:glass ratio on the basis of the depth of the 0.91 μ band alone. The second line of evidence would be the ratio of the Cone Crater curve to that of the surrounding soil. The resulting "spectral type" (McCord et al., 1972) also correlates with a high crystal:glass ratio (later in Fig. 8). If we now add the evidence from the positions of the two pyroxene bands at 0.91 μ and 1.98 μ it would then be apparent that the Cone Crater material does not plot with the more mature soils on Fig. 3, and therefore must be relatively uncontaminated.

There is additional and independent evidence that Cone Crater exposes fresh materials. Burnett et al (1972) report cosmic ray exposure ages of 24 ± 2 m.y. for rocks from the flank of Cone Crater, and ages of 110 m.y. to 590 m.y. for typical rocks near the LM landing area. This is in close agreement with the results of Crozaz et al. (1972) and is further supported by the work of Dran et al. (1972).

COMPARISON OF TELESCOPIC AND LUNAR SAMPLE CURVES

In addition to the laboratory analysis of the spectral reflectivity of the lunar samples, we have been measuring the spectral reflectivity of many 10- to 18-km diameter areas of the moon from earth using ground-based telescopes. In this way we hope to extrapolate from information gained at the Apollo sites to other unvisited areas of the front face of the moon.

The telescope measurements of the spectral reflectivity of 18-km diameter areas containing the Apollo 11 and Apollo 12 landing sites were compared to laboratory measurements of Apollo samples (Adams and McCord, 1970, 1971a, 1971b). Excellent agreement was found for the surface soil samples indicating that (1) the telescope measurements were accurate to a percent or so, (2) the telescope observations determine the soil properties but are little affected by rocks, (3) the surface soil samples taken at Apollo sites are representative on a regional scale, and (4) the absorption bands in the telescope reflectivity curves are a measure of the pyroxene content of the soil. Since the 0.95 μ absorption appears at nearly the same wavelength in all telescope curves measured so far (McCord *et al.*, 1972), the soil must be of nearly uniform average pyroxene content over the entire front surface of the moon, a result in agreement with our hypothesis of the mixing of the lunar soil.

In Fig. 4 telescope measurements of an 18-km and an 8-km diameter area, containing the Apollo 14 and Apollo 15 landing sites respectively, are compared to laboratory measurements of surface soil samples acquired at these sites. The agreement is excellent; the formal errors on the telescope measurements are about the size of

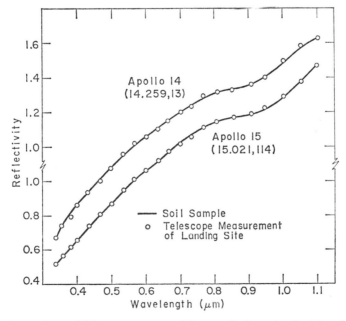

Fig. 4. Comparison of laboratory curves of lunar soils from Apollo 14 and 15 with telescopic curves of 18 km- and 8 km-diameter areas at respective landing sites.

the symbols (<1%). The conclusions from the earlier measurements of this type are confirmed.

The telescopic spectral reflectivity curves for all lunar areas are similar in their shape. However, small but very important differences exist among these curves (McCord *et al.*, 1972). To better display these small differences we have developed what we call the *relative* spectral reflectivity. This quantity is calculated by dividing the spectral reflectivity curve for each area by that for a standard area; in our telescopic case, the standard area is a uniform area of Mare Serenitatis. These ratio curves show the differences between the spectral properties of two lunar areas much more clearly than do the spectral reflectivity curves themselves (see McCord *et al.*, 1972, for detailed presentation).

We have measured and calculated relative spectral reflectivity curves for more than 150 lunar areas, always using the standard area in Mare Serenitatis as the denominator in these ratios. All the curves obtained can be arranged into four non-intersecting sets according to their shape. These four sets—we call them spectral types—are directly correlated with four morphological units: background maria, background uplands, mare bright craters, and upland bright craters (McCord *et al.*, 1972).

In an attempt to understand how these four distinct spectral curve types arise, we have calculated relative spectral reflectivity curves for several of the lunar samples as measured in the laboratory. For these laboratory calculations the Apollo 12 soil sample curve was used as a standard by which all other sample curves were divided.

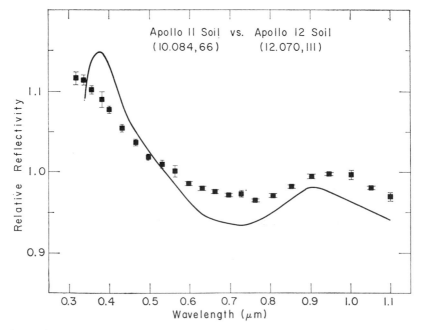

Fig. 5. Comparison of laboratory relative reflectivity curve of Apollo 11 soil (using Apollo 12 soil as standard) with telescopic relative curve of background mare (filled squares).

The Apollo 12 soil sample has a spectral reflectivity very similar to that for the Mare Serenitatis standard area used for the telescope measurements.

Figure 5 shows the relative spectral reflectivity curve for the Apollo 11 soil (determined in the laboratory) along with the relative spectral reflectivity curve for the Apollo 11 site in Tranquilitatis. The similarity is clear; the Apollo 11 soil has a background-mare spectral type.

Figure 6 shows the relative spectral reflectivity curve for a powdered Apollo 12 basalt. Notice how well this spectral curve type matches those observed telescopically for fresh mare bright craters.

The soil from the Apollo 14 landing site yields a background uplands spectral curve type, as can be seen by comparing it with the telescopic background uplands spectral curve type shown in Fig. 7.

And finally, the relative spectral curve type for upland bright craters is duplicated in the laboratory by soil from Cone Crater (Fig. 8). Note that Aristarchus shows an upland bright crater curve even though it is located in a mare area. Apparently Aristarchus has punched through the mare fill, exposing relatively fresh upland-like material from below the mare (McCord *et al.*, 1972).

We have demonstrated that the shapes of the relative spectral reflectivity curves measured using telescopes (observing relatively large lunar areas) can be duplicated using Apollo samples measured in the laboratory. In summary (1) the background mare spectral type corresponds to mature (glassy) mare soil, (2) the background upland spectral type corresponds to mature (glassy) upland soil, (3) the mare bright crater spectral type corresponds to powdered crystalline mare rock, and (4) the

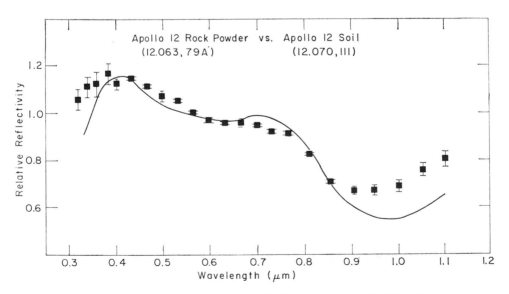

Fig. 6. Comparison of laboratory relative reflectivity curve of Apollo 12 rock powder (using Apollo 12 soil as standard) with telescopic relative curve of mare bright craters (filled squares).

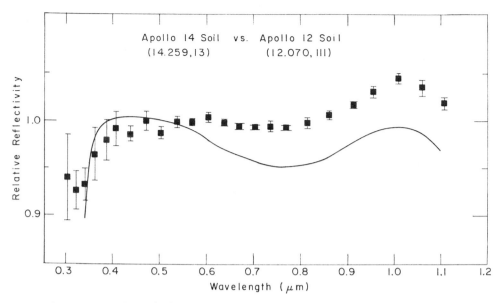

Fig. 7. Comparison of laboratory relative reflectivity curve of Apollo 14 soil (using Apollo 12 soil as standard) with telescopic relative curve of background uplands (filled squares).

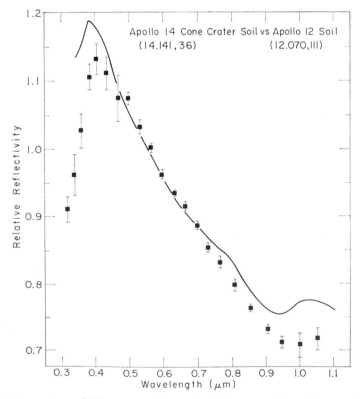

Fig. 8. Comparison of laboratory relative reflectivity curve of Cone Crater soil (using Apollo 12 soil as standard) with telescopic relative curve of Aristarchus (filled squares).

upland bright crater spectral type corresponds to immature (crystalline) soil formed in upland material.

From the comparison of telescope and laboratory measurements of lunar material several conclusions can be drawn:

(1) Telescope measurements are accurate.

(2) The lunar-spectral-type classification derived from telescope observations is strongly supported by laboratory measurements.

(3) The correlation of lunar spectral classifications with lithologic units is verified.

(4) The features in the relative spectral reflectivity curves, from which the lunar spectral type classification is made, can now be explained in terms of the specific mineralogy and composition of the lunar surface material.

(5) The uniformity of spectral properties to this high precision over the background mare and over the background upland regions is further support for the mixing hypothesis. However, the uplands and mare do have slightly different spectral characteristics, indicating that the mixing of the soils is not complete.

(6) With the explanations and confirmations of the telescope results we have obtained from the laboratory measurements of the samples, we can now proceed with meaningful geologic exploration of the front face of the moon down to about 1 km spatial resolution using groundbased telescopes. As before, we can map differences between units on the lunar surface. But it is now possible to interpret these differences in terms of the compositional and mineralogical properties of the surface material. It appears likely that we eventually can specify quantitatively several compositional and mineralogical properties of individual units on the lunar surface including (a) crystal to glass ratio, (b) amount of Ti in glass, and (c) the average pyroxene composition.

Acknowledgments—We thank Dr. Peter Bell of the Geophysical Laboratory, Washington D.C., for providing us with an absorption spectrum of a pyroxene grain from our sample 12063. Mr. Michael Charette of M.I.T. assisted with laboratory measurements of terrestrial pyroxenes and lunar samples. Dr. M. Maurette (Orsay, France) kindly provided irradiation data in advance of publication. This work was supported by NASA grants NGR-22-009-350 and NGR 52-083-003.

REFERENCES

Adams J. B. and Jones R. L. (1970) Spectral reflectivity of lunar samples. *Science* **167**, 737–739.

Adams J. B. and McCord T. B. (1970) Remote sensing of lunar surface mineralogy: Implications from visible and near-infrared reflectivity of Apollo 11 samples. *Proc. Apollo 11 Lunar Sci. Conf.*, *Geochim. Cosmochim. Acta* Suppl. 1, Vol. 3, pp. 1937–1945. Pergamon.

Adams J. B., and McCord T. B. (1971a) Alteration of lunar optical properties: Age and composition effects. *Science* **171**, 567–571.

Adams J. B. and McCord T. B. (1971b) Optical properties of mineral separates, glass, and anorthositic fragments from Apollo mare samples. *Proc. Second Lunar Sci. Conf.*, *Geochim. Cosmochim. Acta* Suppl. 2, Vol. 3, pp. 2183–2195. MIT Press.

Adams J. B. and McCord T. B. (1972) Optical evidence for regional cross-contamination of highland and mare soils. In *Lunar Science—III* (editor C. Watkins), pp. 1–3, Lunar Science Institute Contr. No. 88.

Bell P. M. and Mao H. K. (1972) Initial findings of a study of chemical composition and crystal field spectra of selected grains from Apollo 14 and 15 rocks, glasses and fine fractions. In *Lunar Science—III* (editor C. Watkins), pp. 55–57, Lunar Science Institute Contr. No. 88.

Burnett D. S., Huneke J. C., Podosek F. A., Russ G. P. III, Turner G., and Wasserburg G. J. (1972) The irradiation history of lunar samples. In *Lunar Science—III* (editor C. Watkins), pp. 105–107, Lunar Science Institute Contr. No. 88.

Burns R. G., Huggins F. E., and Abu-Eid R. (1971) Polarized absorption spectra of single crystals of lunar pyroxenes and olivines. *Conference on Lunar Geophysics, Lunar Sci. Inst.*

Burns R. G., Abu-Eid R., and Huggins F. E. (1972) Crystal field spectra of lunar silicates. In *Lunar Science—III* (editor C. Watkins), pp. 108–109, Lunar Science Institute Contr. No. 88.

Carr M. H. and Meyer C. E. (1972) Petrologic and chemical characterization of soils from the Apollo 14 landing site. In *Lunar Science—III* (editor C. Watkins), pp. 116–118, Lunar Science Institute Contr. No. 88.

Chao E. C. T., Boreman J. A., and Minkin J. A. (1972) Apollo 14 glasses of impact origin. In *Lunar Science—III* (editor C. Watkins), pp. 133–134, Lunar Science Institute Contr. No. 88.

Crozaz G., Drozd R., Hohenberg C. M., Hoyt H. P. Jr., Ragan D., Walker R. M., and Yuhas D. (1972) Solar flare and galactic cosmic ray studies of Apollo 14 samples (abstract). In *Lunar Science—III* (editor C. Watkins), pp. 167–169, Lunar Science Institute Contr. No. 88.

Dran J. C., Duraud J. P., Maurette M., Durrieu L., Jouret C., and Legressus C. (1972) Track metamorphism in extraterrestrial breccias. *Proc. Third Lunar Sci. Conf., Geochim. Cosmochim. Acta* Suppl. 3, Vol. 3.

Glass B. P. (1972) Apollo 14 glasses. In *Lunar Science—III* (editor C. Watkins), pp. 312–314, Lunar Science Institute Contr. No. 88.

Hafner S. S., Virgo D., and Warburton D. (1971) Cation distributions and cooling history of clinopyroxenes from Oceanus Procellarum. *Proc. Second Lunar Sci. Conf., Geochim. Cosmochim. Acta* Suppl. 2, Vol. 1, pp. 91–108. MIT Press.

Lewis J. F. and White W. B. (1972) Electronic spectra of iron in pyroxenes. *Jour. Geophys. Res.* (in press).

LSPET (Lunar Sample Preliminary Examination Team) (1971a) Preliminary examination of the lunar samples from Apollo 14, *Science* **173**, 681–693.

LSPET (Lunar Sample Preliminary Examination Team) (1971b) The Apollo 15 lunar samples: A preliminary description. *Science* **175**, 363–375.

Marvin U. B., Wood J. A., Taylor G. J., Reid J. B. Jr., Powell B. N., Dickey J. S. Jr., and Bower J. F. (1971) Relative proportions and probable sources of rock fragments in the Apollo 12 soil samples. *Proc. Second Lunar Sci. Conf., Geochim. Cosmochim. Acta* Suppl. 2, Vol. 1, pp. 679–699. MIT Press.

McCord T. B., Charette M., Johnson T. V., Lebofsky L., Pieters C., and Adams J. B. (1972) Lunar spectral types. *Jour. Geophys. Res.* **77**, 1349–1359.

Meyer C. Jr., Brett R., Hubbard N. J., Morrison D. A., McKay D. S., Aitken F. K., Takeda H., and Schonfeld E. (1971) Mineralogy, chemistry, and origin of the KREEP component in soil samples from the Ocean of Storms. *Proc. Second Lunar Sci. Conf., Geochim. Cosmochim. Acta* Suppl. 2, Vol. 1, pp. 393–411. MIT Press.

Nash D. B. and Conel J. A. (1972) Further studies of the optical properties of lunar samples, synthetic glass, and mineral mixtures. In *Lunar Science—III* (editor C. Watkins), pp. 576–577, Lunar Science Institute Contr. No. 88.

Reid A. M., Ridley W. I., Warner J., Harmon R. S., Brett R., Jakes P., and Brown R. W. (1972) Chemistry of highland and mare basalts as inferred from glasses in the lunar soil. In *Lunar Science—III* (editor C. Watkins), pp. 640–642, Lunar Science Institute Contr. No. 88.

Steele I. M. and Smith J. V. (1972) Mineralogy, petrology, bulk electron-microprobe analyses from Apollo 14, 15, and Luna 16. In *Lunar Science—III* (editor C. Watkins), pp. 721–723, Lunar Science Institute Contr. No. 88.

Wood J. A., Dickey J. S., Marvin U. B., and Powell B. N. (1970) Lunar anorthosites and a geophysical model of the moon. *Proc. Apollo 11 Lunar Sci. Conf., Geochim. Cosmochim. Acta* Suppl. 1, Vol. 1, pp. 965–988. Pergamon.

Proceedings of the Third Lunar Science Conference
(Supplement 3, *Geochimica et Cosmochimica Acta*)
Vol. 3, pp. 3035–3045
The M.I.T. Press, 1972

Far infrared properties of lunar rock

P. E. Clegg, S. J. Pandya, S. A. Foster, and J. A. Bastin

Physics Department, Queen Mary College, Mile End Road,
London, E1 4NS, England

Abstract—This paper describes the continuation of observations of the interaction of far infrared electromagnetic radiation with lunar samples.

In the case of the fines an observed increase of *mass* electromagnetic attenuation coefficient with density indicates that part of the attenuation results from scattering: at 338 μm the pure absorption and scattering coefficients are equal at a porosity of $((\rho_0 - \rho)/\rho_0) = 0.7$. Within the wavelength range 300–3000 μm the scattering coefficient decreases with increasing wavelength although the *ratio* of the scattering to pure absorption terms becomes greater as the wavelength is increased. Within the same wavelength range, the attenuation coefficient shows a marked temperature dependence; the relevance of this effect to absorption mechanisms centered in the infrared is discussed. Previous attenuation measurements in the range 1000–3000 μm have been confirmed using a new filter spectroscopic technique.

The results are discussed in connection with measurements of the thermal conductivity of lunar fines and direct thermal measurements on the lunar surface. It is concluded that there is a relatively sharp variation of conductivity with depth in the upper layers of the lunar regolith.

Far Infrared Properties of Lunar Rock

Although research in the far infrared wavelength region requires complex and sophisticated techniques, its application to the spectroscopic study of lunar samples is of considerable value especially in making more precise our understanding of the regolith layer.

Measurements of the absorption coefficient of the lunar fines may be employed in conjunction with far infrared and microwave telescopic observations of the lunar surface to determine not only the gradient resulting from the internal lunar heat flux but also the effective conductivity of the uppermost centimeters of the regolith layer. A study of the absorption and scattering coefficients at these wavelengths is also desirable, since it is relevant to the radiative transfer processes at shorter wavelengths that are thought to dominate the effective net thermal transfer coefficient in the upper ranges of temperature to which the top layers of the lunar surface are subject.

It is the purpose of this note both to report recent studies that distinguish between scattering and absorption processes in the far infrared and also to describe some measurements of the variation of absorption with temperature. In conclusion, we discuss the relevance of these measurements to the study of the lunar surface layer.

Experimental Technique

Low temperature measurements

To investigate the absorption and scattering properties of rock samples at various temperatures, and in the far infrared wavelength range, the arrangement shown in

3035

Fig. 1 was used. For work between 70K and ambient temperature, the sample was located at S, the holder being in vacuum but attached to a cold finger F in which various refrigerant liquids at ambient pressure were contained. For measurements at 1.4K the sample was placed at S' in the light pipe P just above the detecting element E which in this case was a germanium bolometer. This bolometer, whose action depends on the temperature variation of the resistance of germanium, is sensitive in the wavelength range up to 1000 μm. The sample was prepared by sandwiching a measured mass of lunar fines between two polythene windows separated by a brass ring spacer of known thickness. An identical empty sample container was prepared to determine the background spectrum. The polarizing interferometer used in these measurements has been described by Martin and Puplett (1969). The interferogram was apodized and was transformed using a fast Fourier transform procedure on an I.C.L.1905E computer and the sample spectrum ratioed against the background. Some results are shown in Fig. 2.

Laser measurements

To determine the scattering contribution to the attenuation coefficient for samples of differing density, measurements were made using the 338 μm laser described by Gebbie (1964). A compound sample holder was used: this consisted of two hard plastic discs ("Rigidex 100" I.C.I.) which is known to have very low absorption at far infrared wavelengths. Between the discs of the holder measured amounts of lunar fines were included, separated by a given distance, again using an annulus as a spacer. The discs and spacer were held rigidly together between two annular metal plates by

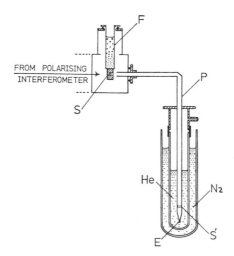

Fig. 1. Measurement of attenuation of lunar fines at low temperatures. For temperatures close to the absolute zero of temperature, the sample is placed at S' while for temperatures between ambient and that of liquid air ($\sim$70K) it is placed in the sample container at S. The beam from the polarizing interferometer passes through the light pipe P to the detecting crystal E.

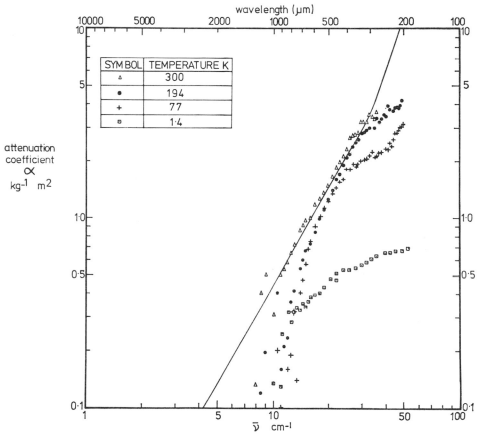

Fig. 2. Mass attenuation coefficient α for the lunar fines measured with the polarizing interferometer at various temperatures. The continuous curve represents the mean of previously reported results (Ade *et al.*, 1971).

three equidistant screws. An identical empty sample holder was placed next to the sample holder so that they could be rapidly interposed at a position that intercepted the laser beam: This rapid interchange is required to reduce the effects of slow drift in the laser intensity. The absolute value of this intensity is, however, sufficiently high for a golay cell to be used as a convenient detector of adequate sensitivity. The average fraction of 338 μm radiation transmitted by the sample was determined and from this fraction the attenuation coefficient was calculated. The experiment was repeated with increasing amounts of Type D material until it was no longer possible to compress the sample to the thickness of the annular spacer. A thin chip of Type C microbreccia (10065,15) was used to determine the mass attenuation coefficient at a higher density but essentially using the same techniques. The results of all these measurements are summarized in Fig. 3.

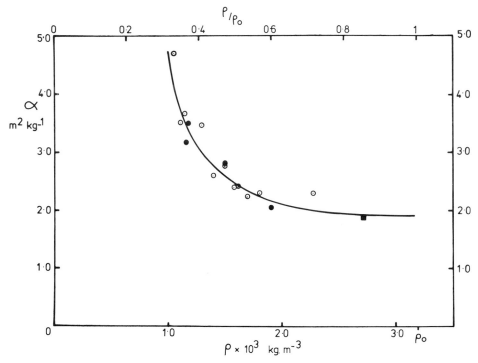

Fig. 3. Mass attenuation coefficient α shown as a function of density ρ at a wavelength of 338 μm for lunar fines (10084,111), circular symbols, and microbreccia (10065,15), square symbol. Filled symbols refer to measurements made with the polarizing interferometer; open symbols refer to measurements with the 338 μm CN laser. The extrapolated absorption coefficient for material without voids, $(\rho/\rho_0) = 1$, is 1.88 $kg^{-1}\ m^2$.

Solar source measurements

Measurements in relatively narrow wavelength bands centered at wavelengths above 1000 μm were made using a 1.6 m aperture Cassegrain telescope as a flux collector and the sun as a source. The lunar fines were held in the device described in the previous section and placed so as to intercept the beam at the secondary focus of the telescope. The radiation was detected by a liquid-helium cooled Indium Antimonide element sensitive in the wavelength range 200–2000 μm. A 6 mm thickness of carbon impregnated polythene together with a wire mesh etalon (Ade and Bastin, 1971) was used to define a pass band centered at 1200 μm: another pass band centered at 1700 μm was produced by the use of carbon impregnated polythene of 6 mm thickness without an ancillary etalon. The telescope was driven so as to point continuously at the same hour-angle as the sun: At the same time scans in declination were made across the solar disc. The sample and the blank container were interposed into the beam during alternate scans and, from the ratio of recorded intensities, the attenuation coefficient of the Type D material was determined. The results are shown in Fig. 4.

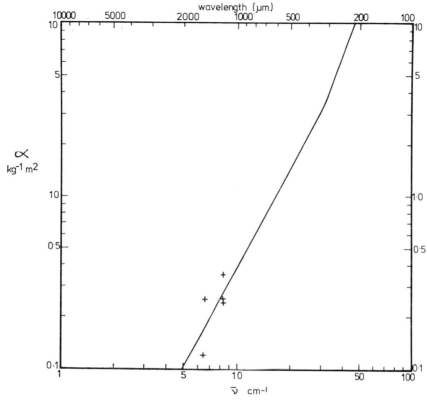

Fig. 4. Mass attenuation coefficient α for lunar fines. The continuous curve represents the mean of previously reported results (Ade *et al.*, 1971). Crosses refer to measurements made with filters using sun as a mm source. These results have been corrected for scattering so that they apply to the same density as the continuous curve.

FAR INFRARED SCATTERING BY LUNAR FINES

As yet no direct observations have been made of the scattering of far infrared radiation by lunar fines. However, the results shown in Fig. 4 suggest that scattering is an important mechanism in the attenuation of a beam passing through a sample of fines.

If the coefficient of mass attenuation at wavelength λ is α_λ then this may be written as the sum of mass coefficients resulting respectively from pure absorption (κ_λ) and scattering (σ_λ) processes:

$$\alpha_\lambda = \kappa_\lambda + \sigma_\lambda \tag{1}$$

The decrease of α_λ with density shown in Fig. 3 is readily understandable if we assume that σ_λ decreases with increasing density. On a simple basis the mass scattering coefficient would be expected to decrease as the sample is compressed. Consider, for example, two particles of the same refractive index separated by a distance d. If d is the same or greater than λ, reflection losses occur at each of the two boundaries when a beam passes from one particle to the next. However, as soon as the sample is compressed to such an extent that d becomes an order of magnitude less than λ, the electromagnetic reflected amplitudes destructively interfere and the intensity of the reflected beam (which constitutes an element of scattered radiation) is greatly reduced.

On this basis the scattering should be a minimum for crystalline basaltic rock. We should not expect the scattering to be zero within such crystalline samples, since they are composed of different minerals each with a different refractive index. Nevertheless, since the refractive indices of the different mineral types are not greatly different, the scattering coefficient for the igneous rock should be very much less than the corresponding coefficient for fine material with an appreciable porosity.

A re-examination of the results reported by Ade *et al.* (1971) show a very definite correlation with density in agreement with the arguments given above. We have made some fresh measurements with more widely differing densities and the results of these are shown in Fig. 5. From equation (1) we have, for two different densities ρ_1 and ρ_2,

$$\frac{\alpha_\lambda(\rho_1)}{\alpha_\lambda(\rho_2)} = \frac{\kappa_\lambda + \sigma_\lambda(\rho_1)}{\kappa_\lambda + \sigma_\lambda(\rho_2)} \tag{2}$$

where we have assumed that κ_λ is independent of density. From Fig. 3 we see that,

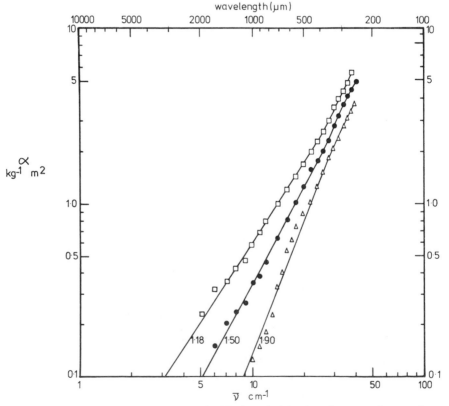

Fig. 5. Attenuation coefficient measured with the polarizing interferometer for samples of fines of different macroscopic density. The figure by each curve refers to the specific gravity, or the density in units $10^3 \mathrm{Kg\ m^{-3}}$.

if our assumption about scattering is correct,

$$\sigma_{337\mu}(1.90) \approx 0,$$

so that, on the physically reasonable assumption that σ_λ is a decreasing function of wavelength (cf. discussion below), we have from equation (2)

$$\frac{\alpha_\lambda(\rho)}{\alpha_\lambda(1.90)} \approx 1 + \frac{\sigma_\lambda(\rho)}{\kappa_\lambda} \qquad (3)$$

for wavelengths greater than $\sim 300 \ \mu m$.

In Fig. 6 we have plotted $\alpha_\lambda(\rho)/\alpha_\lambda(1.90)$ for densities of 1.18 and 1.50×10^3 kg m^{-3}. It can be seen that $\alpha_\lambda(\rho)/\alpha_\lambda(1.90) \approx 1$, for $\lambda \sim 300 \ \mu m$, in agreement with our assumption above, and that $\sigma_\lambda/\kappa_\lambda$ is an increasing function of wavelength. This implies that κ_λ decreases more rapidly with λ than does σ_λ.

The discovery of an appreciable scattered term in the far infrared electromagnetic attenuation within lunar fines has interesting implications in our understanding of the heat balance of the regolith. In the experimental arrangements we have described, a negligible fraction of the radiation scattered through appreciable angles is incident

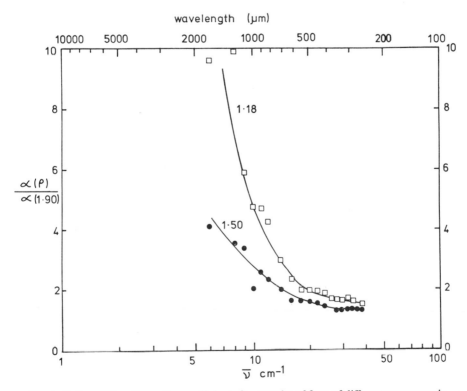

Fig. 6. Ratios of the attenuation coefficients for samples of fines of different macroscopic densities to the coefficient for a sample of macroscopic density 1.90×10^3 Kg m^{-3}. The figure by each curve is the specific gravity, or the density in units 10^3 Kg m^{-3}.

on the detector, and in our analysis we have assumed this fraction to be zero. In the case of far infrared telescopic observations of the moon a different geometry exists and it seems likely that an appreciable fraction (perhaps even the majority) of photons that reach the observing telescope have been scattered through an appreciable angle within the lunar regolith.

If, for some wavelength λ, the scattering coefficient σ_λ is comparable with, or greater than, the true absorption coefficient κ_λ, and if significant thermal gradients occur within the characteristic depth of penetration of radiation at this wavelength, then the theory must be modified. Quantitatively, the two conditions are

$$\frac{\sigma_\lambda}{\kappa_\lambda} \gtrsim 1$$

and

$$\delta_\lambda \gtrsim 1$$

where δ_λ is the ratio of the penetration depth of an electromagnetic wave of wavelength λ to the penetration depth of thermal waves (cf., for example, Clegg et al., 1966). From Fig. 6 it is seen that the first condition is satisfied for $\lambda \gtrsim 700$ μm, while the condition on δ_λ is satisfied for $\lambda \gtrsim 400$ μm (Bastin and Gear, 1967). We therefore expect the theory to require modification for $\lambda \gtrsim 700$ μm. The scattering phenomena we have observed can also to some extent solve the discrepancy between the values of the thermal conductivity determined by microwave telescopic observations of the moon, on the one hand, and on the other from deductions based both on midinfrared measurements (8–20 μm) and direct laboratory thermal measurements with lunar fines. A simple analysis, taking into account the scattered term, now shows that the effective conductivity of the top few centimeters of the regolith may be as high as 3×10^{-4} $Wm^{-1}K^{-1}$.

Temperature Dependence of the Attenuation Coefficient for Lunar Fines

In order to examine the results shown in Fig. 2, we first consider equation (1) in greater detail. We shall now write the coefficients with, in parentheses, those variables on which we think the attenuation may depend in the case of a given powdered material:

$$\alpha(\lambda, \rho, T) = \kappa(\lambda, T) + \sigma(\lambda, \rho). \tag{4}$$

Here, as indicated, we expect the true absorption coefficient κ to be independent of ρ in a first approximation but dependent on T. On the other hand, we expect σ, the scattering coefficient, to be roughly independent of T but to be a decreasing function of ρ since, as discussed above, the number of effective scattering centers will decrease as ρ increases. A similar argument shows that σ should also decrease with increasing wavelength.

As a simple model for the true absorption, we may suppose this to be the result of a single Lorentz type absorption for which the absorption coefficient takes the form:

$$\kappa(\lambda, T) = \frac{M\gamma}{\left(\dfrac{1}{\lambda} - \dfrac{1}{\lambda_0}\right)^2 + \dfrac{\gamma^2}{4}} \tag{5}$$

where λ_0 is the wavelength at the absorption maximum, γ is the damping term and M the transition probability. Equation (5) shows explicitly the wavelength dependence of the process; the temperature dependence is implicit since both γ and M may be temperature dependent. The damping term γ is essentially the result of multiphonon processes and will therefore depend upon the phonon population which is, of course, strongly temperature dependent. The M will be independent of temperature for single phonon processes but will be dependent on temperature for multiphonon processes since these depend on phonon population.

The results shown in Fig. 2 show considerable temperature dependence of the coefficient of attenuation. This we attribute to a decrease in the true absorption coefficient with decrease of temperature. Here γ (and possibly M) will depend upon temperature very roughly as $e^{-h\nu/kT}$, where ν is the frequency of the phonons involved. Analysis of Fig. 2 shows the results to be consistent with phonons at a few tens of wave numbers. However, it is felt that this model is too crude to be pursued profitably. In particular, the effects of particle size and shape on the absorption spectrum (Ruppin and Englman, 1970) renders the problem complex.

Two further points may be made in connection with the results shown in Fig. 2. In the first place, the curvature of the graphs at room temperature is different and indeed in the opposite sense both to that shown in Figs. 4 and 5, and to previously reported results (Ade et al., 1971). This effect in Fig. 2 we feel is purely instrumental, resulting from the insensitivity of the detector at low frequencies and the choice of sample thickness that makes accurate measurement difficult at the highest frequencies. Second, it seems possible that the curve for 1.4K may have no measurable residual absorption, the attenuation resulting purely from the scattering process. We are encouraged in this view by the fact that the magnitude of the attenuation coefficient is equal, within experimental error, to that which may be deduced for the scattering coefficient from the results shown in Fig. 3; but the view is also supported by a plot of attenuation coefficient as a function of reciprocal absolute temperature. More measurements of the attenuation coefficient in the range 0–20°K are clearly required.

Finally our results have implications in the study of the transfer of energy by radiative processes in the regolith. A simple treatment in the case of a pure absorbing medium (see, for example, Clegg et al., 1966) shows that we can define a radiative component k_r to the effective conductivity:

$$k_r = \frac{16\sigma}{3} \frac{T^3}{\kappa(T)} \tag{6}$$

where σ is Stefan's constant and $\overline{\kappa(T)}$ is the Rosseland mean absorption coefficient, a mean which is weighted by the Planck function at temperature T. Although the most important contributions to the radiative transfer probably occur at wavelengths shorter than those covered by our measurements, it seems most unlikely that the Rosseland mean coefficient can do other than increase with increasing temperature. We see thus from equation (4) that the effective radiative term increases less rapidly with temperature than T^3 and is thus probably more important at lower temperatures than had previously been assumed.

Conclusions

The measurements and results of this paper may be summarized as follows:

(i) In the far infrared wavelength range there is, in the case of lunar fines, appreciable attenuation of an electromagnetic beam by scattering processes.

(ii) The ratio of the relative attenuations resulting from scattering and pure absorption decreases with increasing frequency in the range 5–50 cm^{-1}.

(iii) The infrared absorption processes in lunar fines are temperature dependent; this indicates that the radiative contribution to the effective conductivity will probably vary less rapidly with temperature than T^3.

(iv) The *in situ*, laboratory, infrared and microwave determinations of the conductivity of lunar soil can probably all be reconciled if we allow for scattering effects in the microwave region and assume a thermal conductivity which increases rapidly with depth. Such a conclusion is in good agreement with the results of direct measurements at depths of the order of 1 m (Langseth, 1972) and infrared observations during eclipses which relate to the top few millimeters (see, for example, Winter and Saari, 1969).

Acknowledgments—We wish to thank the National Physical Laboratory for the use of the 338 μm CH laser. The interferometric absorption measurements reported here were obtained with a Beckman–R.I.I.C. FS 720 Michelson interferometer converted to operate as a polarizing interferometer. We are most grateful to Dr. S. O. Agrell for his kind loan of microbreccia (10065,15) and are indebted to Mr. P. Ade, Mr. A. C. Marston, Mr. E. Puplett, and Mr. D. G. Vickers for experimental advice and help. One of us (S.J.P.) wishes to thank the World University Service for financial support, and we wish to thank Prof. D. H. Martin and the College in general for its support and encouragement with this work.

References

Ade P. A., Bastin J. A., Marston A. C., Pandya S. J., and Puplett E. (1971) Far Infrared Properties of Lunar Rock. *Proc. Second Lunar Sci. Conf., Geochim. Cosmochim. Acta* Suppl. 2, Vol. 3, 2203–2211, MIT Press.

Bastin J. A. and Gear A. E. (1967) Observations in the wavelength range 1 to 3 mm. *Proc. Roy. Soc. A* **296**, 348–353.

Clegg P. E., Bastin J. A., and Gear A. E. (1966) Heat transfer in lunar rock. *Mon. Not. Royal. Astron. Soc.* **133**, 63–66.

Gebbie H. A., Stone N. W. B., and Findlay F. D. (1964) Interferometric observations on far infrared stimulated emission sources. *Nature* **202**, 169–170.

Langseth M. (1972) *The Moon* (to be published March 1972; October 1971 Lunar Geophysics Conference. Lunar Science Institute).

Martin D. H. and Puplett E. (1969) Polarized interferometric spectrometry for the millimeter and submillimeter spectrum. *Infrared Physics* **10,** 105–109.

Piddington J. H. and Minnet H. C. (1949) Microwave thermal radiation from the moon. *Aust. J. Sci. Res.* **2,** 63–77.

Ruppin R. and Englman R. (1970) Optical phonons of small crystals. *Rep. Prog. Phys.* **33,** 149–196.

Troitski V. S. (1954) Towards the theory of lunar thermal radiation. *Astron. Zhurnal* **31,** 511–523.

Winter D. F. and Saari J. M. (1969) A particulate thermophysical model of the lunar soil. *J. Appl. Chem. (London)* **156,** 1135–1151.

Proceedings of the Third Lunar Science Conference
(Supplement 3, *Geochimica et Cosmochimica Acta*)
Vol. 3, pp. 3047–3067
The M.I.T. Press, 1972

Infrared and Raman spectroscopic studies of structural variations in minerals from Apollo 11, 12, 14, and 15 samples

P. A. Estep, J. J. Kovach, P. Waldstein, and C. Karr, Jr.

Morgantown Energy Research Center,
U.S. Department of the Interior, Bureau of Mines,
Morgantown, West Virginia 26505

Abstract—Infrared and Raman vibrational spectroscopic data, yielding direct information on molecular structure, have been obtained for single grains ($> 150 \mu$) of minerals, basalts, and glasses isolated from Apollo 11, 12, 14, and 15 rock and dust samples, and for grains in Apollo 14 polished butt samples. From the vibrational data, specific cation substitutions were determined for the predominant silicate minerals of plagioclase ($An_{/1}$ to An_{95}), pyroxene (Fs_6 to Fs_{53}), and olivine (Fa_{14} to Fa_{45}). Unique spectral variations for grains of K-feldspar, orthopyroxene, pyroxenoid, and ilmenite were observed to exceed the ranges of terrestrial samples, and these variations may be correlatable with formation histories. Alpha-quartz was isolated as pure single grains, in granitic grains composited with sanidine, and in unique grains that were intimately mixed with varying amounts of glass. Accessory minerals of chromite and ülvospinel were isolated as pure grains and structurally characterized from their distinctive infrared spectra. Fundamental vibrations of the SiO_4 tetrahedra in silicate minerals were used to classify bulk compositions in dust sieved fractions, basalt grains, and glass particles and to compare model characteristics for maria, highland, and rille samples. No hydrated minerals were found in any of the samples studied, indicating anhydrous formation conditions.

INTRODUCTION

PREVIOUSLY PRESENTED VIBRATIONAL SPECTROSCOPIC DATA for Apollo 11 and 12 samples (Estep *et al.*, 1971) were obtained by macro-infrared methods, which required frequent grain combinations to obtain 1 mg of sample for analysis. The application of microsampling techniques in the present work has eliminated the possibility of contamination from such combinations, has furnished new data on grain-to-grain structure variations, has made possible the identification of some trace accessory minerals, and has made it possible to obtain infrared and Raman vibrational data on the same single grain. The improvement in sampling techniques thus necessitated an examination of mineral grains from some Apollo 11 and 12 samples that we previously studied, for an assessment of the technique and in order to compare grains with those from Apollo 14 and 15 samples. We found that single-grain analyses on the Apollo 11 and 12 samples broadened the compositional ranges for pyroxenes (Fs_{15}–Fs_{51}) and olivines (Fa_{26}–Fa_{38}) considerably, compared to ranges previously obtained by grain-combination analyses.

The spectral data show both similarities and differences among the lunar maria, highland, and rille samples studied in this work: three crystalline rocks (12018,26, 12021,24, 14310,93); six dusts (10085,46, 12070,24, 14163,80, 14259,15, 15301,83, 15601,72); one breccia (14321,108); and two potted butts (14310,2, 14321,97). We found that one of the Apollo 15 dusts (15301,83) had some mineral structural features (in ilmenite and bulk compositions) similar to those of highland samples, while these same features in the other Apollo 15 dust (15601,72) were more similar to those in maria samples. No hydrated minerals were found in any of the samples studied, indicating anhydrous formation conditions.

EXPERIMENTAL TECHNOLOGY

Preparation of single lunar grains for infrared and Raman spectroscopy

Single grains for analysis were isolated from the + 100 mesh dry-sieved dust fractions and crushed rock fragments in a dry-nitrogen-filled chamber. For a complete mineral structure characterization, it was necessary to obtain both infrared and Raman spectra on the same individual grain. We therefore first obtained a Raman spectrum on a weighed grain, typically 400 to 200 μ, that was sealed under helium in a 0.5 mm diameter fused-quartz capillary (Uni-mex Company, Griffith, Indiana) to prevent sample oxidation by the laser beam in the Raman instrument. Raman spectra were obtained on a Spex Ramalog (Model 1401) spectrometer,* utilizing a Coherent Radiation argon-krypton (Model 52) laser source, and the Raman scattered light was observed at 90° to the incident beam. The laser beam could be focused to an area of 100 μ^2 for measurements on 150-μ diameter single grains. The thin-walled capillaries (10 μ) in which the samples were enclosed gave no interfering Raman lines when the laser beam was focused on the sample. After Raman data were obtained, the grain was recovered and ground in air to a fine powder for infrared microsampling analysis, utilizing a specially designed microcrushing mortar, similar to the macrocrushing mortar previously described (Estep *et al.*, 1971). Cesium iodide power (Harshaw Chemical Company) was added directly to the preground sample in the mortar to give a sample concentration of approximately 0.33 wt.%. After blending, about 10 μg of this mixture were transferred to a Perkin-Elmer ultra-micro die to prepare a 1.5-mm diameter micropellet at 500 lb total load and under vacuum. Infrared pellets were scanned with a reflecting-type 6× ultra-micro beam condenser mounted on a Model 621 Perkin-Elmer grating spectrophotometer purged with dry air. Both infrared and Raman data were obtained as described on 40 selected grains from only Apollo 12 and 14 samples, while 570 individual grains were isolated and examined by infrared from all the lunar rock chips and dust samples.

Raman and infrared reflectance spectra of grains in polished rock samples

To obtain *in situ* and orientation information, infrared and Raman spectra of single grains were obtained directly from polished slabs of two Apollo 14 lunar rock samples. The laser-Raman beam was focused on selected grains embedded in the polished rock surface (under helium purge) by means of a specially designed aperture-mask device (Makovsky, 1972), using masks with apertures of 0.5, 0.75, 1.0, and 1.5 mm diameters. Infrared specular reflectance spectra (13° angle of incidence) were subsequently obtained on the same selected grains (under dry air purge), utilizing a Perkin-Elmer microspecular reflectance attachment on the infrared instrument, and masks with apertures that were 1.0 and 2.0 mm in diameter. To obtain bulk compositional information from the polished rock samples, similar to the infrared reflectance measurements made by Perry *et al.* (1972), we utilized a mask with a 5.0 mm diameter aperture. Techniques were developed for both Raman and infrared reflectance measurements that allowed good spectra to be obtained from pressed powders of terrestrial standards—a capability particularly necessary for obtaining data from powdered synthetic samples.

---

* Equipment is named in this report for identification only and does not necessarily imply endorsement by the U.S. Bureau of Mines.

RESULTS AND DISCUSSION

Structure determination of mineral grains

The infrared spectral correlations previously developed for determining specific cation substitutions in the predominant silicate minerals of Apollo 11 and 12 samples (Estep *et al.*, 1971) were applied to obtain a comparison with the Apollo 14 and 15 samples. These involved the use of determinative curves derived from studies on synthetic and terrestrial standards. The composition-dependent infrared frequency shifts on which these curves were based allowed a determination of anorthite content in plagioclase feldspars, ferrosilite in pyroxenes, and fayalite in olivines. It has recently been demonstrated that in Raman spectra, chain vibrations are sensitive to the population of the cation sites and reflect the crystal environment of the Si–O groups in a variety of silicate minerals (Griffith, 1969). We observed composition-dependent frequency shifts, specifically in Raman spectra of plagioclase, pyroxenes, and olivines, and from these constructed determinative curves which provided information on the same cation substitutions as obtained from the infrared data. We obtained similar results with infrared reflectance and Raman spectra gathered directly on polished rock or pressed powder mineral samples. Data collected from studies on synthetic and terrestrial standards by these techniques also provided the same composition-dependent frequency shifts which could be utilized to determine cation substitutions in the lunar silicate minerals.

Table 1 presents compositional data obtained for Apollo 11, 12, 14, and 15 silicate minerals. The table includes data only for single grains, whether isolated or measured *in situ* by Raman or infrared reflection techniques in the polished rock samples.

Feldspars

Plagioclase. The abundant Apollo 14 and 15 plagioclase grains occurred in a wide range of morphologies and were predominantly calcium-rich (An_{71}–An_{95}). Compositions shown in Table 1 were derived from both infrared and Raman data, utilizing low-frequency absorption bands that are linearly dependent on calcium content (Estep *et al.*, 1971).

K-feldspars. From the Apollo 14 breccia we isolated seven white granular composite grains of K-feldspar and alpha-quartz with compositions ranging from predominantly K-feldspar to predominantly alpha-quartz. Such granitic components were reported in Apollo 12 breccias (Sclar, 1971; Drake *et al.*, 1970, for rock 12013) and led to the suggestion that these constituents were derived from the lunar highlands where large masses of primary granitic rock could exist in addition to igneous rocks of gabbroic, noritic, and anorthositic composition. Because these grains were recovered in small amounts from the Apollo 14 samples, it is still speculative whether such siliceous material represents any widespread geological formation on the lunar surface. Figure 1 demonstrates the detection of alpha-quartz (curve a) and K-feldspar (curve b) in one of the granitic grains (curve c) in which these minerals are of approximately equal abundance. We further isolated five white granular grains of pure K-feldspar from the same Apollo 14 breccia. Isolation of these pure grains

Table 1. Compositional data for single grains of Apollo 11, 12, 14, and 15 silicate minerals.

Mineral	Sample[1]	Description of mineral grains[2]	Number of analyses[3]	Analytical frequencies,[4] cm^{-1}	Derived composition
Silica					
α-Quartz	10085,46	Black lustrous opaque, from a 2 mm conglomerate	1		
	12070,24	Mixed with glass; medium amber opaque	1		
	14259,15	Deep green, glossy, flat, 600 × 300 μ; Colorless transparent, 400 × 400 μ	2	1075, 791, 773, 454, 392, 368	
		Mixed with glass; light yellow to deep amber transparent, yellow to green to brown opaque	17		
	14321,108	Composited with K-feldspar; white granular, 500–200 μ	7		
			1	467	
	15601,72	Mixed with olivine; medium yellow, honey transparent	2		
		Composited with other lunar silicates; black, smooth surfaced, blocky	1		
Feldspars					
K-Feldspar	14321,108	White granular, 800–200 μ	5	638–631, 542–548	
		Composited with α-quartz; white granular, 500–200 μ	7	637–625, 542–545	
			1	516, 488	
Plagioclase	14163,80	Colorless transparent, white opaque, white granular	10	225–228	An$_{79}$–An$_{85}$
			2	507	>An$_{90}$
	14310,2	Collection of 200 μ colorless to grey transparent grains, 1200 × 1200 μ area	2	(560)	An$_{95}$
			2	508	>An$_{90}$
	14310,93	Colorless transparent	2	225, 230	An$_{79}$, An$_{89}$
	14259,15	Colorless transparent, tinted translucent, white opaque to colored opaque, white granular	20	225–230	An$_{79}$–An$_{89}$
	14321,97	Light grey transparent to chalky white, 1500 × 600 μ area	1	(565)	An$_{90}$
			7	508	>An$_{90}$
	14321,108	Colorless transparent, tinted translucent, white opaque	9	220–225	An$_{71}$–An$_{79}$
	15301,83	Colorless transparent, tinted translucent, white granular	9	225	An$_{79}$
	15601,72	Colorless transparent, white granular	7	225	An$_{79}$
Pyroxenes					
Orthopyroxenes					
Enstatite	14321,108	Light yellow opaque	6	1060, 855	Fs$_6$–Fs$_{10}$
			1	1018, 680, 668	n.d.[5]
Bronzite[6]	14321,108	Light yellow transparent	1	396	Fs$_{11}$
	15301,83	Light yellow opaque	2	396, 393	Fs$_{11}$, Fs$_{16}$
Bronzite-hypersthene	10085,46	Medium amber transparent, from a 1.6 mm conglomerate	1	391	Fs$_{19}$
	14163,80	Yellow transparent	1	392	Fs$_{17}$
	14259,15	Light green opaque, light yellow opaque	2	395, 389	Fs$_{13}$, Fs$_{22}$
	14310,93	Light yellow transparent to translucent	5	393–392	Fs$_{16}$–Fs$_{17}$
	14321,97	Honey transparent, 600 × 600 μ area	1	(877)	Fs$_{21}$
	14321,108	Dark brown transparent	1	383	Fs$_{32}$
	15301,83	Light yellow transparent, light yellow to honey translucent	4	395–392	Fs$_{13}$–Fs$_{17}$
	15601,72	Light yellow transparent	1	390	Fs$_{20}$
Clinopyroxenes					
Pigeonite (Wo$_5$–Wo$_{15}$)	12018,26	Yellow transparent	1	390	Fs$_{24}$
	12021,24	Yellow transparent	17	393–390	Fs$_{20}$–Fs$_{24}$
			1	1008, 676, 652	Fs$_{33}$
	12070,24	Honey transparent	2	390	Fs$_{24}$
	14259,15	Medium yellow transparent	1	390	Fs$_{24}$
	14310,93	Light yellow transparent	3	342–341	Fs$_9$–Fs$_{11}$
			1	1010, 682, 665	Fs$_{30}$
	14321,108	Honey transparent	3	385–383	Fs$_{32}$–Fs$_{35}$
	15301,83	Light yellow transparent	1	390	Fs$_{24}$
	15601,72	Medium yellow to medium amber transparent	5	393–380	Fs$_{20}$–Fs$_{40}$
Subcalcic augite (Wo$_{15}$–Wo$_{25}$)	10085,46	Light and medium amber transparent, from a 2 mm conglomerate	2	388, 394	Fs$_{27}$, Fs$_{18}$
	12018,26	Light medium amber transparent	5	390–383	Fs$_{24}$–Fs$_{35}$
	12021,24	Light amber transparent	3	380	Fs$_{40}$
	12070,24	Honey to dark amber transparent	5	396–380	Fs$_{15}$–Fs$_{40}$
	14163,80	Light yellow transparent	1	392	Fs$_{21}$
	14259,15	Light amber translucent	1	380	Fs$_{40}$
			1	1013, 679, 663	Fs$_{22}$
	14321,97	Honey transparent	1	1004, 679, 663	n.d.

Table 1——*continued*.

Mineral	Sample	Description of mineral grains	Number of analyses	Analytical frequencies, cm^{-1}	Derived composition
Augite $(Wo_{25}-Wo_{45})$	14321,108	Honey to dark amber transparent	3	390–385	$Fs_{24}-Fs_{32}$
	15301,83	Golden yellow to medium amber transparent	3	387–375	$Fs_{29}-Fs_{48}$
	15601,72	Honey to medium amber transparent	15	390–372	$Fs_{24}-Fs_{53}$
	10085,46	Medium to dark amber transparent, from a 2 mm conglomerate	9	390–385, 325–322	$Fs_{24}-Fs_{32}$, $Fs_{35}-Fs_{41}$
	12021,24	Medium to dark amber transparent	4	387–383, 322	$Fs_{29}-Fs_{35}$, Fs_{41}
	12070,24	Honey to reddish amber transparent	8	395–386, 325	$Fs_{16}-Fs_{30}$, Fs_{45}
	14259,15	Medium to reddish amber transparent	2	387	Fs_{29}
	15301,83	Light to dark amber transparent	5	387–383, 321–319	$Fs_{29}-Fs_{35}$, $Fs_{43}-Fs_{46}$
	15601,72	Yellow to dark amber transparent	8	394–385, 320–315	$Fs_{18}-Fs_{32}$, $Fs_{44}-Fs_{53}$
Pyroxenoids	12021,24	Golden yellow transparent, subhedral equant, 500–300 μ	3	728, 705, 672, 660, 636, 575, 560	
	15601,72	Honey transparent, 400 × 400 μ	1	726, 705, 670, 659, 632, 572, 556	
Olivines	12070,24	Honey, opaque to transparent	4	402–394	$Fa_{26}-Fa_{38}$
	14163,80	Light yellow transparent	1	402	Fa_{26}
			1	855, 821	Fa_{27}
	14259,15	Light yellow to amber transparent, light yellow translucent	8	405–390	$Fa_{22}-Fa_{45}$
	14321,97	Gray translucent, with fractures, 1050 × 700 μ area	1	850, 823	Fa_{40}
	14321,108	Light to dark yellow transparent, yellow to green opaque	15	410–390	$Fa_{14}-Fa_{45}$
		Light to medium yellow transparent	2	851, 819	Fa_{20}
	15301,83	Light yellow to honey, green, transparent to opaque; some rounded	8	402–391	$Fa_{26}-Fa_{43}$
	15601,72	Light yellow to honey, green, transparent to opaque; some rounded	21	402–390	$Fa_{26}-Fa_{45}$

[1] Lunar samples were classified as follows: 10085,46, fines to coarse fines; 12018,26, crystalline rock; 12021,24, crystalline rock; 12070,24, < 1 mm contingency fines; 14163,80, < 1 mm bulk fines 14259,15, < 1 mm comprehensive fines; 14310,2, potted butt of basaltic crystalline rock; 14310,93, basaltic crystalline rock; 14321,97, potted butt of coarse grained breccia; 14321,108, coarse grained breccia; 15301,83, < 1 mm comprehensive fines; 15601,72, < 1 mm comprehensive fines.
[2] Isolated single grains ranged 1500–150 μ and were typically 400–200 μ; single grains in polished samples examined by Raman and infrared reflectance techniques ranged 1500–200 μ.
[3] Number of grains isolated for infrared absorption analysis; number of areas examined by Raman and infrared reflectance techniques.
[4] Analytical Raman frequencies are underlined; infrared reflectance frequencies are in parentheses; only selected diagnostic frequencies are listed.
[5] n.d. = not determined.
[6] These unique bronzite grains (shown for example by Fig. 3, curve b) are listed separately from the other bronzites in this table because the Si–O stretching region pattern more closely matched pure enstatite (Fig. 3, curve a). All other bronzite grains isolated from the lunar samples more closely matched hypersthene (Fig. 3, curve c) in this spectral region.

allowed a determination of the distribution of Al and Si cations over the tetrahedral lattice positions, which has been difficult to obtain from x-ray diffraction data because of the similarities of Al and Si x-ray scattering factors (Brown, 1971). To quantify the varying degrees of disorder that are possible in K-feldspars, we applied the infrared method described by Laves and Hafner (1956), Hafner and Laves (1957), and Bahat (1970), to a number of terrestrial samples of the three possible polymorphs, with examples of spectra shown in Fig. 2. In going from ordered microcline to orthoclase to disordered sanidine (curves a, b, c), a general band broadening can be observed, and the frequency of the absorption band A can be seen to decrease from 642 to 632 cm^{-1} while that of band B increases from 529 to 543 cm^{-1}. The frequency difference between A and B can thus be used as an indication of relative degree of disorder in the feldspar lattice. Studies on the terrestrial samples show typical frequency difference decreases from 113 to 98 to 88 cm^{-1} in going from microcline to orthoclase to sanidine, respectively. Spectra of the lunar K-feldspar grains (e.g., Fig. 2, curve d) showed

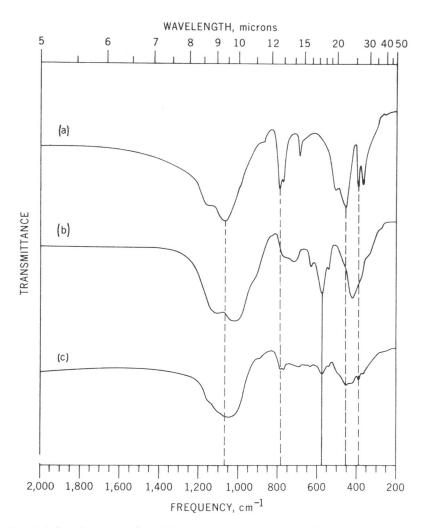

Fig. 1. Infrared spectra of granitic constituents found in Apollo 14 samples. (a) α-quartz from dust 14259,15, 400 μ colorless transparent rounded grain, (b) sanidine from breccia 14321,108, 200 μ white granular grain, (c) granitic grain from breccia 14321,108, 200 μ white granular grain.

A-B frequency differences in the range of 93 to 83 cm^{-1}, indicating for some grains a lower degree of Al/Si ordering than that observed in any terrestrial sanidine samples, except for that from a submarine Hawaiian basalt (77 cm^{-1}) (Andrews, 1971). Since sanidine is the more stable polymorph above about 700°C, its identification in lunar breccia suggests processes on the lunar surface involving high temperatures. The lower degree of Al/Si ordering in the lunar breccia and Hawaiian basalt sanidines may indicate exposures to higher temperatures and/or more rapid quenching.

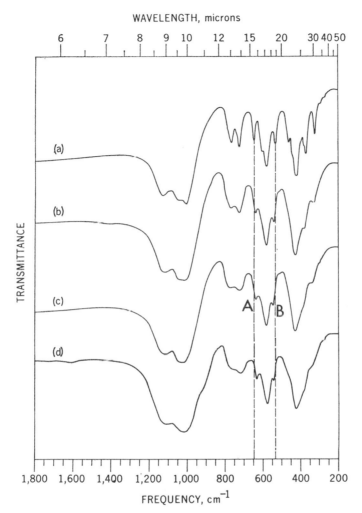

Fig. 2. Infrared spectra of K-feldspars. (a) microcline (Custer County, North Dakota), (b) orthoclase (Ray, Arizona), (c) sanidine (Drachenfels, Germany), (d) sanidine from breccia 14321,108, 200 μ white granular grain.

Pyroxenes

Seven different structural types of pyroxenes, characterized by their distinctive infrared spectra, were isolated from the lunar samples; these are listed in Table 1. Classification criteria were based on unique variations in band frequencies, resolution, splittings, and relative intensities. Changes in these spectral parameters can be due to compositional variations (cation substitution), changes in crystal symmetry, or degree of order in the pyroxene lattice. Whereas phase designations derived by many other techniques are based on chemical composition, the vibrational data yield direct knowledge on structural states. Studies on terrestrial and synthetic pyroxene standards allowed a specific determination of the ferrosilite content and a semiquantitative determination of wollastonite content, as shown in Table 1.

Orthopyroxenes. Orthopyroxenes were isolated from the Apollo 11 dust and from all the Apollo 14 and 15 rock chips and dusts made available to us, as shown in Table 1. Specific ferrosilite contents for these were derived from a determinative curve prepared from 24 analyzed terrestrial samples of orthopyroxenes, for the absorption band which shifts linearly from 400 to 350 cm^{-1} for Fs_5 to Fs_{88}. Figure 3 shows the infrared spectral distinctions for grains of enstatite, bronzite, and hypersthene orthopyroxenes isolated from the Apollo 14 breccia. In lunar enstatite spectra (curve a), a change in band structure in the region of the 400 cm^{-1} analytical band prevented a

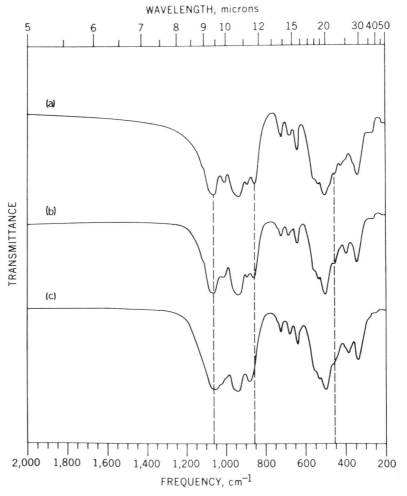

Fig. 3. Infrared spectra of orthopyroxenes from breccia 14321,108. (a) enstatite (Fs_6), 400 μ yellow opaque grain, (b) bronzite (Fs_{11}), 500 μ light yellow transparent grain. (c) hypersthene (Fs_{32}), 400 μ brown transparent grain.

determination of ferrosilite content utilizing this band. The spectral changes (more fine structure) in the 500–400 cm^{-1} region were intermediate between those in ortho-enstatite (from the Norton County achondrite, the Cumberland Falls achondrite and a synthetic from Tem-Pres Research Division, The Carborundum Co., State College, Pa.) and clinoenstatite (a natural sample from Cape Vogel, Australia, a Tem-Pres synthetic, and from the Bencubbin mesosiderite). The close comparison of breccia enstatite spectra with spectra of enstatite from these sources, and the fact that the changes did not occur in the natural terrestrial magnesium-rich orthopyroxenes we examined, suggest that the spectral changes may be the result of shock or high temperatures, rather than a simple composition effect. To obtain ferrosilite content for the breccia enstatite, we utilized the magnitude of the splitting in the Si–O stretching vibrations at 1064 and 854 cm^{-1} (Fs_6), which decreases regularly with increasing ferrosilite content, as demonstrated in Fig. 3.

In a further comparison of lunar orthopyroxene spectra with those obtained from terrestrial metamorphic sources (Howie, 1963; Sadashivaiah and Subbarayudu, 1970), it was noted that a band at 450 cm^{-1} which was of medium intensity in spectra of the terrestrial samples, was considerably weaker in spectra of lunar grains with equivalent ferrosilite contents (Figs. 3 and 4). Further, the intensity of this absorption band in spectra of the terrestrial metamorphic orthopyroxenes can be seen in Fig. 4 to be dependent on ferrosilite content for compositions from Fs_{11} to Fs_{88}. Recent Mössbauer studies on terrestrial metamorphic orthopyroxenes (Virgo and Hafner, 1970) have shown that an increasing ferrosilite content produces a more disordered Fe^{2+} distribution over the nonequivalent octahedrally coordinated sites M1 and M2. Additional Mössbauer studies have demonstrated that this cation distribution is both temperature (Virgo and Hafner, 1969; Dundon and Hafner, 1971) and shock dependent (Dundon and Hafner, 1971). We therefore examined samples of orthopyroxenes for which Mössbauer data are reported in an attempt to determine if the intensity of the 450 cm^{-1} band is related to the same structural variation. Figure 5 demonstrates the effects of shock on the infrared spectrum of orthopyroxene from Bamle, Norway. Samples shocked to 250 kilobars, 450 kilobars, and 1 megabar show a systematically decreasing intensity for the 450-cm^{-1} band and this correlates with the cation disordering detected by Mössbauer studies on these same samples (Dundon and Hafner, 1971). It can also be noted in Fig. 5, curve (d), that the 400-cm^{-1} analytical band nearly disappears with 1 megabar of shock, as observed in the breccia enstatite. We examined a sample of hypersthene (No. 115/30, Fig. 4, curve c) after it was heated to 1000°C for 166 hours and rapidly quenched and found a substantial decrease in the intensity of the 450-cm^{-1} band, correlating well with the cation disordering measured by Mössbauer studies on this same heated sample (Virgo and Hafner, 1969). We further observed the 450-cm^{-1} band to be considerably reduced in intensity in spectra of orthopyroxene grains that we isolated from several chondrites and mesosiderites, and in particular those from the Rose City and Farmington chondrites which have been shown to be cation disordered by Mössbauer studies (Dundon and Hafner, 1971; Dundon and Walter, 1967). The correlation with Mössbauer data supports the interpretation that the intensity of this absorption band might be related to cation disorder in pyroxenes, rather than the lattice-stacking disorder detectable by x-ray

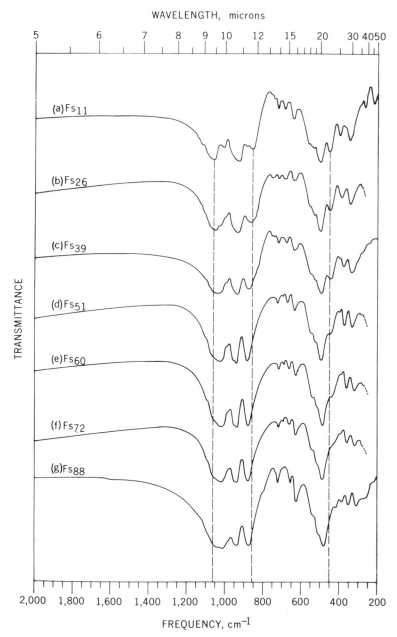

Fig. 4. Infrared spectra of terrestrial metamorphic orthopyroxenes. (a) bronzite (Jackson County, North Dakota, USNM No. 47530), (b) bronzite (Madras, Howie No. 3709), (c) hypersthene (Virgo No. 115/30), (d) ferrohypersthene (Madras, Howie No. 137), (e) ferrohypersthene (Baffin Island, Howie No. 400), (f) eulite (Sudan, Howie No. 1002), (g) eulite (Virgo XYZ). Samples from Howie (1963) and Virgo and Hafner (1970). Spectra for Howie samples Nos. 3701, 137, 400, and 1002 were previously published by Lyon (1963). (Spectra shown as curves b, d, e, f were obtained in KBr.)

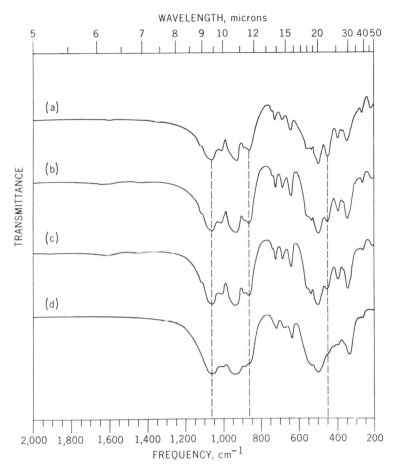

Fig. 5. Infrared spectra of orthopyroxene from Bamle, Norway (Fs_{14}). (a) Natural,
(b) shocked to 250 kilobars, (c) shocked to 450 kilobars, (d) shocked to 1 megabar.

diffraction (Pollack and DeCarli, 1969). The 450-cm^{-1} absorption band appears in a frequency region typical for cation-oxygen stretching vibrations, bending vibrations of Si–O–Si groups in silicate chains, and lattice modes. It is expected that for any of these possible assignments, the distorted M2 octahedron would produce vibrational modes differing from those of the more regular M1 octahedron in the pyroxene structure. Thus, the reduced intensity of the 450-cm^{-1} band in spectra of the ortho-pyroxene grains we isolated from the lunar samples and from meteorites may indicate reduced occupancy of the preferred M2 sites, resulting from exposure to either high temperature or shock events. Spectra of orthopyroxene grains from the Apollo 14 breccia and dust 14259,15 exhibited a 450-cm^{-1} band relatively weaker than that in spectra of dust 10085,46, dust 14163,80, rock 14310,93, and the two Apollo 15 dusts; this suggests differences in their formation histories. Further studies of this spectral variation in natural, heated, and shocked samples could lead to estimates of specific temperatures and pressures, which in turn could be usefully applied in deducing sample history.

Clinopyroxenes. The three distinctive clinopyroxene spectra that we obtained from the lunar samples are shown in Fig. 6. Systematic studies on four pyroxenes synthesized by Turnock, University of Manitoba, Canada (Wo_{20} with Fs_{14}, Fs_{40}, Fs_{56}, and $Wo_{40}Fs_{30}En_{30}$) and eight pyroxenes synthesized by Tem-Pres Research (Wo_{10}, Wo_{20}, Wo_{30}, and Wo_{40}, each with Fs_{22} and Fs_{45}) allowed us to assess the

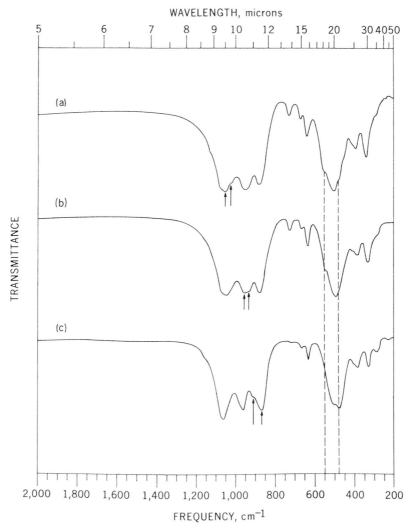

Fig. 6. Infrared spectra of lunar clinopyroxenes. (a) Pigeonite (Fs_{20}) phenocryst from rock 12021,24, 200 μ yellow transparent grain, (b) subcalcic augite (Fs_{15}) from dust 15301,83, 800 × 400 μ golden yellow transparent grain, (c) augite (Fs_{27}) from dust 15301,83, 200 μ light amber transparent grain.

effects of calcium substitution on infrared spectra. Unique variations in band splittings of the Si–O stretching vibrations in the range 1060–850 cm^{-1} were observed, and the strong Si–O–Si bending vibration near 500 cm^{-1} shifts to lower frequencies, shows less fine structure, and splits as calcium increases. Specific ferrosilite contents for clinopyroxenes were determined from the same infrared analytical absorption band near 400 cm^{-1} as used for orthopyroxenes and for Apollo 11 and 12 clinopyroxenes (Estep et al., 1971). Because in iron-rich augites ($>Fs_{40}$) the 400-cm^{-1} analytical band splits and is weakened, an absorption band near 330 cm^{-1} (also varying regularly with ferrosilite content) was used to determine iron substitution. For both analytical bands we observed a general correlation of deepening pyroxene grain color with shifts to lower frequencies, as previously reported for Apollo 11 and 12 clinopyroxenes (Estep et al., 1971). The dark amber iron-rich augites (Fs_{45}–Fs_{53}) gave good spectral matches with ferroaugite ($Wo_{41}Fs_{51}En_8$) from the Skaergaard Intrusion of East Greenland (Williams et al., 1971). A Raman determinative curve for ferrosilite content in lunar pigeonites was derived by plotting the center of gravity of an intense doublet that shifted from 1014 to 1002 cm^{-1} as ferrosilite content varied from Fs_{15} to Fs_{60}. The curve exhibited an unusually abrupt frequency shift near $Fs_{33.5}$, suggesting a possible structural change at this composition where several Apollo 14 grains fell.

Pyroxenoids. We isolated grains of pyroxenoids from rock 12021,24 and dust 15601,72 and used the infrared method described by Lazarev and Tenisheva (1961b) to confirm the pyroxmangite structure and to show that there are seven silicon tetrahedra repeat units in the silicate chains (*Siebenerketten* structure), as determined by Burnham (1971) for the pyroxferroite from an Apollo 11 rock. The method utilizes a series of absorption bands in the 750–550 cm^{-1} region (see Fig. 7), which correspond to that produced by a totally symmetric stretching vibration in SiO_4 tetrahedra. As coupling takes place to form chains, these split to give more absorption bands. Lazarev showed that there is a direct correspondence between the number of bands appearing in this region and the number of silicon-oxygen tetrahedra in the repeating unit of the silicate chain. He demonstrated, for example, that a structure with a repeat unit of seven tetrahedra indicates a chain of the type $[(SiO_3)_n]_\infty$, where $n = 7$. We applied Lazarev's correlations to the three synthetic pyroxenoids of Fig. 7 (curves a, b, e) and verified Liebau's (1956) general correlation that, as the mean octahedral cation size decreases (Ca^{2+} to Fe^{2+}), the number of silicon tetrahedra in the repeat unit of the chain increases. For example, the composition of $Ca_{0.5}Fe_{0.5}SiO_3$ (curve a) shows three absorption bands in the 750–550 cm^{-1} region, indicating that $n = 3$ and verifying that this pyroxenoid has the bustamite structure. The synthetic pyroxenoid $Ca_{0.15}Fe_{0.85}SiO_3$ (curve b) showed seven absorption bands in the scale expanded spectrum of this same region, indicating the pyroxmangite structure. Similarly, scale expansion of the spectrum for $FeSiO_3$ (Lindsley et al., 1964) (curve e) shows the presence of eight absorption bands. Since Lazarev studied only structures for $n = 2$ to $n = 7$, a ninth absorption band may be out of the 750–550 cm^{-1} range in ferrosilite III, or may be accidentally degenerate. Thus we presume that $n = 9$ for the composition $FeSiO_3$, in agreement with the *Neunerketten* chain configuration assigned by Burnham (1966) to this same sample. Spectra of the pyroxenoid grains from both lunar samples (Fig. 7, curves c and d) were scale expanded in the region 750–550 cm^{-1}

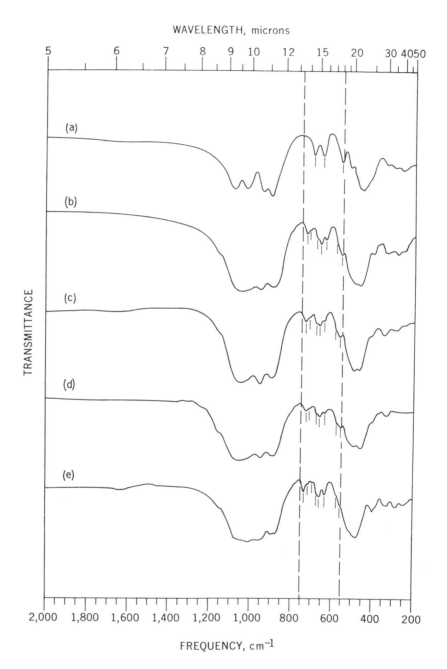

Fig. 7. Infrared spectra of pyroxenoids. (a) Synthetic, $Ca_{0.5}Fe_{0.5}SiO_3$, (b) synthetic, $Ca_{0.15}Fe_{0.85}SiO_3$, (c) lunar, from rock 12021,24, 300 μ golden yellow transparent grain, (d) lunar, from dust 15601,72, 400 μ honey transparent grain, (e) synthetic, $FeSiO_3$ (ferrosilite III), (Synthetic samples synthesized by Lindsley and Dowty.)

and a total of seven absorption bands could be counted in all spectra (as listed in Table 1), indicating the pyroxmangite structure with $n = 7$. All three grains from rock 12021,24 gave identical infrared spectra (for example, curve c) and corresponded closely with that of the synthetic sample $Ca_{0.15}Fe_{0.85}SiO_3$ (curve b), which is near the composition reported by Weill (1971) for pyroxferroite from this rock. However, the spectrum of the pyroxenoid grain from dust 15601,72 (Fig. 7, curve d) showed shifts to lower frequencies for some absorption bands and more closely matched that of natural pyroxmangite ($Ca_{0.056}Mg_{0.054}Fe_{0.38}Mn_{0.51}SiO_3$) from Idaho (USNM 102794, Henderson, 1936). This spectral match, and the lighter color of the Apollo 15 grain suggest less iron in the pyroxmangite structure. These studies indicate that infrared spectroscopy could be applied in determining the compositional limits of the various pyroxenoid structures, which remain still largely unknown (Burnham, 1971).

Olivines

The olivine grains isolated in this work (Fa_{14}–Fa_{45}) were all in the forsterite-fayalite olivine series, and their spectra compared well with that previously published (Estep *et al.*, 1971). For a given composition, lunar olivine spectra compared best with spectra of synthetic equivalents. Some grains from the Apollo 14 and 15 samples gave spectra of poor quality, showing a band broadening particularly in the region 1150–1000 cm^{-1} that suggested a glass or silica association. In two grains (Fa_{41}) from dust 15601,72, alpha-quartz bands at 1075, 791, and 773 cm^{-1} were detected. Determinative curves for fayalite contents were used for an infrared absorption band which shifts linearly from 418 to 356 cm^{-1} for Fa_0 to Fa_{100}, and for a Raman doublet which shifts 865–845 cm^{-1} and 832–818 cm^{-1} for the same fayalite contents.

Opaque oxides

Ilmenite grains were isolated as black, lustrous grains from the maria, highland, and rille samples listed in Table 2. Infrared spectra of some single grains showed weak

Table 2. Single grains of ilmenite isolated from lunar rocks and dusts.

Sample source	Sample type	Number of grains	Size, μ	Analytical frequency, cm^{-1}	Type classification[1]
10085,46	Dust, coarse fines	324[2]	100 avg.	285	Intermediate I–II
		10[3]	200–150	285–275	Intermediate I–II, Type II
12018,26	Crystalline rock	360[2]	100 avg.	290	Intermediate I–II
		3	150	290	Intermediate I–II
		2	200, 150	305	Type I
12021,24	Crystalline rock	349[2]	100 avg.	284	Intermediate I–II
		5	400–200	280–275	Intermediate I–II, Type II
14259,15	Dust	2	200,150	310, 305	> Type I, Type I
14321,108	Breccia	6	300–150	305	Type I
		1	200	320	> Type I
15301,83	Dust	2	150	310	> Type I
15601,72	Dust	2	300, 400	280, 270	Intermediate I–II, < Type II

[1] For terrestrial ilmenites, Type I = 305 cm^{-1} and Type II = 275 cm^{-1} (see text); the closest type is underlined.

[2] These grains were combined for a single infrared macro-analysis (Estep *et al.*, 1971), and are shown here for comparison. They were isolated from the –100 mesh sieved fractions of crushed fragments.

[3] Grains from a 2 mm and a 1.6 mm conglomerate.

Si–O stretching frequencies in the range 1100–1000 cm^{-1}, suggesting silicate over-growths. In spectra of terrestrial ilmenites from a variety of sources and with different formation histories, we have observed gradational frequency shifts for the pre-dominant absorption bands, indicating subtle and continuously variable structural differences. To describe these variations, we have distinguished two end-members, Type I and Type II, according to the position of the lowest frequency-absorption band for terrestrial samples. This diagnostic band varies, as seen in Fig. 8, from 305 cm^{-1} in Type I (curve b) to 275 cm^{-1} in Type II (curve d), and its frequency in spectra of all other terrestrial samples is intermediate between these two values. Similarly, spectra of lunar ilmenites exhibited these same frequency variations and could be classified according to the terrestrial scheme, as shown in Table 2. There were both sample-to-

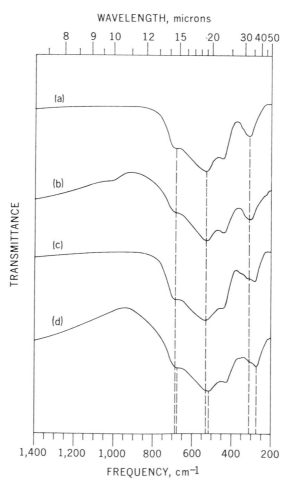

Fig. 8. Infrared spectra of ilmenites. (a) Lunar, from dust 15301,83, 100 μ black lustrous grain, (b) terrestrial, from Canada, Type I, (c) lunar, from dust 15601,72, 300 μ black lustrous grain, (d) terrestrial, from South Carolina, Type II.

sample spectral variations within the maria, highland, and rille (Fig. 8, curves a and c) samples and grain-to-grain spectral variations within single samples. However, we observed a distinct general trend that absorption bands in the spectra of ilmenite from the highland samples and dust 15301,83 occurred at higher frequencies (closer to Type I) than those from the maria samples and dust 15601,72. The presumed shock histories of the samples showing the higher frequencies suggest that lattice disorder may be producing the spectral differences. As shown in Table 2, some ilmenite grains from the Apollo 14 and 15 samples showed frequencies higher (>305 cm^{-1}) and lower (<275 cm^{-1}) than those observed in any terrestrial sample.

From the groundmass of rock 12021,24 we isolated a single anhedral grain of a spinel-group mineral, with a black submetallic luster. Apollo 12 spinels have been shown to vary from chrome spinels to chrome ulvöspinels (Haggerty and Meyer, 1970), which are members of the chromite ($FeCr_2O_4$) to ulvöspinel (Fe_2TiO_4) continuous solid solution series. We compared the spectrum of the grain (Fig. 9, curve c) that we isolated with spectra of a series of 13 USGS standard chromite samples covering a wide range of elemental substitutions, and found that it matched best with sample 148005, shown as curve (b), Fig. 9. The composition for chromite 148005 closely approximates the range of compositions found for lunar chromites in Apollo 12 samples by Haggerty and Meyer (1970), except for TiO_2.

The infrared spectrum of an opaque oxide grain isolated from dust 15601,72 (Fig. 9, curve e) can be seen to correspond better with that of synthetic ulvöspinel (curve d) than with those of Al or Cr-rich spinels (curves a and b), indicating a similar composition.

BULK COMPOSITION OF COMPOSITE LUNAR SAMPLES

Basalt grains and sieved fractions

The frequency of the strong infrared Si–O stretching vibration near 1000 cm^{-1}, previously correlated with silica content (Estep et al., 1971), was used to compare spectra of basalt grains and sieved fractions from the lunar dusts and crushed rock chips. Spectra of basalt grains (1000–300 μ) isolated from the $+100$ mesh sieved fraction of the Apollo 14 samples (two dusts and a crushed breccia) and dust 15301,83, with Si–O frequencies ranging from 1010 to 1000 cm^{-1} indicate a predominance of plagioclase (1010 cm^{-1}) over pyroxene (980 cm^{-1}). Data from our infrared reflectance measurements also indicate a predominance of plagioclase (An_{90}) in the Apollo 14 samples, as ascertained from measurements of 20 mm^2 areas on the two polished butt samples. Absorption spectra of basalt grains from dust 15601,72, with Si–O stretching frequencies of 990–980 cm^{-1}, on the other hand, showed a predominance of pyroxene and were more basic in composition, similar to the maria basalt grains previously studied. These same modal trends were observed in spectra of the -100 mesh sieved fractions of the dusts, obtained by using 200, 325, and 400 mesh screens. Spectra of the Apollo 14 dust sieved fractions (1005–1000 cm^{-1}) and dust 15301,83 sieved fractions (1000 cm^{-1}) showed a predominance of plagioclase, while spectra of dust 15601,72 fractions (980 cm^{-1}) showed a predominance of pyroxene, similar to the -100 mesh sieved fractions of the maria dusts previously studied. From the correla-

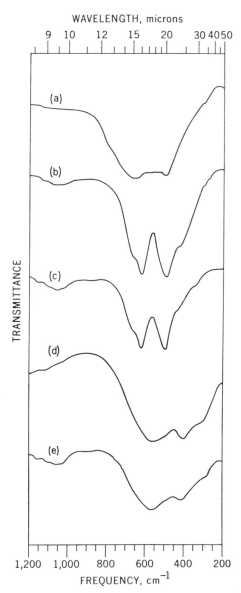

Fig. 9. Infrared spectra of spinels. (a) Synthetic hercynite (FeAl$_2$O$_4$), Tem-Pres Research, (b) chromite (FeCr$_2$O$_4$), Maryland (USGS No. 148005, 7.0% Al$_2$O$_3$, 52.2% Cr$_2$O$_3$, 27.6% FeO, 0.35% TiO$_2$, 9.9% MgO), (c) lunar chromite from rock 12021,24, 200 μ black grain, (d) synthetic ülvospinel (Fe$_2$TiO$_4$), Tem-Pres Research, (e) lunar ülvospinel from dust 15601,72, 200 μ black anhedral grain.

tion plot relating Si–O stretching frequencies to silica content, the bulk compositions of all basaltic grains and sieved fractions (1010–980 cm^{-1}) from Apollo 14 and 15 samples were seen to be a basaltic range (54–42% SiO$_2$), as were those from the maria dusts (52–40% SiO$_2$).

Glasses

Overall infrared spectral characteristics of the glasses isolated from Apollo 14 and 15 samples, shown in Table 3, were similar to those from Apollo 11 and 12 samples, exhibiting three principal bands in the region of Si O stretching, Si–O–Si stretching, and Si–O–Si bending vibrations, near 1000 (s), 700 (w), and 460 (m) cm^{-1}, respectively. Raman spectra of glasses from Apollo 14 samples exhibited the same three principal absorption bands, with similar frequencies, band widths, and relative intensities. Infrared spectra of many glass particles, both beads and irregular fragments, and in particular those of irregular fragments from dust 14259,15, showed a fine structure characteristic of feldspars and pyroxenes, indicating partial devitrification or incomplete melting of these minerals. We estimated silica contents of the glasses from the same correlation plot as used for the basalt grains and sieved fractions. Compositions of true glasses from Apollo 14 samples (1010–980 cm^{-1}, 54–42 wt.% SiO$_2$) were generally less basic than those from Apollo 15 (985–970 cm^{-1}, 44–38 wt.% SiO$_2$), and probably reflect the higher feldspar and lower iron contents in the highland samples. In both infrared and Raman spectra of individual grains of Apollo 14 glasses and in the infrared spectra of Apollo 15 glasses we noted, as with Apollo 11 and 12 glasses, a general trend that, as the frequency of the Si–O stretching vibration decreased, the frequency of the Si–O–Si bending vibration increased. We have not

Table 3. Glass grains isolated from lunar dusts.

Sample source	Glass description[1]	Number analyzed	Frequencies, cm^{-1}	
			Si–O stretch	Si–O–Si bend
12070,24	Mixed with α-quartz; medium amber opaque	1	1000	458
14163,80	True glass; amber transparent, bead and fragment	2	1010, 990	462, 465
	Partially crystalline; light yellow to amber, green transparent fragments	15	1020–980	462–488
14259,15	True glass; yellow to amber, green transparent, beads, fragments, vesicular and conglomerates	11	1010–980	460–495
	Mixed with α-quartz; light yellow to dark amber transparent, yellow to green opaque, beads and fragments	20	1020–980	458–472
	Partially crystalline; light yellow to green, amber transparent, beads and fragments	29	1020–970	458–490
15301,83	True glass; green transparent, beads and fragments	9	980–970	500–510
	Partially crystalline; yellow, green and amber transparent, beads and fragments	14	1000–980	460–501
15601,72	True glass; dark amber transparent fragment, green transparent bead	2	985, 970	470, 500
	Partially crystalline; light yellow to green, brown transparent to opaque, beads and fragments	4	990–980	475–500

[1] Size of beads ranged from 800 to 200 μ; irregular fragments 1500 to 200 μ.

observed this trend in spectra of synthetic glasses or tektites; the structural reason for this unique behavior has not yet been determined.

Several glass grains containing varying amounts of alpha-quartz were isolated from the three Apollo 11, 12, and 14 dusts listed in Table 3, and the distinctive infrared spectra of some of these grains matched well with those of glass grains (also containing quartz) that we isolated from the Stannern eucrite. Their identification in lunar dusts suggests that the quartz previously found in meteorites is not due to terrestrial contamination. In addition, the occurrence of these unique glass grains containing variable amounts of the low-temperature high-pressure phase of silica is geochemically significant. It suggests similar formation processes for grains in lunar dusts and eucrites and supports the hypothesis of a similar origin (Duke and Silver, 1967).

Acknowledgments—We thank the following persons from this laboratory for aiding in the collection of infrared and Raman spectra: A. A. Angotti, R. C. Berkshire, Jr., E. E. Childers, B. D. Stewart, L. E. Makovsky, and W. H. Edwards. We are grateful to J. Dinnin, U.S. Geological Survey, for supplying analyzed chromite samples; to D. Virgo, Carnegie Institution of Washington, R. W. Dundon, Marquette University, and S. S. Hafner, University of Chicago, for samples of pyroxenes analyzed by Mössbauer spectroscopy; to E. Dowty, University of New Mexico, for samples of synthetic pyroxenoids synthesized by Lindsley and Dowty; to J. D. Stevens, Kennecott Copper Corporation, and R. J. P. Lyon, Stanford University, for supplying pellets and data for samples studied by Howie; to G. Kurat, Naturhistorisches Museum, Vienna, Austria, and E. J. Olsen, Chicago Natural History Museum, for meteorite samples; to M. Morgenstein, University of Hawaii, for samples of Hawaiian basalt; and to the following persons for supplying analyzed pyroxenes for our comparisons: R. A. Howie, University of London, King's College, England, I. D. Muir, University of Cambridge, S. S. Ghose, University of California, G. M. Bancroft, University of Western Ontario, Canada, R. V. Fodor, University of New Mexico, J. S. White, Smithsonian Institution, G. V. Subbarayudu, State University of New York at Buffalo, and S. S. Pollack, Carnegie-Mellon University.

This research was sponsored by NASA under Contract No. T-1760A.

References

Andrews J. E. (1971) Abyssal hills as evidence of transcurrent faulting on north Pacific fracture zones. *Geol. Soc. Amer. Bull.* **82**, 463–470.

Bahat D. (1970) Optical and infrared studies on high-temperature alkali feldspars. *J. Geol. Soc. Aust.* **17**, 93–102.

Brown G. E. (1971) Neutron diffraction of Al/Si ordering in sanidine. *Geol. Soc. Amer., Abstracts of 1971 Annual Meeting* **3** (7), 514.

Burnham C. W. (1966) Ferrosilite III: A triclinic pyroxenoid-type polymorph of ferrous metasilicate. *Science* **154**, 513–516.

Burnham C. W. (1971) The crystal structure of pyroxferroite from Mare Tranquillitatis. *Proc. Second Lunar Sci. Conf., Geochim. Cosmochim. Acta* Suppl. 2, Vol. 1, pp. 47–57. MIT Press.

Drake M. J., McCallum I. S., McKay G. M., and Weill D. F. (1970) Mineralogy and petrology of Apollo 12 sample 12013: A progress report. *Earth Planet. Sci. Lett.* **9**, 103–123.

Duke M. B. and Silver L. T. (1967) Petrology of eucrites, howardites, and mesosiderites. *Geochim. Cosmochim. Acta* **31**, 1637–1666.

Dundon R. W. and Hafner S. S. (1971) Cation disorder in shocked orthopyroxene. *Science* **174**, 581–583.

Dundon R. W. and Walter L. S. (1967) Ferrous ion order-disorder in meteoritic pyroxenes and the metamorphic history of chondrites. *Earth Planet. Sci. Lett.* **2**, 372–376.

Estep P. A., Kovach J. J., and Karr C. Jr. (1971) Infrared vibrational spectroscopic studies of minerals from Apollo 11 and Apollo 12 lunar samples. *Proc. Second Lunar Sci. Conf., Geochim. Cosmochim. Acta* Suppl. 2, Vol. 3, pp. 2137–2151. MIT Press.

Griffith W. P. (1969) Raman spectroscopy of minerals. *Nature* **224**, 264–266.

Hafner S. and Laves F. (1957) Order-disorder and infrared absorption, II. Variation in the position and intensity of certain absorption lines of feldspars on the structure of orthoclase and adularia. *Z. Kristallogr.* **109**, 204–225.

Haggerty S. E. and Meyer H. O. A. (1970) Apollo 12: Opaque oxides. *Earth Planet. Sci. Lett.* **9**, 379–387.

Henderson E. P. and Glass J. J. (1936) Pyroxmangite, new locality: Identity of sobralite and pyroxmangite. *Amer. Mineral.* **21**, 273–294.

Howie R. A. (1963) Cell parameters of orthopyroxenes. *Mineral. Soc. Amer. Spec. Paper 1*, 213–222.

Laves F. and Hafner S. (1956) Order-disorder and its effect on infrared absorption spectra, I. (Al, Si) Distribution in feldspars. *Z. Kristallogr.* **108**, 52–63.

Lazarev A. N. and Tenisheva T. F. (1961b) Vibrational spectra of silicates, III. Infrared spectra of the pyroxenoids and other chain metasilicates. *Optics and Spectros.* (USSR) **11**, 316–317.

Liebau F. (1956) Systematology of crystal structures of silicates with highly condensed anions. *Z. Phys. Chem.* **206**, 73–92.

Lindsley D. H. and Burnham C. W. (1970) Pyroxferroite: Stability and x-ray crystallography of synthetic $Ca_{0.15}Fe_{0.85}SiO_3$ pyroxenoid. *Science* **168**, 364–367.

Lindsley D. H., Davis B. T. C., and MacGregor I. D. (1964) Ferrosilite ($FeSiO_3$): Synthesis at high pressures and temperatures. *Science* **144**, 73–74.

Lyon R. J. P. (1963) Evaluation of infrared spectrophotometry for compositional analysis of lunar and planetary soils. Stanford Res. Inst., Final Rept. under contract NASr-49(04), pub. by NASA as Tech. Note D-1871.

Makovsky L. E. (1972) Sampling technique for Raman spectroscopy of minerals. In preparation.

Perry C. H., Agrawal D. K., Anastassakis E., Lowndes R. P., and Tornberg N. E. (1972) Far infrared and Raman spectroscopic investigations of lunar materials from Apollo 11, 12, 14, and 15 (abstract). In *Lunar Science— III* (editor C. Watkins), pp. 605–607, Lunar Science Institute Contr. No. 88.

Pollack S. S. and DeCarli P. S. (1969) Enstatite: Disorder produced by a megabar shock event. *Science* **165**, 591–592.

Sadashivaiah M. S. and Subbarayudu G. V. (1970) Orthopyroxenes from the Kondavidu charnockites, Guntur District, Andhra Pradesh. *Proc. Indian Acad. Sci.* **1970**, 139–148.

Sclar C. B. (1971) Shock-induced features of Apollo 12 microbreccias. *Proc. Second Lunar Sci. Conf.*, Geochim. Cosmochim. Acta Suppl. 2, Vol. 1, pp. 817–832. MIT Press.

Virgo D. and Hafner S. S. (1969) Fe^{2+}, Mg order-disorder in heated orthopyroxenes. *Mineral. Soc. Amer. Spec. Paper 2*, 67–81.

Virgo D. and Hafner S. S. (1970) Fe^{2+}, Mg order-disorder in natural orthopyroxenes. *Amer. Mineral.* **55**, 201–223.

Weill D. F., Grieve R. A., McCallum I. S., and Bottinga Y. (1971) Mineralogy-petrology of lunar samples. Microprobe studies of samples 12021 and 12022; viscosity of melts of selected lunar compositions. *Proc. Second Lunar Sci. Conf.*, Geochim. Cosmochim. Acta Suppl. 2, Vol. 1, pp. 413–430. MIT Press.

Williams P. G. L., Bancroft G. M., Brown M. G., and Turnock A. C. (1971) Anomalous Mössbauer spectra of C2/c clinopyroxenes. *Nature* **230**, 149–151.

Proceedings of the Third Lunar Science Conference
(Supplement 3, *Geochimica et Cosmochimica Acta*)
Vol. 3, pp. 3069–3076
The M.I.T. Press, 1972

Midinfrared emission spectra of Apollo 14 and 15 soils and remote compositional mapping of the moon

Lloyd M. Logan, Graham R. Hunt, Salvatore R. Balsamo,
and John W. Salisbury

Terrestrial Sciences Laboratory, Air Force Cambridge Research Laboratories,
L. G. Hanscom Field, Bedford, Mass. 01730

Abstract—Laboratory measurements of the midinfrared spectral behavior of Apollo 14 and 15 soils show that compositionally diagnostic spectral information is present in their infrared emission. Soil spectra are compared to lunar spectra obtained with a balloon-borne telescope system and it is shown that the remote observations are consistent with the laboratory results. We conclude that the moon does not radiate as a black body in the midinfrared; that the spectral information present in its emission can be used to map lunar surface composition remotely; and that sufficient observational data are already in hand to draw conclusions about regional compositional variations. The problem of regional cross contamination is one that warrants further study.

INTRODUCTION

In principle, a midinfrared emission spectrum of a planetary surface offers information diagnostic of bulk mineralogical composition, or rock type. This is because features which occur in midinfrared spectra are caused by vibrations within the basic repeating units of the material. Thus, the number and location of the bands depends directly upon the molecular structure and, hence, on mineralogical composition.

There has, however, been serious disagreement concerning the amount of such information available in lunar infrared emission. Early laboratory work showed that the molecular vibration bands (reststrahlen bands) of silicate minerals become more difficult to detect as the particle size of the minerals becomes small (Lyon, 1963). Because the lunar surface layer has long been known to be composed primarily of fine particulate material, it has been generally assumed that it radiated essentially as a black body in the midinfrared.

Yet, early observations of lunar emission did detect differences in spectral behavior from place to place on the surface (Hunt and Salisbury, 1964), and one observer determined that lunar emission departed significantly from black body behavior (Murcray, 1965). On the other hand, later observations indicated that, although the emissivity of some areas differed from that of their surroundings, the emissivity of most areas did not (Goetz, 1968). The most recent measurements, which have the advantage of being made with a balloon-borne telescope above most of the infrared-absorbing constituents of the atmosphere, appear to show that the moon does depart from black body behavior, and that its emission contains spectral information (Murcray *et al.*, 1970).

It is the purpose of this paper to demonstrate by laboratory measurements of lunar soil samples that their infrared emission spectra contain diagnostic information, that significant spectral differences exist between soils from different localities, and that these differences are consistent with spectra obtained by Murcray *et al.* (1970) with the balloon-borne telescope system. We shall also examine what predictions can be made concerning the composition of the as yet unexplored lunar highlands.

LABORATORY MEASUREMENTS

For fine particulate silicates, such as are found in the lunar soil, Conel (1969) found that compositionally diagnostic information is available in the form of either a transmission or emission maximum, which occurs on the short wavelength side of the reststrahlen features. This maximum is found only in finely particulate samples where scattering plays a dominant role in the determination of the spectral properties of the material. The transmission or emission maximum corresponds to the minimum in scattering that occurs at the wavelength (the so-called Christiansen frequency) where the refractive index of the material equals unity in the anomalous dispersion that accompanies an absorption maximum.

Logan and Hunt (1970) found that this emission peak becomes a very prominent feature in the spectra of terrestrial silicates under simulated lunar conditions. As described in Logan and Hunt (1970), the lunar thermal environment was simulated by evacuating the sample space to a pressure of less than 10^{-4} Torr, surrounding the sample with a cooled (77°K) radiation shield to simulate lunar surface radiation to cold space, and heating it by illumination with the visible energy from a quartz iodine lamp to simulate solar radiation. They showed that the steep thermal gradient induced near the surface of a sample by lunar thermal conditions is the cause of this large departure from black body behavior. Figure 1 shows that the positions of the emission peaks, like the positions of the contrast-diminished reststrahlen minima, are related to composition, i.e., the more basic the rock, the longer will be the wavelength at

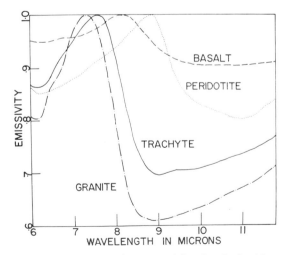

Fig. 1. Emission spectra of representative terrestrial rocks obtained in a simulated lunar environment.

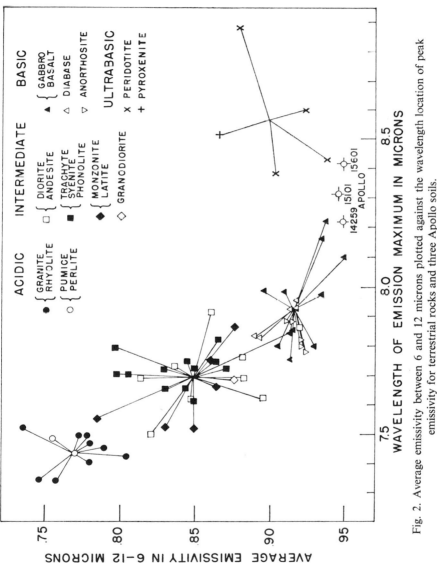

Fig. 2. Average emissivity between 6 and 12 microns plotted against the wavelength location of peak emissivity for terrestrial rocks and three Apollo soils.

which the peak falls. Figure 1 also illustrates the tendency for the spectral contrast, here defined as the average departure of a spectral curve from unit emissivity over the 6–12 μ wavelength range, to vary systematically from acidic to ultrabasic. In going from acidic to basic rocks there is a progressive increase in average emissivity, followed by a slight decrease again for the ultrabasics. The peak wavelength and average emissivity can be used to determine rock type as illustrated in Fig. 2.

Laboratory measurement of the emission spectra of Apollo soils in the simulated lunar environment described above show that they display emission peaks similar to those of particulate terrestrial rocks (Fig. 3). The average emissivities and wavelengths of the emission maxima of the lunar soils are also plotted in Fig. 2.

Adams and McCord (1972) have raised the question of cross contamination of soils from different regions on the moon. Apollo 15 soils offer an excellent test of the effects of cross contamination, because preferential down-slope movement of material along the Apennine front maximizes soil mixing. Figure 4 shows spectra of two Apollo 15 soils, one (15101) obtained at Station 2 on the Apennine front and one (15601)

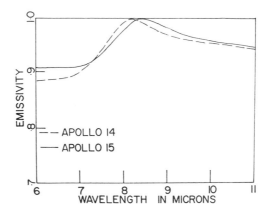

Fig. 3. Emission spectra of Apollo 14259 and 15601 soils.

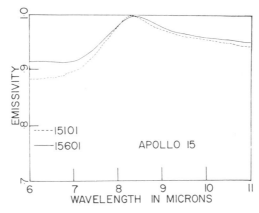

Fig. 4. Emission spectra of Apollo 15101 and 15601 soils.

obtained at Station 9 on the margin of Hadley Rille about 3.5 km north of the front. These soils display differences in spectral contrast and a shift in the wavelength of their emissivity peaks. Thus, it can be seen that for our two Apollo 15 samples, separated by only a few kilometers, cross contamination has not been sufficient to eliminate their spectral differences. Because of the presence of Hadley Rille, which complicates the soil migration picture, it is not possible to draw a more comprehensive conclusion regarding the importance of cross contamination. However, we are encouraged by this one result and plan to examine our other samples in this light.

Comparison of Laboratory Spectra with Lunar Observations

The best available observations of lunar spectral emissivity are those made by Murcray *et al.* (1970) with a balloon-borne telescope. The spot size in these early observations was large (180 arc sec), and the target areas were therefore regional in size. However, definite emissivity peaks were obtained, the wavelength of which differed from region to region, as illustrated in Fig. 5. It is also interesting that, despite the large spot size, these spectra correlate well with laboratory spectra of lunar soil samples. For example, the emissivity peak displayed by the center of Mare Imbrium

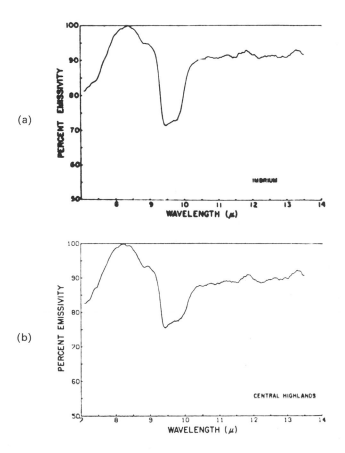

(c)

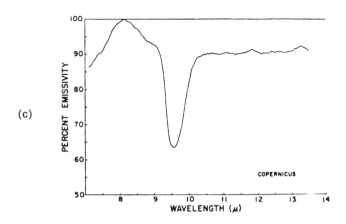

(d)

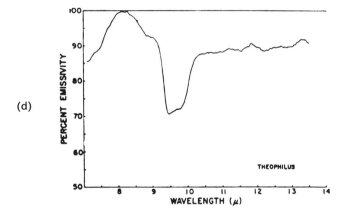

(e)

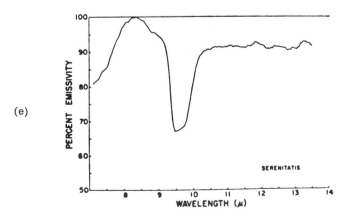

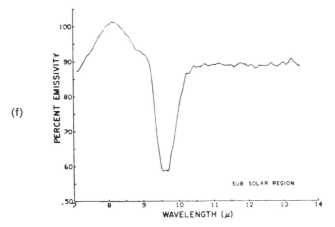

Fig. 5. Emission spectra of six different regions on the moon obtained by Murcray *et al.*
(1970). The strong minimum at 9.6 μ is due to absorption by residual atmospheric ozone
above the balloon system from which these spectra were acquired.

(Fig. 5a) at 8.37 μ is very close to that of the 15601 sample (8.42 μ) taken from Mare
Imbrium margin. The maximum obtained in the Central Highlands (Fig. 5b) spectrum
at 8.24 μ is virtually identical with that displayed by Apollo 14 soil at 8.22 μ, and this
result is reasonable if the regolith in the Central Highlands and the debris ejected from
the Imbrium Basin to form the Fra Mauro formation are both derived from a primeval
lunar crust. An emissivity peak at this long a wavelength for the Central Highlands
implies that the regolith there cannot be "anorthositic," by which we mean a material
composed almost entirely of calcic plagioclase, because the emissivity peaks for such
materials are always shorter than 7.90 μ. On the contrary, the regolith must contain
abundant pyroxene and/or olivine, as does the Apollo 14 soil. Of course, the composition of the Central Highlands target area observed by Murcray *et al.* (1970) may not
be representative. We hope that additional lunar spectra obtained with the balloon-
borne telescope system at higher spatial resolution will soon be published, so that the
heterogeneity of the lunar highlands can be determined. In the currently available
data, the least basic areas are associated with the relatively recent craters, Copernicus
(Fig. 5c) and Theophilus (Fig. 5d). The spectra shown in Fig. 5 were obtained from
areas which encompassed primarily the crater and its ejecta blanket. It may be that
these craters have excavated material that is less basic than that found in the Central
Highlands area that lies between them. In the light of Adam's and McCord's (1972)
work on cross contamination, however, it may be more likely that regolith in and
around these recent craters represents relatively uncontaminated highlands material.

CONCLUSIONS

(1) Our laboratory measurements of the spectral emissivity of lunar soil samples
confirms the conclusion reached from ground-based and balloon-borne measurements
that the moon does not emit in the midinfrared as a black body.

(2) Compositionally diagnostic spectral information is present in infrared emission from the moon, and there is good agreement between spectra of large areas of the surface obtained with a balloon-borne telescope system and spectra measured in the laboratory of small lunar soil samples.

(3) Spectral differences between different lunar soil samples confirm the conclusion reached from the balloon-borne observations, that this spectroscopic remote sensing technique can be used to map compositional variations on the lunar surface.

(4) From the remote sensing data published so far, we can conclude that the maria are more basic than the highlands. However, to the extent that the Central Highlands target area is representative of the highlands as a whole, they appear to be gabbroic rather than anorthositic.

(5) The spectral effect of regional cross contamination is difficult to evaluate. It is not great enough to disguise compositional differences between our Apollo 15 soil samples. It may, however, explain apparent compositional differences between the Central Highlands and the two craters, Copernicus and Theophilus. If so, the age of surface must be taken into account in any attempt to determine its composition remotely.

References

Adams J. B. and McCord T. B. (1972) Optical evidence for regional cross contamination of highlands and mare soils (abstract). In *Lunar Science—III* (editor C. Watkins), p. 1, Lunar Science Contr. No. 88.

Conel J. E. (1969) Infrared emissivities of silicates: Experimental results and a cloudy atmosphere model of spectral emission from condensed particulate mediums. *J. Geophys. Res.* **74**, 1614–1634.

Goetz A. F. H. (1968) Differential infrared lunar emission spectroscopy. *J. Geophys. Res.* **73**, 1455–1466.

Hunt G. R. and Salisbury J. W. (1964) Lunar surface features: Midinfrared spectral observations. *Science* **146**, 641–642.

Logan L. M. and Hunt G. R. (1970) Infrared emission spectra: Enhancement of diagnostic features by the lunar environment. *Science* **169**, 865–866.

Lyon R. J. P. (1963) Evaluation of infrared spectrophotometry for compositional analysis of lunar and planetary soils. NASA Technical Note TND-1871.

Murcray F. H. (1965) The spectral dependence of lunar emissivity. *J. Geophys. Res.* **70**, 4959–4962

Murcray F. H., Murcray D. G., and Williams W. J. (1970) Infrared emissivity of lunar surface features, I. Balloon-borne observations. *J. Geophys. Res.* **75**, 2662–2669.

Proceedings of the Third Lunar Science Conference
(Supplement 3, *Geochimica et Cosmochimica Acta*)
Vol. 3, pp. 3077–3095
The M.I.T. Press, 1972

Far infrared and Raman spectroscopic investigations of lunar materials from Apollo 11, 12, 14, and 15

Clive H. Perry, D. K. Agrawal, E. Anastassakis,
R. P. Lowndes, and N. E. Tornberg

Solid State Spectroscopy Laboratory, Physics Department,
Northeastern University, Boston, Massachusetts 02115

Abstract—We have studied the elastic and inelastic light scattering of twelve lunar surface rocks and eleven lunar soil samples from Apollo 11, 12, 14, and 15, over the range 20–2000 cm^{-1}. The phonons occurring in this frequency region have been associated with the different chemical constituents and are used to determine the mineralogical abundances by comparison with the spectra of a wide variety of terrestrial minerals and rocks. Kramers-Kronig analyses of the infrared reflectance spectra provided the dielectric dispersion (ε' and ε'') and the optical constants (n and k). The dielectric constants at $\sim 10^{11}$ Hz have been obtained for each sample and are compared with the values reported in the 10^2–10^6 Hz range. The emissivity peak at the Christianson frequencies for all the lunar samples lie within the range 1195–1250 cm^{-1}; such values are characteristic of terrestrial basalts. The Raman light scattering spectra provided investigation of small individual grains or inclusions and gave unambiguous interpretation of some of the characteristic mineralogical components.

Introduction

We have studied the interaction of electromagnetic radiation with lunar materials using elastic and inelastic light scattering techniques. These investigations have many important applications as they identify the vibrational modes associated with the different chemical constituents and can be used to determine mineralogical abundances and give structural information. The data can be used for compositional mapping of the lunar surface and the uniqueness of the method suggests the possibility of performing remote infrared petrology using spectral matching techniques (Lyon, 1963; Aronson *et al.*, 1967; Perry and Lowndes, 1970). The establishment of the relative chemical constituents and their relation to the different rock and soil types on the lunar surface is made by comparison with spectra of a wide variety of terrestrial minerals and rocks.

The infrared studies were accomplished from measurements (mainly conducted at 300°K) of the specular component of the reflectivity over the thermal frequency range 20–2000 cm^{-1} (500–5 μ wavelength). The data were subjected to a Kramers-Kronig analysis as described previously (Perry *et al.*, 1971) and the dispersion of the real and imaginary parts of the dielectric function (ε', ε''), the refractive index n, and the absorption coefficient α were obtained. These fundamental optical constants are independent of the method of measurement and can be used directly to simulate emittance spectra of lunar materials for comparison with future remote sensed infrared spectra and radiometric measurements of the lunar surface (Mendell and Low, 1971).

The far-infrared data yield values of the dielectric constant and loss tangent at $\sim 10^{11}$ Hz. It is of interest to extrapolate these values of the dielectric function to lower frequencies to provide useful information for electrical prospecting of subsurface layers in the 10^6–10^9 Hz range (Simmons *et al.*, 1971; Gold *et al.*, 1972; Howard and Tyler, 1972). Comparison can then be made with the values obtained in the 10^2–10^6 Hz range using standard capacitance techniques provided that these measurements have been made on dry material (Collett and Katsuba, 1971; Chung *et al.*, 1971; Chung, 1971; Chung and Westphal, 1972; Lowndes *et al.*, 1972).

The inelastic (Raman) light scattering techniques for determining the frequency, structural, and chemical properties of the constituents of lunar samples offers a complementary tool. In contrast to the infrared measurements which provide macroscopic information over large areas, this technique, by virtue of the small size ($< 50\ \mu$ diameter) of the focused laser beam, allows the investigation of small individual grains or inclusions and has provided unambiguous interpretation of some of the characteristic mineralogical components (Perry *et al.*, 1971; White *et al.*, 1971; Estep *et al.*, 1972; Fabel *et al.*, 1972).

EXPERIMENTAL

Infrared measurements

We have obtained the infrared specular reflectance spectra of the following bulk samples which were in the form of polished butt ends: 10058,56, 12002,186, 12008,23, 12009,48, 12065,115, 12073,42, 14301,20, 14307,18, 14310,184, 14313,51, 14321,98, and 15426,2. The samples were polished to obtain one smooth flat face having an area of at least 0.25 cm².

Pressed disks ~ 0.1 cm thick of the fines samples were examined in a similar manner. The samples were prepared in a standard 0.5 cm diameter die with pressures of about 5×10^9 dynes/cm² being applied for approximately 5 minutes. The densities of all the samples ranged from 2.4–2.6 g/cm³. The following fines samples were examined: 14141,37, 14161,36, 14163,31, 14230,95, 14259,38, 15021,159, 15071,60, 15091,65, 15221,71, 15471,68, and 15531,63.

Dry nitrogen flushed grating spectrometers (PE-301, PE-180, and PE-521) with specular reflectance attachments (angle of incidence 10°–15°) were used for the studies over the frequency range 200–4000 cm⁻¹ (2.5–50 μ). From 20–250 cm⁻¹ a Fourier transform Michelson type interferometer with a liquid-helium-cooled germanium detector was used (Perry *et al.*, 1966).

Raman measurements

Identifiable Raman spectra of a wide variety of individual grains and glassy inclusions were obtained on a number of lunar samples. Various excitation radiations were used, namely Ar⁺ 4880 and 5145 Å and He/Ne 6328 Å, and some resonance effects were noted. The Raman scattered light was analyzed using a Spex 1401 double spectrometer and detected using photon counting techniques. Both oblique incidence and back-scattering sample geometries were employed. The power absorbed by the samples often represented a restriction on the Raman technique as the opaque regions showed a definite tendency to decompose under the influence of the focused laser beam ($< 50\ \mu$ diameter with a power density of 5×10^4 W/cm²). Both Stokes and anti-Stokes components of the spectrum were examined at room temperature.

Generally, the Raman spectra were of lesser complexity than their infrared counterparts and comparison with terrestrial mineral spectra provided unambiguous interpretation. The effects due to the lowering of site symmetry in the $(SiO_n)^{x-}$ group (Griffith, 1969), variation of Fs content in the pyroxenes, and Fa content in the olivines were demonstrated by frequency shifts in the spectra (Perry *et al.*, 1971; Estep *et al.*, 1971; 1972).

Dielectric studies

The low-frequency dielectric constants were determined through three terminal capacitance measurements in the frequency range 10^2–10^5 Hz. The capacitance measurements were recorded on a

General Radio 1615A transformer ratio arm bridge used in conjunction with a General Radio 1232A tuned amplifier and null detector. The capacitances were determined to an accuracy of $\pm 0.1\%$ at the lower frequencies and $\pm 0.4\%$ at 10^5 Hz. The variable temperature cell used to measure the dielectric constants has been described previously (Lowndes and Martin, 1969).

RESULTS AND DISCUSSION

The infrared reflectance spectra from 10–1700 cm^{-1} of some representative lunar rock samples from Apollo 11, 12, and 14 are shown in Fig. 1. From these spectra the real and imaginary parts of the dielectric constant were calculated and are plotted in Fig. 2. In Fig. 3 the infrared reflectance spectra of three Fra Mauro breccias 14301,20; 14307,18 and 14313,51, and a typical Apollo 14 soil 14161,36 (bulk sample) are displayed. The spectra of these breccias can be compared with those of an Apollo 12 mare breccia (12073,42) and an Apollo 15 breccia (15426,2) from the Apennine Front

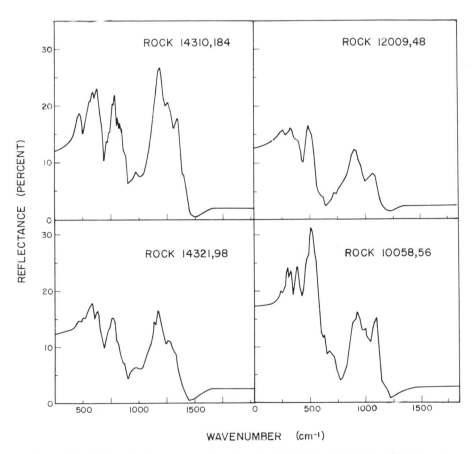

Fig. 1. The infrared reflectance spectra of rock 14310,184, 12009,48, 14321,98, and 10058,56, from 0 to 1700 cm^{-1}.

C. H. PERRY *et al.*

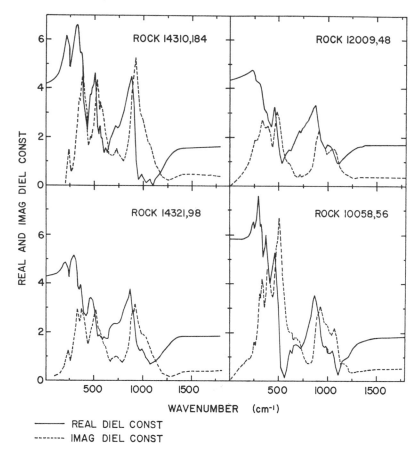

Fig. 2. The real and imaginary parts of the dielectric constant for rock 14310,184, 12009,48, 14321,98, and 10058,56, from 0 to 1700 cm^{-1}.

shown in Fig. 4. Figure 5 comprises the spectra of five Apollo 14 fines (14163,31; 14161,36; 14259,38; 14230,95 (core sample); and 14141,37 (Cone Crater)) and one powdered rock sample 14321,263 (Cone Crater). The reflectance spectra of six Apollo 15 fines are shown in Fig. 6. The samples investigated were 15091,65 and 15221,71 from Station 2 at the foot of the Apennine Front, 15071,60 from Station 1 (Elbow Crater ejecta); 15021,159 from the LM area, 15471,68 from Dune Crater (Station 4), and 15531,63 from Station 9a at the rim of Hadley Rille.

The spectra shown in the first six figures represent the bulk mineralogical and chemical properties of the materials averaged over the area of the sample examined (25–100 mm^2). They form a basis for the comparison of these samples with each other and with known terrestrial materials. In Figs. 7 and 8 are shown some examples of the results of the Raman investigations. Because of differing selection rules in the scattering process, this technique is complementary to that of infrared spectroscopy in the sense that it provides information unobtainable by the latter. Furthermore, be-

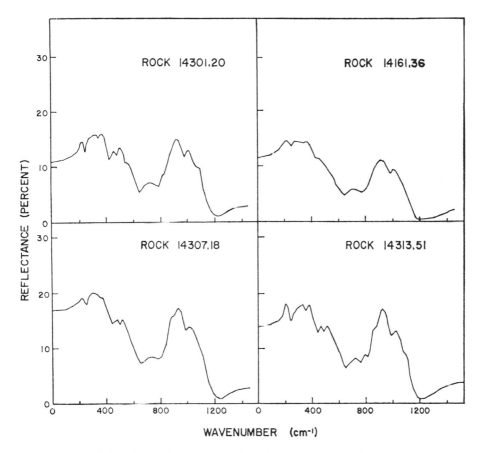

Fig. 3. The infrared reflectance spectra from 0 to 1400 cm^{-1} of three Fra Mauro breccias 14301,20, 14307,18, and 14313,51, and a typical Apollo 14 soil 14161,36 (bulk sample).

cause of the small size of the sample area investigated ($\sim 2 \times 10^{-5}$ cm^2), individual grains and inclusions may be studied, and the resulting spectra are usually simpler and the mineralogical identification is more straightforward. In Fig. 7, for example, the Raman spectra show some of the variations of lunar pyroxenes in samples 10058,56, 12002,186, 12065,115, and 14310,76 (Perry *et al.*, 1971, and as discussed by Estep *et al.*, 1972). The systematic frequency shift of the stretching vibration at ~ 1015 cm^{-1} (e.g., Estep *et al.*, 1972) and the bending vibration at ~ 650 cm^{-1} in these samples can be correlated with the ferrosilite content. In 14310,76 (with Fs$_{25}$–Fs$_{30}$) the latter band is almost a resolved doublet indicating the presence of both ortho- and clinopyroxene. Sample 10058,56 contains mostly clinopyroxene ranging from pigeonite to augite (Fs$_{20}$–Fs$_{40}$). Samples 12002,186 and 12065,115 show broader, weaker spectra and the evaluation is less well defined but both again probably contain clino- and ortho-pyroxene. The Raman spectrum of a typical green glass sphere in sample 15426,2

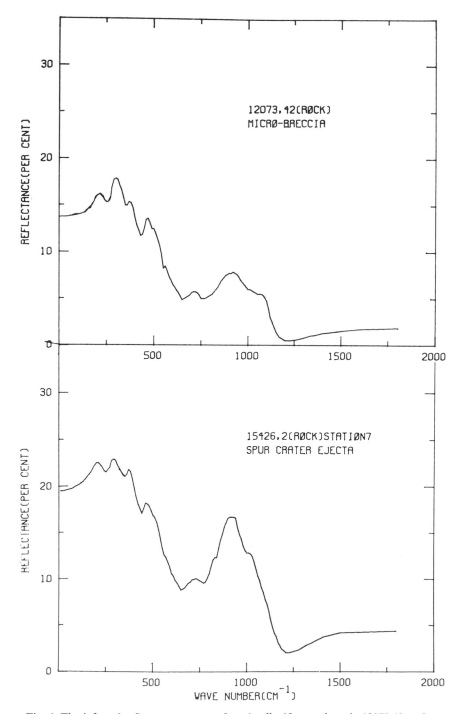

Fig. 4. The infrared reflectance spectra of an Apollo 12 mare breccia 12073,42 and an
Apollo 15 breccia 15426,2 from the Apennine Front.

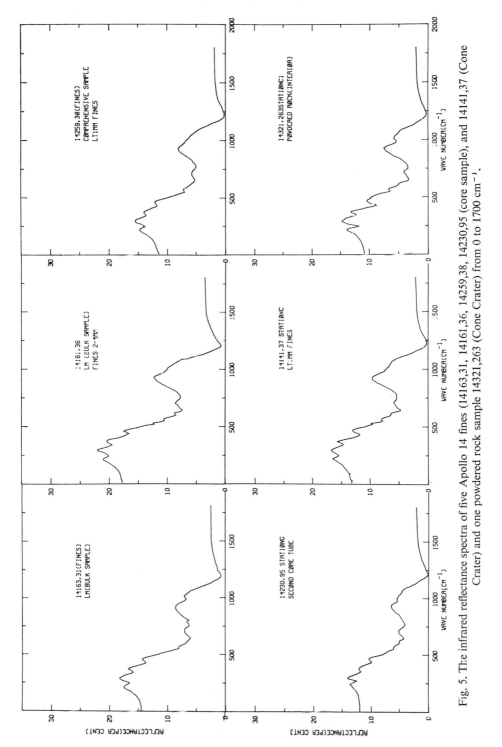

Fig. 5. The infrared reflectance spectra of five Apollo 14 fines (14163,31, 14161,36, 14259,38, 14230,95 (core sample), and 14141,37 (Cone Crater) and one powdered rock sample 14321,263 (Cone Crater) from 0 to 1700 cm⁻¹.

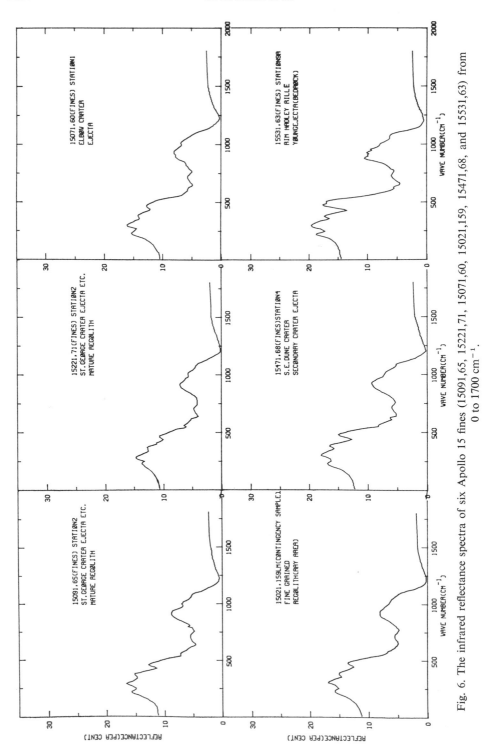

Fig. 6. The infrared reflectance spectra of six Apollo 15 fines (15091,65, 15221,71, 15071,60, 15021,159, 15471,68, and 15531,63) from 0 to 1700 cm^{-1}.

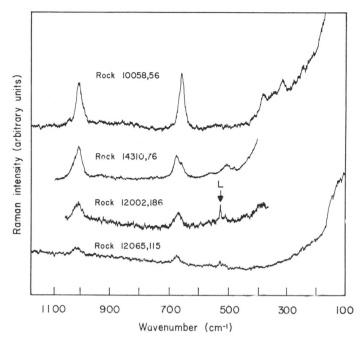

Fig. 7. The Raman spectra of lunar pyroxene inclusions in samples 10058,56, 14310,76, 12002,186, and 12065,115 showing the variation in the frequencies of bands that can be related with the ferrosilite content. The L refers to an extraneous laser line.

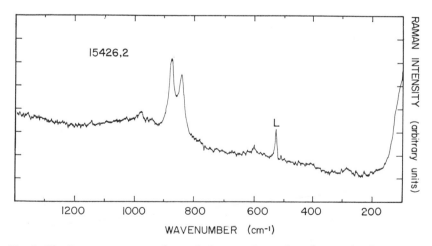

Fig. 8. The Raman spectrum of a typical green glass sphere in sample 15426,2. The strong doublet in the region of 850 cm^{-1} shows the presence of a particle that has undergone partial devitrification to olivine. The L refers to an extraneous laser line.

Table 1. A summary of the lunar rock samples investigated from the Apollo 11, 12, 14, and 15 missions. ω_c is the Christianson frequency corresponding to the emissivity peak and ε_0 is the dielectric constant at far infrared frequencies ($\sim 10^{11}$ Hz). The major chemical abundances are taken from the references cited.

Sample Number	Rock Classification	Spectra (Lattice Modes / M-O-M cation-bend oxygen vibs. / M-O stretch / Combinations)	ω_c (cm^{-1})	ε_0 at 10^{11} Hz	SiO$_2$	Al$_2$O$_3$	MgO	FeO	CaO	TiO$_2$	
14310,184	Basaltic cryst. rock		1250	5.2	50	20	8	8	11	1	§
14321,98	Breccia (cone crater)		1205	4.3	49	16	12	11	8	1	
12002,186	Basalt		1220	3.9							
14301,20	Coherent Breccia		1230	4.0	48	16	10	12	10	2	*
14307,18	Coherent Breccia		1230	5.5	47	16.5	10	12	11	2	**
14313,51	Coherent Breccia		1215	4.8							
15426,2	Breccia		1220	6.5							
12073,42	Microbreccia		1240	2.7	41	15	11	17	11.5	3	†
12008,23	Cumulate		1200	6.7	Major phase, ilmenite						
12009,48	Prophyritic Basalt		1240	4.3	41	11	12.5	20	10	3	†
10058,56	Ophitic Basalt		1220	5,8	41	11	6	17	12	11	×
12065,115	Variolitic Basalt		1200	3.8	39	12	9	22	12	4	†

Major chemical abundance (%)

WAVE NUMBERS (cm^{-1}): 200   600   1000

x Charles et al. (1971).   † LSPET (1970).   § LSPET (1972).   † LSPET (1972).   * Hubbard and Gast (1972).   ** Agrell et al. (1972).

is shown in Fig. 8. The strong doublet in the region of 850 cm^{-1} indicates devitrification to olivine (with Fa$_{20}$–Fa$_{40}$) of a contaminating particle either on or imbedded in the sphere. Raman spectra of lunar glasses have also been reported by White *et al.* (1971) and Fabel *et al.* (1972). However, Fabel *et al.* (1972) report a large number of closely spaced bands at low frequencies in some of their samples. We have observed similar bands in some of our samples but substitution of different laser excitation radiation revealed them to be grating ghosts.

Tables 1, 2, and 3 summarize all the lunar rock and soil samples that we have investigated from the Apollo 11, 12, 14, and 15 missions. A pictorial comparison is shown in Table 1 between the frequencies and oscillator strengths of the stretching, bending, and lattice modes for each sample. For more detailed information reference should be made to Perry *et al.* (1971, Figs. 3 and 4), and Tables 2 and 3.

Several researchers (Conel, 1969; Salisbury *et al.*, 1970) have suggested that the frequency of the peak in the emissivity can be used as a diagnostic feature for determining the silica abundance in fine-powder emissivity spectra. This peak is directly related through the dielectric function ($\varepsilon' = n^2 - k^2$ and $\varepsilon'' = 2nk$, where n is the real part of the refractive index and k is the extinction coefficient) to the high-frequency minimum in the reflectivity and in fact our reflectance spectra are simply inverted emission spectra referenced to blackbody emissivity. For powder spectra both curves are complicated by particle size and packing fraction because of additional diffuse scattering mechanisms (Aronson, 1972). This frequency (known as the Christianson frequency and referred to as ω_c in Table 1) occurs in the region where $n \approx 1$ and $k \approx 0$. Perry *et al.* (1971) have shown that for lunar materials ω_c varies over the narrow range from about 1195 cm^{-1} ($\sim 40\%$ SiO$_2$) to about 1260 cm^{-1} ($\sim 50\%$ SiO$_2$). These results for terrestrial materials correspond to the intermediate to ultrabasic range as reported by Conel (1969) and Salisbury *et al.* (1970). The emissivity peak for 14259 soil by Logan *et al.* (1972) at 1220 cm^{-1} is in good agreement with the ω_c value

Table 2. Comparison of infrared transverse optical phonon frequencies for lunar rock samples from Apollo 14 and 15.

Suggested assignment Combination	14310,184 1065(w)	14321,198	14301,20	14307,18	14313,51	15426,2 1040(w)
M–O stretch					1015(m)	
v_3''	990(m)	1010(w) 980(w)	1010(w)	1005(m)		
v_3'	920(s)	920(m)	940(m)	930(s)	925(s)	
v_3'		890(m)				850(m)
v_1'						810(w)
Cation-oxygen vibrations	725(w)		745(w)	745(w)	730(w)	710(w)
		620(w)				640(w)
	580(w)		570(w)			610(w)
	560(w)					570(w)
M–O–M bend	530(m)		530(m)			
v_4'		520(m)		525(m)	520(m)	
v_4''	470(m)		475(m)	485(m)	470(s)	450(w)
v', v_4'''	390(m)	390(m)	390(m)	380(m)	380(m)	370(w)
v_2''	345(m)	335(m)	345(w)		345(m)	340(w)
Lattice modes	240(w)	240(w)	240(w)		225(w)	280(s)

C. H. PERRY et al.

Table 3. Comparison of the infrared optical phonon frequencies, ω_c (Christianson frequency) and ε_0 (dielectric constant for sample densities 2.4–2.6 g/cm³) for Apollo 14 and 15 lunar fines.

	Fines (LM)	2-4mm Fines (LM)	Comp. <1 mm fines	Core sample	Station C <1 mm fines	Station C1 Powdered rock
Suggested assignment Combinations	14163,31	14161,36	14259,38	14230,95 1060	14141,37 1050	14321,263
M–O stretch	1020 860	1020 930 810	1010 930 870	1010 900	 900	1020 920
Cation-oxygen lattice vibrations	780 690 560	 690 610 560 540	 700 570 540	780 660 610 560	 660 600 550 530 510	800 690 610 560 490
M–O–M bend	450 370	450 360	440 370	460 370	430 360	450 360 330
Lattice modes	290	290 230	300 200	280 230	260 220	270 220
ω_c (cm⁻¹)	1195	1210	1225	1215	1230	1225
ε_0	4.9	5.9	4.0	4.2	4.5	3.8

	Station 2 St. George Crater ejecta		Station 1 Elbow Crater ejecta	Ray area (LM)	Station 4 Dune Crater	Station 9a Hadley Rille
Suggested assignment Combinations	15091,65 1090	15221,71 1100	15071,60	15021,159 1040	15471,68 1100	15531,63 1070
M–O stretch	1000 880	1000 890	1000 920 870 820	 890	1000 950 860	 960 930 910 870
Cation-oxygen lattice vibrations	790 670 600 550	790 670 610 550	780 670 610 580	 660	770 680 620 560	790 700 630 510
M–O–M bend	440 360	440 370	460 370	470 370	450 370	450 380
Lattice modes	280 220	280 230	290 230	300 210	280 230	280 220
ω_c (cm⁻¹)	1220	1220	1225	1220	1210	1205
ε_0	4.0	4.2	3.8	4.0	4.2	4.9

(1225 cm^{-1}) obtained from our reflectance measurements as given in Table 3. Birkebak and Dawson (1972) obtained a value of 1210 cm^{-1} for 14163 soil having a density of 1.6 g/cm^3 which is within experimental error of our values of 1205 cm^{-1} for 14163,31 (LM sample) with a density of $\sim 2.5 \text{ g/cm}^3$. It is interesting to note that the infrared emissivity measurement of surface features in various lunar regions from balloon observations by Salisbury et al. (1970) and Murcray et al. (1970) vary from 1195 to 1235 cm^{-1} again within this narrow region. The 14310,184 sample, which we estimate to have a high anorthite content, compares with noritic gabbros and anorthitic basalts in terms of the Christianson frequency ($\omega_c \approx 1250 \text{ cm}^{-1}$), but there are marked anomalies in the series, as can be seen in Table 1, where ω_c is compared with the major chemical abundances. However, portions of the specimens used to obtain the elemental analyses are not the same as the ones used to measure the ω_c values, and variations can be expected within the same parent sample. Consequently, correlation of the ω_c values with SiO_2 wt.% may be better than the data presented in Table 1, although particle size and packing of the material also effect the scattering components and can cause variations in the measured Christianson frequency.

In Tables 1 and 3 we have also shown the values of the far-infrared dielectric constant at $\sim 10^{11}$ Hz which were calculated from the Kramers-Kronig analysis. Our values for the lunar rocks appear to be in general agreement with Ade et al. (1971) far-infrared measurements at 337 μ where a cyanogen laser was used. For sample 12063,96 (type A basalt) and for 10065,30 (breccia) they obtain values of 6.25 and 3.7, respectively. Based on the results of Perry et al. (1971) one may estimate that both these samples contain 25–30 mole percent of titani-ferrous oxide. Sample 10017,64, another Apollo 11 type A basalt, has a high dielectric constant value of 7 which compares with our measured value of 6.7 for 12008,23, and consequently it probably has a comparable or slightly higher ilmenite content (i.e., 30–35%). Comparison can be made with the dielectric response determined by capacitance measurements at low frequencies (10^2–10^6 Hz) by Chung et al. (1971), Collett and Katsuba (1971), Chung and Westphal (1972), and Lowndes et al. (1972), and at 430 MHz by Gold et al. (1972). The disagreement with previous values of Chung et al. (1971) appears to be due to their samples not being dry (Chung et al., 1972). Our value for a 14310 sample of 5.2 (10^{11} Hz) and 5.99 (10^2–10^6 Hz) compares favorably with Chung and Westphal's (1972) value of 6–9 and the Gold et al. (1972) value of 6.6. However, there are still striking differences in sample 12002 where we obtained a dielectric constant (at 10^{11} Hz) of 4 which is to be compared with Chung et al.'s (1971) result of 8–10 and Collett and Katsuba's (1971) value of 9 at low frequencies. Similar differences were observed in our own measurements in 12008,23 where the dielectric constant varies between 6.7 at 10^{11} Hz and 33 at 10^6–10^2 Hz. As both these samples contain free iron and other high-conductivity materials, they may give rise to additional relaxation mechanisms or free carrier absorption which could account for this strong dispersion of the dielectric constant.

In an earlier report, Perry et al. (1971) have found that sample 12008,23 has a strong vibration at $\sim 890 \text{ cm}^{-1}$. The high value of the measured static dielectric constant of this sample indicates that it has a very high ($> 30\%$) FeO content and it has also been reported that the major phase of the parent rock may be ilmenite (Warner,

1970). Sample 12009,48 has a lower FeO content ($\sim 20\%$) and the stretching vibration is now situated at ~ 920 cm^{-1}. With the increase in plagioclase content, as for example in 12002,186 ($\sim 30\%$ plagioclase), there is a continued shift of this band to higher frequencies (~ 930 cm^{-1}), and in 12065,115 (breccia), with about 50-50 plagioclase-pyroxene content, the trend is further indicated (~ 960 cm^{-1}).

In Fig. 1 we show some of the greatest variations that we have observed in the reflection spectra of lunar samples. The 14310,184 spectrum has features that clearly demonstrate that it is a high calcium oxide, alumina, and silica basalt and it has the highest ω_c frequency compared with the other lunar samples examined. It has high stretching vibrations at ~ 990 cm^{-1} and ~ 920 cm^{-1} and bending vibrations at ~ 470 cm^{-1} and 530 cm^{-1}. The lattice band at 240 cm^{-1} is characteristic of a high plagioclase material and is indicative of its high anorthite (An$_{90}$–An$_{100}$) content. For comparison the Cone Crater sample (14321,98) from Station C1 (which has comparable silica, alumina, and calcium oxide content) shows a lowering of the frequency of the band at ~ 500 cm^{-1}. This band appears to be related to the slightly higher orthopyroxene content. The spectrum of 12009,48 with its reduced silica and alumina content shows a further lowering of the bending vibration to ~ 490 cm^{-1} which appears associated with the increase in FeO content (see Table 1). Generally, as the dominance of the framework silicates change gradually to the more open chain silicates of the pyroxenes, a lowering of the internal frequencies may be expected and is apparently observed. Finally in Fig. 1, the new features in the spectrum of sample 10058,56, which has FeO, SiO$_2$, Al$_2$O$_3$, and CaO contents comparable with those of 12009,48, indicate that they can be directly associated with the increasing TiO$_2$ present and the subsequent reduction in MgO.

The three Apollo 14 breccias (or fragmental rocks) examined, namely, 14307,18, 14313,51, and 14301,20, show unmistakable spectral similarities, as seen in Fig. 3. All are typical of the Fra Mauro feldspar-rich breccias and their spectra, although slightly enhanced in intensity, possess features that correspond closely with those observed in the Apollo 14 fines sample of which the bulk sample 14161,36 from the LM vicinity is a typical example.

From Table 1 it is seen that these Fra Mauro breccias have lower silica and alumina contents than sample 14310 which is more likely of highland origin since it is rich in plagioclase (anorthosites). However, they are intermediate between the highland material and the mare material of Apollo 11 and 12 and these findings have been confirmed in the x-ray fluorescence experiment by Adler *et al.* (1972). These breccias are presumably representative of the ejecta blanket deposited by the impact that formed the Imbrium Basin and contain materials from the upper levels of the crust. The Apollo 14 soils also show considerable homogeneity in the spectra (see Fig. 5) and this is confirmed in the analyses (Compston *et al.*, 1972), where the average soil content is 48% SiO$_2$, 17% Al$_2$O$_3$, 9% MgO, 10% FeO, 11% CaO, and 2% TiO$_2$. These major chemical abundances are very similar to those found in the Apollo 14 breccias (see Table 1) and the spectral similarities between Fig. 3 and Fig. 5 indicate that the Fra Mauro soils were mainly produced by local impacts and consist of particulate debris from the rocks with a small admixture of other material (see, for example, Adams and McCord, 1972). From the infrared and Raman frequencies it would appear that these samples contain plagioclase An_{70}–An_{90}, olivine Fa_{20}–Fa_{35}, and pyroxene Fs_{25}–Fs_{35}, and are in general agreement with the report on 14313 (Floran *et al.*, 1972).

The Cone Crater breccia sample from Station C, 14321,98, has spectral similarities that lie between 14310 and the other three Apollo 14 rock samples. It has the highest magnesium content and lowest calcium content and although the silica abundance is similar to 14310 its alumina and FeO contents are more like the Fra Mauro breccias. It may in fact be part of an earlier regolith that suffered from the Imbrian ejecta and was later ejected during the Cone Crater event. The increase in iron content in 14321 over the 14310 samples is exhibited in a shifting to lower frequencies of the band ~500 cm^{-1}. Its lower Christianson frequency probably results from its high MgO content. This reverse behavior of ω_c is apparent in 12009,48 which again has a comparable abundance of MgO even though the SiO_2 content is less and the FeO content is increased.

In Figs. 5 and 6 we can see a direct comparison between the Apollo 14 and 15 soil samples. Samples 14259,38, which is representative of the regolith, and 14163,31, which is from a depth of a few centimeters, both gave very similar spectra. Sample 14161,36, which consists of 2–4 mm fines, showed some slight sharpening of the reflectivity peaks. The sample 14230,95 from the second core tube again showed general overall spectral similarities and it was concluded that this sample was also representative of the soil in the area and confirmed its homogeneity. Some small spectral differences were observed with the <1 mm fines sample from Station C (14141,37) in the vicinity of Cone Crater. The most noticeable feature was the slight strengthening of the band about 450 cm^{-1} which showed that the Fra Mauro soil in this area was contaminated with the more pyroxene-rich Cone Crater soil ejecta. Some rock fragments from the interior of 14321 (associated with piece 14321,263) again showed some spectral enhancement of the features but they were not as pronounced as 14321,98 which was a polished butt end (presumably from the exterior). Again it would appear in fact that the interior pieces of this rock investigated more closely resembled the surrounding soil. In Fig. 6 the Apollo 15 soils show a larger variety of spectral differences that can be associated with their geographic locations and the corresponding mineralogical compositions. The two samples from the Apennine Front at Station 2 (15091,65 and 15221,71) comprise St. George Crater ejecta and other ejecta materials. The two showed some very slight spectral variations but in fact more closely resembled the Apollo 14 soils. Consequently, it would appear that the soils in the vicinity of the St. George Crater rim are substantially intermixed. To some extent this was also true of 15071,60 from Station 1 which might be expected to be more representative of Elbow Crater ejecta. Evidently, this is not the case and again here the soil is well intermixed. In the vicinity of the LM on the mare surface of Palus Putredinis are the faint rays from either of the craters Aristillus and Autolycus. The spectrum of the contingency sample 15021,159 is presumably representative of the fine-grain regolith and although the spectral features were not significantly different there was a lowering of some frequencies which is usually indicative of an increasing iron content and a corresponding slight decrease in the silica and alumina content. The surface layer here may be more representative of Imbrium or other ejecta from deeper crust levels that would not have traveled as far to the Apollo 15 site compared to the Apollo 14 site. The spectrum of the Dune Crater sample 15471,68 from Station 4 shows further enhancement of the spectral feature at ~ 450 cm^{-1} which can be associated with a slight increase in the pyroxene content and this trend is further enhanced, as seen in Fig. 6, in 15531,63 from the rim of Hadley Rille. The spectra are more similar to some of the Apollo 12 rocks such as 12002,186 or 12009,48 which have high pyroxene contents. It may be expected that these young ejecta materials are more representative of bedrock whereas the material from the Apennine Front contains largely gardened mixtures.

The 15426,2 breccia from Station 7 at Spur Crater does not resemble the Fra Mauro breccias as it is obviously richer in pyroxene but it also does not show the same spectroscopic features as the Apollo 12 samples. In Fig. 4 it can be compared with the microbreccia 12073,42 and it also bears some resemblance to the Cone Crater sample 14321,263 in Fig. 5. However, we have not observed any exact spectroscopic analog from any of the soil samples.

SUMMARY

In this paper we have shown that the infrared and Raman spectra provide all the vibrational frequencies of the materials. These comprise the internal stretching and bending modes associated with the $(SiO_n)^{x-}$ groups, the weak cation-oxygen coupling between adjacent groups and the low-frequency vibrations that can be associated with translational and librational lattice modes. These groups may be isolated, as for example in the olivines, in chains or framework structures which are associated with the pyroxene and plagioclase classes, respectively. Generally as the structure changes from one that is more open to one that has considerable cross coupling, the frequencies of the vibrational modes increase.

The infrared measurements have the advantage that they provide information on the macroscopic properties that may be expected to be more representative of large areas. In consequence, the spectra are more difficult to interpret uniquely but they do indicate spectral differences. The complementary Raman techniques offer a non-destructive tool for determining mineralogical compositions of grains and inclusions and usually provide positively identifiable results.

The Christianson frequencies can be obtained concurrently with the infrared vibrational frequencies from a single infrared reflectivity measurement and a Kramers-Kronig analysis of the spectrum. This frequency, corresponding to the peak in the emittance spectrum, has been shown by others and on a restricted basis by ourselves to be a useful feature for diagnostic applications.

However, we believe that it is necessary to obtain the complete infrared spectrum from 100–2000 cm^{-1} in order to correlate the molecular vibrational frequencies and the Christianson frequency with the chemical constituents and to then deduce the major mineral abundances (Perry and Lowndes, 1970). This is of course done by comparison with terrestrial counterparts and an excellent background for this work has been provided by Estep *et al.* (1971 and 1972). Their comprehensive measurements on a large number of terrestrial materials are an important part of the program as the uniqueness of the vibrational frequencies applied to mineralogical studies has been relatively unexploited by the geologist.

It is apparent from our data that differences in the chemical and mineralogical compositions of the lunar samples can be spectroscopically distinguished. A spectral difference can also be observed between the samples gathered from the different geographical locations, e.g., the spectra of lunar highland material indicate that they are less basic than the Fra Mauro breccias. These materials in turn are less basic than some of the crater ejecta, the Hadley Rille samples, and some of the mare materials from Apollo 11 and 12. These results in consequence have a direct bearing on the origin of the samples and comparison of spectra from various locations (either obtained from the Apollo missions or using remote sensing techniques) will be directly applicable as a petrological tool and will contribute to unraveling the history of the lunar surface.

The reflectance data provide the infrared dielectric dispersion (ε', ε'') and the optical constants, n, the refractive index and α, the absorption coefficient, over the range 10–5000 cm^{-1}. These properties do not depend on the type of measurement and they provide fundamental parameters for synthesizing spectra (e.g., emission spectra) for spectral matching models. Another product of the Kramers-Kronig analysis are values of the far-infrared dielectric constant at $\sim 10^{11}$ Hz. These results are less sensitive to sample dryness compared with the low-frequency dielectric constant obtained from capacitance techniques. The temperature dependence and pressure (density) dependence of the dielectric constants extrapolated to lower frequencies have important applications in radar studies of the depth and composition of the various surface layers. Our far-infrared and dielectric studies on lunar basalts are in agreement, but the difference in the results for the samples containing high abundances of iron and titanium indicate other relaxation processes are present at lower frequencies.

Acknowledgments—We wish to thank Dr. J. W. Salisbury and Dr. G. R. Hunt, Air-Force Cambridge Research Laboratories, Hanscom Field, for their continued interest in this work and for the use of their Perkin-Elmer Model 180 Spectrophotometer in measuring all the Apollo 14 and 15 samples. We are also indebted to Dr. R. Hannah, Perkin Elmer Corporation, Norwalk, Conn., for the use of a similar instrument for the Apollo 11 and 12 samples. This spectrometer provided the investigators with a vast improvement in signal-to-noise and reliability of the reflectance data over those obtained on their own instrumentation. This work was supported by NASA Grant NGR 22-011-069 and by a Northeastern University Grant for basic research. Partial equipment support was provided under NASA Cooperative Agreement NCAw 22-011-079.

References

Adams J. B. and McCord T. B. (1972) Optical evidence for regional cross-contamination of highland and mare soils (abstract). In *Lunar Science—III* (editor C. Watkins), pp. 1–3, Lunar Science Institute Contr. No. 88.

Ade P. A., Bastin J. A., Marston A. C., Pandya S. J., and Puplett E. (1971) Far infrared properties of lunar rock. *Proc. Second Lunar Sci. Conf., Geochim. Cosmochim. Acta* Suppl. 2, Vol. 3, pp. 2203–2211. MIT Press.

Adler I., Trombka J., Gerard J., Lowman P., Yin L., Blodgett H., Gorenstein P., and Bjorkholm P. (1972) Preliminary results from the S-161 x-ray fluorescence experiment (abstract). In *Lunar Science—III* (editor C. Watkins), pp. 4–6, Lunar Science Institute Contr. No. 88.

Agrell S. O., Scoon J. H., Long J. V. P., and Coles J. N. (1972) The occurrence of goethite in a microbreccia from the Fra Mauro formation (abstract). In *Lunar Science—III* (editor C. Watkins), pp. 7–9, Lunar Science Institute Contr. No. 88.

Aronson J. R. (1972) The prospects for remote infrared petrology (abstract). Proceedings of Conference on Lunar Geophysics, *The Moon* (in press).

Aronson J. R., Emslie A. G., Allen R. V., and McLinden H. G. (1967) Studies of the middle and far infrared spectra of mineral surfaces for application in remote compositional mapping of the moon and planets. *J. Geophys. Res.* **72**, 687.

Birkebak R. C. and Dawson J. P. (1972) Thermal radiation properties of Apollo 14 fines (abstract). In *Lunar Science—III* (editor C. Watkins), pp. 83–85, Lunar Science Institute Contr. No. 88.

Charles R. W., Hewitt D. A., and Wones D. R. (1971) H_2O in lunar processes: Stability of hydrous phases in lunar samples 10058 and 12013. *Proc. Second Lunar Sci. Conf., Geochim. Cosmochim. Acta* Suppl. 2, Vol. 1, pp. 645–664. MIT Press.

Chung D. H. (1971) Laboratory studies on seismic and electrical properties of the moon (abstract). Proceedings of Conference on Lunar Geophysics, *The Moon* (in press).

Chung D. H., Westphal W. B., and Simmons G. (1971) Dielectric behavior of lunar samples. *Proc. Second Lunar Sci. Conf., Geochim. Cosmochim. Acta* Suppl. 2, Vol. 3, pp. 2381–2390. MIT Press.

Chung D. H. and Westphal W. B. (1972) Dielectric properties of Apollo 14 lunar samples (abstract). In *Lunar Science—III* (editor C. Watkins), pp. 139–140, Lunar Science Institute Contr. No. 88.

Collett L. S. and Katsuba T. J. (1971) Electrical properties of Apollo 11 and Apollo 12 lunar samples. *Proc. Second Lunar Sci. Conf., Geochim. Cosmochim. Acta* Suppl. 2, Vol. 3, pp. 2367–2379. MIT Press.

Compston W., Vernon M. J., Berry H., Rudowski R., Gray C. M., Ware N., Chappell B. W., and Kaye M. (1972) Age and petrogenesis of Apollo 14 basalts (abstract). In *Lunar Science—III* (editor C. Watkins), pp. 151–153, Lunar Science Institute Contr. No. 88.

Conel J. E. (1969) Infrared emissivities of silicates: Experimental results and a cloudy atmosphere model of spectral emission from condensed particulate mediums. *J. Geophys. Res.* **74**, 1614.

Estep P. A., Kovach J. J., and Karr C. Jr. (1971) Infrared vibrational spectroscopic studies of minerals from Apollo 11 and Apollo 12 lunar samples. *Proc. Second Lunar Sci. Conf., Geochim. Cosmochim. Acta* Suppl. 2, Vol. 3, pp. 2137–2151. MIT Press.

Estep P. A., Kovach J. J., Waldstein P., and Karr C. Jr. (1972) Infrared and Raman spectroscopic studies of structural variations in minerals from Apollo 11, 12, and 14 samples (abstract). In *Lunar Science—III* (editor C. Watkins), pp. 244–246, Lunar Science Institute Contr. No. 88.

Fabel G. W., White W. B., White E. W., and Roy R. (1972) Structure of lunar glasses by Raman and soft x-ray spectroscopy (abstract). In *Lunar Science—III* (editor C. Watkins), pp. 250–251, Lunar Science Institute Contr. No. 88.

Floran R. J., Cameron K., Bence A. E., and Papike J. J. (1972) The 14313 consortium: A mineralogic and petrologic report (abstract). In *Lunar Science—III* (editor C. Watkins), pp. 268–269, Lunar Science Institute Contr. No. 88.

Gold T., Bilson E., and Yerbury M. (1972) Grain size analysis, optical reflectivity measurements and determination of high frequency electrical properties for Apollo 14 lunar samples (abstract). In *Lunar Science—III* (editor C. Watkins), pp. 318–320, Lunar Science Institute Contr. No. 88.

Griffith W. P. (1969) Raman studies on rock-forming minerals. Part I. Orthosilicates and cyclosilicates. *J. Chem. Soc.*, Ser. A, 1372.

Howard H. T. and Tyler G. L. (1972) Bistatic-radar observations of the lunar surface with Apollos 14 and 15 (abstract). In *Lunar Science—III* (editor C. Watkins), p. 398, Lunar Science Institute Contr. No. 88.

Hubbard N. J. and Gast P. W. (1972) Chemical composition of Apollo 14 materials and evidence for alkali volatilization (abstract). In *Lunar Science—III* (editor C. Watkins), pp. 407–409, Lunar Science Institute Contr. No. 88.

Logan L. M., Hunt G. R., Balsamo S. R., and Salisbury J. W. (1972) Mid infrared emission spectrum of Apollo 14 soil: Significance for compositional remote sensing (abstract). In *Lunar Science—III* (editor C. Watkins), pp. 490–492, Lunar Science Institute Contr. No. 88.

Lowndes R. P. and Martin D. H. (1969) Dielectric dispersion and the structure of ionic lattices. *Proc. Roy. Soc.*, Ser. A, **308**, 473.

Lowndes R. P., Rastogi A., and Perry C. H. (1972) Dielectric properties of some lunar samples. To appear.

LSPET (Lunar Sample Preliminary Examination Team) (1970) Preliminary examination of the lunar samples for Apollo 12. *Science* **167**, 1334.

LSPET (Lunar Sample Preliminary Examination Team) (1971) Preliminary examination of lunar samples from Apollo 14. *Science* **173**, 688.

Lyon R. J. P. (1963) Evaluation of infrared spectrophotometry for compositional analysis of lunar and planetary solids. Stanford Res. Inst. Final Report under contract NASr-49(04), NASA Tech. Note, TND-1871.

Mendell W. W. and Low F. J. (1971) Differential flux scans of the moon at $\lambda = 22\ \mu$ (abstract). Proceedings of Conference on Lunar Geophysics, *The Moon* (in press).

Murcray F. H., Murcray D. G., and Williams W. J. (1970) Infrared emissivity of lunar surface features, I. Balloon-borne observations. *J. Geophys. Res.* **75**, 2662.

Perry C. H., Geick R., and Young E. F. (1966) Solid state studies by means of Fourier transform spectroscopy. *Applied Optics* **5**, 1171.

Perry C. H. and Lowndes R. P. (1970) The study of the reflectivity of inorganic materials for remote sensing applications. Final Report AFCRL-70-0512.

Perry C. H., Agrawal D. K., Anastassakis E., Lowndes R. P., Rastogi A., and Tornberg N. E. (1971) Infrared and Raman spectra of lunar samples from Apollo 11, 12, and 14 (abstract). Proceedings of Conference on Lunar Geophysics, *The Moon* (in press).

Salisbury J. W., Vincent R. K., Logan L. M., and Hunt G. R. (1970) Infrared emissivity of lunar surface features, II. Interpretation. *J. Geophys. Res.* **75**, 2671.

Simmons G., Strangway D., Bannister L., Cubley D., La Torraca G., Rossiter J., Watts R., and Annan A. (1971) Surface electrical properties experiment (abstract). Proceedings of Conference on Lunar Geophysics, *The Moon* (in press).

Warner J. (1970) Apollo 12 lunar-sample information. *NASA Technical Report R-353*, p. 75.

White W. B., White E. W., Görz H., Henisch H. K., Fabel G. W., and Roy R. (1971) Examination of lunar glass by optical, Raman and x-ray emission spectroscopy and by electrical measurements. *Proc. Second Lunar Sci. Conf., Geochim. Cosmochim. Acta* Suppl. 2, Vol. 3, pp. 2213–2221. MIT Press.

Proceedings of the Third Lunar Science Conference
(Supplement 3, *Geochimica et Cosmochimica Acta*)
Vol. 3, pp. 3097–3101
The M.I.T. Press, 1972

Reflectance and absorption spectra of Apollo 11 and Apollo 12 samples

I. I. Antipova-Karataeva, Ju. I. Stacheev, and L. S. Tarasov

V. I. Vernadsky Institute of Geochemistry and Analytical Chemistry,
Moscow, U.S.S.R. Academy of Sciences

Abstract—Results are presented from an investigation of the optical parameters (albedo, spectra of diffuse reflectance, transmittance spectra and electronic absorption spectra) of regolith powder returned by Apollo 11 and Apollo 12 and of a surface section of a lunar rock. The diffuse-reflectance spectra of Apollo 11 material is similar to but somewhat higher than that of the Apollo 12 regolith. Apollo 12 fines have an enhanced ferrous band. The lunar rock reflectance coefficient is more than twice as high as that of the regolith. The rock spectra have a strong band centered at 1 μ and structure near 0.3–0.6 μ. This difference between rock and regolith diffuse-reflectance spectra is accounted for by the morphology of a surface of second origin of the regolith particles. The absorption spectra of different areas of the same rock section are unlike and are related to mineral composition. The position difference of the ferrous band can be caused by the Fe^{2+} coordination sphere difference in rocks. Occurrence of the absorption band near 0.56 μ apparently must be attributed to either color centers in solar irradiated plagioclase or to Ti^{3+} ions in octahedrally coordinated sites.

INTRODUCTION

THE RESULTS OF an optical parameters (albedo, spectra of diffuse reflectance, and electronic absorption spectra) investigation of regolith powder returned by Apollo 11 and 12 and of the surface of a rock section are presented in this work.

The diffuse-reflectance spectra (albedo) were obtained for all specimens between 0.24 μ and 1.8 μ relative to an MgO-smoke standard. A Hitachi Model EPS-3T spectrophotometer with an integrating sphere attachment was used. The absorption spectra of the transparent sections of rocks were obtained between 0.34 and 2.4 μ relative to air or to the sectionholder using the same spectrophotometer but without the integrating sphere.

Regolith powders from the Apollo 11 core at depths of ~1.5 cm (sample N10005,34) and ~10–11 cm (10005,35) and from the Apollo 12 core at depths of ~16 cm (12028,229) and ~33 cm (12028,233) were investigated.

Diffuse-reflectance spectra of all specimens were similar and had no evident structure. The spectral features of Apollo 11 and 12 core samples were similar to those described previously (Antipova-Karataeva *et al.*, 1971) for regolith powder returned by Luna 16. The reflectance increased continually between 0.25 and 1.8 μ from blue to red. At 1 μ a broad, weak absorption peak was present which is believed to be due to ferrous ions in octahedrally coordinated sites containing a crystal lattice of regolith-formed minerals (see Fig. 1). The reflectance of Luna 16 regolith powder is close to the reflectance of Apollo 11 fines. The Apollo 12 fines have a higher albedo and an enhanced ferrous band. These albedo differences increase with increasing wavelength. The average reflectance does not vary systematically with depth in the Apollo 12 core tube for the samples examined. By contrast, the reflectivity of Apollo 11 fines from

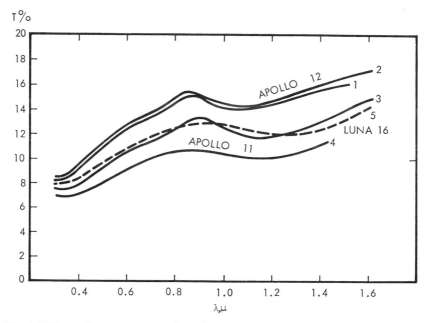

Fig. 1. Diffuse reflectance spectra of regolith: 1—sample N12028,233; 2—N12028,229; 3—N10005,35; 4—N10005,34; 5—Luna 16 regolith.

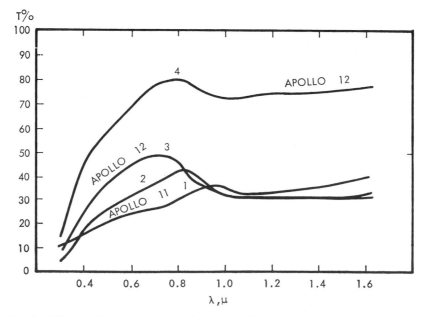

Fig. 2. Diffuse reflectance spectra of rock sections: 1—N10048,53, breccia; 2—N10047,25, coarse-grained basalt; 3—N12018,80, fine-grained basalt; 4—N12034,34 breccia.

~1.5 cm depth (curve 4) is slightly lower than those from ~10–11 cm depth (curve 3). All diffuse reflectance data are in agreement with albedo measurements of regolith powder reported by Hapke *et al.* (1970), Birkebak *et al.* (1971), Garlick *et al.* (1971), Nash and Conel (1971), and Dollfus *et al.* (1972).

Using the same device the diffuse reflectance spectra of four sections of Apollo 11 rocks (sample N10047,25 and N10048,53) and of Apollo 12 (N12034,34 and N12018,80) were obtained. Spectra of these samples are shown on Fig. 2. They are similar in general to those of regolith: the reflectance increased continually between 0.25 and 0.9 μ, and at $\simeq 1$ μ a weak absorption band exists. However, the spectra of lunar crystalline rocks are characterized by an albedo more than twice as high as that of regolith, the spectra are relatively flat between 1.2 and 1.7 μ, have a pronounced absorption band in the ferrous position at 1 μ, and weak bands at 0.5 and 0.3 μ. In addition, the absorption curves of lunar rocks sections between 1 and 1.7 μ are relatively flat, but between 0.25 and 0.8 μ they are much steeper than those of fines. The reflectance difference between measured Apollo 11 and 12 crystalline rocks is lower than those of regolith. Thus, for example, the diffuse reflectance spectra of section N10048,53, N10047,25, and N12018,80 are in general similar, especially between 1 μ and 1.7 μ, but the intensity of section N12074,34 (curve 4) is considerably greater. Wavelength position of the ferrous band changes from one rock sample to another. Diffuse reflectance spectra of rocks and regolith are readily distinguishable from one another by the above-mentioned properties.

The transmittance spectra between 0.36 and 2.4 μ of two transparent sections N10047,25 and N12018,80 were obtained in order to investigate the interconnection between the spectral properties and mineral composition of the rocks. The transmittance spectra of 8×8 mm^2 areas are presented on Fig. 3. The section N12018,80 (curve 2) belonged to fine-grained basalt rock, the grains—pyroxene, ilmenite, and

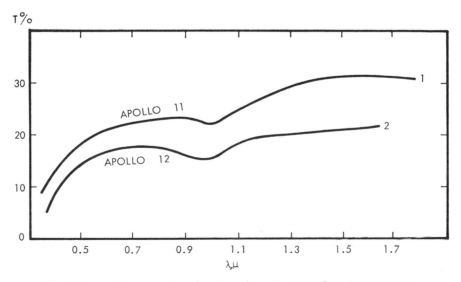

Fig. 3. Transmittance spectra of rock sections: 1—N10047,25; 2—N12018,80.

plagioclase—are distributed regularly. Apollo 11 section N10047,25 (curve 1) is composed of relatively large crystals of transparent feldspar, brownish pyroxene, and untransparent ilmenite. The transmittance curves rise steeply between 0.6 and 2.4 μ, have pronounced absorption bands in the ferrous position at $\sim 1\ \mu$, and weak structure at 0.5 μ. The ferrous band in the Apollo 12 section spectra is more pronounced than that in the Apollo 11 section, and the maximum is shifted toward short wavelengths between 1 μ to 0.94 μ. The position difference of the ferrous band in these two sections can be caused by a Fe^{2+} coordination sphere difference in the rocks.

Relatively large regions were present in the Apollo 11 section consisting of the same crystals. Measurements were made on these homogeneous areas. Absorption spectra of six such areas of section N10047,25 were examined between 0.35 and 2.4 μ relative to a glue-coated glass plate holder (Fig. 4). Dimensions of these areas were $\sim 2 \times 2\ mm^2$. The absorption curves divided into two groups. The curves 2, 4, and 6 have pronounced ferrous absorption bands. In contrast, on curve 1 this band is very weak in intensity, and is absent on curves 3 and 5.

A possible explanation of these absorption spectra properties is that areas 2, 4, and 6 consist mainly of pyroxene and ilmenite crystals in different ratios. The band near 1 μ may be attributed principally to Fe^{2+} ions in pyroxene. The change of wavelength position of the ferrous band from 1 μ to 0.96 μ and to 0.93 μ in the spectra of areas 6, 4, and 2, respectively, may be caused by possible differences in composition of the pyroxenes. Adams (1972) and Burns *et al.* (1972) have also indicated a connection between pyroxene composition and the wavelength position of the Fe^{2+} band. Rock spectra display, in addition to the band near 1 μ, a weak wide band

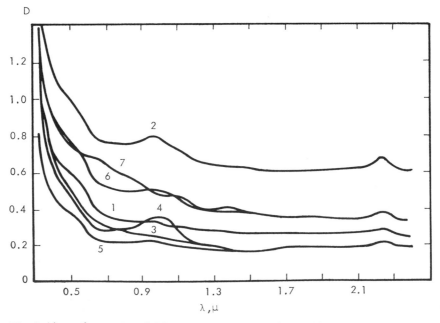

Fig. 4. Absorption spectra of different areas of rock section N10047,25: 1—plagioclase; 2—pyroxene, ilmenite; 3—plagioclase; 4—pyroxene, ilmenite; 5—plagioclase, pyroxene; 6—pyroxene, ilmenite; 7—section of earth basalt.

near 0.56 μ which has a maximum intensity in the spectra of areas 1, 3, 5, and 6 and seems to arise principally from plagioclase. This is true because the band is very weak or absent on the absorption curves of areas 2 and 4 which have very low abundances of plagioclase. The low intensity of this band allows one to attribute it to the "d–d" forbidden transition in metal ions or to color centers in solar irradiated crystals of plagioclase. This band has been attributed to the Ti^{3+} ion in pyroxene (Hapke *et al.*, 1970; Nash and Conel, 1971). However, the absorption band of Ti^{3+} in octahedrally coordinated sites usually has a maxima near 0.5 μ and 0.58 μ; therefore this assumption does not appear to account for this result.

Conclusions

By measurements of diffuse reflectance spectra of regolith and rock sections and absorption spectra of rock sections of Apollo 11 and 12 samples it was found:

(1) Diffuse-reflectance spectra of regolith of Apollo 11, 12, and Luna 16 samples are similar, but the Apollo 12 fines have a somewhat higher albedo.

(2) The rock reflectance coefficient is more than twice as high as that of regolith. The rock spectra have a strong band centered at 1 μ and structure near 0.3–0.6 μ. This difference between rock and regolith spectra is accounted for by the morphology of this regolith particles which constitute a surface of second origin.

(3) The absorption spectra of different areas of the same rock section are different. The differences are related to the mineral composition of the sections. Differences of the Fe^{2+} band can be caused by the differences of the Fe^{2+} ion coordination sphere in rocks, and the occurrence of the absorption band at 0.56 μ must apparently be attributed to color centers in solar irradiated plagioclase or to the Ti^{3+} ions in octahedrally coordinated sites.

References

Adams J. B. (1972) Optical evidence for regional crosscontamination of highland and mare soils (abstract). In *Lunar Science—III* (editor C. Watkins), pp. 1–3, Lunar Science Institute Contr. No. 88.

Antipova-Karataeva I. I., Stacheev Yu. I., and Florensky K. P. (1971) Paper 19. The optical parameters of Mare Foecunditatis regolith. *Space Research XII. Life Science and Space Research X.* Publisher not available.

Birkebak R. C., Cremers C. J., and Dawson J. P. (1971) Spectral directional reflectance of lunar fines as a function of bulk density. *Proc. Second Lunar Sci. Conf., Geochim. Cosmochim. Acta* Suppl. 2, Vol. 3, pp. 2197–2202. MIT Press.

Burns R. G., Rateb M. A., and Huggins F. E. (1972) Crystal field spectra of lunar samples (abstract). In *Lunar Science—III* (editor C. Watkins), pp. 108–109, Lunar Science Institute Contr. No. 88.

Dollfus A., Bowell E., Geake J. E., and Maurette M. (1972) Optical polarimetric and photometric studies of lunar samples (abstract). In *Lunar Science—III* (editor C. Watkins), pp. 180–182, Lunar Science Institute Contr. No. 88.

Garlick G. F. J., Lamb W. E., Steigman G. A., and Geake J. E. (1971) Thermoluminescence of lunar samples and terrestrial plagioclases. *Proc. Second Lunar Sci. Conf., Geochim. Cosmochim. Acta* Suppl. 2, Vol. 3, pp. 2277–2283. MIT Press.

Hapke B. W., Cohen A. J., Cassidy W. A., and Wells E. N. (1970) Solar radiation effects on the optical properties of Apollo 11 samples. *Proc. Apollo 11 Lunar Sci. Conf., Geochim. Cosmochim. Acta* Suppl. 1, Vol. 3, pp. 2199–2212. Pergamon.

Nash D. B. and Conel J. C. (1971) Luminescence and reflectance of Apollo 12 samples. *Proc. Second Lunar Sci. Conf., Geochim. Cosmochim. Acta* Suppl. 2, Vol 3, pp. 2235–2244. MIT Press.

Proceedings of the Third Lunar Science Conference
(Supplement 3, *Geochimica et Cosmochimica Acta*)
Vol. 3, pp. 3103–3126
The M.I.T. Press, 1972

Polarimetric properties of the lunar surface and its interpretation.
Part 5: Apollo 14 and Luna 16 lunar samples*

E. Bowell and A. Dollfus

Paris Observatory,
Meudon, France

and

J. E. Geake

University of Manchester, Institute of Science and Technology,
Manchester, England

Abstract—Optical polarization and albedo measurements have been made in five colors spanning the wavelength range 3520 Å–5800 Å. All fines samples examined have polarization characteristics that agree extremely well with telescopic measurements on 50 km² or larger lunar regions. The range in maximum polarization and albedo is much more restricted for the fines than for selected terrestrial simulators, suggesting that the lunar material has been altered *in situ*. Production of fines and glasses by meteoroidal bombardment and perhaps alteration by solar wind action contribute to the optical uniformity of the lunar regolith. SEM studies show that the negative branches of polarization for fines, rocks, and breccias are closely related to surface microstructure. The well-developed negative branches exhibited by fines are characteristic of great microroughness, whereas the shallower negative branches of rock and breccia surfaces are due to the presence of smooth facets large compared to the wavelength of light. From telescopic observations on lunar regions larger than 50 km², we conclude that nowhere on the moon can exposed rock and breccia surfaces be abundant, for they would then show up by anomalous polarization.

INTRODUCTION

Previous parts in this series of papers were: Part 1, "Telescopic Observations", Dollfus and Bowell (1971); Part 2, "Terrestrial Samples in Orange Light", Dollfus et al. (1971a); Part 3, "Volcanic Samples in Several Wavelengths", Dollfus and Titulaer (1971); Part 4,* "Apollo 11 and 12 Lunar Samples", Dollfus et al. (1971b). The present paper and Part 4 describe work carried out in collaboration between Meudon Observatory, Laboratory "Physique du Système Solaire," France, and the University of Manchester Institute of Science and Technology (U.M.I.S.T.), U.K. J. E. Geake, NASA Principal Investigator at Manchester, also collaborated with A. Dollfus, Meudon, on work relating to Apollo 11 samples, reported by Geake et al. (1970). The present paper extends the investigations to Apollo 14 and Luna 16 samples, and enlarges the interpretation of previous Apollo samples.

---

* Part 4: Apollo 11 and Apollo 12 Lunar Samples, *Proc. Second Lunar Sci. Conf.*, **3**, 2285–2300, was incorrectly titled Part 3.

Sample Measurements

Polarization measurements were carried out with the photoelectric polarimeter at Meudon Observatory. Five filters were used to select colors in the spectral range 3540 Å to 5800 Å. The form of the negative branch, the value of maximum polarization, and the albedo of all samples were determined. Table 1 lists these samples

Table 1. Luna 16 and Apollo 14 samples measured.

Sample	Allocated to	Albedo A at 5800 Å (%)	P_m at 5800 Å (thousandths)	Type
L-16-19-1-116	Dollfus	8.5	173	Fines
14003,12	Dollfus	12.5	78	Fines
14083,5	Maurette	52.5	14	Breccia surface
14163,29	Dollfus	14.0	61	Fines
14259,39	Dollfus	10.6	85	Fines
14267A	Eglinton	12.0	364	Breccia surface
14321,151	Dollfus	18.4	132	Breccia interior
14321,176	Dollfus	12.7	199	Breccia interior

together with albedos and maximum degrees of polarization in orange light. The breccia 14267A was studied in collaboration with European Consortium members from U.K., France, and Switzerland under a program unified by G. Eglinton, Bristol University, U.K. Samples were once again exchanged with M. Maurette of CNRS, Orsay, France.

Figures 1, 2, and 3 show polarization curves for Luna 16 fines L-16-19-1-116, the European Consortium breccia 14267A, and three Apollo 14 breccias, respectively. The degree of linear polarization P (in thousandths) is plotted against the phase angle V (the angle between the source and the direction of observation as seen from the sample). Measurements were made in the specular direction, that is with equal incident and emergent angles and, as far as possible, with the normal to the sample surface contained in the plane of vision. The spectral variation for the Luna 16 fines is well shown by the series of curves in five colors (Fig. 1, negative branch in Fig. 16, also later section). Inset in Fig. 1 is the spectral variation of the albedo A, measured in the colors used for polarimetry. We define albedo as the reflectance with respect to a screen of MgO measured at a phase angle of 5° (Dollfus et al., 1971a). The polarization curve and negative branch for 14267A (Fig. 2) is shown for orange light only; inset is the spectral variation of the maximum degree of polarization P_m. Curves for breccias 14083,5, 14321,151, and 14321,176 are given in Fig. 3a; negative branches are detailed in Fig. 3b.

In this paper we have divided the study of the polarization characteristics of lunar samples into two parts: first, we consider the positive branch, and in particular P_m and its relationship with A; then we consider the negative branch and relate it with the microstructure.

Maximum Polarization and Albedo

Throughout this series of papers stress has been laid on the linear relationship between the maximum degree of polarization and albedo, both for telescopic lunar

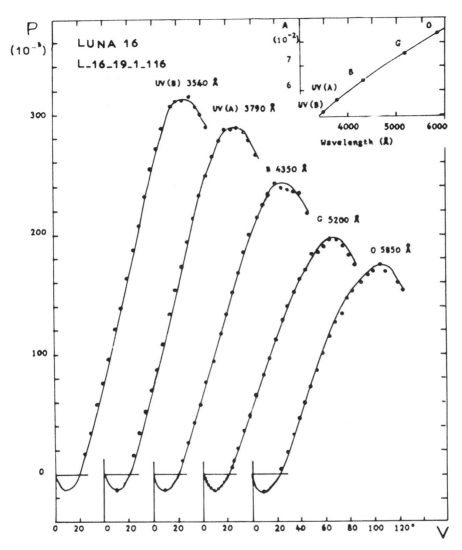

Fig. 1. Polarization curves for Luna 16 fines L-16-19-1-116. The proportion of plane polarized light (*P*, in thousandths) versus phase angle *V*, at five wavelengths, with the curves staggered horizontally to avoid overlap. Inset gives spectral variation of albedo *A*.

measurements and for laboratory measurements on lunar and terrestrial samples. A theoretical explanation of the relationship was given by Dollfus and Titulaer (1971).

Fines

Figure 4 is a plot of log *A* versus log P_m for all the lunar fines samples measured at 5800 Å. Also included is the domain defined by lunar telescopic data (corrected in albedo for reasons given in Dollfus and Titulaer, 1971).

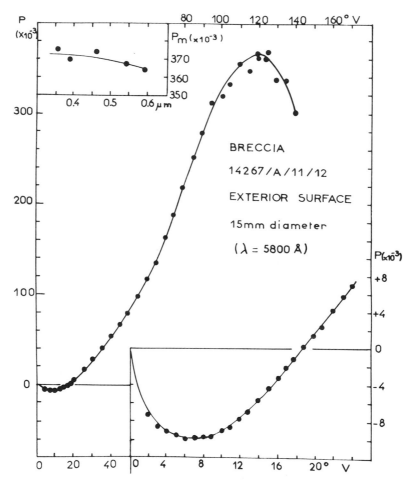

Fig. 2. Polarization curve of a 15 mm diameter region on the outer surface of breccia 14267A. Insets detail the negative branch of polarization and the spectral variation of the maximum degree of polarization (P_m).

The domain in which points for the fines lie is not significantly broader than that given by the lunar telescopic data. This result was not predicted since: (1) telescopic observations take in at least 20 km² of lunar terrain, whereas laboratory sample areas are of the order of 0.1 cm²; evidently the average polarization of a large number of types of material is being observed at the telescope; (2) the domains defined by suites of terrestrial rocks are much more extensive; for example, crushed basalts show a dispersion of up to 0.3 in log A for a given P_m, and extend over greater ranges in A and P_m than do the lunar fines (see Fig. 9 in Dollfus and Titulaer, 1971).

Apollo 14 points i, j, k all lie in the "intermediate" portion of the lunar domain; this corresponds well with the overall optical properties of the Fra Mauro region.

Figure 5 (adapted from Fig. 12, Dollfus et al., 1971b) shows the spectral variation

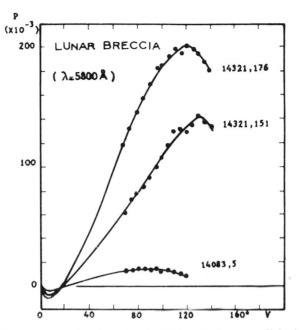

Fig. 3a. Polarization curves for three Apollo 14 breccias in orange light (curves interpolated between $V = 25°$ and $V = 60°$).

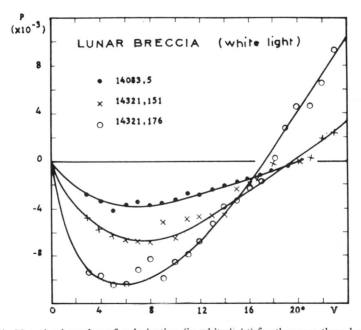

Fig. 3b. Negative branches of polarization (in white light) for the same three breccias.

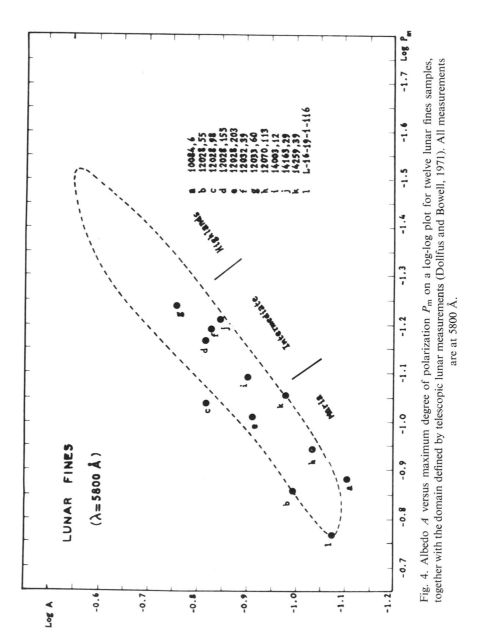

Fig. 4. Albedo A versus maximum degree of polarization P_m on a log-log plot for twelve lunar fines samples, together with the domain defined by telescopic lunar measurements (Dollfus and Bowell, 1971). All measurements are at 5800 Å.

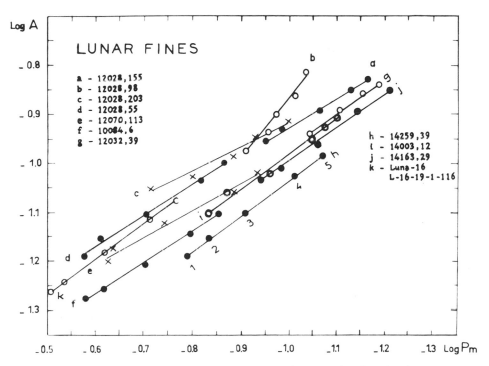

Fig. 5. Albedo versus P_m on a log-log plot for the fines samples given in Fig. 4. For each sample the measurements in five colors lie roughly on a straight line. Numbers on segment *h* refer to wavelengths used: (1) 3540 Å; (2) 3790 Å; (3) 4350 Å; (4) 5200 Å; (5) 5850 Å. Wavelength increases from left to right for the other samples.

of A and P_m for the fines. Most of the segments are parallel and of similar length, indicating high optical opacity and small range in colors. Only core-tube sample 12028,98 departs from this trend. Once more, such great uniformity is not maintained in the polarimetric properties of suites of particulate terrestrial minerals, where spectral variation in optical opacity is encountered (Dollfus and Titulaer, 1971).

Figure 6 shows lunar measurements (data adapted from Dollfus and Bowell, 1971, and Adams and McCord, 1971) for a region in Mare Serenitatis (low albedo) and a region centered on Tycho (high albedo). The segment lengths and slopes agree very well with those for the fines (also plotted).

Figure 7 presents results of albedo measurements as a function of wavelength for 12 fines samples. Figure 8 presents, for the same samples, the maximum degree of polarization P_m as a function of wavelength.

Rocks and breccias

Measurements on rocks 10020,42, 12002,102, and 12051,51 and the breccia 10059,36 were reported by Dollfus *et al.*, 1971b. In all cases P_m exceeded 300×10^{-3}, and albedos ranged from about 10 to 25%. We have increased the data available by measuring four Apollo 14 breccias.

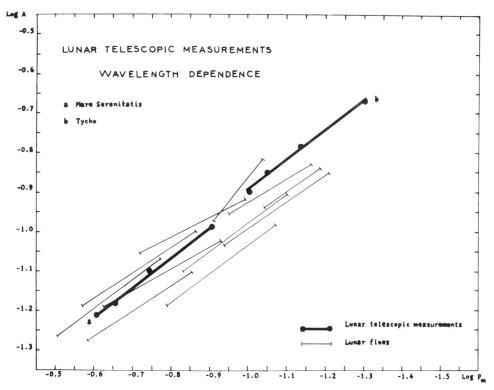

Fig. 6. Wavelength dependence of A and P_m for lunar telescopic measurements. Also shown are laboratory measurements on fines samples. Data adapted from Dollfus and Bowell (1971) and Adams and McCord (1971).

Figure 9 is a log-log plot of A versus P_m for the rocks and breccias. Segments show the spectral variation of log A and log P_m; the left end of each segment corresponds to a measurement made at 3540 Å, and the right end (with symbol) to 5800 Å. Domains indicating the positions of lunar telescopic and terrestrial mineral measurements are also shown for 5800 Å.

The measurements at 5800 Å reveal that, excepting 14083,5, the optical characteristics of rocks and breccias are similar to those of crushed, coarse-grained terrestrial rocks (gabbros and granites). All polarize much more strongly than lunar fines of corresponding albedo, and hence a difference of surface texture is indicated. The consolidated surfaces of the rocks and breccias are generally morphologically less complex than those of the fines, having many large (compared to the wavelength of light) facets that specularly reflect light at all phase angles. This specular component is strongly (positively) polarized. Few consolidated lunar samples possess surface textures comparable in complexity with those of the fines, and in no case does the positive branch of the polarization curves accord with either the fines or the lunar telescopic data.

As expected, the lengths and slopes of segments (relating to the sample's color,

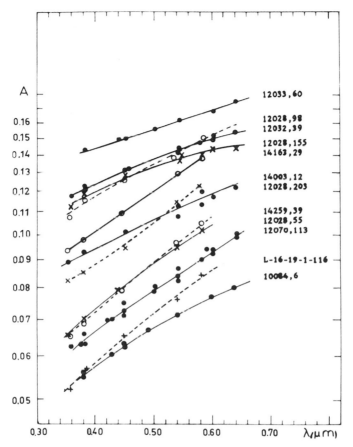

Fig. 7. Albedo (at $V = 5°$) versus wavelength for 12 lunar fines samples. The albedo is plotted on a logarithmic scale.

refractive index, and variation of optical opacity with wavelength) vary much more widely than for the fines (Fig. 5). Such a range is typical of terrestrial coarse-grained rocks (Dollfus and Titulaer, 1971). The segment for the light-hued breccia 14083,5 falls in the domain defined by terrestrial ignimbrites; it has polarization characteristics similar to a Chilean rhyolitic ignimbrite, JG/C354 (see Dollfus *et al.*, 1971a, Table 1), though the spectral variation in P_m for this ignimbrite is somewhat less owing to the presence of partly devitrified glassy shards.

It may be remarked that the two chips from 14321, a 9 kg coarse-grained breccia from Station C1, have almost parallel horizontal segments of similar lengths; this indicates that although variations in albedo and maximum polarization occur from place to place in this complex sample, the optical parameters relating to color and opacity characterized by the log A − log P_m segments do not vary greatly. We are currently making localized measurements to see whether this holds true for small regions on other rocks and breccias.

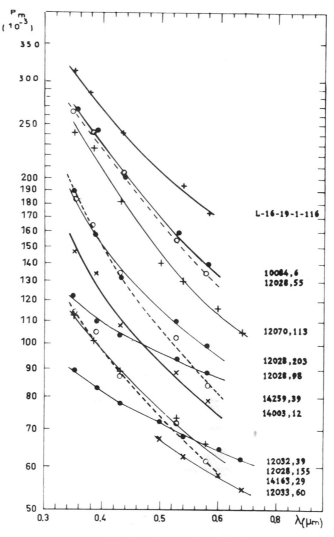

Fig. 8. Maximum degree of polarization (P_m, on a logarithmic scale) versus wavelength, for lunar fines samples.

The Negative Branch

Lunar telescopic measurements

Telescopic measurements show that the moon's polarization curve possesses a negative branch for small phase angles. At $V = 0°$ the degree of polarization is very nearly zero, and descends to about -12×10^{-3} at $V = 10°$. Thereafter it rises and becomes positive at a phase angle of about 22°. Lyot (1929) discovered this phenomenon. He was able to ascertain that all features on the moon possess identical polarization characteristics at small phase angles.

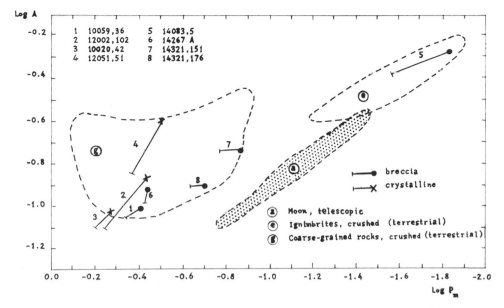

Fig. 9. Log A − log P_m plot for lunar rocks and breccias. Segments show spectral variation in A and P_m: left end refers to 3540 Å; right end (with symbol) to 5800 Å. Domains
for lunar telescopic and terrestrial mineral measurements are shown for 5800 Å.

Numerous observations on small lunar regions were later made by, among others,
Gehrels *et al.* (1964), Pellicori (1969), and Dollfus and Bowell (1971). All results
substantiate Lyot's work and show additionally that the form of the negative branch
is identical over a considerable wavelength range, except that the inversion angle
(for which $P = 0$) increases from about $V = 20.7°$ at 3300 Å to 26.1° at 10500 Å.
The minimum degree of polarization P_{min} remains constant at -12×10^{-3}. Figure
10 shows the lunar negative branch defined by telescopic measurements made by
Lyot (1929) and the present authors. The accuracy of measures, indicated by the
dispersion of points, is generally better than 1×10^{-3}

TERRESTRIAL MINERAL SAMPLES

The degree of polarization exhibited by a substance oriented and illuminated in
a specified way is not usually amenable to calculation. In certain cases the formulas of
Fresnel for the polarization due to a plane dielectric surface and the theory of scattering due to Mie may be employed, but the surfaces of most consolidated and particulate materials are far too complex to admit of theoretical treatment at present.

Two critical experiments demonstrate the conditions necessary for the formation
of a negative branch of polarization: (1) Carbon particles ascending from a smoky
flame were strongly horizontally illuminated with an arc lamp. Viewed in the horizontal plane it can be taken that the arc lamp is the principal source of illumination.
The degree of polarization as a function of phase angle almost obeyed the molecular
scattering law $P = \sin^2 V/(1 + \cos^2 V)$. Maximum polarization was almost total

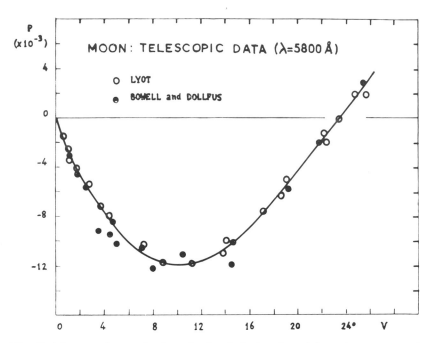

Fig. 10. The moon's negative branch of polarization from telescopic measurements.
Data from Lyot (1929) and Dollfus and Bowell (1971).

and occurred at $V = 90°$. When similar carbon particles were deposited on a metal plate, they exhibited a very pronounced negative branch, for which $P_{min} = -19 \times 10^{-3}$ at $V = 4°$ and $V(0) = 9°$. Experiments using other samples confirm this finding provided that the grains are small ($\lesssim 1$ mm). (2) When blue crystals of $CuSO_4$ were illuminated with red light a well-developed negative branch was measured; but when illuminated with white light and observed through a cell containing $CuSO_4$ solution, the negative branch was completely absent. $CuSO_4$ is highly opaque in the red and partially transparent in the blue.

The negative branch therefore results from the juxtaposition of small grains (experiment (1)) which must be opaque (experiment (2)). From these and other results (see, for example, Dollfus, 1956), we infer that the negative branch owes its origin to multiple scattering between touching submicron sized particle facets.

Great variety in the extent of the negative branch is exhibited by terrestrial minerals (Lyot, 1929; Dollfus, 1956). Such diversity is not reminiscent of telescopic observations of the lunar surface (Dollfus and Bowell, 1971).

Consolidated and partially transparent particulate materials possess negative branches which vary with the inclination of the sample with respect to the plane of vision, in contrast with the orthotropic behavior of the lunar surface. Particulate materials with opaque submillimeter-sized grains may show negative polarization independent of the macrosurface inclination, because light is reflected and scattered more or less orthotropically at small phase angles. Particle size, optical opacity and surface roughness are the principal parameters that control the form of the negative branch.

It was shown in Parts 2 and 3 of this series of papers (Dollfus *et al.*, 1971a; Dollfus and Titulaer, 1971) that the best terrestrial simulators of the lunar negative branch are crushed basaltic rocks and volcanic ashes. These generally contain highly opaque rocks and are capable of forming the complex so-called "fairy castle" structures of lunar fines when reduced to small ($\lesssim 25 \mu$) grain sizes.

Fines

Negative branches for two fines samples have been reported previously (10084,6 in Fig. 7, Geake *et al.*, 1970; and 12070,113 in Fig. 20, Dollfus *et al.*, 1971b). Both showed negative polarization characteristics similar to the lunar telescopic data within the limits of experimental error.

For laboratory measurements great care was taken to reproduce the rough "open" structure that the fines must have on the lunar surface (see description in Dollfus *et al.*, 1971b, p. 2288 *et seq.*). The effect of flattening and compressing a fines sample is to change the minimum degree of polarization P_{min} from -12×10^{-3} to about -9×10^{-3}, and to reduce the inversion angle $V(0)$ from 22° to about 19°; P_m and A are also affected (Geake *et al.*, 1970, p. 2140).

We have since measured one Luna 16 (L-16-19-1-116) and three Apollo 14 (14003,12; 14163,29; 14259,39) fines samples, measurements for which are plotted in Figs. 1 and 16 (Luna 16) and Fig. 11 (Apollo 14). For most samples it is sufficient to make detailed measurements of the negative branch in only one color. In other colors we generally make a single polarization measurement at $V(P_{min})$ (the phase angle at which P_{min} occurs) since, if there is no wavelength variation in P_{min}, it can be taken that the entire negative branch is wavelength independent. Once again there is very little difference between the negative branches of the fines and that given by lunar telescopic data (Fig. 10). For the purposes of numerical comparison, three parameters may be used to characterize the form of negative branches: P_{min}, $V(P_{min})$, and $V(0)$. Table 2 collects results for the fines and shows the average lunar values of these quantities.

Averages of P_{min}, $V(P_{min})$, and $V(0)$ for the fines agree very well with the lunar values, though some individual samples do depart slightly: 12070,113 and 14259,39 certainly have less developed negative branches, and L-16-19-1-116 shows slightly

Table 2. Negative branches of fines (white light).

Sample	P_{min} $(\times 10^{-3})$	$V(P_{min})$ (deg)	$V(0)$ (deg)
10084,6	-11.5	10	22
12070,113	-10	10	21.5
L-16-19-1-116	-11.5	8	20
14003,12	-12.5	9	22
14163,29	-12	9.5	22
14259,39	-10.5	9	20
Average fines	-11	9.5	21.5
Average moon	-12	10	22

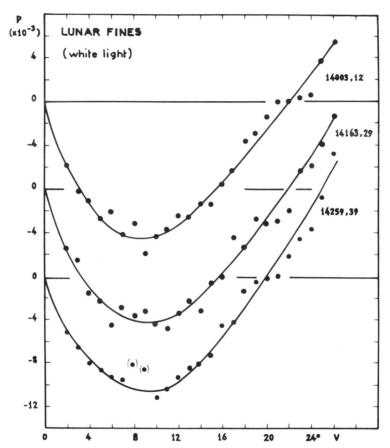

Fig. 11. Negative branches for Apollo 14 fines 14003,12; 14163,29; and 14259,39 in white light.

low $V(P_{min})$. The telescopic negative branch forms an envelope to the laboratory fines measurements, suggesting that the lunar surface bulk density is less than can be obtained in laboratory simulation. However, the uniformity of all these values when compared with a suite of terrestrial rocks is quite striking.

Rocks and breccias

Geake *et al.* (1970) and Dollfus *et al.* (1971b) have reported measurements on five rocks and one breccia. The forms of the negative branches varied greatly between samples.

Laboratory measurement is made difficult by the inhomogeneity of most sample surfaces; generally we have measured 0.1 cm² regions, and have attempted to choose those which are representative of the entire surface.

In the present study we have measured three breccias. Of one, 14321, we were allocated two chips which we have treated independently. 14267A, the European

Consortium breccia of 14 g, is the only sample having a carefully preserved exterior surface, and we have been able to make detailed measurements on localized regions. The third, 14083,5, is an extremely friable light-hued sample with no indication as to exterior surface. Negative branches of polarization in white light are shown for 14083,6, 14321,151, and 14321,176 in Fig. 3b, and for 14267A/11/12, a 15 mm diameter region taking in most of the breccia's exterior surface, in Fig. 2.

We find once more that negative branches differ widely from sample to sample and also from point to point on a sample. In no case does a negative branch exactly resemble those of the fines or of the telescopically observed moon; all are less developed.

Complete data for negative branches of consolidated materials, including regional studies of 14267A, are grouped in Table 3.

Scanning electron microscope (SEM) studies

Preliminary SEM studies (Geake et al., 1970; Dollfus et al., 1971b) were aimed at relating the physical nature of sample surfaces to the form of the negative branch of polarization. It appeared, en gros, that the more complex the surface appearance the more developed was the negative branch. This is the result expected from polarization studies of terrestrial minerals.

We report here further results relating to regions localized on the breccia 14267A and to Luna 16 fines L-16-19-1-116. Figs. 12 to 16 show stereoscan photomicrographs of these samples, together with the measured negative branches of polarization and spectral variation of albedo where available. Commentary on these is as follows:

Breccia 14267A. Figure 12 shows the aspect of the outer surface of breccia 14267A at center, with aluminum pointers and region locations. Surrounding photographs show typical low-magnification stereoscan views of all regions investigated together with a subjective region description. All regions show a superficial coating of small particles.

Figure 13 refers to the dust-covered region *a*. The mosaic (top) shows densely agglomerated small particles. Complex surface structures are built up by interfacial adhesion, and most individual grains are rough on a micron scale. The negative branch of polarization descends to -11×10^{-3} near $V = 9°$; inversion angle

Table 3. Negative branches of rocks and breccia (white light).

Sample	Type	P_{min} ($\times 10^{-3}$)	$V(P_{min})$ (deg)	$V(0)$ (deg)
10020,42	Rock	-6	5	12
10057,54	Rock	-8	8	20
10058,37	Rock	-0.5	4	8
12002,102	Rock	-2	5	13
12051,51	Rock	-4	7	16
10059,36	Breccia	-11	8	21
14083,5	Breccia	-4	7	20
14267A/11/12	Breccia	-10	7	19
14267A/11/c,e,f,i	Breccia	-4	10	18
14267A/11/h	Breccia	-4	4.5	13.5
14267A/11/a	Breccia	-11	8	20.5
14321,151	Breccia	-7	7	19.5
14321,176	Breccia	-10.5	6	17.5

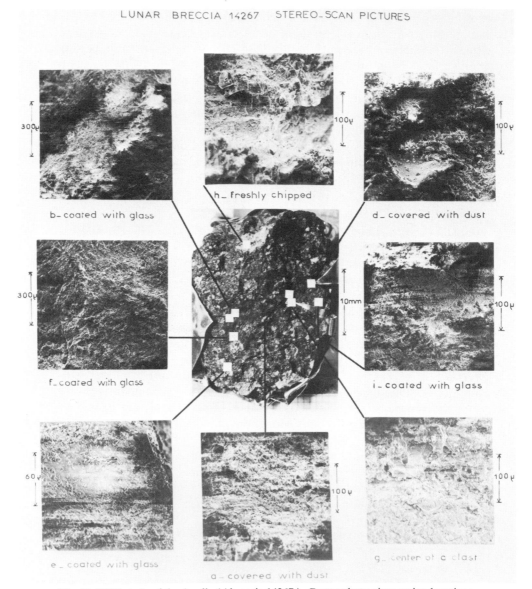

Fig. 12. SEM study of the Apollo 14 breccia 14267A. Center photo gives region locations.
Surrounds show typical photomicrographs of regions studied.

$V(0) = 20°$. Telescopic lunar values of -12×10^{-3}, $10°$, and $23°$, respectively, indicate that region a quite closely resembles the large-scale lunar surface; this is due to the dust covering. If the dust were removed, the underlying surface would not show moon-like polarization.

Figure 14 shows regions c, e, f, i: glass-coated. The top four views at high magnification show small particles lying on a relatively smooth surface; c and e show

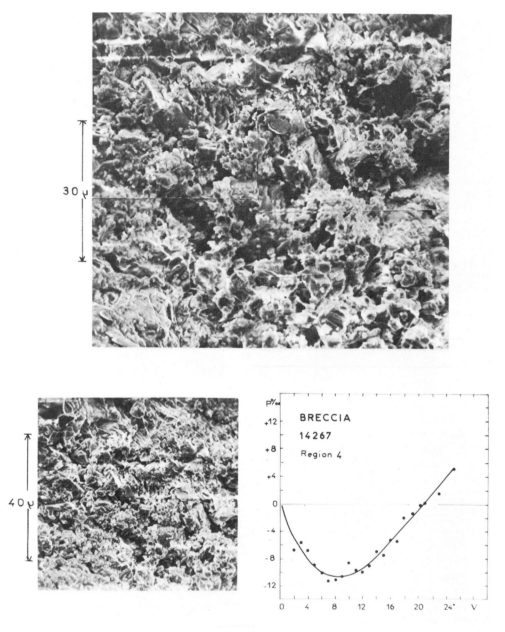

Fig. 13. SEM views of 14267A, dust covered region *a*. Mosaic (top) is centered on view in Fig. 12. At bottom is the negative branch of polarization.

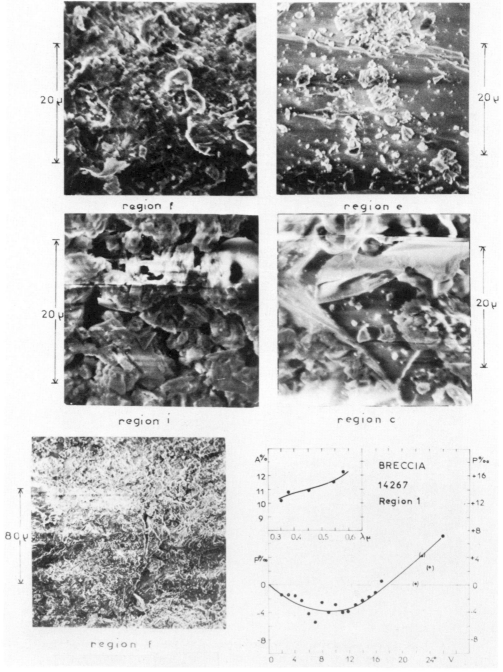

Fig. 14. SEM views of 14267A, glass coated region *f* (see also Fig. 12). Four top views are typical, showing varying amounts of overlying fine particles. At bottom is the negative branch with spectral variation of albedo inset.

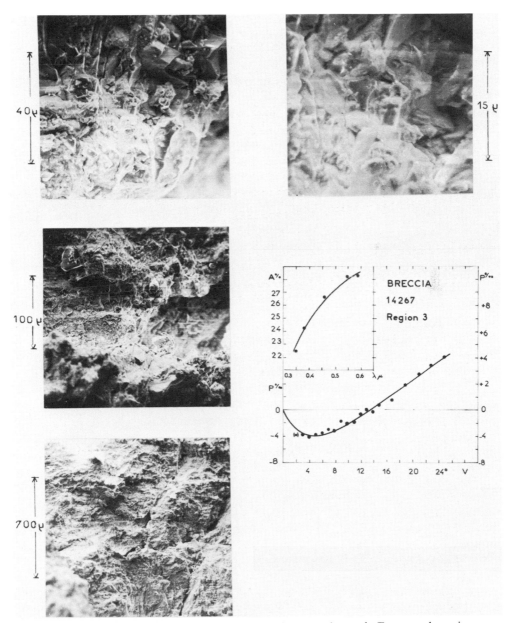

Fig. 15. SEM views of 14267A, chipped freshly exposed area *h*. Fracture planes in crystalline material are well shown, and there are few overlying particles. Negative branch is very unlike the moon's.

individual particles on a very smooth glassy surface; i is an agglomeration of relatively smooth particles. The negative branch of a glass-coated region is quite unlike the moon's. It descends only to $P_m = -4 \times 10^{-3}$ at $V = 9°$, with $V(0) = 17°$. The effect of multiple scattering is much reduced in this instance since complex particle structures are few or absent, and the polarization curve approaches that of a plane dielectric material.

Figure 15 shows region h; here a portion of the outer surface has been chipped off, revealing a freshly exposed interior. Stereoscan views show a fractured solid surface sparsely covered with small particles. The negative branch shows that such a surface cannot reproduce the overall lunar polarization characteristics.

Luna 16, L-16-19-1-116. Figure 16 shows views of this core-tube fines sample, taken from a depth of 20 to 22 cm. Once more, complex interfacial adhesion and clumping of small particles give rise to sufficient multiple scattering to produce a well-developed negative branch which accords with lunar telescopic measurements. The inset albedo curve (see also Fig. 1) is mare-like.

Conclusions and Discussion

Our principal results may be summarized as follows:

Fines

(1) Polarization characteristics of fines material in all cases agree very well with telescopic measurements on 50 km^2 or larger lunar regions.

(2) The range in maximum polarization and albedo is very much more restricted than that of selected terrestrial simulators.

(3) The form of the negative branch is moon-like for all fines samples measured; significant variations do not occur. Microstructure revealed by SEM studies is as predicted by previous laboratory studies of terrestrial minerals.

Rocks and breccias

(1) Hardly any rock or breccia surfaces have a moon-like polarization curve. Most polarize much too strongly at large phase angles.

(2) SEM studies show that most rocks and breccias have smooth microsurfaces relative to the fines material. These are responsible for the poorly developed negative branches.

The similarity between the negative branches of all fines samples measured and the agreement with the lunar telescopic results is quite clear. No fines sample possesses a negative branch more developed than that of the moon, and this suggests that the polarimetric properties of large-scale lunar terrains represent a limiting case, corresponding to a ubiquitous complex structure of highly absorbent, fine particles.

If *exposed* rock or breccia surfaces were abundant anywhere on the moon at a scale measured by telescopes, then the negative branch of polarization would be weakened. We have searched for variations in P_{min} by scanning large areas of the moon near $V(P_{min})$ with the photoelectric polarimeter attached to the Meudon 60 cm reflector.

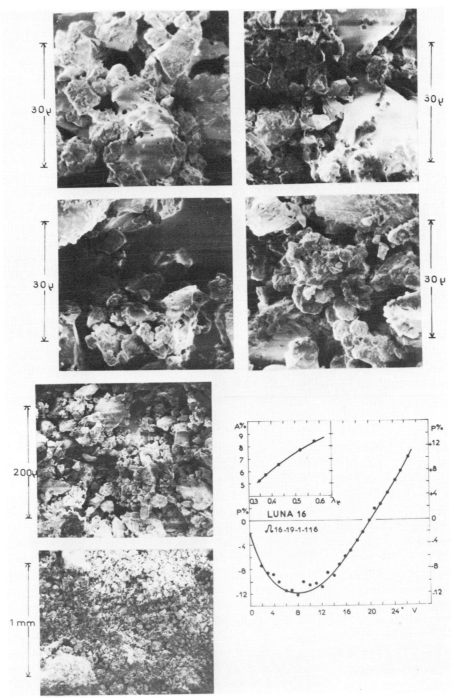

Fig. 16. SEM views of Luna 16 fines L-16-19-1-116. Negative branch and spectral variation of albedo are typical of lunar maria.

No definite variation was detected. There are several possible explanations of this result: (1) the exposed surfaces of rocks and breccias comprise a small proportion of the regolith all over the moon (Bowell, 1971); (2) rocks and breccias in extended fields ($\gtrsim 100$ km^2) are largely covered with fines, or (3) are altered by space exposure so as to have a very rough microtexture and exhibit fines-like polarization. Our laboratory data do not support this last possibility.

The great uniformity in the polarimetric properties of fines must come about by alteration on the lunar surface (even limited suites of terrestrial minerals show far greater diversity). Fragmentation of parent rocks by meteoroidal bombardment may suffice to produce the observed grain-size distribution in lunar fines, but it alone cannot give rise to the observed polarimetric properties (Dollfus et al., 1971a). In particular, the range in albedo and P_m would be too great. We have examined the polarimetric implications of three processes put forward to explain the uniformity of the regolith. Each of these processes could lead to polarization properties according with our telescopic polarization measurements.

Maurette (Borg et al., 1971) has shown the presence of metamictized coatings on fines particles, and has suggested that a "simple amorphous layer on a silicate grain, composed of the same material as the grain itself, will have a smaller index of refraction and could act to trap incident radiation by internal reflection." Evidently, such a process might make particles optically opaque, limit the range in the albedo, and impart them with moon-like polarization. Moreover, calculation for a semi-infinite slab of crystalline material coated with a metamictized layer of about 350 Å thickness qualitatively reproduces the spectral variation in lunar albedo (Bibring et al., 1972). In collaboration with M. Maurette, we have made measurements on samples whose surfaces have been etched in HF, and these show that the albedo rises as the amorphous coating is removed.

Hapke et al. (1971) suggest that fines particles are coated by sputtering deposition due to solar wind bombardment, and possibly by low vaporization-point materials escaping from below the lunar surface. The effect is to reduce the albedo (see also Gold et al., 1971). Measurements on proton-bombarded fines samples (Dollfus et al., 1971) showed that two out of four samples darkened and had altered polarization characteristics, but the other two samples were optically unaffected. It was pointed out that if proton irradiation alters the lunar surface, the effect may vary from place to place and be complicated by other effects, such as heating and uv bleaching.

Adams and McCord (1972) find that both the albedo and color of fines are dominated by the proportion of glass present: the more the glass the lower the albedo. Glass, more or less dark and opaque according to its iron and titanium content, is formed in fines by meteoroid impact vitrification, so that the regolith progressively darkens with time. Three crushed synthetic glasses, prepared from melted and vacuum upwelled terrestrial basalts (Fielder et al., 1967) were polarimetrically examined (Dollfus et al., 1971a, Fig. 12). These had maximum polarization up to five times the lunar values in orange light and albedos less than 5%, for all grain-size fractionations greater than 60 μ. For a mean particle size of 25 μ, however, maximum polarization and albedo abruptly changed to moon-like values corresponding to maria or "intermediate" terrains (of albedo intermediate between maria and highlands). Calculations of maximum polarization and albedo for mixtures of these glasses (grain size less than 160 μ diameter) and suites of particulate opaque terrestrial rocks, assuming that the surface area of each component is proportional to its abundance, show that lunar polarization characteristics can be reproduced (Bowell, 1971, Fig. 12). Moreover, admixture of such glasses to light-hued material is a means of limiting the range of maximum polarization and albedo provided that the glasses are not so finely divided as to have high albedo. Our calculations must be regarded as somewhat schematic, for if very small glass particles adhere to the light-hued component (as suggested by Adams and McCord, 1971), they will dominate the mixture. In this case an even greater reduction in the range of maximum polarization and albedo is to be expected. If this mechanism of mixing is applicable to the entire lunar surface, then it is unnecessary to invoke solar wind irradiation to explain the restricted range of polarimetric properties (Nash and Conel, 1971; Adams and McCord, 1971, reached this conclusion from reflectance studies).

Acknowledgments—We thank M. Maurette for the loan of Apollo 14 samples and G. Eglinton and S. O. Agrell for their arrangement of the European Consortium meetings for the Apollo 14 breccia 14267A. We are also grateful to the Textile Technology Department, UMIST, for the use of their stereoscan microscope and to B. Lomas for operating it. Technical assistance was provided at Meudon by J. Advielle, F. Berling, G. Colin-Morel, F. Colson, J. Cosme, and M.-T. Leveque. Supporting grants were given by ESRO (E.B.) and the Science Research Council (J.E.G.).

References

Adams J. B. and McCord T. B. (1971) Optical properties of mineral separates, glass, and anorthositic fragments from Apollo mare samples. *Proc. Second Lunar Sci. Conf., Geochim. Cosmochim. Acta* Suppl. 2, Vol. 3, pp. 2183–2195. MIT Press.

Adams J. B. and McCord T. B. (1972) Optical evidence for regional crosscontamination of highland and mare soils (abstract). In *Lunar Science—III* (editor C. Watkins), pp. 1–3, Lunar Science Institute Contr. No. 88.

Bibring J. P., Maurette M., Meunier R., Durrieu L., Jouret L., and Eugster O. (1972) Solar wind implantation effects in the lunar regolith (abstract). In *Lunar Science—III* (editor C. Watkins), pp. 71–73, Lunar Science Institute Contr. No. 88.

Borg J., Maurette M., Durrieu L., and Jouret C. (1971) Ultramicroscopic features in micron-sized lunar dust grains and cosmophysics. *Proc. Second Lunar Sci. Conf., Geochim. Cosmochim. Acta* Suppl. 2, Vol. 3, pp. 2027–2040. MIT Press.

Bowell E. (1971) Polarimetric studies. In *Geology and Physics of the Moon* (editor G. Fielder), Chap. 9, pp. 125–133, Elsevier.

Dollfus A. (1956) Polarisation de la lumière renvoyée par les corps solides et les nuages naturels. *Ann. Astrophys.* **19**, 83–113.

Dollfus A. and Bowell E. (1971) Polarimetric properties of the lunar surface and its interpretation, Part 1. Telescopic observations. *Astron. Astrophys.* **10**, 29–53.

Dollfus A. and Titulaer C. (1971) Polarimetric properties of the lunar surface and its interpretation, Part 3. Volcanic samples in several wavelengths. *Astron. Astrophys.* **12**, 199–209.

Dollfus A., Bowell E., and Titulaer C. (1971a) Polarimetric properties of the lunar surface and its interpretation, Part 2. Terrestrial samples in orange light. *Astron. Astrophys.* **10**, 450–466.

Dollfus A., Geake J. E., and Titulaer C. (1971b) Polarimetric properties of the lunar surface and its interpretation, Part 4. Apollo 11 and Apollo 12 lunar samples. *Proc. Second Lunar Sci. Conf., Geochim. Cosmochim. Acta* Suppl. 2, Vol. 3, pp. 2285–2300. MIT Press.

Fielder G., Guest J. E., Wilson L., and Rogers P. (1967) New data on simulated lunar material. *Planet. Space Sci.* **15**, 1653–1666.

Geake J. E., Dollfus A., Garlick G., Lamb W., Walker G., Steigmann G., and Titulaer C. (1970) Luminescence, electron paramagnetic resonance and optical properties of lunar material from Apollo 11. *Proc. Apollo 11 Lunar Sci. Conf., Geochim. Cosmochim. Acta* Suppl. 1, Vol. 3, pp. 2127–2147. Pergamon.

Gehrels T., Coffeen D., and Owings D. (1964) The wavelength dependence of polarization, III. The lunar surface. *Astron. J.* **69**, 826–852.

Gold T., Campbell M. J., and O'Leary B. T. (1971) Optical and high-frequency electrical properties of the lunar sample. *Proc. Apollo 11 Lunar Sci. Conf., Geochim. Cosmochim. Acta* Suppl. 1, Vol. 3, pp. 2149–2154. Pergamon.

Hapke B. W., Cohen A. J., Cassidy W. A., and Wells E. N. (1971) Solar irradiation effects on the optical properties of Apollo 11 samples. *Proc. Apollo 11 Lunar Sci. Conf., Geochim. Cosmochim. Acta* Suppl. 1, Vol. 3, pp. 2199–2212. Pergamon.

Lyot B. (1929) Recherches sur la polarisation de la lumière des planètes et de quelques substances terrestres. *Ann. Obs. Paris* **8**, No. 1; Research on the polarization of light from planets and from some terrestrial substances. *NASA Tech. Transl.* TT F-187, 1964.

Nash D. B. and Conel J. E. (1971) Luminescence and reflectance of Apollo 12 samples. *Proc. Second Lunar Sci. Conf., Geochim. Cosmochim. Acta* Suppl. 2, Vol. 3, pp. 2235–2244. MIT Press.

Pellicori S. F. (1969) Wavelength dependence of polarization, XIX. Comparison of the lunar surface with laboratory samples. *Astron. J.* **74**, 1066.

Proceedings of the Third Lunar Science Conference
(Supplement 3, *Geochimica et Cosmochimica Acta*)
Vol. 3, pp. 3127–3142
The M.I.T. Press, 1972

Lunar surface properties as determined from earthshine and near-terminator photography

D. D. LLOYD and J. W. HEAD

Bell Telephone Laboratories, Inc.,
955 L'Enfant Plaza North, S.W., Washington, D.C. 20024

Abstract—*First earthshine photography.* During Apollo 15, the first earthshine photographs were taken from lunar orbit. The photographs are of photometric interest, particularly as they involve double reflection of sunlight, by the earth, then the moon, prior to photographic exposure. The apparent albedo values obtained for the floor of the crater Aristarchus were anomalously higher than those obtained of the surrounding maria.

Preliminary analysis suggests that the special earthshine illumination conditions make possible new determinations of certain photometric properties of geological areas. New brightness and contrast patterns not seen before include bright and dark bands that are oriented radially on the walls and rim of the crater Aristarchus. The relative brightness of a portion of the crater rim on the eastern side of Aristarchus compared to its relative brightness as shown by direct sunshine photography suggests a physical properties explanation.

Near-terminator photography. A sequence of four photographs was obtained on Apollo 14 using high-speed film that permitted photography within $\frac{1}{2}°$ of the terminator. It is possible to recognize small (considerably less than $\frac{1}{2}°$) variations in a slope of those portions of the surface that are nearly tangential to the local sun. Many geological features stand out in a manner that is not normal in conventional lunar photography.

During Apollo 15 ten sequences of photographs were taken near the terminator using high-speed film.

The preliminary scientific results apply to three general areas:

(1) A sequence of photographs obtained in Mare Vaporum showing mare structure in detail.

(2) Four sequences in the Aristarchus plateau area and in Oceanus Procellarum area west of the Aristarchus plateau.

(3) Five sequences in highland areas.

Selected pictures are reproduced and discussed.

FIRST EARTHSHINE PHOTOGRAPHY

ON REV 34 OF THE APOLLO 15 MISSION the first earthshine pictures of the moon were taken from lunar orbit. A series of 15 photographs was taken, and Fig. 1 shows the location of the four of them which had reasonable photographic images. Most significant are two photographs (Figs. 2 and 3) of Aristarchus (AS-15-101-13591 and AS-15-101-13592). Another photograph (Fig. 4) provides a view of Schroeter's Valley (AS-15-101-13593), and the fourth (Fig. 5) shows the crater Herodotus (AS-15-101-13594). The other photographs were of maria and were significantly underexposed.

A 35-mm Nikon camera with an *f* number of 1.2 was used. (This camera was required by the Gegenschein experiment, S-178.) The low *f* number lens provided more capability for low light level photography than had been available on earlier missions.

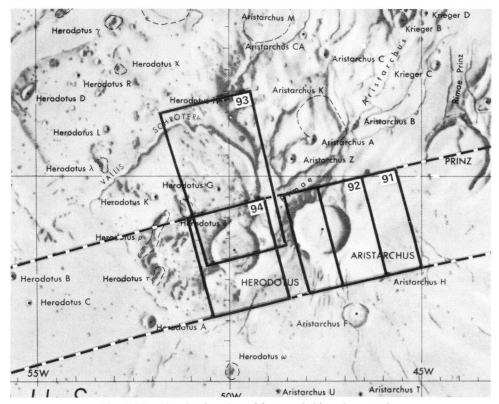

Fig. 1. Map showing location of four earthshine photographs.

The film selected and used for all photographs discussed in this paper was EK 2485, a high-speed black and white recording film (Eastman Kodak, 1970). The predicted response characteristic curve ($D/\log E$ curve) for this film, as used for preflight exposure estimation, is shown in Fig. 6.

The earthshine photographs were taken on 1 August 1971 at 1345 GMT. At that time the moon, earth, sun relationship was as shown in Fig. 7. The eastern limb of the moon is in sunlight; the subsolar point is at $+60.6°E$. The target is in earthshine. Earthshine comes from the portion of the earth that is both sunlit and visible from the moon.

The primary factor that determines the magnitude of earthshine illumination incident on the moon is the angle between the sun-earth and the moon-earth vectors. This angle can be called the moon-earth phase. The moon-earth phase affects the extent of the sunlit earth visible from the moon. For a subsolar point of $+60.6°E$ and a libration such that the earth is at $-6°W$, the moon-earth phase angle is $180° - 66.6°$, or $113.4°$. At large moon-earth phase angles the magnitude of the illumination is significantly reduced. For a moon-earth phase angle of $113.4°$, the mean illumination of the moon by the crescent earth is 1.35 lumens/m² as based on premission data and calculations (NASA, 1969).

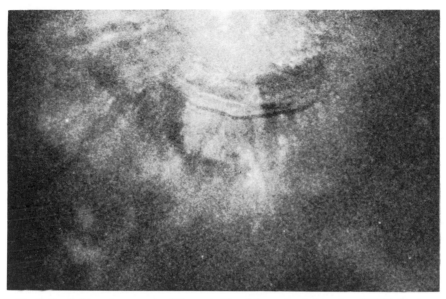

Fig. 2. Earthshine photograph of the east side of the crater Aristarchus (NASA Photograph Number AS-15-101-13591).

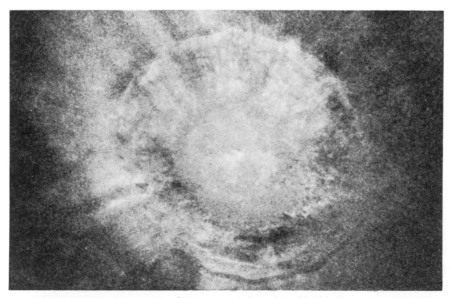

Fig. 3. Earthshine photograph of the crater Aristarchus (NASA Photograph Number AS-15-101-13592).

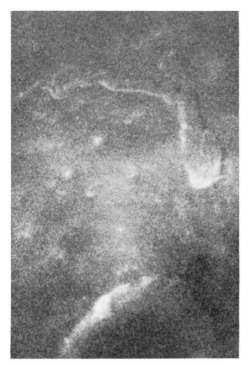 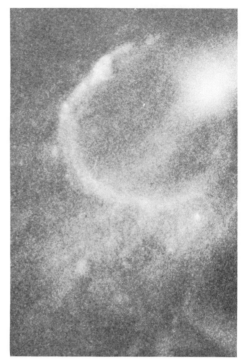

Fig. 4. Earthshine photograph of Schroter's Valley and Cobra Head (NASA Photograph Number AS-15-101-13593).

Fig. 5. Earthshine photograph of the crater Herodotus (NASA Photograph Number AS-15-101-13594).

Albedo data based on photoelectric-photographic measurements from earth are available. The preflight albedo values used for preflight exposure prediction purposes were obtained from Pohn and Wildey (1970). A value of 0.09 was used for the maria area and 0.18 for Aristarchus. A shutter speed of $\frac{1}{8}$ sec produces a slight underexposure of a target of an albedo of 0.18. The predicted density is shown as A on Fig. 6. For an albedo of 0.09, such as the maria, the photograph can be expected to be significantly underexposed. The predicted density is shown as M on Fig. 6.

Results

The density and exposure values obtained for general maria and for the floor of Aristarchus are shown in Fig. 6 (at M and A, respectively).

If the measured exposure values are compared to the predicted values, the following results are obtained:

(1) For the maria area there is reasonably close agreement such that future photographic results can be predicted with reasonable confidence. However, the difference between theory and results is such that it is highly desirable that further analysis be performed to clarify the cause(s) of the difference.

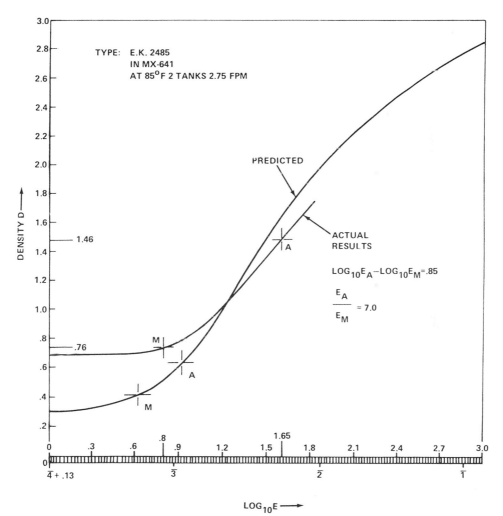

Fig. 6. Comparison of predicted $D/\log E$ Curve for EK 2485 and actual exposures of maria and Aristarchus.

(2) For the floor of Aristarchus the value of exposure obtained is far greater than predicted.

If only the measured exposures are examined, no reference being made to the predicted value, there is a large ratio between the exposure for the floor of Aristarchus and the exposure for the maria. This ratio is about 7, which is a difference in $\log_{10}E$ of about 0.85. This would suggest an "apparent albedo" of 7 × 0.09 or 0.63—a very high albedo value if it can be accepted as such. It should be noted that the measured ratio is independent of any possible errors in estimates of factors used in predicting exposure.

There is another unusual result in the brightness of certain sections of the earthshine photographs. The measurements discussed above suggest that the floor of the

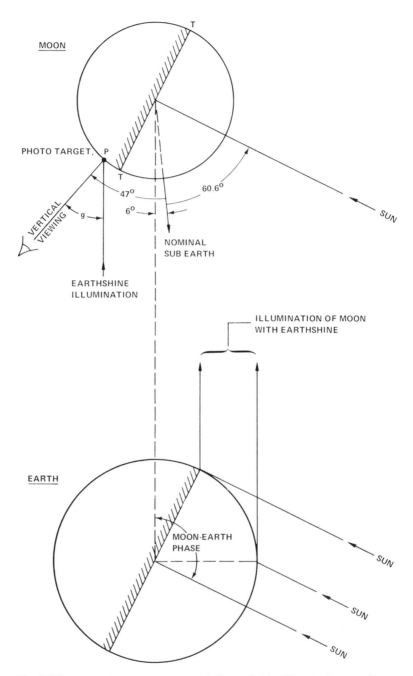

Fig. 7. The sun-earth-moon geometry of the earthshine illumination conditions.

crater Aristarchus has unusual physical properties. We turn next to some unusual albedo patterns in the eastern section of the crater Aristarchus that extend outside the crater to the east.

If we examine Fig. 2 (AS-15-101-13591), the earthshine photograph shows a clearly distinguishable high albedo area located on the western crater wall and rim covering about 40° of the eastern rim and extending eastward. This bright region was also clearly visible to Command Module pilot Al Worden (personal communication).

The area of higher albedo does not stand out in photographs of this area taken under direct sunlight illumination conditions, as illustrated in Fig. 8, a Lunar Orbiter

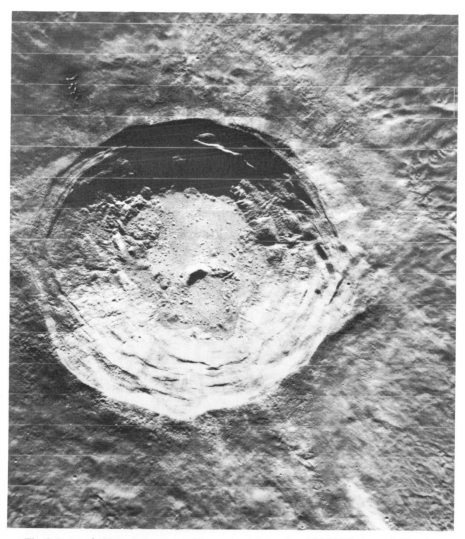

Fig. 8. Lunar Orbiter photograph of the crater Aristarchus (NASA Photograph Number L/O V-198M).

photograph of the crater Aristarchus (LOV-198M). Additional sunlight illumination photographs include an Apollo 15 mapping camera frame covering Aristarchus and the surrounding plateau and maria (AS-15-2609) and a panoramic camera photograph (Fig. 9) looking south toward the east rim of Aristarchus (AS-15-324). The high albedo region seen under earthshine illumination (Figs. 2 and 3) is not seen as a distinctly bright region on these photographs.

Definition and mapping of geologic units associated with the rim of Aristarchus (Head, 1969) has shown that several distinct subunits can be recognized on the crater rim. Surrounding the rim are three sequentially concentric facies. In sequence away from the crater edge these are: (1) a concentric or transverse, dunelike facies immediately surrounding the crater rim (0–8 km) and characterized by a series of concentric dunelike structures of 15–25 m width, dark pools and flow structures, block fields, and a general lack of craters, (2) a radial or longitudinal dune facies (~0–4 km) characterized by radially arrayed dunelike structures transitional between the concentric

Fig. 9. Panoramic camera photograph showing area to the east and north of the crater Aristarchus (NASA Photograph Number AS-15-324).

and rhomboidal facies, and (3) a rhomboidal facies characterized by a coarse and fine rhomboidal pattern formed by barchanlike crescent ridges and finer chevron-shaped ridges whose apices are radially arrayed around and point toward the center of Aristarchus.

Mapping of the rim in the area showing the high albedo in earthshine shows that this region corresponds to the best development of the concentric facies. In this particular area, the dunelike structures are more distinct and less modified by filling of interridge areas. Thus, it appears that the area which has a high albedo under earthshine conditions can be correlated with a geologic subunit defined by distinctive surface morphologic features. This same subunit, however, does not show a distinctly different albedo under sunlight illumination. The possibility also exists that the areas which show high albedo under earthshine may be unmantled by the apparently darker material which fills in the lows between the crater rim ridges, and which partly mantles the crater walls.

The mantle material may be a different rock type or may simply differ from the lighter ejecta material in physical properties. Reference to Fig. 3 and Fig. 9 indicates that the albedo of the predominantly mantled portion of the crater wall (about 330° to 60°) is lower than the dominantly unmantled portion of the crater wall under both earthshine and sunshine illumination. This is not as obvious, of course, for the high albedo portion on the crater rim (Figs. 2 and 9). At the present time, physical property differences possibly relatable to different units of varied geologic origin seem to best account for the earthshine observations of Aristarchus.

There is clearly a need to explain in terms of physical properties not just the differences in albedo in earthshine but also why these differences are not observed under direct sunlight conditions.

NEAR-TERMINATOR PHOTOGRAPHY

For many years, it has been widely accepted that an examination of the lunar surface under near-terminator lighting conditions is extremely valuable to geologists and other scientists. Before the era when photography could be obtained from spacecraft in lunar orbit, a large percentage of the telescopic observations of the moon (both direct and photographic) was conducted when the feature of interest was under near-terminator lighting conditions. When unmanned spacecraft were flown to obtain lunar photographs (e.g., Lunar Orbiter series), the mission parameters were selected so that photographs could be taken near the terminator. One reason is that under near-terminator conditions (low sun elevation), small changes in slope produce greater contrast changes than at high sun-elevation angles. A related desirable phenomenon is that, at low sun elevation, the shadow is longer than the object is high, thus increasing certain information about the object. For example, at 20° the exaggeration is 2.75 (cotan 20°), and detectability and morphologic identification are consequently enhanced.

Historically it has been difficult to obtain photography nearer the terminator than approximately 8° without severe underexposure. The Lunar Orbiter and the Panoramic camera photographs were both optimized for photography at sun elevations of

20° or above when photographing maria (although these cameras can be used to photograph lunar highlands at lower sun elevations because of the higher albedo). Neither camera can produce the desired midrange exposures when operating nearer the terminator than 8°.

Faster films could be used to obtain photography nearer the terminator than 8°, but such fast films entail a resolution penalty which, for most unattended camera systems, would affect all the photography obtained during the mission; that is, the film selected for photography at, for instance, 1° from the terminator would also have to be used for all the other photography.

The ability of an astronaut to change films in the Hasselblad cameras provided an opportunity for use of a very high-speed film. Despite the fact that no image motion compensation is normally available for this camera (although such compensation can be obtained by rotation of the spacecraft), a preliminary set of four photographs from the Apollo 14 mission showed that photography within 0.5° of the terminator could provide lunar photography of special geological interest. The shadow length

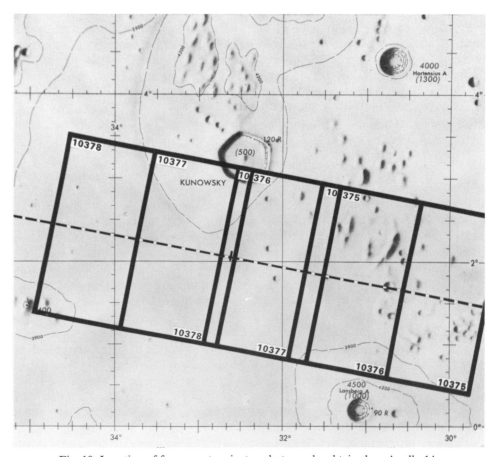

Fig. 10. Location of four near-terminator photographs obtained on Apollo 14.

at 0.5° from the terminator is greater than the height of the corresponding object by a factor of 114.6 (cotan 0.5°). Slight variations (less than 0.5°) in nearly horizontal slopes produced significant variations in scene contrast. The location of the photographs obtained on Apollo 14 is shown in Fig. 10.

During the Apollo 15 mission, 10 sequences of photographs were taken starting a few minutes before crossing of the terminator and continuing past the terminator.

The Hasselblad data camera was used with the 250 mm and 80 mm lenses at the maximum aperture. Photographic operations were conducted by the command module pilot, as specified in the mission requirements document. Ten sequences (sets of approximately six photographs) were taken using magazine R. (This magazine also was used for other photography.)

Discussion

The preliminary scientific results apply to four general areas:

(1) A sequence in Oceanus Procellarum near the crater Kunowsky showing mare flow fronts and craters of internal origin.

(2) A sequence of photographs of the Mare Vaporum area showing mare structure in detail.

(3) Four sequences of the Aristarchus Plateau area and the Oceanus Procellarum area west of the Aristarchus Plateau.

(4) Five sequences of highland areas.

In Oceanus Procellarum near the crater Kunowsky, mare features are particularly enhanced. A single flow front in the mare (center of Fig. 11; 5.5, L-G) can be traced for over 50 km in the area southwest of Kunowsky. This feature is not visible on Lunar Orbiter IV photography of the same area (Fig. 12).

The area of the flow has a lower crater density than the surrounding areas of mare and the low sun elevation enhances the detection of numerous rimless depressions which occur on top of the flow surface (6.5, K.2 and 6.5, I.2). These are interpreted as collapse depressions which are believed to be caused by shrinkage and drainage of the flow material. They range up to 400 m in their longest dimension. On Lunar Orbiter photography these are indistinguishable from raised-rim craters which are interpreted to be of impact origin.

The flow front itself ranges from 3 to 5 m in height and is extremely sinuous in nature (Fig. 11). This height lies in contrast to flow-front elevations in the well known Imbrium Basin flow lobes. There are flow-front scarps ranging from 50 to 150 m in elevation, rather than the 3–5 m seen here. Obviously, the low sun angle considerably enhances the detection of these small flow fronts, and the near-terminator photography provides important data for working out stratigraphic relationships and for understanding the processes of mare deposition.

Crater rims are enhanced in the near-terminator photography, and rimless craters are particularly evident. The ease with which the lack of rims can be detected is shown by the contrast between the two craters to the south-east of Kunowsky (Fig. 11; 4.2, K.3 and 4.7, L.3). Each is approximately 1 km in diameter, but the one to the south has an upraised rim which is 50 m above the mare level. In striking contrast, the crater to the north, although it is sharp rimmed, has no raised rim and thus appears to be of internal origin.

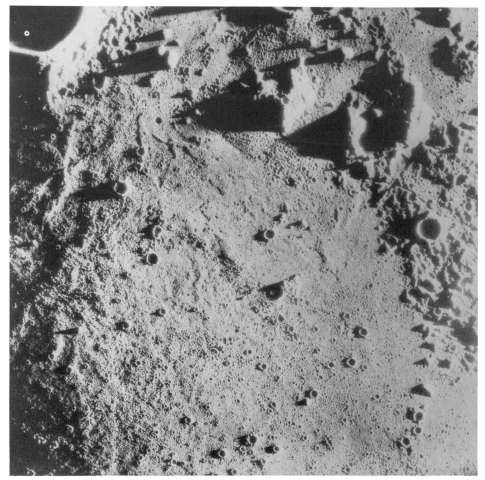

Fig. 11. Near-terminator photography of the Kunowsky region (2°N and 32.5°W). Frame boundaries are approximately 80 km (NASA Photograph Number AS-14-78-10376).

Variations in crater density in mare units are seen in Fig. 11 (8.5, K.5 and 6.5, K.0), as are flow features associated with mare deposits but which are not usually seen on Lunar Orbiter photography.

Apollo 15 near-terminator photography of the Mare Vaporum area shows numerous features made visible or enhanced in the low sun angle view (Fig. 13). These include a 25 km diameter mare dome (upper left) and numerous mare ridges of varying widths and orientations (lower right). Many structures such as flow fronts, elongate craters, and depressions are visible in the near-terminator photography, which provides additional data regarding the surface and subsurface evolution of the maria.

In the western portion of the Aristarchus Plateau, higher crater density on the

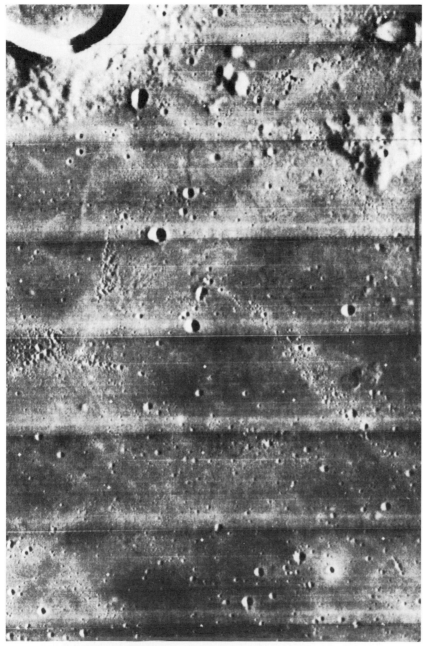

Fig. 12. Lunar Orbiter photograph of the Kunowsky region. Framelet width is approximately 10 km (from NASA Photograph Number L/O IV-133H3).

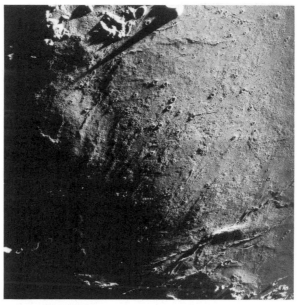

Fig. 13. Apollo 15 near-terminator photograph of Mare Vaporum. Width of frame is approximately 200 km (NASA Photograph Number AS-15-98-13302).

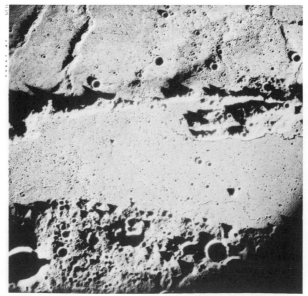

Fig. 14. Apollo 15 near-terminator photography of the western edge of the Aristarchus Plateau. Width of frame is approximately 80 km (NASA Photograph Number AS-15-98-13329).

Fig. 15. Apollo 15 near-terminator photography of the western edge of the Aristarchus Plateau. Width of frame is approximately 30 km. (NASA Photograph Number AS-15-98-13345).

plateau and the highly sinuous rilles at the mare edge are evident (Fig. 14). Mare ridges in the upper left continue northwest into Oceanus Procellarum, and numerous smaller sinuous rilles are seen that are barely visible in photography taken at higher sun angles.

On the western edge of the Aristarchus Plateau, several straight and slightly sinuous rilles originate in the plateau and are obscured in the mare (Fig. 15). Three units of varying crater density can be seen. The most heavily cratered includes the plateau; the intermediate unit is more marelike and appears to be the unit in which the rilles originate; and the least cratered area is in the mare and embays the rilles.

Many photographs were obtained that show lunar-surface areas within a few degrees of the terminator; these photographs are of significant geologic interest. Many geological features stand out in a distinct manner not normal in conventional lunar photography, thus providing additional data on the surface morphology and the configuration of a large number of lunar-surface structures.

Editor's Note: The interested reader should refer to Rennilson J. J. and Holt H. H. (1969) Photometry and polarimetry of the earth. JPL Tech. Report 33-1443, *Analysis of Surveyor Data*, 135–140. The variation of illuminance (earthshine) is large depending on cloud cover and may be as high as 1.9 lm/m² at the phase angle discussed in this paper. This may explain part of the difference between the predicted and the actual results. It does, of course, have no influence on the primary results obtained by examination of *relative* brightness within the actual photographs obtained.

REFERENCES

Eastman Kodak Co. (1970) Kodak 2485 high speed recording film. Data release.

Head J. W. (1969) Distribution and interpretation of crater-related facies: The lunar crater Aristarchus. *Trans. Amer. Geophys. Union* **50,** No. 11, 8.

NASA (1969) Natural environment and physical standards for the Apollo program. NASA M-DE8020.008C, p. 5-2.

Pohn H. A. and Wildey R. L. (1970) Photoelectric-photographic map of the normal albedo of the moon. Professional Paper 599-E, Plate 1, Department of the Interior Geological Survey, 1970-G69152.

Proceedings of the Third Lunar Science Conference
(Supplement 3, *Geochimica et Cosmochimica Acta*)
Vol. 3, pp. 3143–3155
The M.I.T. Press, 1972

Optical properties of lunar glass spherules from Apollo 14 fines

K. J. RAO and A. R. COOPER

Division of Metallurgy and Materials Science, Case Western Reserve University,
Cleveland, Ohio 44106

Abstract—Optical spectra of lunar glass spherules are similar to observed spectra of silicate glasses containing both Fe^{2+} and Fe^{3+}. Increased absorption in the visible region upon heat treatment in the laboratory is attributed to an increased fraction of Fe^{3+}. From previous thermodynamic studies in sodium disilicate glass (Johnston, 1964) it is observed that below a fixed oxygen partial pressure the Fe^{2+}/Fe^{3+} ratio, at a given temperature no longer depends on oxygen pressure but is controlled by the disproportionation reaction $3FeO = Fe_2O_3 + Fe$.

It is suggested that among the consequences of this disproportionation reaction are (a) that the Fe^{2+}/Fe^{3+} ratio in lunar glass may be a measure of the thermal history of the glass—this is supported by the fact that glasses of quite similar chemical composition show significantly different absorption coefficients in the visible—and (b) that the ability for the disproportionation to occur in the liquid and supercooled liquid phases may explain why iron metal is frequently found to be associated with the glassy phase.

Birefringence which has been observed in rapidly heat-treated and quenched lunar glass spherules is shown to be explicable only on the basis of a chemical alteration, such as disproportionation or oxidation. Index of refraction changes which result on heat treatment of lunar spherules are not reconcilable with usual glass transition phenomena and hence also suggest chemical change.

INTRODUCTION

ABOUT ONE PERCENT of glasses in lunar fines occurs as spherules (Glass, 1972). The colors of these glass spheres are surely caused by the presence of transition metal ions (principally Fe and Ti) and vary from green to yellow to brown (Keil, Bunch, and Prinz, 1970). Finely divided Fe inclusions are often present (Housley *et al.*, 1970). The production of these glassy spherules is no doubt due to important events on the lunar crust (meteoritic or volcanic). A study of the properties of spherules provides an opportunity to deduce their thermal history and also, hopefully, to provide useful information regarding more general questions of lunar history.

Optical properties of glasses, such as refractive index and optical absorption, are sensitive to thermal history. Typically, the higher the quenching temperature and quenching rate, the lower the index and the valence states of the transition metal ions which cause absorption in the visible region. Birefringence in an isotropic material such as glass is due to strains which conventionally are related to cooling rates.

Observations on the optical properties of lunar glass spherules selected from sample 14163 and spherules selected from the same sample which have been subjected to laboratory heat treatments, particularly those on birefringence in particles subjected to severe heat treatment, will be reported and discussed. While the relevance of this study to thermal history of the spherules is considered, particular attention has been given to the ferrous-ferric equilibrium in the belief that it is of general interest to lunar petrology.

Experimental

Optical absorption of individual spherules immersed in oils of matching refractive index were measured in the visible region using a Reichert Microspectrophotometer attached to a Reichert Microscope. Details have been discussed elsewhere (Sarkar and Cooper, 1972). These measurements, which are easy to perform, permit a rapid quantitative assessment of "darkening" which was previously observed (Cooper *et al.* 1972) on heat treatment of lunar spherules. However, as will be discussed later, the limitation of this technique to the visible region of the spectrum caused a serious ambiguity in the interpretation of the results. To permit absorbance measurements of a single spherule that include the near-IR region, a special light condensing spectrophotometer cell, to be described elsewhere (Rao, 1972), was therefore designed and utilized with a Cary 14 recording spectrometer.

After completion of the absorption measurements, the spherules were mounted, ground to nearly hemispherical shape, polished and carbon-coated for electron-microprobe analysis. The chemical composition was based on linear interpolation and extrapolation of count rates of five standard glasses of similar composition.

Refractive index, n_e, of the spherules was measured by the Jamin-Lebedeff interference technique (Carl Zeiss, 1967), with a Zeiss Ultraphot II microscope at a wavelength of 546.1 nm. The index measurements are typically reproducible within 0.0002. However, in the case of darker (absorbing) and larger (> 200 μ diameter) spherules, the precision diminishes to 0.0005. Birefringence was measured on a Leitz microscope using an Ehringhaus compensator with a quartz plate.

Very rapid quenching rates were achieved by inserting the lunar spherule into a cavity drilled into a 1-mm diameter Pt–Pt/Rh thermocouple junction. The cavity was plugged and the thermocouple inserted into a horizontal annular furnace mounted on tracks. When the thermocouple reached the appropriate temperature, the furnace was rapidly moved away. Maximum cooling rates ($> 200°$/sec of the thermocouple) were achieved by pouring liquid nitrogen onto the thermocouple.

Results and Discussion

Typical absorption coefficients, $(\ln (I_0/I))/p$ versus wavelength curves of as-received glassy lunar spherules are presented in Fig. 1. I_0, I, and p are, respectively, the incident light intensity, the transmitted light intensity, and the length of the light path through the particle. Additional measurements reported elsewhere (Sarkar and Cooper, 1972) showed similar spectra with evidence for broad peaks centered at ~ 440 nm and ~ 660 nm superimposed on a nearly monotonically decreasing absorption with increasing wavelength.

Chemical composition of the major species along with the nominal diameters of the spherules whose spectra are shown in Fig. 1 are given in Table 1. Although the chemical compositions are very similar, especially in iron oxide and titanium oxide content, the absorption coefficients are distinctly different. Particularly noteworthy is the threefold difference between particle 2 and 7 over the region 400 to ~ 600 nm.

Table 1. Composition of lunar spherules.

Sphere No.	SiO_2	Al_2O_3	FeO	CaO	MgO	TiO_2	K_2O	Na_2O	MnO_2	Cr_2O_3	Diameter, p
2	52.5	13.50	11.8	9.7	6.0	2.8	1.4	1.1	0.1	0.2	~ 45 μ
4	48.5	15.10	13.0	11.1	9.5	2.4	0.1	0.1	0.1	0.1	~ 90 μ
7	42.3	17.50	11.5	14.4	10.3	2.4	0.1	0.3	0.1	0.1	~ 100 μ

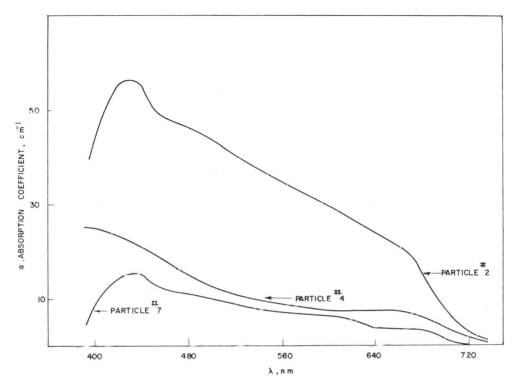

Fig. 1. Absorption coefficient α versus wavelength λ for typical lunar spherules (Apollo 14163).

Since the effect of difference in optical path length has been normalized out, the reason for the difference in absorption coefficient must lie elsewhere.

As is well known, Fe^{2+} causes a broad absorption band centered at ~ 1100 nm. In silicate glass (Varguine and Weinberg, 1955) the absorption at 700 nm is about one-quarter of the peak value. This information coupled with the low (almost vanishing) absorption in the 700 to 720 nm region in lunar spherules (Fig. 1) had led to the suspicion (Sarkar and Cooper, 1972) that Fe^{2+} was present only in small quantities. To confirm this, the spectra from 360 to 1400 nm were measured on a single particle using the special light-condensing spectrophotometer cell. Contrary to this suspicion the results, shown in Fig. 2 where the absorbance is plotted versus wavelength, show instead that there is a large amount of Fe^{2+} in the lunar glass. In this measurement it is not possible to determine the absorption coefficients precisely.

Table 2 describes the various laboratory heat treatments in air given to a single lunar spherule along with the index n_e and the sign and magnitude of birefringence following each heat treatment. (Positive birefringence indicates a higher index for light whose electric vector vibrates in the radial direction than for light whose electric vector is in the tangential direction.) A photomicrograph of a lunar spherule before and after heat treatment under crossed nicols is shown in Fig. 3. The pattern after

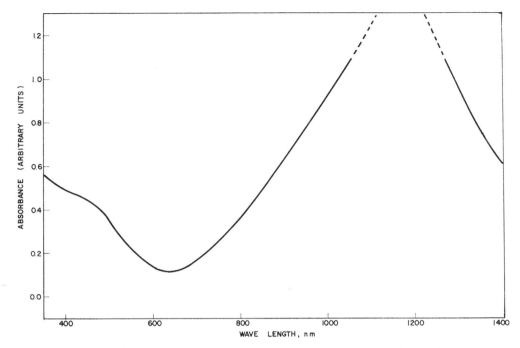

Fig. 2. Absorbance (arbitrary units) versus wavelength λ of lunar spherule (Apollo 14163).

Table 2. Laboratory thermal history of lunar glass spherule
(initial diameter, 176 μ; color, pale yellow).

Heat treatment no.	0	1	2	3	4
Time elapsed on heating between 600°C and maximum temperature	–	$4\frac{1}{2}$ sec	3 sec	7 sec	2 hours
Maximum temperature	–	700°C	800°C	1050°C	625°C
Time elapsed on cooling between maximum temperature and 600°C	–	1.0 sec	0.8 sec	0.6 sec	10 hours
Refractive index n_e	1.6132	1.6133	1.6132	1.6116	1.6126
Birefringence: magnitude sign	–	0.00034 positive	0.0007 positive	Could not be observed clearly*	

* The particle became too dark.

heat treatment is typical of that for a spherically symmetric strain. The symmetry was confirmed by the observation that the pattern did not change upon rotation of the spherules.

Figure 4 shows the influence of the heat treatments given in Table 2 on the visible absorption spectra.

From an application of crystal field theory (Burns, 1970; Bates, 1962), from the empirical results of optical spectra in silicate glasses, from the results of numerous other workers who have studied lunar materials spectroscopically, and from a knowledge of the Fe^{2+}-Fe^{3+} equilibria, especially in silicate glasses (Johnston, 1964), it is

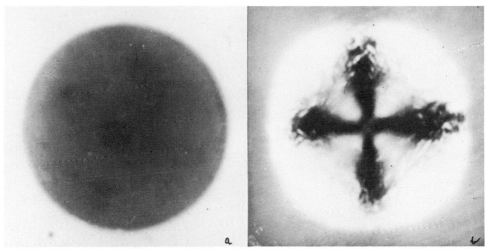

Fig. 3. Birefringence in quenched lunar spherule (a) before heat treatment (b) after heat treatment.

possible to present the following rationalization of the results which is consistent with all our findings:

The spectra in Fig. 3 may be explained as being produced by the presence of both Fe^{2+} and Fe^{3+} ions in the lunar glasses.

There is general agreement with respect to the presence of Fe^{2+} in most lunar materials and glasses because of the intense broad optical absorption peak centered about 1100 nm. There is however no general acceptance (Housley et al., 1970, 1971; Herzenberg and Riley, 1972; Burns et al., 1972) of the presence of significant ($\sim 10\%$ of total iron) quantities of Fe^{3+}. Absorption of pyroxene and olivine minerals in the visible (and near ultraviolet) has been attributed (e.g., Burns et al., 1972) to Ti^{3+}. Among the reasons for the tendency to reject the hypothesis of significant quantities of Fe^{3+} in lunar glasses are the reducing atmosphere, $Po_2 \sim 10^{-13}$ bars (Brown et al., 1970), the high temperatures, and the rapid cooling associated with lunar sphere formation. Also, chemical anaylsis of lunar samples has failed to reveal the presence of Fe^{3+} (Maxwell and Wiik, 1971; Scoon, 1971).

Comparison of the spectra in Fig. 1 with earlier spectral studies in silicate glasses (Weyl, 1951; Varguine and Weinberg, 1954; Bamford, 1961), however, indicates that the absorptions are likely to be due to Fe^{3+}.

In addition, various workers have recently interpreted their results to indicate the presence of Fe^{+3} in lunar glasses and minerals. The results reported include: (a) crystal field spectral studies (Bell and Mao, 1972) in which the ~ 400 nm region absorption is attributed, in part at least, to Fe^{3+}; (b) Mössbauer studies in plagioclase (Finger et al., 1972), in which it is suggested that $\sim 10\%$ of the total iron is present as Fe^{3+}; (c) luminescence studies in plagioclase (Geake, 1972), in which the presence of considerable Fe^{3+} is indicated; (d) ESR studies of glasses (Griscom and Marquardt 1972) in which characteristic resonance due to Fe^{3+} was observed and Fe^{3+} was

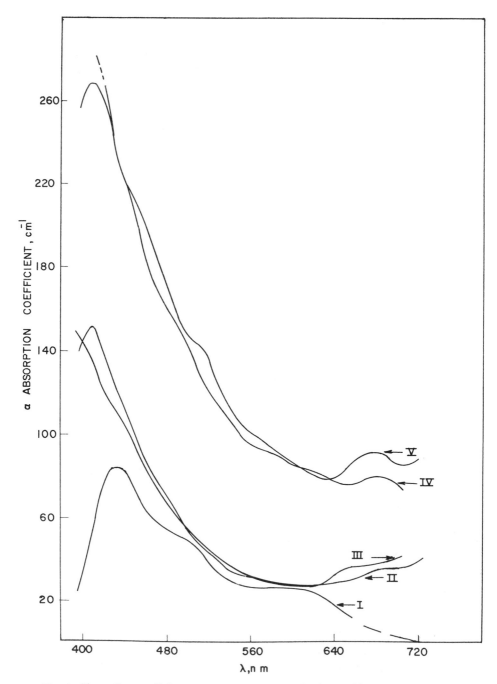

Fig. 4. Absorption coefficient α versus wavelength λ for lunar spherules after various
heat treatments. (See Table 1 for description of heat treatment.)

estimated to be 1 to 5% of the total iron. (Their conclusions which were confirmed by annealing studies in air which supported the conclusion arrived at in earlier ESR studies (Kolopus *et al.*, 1971; Weeks *et al.*, 1972)); (e) absorption studies (Sarkar and Cooper, 1972) in which the spectra in the visible were attributed predominantly to Fe^{3+}; and (f) absorption studies (Nash and Connel, 1972) in which absorption in the 420 to 480 nm region was attributed to Fe^{3+} rather than to Ti^{3+}.

While we do not completely reject the possibility that Ti^{3+} may contribute to the visible absorption spectra, phase-equilibrium studies (MacChesney and Muan, 1961) in the system Ti–Fe–O have revealed that when $FeO/TiO_2 > 1$, Ti^{3+} was not detectable, while Fe^{3+} occurred in considerable proportion. Typically (see Table 2) $FeO/TiO_2 \cong 5$ in these lunar glasses. Furthermore, our belief that most of the absorption of these glasses in the visible region is due to Fe^{3+} is based in part on the fact that the measured absorption increases on heat treatment in air (Fig. 4); such an increase cannot be attributed to further production of Ti^{3+} from Ti^{4+}. Production of Fe^{3+} is, however, feasible, either from oxidation of Fe^{2+} or from chemical disproportionation.

An important basis for our conviction that Fe^{3+} is present is found from the Fe^{2+}-Fe^{3+} equilibrium studies in sodium disilicate melts (Johnston, 1964). These studies, as illustrated by Fig. 5 (taken from the original paper), reveal that at a given

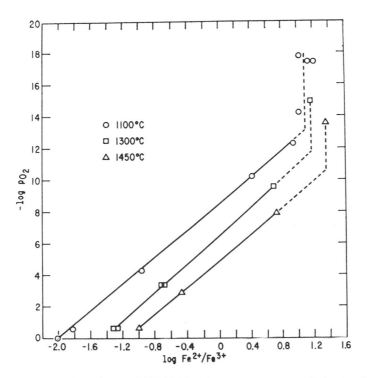

Fig. 5. Equilibrium dependence of Fe^{2+}/Fe^{3+} ratio on oxygen pressure (Johnston, 1964).

temperature below a fixed p_{O_2} the Fe^{2+}/Fe^{3+} ratio no longer depends on oxygen partial pressure but is governed by the disproportionation* reaction

$$3FeO = Fe_2O_3 + Fe \qquad (1)$$

Figure 5 shows that, as temperature is reduced, equilibrium shifts to the right, i.e., to favor Fe^{3+} and Fe^0. This is consistent with the evidence that wüstite is stable only above 570°C; below this temperature it decomposes into Fe and Fe_3O_4 (Barnett, 1953; Remy, 1956; Cotton and Wilkinson, 1962; Reed, 1969). Decomposition of wüstite below 570°C thus gives $\log (Fe^{2+}/Fe^{3+}) = 0$ after complete disproportionation.

From the temperature dependence of $\log (Fe^{2+}/Fe^{3+})$ in the region where $\log (Fe^{2+}/Fe^{3+}) < 1$, Johnston determined that the free energy change ($\Delta F = 52.5$ K cal/mol) in sodium disilicate glass for the reaction

$$2FeO + \tfrac{1}{2}O_2 = Fe_2O_3 \qquad (2)$$

is similar to that for the oxides in the free state which indicates that the Fe^{2+}–Fe^{3+} equilibrium may not depend significantly on the glass composition.

Finally, from Fig. 5 we see that even when disproportionation does not occur, the Fe^{2+}/Fe^{3+} ratio is less than 100 at all temperatures below 1450°C at oxygen pressures ($\sim 10^{-13}$ bars) estimated for the lunar atmosphere. While a quantitative description of how the Fe^{2+}/Fe^{3+} ratio describes thermal history awaits a study of the kinetics, the work of Johnston reveals that the reactions are slow and presumably controlled by diffusion of oxygen or iron ions.

Equations 1 and 2 both favor Fe^{2+} at high temperatures. It is likely that the Fe^{2+}/Fe^{3+} ratio of a lunar sample is determined by its thermal history. In the language of glass science (Davis and Jones, 1953) there is a "valence fictive temperature" (temperature at a given oxygen partial pressure at which the observed Fe^{2+}/Fe^{3+} ratio would be in thermal equilibrium). The lower the "soak temperature" and the slower the cooling rate, the lower the valence fictive temperature.

This background explains the results of our studies. The as-received lunar spherules were presumably molten at temperatures > 1400°C. Depending on their thermal history, an Fe^{2+}/Fe^{3+} ratio was established. All other variables being equal, the higher the maximum temperature and the faster the cooling rate, the lower the expected Fe^{2+}/Fe^{3+} ratio. This provides an explanation for the marked difference in absorption coefficient between particles 2 and 7 (Fig. 1). Particle 2 has a lower valence fictive temperature than particle 7.

The effect on optical properties of the heat treatments given in Table 1 are consistent with Fe^{3+} being formed during heat treatment. What portion is due to oxidation and what portion of it is from disproportionation, is not entirely clear. The fact that the absorption increase occurs so rapidly (the total heat treatment above 600°C lasts only a few seconds) seems to rule out oxidation from the atmosphere, but the small size of the spherules means that diffusion distances are very small. (Suppose $D = 10^{-8}$

---

* Rawcliffe and Rawson (1969) state that "disproportion describes a reaction in which an atom or ion undergoes self-oxidation and reduction simultaneously. . . ."

cm^2/sec and time, $t = 1$ sec. Then the penetration distance, which can be estimated as $\sqrt{Dt}$, is equal to 1 μ; for a 100-μ diameter particle this represents 4% of the volume.) The laboratory heat treatments are fast and, while there are no kinetic data, it is doubtful that they are of sufficient duration to shift the equilibrium of equation 2 significantly. Longer heat treatments in sealed platinum enclosures (Roy et al., 1971) caused changes in valence state, indicating that disproportionation has occurred. It is interesting to examine some of the consequences of equation (1), which states that in the absence of significant Fe vapor pressure there should be an equivalent amount of Fe^0 to Fe^{3+} if the Fe^0 has resulted from disproportionation.

Typically, large spherules are much darker in transmission. This is of course to be expected in view of the longer light path. It also is to be expected (Izard, 1971) that larger particles will cool more slowly and hence have lower valence fictive temperatures and thus more Fe^{3+}.

The evidence (Housley et al., 1970, 1971) that iron inclusions in glassy lunar spherules are often found within the spherule is consistent with their formation by disproportionation.

Figure 5 suggests that we should expect disproportionation of FeO in a lunar magma at temperatures above the liquidus of the common lunar silicate crystalline phases. When crystallization from this magma is slow (Roedder and Wieblen, 1971) the crystals reject the metallic phase. (Rapid crystallization, however, may include some metallic iron (Brett et al., 1971). Possibly the thermodynamics is such that the crystalline silicate phases cannot disproportionate during cooling until temperatures are reached where the kinetics are infinitesimally slow. The liquid or supercooled liquid phase, however, continues to disproportionate on cooling until the kinetics become prohibitively slow. This is consistent with observations (Housley et al., 1970) that most of the metallic iron is present in the glassy portions of lunar dust. We recognize that iron particles whose source is from an impacting meteorite might also be likely to segregate to the glassy phase, but we believe that the disproportionation reaction (equation 1) must be considered as a source of the free iron (Goldstein and Yakowitz, 1971) which is outside usual meteoric compositions.

We also recognize that in the immediate vicinity of a metallic meteorite impact there may be sufficient Fe vapor pressure (Carter, 1971) to drive equation 1 to the left, thus increasing the Fe^{2+}/Fe^{3+} ratio. It may be that anomalously high Fe^{2+}/Fe^{3+} ratios will serve as evidence that a mineral or glass was in the immediate vicinity of such an event.

Spherules heated to temperatures greater than 700°C and quenched developed spherically symmetric birefringence. As shown in Table 2, the extent of birefringence increases with increasing quenching rates and temperatures. Such birefringence as observed in Fig. 4 could originate from radially symmetric crystallites (Hartshorne and Stuart, 1970; Morse and Donnay, 1936). The birefringence however, vanishes on annealing at 600°C, revealing that crystallization is not the cause of the phenomenon. If, on the other hand, the birefringence originates from the failure to relax thermal stresses during rapid quenching (so-called tempering), the following approximate analysis can be made: Let σ_r and σ_T be, respectively, the radial and tangential stresses developed from quenching at a rate $\dot{w}$, of a spherical particle of radius r, linear thermal expansion coefficient α, Young's modulus E, Poisson's ratio v, and thermal diffusivity K. Then the difference of thermal stresses $(\sigma_r - \sigma_T)$ at r may be equated to (Williamson, 1919)

$$\sigma_r - \sigma_T = \frac{\alpha \dot{w} E}{15(1 - v)K} \cdot r^2. \tag{3}$$

The birefringence B is then given by

$$B = \frac{(\sigma_r - \sigma_T)p}{N}, \tag{4}$$

where N is the stress optical coefficient (typically 800 psi/fringe inch) and p is the light path. On substituting typical values for glasses (using the fastest laboratory cooling rate of 250°/sec) and assuming that the light path where the birefringence is greatest has the length $p \cong r$, one finds from equations 3 and 4, $B \cong 1.1 \times 10^4 r^3$ fringes. For a typical lunar spherule of radius 60 μ, therefore, $B = 3 \times 10^{-4}$ fringes. The measured values, however, are about one-tenth of a fringe, which is at least $\sim 10^2$ times greater than the calculated birefringence.

Since stresses from rapid cooling cannot account for the observed birefringence, we believe that the phenomenon has a chemical origin, because in this case the birefringence will only depend on the first power of the radius, as compared with radius cubed in the thermal stress case. If, for example, the outer layer is chemically altered so as to change its specific volume significantly, stresses may be expected. From the preceding discussion of the origin of the Fe^{3+} ions, it is clear that chemical change does occur on heat treatment. If the outer layer which is heated most in the rapid heat treatment, alters to a lower specific volume material, tangential compression will result at the periphery, this can explain the observed birefringence. This explanation is consistent with the recent observation in our laboratories that rapid heat treatments similar to those shown in Table 2 produce significant density increases of lunar spherules. The disappearance of birefringence on annealing is then explained by a uniform alteration throughout the spherule. An alternate explanation based on thermal expansion decrease of the surface layers is possible but is not as appealing.

The variation of refractive indices from the first two heat treatments is not significant. Either these treatments gave cooling rates equivalent to those the spherules received on the lunar surface, or the bulk of the spherule never reached a temperature at which volume relaxation could occur (Davis and Jones, 1953). A marked decrease occurs on quenching from $\sim 1050°$ which may be due to laboratory quenching being more rapid than lunar cooling. On the other hand, the possibility that the index change was in part due to chemical change cannot be discounted. On annealing at 625°C in air, the index increases, but to a value smaller than the original index. The fact that the original index was not regained seems to confirm that chemical change is at least partly responsible for the behavior. Since $\frac{1}{2}Fe_2O_3$ has a lower specific refraction than FeO (Young and Finn, 1940) it is not surprising that the index decreases.

CONCLUSION

The $Fe^{2+}-Fe^{3+}$ equilibrium and its variation with temperatures is crucial to an understanding of the optical properties of lunar glass and minerals. The equilibrium reveals that, except in the presence of Fe vapor, Fe^{3+} ions will be present as a significant fraction of the total iron.

A self-consistent qualitative understanding of observed measurements of optical properties of lunar spherules before and after a variety of laboratory heat treatments is possible on the basis of the chemical change hypothesis. However a quantitative understanding awaits a study of the kinetics of the path toward equilibrium. Controlled heat treatment experiments in low p_{O_2} atmosphere as well as in air may be combined with the measurements described in this paper along with magnetic property measurements (Mössbauer, EPR, and susceptibility) in order to attain this goal.

Acknowledgments—The authors are thankful to Professor A. H. Heuer, Metallurgy and Materials Science Division, Case Western Reserve University, for helpful discussions on several aspects of this paper.

REFERENCES

Bamford C. R. (1961) A study of the magnetic properties of iron in relation to its coloring action in glass. *Phys. Chem. Glasses* **2**, 163–168.

Barnett E. deB. and Wilson C. L. (1953) *Inorganic Chemistry*, p. 197, Longmans Green and Company.

Bates T. (1962) Ligand field theory and absorption spectra of transition metal ions in glasses. In *Modern Aspects of the Vitreous State* (editor J. D. MacKenzie), Vol. 2, Chap. 5, Butterworths.

Bell P. M. and Mao H. L. (1972) Initial findings of a study of chemical composition and crystal field spectra of selected grains from Apollo 14 and 15 rocks, glasses, and fine fractions (abstract). In *Lunar Science—III* (editor C. Watkins), pp. 55–57, Lunar Science Institute Contr. No. 88.

Brett R., Butler P. Jr., Meyer C. Jr., Reid A. M., Takeda H., and Williams R. (1971) Apollo 12 igneous rocks 12004, 12008, 12009, and 12022. A mineralogical and petrological study. *Proc. Second Lunar Sci. Conf.*, *Geochim. Cosmochim. Acta* Suppl. 2, Vol. 1, pp. 301–317. MIT Press.

Brown G. M., Emelius C. H., Holland J. A., and Phillips R. (1970) Mineralogical, chemical, and petrological features of Apollo 11 rocks and their relationship to igneous processes. *Proc. Apollo 11 Lunar Sci. Conf.*, *Geochim. Cosmochim. Acta* Suppl. 1, Vol. 1, pp. 195–219. Pergamon.

Burns R. G. (1970) *Mineralogical Applications of Crystal Field Theory*. Cambridge University Press.

Burns R. G., Abu-Eid R. M., and Huggins F. E. (1972) Crystal field spectra of lunar silicates (abstract). In *Lunar Science—III* (editor C. Watkins) Lunar Science Institute Contr. No. 88, pp. 108–109.

Carl Zeiss (1967) Manual G-41-541-c, Microinterference refractometer.

Carter J. L. (1971) Chemistry and surface morphology of fragments from Apollo 12 soil. *Proc. Second Lunar Sci. Conf.*, *Geochim. Cosmochim. Acta* Suppl. 2, Vol. 1, pp. 873–892. MIT Press.

Cooper A. R., Varshneya A. K., Sarkar S. K., Swift J., Klien L., and Yen F. (1972) Some aspects of thermal history of lunar glass. *J. Am. Ceram. Soc.* (to appear),

Cotton F. A. and Wilkinson G. (1962) *Advanced Inorganic Chemistry*, p. 708, Interscience.

Davies R. O. and Jones G. O. (1953) Thermodynamic and kinetics properties of glasses. *Adv. Phys.* **2**, 370–410.

Finger L. W., Hafner S. S., Schurmann K., Virgo D., and Warburton D. (1972) Distinct cooling histories and reheating of Apollo 14 rocks (abstract). In *Lunar Science—III* (editor C. Watkins) Lunar Science Institute Contr. No. 88, pp. 259–261.

Geake J. E., Walker G., Mills A. A., and Garlick G. F. J. (1972) Luminescence of lunar material excited by protons or electrons (abstract). In *Lunar Science—III* (editor C. Watkins) Lunar Science Institute Contr. No. 88, pp. 294–296.

Glass B. P. (1972) Investigations of glass particles recovered from Apollo 11 and 12 fines. (Private communication.)

Goldstein J. I. and Yakowitz H. (1971) Metallic inclusions and metal particles in the Apollo 12 lunar soil. *Proc. Second Lunar Sci. Conf., Geochim. Cosmochim. Acta* Suppl. 2, Vol. 1, pp. 177–191. MIT Press.

Griscom D. L. and Marquardt C. L. (1972) Electron spin resonance studies of iron phases in lunar glasses and simulated lunar glasses (abstract). In *Lunar Science—III* (editor C. Watkins) Lunar Science Institute Contr. No. 88, pp. 341–343.

Hampe W. (1958) Beitrag zur deutung der anomalen optichan eigenschaften feinstteihger metalkolloide in grober konzentration. *Z. Physik* **152**, 476–494.

Hartshorne N. H. and Stuart A. (1970) *Crystals and the Polarizing Microscope.* Edward Arnold, London.

Herzenberg C. L., Moler R. B., and Riley D. L. (1972) Mössbauer instrumental analysis of Apollo 12 lunar rock and soil samples. *Proc. Second Lunar Sci. Conf., Geochim. Cosmochim. Acta* Suppl. 2, Vol. 3, pp. 2103–2123. MIT Press.

Housley R. M., Blander M., Abdel-Gawad M, Grant R. W., and Muir A. H. Jr. (1970) Mössbauer spectroscopy of Apollo 11 samples. *Proc. Apollo 11 Lunar Sci. Conf., Geochim. Cosmochim. Acta* Suppl. 1, Vol. 3, pp. 2251–2268. Pergamon.

Housley R. M., Grant R. W., Muir A. H. Jr., Blander M., and Abdel-Gawad M. (1971) Mössbauer studies of Apollo 12 samples. *Proc. Second Lunar Sci. Conf., Geochim. Cosmochim. Acta* Suppl. 2, Vol. 3, pp. 2125–2136. MIT Press.

Isard J. (1971) The formation of spherical glass particles on the lunar surface. *Proc. Second Lunar Sci. Conf., Geochim. Cosmochim. Acta* Suppl. 2, Vol. 3, pp. 2003–2008. MIT Press.

Johnston W. D. (1964) Oxidation-reduction equilibrium in iron-containing glass. *J. Am. Ceram. Soc.* **47**, 198–201.

Keil K., Bunch T. E., and Prinz M. (1970) Mineralogy and composition of Apollo 11 lunar samples. *Proc. Apollo 11 Lunar Sci. Conf., Geochim. Cosmochim. Acta* Suppl. 1, Vol. 1, pp. 561–598. Pergamon.

Kolopus J. L., Kline D., Charelain A., and Weeks R. A. (1971) Magnetic resonance properties of lunar samples. *Proc. Second Lunar Sci. Conf., Geochim. Cosmochim. Acta* Suppl. 2, Vol. 3, pp. 2501–2514. MIT Press.

MacChesney J. B. and Maun A. (1961) Phase equilibria in the system iron oxide-titanium oxide. *Amer. Mineral.* **46**, 572–582,

Maxwell J. A. and Wilk H. B. (1971) Chemical composition of Apollo 12 lunar samples, 12004, 12033, 12051, 12052, and 12065. Abstracts of Second Lunar Science Conference.

Morse H. W. and Donnay J. D. H. (1936) Optics and structure of three dimensional spherulites. *Amer. Mineral.* **21**, 391–426.

Nash D. and Conel J. (1972) Further studies of optical properties of lunar samples, synthetic glass and mineral mixtures (abstract). In *Lunar Science—III* (editor C. Watkins) Lunar Science Institute Contr. No. 88, pp. 576–577.

Rao K. J. (1972) A simple light condensing technique to study the spectra of microscopic particles in visible and near IR. *J. Sci. Instruments* (communicated).

Rawcliffe C. T. and Rawson D. H. (1969) *Principles of Inorganic and Theoretical Chemistry.* Heinemann Ltd., London.

Reed T. B. (1966) The role of oxygen pressure in the control and measurement of composition in 3d metal oxides. In *The Chemistry of Extended Defects in Non-Metallic Solids* (editors L. Eyring and M. O'Keefe), p. 21, North Holland.

Remy H. (1956) *Treatise on Inorganic Chemistry*, p. 271, Elsevier.

Roedder E. and Weiblen P. W. (1971) Lunar petrology of silicate melt inclusions, Apollo 11 rocks. *Proc. Apollo 11 Lunar Sci. Conf, Geochim. Cosmochim. Acta* Suppl. 1, Vol. 1, pp. 801–837. Pergamon.

Roy R., Roy D. M., Kurtossy S., and Faile S. P. (1971) Lunar glass. I. Densification and relaxation studies. *Proc. Second Lunar Sci. Conf., Geochim. Cosmochim. Acta* Suppl. 2, Vol. 3, pp. 2069–2078. MIT Press,

Sarkar S. K. and Cooper A. R. (1972) Optical absorption and compositions of lunar glassy spheres. (Communicated.)

Scoon J. H. (1971) Chemical analysis of lunar samples 12040 and 12064. *Proc, Second Lunar Sci. Conf., Geochim. Cosmochim. Acta* Suppl. 2, Vol. 2, pp. 1259–1260. MIT Press.

Varguine V. V. and Wienberg T. I. (1955) IVe Congres International de Verre.

Weeks R. A., Kolopus J. L., and Kline D. (1972) Magnetic phases in lunar material and their electron magnetic resonance spectra: Apollo 14 (abstract). In *Lunar Science—III* (editor C. Watkins) Lunar Science Institute Contr. No. 88, pp. 791–793.

Weyl W. A. (1951) *Coloured Glasses.* Society of Glass Technology, Sheffield, England.

Williamson E. D. (1919) Strains due to temperature gradients with special reference to optical glass. *J. Washington Acad. Sci.* **9,** 209–217.

Young J. C. and Finn A. N. (1940) Effect of composition and other factors on the specific refraction and dispersion of glasses. *J. Res. Natl. Bur. Standards* **25,** 759–782.

Proceedings of the Third Lunar Science Conference
(Supplement 3, *Geochimica et Cosmochimica Acta*)
Vol. 3, pp. 3157–3160
The M.I.T. Press, 1972

Dielectric properties of Apollo 14 lunar samples at microwave and millimeter wavelengths

H. L. BASSETT and R. G. SHACKELFORD

Georgia Institute of Technology, Atlanta, Georgia 30332

Abstract—The relative dielectric constant and loss tangent of lunar sample 14163,164 (fine dust) have been determined as a function of density at 9.375, 24, 35, and 60 GHz. In addition, such measurements have also been performed on lunar sample 14310,74 (solid rock) at 9.375 GHz. The loss tangent was found to be frequency independent at these test frequencies and had a value of 0.015 for the lunar dust sample.

INTRODUCTION

THE DIELECTRIC CONSTANT and loss tangent data of lunar sample 14163,164 (fine dust) have been determined as a function of density at 9.375, 24, 35, and 60 GHz. In addition, the dielectric constant and loss tangent of lunar sample 14310,74 (solid rock) have been measured at 9.375 GHz.

MEASUREMENT TECHNIQUE

The short-circuit wave guide technique (Redheffer *et al.*, 1952; Dakin and Works, 1947), where the rectangular test sample is formed by compressing the powdered sample (14163,164) in the short-circuited wave guide test cell, was used to determine the dielectric properties. A block diagram of the 9.375 GHz test setup is presented in Fig. 1. As indicated, the material undergoing test rests in the sample holder against the metal short circuit. In the case of the powder sample measurement, care was taken to keep the front surface of the lightly compressed sample parallel with the plane of the short circuit. Careful alignment of the compression mandrel and placement of the slotted section and the sample holder in a vertical plane resulted in more than adequate parallelism.

Sample holders and mandrels were fabricated for each frequency band of interest so that microwave measurements could be made as a function of sample density. The mandrels were machined from high carbon steel to withstand the pressures required to compress the powdered material to 77% integral density.

RESULTS

The dielectric constant versus sample density (lunar sample 14163,164) is presented in Figs. 2 and 3. The 9.375 measurement data are plotted in Fig. 2 and are compared to the 24 GHz and 60 GHz measurements in Fig. 3. The straight lines on the density-dielectric curves are the linear least mean square estimates. For example, at 9.375 GHz, the dielectric constant is 3.59 at a density of 1.71 gm/cm^3 and increases linearly to a value of 4.45 at a density of 2.06 gm/cm^3. The loss tangent remains relatively constant at a value of 0.015 at each density and at each test frequency.

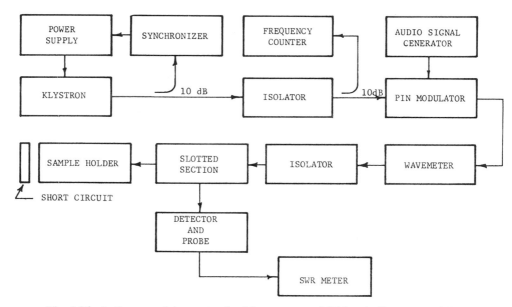

Fig. 1 Block diagram of Apparatus for Measurement of Dielectric Constant and Loss Tangent (Short-Circuit Waveguide Technique).

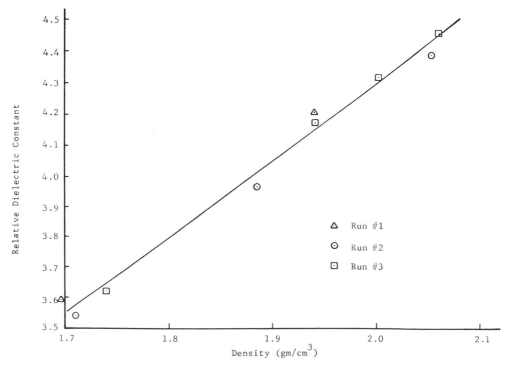

Fig. 2 Dielectric Constant vs. Density at 9.375 GHz for Sample 14163,164.

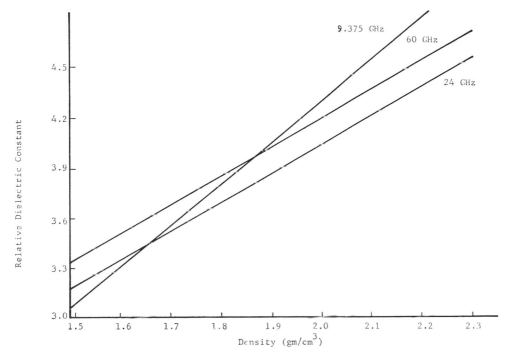

Fig. 3

The solid sample relative dielectric constant was found to be 6.46 at a frequency of 9.375 GHz and the loss tangent is 0.0075. The density of this particular sample (14310,74) is 2.814 gm/cm³. The relative dielectric constant of 6.46 is interesting, since it was noted that one would obtain the same value if the curve of Fig. 2 were extended to include the actual density, 2.814 gm/cm³. Thus, solid sample 14310,74 is the 100% dense equivalent of the lunar fine sample 14163,164. It would now be possible to change the abscissa of Fig. 2 to "percent theoretical density" as opposed to density, per se.

CONCLUSIONS

Although it was stated in the above results that the dielectric constant versus density curve could be extrapolated linearly to obtain the 100% dense value, the number of data points is not sufficient to categorize this variation as being linear over all densities. Usually, the variation of dielectric with density of a base material uniformly mixed with air conforms to the empirical relation (Jasik, 1961; Breeden and Shackelford, 1971):

$$\log \frac{\varepsilon'}{\varepsilon_0} = \frac{\rho'}{\rho} \log \frac{\varepsilon}{\varepsilon_0},$$

where $\varepsilon'/\varepsilon_0$ is the relative dielectric constant of the sample with air, $\varepsilon/\varepsilon_0$ is the relative dielectric constant of the original solid, ρ' is the density of the material with air, and

ρ is the density of the original solid. The data do not indicate that this relation holds for the lunar dust. Also, the linear variation of dielectric constant with density is valid only over the limited measurement range. It is possible that the relative dielectric constant varies in a logarithmic manner at very low densities and varies linearly at the higher densities.

The loss tangent of sample 14163,164 is independent of frequency and appears to be independent of density. As indicated by Chung and Westphal (1972), the frequency independence of the loss tangent exists over a very broad range of frequencies, and these data support their conclusion.

References

Breeden K. H. and Shackelford R. G. (1971) A Novel Technique for Determining the Dielectric Constants of Solids at Millimeter Wavelengths, Georgia Institute of Technology Report under Contracts USDA 12-14-100-9071(33) and Nonr-991(13).

Chung D. H. and Westphal W. B. (1972) Dielectric Properties of Apollo 14 Lunar Samples, in *Lunar Science—III* (editor C. Watkins), p. 139, Lunar Science Institute Contribution No. 88.

Dakin T. W. and Works C. N. (1947) Microwave Dielectric Measurements. *J. Appl. Phys.* **18**, No. 9, 789–796.

Jasik H. (editor) (1961) *Antenna Engineering Handbook*, pp. 32–30. McGraw-Hill.

Redheffer R. M., Wildman R. C., and O'Gorman V. (1952) The Computation of Dielectric Constants. *J. Appl. Phys.* **23**, No. 4, 505–508.

Proceedings of the Third Lunar Science Conference
(Supplement 3, *Geochimica et Cosmochimica Acta*)
Vol. 3, pp. 3161–3172
The M.I.T. Press, 1972

Dielectric properties of Apollo 14 lunar samples

D. H. Chung, W. B. Westphal, and G. R. Olhoeft

Massachusetts Institute of Technology,
Cambridge, Massachusetts 02139

Abstract—Laboratory characterization of dielectric properties of lunar samples 14301,41, 14310,75, 14318,30 and 14321,163 is made. Our measurements of dielectric constants and losses were made over a range of frequency from 100 Hz to 10 MHz and temperature from 77° to 473°K by two-terminal capacitance substitution methods. The dielectric behavior of these samples is generally similar to other lunar samples from Apollo 11 and Apollo 12 sites when these samples are free from absorbed moisture. As did sample 12002,58, sample 14310,75 showed a distinctive dispersion that may be associated with the presence of water, and different values of the activation energy for conduction as the temperature was varied. The activation energies range from about 0.03 to 0.5 eV. Values of the dielectric constant for lunar basalts like sample 14310 seem to run a few percent higher than those for terrestrial basalts with similar chemical composition. Our lunar samples contained metallic free iron and other high-conductivity materials in minute amounts; we believe our observed values of relatively high "apparent" dielectric constants for these lunar samples appear consistent with the presence of the high-conductivity materials in the samples.

INTRODUCTION

LABORATORY STUDIES on electrical properties of returned lunar samples from Apollo missions become of fundamental importance to the analysis of signals for interplanetary communications and also to the planning and subsequent interpretation of future in situ electrical experiments on the moon. For the last two years, since the receipt of Apollo 11 lunar samples and to date with Apollo 15, we have been concerned with laboratory characterization of electrical properties of returned lunar samples; much of our effort has been reported in various publications. As part of our continuing studies on the physical properties of lunar samples, we report in this paper laboratory measurements of dielectric constants and losses of lunar samples 14301, 14310, 14318, and 14321 returned from Apollo 14 mission.

Our measurements of dielectric constants and losses were made over a range of frequency from 100 Hz to 10 MHz and temperature from 77° to 473°K by two-terminal capacitance substitution methods. The methods have been described by von Hippel (1954) and are those adopted by the American Society for Testing Materials as the standard technique under ASTM Specification D150-68. Laboratory measuring equipment includes wide-range bridge devices, resonant cavities, and standing-wave equipment available at the MIT Laboratory for Insulation Research. A laboratory-built capacitance bridge covers a wide range of frequencies; the resolution is 0.002 pF or about 2% in dielectric constant and a few microradians in loss angle. Effects of some imperfect geometries of lunar samples and parallel capacitance of sample holders limit the accuracy of the dielectric constant values to about 5% and of the loss tangent measurements to about 10%.

Lunar Samples

The chemical and mineralogical description of samples 14301, 14310, 14318, and 14321 may be found in a report of the Lunar Sample Preliminary Examination Team (LSPET, 1971); readers are referred to this report for detailed information about these samples. A brief description of their character and density is made in Table 1.

Methods of transportation, handling, heating, and drying of lunar samples, along with other experimental details utilized in the present work were exactly the same as ones described by Chung et al. (1970, 1971), and this same description will not be repeated here. In the present work, our special attention was given to drying and baking of lunar samples at 570°K in vacuum oven so that our samples would be free from absorbed moisture.

Dielectric Properties of Lunar Samples

Values of the dielectric constant κ', loss tangent, and conductivity for sample 14321,163 are shown in Figs. 1 through 3. We note the very flat frequency-dependence of κ', indicating the successful minimization of water contamination as discussed earlier by Chung et al. (1971) and Katsube et al. (1971). In their measurements of κ' for some terrestrial basalts, Saint-Amant and Strangway (1970) made similar observations. The effects of absorbed water on κ' were discussed by Chung (1972). In characterizing the dielectric constant spectra, it is convenient to define the "Percent Frequency Effect (PFE)" as:

$$\text{PFE (in \%)} = 100(\kappa'_{10^2} - \kappa'_{10^6})/\kappa'_{10^2} \qquad (1)$$

where κ'_{10^2} and κ'_{10^6} refer to the measured κ' at 10^2 and 10^6 Hz, respectively. Large values of PFE imply large dispersion at low frequencies; for lunar samples these

Table 1. Lunar samples and their material characteristics

Sample number	Density (gm/cm³)	Classification	Characteristics
14301	2.30	Clastic	Feldspathic clasts with plagioclase as the major phase. Other phases seen are pyroxene, olivine, ilmenite, metallic iron and zircon crystals.
14310	2.85	Crystalline	Fine-grained crystalline rock with scattered small cognate inclusions. Plagioclase and anhedral pyroxenes are the major phases; about 70 volume percent plagioclase and a 30% clinopyroxene. Minor phases are ilmenite, troilite, metallic iron and some spinels with iron-rich compositions.
14318	2.30	Clastic	Feldspathic clasts, very similar to sample 14301.
14321	2.35	Fragmented	Complex texture showing a variety of rock and mineral clasts set in matrices. Anorthitic plagioclase and clinopyroxene are the major phases. Orthopyroxene, olivine, ilmenite, troilite, and metallic iron are the minor phases.

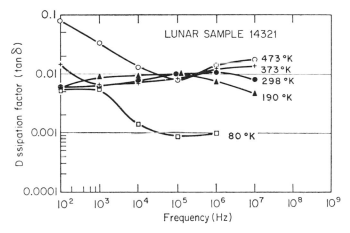

Fig. 1. Dielectric losses in sample 14321,163 as a function of frequency and temperature.

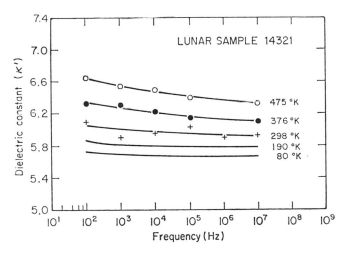

Fig. 2. Dielectric constant of sample 14321,163 as a function of frequency and temperature.

values are tabulated in Table 2. The flat frequency dependencies as observed earlier by Katsube and Collett (1971) on Apollo 11 and 12 lunar samples have PFE values in the range of 0 to about 17. Our earlier data on lunar samples containing small amounts of absorbed water show a PFE value in excess of 80%. With baking the lunar samples in a vacuum oven at 570°K for about 10 hours and then the measurement of κ' in the dry nitrogen atmosphere, the PFE values dropped down to 1 to 8%. As is seen in Table 3, sample 14321,163 has dielectric properties that are very similar to those of other lunar samples studied by Chung et al. (1972, 1971) and Katsube and Collett (1971).

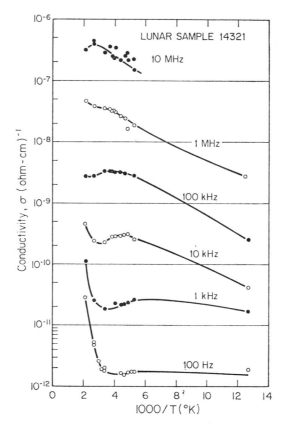

Fig. 3. Electrical conductivity of sample 14321,163 as a function of frequency and temperature.

Table 2. Extrema of loss tangents and percent frequency effect (PFE) of lunar samples.

Sample number	tan δ ($10^3 - 10^6$ Hz)	PFE* (%)	Sources and references
10020	0.1–0.4	82	Chung et al. (1970)
10046	0.06–0.2	79	Chung et al. (1970)
10017	0.02–0.09	0.3	Katsube and Collett (1971)
10065	0.02–0.1	17	Katsube and Collett (1971)
10084	0.01–0.11	0	Katsube and Collett (1971)
12070	0.005–0.13	3	Katsube and Collett (1971)
12002,84	0.02–0.05	6	Katsube and Collett (1971)
12002,58	0.03–0.05	30	Chung et al. (1971)
12022	0.2–0.3	81	Chung et al. (1971)
14301	0.03–0.06	8	This work
14310	0.007–0.02	7	This work
14318	0.025–0.06	8	This work
14321	0.008–0.01	1	This work

* High value of PFE may represent low-frequency dispersion associated with the presence of the absorbed water in the sample (see text).

Table 3. Dielectric constants and losses and conductivities of lunar samples and their comparison with some terrestrial materials.

Material and sample no.	Density (gm/cm³)	T (°K)	Freq. (Hz)	κ'	tan δ	Conductivity T (°K)	Conductivity (mho/cm)	Refer-ence*
CN Basalt-A	2.55	300	10^6	10.1	0.03	473–573	3×10^{-12}	1
CN Basalt-B	2.60	77–473	10^6	6	0.05	77–473	$\sim 10^{-10}$	2
Andesite Basalt-A	?	—	—	—	—	573–720	$\sim 10^{-6}$	3
Andesite Basalt-B	?	—	—	—	—	467–850	$\sim 10^{-6}$	4
Diabase	?	—	—	—	—	473–900	$\sim 10^{-6}$	4
LMT-J Basalt	3.01	77–473	10^6	6	0.002	77–473	10^{-10}	2
10020	3.18	77–473	10^6	9–16	0.002–0.3	278–473	10^{-10}	2
10046	2.21	77–250	10^6	9–13	0.09–0.2	77–250	10^{-10}	2
10017	3.10	300	10^6	8.8	0.075	300	4.2×10^{-10}	5
10065	2.45	300	10^6	7.3	0.053	300	6.7×10^{-10}	5
10084	?	300	10^6	3.8	0.0175	300	2.2×10^{-10}	5
12070	?	300	10^6	3.0	0.025	300	2.2×10^{-10}	5
12002,84	3.10	300	10^6	8.3	0.051	300	2.4×10^{-10}	5
12002,85	3.04	300	10^6	7.8	0.056	300	1.9×10^{-10}	5
12002,58	3.30	78–473	10^6	8–10	0.02–0.09	300–473	10^{-10}	6
12022,60	3.32	78–473	10^6	7–14	0.002–0.2	77–300	10^{-10}	6
14301,41	2.30	77–340	10^6	5	0.001–0.6	77–340	—	7
14310,37	2.17	300	10^6	4.8	0.05	300	1.5×10^{-11}	7
14321,163	2.40	77–376	10^6	6	0.001 0.15	77–200	—	7
14321,228	2.35	300	10^6	5.9	0.01	300	2×10^{-12}	7
14310,75	3.30	200–374	10^6	6	0.006–0.015	77–200	—	7
14310,75	3.30	200–374	10^6	7	0.006–0.015	200–300	—	7
14310,72	2.86	300	10^6	6	0.02	300	2×10^{-12}	7
14318,30	2.30	300	10^6	5.1	0.05	300	1.5×10^{-12}	7

* References: (1) St. Amant and Strangway (1970). (2) Chung, Westphal, and Simmons (1970). (3) Noritomi (1961). (4) Parkhomenko (1967). (5) Katsube and Collett (1971). (6) Chung, Westphal, and Simmons (1971). (7) This work.

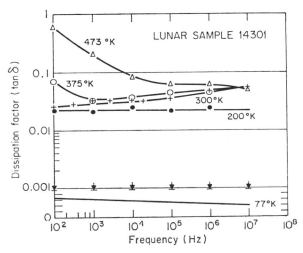

Fig. 4. Dielectric losses in sample 14301,41 as a function of frequency and temperature.

The dielectric spectra measured on sample 14301,41 are shown in Figs. 4 through 6. We observed a slight dispersion at low frequencies as is shown in Fig. 5. We also observed a Maxwell-Wagner type relaxation at high frequencies and high temperatures. Limited work on sample 14318 shows almost the same dielectric spectra as those observed for sample 14301. The dielectric properties measured on sample 14310,75 are shown in Figs. 6 through 10. The sample shows a strong relaxation effect of the Maxwell-Wagner type; an analysis of this relaxation is made in Fig. 11.

From the dielectric theory of solids, we have (see, for example, Poole and Farach, 1971)

$$\kappa = \frac{\varepsilon}{\varepsilon_0} = \kappa' - i\kappa'' = \kappa_\infty + \frac{\kappa_0 - \kappa_\infty}{1 + (i\omega\tau_0)^\lambda}; \qquad \lambda = (1 - \alpha) \qquad (2)$$

where all the terms are exactly as described by Poole and Farach (1971); we have neglected contributions from the dc conductivity effects. For a single relaxation of the Debye type, the coefficient α is zero. An analysis of our data on sample 14310 in accordance with equation (2) is shown in Fig. 11, and we obtained a value of $\alpha = 0.345$ to give the best fit to our data. The single point at 100 Hz does not fit the Cole-Cole plot of Fig. 11; we believe that either a second distribution which may be appearing at lower frequencies, as noted by Khalafalla (1971), or an effect arising from the electrode contributions to the dispersion at this frequency range is associated with this datum point.

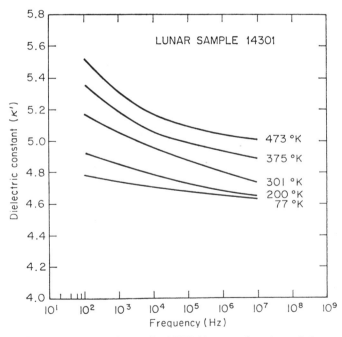

Fig. 5. Dielectric constant of sample 14301,41 as a function of frequency and temperature.

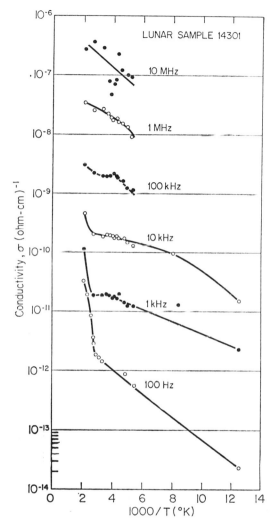

Fig. 6. Electrical conductivity of sample 14301,41 as a function of frequency and temperature.

An expression for loss tangent corresponding to equation (2) is

$$\tan \delta = \frac{\kappa''}{\kappa'} = \frac{(\kappa_0 - \kappa_\infty)(\omega\tau_0)^{1-\alpha} \sin (1 - \alpha)(\pi/2)}{\kappa_0 + (\kappa_0 + \kappa_\infty)(\omega\tau_0)^{1-\alpha} \cos (1 - \alpha)(\pi/2) + \kappa_\infty(\omega\tau_0)^{2(1-\alpha)}} \quad (3)$$

where ω is the frequency and τ the relaxation time. When $\alpha = 0$, equation (2) reduces to the well-known equation of the Debye case, and when $\alpha = 1$, equation (3) results in an expression describing a frequency-independent, constant loss tangent. Thus, for sample 14310,75, the relaxation frequency is spread over a range of τ (about the τ_0 point) as implied by $\alpha = 0.345$.

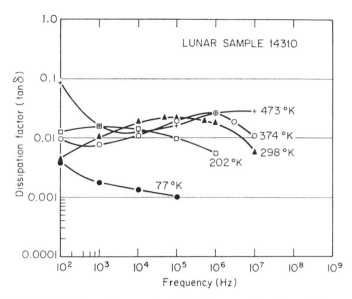

Fig. 7. Dielectric losses in sample 14310,75 as a function of frequency and temperature.

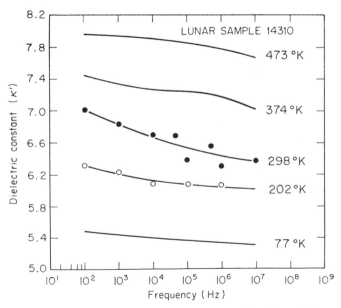

Fig. 8. Dielectric constant of sample 14310,75 as a function of frequency and temperature.

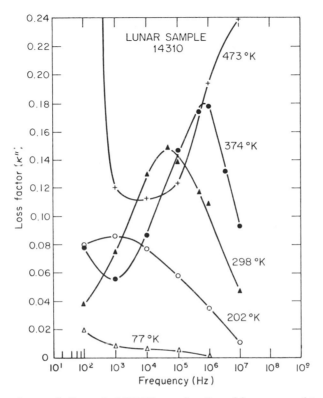

Fig. 9. Loss factor κ'' of sample 14310,75 as a function of frequency and temperature; illustration of the Maxwell-Wagner type relaxation due possibly to H_2O residue.

The loss tangent shown in Fig. 9 is typical of the dielectric losses seen in most igneous lunar basalts we studied. The activation energies associated with these relaxations can be estimated either from the temperature dependence of the frequency of peak loss, or from the peak value of κ'' varying with temperature in the usual way (see, for example, Saint-Amant and Strangway, 1970). For sample 14310,75 we estimate that the activation energy E_0 is in the neighborhood of 0.03 eV as obtained from the temperature dependence of κ'' data. The dielectric E_0 calculated from the temperature dependence of the relaxation frequency as inferred by the peak loss is in the range of about 0.2 to 0.5 eV.

The dielectric constants and losses, conductivities, and the activation energies for the lunar samples studied thus far are summarized in Table 4, and a comparison with the properties of similar materials of terrestrial origins is made in the table. An attempt to correlate the measured properties with the chemical composition of lunar samples is made also in Table 4; as is seen from this table, however, no generalization can be made at this time.

In summary, then, the following observations are listed:

(1) The dielectric behavior of samples 14301, 14310, 14318, and 14321 is generally similar to other lunar samples from Apollo 11 and Apollo 12 sites when these samples

Table 4. Electrical parameters of lunar samples and their possible relation to concentration of high-conductivity phases (all the parameters are evaluated at 300°K).

Sample no.	Sample classification	FeO (mole %)	TiO₂ (mole %)	Ilmenite (mole %)	κ'	tan δ	E_0(diel) (eV)	E_0(cond) (eV)
10020	Crystalline	19.35	10.72	15	11	0.15	—	0.22
10046	Breccia	19.22	10.35	9	13	0.07	—	—
10017	Crystalline	19.79	11.82	15–20	8.8	0.075	—	—
10065	Breccia	—	—	—	7.3	0.053	—	—
10084	Fines	15.62	7.5	12–14	3.8	0.0175	—	—
12070	Fines	17.0	3.1	3–6	3.0	0.025	—	—
12002,84	Igneous	13.0	4.3	3–4	8	0.05	—	—
12002,58	Igneous	13.1	3.9	4	8	0.05	0.04	0.23
12022	Crystalline	22.0	5.1	9–11	7–14	0.2	—	0.16
14301	Clastic	9.8	1.7	2	4.8	0.05	—	—
14321	Breccia	13.0	2.4	4	5.9	0.01	0.07	—
14310	Crystalline	7.7	1.3	2	6.4	0.02	0.3	0.24

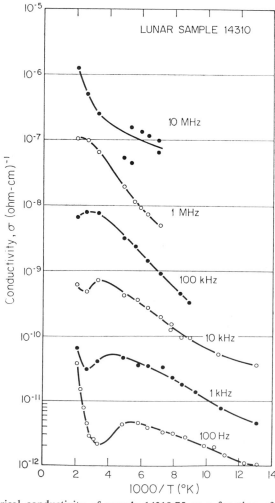

Fig. 10. Electrical conductivity of sample 14310,75 as a function of frequency and temperature.

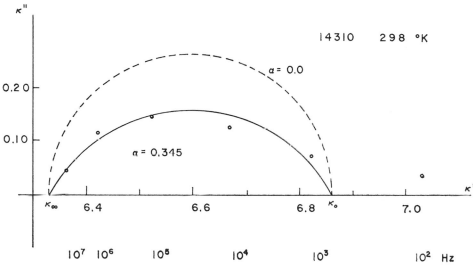

Fig. 11. Cole-Cole plot of sample 14310,75.

are free from absorbed moisture. As did sample 12002, sample 14310 showed a distinctive dispersion which may be associated with the presence of water, and different values of the activation energy as the temperature was varied. The activation energies range from about 0.03 to 0.5 eV and they seem to increase with increasing temperature.

(2) An abrupt change in the activation energy between 200 and 300°K from 0.02 to 0.3 eV, as seen in the conductivity versus temperature, is observed for all the lunar samples we studied. The mechanism of this change is not understood at the present time; it is suggestive of freezing effects of residual water vapor.

(3) The dielectric properties of the lunar samples fall within the range of terrestrial basalts; the lunar samples, however, exhibit slightly lower activation energies for conduction and slightly higher losses than the terrestrial basalt.

Acknowledgments—We are grateful to Professor A. R. von Hippel for his interest and encouragement during our work on lunar samples. We thank Drs. D. W. Strangway and L. S. Collett for reviewing the present paper. This study was supported by NASA under Grant NGR-22-009-597.

REFERENCES

Chung D. H. (1972) Laboratory studies on seismic and electrical properties of the moon. *The Moon* (in press).

Chung D. H., Westphal W. B., and Simmons G. (1970) Dielectric properties of Apollo 11 lunar samples and their comparison with earth materials. *J. Geophys. Res.* **75,** 6524–6531.

Chung D. H., Westphal W. B., and Simmons G. (1971) Dielectric behavior of lunar samples: Electromagnetic probing of the lunar interior. *Proc. Second Lunar Sci. Conf., Geochim. Cosmochim. Acta* Suppl. 2, Vol. 3, pp. 2381–2390. MIT Press.

Katsube T. J. and Collett L. S. (1971) Electrical properties of Apollo 11 and 12 rock samples. *Proc. Second Lunar Sci. Conf., Geochim. Cosmochim. Acta* Suppl. 2, Vol. 3, pp. 2367–2379. MIT Press.

Khalafalla A. S. (1971) Dielectric relaxation and rock geophysical characteristics. *Tech. Rept. 12288-IRI* (ARPA Contract H0210026). Honeywell, Inc.

LSPET (Lunar Sample Preliminary Examination Team) (1971) Preliminary examination of the lunar samples from Apollo 14. *Science* **173**, 681–693.

Noritomi K. (1961) The electrical conductivity of rocks and the determination of the electrical conductivity of the Earth's interior. *J. Min. Coll. Akita Univ.*, Ser. A, Vol. 1.

Parkhomenko E. I. (1967) *Electrical Properties of Rocks*, Plenum Press.

Poole C. P. and Farach H. A. (1971) *Relaxation in Magnetic Resonance*, pp. 319–348. Academic Press.

Saint-Amant M. and Strangway D. W. (1970) Electrical properties of dry, geologic materials. *Geophysics* **35**, 624–645.

von Hippel A. R. (1954) *Dielectric Materials and Applications*, MIT Press.

Proceedings of the Third Lunar Science Conference
(Supplement 3, *Geochimica et Cosmochimica Acta*)
Vol. 3, pp. 3173–3185
The M.I.T. Press, 1972

Electrical conductivity and Mössbauer study of Apollo lunar samples

F. C. Schwerer, G. P. Huffman, and R. M. Fisher

U. S. Steel Research Laboratory, Monroeville, Penna. 15146

and

Takesi Nagata

Geophysical Institute, University of Tokyo, Japan

Abstract—Electrical conductivity data are presented for a sample from lunar crystalline rock 12053 and from a terrestrial augite that represent the preliminary phase of a study of the effects of oxidizing and reducing atmospheres. Detailed Mössbauer analyses are presented for samples from lunar rocks Apollo Nos. 10048, 12053, 14053, 14063, 14047, and 14303. Discussions of the Mössbauer spectra emphasize (1) changes that occur during the heating of lunar conductivity samples including increases in the ilmenite content of 10048, (2) phase distribution of Fe in Apollo 14 samples, and (3) distinctive features unique to spectra obtained at low temperatures and in applied magnetic fields.

Introduction

THERE HAS BEEN considerable interest in obtaining the electrical conductivity profile of the lunar interior from direct probings by various electromagnetic experiments. It is of further interest to compare such profiles with those obtained from known physical properties of lunar or lunar-like materials and theoretical temperature-pressure profiles. It is well known that the electrical conductivity of poorly conducting solids is extremely sensitive to chemical defects. Recent work on terrestrial olivines, for example, has emphasized this point (see Housley and Morin, 1972). An earlier conductivity study of lunar surface rocks (Schwerer *et al.*, 1971) indicated large hysteresis in electrical conductivity after heating to moderate temperatures. It is now apparent that to obtain conductivity data for lunar materials that are relevant for the purposes discussed above, it is necessary to understand the effects of the chemical environment as well as of thermal history. Electrical conductivity and Mössbauer studies have recently been initiated to obtain a more detailed understanding of several of these effects. At the present time the difficulties associated with the conductivity measurements have been only partially resolved. Consequently, this report appears in two sections. The first section describes progress in the study of electrical conductivity of lunar samples including new data for an Apollo 12 sample and for a terrestrial augite. The second section contains detailed Mössbauer analyses of Apollo 11 and 12 conductivity samples and of Apollo 14 samples.

Discussed in this section are (1) the changes that occur in the Mössbauer spectra of the conductivity samples on heating and their connection with the observed hysteresis effect, (2) the phase distribution of Fe in Apollo 14 samples, and (3) Mössbauer spectra obtained at very low temperatures and in applied magnetic fields.

Electrical Conductivity

Previous measurements of electrical conductivity of samples from Apollo 11 and 12 lunar rocks showed large increases by factors of as much as 10^5 in the conductivity measured at room temperature $\sigma(0)$ after heating to temperatures above approximately 500°C (Schwerer et al., 1971). This effect was observed for samples heated in vacua or in flowing atmospheres of Ar, He, or He–2% H_2 for both crystalline and fragmental rocks. Attempts to recover the original conductivity values by heating in neutral or reducing environments were unsuccessful. New studies of the nature of this hysteresis in Apollo 11 and 12 samples and in a terrestrial augite using improved experimental procedures have been initiated recently and the preliminary results are presented in this paper.

Current experimental practice involves dry-cutting specimens of sizes $6 \times 6 \times 2$ (mm). Lakeside 70C bonding compound is used to hold the samples during cutting and is removed by several washings in xylene and ethanol. Samples are given a final cleaning with a freon degreasing spray and mounted in a modified three-terminal, guard-ring conductivity apparatus (see Dunlap, 1959). Electrical contacts are painted on the sample with air-drying Silver Conductive Paint, 6C-21-1 (G. C. Electronics, Rockford, Ill.). A working voltage of 10 V is used and linearity of current-voltage is checked with regard to polarity and when feasible with regard to voltage from 1 to 10 V. The sample chamber is evacuated to approximately 2×10^{-7} Torr and pumped continuously. The sample is tempered for at least 12 hours at 150°C before continuing measurements.

The dc electrical conductivity of lunar samples should be dominated by the two abundant silicate phases—plagioclases and pyroxenes. However, it has been reported that the conductivities of terrestrial plagioclases (of Noritomi and Asada, 1957) are low compared to lunar samples, suggesting that the conductivity of lunar samples is predominantly due to pyroxene phases. For this reason, studies of terrestrial pyroxenes have been initiated and the results are available for an augite. The composition expressed as weight percent oxides is 48.4 SiO_2, 9.0 Al_2O_3, 7.6 FeO, 13.6 MgO, and 17.1 CaO. The sample is predominantly single phase with sparse internal boundaries and sparse inclusions (20–50 μm diameter) of FeS and an iron oxide. Conductance was initially measured in the range 165°C to 325°C in vacua and in flowing He gas. (To facilitate comparison of surface and bulk effects, data have been reported as conductance. If the surface layer is thin and highly conducting, a geometrical factor can be estimated from the effective area of the guarded electrode (American Society for Testing and Materials, Publ. D257-61). Under these conditions the bulk conductivity in $(\Omega cm)^{-1}$ can be obtained by multiplying the conductance values by 5.3.) At constant temperature, conductance increased in vacua and decreased in He. At 165°C, the conductances were 6×10^{-11} mho and 11×10^{-11} mho in He and vacua, respectively. Heat treatments at higher temperatures in vacua resulted in substantially larger increases in conductance, e.g., $G(165°C) = 3 \times 10^{-9}$ mho after heating to 325°C in vacua. This value decreased to 2.5×10^{-10} in a He atmosphere and to 6×10^{-11} in air. Apparently these increases in conductance are due to a reduction of the sample in vacua and the decreases in conductance are due to oxidation in air and He (presumably from residual water vapor in the latter).

The dependence of the conductance near 200°C on the O_2 or H_2 content of the environment is illustrated in more detail in Fig. 1. At this temperature, reducing conditions cause increases in conductance whereas oxidation results in decreased conductances. At higher temperatures, however, both reducing and oxidizing atmospheres cause decreases in bulk conductance (for example, see data for 407°C in Fig. 1). At the higher temperatures, the conductance is greatest near the conditions for a neutral atmosphere and is close to the value obtained after initial heating in vacua. This maximum is most easily observed during the approach to equilibrium in a slightly reducing atmosphere for an initially oxidized sample. Although the data are generally reproducible for repeated cycles there is hysteresis associated with going from high H_2 (or O_2) concentrations to more neutral conditions. In general, the times to reach near steady-state are long and the sample was equilibrated for times of 5 to 30 hours depending on the atmosphere.

The results discussed above are for the "bulk" conductance as measured between the guarded electrodes. The "surface" conductances are also shown in Fig. 1. It is difficult to be certain that the guard ring prevents surface leakage. Furthermore,

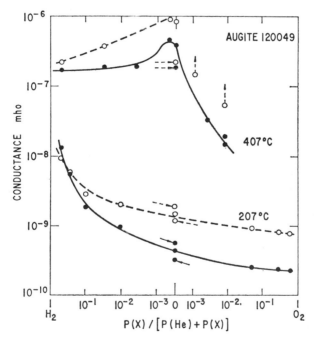

Fig. 1. Variation of conductivity of terrestrial augite sample with O_2 or H_2 content of gas environment. Base atmosphere is He. Solid curves and points are conductances between guarded electrodes; broken curves and open points are conductances between guard electrodes.

cracks and internal boundaries may provide additional, unguarded leakage paths. The behavior observed at 200°C in Fig. 1 makes the situation more obscure. However, at higher temperatures, "surface" conductance increases by three or four orders of magnitude after long times in an oxidizing atmosphere, whereas the "bulk" conductance shows less dramatic behavior, see Fig. 1. This would seem to indicate that conduction on external surfaces is being guarded effectively.

As a result of these observations, in particular the decrease in conductivity after heating in an oxidizing atmosphere at low temperatures, an Apollo 12 sample which had previously been studied was given further heat treatments. Because of size limitations these measurements were made with a two-probe arrangement which did not eliminate surface conduction. The initial heat treatments are indicated in Fig. 2 by solid dots. The sample was heated to successively higher temperatures in a He–2% H_2 atmosphere. After heating to temperatures above approximately 450°C, the room temperature conductivity, $\sigma(0)$, increased to large values. After heating to 800°C the conductivity had the value indicated by point 7. Subsequent to the initial part of the augite study described above, the sample was heated in air at 500°C for about one hour, whereupon the value of $\sigma(0)$ was observed to have decreased to within a factor of 4 of the original value (point 8 in Fig. 2). Subsequent heating in air

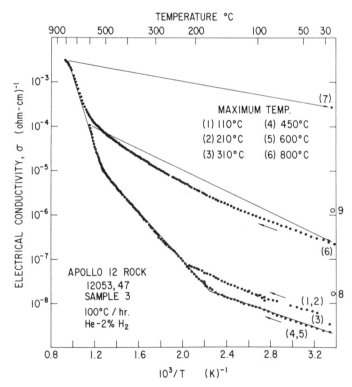

Fig. 2. Conductivity of sample from lunar crystalline rock Apollo 12053 during runs to successively higher temperatures (curves 1–6) in an atmosphere of He–2% H_2. Point 7 is value after heating to 800°C. Points 8 and 9 are values after heating in air to 490°C and 800°C, respectively.

at 800°C resulted in an increase in $\sigma(0)$ (point 9, Fig. 2). These preliminary observations are qualitatively similar to those for the terrestrial augite.

Two recent reports relating to the electrical conductivity of lunar rocks are of particular interest in view of these findings. First, Housley and Morin (1972) have found that olivine single crystals show very large decreases in conductivity when fired in reducing atmospheres at temperatures near 600°C. This effect is larger but in qualitative agreement with the present observations for the terrestrial augite (for comparison see data for 400°C in Fig. 1). Second, Wright (1971) has suggested that nonstoichiometries due to oxygen deficiencies may lead to increased conductivity values in lunar samples with attendant anomalies in the lunar conductivity profile. Insofar as the terrestrial augite can be considered to be an analog of lunar material, the low temperature data illustrated in Fig. 1 support this suggestion.

Obviously the measurement of conductivity in lunar samples still involves several difficulties. In particular, it is necessary either to select and control appropriate chemical environments for the sample or to eliminate leakage conduction paths in these multiphase rocks. Work is continuing in these areas.

Mössbauer Analysis

Conductivity samples

In an attempt to understand the heating effects observed in conductivity studies, Mössbauer spectra were obtained at several temperatures for crystalline rock 12053 and breccia 10048 in their virgin state and after heating to 800°C in the conductivity experiments. The Mössbauer samples were taken from the actual conductivity specimens before and after the conductivity measurements shown in Fig. 2 of this paper and Fig. 3 of a previous paper (Schwerer *et al.*, 1971). Spectra of 12053 are shown in Fig. 3 of this paper. The solid curves are computer fits assuming five quadrupole doublets associated with the M1 and M2 sites in the augite and pigeonite fractions of the pyroxene phase, and with the ilmenite phase.* As seen in Table 1, an expected redistribution of the pyroxene Fe occurs on heating (Hafner *et al.*, 1971) with approximately 2% of the M2 Fe being transferred to M1 sites. Additionally, small changes occur in the quadrupole splittings, isomer shifts, and peak widths as shown in Table 2. These changes and the increase in absorption near zero velocity may be indicative of some oxidation of Fe^{2+} to Fe^{3+}.

In Fig. 4, spectra of the breccia 10048 in the heated and virgin states are shown. Again, as indicated in Table 1, approximately 2% of the pyroxene Fe is transferred from M2 to M1 sites by heating. More striking is the fact that the amount of Fe in

---

* It is perhaps worthwhile to note the constraints used in fitting the various spectra appearing in this paper. As mentioned, for 12053, five quadrupole doublets were used, with all peak widths constrained to be equal but with no constraints on peak heights or positions; the fit obtained in this case was significantly better than that found using the conventional three quadrupole doublets. The silicate and ilmenite peaks of all other spectra (excepting 14047, Fig. 8, which is discussed later in the text) were fit with three quadrupole doublets; the height, width and position of each peak was generally independently varied, but in some cases, the widths of the two peaks in a single doublet were constrained to be equal. For magnetic phases, the peak heights were constrained to the conventional 3 : 2 : 1 ratio.

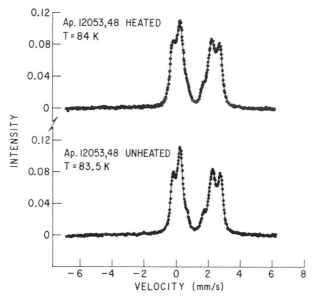

Fig. 3. Mössbauer spectra of 12053 in the heated and virgin states. All spectra shown in this paper were obtained in transmission, the absorber temperatures are given in the figures, velocities are measured with respect to metallic Fe at room temperature and the light solid curves are the results of a least squares computer analysis.

the ilmenite phase is increased by about 30% on heating, with corresponding decreases in the other phases. Since the spectra of Fig. 4 were obtained from two different samples of 10048, the possibility of sampling errors must be ruled out. This is done to some extent by magnetic susceptibility data obtained from several pieces of the conductivity sample before and after heating to 800°C and illustrated in Fig. 5 by

Table 1. Distribution of Fe in conductivity samples before and after heating and fractional occupancy of M1 and M2 sites in pyroxene. (In addition to pyroxene, 10048 contains some glass and olivine, but the M1, M2 terminology is retained for simplicity.)

Sample no.	Silicate			Ilm.	Fe	FeS
	Total	M1	M2			
12053, unheated	0.925	0.460	0.540	0.075	0	0
12053, heated	0.920	0.481	0.519	0.080	0	0
10048, unheated	0.733	0.272	0.728	0.177	0.052	0.039
10048, heated	0.700	0.291	0.709	0.231	0.045	0.025

Table 2. Quadrupole splittings (ε) and isomer shifts (δ) at 84°K of conductivity sample 12053 before and after heating. Here p and α denote pigeonite and augite components of pyroxene.

Site	Before ($\Gamma = 0.31$ mm/sec)		After ($\Gamma = 0.33$ mm/sec)	
	ε (mm/sec)	δ (mm/sec)	ε (mm/sec)	δ (mm/sec)
M1$_p$	3.04	1.27	3.02	1.25
M1$_\alpha$	2.60	1.28	2.57	1.27
M2$_p$	2.11	1.24	2.08	1.24
M2$_\alpha$	1.79	1.21	1.71	1.24
Ilm.	1.05	1.18	1.06	1.19

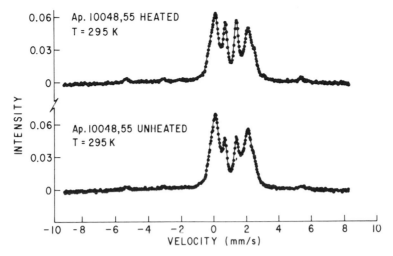

Fig. 4. Mössbauer spectra of 10048 in the heated and virgin states.

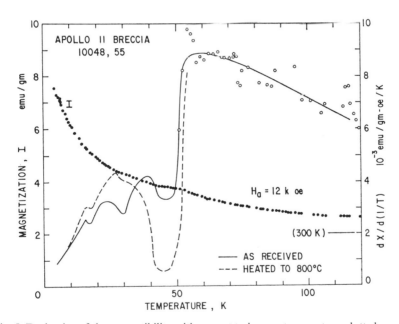

Fig. 5. Derivative of the susceptibility with respect to inverse temperature plotted versus temperature for 10048 before (solid curve) and after (dashed curve) heating. Scatter of derivative values is large only at high temperatures and is illustrated for the solid curve by open circles.

derivative curves ($d\chi/dT^{-1}$). The peak near 56°K is due to the antiferromagnetic transition of ilmenite (Nagata *et al.*, 1972) and is clearly stronger for the heated sample. Mössbauer spectra on a separate single sample of 10048 in the virgin state and after heating for $\frac{1}{2}$ hour at 800°C in a He–2% H$_2$ atmosphere show an increase of about 24% in the amount of ilmenite; however, the sample was heated in a powder form, and a large part of the metallic Fe apparently became volatile, coating the enclosing tube. It is not clear what effect this loss in metallic Fe might have on reactions capable of producing ilmenite (Housley *et al.*, 1970; El Goresy *et al.*, 1972).*

As previously discussed, it is probable that part of the conductivity increase which occurs on heating arises from the reduction of surface layers and part from changes in the bulk material. Mössbauer results offer at least a qualitative explanation of the bulk part of the conductivity change. If, below about 400°C, conduction in the pyroxene phase is due to a transfer of d holes among Fe ions, higher conductivity should be favored by a more equal distribution of Fe atoms among M1 and M2 sites, since the average Fe–Fe distance along the metal chains in the c-axis direction should then be smaller. The transfer of Fe from M2 to M1 sites is in accord with this idea. 10048, which showed the largest thermal hysteresis in conductivity, also showed an apparent increase in ilmenite content of about 30%. The conductivity of oxidized ilmenite at room temperature is about 10^{-2} $(\Omega cm)^{-1}$ (Ishikawa and Sawada, 1956), just slightly greater than that of 10048 after heating to 800°C (Schwerer *et al.*, 1971). Thus, it is possible that creation of nonstoichiometric ilmenite may account for part of the observed conductivity increase.

Apollo 14 samples

Mössbauer spectra have been obtained and analyzed for a number of Apollo 14 samples. Generally they are more heterogeneous than samples returned on previous Apollo missions. The distribution of Fe in the various mineral phases is presented in Table 3. Spectra of several of the more interesting samples are shown in Figs. 6 to 9 and discussed in more detail below.

Table 3. Fractional distribution of Fe in Apollo 14 samples. P, G, and O refer to pyroxene, glass and olivine. If underlined they are major constituents of the silicate phase. M1 and M2 values indicate relative contributions to the outer and inner silicate doublets and have crystallographic meaning only if the major silicate phase is pyroxene.

Sample no.	Total	Silicate M1	Silicate M2	Ilm.	Fe	FeS	Other
14047,47	0.849 (P, G, O)	0.227	0.773	0.03	0.06	≲0.01	0.06 (broad, para.)
14053,48	0.807 (P)	0.314	0.686	0.065	0.078	0.03	0.02 (magnetic)
14063,47	0.85 (~0.75 P, 0.25 O)	0.369 (295°K)	0.631 (295°K)	0.145	≲0.01	—	—
		0.436 (84°K)	0.564 (84°K)				
14259,69 (Fines)	0.86 (P, G, O)	0.22	0.78	0.02	0.04	≲0.01	0.07 (broad, para.)
14301,65	0.909 (P, O)	0.209	0.791	0.06	0.017	0.015	—
14303,35	0.842 (P, O)	0.185	0.815	0.113	0.025	0.02	—

* One of our reviewers has suggested that the increase in ilmenite and other changes which occur for 10048 on heating may result primarily from devitrification of glass.

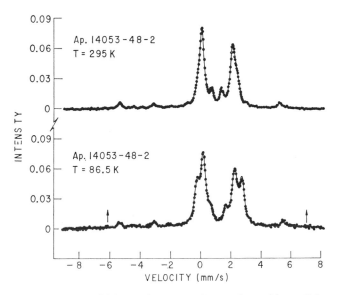

Fig. 6. Mössbauer spectra of 14053. The arrows denote the positions of the outermost peaks for the weak magnetic phase discussed in the text.

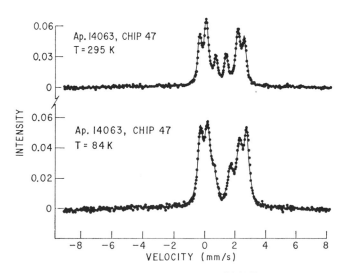

Fig. 7. Mössbauer spectra of 14063.

14053. This rock has a very high content of nearly pure metallic Fe; its magnetic properties have been reported elsewhere (Nagata *et al.*, 1972). At 86°K, a weak magnetic phase believed to be a spinel appears (El Goresy *et al.*, 1972; Haggerty, 1972); its hyperfine field is about 400 kG.

14063. The Mössbauer spectra of Fig. 7 indicate a high olivine content for this sample. This is seen from the large quadrupole splitting (2.92 mm/sec) of the outer silicate

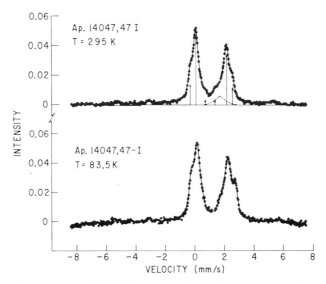

Fig. 8. Mössbauer spectra of 14047. The arrows denote the positions and intensities of the usual silicate and ilmenite peaks and the dashed curve is the additional peak discussed in the text.

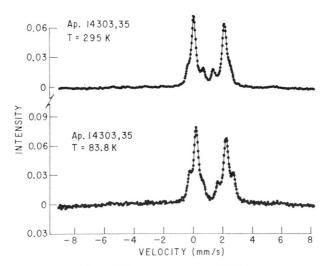

Fig. 9. Mössbauer spectra of 14303.

doublet at room temperature and the increase in intensity of this doublet which occurs when the pyroxene M1 peaks shift outward at 84°K.

14047. As indicated in Table 3, this is perhaps the most heterogeneous of the 14 samples studied. The silicate phase is quite complex, containing a large amount of glass, and the metallic Fe peaks are broad, indicating a high Ni content. Additionally, a broad ($\Gamma = 0.72$ mm/sec) paramagnetic peak occurs at 1.68 mm/sec, presumably arising from a glassy phase. A satisfactory fit to the spectral data could not be obtained without the inclusion of this peak. This was also the case for the fines, 14259, which are very similar to 14047.

14303. The Fe occupancy of M2 sites in the pyroxene phase is highest for this sample. Conductivity measurements on this sample may therefore provide a valuable test of the effect of site redistribution of the pyroxene Fe, discussed in the previous section.

Low temperature Mössbauer results

At liquid helium temperatures, the spin relaxation time in the pyroxene phase is long and magnetic hyperfine splitting is observed in the Mössbauer spectra. Typical spectra (samples 12053 and 14303) are shown in Figs. 10 and 11. For 12053, application of a 900 oer. field produces a considerable decrease in the complexity of the spectrum. In the simplest spin relaxation phenomena, where the spins are dilute and the hyperfine interaction and g tensors are isotropic, such a result could be interpreted in terms of a decoupling of nuclear and electronic spins (Lang and Oosterhuis, 1969; Johnson, 1971). However, the cations sites in pyroxene, particularly the M2 site, are not expected to satisfy the isotropy requirement. Additionally, Hafner has given an average composition of $En_{57}Fs_{32}Wo_{11}$ for pigeonite separated from 12053 (Hafner *et al.*, 1971), and it seems likely that only a small fraction of the Fe spins would be isolated enough to satisfy the dilution requirement. An alternative possibility is that groups of spins, behaving in a superparamagnetic fashion, are slowly relaxing. The applied field might then have the effect of stabilizing the smaller spin clusters. Further experiments in which the applied field is varied should clarify this point.

As seen in Fig. 11, the applied field has little effect on the spectrum of 14303. Two possible explanations suggest themselves. (1) A large portion of the pyroxene

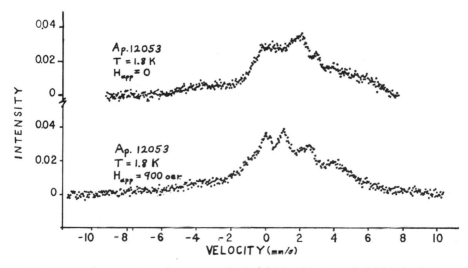

Fig. 10. Mössbauer spectra of 12053 at 1.8°K in 0 field and in an applied field of 900 oer.

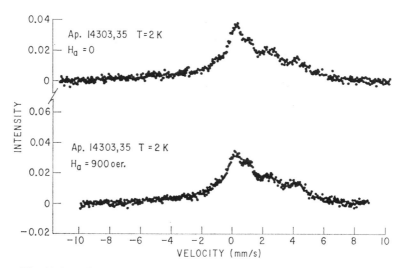

Fig. 11. Mössbauer spectra of 14303 at 2°K in 0 field and in a 900 oer field.

Fe in 14303 is magnetically ordered at 2°K. (2) Since 82% of the pyroxene Fe occupies M2 sites in 14303 (Table 3), the applied field should have little effect on the Mössbauer pattern if only one component of the M2 site g tensor is large (Lang and Oosterhuis, 1969), giving a strongly preferred crystal direction for M2 site spins. For several reasons we favor explanation (2) over (1). First, the hyperfine fields have a typical relaxation mechanism temperature dependence, first causing a slight broadening of the pyroxene peaks at about 20 to 30°K and gradually increasing over the whole temperature range down to 2°K (Huffman *et al.*, 1971); second, results for ortho-pyroxenes indicate that they must contain at least 50 to 60% Fs to be magnetically ordered at any temperature (Shenoy *et al.*, 1969), and; third, our own susceptibility measurements give no evidence for magnetic ordering in pyroxene for either 14303 or 12053 down to 4°K; if anything, the results indicate a stronger interaction between spins for 12053 than for 14303. Explanation (2), on the other hand, would suggest a large separation between d electron states of $|xy\rangle$ and $|xz\rangle|yz\rangle$, symmetry at M2 sites. The small temperature dependence of the M2 quadrupole splitting (Shenoy *et al.*, 1969) supports this suggestion.

Finally, we note that the tentative conclusions of this section, while interesting, must be viewed as being extremely qualitative at this point. Further experiments on well characterized terrestrial pyroxenes and pyroxene separates from lunar samples will be required to obtain more quantitative results.

Acknowledgments—The authors gratefully acknowledge several discussions with R. M. Housley of North American Rockwell Science Center and W. T. Oosterhuis of Carnegie-Mellon University, and correspondence with D. A. Wright of the University of Durham. The sample of terrestrial augite was provided by the Smithsonian Institution through the courtesy of J. S. White, Jr. J. L. Bomback of this laboratory provided microprobe analyses of the terrestrial augite and J. R. Anderson, J. W. Conroy, and G. R. Dunmyre provided valuable technical assistance. This work was largely supported by NASA Contract NAS 9-12271.

REFERENCES

Dunlap W. C. (1959) Conductivity Measurements on Solids. *Methods of Experimental Physics*, Vol. 6B (editors K. Lark-Horovitz and V. A. Johnson). Academic Press.

El Goresy A., Ramdohr P., and Taylor L. A. (1972) Fra Mauro Crystalline Rocks: Petrology, Geochemistry, and Subsolidus Reduction of Opaque Minerals (abstract). In *Lunar Science—III* (editor C. Watkins), pp. 224–226, Lunar Science Institute Contr. No. 88.

Hafner S. S., Virgo D., and Warburton D. (1971) Cation Distributions and Cooling History of Clinopyroxenes from Oceanus Procellarum. *Proc. Second Lunar Sci. Conf., Geochim. Cosmochim. Acta* Suppl. 2, Vol. 1, pp. 91–108. MIT Press.

Haggerty S. E. (1972) Lunar Spinels: Reduction and Compositions (abstract). In *Lunar Science—III* (editor C. Watkins), pp. 348–349, Lunar Science Institute Contr. No. 88.

Housley R. M., Blander M., Abdel-Gawad M., Grant R. W., and Muir A. H. Jr. (1970) Mössbauer Spectroscopy of Apollo 11 Samples. *Proc. Apollo 11 Lunar Sci. Conf., Geochim. Cosmochim. Acta* Suppl. 1, Vol. 3, pp. 2251–2268. Pergamon.

Housley R. M. and Morin F. J. (1972) Electrical Conductivity of Olivine and the Lunar Temperature Profile. Proceedings of Conference on Lunar Geophysics, *The Moon* (in press).

Huffman, G. P., Dunmyre G. R., Fisher R. M., Wasilewski P. J., and Nagata T. (1971) Mössbauer and Supplementary Studies of Apollo 11 Lunar Samples. *Mössbauer Effect Methodology*, Vol. 6, pp. 209–224. Plenum Press.

Ishikawa Y. and Sawada S. (1956) Study on Substances having the Ilmenite Structure I; Physical Properties of Synthesized $FeTiO_3$ and $NiTiO_3$, *J. Phys. Soc. Japan* **11**, 496–501.

Johnson C. E. (1971) Mössbauer Spectroscopy and Biophysics. *Physics Today*, Feb. 1971, p. 35.

Lang G. and Oosterhuis W. J. (1969) Calculated Paramagnetic Mössbauer Spectra of Spin-1/2 Iron Salts. *J. Chem. Phys.* **51**, 3608.

Nagata T., Fisher R. M., and Schwerer F. C. (1972) Lunar Rock Magnetism. Proc. Conf. on Lunar Geophysics, *The Moon* (in press).

Noritomi K. and Asada A. (1957) Studies of the Electrical Conductivity of a Few Samples of Granite and Andesite. *Sci. Rept. Tohoku Univ.* **7**, 201–207.

Schwerer F. C., Nagata T., and Fisher R. M. (1971) Electrical Conductivity of Lunar Surface Rocks and Chondritic Meteorites. *The Moon* **2**, 408–422.

Shenoy G. K., Kalvius G. M., and Hafner S. S. (1969) Magnetic Behavior of the $FeSiO_3$–$MgSiO_3$ Orthopyroxene System from NGR in Fe. *J. Appl. Phys.* **40**, 1314–1316.

Wright D. A. (1971) Electrical Conductivity of Lunar Rock. *Nature* **231**, 169–170.

Proceedings of the Third Lunar Science Conference
(Supplement 3, *Geochimica et Cosmochimica Acta*)
Vol. 3, pp. 3187–3193
The M.I.T. Press, 1972

Grain size analysis, optical reflectivity measurements, and determination of high-frequency electrical properties for Apollo 14 lunar samples

T. Gold, E. Bilson, and M. Yerbury

Center for Radiophysics and Space Research,
Cornell University, Ithaca, New York 14850

Abstract—The particle size distribution is measured for the Apollo 14 bulk and contingency fines as well as for two subsamples from the 14230 core sample. Among these samples there seems to be no significant variation in grain size distribution. Reflectivity measurements on lunar fines from different Apollo missions show that their albedo decrease significantly after being subjected to a dose of proton bombardment which would be equivalent to approximately 1.5×10^4 years of solar wind. Results of dielectric constant and power absorption length measurements are reported for Apollo 14 fines and an Apollo 14 rock sample. A strikingly long absorption length 28 wavelengths is observed for the rock sample 14310,161 at 450 MHz.

Grain Size Analysis

The particle size distribution has been determined for several samples of Apollo 14 lunar fines by the sedimentation rate analysis in a column of water. This method has been described earlier (Gold *et al.*, 1971); it is accurate for comparative measurements, and together with earlier calibration measurements it gives definitive results. It is restricted to the particle size range 1–100 μ. For the purpose of comparing the size distributions of fines collected at different locations and/or at different depths the method is very useful and simple. Our main interest has been the comparison of surface material at the different sites, and the search for physically distinct layers in the core samples, such layers being clearly present in the Apollo 12 double core sample (Gold *et al.*, 1971).

Figure 1a compares the differential particle size distribution of the Apollo 14 bulk and contingency fines (both from locations near the LM) with that of two subsamples from the 14230 core sample. Among these samples there seems to be no significant variation in grain size distribution. This is true despite the existence in this core tube of three distinct morphologic units, as reported by Fryxell and Heiken (1971). Thus the sample was probably not mixed, either on the lunar surface or in the collection process, and a distinct origin has to be envisioned for each of the units in which, nevertheless, the same size distribution was generated. The study of the Apollo 14 double core sample, when available, would probably prove most interesting.

Figure 1b compares the differential particle size distribution of three surface fine samples: one from each Apollo mission. These curves show that A14163 is appreciably finer grained than either the Apollo 11 or 12 fines.

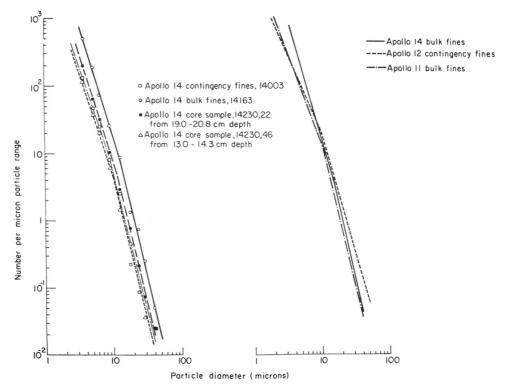

Fig. 1. (a) The differential particle size distribution for the Apollo 14 bulk and contingency fines and two of the 14230 core subsamples. (b) The differential particle size distribution for the Apollo 11 and 14 bulk fines and the Apollo 12 contingency fines.

REFLECTIVITY MEASUREMENTS AND SURFACE DARKENING BY SIMULATED SOLAR WIND

The results of the laboratory determination of the optical properties of Apollo 11 and 12 dust samples were in close agreement with the earlier telescopic measurements on the lunar maria (Gold *et al.*, 1970, 1971). A number of investigations have been thus conducted in order to explain the low albedo, the spectrum, and other optical properties of these dust samples which seem to be characteristic of the lunar lowland material. There has been strong evidence that the high percentage of dark glass content in the mare soil could account for the low albedo of the powder (Adams and McCord, 1971); however, it has not been demonstrated that it also accounts for the polarization properties, which demand high opacity.

It has been proposed (Gold, 1960) that the low albedo is due to the darkening effect of the solar wind. Extensive experiments with terrestrial rock powders irradiated with approximately 100 coulomb/cm^2, of 2 keV protons (Hapke, 1965) showed that the entire range of optical properties of the moon (color, albedo, polarization) could be reproduced with a variety of powders.

Evidence of solar wind sputtered coating on lunar fines was found by Gold *et al.* (1970) and Hapke *et al.* (1970). Irradiation experiments have been also performed on lunar material. Hapke *et al.* found that the albedo of a pulverized sample of crystalline rock 10022 decreased from 0.18 to 0.12 when irradiated by approximately 20 coulomb/cm^2 of 2 keV hydrogen ions. A similar proton irradiation of the Apollo 11 fines resulted in a very small decrease of the albedo only (0.097 to 0.094), perhaps because lunar fines had already suffered near saturation darkening on the moon. Dollfus *et al.* (1971) irradiated two Apollo 11 and two Apollo 12 surface and core samples of fines with 4 μamp/cm^2 hr of 60 keV protons. Two of these samples had a slightly lower albedo after irradiation and their polarization characteristics were also changed. Two samples showed no change in these optical properties. Nash and Conel (1971) observed no change in reflectance properties of any of the Apollo 12 double core samples they examined, as a result of proton bombardment by approximately 3.2×10^{-2} coulomb of H ions per cm^2 at 4 keV.

Lunar fines from different Apollo missions have been subjected to simulated solar wind also in this laboratory. The proton irradiation chamber used in these experiments was described by Hapke (1964). Although the vacuum system makes use of an oil diffusion pump, we believe that the darkening induced by proton bombardment cannot be due to organic contamination (i.e., cracked pump oil) since along with each sample a control MgO sample was always placed in the target area, and the albedo of the latter remained consistently unchanged after irradiation. As can be seen in Fig. 2, the albedo of both the Apollo 11 bulk fines and the more reflectant Apollo

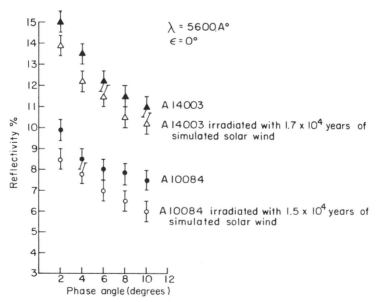

Fig. 2. Reflectivity of an Apollo 11 and an Apollo 14 powder sample before and after being subjected to simulated solar wind (2 keV, 0.16 mA/cm^2 proton beam, irradiation times: 13.5 hours and 14.75 hours respectively).

14 bulk fines decreased significantly after being subjected to 8.4 coulomb/cm^2 of 2 keV proton bombardment, the equivalent of approximately 1.5 × 10^4 yr of solar wind. The albedo of a finely ground terrestrial olivine basalt sample decreased from 34.3% (at 10° phase angle and $\varepsilon = 0°$ viewing angle) to 22.8% after being irradiated with approximately 3 coulomb/cm^2 or 5 × 10^3 yr of solar wind in the same experimental conditions.

In these experiments we used a larger dose of irradiation than was used by the other investigators (except Hapke) quoted earlier. This might be responsible for the more pronounced changes in albedo observed with our samples. Since pulverized lunar rock appears to have almost invariably a much higher albedo than lunar fines, one can conclude that either the bulk of the fines are not derived from the pulverization of lunar rocks, or a subsequent surface treatment on the moon, such as by sputtering, greatly diminishes the albedo. We therefore plan to undertake more extensive investigations involving irradiation of different pulverized lunar rock samples. The comparison of the chemical composition of the surfaces of the grains of irradiated rock powder samples and of the fines may demonstrate the process that has been active on the moon.

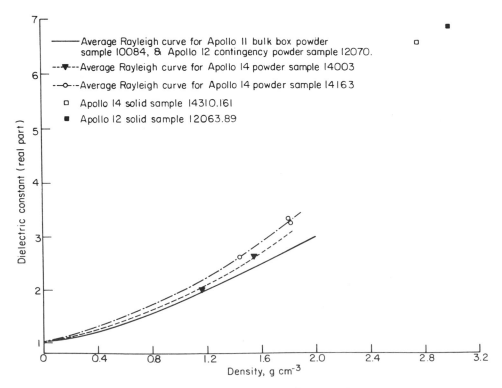

Fig. 3. Dielectric constant measurements for two Apollo 14 powder samples as a function of bulk powder density, compared to the average Rayleigh curve which fits the measured data for the Apollo 11 bulk fines and Apollo 12 contingency fines. Dielectric constant versus density points sample are also shown.

ELECTRICAL PROPERTIES

Two surface fine samples (A14163 and A14003) and one rock sample (A14310,161) have been analyzed by techniques described in detail earlier (Campbell and Ulrichs, 1969). Moisture effects were avoided by cutting the solid samples dry in laboratory atmosphere and all the samples were vacuum baked at 120°C for 1 day. The dielectric constant measurements are shown in Fig. 3, the power absorption length in these samples is shown in Fig. 4. (Some earlier Apollo 11 and 12 data are also shown in these figures.)

The variation of the dielectric constant as a function of the bulk powder density follows the Rayleigh formula (Campbell and Ulrichs, 1969) and the curves for the Apollo 14 samples differ only slightly from the average Rayleigh curve for the Apollo 11 bulk box powder and the Apollo 12 contingency powder sample. The single point of dielectric constant versus density for the only Apollo 14 rock sample analyzed is close to the corresponding data point of a typical Apollo 12 rock sample. The ground-based radar determinations of the dielectric constant (see Evans and Hagfors, 1968) are in good agreement with measurements on all our lunar powder samples from the three Apollo missions so far if one assumes a density of about 1.5–1.7 g/cm^3 for the soil at a depth of 20 cm. (According to the latest available data, the bulk density

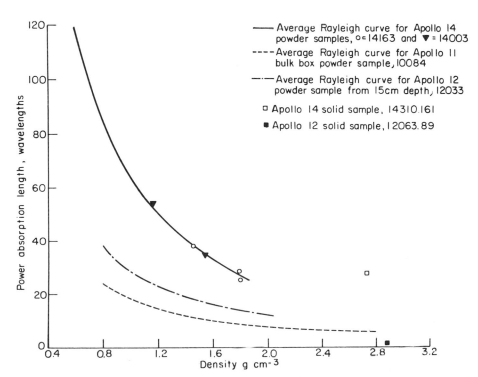

Fig. 4. The variation with density of the absorption length in two Apollo 14 powder samples, in the Apollo 11 bulk box powder samples, in the Apollo 12 powder sample from 15 cm depth. Points for an Apollo 14 and an Apollo 12 solid sample are also shown.

ranges for the Apollo 12 and 14 core tube samples were 1.74 to 1.98 g/cm^3 and 1.60 to 1.75 g/cm^3, respectively (Carrier III, 1972).

A new and striking result is the very long absorption length, 28 wavelengths, observed for the Apollo 14 rock sample, 14310,161 at 450 MHz. This can be compared with only 1.7 wavelengths for the Apollo 12 rock sample, 12063,89 (the power absorption of this latter rock being typical for other Apollo 11 and 12 rock samples as well). The densities and dielectric constants of these two rock samples were similar.

Apollo 14 dust samples are also less absorbent than the Apollo 11 and 12 samples. At a packing density of 1.6 g/cm^3 the absorption length is approximately 34 wavelengths or 3.5 times greater than the absorption length in A10084 or 2.2 times greater than that in A12033.

Materials of such low absorption are not common constituents of the crust of the earth, and such behavior cannot be predicted on the basis of chemical analysis. Very small effects such as small impurity levels in particular crystal sites probably dominate in the definition of the power absorption. The possibility that rocks such as 14310 are typical in certain regions of the lunar surface adds new significance to lunar radar observations. At a wavelength of 10 meters reflections from a depth of over 100 meters may well contribute very significantly to the radar echoes. If any soil exists which is the pulverized version of rock 14310, then radio wave penetrations of 70 wavelengths would be expected, and radar echoes from several hundred meters depth could be observed.

Acknowledgments—We wish to thank Mr. H. J. Eckelmann and G. A. Smith for technical assistance. Work on lunar samples was carried out under NASA Grant NGR-33-010-137.

References

Adams J. B. and McCord T. G. (1971) Optical properties of mineral separates, glass, and anorthositic fragments from Apollo mare samples. *Proc. Apollo 12 Lunar Sci. Conf., Geochim. Cosmochim. Acta* Suppl. 2, Vol. 3, pp. 2183–2195. MIT Press.

Campbell M. J. and Ulrichs J. (1969) Electrical properties of rocks and their significance for lunar radar observations. *J. Geophys. Res.* **74**, 5867–5881.

Carrier W. D. III (1972) Core sample depth relationships: Apollo 14 and 15. In *Lunar Science—III* (editor C. Watkins), pp. 122–124, Lunar Science Institute Contr. No. 88.

Dollfus A., Geake J. E., and Titulaer C. (1971) Polarimetric properties of the lunar surface and its interpretation. Part 3: Apollo 11 and Apollo 12 lunar samples. *Proc. Apollo 12 Lunar Sci. Conf., Geochim. Cosmochim. Acta* Suppl. 2, Vol. 3, pp. 2285–2300. MIT Press.

Evans J. V. and Hagfors T. (1968) *Radar Astronomy*, McGraw-Hill.

Fryxell R. and Heiken G. (1971) Description, dissection, and subsampling of Apollo 14 core sample 14230. NASA TMX-58070.

Gold T. (1960) Processes on the lunar surface. Proc. of Symposium No. 14 of the I.A.U. *The Moon*, pp. 433–439. Academic Press.

Gold T., Campbell M. J., and O'Leary B. T. (1970) Optical and high-frequency electrical properties of the lunar sample. *Proc. Apollo 11 Lunar Sci. Conf., Geochim. Cosmochim. Acta* Suppl. 1, Vol. 3, pp. 2149–2154. Pergamon.

Gold T., O'Leary B. T., and Campbell M. (1971) Some physical properties of Apollo 12 lunar samples. *Proc. Apollo 12 Lunar Sci. Conf., Geochim. Cosmochim. Acta* Suppl. 2, Vol. 3, pp. 2173–2181. MIT Press.

Hapke B. W. (1964) Effects of a simulated solar wind on the photometric properties of rocks and powders. Cornell University, CRSR 169.

Hapke B. W. (1965) Optical properties of the Moon's surface. Proc. of the 1965 IAU-NASA symposium. *The Nature of the Lunar Surface*, pp. 141–154. Johns Hopkins.

Hapke B. W., Cohen A. J., Cassidy W. A., and Wells E. N. (1970) Solar radiation effects on the optical properties of Apollo 11 samples. *Proc. Apollo 11 Lunar Sci. Conf., Geochim. Cosmochim. Acta* Suppl. 1, Vol. 3, pp. 2199–2212. Pergamon.

Nash D. B. and Conel J. E. (1971) Luminescence and reflections of Apollo 12 samples. *Proc. Apollo 12 Lunar Sci. Conf., Geochim. Cosmochim. Acta* Suppl. 2, Vol. 3, pp. 2235–2244. MIT Press.

Proceedings of the Third Lunar Science Conference
(Supplement 3, *Geochimica et Cosmochimica Acta*)
Vol. 3, pp. 3195–3200
The M.I.T. Press, 1972

CESEMI studies of Apollo 14 and 15 fines

Herta Görz, E. W. White, G. G. Johnson, Jr., and Mary W. Pearson

Materials Research Laboratory, The Pennsylvania State University,
University Park, Pennsylvania 16802

Abstract—Quantitative size and shape analyses have been carried out on six lunar fines from Apollo 14 and 15 by computer evaluation of scanning electron microscope images (CESEMI). For the size range of 0.50 to 30 μm diameter, the distributions are log-normal on a number count basis. This finding is in contrast with results for coarser materials in the fines reported by King *et al.* suggesting the operation of a different process. Aspect ratios range from 0.1 to 1.0 with an average value of about 0.6. Only slight differences are observed among samples. A shape complexity factor (ratio of particle perimeter to ellipse perimeter) also varies only slightly among the six samples.

INTRODUCTION

The CESEMI technique (Computer Evaluation of Scanning Electron Microscope Images) permits the size, shape, and chemical analysis of fine particles in the micron and submicron range (McMillan *et al.*, 1969; White *et al.*, 1970; Görz *et al.*, 1971). This paper presents results on the size and shape analysis of six lunar fines including samples 14163,158, 15031,44, 15041,50, 15231,49, 15501,25, and 15531,40. The purpose of this study is to determine the shape of the grain size distribution curve in the lower particle size range (0.5–30 μm) that cannot be readily done by other means. Several authors (Frondel *et al.*, 1971; Heywood, 1971; King *et al.*, 1971; Sellers *et al.*, 1971) have published size distribution curves derived from weight determinations by sieving and Coulter counting techniques. The lower cut-off of those curves occurs at around 2–10 μm, a size range where the scanning electron microscope technique is helpful in describing the finer grain sizes. [Also see Gold *et al.*, 1971, for additional lunar grain size data using SEM—Ed.]

Microscopy studies of the six fines indicate an upper size limit of around 600 μm, the shape of the particles ranging from spherical (brown glass, seldom) to angular with sharp edges. The minerals identified were pyroxene (greenish brown, sharp edges, sometimes isotropic), olivine (green to brown, rounded), plagioclase (sometimes twins, optical character negative), opaques, and glasses with various colors, shapes, and refractive indices. The amount of glasses in these samples is higher than in the samples of Apollo 11 and 12 studied. It is remarkable that the appearance of the glass fragments regarding color and shape resembles very often that of clinopyroxene. Only very few glass spheres or dumbbells have been found.

SPECIMEN PREPARATION AND RECORDING PARAMETERS

The medium for dispersing the fines on a substrate for the CESEMI recordings is a eutectic mixture of camphor and naphthalene (Thaulow and White, in press). For this study, highly polished metallic Be was chosen as a substrate. The samples were not given a conductive coating. The secondary electron images were recorded using a voltage of 25 keV and 10^{-9} amp specimen current. The secondary electron detector was optimized to suppress shadows. All images were recorded at a picture point density of 256 points per line and 256 lines. Six images were recorded for each of the six samples; three were taken at 300×, three at 1250× magnification.

RESULTS AND DISCUSSION

In order to accurately measure the size distribution over the range of 0.5 to 30 μm, data from the two magnification settings had to be combined, taking into account the relative areas represented by the two magnifications. The data were merged with no overlap between magnifications. For the magnifications 300× and 1250× the following limits in Table 1 were used:

Table 1. Limits for merging.

Magnification	Lower Size Limit	Upper Size Limit
1250×	0.5 μm	5.0 μm
300×	5.0 μm	30.0 μm

The number of counts must be weighted to account for the relative areas sampled. The area sampled at 1250× is approximately $\frac{1}{17}$ that of the area sampled at 300×, hence the number of counts are adjusted accordingly.

Results of the particle size analyses are summarized in Figs. 1 and 2. On the logarithmic scale is recorded the equivalent circular diameter of the particles; on the probability scale the cumulative frequency of the number of grains. The size distributions are essentially log-normal. Because of the selected cut-offs discussed above, the smallest grain size counted is 0.5 μ. A slight uncertainty due to statistics of the counting is evidenced in the upper size range (above 98%).

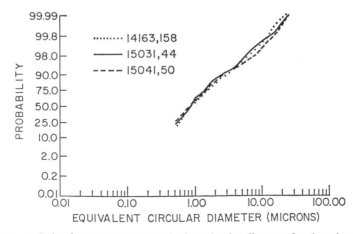

Fig. 1. Cumulative frequency versus equivalent circular diameter for three lunar fines.

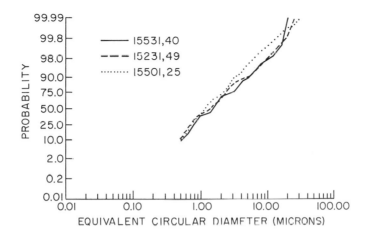

Fig. 2. Cumulative frequency versus equivalent circular diameter for three lunar fines.

Table 2 presents the total number of grains processed for establishing the frequency curves, the median size of the equivalent circular diameter, and the slope of the distribution curves mathematically fitted for data between 5% and 95% probability.

The shapes of the particles in the size range of 1.25–30 μm equivalent circular diameter were characterized from ellipses calculated by least square fits to each particle perimeter (Matson *et al.*, 1970). Histograms of the aspect ratios (minor/major axes of calculated ellipses) are given in Fig. 3. An aspect ratio of 1.0 indicates an equant particle, while a ratio of 0.1 indicates an elongate shape. Most grains seem to be slightly-to-medium elongated with a ratio between 0.4–0.7. Samples 15501 and 15531 show bimodal distributions.

Figure 4 summarizes the histograms of the shape complexity factor (particle perimeter/fitted ellipse perimeter). Around 80% of the grains measured have a shape complexity factor in the range 0.80–1.00. If a particle has a shape complexity factor greater than 1, then the particle is highly angular and complex with possible reentrant perimeter. If the factor is equal to 1, the particle is smooth and not reentrant, while a factor less than 1 shows an extremely regular particle but not elliptical in shape.

Table 2. Number of particles, median size of merged equivalent circular diameter, and slope of distribution curves for six lunar fines.

Sample	Number of Particles Measured	Median Size of Equivalent Circular Diameter (μm)	Slope of Distribution Curve
14163,158	471	0.90	0.97
15031,44	603	0.86	0.97
15041,50	220	0.84	0.91
15231,49	365	1.37	0.89
15501,25	195	1.17	1.09
15531,40	157	1.49	0.92

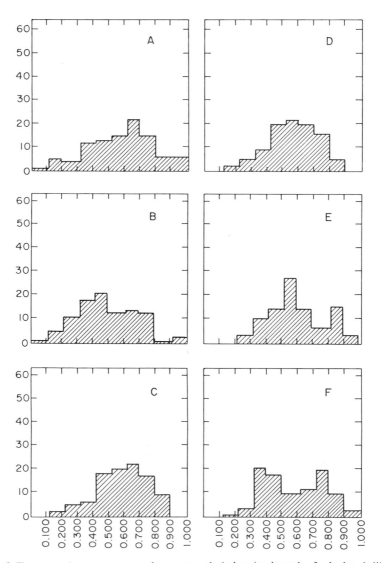

Fig. 3. Frequency in percent versus the aspect ratio (minor/major axis of calculated ellipse) for six lunar fines.

A. Sample 14163,158   D. Sample 15231,49
B. Sample 15031,44   E. Sample 15501,25
C. Sample 15041,50   F. Sample 15531,40

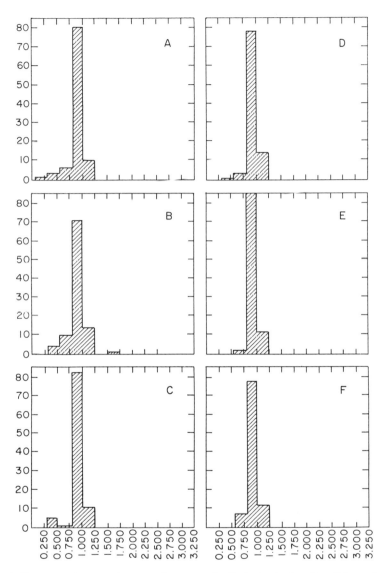

Fig. 4. Frequency in percent versus shape complexity factor (ratio of particle perimeter to ellipse perimeter) for six lunar fines.

A. Sample 14163,158   D. Sample 15231,49

B. Sample 15031,44   E. Sample 15501,25

C. Sample 15041,40   F. Sample 15531,40

CONCLUSIONS

The six samples of lunar fines used in this study all have size distributions that are log-normal on a number count basis over the size range from 0.5 to 30 μm. All six distribution curves have the same slope. As the distributions are log-normal on a number count basis they are, of necessity, nonlog-normal on a weight basis. This finding is quite different from King et al. found for the coarser fractions where they find log-normal behavior on a weight basis.

Sellers et al. interpret the log-normal behavior (weight basis) to be indicative of communition processes. Since we observe a different distribution function for the finer fraction (0.5–30 μm) one can only assume that different processes are operative in generating these fine particles.

Acknowledgment—This work was supported under NASA Grant NGR-39-009-183.

REFERENCES

Frondel C., Klein C. Jr., and Ito J. (1971) Mineralogical and chemical data on Apollo 12 lunar fines. *Proc. Second Lunar Sci. Conf., Geochim. Cosmochim. Acta* Suppl. 2, Vol. 1, pp. 719–726. MIT Press.

Gold T., O'Leary B. T., and Campbell M. (1971) Some physical properties of Apollo 12 lunar samples. *Proc. Second Lunar Sci. Conf., Geochim. Cosmochim. Acta* Suppl. 2, Vol. 3, pp. 2173–2181. MIT Press.

Görz H., White E. W., Roy R., and Johnson G. G. Jr. (1971) Particle size and shape distributions of lunar fines by CESEMI. *Proc. Second Lunar Sci. Conf., Geochim. Cosmochim. Acta* Suppl. 2, Vol. 3, pp. 2021–2025. MIT Press.

Heywood H. (1971) Particle size and shape distribution for lunar fines sample 12057,72. *Proc. Second Lunar Sci. Conf., Geochim. Cosmochim. Acta* Suppl. 2, Vol. 3, pp. 1989–2001. MIT Press.

King E. A., Butler J. C., Carman M. F. (1971) The lunar regolith as sampled by Apollo 11 and Apollo 12: Grain size analyses, and origins of particles. *Proc. Second Lunar Sci. Conf., Geochim. Cosmochim. Acta* Suppl. 2, Vol. 1, pp. 737–746. MIT Press.

Kim Y. K., Lee S. M., Yang T. H., Kim T. H., and Kim C. K. (1971) Mineralogical and chemical studies of lunar fines 10084,148 and 12070,98. *Proc. Second Lunar Sci. Conf., Geochim. Cosmochim. Acta* Suppl. 2, Vol. 1, pp. 741–755. MIT Press.

Matson W. L., McKinstry H. A., Johnson G. G. Jr., White E. W., and McMillan R. E. (1970) Computer processing of SEM images by contour analyses. *Pattern Recognition* **2,** 303–312.

McKay D. S., Morrison D. A., Clinton U. S., Ladle G. H., and Lindsay J. F. (1971) Apollo 12 soil and breccia. *Proc. Second Lunar Sci. Conf., Geochim. Cosmochim. Acta* Suppl. 2, Vol. 1, pp. 755–776. MIT Press.

McMillan R. E., Johnson G. G. Jr., and White E. W. (1969) Computer processing of binary maps of SEM images. Second Annual Scanning Electron Microscope Symposium, IITRI, 439–444.

Sellers G. A., Woo C. C., and Bird M. L. (1971) Composition and grain-size characteristics of fines from the Apollo 12 double-core tube. *Proc. Second Lunar Sci. Conf., Geochim. Cosmochim. Acta* Suppl. 2, Vol. 1, pp. 665–678. MIT Press.

Thaulow N and White E. W. General method for dispersing and disaggregating particulate samples for quantitative SEM and optical microscope studies (to be published in *Powder Technology*).

White E. W., Görz H., Johnson G. G. Jr., and McMillan R. E. (1970) Particle size distributions of particulate aluminas from computer processed SEM images. Third Annual Scanning Electron Microscope Symposium, IITRI, 57–64.

White E. W., Mayberry K., and Johnson G. G. Jr. Computer analysis of multichannel SEM and x-ray images from fine particles (to be published in *Pattern Recognition*).

Proceedings of the Third Lunar Science Conference
(Supplement 3, *Geochimica et Cosmochimica Acta*)
Vol. 3, pp. 3201–3212
The M.I.T. Press, 1972

Scanning electron microscope and energy dispersive x-ray analysis of the surface features of Surveyor III television mirror

J. C. Mandeville* and H. Y. Lem

Space Science Division, NASA-Ames Research Center,
Moffett Field, California 94035

Abstract—The mirror of the Surveyor III television camera has been searched for micrometeoroid impact sites. Optical and scanning electron microscope analysis shows that the mirror is covered with a homogeneous layer of dust whose size and composition are consistent with a lunar origin. Further, the surface is pitted and the concentration of the pits increases from the bottom to the top of the mirror. Impact features are found at a concentration of $10^6/cm^2$ and nearly 75% are smaller than 4 μm. The shape and the orientation of most pits seem to show that they were caused by rather low-velocity particles striking the mirror with an oblique or grazing angle and were coming from some source below the mirror. No definite hypervelocity impacts were found and a flux limit of 1×10^{-2} m^{-2} sec^{-1} (2π steradians)$^{-1}$ can be assumed for particles with masses larger than 1×10^{-13} g.

Introduction

The Planetology Branch, Ames Research Center, received the mirror of the Surveyor III television camera from the Jet Propulsion Laboratory to search for micrometeoroid impact sites. The Surveyor III mirror was exposed to the lunar environment for 944 days. Previous investigators (Hallgren *et al.*, 1972) had already studied this mirror but we had the first opportunity to observe it with a scanning electron microscope (SEM). Of the various television camera surfaces this one seemed the most appropriate for the detection of small impact craters. We were also allowed to remove the two trunnions (aluminum pieces on each side bearing the mirror) for SEM analysis. In this paper we show the results of this examination.

The Mirror

The mirror was manufactured by electrode deposition of nickel plating 80 μm thick on a beryllium block. This plating was polished and then vacuum coated with a 0.2 μm thin film of aluminum. Finally, a 0.1 μm thin coating of silicon oxide was deposited on top of the aluminum film. When the mirror was first removed from the camera, it was totally covered with dust except for the portion wiped off by the astronaut's finger. During later analysis by other investigators a 2 cm wide strip of dust along its main axis was taken off by acetate film stripping. Also, an area was rubbed off by a cotton tipped swab (Q-Tip). In our investigations we examined three

* On leave from the DERTS, Departement de l'Office National d'Etudes et de Recherches Aerospatiales, Toulouse, France, under a grant from the European Space Research Organization.

areas. The first was an untouched area covered with debris; the second, the area where the dust had been stripped off; and the third, the area where the dust had been rubbed off with the Q-Tip (Fig. 1).

Prior to SEM examination, we scanned the surface with an optical microscope along the main axis of the mirror. From this examination we found that the untouched areas of the mirror were covered with a homogeneous layer of dust. Furthermore, the surface was pitted and the concentration of pits increased from the bottom to the top of the mirror. A significant change occurred across an oblique line in the lower part of the mirror near the rotation axis. The bottom seemed to have been shielded from impacting particles and could be used as a control surface. The optical scan

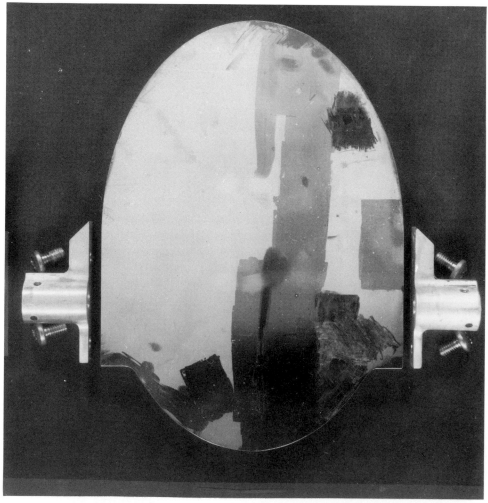

Fig. 1. Surveyor III television mirror and dismantled aluminum trunnions, showing areas stripped and rubbed free of contaminating dust.

also showed that the mirror was covered with a contaminant that may have been deposited by the combustion of the vernier engines. This contaminant could not be found in the lower part of the mirror; it was more adherent than the dust and was only removed by Q-Tip rubbing.

This coating caused interference bands and brown coloration of the mirror. In the optical microscope it looked like a deposit of irregular shaped, micro-sized blisters. The x-ray attachment to the SEM did not give any information about its composition. Cour-Palais *et al.* (1971) encountered the same material on a Surveyor polished aluminum tube. Finally, we found splatters (at the average density of 15/mm²) that looked like impact craters (Hallgren *et al.*, 1972). However, a closer view with the SEM showed that these features were probably contamination which was on the mirror before the Surveyor launching.

The small sample chamber on the SEM confined our examination to the central portion of the mirror. The three areas of interest (untouched region, stripped area, and Q-Tip rubbed zone) were visible in this central 3 × 3 cm area.

In the untouched area the search for impact sites was difficult due to the great amount of dust on the surface (Fig. 2). This dust consisted mainly of angular particles with 1% to 2% spherical particles. The concentration of the debris was about 3×10^7 particles/cm²; few particles are larger than 4 μm and 80% are smaller than 2 μm (Fig. 3). The size and the concentration of these particles are in agreement with similar measurements performed by others on this mirror (Carr and Proudfoot,

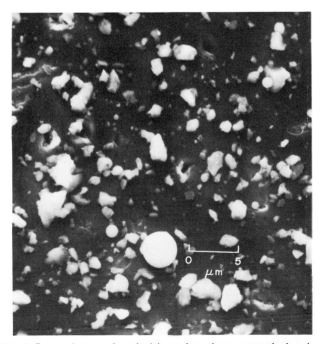

Fig. 2. Lunar dust on the television mirror in an untouched region.

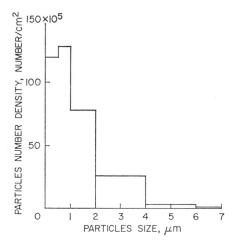

Fig. 3. Size distribution of the particles on the untouched area of the mirror.

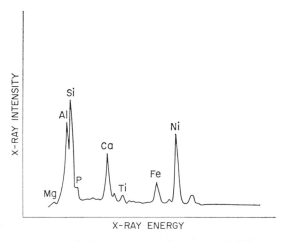

Fig. 4. X-ray spectrum of a spherical particle on the mirror (the Ni peak corresponds to the substrate).

private communication, 1971) and on the Surveyor III optical filters (Brownlee *et al.*, 1971). The composition of the debris (Mg, Al, Si, Ca, Ti, Fe) as given by the x-ray spectrometer is consistent with a lunar origin (Fig. 4).

In the stripped area and the Q-Tip zone of the mirror where lunar dust did not obstruct the view, impact features were found at the average rate of $10^6/cm^2$ (Fig. 5). The largest of the impact features were elongated gouges; the smallest were nearly circular dents. Their sizes ranged from 25 μm length and 8 μm width, down to 1 μm diameter. Nearly 75% of the pits were smaller than 4 μm (Fig. 6). We found spherical or irregular particles imbedded in several pits. These particle compositions were consistent with a lunar origin (Figs. 7, 8). The shape and the orientation of most pits seem to show that they were caused by rather low-velocity particles (some hundred

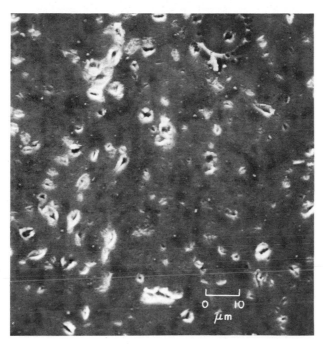

Fig. 5. Impact pits on the mirror in the region where dust has been removed by stripping.

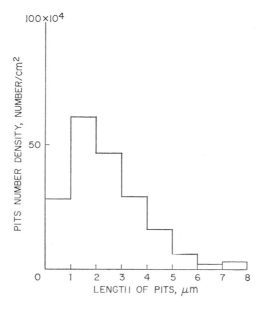

Fig. 6. Size distribution of the impact pits on the mirror.

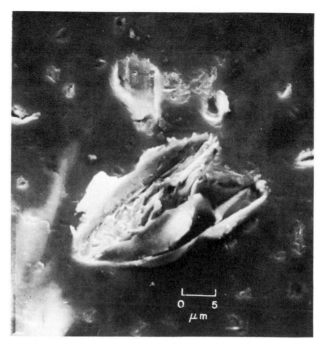

Fig. 7. Impact pit showing an imbedded particle of irregular shape.

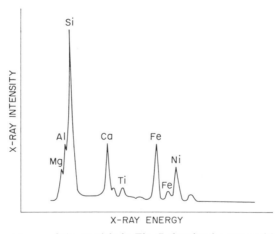

Fig. 8. X-ray spectrum of the particle in Fig. 7 showing its composition is typical of lunar grain composition.

meters per second) striking the mirror with an oblique or grazing angle and coming from some source below the mirror.

Of particular interest were the splatters scattered on the surface of the mirror. Two kinds of features could be distinguished; the first were small spall-like features with a central hole, about 5 to 10 μm in diameter, the second were circular features, with cracked rims raised above the surface of the mirror. Their diameters ranged from 8 to 40 μm (Fig. 9). The energy-dispersive spectrometer attached to the SEM

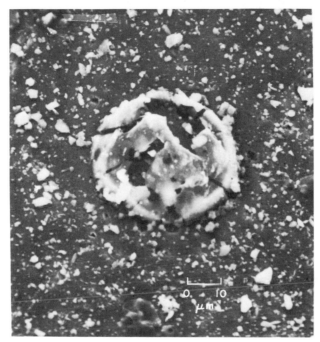

Fig. 9. Contamination splatter which was present on the mirror surface before its flight
to the moon.

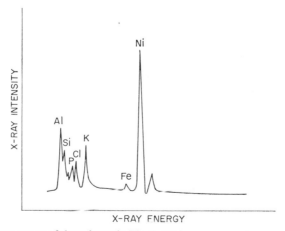

Fig. 10. X-ray spectrum of the splatter in Fig. 9. Major components correspond to those
of thermal paint used on the spacecraft.

showed that these splatters were mainly composed of potassium, aluminum, and silicon which is the composition of the thermal paint used on the spacecraft (Fig. 10). This contaminant was on the mirror before the Surveyor launching and landing because the splatters were covered by lunar dust.

The Trunnions

When the trunnions were removed from the mirror they showed a brown coloration on the part exposed to the lunar environment. The left trunnion was colored more than the right one (Fig. 11). This coloration was similar to that on the mirror and was probably caused by the effluent from the vernier engines. A portion of each trunnion was fastened tightly against the mirror by screws and was completely protected from the lunar environment. This served as an excellent control area free of any features caused by lunar exposure (Fig. 12). Under SEM examination, all parts of the aluminum trunnions were found to be rather rough, with parallel polishing grooves and small, irregular scratches. Careful scanning of the areas which had been exposed to the moon's environment revealed a great amount of lunar debris and numerous impact-looking features (Fig. 13). Most of the pits were hemispherical or elliptical gouges crossing the polishing grooves with, often, the crater lip folded out away from the direction of impact. Their sizes ranged from 0.5 to 10 μm. Spherical lunar particles were found imbedded in many pits (Figs. 14 and 15). An average pit concentration of $3 \times 10^4/\text{cm}^2$ was measured on the left trunnion and $1 \times 10^4/\text{cm}^2$ on the right. These impact features, whose morphology strongly differs from polishing scratches, were not found on the unexposed areas. Comparing these with simulated impacts shows that they were caused by particles striking the trunnions with a velocity of several hundred meters per second.

Fig. 11. Left and right trunnions removed from the mirror showing protected and exposed areas. The small projected exposed area on the left trunnion is darker than the right because of denser dust deposits.

Fig. 12. Right trunnion, protected area magnified in the scanning electron microscope showing no impact pits.

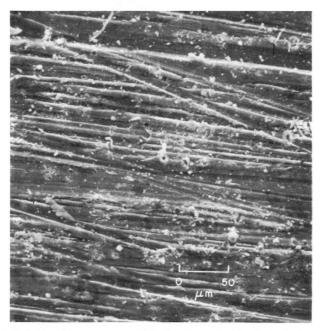

Fig. 13. Left trunnion, area exposed to the moon showing impact pits and dust.

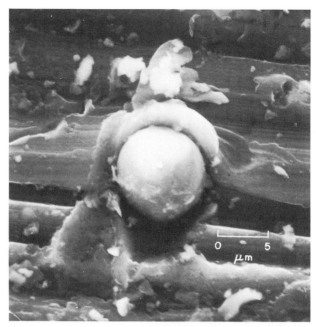

Fig. 14. Left trunnion, impact pit with imbedded particle. Note the crater lip folded
out away from the direction of impact and across the polishing grooves.

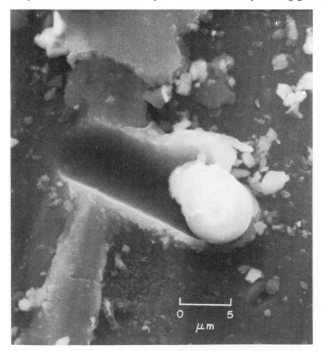

Fig. 15. Left trunnion, another pit with particle imbedded. Observe the shallow angle
of impact shown by the gouged track across a polishing groove.

DISCUSSION

No obvious hypervelocity impacts were found on the 5 mm² scanned area on the mirror or on the trunnions. Because of changes in elevation of the mirror during its operation on the moon, it is difficult to compute its actual exposure to direct impacting micrometeoroids. However, a flux limit of 1×10^{-2} m^{-2} sec^{-1} (2 π steradians)$^{-1}$ for particles with masses larger than 1×10^{-13} g is calculated.

Three possibilities are considered for the origin of the dust and the numerous impacts found on the trunnions and the mirror:

(1) Lunar secondary ejecta produced by primary meteoroid impacts might have bombarded the mirror. The location of the camera and the orientation of the mirror are suitable for this hypothesis. However, studies on lunar secondary ejecta (Gault *et al.*, 1963; Zook, 1970) and the small flux of meteoroids reject this possibility.

(2) The pitting may have been due to lunar material blown toward the Surveyor III spacecraft by the Lunar Module (LM) as it landed. The approaching LM passed beyond the north rim of the Surveyor crater and landed 155 m NW of the Surveyor spacecraft. Dust clouds were caused by the LM before it landed while it was at an altitude of about 175 feet (Bean *et al.*, 1970). The back of the camera hood on the Surveyor III camera was pointing toward the LM during the landing and showed evidence of pitting from that source (Cour-Palais *et al.*, 1971). A heavier layer of dust was found on the mirror after it was returned to earth than was calculated to be on the mirror during the Surveyor operation in 1967 (Caroll *et al.*, 1971). However, the protecting hood of the camera shielded the mirror, and especially the trunnions, from direct impacts from the LM direction (Brownlee *et al.*, 1971). Therefore, pitting from this source is improbable.

(3) Most likely, however, the pitting was due to lunar material being blasted toward the camera by the Surveyor's own vernier engines. This is apparent because the mirror was facing the No. 3 vernier thruster upon landing; the lower part of the mirror was shielded by the housing and is relatively free of pits; the direction of the pitting material as shown by the orientation and the morphology of impacts was from below the mirror. The material probably came from the area under the crushable block and the vernier engine located near leg 3. This area was found by simulation tests conducted at JPL (Christensen *et al.*, 1967). A light was placed on the ground near leg 3 to determine the position where the light would shine on the camera mirror positioned as it was during the landing. The illuminated area on the mirror was the location of heaviest pitting. A glare pattern observed in the early photographic images taken by the Surveyor camera is also consistent with such pitting and dust deposition (Gault *et al.*, 1967). Computer simulation of the landing dynamics, carried out at JPL (Christensen *et al.*, 1967) has provided estimates of the vernier engine thrust levels and the vernier engine height above the surface during the landing. This showed the dust particles could travel along a trajectory that eventually caused them to hit the mirror and contaminate and abrade its surface. Studies on soil erosion by landing rockets has shown that the terminal velocity of particles entrained in rocket plumes can be as great as 500 m/sec for 10 μm diameter particles (Hutton, private communication, 1971). Such a velocity could cause the observed damage.

The study of the Surveyor mirror as a means for defining the lunar micrometeoroid environment was difficult due to the heavy deposition of dust and the pitting of the surface. Nevertheless this mirror provided an excellent opportunity to study the low-velocity effects of erosive material thrown up by descent engines of landing spacecraft. The x-ray attachment to the SEM was a very useful tool in determining the possible source of impacting particles or contamination.

Acknowledgments—We would like to acknowledge the help given by D. E. Gault, N. H. Farlow, and all the Ames staff.

REFERENCES

Bean A. L., Conrad C. Jr., and Gordon R. F. (1970) Crew observations. Apollo 12 Preliminary Science Report, NASA SP-235, 29–38.

Brownlee D., Bucher W., and Hodge P. (1971) Micrometeoroid flux from Surveyor glass surfaces. *Proc. Second Lunar Sci. Conf., Geochim. Cosmochim. Acta* Suppl. 2, Vol. 3, pp. 2781–2789. MIT Press.

Caroll W. F., Blair P. M., Jacob S., and Leger L. (1971) Discoloration and lunar dust contamination of Surveyor surfaces. *Proc. Second Lunar Sci. Conf., Geochim. Cosmochim. Acta* Suppl. 2, Vol. 3, pp. 2735–2742. MIT Press.

Christensen E. M., Batterson S. A., Benson H. E., Choate R., Jaffe L. D., Jones R. H., Ko H. Y., Spencer R. L., Sperling F. B., and Sutton G. H. (1967) Lunar surface mechanical properties. *Surveyor III Mission Report Part II*, JPL TR 32-1177, 111–153.

Cour-Palais B. G., Flaherty R. E., High R. W., Kessler D. J., McKay D. S., and Zook H. A. (1971) Results of the Surveyor III sample impact examination conducted at the Manned Spacecraft Center. *Proc. Second Lunar Sci. Conf., Geochim. Cosmochim. Acta* Suppl. 2, Vol. 3, pp. 2767–2780. MIT Press.

Gault D. E., Collins R., Gold T., Green J., Kuiper G. P., Masursky H., O'Keefe J. A., Phinney R., and Shoemaker E. M. (1967) Lunar theory and processes. *Surveyor III Mission Report Part II*, JPL TR 32-1177, 193–213.

Gault D. E., Shoemaker E. M., and Moore H. J. (1963) Spray ejected from the lunar surface by meteoroid impact. NASA Technical Note TN D-1767.

Hallgren D. S., Laudate A. T., Wachtel D., and Hemenway C. L. (1972) Study of Surveyor III parts for micrometeorite impact craters. *Space Research XII* (in press). Akademie-Verlag.

Zook H. A. (1970) The problem of secondary ejecta near the lunar surface. Trans. of the 1967 National Symposium on Saturn V/Apollo and Beyond, Vol. 1; *Paper EN-8* American Astronautical Society.

Proceedings of the Third Lunar Science Conference
(Supplement 3, *Geochimica et Cosmochimica Acta*)
Vol. 3, pp. 3213–3221
The M.I.T. Press, 1972

Core sample depth relationships: Apollo 14 and 15

W. David Carrier, III

Manned Spacecraft Center, NASA,
Houston, Texas 77058

Stewart W. Johnson

Air Force Institute of Technology, Wright-Patterson Air Force Base,
Ohio 45433

and

Lisimaco H. Carrasco and Ralf Schmidt

Lockheed Electronics Company,
Houston, Texas 77058

Abstract—The depth relationships for the Apollo 14 and 15 core tubes and the Apollo 15 drill core are presented, as determined from laboratory simulation studies. Sample at a depth of 40 cm in the Apollo 14 double core tube (virtually the same as the Apollo 12 tubes) represents material from a depth of approximately 58 cm in the lunar surface. The new design of the Apollo 15 core tube results in much less sample disturbance and the depth relationship is practically one-to-one, with sample recovery approaching 100%. The depth relationship for the drill core is also probably close to one-to-one, and its recovery ratio was also 100%. The *in situ* bulk density at the Apollo 14 core tube sites was 1.45 to 1.6 g/cm³. The Apollo 15 densities ranged from 1.36 to 1.93 g/cm³.

Introduction

The Apollo 11, 12, 14, and 15 missions have returned to earth 15 core tube samples (7 singles and 4 doubles) and 1 drill sample (6 sections) representing a total of 5578 g of regolith with well-preserved stratigraphy. The core tubes were driven into the lunar surface to various depths ranging from 25 cm for an Apollo 11 single-core tube to 70 cm for an Apollo 15 double-core tube. The drill core was first used on Apollo 15 and penetrated to a depth of approximately 236 cm. Information concerning the true depth from which a core sample originated is of particular importance to investigators studying gradients of such properties as solar wind composition, cosmic-ray track densities, and thermoluminescence. Depth relationships for the Apollo 11 and 12 core tube samples were presented previously (Carrier *et al.*, 1971); the present paper reports additional results for the Apollo 14 and 15 core tubes and for the Apollo 15 drill core. The simulant and procedures used in obtaining these results are in general as described in our earlier paper; exceptions are pointed out in applicable sections of the paper.

The shapes and dimensions of the bit ends of the Apollo 14 and 15 core tubes and the Apollo 15 drill core are presented in Fig. 1. The Apollo 15 core tube has a much greater diameter and a thinner wall than did the Apollo 14 core tube. As a result,

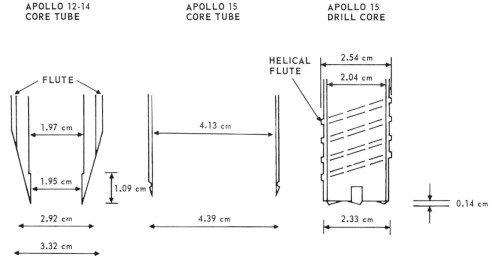

Fig. 1. Comparison of Apollo core bits. Shown are cut-away views giving diameters at various sections. The Apollo 12–14 and 15 core tubes are advanced into the lunar soil by pushing and, if necessary, by hammering. The Apollo 15 drill is battery powered, and the core is advanced by a rotary-percussion action.

Table 1. Apollo 14 returned core tube sample data.

Core tube serial no.		LRL sample no.	Returned sample weight (g)	Returned sample length (cm)	Bulk[a] density (g/cm³)	Total[b] sample length (cm)	Core tube depth (cm)	Percent core recovery
Double	⎧ 2045	14211	39.5	7.5[g]	1.73	40.5	13[c]; 64[d]	63
	⎩ 2044	14210	169.7	31.9	1.75			
	2022[e]	14220	80.7	16.5[g]	1.60	17.6	15[c]; <36[d]	>49
	2043[f]	14230	76.0	~12.5[g]	—	—	23/45[f]	—

[a] Based on a sample diameter of 1.97 cm.
[b] Adjusted to include the length of the sample in the bit; for Apollo 14, an additional 1.09 cm.
[c] Depth before final driving (maximum depth that was photographed).
[d] Crew estimate: no photograph taken.
[e] This core tube struck a rock during the first attempt at a triple at North Triplet.
[f] This core tube was driven twice: first on Cone Crater where some or all of the sample fell out; second at North Triplet during the second attempt at a triple core where some of the sample fell out.
[g] Heiken (1971) determined the sample length by means of x-radiography of the core tube.

depth relationships established earlier for the Apollo 12 and 14 core tubes do not hold for the Apollo 15 core tube. The drill core is emplaced by a rotary-percussion action. There has previously been some speculation that this motion would cause considerable disturbance of the sample gathered with the drill. One purpose of our investigation has been to evaluate the degree of disturbance.

Apollo 14 Core Tube Samples

The Apollo 14 mission returned four core tube samples: two singles and a double (two tubes connected end-to-end). The basic data concerning the samples are summarized in Table 1. The locations of these core tubes at the Fra Mauro landing site are shown in Swann *et al.* (1971).

To date, only one of the Apollo 14 core tubes (S/N2043; LRL No. 14230) has been dissected and distributed for analysis. The description of this core sample is presented in Fryxell and Heiken (1971). Unfortunately, this core sample was severely disturbed, as discussed in Mitchell *et al.* (1971), and determination of its depth relationship is highly speculative.

Heiken (1971) has determined the sample lengths in the remaining unopened core tubes by means of x-radiography; the percent core recoveries agree very well with the recovery ratios for the Apollo 12 samples and for the simulated core samples:

	Percent core recovery (total sample length ÷ core tube depth)	
	Single	Double
Simulation	50	63
Apollo 12	~50; 56	61
Apollo 14	>49	63

One of the interesting aspects of the Apollo 14 core samples is that the bulk densities in the tubes are significantly less than in the Apollo 12 core tubes, despite the fact that the Apollo 14 crew encountered the same or slightly greater resistance to driving and despite the fact that the core recoveries are similar. The bulk densities in the upper and lower halves of the Apollo 14 double are 1.73 and 1.75 g/cm^3 versus 1.98 and 1.96 g/cm^3, respectively, for the Apollo 12 double (Carrier *et al.*, 1971). Similarly, the bulk density for the 14 single is 1.60 g/cm^3 versus 1.74 g/cm^3 for the 12 single. These differences were first noted in Mitchell *et al.* (1971) and several possible explanations were suggested. Of these, the most likely is a difference in the specific gravity of the soils at the two sites. A specific gravity of 3.1 was determined for one sample of Apollo 12 soil (Carrier, 1970); however, no measurements have been reported for the Apollo 14 soil. Nonetheless, LSPET (1971) has found that the Apollo 14 soils consist of far more microbreccias and far less crystalline fragments than the Apollo 12 soils. Furthermore, Cremers (1972) has reported that the minimum density at which the Apollo 14 soil could be placed for his thermal conductivity experiments was 1.1 g/cm^3, as compared with a value of 1.3 g/cm^3 for the Apollo 12 soil. Consequently, the Apollo 14 soil particles are probably less dense than the Apollo 12 soil particles and may contain a significantly greater amount of intragranular porosity; based on Cremers' data, the effective specific gravity of the Apollo 14 soil must be approximately 15% less than that of the Apollo 12 soil. In which case, it can be calculated that the intergranular porosity of the soil in the Apollo 12 and 14 core tubes are quite similar, even though the bulk densities are significantly different. This result has several interrelated implications: (1) Since the percent core recoveries and the intergranular porosities of the core samples are similar for the two sites, then the *in situ* intergranular porosities must also be similar. (2) Since the *in situ* bulk density for the top 30 to 60 cm at the Apollo 12 core tube sites has been estimated to be 1.7 to 1.9 g/cm^3 (Carrier *et al.*, 1971), then the *in situ* bulk density at the Apollo 14 core tube sites must be 1.4^5 to 1.6 g/cm^3. (3) Finally, inasmuch as the intergranular porosity of a soil is one of the most important factors controlling its core sampling behavior (Houston and Mitchell, 1971), and because the percent core recoveries on Apollo 14, 12, and in our simulant are in such good agreement, it has been concluded that the simulation data that were previously obtained are valid for determination of the depth relationships for the Apollo 14 double and single. These depth relationships are shown in Fig. 2.

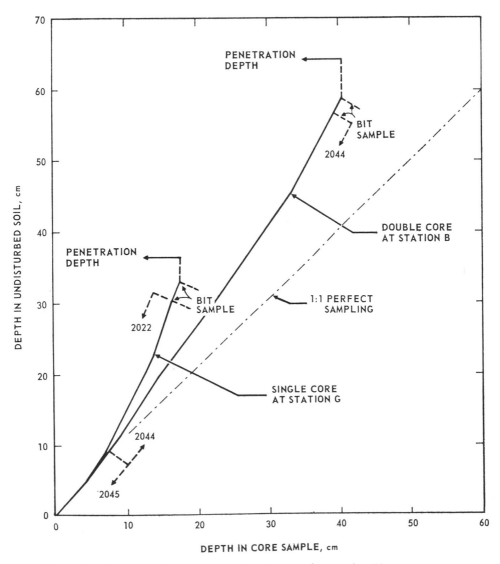

Fig. 2. Depth relationships for the Apollo 14 core tube samples. These curves were prepared by driving core tubes into a lunar soil simulant containing colored layers. The bit sample for the double was returned to earth in Bag 2N and has been designated LRL Sample 14411; the bit sample for the single was returned in Bag 18N and is LRL Sample 14414.

The penetration depth for these two cores has been estimated on the basis of Crew comments; no photographs were taken of the tubes after they had been driven into the lunar surface. The penetration depth for the single core indicated in Fig. 2 is within ± 1.5 cm and that of the double within ± 2 cm. In addition, there is an experimental error of $\pm 5\%$ of the depth in the core sample: that is, at a depth of 10 cm in the sample, the depth in the undisturbed soil is accurate to ± 0.5 cm; and at 40 cm, ± 2 cm.

The departure of the actual depth-relationship curves in Fig. 2 from the linear curve for perfect sampling is due to a combination of two factors: densification of the soil in the core tube; and incomplete sampling of the soil in the path of the tube. In regard to the latter, our simulations have shown that although a continuous stratigraphy was recovered, approximately 34% of the mass of the soil that could have been sampled by the Apollo 14 single core was pushed away as the tube was advanced into the soil; and the mass loss from the double core was approximately 20%. This phenomenon is due to the bit geometry and the presence of a teflon follower inside the tube (see Carrier *et al.* (1971) for additional information).

Apollo 15 Core Tube Samples

The Apollo 15 mission returned five core tube samples: one single and two doubles. The basic data concerning the samples are summarized in Table 2. The locations of these core tubes at the Hadley-Apennine landing site are shown in Swann *et al.* (1972).

To date, the Apollo 15 core tubes have only been weighed and x-rayed in the Lunar Receiving Laboratory. The x-radiographs reveal considerable stratigraphy and a detailed description is presented in LSPET (1972).

The new core tubes used for the first time on Apollo 15 induce considerably less disturbance in the samples than the earlier core tubes. The significance of this improvement in regard to the present study is that the depth relationship is nearly one-to-one and is virtually independent of the simulant. That is, the depth relationship is

Table 2. Apollo 15 returned core tube sample data.

Core tube serial no.	LRL sample no.	Returned sample weight (g)	Returned sample length (cm)	Bulk[a] density (g/cm^3)	Core tube depth (cm)	Percent core recovery
Double {2003	15008	510.1	28 $\pm$ 1	1.36 $\pm$ 0.05	70.[1]	87–90
2010	15007	768.7	33.9 (34.9[b])	1.69 (1.64)		(88–91)
2007	15009	622.0	36.2[d] (32.9–34.6[c,d])	1.30 (1.36–1.43)	34.[6]	105 (95–100)
Double {2009	15011	660.7	29.2 $\pm$ 0.5	1.69 $\pm$ 0.03	67.[6]	91–93
2014	15010	740.4	32.9[e] (34.9[b,e])	1.91 (1.79)		(94–96)

[a] Based on a sample diameter of 4.13 cm.
[b] Nominal full length is 34.9 cm: sample either fell out of top of lower half, or sample was compressed when keeper inserted.
[c] 100% core recovery length would be 34.6 cm: keeper slipped out of position.
[d] Approximately 6 cm^3 of sample apparently fell out of bottom of core before it was capped.
[e] Approximately 54 cm^3 of sample apparently fell out of bottom of core before it was capped.

practically the same over a wide range of soil properties. This is important because, as discussed by Mitchell *et al.* (1972), there was a wider spectrum of soils at the Apollo 15 site than has previously been encountered.

The depth relationships for the Apollo 15 core tube samples are shown in Fig. 3. A few comments are in order. The sample lengths in the upper halves of the two doubles cannot be determined accurately from the x-radiographs, due to image distortion caused by parallax. This introduces an uncertainty in the sample length of ± 1 cm in the case of the first double and ± 0.5 cm in the case of the second double. When the core tubes are eventually opened, this uncertainty will be eliminated. There is, how-

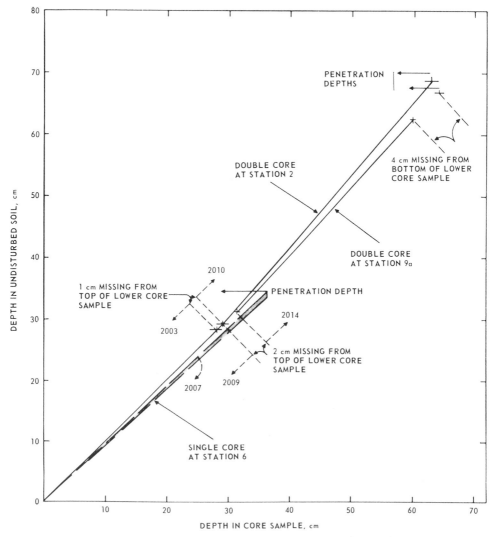

Fig. 3. Depth relationships for the Apollo 15 core tube samples. The new core tubes induce very little sample disturbance and the curves are nearly one-to-one.

ever, an experimental error of $\pm 3\%$ for the depth in the undisturbed soil for all the Apollo 15 core tube samples.

Although the Apollo 15 crew did not observe any sample falling out of the core tubes, the x-radiographs reveal that a small portion of the sample is apparently missing from the tops and bottoms of some of the cores. This is indicated in Fig. 3. In addition, the sample in the single core evidently expanded when the keeper failed to function properly (see Mitchell *et al.*, 1972). The original percent core recovery for the single was probably 95 to 100%, corresponding to a sample length of 32.9 to 34.6 cm. As the present sample length is 36.2 cm, this means the sample expanded anywhere from 1.6 to 3.3 cm. The curve for the single core sample in Fig. 3 presents the range of possibilities.

APOLLO 15 DRILL CORE SAMPLE

The drill core sample recovered on Apollo 15 consists of six sections, the basic data for which are shown in Table 3. The sample was taken at Station 8 near the ALSEP, as shown in Swann *et al.* (1972). The drill stems are presently being opened and samples distributed to various investigators.

Premission drill tests were conducted in the same simulant used in the core tube studies. The results of one of these tests is shown in Fig. 4. These tests demonstrated that despite the agitation of the core stem as it is advanced into the soil (~ 280 rpm and 2270 blows/min at 40 in-pounds/blow), preservation of the *in situ* stratigraphy is excellent. This has been confirmed with the lunar core sample, in which over 50 individual layers have been identified on the basis of x-radiographs (LSPET, 1972).

However, the observed penetration rates for the Apollo 15 drill core and heat flow bores, as well as the penetration resistance measured by the self-recording penetrometer (Mitchell *et al.*, 1972), indicated a lower inter-granular porosity than had previously been assumed for the lunar surface. Thus, a different lunar soil simulant has been prepared, consisting of a ground basalt placed at a high relative density. Preliminary tests with this simulant permit the following tentative conclusions:

(1) A high percentage core recovery, as was obtained in the Apollo 15 core, can only be achieved when the soil is placed at a low porosity.

Table 3. Apollo 15 returned drill core sample data.

	Drill stem serial no.	LRL sample no.	Returned sample weight (g)	Returned sample length (cm)	Bulk[a] density (g/cm^3)	Drill stem depth (cm)	Percent core recovery
Top	022	15006	210.6	39.9[b]	1.62		
	023	15005	239.1	39.9	1.84		
	011	15004	227.9	39.9	1.75		
	020	15003	223.0	39.9	1.79	~ 236	~ 102
	010	15002	210.1	39.9	1.62		
Bottom	027	15001	232.8	37.0[b] (42.5)[c]	1.93		

[a] Based on a sample diameter of 2.04 cm.

[b] The drill stems have been opened and the previously reported sample lengths have been corrected.

[c] Nominal length is 42.5 cm: ~ 5.5 cm fell out of bottom of drill stem.

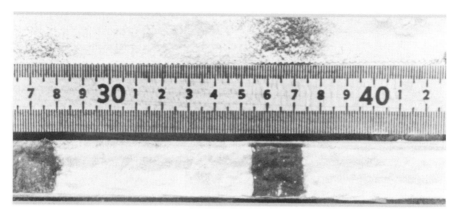

Fig. 4. Drill core tests were performed in the same simulant used to prepare the curves shown in Figs. 2 and 3. It had been anticipated before the tests that the drill would "barber-pole" the stratigraphy, but it can be seen that the stratigraphy is very well preserved.

(2) The average penetration rate at this low porosity is similar to that observed during the Apollo 15 mission.

(3) The percentage densification at this low porosity is quite small: $\pm 2\%$. Thus, the densities in the drill stems may be very close to the *in situ* values.

(4) Astronaut Scott reported that he was able to feel hard and soft layers as he was drilling with the core. The different densities in the six core stems (see Table 3) probably reflect these layers.

(5) As a consequence of the above observations, the depth relationship for the drill core sample is probably close to one-to-one.

Although considerably more drill tests are required, it is recommended that a one-to-one depth relationship be assumed for the present. The absolute depth of drill core sample is probably accurate to within $\pm 5\%$.

CONCLUSIONS

(1) The depth relationships for two of the three Apollo 14 core tube samples have been determined and are shown in Fig. 2. The third sample is too disturbed for analysis. In single-core sample 2002, material from a depth-in-core of 16 cm came from a depth of nearly 29 cm beneath the lunar surface. In the double core (2045 and 2044), the bit sample was 40 cm deep as measured on the recovered core but represented material from 58 cm beneath the surface. Individual layers in the Apollo 14 cores are quite well preserved but the sampling process is such that there is some compaction and some material is pushed away from the driven core tube, such that the recovery ratio is generally between 49% and 63%.

(2) The Apollo 15 core tube has a greater diameter and a thinner wall than the Apollo 11, 12, and 14 core tubes. The depth-in-core-tube versus true depth in the regolith for the Apollo 15 core tube samples is practically one-to-one, as shown in Fig. 3. Sample recovery with this core tube approaches 100%.

(3) The Apollo 15 returned drill core, which penetrated to a depth of about 236 cm, contains sample in which the preservation of *in situ* stratigraphy is excellent. Preliminary drill tests indicate that the densities in the drill stem may be very close to the *in situ* values, and that the depth relationship is probably one-to-one. The excellent quality of the drill core sample is primarily due to the low intergranular porosity of the soil encountered at the drill site.

(4) Bulk density calculations for Apollo 15 core tube and drill stem samples reveal a range of values from 1.36 g/cm^3 for the top tube of an Apollo 15 double to 1.93 g/cm^3 for the bottom section of the Apollo 15 six-section drill stem. These values for the returned samples are probably close to *in situ* densities. Bulk density ranges for the Apollo 12 and 14 core tube samples were 1.74 to 1.98 g/cm^3 and 1.6 to 1.75 g/cm^3, respectively, corresponding to *in situ* values of 1.7 to 1.9 g/cm^3 and 1.4^5 to 1.6 g/cm^3. More investigation is needed to ascertain the reason for the bulk density variations with depth and between sites. Variations in both intergranular and intragranular porosity as well as compositional differences could account for the variations in density.

REFERENCES

Carrier W. D. III (1970) Lunar soil mechanics on the Apollo missions. Spring Meeting, Texas Section, ASCE, Galveston, Texas.

Carrier W. D. III, Johnson S. W., Werner R. A., and Schmidt R. (1971) Disturbance in samples recovered with the Apollo core tubes. *Proc. Second Lunar Sci. Conf., Geochim. Cosmochim. Acta* Suppl. 2, Vol. 3, pp. 1959–1972. MIT Press.

Cremers C. J. (1972) Thermal conductivity of Apollo 14 fines at low density (abstract). In *Lunar Science—III* (editor C. Watkins), pp. 160–162, Lunar Science Institute Contr. No. 88.

Fryxell R. and Heiken G. (1971) Description, dissection and sub-sampling of lunar core 14230, Triplet Crater, Apollo 14 mission. NASA Technical Memorandum TMX-58070.

Heiken G. (1971) Apollo 14 core tubes. NASA Manned Spacecraft Center, Houston, Texas, Memorandum to TN12/Curator, Planetary and Earth Sciences, April 29.

Houston W. N. and Mitchell J. K. (1971) Lunar core tube sampling. *Proc. Second Lunar Sci. Conf., Geochim. Cosmochim. Acta* Suppl. 2, Vol. 3, pp. 1953–1958. MIT Press.

LSPET (Lunar Sample Preliminary Examination Team) (1971) Preliminary examination of the lunar samples from Apollo 14. *Science* **173**, 681–693.

LSPET (Lunar Sample Preliminary Examination Team) (1972) The Apollo 15 lunar samples: A preliminary description. *Science* **175**, 363–375.

Mitchell J. K., Bromwell L. G., Carrier W. D. III, Costes N. C., and Scott R. F. (1971) Soil mechanics experiment. Apollo 14 Preliminary Science Report, NASA SP-272, pp. 87–108.

Mitchell J. K., Bromwell L. G., Carrier W. D. III, Costes N. C., Houston W. N., and Scott R. F. (1972) Preliminary analysis of soil behavior. Apollo 15 Preliminary Science Report, NASA SP-289, pp. 7-1 to 7-28.

Swann G. A., Bailey N. G., Batson R. M., Eggleton R. G., Hait M. H., Holt H. E., Larson K. B., McEwen M. C., Mitchell E. D., Schaber G. G., Schafer J. B., Shepard A. B., Sutton R. L., Trask N. J., Ulrich G. E., Wilshire H. G., and Wolfe E. W. (1971) Preliminary geologic investigations of the Apollo 14 site. Apollo 14 Preliminary Science Report, NASA SP-272, pp. 39–86.

Swann G. A., Bailey N. G., Batson R. M., Freeman V. L., Hait M. H., Head J. W., Holt H. E., Howard K. A., Irwin J. B., Larson K. B., Muehlberger W. R., Reed V. S., Rennilson J. J., Schaber G. G., Scott D. R., Silver L. T., Sutton R. L., Ulrich G. E., Wilshire H. G., and Wolfe E. W. (1972) Preliminary geologic investigation of the Apollo 15 landing site. Apollo 15 Preliminary Science Report, NASA SP-289, pp. 5-1 to 5-112.

Proceedings of the Third Lunar Science Conference
(Supplement 3, *Geochimica et Cosmochimica Acta*)
Vol. 3, pp. 3223–3234
The M.I.T. Press, 1972

Strength and compressibility of returned lunar soil

W. David Carrier, III

Manned Spacecraft Center, NASA,
Houston, Texas 77058

and

Leslie G. Bromwell and R. Torrence Martin

Massachusetts Institute of Technology,
Cambridge, Massachusetts 02139

Abstract—Two oedometer and three direct shear tests have been performed in vacuum on a 200 g sample of lunar soil from Apollo 12 (12001,119). The compressibility data have been used to calculate bulk density and shear wave velocity versus depth on the lunar surface. The shear wave velocity was found to increase approximately with the one-fourth power of the depth and the results suggest that the Apollo 14 Active Seismic Experiment may not have detected the Fra Mauro formation at a depth of 8.5 m, but only naturally consolidated lunar soil. The shear data indicate that the strength of the lunar soil sample is about 65 % that of a ground basalt simulant at the same void ratio. This difference has been tentatively attributed to mechanical breakdown of fragile lunar soil particles such as microbreccias and agglutinates. The shear strength of other lunar soils at the same void ratio will probably vary according to the proportion of these particles. Thus, the Apollo 14 soils, which contain a higher percentage of microbreccias, are likely to have an even higher compressibility and a lower shear strength. Gas pressure bursts consisting primarily of hydrogen and helium, apparently derived from solar wind, occurred during both compression and shear and are also attributed to the crushing of these fragile grains. Detection of these gases at the lunar surface could indicate that a soil mass was consolidating or creeping downslope.

Introduction

The writers were allotted just over 200 g of vacuum soil from Apollo 12 (sample 12001,119) for the purpose of performing vacuum oedometer (one-dimensional compression) and direct shear tests; two oedometer and three direct shear tests have been performed. This sample was originally returned to the Lunar Receiving Laboratory (LRL) at a pressure of 10^{-2} Torr and was stored at 10^{-9} Torr for over a year before these experiments were performed. The test specimens were prepared at a pressure of less than 2×10^{-6} Torr; and they were tested at a pressure of less than 5×10^{-8} Torr. The purpose of these measures was to maintain the soil particle surfaces as free of terrestrial contamination as possible. Previous vacuum work on terrestrial soils (c.f. Bromwell, 1966; Nelson and Vey, 1968; Vey and Nelson, 1965; Johnson *et al.*, 1970) has shown that the level of cleanliness has an important influence on the stress-strain behavior. These tests are state-of-the-art in terms of the vacuum system, as no earlier vacuum terrestrial soil testing has included preparing the sample in a vacuum. Until now, the sample had been prepared in air or in an inert atmosphere and then pumped down. In these tests, the sample has not been exposed to a pressure above 2×10^{-6} Torr since it was returned to the LRL.

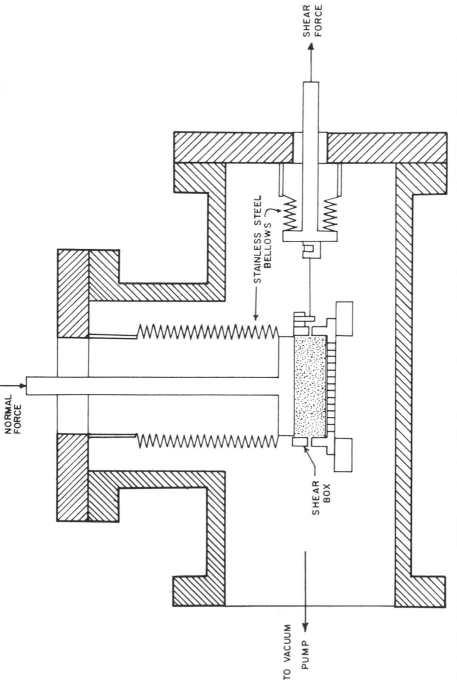

Fig. 1. Ultra-high vacuum (UHV) test chamber. The lunar sample was placed in the shear box and inserted into the UHV chamber at a pressure of less than 2×10^{-6} Torr. The UHV chamber was then sealed and transferred to a test bench where the sample was compressed and sheared at a pressure of less than 5×10^{-8} Torr.

Test Equipment

The test specimens were prepared in the F-401 glove test chamber in the LRL. The F-401 chamber is approximately 75 cm in diameter and 150 cm long; all of the operations were performed one-handed through a spacesuit-type arm and glove. The samples were placed in a shear box which was then inserted into a small ultra-high vacuum (UHV) test chamber (shown in Fig. 1), which was then sealed before removing it from F-401. An auxiliary ion pump, mounted directly on the UHV chamber, was used to maintain the vacuum until the chamber could be transferred to an external test stand and pumping station.

After the pressure in the UHV chamber had been reduced to less than 5×10^{-8} Torr, the oedometer and direct shear tests were performed. The normal and shear forces were transmitted to the sample by means of the bellows shown in Fig. 1. A detailed description of the UHV chamber may be found in Bromwell (1966).

Compressibility

The data from the two vacuum oedometer tests are presented in Fig. 2 and Table 1. The void ratio has been calculated assuming the specific gravity for this sample is

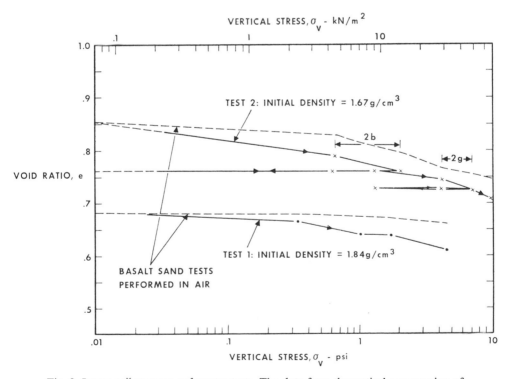

Fig. 2. Lunar soil vacuum oedometer tests. The data from the vertical compression of lunar soil in the shear box is shown and compared with the results from tests on a ground basalt simulant.

Table 1. Lunar soil vacuum oedometer data (see Fig. 2).

	Void ratio,* e	Vertical stress, σ_v (psi)	Vertical stress, σ_v (kN/m²)	Increment label
Test 1	0.684	0	0	1a
	0.667	0.33	2.27	1b
	0.641	1.01	6.96	1c
	0.638	1.72	11.85	1d
	0.612	4.53	31.21	
Test 2	0.854	0	0	2a
	0.789	0.64	4.41	2b
	0.759	2.03	13.98	2c
	0.763	0	0	2d
	0.762	0.61	4.20	2e
	0.761	1.32	9.09	2f
	0.746	4.15	28.59	2g
	0.725	6.97	48.02	2h
	0.725	4.13	28.45	2i
	0.728	1.29	8.89	2j
	0.727	4.13	28.45	2k
	0.724	6.97	48.02	2l
	0.708	9.80	67.51	

* Specific gravity of 3.1 assumed: $e = (3.1/\rho) - 1$ (ρ in g/cm³).

Table 2. Estimated vacuum oedometer data for test 1B.

Void ratio, e	Vertical stress σ_v (psi)	Vertical stress σ_v (kN/m²)	Increment label
0.622	0.28	1.93	1Ba
0.597	1.68	11.56	1Bb
0.575	4.49	30.95	1Bc
0.561	7.32	50.40	1Bd
0.551	10.15	69.92	

3.1, as was measured for another Apollo 12 soil sample (Carrier, 1970). The first sample was placed in a medium-dense condition ($\rho = 1.84$ g/cm³; $e = 0.684$), compressed vertically in four increments to 31.21 kN/m², and then sheared (Test 1). Only a small normal load was left on the sample (1.93 kN/m²) while the top half of the shear cup was moved back into position. The sample was then recompressed in four increments to a normal stress more than twice the previous load, and then sheared (Test 1B). The void ratio during compression could not be accurately determined for this test and has not been included in Fig. 2; estimated void ratios are presented in Table 2.

The sample test chamber was then moved back into F-401 and the sample removed from the shear box. The soil was then re-placed in the shear box as loosely as could be attained within the constraints of the F-401 vacuum system ($\rho = 1.67$ g/cm³; $e = 0.854$). After that, it was compressed in a series of increments and then sheared (Test 2).

The *in situ* density of the lunar soil at the Apollo 12 site has been estimated at 1.6 to 2 g/cm³ (Scott, *et al.*, 1971), and more recently, based on Apollo 12 core tube simulations, at 1.7 to 1.9 g/cm³ (Carrier *et al.*, 1971). Thus, the first test represents the mid-range of the predicted *in situ* density and the second test is near the lower bound. (Under less difficult working conditions, the lunar soil can be placed at a much lower density; a value of 1.36 g/cm^3 (Costes *et al.*, 1970) was achieved with Apollo 11 soil in a nitrogen atmosphere.)

For comparison, two oedometer tests were performed on a basalt sand simulant and the results are plotted in Fig. 2. The simulant has a similar grain size distribution as that of the lunar soil sample and it was placed at the same initial void ratios as in vacuum oedometer Tests 1 and 2. At the medium-dense initial density, the compressibility of the lunar soil is slightly higher than that of the basalt sand. At the lower density, the large decrease in the lunar soil void ratio during the initial load increment causes the lunar soil to have a higher compressibility than the simulant; but a portion of this difference may have been due to a seating error with the normal force bellows. The loose lunar soil also showed higher compressibility at the higher load increments. Additional tests are needed to clarify this point. The higher compressibility of the lunar soil is probably due to the crushing of fragile particles; this interpretation is discussed further in the sections on shear strength and gas composition. It should also be noted that the e-log σ_v curves do not merge at a vertical stress of less than 70 kN/m^2 and this implies that lunar soil in $\frac{1}{6}$ g can exist at significantly different densities at depths greater than 25 m.

Two rebound cycles were also performed as part of Test 2; it can be seen in Fig. 2 and Table 1 that the lunar soil skeleton did not rebound to any significant extent. Thus, the elastic (or recoverable) portion of one-dimensional settlement is negligible compared to the plastic (or irrecoverable) portion. Furthermore, compression during each increment of vertical load is virtually instantaneous, as would be anticipated, with 50% of the compression occurring in less than 2 to 3 sec and 90% in less than 7 to 8 sec.

Density of lunar soil versus depth

Scientists and engineers alike are interested in determining the density of the lunar soil with depth: the scientist for interpretation of data from the core samples, the seismometers, the heat-flow probe, and the radar experiments, as well as for understanding the mode of deposition of the lunar soil; the engineer for design of foundations for observatories, design of excavating equipment for tunnels and mines, and design of underground living quarters. Over the years, many estimates have been made of the density of lunar soil; see Mitchell *et al.* (1972) for a comprehensive review of this topic. The compression data that we now have permit a new look at this question.

By assuming that the soil densifies with depth only due to self-weight in lunar gravity, the vacuum oedometer data in Table 1 can be used for preliminary calculations of density versus depth, settlements due to imposed loads, and shear wave velocity versus depth. For a given surface density, this assumption yields a lower bound estimate of the density versus depth; that is, the actual density at a certain depth on the lunar surface can be higher than the calculated value, but it cannot be lower. If it is assumed that the lunar surface density is 1.67 g/cm^3 (i.e., the initial density of compression Test 2), density versus depth curves can be calculated for both

lunar and earth gravity conditions as shown in Fig. 3, along with similar curves for
the Test 1 data. In the lunar case for Test 2, the density increases very rapidly to a value
of 1.72 g/cm^3 at a depth of 30 cm and very slowly thereafter to a value of 1.77 g/cm^3
at a depth of 8 m. The six-fold increase in self-weight for the earth case increases the
density with depth considerably. Below about 3.9 m the stress in earth gravity exceeds
the maximum reached in Test 2 (67.5 kN/m^2) and the data were extrapolated for the
greater depths (indicated by the broken line). The density calculated from Test 1 is far
less dependent on depth in the top 30 cm than for Test 2, owing to the much higher
initial density in Test 1 and the consequent decrease in compressibility. The density
profile is also less affected by the change in gravity.

Previously, the thick-walled core tubes produced too much sample disturbance
for determination of the *in situ* density except within broad bounds. The relatively
undisturbed samples returned in the new thin-walled core tubes on Apollo 15 (also
planned for Apollo 16 and 17), permit a far more accurate determination of the average
density for the top 30 to 60 cm of the lunar surface. However, exactly how the density
varies with depth in the top 30 cm of the lunar surface probably cannot be determined
on returned lunar samples. The samples experience at least six times earth gravity
(36 times lunar gravity) during re-entry into the earth's atmosphere and that is more
than sufficient to obliterate a density change such as that predicted by Test 2 in Fig. 3.
The actual *in situ* density profile is undoubtedly far more complex than the simple
profiles that have been calculated. Constant meteorite bombardment of the lunar

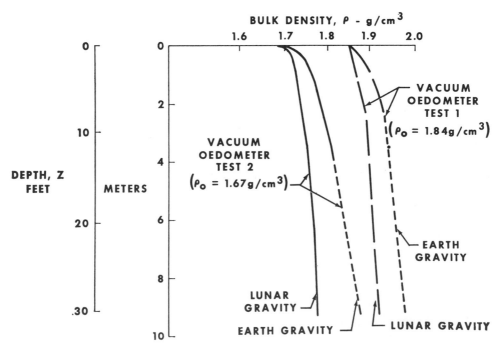

Fig. 3. Lunar soil density versus depth. The vacuum oedometer data have been used to
calculate densification due to the self-weight of the soil in lunar and earth gravity.

surface undoubtedly changes the shape of the profile vertically, laterally, and temporally. *In situ* testing devices are an absolutely necessary complement to the testing of returned samples; the Apollo 15 Self-Recording Penetrometer returned very important data (Mitchell *et al.*, 1972).

Settlements in lunar soil

The oedometer data can also be used to estimate the settlement of a lunar structure which applies a load over a large area, such as an observatory. For a surface load of 10 kN/m², the settlement due to one-dimensional compression would be greater than 6 cm for the Test 1 soil and greater than 9 cm for the Test 2 soil. These are large settlements and careful foundation design would be required to maintain differential settlements within acceptable limits.

Shear wave velocity in lunar soil

In the design of vibrating machinery, it is essential to have an estimate of the *in situ* dynamic properties of the foundation soil. It is also required to know how these properties will be changed by the application of the static load of the equipment. Although no tests have been made on returned lunar soil in vacuum, nor at the very low confining stresses corresponding to the lunar surface, the vacuum oedometer data may be used to estimate the shear wave velocity in the lunar regolith.

Hardin and Black (1966) experimentally determined the following relationship for sands in air:

$$G = 3228 \frac{(2.973 - e)^2}{(1 + e)} \sigma^{1/2},$$

where G = shear modulus, in kN/m²,

σ = average normal effective stress, in kN/m²

$= \dfrac{\sigma_v + 2\sigma_h}{3}$, where σ_v = vertical stress and σ_h = horizontal stress.

This relationship is known to be valid for stresses on the order of 140 kN/m² and above, whereas the stresses in the lunar surface are significantly lower: less than 20 kN/m² at 10 m. Furthermore, the lunar soil is not entirely cohesionless. Despite these drawbacks, Hardin and Black's formula does provide a starting point.

The shear modulus G was calculated according to the above equation for the Test 1 and 2 soil profiles (σ_h was assumed to equal $\frac{1}{2}\sigma_v$; thus, $\sigma = \frac{2}{3}\sigma_v$) and the shear wave velocity V_s, then calculated from the relationship

$$V_s = \sqrt{G/\rho}.$$

These shear wave velocity profiles are shown in Fig. 4.

It can be seen that the shear wave velocity increases very rapidly from 0 at the surface to 61 to 66 m/sec at a depth of 30 cm, and increases very slowly with depth thereafter, the velocity being approximately proportional to the one-fourth power of

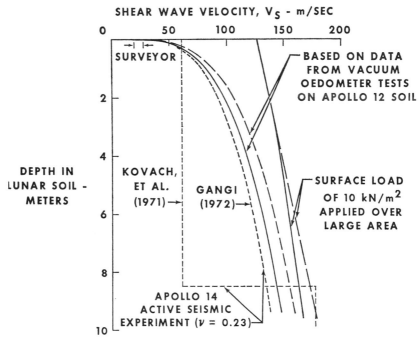

Fig. 4. Shear wave velocity versus depth. The oedometer data suggest that the Apollo 14 Active Seismic Experiment may not have detected the Fra Mauro formation at a depth of 8.5 m, but only lunar soil consolidated under its own weight.

the depth. The estimated velocity profiles for the Test 1 and 2 soil profiles of Fig. 4 do not vary significantly and indicate that the shear wave velocity is more dependent on the stress than on the void ratio.

The two-layer compression wave velocity model based on the data from the Apollo 14 active seismic experiment (Kovach *et al.*, 1971) may be converted to a two-layer shear wave velocity model using a value of Poisson's ratio of 0.23 (Kovach and Watkins, 1972). This is also shown in Fig. 4, as well as Surveyor data for the upper few centimeters from Sutton and Duennebier (1970). Although not immediately apparent in Fig. 4, work by Gangi (1972) has shown that the $\frac{1}{4}$-power type of relationship fits the data from the active seismic experiment very well. Thus, although Kovach *et al.* have interpreted their data as indicating the presence of the Fra Mauro formation at a depth of 8.5 m at the Apollo 14 landing site, in fact, their seismic data may only reflect the natural consolidation of lunar soil.

For the case of a uniform load of 10 kN/m² acting over a large area, the shear wave velocity profile would be altered considerably, as shown in Fig. 4. The profile would become nearly homogeneous, with a value of 125 m/sec at the surface, increasing nearly linearly to a value of 168 to 180 m/sec at 10 m. This alteration of the velocity profile must be taken into account in the design of vibrating machinery for the lunar surface.

Shear Strength

The results of the three vacuum direct shear tests are summarized in Table 3. Test 1 was performed after the sample had been consolidated to a vertical stress of 31.21 kN/m^2. After shear Test 1 was completed, the shear force was removed and the shear box returned to its initial position. The normal force was then increased to 69.92 kN/m^2, and a second shear test (Test 1B) was run. Shear Test 2 was run after the sample had been returned to the F-401 chamber and reconstituted at a higher void ratio. The vertical stress for Test 2 was 67.51 kN/m^2.

In all of the tests, the sample reached failure at a shear displacement of approximately 0.27 cm. This represents a strain (expressed as $\Delta L/L_0 \times 100\%$) of about 4%. The change in height of the sample was small in all of the tests, ranging from a vertical compression of about 0.013 cm ($\Delta H/H_0 = 1\%$) in Test 2 to a vertical expansion of 0.005 cm ($\Delta H/H_0 = 0.3\%$) in Test 1B.

The three shear tests do not provide sufficient data to determine the values of ϕ (friction angle) and c (cohesion) independently, since all three tests were run at different initial void ratios. However, both the test data and visual observation of the soil during the placement and testing indicate that the cohesion was small, probably ranging from 0 to 0.7 kN/m^2. For cohesion values in this range, the friction angles indicated by the test data are 28° for the loose sample (Test 2) and 34° to 35° for the medium-dense samples (Tests 1 and 1B).

In Fig. 5, the strength of the lunar soil sample is compared with the strength for the ground basalt simulant measured in air (Mitchell and Houston, 1970). Only the frictional component (tan ϕ) of the basalt strengths has been plotted, as the cohesion can be varied over a wide range for this material by controlling the moisture content. Hence, the actual strength of the basalt would be somewhat higher than shown in Fig. 5. Also, the strength of the basalt under high vacuum conditions should be higher than in air.

Figure 5 indicates that the lunar soil sample had considerably less strength than the ground basalt at the same void ratio, the lunar soil strength being only about 65% of the simulant strength. The reason for the lower strength of the lunar sample is not known. However, it may be due to a difference in the particle composition of the two soils. The basalt simulant consists of strong, coherent rock fragments produced by grinding. The lunar sample, on the other hand, contains many microbreccias, agglutinates, and other weakly cemented particles. During shear, these particles could easily have been broken down into smaller particles.

Table 3. Lunar soil shear test results.

Test no.	Void ratio,* e_0	Vertical stress, σ_v (kN/m^2)	Peak shear stress, τ (kN/m^2)	τ/σ_v	ϕ (for $c = 0$ to 0.7 kN/m^2)
1	0.612	31.21	21.22	0.680	34°
1B	0.55	69.92	50.15	0.717	~35°
2	0.708	67.51	36.30	0.538	28°

* At end of consolidation, but prior to shear; specific gravity of 3.1 assumed.

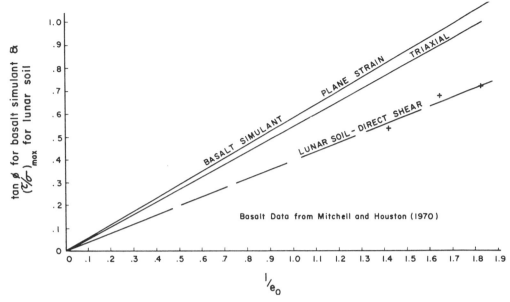

Fig. 5. Comparison of strengths of lunar soil and ground basalt simulant. The shear strength of the lunar soil is about 65 % of that of the basalt simulant at the same void ratio.

Mitchell and Houston (1970) showed that decreasing the average particle size of the basalt simulant resulted in a significant decrease in friction angle. For example, as the percentage finer than sand size (0.06 mm) was increased from 0 to 50%, the friction angle decreased from about 40° to 30°. Hence, particle breakdown of the lunar soil could have resulted in a lower friction angle. Whether or not particle degradation actually occurred during the oedometer and shear tests on the lunar sample is not known. If it is a factor, then the compressibility and the shear strength of other lunar soils at the same void ratio will probably vary according to the percentage of fragile particles. Thus, the Apollo 14 soils, which contain a greater proportion of microbreccias than the Apollo 12 soils, are likely to have an even higher compressibility and lower shear strength. Further study of this phenomenon and its implications for lunar soil behavior is needed.

Gas Composition

During consolidation, gas was released from the lunar soil each time the applied load exceeded the maximum load previously applied. These short duration pressure bursts ranged from $\Delta P = 0.6$ to 3.2×10^{-8} Torr, with the larger ΔP's occurring at the higher total loads. From a comparison of the soil pore volume to test chamber volume, it is estimated that the pressure in the soil pores was 250 times larger than the observed pressure in the chamber at the time of the bursts.

Because of the short duration of the pressure bursts, it was possible to measure only one atomic mass unit (amu) during any single burst. The following gases contributed to the bursts: amu = 2 (H_2), amu = 4 (He), amu = 28 (unidentified). Water (amu = 18), which was a major component of the background spectra, did not contribute to the observed pressure bursts. Other major gas components in the background were hydrogen and amu = 28. Helium was not detected in the background spectra. Assuming that all pressure bursts were similar in composition, the gas released from the lunar soil was 60% hydrogen, 35% helium, and 4.6% amu = 28.

During shear, the total pressure in the test chamber rapidly rose to 4% above the base pressure prior to shear. During Test 2, helium, which was monitored throughout shear, increased rapidly to 50% above the helium base remaining after consolidation. Both total pressure and helium partial pressure showed periodic very small bursts until near the maximum shear stress, at which time there was a large burst in both total pressure and helium partial pressure. The total pressure during this burst went from 8.2 to greater than 16×10^{-8} Torr, and the helium partial pressure increased from 0.6 to greater than 2.6×10^{-9} Torr. Simultaneously, a large drop in the shear force occurred.

The observed gas burst composition during compression and shear is consistent with the particle-crushing studies of Funkhouser et al. (1971) and the pyrolysis studies of Gibson and Moore (1972). In these studies, it was concluded that the majority of the gas (H_2 and He components) was derived from solar wind, while other gases (amu = 28 component) may be indigenous to the soil grains in small gas-rich inclusions.

These gas pressure bursts are undoubtedly related to the higher compressibility and the lower shear strength of the lunar soil as discussed in the preceeding sections. The gas bursts and the lower strengths are probably the result of crushing of fragile grains when stress is applied to the soil sample.

CONCLUSIONS

(1) The compressibility of returned lunar soil in vacuum is the same or greater than that of ground basalt of similar grain size in air.

(2) Lunar soil can exist at significantly different porosities at depths greater than 25 m below the lunar surface.

(3) The rate of densification of lunar soil due to self-weight is small below a depth of about 30 cm. However, other factors, such as meteorite impact, may cause the in situ density to be altered significantly.

(4) The lunar shear wave velocity probably increases approximately with the one-fourth power of the depth. As a consequence, the Apollo 14 Active Seismic Experiment may not have detected the Fra Mauro formation at a depth of 8.5 m but only naturally consolidated lunar soil.

(5) The frictional strength (tan ϕ) of this lunar soil sample is about 65% that of ground basalt at the same void ratio. This is probably due to the mechanical crushing of fragile grains such as microbreccias and agglutinates. The compressibility and shear strength of other lunar soils will probably vary according to the percentage of these particles.

(6) The breakdown of these grains is also probably the source of the hydrogen and helium gas pressure bursts that were observed when the soil was compressed or

sheared. Detection of these gases at the lunar surface could indicate that a soil mass was consolidating or moving downslope.

Acknowledgments—The authors owe a debt of gratitude to the NASA and BRN staff of the LRL and especially to Ray Ayres and Bob Eason for their invaluable assistance in conducting these experiments.

REFERENCES

Bromwell L. G. (1966) The friction of quartz in high vacuum. Research Report R66-18, Department of Civil Engineering, MIT.

Carrier W. D. III (1970) Lunar soil mechanics on the Apollo missions. Spring Meeting of the Texas Section, ASCE, Galveston, Texas.

Carrier W. D. III, Johnson S. W., Werner R. A., and Schmidt R. (1971) Disturbance in samples recovered with the Apollo core tubes. *Proc. Second Lunar Sci. Conf., Geochim. Cosmochim. Acta* Suppl. 2, Vol. 3, pp. 1959–1972. MIT Press.

Costes N. C., Carrier W. D. III, Mitchell J. K., and Scott R. F. (1970) Apollo 11: Soil mechanics results. ASCE *J. Soil Mech. Foundations Div.* **96,** 2045–2080.

Funkhouser J., Jessberger E., Müller O., and Zähringer J. (1971) Active and inert gases in Apollo 12 and Apollo 11 samples released by crushing at room temperature and by heating at low temperatures. *Proc. Second Lunar Sci. Conf., Geochim. Cosmochim. Acta* Suppl. 2, Vol. 2, pp. 1381–1396. MIT Press.

Gangi A. F. (1972) The lunar seismogram. *The Moon* **4,** 40–48.

Gibson E. K. Jr. and Moore G. W. (1972) Inorganic gas release and thermal analysis study of Apollo 14 and 15 soils. *Proc. Third Lunar Sci. Conf., Geochim. Cosmochim. Acta* Suppl. 3, Vol. 2. MIT Press.

Hardin B. O. and Black W. L. (1966) Sand stiffness under various triaxial stresses. ASCE *J. Soil Mech. Foundations Div.* **92,** 27–42.

Houston W. N. and Mitchell J. K. (1971) Lunar core tube sampling. *Proc. Second Lunar Sci. Conf., Geochim. Cosmochim. Acta* Suppl. 2, Vol. 3, pp. 1953–1958. MIT Press.

Johnson S. W., Pyrz A. P., Lee D. G., and Thompson J. E. (1970) Simulating the effects of gravitational field and atmosphere on behavior of granular media. *Spacecraft and Rockets* **7,** 1311–1317.

Kovach R. L., Watkins J. S., and Landers T. (1971) Active seismic experiment. Apollo 14 Preliminary Science Report, NASA SP-272, pp. 163–174.

Kovach R. L. and Watkins J. S. (1972) The near-surface velocity structure of the moon (abstract). In *Lunar Science—III* (editor C. Watkins), pp. 461–462, Lunar Science Institute Contr. No. 88.

Mitchell J. K. and Houston W. N. (1970) Lunar surface engineering properties experiment definition. Final Report (June 20, 1968 to July 19, 1969), Vol. I, Mechanics and Stabilization of Lunar Soils, University of California at Berkeley, Space Sciences Laboratory, NASA CR-102963.

Mitchell J. K., Bromwell L. G., Carrier W. D. III, Costes N. C., and Scott R. F. (1971) Soil mechanics experiment. Apollo 14 Preliminary Science Report, NASA SP-272, pp. 87–108.

Mitchell J. K., Scott R. F., Houston W. N., Costes N. C., Carrier W. D. III, and Bromwell L. G. (1972) Mechanical properties of lunar soil: Density, porosity, cohesion, and angle of internal friction. *Proc. Third Lunar Sci. Conf., Geochim. Cosmochim. Acta* Suppl. 3, Vol. 3. MIT Press.

Nelson J. D. and Vey E. (1968) Relative cleanliness as a measure of lunar soil strength. *J. Geophys. Res.* **73,** 3747–3764.

Scott R. F., Carrier W. D. III, Costes N. C., and Mitchell J. K. (1971) Apollo 12 soil mechanics investigation. *Geotechnique* **21,** 1–14.

Sutton G. H. and Duennebier F. K. (1970) Elastic properties of the lunar surface from surveyor spacecraft data. *J. Geophys. Res.* **75,** 7439–7444.

Vey E. and Nelson J. D. (1965) Engineering properties of simulated lunar soils. ASCE *J. Soil Mech. Foundations Div.* **91,** 25–52.

Proceedings of the Third Lunar Science Conference
(Supplement 3, *Geochimica et Cosmochimica Acta*)
Vol. 3, pp. 3235–3253
The M.I.T. Press, 1972

Mechanical properties of lunar soil:
Density, porosity, cohesion, and angle of internal friction

J. K. Mitchell and W. N. Houston

University of California,
Berkeley, California

R. F. Scott

California Institute of Technology,
Pasadena, California

N. C. Costes

Marshall Space Flight Center,
Huntsville, Alabama

W. D. Carrier, III

Manned Spacecraft Center,
Houston, Texas

and

L. G. Bromwell

Massachusetts Institute of Technology,
Cambridge, Massachusetts

Abstract—The mechanical properties of lunar soils are remarkably similar to those of terrestrial soils of comparable gradation (silty fine sand), even though the two soil types are compositionally dissimilar. Particle size distribution, density, and particle shape control physical behavior.

A variety of data sources indicate that density and strength characteristics vary locally and with depth. Density may be low (1.0 g/cm³) at the surface in some areas but may be as high as 2.0 g/cm³ at depths of a few centimeters in others. Densities greater than 1.5 g/cm³ are probable at depths of 10 to 20 cm.

For a given lunar soil, porosity appears to be the most important single variable controlling cohesion and friction angle. Most probable values of cohesion appear to be in the range of 0.1 to 1.0 kN/m². The most probable range of lunar soil friction angle is about 30° to 50° with the higher values associated with lower porosities. Data from the Soviet Lunokhod I show specific indication of an increase in strength parameters, and therefore also density, with depth. Other data indicate that soil on slopes is less dense and weaker than the soil covering level areas.

Introduction

THE PHYSICAL AND MECHANICAL PROPERTIES of lunar soil *in situ* have been under study since well before the first lunar landings because of their importance to the interpretation of lunar history and processes, their relevance to the analysis of data from several lunar surface, orbital, and terrestrially based experiments and observations, and because of the need for data to solve engineering and operational problems. Density,

porosity, strength, compressibility, and stress-strain characteristics, and their variations both regionally and locally are of particular importance.

The current state of knowledge (through the Apollo 15 mission) of density (ρ),
porosity (n), interparticle cohesion (c), and frictional resistance between grains (as
reflected by angle of internal friction, ϕ) is reviewed in this paper. Emphasis is on the
unconsolidated fine-grained lunar regolith, i.e., soils whose particles are predominantly
smaller than 1 mm. Ranges in grain-size distribution for several soil samples returned
from the four Apollo landing sites are shown in Fig. 1. With the exception of the curves
for the coarse layer in the Apollo 12 double core tube and the coarse layer in the
Apollo 14 trench, all samples define a band that is typical of well-graded terrestrial silty
fine sands.

METHODS

Pre-Apollo studies

A number of early (pre-Surveyor) estimates of the physical properties of lunar soils
are summarized by Mitchell and Smith (1969), and will not be reviewed here, except
to note that density estimates centered on values less than 1 g/cm³. Choate (1966)
determined lunar slope angles from Ranger photographs. This information made
possible estimation of some lower bound values for soil strength parameters. Jaffe
(1964, 1965) estimated a lower bound bearing capacity based on stability analyses of
crater walls photographed by Ranger.

A number of boulder tracks and their associated boulders were observed on
slopes visible in lunar Orbiter high-resolution photographs. By using assumptions

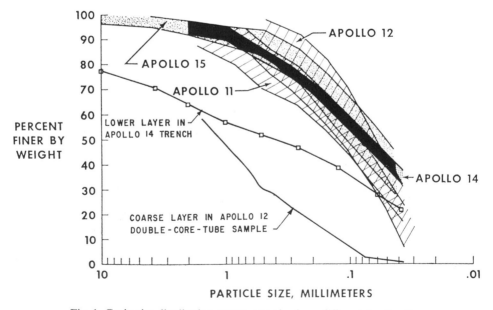

Fig. 1. Grain size distribution ranges, samples from different Apollo sites.

relative to boulder and soil density in conjunction with bearing capacity theory, estimates have been made (Moore, 1970; Hovland and Mitchell, 1971) of the strength characteristics of the soils over which the boulders rolled.

The results of the Surveyor program provided the first conclusive evidence that the lunar surface materials are predominantly fine grained and slightly cohesive and behave in a manner comparable to terrestrial soils of similar gradation. Semiquantitative and quantitative estimates of density, cohesion, and friction angle were made in several ways (Christensen *et al.*, 1967, 1968a, 1968b, 1968c; Scott and Roberson, 1967, 1968a, 1968b, 1969; Scott, 1968)

Simulants for study of lunar soil properties

The definitive information on the texture, particle size, and general behavior of lunar soil provided by the Surveyor results made possible the preparation of lunar soil simulants. Ground basalt has been most widely used, and it has been possible to duplicate closely the range of soil properties believed to exist on the moon (Costes *et al.*, 1969a, 1971; Green and Melzer, 1970; Mitchell *et al.*, 1969, 1971a).

Figure 2 shows the relationship between friction angle and porosity for a lunar soil simulant studied by Mitchell *et al.* (1971a), and Fig. 3 illustrates the variation of cohesion with porosity for the same soil.

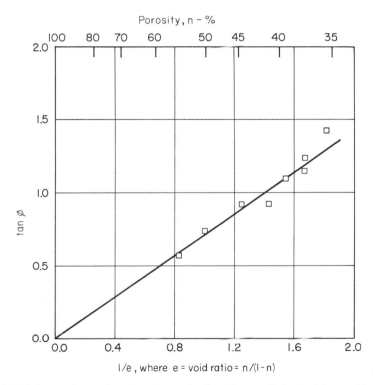

Fig. 2. Friction angle as a function of porosity for a lunar soil simulant (ground basalt).

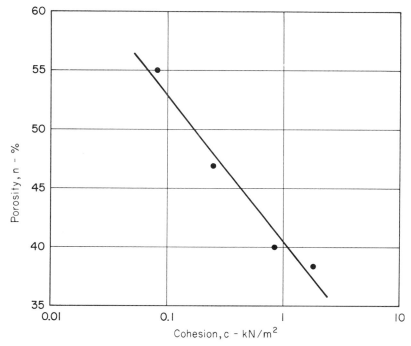

Fig. 3. Cohesion as a function of porosity for a lunar soil simulant (ground basalt).

The application of mechanics analyses

It has been possible to estimate the friction angle and cohesion for soil on the moon using mechanics analyses (usually developed from plasticity theory) of observed interactions; e.g., penetrometers, footprints, boulder and vehicle tracks. With the friction angle and cohesion known, it is then possible to estimate porosity. Conversion of porosity to the *in situ* density on the moon requires a knowledge of the average specific gravity of the soil particles. Only two determinations of specific gravity have been made to the authors' knowledge—on single samples of Apollo 11 and 15 soil. In each case a value of 3.1 was obtained. Other evidence suggests that this value may not be generally applicable and that lower values may hold in many cases.

Core tube samples

Core tube samples returned by each of the Apollo missions and by Luna 16 provide data on soil density. Because of disturbance during sampling, earth return, and handling, the core sample density may differ significantly from the actual *in situ* density (Carrier *et al.*, 1971; Houston and Mitchell, 1971). The larger diameter and reduced wall thickness used for the Apollo 15 core tubes resulted in the acquisition of much less disturbed samples than in previous missions, and the densities of these samples can be considered much more representative of *in situ* values than directly measured densities from core tube samples obtained previously. One drill core sample was returned by the Soviet Luna 16 (Vinogradov, 1971).

Soil mechanics trench

A shallow trench experiment was carried out as part of the Apollo 14 and 15 lunar surface activities for the purposes of observing soil profiles and determining the soil cohesion. Analysis of the failure conditions yields information on the strength parameters, c and ϕ. Two important features of the trench experiment are that the computed cohesion is not a sensitive function of the friction angle, and the calculation is virtually independent of the value used for soil density.

Penetrometer measurements

A self-recording penetrometer was used for the first time during Apollo 15. This device can penetrate to a maximum depth of 76 cm, and could measure penetration force to a maximum of 111 N. The record of each penetration was scribed on a recording drum which was returned to earth for analysis. Two penetration records obtained in the vicinity of the Apollo 15 ALSEP site (Station 8) are shown in Fig. 4. These

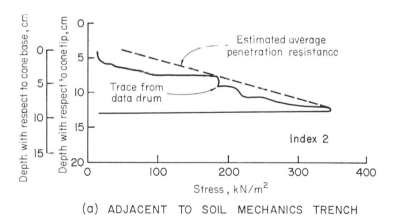

(a) ADJACENT TO SOIL MECHANICS TRENCH

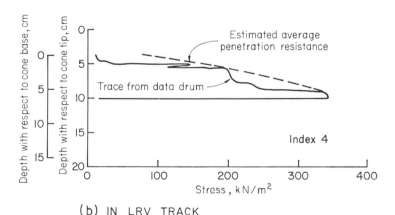

(b) IN LRV TRACK

Fig. 4. Stress versus penetration records, Apollo 15.

data were used in conjunction with the trench data to determine porosity, cohesion and friction angle from direct comparison of behavior with that of terrestrial simulants and from theoretical analyses.

Soviet measurements by Lunokhod I

The most systematic and extensive set of quantitative measurement of surface soil mechanical properties to date has been obtained by the Soviet rover, Lunokhod I, delivered to the western part of Mare Imbrium by Luna 17. A cone penetrometer device, configured with vanes, as shown in the upper right of Fig. 5, was used for a total of 327 measurements along a 5224 m traverse (Leonovich *et al.*, 1971). This device could penetrate to a maximum depth of 10 cm and could be twisted in the ground causing the soil to fail in the manner of a conventional vane shear test. Four penetration curves representing different surface conditions are shown in Fig. 5.

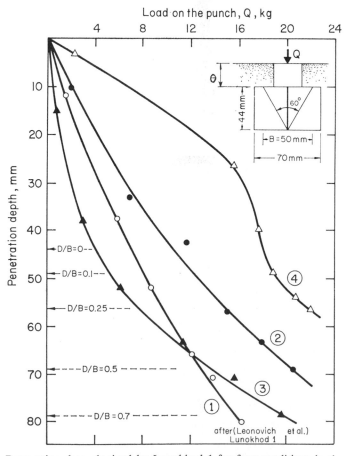

Fig. 5. Penetration data obtained by Lunokhod 1 for four conditions in the western part of Mare Imbrium (data from Leonovich *et al.*, 1971). 1—Level intercrater region; 2—Crater slope; 3—Crater wall; 4—Sector covered by small rocks.

Tests on returned samples

Limited testing of the mechanical properties of the less than 1 mm fraction of the Apollo 11 bulk soil sample was done (Costes *et al.*, 1970), and three direct shear tests on a 200 g sample from Apollo 12 that had not been exposed to a pressure above 2×10^{-6} Torr were conducted by Carrier *et al.* (1972).

DENSITY AND POROSITY

Background

Table 1 summarizes some of the density estimates that have been made since early in the lunar exploration program.

A density of 0.3 g/cm³ (corresponding to a porosity of 90%) was assumed by Jaffe (1964, 1965) in an effort to calculate lower bound bearing capacities. Halajian (1964) also assumed a very low density, 0.4 g/cm³, but believed that the strength of the lunar surface was similar to that of pumice. The grain size distribution and the lunar soil-footpad interaction observed on Surveyor I suggested a value of 1.5 g/cm³ (Christensen *et al.*, 1967). The Russian probe, Luna 13, provided the first in-place (December 1966) measurement of soil density on the moon by means of a gamma-ray device. The calibration curve for this device was double-valued, and it was necessary to choose between a value of 0.8 and 2.1 g/cm³. Cherkasov *et al.* (1968) chose the lesser value. Based on the results from the soil mechanics surface sampler experiments on Surveyors III and VII, Scott and Roberson (1967, 1968) confirmed the Surveyor I value of 1.5 g/cm³ and argued (Scott, 1968) that the Russian investigators had chosen the wrong portion of their calibration curve.

The drive tube data from Apollo 11 were also ambiguous because of the shape of the bit. The bulk densities of the soil in the two core tubes were 1.59 and 1.71 g/cm³ (Costes *et al.*, 1969b), or 1.54 g/cm³ and 1.75 g/cm³ as later reported by Costes and

Table 1. Estimates of lunar soil density.

Bulk density ρ (g/cm³)	Investigator	Mission
0.3	Jaffe (1964, 1965)	
0.4	Halajian (1964)	
1.5	Christensen *et al.* (1967)	Surveyor I
0.8	Cherkasov *et al.* (1968)	Luna 13
1.5	Scott and Roberson (1967, 1968) and Scott (1968)	Surveyor III & VII
1.54 to 1.75	Costes and Mitchell (1970)	Apollo 11
0.74 to > 1.75	Scott *et al.* (1971)	Apollo 11
1.81 to 1.92*	Costes *et al.* (1971)	Apollo 11
1.6 to 2.0	Scott *et al.* (1971)	Apollo 11
1.80 to 1.84*	Costes *et al.* (1971)	Apollo 12
1.55 to 1.90	Houston and Mitchell (1971)	Apollo 12
1.7 to 1.9	Carrier *et al.* (1971)	Apollo 12
1.2	Vinogradov (1971)	Luna 16
1.5 to 1.7	Leonovich *et al.* (1971)	Lunokhod I
1.45 to 1.6	Carrier *et al.* (1972)	Apollo 14
1.35 to 2.15	Mitchell *et al.* (1972)	Apollo 15

*Upper bound estimates.

Mitchell (1970) taking into account possible differences in core tube diameter. These densities could have indicated an *in situ* density anywhere from 0.75 g/cm^3 to more than 1.75 g/cm^3 (Scott *et al.*, 1971).

The shape of the Apollo 12 drive tube bits reduced the uncertainty, and the density at this site was estimated to be 1.6 to 2 g/cm^3 (Scott *et al.*, 1971). Core tube simulations performed later by Houston and Mitchell (1971) and Carrier *et al.* (1971), yielded additional estimates of 1.55 to 1.9 g/cm^3 and 1.7 to 1.9 g/cm^3, respectively. Based on penetration resistance data from the Apollo 11 and 12 landing sites, Costes *et al.* (1971) gave upper bound estimates of the density at the two sites of 1.81 to 1.94 g/cm^3 and 1.81 to 1.84 g/cm^3, respectively. Carrier *et al.* (1972) have determined *in situ* densities of 1.45 to 1.6 g/cm^3 for the Apollo 14 core tube samples. Vinogradov (1971) estimated a value of 1.2 g/cm^3 from a rotary drill sample returned by Luna 16. By comparison of Lunokhod data with the results of studies of the Luna 16 sample, Leonovich *et al.* (1971) estimated densities in the range of 1.5 to 1.7 g/cm^3 for the areas traversed by Lunokhod.

Density of the lunar soil at the Apollo 15 site

A preliminary estimate has been made of density versus depth at the three Apollo 15 core tube locations, as shown in Fig. 6. The top 25 to 35 cm of soil along the

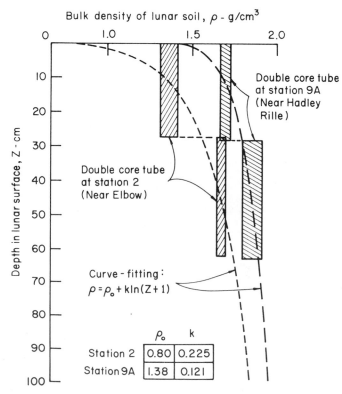

Fig. 6. Density of lunar soil at Apollo 15 core tube sites.

Apennine Front (Stations 2 and 6) have very similar, low average values, 1.35 g/cm³, so only the data for Station 2 are shown. The soil density evidently increases fairly rapidly with depth. The soil density at the Front is approximately 10% less than observed at any previous Surveyor or Apollo site and approaches that of the Luna 16 site (1.2 g/cm³). The average soil density at the edge of Hadley Rille (Station 9A) is significantly higher in the top 30 cm (1.69 g/cm³) and increases less rapidly with depth than at Station 2.

If the density of the lunar soil is assumed to increase with depth primarily because of self-weight, a monotonic curve may be fitted to the data for the two double-core tubes, as shown in Fig. 6. The surface density at the Front would then be 0.80 g/cm³; the surface density at the Rille is 1.38 g/cm³, or more than 70% greater. At a depth of 2.8 m, the density at both locations would be 2.07 g/cm³.

The fact that the density of the soil on the slopes of the Apennine Front is much less than that in the mare area suggests that the soil on the slopes is considerably weaker, although quantitative comparisons are not available. If it can be shown that the soil covering slopes is generally much weaker, then development of reasonable hypotheses for downslope movement of material will be greatly facilitated.

The *in situ* density at the soil mechanics trench (Station 8, near the ALSEP site) has been estimated to be in the range of 1.92 to 2.01 g/cm³ based on penetration test results. A density range of 1.62 to 1.93 g/cm³ has been estimated (Carrier *et al.*, 1972) for the samples in the deep drill stem obtained from the same area. Average density of these samples is of the order of 1.8 g/cm².

COHESION AND FRICTION

Introduction

Most estimates of lunar soil cohesion and friction angles have been based on the results of analyses of soil failure conditions, e.g., under rolling boulders, by penetration, and by failure of a trench wall. Both soil cohesion (c) and internal friction (ϕ) are important in resisting the applied loads, since soil shear strength (s) is given by

$$s = c + \sigma \tan \phi, \qquad (1)$$

where σ is the stress normal to the failure plane. Thus, in many cases it is impractical to discuss the magnitude and variation of one parameter independently of the other. A number of estimates of lunar soil cohesion and friction angle developed from data available prior to the Apollo and Lunokhod missions is summarized in Table 2.

Regional variability as indicated by boulder track records

A large range of friction angle and cohesion values is indicated by the data in Table 2. Because of the assumptions and uncertainties associated with most of the analyses, it is difficult to establish whether or not such variations really exist on the lunar surface.

Some light is shed on this question by the results of boulder track analyses reported by Hovland and Mitchell (1971). In this study 69 boulder tracks from 19 different

J. K. MITCHELL *et al.*

Table 2. Estimates of lunar soil cohesion and friction angle based on pre-Apollo data.

Basis	Cohesion c (kN/m²)	Friction angle ϕ (deg)	Reference
(1) Boulder track analysis—Orbiter data	0.35	33	Nordmeyer (1967)
(2) Surveyor I strain gage and TV data	0.15–15	55	Jaffe (1967)
(3) Surveyor I	0.13–0.4	30–40	Christensen *et al.* (1967)
(4) Surveyor III, soil mechanics surface samples		> 35	Scott and Roberson (1968)
(5) Surveyor III, landing data	0 for 10 for	45–60 0	Christensen *et al.* (1968c)
(6) Surveyor VI, vernier engine firing	> 0.07 for	35	Christensen *et al.* (1968a)
(7) Surveyor VI, attitude control jets	0.5–1.7		
(8) Surveyor III and VII, soil mechanics surface samples	0.35–0.70	35–37	Scott and Roberson (1969)
(9) Lunar Orbiter boulder track records	0.1	10–30	Moore (1970)
(10) Lunar Orbiter boulder track records	0.5*	21–55 39†	Hovland and Mitchell (1971)

* Assumed.
† Mean of 69 values.

locations were selected from Lunar Orbiter high resolution photography. These tracks were formed on slopes estimated to range from 0° to 30°. Analyses for friction angle values were made using bearing capacity theory for footings on slopes (Meyerhof, 1951) modified for application to the boulder track formation mechanisms (Hovland and Mitchell, 1971). The range in values of friction angle computed in this study is shown in Table 2, together with the value of cohesion assumed for the analyses. These friction angles were converted to porosity by Houston *et al.* (1972) and analyzed statistically. A mean porosity of about 44% was obtained with a standard deviation of 6.6%. Additional evidence of variability in lunar soil properties was obtained from analyses of astronaut footprint depth (Houston *et al.*, 1972).

Cohesion and friction angle values from Apollo 11, 12, and 14 data

During the first three Apollo landings, no force or deformation measuring devices were utilized to determine directly the in place mechanical properties of the lunar soil. Consequently, inferences on these properties were made from (a) observed soil deformations resulting from the interaction of the soil with objects of known geometry and weight; (b) assumptions on the ranges of loads applied by the astronauts in pushing the Apollo simple penetrometer (Apollo 14 mission), core tubes or other shafts and poles into the soil; (c) slope stability analyses applied to natural crater slopes, incipient slope failures of soft-rimmed craters due to loads imposed by walking astronauts, and to the collapse of the soil mechanics trench during the Apollo 14 mission; (d) LM landing dynamics and soil erosion caused by the LM engine exhaust; (e) penetration tests on loose and densely compacted Apollo 11 soil bulk sample; and (f) studies on simulated lunar soils. Values of c and ϕ obtained on the basis of these analyses are listed in Table 3.

Table 3. Estimates of lunar soil cohesion and friction angle based on Apollo 11, Apollo 12, and Apollo 14 data.

Mission	Basis	Cohesion c (kN/m²)	Friction angle ϕ (deg)	Reference
Apollo 11	Astronaut footprints, LM landing data, crater slope stability	Consistent with lunar soil model from Surveyor data		Costes *et al.* (1969)
Apollo 11	Penetrometer tests in LRL on bulk soil sample	0.3–1.4	35–45	Costes *et al.* (1970)
Apollo 11	Penetration of core tubes, flagpole, SWC shaft	0.8–2.1	37–45	Costes *et al.* (1971)
Apollo 12	Astronaut footprints, LM landing data, crater slope stability	Consistent with lunar soil model from Surveyor data		Scott *et al.* (1970)
Apollo 12	Penetration of core tubes, flagpole, SWC shaft	0.6–0.8	38–44	Costes *et al.* (1971)
Apollo 14	Soil mechanics trench	< 0.03–0.3	35–45	Mitchell *et al.* (1971b)
Apollo 14	Apollo simple penetrometer	Soil shear strength equal to or greater than that of soil model from Surveyor data		Mitchell *et al.* (1971b)
Apollo 14	MET tracks		37–47	Mitchell *et al.* (1971b)

Variability of lunar soil properties within the Fra Mauro landing site from MET track analysis

Analysis of tracks left by the Modular Equipment Transporter (MET) at various points of the geological traverses during the Apollo 14 mission yielded information on the variability of lunar soil properties near the surface on a landing site scale (maximum distance from the LM was about 1450 m).

Tracks left by the MET were analyzed using a dimensional analysis relating to the interaction of pneumatic tires with granular, primarily cohesionless soils, results of studies on lunar soil simulants under terrestrial and lunar gravity levels (Green and Melzer, 1970; Costes *et al.*, 1971), and bearing capacity theory as applied to cone penetrometers. Details of this analysis are given by Mitchell *et al.* (1971b).

The results showed that soil located in intercrater areas on firm level ground was weaker than soil located in soft pockets and on fresh crater rims and slopes. No appreciable differences in lunar soil properties between regions of different geologic age were discernible, however, on the basis of the computed values.

Cohesion and friction angle values at the Hadley-Apennine landing site

The Apollo 15 mission provided the first U.S. opportunity for definitive evaluation of cohesion and friction at a given location. The results of simulation studies, drill stem examination, the soil mechanics trench analysis, and measurements using the self-recording penetrometer have been used collectively to determine properties at Station 8 (ALSEP site).

Using the penetration test data in Fig. 4 in conjunction with the results of simulation studies (Houston and Namiq, 1971; Costes *et al.*, 1971; Green and Melzer, 1970), the ranges in, and best estimates of, porosity, density, and friction angle for the near surface soil in a level intercrater region near the Apollo 15 ALSEP site have been estimated as indicated in Table 4. It may be noted that these values of density and friction angle are near the high end of the ranges associated with previous estimates. The firmness of the soil at this location as indicated by its resistance to drilling during installation of the heat flow experiment is consistent with this result.

Soil mechanics analysis of the trench wall failure and the measured values of penetration resistance (Mitchell *et al.*, 1972; Durgunoglu, 1972) give values of cohesion and friction angle which are consistent with the values obtained by comparison of observed behavior with simulants. The dashed line in Fig. 7 shows the values of cohesion and friction angle which satisfy equilibrium equations at incipient failure of the

Table 4. Soil characteristics at Station 8 (near the ALSEP site), Apollo 15 mission.

	Porosity n (%)	Void ratio e	Density* ρ (g/cm³)	Friction angle ϕ
Range	35–38	0.54–0.61	1.92–2.01	47.5–51.5
Best estimate	36.5	0.58	1.97	49.5

* Assumes average specific gravity of soil grains to be 3.1.

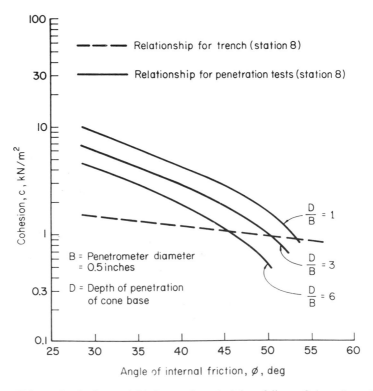

Fig. 7. Values of cohesion and friction angle at incipient failure of the soil mechanic trench wall and for a 25-pound force applied to the self-recording penetrometer at different penetration depths, Apollo 15 ALSEP site.

trench wall. Also shown in Fig. 7 (solid curves) are values of cohesion and friction angle which correspond to loading of the 0.5 sq in. cone penetrometer to its maximum recordable capacity (25 pounds) and three penetration depths.

The depth-to-cone base diameter (D/B) ratio for tests at Station 8 fell in the range of about 2.5 to 4.1. Thus the curve in Fig. 7 for $D/B = 3$ is appropriate. The interaction of this curve with that for the trench wall failure gives conditions that satisfy both the trench and penetration test simultaneously. All the data taken collectively are quite consistent and indicate values of properties that can be summarized as follows for the soil at the soil mechanics trench in the Hadley-Apennine landing area:

Porosity	36.5%
Void ratio	0.58
Density	2 g/cm³ or slightly less
	(for a specific gravity = 3.1)
Cohesion	1 kN/m²
Angle of internal friction	50°

This value of angle of internal friction may be compared with that to be expected for a terrestrial soil of similar characteristics. Koerner (1970) has proposed relationships for angle of internal friction in terms of particle shape, particle size, gradation, relative density, and mineral type. When applied to the lunar soil *in situ* at Station 8, these relationships give an estimated friction angle of 50° to 52°. This compares well with the value of 50° deduced from simulation studies and theoretical analyses.

Analysis of Lunokhod I data

It was indicated earlier in this paper that the Soviet rover, Lunokhod I, was equipped with a cone penetrometer and vane shear test device and that a large number (327) of measurements to depths up to 10 cm had been made along a traverse exceeding 5 km in length in the western part of Mare Imbrium. An analysis of some of the data reported by Leonovich *et al.* (1971) has been made which provides quantitative indication of soil property variations between different areas and with depth.

In the analysis of vane shear test results the usual assumptions are that the soil fails as a cylinder defined by the vane dimensions (see upper right of Fig. 5) when it rotates and that the resistance to the torque applied to the vane is provided by soil cohesion; i.e., a $\phi = 0$ soil is assumed. This approach appears to have been used by Leonovich *et al.* (1971), who report values of "torque resistance" in the range of 2–9 kN/m², with a greatest frequency of 4.5 kN/m². These values are of a magnitude several times greater than have been estimated previously for lunar soil cohesion. The Lunokhod data have been reanalyzed using an expression derived by Farrent (1960) for vane shear tests in soils exhibiting both cohesive and frictional resistance.

Values of c and ϕ corresponding to the maximum and minimum measured torque are shown by the upper and lower curves in Fig. 8, and the middle curve corresponds to the highest frequency torque resistance.* It may be seen that for friction angles in the probable range of 35° to 50° the corresponding values of cohesion are well within

---

* Since both c and ϕ are unknown, they cannot be determined uniquely by Farrent's (1960) equation.

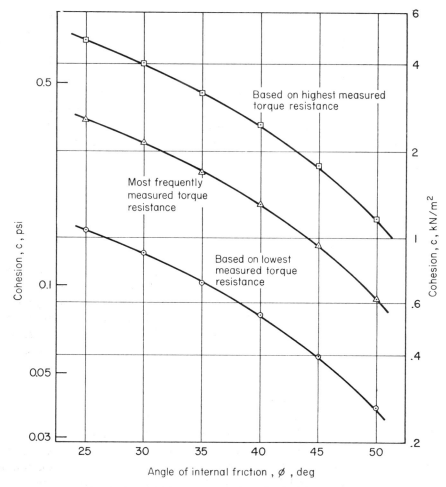

Fig. 8. Range of cohesion and friction angle values obtained from Lunokhod vane shear test data (data from Leonovich *et al.*, 1971).

the range established by previous studies. For an assumed ϕ of 50° one would anticipate a high density and high torque resistance which, according to Fig. 8 would indicate a cohesion of about 1 kN/m² , as was the case also at the Apollo 15 ALSEP site.

An analysis of the penetration data shown in Fig. 5 has been made using the method developed by Durgunoglu (1972). According to classical plasticity theory, unit penetration resistance q is given by

$$q = cN_c\xi_c + B\rho g N_{\gamma q}\xi_{\gamma q}, \tag{2}$$

where c = cohesion; ρ = soil density; g = acceleration due to gravity; B = cone base diameter = 50 mm; N_c, $N_{\gamma q}$ = bearing capacity factors = $f(\phi, \alpha, \delta/\phi, D/B)$; ξ_c, $\xi_{\gamma q}$ = shape factors; ϕ = soil friction angle; δ = soil to cone friction angle; $\alpha = \frac{1}{2}$ cone apex angle; D = penetration depth (cone base). For the present analysis

$\alpha = 30°$ and δ/ϕ was assumed as 0.5. The computed results are insensitive to soil density, and it was assumed to be 1.7 g/cm³.

From a histogram of "carrying capacity" (i.e., q for $D/B = 0$) values, upper and lower bounds, as well as the most frequently measured values, were taken, and corresponding values of c and ϕ were computed. The penetrometer and vane shear device gave comparable results.

It is important to note in connection with plots of the type shown in Figs. 7 and 8 that increases in friction angle do not, in general, imply decreases in soil cohesion. The curves simply indicate corresponding values of c and ϕ that could account for the measured values of strength or penetration resistance. For a given soil—and in the absence of significant cementation between particles, for which there is no evidence thus far—variations in c and ϕ are closely related to variations in density. As density increases both c and ϕ tend to increase.

Combinations of c and ϕ corresponding to the penetration resistance at several depths according to Curve 1 of Fig. 5 (level intercrater region) were computed. The results showed clearly that strength, and probably also density, increase significantly with depth. Similar results were obtained from analysis of Fig. 5, Curve 2 for penetration into a crater slope and Curve 3 for penetration into a crater wall. The results for the crater wall show a particularly marked increase in strength with depth.

Curves corresponding to $D/B = 0$ for the level ground, crater slope, and crater wall are compared in Fig. 9. Unfortunately, Leonovich et al. (1971) do not describe the nature or size of the crater in which the curves shown in Fig. 5 were obtained. Because of trafficability considerations, however, it is not likely that the crater was sharp, blocky, or deep. The curves in Fig. 9 would appear compatible with a subdued shallow crater where soil on the inside had a long history of downslope movement and could be expected to be in a relatively loose and weak condition.

CONCLUSIONS

The fine-grained soil that blankets the lunar surface has a grain size distribution that corresponds to terrestrial silty fine sands, although coarser material may be found locally. Available evidence from a variety of sources—photographs, simulation studies, core tube samples, trenching experiments, and penetrometer measurements—indicate that the mechanical properties of the soil are remarkably similar to those of terrestrial soils of comparable gradation, even though the two types of soil are compositionally dissimilar. This is a direct reflection of the fact that for soils in this particle size range, density and particle size and shape distribution exert a larger influence on mechanical properties than does composition.

Lunar soils do differ from terrestrial soils, however, in that the lunar materials are somewhat more cohesive. Whether this is due to compositional or environmental differences is not yet known.

The density may be less than 1 g/cm³ at the surface in some areas and as high as 2 g/cm³ at very shallow (a few centimeters) depth in others. There is strong evidence that density and strength increase with depth, rapidly in the case of low surface densities, and gradually in the case of high surface densities. Densities greater than 1.5 g/cm³ are probable below depths of 10 to 20 cm.

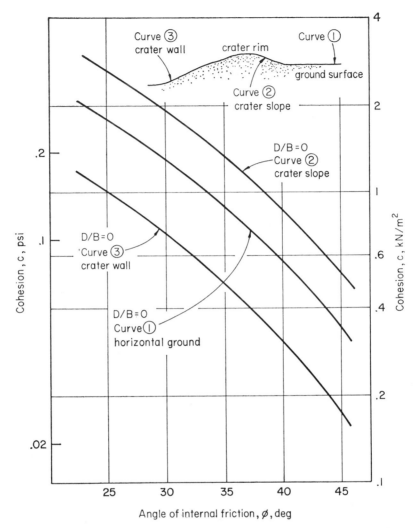

Fig. 9. Comparison of strength characteristics for level ground, crater wall, and crater slope (data from Leonovich *et al.*, 1971).

The soil on slopes may be significantly less dense and weaker than the soil on level areas. This finding may be important in understanding mechanisms of downslope movement of soil, since the relationship between strength and slope angle will influence the extent to which different driving forces can cause movement.

For a given grain size distribution, porosity appears to be the most important single variable controlling cohesion and friction angle, and relationships similar to those in Figs. 2 and 3 are probable, although present data are insufficient to define them uniquely. Most probable values of cohesion appear to be in the range of 0.1 to 1 kN/m².

Evidence is accumulating that the angle of internal friction of lunar soils varies in a manner similar to that of terrestrial soils and that its most probable range is from about 30° to 50°.

The most comprehensive set of data thus far available are for the soil at shallow depth near the soil mechanics trench (Station 8) at the Hadley-Apennine site. From these data the determined soil properties are:

Porosity	36.5%
Void ratio	0.58
Density	2 g/cm^3
Cohesion	1 kN/m^2
Angle of internal friction	50°

These results indicate that the soil at this location is near the lower end of the range of porosities likely to be encountered (Houston *et al.*, 1972).

Analysis of penetration and vane shear data obtained by the Soviet Lunokhod 1 has yielded cohesion and friction angle estimates that are consistent with other information gathered to date. Lunokhod data provide specific indication of an increase in strength parameters, and therefore also density, with depth below the surface. The results show further that the soil properties differ between level intercrater areas, crater walls, and crater slopes.

Acknowledgments—This paper was prepared under NASA Contract NAS 9-11266, Principal Investigator Support for the Soil Mechanics Experiment. H. Turan Durgunoglu, Department of Civil Engineering, University of California, Berkeley, assisted in the analysis of Lunokhod data.

REFERENCES

Carrier W. D. III, Bromwell L. G., and Martin R. T. (1972) Strength and compressibility of returned lunar soil (abstract). In *Lunar Science—III* (editor C. Watkins), pp. 119–121, Lunar Science Institute Contr. No. 88.

Carrier W. D. III, Johnson S. W., Carrasco L. H., and Schmidt R. (1972) Core sample depth relationships: Apollo 14 and 15 (abstract). In *Lunar Science—III* (editor C. Watkins), pp. 122–124, Lunar Science Institute Contr. No. 88.

Carrier W. D. III, Johnson S. W., Werner R. A., and Schmidt R. (1971) Disturbance in samples recovered with the Apollo core tubes. *Proc. Second Lunar Sci. Conf., Geochim. Cosmochim. Acta* Suppl. 2, Vol. 3, pp. 1959–1972. MIT Press.

Cherkasov I. I., Vakhnin V. M., Kemurjian A. L., Mikhailov L. N., Mikheyev V. V., Musatov A. A., Smorodinov M. I., and Shvarey V. V. (1968) Determination of the physical and mechanical properties of the lunar surface layer by means of Luna 13 automatic station. *Moon and Planets II* (editor A. Dollfus), pp. 70–76, North-Holland.

Choate R. (1966) Lunar slope anales and surface roughness from Ranger photographs. Jet Propulsion Lab. Tech. Report 32-994, pp. 411–432.

Christensen E. M., Batterson S. A., Benson H. E., Chandler C. E., Jones R. H., Scott F. R., Shipley E. N., Sperling F. B., and Sutton G. H. (1967) Lunar surface mechanical properties—Surveyor I. *J. Geophys. Res.* **72**, 801–813.

Christensen E. M., Batterson S. A., Benson H. E., Choate R., Hutton R. E., Jaffe L. D., Jones R. H., Ko H. Y., Schmidt F. N., Scott R. F., Spencer R. L., Sperling F. B., and Sutton G. H. (1968a) Lunar surface mechanical properties, in Surveyor VI, a preliminary report. NASA SP-166, pp. 41–95.

Christensen E. M., Batterson S. A., Benson H. E., Choate R., Hutton R. E., Jaffe L. D., Jones R. H., Ko H. Y., Schmidt F. N., Scott R. F., Spencer R. L., and Sutton G. H. (1968b) Lunar surface mechanical properties. *J. Geophys. Res.* **73**, 7169–7192.

Christensen E. M., Batterson S. A., Benson H. E., Choate R., Jaffe L. D., Jones R. H., Ko H. Y., Spencer R. L., Sperling F. B., and Sutton G. H. (1968c) Lunar surface mechanical properties at the landing site of Surveyor III. *J. Geophys. Res.* **73**, 4081–4094.

Costes N. C., Hadjidakis C. G., Holloway D. M., Olson J. P., and Smith R. E. (1969a) Lunar soil simulation studies in support of the Apollo 11 mission. Technical Memo, Geotechnical Research Laboratory, NASA Marshall Space Flight Center, Huntsville, Alabama.

Costes N. C., Carrier W. D., Mitchell J. K., and Scott R. F. (1969b) Apollo 11 soil mechanics investigation. Apollo 11 preliminary science report, NASA SP-214, pp. 85–122.

Costes N. C., Carrier W. D. III, Mitchell J. K., and Scott R. F. (1970) Apollo 11: Soil mechanics results. *J. Soil Mechanics and Foundations Div.*, ASCE **96**, 2045–2080.

Costes N. C., Cohron G. T., and Moss D. C. (1971) Cone penetration resistance test—An approach to evaluating the in-place strength and packing characteristics of lunar soils. *Proc. Second Lunar Sci. Conf., Geochim. Cosmochim. Acta* Suppl. 2, Vol. 3, pp. 1973–1987. MIT Press.

Costes N. C. and Mitchell J. K. (1970) Apollo 11 soil mechanics investigation. *Proc. Apollo 11 Lunar Sci. Conf., Geochim. Cosmochim. Acta* Suppl. 1, Vol. 2, pp. 2025–2044. Pergamon.

Durgunoglu H. T. (1972) The static penetration resistance of soils, Ph.D. Dissertation, in preparation. Dept. of Civil Engineering, University of California, Berkeley.

Farrent T. A. (1960) The interpretation of vane tests in soils having friction. *Proc. Third Australia–New Zealand Conf. on Soil Mechanics and Foundation Engineering*, pp. 81–88.

Green A. J. and Melzer K. J. (1970) Performance of Boeing-GM wheels in a lunar soil simulant (basalt). Technical Report M-70-15, USAE WES, Vicksburg, Miss.

Halajian J. D. (1964) The case for a cohesive lunar surface model. Grumman Research Dept. Report ADR 04-40-64.2, Grumman Aircraft Engineering Corp., Bethpage, New York.

Houston W. N. and Mitchell J. K. (1971) Lunar core tube sampling. *Proc. Second Lunar Sci. Conf., Geochim. Cosmochim. Acta* Suppl. 2, Vol. 3, pp. 1953–1958. MIT Press.

Houston W. N. and Namiq L. I. (1971) Penetration resistance of lunar soils. *Journal of Terramechanics* **8**, 59–69.

Houston W. N., Hovland H. J., Mitchell J. K., and Namiq L. I. (1972) Lunar soil porosity and its variation as estimated from footprints and boulder tracks (abstract). In *Lunar Science—III* (editor C. Watkins), pp. 395–397, Lunar Science Institute Contr. No. 88.

Hovland H. J. and Mitchell J. K. (1971) Mechanics of rolling sphere-soil slope interaction. Final Report, Vol. II of IV, NASA Contract 8-21432, Space Sciences Laboratory, University of California, Berkeley.

Jaffe L. D. (1964) Depth and strength of lunar dust. *Trans. Amer. Geophys. Union* **45**, 628.

Jaffe L. D. (1965) Strength of the lunar dust. *J. Geophys. Res.* **70**, 6139–6146.

Jaffe L. D. (1967) Surface structure and mechanical properties of the lunar maria. *J. Geophys. Res.* **72**, 1727–1731.

Koerner R. M. (1970) Effect of particle characteristics on soil strength. *J. Soil Mechanics and Foundations Div.*, ASCE **96**, 1221–1234.

Leonovich A. K., Gromon V. V., Rybakov A. V., Petrov V. K., Pavlov P. S., Cherkasov I. I., and Shvarav V. V. (1971) Studies of lunar ground mechanical properties with the self-propelled Lunokhod-1. In *Peredvizhnaya Laboratoriya na Luna-Lunokhod-1*, Chap. 8, pp. 120–135.

Meyerhof G. G. (1951) The ultimate bearing capacity of foundations. *Geotechnique* **2**, 301.

Mitchell J. K., Bromwell L. G., Carrier W. D., Costes N. C., Houston W. N., and Scott R. F. (1972) Soil mechanics experiment. Apollo 15 Preliminary Science Report, NASA SP-289, pp. 7-1–7-28.

Mitchell J. K., Houston W. N., Vinson T. S., Durgunoglu T., Namiq L. I., Thompson J. B., and Treadwell D. D. (1971a) Lunar surface engineering properties experiment definition. Final report. Vol. I of IV, Mechanics, properties and stabilization of lunar soils. NASA Contract NAS 8-21432, Space Sciences Laboratory, University of California, Berkeley.

Mitchell J. K., Bromwell L. G., Carrier W. D., Costes N. C., and Scott R. F. (1971b) Soil mechanics experiment. Apollo 14 Preliminary Science Report, NASA SP-272, pp. 87–108.

Mitchell J. K. and Smith S. S. (1969) Engineering properties of lunar soils. Vol. I, Chap. 2, of Materials studies related to lunar surface exploration. Final Report under Contract NAS 05-003-189, Univ. of Calif., Berkeley.

Moore H. J. (1970) Estimates of the mechanical properties of lunar surface using tracks and secondary impact craters produced by blocks and boulders. Interagency Report: Astrogeology 22. United States, Dept. of the Interior, Geological Survey.

Nordmeyer E. F. (1967) Lunar surface mechanical properties derived from track left by nine meter boulder. MSC Internal Note No. 67-TH-1, NASA.

Scott R. F. (1968) The density of lunar surface soil. *J. Geophys. Res.* **73**, 5469–5471.

Scott R. F., Carrier W. D., Costes N. C., and Mitchell J. K. (1970) Mechanical properties of the lunar regolith. Apollo 12 Preliminary Science Report, NASA SP-235, pp. 161–182.

Scott R. F., Carrier W. D. III, Costes N. C., and Mitchell J. K. (1971) Apollo 12 soil mechanics investigation. *Geotechnique* **21**, 1–14.

Scott R. F. and Roberson F. I. (1967) Soil mechanics surface sampler: Lunar tests, results, and analyses —Surveyor III Mission Report, Part II: Scientific Data and Results. Technical Report 32-1177, pp. 69–110, JPL, Pasadena, Calif.

Scott R. F. and Roberson F. I. (1968a) Soil mechanics surface sampler. Surveyor Project Final Report, Part II: Science Results. Technical Report 32-1265, pp. 195–206, JPL, Pasadena, Calif.

Scott R. F. and Roberson F. I. (1968b) Surveyor III—Soil mechanics surface sampler: Luna surface tests, results, and analysis. *J. Geophys. Res.* **73**, 4045–4080.

Scott R. F. and Roberson F. I. (1969) Soil mechanics surface sampler. In *Surveyor Program Results*, NASA SP-184, pp. 171–179.

Vinogradov A. P. (1971) Preliminary data on lunar ground brought to earth by automatic probe Luna 16. *Proc. Second Lunar Sci. Conf., Geochim. Cosmochim. Acta* Suppl. 2, Vol. 1, pp. 1–16. MIT Press.

Proceedings of the Third Lunar Science Conference
(Supplement 3, *Geochimica et Cosmochimica Acta*)
Vol. 3, pp. 3255–3263
The M.I.T. Press, 1972

Lunar soil porosity and its variation as estimated from footprints and boulder tracks

W. N. HOUSTON, H. J. HOVLAND, and J. K. MITCHELL

University of California, Berkeley, California

and

I. I. NAMIQ

Baghdad, Iraq

Abstract—Statistical variation in lunar soil porosity has been assessed through analyses of astronaut footprints and imprints made by rolling boulders. The mean porosities and standard deviations were found to be about the same for the level, intercrater areas for each of the four Apollo sites. For 273 observations of footprint depth at the four sites combined, a mean porosity of 43.3% and a standard deviation of 2.8% were obtained for the upper 15 cm. The analyses indicate a gradually decreasing porosity with depth, although local deviations from this pattern are known to occur. The equivalence of mean porosity and standard deviation for the four Apollo sites suggests that the lunar processes affecting the porosity of the upper few cm may have reached a steady state.

Porosities of crater rim material and crater and rille slopes were found to be somewhat higher, with an average value of about 46 to 47%, but occasionally much higher.

The variations in porosity were found to occur on a relatively small scale—probably of the order of a few meters. This result indicates that it is very unlikely that the porosity at a given lunar traverse station stop could be characterized by one or two core tube samples.

INTRODUCTION

SEVERAL CORE TUBE DENSITY VALUES have been obtained during the Apollo 11–15 missions (Mitchell *et al.*, 1972), but the number of values is typically too small to yield a statistical variation, especially at any one landing site. The wide range in porosity which may be inferred from the core tube results suggests that lunar soil porosity may best be described in terms of statistical parameters.

Statistical variations in lunar soil porosity have been derived from study of two sources of lunar data: (1) astronaut footprints—Apollo 11, 12, 14, and 15, and (2) boulder tracks—Lunar Orbiter Missions II, III, and V.

Photographic coverage of a large number of footprints is available for each landing site, and these footprints have been analyzed as indicators of average soil porosity within the top 15 cm of lunar soil. These data were derived primarily from photographs of generally level intercrater regions, plus some crater rims.

Variation in porosity of lunar soil has also been derived from the study of tracks (imprints) made by boulders rolling down slopes. These analyses have led to estimates of the average lunar soil porosity in the upper few meters on crater and rille slopes.

Analysis of Footprints

Depth estimates from photographs

The first step in the analysis of footprints was the measurement of footprint depths from the photographs from each Apollo mission. The known length and width of the astronaut boot and the treads on the bottom of the boot were used to establish scale for the measurement of the footprint depth. The depth of each footprint was measured in several places and averaged. In cases where the depth could not be measured directly, the length of the shadow cast by the up-sun wall of the imprint and the sun angle were used to compute the depth. Footprints indicating that the astronaut did not apply his weight evenly, e.g., where the heel dug in deeply and the toe did not touch, were not analyzed.

Calibration of depth estimates

After all footprint depth estimates had been made for the Apollo missions, a series of tests was made for calibration purposes. Footprints were made in a lunar soil simulant and photographed and the depths were estimated from the photographs. The footprint depths were subsequently measured directly with a scale in the laboratory. The average difference between the values measured directly and those estimated from the photographs was about 15%, and a correction of this magnitude was then applied to the estimated footprint depths for the Apollo missions.

Conversion of footprint depth to porosity

An extensive model testing program using a lunar soil simulant and finite element analyses of footprints as plateload tests (including an accounting for the effect of reduced gravity) was used to develop a correlation between astronaut footprint depth and average soil porosity in the top 15 cm. The lunar soil simulant (designated LSS No. 2) is described by Houston and Namiq (1971). Its grain size distribution curve is very near to the average distribution curve for returned samples from the Apollo missions, and the simulant has been found to represent the mechanical properties of the lunar soil well.

The model tests and theoretical analyses are described in detail by Namiq (1970). The correlation curve, presented in Fig. 1, represents the final result of these analyses. This curve was used to convert each footprint depth to porosity.

Histograms and statistical parameters

The porosity values were grouped initially by Apollo mission and, in addition, subgroupings were made for each set of data to see if local and regional variations could be distinguished. Porosity values for crater rims were also separated from values for intercrater areas.

Subsequently, various combinations of data groupings were made for additional analyses. For each set of data the mean and sample standard deviation were computed, and all the results are given in Table 1. Shown also in Table 1 are the results from the Boulder Track Analyses, described in a later section.

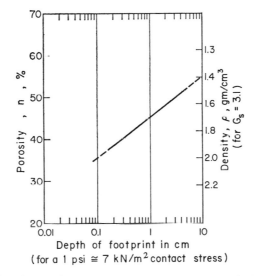

Fig. 1. Predicted variation of footprint depth with average porosity in the top 15 cm for the lunar surface.

Table 1. Statistical parameters for all data sets.

Data source	Description of data source	Number of observations, N	Mean porosity $\bar{n}$ in %	Standard deviation of sample, s
	All data	33	43.8	2.2
Apollo 11	All data less crater rims	30	43.3	1.8
	Crater rims only	3	48.4	2.0
	All data	119	43.8	3.6
Apollo 12	All data less crater rims	88	42.8	3.1
	Crater rims only	31	46.5	4.8
	Near LM less crater rims	28	43.9	2.5
	All data	42	43.9	2.3
Apollo 14	All data less crater rims	38	43.3	2.2
	Crater rims only	4	50.1	2.9
	All data	129	43.6	2.8
	All data less crater rims	117	43.4	2.9
Apollo 15	Crater rims only	12	46.0	1.8
	ALSEP Pan less crater rims	35	43.0	2.2
	Near LM less crater rims	35	43.8	2.2
	Near Soil Mechanics Trench, Station 8	13	44.1	2.2
	Apollos 11, 12, 14, & 15—all data	323	43.8	3.0
	Apollos 11, 12, 14, & 15—all data less crater rims	273	43.3	2.8
Combinations as noted	Apollos 11, 12, 14, & 15—crater rims only	50	46.7	4.0
	Apollos 11, 14, & 15—all data	204	43.7	2.6
	Apollos 11, 14, & 15—all data less crater rims	185	43.3	2.6
	Apollos 11, 14, & 15—crater rims only	19	47.2	2.1
Boulder track analyses	19 locations on lunar surface as noted in Hovland (1970)	69	43.9	6.6

As examples to show the shapes of the porosity distributions, histograms were prepared for each mission using all data points and are shown in Fig. 2. These histograms show a slight but definite skewness toward lower porosity, but a bell-shaped normal distribution curve provides a fair fit for these data. A normal distribution was assumed for the statistical analyses which follow.

Comparative analyses

A comparison of data derived from Apollo 11, 14, and 15 (see Table 2) was made using Student's *t* test and the *F* distribution, and no statistical difference in either the means or sample variances was found at the 5% significance level.

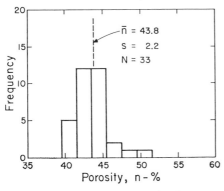
Fig. 2a. Apollo 11 footprint data.

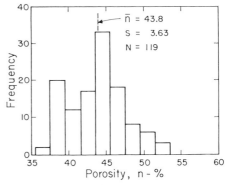
Fig. 2b. Apollo 12 footprint data.

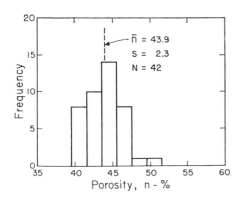
Fig. 2c. Apollo 14 footprint data.

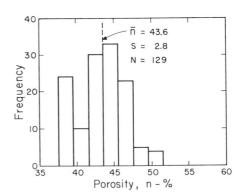
Fig. 2d. Apollo 15 footprint data.

Table 2. Comparative statistics for all data points for all Apollo missions

Data source	Number of observations N	Mean porosity $\bar{n}$ in %	Standard deviation of sample, s
Apollo 11, all data	33	43.8	2.2
Apollo 12, all data	119	43.8	3.6
Apollo 14, all data	42	43.9	2.3
Apollo 15, all data	129	43.6	2.8
Apollos 11, 12, 14, & 15, all data	323	43.8	3.0

A similar comparison was made between the Apollo 12 data ($s = 3.6$, Table 2) and the collection of data from Apollo 11, 14, and 15 ($s = 2.6$, Table 3). Although the means were not significantly different, the standard deviation was found to be significantly higher for the Apollo 12 data.

Closer examination of the data in Table 1, however, shows that the relatively high standard deviation for the Apollo 12 data is due almost entirely to the high standard deviation of the Apollo 12 crater rim porosities ($s = 4.8$, Table 1). When the standard deviations for all data exclusive of crater rims for Apollo 12 ($s = 3.1$, Table 1) and Apollo 11, 14, and 15 ($s = 2.6$, Table 1) were compared, there was no difference at the 5% significance level.

Furthermore, a comparison of the standard deviation for the Apollo 11, 14, and 15 crater rim porosities ($s = 2.1$, Table 4) with the Apollo 11, 14, and 15 data without crater rims ($s = 2.6$, Table 3) showed that the crater rims had slightly less variance, but not significantly less at the 5% significance level.

To study further the possible differences between the variation in porosity on crater rims and in generally level intercrater areas, a comparison was made between the standard deviations for crater rims for all Apollo missions ($s = 4.0$, Table 4) and for intercrater areas for all Apollo missions ($s = 2.8$, Table 3). The crater rims were found to have a significantly higher standard deviation; however, as explained above, the relatively high standard deviation for the crater rims for the collection of all Apollo missions is due to the strong influence of the high value for Apollo 12. Thus the Apollo 12 data suggest that crater rims have higher variation than intercrater (generally level) areas, but Apollo 11, 14, and 15 data suggest that the variation is about the same or slightly less for crater rims than in intercrater areas.

These differences may well arise out of the difficulty in arbitrarily classifying a footprint as being either on a crater rim or in between craters. Some judgment is required

Table 3. Comparisons showing effect of removing crater rim values.

Data source	Number of observations N	Mean porosity $\bar{n}$ in %	Standard deviation of sample, s
Apollos 11, 14, & 15, all data	204	43.7	2.6
Apollos 11, 14, & 15, all data minus crater rims (intercrater areas)	185	43.3	2.6
Apollos 11, 12, 14, & 15, all data minus crater rims (intercrater areas)	273	43.3	2.8

Table 4. Crater rim statistics.

Data source	Number of observations N	Mean porosity $\bar{n}$ in %	Standard deviation of sample, s
Apollos 11, 12, 14, & 15, crater rims only	50	46.7	4.0
Apollo 12, crater rims only	31	46.5	4.8
Apollos 11, 14, & 15, crater rims only	19	47.2	2.1

in making this decision, and it is possible that these differences may be due to this factor. Therefore the conclusion that the porosity variability is higher for crater rims than for intercrater areas is only tentative and is not yet strongly supported.

After establishing equality of variance for the Apollo 11, 14, and 15 crater rims and intercrater areas, equality of means was checked using Student's t test. The mean for the crater rims ($\bar{n} = 47.2\%$, Table 4) was found to be significantly greater than the mean for the intercrater areas ($\bar{n} = 43.3\%$, Table 3) at the 5% level—as well as at the 0.5% level. Thus these two sets of data were combined as shown in Tables 1 and 3 and Fig. 3.

Except for the significantly higher mean porosity values on crater rims than for intercrater areas noted above, no statistical difference was found between the mean porosities for any of the groupings, subgroupings, or collections reported in Table 1. As a matter of fact, the mean values were remarkably similar in magnitude, covering a range of only 1.3%, excluding crater rim groupings, with most values falling in an 0.5% band. These data suggest that, to the extent that the four Apollo sites represent the lunar surface, the mean porosity in generally lever intercrater areas does not vary significantly over the lunar surface.

Other interesting comparisons may be made by looking at the variations in porosity within the Apollo 12 and 15 sites. The Apollo 15 subgroupings, for example (see Table 1), show that both the mean and standard deviations for the relatively localized areas associated with the ALSEP Pan, the LM, and Station 8 are not significantly different from the corresponding values for the entire Apollo 15 site, nor for the entire collection of Apollo sites. This comparison indicates that, although variations in porosity exist on the lunar surface (a standard deviation of 2.5 to 3% in porosity is appreciable), these variations apparently occur on a scale of only a few meters—very likely less than a few tens of meters. Otherwise, porosity estimates from within a relatively small area, such as near the Soil Mechanics Trench at Station 8, would show smaller variation than for the Apollo 15 site as a whole.

The mean porosity at Station 8 ($\bar{n} = 44.1\%$) does not differ significantly from the mean values determined for the Apollo 15 site ($\bar{n} = 43.6\%$), but it is considerably higher than the value estimated at the Soil Mechanics Trench. Mitchell *et al.* (1972) estimated $n = 35\text{--}38\%$ for the Soil Mechanics Trench from analysis of measured

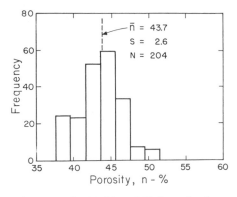

Fig. 3. Apollo 11, 14, and 15 footprint data.

penetration resistance. It is pertinent to note, however, that because of surface disturbance very few footprints from immediately around the trench could be analyzed.

Using the normal distribution assumption and only the porosity data gathered at Station 8 from footprint analysis, 90% confidence limits for the true mean and standard deviation may be computed to be 43.0 and 45.2% and 1.74 and 3.46%, respectively. These values make it possible to conclude that the probability is at least 0.8 that 95% of the porosity values around Station 8 fall within the range 36 to 52%. Therefore, it appears that the porosity at the Soil Mechanics Trench is unusually low—even with respect to the surrounding area—and that the porosity may change appreciably over very short distances, perhaps as small as a meter or two.

A density range of 1.62 to 2.15 g/cm^3 with an average value of about 1.8 g/cm^3 has been estimated by Carrier et al. (1972) for the samples in the deep drill stem obtained from the Station 8 area. Using an assumed value of specific gravity of 3.1, this density range corresponds to a porosity range of about 31 to 48% with an average of about 42%.

ANALYSES OF BOULDER TRACKS

Determination of ϕ values

Theoretical analyses (Hovland and Mitchell, 1972; Hovland, 1970) of the deformation mechanism associated with rolling boulders have led to the development of a relationship between the boulder track geometry and the mechanical properties of the soil, including porosity, through correlation. Sixty-nine lunar boulder tracks from 19 different locations on the moon were examined using lunar orbiter photography. Measurements of the track widths show that some boulders sank considerably deeper than others.

Using bearing capacity theory and an average value of cohesion of 0.5 kN/m^2, the average friction angle, ϕ, of the lunar soil was estimated for each of the 69 boulder tracks.

Conversion of ϕ values to porosity

The relationship between ϕ and porosity for the lunar soil simulant (LSS No. 2) as reported by Mitchell et al. (1972) was used to obtain porosity values which were compared with porosity values determined from the analysis of footprints. It was necessary to make a correction in the ϕ versus porosity curve for the lunar soil simulant before the conversion to account for the higher average confining pressure to which the soil under the rolling boulders was subjected.

Reduction of data and comparison with footprint results

The mean and standard deviation for the boulder tracks were computed to be 43.9 and 6.6% respectively, and the histogram is shown in Fig. 4.

The mean value is essentially the same as was obtained from the footprints described in the preceding section, but the standard deviation is considerably higher—suggesting that soil porosity is more variable on slopes and crater walls than on generally level intercrater areas.

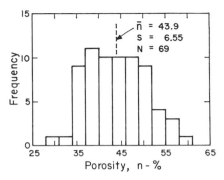

Fig. 4. Boulder track data from Orbiter photographs.

The apparent agreement between the mean values for the boulder track and footprint data may be misleading. Due to the large size of the boulders, the porosity estimates obtained represent averages for the upper few meters, whereas the footprint data represent average values for the upper 15 cm. It has been shown that porosity generally decreases with depth (Mitchell *et al.*, 1972; Houston and Namiq, 1971), although local exceptions occur. Therefore, agreement between the average values for boulder track and footprint data implies that the average porosity of the top 15 cm of the boulder track slopes must be higher than the value obtained from the footprints for level intercrater areas. It is difficult to estimate precisely how much higher, but reasonable estimates of porosity variation with depth as caused by compression under self-weight (Houston and Namiq, 1971) suggest 2 or 3% higher. This means that the average value for the top 15 cm on the boulder track slopes may be of the order of 46 to 47%, or about the same as was obtained for crater rims.

CONCLUSIONS

(1) Analyses of footprint depths within generally level intercrater areas indicate that both the mean porosity and the standard deviation (for the top 15 cm) are essentially the same for the four Apollo sites and are equal to 43.3 and 2.8%, respectively, for 273 observations. The corresponding 95% confidence limits are 42.9 to 43.7% and 2.6 to 3%, respectively, if normal distribution for porosity is assumed. Assuming the four Apollo sites are representative of the lunar surface, it may be concluded that the probability is at least 0.9 that the indicated porosity ranges in the table below include 67 and 95% of the porosity values for level intercrater areas. The corresponding values of density, angle of internal friction ϕ, and cohesion c, as estimated from lunar soil simulant properties (Mitchell *et al.*, 1972) are shown also for comparison.

	Mean	Standard deviation	Range which includes 67% of porosity values	Range which includes 95% of porosity values
Porosity, %	43.3	2.8	40–46.6	36.8–49.8
Density, assuming $G_s = 3.1$, g(cm^3	1.76		1.66–1.86	1.55–1.96
Corresponding ϕ values	43°		19.5–47	36–51
Corresponding c values, kN/m^2	0.55		0.35–1.10	0.18–2.00

(2) Crater rims are significantly softer than level intercrater areas, and footprint data suggest that the average crater rim porosity may be about 46.7% with a standard deviation of about 4%. The variation in crater rim porosity may be slightly higher than for intercrater areas, but this conclusion is only tentative and may depend on the arbitrary selection of the dividing line between crater rim and intercrater areas.

(3) The observed variations in porosity occur on a relatively small scale, probably of the order of a few meters. The computed porosity at the Apollo 15 Station 8 Soil Mechanics Trench was probably unusually low—even with respect to the immediately surrounding area.

(4) Because variations in porosity probably occur on a relatively small scale, it is very unlikely that one core tube, or even two, would be sufficient to characterize the porosity at a given lunar traverse station stop, except for a very small area immedi-around the core tube.

(5) The analysis of boulder track data indicates that the average porosity for the top few meters for the 19 slope locations examined is about the same as the average porosity for the top 15 cm at the Apollo sites, although the variability in boulder track porosities is much higher. In view of the general decrease in porosity with depth, the comparison indicates that the average porosity for the top 15 cm on the slopes and crater walls is higher than for the corresponding depth range at the Apollo sites. This difference has been roughly estimated to be about 2 or 3%. This indicates that soil slopes and crater walls are softer than intercrater areas, and may have near-surface average porosities close to that of crater rims. These findings are consistent with the tentative conclusion that many slope and crater wall soil layers have been loosened somewhat by downslope movements.

(6) The equivalence of porosity means and standard deviations for the intercrater areas of the four Apollo sites is consistent with the results of the MET track studies (Mitchell et al., 1972). This equivalency suggests that the lunar processes affecting the porosity of the upper 15 cm may have reached a steady state.

References

Carrier W. D. III, Johnson S. W., Carrasco L. H., and Schmidt R. (1972) Core sample depth relationships: Apollo 14 and 15 (abstract). In *Lunar Science—III* (editor C. Watkins), pp. 122–124, Lunar Science Institute Contr. No. 88.

Houston W. N. and Namiq L. I. (1971) Penetration resistance of lunar soils. *J. Terramechanics* **8**, 59–69.

Hovland H. J. (1970) Mechanics of rolling sphere-soil slope interaction. Ph.D. dissertation, Department of Civil Engineering, University of California, Berkeley, California.

Hovland H. J. and Mitchell J. K. (1972) Variability of lunar surface materials as indicated by boulder tracks. *Science*, submitted for publication.

Mitchell J. K., Scott R. F., Houston W. N., Costes N. C., Carrier W. D. III, and Bromwell L. G. (1972) Mechanical properties of lunar soils: Density, porosity, cohesion, and angle of internal friction. *Proc. Third Lunar Sci. Conf., Geochim. Cosmochim. Acta* Suppl. 3, Vol. 3. MIT Press.

Namiq L. I. (1970) Stress-deformation study of a simulated lunar soil. Ph.D. dissertation, Department of Civil Engineering, University of California, Berkeley, California.

Lunar Sample Cross Reference
(1972 Proceedings)

Two sample cross references are presented in this Proceedings set. This sample cross reference refers to the *Proceedings of the Third Lunar Science Conference* (set 3). The sample cross reference to the *Proceedings of the Apollo 11 Lunar Science Conference* (set 1) and the *Proceedings of the Second Lunar Science Conference* (set 2) is presented at the end of Volume 2. The format of the sample cross reference is as follows:

Sample Number
   (set number, volume number) page numbers . . . .

Luna 16
   (3, 2)   1161, 1201, 1343, 1361, 1377, 1397, 1421, 1455, 1515, 1569, 1821
   (3, 3)   2735, 2845, 2867, 2905, 2919, 3103
10002
   (3, 2)   1479, 1703, 2059
10003
   (3, 2)   1531, 1821
10004
   (3, 2)   1719
10005
   (3, 2)   1421, 1719
   (3, 3)   2845
10017
   (3, 1)   569
   (3, 2)   1251, 1269, 1531, 1659, 1693, 1733, 1763, 1821, 1891, 2003
10018
   (3, 2)   1251, 1397
10019
   (3, 1)   673
   (3, 2)   1397, 1473
   (3, 3)   2793
10020
   (3, 1)   251, 819
   (3, 2)   1531
   (3, 3)   2343, 2527, 2599, 3103
10021
   (3, 2)   1397
10022
   (3, 3)   3187
10044
   (3, 1)   615
   (3, 2)   1251, 1787
   (3, 3)   2981
10045
   (3, 1)   251, 533
10046
   (3, 1)   673, 1037
   (3, 2)   1821, 2059
   (3, 3)   2599, 2883

10047
  (3, 1)   295, 533
  (3, 3)   2933
10048
  (3, 3)   2363, 2423, 2981, 3173
10049
  (3, 2)   1479
  (3, 3)   2343
10050
  (3, 1)   555
  (3, 2)   1531
10057
  (3, 2)   1251, 1479, 1623, 1703, 1787, 1891, 2059
  (3, 3)   2527, 2599, 2919, 3103
10058
  (3, 1)   533
  (3, 3)   2919, 3077, 3103
10059
  (3, 2)   1397, 2069
  (3, 3)   2883, 3103
10060
  (3, 2)   1397
10061
  (3, 1)   819
  (3, 2)   1397
10065
  (3, 3)   2527, 3085
10071
  (3, 2)   1821, 1891
10072
  (3, 1)   533
10073
  (3, 2)   1397, 2003
10084
  (3, 1)   753, 939, 1065, 1077
  (3, 2)   1133, 1201, 1251, 1269, 1361, 1377, 1397, 1637, 1763, 1779, 1821, 1947, 1981, 2059, 2069
  (3, 3)   2343, 2465, 2495, 2681, 2883, 2905, 2933, 2971, 3085, 3103, 3187
10085
  (3, 1)   753
  (3, 2)   1361, 1377, 1515, 1623
  (3, 3)   2397, 2465, 3047
10086
  (3, 2)   1479, 2069, 2091, 2109
10087
  (3, 3)   2243, 2259
10089
  (3, 2)   2025
12001
  (3, 1)   753, 1029, 1077
  (3, 2)   1251, 1693, 1821, 2109
  (3, 3)   2243, 2259, 2397, 2503, 3225
12002
  (3, 1)   171

(3, 2) 1269, 1479, 1659, 1693, 1719, 1733, 1779, 1787, 1891
(3, 3) 2449, 2527, 2981, 3077, 3103, 3161
12003
(3, 2) 1161
12004
(3, 1) 865
(3, 2) 1149
12006
(3, 3) 2793
12008
(3, 3) 3077
12009
(3, 1) 171, 865
(3, 2) 1531, 1891
(3, 3) 2587, 3077
12010
(3, 2) 1181, 1201, 1671, 1899
12013
(3, 1) 281, 591, 739
(3, 2) 1161, 1181, 1201, 1361, 1377, 1515, 1531, 1637, 1651, 2003
(3, 3) 2157
12017
(3, 3) 2735, 2767
12018
(3, 2) 1479, 1891
(3, 3) 2343, 2527, 2811, 3047
12020
(3, 2) 1787
(3, 3) 2337, 2343, 2811, 2981
12021
(3, 1) 171, 251, 481, 493, 507, 533, 569, 591, 615
(3, 2) 1149, 1479, 1531, 1763, 1779, 1891, 1917
(3, 3) 2495, 2735, 2767, 3047
12022
(3, 2) 1149, 1479, 1531, 1763, 1787, 2003
(3, 3) 2527, 2599, 2905, 2919
12023
(3, 2) 2091, 2119
12024
(3, 3) 2793
12025
(3, 1) 983, 1065
(3, 2) 1569, 2069
(3, 3) 2845, 2905, 2919, 2997
12028
(3, 1) 983, 1065
(3, 2) 1181, 1361, 1377, 1397, 1569, 1623, 2069
(3, 3) 2811, 2905, 2919, 2997, 3103
12030
(3, 1) 983
(3, 2) 1857, 1967
(3, 3) 2503, 2811, 2919
12031
(3, 2) 1397, 1671

12032
   (3, 1)   379, 983
   (3, 2)   1149, 1181, 1269, 1315, 1397, 1569, 2051, 2069
   (3, 3)   2681, 2919, 2971, 3103
12033
   (3, 1)   379, 423, 753, 983
   (3, 2)   1149, 1161, 1181, 1201, 1315, 1343, 1377, 1397, 1515, 1531, 1569, 1857, 1967, 1981, 2003, 2051, 2069, 2109
   (3, 3)   2259, 2503, 2681, 2919, 2971, 2997, 3103, 3187
12034
   (3, 1)   739, 753, 819
   (3, 2)   1181, 1315, 1397, 1531, 1569
12035
   (3, 1)   251
   (3, 2)   1149, 1531
12036
   (3, 1)   251
   (3, 3)   2919
12037
   (3, 1)   983
   (3, 2)   1315, 1377
   (3, 3)   2811
12038
   (3, 1)   243, 401, 507, 555, 1065
   (3, 2)   1161, 1269, 1531, 1779
   (3, 3)   2565, 2735, 2767, 2811
12039
   (3, 1)   591
12040
   (3, 1)   243, 251, 507, 545
   (3, 2)   1623
12041
   (3, 1)   983
12042
   (3, 1)   939, 1065
   (3, 2)   1569, 1857, 1967
   (3, 3)   2811, 2919
12044
   (3, 1)   983
   (3, 2)   1149, 1315, 1857, 1967
   (3, 3)   2919
12047
   (3, 3)   2793
12051
   (3, 1)   581
   (3, 2)   1149, 1787
   (3, 3)   2735, 2955, 3103
12052
   (3, 1)   171, 533, 1065
   (3, 2)   1531, 1891
   (3, 3)   2527
12053
   (3, 1)   171, 401, 493, 507, 1077

   (3, 2)   1251, 1269, 1479, 1693, 1763, 1917
   (3, 3)   2793, 3173
12054
   (3, 3)   2735, 2767, 2793
12055
   (3, 3)   2793
12057
   (3, 1)   739
   (3, 2)   1149, 1361, 2059
   (3, 3)   2465
12062
   (3, 2)   1659
12063
   (3, 1)   295, 545, 591
   (3, 2)   1149, 1251, 1269, 1531, 1623, 1703, 1891, 2059
   (3, 3)   2337, 2363, 2527, 2565, 2767, 2793, 2919, 3021, 3187
12064
   (3, 2)   1149, 1531, 1623
12065
   (3, 1)   171
   (3, 2)   1703, 1787
   (3, 3)   2527, 3077
12070
   (3, 1)   379, 753, 939, 983
   (3, 2)   1133, 1149, 1181, 1251, 1269, 1315, 1343, 1361, 1397, 1479, 1531, 1569, 1637, 1763,
              1779, 1899, 1981, 2059, 2069
   (3, 3)   2343, 2465, 2503, 2955, 2971, 3047, 3103
12073
   (3, 1)   281, 819
   (3, 2)   1315, 1637, 1857, 1967
   (3, 3)   2767, 3077
12075
   (3, 2)   2059
14001
   (3, 1)   739, 983
   (3, 2)   1487, 1651, 1865
   (3, 3)   2811, 2919
14002
   (3, 1)   971, 983, 995
   (3, 2)   1377
   (3, 3)   2811
14003
   (3, 1)   141, 673, 753, 887, 983, 1015, 1029, 1037, 1065
   (3, 2)   1149, 1181, 1215, 1231, 1293, 1335, 1343, 1377, 1455, 1473, 1503, 1531, 1797, 1857,
              1865, 1967, 2051, 2059, 2091, 2109, 2131
   (3, 3)   2465, 2479, 2503, 2811, 2981, 3103, 3187
14006
   (3, 1)   623, 645, 739, 753, 771
   (3, 2)   1515, 1797, 2003
   (3, 3,)   2883
14034
   (3, 1)   771

14036
  (3, 1)   251
14042
  (3, 1)   645
  (3, 2)   2051
14045
  (3, 2)   1671
14047
  (3, 1)   645
  (3, 2)   1149, 1181, 1215, 1231, 1327, 1377, 1899, 2051, 2131
  (3, 3)   2363, 2387, 2423, 2845, 3173
14049
  (3, 1)   645, 753
  (3, 2)   1201, 1215, 1293, 1645, 2041, 2051, 2059, 2131
  (3, 3)   2387, 2449, 2465, 2883
14053
  (3, 1)   115, 131, 207, 251, 281, 295, 305, 333, 379, 493, 533, 581, 603, 615, 645, 771, 865
  (3, 2)   1149, 1161, 1201, 1215, 1269, 1275, 1361, 1377, 1397, 1455, 1487, 1515, 1531, 1557, 1589, 1651, 1671
  (3, 3)   2343, 2363, 2423, 2767, 2919, 3173
14055
  (3, 1)   753, 771, 887
  (3, 3)   2845
14059
  (3, 1)   887
14063
  (3, 1)   305, 471, 645, 753, 771, 819, 887, 907, 971
  (3, 2)   1161, 1181, 1215, 1231, 1275, 1361, 1377, 1429, 1531, 1651, 2051, 2069
  (3, 3)   2343, 2387, 2423, 2883, 3173
14064
  (3, 1)   753
14065
  (3, 2)   1181, 1377
  (3, 3)   2587
14066
  (3, 1)   305, 623, 645, 687, 753, 771, 865, 887
  (3, 2)   1181, 1215, 1251, 1343, 1487, 1651, 1671, 2051, 2131
  (3, 3)   2343, 2845
14068
  (3, 1)   379, 623, 645, 723, 739, 865, 895
  (3, 2)   1161, 1429, 1515, 1797, 2003
  (3, 3)   2811
14072
  (3, 1)   131, 281, 333, 771, 971
  (3, 2)   1161, 1201, 1231, 1275, 1455, 1487, 1613
14073
  (3, 1)   141, 281, 333
  (3, 2)   1181, 1455, 1487, 1589
  (3, 3)   2767, 2919
14082
  (3, 1)   27, 645, 753
  (3, 2)   1671
14083
  (3, 2)   1161, 1181, 1215, 2131

    (3, 3)   2883, 2919, 3103
14130
    (3, 3)   2243
14141
    (3, 1)   141, 673, 983
    (3, 2)   1201, 1377, 2051, 2069, 2131
    (3, 3)   2811, 2845, 2867, 2883, 2905, 2919, 2997, 3021, 3077
14142
    (3, 1)   983, 995
    (3, 2)   1377
14143
    (3, 1)   739, 983
14146
    (3, 1)   983, 995
    (3, 2)   1377
14147
    (3, 1)   983
14148
    (3, 1)   673, 723, 907, 927, 983
    (3, 2)   1201, 1293, 1377, 1429, 1651, 1671, 1797, 1989, 2051, 2069, 2131
    (3, 3)   2503, 2811, 2845, 2867, 2883, 2905, 2919, 2997
14149
    (3, 1)   673, 723, 907, 927, 983
    (3, 2)   1201, 1293, 1377, 1429, 1503, 1569, 1651, 1671, 1797, 1989, 2051, 2069, 2131
    (3, 3)   2503, 2845, 2867, 2883, 2905, 2919, 2997
14150
    (3, 2)   1651
14151
    (3, 1)   983, 995
    (3, 2)   1377, 1651
    (3, 3)   2811
14152
    (3, 1)   983
    (3, 2)   1557, 1651
    (3, 3)   2811
14153
    (3, 1)   983
14154
    (3, 1)   983, 995
    (3, 2)   1377
14156
    (3, 1)   363, 673, 723, 907, 927, 983
    (3, 2)   1201, 1293, 1377, 1429, 1651, 1671, 1797, 1989, 2051, 2069, 2091, 2131
    (3, 3)   2503, 2845, 2867, 2883, 2905, 2919, 2997
14160
    (3, 2)   1671, 1891
14161
    (3, 1)   141, 401, 645, 739, 837, 907, 983, 1077
    (3, 2)   1161, 1515, 1671, 1797, 1865, 1899, 2003
    (3, 3)   2423, 2811, 2883, 3077
14162
    (3, 1)   207, 251, 569, 673, 753, 837, 895, 939, 983, 995
    (3, 2)   1671, 1797
    (3, 3)   2479, 2811, 2883

14163
  (3, 1)   141, 251, 545, 603, 673, 723, 771, 837, 887, 907, 927, 983, 995, 1015, 1029, 1037,
           1065, 1077, 1085
  (3, 2)   1133, 1149, 1161, 1181, 1201, 1215, 1231, 1251, 1269, 1275, 1307, 1327, 1337,
           1343, 1377, 1397, 1455, 1465, 1479, 1503, 1515, 1531, 1637, 1645, 1671, 1681,
           1693, 1703, 1763, 1771, 1779, 1797, 1857, 1899, 1917, 1947, 1967, 1989, 2003,
           2015, 2025, 2029, 2041, 2051, 2059, 2069, 2109, 2149
  (3, 3)   2243, 2479, 2495, 2503, 2611, 2681, 2811, 2845, 2867, 2883, 2919, 2949, 2971, 2981,
           2997, 3009, 3047, 3077, 3103, 3157, 3187, 3195
14166
  (3, 1)   645
  (3, 2)   1857, 1967
14167
  (3, 1)   379, 895
  (3, 2)   1557, 1613, 1651, 1857, 1967
  (3, 3)   2919
14168
  (3, 1)   1077
14169
  (3, 2)   1651
14170
  (3, 2)   1651
14171
  (3, 1)   105, 645, 887
14190
  (3, 1)   141
14191
  (3, 1)   305
14192
  (3, 2)   1557
14193
  (3, 2)   1557
14210
  (3, 3)   3213
14211
  (3, 3)   3213
14220
  (3, 3)   3213
14230
  (3, 1)   673, 753, 927, 983, 1015
  (3, 2)   1181, 1201, 1293, 1797, 2051
  (3, 3)   2397, 2883, 3077, 3187, 3213
14233
  (3, 3)   2845
14240
  (3, 2)   1181, 1215, 1429, 1947, 2051, 2069, 2091, 2109, 2119, 2131
14257
  (3, 1)   251, 983, 1095
  (3, 2)   1557, 1865, 2059
  (3, 3)   2793, 2811
14258
  (3, 1)   305, 379, 673, 739, 837, 971, 983, 995
  (3, 2)   1201, 1377
  (3, 3)   2811

14259
  (3, 1)   141, 185, 251, 363, 645, 673, 723, 753, 771, 907, 927, 983, 995, 1015
  (3, 2)   1149, 1181, 1201, 1215, 1251, 1269, 1337, 1343, 1377, 1397, 1455, 1473, 1503, 1531,
           1569, 1645, 1651, 1671, 1693, 1703, 1719, 1763, 1857, 1917, 1967, 1981, 1989, 2051
  (3, 3)   2259, 2343, 2479, 2619, 2627, 2655, 2665, 2811, 2845, 2867, 2905, 2919, 2971, 2981,
           2997, 3047, 3069, 3077, 3103
14260
  (3, 1)   723, 771, 907, 983
14261
  (3, 1)   739
14262
  (3, 1)   995
  (3, 2)   1201, 1377
14264
  (3, 1)   739
14265
  (3, 2)   1651
14267
  (3, 2)   2069
  (3, 3)   3103
14268
  (3, 1)   739
14270
  (3, 1)   645, 819
  (3, 2)   1201
  (3, 3)   2845
14271
  (3, 2)   1651
14272
  (3, 2)   1651
14273
  (3, 2)   1651
14276
  (3, 2)   1133, 1215, 1455
14298
  (3, 2)   1429, 2051, 2069, 2109, 2119
14301
  (3, 1)   379, 603, 645, 673, 895, 907
  (3, 2)   1161, 1215, 1327, 1623, 1671, 1797, 1899, 2051, 2131
  (3, 3)   2363, 2387, 2423, 2479, 2503, 2767, 2811, 2845, 2919, 2981, 3077, 3161
14302
  (3, 2)   1733
14303
  (3, 1)   251, 471, 645, 887
  (3, 2)   1133, 1149, 1215, 1865, 2059
  (3, 3)   2363, 2387, 2423, 2479, 2503, 2793, 2811, 3173
14304
  (3, 1)   645, 895, 1037
14305
  (3, 1)   105, 141, 251, 281, 379, 645, 673, 687, 887, 895, 1095
  (3, 2)   1149, 1251, 1269, 1293, 1637, 1651, 1671, 1733, 1763, 1917, 1989, 2041, 2051, 2059
  (3, 3)   2343, 2713, 2767, 2793, 2811, 2883

14306
- (3, 1)   27, 545, 645, 673, 771, 819, 853
- (3, 2)   1231, 1455, 1623
- (3, 3)   2343, 2767, 2919

14307
- (3, 1)   623, 645, 753, 895
- (3, 2)   1161, 1429, 1515, 1531, 1651, 1797
- (3, 3)   3077

14308
- (3, 1)   645

14310
- (3, 1)   115, 131, 141, 159, 171, 185, 197, 207, 231, 251, 281, 305, 333, 351, 379, 401, 423, 493, 507, 533, 555, 569, 581, 591, 603, 615, 645, 687, 771, 865, 971
- (3, 2)   1133, 1149, 1161, 1181, 1215, 1231, 1269, 1275, 1293, 1307, 1327, 1343, 1361, 1377, 1397, 1455, 1487, 1503, 1515, 1531, 1557, 1589, 1613, 1623, 1651, 1659, 1671, 1681, 1719, 1891, 1989, 2015, 2041, 2051
- (3, 3)   2243, 2337, 2387, 2423, 2449, 2479, 2503, 2527, 2565, 2577, 2619, 2627, 2713, 2735, 2755, 2811, 2845, 2867, 2919, 2941, 2971, 2981, 3021, 3047, 3077, 3157, 3161, 3187

14311
- (3, 1)   379, 623, 645, 673, 739, 753, 771, 1065
- (3, 2)   1149, 1327, 1335, 1623, 2051, 2069, 2091
- (3, 3)   2387, 2423, 2479, 2503, 2557, 2767, 2811, 2845, 2919

14312
- (3, 1)   645, 753, 887
- (3, 3)   2387, 2449, 2971

14313
- (3, 1)   379, 645, 661, 687, 771, 895
- (3, 2)   1275, 1455, 2051, 2069
- (3, 3)   2363, 2387, 2449, 2557, 2767, 3077

14314
- (3, 1)   379, 645, 753, 887, 895

14315
- (3, 1)   379, 645, 723, 753, 895
- (3, 2)   1215, 1671
- (3, 3)   2811

14318
- (3, 1)   379, 623, 645, 673, 707, 723, 753, 771, 819, 887
- (3, 2)   1181, 1215, 1531, 1671, 2051, 2131
- (3, 3)   2343, 2479, 2503, 2599, 2767, 3161

14319
- (3, 1)   251, 569, 645, 753, 887, 895
- (3, 2)   1293

14320
- (3, 1)   141, 251, 645, 753, 771, 887

14321
- (3, 1)   27, 115, 207, 251, 305, 351, 401, 645, 723, 753, 771, 819, 865, 887, 895, 907, 1077
- (3, 2)   1149, 1161, 1201, 1231, 1251, 1307, 1327, 1335, 1337, 1343, 1361, 1377, 1429, 1455, 1465, 1479, 1487, 1613, 1623, 1645, 1651, 1659, 1671, 1681, 1693, 1719, 1747, 1771, 1779, 1891, 1899, 1917, 1989, 2041, 2051, 2059, 2069, 2131
- (3, 3)   2243, 2337, 2363, 2387, 2417, 2449, 2479, 2495, 2503, 2565, 2767, 2811, 2845, 2867, 2883, 2919, 2949, 2981, 3047, 3077, 3103, 3161

14359
- (3, 3)   2997

14411
 (3, 3)   3213
14414
 (3, 3)   3213
14421
 (3, 1)   995
 (3, 2)   1215, 1293, 2051, 2131
 (3, 3)   3021
14422
 (3, 2)   1429, 2051, 2091
14923
 (3, 2)   1397
15001
 (3, 3)   2905, 3213
15002
 (3, 3)   2905, 3213
15003
 (3, 3)   2905, 3213
15004
 (3, 3)   2905, 3213
15005
 (3, 3)   2905, 3213
15006
 (3, 3)   2905, 3213
15007
 (3, 3)   3213
15008
 (3, 3)   3213
15009
 (3, 3)   3213
15010
 (3, 3)   3213
15011
 (3, 3)   3213
15015
 (3, 1)   939
 (3, 3)   2243, 2577, 2599, 2981
15016
 (3, 2)   1455, 1659
15017
 (3, 1)   939
15021
 (3, 1)   723
 (3, 2)   1181, 1251, 1327, 1397, 1429, 1671, 1747, 1797, 1865, 2029, 2131
 (3, 3)   3077
15022
 (3, 1)   723
15023
 (3, 2)   1161
15031
 (3, 2)   1659, 1671, 1681, 1703, 2069
 (3, 3)   3195

Author Index

Subject Index

(Where an index entry refers to the opening page of an article, the entry includes the entire article.)